PHYSICS OF PARTICLE ACCELERATORS

Particle Accelerator School
(1987 and 1988)

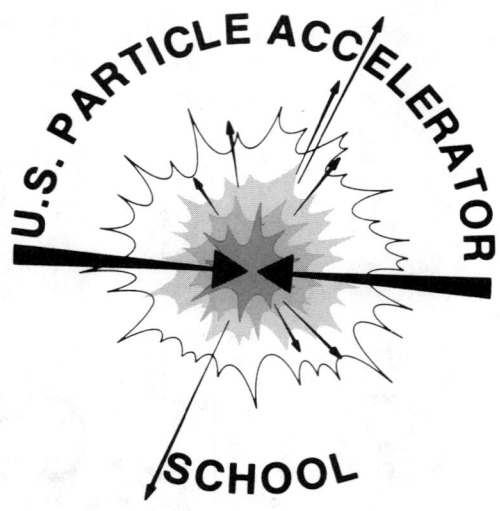

The U.S. Particle Accelerator School
is dedicated to the future of
Accelerator-based Science and Technology

Fermi National Accelerator Laboratory
July 20–August 14, 1987
and
Cornell University
August 1–12, 1988

AIP CONFERENCE PROCEEDINGS 184

RITA G. LERNER
SERIES EDITOR

VOLUME TWO

PHYSICS OF PARTICLE ACCELERATORS

FERMILAB SUMMER SCHOOL, 1987
CORNELL SUMMER SCHOOL, 1988

EDITORS:
**MELVIN MONTH &
MARGARET DIENES**
BROOKHAVEN NATIONAL
LABORATORY

AMERICAN INSTITUTE OF PHYSICS   NEW YORK 1989

Authorization to photocopy items for internal or personal use, beyond the free copying permitted under the 1978 US Copyright Law (see statement below), is granted by the American Insitute of Physics for users registered with the Copyright Clearance Center (CCC) Transactional Reporting Service, provided that the base fee of $3.00 per copy is paid directly to CCC, 27 Congress St., Salem, MA 01970. For those organizations that have been granted a photocopy license by CCC, a separate system of payment has been arranged. The fee code for users of the Transactional Reporting Service is: 0094-243X/87 $3.00.

Copyright 1989 American Institute of Physics.

Individual readers of this volume and non-profit libraries, acting for them, are permitted to make fair use of the material in it, such as copying an article for use in teaching or research. Permission is granted to quote from this volume in scientific work with the customary acknowledgment of the source. To reprint a figure, table or other excerpt requires the consent of one of the original authors and notification to AIP. Republication or systematic or multiple reproduction of any material in this volume is permitted only under license from AIP. Address inquiries to Series Editor, AIP Conference Proceedings, AIP, 335 E. 45th St., New York, NY 10017.

L.C. Catalog Card No. 89-83575
ISBN 0-88318-384-6
DOE CONF 8707208

Printed in the United States of America.

Contents

VOLUME ONE
Preface ix
Acknowledgments xvii
1987/1988 School Organization xvii

ACCELERATOR PHYSICS AND TECHNOLOGY

An Introduction to the Physics of Particle Accelerators 2
    D. A. Edwards and M. J. Syphers
Transverse Motion of Single Particles in Accelerators 190
    R. Talman
Fundamentals—Longitudinal Motion 243
    W. T. Weng
RF System Considerations for a Large Hadron Collider 288
    E. Raka
Beam Observation and the Nature of Instabilities 343
    J. Gareyte
Bunched Beam Diagnostics 430
    R. H. Siemann
Fields, Impedances, and Structures 472
    K. Y. Ng
Introduction to Wake Fields and Wake Potentials 525
    P. B. Wilson
Characteristics of Synchrotron Radiation 565
    K. J. Kim
Review of Linear Collider Beam–Beam Interaction 633
    P. Chen
The Linear Collider Beam–Beam Problem 680
    R. Hollebeek
A Quantum Treatment of Beamstrahlung and its Application
to Ribbon Pulses 711
    R. Blankenbecler and S. D. Drell
Methods of Beam Optics 758
    F. Willeke and G. Ripken
Nonlinear Dynamics 820
    G. Guignard
Phase Space Concepts 891
    L. Michelotti
Comments on Nonlinear Dynamics Studies in Storage Rings 946
    A. W. Chao
The Description of Particle Accelerators Using High-Order
Perturbation Theory on Maps 961
    M. Berz
Methods of Stability Analysis in Non-linear Mechanics 995
    R. L. Warnock, R. D. Ruth, W. Gabella, and K. Ecklund

Advanced Nonlinear Theory: Long-Term Stability at the SSC 1015
   S. Heifets
A Review on the Lattice Design of Large Hadron Colliders 1062
   S. Y. Lee
The Physics of Codes 1084
   R. K. Cooper and M. E. Jones
Canonical Integrators as Tracking Codes 1106
   E. Forest

VOLUME TWO

A Compendium of Computer Codes Used in Particle Accelerator
Design and Analysis 1137
   Los Alamos Accelerator Code Group
Superconducting Magnet System 1327
   P. Dahl
Superconductor Developments 1377
   D. Larbalestier
Radio-Frequency Acceleration 1393
   M. Puglisi
Principles and Technology of RFQs 1477
   A. Schempp
Photocathode RF guns 1500
   R. L. Sheffield
Low Emittance Thermionic Electron Guns 1532
   W. B. Herrmannsfeldt
Intense, Low Emittance Injectors for RF Electron Linacs 1543
   T. I. Smith
Final Focus Systems for Linear Colliders 1553
   R. A. Erickson
Positrons for Linear Colliders 1592
   S. Ecklund
Free-Electron Lasers 1615
   C. A. Brau
Free-Electron Lasers 1707
   R. H. Pantell
Fundamentals of Intense Relativistic Electron Beams 1729
   R. B. Miller
Application of Pulse Power Technology to Ultra High Energy
Electron Accelerators 1759
   J. A. Nation
Coherence and Statistical Properties of Photon Beams
with Application to the Free-Electron Laser 1776
   A. Bhattacharjee and I. Gjaja
Plasma Acceleration of Particle Beams 1798
   T. Katsouleas and J. M. Dawson

THE TEVATRON

Introduction	1830
D. A. Edwards	
Antiproton Source	1845
J. Peoples	
Primer on Beam Dynamics in Synchrotrons	1878
L. C. Teng	
Emittances Through the Fermilab Accelerator Chain	1910
D. Finley	
Detection and Compensation of Transverse Resonances	1928
R. Gerig	
Longitudinal Phase Space in Circular Accelerators	1941
P. S. Martin and S. Ohnuma	
Longitudinal Instabilities and Stability Criteria	1969
K. Y. Ng	
Resonant Extraction at the Tevatron	2009
M. Harrison	
Beam Loss	2033
A. Van Ginneken, D. Edwards, and M. Harrison	
Design and Operation of the Quench Protection System	
for the Fermilab Tevatron	2073
P. S. Martin	

PERSPECTIVES

Science in the Age of Accelerators	2098
M. L. Perl	
Accelerator Projects, Worldwide	2116
L. C. Teng	
Technology and Particle Beams	2127
D. L. Hartill	
Synchrotron Radiation	2138
H. Winick	
Induction Accelerators and Free-Electron Lasers at LLNL	2183
R. J. Briggs	
Progress on Next Generation Linear Colliders	2208
R. D. Ruth	
The SSC Project	2227
J. Peoples	
Parameter Choices for the SSC	2240
J. M. Peterson	
Supercollider Physics	2255
C. Quigg	
Particle Accelerators and Discoveries in Elementary	
Particle Physics	2280
L. W. Jones	
Federal Science Policy	2300
Harold P. Hanson	

Preface

This volume is based on lectures presented at the 1987 U.S. Particle Accelerator School, held at Fermi National Accelerator Laboratory in July and August 1987, in addition to selected material from the school at Cornell University in August 1988. As is our custom, review articles are also included for completeness and to enhance the book's quality.

The schools, sponsored by the U.S. Department of Energy and the National Science Foundation and supported by national laboratories across the United States, are part of a continuing series at different locations, as shown in Table I.

Table I

Fermilab	July 13–24, 1981	AIP Conf. Proc. 87
SLAC	August 2–13, 1982	AIP Conf. Proc. 105
BNL/SUNY	July 6–16, 1983	AIP Conf. Proc. 127
Fermilab	August 13–24, 1984	AIP Conf. Proc. 153
SLAC	July 15–26, 1985	
Fermilab	July 20–August 14, 1987	AIP Conf. Proc. 184
Cornell	August 1–12, 1988	
Sardinia*	January 31–February 5, 1985	Springer-Verlag Lecture Notes in Physics 247
Texas*	October 23–29, 1986	Springer-Verlag Lecture Notes in Physics 296
Capri*	October 20–26, 1988	Springer-Verlag Lecture Notes in Physics

*Organized jointly with the CERN Accelerator School.

The purpose of these schools derives from a strong recommendation by a subpanel of the High Energy Physics Advisory Panel (HEPAP), convened in 1979–1980 to assess the current state of accelerator and detector R&D, that the high energy physics community encourage more scientists and students to work in the field of high energy particle accelerators. These national summer schools constitute one response to that recommendation; indeed, their main purpose is to attract scientists and students to this field and to enhance their education in it.

The school is guided in its operation by the following objectives: (i) to present the entire spectrum of current knowledge about particle accelerators; (ii) to help train scientists planning to work in accelerator physics, in order to strengthen the base of particle accelerator specialists; (iii) to encourage accelerator physics programs in laboratories and universities by providing texts and training for their potential faculties; (iv) to foster dialogue between accelerator physicists and those working in accelerator-based sciences; and (v) to enhance the cultural side of accelerator technology by supporting symposia on the role of accelerators and particle beams in society, by fostering international collaboration on accelerator education, and by granting awards for excellence. Success in achieving these goals could be an important factor in continuing the advances in accelerator development necessary for vigorous programs in accelerator-based areas of science, medicine, industry, and defense.

The U.S. Particle Accelerator School has been operating since 1981. The standard symposium-style summer schools were conducted at Fermilab (1981), SLAC (1982), BNL/SUNY (1983), Fermilab (1984), SLAC (1985), and Fermilab (1987). The 7th will be held at BNL in 1989. These are two-week schools of intensive courses on the physics and technology of particle accelerators. In 1987, the U.S. School began operating on a biyearly schedule. In odd years it is comprised of two 2-week sessions of basic accelerator physics plus advanced subjects, one in the university format and the other in the symposium format. Introducing the university style allows the courses to be presented in greater depth, promotes student interaction and feedback, and encourages younger students to attend, especially since university credit is earned for courses successfully completed. There have been two university-related schools, at the University of Chicago (1987) and at Cornell University (1988), and they will now be held yearly at selected universities. The third will be held at the University of California at Berkeley in 1989. The proceedings of these schools are structured to be read as comprehensive textbooks and have been published (with the overall title *The Physics of Particle Accelerators*) by the American Institute of Physics as part of its Conference Proceedings Series, as listed above.

In 1985 the U.S. Particle Accelerator School and the CERN Accelerator School agreed to initiate a series of joint U.S.–CERN Topical Courses. The first of these international ventures was held in Santa Margherita di Paula, Sardinia, January 31–February 5, 1985. Since then, two others have taken place, in South Padre Island, Texas, October 23–29, 1986, and in Anacapri, Italy, October 20–26, 1988, with a fourth being planned for Hilton Head, South Carolina, in November 1990. The proceedings of these joint courses are being published by Springer-Verlag as part of its Lecture Notes in Physics Series. The Sardinia book, No. 247 in the series, was entitled *Nonlinear Dynamics Aspects of Particle Accelerators*, and the Texas book, No. 296, *Frontiers of Particle Beams*. The Capri proceedings will be published in 1989.

The school organization consists of a steering committee, which develops school policy, a school office, which carries out the administration, a program committee, and a local arrangements group. The steering committee is chaired by Burton Richter, and the school office director is Melvin Month. The 1987 program committee consisted of A. W. Chao (URA/SSC), G. Dugan (Fermilab), D. A. Edwards (Fermilab), G. Fisher (SLAC), S. Humphries, Jr. (U. of New Mexico), L. Michelotti (Fermilab), M. Month (BNL), chairman, C. Pellegrini (BNL), M. P. Reiser (U. of Maryland), A. G. Ruggiero (BNL), R. Ruth (SLAC), R. H. Siemann (Cornell), T. P. Wangler (LANL), and W. T. Weng (BNL). The 1988 program committee was essentially the set of lecturers for the university-type courses in 1987 (U. of Chicago) and 1988 (Cornell); the committee was comprised of C. A. Brau (UC Santa Barbara/LANL), K. L. Brown (SLAC), D. A. Edwards (Fermilab), V. L. Granatstein (U. of Maryland), M. Month (BNL), chairman, C. Pellegrini (BNL), R. H. Siemann (Cornell), R. Talman (Cornell), and P. B. Wilson (SLAC). A complete organization chart for 1987 and 1988 appears after the preface.

The texts produced by the school are compiled for the most part from the formal lectures. Each course is designed to introduce the concepts at a basic level and to include state-of-the-art material. The courses cover a broad range, and the list continues to grow as the school matures, as seen in Table II.

Table II

Principles of Linear and Circular Accelerators and Storage Rings	Particle Tracking
	Space–Charge Phenomena
Design of Circular Colliders	Coherent Phenomena and Instabilities
Design of Linear Colliders	Polarized Beams
Commissioning and Operation of Accelerators	Synchrotron Radiation
	Free-Electron Laser
Beam Optics and Magnetic Lattices	Scattering Phenomena
Magnet Design	Noise
Field Errors, Beam Measurements, and Controls	Beam Cooling
	Beam Simulation
Measurement Techniques	Statistical Phenomena in Particle Beams
Radiofrequency Systems	
Vacuum Systems	Concepts and Architecture of Control Systems
Superconducting Magnets and Radiofrequency Cavities	Computer Modeling
The Beam–Beam Interaction	New Accelerator Ideas
Methods in Nonlinear Theory	Advanced Accelerator Techniques

Participation in the school both by lecturers and by students continues to be excellent. For the symposium-style schools, the major U.S. accelerator laboratories (SLAC, Fermilab, and BNL) provide about 50% of the students. Efforts have been made to improve university and foreign participation, primarily by greater interaction by the school with U.S. universities and foreign institutions. Table III shows the success achieved in raising university participation, with significant increases continuing through 1988; foreign participation showed some improvement in 1987. Enhancement of participation by non-U.S. students has also been achieved in the joint U.S.–CERN topical courses, as seen in the numbers for Sardinia, Texas, and Capri. This is expected to continue at our joint school in South Carolina in 1990.

In 1985 the U.S. Particle Accelerator School introduced its program of prizes and awards to recognize excellence in the field of particle beam physics. Two prizes for Achievement in Accelerator Physics and Technology are awarded each year for significant contributions to the advancement of particle accelerator technology and to the increased understanding of the physics of particle beams.

Table III U.S. Particle Accelerator School Participation

Symposium Schools

Source of Students	1981 Fermilab	1982 SLAC	1983 BNL/SUNY	1984 Fermilab	1985 SLAC	1987 Fermilab
Major Labs. (U.S.)	65	92	78	94	138	67
Universities (U.S.)	32	14	37	55	67	36
Other (U.S.)	13	24	17	15	33	29
Foreign	10	19	16	31	21	22
Total Students	120	149	148	195	259	154
Total Lecturers	24	19	33	50	47	32

Source of Students	University Schools		Joint U.S.–CERN Topical Courses		
	1987 Chicago	1988 Cornell	1985 Sardinia	1986 Texas	1988 Capri
Major Labs. (U.S.)	19	15	16	14	7
Universities (U.S.)	38	47	8	28	6
Other (U.S.)	29	37	14	35	16
Foreign	17	26	58	33	102
Total Students	103	125	96	110	131
Total Lecturers	9	8	23	27	25

They are awarded on a competitive basis, with nominations open to all. The winners are chosen by a prize committee, and the awards are presented at a ceremony, generally during a school event. Each prize consists of a certificate of achievement and a cash award of $2000. Much gratitude is extended to the Universities Research Association, the Houston Area Research Center, the Continuous Electron Beam Accelerator, The Vacuum Products Division of Varian, Intermagnetics General Corporation, and the Westinghouse Electric Company for their support.

Table IV lists the prize committees, including the recently selected 1989 committee.

Table IV   Prize Committees

1985	1986	1987	1988	1989
B. Richter	R. L. Gluckstern	H. T. Edwards	R. Billinge	J. E. Leiss
A. E. Sessler	E. Picasso	E. A. Knapp	J. E. Leiss	W. K. H. Panofsky
M. Tigner	B. Richter	C. Pellegrini	C. Pellegrini	R. H. Siemann
	M. Sands	G.-A. Voss	J. Rees	S. Van der Meer

The prizes represent a high level of achievement, as can be seen from the following list of the winners and their citations.

1985   Presented at Stanford Linear Accelerator Center, Palo Alto, CA, July 25: Helen T. Edwards, for essential contributions in making the world's first superconducting synchrotron a reality;
John M. J. Madey, for the invention and demonstration of the free-electron laser.

1986   Presented at South Padre Island, TX, October 28:
Helmut Piel and Maury Tigner, for their contributions to making rf superconductivity a practical reality;
Thomas Weiland, for the development of novel methods for calculating electromagnetic fields in complex structures.

1987   Presented at Fermi National Accelerator Laboratory, Batavia, IL, August 13:
Klaus Halbach, for making high field permanent magnets practical tools for accelerator technology;
Lars Thorndahl, for essential theoretical and experimental contributions to the stochastic cooling of particle beams.

1988   Presented at Snowmass, CO, July 13:
I. M. Kapchinskii and V. A. Teplyakov, for the invention and early development of the radiofrequency quadrupole;
Andrew M. Sessler, for pioneering work on collective beam phenomena, in particular the negative mass and resistive wall instabilities.

The schools in 1987 and 1988 have developed a new format. Intensive courses over a two-week period, each comprising 45 lecture hours, are offered to students. By successfully completing one of these courses, a student may earn up to 3 credit hours from the sponsoring university. In the first two years, at the University of Chicago and Cornell University, eight courses were given, and five more are being offered for the 1989 school to be held at the University of California at Berkeley, June 19–30, 1989. Course outlines for these 13 courses are described below.

THE UNIVERSITY OF CHICAGO and THE U.S. PARTICLE ACCELERATOR SCHOOL 1987 COURSE DESCRIPTIONS

PHYSICS 700 HIGH ENERGY STORAGE RINGS—R. Talman, Cornell

Equations of motion. Beam Concepts. Betatron and synchrotron oscillations. Periodicity. Stability. Nonlinear behavior, including resonances and introduction to chaos. Collective Behavior. Statistical phenomena. Computer modeling. (1 credit = 3. 33 semester hours = 5. 0 quarter hours).

PHYSICS 701 RELATIVISTIC ELECTRONICS—V. L. Granatstein, U. of Maryland

Electromagnetic wave theory. Resonators. Reentrant cavities. Electron beam fundamentals. Space–charge waves. Klystrons. Gyrotrons. Free-electron lasers. Emphasis on understanding the principles of rf power sources. (1 credit).

PHYSICS 702 PARTICLE BEAMS: MAGNETIC OPTICS—K. L. Brown, SLAC

Fundamentals of charged-particle optics. Differential equations. Matrix methods. Description of first-order optics and optics modules. Higher-order optics. Geometric and chromatic abberrations. Optical symmetries. Design of optical systems. New techniques. (1/2 credit).

PHYSICS 703 PARTICLE BEAMS: ACCELERATION—P. B. Wilson, SLAC

Basic principles of acceleration of particle beams. Equations of motion. Phase stability. Introduction to beam loading and beam–cavity interactions. Wake-field formalism and phasor diagrams. Emphasis on developing a physical understanding of the interactions between particle beams and rf structures. (1/2 credit).

CORNELL UNIVERSITY and THE U.S. PARTICLE ACCELERATOR SCHOOL 1988 COURSE DESCRIPTIONS

PHYSICS 686 THEORY, SIMULATION, AND OBSERVATION OF BEAMS IN STORAGE RINGS—Robert H. Siemann, Cornell

The phenomenology of particles and beams in storage rings from an experimental viewpoint. Topics include single-particle motion and effects of magnetic field nonlinearities, coherent motion, including instabilities and wake-field calculations, and beam–beam interactions. (1 credit = 3 semester hours).

PHYSICS 687 PHYSICS OF PARTICLE ACCELERATORS—Donald A. Edwards, Fermilab

Topics in the design and performance of particle accelerators, with emphasis on the basic physics. Single-particle dynamics of linacs and synchrotrons, including nonlinear motion. Intensity dependence, including space–charge effects, coherent instabilities, and statistical phenomena. Examples of current accelerator research and development, chosen from large synchrotron (SSC/LHC) and linear collider activity. (1 credit).

PHYSICS 688 INTRODUCTION TO SYNCHROTRON RADIATION SOURCES—Claudio Pellegrini, BNL

The small beam emittance characteristic of synchrotron radiation sources dominates their design and performance. Description of the basic linear and nonlinear electron dynamics in a storage ring optimized for the generation of synchrotron radiation. Discussion of the basic properties of synchrotron radiation production from bending magnets and undulators. (1 credit).

PHYSICS 689 INTRODUCTION TO FREE-ELECTRON LASERS—Charles A. Brau, UC Santa Barbara/Los Alamos

The theory and experimental development of free-electron lasers, including time-dependent and three-dimensional effects. The basic physics of wigglers and undulators, laser optics, rf and induction linacs, storage rings, and other accelerators. (1 credit).

THE UNIVERSITY OF CALIFORNIA AT BERKELEY and THE U.S. PARTICLE ACCELERATOR SCHOOL 1989 COURSE DESCRIPTIONS

PHYSICS 401 THEORY AND DESIGN OF PARTICLE BEAMS—M. P. Reiser, U. of Maryland

The physical characteristics and theoretical descriptions of charged-particle beams in particle accelerators, microwave tubes, free-electron lasers, and other devices. Topics include review of classical mechanics for charged-particle motion in electromagnetic fields; beam optics without space charge, beam optics with space charge (paraxial theory); self-consistent models of beams with space charge; transport of intense, high-brightness beams in uniform and periodic focusing channels; nonlinear effects and emittance growth; charge neutralization in background gas. (1 course credit = 3 semester hours).

PHYSICS 402 INTRODUCTION TO ACCELERATOR PHYSICS—H. Wiedeman, Stanford

History and principles of charged-particle accelerators. Physics of linear accelerators, particle sources, synchrotrons, and storage rings. Beam dynamics in electromagnetic fields. Optics, emittance, and acceleration. Lifetime, instabilities, and other limitations. Injection and beam accumulation. Synchrotron radiation. Scaling laws for colliding beams and for beams for synchrotron radiation sources. Applications of accelerators for basic and applied research, technology, and medicine. (1 course credit).

PHYSICS 403 INTRODUCTION TO FREE-ELECTRON LASERS—W. B. Colson, Berkeley Research Associates

History of free-electron lasers (FELs). Characteristics of FEL configurations. Introductory topics include the undulator field, the electron trajectories, and the properties of the radiation. Electron phase-space dynamics. Classical bunching and coherent radiation. The self-consistent pendulum equation and the slowly varying wave equation. Single-mode analysis includes gain mechanisms, coherence development, and energy extraction. Alternative undulator designs like the FEL klystron and the tapered undulator. Advanced topics include coherence theory and optical self-guiding. Classical and quantum descriptions of the fundamental FEL mechanism are compared. Numerical methods. Selected experimental results are described and interpreted. (1 course credit).

PHYSICS 404 PRINCIPLES OF ACCELERATION—R. B. Palmer, BNL

The mechanisms of acceleration are derived from first principles. Real devices are discussed. Practical limitations are illustrated with examples. General theorem for the acceleration of charged particles in electromagnetic fields. Radiation pressure and quantum mechanical considerations. Ponderomotive force. Plasma wave acceleration. The inverse free-electron laser. The inverse Cerenkov effect. The grating accelerator. A two-dimensional linac. The Panofsky–Wenzel theorem. Alvarez and traveling-wave structures. (2/3 course credit).

PHYSICS 405 INTRODUCTION TO BEAM INSTABILITIES—A. W. Chao, Universities Research Association

Review of relativistic single-particle motion in synchrotrons. Electromagnetic fields due to the beam–cavity interaction and the resistive wall. Wakefields and impedance. Coasting-beam instabilities. Introduction to dispersion relations and Landau damping. Physical models of bunched-beam instabilities. Coherent modes and beam signals. The course includes an introduction to various mechanisms of beam instability and a brief survey of experimental observations. (2/3 course credit).

The field of particle accelerators or beam physics has made great strides in the last five decades and this progress will surely continue into the foreseeable future. The U.S. Particle Accelerator School was formed in large measure because of the recognition of the fact that beam physics is a field of science requiring an academic tradition but not having one. The school hopes to help this new science become established as an academic discipline. A new organization in the American Physical Society, the Particle Beam Physics Topical Group, has essentially the same goal and recently has prepared a White Paper which describes both the field and its relationship to the academic matters of education and publication. It seems appropriate to print this White Paper here and to wish this new group well in its efforts.

BEAM PHYSICS WHITE PAPER
Particle Beam Physics Topical Group, American Physical Society

Beam physics has progressed in the last 60 years from a sideline practiced by nuclear and particle physicists in building the instruments they used in their research studies to a highly specialized sub-discipline of science with a core body of knowledge and extensive archived publications. However, because of a variety of factors, this field has not developed an academic tradition as have other sub-fields of physics. This White Paper addresses one aspect of this problem, the status of beam physics within the American Physical Society.

What Is Beam Physics?

Beam physics is the study of particle and photon beams, their nature, their behavior, and their interactions, including the interaction of beams with matter, of beams with beams, and of particle beams with radiation. Evolving from concepts and ideas derived from classical mechanics, electromagnetism, statistical physics, and quantum physics, the study of beams is opening up a very rich field, with new effects being discovered and new types of beams with novel characteristics being realized. There are in fact a growing number of research areas in beam physics that need theoretical understanding and experimental verification and these represent challenging opportunities to present and future generations of scientists and engineers.

In the process of describing and understanding beams, there is a growing interdependence on other branches of physics. For example, nonlinear dynamical theories play an important part in beam physics and a number of advances in this field have been spurred by research on particle beams.

Physicists have created, in beams, a novel state of matter of remarkable scope. During the past few decades, the energy of particle beams has been extended from keV to MeV to GeV to TeV; beam currents have gone from microamps to milliamps to amps to mega amps; beam pulse durations have ranged from nanoseconds to continuous; beam lifetimes have been extended from microseconds to a week; species have multiplied from electron and proton beams to atomic, molecular, laser, and neutron beams, and then to pi-meson, K-meson, muon, neutrino, and antiproton beams; in addition beams are bunched, squeezed, expanded, modulated, and chopped.

The growth of beam physics is intimately tied to an expanding base of accelerator technology, including magnet systems, superconductivity, radiofrequency systems, particle sources, and others. The development of new types of particle beams is often tied to an extension of existing technology, for example, superconducting magnets for high energy accelerators, intense sources to produce very low emittance beams, and very high gradient accelerating fields to produce high energy beams in short distances. As a consequence, a substantial part of the R&D is directed toward advances in associated technologies.

The invention and continued development of particle accelerators and the associated technologies have had a profound impact on many sub-fields of pure and applied science and on our overall technological capabilities.

In the fields of high energy and nuclear physics, a marvelous complement of accelerator facilities exists around the world. In other areas of science, synchrotron light sources and accelerator-driven pulsed-neutron sources have opened up revolutionary new research opportunities in materials, chemistry, and biological research. Hardly a field of science is not beneficially affected. In industry and medicine there are literally thousands of accelerators in use in health care treatment, radiation sterilization, radiation processing, ion implantation, microchip production, etc.

Even with all these industrial applications we have only touched the surface. Research now going on has demonstrated that we can manufacture specialty materials, laying down controlled numbers of atom layers of different elements by ion implantation, and producing materials of totally new physical properties. Free-electron lasers offer the promise of high power, inexpensive, tunable photon sources which could revolutionize many areas of science.

In the area of defense, accelerators have in the past played an important but peripheral role. This situation is changing. The historical development of military ordnance has been characterized by the development of new methods to transfer energy to a target. At present, research is directed toward the use of particle beams and photon beams to take advantage of their very small transit times and potential ability to transmit appreciable energy density over large distances.

There is every reason to believe that the field of beam physics is still in its infancy. Many new developments have only begun to have general application in accelerators. They will make possible new instruments for a variety of applications in the future.

Problems

Despite the success and importance of beam physics to the practice of a variety of branches of science and modern technology, several important shortcomings of the field are apparent. The field has not developed a strong academic tradition, with relatively few academic physics departments giving courses in beam physics or granting degrees with work in this sub-discipline. The field has not developed a tradition of publication in refereed professional journals, perhaps out of convenience, because archived conference reports have recorded, perhaps inadequately, progress in the field. Adequate funding for generic research in fundamental beam physics has not been generated from the federal funding agencies, and little of the generic funding which has been forthcoming has gone to academically based work. Almost all funding has been associated with specific projects which typically undertake research only until solutions adequate to instrument construction are found. Finally, the practitioners of the field have not organized into a professional society subgroup, such as an APS division, to increase representation within the physics community.

What Can Be Done?

A small step which might be taken would be to petition the American Physical Society to consider forming a Division of Beam Physics to help its members solve some of the problems enumerated above. However, such an organization involves both benefits and responsibilities. Benefits of a Beam Division in the APS would include an expanded forum for discussion of sub-discipline problems with other members of the physics community, expanded access to journals for publication, and expanded opportunities for meetings and other professional gatherings. With these benefits go the responsibilities of disciplining our field to shift publication to the refereed journals, to organize consistently interesting and timely meetings, to participate in the affairs of the APS in all aspects, and to utilize that membership to try to professionalize, particularly in developing an academic tradition. It may be that such an organization can significantly help us to achieve these goals.

1987 Executive Committee

Chair	James E. Leiss	(4/88)
Vice Chair	Edward A. Knapp	(4/88)
Secy-Treas.	Melvin Month	(4/90)
Members	Richard Briggs	(4/88)
	Alexander W. Chao	(4/89)
	Denis Keefe	(4/89)
	Claudio Pellegrini	(4/89)
	John Peoples	(4/88)
	Robert L. Gluckstern	White Paper Committee Member
	Phillip A. Sprangle	White Paper Committee Member
	Richard Talman	White Paper Committee Member

The U.S. Particle Accelerator School is dedicated to this new science of beam physics. It is our hope that these continuing schools and the texts resulting from them will stimulate students, scientists, and engineers to enter this exciting field and become part of the bold venture to conceive and build new instruments of science and technology destined to push back the frontiers of knowledge.

Melvin Month, Director
U.S. Particle Accelerator School
January 30, 1989

ACKNOWLEDGMENTS

On behalf of the U.S. Particle Accelerator School, we are pleased to extend our thanks to the following people whose hard work enabled the participants to enjoy the intense schedule of courses in accelerator physics and who helped in the completion of these proceedings.

For 1987, at Fermilab, thanks go to Barbara Burwell, Peggy McAuliff, and the cafeteria staff; to Fred Ullrich, Al Johnson, and Jackie Stevens for duplicating; to Angela Gonzales for artwork; and to the personnel in building management, grounds, housing, custodial, and accounting.

For 1988, at Cornell, thanks go to Judith Eger, Diane Sheridan, Vallerie Sellers, and Kathy Kopsa of the Cornell Summer Sessions Office, and to the Housing staff for their help and cooperation.

In the preparation of this volume we received invaluable help from Mary Wigger and the staff of the BNL Technical Publishing Center, for which we express our thanks.

U.S. PARTICLE ACCELERATOR SCHOOL
1987/1988 Organization

Steering Committee
M. Blume (BNL)
H. T. Edwards (Fermilab)
H. A. Grunder (CEBAF)
B. McDaniel (Cornell)
B. Richter (SLAC), Chairman

Central Office
B. Chrisman (Fermilab), Advisor
M. Dienes (Consultant), Editor
H. T. Edwards (Fermilab), Advisor
A. Gonzales (Fermilab), Design and Art
M. Month (BNL), Director
M. Paul (Fermilab), Administrator
S. Winchester (Fermilab), Secretary

1987 Program Committee
A. W. Chao (URA/SSC)
G. Dugan (Fermilab)
D. A. Edwards (Fermilab)
G. Fischer (SLAC)
S. Humphries, Jr. (U. of New Mexico)
L. Michelotti (Fermilab)
M. Month (BNL), Chairman
C. Pellegrini (BNL)
M. P. Reiser (U. of Maryland)
A. G. Ruggiero (BNL)
R. Ruth (SLAC)
R. H. Siemann (Cornell)
T. P. Wangler (LANL)
W. T. Weng (BNL)

1988 Program Committee
C. A. Brau (UC Santa Barbara/LANL)
K. L. Brown (SLAC)
D. A. Edwards (Fermilab)
V. L. Granatstein (U. of Maryland)
M. Month (BNL), Chairman
C. Pellegrini (BNL)
R. H. Siemann (Cornell)
R. Talman (Cornell)
P. B. Wilson (SLAC)

1988 Local Administration (Cornell)
T. Allen
K. Berkelman, LNS Director
D. Cassel
J. Eger, Dir. of Spec. Programs
D. Hartill
D. Held, LNS Exec. Officer
B. McDaniel
N. Mistry
M. Norris, Exec. Secretary
M. Wright, LNS Admin. Manager

A COMPENDIUM OF COMPUTER CODES USED IN PARTICLE ACCELERATOR DESIGN AND ANALYSIS

by the
Los Alamos Accelerator Code Group

TABLE OF CONTENTS

1. Introduction 1138
2. Subject Index 1141
3. List of Codes 1146
4. Data Sheets on Codes 1147

By acceptance of this article, the publisher recognizes that the U.S. Government retains a nonexclusive, royalty-free license to publish or reproduce the published form of this contribution, or to allow others to do so, for U.S. Government purposes.

Work performed under the auspices of the U.S. Department of Energy.

A COMPENDIUM OF COMPUTER CODES USED IN PARTICLE ACCELERATOR DESIGN AND ANALYSIS

by the
Los Alamos Accelerator Code Group

1. INTRODUCTION

Support for this compilation has been provided by the Offices of High Energy and Nuclear Physics, U.S. Department of Energy. We are extremely grateful for their foresightedness.

In searching the accelerator literature, we have come across only two previous comprehensive surveys of useful accelerator codes. The first was a book by John Colonias, "Particle Accelerator Design: Computer Programs," Academic Press (1974). The second was a review article by Eberhard Keil, "Computer Programs in Accelerator Physics," in "Physics of High Energy Particle Accelerators," (SLAC Summer School, 1982) edited by Melvin Month, American Institute of Physics, AIP Conf. Proc. No.105 (1983). Colonias gives a comprehensive discussion of 35 codes. Keil mentions 21 codes. There are only 3 codes that are mentioned in both surveys. This is perhaps an indication of how rapidly codes become obsolete and new codes are written.

In addition to these surveys, there are four other sources worth mentioning:
1. "Computing in Accelerator Design and Operation, Proceedings of the 1983 Berlin Conference," Springer-Verlag, Berlin, 1984.
2. "Nonlinear Dynamics Aspects of Particle Accelerators, Proceedings of the 1985 Sardinia Conference," Springer-Verlag, Berling, 1986.
3. "Proceedings of the Workshop on Accelerator Orbit and Particle Tracking Programs," Brookhaven National Laboratory Report BNL-31761.
4. "Workshop on Orbital Dynamics and Applications to Accelerators," March 7-12, 1985 at Lawrence Berkeley Lab. published in Part. Accel. 19(1986)1-262.

In preparing this compilation, we came across the names of more than 150 programs that have been used in the design and analysis of accelerators. Many are obsolete and some are not easily transported from the institution where they were created. All are included in this compilation and filed at Los Alamos, but the obsolete codes have been removed from this article. Other than the judgement as to whether or not a particular code is obsolete, we have not made any critical evaluations.

© 1989 American Institute of Physics

Computer codes and code compilations share the common problem of obsolesence. This compilation will probably be almost useless in three years or less. Useful codes become widely distributed. Users make improvements in distributed code, and the original author loses control over the evolution of the code as variations proliferate. Many times authors tire of maintaining, documenting, and distributing their codes. Every generation of accelerator physicists produces code-builders; persons who feel that they can design more comprehensive or easier-to-use codes to do the tasks done by previous codes.

On the whole this environment of change is healthy, but when the number of codes and the variation on the same code become too large, there is a legitimate concern about duplicaton of effort and confusion in comparing the ouputs of different codes that supposedly do nearly the same calculations.

It might be useful to establish a mechanism for evaluating and making comparisons between codes. Such a mechanism would also be useful in maintaining and guiding the evolution of codes abandoned by the original authors and in setting standards for types of input, output, and documentation.

The design of the next generation of high-energy accelerators will probably be done as an international collaborative effort and it would make sense to establish, either formally or informally, an international center for accelerator codes with branches for maintenance, distribution, and consultation at strategically located accelerator centers around the world.

This arrangement could have at least three beneficial effects. It would cut down duplication of effort, provide long-term support for the best codes, and provide a stimulating atmosphere for the evolution of new codes. It does not take much foresight to see that the natural evolution of accelerator design codes is toward the development of so-called Expert Systems; systems capable of taking design specifications of future accelerators and producing specifications for optimized magnetic trasport and acceleration components, taking present-day programs such as TRANSPORT, POISSON, and SUPERFISH as tools in the optimization process. Such a program would also serve to codify the experience of two generations of accelerator designers before it is lost as these designers reach retirement age.

It is our hope that this compilation will stimulate some thought in this direction. This compilation was assembled by first sending a questionnaire to everyone that we could find who had written a code that might be useful in accelerator design and analysis. About one-third of the questionnaires were returned. We then set about searching the literature for descriptions of the remaining codes. We also telephoned authors when we could not get sufficient information from the literature. Certainly there are useful codes that we did not find. We solicit the readers of this document to write to us about our omissions as well as any errors in content. We are planning to update this document at least once.

This document is organized so that each code is described on a one- or two-page data sheet. The data sheets are arranged alphabetically by code name but are not numbered. In this way, it will be easy to insert new codes as they are discovered.

There are a number of simulation codes that have no names, and we have not taken the time to obtain detailed information on all of them. However, there are two fairly current codes that we thought worthwhile to include here. Therefore we have arbitrarily assigned them names. One of these is a CERN code by Myers, which we have called BEAMBEAM and the other is a DESY code by Piwinski, which we have called BMBMI.

The code data sheets are preceded by three indexes: 1. subject, 2. person-to-contact, and 3. code acronym. It was not useful to list codes by authors because, in many cases, the original author is no longer associated with the code and many other persons have contributed to maintaining and improving the code.

We would like to thank those who replied to our request for information. Special thanks goes to Roger Peng, who did most of the organization and typing, and to Gary Benson, who wrote the T_EX-formatting macro for the data sheets.

It is our sincere hope that this document will be helpful to persons entering the accelerator field. It has certainly been a revelation to us.

Before closing this introduction, I would like to bring up a very important point. This compilation is continuously updated by the Los Alamos Accelerator Code Group. If you know of computer codes for particle accelerator design and analysis which are not included in the compilation here, please inform us by sending the name of the code, a brief description, and an address for further contact. Your effort will help in creating a more comprehensive compilation and will be appreciated by the accelerator community. A copy of this compendium and futher updates are also available from us. Our address is:

> Los Alamos Accelerator Code Group
> AT-6, Mail Stop H-829
> Los Alamos National Laboratory
> Los Alamos, NM 87545
>
> Telephone: (505)667-2839 or
>   FTS 843-2839

2. SUBJECT INDEX

ANALYSIS-Impedances
    KN7C
    MAFIA
ANALYSIS-MISALIGNMENTS/ORBIT CORRECTIONS
    ALIGN
    CODINV
    MICADO
    PETROC
    PETROS
ANALYSIS-Space Charge Effects
    KOBRA
    ZFIELD
ANALYSIS-Spin Depolarization
    SLIM
ANALYSIS-Stability
    AZTEC
    BBI
    MARYLIE
    MAFIA
    PETROS
    SLIM
    SCHAR
    SYNCH
    TRANSVRS
    ZAP
ANALYSIS-Wakefield Effects
    BCI
    MAFIA
    SIMTRAC
    TBCI
    TRANSVRS
ANALYSIS-Other
    BIM2D (General magnetic field calculations in 2D)
    CARMEN (General magnetic field calculations in 3D)
    EBQ   (Particle Distributions)
    SLIM (Depolarization of Electron Beams)
    ZAP (Intrabeam Scatter, Gas Scatter, Touschek Effect)
COMPONENTS-Ion Sources/Electron Guns
    AXCEL-GSI
    BEAM
    CARMEN
    EBQ
    EGUN
    KOBRA
    MASK
    RAY
    SCHAR
    SNOW
    TOSCA
    WOLF

COMPONENTS-Magnets
    CARMEN
    DE2D
    DIFDIRA
    EDDYNET
    EFFI(3D)
    EMD
    FATIMA
    FORGY (See TRIM)
    GFUN3D
    LINDA
    MADEST
    MAFCO
    MAFCO-W
    MAGFOR
    MAGNET
    MAGNUS
    PANDIRA GROUP CODES
    PAR2DOPT
    PE2D
    POISCR
    POISSON GROUP CODES
    POISSON-BNL
    POISSON-LBL
    POISSON-TAC
    PROFI
    SATDSK
    TOSCA
    TRIDIF
    TRIM(ANL)
COMPONENTS-RF Cavities
    AZTEC
    BCI
    CAV3D
    CAVIT
    CURE
    DISPER
    DISPERSION
    HAX
    H2DB
    LACC
    LALA
    LALAGE
    LANS
    LILA
    LOOPER
    MAFIA
    MESSYMESH
    MULTIMODE
    OSCAR2D
    PISCES
    PRUD-M
    PRUD-O
    PRUD-OB
    SHRIMP
    SUPERFISH GROUP CODES
    TBCI
    TRANSVRS
    ULTRAFISH
    URMEL
    URMEL-T

```
COMPONENTS-Other
        JASON (Electrostatics)
        RELAX-3D (3D Electrostatics)
        RMKT (Klystron)
OPTIMIZATION-Cyclotrons
        BEAMTRACE
        COSY 5.0
        GIOS
        GOBLIN
        SATDSK
        SINAC
        TRAJECTORY
OPTIMIZATION-LINACS
        EBQ
        HOPI
        PARMILA
        PARMTEQ
        SCHAR
        TRACE
        TRACE3D
OPTIMIZATION-Spectrometers/Transport lines
        BEAMTRACE
        DIMAD
        GIOS
        HARMON
        MAPPOT
        MARYLIE
        MIRKO
        MOTER
        PARMILA
        PATH
        PATRICIA
        PINWHEEL
        RAY
        SYNCH
        TRAMP
        TRANCO
        TRANSOPTR
        TRANSPORT
        TRANSPORT LBL
OPTIMIZATION-Synchrotrons
        AGS
        BEAMTRACE
        COSY 5.0
        DIMAD
        GIOS
        HARMON
        LATTICE
        MAD
        MARYLIE
        MIRKO
        PAQUASEX
        RACETRACK
        RING
        SYNCH
        TEAPOT
        WIGWAM
```

OPTIMIZATION-Other
    BEAMTRACE
    COSY 5.0 (Optical Systems)
    COMFORT (Insertion lines and circular machines)
    DIMAD (Storage Rings)
    MARYLIE (Beam Lines and Storage Rings)
    OPTIC II (Electrostatic Accelerators)

SIMULATION/TRACKING-Colliding Beams
    (BEAMBEAM)
    (BMBMI)
    SYMP3

SIMULATION/TRACKING-Cyclotrons
    BEAMTRACE
    COSY 5.0
    GIOS
    NAJO
    PINWHEEL
    SINAC
    TRAJECTORY

SIMULATION/TRACKING-LINACS
    BEDLAM
    CCRTRACE
    DECAY-TURTLE
    EBQ
    GENMAP 3.0
    GIANT
    LTRACK
    MAFCO III
    MOTION
    PARMELA (Electron)
    PARMILA (Ion)
    PARMTEQ (RFQ)
    RAYTRACE
    RFQLIB
    SCHAR
    SCOP-2
    SCOP-RZ
    TRACE3D
    ZFIELD

SIMULATION/TRACKING-Synchrotrons
    ARCHSIM
    COSY 5.0
    DECAY-TURTLE
    DIMAD
    EVOL
    GENMAP 3.0
    GIOS
    LATTICE
    LIEPOT
    LILA
    LIMATRA
    MAFCO III
    MARYLIE
    MATRACE
    MIRKO
    PATPET
    PATRAC
    PATRICIA
    PATTV
    PETROC

            PETROS
            RACETRACK
            RING
            SCOP-2
            SCOP-RZ
            SIMTRAC
            SLIM
            SYNCH
            TEAPOT
SIMULATION/TRACKING-Spectrometers
            BEAMTRACE
            PINWHEEL
            TRACK
SIMULATION/TRACKING-Storage Rings
            DIMAD
            LATTICE
            MARYLIE
            SCOP-2
            SCOP-RZ
            SYNCH
SIMULATION/TRACKING-Transport and Beam Lines
            MARYLIE
            MOTION
            REVMOC
            SPEAM VI
            TRAMP
            TRANCO
            TRANSPORT
            TRANSPORT_LBL
            TRIO
            TURTLE
SIMULATION/TRACKING-Other
            COSY 5.0 (General Optics)
            GOC3D (General Magnetic Field)
            KOBRA (Space Charge Effects)
            MAFCO III (General Field Configurations)
            MISAR (Intense Beam Accumulator Rings)
            OPTIC II (Electrostatic Accelerators)
            SINAC (General Magnets)
            SOTRM (Generate Transport Matrices from Magnetic Field)
OTHER APPLICATIONS-
            COMFORT (Control Program)
            GIANT (Control Program)
            GO (Executive Program)
            GRAPHIC (Executive Program)
            HETC (Target and Shielding Design)
            HOPI (Control)
            ISIS (Modeling of Intense Charged Particle Beams)
            ITS (Charged Particle Transport Code)
            MARTUR (Radiation Loading Calculation)
            MEBT (Beam Diagnostics)
            SOTRM (Generate Transport Matrices from Magnetic Field)
            TRANCO (Control)
            WAVE (Laser Beat Wave Acceleration)
            WIGWAM (Electron Storage Ring and Wiggler Performance)

3. LIST OF CODES

AGS
ALIGN
ARCHSIM
AXCEL-GSI
AZTEC
BBI
BEAM
  (BEAMBEAM)
BEAMTRACE
BEDLAM
BIM2D
  (BMBMI)
CARMEN
CAV3D & CAVIT
CCRTRACE
CODINV
COMFORT
COSY 5.0
DECAY-TURTLE
DE2D
DIFDIRA
DIMAD
DISPERSION
EBQ
EDDYNET
EFFI(3D)
EGUN
EMD
EVOL
FATIMA
FORGY
GENMAP 3.0
GIANT
GIOS
GO
GOBLIN
GOC3D
HARMON
HAX
H2DB
HETC
HOPI
ISIS
ITS
JASON
KN7C

KOBRA
LACC
LALAGE
LANS
LATTICE
LIEPOT
LIMATRA
LINDA
LOOPER
LTRACK
MAD
MADEST
MAFCO
MAFCO III
MAFCO-W
MAFIA
MAGFOR
MAGNET
MAGNUS
MAPPOT
MARTUR
MARYLIE
MASK
MATRACE
MEBT
MICADO
MIRKO
MISAR
MOTER
MOTION
MULTIMODE
NAJO
OPTIC II
OSCAR2D
PANDIRA
PAQUASEX
PAR2DOPT
PARMELA
PARMILA
PARMTEQ
PATH
PATPET
PATRAC
PATRICIA
PATTV
PETROC

PETROS
PE2D
PINWHEEL
PISCES
POISCR
POISSON GROUP
  CODES
POISSON-BNL
POISSON-LBL
POISSON-TAC
PROFI
PRUD-M
PRUD-O
PRUD-OB
RACETRACK
RAY
RAYTRACE
RELAX3D
REVMOC
RFQLIB
RING
RMKT
SATDSK
SCHAR
SCOP-2
SCOP-RZ
SHRIMP
SIMTRAC
SINAC
SLIM
SNOW
SOTRM
SPEAM VI
SUPERFISH GROUP
  CODES
SYMP3
SYNCH
TBCI
TEAPOT
TOSCA
TRACE
TRACE3D
TRACK
TRAJECTORY
TRAMP
TRANCO

STRANSOPTR
TRANSPORT
TRANSPORT-LBL
TRANSVRS
TRIDIF
TRIM (ANL) &
  FORGY
TRIO
TURTLE
ULTRAFISH
URMEL
URMEL-T
WAVE
WIGWAM
WOLF
ZAP
ZFIELD

Date of Latest Version: unknown   Program Name: AGS

Person to Contact: E. Keil
          Address: LEP Division
                   CERN
                   1211 Geneva 23
                   Suisse, Switzerland

Telephone Number:

Classification of Computer Code:
    Component Design:
        ☐ Ion Source, ☐ Magnet, ☐ RF-cavity, ☐
    Accelerator Optimization:
        ☐ Linac, ☐ Cyclotron, ☒ Synchrotron, ☐
    Tracking or Simulation:
        ☐ Linac, ☐ Cyclotron, ☐ Synchrotron, ☐
    Analysis:
        ☐ Stability, ☐ Impedances, ☐
    Other:

Short Description: (Purpose, capabilities, algorithms, special features, etc.)

   Program AGS computes the transformation matrices of the elements that make up the synchrotron, computes the betatron and closed orbit functions, the coordinates of the equilibrium orbit, and other pertinent quantities. A typical run is completed in less than one minute. Memory requirements depend on the number of elements that the program can handle. For maximum efficiency the program is overlaid and more than 1500 elements can be accommodated in less than $50 K_8$ memory locations. Program AGS is, in many respects, similar to program SYNCH.

Publications describing the code:

   E. Keil, Y. Marti, B. W. Montague and A. Sudboe, "AGS - The ISR Computer Program for Synchrotron Design, Orbit Analysis and Function Matching" CERN Internal Report CERN 75-13(1975).

Is code documentation available? ☐ Yes ☒ No

How may the code be obtained?
   No longer supported by the authors.

Source language: FORTRAN

Computers it runs on: CDC 6600

It is available as: ☐ Source code, ☐ Executable only

Source Media: ☐ Listing, ☐ Tape, ☐ Diskette, ☐ Cards, ☐ Networks
   Tape format:
   Diskette size & format:

Available through: ☐ DECNET, ☐ ARPANET, ☐ BITNET
                   ☐

Network Address: keil@cernvm

Date of Latest Version: unknown

Program Name: ALIGN

Person to Contact: Elon R. Close MS-50B/2239
    Address: Lawrence Berkeley Laboratory
        1 Cyclotron Rd.
        Berkeley, CA 94720
        USA

Telephone Number: (415) 486-6166, FTS 451-6166

Classification of Computer Code:
    Component Design:
        ☐ Ion Source, ☐ Magnet, ☐ RF–cavity, ☐
    Accelerator Optimization:
        ☐ Linac, ☐ Cyclotron, ☐ Synchrotron, ☐
    Tracking or Simulation:
        ☐ Linac, ☐ Cyclotron, ☐ Synchrotron, ☐
    Analysis:
        ☐ Stability, ☐ Impedances, ☒ Closed Orbit Correction.
    Other:

Short Description: (Purpose, capabilities, algorithms, special features, etc.)
    This code simulates survey misalignments of magnetic elements in a circular accelerator. It constructs a closed orbit corresponding to the misalignment errors and finds corrector strengths needed to sustain particle beam within the vacuum chamber. This is a Monte Carlo type program that performs a collection of misalignments over an ensemble of machines. ALIGN was developed and used for the construction of PEP, has been exported to CERN where it was used in the design of LEP.

Publications describing the code:
    E. Close, et al, "Proposed Orbit and Vertical Dispersion Correction System for PEP," Stanford Linear Accelerator internal report PEP Note 271 and CONF-7903271-151 (1979).

Is code documentation available? ☐ Yes ☒ No

How may the code be obtained?
    Call Elon R. Close.

Source language: FORTRAN

Computers it runs on: CDC

It is available as: ☐ Source code, ☐ Executable only
Source Media: ☐ Listing, ☐ Tape, ☐ Diskette, ☐ Cards, ☐ Networks
    Tape format:
    Diskette size & format:
Available through: ☐ DECNET, ☐ ARPANET, ☐ BITNET
        ☐
Network Address:

Date of Latest Version: July 1986

Program Name: ARCHSIM

Person to Contact: Henry A. Thiessen
Address: MP-14, MS H847
Los Alamos National Lab.,
Los Alamos, NM 87545
USA

Telephone Number: (505)667-8991, FTS 843-8991

Classification of Computer Code:
Component Design:
☐ Ion Source, ☐ Magnet, ☐ RF-cavity, ☐
Accelerator Optimization:
☐ Linac, ☐ Cyclotron, ☐ Synchrotron, ☐
Tracking or Simulation:
☐ Linac, ☐ Cyclotron, ☒ Synchrotron, ☐
Analysis:
☐ Stability, ☐ Impedances, ☐
Other:

Short Description: (Purpose, capabilities, algorithms, special features, etc.)

ARCHSIM simulates the acceleration cycle of a rapid-cycling proton synchrotron. A lattice can consist of up to 100 cells and rf cavities. Transport of the beam in six dimensions includes all second-order optical terms. The rf field and proton velocity are treated exactly. Longitudinal space charge is handled in a self-consistent manner. The fluctuations due to the finite number of particles are handled by a Gaussian smoothing algorithm. The program runs on a VAX 11/780 and can track 100 particles without space charge through the full acceleration cycle from 0.8 to 32 GeV in 49 minutes ($\sim$ 5000 turns). A thousand particles with space charge takes about ten hours of computer time.

The motivation for writing this tracking program was the need to explore the effect of various rf accelerator-cavity parameters on the beam dynamics and stability of a proposed 32-GeV rapid-cycling synchrotron.

Currently the program is under revision to improve the graphics (phase space plots and particle distribution histograms), and the number and type of optical elements available. The program uses second order transport matrices generated by another program like DIMAT. These matrices are checked to see if they are symplectic before doing the simulation.

Publications describing the code:

Henry A. Thiessen and John L. Warren, "ARCHSIM: A Proton Synchrotron Tracking Program Including Longitudinal Space Charge," Computing in Accelerator Design and Operation, Proc., Berlin 1983, Springer-Verlag Berlin (1984) 225-60.

Is code documentation available? ☐ Yes ☒ No

How may the code be obtained?
Contact H. A. Thiessen.

Source language: FORTRAN

Computers it runs on: VAX 11/780

It is available as: ☒ Source code, ☐ Executable only

Source Media: ☐ Listing, ☒ Tape, ☐ Diskette, ☐ Cards, ☒ Networks
   Tape format: 9 Track 1600 bpi
   Diskette size & format:

Available through: ☐ DECNET, ☒ ARPANET, ☒ BITNET
   ☐

Network Address: hks@lanl.arpa

Date of Latest Version: 1986

Program Name: AXCEL-GSI

Person to Contact: P. Spädtke
Address: GSI-Darmstadt
Postfach 11 05 41
6100-Darmstadt
Fed. Rep. Germany

Telephone Number: 6151-359-323

Classification of Computer Code:
Component Design:
[x] Ion Source, ☐ Magnet, ☐ RF–cavity, [x] Electron Gun/DC-Beam Transport

Accelerator Optimization:
☐ Linac, ☐ Cyclotron, ☐ Synchrotron, ☐
Tracking or Simulation:
☐ Linac, ☐ Cyclotron, ☐ Synchrotron, ☐
Analysis:
☐ Stability, ☐ Impedances, ☐
Other:

Short Description: (Purpose, capabilities, algorithms, special features, etc.)

2D-code, cylinder symmetry, plasma boundary, electrons and ions, cathode simulation/plasma simulation, including symmetric magnetic fields(!). Interactive code. Diagnostic: trajectories, emittances, transverse energies ...

Publications describing the code:

Peter Spädtke, "Computer Simulation of High-Current DC-Ion Beams," Proc. 1984 Linac Accelerator Conf. Seeheim May 7-11, 1984, GSI, Darmstadt internal report GSI-84-11.

J. Klabunde, P. Spädtke, and A. Schönlein, "High Current Beam Transport Experiments At GSI", IEEE Trans. NS-32 (1985) 2462.

Is code documentation available? [x] Yes ☐ No

How may the code be obtained?
Contact Peter Spädtke.

Source language: Fortran 77

Computers it runs on: IBM/VAX

It is available as: [x] Source code, ☐ Executable only

Source Media: ☐ Listing, ☐ Tape, ☐ Diskette, ☐ Cards, ☐ Networks
Tape format:
Diskette size & format:

Available through: ☐ DECNET, ☐ ARPANET, ☐ BITNET
☐

Network Address: ul13%DDAGSI3.bitnet

Date of Latest Version: 1972

Program Name: AZTEC

Person to Contact: S. J. Sackett, MS L-122
Address: Applied Mechanics Group
Lawrence Livermore Laboratory
Box 808
Livermore, CA 94550
USA
Telephone Number: (415) 422-8709, FTS 532-8709

Classification of Computer Code:
Component Design:
☐ Ion Source, ☐ Magnet, ☒ RF-cavity, ☐
Accelerator Optimization:
☐ Linac, ☐ Cyclotron, ☐ Synchrotron, ☐
Tracking or Simulation:
☐ Linac, ☐ Cyclotron, ☐ Synchrotron, ☐
Analysis:
☒ Stability, ☐ Impedances, ☐
Other:

Short Description: (Purpose, capabilities, algorithms, special features, etc.)
AZTEC calculates the fields due to an azimuthally bunched beam in cylindrically symmetric structures of arbitrary geometry. The computed fields are then used to calculate the self-impedance of the beam for stability studies. Any combination of dielectric, magnetic, and conducting materials is allowed. Material properties, however, are assumed to be isotropic and linear. Boundaries between different materials within the problem space may be arbitrary curves, as may the contour defining the beam region.

Publications describing the code:
S. T. Sackett and A. A. Garren, "AZTEC — A Code for Calculating the Impedance of an Azimuthally Bunched Beam in a Cylindrically Symmetric Structure," Lawrence Berkeley Laboratory Internal Report LBL-774(1972).

John S. Colonias, "Particle Accelerator Design: Computer Programs," Academic Press, New York (1974) 281.

Is code documentation available? ☐ Yes ☒ No

How may the code be obtained?
No longer supported by the authors.

Source language: FORTRAN & Compass.

Computers it runs on: CDC 7600

It is available as: ☐ Source code, ☐ Executable only
Source Media: ☐ Listing, ☐ Tape, ☐ Diskette, ☐ Cards, ☐ Networks
Tape format:
Diskette size & format:
Available through: ☐ DECNET, ☐ ARPANET, ☐ BITNET
☐
Network Address:

Date of Latest Version: June 1984   Program Name: BBI

Person to Contact: Mme. Monica Gygi/B. Zotter
Address: CERN
LEP Theory Div.
1211 Geneva 23
Switzerland

Telephone Number: 83 2951/83 6637

Classification of Computer Code:
    Component Design:
       ☐ Ion Source, ☐ Magnet, ☐ RF–cavity, ☐
    Accelerator Optimization:
       ☐ Linac, ☐ Cyclotron, ☐ Synchrotron, ☐
    Tracking or Simulation:
       ☐ Linac, ☐ Cyclotron, ☐ Synchrotron, ☐
    Analysis:
       ☒ Stability, ☐ Impedances, ☐
    Other:

Short Description: (Purpose, capabilities, algorithms, special features, etc.)
    Calculation of bunched beam instabilities:

       Longitudinal/transverse coupled bunch modes.

       Bunch lengthening, transverse mode-coupling.

       Tune-Shifts, intra beam scattering etc.

Publications describing the code:
    Internal CERN Report, LEP Theory Note 83-2.

    Albert Hofmann, Kurt Hübner and Bruno Zotter, "A Computer Code For The Calculation Of Beam Stability In Circular Electron Machines," IEEE Trans. NS-26 (1979) 3514

Is code documentation available? ☒ Yes ☐ No

How may the code be obtained?

Source language: FORTRAN

Computers it runs on: CDC

It is available as: ☒ Source code, ☐ Executable only

Source Media: ☐ Listing, ☒ Tape, ☐ Diskette, ☐ Cards, ☐ Networks
Tape format:
Diskette size & format:

Available through: ☐ DECNET, ☐ ARPANET, ☒ BITNET
                     ☐

Network Address:

Date of Latest Version: Jan. 1985   Program Name: BEAM

Person to Contact: Murray Shubaly
    Address: Group AT-1, MS H817
             Los Alamos National Laboratory
             Los Alamos, NM 87545
             USA

Telephone Number: (505) 667-9124, FTS 843-9124

Classification of Computer Code:
    Component Design:
        ☒ Ion Source, ☐ Magnet, ☐ RF–cavity, ☐
    Accelerator Optimization:
        ☐ Linac, ☐ Cyclotron, ☐ Synchrotron, ☐
    Tracking or Simulation:
        ☐ Linac, ☐ Cyclotron, ☐ Synchrotron, ☐
    Analysis:
        ☐ Stability, ☐ Impedances, ☐
    Other:

Short Description: (Purpose, capabilities, algorithms, special features, etc.)

BEAM (Beam Extraction and Acceleration Modeling) is a second generation 2D ion source code based on AXCEL with some features from SNOW. It has the following capabilities: Starting from the unperturbed plasma, the code calculates ion trajectories and electrostatic potentials, the electrode boundaries need not coincide with mesh lines, the mesh density is variable to permit finer resolution in critical regions. Both extraction and injection calculations are possible, with variable current density in the source plasma, variable ion injection energy and angle, and finite ion temperature effects. Space charge neutralization is included. The output of the code gives values of rms emittance, divergence and radius, maximum divergence and radius, and overlaid equipotential and trajectory plots. Both rectangular and cylindrical geometry are treated. It does not at present handle axial magnetic fields, but this improvement is planned in the near future. Only partially complete documentation is available, but the coding is well commented.

Publications describing the code:
    M. R. Shubaly, R. A. Judd and R. W. Hanam, "BEAM, An Improved Beam Extraction and Acceleration Modeling Code," IEEE Trans. NS-28 (1981) 2655.

Is code documentation available? ☒ Yes ☐ No

How may the code be obtained?
    Call Murray Shubaly or The Los Alamos Accelerator Code Group (505) 667-6677 or 667-2839.

Source language: FORTRAN IV

Computers it runs on: CDC 7600, CYBER 175, VAX, IBM.

It is available as: ☒ Source code, ☐ Executable only

Source Media: ☐ Listing, ☒ Tape, ☐ Diskette, ☐ Cards, ☒ Networks
    Tape format: 1600 bpi, 9 track.
    Diskette size & format:

Available through: ☐ DECNET, ☒ ARPANET, ☒ BITNET
                   ☐
Network Address: hks@lanl.arpa

Date of Latest Version: unknown   Program Name: (BEAMBEAM)

Person to Contact: Steve Myers
          Address: CERN
                   LEP-Divsion
                   1211 Geneva 23
                   Switzerland

Telephone Number:

Classification of Computer Code:
    Component Design:
        ☐ Ion Source, ☐ Magnet, ☐ RF–cavity, ☐
    Accelerator Optimization:
        ☐ Linac, ☐ Cyclotron, ☐ Synchrotron, ☐
    Tracking or Simulation:
        ☐ Linac, ☐ Cyclotron, ☐ Synchrotron, ☒   Colliding Beams.
    Analysis:
        ☐ Stability, ☐ Impedances, ☐
    Other:

Short Description: (Purpose, capabilities, algorithms, special features, etc.)

A multi-particle two-beam (strong-strong) simulation program has been written for investigation of the beam-beam effect in 'LEP'. The motion of the superparticles is treated in six-dimensional phase space and the effects of quantum excitation and radiation damping are included. The effects of perturbations to the superperiodicity (errors) are also included. Non-zero dispersion at the RF cavities allows computation of single beam synchro-betatron resonances.

After each revolution the parameters influencing the beam-beam force (e.g. the beam dimensions and the beam current) are reevaluated in order to simulate a real situation. For the beam-beam force an elliptical beam with Gaussian charge distribution has been assumed. The computation of this force is speeded up by using tabulated values of the complex error function and a fast interpolation procedure.

The initial distribution of a large number of particles (typically 200) in the three phase planes are random with pre-specified variances. Each particle in each beam is 'tracked' through (i) an RF cavity, (ii) a beam-beam interaction and (iii) a traversal of a machine arc. This procedure is repeated until each beam has completed one run. The position of each particle is then compared with aperture limitations (typically 10 σ) and those particles which fall outside are excluded from further tracking. The remaining particles are then used to recalculate the beam current, the specific luminosity, the beam variances and hence the new beam-beam kick parameters. This cycle is repeated until the 'beam' has been circulating for about 1.5 damping times.

Publications describing the code:

S. Myers, "Beam-Beam Simulation for LEP," IEEE Trans NS-28 (1981) 2503. (See also IEEE Trans NS-30 (1983) 2466, and Nucl. Insts. Meth. 211 (1983) 263.)

Is code documentation available? ☐ Yes ☐ No

How may the code be obtained?
    unknown

Date of Latest Version: 1985

Program Name: BEAMTRACE

Person to Contact: M. Berz or H. Wollnik
Address: II. Physikal Institut
Heinrich Buffring 14–16
6300 Giessen,
Fed. Rep. Germany

Telephone Number: 641-702-2770

Classification of Computer Code:
Component Design:
☐ Ion Source, ☐ Magnet, ☐ RF–cavity, ☐
Accelerator Optimization:
☐ Linac, ☒ Cyclotron, ☒ Synchrotron, ☒ magnetic optical system

Tracking or Simulation:
☐ Linac, ☒ Cyclotron, ☐ Synchrotron, ☒ mass spectrometers

Analysis:
☐ Stability, ☐ Impedances, ☐
Other:

Short Description: (Purpose, capabilities, algorithms, special features, etc.)
PIC-code, usual beam elements, space charge, transport beam by 2nd order matrices.

Publications describing the code:
H. Wollnik, J. Brezina and M. Berz, "GIOS-Beam Trace, a program for the design of high resolution mass spectrometer", 2nd Intl. Conf. on Charg. Particle Optics, Albuquerque, May 19–23, 1986. (To be published in Nucl. Inst. Meth.).

GSI-report THD 26, Darmstadt (1984).

Is code documentation available? ☒ Yes ☐ No

How may the code be obtained?

Source language: FORTRAN

Computers it runs on: VAX, Cyber

It is available as: ☒ Source code, ☐ Executable only
Source Media: ☐ Listing, ☒ Tape, ☒ Diskette, ☐ Cards, ☒ Networks
Tape format: as desired
Diskette size & format: as desired
Available through: ☐ DECNET, ☐ ARPANET, ☒ BITNET
☐
Network Address: ug21%ddagsi3@bitnet

Date of Latest Version: Aug. 1985   Program Name: BEDLAM

Person to Contact: Walter P. Lysenko
         Address: Group AT-6, MS H829
                  Los Alamos National Laboratory
                  Los Alamos, NM 87545

Telephone Number: (505) 667-7431, FTS 843-7431

Classification of Computer Code:
    Component Design:
        ☐ Ion Source, ☐ Magnet, ☐ RF–cavity, ☐
    Accelerator Optimization:
        ☐ Linac, ☐ Cyclotron, ☐ Synchrotron, ☐
    Tracking or Simulation:
        ☒ Linac, ☐ Cyclotron, ☐ Synchrotron, ☐
    Analysis:
        ☐ Stability, ☐ Impedances, ☐
    Other:

Short Description: (Purpose, capabilities, algorithms, special features, etc.)

   BEDLAM is a fourth-order moment simulation code. The beam at the input to a linear accelerator is specified as a collection of moments of the phase-space distribution. Then the moment equations, which describe the time evolution of the moments, are numerically integrated. No particles are traced in this approach. The computed distribution and the external forces are computed consistently to a given order of accuracy. Although BEDLAM includes moments to fourth order only, it could be systematically extended to any order. Another feature of this method is that physically interesting and intuitive quantities, such as beam sizes and rms emittances, are computed directly. This code is still under development to include space charge effects

Publications describing the code:

   P. J. Channell, L. M. Healy and W.P. Lysenko, "The Moment Code BEDLAM," IEEE Trans. NS-32 (1985) 2565.

Is code documentation available? ☐ Yes ☒ No

How may the code be obtained?

   Los Alamos Accelerator Code Group, (505) 667-6677 or 667-2839.

Source language: FORTRAN

Computers it runs on: CRAY, SUN

It is available as: ☒ Source code, ☐ Executable only

Source Media: ☐ Listing, ☒ Tape, ☐ Diskette, ☐ Cards, ☒ Networks
    Tape format: whatever
    Diskette size & format:

Available through: ☐ DECNET, ☒ ARPANET, ☒ BITNET
                   ☐

Network Address: hks@lanl.arpa

Date of Latest Version: Dec. 1984

Program Name: BIM2D

Person to Contact: John S. Whitney
Address: Vector Fields, Ltd.
Osney Mead
Oxford OX2 OEE
England

Telephone Number: 0865 248236

Classification of Computer Code:
Component Design:
☐ Ion Source, ☒ Magnet, ☐ RF–cavity, ☐
Accelerator Optimization:
☐ Linac, ☐ Cyclotron, ☐ Synchrotron, ☐
Tracking or Simulation:
☐ Linac, ☐ Cyclotron, ☐ Synchrotron, ☐
Analysis:
☐ Stability, ☐ Impedances, ☒   General purpose magnetic field calculation in 2D

Other:

Short Description: (Purpose, capabilities, algorithms, special features, etc.)

BIM2D is an interactive computer program for solving linear magnetostatic problems using boundary integral methods.

The computer model assumes that the magnet can be represented by a two-dimensional (x-y) cross-section of iron and conductor regions. The permeability in each iron region and the current density in each conductor region are specified constants which can be changed between repeated analyses. The limitation to fixed permeability allows an integral formulation where only the edges of the iron cross-section need to be meshed; the analysis is performed very quickly on-line so that the system is truly interactive.

Input to the program is via a graphics terminal and is completely interactive. As the magnet data are input, the model is displayed on the terminal screen. Any errors can be seen and corrected immediately.

Magnetic fields can be calculated at points, along a line or over a grid, displayed as contour maps or graphs and consequent forces can be obtained. Listings, graphs and plots are available as output.

Publications describing the code:
Rutherford Report RL-79-088(1979)

Data Sheet Ref: 028631

Is code documentation available? ☒ Yes ☐ No

How may the code be obtained?
By license agreement with Vector Fields Ltd

Source language: FORTRAN 77

Computers it runs on: PRIME, VAX

It is available as: ☒ Source code, ☐ Executable only

Source Media: ☐ Listing, ☒ Tape, ☐ Diskette, ☐ Cards, ☐ Networks
  Tape format: As required
  Diskette size & format:

Available through: ☐ DECNET, ☐ ARPANET, ☐ BITNET
  ☐

Network Address:

Date of Latest Version: unknown

Program Name: (BMBMI)

Person to Contact: A. Piwinski
    Address: DESY
                Notkestrasse 85
                2000 Hamburg 52
                Fed. Rep. Germany

Telephone Number:

Classification of Computer Code:
    Component Design:
        ☐ Ion Source, ☐ Magnet, ☐ RF-cavity, ☐
    Accelerator Optimization:
        ☐ Linac, ☐ Cyclotron, ☐ Synchrotron, ☐
    Tracking or Simulation:
        ☐ Linac, ☐ Cyclotron, ☐ Synchrotron, ☒ Colliding Beams
    Analysis:
        ☐ Stability, ☐ Impedances, ☐
    Other:

Short Description: (Purpose, capabilities, algorithms, special features, etc.)

The beam-beam interaction was simulated on a digital computer taking into account the horizontal and vertical betatron oscillation, the synchrotron oscillation, a horizontal dispersion at the interaction point, quantum fluctuation, damping and the exponential decay of the voltage at the separators. The space charge forces were derived from the exact potential of a time-independent gaussian bunch, were calculated for 10000 points and then quadratically interpolated for each passage. The simulation was done for 225 particles starting with a gaussian distribution. The beam height and the beam width were calculated as the root mean square of the coordinates of the particles at the interaction point during 20 revolutions. The collision of two strong bunches was also simulated.

Publications describing the code:

A. Piwinski, "Computer Simulation of the Beam-Beam Interaction," 11th Int'l Conf. on High Energy Accelerators, Geneva, July 7-11, 1980, Berkhaüser Verlag, Basel.

A. Piwinski, "Computer Simulation of the Beam-Beam Interaction at a Crossing Angle," IEEE Trans. NS-32 (1985) 2240.

Is code documentation available? ☐ Yes ☐ No

How may the code be obtained?
    unknown

Source language:
Computers it runs on:
It is available as: ☐ Source code, ☐ Executable only
Source Media: ☐ Listing, ☐ Tape, ☐ Diskette, ☐ Cards, ☐ Networks
    Tape format:
    Diskette size & format:
Available through: ☐ DECNET, ☐ ARPANET, ☐ BITNET ☐

Network Address:

Date of Latest Version: Version 1.0, 1986

Program Name: CARMEN

Person to Contact: John S. Whitney
Address: Vector Fields, Ltd.
Osney Mead
Oxford OX2 OEE
England

Telephone Number: 0865 248236

Classification of Computer Code:
Component Design:
[x] Ion Source, [x] Magnet, ☐ RF–cavity, ☐
Accelerator Optimization:
☐ Linac, ☐ Cyclotron, ☐ Synchrotron, ☐
Tracking or Simulation:
☐ Linac, ☐ Cyclotron, ☐ Synchrotron, ☐
Analysis:
☐ Stability, ☐ Impedances, [x] General purpose magetic field, trajectory and eddy current in 3D.

Other:

Short Description: (Purpose, capabilities, algorithms, special features, etc.)

CARMEN is an advanced program for the calculation of eddy currents in three dimensions. The algorithm used in CARMEN gives nearly optimal economy for magnetic field computation and it is implemented using state-of-the-art numerical methods. In addition to magnetic fields the program can also be used to model any system governed by Poisson's equation. This includes electrostatics and current flow.

Applications include fusion magnets, particle accelerators, MRI gradient field eddy currents, nondestructive testing, electrical machines, eddy current heating, electron lenses and deflection magnets.

CARMEN uses a discrete finite element model in order to solve the partial differential equations governing the behavior of a system.

The mesh is formed from hexahedra with 'ruled' faces which are automatically subdivided into elements. A 2D grid is created initially and this can then be swept through space thus creating 3D volumes. The sweep operations include translation, rotation and projection.

The mesh primitive blocks are assigned material names and geometric properties, for example orientation.

CARMEN uses 8 and 20 node isoparametric 'brick' elements. These can be mixed together; the program will enforce inter-element continuity. The type of element created in each primitive may be selected by the user. This allows the higher-order elements to be used where solution accuracy is important. Two result-evaluation modes are provided to give a choice between speed and accuracy.

The suite of programs was designed to be used in a distributed computing environment. Data files created for CARMEN can be easily transferred between computers and result files from CARMEN can be returned. CARMEN provides full check point, drop file and restart facilities to maximize the efficient use of computer resources. The PCARMEN program allows results to be displayed graphically and further calculations can be performed, e.g. particle trajectories.

Publications describing the code:
> CRI Emson, J. Simkin and CW Trowbridge, "Further Developments in Three Dimensional Eddy Current Analysis", IEEE Trans, MAG-21(1985) 2231.

> Data Sheet Ref: 028671. From Vector Fields.

Is code documentation available? ☒ Yes ☐ No

How may the code be obtained?
> By licence agreement with Vector Fields, Ltd.

Source language: FORTRAN 77

Computers it runs on: PRIME, VAX IBM

It is available as: ☒ Source code, ☐ Executable only

Source Media: ☐ Listing, ☒ Tape, ☐ Diskette, ☐ Cards, ☐ Networks
    Tape format: As required
    Diskette size & format:

Available through: ☐ DECNET, ☐ ARPANET, ☐ BITNET
☐

Network Address:

Date of Latest Version: Apr. 1985   Program Name: CAV3D, CAVIT

Person to Contact: Dr. Wolfgang Wilhelm
    Address: Physik Department E12
             Technische Universität München
             Fed. Rep. Germany

Telephone Number: 089/3209 2435

Classification of Computer Code:
    Component Design:
        ☐ Ion Source, ☐ Magnet, ☒ RF–cavity, ☐
    Accelerator Optimization:
        ☐ Linac, ☐ Cyclotron, ☐ Synchrotron, ☐
    Tracking or Simulation:
        ☐ Linac, ☐ Cyclotron, ☐ Synchrotron, ☐
    Analysis:
        ☐ Stability, ☐ Impedances, ☐
    Other:

Short Description: (Purpose, capabilities, algorithms, special features, etc.)
    CAVIT is a 2D-code for cavities with constant cross-section, the calculation is performed in a square mesh; accuracy is better than 1 percent.

    CAV3D calculates low frequency modes of any cavity with a 3D cubic mesh, and accuracy of about 5 percent.

Publications describing the code:
    W. Wilhelm, Particle Accelerators 12 (1984) 139

Is code documentation available? ☒ Yes ☐ No

How may the code be obtained?
    From the author

Source language: FORTRAN

Computers it runs on: Cyber, DEC-10

It is available as: ☒ Source code, ☐ Executable only

Source Media: ☐ Listing, ☒ Tape, ☐ Diskette, ☐ Cards, ☐ Networks
    Tape format:
    Diskette size & format:

Available through: ☐ DECNET, ☐ ARPANET, ☐ BITNET
                   ☐

Network Address:

Date of Latest Version: Dec. 1985

Program Name: CCRTRACE

Person to Contact: Roger Cole MS-810
Address: MS-H810, Group MP-1
Los Alamos National Laboratory
Los Alamos, NM 87545
USA

Telephone Number: (505) 667-7193, FTS 843-7193

Classification of Computer Code:
Component Design:
☐ Ion Source, ☐ Magnet, ☐ RF–cavity, ☐
Accelerator Optimization:
☐ Linac, ☐ Cyclotron, ☐ Synchrotron, ☐
Tracking or Simulation:
☒ Linac, ☐ Cyclotron, ☐ Synchrotron, ☐
Analysis:
☐ Stability, ☐ Impedances, ☐
Other:

Short Description: (Purpose, capabilities, algorithms, special features, etc.)
CCRTRACE is a first-order envelope tracing code, currently for transverse effects only. The "guts" of CCRTRACE is a subroutine library with powerful, simple interface. The operator interface runs on color CRT in CCR or on VT640. It was designed for use on the Los Alamos Meson Physics Facility (LAMPF) accelerator.

Publications describing the code:
MP-1-3563-2 (A Los Alamos National Laboratory internal report).

Is code documentation available? ☒ Yes ☐ No

How may the code be obtained?
Not readily available.

Source language: FLECS

Computers it runs on: VAX/VMS

It is available as: ☐ Source code, ☐ Executable only

Source Media: ☐ Listing, ☐ Tape, ☐ Diskette, ☐ Cards, ☐ Networks
Tape format:
Diskette size & format:

Available through: ☐ DECNET, ☐ ARPANET, ☐ BITNET
☐

Network Address:

Date of Latest Version: July 1985

Program Name: CODINV

Person to Contact: John L. Warren
Address: Group AT-6, MS H-829
Los Alamos National Lab.
Los Alamos, NM 87545
USA

Telephone Number: (505) 667-6677, FTS 843-6677

Classification of Computer Code:
Component Design:
☐ Ion Source, ☐ Magnet, ☐ RF–cavity, ☐
Accelerator Optimization:
☐ Linac, ☐ Cyclotron, ☐ Synchrotron, ☐
Tracking or Simulation:
☐ Linac, ☐ Cyclotron, ☐ Synchrotron, ☐
Analysis:
☐ Stability, ☐ Impedances, ☒ Closed Orbit Distortion Corrections

Other:

Short Description: (Purpose, capabilities, algorithms, special features, etc.)

A certain class of magnet misalignments in storage rings and other accelerators produces closed-orbit distortions (CODs). Quite often the CODs are measured at a fewer number of locations (N) than the number of misalignment parameters (M). There is a linear relation between COD measurements, u(j), j=1, ..., N and the misalignment parameters e(k), k=1, ..., M. Hence the e(k)'s are underdetermined. If M < 2N, one can obtain an overdetermined set of equations by measuring the COD at two different quadrupole settings. There are several ways of inverting the COD measurements to get misalignment parameters that are fairly insensitive to errors in the measured CODs. A computer program called CODINV has been written to test some of these schemes. Two schemes give fairly good results when applied to the lattice of the Los Alamos Proton Storage Ring (PSR). The first scheme requires measurements at two nearby tunes and the use of singular-value decomposition methods. The second scheme requires measurements of the CODs in the FODO and DOFO cell arrangements but is easier mathematically.

Publications describing the code:

J. L. Warren and P. J. Channell, "New Method for Inverting the Closed-Orbit-Distortion Problem," Part. Accel. Conf. Santa Fe, IEEE Trans NS-30(1983) 2415.

John L. Warren, "Determination of Magnet Misalignments from Measurement of Closed Orbit Distortion," Los Alamos National Laboratory Internal Report AT-6: ATN-83-13 (1983).

Is code documentation available? ☒ Yes ☐ No

How may the code be obtained?
Contact Barbara Blind, AT-3, MS H808, Los Alamos National Laboratory (505) 667-9130.

Source language: FORTRAN IV

Computers it runs on: VAX

It is available as: ☒ Source code, ☐ Executable only

Source Media: ☐ Listing, ☒ Tape, ☐ Diskette, ☐ Cards, ☒ Networks
　　Tape format: 9 track, 1600 bpi
　　Diskette size & format:

Available through: ☐ DECNET, ☒ ARPANET, ☒ BITNET
　　　　　　　　　　☐

Network Address: hks@lanl.arpa

Date of Latest Version: unknown

Program Name: COMFORT

Person to Contact: Hamid Shoaee
Address: SLAC Bin 26
P.O. Box 4349
Stanford, CA 94305

Telephone Number: (415) 854-3300 x 2954, FTS 461-9300 x 2954

Classification of Computer Code:
Component Design:
☐ Ion Source, ☐ Magnet, ☐ RF–cavity, ☐
Accelerator Optimization:
☐ Linac, ☐ Cyclotron, ☐ Synchrotron, ☒ Insertion lines and circular machines.

Tracking or Simulation:
☐ Linac, ☐ Cyclotron, ☐ Synchrotron, ☐
Analysis:
☐ Stability, ☐ Impedances, ☐
Other: control

Short Description: (Purpose, capabilities, algorithms, special features, etc.)
A lattice program for analysis and control of charged particle beam transport systems. Main features include:

1) MAD style input format

2) Linear lattice parameter

3) Linear lattice matching

4) Chromaticity correction

5) Beam and RF parameter calculations for rings.

Publications describing the code:
M. D. Woodley, M. J. Lee, J. Jäger, A. S. King, "Control of Machine Functions or Transport Systems," IEEE Trans NS-30 (1983) 2367.

Is code documentation available? ☒ Yes ☐ No

How may the code be obtained?
Call Hamid Shoaee

Source language: FORTRAN 77

Computers it runs on: VAX (vms), IBM 3081 (vm)

It is available as: ☐ Source code, ☐ Executable only
Source Media: ☐ Listing, ☐ Tape, ☐ Diskette, ☐ Cards, ☐ Networks
Tape format:
Diskette size & format:
Available through: ☐ DECNET, ☐ ARPANET, ☐ BITNET
☐
Network Address:

Date of Latest Version: 1986   Program Name: COSY 5.0

Person to Contact: Martin Berz or H. Wollnik
Address: II Physikal Institut
Heinrich Buffring 14–16
6300 Giessen,
Fed. Rep. Germany

Telephone Number: 641-702-2770

Classification of Computer Code:
Component Design:
☐ Ion Source, ☐ Magnet, ☐ RF-cavity, ☐
Accelerator Optimization:
☐ Linac, ☒ Cyclotron, ☒ Synchrotron, ☒   Optical Systems

Tracking or Simulation:
☐ Linac, ☒ Cyclotron, ☒ Synchrotron, ☒   optics

Analysis:
☐ Stability, ☐ Impedances, ☐
Other:

Short Description: (Purpose, capabilities, algorithms, special features, etc.)
Fifth order matrix transport and method of power series tracking, usual beam line elements, arbitrary field distributions and fringe fields, fitting capabilities, space charge (under development), very general input language.

Publications describing the code:
M. Berz, H.C. Hofmann and H. Wollnik, "COSY 5.0", Proceedings of Second Charged Particle Optics Conference, Alburquerque, 1986. To be published in "Nuclear Instrument and Methods".

Is code documentation available? ☒ Yes ☐ No

How may the code be obtained?
Contact above address

Source language: FORTRAN

Computers it runs on: VAX, Cyber, CRAY

It is available as: ☒ Source code, ☐ Executable only

Source Media: ☐ Listing, ☒ Tape, ☒ Diskette, ☐ Cards, ☒ Networks
Tape format: as desired
Diskette size & format: as desired

Available through: ☐ DECNET, ☐ ARPANET, ☒ BITNET
☐

Network Address: ug21%ddagsi3@bitnet

Date of Latest Version: 1974   Program Name: DECAY TURTLE

Person to Contact: Program Library, or C. Iselin
Address: CERN
DD Div., LEP Theory Group
CH-1211 Geneva 23
Switzerland

Telephone Number: (22) 83 23 77 or (22) 83 36 57

Classification of Computer Code:
Component Design:
☐ Ion Source, ☐ Magnet, ☐ RF–cavity, ☐
Accelerator Optimization:
☐ Linac, ☐ Cyclotron, ☐ Synchrotron, ☐
Tracking or Simulation:
☒ Linac, ☐ Cyclotron, ☒ Synchrotron, ☒   Beam Transport

Analysis:
☐ Stability, ☐ Impedances, ☐
Other:

Short Description: (Purpose, capabilities, algorithms, special features, etc.)
Tracking of particles with first- and second-order matrix formalism. Optional Decay included.

Publications describing the code:
CERN 74-02

Is code documentation available? ☒ Yes ☐ No

How may the code be obtained?
Contact C. Iselin. (You can also contact Dave Carey at Fermilab (312) 840–3639)

Source language: FORTRAN 66

Computers it runs on: IBM/CDC

It is available as: ☒ Source code, ☐ Executable only

Source Media: ☐ Listing, ☒ Tape, ☐ Diskette, ☐ Cards, ☐ Networks
Tape format: 9 track 1600 bpi
Diskette size & format:

Available through: ☐ DECNET, ☐ ARPANET, ☐ BITNET
☐

Network Address: FCI @ CERN VM

Date of Latest Version: 1986

Program Name: DE2D

Person to Contact: Fan Mingwu
 Address: Institute of Atomic Energy
 P. O. Box 275
 Beijing
 P. R. of China

Telephone Number: Beijing 868221 ext.341

Classification of Computer Code:
 Component Design:
  ☐ Ion Source, ☒ Magnet, ☐ RF-cavity, ☐
 Accelerator Optimization:
  ☐ Linac, ☐ Cyclotron, ☐ Synchrotron, ☐
 Tracking or Simulation:
  ☐ Linac, ☐ Cyclotron, ☐ Synchrotron, ☐
 Analysis:
  ☐ Stability, ☐ Impedances, ☐
 Other:

Short Description: (Purpose, capabilities, algorithms, special features, etc.)
 Calculate static magnetic field, electric field and eddy current problems in two cartesian dimensions or cylindrically symmetric configuration with permeable iron. It is based on Finite Element Methods and consists of three codes: MESH2D, DE2D, and DE2DD.

Publications describing the code:
 Fan Mingwu, Maio Yixin and Yan Weili, "DE2D - Interactive Software Package for 2D Magnetostatic, Electrostatic and Eddy Current Field Computations," IEEE Trans. Mag-21(1985)2539.

Is code documentation available? ☒ Yes ☐ No

How may the code be obtained?
 Contact Institute of Atomic Energy at above address

Source language: FORTRAN 77

Computers it runs on: VAX, PDP, IBM-PC/XT

It is available as: ☒ Source code, ☐ Executable only
Source Media: ☒ Listing, ☒ Tape, ☒ Diskette, ☐ Cards, ☐ Networks
 Tape format:
 Diskette size & format:
Available through: ☐ DECNET, ☐ ARPANET, ☐ BITNET
 ☐
Network Address:

Date of Latest Version: 1973

Program Name: DIFDIRA

Person to Contact: Herbert F. Vogel
Address: MS B220, Group X-2
Los Alamos National Laboratory
Los Alamos, NM 87545
USA

Telephone Number: (505) 667-8949, FTS 843-8949

Classification of Computer Code:
Component Design:
☐ Ion Source, ☒ Magnet, ☐ RF–cavity, ☐
Accelerator Optimization:
☐ Linac, ☐ Cyclotron, ☐ Synchrotron, ☐
Tracking or Simulation:
☐ Linac, ☐ Cyclotron, ☐ Synchrotron, ☐
Analysis:
☐ Stability, ☐ Impedances, ☐
Other:

Short Description: (Purpose, capabilities, algorithms, special features, etc.)

2-d eddy currents and their magnetic field, driven from a current source with arbitrary pulse shape. Implementation by modification of the Poisson code, i.e., current normal to the (B_x, B_y) or (B_r, B_z) plane.

Publications describing the code:
None

Is code documentation available? ☐ Yes ☒ No

How may the code be obtained?

Source language: FORTRAN

Computers it runs on: CDC 7600

It is available as: ☒ Source code, ☐ Executable only

Source Media: ☐ Listing, ☒ Tape, ☐ Diskette, ☐ Cards, ☐ Networks
Tape format:
Diskette size & format:

Available through: ☐ DECNET, ☐ ARPANET, ☐ BITNET ☐

Network Address:

Date of Latest Version: Jan. 1986

Program Name: DIMAD

Person to Contact: R. Servranckx
Address: SLAC
P.O. Box 4349
Stanford, CA 94025
USA

Telephone Number: (415) 854-3300 ext. 2741 or (306) 966-6054

Classification of Computer Code:
Component Design:
☐ Ion Source, ☐ Magnet, ☐ RF–cavity, ☐
Accelerator Optimization:
☐ Linac, ☐ Cyclotron, ☒ Synchrotron, ☒ Rings, Beam Lines

Tracking or Simulation:
☐ Linac, ☐ Cyclotron, ☒ Synchrotron, ☒ Rings, Beam Lines

Analysis:
☐ Stability, ☐ Impedances, ☐
Other:

Short Description: (Purpose, capabilities, algorithms, special features, etc.)

1) General design fittings

2) Basis: Transport, 2nd order formalism

3) Closed orbit studies: stable or unstable special resonance extraction studies.

4) Extensive error and misalignment handling capabilities.

Publications describing the code:
SLAC Report 285 UC-28(A)

Is code documentation available? ☒ Yes ☐ No

How may the code be obtained?
Contact Roger Servranckx

Source language: FORTRAN

Computers it runs on: VAX, IBM, CDC

It is available as: ☒ Source code, ☐ Executable only

Source Media: ☐ Listing, ☒ Tape, ☐ Diskette, ☐ Cards, ☒ Networks
Tape format: 9T 1600 BPI
Diskette size & format:

Available through: ☐ DECNET, ☐ ARPANET, ☒ BITNET
☐

Network Address: RVS @ SLACVM

Date of Latest Version: 1981   Program Name: DISPER

Person to Contact: S. O. Schriber
           Address: AT Division, MS H811,
                    Los Alamos National Laboratory
                    Los Alamos, NM 87545
                    USA

Telephone Number: (505) 667-7634, FTS 843-7634

Classification of Computer Code:
    Component Design:
        ☐ Ion Source, ☐ Magnet, ☒ RF–cavity, ☐
    Accelerator Optimization:
        ☐ Linac, ☐ Cyclotron, ☐ Synchrotron, ☐
    Tracking or Simulation:
        ☐ Linac, ☐ Cyclotron, ☐ Synchrotron, ☐
    Analysis:
        ☐ Stability, ☐ Impedances, ☐
    Other:

Short Description: (Purpose, capabilities, algorithms, special features, etc.)

   Program DISPER does a weighted, non-linear, least squares fit to experimentally measured frequencies
   of mode spectra from arrays of rf cavities. The fit can be for singly or doubly periodic systems with
   up to second neighbor coupling constants and various end cavity terminations. The fit is to a model
   consisting of coupled RLC circuits.

Publications describing the code:

   S. O. Schriber, "Fitting of an Ordered Set of Mode Frequencies," Atomic Energy of Canada Limited,
   Report No. AECL–3669 (1970).

Is code documentation available? ☒ Yes ☐ No

How may the code be obtained?

   From the author/see above; H. Euteneur, Inst. für Kernphysik, Universität Mainz, Postfach 3980, 6500
   Mainz, BRD; S. Inagaki, KEK – National Laboratory for High Energy Physics, Oho-Machi, Tskuba-
   Gun, Ibaraki-Ken, JAPAN 300-32

Source language: FORTRAN

Computers it runs on: CDC, CYBER

It is available as: ☒ Source code, ☐ Executable only

Source Media: ☐ Listing, ☒ Tape, ☐ Diskette, ☐ Cards, ☐ Networks
    Tape format:
    Diskette size & format:

Available through: ☐ DECNET, ☐ ARPANET, ☐ BITNET
                   ☐

Network Address:

Date of Latest Version: July 1985

Program Name: DISPERSION

Person to Contact: Eric N. Opp
    Address: MRJ Inc., Suite 200
        10455 White Granite Dr.
        Oakton, VA 22124
        USA

Telephone Number: (202) 385-0818

Classification of Computer Code:
    Component Design:
        ☐ Ion Source, ☐ Magnet, ☒ RF–cavity, ☐
    Accelerator Optimization:
        ☐ Linac, ☐ Cyclotron, ☐ Synchrotron, ☐
    Tracking or Simulation:
        ☐ Linac, ☐ Cyclotron, ☐ Synchrotron, ☐
    Analysis:
        ☐ Stability, ☐ Impedances, ☐
    Other:

Short Description: (Purpose, capabilities, algorithms, special features, etc.)

   DISPERSION calculates the frequency dispersion relation $\omega(k)$ for the azimuthally symmetric modes in a periodic array of rf cavities. The code is a modification of the SUPERFISH package. The codes LATTICE (mesh generator) and SUPERFISH were modified. LATTICE was modified to set up the appropriate periodic boundary conditions for the basic structure cell.

Publications describing the code:

   R. L. Gluckstern and E. N. Opp, "Calculation of Dispersion Curves in Periodic Structures," IEEE Trans. MAG-21 (1985) 2344.

   E. N. Opp, "Calculation of Dispersion Curves in Periodic Structures Using SUPERFISH," Los Alamos National Lab. Internal Report AT-6: ATN-85–4, (1985).

Is code documentation available? ☒ Yes ☐ No

How may the code be obtained?
   Contact the Los Alamos Code Group, MS H829, Los Alamos National Laboratory, Los Alamos, NM 87545, phone (505) 667–6677 or 667–2839.

Source language: FORTRAN

Computers it runs on: CRAY, VAX

It is available as: ☒ Source code, ☐ Executable only

Source Media: ☐ Listing, ☒ Tape, ☐ Diskette, ☐ Cards, ☒ Networks
    Tape format: 9 Track, 1600 bpi
    Diskette size & format:

Available through: ☐ DECNET, ☒ ARPANET, ☒ BITNET
                   ☐

Network Address: hks@lanl.arpa

Date of Latest Version: Nov. 1982   Program Name: EBQ

Person to Contact: Arthur C. Paul L-626
         Address: Lawrence Livermore National Laboratory
                  Livermore, CA 94550
                  USA

Telephone Number: (415) 423-3183, FTS 543-3183

Classification of Computer Code:
   Component Design:
      [x] Ion Source, ☐ Magnet, ☐ RF-cavity, [x]   Ion and Electron Guns
   Accelerator Optimization:
      [x] Linac, ☐ Cyclotron, ☐ Synchrotron, ☐
   Tracking or Simulation:
      [x] Linac, ☐ Cyclotron, ☐ Synchrotron, ☐
   Analysis:
      ☐ Stability, ☐ Impedances, [x]   Particle Distribution
   Other:

Short Description: (Purpose, capabilities, algorithms, special features, etc.)

   The EBQ (electric field E, magnetic fields B, and space charge Q) code simulates steady state problems involving space charge transport of charged particles in cylindrically symmetric devices, providing a fairly flexible and forgiving data input structure.

   This two-dimensional program accepts data specifying the externally applied electric and magnetic fields. The electric and magnetic self-fields of the particles are used to obtain self-consistent azimuthally symmetric charge and current distributions. The code follows particle trajectories and employs a unique method of assigning values of the charge density to grid points. This method provides sufficient accuracy to model the cancellation that occurs between radial electric and magnetic self-forces in a relativistic beam.

   The orbits are treated in Cartesian geometry (position and momentum) with z as the independent variable. Poisson's equation is solved in cylindrical geometry on an orthogonal rectangular mesh.

   EBQ can also handle problems involving multiple ion species where the space charge forces from each must be included. Such problems arise in the design of ion sources where different charge and mass states are present.

Publications describing the code:
   Lawrence Berkeley Laboratory Internal Report LBL-13241

Is code documentation available? [x] Yes ☐ No

How may the code be obtained?
   Contact Arthur Paul.

Source language: FORTRAN IV

Computers it runs on: CDC 7600

It is available as: ☒ Source code, ☐ Executable only

Source Media: ☐ Listing, ☒ Tape, ☐ Diskette, ☐ Cards, ☐ Networks
  Tape format: 7 track BCD,
  Diskette size & format:

Available through: ☐ DECNET, ☐ ARPANET, ☐ BITNET
  ☐

Network Address:

Date of Latest Version: 1986   Program Name: EDDYNET

Person to Contact: Larry Turner
            Address: Argonne National Laboratory
                     9700 S. Cass Ave.
                     Argonne, IL 60439-4814
                     U.S.A.

Telephone Number: (312)972-6257, FTS 972-6257

Classification of Computer Code:
    Component Design:
        ☐ Ion Source, ☒ Magnet, ☐ RF–cavity, ☐
    Accelerator Optimization:
        ☐ Linac, ☐ Cyclotron, ☐ Synchrotron, ☐
    Tracking or Simulation:
        ☐ Linac, ☐ Cyclotron, ☐ Synchrotron, ☐
    Analysis:
        ☐ Stability, ☐ Impedances, ☒   Eddy Currents

    Other:

Short Description: (Purpose, capabilities, algorithms, special features, etc.)
    EDDYNET uses a wire grid approach to solve eddy current problems. The conducting surface is approximated by a quadrilateral mesh of conducting lines. Line resistances and loop inductances are defined in a manner consistent with the approximation. The system of loop equations, with a dense matrix, is solved repeatedly to give the time development of the eddy currents, magnetic field, and dissipated power.

Publications describing the code:
    L. R. Turner and R. J. Lari, "Developments in Eddy Current Computations with EDDYNET," IEEE Trans. MAG-19(1983)2577-80.

    L. R. Turner, "Eddy Current Analysis of the ZT-P Shells with EDDYNET," Computational Electronics, Elsevier Science Publishers (1986) pp.181-89.

Is code documentation available? ☐ Yes ☒ No

How may the code be obtained?
    Contact Larry Turner

Source language: FORTRAN

Computers it runs on: IBM, CRAY

It is available as: ☐ Source code, ☐ Executable only

Source Media: ☐ Listing, ☐ Tape, ☐ Diskette, ☐ Cards, ☐ Networks
    Tape format:
    Diskette size & format:

Available through: ☐ DECNET, ☐ ARPANET, ☐ BITNET
                    ☐

Network Address:

Date of Latest Version: unknown

Program Name: EFFI(3D)

Person to Contact: S. J. Sackett
  Address: L-122
    Lawrence Livermore National Laboratory
    Livermore, CA 94550
    USA

Telephone Number: (415) 422-8709, FTS 532-8709

Classification of Computer Code:
  Component Design:
    ☐ Ion Source, ☒ Magnet, ☐ RF–cavity, ☐
  Accelerator Optimization:
    ☐ Linac, ☐ Cyclotron, ☐ Synchrotron, ☐
  Tracking or Simulation:
    ☐ Linac, ☐ Cyclotron, ☐ Synchrotron, ☐
  Analysis:
    ☐ Stability, ☐ Impedances, ☐
  Other:

Short Description: (Purpose, capabilities, algorithms, special features, etc.)
  EFFI calculates the electromagnetic field and vector potential in coil systems of arbitrary geometry. The coils are made from circular arc and/or straight segments of rectangular cross-sectional conductor. EFFI can also calculate magnetic flux lines, magnetic force, and inductance. The methods used for the calculations are based on a combination analytical and numerical integration of the Biot-Savart law for a volume distribution of current. These methods yield accurate field values inside and outside the conductor.

Publications describing the code:
  Steven J. Sackett, "EFFI—A Code for Calculating the Electromagnetic Field, Force and Inductance in Coil Systems of Arbitrary Geometry," Lawrence Livermore Laboratory, Internal Report UCRL-52402, (1978).

Is code documentation available? ☐ Yes ☐ No

How may the code be obtained?
  unknown

Source language: unknown

Computers it runs on: CDC

It is available as: ☐ Source code, ☐ Executable only
Source Media: ☐ Listing, ☐ Tape, ☐ Diskette, ☐ Cards, ☐ Networks
  Tape format:
  Diskette size & format:
Available through: ☐ DECNET, ☐ ARPANET, ☐ BITNET
  ☐
Network Address:

Date of Latest Version: May, 1986

Program Name: EGUN

Person to Contact: W. B. Herrmannsfeldt
Address: SLAC
SLAC Bin 26
Stanford University
Stanford, CA 94305

Telephone Number: (415) 854-3300, FTS 461-9300, ext 3342

Classification of Computer Code:
Component Design:
[x] Ion Source, ☐ Magnet, ☐ RF-cavity, [x]   Gun Design

Accelerator Optimization:
☐ Linac, ☐ Cyclotron, ☐ Synchrotron, ☐
Tracking or Simulation:
☐ Linac, ☐ Cyclotron, ☐ Synchrotron, ☐
Analysis:
☐ Stability, ☐ Impedances, ☐
Other:

Short Description: (Purpose, capabilities, algorithms, special features, etc.)

The program is specifically written to compute trajectories of charged particles in electrostatic and magnetostatic focusing systems including the effects of space charge and self-magnetic fields. Starting options include Child's Law conditions on cathodes of various shapes. Either rectangular or cylindrically symmetric geometry may be used. Magnetic fields may be specified using arbitrary configurations of coils, or the output of a magnet program such as Poisson or by an externally calculated array of the axial fields.

Publications describing the code:
SLAC-226.

Is code documentation available? [x] Yes ☐ No

How may the code be obtained?
Write or call W. B. Herrmannsfeldt.

Source language: FORTRAN

Computers it runs on:
It is available as: [x] Source code, ☐ Executable only
Source Media: ☐ Listing, [x] Tape, ☐ Diskette, ☐ Cards, [x] Networks
Tape format:
Diskette size & format:
Available through: ☐ DECNET, ☐ ARPANET, [x] BITNET
☐
Network Address: SLACVM@WBHAP

Date of Latest Version: Oct. 1985 Program Name: EMD (Expert Magnet Design)

Person to Contact: Ann Aldridge
    Address: Group C-3, MS B265
             Los Alamos National Laboratory
             Los Alamos, NM 87545
             USA

Telephone Number: (505) 667-7047, FTS 843-7047

Classification of Computer Code:
    Component Design:
        ☐ Ion Source, ☒ Magnet, ☐ RF–cavity, ☐
    Accelerator Optimization:
        ☐ Linac, ☐ Cyclotron, ☐ Synchrotron, ☐
    Tracking or Simulation:
        ☐ Linac, ☐ Cyclotron, ☐ Synchrotron, ☐
    Analysis:
        ☐ Stability, ☐ Impedances, ☐
    Other:

Short Description: (Purpose, capabilities, algorithms, special features, etc.)
    An AI (Expert) program for designing "H" type dipole bending magnets. Given particle, energy, bend angle, etc., it designs the coil using LAMPF Standard conductor, determines ΔP, ΔT, V, I, power and prints out cross section coordinates in Poisson format for advanced field quality design.

Publications describing the code:
    Internal notes; For technical info contact Ed Bush, MP-8 LANL, (505) 667-5968

Is code documentation available? ☒ Yes ☐ No

How may the code be obtained?
    See Ann Aldridge

Source language: LISP

Computers it runs on: VAX or VMS with UNIX

It is available as: ☒ Source code, ☐ Executable only
Source Media: ☐ Listing, ☒ Tape, ☐ Diskette, ☐ Cards, ☐ Networks
    Tape format: as desired
    Diskette size & format:
Available through: ☐ DECNET, ☒ ARPANET, ☐ BITNET
                   ☐
Network Address:

Date of Latest Version: May 1986

Program Name: EVOL

Person to Contact: Steve Peggs MS 90/4040
Address: URA Design Center
c/o UCLBL
Berkeley, CA 94720
USA

Telephone Number: (415) 486–4772, 486–6559, FTS 451–4772

Classification of Computer Code:
Component Design:
☐ Ion Source, ☐ Magnet, ☐ RF–cavity, ☐
Accelerator Optimization:
☐ Linac, ☐ Cyclotron, ☐ Synchrotron, ☐
Tracking or Simulation:
☐ Linac, ☐ Cyclotron, ☒ Synchrotron, ☐
Analysis:
☐ Stability, ☐ Impedances, ☐
Other: Beam-Beam Interaction.

Short Description: (Purpose, capabilities, algorithms, special features, etc.)

EVOL is a tracking program that includes sextupoles, multiple beam-beam collisions, external tune modulation, and other effects. Its construction emphasizes operational speed and the nested scanning of two or three configuration variables, such as betatron tune, amplitude and chromaticity, at the expense of simplification in the physical model.

EVOL was originally written at CERN to simulate nonlinearities in the SPS collider. It is now being used and developed further, in round and flat beam versions, at the Cornell Electron Storage Ring, CESR. Single particles are tracked for many turns, for example 10^5, around a nonlinear lattice in the presence of a set of physical effects chosen by the user from a "library". These effects interact with each other strongly or weakly, in ways that are theoretically understood to a greater or lesser degree.

Publications describing the code:
S. Peggs "Hadron Collider Behavior in the Nonlinear Numerical Model EVOL," Part. Acc. 17 (1985) 11.

Is code documentation available? ☒ Yes ☐ No

How may the code be obtained?
Contact Steve Peggs.

Source language: FORTRAN 77

Computers it runs on: VAX-VMS, IBM

It is available as: ☒ Source code, ☐ Executable only

Source Media: ☐ Listing, ☒ Tape, ☐ Diskette, ☐ Cards, ☒ Networks
Tape format: whatever
Diskette size & format:

Available through: ☒ DECNET, ☒ ARPANET, ☒ BITNET
☐

Network Address: sgp@lbl

Date of Latest Version: 1970

Program Name: FATIMA

Person to Contact: C. Iselin
Address: LEP Theory Group
CERN
CH-1211 Geneva 23
Switzerland

Telephone Number: (22) 83 36 57

Classification of Computer Code:
Component Design:
☐ Ion Source, ☒ Magnet, ☐ RF–cavity, ☐
Accelerator Optimization:
☐ Linac, ☐ Cyclotron, ☐ Synchrotron, ☐
Tracking or Simulation:
☐ Linac, ☐ Cyclotron, ☐ Synchrotron, ☐
Analysis:
☐ Stability, ☐ Impedances, ☐
Other:

Short Description: (Purpose, capabilities, algorithms, special features, etc.)
2nd order Finite Element

Publications describing the code:
None

Is code documentation available? ☐ Yes ☐ No

How may the code be obtained?
Obsolete. Code not obtainable.

Source language:
Computers it runs on:
It is available as: ☐ Source code, ☐ Executable only
Source Media: ☐ Listing, ☐ Tape, ☐ Diskette, ☐ Cards, ☐ Networks
Tape format:
Diskette size & format:

Available through: ☐ DECNET, ☐ ARPANET, ☐ BITNET
☐

Network Address: fci@cernvm

Date of Latest Version: Dec. 1986   Program Name: GENMAP 3.0

Person to Contact: Alex J. Dragt
Address: Dept. of Physics
University of Maryland
College Park, MD 20742
USA

Telephone Number: (301)454-7324

Classification of Computer Code:
Component Design:
☐ Ion Source, ☐ Magnet, ☐ RF–cavity, ☐
Accelerator Optimization:
☐ Linac, ☐ Cyclotron, ☐ Synchrotron, ☐
Tracking or Simulation:
☒ Linac, ☐ Cyclotron, ☒ Synchrotron, ☐
Analysis:
☐ Stability, ☐ Impedances, ☐
Other:

Short Description: (Purpose, capabilities, algorithms, special features, etc.)

GENMAP 3.0 is a program to compute numerically transfer maps (through third order) for arbitrary beamline elements using Lie algebraic methods. The code uses canonical variables (x, p_x, y, p_y, t, p_t). To study a particular beam element or transport system, the user must modify a subroutine to specify the Hamiltonian as a power series around the design trajectory. This involves specifying the electromagnetic fields in question (and various derivatives) by analytic approximations or interpolation tables. The output of GENMAP is the Lie algebraic transfer map for the system under study. GENMAP is generallly used to compute transfer maps for beamline elements with fringe fields. These maps can be utilized by MARYLIE 3.0 for tracking and analysis of transport systems with realistic fringe fields. GENMAP can also be incorporated as a subroutine in MARYLIE 3.0 for optimization of systems with fringe fields.

Publications describing the code:

R. Ryne, "Numerical Computation of Transfer Maps using Lie Algebraic Methods," Los Alamos National Lab. internal report AT-6:ATN-86-21 (June 1986).

R. Ryne, "Numerical Computation of the Transfer Map for a Magnetic Dipole with Mid-Plane Symmetry using Lie Algebraic Methods," Los Alamos National Lab. internal report AT-6:ATN-86-25 (August 1986).

A. Dragt et al, "MARYLIE 3.0, A Program for Charge Particle Beam Transport Based on Lie Algebraic Methods," 1985 User's Guide; unpublished.

A. Dragt and E. Forest, J. Math. Phys. 24(1984)2734.

Is code documentation available? ☐ Yes ☒ No

How may the code be obtained?

Two versions are on mass storage at Los Alamos National Laboratory:
(1) /095680/dec86/sgenrec :The version for a Rare Earth Cobalt quadrupole magnet
(2) /095680/dec86/sgendip :The version for a magnetic dipole with midplane symmetry.
These are "standard text" versions of the source file. The code can also be obtained form the Los Alamos Accelerator Code Group; contact Helen K. Stokes at (505)667-9131 or -2839, (FTS 843-9131).

Source language: FORTRAN 77

Computers it runs on: Any supporting FORTRAN 77

It is available as: ☒ Source code, ☐ Executable only

Source Media: ☐ Listing, ☒ Tape, ☐ Diskette, ☐ Cards, ☒ Networks
  Tape format:
  Diskette size & format:

Available through: ☐ DECNET, ☒ ARPANET, ☒ BITNET
  ☐

Network Address: hks@lanl.arpa

Date of Latest Version: unknown   Program Name: GIANT

Person to Contact: Hamid Shoaee
    Address: Stanford Linear Accelerator Center
        P.O. Box 4349, SLAC Bin 26
        Stanford, CA 94305
        USA

Telephone Number:

Classification of Computer Code:
    Component Design:
        ☐ Ion Source, ☐ Magnet, ☐ RF–cavity, ☐
    Accelerator Optimization:
        ☐ Linac, ☐ Cyclotron, ☐ Synchrotron, ☐
    Tracking or Simulation:
        ☒ Linac, ☐ Cyclotron, ☐ Synchrotron, ☐
    Analysis:
        ☐ Stability, ☐ Impedances, ☐
    Other: Control Program.

Short Description: (Purpose, capabilities, algorithms, special features, etc.)

Many model-driven diagnostic and correction procedures have been developed at SLAC for the on-line computer controlled operation of SPEAR, PEP, the LINAC, and the Electron Damping Ring. In order to facilitate future applications and enhancements, these procedures are being collected into a single program, GIANT. The program allows interactive diagnosis as well as performance optimization of any beam transport line or circular machine. The test systems for GIANT are those of the SLC projects.

Publications describing the code:

J. Jäger, M. Lee, R. Servranckx and H. Shoaee, "GIANT—A Computer code for General Interactive Analysis of Trajectories," IEEE Trans. NS 32 (1985) 1877–82.

Is code documentation available? ☐ Yes ☐ No

How may the code be obtained?
    Unknown.

Source language: FORTRAN 77

Computers it runs on:
It is available as: ☐ Source code, ☐ Executable only
Source Media: ☐ Listing, ☐ Tape, ☐ Diskette, ☐ Cards, ☐ Networks
    Tape format:
    Diskette size & format:
Available through: ☐ DECNET, ☐ ARPANET, ☐ BITNET
    ☐
Network Address:

Date of Latest Version: 1986

Program Name: GIOS

Person to Contact: H. Wollnik
Address: II. Physikal Institut
Heinrich Buffring 14–16
6300 Giessen,
Fed. Rep. Germany

Telephone Number: 641-702-2770

Classification of Computer Code:
Component Design:
☐ Ion Source, ☐ Magnet, ☐ RF–cavity, ☐
Accelerator Optimization:
☐ Linac, ☒ Cyclotron, ☒ Synchrotron, ☐
Tracking or Simulation:
☐ Linac, ☒ Cyclotron, ☒ Synchrotron, ☐
Analysis:
☐ Stability, ☐ Impedances, ☐
Other: Design of Mass Spectrometers and electromagnetic transport lines

Short Description: (Purpose, capabilities, algorithms, special features, etc.)

Third order matrix method, usual beam line elements + fringe field approximation, fitting capabilities, space charge approximation.

Publications describing the code:

H. Wollnik, J. Brezina and M. Berz, "GIOS – BEAMTRACE, a Program for the Design of High Resolution Mass Spectrometer," 2nd Intl. Conf. on Char. Part. Optics, Albuquerque, May 19–23, 1986. NIM (To be published).

(GSI Report THD 26 Darmstadt (1984)).

Is code documentation available? ☒ Yes ☐ No

How may the code be obtained?
Contact above address

Source language: Fortran

Computers it runs on: VAX, Cyber

It is available as: ☒ Source code, ☐ Executable only
Source Media: ☐ Listing, ☒ Tape, ☒ Diskette, ☐ Cards, ☒ Networks
Tape format: as desired
Diskette size & format: as desired
Available through: ☐ DECNET, ☐ ARPANET, ☒ BITNET
☐
Network Address: ug21%ddagsi3@bitnet

Date of Latest Version: unknown   Program Name: GO

Person to Contact: Hamid Shoaee
Address: SLAC
P.O. BOX 4349, SLAC Bin 26
Stanford, CA 94305
USA

Telephone Number: (415) 854-3300 ext. 2954, FTS 461-9300 ext. 2954

Classification of Computer Code:
Component Design:
☐ Ion Source, ☐ Magnet, ☐ RF–cavity, ☐
Accelerator Optimization:
☐ Linac, ☐ Cyclotron, ☐ Synchrotron, ☐
Tracking or Simulation:
☐ Linac, ☐ Cyclotron, ☐ Synchrotron, ☐
Analysis:
☐ Stability, ☐ Impedances, ☐
Other: An executive program.

Short Description: (Purpose, capabilities, algorithms, special features, etc.)
GO is an executive program placed on the PEP group's public disk (PUBRL 192) to facilitate the use of several PEP related computer programs available on VM. The exec's program list currently includes: CELL, COLLIDER, MAGIC, PATRICIA, PETROS, TRANSPORT, and TURTLE. In addition, provisions have been made to allow addition of new programs to this list as they become available. The GO exec is directly callable from inside the Wylbur editor (in fact, currently this is the only way to use the GO exec.) It provides the option of running any of the above programs in either interactive or batch mode. In the batch mode, the GO exec sends the data in the Wylbur active file along with the information required to run the job to the batch monitor (BMON, a virtual machine that schedules and controls execution of batch jobs). This enables the user to proceed with other VM activities at his/her terminal while the job executes, thus making it of particular interest to the users with jobs requiring much CPU time to execute and/or those wishing to run multiple jobs independently. In the interactive mode, useful for small jobs requiring less CPU time, the job is executed by the user's own Virtual Machine using the data in the active file as input. At the termination of an interactive job, the GO exec facilitates examination of the output by placing it in the Wylbur active file.

Publications describing the code:
Shoaee, H., "GO, an Exec for Running the Programs: CELL, COLLIDER, MAGIC, PATRICIA, PETROS, TRANSPORT And TURTLE," SLAC internal report No. PEP-NOTE-369 (1982) 8p.

Is code documentation available? ☐ Yes ☐ No

How may the code be obtained?

Source language:
Computers it runs on:
It is available as: ☐ Source code, ☐ Executable only
Source Media: ☐ Listing, ☐ Tape, ☐ Diskette, ☐ Cards, ☐ Networks
   Tape format:
   Diskette size & format:
Available through: ☐ DECNET, ☐ ARPANET, ☐ BITNET
☐
Network Address:

Date of Latest Version: unknown

Program Name: GOBLIN

Person to Contact: C. Kost
    Address: TRIUMPH, Univ. B. C.
                4004 Wesbrook Mall
                Vancouver, B.C.
                Canada V6T 2A3

Telephone Number: 604 2221047, ext. 310

Classification of Computer Code:
    Component Design:
        ☐ Ion Source, ☐ Magnet, ☐ RF–cavity, ☐
    Accelerator Optimization:
        ☐ Linac, ☒ Cyclotron, ☐ Synchrotron, ☐
    Tracking or Simulation:
        ☐ Linac, ☐ Cyclotron, ☐ Synchrotron, ☐
    Analysis:
        ☐ Stability, ☐ Impedances, ☐
    Other:

Short Description: (Purpose, capabilities, algorithms, special features, etc.)
    None available

Publications describing the code:
    None available

Is code documentation available? ☐ Yes ☐ No

How may the code be obtained?
    This code is not "portable" to any other institution.

Source language:

Computers it runs on:

It is available as: ☐ Source code, ☐ Executable only

Source Media: ☐ Listing, ☐ Tape, ☐ Diskette, ☐ Cards, ☐ Networks
    Tape format:
    Diskette size & format:

Available through: ☐ DECNET, ☐ ARPANET, ☐ BITNET
                    ☐

Network Address:

Date of Latest Version: unknown   Program Name: GOC3D

Person to Contact: A. C. Paul
        Address: MS L-626
                 Lawrence Livermore National Laboratory
                 Livermore, CA 94550
                 USA

Telephone Number: (415) 423-3183, FTS 543-3183

Classification of Computer Code:
    Component Design:
        ☐ Ion Source, ☐ Magnet, ☐ RF–cavity, ☐
    Accelerator Optimization:
        ☐ Linac, ☐ Cyclotron, ☐ Synchrotron, ☐
    Tracking or Simulation:
        ☐ Linac, ☐ Cyclotron, ☐ Synchrotron, ☒   General Magnetic Field

    Analysis:
        ☐ Stability, ☐ Impedances, ☐
    Other:

Short Description: (Purpose, capabilities, algorithms, special features, etc.)
    GOC3D is a general orbit code incorporating a flexible selection of magnetic field input geometries encountered in polar accelerator design. The code calculates the field as the sum of fields obtained from up to two independent field arrays. The main field array can be either radial (1 dimensional), median plane polar (2 dimensional) non-median plane (two dimensional r, z, input of B_z and B_r) or three dimensional with input of B_z, B_r, and B_θ. For median plane expansion the code is accurate to z^2. The code takes z as the independent variable and will trace rays, track phase space, or determine equilibrium orbit properties of the magnetic field.

Publications describing the code:
    Unknown.

Is code documentation available? ☐ Yes ☐ No

How may the code be obtained?
    Unknown.

Source language:
Computers it runs on:
It is available as: ☐ Source code, ☐ Executable only
Source Media: ☐ Listing, ☐ Tape, ☐ Diskette, ☐ Cards, ☐ Networks
    Tape format:
    Diskette size & format:
Available through: ☐ DECNET, ☐ ARPANET, ☐ BITNET
                    ☐
Network Address:

Date of Latest Version: Oct. 1985

Program Name: HARMON

Person to Contact: Martin Donald
Address: SLAC
P.O. Box 4349
Stanford, CA 94305
USA

Telephone Number: 415-854-3300, ext. 3205

Classification of Computer Code:
Component Design:
☐ Ion Source, ☐ Magnet, ☐ RF-cavity, ☐
Accelerator Optimization:
☐ Linac, ☐ Cyclotron, ☒ Synchrotron, ☐
Tracking or Simulation:
☐ Linac, ☐ Cyclotron, ☐ Synchrotron, ☐
Analysis:
☐ Stability, ☐ Impedances, ☐
Other: Beam Line Optimizer

Short Description: (Purpose, capabilities, algorithms, special features, etc.)

Purpose: To optimize the strength of sextupole correction magnets to improve beam dynamics.

Algorithms: Least Squares minimization of a set of nonlinear functions including tune shift versus momentum, tune shift versus amplitude and distortion of phase space.

A new version will be released with the data-based version of MAD.

This version will have data input closely resembling that of MAD.

The code at present will only handle closed machines. It is hoped to extend it to handle beam lines as well.

Publications describing the code:

M. Donald, "Chromaticity Correction in Circular Accelerators and Storage Rings. A User's Guide to the HARMON Program." PEP Note 311 (1979).

M. Donald and D. Schofield, "A User's Guide to the HARMON Program," CERN LEP Note 420

Is code documentation available? ☐ Yes ☐ No

How may the code be obtained?

Code is a module of the program MAD by F. C. Iselin (CERN).

Code is presently only documented by publications above.

Source language: FORTRAN 77

Computers it runs on: CDC, IBM, VAX

It is available as: ☒ Source code, ☐ Executable only

Source Media: ☐ Listing, ☒ Tape, ☐ Diskette, ☐ Cards, ☒ Networks
   Tape format: ASCII or EBCDIC
   Diskette size & format:
Available through: ☒ DECNET, ☐ ARPANET, ☒ BITNET
                               ☐
Network Address: MHD@SLACVM (BITNET). MHD@SLACPCR (BITNET). PCR::MHD (DECNET).

Date of Latest Version: Jan. 1984   Program Name: HAX

Person to Contact: Masahiro Hara
     Address: Cyclotron Laboratory
                RIKEN (The Institute of Physical and Chemical Research)
                WAKO, SAITAMA, 351–01.
                Japan

Telephone Number: 0484-62-1111 ext. 4011

Classification of Computer Code:
    Component Design:
        ☐ Ion Source, ☐ Magnet, ☒ RF–cavity, ☐
    Accelerator Optimization:
        ☐ Linac, ☐ Cyclotron, ☐ Synchrotron, ☐
    Tracking or Simulation:
        ☐ Linac, ☐ Cyclotron, ☐ Synchrotron, ☐
    Analysis:
        ☐ Stability, ☐ Impedances, ☐
    Other:

Short Description: (Purpose, capabilities, algorithms, special features, etc.)
    To calculate resonant frequencies, electric lines of force, magnetic lines of force, electric fields, and magnetic field for axi-symmetric modes (TM or TE). Based on finite element method.

Publications describing the code:
    M. Hara, T. Wada, A. Toyama, and F. Kikuchi, "Calculation of RF Electromagnetic Field by Finite Element Method," Scientific Papers of the Institute of Physical and Chemical Research, 75 (1981) 143–75

Is code documentation available? ☒ Yes ☐ No

How may the code be obtained?
    contact author

Source language: FORTRAN 77

Computers it runs on: FACOM M380

It is available as: ☒ Source code, ☐ Executable only

Source Media: ☐ Listing, ☒ Tape, ☐ Diskette, ☐ Cards, ☐ Networks
    Tape format: NL/SL 1600 bpi
    Diskette size & format:

Available through: ☐ DECNET, ☐ ARPANET, ☐ BITNET
                    ☐

Network Address:

Date of Latest Version: Jan. 1984   Program Name: H2DB

Person to Contact: Masahiro Hara
       Address: Cyclotron Laboratory
                  Riken (The Institute of Physical and Chemical Research)
                  WAKO, SAITAMA, 351-01,
                  Japan

Telephone Number: 0434-62-1111 ext. 4011

Classification of Computer Code:
    Component Design:
        ☐ Ion Source, ☐ Magnet, ☒ RF-cavity, ☐
    Accelerator Optimization:
        ☐ Linac, ☐ Cyclotron, ☐ Synchrotron, ☐
    Tracking or Simulation:
        ☐ Linac, ☐ Cyclotron, ☐ Synchrotron, ☐
    Analysis:
        ☐ Stability, ☐ Impedances, ☐
    Other:

Short Description: (Purpose, capabilities, algorithms, special features, etc.)

    To calculate cut-off frequencies, electric lines of force, and magnetic lines of force for waveguide with arbitrary cross section. Mesh generator and graphic display codes are included. Based on finite element method.

Publications describing the code:

    M. Hara, T. Wada, A. Toyama and F. Kikuchi, "Calculation of RF Electromagnetic Field by Finite Element method," Scientific Papers of the Institute of Phyical and Chemical Research, 75 (1981) 143–175.

Is code documentation available? ☒ Yes ☐ No

How may the code be obtained?
    Contact author.

Source language: FORTRAN 77

Computers it runs on: FACOM M380

It is available as: ☒ Source code, ☐ Executable only

Source Media: ☐ Listing, ☒ Tape, ☐ Diskette, ☐ Cards, ☐ Networks
    Tape format: NL/SL 1600 bpi
    Diskette size & format:

Available through: ☐ DECNET, ☐ ARPANET, ☐ BITNET ☐

Network Address:

Date of Latest Version: Oct. 1986   Program Name: HETC

Person to Contact: Richard E. Prael
Address: MS B266, X-6
Los Alamos National Laboratory
Los Alamos, NM 87545
USA

Telephone Number: (505)667-7283, FTS 843-7283

Classification of Computer Code:
    Component Design:
        ☐ Ion Source, ☐ Magnet, ☐ RF-cavity, ☐
    Accelerator Optimization:
        ☐ Linac, ☐ Cyclotron, ☐ Synchrotron, ☐
    Tracking or Simulation:
        ☐ Linac, ☐ Cyclotron, ☐ Synchrotron, ☐
    Analysis:
        ☐ Stability, ☐ Impedances, ☐
    Other: Target and shielding design

Short Description: (Purpose, capabilities, algorithms, special features, etc.)
    Transport of nucleons, pions, and muons in general 3D geometry. Interaction physics by Bertini intranuclear cascade model with evaporation model. Couples to MCNP for transport of low energy neutrons and photons.

Publications describing the code:
    R. E. Prael, "High-Energy Particle Monte Carlo at Los Alamos," Los Alamos National Lab. internal report LA-UR-85-1243.

Is code documentation available? ☒ Yes ☐ No

How may the code be obtained?
    Original ORNL version is available from Radiation Shilding Information Center (RSIC); Los Alamos version available upon special request from R. Prael.

Source language: FORTRAN

Computers it runs on: CDC7600 and CRAY

It is available as: ☒ Source code, ☐ Executable only

Source Media: ☐ Listing, ☐ Tape, ☐ Diskette, ☐ Cards, ☐ Networks
    Tape format:
    Diskette size & format:

Available through: ☐ DECNET, ☐ ARPANET, ☐ BITNET
    ☐

Network Address:

Date of Latest Version: unknown

Program Name: HOPI

Person to Contact: J. L. Le Maire
Address: Brookhaven National Laboratory
Upton, L.I., NY 11983
USA

Telephone Number:

Classification of Computer Code:
  Component Design:
    ☐ Ion Source, ☐ Magnet, ☐ RF–cavity, ☐
  Accelerator Optimization:
    ☒ Linac, ☐ Cyclotron, ☐ Synchrotron, ☐
  Tracking or Simulation:
    ☐ Linac, ☐ Cyclotron, ☐ Synchrotron, ☐
  Analysis:
    ☐ Stability, ☐ Impedances, ☐
  Other: Control

Short Description: (Purpose, capabilities, algorithms, special features, etc.)

HOPI, an on-line computer program written on the PDP 10, matches the beam from the 200 Mev linac to the AGS without the necessity of making emittance measurements; it performs the matching by modifying independently the horizontal and vertical emittance. Experimental results show success with this method, which can be applied to any matching section.

Publications describing the code:

Le Maire, J. L., "HOPI: On-line Injection Optimization Program," Brookhaven National Laboratory internal report no. BNL-50741 (1977) 30p.

Is code documentation available? ☐ Yes ☐ No

How may the code be obtained?

Source language:
Computers it runs on: PDP 10

It is available as: ☐ Source code, ☐ Executable only
Source Media: ☐ Listing, ☐ Tape, ☐ Diskette, ☐ Cards, ☐ Networks
  Tape format:
  Diskette size & format:
Available through: ☐ DECNET, ☐ ARPANET, ☐ BITNET
  ☐
Network Address:

Date of Latest Version: June 1986

Program Name: ISIS

Person to Contact: Michael E. Jones
Address: MS H829, Group AT-6
Los Alamos National Laboratory
Los Alamos, NM 87545
USA

Telephone Number: (505) 667-7760, FTS 843-7760

Classification of Computer Code:
Component Design:
☐ Ion Source, ☐ Magnet, ☐ RF-cavity, ☐
Accelerator Optimization:
☐ Linac, ☐ Cyclotron, ☐ Synchrotron, ☐
Tracking or Simulation:
☐ Linac, ☐ Cyclotron, ☐ Synchrotron, ☐
Analysis:
☐ Stability, ☐ Impedances, ☐
Other: Modeling of intense charged particle beams.

Short Description: (Purpose, capabilities, algorithms, special features, etc.)
ISIS is a fully relativistic PIC code that can handle time-dependent electromagnetic fields. It has been applied to wakefield problems in plasmas and photocathode electron source design for FEL's. There are a wide variety of options, including 1 dimensional and 2-1/2 dimensional geometry, cylindrical and Cartesian coordinates, multiple internal boundary, emission and particle creation, and optimization on CRAY computers.

Publications describing the code:
None

Is code documentation available? ☐ Yes ☒ No

How may the code be obtained?
Contact M. Jones. Note: This code is not easily transportable because it has been optimized for the Los Alamos CRAY's.

Source language: CFT1.14,CAL

Computers it runs on: CRAY'S at LANL

It is available as: ☒ Source code, ☐ Executable only

Source Media: ☐ Listing, ☐ Tape, ☐ Diskette, ☐ Cards, ☐ Networks
Tape format:
Diskette size & format:

Available through: ☐ DECNET, ☐ ARPANET, ☐ BITNET
☐

Network Address:

Date of Latest Version: Nov. 1984   Program Name: ITS (INTEGRATED TIGER SERIES)

Person to Contact: H. Grady Hughes III
Address: MS B-226, X-6
Los Alamos National Laboratory
Los Alamos, NM 87545
USA

Telephone Number: (505)667-3926, FTS 843-3926

Classification of Computer Code:
Component Design:
☐ Ion Source, ☐ Magnet, ☐ RF–cavity, ☐
Accelerator Optimization:
☐ Linac, ☐ Cyclotron, ☐ Synchrotron, ☐
Tracking or Simulation:
☐ Linac, ☐ Cyclotron, ☐ Synchrotron, ☐
Analysis:
☐ Stability, ☐ Impedances, ☐
Other: Charged Particle Transport Code.

Short Description: (Purpose, capabilities, algorithms, special features, etc.)
Electron/photon Monte Carlo transport code. ITS is actually a system of eight codes: TIGER, TIGERP, CYLTRAN, CYLTRANP, CYLTRANM, ACCEPT, ACCEPTP, AND ACCEPTM. Special features include:

1) Generalized 1-d, 2-d and 3-d analytic geometry

2) Complete particle cascade

3) External EM fields

4) Cerenkov & transition radiation

5) Well verified by experiment

These codes have been used to simulate the interaction of electron beams, generated by pulsed-power accelerators, with various target materials. They are based on the ETRAN system, which was developed for an energy range from 10 keV up to a few tens of MeV. Modifications have extended their applicability up to 1 GeV. Physical theories used in the code are equivalent to those employed in the SANDYL code.

Publications describing the code:
M. J. Berger, "Monte Carlo Calculation of the Penetration and Diffusion of Fast Charged Particles," Methods in Computational Physics, Vol. 1, Academic Press, New York (1963).

J. A. Halbleib and T. A. Mehlhorn, Sandia Report SAND84-0573, November 1984.

T. A. Mehlhorn and J. A. Halbleib, "Monte Carlo Benchmark Calculations of Energy Deposition by Electron/Photon Showers Up to 1 GeV," Proc. of Amer. Nuc. Soc. Top. Conf. on Computational Methods, Sandia National Labs. Internal Report No.: SAND–82–2230C; CONF-830304-2 (1983)7p.

J. M. Peek and J. A. Halbleib, "Improved Atomic Data for Electron-Transport Predictions by the Codes TIGER and TIGERP: II Electron Stopping and Range Data," Sandia National Lab. Internal Report No.: SAND-83-0481(l983)43p.

Is code documentation available? ☒ Yes ☐ No

How may the code be obtained?
    Contact Groups X-6 at Los Alamos or J.A. Halbleib at Sandia National Laboratory, Org.1231, Bldg.960, Albuquerque, NM 98123; telephone (505)844-1575 or FTS 884-1575.

Source language: FORTRAN 77

Computers it runs on: CRAY, VAX, CDC 7600, IBM

It is available as: ☒ Source code, ☐ Executable only

Source Media: ☒ Listing, ☒ Tape, ☐ Diskette, ☐ Cards, ☐ Networks
    Tape format: VAX BACKUP
    Diskette size & format:

Available through: ☐ DECNET, ☐ ARPANET, ☐ BITNET
    ☐

Network Address:

Date of Latest Version: unknown   Program Name: JASON

Person to Contact: S. J. Sackett
       Address: MS L-122
                Lawrence Livermore National Laboratory
                Livermore, CA 94550
                USA

Telephone Number: (415) 422-8709 or FTS (415) 532-8709

Classification of Computer Code:
   Component Design:
      ☐ Ion Source, ☐ Magnet, ☐ RF–cavity, ☒ Electrostatics
   Accelerator Optimization:
      ☐ Linac, ☐ Cyclotron, ☐ Synchrotron, ☐
   Tracking or Simulation:
      ☐ Linac, ☐ Cyclotron, ☐ Synchrotron, ☐
   Analysis:
      ☐ Stability, ☐ Impedances, ☐
   Other:

Short Description: (Purpose, capabilities, algorithms, special features, etc.)
   JASON solves general electrostatics problems having either slab or cylindrical symmetry. More specifically, it solves the self-adjoint elliptic equation, $\nabla \cdot (K \nabla V) - \gamma V + \rho = 0$ in an arbitrary two-dimensional domain. For electrostatics, V is the electrostatic potential, K is the dielectric tensor, and ρ is the free-charge density. The parameter γ is identically zero for electrostatics but may have a positive nonzero value in other cases (e.g., capillary surface problems with gravity loading). The system of algebraic equations used in JASON is generated by the finite element method. Four-node quadrilateral elements are used for most of the mesh. Triangular elements, however, are occasionally used on boundaries to avoid severe mesh distortions.

Publications describing the code:
   S. J. Sackett, "JASON—A Code for Solving General Electrostatics Problems. User's Manual," LLNL internal report (1978), also available from US-NTIS (Natl. Tech. Info. Serv.) as report UCID-17814.

Is code documentation available? ☒ Yes ☐ No

How may the code be obtained?
   unknown

Source language:
Computers it runs on:
It is available as: ☐ Source code, ☐ Executable only
Source Media: ☐ Listing, ☐ Tape, ☐ Diskette, ☐ Cards, ☐ Networks
   Tape format:
   Diskette size & format:
Available through: ☐ DECNET, ☐ ARPANET, ☐ BITNET
                   ☐
Network Address:

Date of Latest Version: 1984

Program Name: KN7C

Person to Contact: E. Keil
        Address: LEP Division
                CERN
                1211 GENEVA 23
                Switzerland

Telephone Number: 41-22-83.34.26

Classification of Computer Code:
    Component Design:
        ☐ Ion Source, ☐ Magnet, ☐ RF–cavity, ☐
    Accelerator Optimization:
        ☐ Linac, ☐ Cyclotron, ☐ Synchrotron, ☐
    Tracking or Simulation:
        ☐ Linac, ☐ Cyclotron, ☐ Synchrotron, ☐
    Analysis:
        ☐ Stability, ☒ Impedances, ☐
    Other:

Short Description: (Purpose, capabilities, algorithms, special features, etc.)

Finds the resonant frequencies, the field pattern and the longitudinal loss factors for axisymmetric waveguide modes propagated at $v = \beta c$ in disk-loaded waveguide by field matching, using matrix techniques.

Publications describing the code:

E. Keil, Nucl. Instr. Meth. 100 (1972) 419.

E. Keil, in Phys. of High Energy Particle Accelerators, AIP Conf. Proc. No. 105 (1983).

E. Keil, CERN 84-01 (1984).

Is code documentation available? ☒ Yes ☐ No

How may the code be obtained?
    Contact E. Keil

Source language: FORTRAN 5

Computers it runs on: CDC

It is available as: ☒ Source code, ☐ Executable only

Source Media: ☒ Listing, ☒ Tape, ☐ Diskette, ☐ Cards, ☒ Networks
    Tape format: As desired
    Diskette size & format:

Available through: ☐ DECNET, ☐ ARPANET, ☒ BITNET ☐

Network Address: KEIL @ CERNVM

Date of Latest Version: 1986

Program Name: KOBRA

Person to Contact: P. Spädtke
Address: GSI-DARMSTADT
Postfach 11 05 41
6100-DARMSTADT
Fed. Rep. Germany

Telephone Number: 6151-359-323

Classification of Computer Code:
Component Design:
☒ Ion Source, ☐ Magnet, ☐ RF-cavity, ☒ low energy (DC) beam.

Accelerator Optimization:
☐ Linac, ☐ Cyclotron, ☐ Synchrotron, ☐
Tracking or Simulation:
☐ Linac, ☐ Cyclotron, ☐ Synchrotron, ☐
Analysis:
☐ Stability, ☐ Impedances, ☐
Other: Transport/space charge effects.

Short Description: (Purpose, capabilities, algorithms, special features, etc.)

KOde zur Berechnung RAumladungsbehafteter teilchenbahnen im-d raum.

KOBRA3 calculates the trajectories of charged particles in static electro-magnetic fields in three dimensions, including extraction problems. The electric field is determined by the potential distribution from the given electrode arrangement. Space charge is taken into account by an iterative process. The self-consistent formation of a plasma meniscus can be calculated. The magnetic field distribution is either analytically determined or read in from a table; six types of distribution are offered. The potential fields of any plane may be displayed as equipotential line drawings or by 3D diagrams, in which the potential is represented as the third coordinate. The calculated particle trajectories can be displayed by emittance diagrams, represented in two dimensions by projection onto any plane, or as three dimensional diagrams.

KOBRA is partitioned into eight programs, KOBRA1, KOBRA2, KOBRA3 and KOBRA4, which calculate the mesh, the potential, the magnetic field and the particle trajectories respectively; KOBRA5, KOBRA6, KOBRA7, and KOBRA8, which display the results.

Publications describing the code:

P. Spädtke, "KOBRA3 – Three Dimensional Raytracing Including Space-Charge Effects," IEEE Trans. NS-32 (1982) 2465.

Examples in Proceedings of Linac Conference/Seeheim 1984. GSI, Darmstadt internal report GSI-84-11 (see conference index under Spädtke).

Is code documentation available? ☒ Yes ☐ No

How may the code be obtained?
Implementation on request.

Source language: FORTRAN 77

Computers it runs on: IBM, VAX

It is available as: ☒ Source code, ☐ Executable only

Source Media: ☐ Listing, ☐ Tape, ☐ Diskette, ☐ Cards, ☐ Networks
　Tape format:
　Diskette size & format:

Available through: ☐ DECNET, ☐ ARPANET, ☒ BITNET
　　　　　　　　　　☐

Network Address: ul13%DDAGSl3.BITNET

Date of Latest Version: 1977   Program Name: LACC

Person to Contact: A. Konrad
Address: General Electric
Corporate Research and Development
Building 37, Room 355
Schenectady, NY 12301
USA
Telephone Number: (518) 387-5083

Classification of Computer Code:
Component Design:
☐ Ion Source, ☐ Magnet, ☒ RF-cavity, ☐
Accelerator Optimization:
☐ Linac, ☐ Cyclotron, ☐ Synchrotron, ☐
Tracking or Simulation:
☐ Linac, ☐ Cyclotron, ☐ Synchrotron, ☐
Analysis:
☐ Stability, ☐ Impedances, ☐
Other:

Short Description: (Purpose, capabilities, algorithms, special features, etc.)

Linear Accelerator Cavity Code solves the classical electromagnetic field problem of the empty, axially symmetric resonator with conducting walls. The program algorithm is based on a variational formulation coupled with the high-order polynomial, triangular finite element method for the magnetic field calculation. Various other numerical methods such as one- and two-dimensional Newton-Cotes integration are used to obtain the performance measuring quantities (e.g. transit time factor, stored energy, power loss, shunt impedance, Q-factor). This is a modification of a 1973 program called AXISYMM - VECTOR - HELMHOLTZ - FINTEL6.

Publications describing the code:

A. Konrad, "A Linear Accelerator Cavity Code Based on the Finite Element Method," Comput. Phys. Commun. v 13, (1978) 349-362.

A. Konrad, "A Linear Accelerator Cavity Field Calculation by the Finite Element Method," IEEE Trans. NS-20 (1973) 802-808.

A. Konrad, "Evaluation of the LACC Program," Comput. Phys. Commun. (Netherlands) v 14:3. (1978) 177-184 [1].

Is code documentation available? ☐ Yes ☐ No

How may the code be obtained?
CPC Program Library, Queen's University of Belfast, N. Ireland.

Source language: FORTRAN IV

Computers it runs on: IBM 360/75

It is available as: ☒ Source code, ☐ Executable only
Source Media: ☐ Listing, ☐ Tape, ☐ Diskette, ☐ Cards, ☐ Networks
Tape format:
Diskette size & format:
Available through: ☐ DECNET, ☐ ARPANET, ☐ BITNET
☐
Network Address:

Date of Latest Version: unknown   Program Name: LALAGE

Person to Contact: P. Fernandes
          Address: Institute per la Matematica Applicata
                   del C.N.R.
                   Genova,
                   Italy

Telephone Number:

Classification of Computer Code:
   Component Design:
      ☐ Ion Source, ☐ Magnet, ☒ RF–cavity, ☐
   Accelerator Optimization:
      ☐ Linac, ☐ Cyclotron, ☐ Synchrotron, ☐
   Tracking or Simulation:
      ☐ Linac, ☐ Cyclotron, ☐ Synchrotron, ☐
   Analysis:
      ☐ Stability, ☐ Impedances, ☐
   Other:

Short Description: (Purpose, capabilities, algorithms, special features, etc.)
   LALAGE is an improved version of the LALA program to compute resonant frequencies and fields for all the modes of the lowest TM_{01} band-pass of multicell structures.

Publications describing the code:
   P. Fernandes, R. Parodi, "LALAGE – A Computer Program to Calculate the TM_{01} Modes of Cylindrically Symmetrical Multicell Resonant Structures," Part. Accel. 12(1982)131-7.

Is code documentation available? ☐ Yes ☐ No

How may the code be obtained?
   unknown.

Source language:

Computers it runs on:

It is available as: ☐ Source code, ☐ Executable only

Source Media: ☐ Listing, ☐ Tape, ☐ Diskette, ☐ Cards, ☐ Networks
   Tape format:
   Diskette size & format:

Available through: ☐ DECNET, ☐ ARPANET, ☐ BITNET
                    ☐

Network Address:

Date of Latest Version: unknown   Program Name: LANS

Person to Contact: B. M. Fomel
          Address: USSR Academy of Sciences
                   Siberian Division
                   Institute of Nuclear Physics
                   Novosibirsk, 90
                   USSR
Telephone Number:

Classification of Computer Code:
   Component Design:
      ☐ Ion Source, ☐ Magnet, ☒ RF-cavity, ☐
   Accelerator Optimization:
      ☐ Linac, ☐ Cyclotron, ☐ Synchrotron, ☐
   Tracking or Simulation:
      ☐ Linac, ☐ Cyclotron, ☐ Synchrotron, ☐
   Analysis:
      ☐ Stability, ☐ Impedances, ☐
   Other:

Short Description: (Purpose, capabilities, algorithms, special features, etc.)
   LANS is a code developed for calculation of axisymmetric cavities. The mathematical basis of this code is the method of inverse iterations with a shift, which is the most adequate for the problem of finding the eigenfrequencies and fields for the cavities. This code has some advantages compared with SUPERFISH; it requires a smaller number of operations necessary for calculations and it gives better resolution of resonance modes with close frequencies.

Publications describing the code:
   B. M. Fomel; V. P. Jackowley; M. M. Karliner; P. B. Lysyansky, "LANS – A New Code for Evaluation of the Electromagnetic Fields and Resonance Frequencies of Axisymmetrical RF Cavities," Part. Accel. (United Kingdom) 11 (1981.) 173-9.

Is code documentation available? ☐ Yes ☐ No

How may the code be obtained?
   unknown

Source language:
Computers it runs on:
It is available as: ☐ Source code, ☐ Executable only
Source Media: ☐ Listing, ☐ Tape, ☐ Diskette, ☐ Cards, ☐ Networks
   Tape format:
   Diskette size & format:
Available through: ☐ DECNET, ☐ ARPANET, ☐ BITNET
                    ☐
Network Address:

Date of Latest Version: unknown

Program Name: LATTICE

Person to Contact: John Staples
    Address: Lawrence Berkeley Laboratory
               Bldg. 64, Room 224A
               1 Cyclotron Rd.
               Berkeley, CA 94720

Telephone Number:

Classification of Computer Code:
    Component Design:
        ☐ Ion Source, ☐ Magnet, ☐ RF–cavity, ☐
    Accelerator Optimization:
        ☐ Linac, ☐ Cyclotron, ☒ Synchrotron, ☐
    Tracking or Simulation:
        ☐ Linac, ☐ Cyclotron, ☐ Synchrotron, ☐
    Analysis:
        ☐ Stability, ☐ Impedances, ☐
    Other:

Short Description: (Purpose, capabilities, algorithms, special features, etc.)

    LATTICE is a computer code which enables an interactive user to calculate the functions of a synchrotron lattice. This program satisfies the requirements at LBL for a simple interactive lattice program by borrowing ideas from both TRANSPORT and SYNCH. A fitting routine is included.

    A version of LATTICE exists that is written in BASIC and runs on HP 9845.

    John Staples and Arthur C. Paul have diverging versions in Pascal which run on the IBM PC. The latter version is self-booting with complete on line documentation. Address: Arthur C. Paul, MS L-626, Lawrence Livermore National Laboratory, P.O. Box 808, Livermore, CA 94550.

Publications describing the code:
    John Staples, "LATTICE: An interactive lattice computer code," LBL internal report no. LBL-4843, (1976) 18p.

Is code documentation available? ☒ Yes ☐ No

How may the code be obtained?
    Contact John Staples

Source language: FORTRAN

Computers it runs on: CDC 6600

It is available as: ☐ Source code, ☐ Executable only

Source Media: ☐ Listing, ☐ Tape, ☐ Diskette, ☐ Cards, ☐ Networks
    Tape format:
    Diskette size & format:

Available through: ☐ DECNET, ☐ ARPANET, ☐ BITNET
                     ☐

Network Address:

Date of Latest Version: July 1986   Program Name: LIEPOT

Person to Contact: Etienne Forest
Address: MS 90/4040
URA Design Center
c/o UCLBL
Berkeley, CA 94720
USA
Telephone Number: (415) 486–6580 or FTS 451–6580

Classification of Computer Code:
    Component Design:
        ☐ Ion Source, ☐ Magnet, ☐ RF–cavity, ☐
    Accelerator Optimization:
        ☐ Linac, ☐ Cyclotron, ☐ Synchrotron, ☐
    Tracking or Simulation:
        ☐ Linac, ☐ Cyclotron, ☒ Synchrotron, ☐
    Analysis:
        ☐ Stability, ☐ Impedances, ☐
    Other:

Short Description: (Purpose, capabilities, algorithms, special features, etc.)
    LIEPOT generates a Lie-algebraic map that produces tracking results equivalent those produced by the code TEAPOT. This allows the user to interface with the code MARYLIE and hence calculate auxiliary quantities such as chromaticity and nonlinear invariants. The results are equivalent to 6x6 third order transport matrices.

Publications describing the code:
    E. Forest, "Lie Algebraic Maps and Invariants Produced by Tracking Codes," SSC Design Group internal report no. 78 (1986).

    Lie Algebraic Maps and Invariants Produced by Tracking Codes.

Is code documentation available? ☐ Yes ☒ No

How may the code be obtained?
    Contact Etienne Forest.

Source language: FORTRAN 77

Computers it runs on: CRAY-XMP, VAX

It is available as: ☒ Source code, ☐ Executable only

Source Media: ☐ Listing, ☒ Tape, ☐ Diskette, ☐ Cards, ☐ Networks
    Tape format: As desired.
    Diskette size & format:

Available through: ☐ DECNET, ☐ ARPANET, ☐ BITNET, ☐

Network Address:

Date of Latest Version: 1980

Program Name: LIMATRA

Person to Contact: G. von Holtley
Address: LEP Theory Division
CERN
1211 GENEVA 23
Switzerland

Telephone Number: CERN 41-22-83.53.93

Classification of Computer Code:
Component Design:
☐ Ion Source, ☐ Magnet, ☐ RF–cavity, ☐
Accelerator Optimization:
☐ Linac, ☐ Cyclotron, ☐ Synchrotron, ☐
Tracking or Simulation:
☐ Linac, ☐ Cyclotron, ☒ Synchrotron, ☐
Analysis:
☐ Stability, ☐ Impedances, ☐
Other:

Short Description: (Purpose, capabilities, algorithms, special features, etc.)
A fast and flexible many-particle tracking code in tranverse and longitudinal phase space, using matrix formalism and lumped linear or non-linear perturbations with systematic, harmonic or random azimuthal distributions. In particular it is useful for studies of non-linear resonances, etc.

Publications describing the code:
"User Guide to the Synchrotron Tracking Computer Program LIMATRA," G. Von Holtley, LAB II-DI-PA/Int. 75–3

Is code documentation available? ☒ Yes ☐ No

How may the code be obtained?
From author

Source language: FORTRAN

Computers it runs on: IBM

It is available as: ☒ Source code, ☐ Executable only

Source Media: ☐ Listing, ☒ Tape, ☐ Diskette, ☐ Cards, ☐ Networks
Tape format:
Diskette size & format:

Available through: ☐ DECNET, ☐ ARPANET, ☐ BITNET
☐

Network Address:

Date of Latest Version: 1970   Program Name: LINDA

Person to Contact: Stanley Snowdon
Address: FNAL
P.O. Box 500
Batavia, IL 60510
USA

Telephone Number: (312) 840-3804, FTS 370-3804

Classification of Computer Code:
   Component Design:
      ☐ Ion Source, ☒ Magnet, ☐ RF–cavity, ☐
   Accelerator Optimization:
      ☐ Linac, ☐ Cyclotron, ☐ Synchrotron, ☐
   Tracking or Simulation:
      ☐ Linac, ☐ Cyclotron, ☐ Synchrotron, ☐
   Analysis:
      ☐ Stability, ☐ Impedances, ☐
   Other:

Short Description: (Purpose, capabilities, algorithms, special features, etc.)

   LINDA uses a combination of scalar and vector potentials to model 2-D magnetostatic problems. The code is very accurate and can handle up to 30,000 mesh points. Will handle iron with nonuniform permeability. Main limitation is that there can be only one region of iron in the problem. Input is much simpler than that for POISSON.

Publications describing the code:

   J. S. Colonias, "Particle Accelerator Design Computer Programs," Academic Press, New York (1974) 39–62.

Is code documentation available? ☐ Yes ☐ No

How may the code be obtained?
   Contact Stan Snowdon.

Source language: FORTRAN

Computers it runs on: IBM 360/75, CDC6600.

It is available as: ☒ Source code, ☐ Executable only

Source Media: ☐ Listing, ☒ Tape, ☐ Diskette, ☐ Cards, ☐ Networks
   Tape format:
   Diskette size & format:

Available through: ☐ DECNET, ☐ ARPANET, ☐ BITNET ☐

Network Address:

Date of Latest Version: 1984   Program Name: LOOPER

Person to Contact: S. O. Schriber
         Address: AT Division, MS H811
                  Los Alamos National Laboratory
                  Los Alamos, NM 87545
                  USA

Telephone Number: (505) 667-7634, FTS 843-7634

Classification of Computer Code:
    Component Design:
        ☐ Ion Source, ☐ Magnet, ☒ RF–cavity, ☐
    Accelerator Optimization:
        ☒ Linac, ☐ Cyclotron, ☐ Synchrotron, ☐
    Tracking or Simulation:
        ☐ Linac, ☐ Cyclotron, ☐ Synchrotron, ☐
    Analysis:
        ☒ Stability, ☐ Impedances, ☐
    Other:

Short Description: (Purpose, capabilities, algorithms, special features, etc.)
    Program LOOPER calculates the rf characteristics (as seen by the drive and/or component elements) of coupled rf resonators using coupled RLC circuit theory. Input of cavity Q and impedance permits calculation of power losses and average on-axis fields that agree very well with multicell SUPERFISH calculations. The program handles off-resonance characteristics, all passband modes, beam loading, multi-neighbor coupling, each element different, stability, and bridges between linacs.

Publications describing the code:
    None

Is code documentation available? ☒ Yes ☐ No

How may the code be obtained?
    From the author/see above; H. Euteneur, Inst. für Kernphysik, Universität Mainz, Postfach 3980, 6500 Mainz, BRD; S. Inagaki, KEK – National Laboratory for High Energy Physics, Oho-Machi, Tskuba-Gun, Ibaraki-Ken, JAPAN 300-32

Source language: FORTRAN

Computers it runs on: CDC, CYBER

It is available as: ☒ Source code, ☐ Executable only

Source Media: ☐ Listing, ☒ Tape, ☐ Diskette, ☐ Cards, ☐ Networks
    Tape format:
    Diskette size & format:

Available through: ☐ DECNET, ☐ ARPANET, ☐ BITNET
                   ☐

Network Address:

Date of Latest Version: Nov. 1986

Program Name: LTRACK

Person to Contact: Dominic Chan
Address: AT-6, MS H829
Los Alamos National Lab.
Los Alamos, NM 87545
USA

Telephone Number: (505)665-0376 or FTS 845-0376

Classification of Computer Code:
Component Design:
☐ Ion Source, ☐ Magnet, ☐ RF–cavity, ☐
Accelerator Optimization:
☐ Linac, ☐ Cyclotron, ☐ Synchrotron, ☐
Tracking or Simulation:
☒ Linac, ☐ Cyclotron, ☐ Synchrotron, ☐
Analysis:
☐ Stability, ☐ Impedances, ☐
Other:

Short Description: (Purpose, capabilities, algorithms, special features, etc.)

LTRACK is a first-order-matrix, beam-transport code that takes into account the longitudinal wake-force of the monopole modes and the transverse wake-force of the dipole and quadrupole modes. Provision is made for error analysis, including orbit correction. It has been used in the study of wake effects in the Stanford Linear Collider (SLC) and in the Los Alamos Free Electron Laser Energy Recovery Experiment (FEL ERX). The present version is very similar to that developed at SLAC by Karl Bane, except that bending magnet and edge rotation has been added. A user's package with explanations for code installation has been prepared.

Publications describing the code:

A. W. Chao and R. K. Cooper, "Transverse Quadrupole Wakefield Effects in High Intensity Linacs," Part. Accel. 13(1983)1-12.

Is code documentation available? ☒ Yes ☐ No

How may the code be obtained?

Contact Dominic Chan or The Los Alamos Accelerator Code Group by calling Helen K. Stokes, (505)667-9131 or -2839.

Source language: FORTRAN

Computers it runs on: CRAY

It is available as: ☒ Source code, ☐ Executable only

Source Media: ☐ Listing, ☒ Tape, ☐ Diskette, ☐ Cards, ☒ Networks
Tape format:
Diskette size & format:

Available through: ☐ DECNET, ☒ ARPANET, ☒ BITNET
☐

Network Address: hks@lanl.arpa

Date of Latest Version: 1986

Program Name: MAD

Person to Contact: C. Iselin
Address: LEP Theory Group
CERN
CH-1211 Geneva 23
Switzerland

Telephone Number: (22) 83 36 57

Classification of Computer Code:
Component Design:
☐ Ion Source, ☐ Magnet, ☐ RF–cavity, ☐
Accelerator Optimization:
☐ Linac, ☐ Cyclotron, ☒ Synchrotron, ☐
Tracking or Simulation:
☐ Linac, ☐ Cyclotron, ☐ Synchrotron, ☐
Analysis:
☐ Stability, ☐ Impedances, ☐
Other:

Short Description: (Purpose, capabilities, algorithms, special features, etc.)
Programming system with a common data base for optics and design. Includes survey, linear lattice, matching, tracking.

Publications describing the code:
C. Iselin, "The MAD Program," CERN-LEP/TH-85/15

Is code documentation available? ☒ Yes ☐ No

How may the code be obtained?
Tape or Network.

Source language: FORTRAN 77

Computers it runs on: IBM, CDC, VAX, NORD

It is available as: ☒ Source code, ☐ Executable only

Source Media: ☐ Listing, ☒ Tape, ☐ Diskette, ☐ Cards, ☒ Networks
Tape format: 9 track 1600 bpi
Diskette size & format:

Available through: ☐ DECNET, ☐ ARPANET, ☒ BITNET
☐

Network Address: FCI @ CERNVM

Date of Latest Version: 1987   Program Name: MADEST

Person to Contact: K. M. Thompson
          Address: Argonne National Laboratory
                  9700 S. Cass Ave., Bldg. 360
                  Argonne, IL 60439
                  U.S.A.

Telephone Number: (312)972-6265 or FTS 972-6265

Classification of Computer Code:
    Component Design:
        ☐ Ion Source, ☒ Magnet, ☐ RF-cavity, ☐
    Accelerator Optimization:
        ☐ Linac, ☐ Cyclotron, ☐ Synchrotron, ☐
    Tracking or Simulation:
        ☐ Linac, ☐ Cyclotron, ☐ Synchrotron, ☐
    Analysis:
        ☐ Stability, ☐ Impedances, ☐
    Other:

Short Description: (Purpose, capabilities, algorithms, special features, etc.)
    MADEST is an interactive program used to develop the geometrical designs for the cores and conventional coils of various types of magnets. It does NOT involve magnetic field calculations. From the designs MADEST can be used to develop cost estimates for designing, fabricating, and installing systems of magnets. The code is under development.

Publications describing the code:
    K. M. Thompson,"An interactive Computer Program for the Design and Costing of Magnets," Journal de Physique C1(1984)(January)

Is code documentation available? ☐ Yes ☒ No

How may the code be obtained?
    Contact K. M. Thompson

Source language: The FORTRAN version is still under development; also Hewlett Packard BASIC

Computers it runs on: VAX, HP9845, HP200, and HP300.

It is available as: ☐ Source code, ☐ Executable only

Source Media: ☐ Listing, ☐ Tape, ☐ Diskette, ☐ Cards, ☐ Networks
    Tape format:
    Diskette size & format:

Available through: ☐ DECNET, ☐ ARPANET, ☐ BITNET
                    ☐

Network Address:

Date of Latest Version: unknown

Program Name: MAFCO

Person to Contact: J. C. Brown
Address: MS L561
Lawrence Livermore National Laboratory
P.O. Box 808,
Livermore, CA 94550

Telephone Number: (415) 423–4157

Classification of Computer Code:
Component Design:
☐ Ion Source, ☒ Magnet, ☐ RF–cavity, ☐
Accelerator Optimization:
☐ Linac, ☐ Cyclotron, ☐ Synchrotron, ☐
Tracking or Simulation:
☐ Linac, ☐ Cyclotron, ☐ Synchrotron, ☐
Analysis:
☐ Stability, ☐ Impedances, ☐
Other:

Short Description: (Purpose, capabilities, algorithms, special features, etc.)

Program MAFCO is capable of calculating the magnetic fields resulting from a given set of current-carrying conductors of arbitrary two- or three- dimensional geometry in which no permeable material is present. The elements which comprise the generalized coil geometry are:

1. Circular loops with designated position and orientation in space.

2. Circular arcs with designated position and orientation in space.

3. Helices along the z axis (in the cylindrical coordinate system) with any designated pitch, starting point, and ending point.

4. Straight lines with any arbitrary orientation.

5. General elements specified by a list of points which the program connects with straight lines.

All of these elements are assumed to be infinitely thin.

Arthur C. Paul (see TRANSPORT for address) has a Pascal version which runs on the IBM PC.

Publications describing the code:

J. C. Brown and W. A. Perkins, "MAFCO - A Magnetic Field Code for Handling General Current Elements in 3-D," Univ. of California Internal Report no. UCRL–7744 (1966).

John S. Colonias, "Particle Accelerator Design: Computer Programs," Academic Press (1974) pp. 119–28.

A plotting code with documents also exists: T. N. Haratani, R. W. Moir, "A Code for Viewing MAFCO Conductors from any Angle," Univ. of California Internal Report no. UCRL–51397 (1973).

Is code documentation available? ☐ Yes ☐ No

How may the code be obtained?
   unknown

Source language: FORTRAN

Computers it runs on: CDC 6600/7600

It is available as: ☐ Source code, ☐ Executable only

Source Media: ☐ Listing, ☐ Tape, ☐ Diskette, ☐ Cards, ☐ Networks
   Tape format:
   Diskette size & format:

Available through: ☐ DECNET, ☐ ARPANET, ☐ BITNET
   ☐

Network Address:

Date of Latest Version: unknown   Program Name: MAFCO III

Person to Contact: S. J. Sackett
         Address: MS L-122
                  Lawrence Livermore National Laboratory
                  Livermore, CA 94550
                  USA

Telephone Number: (415) 422-8709, FTS 532-8709

Classification of Computer Code:
   Component Design:
      ☐ Ion Source, ☐ Magnet, ☐ RF-cavity, ☐
   Accelerator Optimization:
      ☐ Linac, ☐ Cyclotron, ☐ Synchrotron, ☐
   Tracking or Simulation:
      ☒ Linac, ☐ Cyclotron, ☒ Synchrotron, ☐
   Analysis:
      ☐ Stability, ☐ Impedances, ☐
   Other:

Short Description: (Purpose, capabilities, algorithms, special features, etc.)
Program MAFCOIII is a combination of MAFCO and ZAM. MAFCO performs the calculation of the magnetic and/or electric fields resulting from the coil configuration specified; while ZAM performs the step-by-step solution of the system of first-order differential equations, by fourth-order Adams-Moulton predictor-corrector method, to obtain the particle trajectories desired. The program is extremely flexible, as evidenced by the generalized coil geometry that MAFCO accepts.

Publications describing the code:
W. A. Perkins and S. J. Sackett, "MAFCOIII – A Code for Calculating Particle Trajectories in Magnetic and Electric Fields," Lawrence Berkeley Laboratory internal report LBL-765 (1972).

J. S. Colonias, "Particle Accelerator Design: Computer Programs," Academic Press (1974) 227-32.

Is code documentation available? ☐ Yes ☐ No

How may the code be obtained?
   unknown

Source language: FORTRAN

Computers it runs on: CDC 6600/7600

It is available as: ☐ Source code, ☐ Executable only

Source Media: ☐ Listing, ☐ Tape, ☐ Diskette, ☐ Cards, ☐ Networks
   Tape format:
   Diskette size & format:

Available through: ☐ DECNET, ☐ ARPANET, ☐ BITNET
                              ☐

Network Address:

Date of Latest Version: unknown   Program Name: MAFCO-W

Person to Contact: T. F. Yang
           Address: University of Wisconsin
                    Madison, WI
                    USA

Telephone Number:

Classification of Computer Code:
    Component Design:
        ☐ Ion Source, ☒ Magnet, ☐ RF–cavity, ☐
    Accelerator Optimization:
        ☐ Linac, ☐ Cyclotron, ☐ Synchrotron, ☐
    Tracking or Simulation:
        ☐ Linac, ☐ Cyclotron, ☐ Synchrotron, ☐
    Analysis:
        ☐ Stability, ☐ Impedances, ☐
    Other:

Short Description: (Purpose, capabilities, algorithms, special features, etc.)

   MAFCO-W has been written for calculating magnetic fields of finite size conductors of general configuration which can be approximated by arc segments or straight segments of rectangular cross section. The magnetic field components were obtained by integrating the Biot-Savart law over the volume of the conductor. Their mathematical expressions were first reduced to single integration analytically and then integrated numerically. The magnetic fields for the conceptual Tokamak fusion reactors UWMAK-I and II were calculated and anlyzed.

Publications describing the code:

   T. F. Yang, "Magnetic Field Code for Handling General Current-carrying conductors in 3-D," 5th Int'l Conf. on Magnet Technology in Frascati (1975), 203–9; published by Comitato Nazionale per. l'Energia Nucleare, Frascati, Italy (1975).

Is code documentation available? ☐ Yes ☐ No

How may the code be obtained?
    unknown

Source language:

Computers it runs on:

It is available as: ☐ Source code, ☐ Executable only

Source Media: ☐ Listing, ☐ Tape, ☐ Diskette, ☐ Cards, ☐ Networks
    Tape format:
    Diskette size & format:

Available through: ☐ DECNET, ☐ ARPANET, ☐ BITNET
                   ☐

Network Address:

Date of Latest Version: Feb. 1986

Program Name: MAFIA

Person to Contact: Thomas Weiland
   Address: Deutches Electronen Synchrotron (DESY)
   Notkestrasse 85 d2000
   Hamburg 52
   Federal Republic of Germany

Telephone Number: 49-40-8998-3196

Classification of Computer Code:
   Component Design:
      ☐ Ion Source, ☐ Magnet, ☒ RF–cavity, ☐
   Accelerator Optimization:
      ☐ Linac, ☐ Cyclotron, ☐ Synchrotron, ☐
   Tracking or Simulation:
      ☐ Linac, ☐ Cyclotron, ☐ Synchrotron, ☐
   Analysis:
      ☒ Stability, ☒ Impedances, ☒ Wake field effects
   Other:

Short Description: (Purpose, capabilities, algorithms, special features, etc.)

MAFIA is a collection of codes (M3, R3, E31, E32, P3 and T3) for calculating the resonant frequencies and transient electromagnetic fields in a fully 3 dimensional geometry. M3 is the mesh generator. R3 generates the eigenvalue matrix. E31 is a "standard" eigenvalue solver. E32 is an eigenvalue solver that uses multigrid methods. P3 is a postprocessor that displays a variety of one and two dimensional plots of the fields in the cavity and also prints a variety of numerical results.

T3 is a 3D version of TBCI and is used for analyzing the electromagnetic interaction between bunched beams of charged particles and vacuum chambers containing rf cavities, bellows, etc. There are two postprocessors used with the code. W3COR subtracts the tube wake field from the total wake. W3OUT reads and prints a wake; it can calculate the gradient impedance, plot the bunch density, and normalize the wakes to ± 1.

Publications describing the code:

T. Weiland, "On the Unique Numerical Solution of the Maxwellian Eigenvalue Problem in Three Dimensions," Part. Accl. 17(1985)227-42.

T. Weiland, Part. Accel. 15(1984)245-91.

T. Weiland, SLAC Linac Conf. (1986)

Is code documentation available? ☒ Yes ☐ No

How may the code be obtained?

The codes are available from T. Weiland on a "friendly user" basis. Executable forms are available on Los Alamos National Laboratory computers. (For more info. contact Therese Barts (505) 667–9385.) The codes are also available on the MFE computer network. For information of the MFE access, contact Carol Tull FTS 532-1556, or Therese Barts FTS 843-9385.

Source language: FORTRAN 77

Computers it runs on: IBM 3081, CRAY, VAX

It is available as: ☒ Source code, ☐ Executable only

Source Media: ☐ Listing, ☒ Tape, ☐ Diskette, ☐ Cards, ☐ Networks
   Tape format: EBCDIC
   Diskette size & format:

Available through: ☐ DECNET, ☐ ARPANET, ☒ BITNET
   ☐

Network Address: MPYWEI %DHIIDESY3.BITNET

Date of Latest Version: unknown
Program Name: MAGFOR

Person to Contact: W. D. Cain
Address: Oak Ridge National Laboratory
Oak Ridge, TN 37830
USA

Telephone Number:

Classification of Computer Code:
Component Design:
☐ Ion Source, ☒ Magnet, ☐ RF–cavity, ☐
Accelerator Optimization:
☐ Linac, ☐ Cyclotron, ☐ Synchrotron, ☐
Tracking or Simulation:
☐ Linac, ☐ Cyclotron, ☐ Synchrotron, ☐
Analysis:
☐ Stability, ☐ Impedances, ☐
Other:

Short Description: (Purpose, capabilities, algorithms, special features, etc.)

MAGFOR calculates electromagnetic fields and forces in coil systems of arbitrary geometry. The coils may be modeled by using 20-node isoparametric hexahedrons; 8-node rectangular cross-sectional straight segments; rectangular cross-sectional circular arcs: and/or filamenting circular loops. A combination of analytical and numerical integration of the Biot-Savart law for a volume distribution of current is used for calculating magnetic fields. Volumetric body forces are calculated for the 20-node isoparametric brick by numerically integrating the vector product J x B over its volume, where the magnetic field at each Gauss point is obtained by interpolating the magnetic field at the node points by using shape functions. The force is distributed to the node points of the element, again using the shape functions in a consistent manner that maintains inter-element torsion. Body forces obtained from MAGFOR were compared with body forces from the computer code EFFI for several coil configurations considered in the design of the Advanced Toroidal Facility (ATF).

Publications describing the code:

W. D. Cain, "MAGFOR: A Magnetics Code to Calculate Field and Forces in Twisted Helical Coils of Constant Cross Section," Fusion Engineering, vol. 2 (1983) 1223–1227.

Is code documentation available? ☐ Yes ☐ No

How may the code be obtained?

Source language:
Computers it runs on: machine-independent

It is available as: ☐ Source code, ☐ Executable only

Source Media: ☐ Listing, ☐ Tape, ☐ Diskette, ☐ Cards, ☐ Networks
Tape format:
Diskette size & format:

Available through: ☐ DECNET, ☐ ARPANET, ☐ BITNET
☐

Network Address:

Date of Latest Version: 1977   Program Name: MAGNET

Person to Contact: Program Library
           Address: CERN,
                    DD Div
                    CH-1211 Geneva 23
                    Switzerland

Telephone Number: (22) 83 2377

Classification of Computer Code:
   Component Design:
      ☐ Ion Source, ☒ Magnet, ☐ RF–cavity, ☐
   Accelerator Optimization:
      ☐ Linac, ☐ Cyclotron, ☐ Synchrotron, ☐
   Tracking or Simulation:
      ☐ Linac, ☐ Cyclotron, ☐ Synchrotron, ☐
   Analysis:
      ☐ Stability, ☐ Impedances, ☐
   Other:

Short Description: (Purpose, capabilities, algorithms, special features, etc.)
   Magnet Design in 2 dimensions. Finite Difference Method with two Potentials.

Publications describing the code:
   CERN Program Library Writeup T600

Is code documentation available? ☒ Yes ☐ No

How may the code be obtained?
   (Contact CERN Program Library)

Source language: FORTRAN 66

Computers it runs on: IBM/CDC

It is available as: ☒ Source code, ☐ Executable only

Source Media: ☐ Listing, ☒ Tape, ☐ Diskette, ☐ Cards, ☐ Networks
   Tape format: 9 track 1600 bpi
   Diskette size & format:

Available through: ☐ DECNET, ☐ ARPANET, ☐ BITNET
                  ☐

Network Address:

Date of Latest Version: Oct. 1986   Program Name: MAGNUS

Person to Contact: Sergio Pissanetzky
           Address: Texas Accelerator Center
                    2319 Timberloch
                    The Woodlands, TX 77380
                    USA

Telephone Number: (713)363-0121

Classification of Computer Code:
  Component Design:
    ☐ Ion Source, ☒ Magnet, ☐ RF–cavity, ☐
  Accelerator Optimization:
    ☐ Linac, ☐ Cyclotron, ☐ Synchrotron, ☐
  Tracking or Simulation:
    ☐ Linac, ☐ Cyclotron, ☐ Synchrotron, ☐
  Analysis:
    ☐ Stability, ☐ Impedances, ☐
  Other:

Short Description: (Purpose, capabilities, algorithms, special features, etc.)

The MAGNUS package for 3-D magnetic field calculations consists of the preprocessor KUBIK, the MAGNUS solver, and the postprocessor EPILOG. The program employs the Finite Element Method (FEM), with a total scalar potential in regions with iron and a partial scalar potential in regions with current. The use of the total potential in iron avoids severe round-off errors that would arise if a partial potential were used (small difference between large numbers). The program is interfaced with the international Graphic Kernel Systems (GKS).

The mesh generator KUBIK prompts the user to input all information describing the geometry of the problem and the desired mesh refinement. Names are also assigned to regions or materials, boundaries, etc. KUBIK is a truly 3-D mesh generator. Modules representing simple bodies, or parts of bodies, are independently defined, each with its own mesh inside. The modules are then assembled into the final 3-D structure of any degree of complexity. KUBIK runs either interactively or in batch, and accepts commands that print a variety of tables or plots.

MAGNUS has a library of solid and filament conductor elements, out of which the user can construct conductors of practically any shape in 3-D. Commands exist that will generate new conductors by displacement, reflection or rotation of existing conductors, or produce tables or plots. The independent program WIRE can be used to calculate the field of the conductors alone (no iron or boundary conditions) at any point of space. The conductors are completely independent of the mesh created by KUBIK, and can be changed without modifying the mesh. MAGNUS also has a library of magnetization tables, including several American steels and Japanese steels at different temperatures, and ideal materials such as pure iron or nickel. The user can input additional tables.

The MAGNUS solver obtains a solution in the mathematical sense: the magnetic potential given as a function of the coordinates at every point in the solution domain. A high efficiency and surprisingly short execution times are achieved by means of sophisticated programming and the use of sparse matrix techniques. The solver runs in batch for the number of iterations specified by the user, or until the desired accuracy is obtained. It generates a drop file and can be restarted if so desired.

Once the solution is available, the user runs the postprocessor EPILOG interactively on VAX. EPILOG accepts commands that will compute and print a variety of derived quantities, like Z-averaged harmonic coefficients, spherical 3-D harmonic coefficients, energy, inductances, line or surface integrals of the field, etc. EPILOG can also generate tables of field, permeability, potential, etc., and a variety of plots. EPILOG is a very useful design tool for the physicist or engineer.

Accuracy is a primary consideration in the MAGNUS package. It is now known that a very important source of large errors is the inappropriate interpolation of magnetization tables. In MAGNUS, interpolation is done using the FKP interpolation relation, the most accurate known rule. Another source of inaccuracy is round-off in the calculation of the field of solid conductors. Careful mathematical/numerical techniques have been used to rewrite the expressions in such a way that round-off will not affect the results. The accuracy, even in single precision, is better than with the usual expressions in double precision. Numerical quadature in MAGNUS is done by Carl de Boor's method, which guarantees the final accuracy, rather than by the usual n-point Gauss formula, which is very inaccurate in many cases but extensively used in other programs.

Publications describing the code:

S. Pissanetzky, "The Interpolation of Magnetization Tables," COMPEL 5(1986)41-56.

S. Pissanetzky, "The Design of Superferric Magnets for the Superconducting Super Collider and the New Program MAGNUS for Three-Dimensional Magnetostatics," IEEE Trans. MAG-21 (1985) 2457.

S. Pissanetzky, "The New Version of the Finite Element 3D Magnetostatics Program MAGNUS," Comp. Electromagnetics (1986) 121-32, Ed. Elsevier Science Publ. B. V. (North Holland).

S. Pissanetzky, "Automatic Three-Dimensional Finite Element Mesh Generation Using the Program KUBIK," Computers Phys. Comm. 32 (1984) 245-65. (See also CPC 32 (1984) 267-74.)

S. Pissanetzky, "Sparse Matrix Technology," Academic Press, London (1984).

S. Pissanetzky, "KUBIK: An Automatic Three-Dimensional Finite Element Mesh Generator," Int. J. Num. Meth. Engng. 17(1981)255-69.

Is code documentation available? ☒ Yes ☐ No

How may the code be obtained?

By license agreement with Ferrari High Technology Products, P.O. Box 1866, Orange Park, FL 32067-1866.

By agreement with the Texas Accelerator Center, the code is also available free of charge to some HEP groups at National Laboratories through the MFE network.

Source language: FORTRAN IV

Computers it runs on: VAX and CRAY

It is available as: ☒ Source code, ☐ Executable only

Source Media: ☐ Listing, ☒ Tape, ☐ Diskette, ☐ Cards, ☐ Networks
Tape format:
Diskette size & format:

Available through: ☐ DECNET, ☐ ARPANET, ☐ BITNET
☒ NMFECC

Network Address:

Date of Latest Version: July 1986

Program Name: MAPPOT

Person to Contact: Etienne Forest
Address: MS 90/4040, URA Design Center
c/o UCLBL
Berkeley, CA 94720
USA

Telephone Number: (415) 486-6580, FTS 451-6580

Classification of Computer Code:
  Component Design:
    ☐ Ion Source, ☐ Magnet, ☐ RF–cavity, ☐
  Accelerator Optimization:
    ☐ Linac, ☐ Cyclotron, ☐ Synchrotron, ☐
  Tracking or Simulation:
    ☐ Linac, ☐ Cyclotron, ☒ Synchrotron, ☐
  Analysis:
    ☐ Stability, ☐ Impedances, ☐
  Other:

Short Description: (Purpose, capabilities, algorithms, special features, etc.)
  MAPPOT generates a Lie-algebraic map that produces tracking results equivalent to those produced by the third-order matrix code RACETRACK. The output can be put into MARYLIE for the calculation of auxilary quantities such as chromaticity and nonlinear invariants.

Publications describing the code:
  None yet.

Is code documentation available? ☐ Yes ☒ No

How may the code be obtained?

Source language: FORTRAN 77

Computers it runs on: CRAY-XMP

It is available as: ☐ Source code, ☐ Executable only

Source Media: ☐ Listing, ☐ Tape, ☐ Diskette, ☐ Cards, ☐ Networks
  Tape format:
  Diskette size & format:

Available through: ☐ DECNET, ☐ ARPANET, ☐ BITNET
                   ☐

Network Address:

Date of Latest Version: unknown   Program Name: MARTUR

Person to Contact: I. S. Baksjev
         Address: Institute of High Energy Physics
                  Serpukhov
                  USSR

Telephone Number:

Classification of Computer Code:
    Component Design:
        ☐ Ion Source, ☐ Magnet, ☐ RF–cavity, ☐
    Accelerator Optimization:
        ☐ Linac, ☐ Cyclotron, ☐ Synchrotron, ☐
    Tracking or Simulation:
        ☐ Linac, ☐ Cyclotron, ☐ Synchrotron, ☐
    Analysis:
        ☐ Stability, ☐ Impedances, ☐
    Other: radiation loading calculations

Short Description: (Purpose, capabilities, algorithms, special features, etc.)
    A set of MARTUR computer codes is created to calculate spatial distributions of high-energy proton losses in accelerator structures and arising energy release, induced radioactivity, radiation loadings on equipment and shields. The set of computer codes consists of three major codes: ESSEPT—the code for simulation of initial interaction of circulating proton beam with the energy E_O with an arbitrary "target"; TURTLM—the code for calculation of particle transport; MARSGT—the code for nuclear-electromagnetic cascade calculation, which appear in magnetic structure and due to the loss of transported fast protons.

Publications describing the code:
    Bajshev, I.S.; Maslov, M.A.; Mokhov, N.V., "MARTUR Set of Computer Codes for Calculation of Particle Interactions and Transport in Proton Accelerators", All-Union Conference on Charged Particle Accelerators Proceedings vol. 2. (1983) 167–170.

Is code documentation available? ☐ Yes ☐ No

How may the code be obtained?

Source language:
Computers it runs on:
It is available as: ☐ Source code, ☐ Executable only
Source Media: ☐ Listing, ☐ Tape, ☐ Diskette, ☐ Cards, ☐ Networks
    Tape format:
    Diskette size & format:
Available through: ☐ DECNET, ☐ ARPANET, ☐ BITNET
                    ☐
Network Address:

Date of Latest Version: Dec. 1986

Program Name: MARYLIE

Person to Contact: Alex J. Dragt
Address: Physics Department
University of Maryland
College Park, MD 20742
USA

Telephone Number: (301) 454-7324

Classification of Computer Code:
Component Design:
☐ Ion Source, ☐ Magnet, ☐ RF–cavity, ☐

Accelerator Optimization:
☐ Linac, ☐ Cyclotron, ☒ Synchrotron, ☒ Beam Lines and Storage Rings.

Tracking or Simulation:
☐ Linac, ☐ Cyclotron, ☒ Synchrotron, ☒ Beam Lines and Storage Rings.

Analysis:
☒ Stability, ☐ Impedances, ☐

Other: Nonlinear Orbit Behavior, aberrations

Short Description: (Purpose, capabilities, algorithms, special features, etc.)

The program employs algorithms based on a Lie-algebraic formulation of charged particle trajectory calculations, and is able to compute transfer maps for and trace rays through single or multiple beam-line elements. This is done for the full 6-dimensional phase space. All nonlinearities, including chromatic effects, through third (octupole) order are included. In addition, MARYLIE is exactly symplectic (canonical) through all orders.

MARYLIE may be used both for particle tracking around or through a lattice and for analysis of linear and nonlinear lattice properties. Tracking can be performed element to element, lump to lump, or any mixture of the two. (A lump is a collection of elements combined together and treated by a single transfer map.) The speed for element to element tracking is comparable to that of other codes. When collections of elements can be lumped together to form single transfer maps, tracking speeds can be orders of magnitude faster.

MARYLIE also has powerful analytic tools. They include the calculation of first, second, and third order dispersion; tunes and first and second order chromaticities; the other linear lattice functions and their energy dependence through second order; the dependence of tune on betatron amplitude; nonlinear lattice functions; nonlinear phase-space distortion; transfer map normal forms; nonlinear resonance driving terms; and nonlinear invariants. Finally, MARYLIE can be used to give an explicit representation for the linear and nonlinear properties of the total transfer map of a system. This information can be used to evaluate or improve the optical quality of a single pass system such as a beam transport line or linear collider. MARYLIE 3.0 running (in double precision) on a VAX 11/785 requires 140K bytes of memory, and can evaluate (track) approximately 25 maps per second.

A vectorized version running on a CRAY X-MP (in single precision and using only one processor) requires 212K words of memory, and can evaluate approximately 4000 maps per second.

Publications describing the code:

D. R. Douglas, "Lie Algebraic Methods for Particle Accelerator Theory," Ph.D. thesis, Univ. Md. (1982) unpublished.

Alex J. Dragt, Robert D. Ryne, Liam M. Healy, Filippo Neri, David R. Douglas, Etienne Forest, "MARYLIE 3.0, A Program for Charged Particle Beam Transport Based on Lie Algebraic Methods."

David R. Douglas and Etienne Forest, "A Program for Nonlinear Analysis of Accelerator and Beamline Lattices," IEEE Trans. NS-32 (1985) 2311.

Is code documentation available? ☒ Yes ☐ No

How may the code be obtained?
From Maryland when released.

Source language: FORTRAN 77

Computers it runs on: CRAY, IBM, VAX UNIVAC, CDC.

It is available as: ☒ Source code, ☐ Executable only

Source Media: ☐ Listing, ☒ Tape, ☐ Diskette, ☐ Cards, ☐ Networks
Tape format:
Diskette size & format:

Available through: ☐ DECNET, ☒ ARPANET, ☒ BITNET
☐

Network Address: dragt@umcincom

Date of Latest Version: unknown   Program Name: MASK

Person to Contact: Adam Drobot
Address: SAI
1710 Goodridge Dr.
McLean, VA 22102
USA

Telephone Number:

Classification of Computer Code:
Component Design:
☒ Ion Source, ☐ Magnet, ☐ RF-cavity, ☐
Accelerator Optimization:
☐ Linac, ☐ Cyclotron, ☐ Synchrotron, ☐
Tracking or Simulation:
☐ Linac, ☐ Cyclotron, ☐ Synchrotron, ☐
Analysis:
☐ Stability, ☐ Impedances, ☐
Other:

Short Description: (Purpose, capabilities, algorithms, special features, etc.)
The MASK code is a 2-1/2 dimensional particle-in-cell code which has been applied to the simulation of a number of microwave devices. It has been used to simulate klystrons and other electron accelerator components.

Publications describing the code:
Eppley, K.; Brandon. S.; Drobot, A.; Hanerfeld, H.; Herrmannsfeldt, W.; Melendez, R.; Nielsen, D.; Yu. S., "Results of Simulations of High-power Klystrons," Proc. of Part. Accl. Conf. published in Trans. IEEE NS-32 (1985) 2903–2905.

Yu, S. S.; Drobot, A.; Wilson, P., "Two and One-half Dimension Particle-in-cell Simulation of High-power Klystrons," Proc. of Part. Accl. Conf. published in Trans. IEEE NS-32 (1985) 2918–2920.

Hanerfeld. H., "Computational Needs for Modelling Accelerator Components," SLAC internal report, SLAC-PUB-3708.

Is code documentation available? ☐ Yes ☐ No

How may the code be obtained?
From Adam Drobot. (The code may also be available from SLAC, for instance from W. Herrmansfeldt.)

Source language:
Computers it runs on:
It is available as: ☐ Source code, ☐ Executable only
Source Media: ☐ Listing, ☐ Tape, ☐ Diskette, ☐ Cards, ☐ Networks
Tape format:
Diskette size & format:
Available through: ☐ DECNET, ☐ ARPANET, ☐ BITNET
☐
Network Address:

Date of Latest Version: July 1986

Program Name: MATRACE

Person to Contact: Etienne Forest
Address: MS 90/4040
URA Design Center
c/o UCLBL
Berkeley, CA 94720

Telephone Number: (415) 486-6580, FTS 451-6580

Classification of Computer Code:
Component Design:
☐ Ion Source, ☐ Magnet, ☐ RF–cavity, ☐
Accelerator Optimization:
☐ Linac, ☐ Cyclotron, ☐ Synchrotron, ☐
Tracking or Simulation:
☐ Linac, ☐ Cyclotron, ☒ Synchrotron, ☐
Analysis:
☐ Stability, ☐ Impedances, ☐
Other:

Short Description: (Purpose, capabilities, algorithms, special features, etc.)

MATRACE can be used as a postprocessor to the code RACETRACK. It will generate up to third-order matrices for variables (x, x', y, y') that give equivalent tracking results for the full ring relative to a given trajectory. This is useful for studying misalignments. The output of MATRACE can be put into MARYLIE, which can generate such quantities as chromaticity and non-linear invariants.

Publications describing the code:

E. Forest, "Lie Algebraic Maps and Invariants Produced by Tracking Codes," SSC Design Group internal report SSC-78 (1986).

Is code documentation available? ☐ Yes ☒ No

How may the code be obtained?
Call Etienne Forest.

Source language: FORTRAN

Computers it runs on: VAX, CRAY-XMP

It is available as: ☐ Source code, ☐ Executable only

Source Media: ☐ Listing, ☒ Tape, ☐ Diskette, ☐ Cards, ☐ Networks
Tape format: As desired.
Diskette size & format:

Available through: ☐ DECNET, ☐ ARPANET, ☐ BITNET ☐

Network Address:

Date of Latest Version: Dec. 1985

Program Name: MEBT

Person to Contact: C. T. Mottershead
Address: MS H829, Group AT-6
Los Alamos National Laboratory
Los Alamos, NM 87545
USA

Telephone Number: (505) 667-9730 FTS 843-9730

Classification of Computer Code:
Component Design:
☐ Ion Source, ☐ Magnet, ☐ RF-cavity, ☐
Accelerator Optimization:
☐ Linac, ☐ Cyclotron, ☐ Synchrotron, ☐
Tracking or Simulation:
☐ Linac, ☐ Cyclotron, ☐ Synchrotron, ☐
Analysis:
☐ Stability, ☐ Impedances, ☐
Other: Beam Diagnostics

Short Description: (Purpose, capabilities, algorithms, special features, etc.)

MEBT = Maximum Entropy Beam Tomography. Intense particle beams require noninterceptive diagnostics. One of these is the light emitted from interaction of the beam with residual gas. If the light is produced by a first-order process linear in the beam density, its profile measured across the beam may be interpreted as a tomographic projection of that density distribution. With a small number of such projections, and appropriate transfer matrices connecting them, Minerbo's maximum entropy (MENT) algorithm may be used to construct an estimate of the beam density distribution in both coordinate and phase space. This MENT algorithm is running as part of an integrated software system on an LSI 11/23 mounted in the same diagnostic node where the data is recorded. The solution usually converges in about 5 iterations, each of which takes a few seconds in the typical case of 3 or 4 views of 25 samples each. The subroutine implementing the MENT algorithms are being rewritten to make them portable.

Publications describing the code:

C. T. Mottershead, "Maximum Entropy Beam Diagnostic Tomography," IEEE Trans. NS-32 (1985) 1970.

G. N. Minerbo, "MENT: A Maximum Entropy Algorithm for Reconstructing a Source from Projection Data," Comp. Graphics Image Proc., 10 (1979) 48.

O. R. Sander, G. N. Minerbo, R. A. Jameson, and D. D. Chamberlin, "Beam Tomography in Two and Four Dimensions," Proc. 1979, Linac Conf. Brookhaven National Laboratory report BNL-51134 (1980).

Is code documentation available? ☐ Yes ☒ No

How may the code be obtained?
Contact Tom Mottershead.

Source language: FORTRAN

Computers it runs on: PDP11 & VAX

It is available as: ☒ Source code, ☐ Executable only

Source Media: ☐ Listing, ☒ Tape, ☐ Diskette, ☐ Cards, ☒ Networks
   Tape format:
   Diskette size & format:

Available through: ☐ DECNET, ☒ ARPANET, ☒ BITNET
                                ☐

Network Address: ctm@lanl.arpa

Date of Latest Version: Dec. 1985

Program Name: MICADO

Person to Contact: Yolande Marti
Address: LEP Theory Division
CERN
1211 Geneva 23
Switzerland

Telephone Number: (022) 832948

Classification of Computer Code:
Component Design:
☐ Ion Source, ☐ Magnet, ☐ RF–cavity, ☐
Accelerator Optimization:
☐ Linac, ☐ Cyclotron, ☐ Synchrotron, ☐
Tracking or Simulation:
☐ Linac, ☐ Cyclotron, ☐ Synchrotron, ☐
Analysis:
☐ Stability, ☐ Impedances, ☒ Error correction and closed orbit correction.

Other:

Short Description: (Purpose, capabilities, algorithms, special features, etc.)

MICADO is a solver of rectangular systems of linear equations. It is recommended for over-determined systems (more equations than unknowns) The solution is iterative and gives at each iteration the most efficient solutions to reduce the norm of the residual vector. It has been extensively tested for orbit correction. A recent application has been made to dynamic aperture correction.

Publications describing the code:

B. Auhn and Y. Marti, "Closed orbit correction of A. G. machines using a small number of magnets," CERN/ISR-MA/73-17.

G. Guignard, Marti, "PETROC users' guide," CERN ISR-BOM-TH/81-32

Is code documentation available? ☒ Yes ☐ No

How may the code be obtained?
Contact LEP Division

Source language: FORTRAN 77

Computers it runs on: IBM

It is available as: ☒ Source code, ☐ Executable only

Source Media: ☒ Listing, ☒ Tape, ☐ Diskette, ☐ Cards, ☒ Networks
Tape format: unlabeled, 1600 Bpi, 3200 char/block, 80 char/record
Diskette size & format:

Available through: ☐ DECNET, ☐ ARPANET, ☒ BITNET
☐

Network Address: MAR at CERNVM.

Date of Latest Version: Jan. 1986

Program Name: MIRKO

Person to Contact: Bernhard J. Franczak
Address: c10 GSI
Postfach 110541
D-6100 Darmstadt-11
Fed. Rep. Germany

Telephone Number: 49 6151-359370

Classification of Computer Code:
Component Design:
☐ Ion Source, ☐ Magnet, ☐ RF-cavity, ☐
Accelerator Optimization:
☐ Linac, ☐ Cyclotron, ☒ Synchrotron, ☒ Beam lines

Tracking or Simulation:
☐ Linac, ☐ Cyclotron, ☒ Synchrotron, ☒ Beam lines

Analysis:
☐ Stability, ☐ Impedances, ☐
Other:

Short Description: (Purpose, capabilities, algorithms, special features, etc.)

Purpose: Design of synchrotrons and beam lines, optimization of focusing elements in first order.

Algorithms: Linear matrix formalism for transformation of ellipses and single particles; tracking of single particles through linear matrices and thin non-linear lenses representing multipoles.

Special features:

Interactive operation employing a command structure.

Graphic output of envelopes, ellipses, and particle distributions.

On-line help available

Interactive graphics using the cursor

Detailed investigation of non-linearities in synchrotrons.

Publications describing the code:

B. Franczak, "MIRKO—An Interactive Program for Beam Lines and Synchrotrons," Conf. on Computing in Accelerator Design and Operation, Berlin (W) (1983) 170.

Springer-Verlag (1984).

Is code documentation available? ☒ Yes ☐ No

How may the code be obtained?
Contact author

Source language: FORTRAN 77

Computers it runs on: VAX, IBM

It is available as: ☒ Source code, ☐ Executable only
Source Media: ☐ Listing, ☒ Tape, ☐ Diskette, ☐ Cards, ☒ Networks
   Tape format: as needed
   Diskette size & format:
Available through: ☐ DECNET, ☐ ARPANET, ☐ BITNET
   ☒ EARN

Network Address: PT01 at DDAGSI3

Date of Latest Version: 1983 Program Name: MISAR

Person to Contact: Donald A. Swenson
Address: SAI
505 Marquette N.W., Suite 1200
Albuquerque, NM 87102
USA

Telephone Number: (505) 247-8787

Classification of Computer Code:
Component Design:
☐ Ion Source, ☐ Magnet, ☐ RF–cavity, ☐
Accelerator Optimization:
☐ Linac, ☐ Cyclotron, ☐ Synchrotron, ☐
Tracking or Simulation:
☐ Linac, ☐ Cyclotron, ☐ Synchrotron, ☒ Accumulator Rings

Analysis:
☐ Stability, ☐ Impedances, ☐
Other:

Short Description: (Purpose, capabilities, algorithms, special features, etc.)
MISAR is a PARMILA-like multiparticle simulation code that follows the transverse coordinates of a collection of macroparticles from the inflector, around the lattice of a circular machine, incorporating the space-charge forces of the beam as modified by the image effects in the walls, for a number of turns. No longitudinal structure in the beam is allowed. After each turn around the machine, the number of particles in the "beam" is increased by the addition of a new quantity of beam from the inflector. Provisions are made for time-dependent pulsed bumps that can move the equilibrium orbit in the vicinity of the inflector to establish the multiturn injection process. As in PARMILA, there are a variety of options for generating the initial coordinates of the inflected beam and a variety of options for displaying the properties of the accumulated beam. The space-charge effects are supplied by a SCHEFF subroutine, which is called one or more times during each basic period of the lattice.

Publications describing the code:
D. A. Swenson and K. R. Crandall, "MISAR: A Particle Tracking Code for Multiturn Injection Studies in Accumulator Rings," Proc. of Workshop on Accelerator Orbit and Particle Tracking Programs, Brookhaven 1982, Brookhaven Internal Report BNL-31761. Also Los Alamos Report LA-UR-82-1585.

Is code documentation available? ☒ Yes ☐ No

How may the code be obtained?
Los Alamos Accelerator Code Group. (Call John Warren at (505) 667-6677 (or 667-2839), FTS 843-6677).

Source language: FORTRAN

Computers it runs on: CDC 7600

It is available as: ☒ Source code, ☐ Executable only

Source Media: ☐ Listing, ☒ Tape, ☐ Diskette, ☐ Cards, ☒ Networks
Tape format:
Diskette size & format:

Available through: ☐ DECNET, ☒ ARPANET, ☒ BITNET
☐

Network Address: hks@lanl.arpa

Date of Latest Version: 1979   Program Name: MOTER

Person to Contact: Edward A. Heighway
        Address: MS H829, Group AT-6
                    Los Alamos National Laboratory
                    Los Alamos, NM 87545
                    USA

Telephone Number: (505) 667-1543, FTS 843-1543

Classification of Computer Code:
    Component Design:
        ☐ Ion Source, ☐ Magnet, ☐ RF–cavity, ☐
    Accelerator Optimization:
        ☐ Linac, ☐ Cyclotron, ☐ Synchrotron, ☒   Beam lines, mass spectrometers

    Tracking or Simulation:
        ☐ Linac, ☐ Cyclotron, ☐ Synchrotron, ☐
    Analysis:
        ☐ Stability, ☐ Impedances, ☐
    Other:

Short Description: (Purpose, capabilities, algorithms, special features, etc.)
    MOTER is a ray tracing program intended for analysis and optimization of a system of magnetic elements. Several features are included in MOTER which are not available in other codes. Among these are Monte Carlo simulation of the beam phase space, a sophisticated definition of the performance including the possibility of computer correction of aberrations based on measurements of the trajectory of each event, the automatic optimization of any parameter of the magnet system, the possibility of the use of field maps for dipoles, quadrupoles, and multipoles, and the availability of several new element types including an ExB separator, an r.f. accelerating gap, a wedge degrader, and various slits and scatterers. To the greatest possible extent, MOTER makes use of the definition of parameters identical to program RAYTRACE from which it evolved. In order to minimize the pitfalls of problem setup, it is suggested that the MOTER user first study his problem with the standard codes TRANSPORT, TURTLE, and RAYTRACE, in that order.

Publications describing the code:
    H. A. Thiessen and M. Klein, "Design of Mass Spectrometer at LASL," NTIS CONF-7209208, Proc. IV Int'l Conf. on Magnet Technology, Brookhaven National Laboratory (1972) 8.

Is code documentation available? ☒ Yes ☐ No

How may the code be obtained?
    Contact the Los Alamos Accelerator Code Group (505) 667-6677 (or 667-2839), FTS 843-6677.

Source language: FORTRAN

Computers it runs on: VAX, CRAY

It is available as: ☒ Source code, ☐ Executable only

Source Media: ☐ Listing, ☒ Tape, ☐ Diskette, ☐ Cards, ☒ Networks
    Tape format:
    Diskette size & format:

Available through: ☐ DECNET, ☒ ARPANET, ☒ BITNET
                                          ☐

Network Address: hks@lanl.arpa

Date of Latest Version: Apr. 1986

Program Name: MOTION

Person to Contact: Klaus Bongardt, ASI
Address: KFA Jülich
Postfach 1913
D-5170 Jülich,
Fed. Rep. Germany

Telephone Number: (02461)61 3544

Classification of Computer Code:
    Component Design:
        ☐ Ion Source, ☐ Magnet, ☐ RF–cavity, ☐
    Accelerator Optimization:
        ☐ Linac, ☐ Cyclotron, ☐ Synchrotron, ☐
    Tracking or Simulation:
        ☒ Linac, ☐ Cyclotron, ☐ Synchrotron, ☒ Beam lines
    Analysis:
        ☐ Stability, ☐ Impedances, ☐
    Other:

Short Description: (Purpose, capabilities, algorithms, special features, etc.)

    Macroparticle tracking code, non-relativistic. Integrates equations of motion through any user-defined external (electric and/or magnetic) field. Built-in components: quadrupoles, solenoids, dipoles, bunching cavities, octupoles, rf gaps. Full 3D space charge. No edge focusing in dipole magnets. Initial distributions available: 4D waterbag, 4D K-V, user defined. Highly specialized and time consuming. Good for design and analysis of low-energy beam lines.

Publications describing the code:

    "MOTION—A Versatile Multiparticle Simulation Code" by K. Mittag and D. Sanitz, Proc. 1981 Linac Conference, Los Alamos Report LA-9234-c, 156–158.

Is code documentation available? ☒ Yes ☐ No

How may the code be obtained?
    Contact Klaus Bongardt.

Source language: FORTRAN 77

Computers it runs on: CRAY, IBM 3081

It is available as: ☐ Source code, ☐ Executable only

Source Media: ☐ Listing, ☒ Tape, ☐ Diskette, ☐ Cards, ☐ Networks
    Tape format:
    Diskette size & format:

Available through: ☐ DECNET, ☐ ARPANET, ☐ BITNET
    ☐

Network Address:

Date of Latest Version: unknown

Program Name: MULTIMODE

Person to Contact: A. I. Fedoseyev or V. V. Gusev
Address: Institute for High Energy Physics,
Serpukhov,
Moscow region,
USSR

Telephone Number:

Classification of Computer Code:
Component Design:
☐ Ion Source, ☐ Magnet, ☒ RF–cavity, ☐
Accelerator Optimization:
☐ Linac, ☐ Cyclotron, ☐ Synchrotron, ☐
Tracking or Simulation:
☐ Linac, ☐ Cyclotron, ☐ Synchrotron, ☐
Analysis:
☐ Stability, ☐ Impedances, ☐
Other:

Short Description: (Purpose, capabilities, algorithms, special features, etc.)
The MULTIMODE code is used for computing the lowest eigenfrequencies and electromagnetic fields in homogeneous waveguides and axially symmetric cavities. Eight-node isoparametric finite elements are used, which give the exact approximation of curvilinear region boundaries and high accuracy in computations of frequencies on a small number of grid nodes. The comparison of MULTIMODE with other programs shows that MULTIMODE attains the same accuracy while running 10-100 times faster. The program allows the computation of both simple and degenerate frequencies.

Publications describing the code:
"MULTIMODE—A Powerful Code for Frequency Spectrum Computation of Electromagnetic Fields in Axially Symmetric Cavities and Longitudinally Homogeneous Waveguides of Arbitrary Shape," Nuclear Instru. & Methd. 227 (1984) 411–19.

Is code documentation available? ☐ Yes ☐ No

How may the code be obtained?

Source language:
Computers it runs on:
It is available as: ☐ Source code, ☐ Executable only
Source Media: ☐ Listing, ☐ Tape, ☐ Diskette, ☐ Cards, ☐ Networks
Tape format:
Diskette size & format:
Available through: ☐ DECNET, ☐ ARPANET, ☐ BITNET
☐
Network Address:

Date of Latest Version: 1980   Program Name: NAJO

Person to Contact: J. Sauret or A. Chambert
Address: GANIL
Boite Postal 5027
F. 14000 CAEN
France

Telephone Number: 31.45.46.47

Classification of Computer Code:
Component Design:
     ☐ Ion Source, ☐ Magnet, ☐ RF–cavity, ☐
Accelerator Optimization:
     ☐ Linac, ☐ Cyclotron, ☐ Synchrotron, ☐
Tracking or Simulation:
     ☐ Linac, ☒ Cyclotron, ☐ Synchrotron, ☐
Analysis:
     ☐ Stability, ☐ Impedances, ☐
Other:

Short Description: (Purpose, capabilities, algorithms, special features, etc.)

This is a general multiparticle code developed for studying particle motion in cyclotrons. Related to the structure of the GANIL separated-sector cyclotrons, it could be adapted to other configurations.

Its main limitation comes from the shape of the accelerating gaps which are presently restricted to radial ones. Accelerating field effects are expressed as kicks applied at the gap centers allowing for a complete decoupling in the treatment of the magnetic and electric fields.

Simplified versions have been derived, restricted either to the median plane (JOAN) or to a single particle in this plane (ANJO). In its most general version the code takes into account space charge effects. A more precise description of these codes including listings and examples is given in the internal report listed below.

Publications describing the code:
     J. Sauret, A. Chabert and M. Prome, "Multiparticle Codes Developed at GANIL," in the Proc. of the Conf. on "Accelerator Design and Operation" Berlin (1983) 164–9, Springer-Verlag, 1984.

     Le Groupe Theorie Parametres, "Les Programmes ANJO, JOAN, NAJO," GANIL Internal Report 80R/132/TP/06.

Is code documentation available? ☒ Yes ☐ No

How may the code be obtained?
     J. Sauret or A. Chambert

Source language: FORTRAN

Computers it runs on: UNIVAC 1108, (MODCOMP in the near future)

It is available as: ☒ Source code, ☐ Executable only
Source Media: ☒ Listing, ☒ Tape, ☒ Diskette, ☐ Cards, ☐ Networks
     Tape format: 1600 BPI
     Diskette size & format: IBM.32.70
Available through: ☐ DECNET, ☐ ARPANET, ☐ BITNET
     ☐
Network Address:

Date of Latest Version: 1983

Program Name: OPTIC II

Person to Contact: Joe Tesmer
   Address: MS K764, Group P-10
   Los Alamos National Laboratory
   Los Alamos, NM 87545
   USA

Telephone Number: (505) 667-6370, FTS 843-6370

Classification of Computer Code:
   Component Design:
      ☐ Ion Source, ☐ Magnet, ☐ RF-cavity, ☐
   Accelerator Optimization:
      ☐ Linac, ☐ Cyclotron, ☐ Synchrotron, ☒ Electrostatic Accelerator

   Tracking or Simulation:
      ☐ Linac, ☐ Cyclotron, ☐ Synchrotron, ☒ Electrostatic Accelerator

   Analysis:
      ☐ Stability, ☐ Impedances, ☐
   Other:

Short Description: (Purpose, capabilities, algorithms, special features, etc.)
   Early program used for beam optics in electrostatic accelerators. Good treatment of accelerating tubes and strippers. Time dispersion treatment of bunched beams.

   Custodian: J. D. Larson, 10011 E 35th Terr., Independence, MO 64052

   Based on code by T. J. Devlin (UCRL-9727)

Publications describing the code:
   T. J. Deubin, Univ. of California Internal Report UCRL 9727

   S. Penner, "Calculations of Properties of Magnetic Deflection Systems," Rev. Sci. Inst. 32 (1961) 150.

   J. D. Larson, "New Developments in Beam Transport Through Tandem Accelerators," Nuclear Inst. and Methods, 122 (1974) 53-63.

Is code documentation available? ☒ Yes ☐ No

How may the code be obtained?
   Contact Joe Tesmer.

Source language: FORTRAN

Computers it runs on: VAX

It is available as: ☒ Source code, ☐ Executable only

Source Media: ☐ Listing, ☒ Tape, ☐ Diskette, ☐ Cards, ☐ Networks
   Tape format: As desired.
   Diskette size & format:

Available through: ☐ DECNET, ☐ ARPANET, ☐ BITNET
   ☐

Network Address:

Date of Latest Version: unknown   Program Name: OSCAR2D

Person to Contact: Paolo Fernandez
        Address: Instituto per la Matematica Applicata
                 Consigilo Nazionale delle Ricerche
                 Via L. B. Alberti, 4
                 16132 Genova, ITALY

Telephone Number: unkown

Classification of Computer Code:
    Component Design:
        ☐ Ion Source, ☐ Magnet, ☒ RF–cavity, ☐
    Accelerator Optimization:
        ☐ Linac, ☐ Cyclotron, ☐ Synchrotron, ☐
    Tracking or Simulation:
        ☐ Linac, ☐ Cyclotron, ☐ Synchrotron, ☐
    Analysis:
        ☐ Stability, ☐ Impedances, ☐
    Other:

Short Description: (Purpose, capabilities, algorithms, special features, etc.)
    This is a 2–D rf cavity code. (No more information available at this time)

Publications describing the code:
    unknown

Is code documentation available? ☐ Yes ☐ No

How may the code be obtained?

Source language:
Computers it runs on:
It is available as: ☐ Source code, ☐ Executable only
Source Media: ☐ Listing, ☐ Tape, ☐ Diskette, ☐ Cards, ☐ Networks
    Tape format:
    Diskette size & format:
Available through: ☐ DECNET, ☐ ARPANET, ☐ BITNET
                   ☐
Network Address:

Date of Latest Version: Nov. 1985   Program Name: PANDIRA GROUP CODES

Person to Contact: Los Alamos Accelerator Code Group
         Address: MS H829, Group AT-6
                  Los Alamos National Laboratory
                  Los Alamos, NM 87545
                  USA

Telephone Number: (505) 667-9131 or 667-2839, FTS 843-9131

Classification of Computer Code:
   Component Design:
      ☐ Ion Source, ☒ Magnet, ☐ RF-cavity, ☐
   Accelerator Optimization:
      ☐ Linac, ☐ Cyclotron, ☐ Synchrotron, ☐
   Tracking or Simulation:
      ☐ Linac, ☐ Cyclotron, ☐ Synchrotron, ☐
   Analysis:
      ☐ Stability, ☐ Impedances, ☐
   Other:

Short Description: (Purpose, capabilities, algorithms, special features, etc.)

   Calculates static magnetic field in two cartesian dimensions or cylindrically symmetric configurations in 3D. Handles permanent magnet materials as well as permeable iron and current carrying coils. Can solve ferroelectric problems also. Uses the "direct" method to solve the 2D, generalized POISSON equation. Included in the group of codes is AUTOMESH, LATTICE, and FORCE.

Publications describing the code:

   K. Halbach, "Design of Permanent Multipole Magnets with Oriented Rare Earth Cobalt Materials," Nucl. Inst. and Meth., 169 (1980) 1-10

Is code documentation available? ☒ Yes ☐ No

How may the code be obtained?

   Send blank tape to above address; specifying version desired (VAX or CRAY). Also available through ARPANET, DECNET or BITNET; telephone for instructions.

Source language: FORTRAN 77

Computers it runs on:
It is available as: ☒ Source code, ☐ Executable only

Source Media: ☐ Listing, ☒ Tape, ☐ Diskette, ☐ Cards, ☒ Networks
   Tape format: 9 track, 1600 bpi, 80 char/line
   Diskette size & format:

Available through: ☐ DECNET, ☒ ARPANET, ☒ BITNET
                    ☐

Network Address: hks@lanl.arpa

Date of Latest Version: unknown Program Name: PAQUASEX

Person to Contact: S. Kheifets
   Address: Stanford Linear Accelerator Center
            Stanford University,
            Stanford, CA 94305
            USA

Telephone Number:

Classification of Computer Code:
   Component Design:
      ☐ Ion Source, ☐ Magnet, ☐ RF–cavity, ☐
   Accelerator Optimization:
      ☐ Linac, ☐ Cyclotron, ☒ Synchrotron, ☐
   Tracking or Simulation:
      ☐ Linac, ☐ Cyclotron, ☐ Synchrotron, ☐
   Analysis:
      ☐ Stability, ☐ Impedances, ☐
   Other:

Short Description: (Purpose, capabilities, algorithms, special features, etc.)

PAQUASEX is essentially a combination of the three codes PATRICIA, QUADS, and MICROSEX. The system is designed to do a configuration survey over a grid of points in the space of main configuration parameters ν_x, ν_y, β_x^*, β_y^* and η_x^* (the star means the value of a parameter at the interaction point).

The system starts by preparing with the help of PATMOD input data decks for QUADS and MICROSEX, i.e. target values of desired parameters. One option prepares a deck for a grid of 5x5 points in ν_x, ν_y space. The other option prepares five sets of five points. Each set of five points are increments in one of the five above-mentioned parameters (keeping all others fixed). These options are selected by means of the control code number KW(16).

Publications describing the code:

S. Kheifets, "Tracking Studies in PEP and Description of the Computer Code PATRICIA," in Proc. Workshop Orbit and Particle Tracking Programs at BNL, (1982) BNL informal report BNL-31761.

Is code documentation available? ☐ Yes ☐ No

How may the code be obtained?
   unknown

Source language:
Computers it runs on:
It is available as: ☐ Source code, ☐ Executable only
Source Media: ☐ Listing, ☐ Tape, ☐ Diskette, ☐ Cards, ☐ Networks
   Tape format:
   Diskette size & format:
Available through: ☐ DECNET, ☐ ARPANET, ☐ BITNET
                   ☐
Network Address:

Date of Latest Version: July 1986   Program Name: PAR2DOPT

Person to Contact: Gerry Morgan
          Address: Brookhaven National Lab.
                   Bldg. 902B
                   Upton, Long Island, NY 11973
                   USA

Telephone Number: (516) 282-4841, FTS 666-4841

Classification of Computer Code:
    Component Design:
        ☐ Ion Source, ☒ Magnet, ☐ RF-cavity, ☐
    Accelerator Optimization:
        ☐ Linac, ☐ Cyclotron, ☐ Synchrotron, ☐
    Tracking or Simulation:
        ☐ Linac, ☐ Cyclotron, ☐ Synchrotron, ☐
    Analysis:
        ☐ Stability, ☐ Impedances, ☐
    Other:

Short Description: (Purpose, capabilities, algorithms, special features, etc.)
   PAR2DOPT is used in the optimization of coil placement of circular cross section magnets of the type used in the design of SSC magnets. It optimizes the azimuthal position and tilt of layered turns and the number of turns per block. The layers of turns need not be fully keystoned, but can be shimmed with wedge-shaped material. The code can be run with or without infinite permeability iron surrounding the coils. The optimizer is based on the CERN code MINUIT. It has some Tektronics 4010-based graphics output. The code is still in the development stage and has no documentation.

Publications describing the code:
   None. The authors of the present code include Richard Fernow, Gerry Morgan and Patrick Thompson. Shlomo Caspi at LBL has a copy.

Is code documentation available? ☐ Yes ☒ No

How may the code be obtained?
    Call Gerry Morgan.

Source language: FORTRAN

Computers it runs on: VAX, CDC 7600

It is available as: ☐ Source code, ☐ Executable only

Source Media: ☐ Listing, ☒ Tape, ☐ Diskette, ☐ Cards, ☐ Networks
    Tape format: whatever
    Diskette size & format:

Available through: ☐ DECNET, ☐ ARPANET, ☐ BITNET
                    ☐

Network Address:

Date of Latest Version: Apr. 1985

Program Name: PARMELA

Person to Contact: Lloyd M. Young
Address: MS H817, Group AT-1
Los Alamos National Laboratory
Los Alamos NM 87545
USA

Telephone Number: (505) 667-1951, FTS 843-1951

Classification of Computer Code:
Component Design:
☐ Ion Source, ☐ Magnet, ☐ RF–cavity, ☐
Accelerator Optimization:
☐ Linac, ☐ Cyclotron, ☐ Synchrotron, ☐
Tracking or Simulation:
☒ Linac, ☐ Cyclotron, ☐ Synchrotron, ☒
Analysis:
☐ Stability, ☐ Impedances, ☐
Other:

Short Description: (Purpose, capabilities, algorithms, special features, etc.)
PARMELA means Phase And Radial Motion in Electron Linear Accelerators. It is a variation of PARMILA that applies to standing wave electron linacs and transport lines. The user must supply the linac structure and the fields in the basic rf cell. Multiparticle tracking is done with space charge forces. The independent variable is time, as opposed to distance along the beam line, which is used in the ion version PARMILA. The code was written by Ken Crandall. The code will generate several types of input electron distributions. The output is a file of particle distribution in 6D phase space at the exit of each rf cell. There is a postprocessor called PARGRAPH which will make "phase space scatter plots". The code is partially documented.

Publications describing the code:
None

Is code documentation available? ☐ Yes ☒ No

How may the code be obtained?
Contact Lloyd Young

Source language: FORTRAN

Computers it runs on: CDC7600, CRAY-1

It is available as: ☒ Source code, ☐ Executable only

Source Media: ☒ Listing, ☒ Tape, ☒ Diskette, ☐ Cards, ☒ Networks
Tape format: as desired
Diskette size & format: 5 1/4" IBM-PC

Available through: ☐ DECNET, ☒ ARPANET, ☒ BITNET
☐

Network Address: hks@lanl.arpa

Date of Latest Version: Jan. 1986

Program Name: PARMILA

Person to Contact: Los Alamos Accelerator Code Group
Address: MS H829, Group AT-6,
Los Alamos National Laboratory
Los Alamos, NM 87545
USA

Telephone Number: (505) 667-6677 or -2839, FTS 843-6677

Classification of Computer Code:
  Component Design:
    ☐ Ion Source, ☐ Magnet, ☐ RF–cavity, ☐
  Accelerator Optimization:
    ☒ Linac, ☐ Cyclotron, ☐ Synchrotron, ☒ transport lines
  Tracking or Simulation:
    ☒ Linac, ☐ Cyclotron, ☐ Synchrotron, ☒ transport lines
  Analysis:
    ☐ Stability, ☐ Impedances, ☐
  Other:

Short Description: (Purpose, capabilities, algorithms, special features, etc.)
PARMILA means Phase And Radial Motion in Ion Linear Accelerators. Given the electric and magnetic fields in one rf cavity and the gap-length-to-cell-length function for the cavity design from a code like SUPERFISH, PARMILA will generate the layout for a multicell DTL. It also does multiparticle tracking with space charge through the linac or through a transport line. There are several choices of input particle distributions: KV, Gaussian, waterbag, uniform, rectangular and experimental data. The default space charge subroutine assumes a circular beam and makes the impulse approximation once per rf cell. Other subroutines can be substituted if desired. The output is a file with the phase space distribution at the exit of each rf cell. There is a postprocessor called OUTPROC that will plot beam profiles as a function the beam direction z and particle distribution for cross sections of phase space, e.g. (x, x'), (y, y'), (ϕ, E), etc. OUTPROC also calculates moments of the distribution. The code is partially documented.

Publications describing the code:
  None

Is code documentation available? ☒ Yes ☐ No

How may the code be obtained?
  Contact The Los Alamos Accelerator Code Group

Source language: FORTRAN

Computers it runs on: CDC7600, CRAY-1

It is available as: ☒ Source code, ☐ Executable only
Source Media: ☒ Listing, ☒ Tape, ☐ Diskette, ☐ Cards, ☒ Networks
  Tape format: 9 track, 1600 bpi
  Diskette size & format:
Available through: ☐ DECNET, ☒ ARPANET, ☒ BITNET
  ☐
Network Address: hks@lanl.arpa

Date of Latest Version: Jan. 1986   Program Name: PARMTEQ (B or C)

Person to Contact: The Los Alamos Accelerator Code Group
Address: MS H829, Group AT-6
Los Alamos National Laboratory
Los Alamos, NM 87545
USA

Telephone Number: (505) 667-6677 or -2839, FTS 843-6677

Classification of Computer Code:
    Component Design:
        ☐ Ion Source, ☐ Magnet, ☐ RF–cavity, ☐
    Accelerator Optimization:
        ☒ Linac, ☐ Cyclotron, ☐ Synchrotron, ☐
    Tracking or Simulation:
        ☒ Linac, ☐ Cyclotron, ☐ Synchrotron, ☐
    Analysis:
        ☐ Stability, ☐ Impedances, ☐
    Other:

Short Description: (Purpose, capabilities, algorithms, special features, etc.)
    PARMTEQ is a version of PARMILA that will generate a design for an RFQ ion accelerator and also do multiparticle tracking with space charge through the linac. To do the design layout the program needs the output of two codes RFQIK and CURLY, described elsewhere. There are several types of input distributions available.

    In PARMTEQ(B) Z (distance along the beam) is the independent variable. In PARMTEQ(C) time is the independent variable. The output is a file of particle distributions over 6D phase space at the end of each RFQ "cell." There is a post processor called OUTPROC that will calculate moments, make phase space plots and beam profiles.

Publications describing the code:
    None

Is code documentation available? ☐ Yes ☒ No

How may the code be obtained?
    Call The Los Alamos Accelerator Code Group

Source language: FORTRAN

Computers it runs on: CDC 7600, CRAY-1

It is available as: ☒ Source code, ☐ Executable only

Source Media: ☐ Listing, ☒ Tape, ☐ Diskette, ☐ Cards, ☒ Networks
    Tape format: 9 Track 1600 bpi in almost any format
    Diskette size & format:

Available through: ☐ DECNET, ☒ ARPANET, ☒ BITNET
    ☐

Network Address: hks@lanl.arpa

Date of Latest Version: 1984   Program Name: PATH

Person to Contact: Los Alamos Accelerator Code Group
       Address: MS H829, Group AT-6
               Los Alamos National Laboratory
               Los Alamos, NM 87545
               USA

Telephone Number: (505) 667–6677 (or 667–2839), FTS 843–6677

Classification of Computer Code:
    Component Design:
       ☐ Ion Source, ☐ Magnet, ☐ RF–cavity, ☐
    Accelerator Optimization:
       ☐ Linac, ☐ Cyclotron, ☐ Synchrotron, ☐
    Tracking or Simulation:
       ☐ Linac, ☐ Cyclotron, ☐ Synchrotron, ☒   Beam Transport.

    Analysis:
       ☐ Stability, ☐ Impedances, ☐
    Other:

Short Description: (Purpose, capabilities, algorithms, special features, etc.)
    PATH is a group of computer programs for simulating charged-particle beam-transport systems. It was developed for evaluating the effects of some aberrations without a time-consuming integration of trajectories through the system. The beam-transport portion of PATH is derived from the well-known program DECAY TURTLE. PATH contains all features available in DECAY TURTLE (including the input format) plus additional features such as a more flexible random-ray generator, longitudinal phase space, some additional beamline elements, and space-charge routines. One of the programs also provides a simulation of an Alvarez linear accelerator. The programs, originally written for a CDC 7600 computer system, also are available on a VAX/VMS system. All of the programs are interactive with input prompting for ease of use.

Publications describing the code:
    John A. Farrell, "PATH – A Lumped Element Beam Transport Simulation Program with Space Charge," Proc. of Berlin Conf. on Computing in Accel. Design and Operation," W. Busse and R. Zelazny ed., Springer-Verlag, Berlin (1984) 267.

Is code documentation available? ☒ Yes ☐ No

How may the code be obtained?
    Contact the Los Alamos Accelerator Code Group.

Source language: FORTRAN

Computers it runs on: VAX, CDC 7600

It is available as: ☒ Source code, ☐ Executable only

Source Media: ☐ Listing, ☒ Tape, ☐ Diskette, ☐ Cards, ☒ Networks
    Tape format: 9 Track 1600 bpi
    Diskette size & format:

Available through: ☐ DECNET, ☒ ARPANET, ☒ BITNET
                     ☐

Network Address: hks@lanl.arpa

Date of Latest Version: July 1986   Program Name: PATPET

Person to Contact: Helmut Wiedemann
    Address: Stanford Synchrotron Radiation Laboratory
                BIN #69
                P.O. Box 4349,
                Stanford, CA 94305

Telephone Number: (415) 497–2503, FTS 461–9300 ext 2503

Classification of Computer Code:
    Component Design:
        ☐ Ion Source, ☐ Magnet, ☐ RF–cavity, ☐
    Accelerator Optimization:
        ☐ Linac, ☐ Cyclotron, ☐ Synchrotron, ☐
    Tracking or Simulation:
        ☐ Linac, ☐ Cyclotron, ☒ Synchrotron, ☐
    Analysis:
        ☐ Stability, ☐ Impedances, ☐
    Other:

Short Description: (Purpose, capabilities, algorithms, special features, etc.)
    PATPET is a combination of PATRICIA and PETROS. It is a tracking program that takes into account multipoles, systematic field errors, and misalignments. It produces dynamic apertures in one run based on tracking 400 particles.

Publications describing the code:
    Users' Guide (draft).

Is code documentation available? ☐ Yes ☐ No

How may the code be obtained?
    Contact Helmut Wiedemann.

Source language: FORTRAN

Computers it runs on: VAX

It is available as: ☒ Source code, ☐ Executable only

Source Media: ☐ Listing, ☐ Tape, ☒ Diskette, ☐ Cards, ☐ Networks
    Tape format:
    Diskette size & format: 8" Floppy.

Available through: ☐ DECNET, ☐ ARPANET, ☐ BITNET
    ☐

Network Address: SSRL750

Date of Latest Version: unknown   Program Name: PATRAC

Person to Contact: A. Hilaire
         Address: LEP Theory Div.
                  CERN
                  1211 Geneva 23
                  Switzerland

Telephone Number:

Classification of Computer Code:
    Component Design:
        ☐ Ion Source, ☐ Magnet, ☐ RF–cavity, ☐
    Accelerator Optimization:
        ☐ Linac, ☐ Cyclotron, ☐ Synchrotron, ☐
    Tracking or Simulation:
        ☐ Linac, ☐ Cyclotron, ☒ Synchrotron, ☐
    Analysis:
        ☐ Stability, ☐ Impedances, ☐
    Other:

Short Description: (Purpose, capabilities, algorithms, special features, etc.)
    PArticle TRACking is a tracking program using a magnet matrix formalism for elements up to quadrupoles and thin lens approximation for multipoles up to 12 poles. The magnet matrices are similar to those used of the code AGS but have been extended to allow rotated magnets and hence coupled motions.

Publications describing the code:
    P. Fougeras, A. Hilaire and A. Warman, "PATRAC: Particle Tracking Program," Proc. of Workshop on Accelerator Orbit and Tracking Programs, Brookhaven, Brookhaven National Laboratory Informal Report BNL-317 (1982).

Is code documentation available? ☐ Yes ☐ No

How may the code be obtained?
    unknown

Source language:
Computers it runs on:
It is available as: ☐ Source code, ☐ Executable only
Source Media: ☐ Listing, ☐ Tape, ☐ Diskette, ☐ Cards, ☐ Networks
    Tape format:
    Diskette size & format:
Available through: ☐ DECNET, ☐ ARPANET, ☐ BITNET
                    ☐
Network Address:

Date of Latest Version: unknown

Program Name: PATRICIA

Person to Contact: S. Kheifets
Address: Stanford Linear Accelerator Center
Stanford, CA 94305
USA

Telephone Number:

Classification of Computer Code:
Component Design:
☐ Ion Source, ☐ Magnet, ☐ RF–cavity, ☐
Accelerator Optimization:
☐ Linac, ☐ Cyclotron, ☐ Synchrotron, ☐
Tracking or Simulation:
☐ Linac, ☐ Cyclotron, ☒ Synchrotron, ☐
Analysis:
☐ Stability, ☐ Impedances, ☐
Other:

Short Description: (Purpose, capabilities, algorithms, special features, etc.)

The program does the following calculations: a) It adjusts horizontal and vertical chromaticities to the values prescribed by the user. b) It calculates Twiss parameters and eta functions of the lattice. c) It calculates emittances of the beam and relevant parameters of the ring. d) It performs harmonic analysis of the particle motion and produces its frequency spectrum. e) It tracks up to four particles simultaneously through up to one thousand revolutions. The oscillations in all three degrees of motion can be included into calculations, but horizontal and vertical motions are treated independently (no coupling is taken into account besides that which appears from the passage of a displaced particle through a sextupole). In all these calculations usual (3x3) matrix formalism is used. Sextupoles are treated in thin lens approximation. PATRICIA does not fit parameters of a linear lattice. The program uses the lattice which is supplied to it and attempts to find a periodic solution for the Twiss parameters and the dispersion function. If no periodic solution can be found for the on-momentum particle, the program stops. To investigate the influence of higher multipole fields in different elements of a machine, an optional version of PATRICIA under the name PNWM can be used. The action of the nonlinear field in a given element is approximated by an effective integrated nonlinear "kick". The longitudinal position of the kick is at the discretion of the user.

Publications describing the code:

S. Kheifets, "Tracking Studies in PEP and Description of the Computer Code PATRICIA," SLAC-PUB-2922 or BNL-31761 (1982) 89.

G. F. Dell, "Studies of the Chromatic Properties and Dynamic Aperture of the BNL Colliding Beam Accelerator," IEEE Trans. NS-30 (1985) 2469.

Is code documentation available? ☐ Yes ☐ No

How may the code be obtained?
unknown

Source language: PASCAL

Date of Latest Version: unknown

Program Name: PATTV

Person to Contact: John M. Jowett
Address: LEP Division
CERN
CH-1211 Geneva 23
Switzerland

Telephone Number: (022) 83 66 43 or 83 50 86

Classification of Computer Code:
Component Design:
☐ Ion Source, ☐ Magnet, ☐ RF–cavity, ☐
Accelerator Optimization:
☐ Linac, ☐ Cyclotron, ☒ Synchrotron, ☐
Tracking or Simulation:
☐ Linac, ☐ Cyclotron, ☒ Synchrotron, ☐
Analysis:
☐ Stability, ☐ Impedances, ☐
Other:

Short Description: (Purpose, capabilities, algorithms, special features, etc.)
PATTV is a version of H. Wiedemann's PATRICIA program which has been used at CERN since 1982. The main modification of the program was to provide high quality graphics via the CERN GD3 package. It is possible to watch animated "movies" of the particle motion on a terminal screen. Some additional analysis of particle power spectra is included.

Publications describing the code:

J. M. Jowett, "A Method for Distinguishing Chaotic from Quasi-periodic Motions in Orbit Tracking Programs," in Computing in Accelerator Design and Operation edited by W. Busse and R. Zelazny, Springer-Verlag, Berlin (1984) pp.261-6.

J. M. Jowett, "A New IBM Version of the Program PATRICIA," CERN LEP Theory Note No.1 (1982).

J. M. Jowett, "An easy way to run PATTV (PATRICIA)," CERN LEP Theory Note No.16 (1983).

Is code documentation available? ☒ Yes ☐ No

How may the code be obtained?
Contact J. M. Jowett

Source language: FORTRAN 77

Computers it runs on: IBM

It is available as: ☒ Source code, ☐ Executable only

Source Media: ☒ Listing, ☒ Tape, ☐ Diskette, ☐ Cards, ☒ Networks
Tape format:
Diskette size & format:

Available through: ☐ DECNET, ☐ ARPANET, ☒ BITNET
☒ EARNET, etc.

Network Address: JOWETT @ CERNVM

Date of Latest Version: Oct. 1985

Program Name: PE2D

Person to Contact: John S. Whitney
Address: Vector Fields, Ltd
Osney Mead
Oxford OX2 OEE
England

Telephone Number: 0865 248236

Classification of Computer Code:
Component Design:
☐ Ion Source, ☒ Magnet, ☐ RF-cavity, ☐
Accelerator Optimization:
☐ Linac, ☐ Cyclotron, ☐ Synchrotron, ☐
Tracking or Simulation:
☐ Linac, ☐ Cyclotron, ☐ Synchrotron, ☒ trajectory calculations in 2D
Analysis:
☐ Stability, ☐ Impedances, ☐
Other:

Short Description: (Purpose, capabilities, algorithms, special features, etc.)
PE2D is a 2D code for the analysis of magnetostatic or electrostatic fields and steady state and transient eddy currents. It can be used for a wide range of applications including fusion and accelerator magnets, electron beam lenses, non-destructive testing, actuators, and MRI magnet shielding.

PE2D enables the solution of the partial differential equation governing a system to be computed using the finite element method. The packaged pre-processor provides powerful tools to aid data input.

The design requiring analysis is defined as an assembly of simple primitives, for example curvilinear quadrilaterals, which are then automatically subdivided by the program. The simple primitives have symmetry properties and may be replicated by rotation, reflection and translation. Using these features together with the copy and modify facilities, it is easy to model even the most complex geometry.

The geometric primitives have assigned material properties. These may be material constants such as permeability, conductivity and current density. Material properties can be specified as tables of function values.

PE2D uses either first or second order triangular finite elements. The first-order solution can be used to obtain a fast test of the model before solving to higher accuracy using second order elements.

There are three analysis programs provided with PE2D:

Static fields (non linear and laminated materials)

Transient fields (non linear)

Steady state alternating current fields (linear)

The post-processor provides extensive facilities for presentation and display of the results. These include potentials, fields and forces.

State-of-the-art error analysis and display provide the user with information necessary for improving the input data to achieve the necessary accuracy in an economical way.

Publications describing the code:
N. J. Diserens, "A Space Charge Beam Option For The PE2D And TOSCA Packages," IEEE Trans. MAG-18 (1982) 362–366.

Data Sheet Ref: 118522 from Vector Fields

Is code documentation available? ☒ Yes ☐ No

How may the code be obtained?
By license agreement with Vector Fields, Ltd

Source language: FORTRAN 77

Computers it runs on: PRIME, VAX, IBM

It is available as: ☒ Source code, ☐ Executable only

Source Media: ☐ Listing, ☒ Tape, ☐ Diskette, ☐ Cards, ☐ Networks
Tape format: As required
Diskette size & format:

Available through: ☐ DECNET, ☐ ARPANET, ☐ BITNET
☒ DOE Network

Network Address: Contact Bob Lari - Argonne National Laboratory (312)972-6632

Date of Latest Version: Dec. 1985   Program Name: PETROC

Person to Contact: Gilbert Guignard or Yolande Marti
Address: LEP Division
CERN
1211 GENEVA 23
Switzerland

Telephone Number: 41-22-83.59.75

Classification of Computer Code:
Component Design:
☐ Ion Source, ☐ Magnet, ☐ RF–cavity, ☐
Accelerator Optimization:
☐ Linac, ☐ Cyclotron, ☐ Synchrotron, ☐
Tracking or Simulation:
☐ Linac, ☐ Cyclotron, ☒ Synchrotron, ☐
Analysis:
☐ Stability, ☐ Impedances, ☒ Closed Orbit Distortion.

Other:

Short Description: (Purpose, capabilities, algorithms, special features, etc.)
It computes betatron and dispersion functions, synchrotron frequency. It calculates the radiation integrals, damping partition numbers, beam emittances, bunch length and energy spread. The effect of given or random distortions (misalignments) on the closed orbit and betatron motion is determined (with or without radiation losses). Different algorithms for correcting the orbit are included (amplitude minimization, successive bumps, iterative method using a small number of correctors).

Publications describing the code:
G. Guignard and Y. Marti, "PETROC Users' Guide," CERN Internal Report, CERN/ISR-BOM-TH/81-32

Is code documentation available? ☒ Yes ☐ No

How may the code be obtained?
Contact G. Guignard or Y. Marti

Source language: FORTRAN 77

Computers it runs on: IBM

It is available as: ☒ Source code, ☐ Executable only

Source Media: ☒ Listing, ☒ Tape, ☐ Diskette, ☐ Cards, ☒ Networks
Tape format: Unlabeled tape. 1600 Bpi, 3200 char/block, 80 char/record
Diskette size & format:

Available through: ☐ DECNET, ☐ ARPANET, ☒ BITNET
☐

Network Address: MAR@CERNVM.

Date of Latest Version: unknown

Program Name: PETROS

Person to Contact: K. Steffen
Address: DESY
Notkestrasse 85
2000 Hamburg 52
Fed. Rep. Germany

Telephone Number:

Classification of Computer Code:
Component Design:
☐ Ion Source, ☐ Magnet, ☐ RF–cavity, ☐
Accelerator Optimization:
☐ Linac, ☐ Cyclotron, ☐ Synchrotron, ☐
Tracking or Simulation:
☐ Linac, ☐ Cyclotron, ☒ Synchrotron, ☐
Analysis:
☒ Stability, ☐ Impedances, ☒ Orbit Correction

Other:

Short Description: (Purpose, capabilities, algorithms, special features, etc.)

A computer program which simulates effects of possible error sources on the beam optics and the improvements due to orbit correction is a necessary tool for the study and design of large electron rings. Such a program called 'PETROS' exists at the DESY laboratory. PETROS can work in two modes: 1. Uncoupled transverse motions are assumed and the usual three-dimensional matrices are used in each plane. 2. Coupled transverse motions are considered and five-dimensional matrices are used throughout. It treats non-linear fields and effects of the radiation losses due to bending. It computes the linear transformation matrices of a ring structure, the corresponding betatron and dispersion functions, betatron and synchrotron frequencies. It calculates the five synchrotron radiation integrals, the damping partition numbers, the damping times, the length deviation of off-momentum orbits, beam emittances, bunch length, relative energy spread and synchrotron lifetime, the effect of prescribed or random distortions, taking into account the radiation losses due to bending. It simulates closed orbit corrections and gives the corresponding kick amplitudes.

Publications describing the code:

K. Steffen and J. Kewish, "Study of Integer Difference Resonance in Distorted PETRA Optics," DESY PET-76/09 (1976).

B. Zotter, "A Short Guide for the Use of Program PETROS at CERN," CERN report LEP-70/37 (1978).

G. Guignard and Y. Marti, "Numerical Simulations of Orbit Correction in Large Electron Rings," Proc. of Conference on Computing in Accelerator Design and Operation, Berlin 1983, Springer Verlag, Berlin, Lecture Notes in Physics No. 215, (1984).

Is code documentation available? ☒ Yes ☐ No

How may the code be obtained?
Unknown (Seems to be available from DESY as PETROS and from CERN as PETROC.)

Date of Latest Version: unknown

Program Name: PINWHEEL

Person to Contact: E. R. Close
Address: 1 Cyclotron Road
Lawrence Berkeley Laboratory
Berkeley, CA 94720
USA

Telephone Number: (415) 486-6166, FTS 451-6166

Classification of Computer Code:
Component Design:
☐ Ion Source, ☐ Magnet, ☐ RF-cavity, ☐
Accelerator Optimization:
☐ Linac, ☐ Cyclotron, ☐ Synchrotron, ☐
Tracking or Simulation:
☐ Linac, ☒ Cyclotron, ☐ Synchrotron, ☒ Spectrometer

Analysis:
☐ Stability, ☐ Impedances, ☐
Other:

Short Description: (Purpose, capabilities, algorithms, special features, etc.)
PINWHEEL was used for tracking orbits of charged particles in a combined electric and magnetic field. Runge-Kutta integration is used to solve the first-order Hamilton's equations of motion. The program has 3 parts: control, plotting and integration.

Publications describing the code:

M. Reiser and J. Kopf, "Electrolytic Tank Facility and Computer Program for Central Region Studies for the MSU Cyclotron," Michigan State University report MSUCP-19 (1964).

E. R. Close, "PINWHEEL — Orbit Tracking in Combined Electric and Magnetic Fields." Lawrence Berkeley Laboratory Report T1-BKY-PINWEL (1964).

John S. Colonias, "Particle Accelerator Design: Computer Programs," Academic Press, New York (1974) 246.

Is code documentation available? ☐ Yes ☐ No

How may the code be obtained?
unknown; probably unavailable.

Source language: FORTRAN IV

Computers it runs on: CDC 6600/7600

It is available as: ☐ Source code, ☐ Executable only
Source Media: ☐ Listing, ☐ Tape, ☐ Diskette, ☐ Cards, ☐ Networks
Tape format:
Diskette size & format:
Available through: ☐ DECNET, ☐ ARPANET, ☐ BITNET
☐
Network Address:

Date of Latest Version: 1985

Program Name: PISCES

Person to Contact: Yoshihisa Iwashita
      Address: Institute for Chemical Research
               Kyoto University
               Torii-cho, Awataguchi
               Sakyo-ku, Kyoto 606,
               JAPAN
Telephone Number: Unknown

Classification of Computer Code:
   Component Design:
      ☐ Ion Source, ☐ Magnet, ☒ RF–cavity, ☐
   Accelerator Optimization:
      ☐ Linac, ☐ Cyclotron, ☐ Synchrotron, ☐
   Tracking or Simulation:
      ☐ Linac, ☐ Cyclotron, ☐ Synchrotron, ☐
   Analysis:
      ☐ Stability, ☐ Impedances, ☐
   Other:

Short Description: (Purpose, capabilities, algorithms, special features, etc.)

A new code, PISCES, has been developed for calculating a complete set of rf electromagnetic modes in an axisymmetric cavity. The finite-element method is used with up to third-order shape functions. Although two components are enough to express these modes, three components are used as unknown variables to take advantage of the symmetry of the element matrix. The unknowns are taken to be either the electric field components $\mathbf{E} = (E_r, E_\phi, E_z)$ or the magnetic field components $\mathbf{H} = (H_r, H_\phi, H_z)$. The zero-divergence condition is satisfied by the shape function within each element.

Publications describing the code:

Y. Iwashita, "Calculation of RF Fields in Axisymmetric Cavities," Los Alamos National Laboratory report LAUR-85-1892.

Is code documentation available? ☐ Yes ☒ No

How may the code be obtained?
   Contact author.

Source language: FORTRAN

Computers it runs on: VAX-780

It is available as: ☒ Source code, ☐ Executable only

Source Media: ☐ Listing, ☒ Tape, ☐ Diskette, ☐ Cards, ☐ Networks
   Tape format:
   Diskette size & format:

Available through: ☐ DECNET, ☐ ARPANET, ☐ BITNET
                    ☐

Network Address:

Date of Latest Version: 1980   Program Name: POISCR

Person to Contact: Program Library
Address: DD Div
CERN
CH - 1211 Geneva
Switzerland

Telephone Number: (22) 83 2377

Classification of Computer Code:
Component Design:
☐ Ion Source, ☒ Magnet, ☐ RF–cavity, ☐
Accelerator Optimization:
☐ Linac, ☐ Cyclotron, ☐ Synchrotron, ☐
Tracking or Simulation:
☐ Linac, ☐ Cyclotron, ☐ Synchrotron, ☐
Analysis:
☐ Stability, ☐ Impedances, ☐
Other:

Short Description: (Purpose, capabilities, algorithms, special features, etc.)
Magnet design in 2 dimensions.

Finite-element method, triangular mesh.

Also for various scalar potential distributions.

Publications describing the code:
CERN Program Library Writeup T602

Is code documentation available? ☒ Yes ☐ No

How may the code be obtained?
CERN Program Librarian

Source language: FORTRAN 77

Computers it runs on: IBM/CDC

It is available as: ☒ Source code, ☐ Executable only

Source Media: ☐ Listing, ☒ Tape, ☐ Diskette, ☐ Cards, ☐ Networks
Tape format: 9 track 1600 bpi
Diskette size & format:

Available through: ☐ DECNET, ☐ ARPANET, ☐ BITNET
☐

Network Address:

Date of Latest Version: Nov. 1985

Program Name: POISSON GROUP CODES

Person to Contact: Los Alamos Accelerator Code Group
Address: MS H829, Group (AT-6)
Los Alamos National Laboratory
Los Alamos, NM 87545
USA

Telephone Number: (505) 667-9131 (or 667-2839), FTS 843-9131.

Classification of Computer Code:
Component Design:
☐ Ion Source, ☒ Magnet, ☐ RF–cavity, ☐
Accelerator Optimization:
☐ Linac, ☐ Cyclotron, ☐ Synchrotron, ☐
Tracking or Simulation:
☐ Linac, ☐ Cyclotron, ☐ Synchrotron, ☐
Analysis:
☐ Stability, ☐ Impedances, ☐
Other:

Short Description: (Purpose, capabilities, algorithms, special features, etc.)

Calculate static magnetic fields in two cartesian dimensions or cylindrically symmetric configurations in 3D. Will handle problems with permeable iron, but not permanent magnets. Can also solve electrostatic problems. Uses over-relaxation method to solve 2D generalized Poisson equation. Included in the group of codes is AUTOMESH, LATTICE, FORCE, and MIRT. MIRT is an optimization code.

Publications describing the code:

K. Halbach. "A Program for Inversion of System Analysis and Its Application to the Design of Magnets." Proc. 2nd Conf. on Magnet Technology, Oxford, England, (1967).

Is code documentation available? ☒ Yes ☐ No

How may the code be obtained?

Send blank tape to above address; specify version desired—VAX or CRAY.

Also available through ARPANET, DECNET or BITNET. Telephone us for instructions.

Source language: FORTRAN 77

Computers it runs on: VAX, CRAY

It is available as: ☒ Source code, ☐ Executable only
Source Media: ☐ Listing, ☒ Tape, ☐ Diskette, ☐ Cards, ☒ Networks
Tape format: 9 Track 1600 bpi
Diskette size & format:
Available through: ☐ DECNET, ☒ ARPANET, ☒ BITNET
☐
Network Address: hks@lanl.arpa

Date of Latest Version: Oct. 86

Program Name: POISSON–BNL

Person to Contact: R. C. Gupta
    Address: Brookhaven National Laboratory
                Building 902-B
                Upton, NY 11973
                U.S.A.

Telephone Number: (516)282-4805, FTS 666-4805

Classification of Computer Code:
    Component Design:
        ☐ Ion Source, ☒ Magnet, ☐ RF–cavity, ☐
    Accelerator Optimization:
        ☐ Linac, ☐ Cyclotron, ☐ Synchrotron, ☐
    Tracking or Simulation:
        ☐ Linac, ☐ Cyclotron, ☐ Synchrotron, ☐
    Analysis:
        ☐ Stability, ☐ Impedances, ☐
    Other:

Short Description: (Purpose, capabilities, algorithms, special features, etc.)
    POISSON–BNL is the modified version of the 1981 Los Alamos National Lab. version of the Poisson Group codes created by Holsinger and Halbach. Major modifications have been made in the AUTOMESH and LATTICE programs. The user now has more control over the type of mesh to be generated and one can use different mesh size at any number of places, anywhere in a model. These improvements allow one to describe the finer details of a complicated geometry with a reasonable number of mesh points. In POISSON one now has access to the intermediate results while the original run is progressing, for a better control of convergence. Also there is a lesser chance that a solution will diverge.

Publications describing the code:
    R. C. Gupta, "Modifications in the AUTOMESH and other POISSON Group Codes," Workshop on Electromagnetic Field Computation, Oct. 20-21, 1986. (To be published) (To be used in addition to the manual available from the Los Alamos Accelerator Code Group for the standard Poisson Group Codes.)

Is code documentation available? ☒ Yes ☐ No

How may the code be obtained?
    Contact R. C. Gupta

Source language: FORTRAN 77

Computers it runs on: VAX

It is available as: ☒ Source code, ☐ Executable only

Source Media: ☐ Listing, ☐ Tape, ☐ Diskette, ☐ Cards, ☒ Networks
    Tape format:
    Diskette size & format:

Available through: ☒ DECNET, ☐ ARPANET, ☒ BITNET
☐

Network Address: BITNET: gupta@bnldag; DECNET(PHYSNET): BNLDAG::GUPTA

Date of Latest Version: July 1986

Program Name: POISSON-LBL

Person to Contact: S. Caspi
Address: MS 46-161
Lawrence Berkeley Laboratory
1 Cyclotron Road
Berkeley, CA   94720
USA
Telephone Number: (415) 486-7244, FTS 451-7244

Classification of Computer Code:
Component Design:
☐ Ion Source, ☒ Magnet, ☐ RF–cavity, ☐
Accelerator Optimization:
☐ Linac, ☐ Cyclotron, ☐ Synchrotron, ☐
Tracking or Simulation:
☐ Linac, ☐ Cyclotron, ☐ Synchrotron, ☐
Analysis:
☐ Stability, ☐ Impedances, ☐
Other:

Short Description: (Purpose, capabilities, algorithms, special features, etc.)

This is a generalization of the standard POISSON code which replaces simple constant Neumann/Dirichlet boundary conditions by more general conditions expressible in the form of a series of harmonic functions giving the physically correct behavior of the potential at large distances. The code can handle the superposition of an externally applied field as well.

Although the original version was written for the HP1000, there exists a version which runs on the MFE CRAY. Documentation exists for the original POISSON codes; changes are described in the publications below.

Publications describing the code:

S. Caspi, M. Helm, and L. J. Laslett, "Incorporation of a Circular Boundary Condition into the Program POISSON," Lawrence Berkeley Internal Report LBL-17064 SSC-MAG-5 (Feb.1984).

S. Caspi, M. Helm, and L. J. Laslett, "The Generalization of a Circular Boundary Condition in the Program POISSON to Include No Symmetry and Axis-symmetry of Revolution," Lawrence Berkeley Internal Report LBL-18063 SSC-MAG-12(Jul.1984).

S. Caspi, M. Helm, and L. J. Laslett, "Incorporation of an Elliptical Boundary Condition into the Program POISSON," Lawrence Berkeley Internal Report LBL-18798 SSC-MAG-28(Dec.1984).

S. Caspi, M. Helm, and L. J. Laslett, "Incorporation of Superposition into the Program POISSON," Lawrence Berkeley Internal Report LBL-19050 SSC-MAG-31(Jan.1985).

S. Caspi, M. Helm, and L. J. Laslett, "The Application of Program POISSON to Axially-Symmetric Problems — Magnetostatic and Electrostatic — with Use of Prolate Spheroidal Boundary," Lawrence Berkeley Internal Report LBL-20893 SSC-MAG-68(Jan.1986).

S. Caspi, M. Helm, and L. J. Laslett, "Numerical Solution of Boundary Condition to Poisson's Equation and Its Incorporation into the Program POISSON," IEEE Trans. NS-32 (1985) 3722.

S. Caspi, M. Helm, and L. J. Laslett, "Incorporation of Toroidal Boundary Condition in the Program POISSON," Lawrence Berkeley Internal Report (in progress)(Dec.1986).

Is code documentation available? ☒ Yes ☐ No

How may the code be obtained?
    Call Shlomo Caspi.

Source language: FORTRAN 77

Computers it runs on: HP1000, CRAY.

It is available as: ☐ Source code, ☐ Executable only

Source Media: ☐ Listing, ☒ Tape, ☐ Diskette, ☐ Cards, ☐ Networks
    Tape format:
    Diskette size & format:

Available through: ☐ DECNET, ☐ ARPANET, ☐ BITNET
    ☐

Network Address:

Date of Latest Version: Apr. 1986

Program Name: POISSON-TAC

Person to Contact: W. Schmidt or S. Pissanetzky
Address: Texas Accelerator Center
2319 Timberloch Place
The Woodlands, TX 77380
U.S.A.

Telephone Number: (713)363-0121

Classification of Computer Code:
Component Design:
☐ Ion Source, ☒ Magnet, ☐ RF–cavity, ☐
Accelerator Optimization:
☐ Linac, ☐ Cyclotron, ☐ Synchrotron, ☐
Tracking or Simulation:
☐ Linac, ☐ Cyclotron, ☐ Synchrotron, ☐
Analysis:
☐ Stability, ☐ Impedances, ☐
Other:

Short Description: (Purpose, capabilities, algorithms, special features, etc.)
This is a modification of the original POISSON code developed by Holsinger and Halbach, where the magnetization table is accurately interpolated, table truncation errors are avoided, and data input is simpler. Imagine the field at a point in a magnet as given by the sum of contributions from each little element of magnetized iron, plus a contribution from the currents. The field can be accurately calculated only if the magnetization is accurately known at each point. In POISSON, the magnetization is calculated by interpolating a table with the assumption that $1/\mu$ is a linear function of B^2 in each interval. It is now known that such an assumption produces errors as large as 5% in some intervals of the POISSON internal table (1010 steel), which are larger than the experimental errors in the measurement of μ.[1] To solve this difficulty, a table with 195 points has been generated for 1008 steel using accurate interpolation techniques.[2] This table has been implemented as the internal table of POISSON-TAC. The points are so close that POISSON's assumption of $1/\mu \propto B^2$ does not introduce any appreciable errors, and good field accuracy can be obtained. POISSON-TAC will issue a warning when truncation of the magnetization table occurs as a consequence of high fields during iteration. It has been shown that convergence to an incorrect solution takes place when truncation errors are present.[2] However, if truncation apppears only during the initial iterations and then stops, convergence is to the correct solution.

POISSON-TAC also has an improved data input scheme. POISSON-TAC is available for VAX or FPS, and documentation exists for the original POISSON code.

Publications describing the code:

1. S. Pissanetaky,"The Interpolation of Magnetization Tables,"COMPEL 5(1986)41-56.

2. R. Carcagno and S. Pissanetzky,"A Smooth Magnetization Table for 1008 Steel at 4.2K," Texas Accelerator Center Report TAC-257/85.

Is code documentation available? ☒ Yes ☐ No

How may the code be obtained?
Contact W. Schmidt or S. Pissanetzky

Source language: FORTRAN 77

Computers it runs on: VAX and FPS

It is available as: ☒ Source code, ☐ Executable only

Source Media: ☐ Listing, ☒ Tape, ☐ Diskette, ☐ Cards, ☐ Networks
   Tape format:
   Diskette size & format:

Available through: ☐ DECNET, ☐ ARPANET, ☐ BITNET
   ☐

Network Address:

Date of Latest Version: Apr. 1986

Program Name: PROFI

Person to Contact: PROFI Engineering
Address: Wilhelminen Straße
D-6100
Darmstadt
Fed. Rep. Germany

Telephone Number: (06151) 26418

Classification of Computer Code:
Component Design:
☐ Ion Source, ☒ Magnet, ☐ RF–cavity, ☒ Electric machines

Accelerator Optimization:
☐ Linac, ☐ Cyclotron, ☐ Synchrotron, ☐

Tracking or Simulation:
☐ Linac, ☐ Cyclotron, ☐ Synchrotron, ☐

Analysis:
☐ Stability, ☐ Impedances, ☐

Other:

Short Description: (Purpose, capabilities, algorithms, special features, etc.)

The computer program PROFI (program for calculation of fields) which calculates 2- or 3-dimensional nonlinear magnetostatics fields, linear electrostatic or stationary electric fields, stationary non-linear 2-dimensional eddy-current fields and stationary temperature distributions. The program uses the finite difference method. The calculations may be carried out in one of five different coordinate systems, two of them being 3-dimensional. A set of service programs for preparing the input data, analysing the results, data handling etc. simplifies the use of the program.

Publications describing the code:

W. Müller et al., "Numerical Solution of 2- or 3D Nonlinear Field problems by means of the Computer Program PROFI," Archiv für Elektrotechnik 65 (1982) 299.

Is code documentation available? ☒ Yes ☐ No

How may the code be obtained?

This code can be bought from PROFI Engineering. Purchase also includes updates and some assistance in learning the code.

Source language: FORTRAN 77

Computers it runs on: IBM, VAX, CDC

It is available as: ☒ Source code, ☐ Executable only

Source Media: ☐ Listing, ☒ Tape, ☐ Diskette, ☐ Cards, ☐ Networks
Tape format: IBM/VAX standard
Diskette size & format:

Available through: ☐ DECNET, ☐ ARPANET, ☐ BITNET
☐

Network Address:

Date of Latest Version: unknown

Program Name: PRUD-M

Person to Contact: A. G. Daikovsky
Address: Insitute for High Energy Physics
Serpukhov
USSR

Telephone Number:

Classification of Computer Code:
Component Design:
☐ Ion Source, ☐ Magnet, ☒ RF–cavity, ☐
Accelerator Optimization:
☐ Linac, ☐ Cyclotron, ☐ Synchrotron, ☐
Tracking or Simulation:
☐ Linac, ☐ Cyclotron, ☐ Synchrotron, ☐
Analysis:
☐ Stability, ☐ Impedances, ☐
Other:

Short Description: (Purpose, capabilities, algorithms, special features, etc.)

A program package for calculating eigenfrequencies and electromagnetic fields with azimuthal variations in axial-symmetric cavities of an arbitrary shape. The method is based on the representation of the equations of electrodynamics in variables ρh_φ, ρe_φ. Apart from frequencies and fields, the accumulated energy, distribution of losses in the metal, and other characteristics important for application are also computed. The package offers wide possibilities for graphic representation of the field topology, facilitating the analysis, and optimization of complicated accelerating structures. The program works in two modes: a mode of estimating the frequency spectrum in the specified interval and a mode of accurate computation of a specific frequency, related fields and derived quantities.

Publications describing the code:

A. G. Daikovsky, Y. I. Portugalov and A. D. Ryabov, "PRUD-code for Calculation of the Nonsymmetric Modes in Axial Symmetric Cavities," Part. Accel. 12 (1982) 59.

Abramov, A. G.; Dajkovskij, A. G.; Ershov, S. Yu.; Portugalov, Yu. I.; Portugalova, L. D., "Method to Find Eigen Electromagnetic Fields in Cavities of Arbitrary Shape. PRUD-M Program Package to Find Azimuthal Nonuniform modes in Axial-Symmetric Cavities. Part 2," Gosudarstvennyi Komitet po Ispol'zovaniyu Atomnoi Energii SSSR, Serpukhov, Inst. Fiziki Vysokikh Energii, Report No. IFVE-OMVT-83-179 (1983), in Russian.

Is code documentation available? ☐ Yes ☐ No

How may the code be obtained?
unknown

Source language:
Computers it runs on:
It is available as: ☐ Source code, ☐ Executable only
Source Media: ☐ Listing, ☐ Tape, ☐ Diskette, ☐ Cards, ☐ Networks
Tape format:
Diskette size & format:
Available through: ☐ DECNET, ☐ ARPANET, ☐ BITNET
☐
Network Address:

Date of Latest Version: unknown

Program Name: PRUD-O

Person to Contact: A. G. Abramov
Address: Institute for High Energy Physics
Serpukhov
USSR

Telephone Number:

Classification of Computer Code:
Component Design:
☐ Ion Source, ☐ Magnet, ☒ RF–cavity, ☐
Accelerator Optimization:
☐ Linac, ☐ Cyclotron, ☐ Synchrotron, ☐
Tracking or Simulation:
☐ Linac, ☐ Cyclotron, ☐ Synchrotron, ☐
Analysis:
☐ Stability, ☐ Impedances, ☐
Other:

Short Description: (Purpose, capabilities, algorithms, special features, etc.)
A program package intended for calculating azimuthal-symmetric modes in axisymmetric cavities as well as critical modes in longitudinally uniform waveguides. The discretization of electrodynamics equations uses eight-node quadrilateral isoparametric elements. The block power method for solving algebraic eigenvalue problems includng estimations of convergence rate is used. To illustrate performance of the program and its separate units and estimate the accuracy, counting time, possibility of calculation oscillations with multiple eigenvalues, the program has been checked on problems having analytical solutions: oscillations in spherical and cylindrical resonators, waves in a rectangular waveguide. It is concluded that frequency by the PRUD-O program is more accurate than by the SUPERFISH program by approximately two orders .

Publications describing the code:
Abramov, A. G.; Dajkovskij, A. G.; Ershov, S. Yu.; Portugalov, Yu. I.; Ryabov, A. D., "PRUD-O Program Package for Accelerating Structure Calculation," Gosudarstvennyi Komitet po Ispol'zovaniyu Atomnoi Energii, Report No. IFVE-OMVT-83-3 (1983), in Russian.

Is code documentation available? ☐ Yes ☐ No

How may the code be obtained?
unknown

Source language:
Computers it runs on:
It is available as: ☐ Source code, ☐ Executable only
Source Media: ☐ Listing, ☐ Tape, ☐ Diskette, ☐ Cards, ☐ Networks
Tape format:
Diskette size & format:
Available through: ☐ DECNET, ☐ ARPANET, ☐ BITNET
☐
Network Address:

Date of Latest Version: unknown   Program Name: PRUD-OB

Person to Contact: A. G. Abramov
            Address: Institute for High Energy Physics
                     Serpukhov
                     USSR

Telephone Number:

Classification of Computer Code:
    Component Design:
        ☐ Ion Source, ☐ Magnet, ☒ RF–cavity, ☐
    Accelerator Optimization:
        ☐ Linac, ☐ Cyclotron, ☐ Synchrotron, ☐
    Tracking or Simulation:
        ☐ Linac, ☐ Cyclotron, ☐ Synchrotron, ☐
    Analysis:
        ☐ Stability, ☐ Impedances, ☐
    Other:

Short Description: (Purpose, capabilities, algorithms, special features, etc.)
    The PRUD-OB program package is intended for calculating the azimuthally homogeneous modes in accelerator periodic axially-symmetrical systems. The problem is reduced to determination of two real functions for the structure half-period. The package is oriented for the problems of determining the periodic structure dispersion characteristics.

Publications describing the code:
    Abramov, A. G.; Dajkovskij, A. G.; Portugalov, Yu. I.; Ryabov, A. D., "Modification of the PRUD-O Program Package for Calculating the Periodic Structures," Gosudarstvennyi Komitet po Ispol'zovaniyu Atomnoi Energii SSSR, Serpukhov, Inst. Fiziki Vysokikh Energii, Report No. IFVE-OMVT-83-178 (1983), in Russian.

Is code documentation available? ☐ Yes ☐ No

How may the code be obtained?
    unknown

Source language:
Computers it runs on:
It is available as: ☐ Source code, ☐ Executable only
Source Media: ☐ Listing, ☐ Tape, ☐ Diskette, ☐ Cards, ☐ Networks
    Tape format:
    Diskette size & format:
Available through: ☐ DECNET, ☐ ARPANET, ☐ BITNET
                   ☐
Network Address:

Date of Latest Version: unknown

Program Name: RACETRACK

Person to Contact: A. Wrulich
Address: DESY
Notkestrasse 85
2000 Hamburg 52
Fed. Rep. Germany

Telephone Number:

Classification of Computer Code:
Component Design:
☐ Ion Source, ☐ Magnet, ☐ RF–cavity, ☐
Accelerator Optimization:
☐ Linac, ☐ Cyclotron, ☒ Synchrotron, ☐
Tracking or Simulation:
☐ Linac, ☐ Cyclotron, ☒ Synchrotron, ☐
Analysis:
☐ Stability, ☐ Impedances, ☐
Other:

Short Description: (Purpose, capabilities, algorithms, special features, etc.)
RACETRACK is a computer code to simulate transverse nonlinear particle motion in accelerators. Transverse magnetic fields of higher order are treated in thin magnet approximation. Multipoles up to 20 poles are included. Energy oscillations due to the nonlinear synchrotron motion are taken into account. Several additional features, as linear optics calculations, chromaticity adjustment, tune variation, orbit adjustment and others are available to guarantee a fast treatment of nonlinear dynamical problems.

Publications describing the code:
A. Wrulich, "RACETRACK — A Computer Code for the Simulation of Nonlinear Particle Motion in Accelerators," DESY Internal Reports 84/07 and 84/026 (1984).

Is code documentation available? ☐ Yes ☐ No

How may the code be obtained?
unknown

Source language: unknown

Computers it runs on:
It is available as: ☐ Source code, ☐ Executable only
Source Media: ☐ Listing, ☐ Tape, ☐ Diskette, ☐ Cards, ☐ Networks
Tape format:
Diskette size & format:
Available through: ☐ DECNET, ☐ ARPANET, ☐ BITNET
☐
Network Address:

Date of Latest Version: 1986   Program Name: RAY

Person to Contact: P. Spädtke
        Address: GSI
                 Postfach 11 05 41
                 6100-Darmstadt
                 Fed. Rep. Germany
Telephone Number: 06151/359-323

Classification of Computer Code:
    Component Design:
        ☒ Ion Source, ☐ Magnet, ☐ RF–cavity, ☒ Electron gun
    Accelerator Optimization:
        ☐ Linac, ☐ Cyclotron, ☐ Synchrotron, ☐
    Tracking or Simulation:
        ☐ Linac, ☐ Cyclotron, ☐ Synchrotron, ☒ Beam transport lines
    Analysis:
        ☐ Stability, ☐ Impedances, ☐
    Other:

Short Description: (Purpose, capabilities, algorithms, special features, etc.)
    Simulation of ions/electrons within electrostatic/magnetostatic fields. Electrostatic potentials are calculated exactly, interactively. Menu-driven program. Including high-resolution colored graphics. 2D-code. Cylindrically symmetric.

Publications describing the code:
    Presentation on low energy ion beams. Conference in GB, 1986.

Is code documentation available? ☒ Yes ☐ No

How may the code be obtained?
    On request.

Source language: Machine Language.

Computers it runs on: Commodore C64, C128

It is available as: ☐ Source code, ☒ Executable only

Source Media: ☐ Listing, ☐ Tape, ☒ Diskette, ☐ Cards, ☐ Networks
    Tape format:
    Diskette size & format: 5 1/4" IBM DOS

Available through: ☐ DECNET, ☐ ARPANET, ☐ BITNET
                    ☐
Network Address:

Date of Latest Version: 1986

Program Name: RAYTRACE

Person to Contact: Stanley Kowalski
Address: Laboratory of Nuclear Science
Bldg. 26-505
MIT
Cambridge, MA 02139
USA
Telephone Number: (617)253-4288

Classification of Computer Code:
  Component Design:
    ☐ Ion Source, ☐ Magnet, ☐ RF–cavity, ☐
  Accelerator Optimization:
    ☐ Linac, ☐ Cyclotron, ☐ Synchrotron, ☐
  Tracking or Simulation:
    ☒ Linac, ☐ Cyclotron, ☐ Synchrotron, ☐
  Analysis:
    ☐ Stability, ☐ Impedances, ☐
  Other:

Short Description: (Purpose, capabilities, algorithms, special features, etc.)

RAYTACE is an ion-optical computer code, which numerically integrates the particle differential equations of motion through real fields and can be used to trace rays one-by-one through a sequence of electromagnetic devices. The main types of elements that are presently supported include dipoles (6 versions), multipoles (4 through 12 poles), electrostatic deflector, velocity filter, lens, and solenoid. For an ion-optical system with a symmetry plane, the accuracy of the trajectory calculations in this plane is comparable to the accuracy of the description of the electric and magnetic fields, i.e., RAYTRACE computes to essentially infinite order. The field components for trajectories off this median plane for dipoles are described by a fourth-order Taylor series; off-axis in multipoles the field is described by a Taylor series carried to at least fifth-order.

Users of RAYTRACE practically always start with TRANSPORT to determine first and second order parameters — in other words the basic layout of the system. RAYTRACE is then used to fine tune the system. First and second order parameters generally have to be readjusted slightly, and when dipoles are involved there are also zeroth order adjustments, i.e., centerline offsets. The major function of RAYTRACE, however, is to calculate higher-order aberrations in the optics, and to aid in correcting these aberrations, whenever possible. The program does not have a built-in automatic fitting routine for minimizing image aberrations, etc., but it has been used as a subroutine for such programs.

Since the program traces one ray at a time, it is not readily adaptable to handle space charge forces as they occur in systems with intense beams.

Publications describing the code:
  S. Kowalski and H. A. Enge, "RAYTRACE," Proc. of Second Int. Conf. on Charged Particle Optics, Albuquerque (1986) to be published. Also there is a MIT internal report.

Is code documentation available? ☒ Yes ☐ No

How may the code be obtained?
  Call Los Alamos Accelerator Code Group, AT-6, Los Alamos National Laboratory (505) 667–6677 (or 667–2839). You can also contact Stan Kowalski directly.

Source language: FORTRAN

Computers it runs on: VAX

It is available as: ☒ Source code, ☐ Executable only

Source Media: ☐ Listing, ☒ Tape, ☐ Diskette, ☐ Cards, ☒ Networks
  Tape format: 9 track, 1600bpi
  Diskette size & format:

Available through: ☐ DECNET, ☒ ARPANET, ☒ BITNET
  ☐

Network Address: hks@lanl.arpa or sk@mitlns

Date of Latest Version: Oct. 1985

Program Name: RELAX3D

Person to Contact: Corrie Kost
       Address: Triumf
                4004 Wesbrook Mall
                Vancouver, B.C.
                Canada V6T-2A3

Telephone Number: 604-2221047 ext. 310

Classification of Computer Code:
   Component Design:
      ☐ Ion Source, ☐ Magnet, ☐ RF–cavity, ☒ Electrostatic devices
   Accelerator Optimization:
      ☐ Linac, ☐ Cyclotron, ☐ Synchrotron, ☐
   Tracking or Simulation:
      ☐ Linac, ☐ Cyclotron, ☐ Synchrotron, ☐
   Analysis:
      ☐ Stability, ☐ Impedances, ☐
   Other:

Short Description: (Purpose, capabilities, algorithms, special features, etc.)

   User-friendly interactive program which solves the Laplace/Poisson equation in 3D Cartesian or 2D cylindrical coordinate system with dielectrics (2D only) by the method of successive over-relaxation (finite difference). Problem cases are dynamically loaded from a user-written subroutine describing the geometry. Contour plots of the potential distribution along any slice can be produced.

Publications describing the code:

   H. Houtman, C. J. Kost, "A FORTRAN Program (RELAX3D) to Solve the 3 Dimensional Poisson (Laplace) Equation", Proc. EPS Conf. on Computing in Accelerator Design and Operation, Berlin 1983, Springer-Verlag (1984).

Is code documentation available? ☒ Yes ☐ No

How may the code be obtained?
   Contact Corrie Kost.

Source language: FORTRAN

Computers it runs on: VAX

It is available as: ☒ Source code, ☐ Executable only

Source Media: ☐ Listing, ☒ Tape, ☐ Diskette, ☐ Cards, ☐ Networks
   Tape format: BACKUP
   Diskette size & format:

Available through: ☐ DECNET, ☐ ARPANET, ☐ BITNET
                   ☐

Network Address:

Date of Latest Version: Nov. 1985

Program Name: REVMOC

Person to Contact: Corrie Kost
Address: Triumf
4004 Wesbrook Mall
Vancouver, B.C.
Canada V6T-2A3

Telephone Number: 604-2221047 ext. 310

Classification of Computer Code:
Component Design:
☐ Ion Source, ☐ Magnet, ☐ RF–cavity, ☐
Accelerator Optimization:
☐ Linac, ☐ Cyclotron, ☐ Synchrotron, ☐
Tracking or Simulation:
☐ Linac, ☐ Cyclotron, ☐ Synchrotron, ☒ beam line transport
Analysis:
☐ Stability, ☐ Impedances, ☐
Other:

Short Description: (Purpose, capabilities, algorithms, special features, etc.)

Second-order Monte Carlo beam transport program which includes the effects of multiple scattering, decay, nuclear scattering, and energy loss. The program cannot optimize beam line elements and is thus primarily used to do detailed checks on a TRANSPORT-designed beam line. Aberration coefficients (transfer matrices) can be calculated for the full beam line.

Publications describing the code:

C. J. Kost, P. A. Reeve, "A Monte Carlo Beam Transport Program, REVMOC," Proc. EPS Conference on Computing in Accelerator Design and Operation, Berlin, 1983, Springer-Verlag (1984).

Is code documentation available? ☒ Yes ☐ No

How may the code be obtained?
Contact Corrie Kost.

Source language: FORTRAN

Computers it runs on: VAX

It is available as: ☒ Source code, ☐ Executable only

Source Media: ☐ Listing, ☒ Tape, ☐ Diskette, ☐ Cards, ☐ Networks
Tape format: BACKUP
Diskette size & format:

Available through: ☐ DECNET, ☐ ARPANET, ☐ BITNET
☐

Network Address:

Date of Latest Version: Jan. 1986

Program Name: RFQLIB

Person to Contact: Walter P. Lysenko
Address: MS H829, Group AT-6
Los Alamos National Laboratory
Los Alamos, NM 87545
USA

Telephone Number: (505) 667-7431

Classification of Computer Code:
Component Design:
☐ Ion Source, ☐ Magnet, ☐ RF–cavity, ☐
Accelerator Optimization:
☐ Linac, ☐ Cyclotron, ☐ Synchrotron, ☐
Tracking or Simulation:
☒ Linac, ☐ Cyclotron, ☐ Synchrotron, ☐
Analysis:
☐ Stability, ☐ Impedances, ☐
Other:

Short Description: (Purpose, capabilities, algorithms, special features, etc.)

The RFQLIB system does particle tracing simulations for RFQ linear accelerators. The particle equations of motion are numerically integrated using time as the independent variable. The forces on the particles are computed in two subroutines. Subroutine FOR computes the external electric forces of the rf field in the RFQ. The RFQ parameters are stored in a table as a function of the longitudinal coordinate. Interpolation is used to get the parameters at given values of the synchronous particle position. Subroutine SCFOR computes the space charge forces by a particle-in-cell method using the electrostatic approximation. An r-z Poisson solver is used with a conducting boundary at r=const and with periodic boundary conditions in the z-direction.

Publications describing the code:

W. P. Lysenko, "An RFQ Simulation Code" in Proc. of 1984 Linear Accel. Conf., ed. by N. Angert, GSI-84-11 (1984) 327.

See also LANL Informal Report AT6: ATN-84-1.

Is code documentation available? ☒ Yes ☐ No

How may the code be obtained?
Call Walter Lysenko.

Source language: FORTRAN

Computers it runs on: CRAY

It is available as: ☒ Source code, ☐ Executable only
Source Media: ☐ Listing, ☒ Tape, ☐ Diskette, ☐ Cards, ☒ Networks
Tape format: whatever
Diskette size & format:

Available through: ☒ DECNET, ☒ ARPANET, ☒ BITNET
☐

Network Address: WPL @ LANL on ARPANET.

Date of Latest Version: Jan. 1985   Program Name: RING

Person to Contact: Eva S. Bozoki
Address: NSLS Dept.
Brookhaven National Laboratory,
Upton, NY 11973
USA

Telephone Number: (516) 282-3701, FTS 666-3701

Classification of Computer Code:
Component Design:
☐ Ion Source, ☐ Magnet, ☐ RF–cavity, ☐
Accelerator Optimization:
☐ Linac, ☐ Cyclotron, ☒ Synchrotron, ☐
Tracking or Simulation:
☐ Linac, ☐ Cyclotron, ☒ Synchrotron, ☐
Analysis:
☐ Stability, ☐ Impedances, ☐
Other:

Short Description: (Purpose, capabilities, algorithms, special features, etc.)

It is a modeling program, which can be used

1) off-line as a design program (like e.g. SYNCH, MAD, etc.) or

2) on-line as a control program. (When used as a control program, the program-microprocessor interface is specific to the installation.)

It has two major modules, one for:

1) tune and chromaticity optimization/control, and one for

2) orbit calculation/correction/control (It uses the MICADO algorithm for minimizing the orbit displacements around the ring).

In addition to the standard lattice elements (drift, bend, quad, sext.) undulators can also be included. Edge focusing is calculated as in TRANSPORT. Chromaticity due to dipoles is calculated as in SYNCH.

Machine and beam parameters, synchrotron integrals, damping partitions, rate of change of dumping partitions, energy spread, spatial beam size with and without coupling, bunch length, quantum and Toushek lifetime, etc. are calculated on demand.

Publications describing the code:

Eva S. Bozoki, "High Level Control Programs at NSLS," Conf. on Computing in Accelerator Design and Operation, 1983, Springer-Verlag (1984) 420.

Brookhaven National Laboratory internal report no. BNL-31361 (1982).

Brookhaven National Laboratory internal report no. BNL-35507 (1984).

Is code documentation available? ☒ Yes ☐ No

How may the code be obtained?
    Contact Eva Bozoki.

Source language: FORTRAN

Computers it runs on: DG

It is available as: ☒ Source code,   ☐ Executable only

Source Media: ☐ Listing, ☒ Tape, ☐ Diskette, ☐ Cards, ☐ Networks
    Tape format:
    Diskette size & format:

Available through: ☐ DECNET,   ☐ ARPANET,   ☒ BITNET
                     ☐

Network Address:

Date of Latest Version: May 1985   Program Name: RMKT

Person to Contact: Bruce Carlsten
Address: MS H825, Group AT-7
Los Alamos National Laboratory
Los Alamos, NM 87545
USA

Telephone Number: (505) 667-5657, FTS 843-5657

Classification of Computer Code:
Component Design:
☐ Ion Source, ☐ Magnet, ☐ RF-cavity, ☒ Klystron simulations.

Accelerator Optimization:
☐ Linac, ☐ Cyclotron, ☐ Synchrotron, ☐

Tracking or Simulation:
☐ Linac, ☐ Cyclotron, ☐ Synchrotron, ☐

Analysis:
☐ Stability, ☐ Impedances, ☐

Other:

Short Description: (Purpose, capabilities, algorithms, special features, etc.)

Ring model, time as the independent parameter, 2 1/2 D, large signal beam-rf cavity interaction simulations, use of so-called "dynamic wavelength" to ensure self-consistent particle motion, iterations to ensure self-consistent cavity gap voltages and space charge.

Publications describing the code:

P. Tallerico and B. Carlsten, "Computer Modeling the Klystron," IEEE Trans. on Nucl. Sci., NS-30 (1983) 2170.

P. Tallerico and B. Carlsten, "Self-Consistent Klystron Simulations," IEEE Trans. on Nucl. Sci., NS-32 (1985) 2837.

Is code documentation available? ☒ Yes ☐ No

How may the code be obtained?
Contact Bruce Carlsten.

Source language: FORTRAN

Computers it runs on: CRAY

It is available as: ☒ Source code, ☐ Executable only

Source Media: ☐ Listing, ☒ Tape, ☐ Diskette, ☐ Cards, ☒ Networks
Tape format: As desired
Diskette size & format:

Available through: ☐ DECNET, ☒ ARPANET, ☒ BITNET
☐

Network Address:

Date of Latest Version: 1981   Program Name: SATDSK

Person to Contact: G. S. McNeilly
         Address: Oak Ridge National Laboratory
                  Building 4500N
                  Oak Ridge, TN 37831-6238
                  U.S.A.

Telephone Number:

Classification of Computer Code:
   Component Design:
      ☐ Ion Source, ☒ Magnet, ☐ RF–cavity, ☐
   Accelerator Optimization:
      ☐ Linac, ☒ Cyclotron, ☐ Synchrotron, ☐
   Tracking or Simulation:
      ☐ Linac, ☐ Cyclotron, ☐ Synchrotron, ☐
   Analysis:
      ☐ Stability, ☐ Impedances, ☐
   Other:

Short Description: (Purpose, capabilities, algorithms, special features, etc.)
   SATDSK calculates the median plane magnetic field due to fully saturated iron poletips. Optionally, SATDSK calculates the magnetic field due to disks of magnetic charge, which can simulate the effect of holes in the iron poletip, or circular trim rods embedded in the poletip. SATDSK is intended for poletip geometries that are both symmetric about the median plane, and have azimuthal sector symmetry. Thus, the program is primarily designed to simulate the magnetic field due to iron poletips in superconducting cyclotrons.

Publications describing the code:
   Gregory S. McNeilly, "SATDSK: A Numerical Simulation of the Magnetic Field Due to Saturated Iron in Cyclotron Poletips," Computer Phys. Comm. 23(1981)199.

Is code documentation available? ☒ Yes ☐ No

How may the code be obtained?
   CPC Program Library, Queen's University of Belfast, N. Ireland; Catalogue number: ABKI

Source language: FORTRAN

Computers it runs on: IBM 360/91

It is available as: ☒ Source code, ☐ Executable only

Source Media: ☐ Listing, ☐ Tape, ☐ Diskette, ☐ Cards, ☐ Networks
   Tape format:
   Diskette size & format:

Available through: ☐ DECNET, ☐ ARPANET, ☐ BITNET
                   ☐
Network Address:

Date of Latest Version: Feb. 1986

Program Name: SCHAR

Person to Contact: Prof. R. J. Hayden
Address: University of Montana
Physics Department
Missoula, MT 59812
U.S.A.

Telephone Number: (406)243-2073

Classification of Computer Code:
Component Design:
☒ Ion Source, ☐ Magnet, ☐ RF-cavity, ☒ Charge Exchange Solenoids

Accelerator Optimization:
☒ Linac, ☐ Cyclotron, ☐ Synchrotron, ☐
Tracking or Simulation:
☒ Linac, ☐ Cyclotron, ☐ Synchrotron, ☐
Analysis:
☒ Stability, ☐ Impedances, ☐
Other:

Short Description: (Purpose, capabilities, algorithms, special features, etc.)

The code traces macrofilaments or macroparticles through electromagnetic fields with or without space charge. Fields may be analytical or tabulated grid values. Input options for the particle distribution include: measured, KV, 4 Vol or 6 Vol. in phase space.

Publications describing the code:

R. J. Hayden and M. J. Jakobson, "The Space Charge Computer Program SHAR," IEEE Trans. NS-30(1983)2540.

R. J. Hayden and M. J. Jakobson, "Macrofilament Simulation of High Current Beam Transport," IEEE Trans. NS-32(1985)2519.

M. J. Jakobson and R. J. Hayden, Proc. of Ion Optics Conf. (May 1986) To be published in Nuc. Instru. & Meth.

Is code documentation available? ☐ Yes ☒ No

How may the code be obtained?
Contact Prof. Hayden or Prof. Jakobson at above address.

Source language: FORTRAN

Computers it runs on: VAX

It is available as: ☒ Source code, ☐ Executable only

Source Media: ☒ Listing, ☒ Tape, ☐ Diskette, ☐ Cards, ☐ Networks
Tape format: 1600BPI-9TRACK
Diskette size & format:

Available through: ☐ DECNET, ☐ ARPANET, ☐ BITNET
☐

Network Address:

(Documentation is in the form of comment lines in the program.)
Date of Latest Version: Apr. 1986

Program Name: SCOP-2

Person to Contact: Ingo Hofmann
Address: GSI
Postfach 110541
D-6100 Darmstadt-11
Fed. Rep. Germany

Telephone Number:

Classification of Computer Code:
Component Design:
☐ Ion Source, ☐ Magnet, ☐ RF-cavity, ☐
Accelerator Optimization:
☐ Linac, ☐ Cyclotron, ☐ Synchrotron, ☐
Tracking or Simulation:
☒ Linac, ☐ Cyclotron, ☒ Synchrotron, ☒ Storage Rings

Analysis:
☐ Stability, ☐ Impedances, ☐
Other:

Short Description: (Purpose, capabilities, algorithms, special features, etc.)

A 2D (x-y Cartesian) PIC code used for space-charge dominated beam transport studies. Recently used for studies of resonance crossing in storage rings under space-charge-dominated conditions. 2D trajectories, tracking.

Publications describing the code:

I. Bozsik and Ingo Hofmann, "Space Charge Effects in the Focusing of Intense Ion Beams," Nucl. Inst. Mtds. 187 (1981) 305-311.

Is code documentation available? ☒ Yes ☐ No

How may the code be obtained?

Source language: FORTRAN

Computers it runs on: IBM 3090

It is available as: ☒ Source code, ☐ Executable only
Source Media: ☐ Listing, ☒ Tape, ☐ Diskette, ☐ Cards, ☒ Networks
Tape format:
Diskette size & format:
Available through: ☐ DECNET, ☒ ARPANET, ☒ BITNET
☐
Network Address: (not given)

Date of Latest Version: Apr. 1986   Program Name: SCOP-RZ

Person to Contact: Ingo Hofmann
Address: GSI
Postfach 110541
D-6100 Darmstadt-11
Fed. Rep. Germany

Telephone Number:

Classification of Computer Code:
Component Design:
☐ Ion Source, ☐ Magnet, ☐ RF–cavity, ☐
Accelerator Optimization:
☐ Linac, ☐ Cyclotron, ☐ Synchrotron, ☐
Tracking or Simulation:
☒ Linac, ☐ Cyclotron, ☒ Synchrotron, ☒ Storage Rings

Analysis:
☐ Stability, ☐ Impedances, ☐
Other:

Short Description: (Purpose, capabilities, algorithms, special features, etc.)

A-particle-in-cell simulation code with 3D trajectories and a 2D Poisson solver (r-z). Conducting cylindrical pipe as boundary condition. A user-defined impedance can be included. Code has been used to study rf bunching and the longitudinal microwave instability.

Publications describing the code:

I. Hofmann and I. Boszik, Proc. of the Sympsouim on Accelerator Aspects of Heavy Ion Fusion, GSI Darmstadt (1982) 181.

Is code documentation available? ☒ Yes ☐ No

How may the code be obtained?
Write to Ingo Hofmann.

Source language: FORTRAN 77

Computers it runs on: IBM 3090 and CRAY

It is available as: ☒ Source code, ☐ Executable only

Source Media: ☐ Listing, ☒ Tape, ☐ Diskette, ☐ Cards, ☒ Networks
Tape format:
Diskette size & format:

Available through: ☐ DECNET, ☒ ARPANET, ☒ BITNET
☐

Network Address:

Date of Latest Version: Aug. 1985   Program Name: SHRIMP

Person to Contact: Robert Ryne / Robert Gluckstern
           Address: Dept. of Physics and Astronomy
                    University of Maryland
                    College Park, MD 20742
                    USA

Telephone Number: (301) 454-7476

Classification of Computer Code:
    Component Design:
        ☐ Ion Source, ☐ Magnet, ☒ RF–cavity, ☐
    Accelerator Optimization:
        ☐ Linac, ☐ Cyclotron, ☐ Synchrotron, ☐
    Tracking or Simulation:
        ☐ Linac, ☐ Cyclotron, ☐ Synchrotron, ☐
    Analysis:
        ☐ Stability, ☐ Impedances, ☐
    Other:

Short Description: (Purpose, capabilities, algorithms, special features, etc.)
    SHRIMP is a post-processor to the program SUPERFISH. SHRIMP computes the cavity frequency by a variational technique, using the fields calculated by SUPERFISH. The frequency computed by SHRIMP has an error $\delta f/f \propto 1/N^2$ (for an N x N mesh). Thus, by using SHRIMP as a post processor, one can make the SUPERFISH mesh coarser, and still obtain comparable accuracy in f.

Publications describing the code:
    R. L. Gluckstern, R. D. Ryne, R. F. Holsinger, "Numerical Programs for Obtaining Accurate Resonant Frequencies of Modes in Azimuthally Symmetric Electromagnetic Cavities," Proceedings of the COMPUMAG Conference, Genoa, Italy, IEEE Trans. MAG-19 (1983) 141.

Is code documentation available? ☐ Yes ☒ No

How may the code be obtained?
    It is located in MASS under /095680/585/SHRIMP (at Los Alamos National Laboratory). Contact the Los Alamos Accelerator Code Center (505) 667-6677 (or 667-2839), FTS 843-6677.

Source language: FORTRAN

Computers it runs on: CRAY

It is available as: ☒ Source code, ☐ Executable only

Source Media: ☐ Listing, ☒ Tape, ☐ Diskette, ☐ Cards, ☐ Networks
    Tape format: 9trk, 1600 bpi
    Diskette size & format:

Available through: ☐ DECNET, ☒ ARPANET, ☒ BITNET
                   ☐

Network Address: hks@lanl.arpa

Date of Latest Version: Apr. 1982   Program Name: SIMTRAC

Person to Contact: Daniel Brandt
Address: LEP Div.
CERN
1211 Geneva 23
Switzerland

Telephone Number:

Classification of Computer Code:
Component Design:
☐ Ion Source, ☐ Magnet, ☐ RF-cavity, ☐
Accelerator Optimization:
☐ Linac, ☐ Cyclotron, ☐ Synchrotron, ☐
Tracking or Simulation:
☐ Linac, ☐ Cyclotron, ☒ Synchrotron, ☐
Analysis:
☐ Stability, ☐ Impedances, ☒ Wakefield Effects

Other:

Short Description: (Purpose, capabilities, algorithms, special features, etc.)
  SIMTRAC is a simulation program for tracking longitudinal and transverse single-bunch effects in a circular electron machine for a number N of superparticles. The program includes damping, collective effects such as transition, beam-loading of rf cavities and wakefields. For a typical run 1000 superparticles can be followed for around 5000 turns. Output includes beam dimensions every NREPR turns, phase space plots, bucket contours, and averages over a given number of turns.

Publications describing the code:
  D. Brandt, "SIMTRAC — A Simulation program for Tracking Longitudinal and Transverse Single Bunch Effects," CERN internal report LEP Note 512, 15 (1984).

Is code documentation available? ☒ Yes ☐ No

How may the code be obtained?
  Contact Daniel Brandt.

Source language: FORTRAN

Computers it runs on: IBM

It is available as: ☐ Source code, ☐ Executable only
Source Media: ☐ Listing, ☐ Tape, ☐ Diskette, ☐ Cards, ☐ Networks
Tape format:
Diskette size & format:

Available through: ☐ DECNET, ☐ ARPANET, ☐ BITNET
☐

Network Address:

Date of Latest Version: Apr. 1986

Program Name: SINAC

Person to Contact: Gerhard Rudolf
    Address: SIN
        CH 5234 Villigen,
        Switzerland

Telephone Number: (059) 99-3394

Classification of Computer Code:
    Component Design:
        ☐ Ion Source, ☐ Magnet, ☐ RF–cavity, ☐
    Accelerator Optimization:
        ☐ Linac, ☒ Cyclotron, ☐ Synchrotron, ☐
    Tracking or Simulation:
        ☐ Linac, ☒ Cyclotron, ☐ Synchrotron, ☒   Magnets
    Analysis:
        ☐ Stability, ☐ Impedances, ☐
    Other:

Short Description: (Purpose, capabilities, algorithms, special features, etc.)
    Orbit calculations from magnetic field measurements; processing of magnetic field measurements in cylindrical coordinates.

Publications describing the code:
    IV Int. Conf. on Isochronous Cyclotrons, Gatlinburg, TN, IEEE/NS-13, (1966) 194–214.

Is code documentation available? ☒ Yes ☐ No

How may the code be obtained?
    Write to Gerhard Rudolf, SIN.

Source language: FORTRAN 77

Computers it runs on: VAX, CDC/NOSVE

It is available as: ☒ Source code, ☐ Executable only

Source Media: ☐ Listing, ☒ Tape, ☐ Diskette, ☐ Cards, ☐ Networks
    Tape format: 80 char/line, 10 lines/block
    Diskette size & format:

Available through: ☐ DECNET, ☐ ARPANET, ☐ BITNET
    ☐

Network Address:

Date of Latest Version: 1985   Program Name: SLIM

Person to Contact: Louis Hand
            Address: Newman Laboratory of Nuclear Science
                     Cornell University
                     Ithaca, NY 14853
                     USA

Telephone Number: (607) 255-6023 (or 1000)

Classification of Computer Code:
    Component Design:
        ☐ Ion Source, ☐ Magnet, ☐ RF–cavity, ☐
    Accelerator Optimization:
        ☐ Linac, ☐ Cyclotron, ☐ Synchrotron, ☐
    Tracking or Simulation:
        ☐ Linac, ☐ Cyclotron, ☒ Synchrotron, ☐
    Analysis:
        ☒ Stability, ☐ Impedances, ☒   depolarization

    Other:

Short Description: (Purpose, capabilities, algorithms, special features, etc.)

SLIM is a tracking code which includes spin orbit interaction to calculate the depolarization of polarized beams due to closed orbit distortions caused by misalignments. It uses an 8x8 matrix formalism, and includes effects of radiation damping. Stability is determined by looking at the eigenvalues of the total transport system.

There exists documentation for the 1984 version of the code as written by Alex Chao, but it is unknown whether D. Barber or L. Hand have updated the documentation.

Publications describing the code:

A. W. Chao, "Evaluation of Beam Distribution Parameters in an Electron Storage Ring." J. Appl. Phys. 50 (1979) 595.

A. W. Chao, "Evaluation of Radiative Spin Polarization in a Electron Storage Ring," Nucl. Inst. Mtds. 180 (1981) 29.

A. W. Chao, "Calculation of Polarization Effects," Computing in Accelerator Design and Operation, Proc. of 1983 Berlin Conf. Springer-Verlag, Berlin, (1984) 59.

H. Mais and G. Ripken, DESY report 83-062 (1983).

Is code documentation available? ☒ Yes ☐ No

How may the code be obtained?
    Desmond P. Barber, DESY, Notkestrasse 85, D-2000, Hamburg 52, Fed. Rep. Germany.

    Louis Hand, Cornell Univ. Ithaca, NY 14853, (607) 255-6023 (or 1000).

Source language: FORTRAN

Computers it runs on: CDC, IBM

It is available as: ☐ Source code, ☐ Executable only

Source Media: ☐ Listing, ☐ Tape, ☐ Diskette, ☐ Cards, ☐ Networks
  Tape format:
  Diskette size & format:

Available through: ☐ DECNET, ☐ ARPANET, ☐ BITNET
  ☐

Network Address:

Date of Latest Version: Mar. 1982

Program Name: SNOW

Person to Contact: James P. Brainard
Address: Org. 2564, Building 891
Sandia National Laboratory
Albuquerque, NM 87185
USA

Telephone Number: (505) 844-6462, FTS 532-6462

Classification of Computer Code:
Component Design:
☒ Ion Source, ☐ Magnet, ☐ RF-cavity, ☐
Accelerator Optimization:
☐ Linac, ☐ Cyclotron, ☐ Synchrotron, ☐
Tracking or Simulation:
☐ Linac, ☐ Cyclotron, ☐ Synchrotron, ☐
Analysis:
☐ Stability, ☐ Impedances, ☐
Other:

Short Description: (Purpose, capabilities, algorithms, special features, etc.)

A digital computer program, SNOW, has been developed for the simulation of dense ion beams. The program simulates the plasma expansion cup (but not the plasma source itself), the acceleration region, and a drift space with neutralization if desired. The ion beam is simulated by computing representative trajectories through the device. The potentials are simulated on a large rectangular matrix array which is solved by iterative techniques. Poisson's equation is solved at each point within the configuration using space-charge densities computed from the ion trajectories combined with background electron and/or ion distributions. (Note that some changes have been made in the code recently that have not been documented. It may be difficult to run the code without personal help from Brainard. Jack Boers is presently writing a new version of the code.)

Publications describing the code:

Jack E. Boers, "SNOW – A Digital Computer Program for the Simulation of Ion Beam Devices," Sandia Laboratory Internal Report no. SAND79-1027 (1980).

Is code documentation available? ☒ Yes ☐ No

How may the code be obtained?

Contact John Brainard at the above address or Jack Boers, Varian Corp., Gloucester, MA 01930, Phone (617) 281-2000, ext. 4344.

Source language: FORTRAN77

Computers it runs on: CRAY, VAX

It is available as: ☒ Source code, ☐ Executable only
Source Media: ☐ Listing, ☒ Tape, ☐ Diskette, ☐ Cards, ☐ Networks
Tape format:
Diskette size & format:
Available through: ☐ DECNET, ☐ ARPANET, ☐ BITNET
☐
Network Address:

Date of Latest Version: unknown

Program Name: SOTRM

Person to Contact: E. R. Close
Address: 1 Cyclotron Road
Lawrence Berkeley Laboratory
Berkeley, CA 94720
USE

Telephone Number: (415) 486-6166, FTS 451-6166

Classification of Computer Code:
Component Design:
□ Ion Source, □ Magnet, □ RF-cavity, □
Accelerator Optimization:
□ Linac, □ Cyclotron, □ Synchrotron, □
Tracking or Simulation:
□ Linac, □ Cyclotron, □ Synchrotron, ☒ Beam Transport

Analysis:
□ Stability, □ Impedances, □
Other: Generate transport matrices from fields

Short Description: (Purpose, capabilities, algorithms, special features, etc.)

In the design of beam transport systems, it is often desirable to generate transformation elements from a magnetic field by numerically integrating the orbits through the field. Such a transformation matrix is needed when only a measured field is available or when the effect of various trial magnetic fields is being investigated. Essentially, program SOTRM produces first- and second-order elements when an arbitrary magnetic field is given. The resulting transformation matrix is readily applicable to beam transport programs such as TRANSPORT.

SOTRM formulates a system of equations which, when integrated, produces the coordinates of the reference particle and of any nearby particle(s) specified. Once this is completed, the program calculates (if requested) the first- and second-order transformation matrix elements using the reference orbit as the origin in a suitably chosen coordinate system.

Publications describing the code:

E. R. Close, "SOTRM — A Program to Generate First and Second-Order Matrix Elements by Tracking Charged Particles in a Specified Magnetic Field," Lawrence Berkeley Laboratory Report UCRL-19823 (1970).

E. R. Close, "Generation of First and Second-Order Transformation Elements from a Given Magnetic Field," Nucl. Inst. Methods 89 (1970) 205.

John S. Colonias, "Particle Accelerator Design: Computer Programs," Academic Press, New York (1974) 194.

Is code documentation available? □ Yes □ No

How may the code be obtained?
unknown

Source language: FORTRAN

Computers it runs on: CDC 6600, 17600

It is available as: ☐ Source code, ☐ Executable only

Source Media: ☐ Listing, ☐ Tape, ☐ Diskette, ☐ Cards, ☐ Networks
   Tape format:
   Diskette size & format:

Available through: ☐ DECNET, ☐ ARPANET, ☐ BITNET
   ☐

Network Address:

Date of Latest Version: Jan. 1986   Program Name: SPEAM VI

Person to Contact: Corrie Kost
Address: TRIUMF
4004 Wesbrook Mall
Vancouver, B.C.
Canada V6T-2A3

Telephone Number: (604) 222-1047 ext. 310

Classification of Computer Code:
   Component Design:
      ☐ Ion Source, ☐ Magnet, ☐ RF–cavity, ☐
   Accelerator Optimization:
      ☐ Linac, ☐ Cyclotron, ☐ Synchrotron, ☐
   Tracking or Simulation:
      ☐ Linac, ☐ Cyclotron, ☐ Synchrotron, ☒ Beam Line
   Analysis:
      ☐ Stability, ☐ Impedances, ☐
   Other:

Short Description: (Purpose, capabilities, algorithms, special features, etc.)
   Calculates and plots the rms beam envelopes of a continuous non-relativistic proton beam through various elements (magnetic and electrostatic). Space charge forces are included. Either the generalized Kapchinsky-Vladimirsky (KV) equations or those of Emigh may be used.

Publications describing the code:
   TRIUMF design notes TRI-DN-73-11, TRI-DN-74-31, TRI-DN-74-32.

Is code documentation available? ☒ Yes ☐ No

How may the code be obtained?
   Contact Corrie Kost.

Source language: FORTRAN

Computers it runs on: VAX

It is available as: ☒ Source code, ☐ Executable only

Source Media: ☐ Listing, ☒ Tape, ☐ Diskette, ☐ Cards, ☐ Networks
   Tape format: BACKUP
   Diskette size & format:

Available through: ☐ DECNET, ☐ ARPANET, ☐ BITNET ☐

Network Address:

Date of Latest Version: Jan. 1985   Program Name: SUPERFISH Group Codes

Person to Contact: Los Alamos Accelerator Code Group
    Address: MS H829, Group AT-6
             Los Alamos National Laboratory
             Los Alamos, NM 87545
             USA

Telephone Number: (505) 667-6677, FTS 843-6677

Classification of Computer Code:
    Component Design:
        ☐ Ion Source, ☐ Magnet, ☒ RF–cavity, ☐
    Accelerator Optimization:
        ☐ Linac, ☐ Cyclotron, ☐ Synchrotron, ☐
    Tracking or Simulation:
        ☐ Linac, ☐ Cyclotron, ☐ Synchrotron, ☐
    Analysis:
        ☐ Stability, ☐ Impedances, ☐
    Other:

Short Description: (Purpose, capabilities, algorithms, special features, etc.)
    The SUPERFISH package evaluates the eigenfrequencies and fields for arbitrary-shaped 2-D waveguides in cartesian coordinates and 3-D axially symmetric rf cavities in cylindrical coordinates. The package contains codes to generate the mesh, plot fields and evaluate auxiliary quantities of interest to drift tube linac design, e.g., transit time factors, power losses, effect of perturbations.

Publications describing the code:
    K. Halbach and R. F. Holsinger, "SUPERFISH — A Computer Program for Evaluation of RF Cavities with Cylindrical Symmetry," Part. Accel. 7 (1976) 213.

    K. Halbach et. al., "Properties of the Cylindrical RF Evaluation Code SUPERFISH," Proc. 1976 Linear Accel. Conf., Chalk River Nuclear Lab Report AECL-5677, 122.

Is code documentation available?   ☒ Yes   ☐ No

How may the code be obtained?
    Send blank tape to above address; specify version desired—VAX or CRAY. Also available through ARPANET, DECNET, OR BITNET. Telephone us for instructions.

Source language: FORTRAN 77

Computers it runs on: VAX, CRAY

It is available as: ☒ Source code, ☐ Executable only

Source Media: ☐ Listing, ☒ Tape, ☐ Diskette, ☐ Cards, ☒ Networks
    Tape format: 9 track, 1600 bpi
    Diskette size & format:

Available through: ☒ DECNET, ☒ ARPANET, ☒ BITNET
                     ☐

Network Address: hks@lanl.arpa

Date of Latest Version: Apr. 1986

Program Name: SYMP3

Person to Contact: Gerry P. Jackson
Address: Fermilab
P.O. Box 500
Batavia, Il 60510
USA

Telephone Number: (312) 840–2317 or (3000)

Classification of Computer Code:
Component Design:
☐ Ion Source, ☐ Magnet, ☐ RF–cavity, ☐
Accelerator Optimization:
☐ Linac, ☐ Cyclotron, ☐ Synchrotron, ☐
Tracking or Simulation:
☐ Linac, ☐ Cyclotron, ☐ Synchrotron, ☒ Colliding Beam

Analysis:
☐ Stability, ☐ Impedances, ☐
Other:

Short Description: (Purpose, capabilities, algorithms, special features, etc.)

A computer program to simulate colliding beam dynamics in $e^+ - e^-$ storage rings has been written. The first version of the program did not incorporate sextupoles but showed some of the characteristics measured in various machines in the past. The focus of the present work is the understanding of the effects of sextupoles on these results. To do this thin sextupoles are added in two ways. The first employs a linear-transfer/nonlinear-kick algorithm for each lattice cell. The second method is to create a symplectic second-order transfer map for the entire machine. While the first method is exact, it is slow for machine lattices with many sextupoles. The luminosities, beam sizes, and tune shifts from these programs are calculated. In addition, the shapes of the time-averaged transverse distributions are obtained. The beam-beam interaction is accomplished each turn by first calculating the beam centroids and rms sizes, and then using this information to determine the transverse kicks received by each test particle. The bunch positions and sizes are output each turn. In addition, the test particle positions are binned and accumulated in 1000 turn intervals.

Radiation excitation and damping are added to each test particle each turn in order to maintain the initial (noncolliding) horizontal and vertical emittances. Since there are no energy oscillations the contribution to the horizontal beam size from the horizontal off-energy function at the interaction region is replaced by additional betatron emittance.

Early versions of the program were in FORTRAN but the latest version has been optimized for the IBM supercomputer at Cornell. It contains FPS assembly language programming which would not be portable to other machines.

Publications describing the code:

G. P. Jackson and R. H. Siemann, "A Computer Simulation Study of e+e- Storage Ring Performance as a Function of Sextupole Distribution," IEEE Trans NS-32 (1985) 2541.

Is code documentation available? ☐ Yes ☐ No

How may the code be obtained?
   Call Gerry Jackson.

Source language: FORTRAN + FPS

Computers it runs on: IBM FPS-264

It is available as: ☐ Source code, ☐ Executable only

Source Media: ☐ Listing, ☐ Tape, ☐ Diskette, ☐ Cards, ☐ Networks
   Tape format:
   Diskette size & format:

Available through: ☐ DECNET, ☐ ARPANET, ☐ BITNET
                   ☐

Network Address:

Date of Latest Version: Jan. 1986

Program Name: SYNCH

Person to Contact: Ardith S. Kenney
Address: Lawrence Berkeley Laboratory
1 Cyclotron Road
Building 46/161
Berkeley, CA 94720
USA
Telephone Number: (415) 486-6631, FTS 451-6631

Classification of Computer Code:
  Component Design:
    ☐ Ion Source, ☐ Magnet, ☐ RF-cavity, ☐
  Accelerator Optimization:
    ☐ Linac, ☐ Cyclotron, ☒ Synchrotron, ☒ Transport lines
  Tracking or Simulation:
    ☐ Linac, ☐ Cyclotron, ☒ Synchrotron, ☐
  Analysis:
    ☒ Stability, ☐ Impedances, ☐
  Other:

Short Description: (Purpose, capabilities, algorithms, special features, etc.)

SYNCH is a computer program for use in the design and analysis of synchrotrons, storage rings and transport lines. Lattices are defined by statements describing beamlines and their components: drifts, dipoles, quadrupoles, sextupoles, other beamlines, etc. Betatron functions and closed orbit distortions due to momentum deviation or misalignments can be obtained. Orbits and beam ellipses can be tracked, and emittances, damping time, etc., calculated. Design of machines is done by versatile fitting algorithms.

Publications describing the code:

A. A. Garren and A. S. Kenney, LBL; E. D. Courant, BNL; M. J. Syphers, FNAL, "A User's Guide to SYNCH," (1985).

Is code documentation available? ☒ Yes ☐ No

How may the code be obtained?
  A. S. Kenney

Source language: FORTRAN

Computers it runs on: VAX, CDC

It is available as: ☒ Source code, ☒ Executable only

Source Media: ☒ Listing, ☒ Tape, ☐ Diskette, ☐ Cards, ☒ Networks
  Tape format:
  Diskette size & format:

Available through: ☒ DECNET, ☒ ARPANET, ☒ BITNET
                   ☐

Network Address: HEPNET/DECNETCSA2::ARDITH BITNETardith@lbl MILNET/ARPANETardith@csa2.arp

Date of Latest Version: Jan. 1986   Program Name: TBCI

Person to Contact: Thomas Weiland
        Address: Deutsches Elektronen Synchrotron/DESY
                 Notkestrasse 85 d2000
                 Hamburg 52
                 Federal Republic of Germany

Telephone Number: 49-40-8998-3196

Classification of Computer Code:
    Component Design:
        ☐ Ion Source, ☐ Magnet, ☒ RF–cavity, ☐
    Accelerator Optimization:
        ☐ Linac, ☐ Cyclotron, ☐ Synchrotron, ☐
    Tracking or Simulation:
        ☐ Linac, ☐ Cyclotron, ☐ Synchrotron, ☐
    Analysis:
        ☒ Stability, ☒ Impedances, ☒ Wakefield Effects

Other:

Short Description: (Purpose, capabilities, algorithms, special features, etc.)

TBCI analyzes the electromagnetic interaction between bunched beams of charged particles moving through cylindrically symmetric cavities by calculating wake fields. The default Gaussian shape function for the bunch may be replaced by a shape of the user's choice.

There are several post processors for TBCI. WAKCOR subtracts a tube wake field from the total wake. WAKFLS reads fields and wakes as saved at every time step and does a fourier transformation. WAKOUT reads the wake field and prints it. It can also calculate the gradient impedance, plot the bunch density and normalize wakes to ± 1.

In addition there are two variations of TBCI. TBCI00 follows the progress of TEM waves launched into a structure, e.g., a series of rf cavities connected by a vacuum pipe, from the left open boundary.

TBCI01 is the same as TBCI00 except that TM01 waves are launched into the structure.

Publications describing the code:

    T. Weiland, Proceedings of the XIth International Conference of High Energy Accelerators, Geneva (1980) 570–5.

    T. Weiland, Nucl. Instr. & Meth. 212(1983)13–34.

Is code documentation available? ☒ Yes ☐ No

How may the code be obtained?

    One must get the source code directly from Thomas Weiland.

    Executable form of the code is installed at Los Alamos and Lawrence Livermore National Laboratories. (For more information on these contact Therese Barts (505) 667-9385, FTS 843-9385 at LANL.)

Source language: FORTRAN 77

Computers it runs on: CRAY, VAX/VMS, IBM 3081.

It is available as: ☒ Source code, ☐ Executable only

Source Media: ☐ Listing, ☐ Tape, ☐ Diskette, ☐ Cards, ☒ Networks
　Tape format: EBCDIC
　Diskette size & format:

Available through: ☐ DECNET, ☐ ARPANET, ☒ BITNET
　　　　　　　　　☐

Network Address: mpywei%dhhdesy3.bitnet

Date of Latest Version: Oct. 1986

Program Name: TEAPOT

Person to Contact: Lindsay Schachinger
Address: SSC-CDG
c/o LBL, MS 90–4000
One Cyclotron Road
Berkeley, CA 94708
U.S.A.
Telephone Number: (415) 486-6590, FTS 451-6590

Classification of Computer Code:
Component Design:
☐ Ion Source, ☐ Magnet, ☐ RF–cavity, ☐
Accelerator Optimization:
☐ Linac, ☐ Cyclotron, ☒ Synchrotron, ☐
Tracking or Simulation:
☐ Linac, ☐ Cyclotron, ☒ Synchrotron, ☐
Analysis:
☐ Stability, ☐ Impedances, ☐
Other:

Short Description: (Purpose, capabilities, algorithms, special features, etc.)

TEAPOT (Thin Element Accelerator Program for Optics and Tracking), developed for design work on the Supercollider (SSC), is a tracking code which treats all elements (aside from drifts) as thin elements. TEAPOT reads a lattice in Standard (MAD) Input Format and converts all thick elements to thin ones. If a quadrupole is of "interaction region" type, it is split into four thin quadrupoles. TEAPOT neglects fringe fields. A Twiss analysis can be performed and the tunes can be adjusted using a thin lens matrix representation of the machine. Magnetic errors and misalignments can be added to elements, and the resulting lattice can be tracked exactly. A full Twiss analysis with errors is also available, which uses tracking to derive the transfer matrices for the machine. The machine can be decoupled using skew quadrupoles, the tunes can be readjusted, and the chromaticity can be fit in the presence of errors. The command format for TEAPOT is a dialect of that used by MAD.

There exists a postprocessor called MATPOT that produces a third order 4x4 matrix representation for the full ring. The output of MATPOT can be put into MARYLIE for the calculation of auxiliary quantities such as chromaticity and nonlinear invariants. MATPOT was written by Etienne Forest.

Publications describing the code:

L. Schachinger and R. Talman, Part. Acc. (to be published)

L. Schachinger and R. Talman, "TEAPOT—A Thin Element Accelerator Program for Optics and Tracking," SSC Central Design Group internal report SSC-52 (1985).

Etienne Forest, "Lie Algebraic Maps and Invariants Produced by Tracking Codes," SSC-78 (1986).

Is code documentation available? ☒ Yes ☐ No

How may the code be obtained?
Contact Lindsay Schachinger.

Source language: FORTRAN 77

Computers it runs on: VAX, CRAY, and SUN

It is available as: ☒ Source code, ☐ Executable only

Source Media: ☐ Listing, ☒ Tape, ☐ Diskette, ☐ Cards, ☒ Networks
  Tape format:
  Diskette size & format:

Available through: ☒ DECNET, ☒ ARPANET, ☒ BITNET
  ☐

Network Address: CSA::LINDSAY (decnet); LINDSAY@LBL-CSA3 (arpa or milnet); LINDSAY@LBL (bitnet).

Date of Latest Version: Dec. 1985

Program Name: TOSCA (Ver. 4.3)

Person to Contact: John S. Whitney
Address: Vector Fields, Ltd.
Osney Mead
Oxford OX2 OEE
England

Telephone Number: 0865 248236

Classification of Computer Code:
Component Design:
☒ Ion Source, ☒ Magnet, ☐ RF–cavity, ☐
Accelerator Optimization:
☐ Linac, ☐ Cyclotron, ☐ Synchrotron, ☐
Tracking or Simulation:
☐ Linac, ☐ Cyclotron, ☐ Synchrotron, ☐
Analysis:
☐ Stability, ☐ Impedances, ☐
Other:

Short Description: (Purpose, capabilities, algorithms, special features, etc.)

TOSCA is a 3D code for magnetostatic and electrostatic fields. It is the most advanced program available for non-linear magnetostatic field computation. It can be used for a wide range of applications including fusion magnets, particle accelerators, electron lenses and deflection magnets, corrosion protection and non-destructive testing. TOSCA uses a discrete finite element model in order to solve the partial differential equations governing the behavior of a system.

The finite element mesh is formed from hexahedra with 'ruled' faces which are automatically subdivided into elements. A 2D grid is created initially and this can then be swept through space thus creating 3D volumes. The sweep operations include translation, rotation and projection.

One of TOSCA's special features is that the finite element mesh does not have to model the conductors. These can slice through the mesh quite arbitrarily. The conductors are modeled using a set of primitive shapes that include arcs, bars, curved-sided hexahedra and more complex complete circuits.

The mesh primitive blocks are assigned material names and geometric properties, for example, orientation. Facilities are provided for input of nonlinear constitutive relationships and for display of the function values and derivatives.

TOSCA uses 8 and 20 node isoparametric 'brick' elements. These can be mixed together; the program will enforce inter-element continuity. The type of element created in each primitive may be selected by the user. This allows the higher order elements to be used where solution accuracy is important. Three result evaluation modes are provided to give a choice between speed and accuracy.

The suite of programs was designed to be used in a distributed computing environment. Data files created by SCARPIA for TOSCA can be easily transferred between computers and result files from TOSCA can be returned. TOSCA provides full check-point, drop-file and restart facilities so that the program maximizes the efficient use of computer resources. The PTOSCA program allows results to be displayed graphically and further calculations can be performed, e.g. particle trajectories.

Publications describing the code:
> IEEE Proc. Vol. 127 Pt. B No. 6 (1980).
>
> Vector Fields, Data Sheet Ref: 018611

Is code documentation available? ☒ Yes ☐ No

How may the code be obtained?
> By license agreement with Vector Fields, Ltd.

Source language: FORTRAN 77

Computers it runs on: PRIME, VAX, IBM

It is available as: ☒ Source code, ☐ Executable only

Source Media: ☐ Listing, ☒ Tape, ☐ Diskette, ☐ Cards, ☐ Networks
  Tape format: As required
  Diskette size & format:

Available through: ☐ DECNET, ☐ ARPANET, ☐ BITNET
  ☒ DOE Network

Network Address: Contact Robert J. Lari, Argonne Natl Lab, (312) 972-6632

Date of Latest Version: Apr. 1986
Program Name: TRACE

Person to Contact: The Los Alamos Accelerator Code Group
Address: MS H829, Group AT-6
Los Alamos National Laboratory
Los Alamos, NM 87545
USA

Telephone Number: (505)667-6677 or -2839, FTS 843-6677

Classification of Computer Code:
Component Design:
☐ Ion Source, ☐ Magnet, ☐ RF–cavity, ☐
Accelerator Optimization:
☒ Linac, ☐ Cyclotron, ☐ Synchrotron, ☐
Tracking or Simulation:
☐ Linac, ☐ Cyclotron, ☐ Synchrotron, ☐
Analysis:
☐ Stability, ☐ Impedances, ☐
Other:

Short Description: (Purpose, capabilities, algorithms, special features, etc.)

TRACE is an interactive, first-order envelope-tracing, beam-dynamics computer code with space charge. It includes some unique features as well as a number of elements not commonly found in other beam-transport programs such as the permanent-magnet quadrupole (PMQ), radio-frequency quadrupole (RFQ), RF gap, accelerator column, and accelerator tank. The code also has a number of fitting capabilities, allowing almost any element parameter in the beamline to be varied, including space charge. TRACE calculations provide immediate graphic display, including the beam envelope and the phase-space ellipses in the transverse dimensions. The program is easy to use and contains its own help package that lists all instructions necessary for input, calculations, and graphic output.

Publications describing the code:

K. R. Crandall; D. P. Rusthoi, "TRACE: An Interactive Beam-Transport Code," Proceedings of the 1984 Linear Accelerator Conference, Darmstadt-Seeheim, FRG (1984) 371-373.

K. R. Crandall, D. P. Rusthoi, "Documentation for TRACE: An Interactive Beam-Transport Code," Los Alamos National Laboratory Internal Report No. LA-10235-MS (1985) 66 pp.

Is code documentation available? ☒ Yes ☐ No

How may the code be obtained?
Contact The Los Alamos Accelerator Code Group or D. Rusthoi(505) 667-2796, FTS 843-2796

Source language: FORTRAN

Computers it runs on: CDC 6600, 7600, VAX-11/750

It is available as: ☒ Source code, ☐ Executable only

Source Media: ☐ Listing, ☒ Tape, ☐ Diskette, ☐ Cards, ☐ Networks
Tape format: VAX: 1600 bpi, ASCII
Diskette size & format:

Available through: ☐ DECNET, ☒ ARPANET, ☒ BITNET
☐

Network Address:

Date of Latest Version: Jan. 1986   Program Name: TRACE3D

Person to Contact: The Los Alamos Accelerator Code Group
Address: MS H829, Group AT-6
Los Alamos National Laboratory
Los Alamos, NM 87545
USA

Telephone Number: (505)667-6677 or -2839, FTS 843-6677

Classification of Computer Code:
Component Design:
☐ Ion Source, ☐ Magnet, ☐ RF–cavity, ☐
Accelerator Optimization:
☒ Linac, ☐ Cyclotron, ☐ Synchrotron, ☐
Tracking or Simulation:
☒ Linac, ☐ Cyclotron, ☐ Synchrotron, ☐
Analysis:
☐ Stability, ☐ Impedances, ☐
Other:

Short Description: (Purpose, capabilities, algorithms, special features, etc.)

TRACE 3-D is an interactive program that calculates the envelopes of a bunched beam (including linear space-charge forces) through a user-defined transport system. The transport system may consist of the the following elements: 1) drift, 2) thin lens, 3) quadrupole, 4) permanent-magnet quadrupole (PMQ), 5) solenoid, 6) doublet, 7) triplet, 8) bending magnet, 9) edge angle (for bend), 10) rf gap, 11)radio-frequency-quadrupole cell (RFQ), 12) rf cavity, 13) coupled-cavity tank, 14) a user-defined element, and 15) a coordinate rotation.

The beam is represented by a 6 x 6 σ-matrix (introduced by the TRANSPORT program) defining a hyperellipsoid in six-dimensional phase space. The projection of this hyperellipsoid on any two-dimensional plane is an ellipse that defines the boundary of the beam in that plane. Using a sequence of matrix transformations, the beam can be "followed" between any two elements. The user can change any parameter and observe the effect on the beam envelopes and on the output beam ellipses. Also, several matching options are available that determine values for the ellipse parameters or for specified transport-system parameters (such as quadrupole gradients) to meet specified objectives.

Publications describing the code:

K. R. Crandall and R. S. Mills, "TRACE 3-D Docmentation," Los Alamos National Laboratory internal document (1985).

Is code documentation available? ☒ Yes ☐ No

How may the code be obtained?

From the Los Alamos Accelerator Code Group. Contact Helen Stokes, AT-6, LANL, (505) 667-9131 or (667-2839); FTS 843-9131.

Source language: FORTRAN

Computers it runs on: CRAY

It is available as: ☐ Source code, ☐ Executable only

Source Media: ☐ Listing, ☒ Tape, ☐ Diskette, ☐ Cards, ☒ Networks
 Tape format:
 Diskette size & format:

Available through: ☐ DECNET, ☒ ARPANET, ☒ BITNET
 ☐

Network Address: hks@lanl.arpa

Date of Latest Version: 1981

Program Name: TRACK

Person to Contact: Robert J. Lari
Address: Argonne National Laboratory
Argonne, IL 60439
USA

Telephone Number:

Classification of Computer Code:
Component Design:
☐ Ion Source, ☐ Magnet, ☐ RF–cavity, ☐
Accelerator Optimization:
☐ Linac, ☐ Cyclotron, ☐ Synchrotron, ☐
Tracking or Simulation:
☐ Linac, ☐ Cyclotron, ☐ Synchrotron, ☒ spectrometers

Analysis:
☐ Stability, ☐ Impedances, ☐
Other:

Short Description: (Purpose, capabilities, algorithms, special features, etc.)

The TRACK/BEAM commands of GFUN-3D are useful for tracking a charged particle through a magnetic field and plotting the path. It was felt the subroutines associated with these two commands would form a useful stand-alone program where the user supplied the magnetic field instead of the GFUN 3-D calculated field. Hence, measured fields could also be used. The user-supplied fields are stored on disk as a field map or as the edge field of a uniform field magnet. The graphic subroutines are written for use on the Tektronix 4012.

Publications describing the code:

R. J. Lari, "TRACK — A Program to Track Charged Particles Through a Magnetic Field and Plot the Path," Proc. of a Workshop on High-resolution Large-acceptance Spectrometers, at Argonne National Laboratory (1981), Argonne National Laboratory report ANL/PHY-81-2 and CONF-8109123.

Is code documentation available? ☒ Yes ☐ No

How may the code be obtained?
Contact Robert J. Lari

Source language: FORTRAN

Computers it runs on: IBM

It is available as: ☒ Source code, ☐ Executable only

Source Media: ☐ Listing, ☒ Tape, ☐ Diskette, ☐ Cards, ☐ Networks
Tape format: LRECL=80, RECFM=FB, BLKSIZE=800.
Diskette size & format:

Available through: ☐ DECNET, ☐ ARPANET, ☐ BITNET
☐

Network Address: None

Date of Latest Version: unknown   Program Name: TRAJECTORY

Person to Contact: A. C. Paul
Address: MS L626
Lawrence Livermore National Laboratory
Livermore, CA 94550
USA

Telephone Number: (415) 423–3183, FTS 543–3183

Classification of Computer Code:
Component Design:
☐ Ion Source, ☐ Magnet, ☐ RF–cavity, ☐
Accelerator Optimization:
☐ Linac, ☒ Cyclotron, ☐ Synchrotron, ☐
Tracking or Simulation:
☐ Linac, ☒ Cyclotron, ☐ Synchrotron, ☐
Analysis:
☐ Stability, ☐ Impedances, ☐
Other:

Short Description: (Purpose, capabilities, algorithms, special features, etc.)

An orbit and ion optic matrix-transport program originally used for the 184-inch LBL cyclotron.

Transport-matrix output can be used as input to TRANSPORT or OPTIC. $\sim 50~K_8$ memory, runs in ~ 1 min. on CDC 6600.

Will track protons and pions in the median plane of the cyclotron.

Publications describing the code:

A. C. Paul, "TRAJECTORY — An Orbit and Ion Optic Matrix Program for the 184-inch Cyclotron," Lawrence Berkeley Laboratory Report UCRL-19407 (1969).

John S. Colonias, "Particle Accelerator Design: Computer Programs," Academic Press, New York (1974) 203.

Is code documentation available? ☐ Yes ☐ No

How may the code be obtained?
Unknown

Source language: FORTRAN

Computers it runs on: CDC 6600/7600, VAX 11/70's

It is available as: ☐ Source code, ☐ Executable only

Source Media: ☐ Listing, ☐ Tape, ☐ Diskette, ☐ Cards, ☐ Networks
Tape format:
Diskette size & format:

Available through: ☐ DECNET, ☐ ARPANET, ☐ BITNET
☐

Network Address:

Date of Latest Version: unknown

Program Name: TRAMP

Person to Contact: J. W. Gardner
    Address: Rutherford-Appleton Laboratories
               Chilton, Didcot
               Oxon OX11 OQX
               England

Telephone Number:

Classification of Computer Code:
    Component Design:
        ☐ Ion Source, ☐ Magnet, ☐ RF–cavity, ☐
    Accelerator Optimization:
        ☐ Linac, ☐ Cyclotron, ☐ Synchrotron, ☒   Beam Transport

    Tracking or Simulation:
        ☐ Linac, ☐ Cyclotron, ☐ Synchrotron, ☒   Beam Transport

    Analysis:
        ☐ Stability, ☐ Impedances, ☐
    Other:

Short Description: (Purpose, capabilities, algorithms, special features, etc.)

Program TRAMP was developed at Rutherford High Energy Laboratory by Gardner and Whiteside to provide solutions to problems encountered in beam transport design. It has been extensively modified by various experimenters to fit the needs and the computer facilities of their respective laboratories.

The LBL version is capable of tracking and matching trajectories, beam profiles, or phase-space ellipses through a given beam transport system. Most beam elements are represented by 2x2 matrices for each plane, but the code handles sextupoles by integration of trajectories. Matching can be done on dispersion.

Publications describing the code:

    J. W. Gardner and D. Whiteside, "TRAMP — Tracking and Matching Program," Rutherford Laboratory Report NIRL/M/21 (1961).

    J. W. Gardner and D. Whiteside, "A FORTRAN version of TRAMP," Rutherford Laboratory Report NIRL/M/41 (1963).

    John S. Colonias, "Particle Accelerator Design: Computer Programs," Academic Press, New York (1974) 176.

Is code documentation available? ☐ Yes ☐ No

How may the code be obtained?
    Unknown (May still be available from LBL).

Source language: FORTRAN

Computers it runs on: CDC 6600

It is available as: ☐ Source code, ☐ Executable only

Date of Latest Version: Mar. 1985

Program Name: TRANCO

Person to Contact: Eva S. Bozoki
Address: NSLS Dept.
Brookhaven National Laboratory
Upton, NY 11973
USA

Telephone Number: (516) 282-3701

Classification of Computer Code:
Component Design:
☐ Ion Source, ☐ Magnet, ☐ RF–cavity, ☐
Accelerator Optimization:
☐ Linac, ☐ Cyclotron, ☐ Synchrotron, ☒ Transport Lines

Tracking or Simulation:
☐ Linac, ☐ Cyclotron, ☐ Synchrotron, ☒ Transport lines

Analysis:
☐ Stability, ☐ Impedances, ☐
Other: Control

Short Description: (Purpose, capabilities, algorithms, special features, etc.)

Purpose: To provide a tool to examine and control of transport lines in terms of beam and machine parameters using the mathematical model of the system.

It is a modeling program, which can be used

1) off-line as a design program (like, e.g., TRANSPORT) or

2) on-line as a control program. (When used as a control program, the program-microprocessor interface is specific to the installation.)

The program can perform

1) ellipse matching — calculate/control the orientation and shape of the phase space ellipses,

2) ellipse positioning — calculate/control the position of the center of the phase space ellipses,

3) beam steering.

A colored graphic display program for GENESCO GCT-300 display system is also available.

Lattice data are read from input files. User-program interface is through screen and it is designed in such a way as to facilitate the input and minimize the effort (using data validation and default options).

Publications describing the code:

Is code documentation available? ☐ Yes ☐ No

How may the code be obtained?
> Eva S. Bozoki, "High Level Control Programs at NSLS," Computing in Accelerator Design and Operation (1983) 420.

Source language: FORTRAN

Computers it runs on: DG

It is available as: ☐ Source code, ☐ Executable only

Source Media: ☐ Listing, ☐ Tape, ☐ Diskette, ☐ Cards, ☐ Networks
Tape format:
Diskette size & format:

Available through: ☐ DECNET, ☐ ARPANET, ☐ BITNET
☐

Network Address:

Date of Latest Version: Jan. 1986 Program Name: TRANSOPTR

Person to Contact: Mark S. de Jong
Address: Accel. Phys. Branch
Chalk River Nuclear Laboratories
Chalk River, Ontario K0J-1-J0
Canada

Telephone Number: (613) 584-3311

Classification of Computer Code:
Component Design:
☐ Ion Source, ☐ Magnet, ☐ RF–cavity, ☐
Accelerator Optimization:
☐ Linac, ☐ Cyclotron, ☐ Synchrotron, ☒   Beam Transport

Tracking or Simulation:
☐ Linac, ☐ Cyclotron, ☐ Synchrotron, ☐
Analysis:
☐ Stability, ☐ Impedances, ☐
Other:

Short Description: (Purpose, capabilities, algorithms, special features, etc.)

A beam transport design code with parametric optimization. The code analyzes the transport of charged particle beams through a user defined magnet system. Space charge effects may be included either in two dimensions, treating transverse forces only, or in three dimensions by treating both transverse and longitudinal forces on the beam. The magnet system parameters are varied (within user defined limits) until the properties of the transported beam and/or the system transport matrix match those properties requested by the user. The code uses matrix formalism to represent the transport elements and optimization is achieved using the variable metric method. For problems without space charge a first or second order matrix formalism can be selected. Any constraints that can be expressed algebraically may be included by the user as part of his design.

Publications describing the code:
R. M. Hutcheon and E. A. Heighway, Nuc. Inst. & Mtds. 187 (1981) 89–95.

E. A. Heighway and M. S. de Jong, "A First Order Space Charge Option for TRANSOPTR," IEEE Trans. NS30 (1983) 2666.

Is code documentation available? ☒ Yes ☐ No

How may the code be obtained?
From Mark de Jong, or from Edward A. Heighway, AT-6, MS H829, Los Alamos National Laboratory, Los Alamos, NM 87545 (505) 667-1543, FTS 843-1543.

Source language: FORTRAN 77

Computers it runs on: CDC, CYBER, CRAY

It is available as: ☒ Source code, ☐ Executable only

Source Media: ☒ Listing, ☒ Tape, ☐ Diskette, ☐ Cards, ☐ Networks
Tape format: as desired
Diskette size & format:

Available through: ☐ DECNET, ☐ ARPANET, ☐ BITNET
☐
Network Address:

Date of Latest Version: June 1985   Program Name: TRANSPORT

Person to Contact: David C. Carey
       Address: Fermilab
                   P.O. Box 500
                   Batavia, IL 60510
                   USA

Telephone Number: (312) 840-3639, FTS 370-3639

Classification of Computer Code:
     Component Design:
         ☐ Ion Source, ☐ Magnet, ☐ RF-cavity, ☐
     Accelerator Optimization:
         ☐ Linac, ☐ Cyclotron, ☐ Synchrotron, ☒   Beam Line
     Tracking or Simulation:
         ☐ Linac, ☐ Cyclotron, ☐ Synchrotron, ☒   Beam Line
     Analysis:
         ☐ Stability, ☐ Impedances, ☐
     Other:

Short Description: (Purpose, capabilities, algorithms, special features, etc.)
     Beam line transfer matrix calculation and fitting program. Calculates floor coordinates, beam matrix, and first, second and some third-order transfer matrices. Elements included arc-bending-magnet, quadrupole, sextupole, octupole, solenoid, and accelerating cavity. Can simulate misalignments, beam steering, and random errors for any beam line parameter. The code will accept input in the MAD format

Publications describing the code:
     (TRANSPORT manual) SLAC-91, or NAL-91, or CERN-80.04

Is code documentation available? ☒ Yes ☐ No

How may the code be obtained?
     Contact Sue McNamara, Program Librarian,

     Fermilab, P.O. Box 500, Batavia, IL 60510.

Source language: FORTRAN

Computers it runs on: CDC, IBM, VAX

It is available as: ☐ Source code, ☐ Executable only

Source Media: ☐ Listing, ☒ Tape, ☐ Diskette, ☐ Cards, ☐ Networks
     Tape format: most anything
     Diskette size & format:

Available through: ☐ DECNET, ☐ ARPANET, ☒ BITNET
                      ☐

Network Address: b90665@fnal or in Europe try FCI@CERNVM

Date of Latest Version: unknown   Program Name: TRANSPORT, LBL Version

Person to Contact: Arthur C. Paul
          Address: MS L-626,
                   Lawrence Livermore National Laboratory
                   P.O. Box 808
                   Livermore, CA 94550
                   USA
Telephone Number: (415) 423-3183, FTS 543-3183

Classification of Computer Code:
    Component Design:
        ☐ Ion Source, ☐ Magnet, ☐ RF-cavity, ☐
    Accelerator Optimization:
        ☐ Linac, ☐ Cyclotron, ☐ Synchrotron, ☒   Beam lines

    Tracking or Simulation:
        ☐ Linac, ☐ Cyclotron, ☐ Synchrotron, ☒   Beam lines

    Analysis:
        ☐ Stability, ☐ Impedances, ☐
    Other:

Short Description: (Purpose, capabilities, algorithms, special features, etc.)
    TRANSPORT is a computer code for calculating properties of charged particle beam transport systems using the matrix method in a six-dimensional phase space and a version of TRANSPORT was translated into FORTRAN from the original BALGOL SLAC TRANSPORT. Some of the important additions are polygon transformation, ray tracing, particle separator, space charge, output plotting, interactive on-line calculations, and flexible data manipulation procedures.

Publications describing the code:
    A. C. Paul, "TRANSPORT: An Ion Optic Program. LBL Version," LBL Internal Report 2697 (1975) 6pp.

Is code documentation available? ☐ Yes ☐ No

How may the code be obtained?

Source language:
Computers it runs on:
It is available as: ☐ Source code, ☐ Executable only
Source Media: ☐ Listing, ☐ Tape, ☐ Diskette, ☐ Cards, ☐ Networks
    Tape format:
    Diskette size & format:
Available through: ☐ DECNET, ☐ ARPANET, ☐ BITNET
                   ☐
Network Address:

Date of Latest Version: unknown

Program Name: TRANSVRS

Person to Contact: Karl Bane
   Address: Stanford Linear Accelerator Center
            SLAC BIN 26
            P.O. Box 4349
            Stanford, CA 94305
            USA
Telephone Number: (415) 497-2026, FTS 461-9300 ext. 2026

Classification of Computer Code:
   Component Design:
      ☐ Ion Source, ☐ Magnet, ☒ RF–cavity, ☐
   Accelerator Optimization:
      ☐ Linac, ☐ Cyclotron, ☐ Synchrotron, ☐
   Tracking or Simulation:
      ☐ Linac, ☐ Cyclotron, ☐ Synchrotron, ☐
   Analysis:
      ☒ Stability, ☐ Impedances, ☒ Wakefield Effects
   Other:

Short Description: (Purpose, capabilities, algorithms, special features, etc.)
   TRANSVRS is a code to calculate the frequencies of a large number of deflecting modes in a periodic array of rf cavities. Using these frequencies one can calculate the transverse wakefield forces on a bunch beam tending to cause beam breakup.

Publications describing the code:
   K. Bane and B. Zotter, "Transverse Modes in Periodic Cylindrical Cavities," Proc. of 11th Int'l Conf. on High Energy Accelerator, Geneva, Switzerland, July 7–11, 1980, Birkhauser Verlag, Basel (1980).

Is code documentation available? ☐ Yes ☐ No

How may the code be obtained?
   Call Karl Bane.

Source language:
Computers it runs on:
It is available as: ☐ Source code, ☐ Executable only
Source Media: ☐ Listing, ☐ Tape, ☐ Diskette, ☐ Cards, ☐ Networks
   Tape format:
   Diskette size & format:
Available through: ☐ DECNET, ☐ ARPANET, ☐ BITNET
                   ☐
Network Address: KBAME@SLACVM.BITNET

Date of Latest Version: unknown

Program Name: TRIDIF

Person to Contact: John R. Freeman
Address: Org.1241, Bldg.980
Sandia National Laboratory
P.O. Box 5800
Albuquerque, NM 87115
USA
Telephone Number: (505)844-5254, FTS 844-5254

Classification of Computer Code:
Component Design:
☐ Ion Source, ☒ Magnet, ☐ RF–cavity, ☐
Accelerator Optimization:
☐ Linac, ☐ Cyclotron, ☐ Synchrotron, ☐
Tracking or Simulation:
☐ Linac, ☐ Cyclotron, ☐ Synchrotron, ☐
Analysis:
☐ Stability, ☐ Impedances, ☐
Other:

Short Description: (Purpose, capabilities, algorithms, special features, etc.)

TRIDIF is a time-dependent diffusion version of the well-known PANDIRA-POISSON-TRIM triangular mesh magnet code. Modifications allow TRIDIF to treat field diffusion in materials with time-varying permeabilities. Good agreement between the measured and computed magnetic fields was found for a simple test experiment.

See also documentation on PANDIRA, POISSON and TRIM.

Publications describing the code:

M. L. Hodgdon; J. R. Freeman, "Transient Magnetic-field Calculations with TRIDIF," Sandia National Laboratory Internal Report No. SAND-81-2001C; CONF-810954-1 (1981) 10pp.

Is code documentation available? ☐ Yes ☐ No

How may the code be obtained?
Contact John Freeman; Hodgdon has left Sandia.

Source language:

Computers it runs on:

It is available as: ☐ Source code, ☐ Executable only

Source Media: ☐ Listing, ☐ Tape, ☐ Diskette, ☐ Cards, ☐ Networks
Tape format:
Diskette size & format:

Available through: ☐ DECNET, ☐ ARPANET, ☐ BITNET
☐
Network Address:

Date of Latest Version: 1975   Program Name: TRIM (ANL) & FORGY

Person to Contact: Robert J. Lari 360
           Address: Argonne National Laboratory
                    9700 S. Cass Ave.
                    Argonne, IL 60439
                    USA

Telephone Number: (312)972-6632, FTS 972-6632

Classification of Computer Code:
    Component Design:
        ☐ Ion Source, ☒ Magnet, ☐ RF–cavity, ☐
    Accelerator Optimization:
        ☐ Linac, ☐ Cyclotron, ☐ Synchrotron, ☐
    Tracking or Simulation:
        ☐ Linac, ☐ Cyclotron, ☐ Synchrotron, ☐
    Analysis:
        ☐ Stability, ☐ Impedances, ☐
    Other:

Short Description: (Purpose, capabilities, algorithms, special features, etc.)
    TRIM — 2D magnetostatic code, FORGY — Force calc on coils and steel

Publications describing the code:
    Alan M. Winslow. UCRL-7784-T (1965).

    R. Lari, TRIM — Unpublished User Guide.

    R. Lari, FORGY — ANL Internal Report TKK/RJL-2 (1972).

Is code documentation available? ☒ Yes ☐ No

How may the code be obtained?
    Contact Robert Lari.

Source language: FORTRAN

Computers it runs on: IBM 370

It is available as: ☒ Source code, ☐ Executable only

Source Media: ☐ Listing, ☒ Tape, ☐ Diskette, ☐ Cards, ☐ Networks
    Tape format: LRECL = 80, RECFM = FB, BLKSIZE = 800
    Diskette size & format:

Available through: ☐ DECNET, ☐ ARPANET, ☐ BITNET
    ☐

Network Address: None

Date of Latest Version: unknown   Program Name: TRIO

Person to Contact: T. Matsuo
Address: College of General Education
Osaka Univ.,
Toyonaka,
Japan

Telephone Number:

Classification of Computer Code:
Component Design:
☐ Ion Source, ☐ Magnet, ☐ RF–cavity, ☐
Accelerator Optimization:
☐ Linac, ☐ Cyclotron, ☐ Synchrotron, ☐
Tracking or Simulation:
☐ Linac, ☐ Cyclotron, ☐ Synchrotron, ☒ Beam Lines
Analysis:
☐ Stability, ☐ Impedances, ☐
Other:

Short Description: (Purpose, capabilities, algorithms, special features, etc.)

TRIO (Third Order Ion Optics) is a computer program for the calculation of ion trajectories; it is applicable to any ion optical system consisting of drift spaces, cylindrical or toroidal electric sector fields, homogeneous or inhomogeneous magnetic sector fields, magnetic and electrostatic Q-lenses. The influence of the fringing field is taken into consideration. The trajectory calculation can execute with accuracy up to third order. Any one of three dispersion bases, momentum, energy, mass and energy, may possibly be selected.

Publications describing the code:

T. Matsuo; H. Matsuda; Y. Fujita; H. Wollnik "Computer program 'TRIO' for Third Order Calculation of Ion Trajectory," Shitsuryo Bunseki (Japan) v. 24:1 (1976) 19-62. Also available as: Proceedings of the Third Symposium on Ion Sources and Application Technology (1979) 25-28.

H. Wollnik and Matsuo, "Addition of Flight Time Calculation to Computer Program TRIO," Mass Spectroscopy 27 (1979) 131-134.

Is code documentation available? ☐ Yes ☐ No

How may the code be obtained?
unknown.

Source language:
Computers it runs on:
It is available as: ☐ Source code, ☐ Executable only
Source Media: ☐ Listing, ☐ Tape, ☐ Diskette, ☐ Cards, ☐ Networks
Tape format:
Diskette size & format:
Available through: ☐ DECNET, ☐ ARPANET, ☐ BITNET
☐
Network Address:

Date of Latest Version: Apr. 1985

Program Name: TURTLE

Person to Contact: David C. Carey
Address: Fermilab
P.O. Box 500
Batavia, IL 60510
USA

Telephone Number: (312) 840-3639, FTS 370-3639

Classification of Computer Code:
　Component Design:
　　☐ Ion Source, ☐ Magnet, ☐ RF-cavity, ☐
　Accelerator Optimization:
　　☐ Linac, ☐ Cyclotron, ☐ Synchrotron, ☐
　Tracking or Simulation:
　　☐ Linac, ☐ Cyclotron, ☐ Synchrotron, ☒　Beam Line
　Analysis:
　　☐ Stability, ☐ Impedances, ☐
　Other:

Short Description: (Purpose, capabilities, algorithms, special features, etc.)

　Simulation of single-pass, charged-particle beam lines and spectrometers. Includes geometric terms and all-order chromatic effects for individual quadrupoles, bending magnets, sextupoles, and solenoids. Accumulates effects regardless of order. Can make specified one- and two-dimensional histograms of particle coordinates at any beam line location.

Publications describing the code:

　National Accelerator Laboratory internal report 64 (TURTLE manual).

Is code documentation available? ☒ Yes ☐ No

How may the code be obtained?
　Sue McNamara, Program Librarian

　Fermilab, P.O. Box 500, Batavia, IL 60510

Source language: FORTRAN

Computers it runs on: CDC, IBM, VAX

It is available as: ☒ Source code, ☐ Executable only

Source Media: ☐ Listing, ☒ Tape, ☐ Diskette, ☐ Cards, ☐ Networks
　Tape format: most anything
　Diskette size & format:

Available through: ☐ DECNET, ☐ ARPANET, ☒ BITNET
　　　　　　　　　☐

Network Address: b90665@fnal

Date of Latest Version: Jun. 1985   Program Name: ULTRAFISH

Person to Contact: Los Alamos Accelerator Code Group
Address: MS H-829, Group AT-6
Los Alamos National Laboratory
Los Alamos, NM 87544
USA

Telephone Number: (505) 667–6677 (or 667–2839), FTS 843–6677

Classification of Computer Code:
    Component Design:
       ☐ Ion Source, ☐ Magnet, ☒ RF–cavity, ☐
    Accelerator Optimization:
       ☐ Linac, ☐ Cyclotron, ☐ Synchrotron, ☐
    Tracking or Simulation:
       ☐ Linac, ☐ Cyclotron, ☐ Synchrotron, ☐
    Analysis:
       ☐ Stability, ☐ Impedances, ☐
    Other:

Short Description: (Purpose, capabilities, algorithms, special features, etc.)
    ULTRAFISH computes the resonant frequencies and fields in an rf cavity which is a figure of revolution for azimuthally assymmetric modes. It will handle regions of different permeability and dielectric constant. It works for some geometries, but not others. The code has essential problems involving boundary conditions that have never been overcome.

Publications describing the code:
    R. L. Gluckstern, R. F. Holsinger, K. Halbach and G. N. Minerbo, "ULTRAFISH — Generalization of SUPERFISH to $m \geq 1$," Proc. 1981 Linear Accl. Conf. in Santa Fe, Los Alamos Report LA-9234-C, p. 102.

Is code documentation available? ☐ Yes ☒ No

How may the code be obtained?
    It is not available for distribution.

Source language:

Computers it runs on:

It is available as: ☐ Source code, ☐ Executable only

Source Media: ☐ Listing, ☐ Tape, ☐ Diskette, ☐ Cards, ☐ Networks
    Tape format:
    Diskette size & format:

Available through: ☐ DECNET, ☐ ARPANET, ☐ BITNET
☐

Network Address:

Date of Latest Version: Jan. 1986   Program Name: URMEL

Person to Contact: Thomas Weiland
          Address: Deutches Elektronen Synchrotron/DESY
                   Notkestrasse 85 d2000
                   Hamburg 52
                   Federal Republic of Germany

Telephone Number: 49-40-8998-3196

Classification of Computer Code:
  Component Design:
    ☐ Ion Source, ☐ Magnet, ☒ RF–cavity, ☐
  Accelerator Optimization:
    ☐ Linac, ☐ Cyclotron, ☐ Synchrotron, ☐
  Tracking or Simulation:
    ☐ Linac, ☐ Cyclotron, ☐ Synchrotron, ☐
  Analysis:
    ☒ Stability, ☒ Impedances, ☒   Wake field effects
  Other:

Short Description: (Purpose, capabilities, algorithms, special features, etc.)

URMEL computes symmetric (m=0) and asymmetric (m>0) resonant modes in cavities and frequencies of longitudinally homogeneous fields in waveguides for cylindrically symmetric accelerating structures. It uses a rectangular mesh. Only the electric field components in the (r,z) plane are used for the calculation of the transverse modes instead of H_ϕ & E_ϕ. The discretization is based on FIT (Finite Integration Techniques) described in the reference. Many modes are found on one pass.

Publications describing the code:

T. Weiland, Electronics & Communication (AEÜ) 31 (1977) 116.

T. Weiland, Nucl. Inst. & Meth. 216 (1983) 329.

Is code documentation available? ☒ Yes ☐ No

How may the code be obtained?
  One must get the code directly from Thomas Weiland.

  Executable form of the code is installed at Los Alamos and Lawrence Livermore National Labratories (For more information on these contact Therese Barts (505) 667-9385 at Los Alamos.)

Source language: FORTRAN 77

Computers it runs on: CRAY, VAX/VMS, IBM 3081

It is available as: ☒ Source code, ☐ Executable only

Source Media: ☐ Listing, ☒ Tape, ☐ Diskette, ☐ Cards, ☒ Networks
  Tape format: as desired
  Diskette size & format:

Available through: ☐ DECNET, ☐ ARPANET, ☒ BITNET
                    ☐

Network Address: mpywei %dhhdesy3.bitnet

Date of Latest Version: Apr. 1986   Program Name: URMEL-T

Person to Contact: Thomas Weiland
         Address: Deutsches Elektronen Synchrotron/DESY
                  Notkestrasse 85 d2000
                  Hamburg 52
                  Federal Republic of Germany

Telephone Number: 49-40-8998-3196

Classification of Computer Code:
    Component Design:
        ☐ Ion Source, ☐ Magnet, ☒ RF–cavity, ☐
    Accelerator Optimization:
        ☐ Linac, ☐ Cyclotron, ☐ Synchrotron, ☐
    Tracking or Simulation:
        ☐ Linac, ☐ Cyclotron, ☐ Synchrotron, ☐
    Analysis:
        ☒ Stability, ☒ Impedances, ☒   Wake field effects

Other:

Short Description: (Purpose, capabilities, algorithms, special features, etc.)

URMEL-T computes symmetric (m=0) and assymetric (m>0) resonant modes in cavities and frequencies of longitudinally homogeneous fields in waveguides for cylindrically symmetric accelerating structures. It uses a triangular mesh instead of the rectangular mesh used by URMEL.

Publications describing the code:

T. Weiland, Electronics & Communication (AEÜ) 31 (1977) 116.

U. Van Rienen and T. Weiland, "Triangular Discretization Method for the Evaluation of RF-fields in Waveguides and Cylindrically Symmetric Cavities," IEEE Trans. MAG-21 (1985) 2317–20.

Is code documentation available? ☒ Yes ☐ No

How may the code be obtained?

One must get the code directly from Thomas Weiland.

Executable form of the code is installed at Los Alamos and Lawrence Livermore National Laboratories. For more information on these contact Therese Barts (505) 667-9385 at Los Alamos National Laboratory.

Source language: FORTRAN 77

Computers it runs on: CRAY, VAX/VMS, IBM 3081

It is available as: ☒ Source code, ☐ Executable only

Source Media: ☐ Listing, ☒ Tape, ☐ Diskette, ☐ Cards, ☒ Networks
    Tape format: EBCDIC
    Diskette size & format:

Available through: ☐ DECNET, ☐ ARPANET, ☒ BITNET
                   ☐

Network Address: mpywei %dhhdesy3.bitnet

Date of Latest Version: June, 1986   Program Name: WAVE

Person to Contact: David W. Forslund
     Address: MS E531, X-DO
               Los Alamos National Laboratory
               Los Alamos, NM 87545
               USA

Telephone Number: (505) 667–4370, FTS 843-4370

Classification of Computer Code:
     Component Design:
         ☐ Ion Source, ☐ Magnet, ☐ RF–cavity, ☐
     Accelerator Optimization:
         ☐ Linac, ☐ Cyclotron, ☐ Synchrotron, ☐
     Tracking or Simulation:
         ☐ Linac, ☐ Cyclotron, ☐ Synchrotron, ☐
     Analysis:
         ☐ Stability, ☐ Impedances, ☐
     Other: Laser beat wave accelerators

Short Description: (Purpose, capabilities, algorithms, special features, etc.)
     WAVE is a 2D, particle-in-cell code for self-consistently solving Newton's equations of motion and Maxwell's equations. It has application to space charge problems, plasmas and analysis of laser-plasma interactions. It is portable to any installation. Partial documentation exists.

Publications describing the code:
     D. W. Forslund, "Fundamentals of Plasma Simulation," Space Science Reviews 42 (1985) 3–16.

     R. L. Morse and C. W. Nielson, "Numerical Simulation of the Weibel Instability in One and Two Dimensions," Phys. Fluids 14 (1971) 830.

     C. Joshi et al, "Ultrahigh Gradient Particle Acceleration by Intense Laser-driven Plasma Density Waves," Nature 311 (1984) 525.

Is code documentation available? ☒ Yes ☐ No

How may the code be obtained?
     Call David Forslund.

Source language: FORTRAN 77

Computers it runs on: PC's, VAX, CRAY, IBM

It is available as: ☐ Source code, ☐ Executable only

Source Media: ☐ Listing, ☒ Tape, ☐ Diskette, ☐ Cards, ☒ Networks
     Tape format: as desired
     Diskette size & format:

Available through: ☐ DECNET, ☒ ARPANET, ☒ BITNET
                     ☐

Network Address: dwf@lanl.arpa

Date of Latest Version: Dec. 1986

Program Name: WIGWAM

Person to Contact: John M. Jowett
Address: LEP Division
CERN
CH-1211 Geneva 23
Switzerland

Telephone Number: (022) 83 66 43 or 83 50 86

Classification of Computer Code:
   Component Design:
      ☐ Ion Source, ☐ Magnet, ☐ RF-cavity, ☐
   Accelerator Optimization:
      ☐ Linac, ☐ Cyclotron, ☒ Synchrotron, ☐
   Tracking or Simulation:
      ☐ Linac, ☐ Cyclotron, ☐ Synchrotron, ☐
   Analysis:
      ☐ Stability, ☐ Impedances, ☐
   Other: Electron storage ring performance, wigglers

Short Description: (Purpose, capabilities, algorithms, special features, etc.)

   WIGWAM evaluates the parameters and performance of electron-positron storage rings in a very fexible way. It includes the effects and will calculate the excitations for normal dipole wigglers and nonlinear wigglers (combined function quadrupole-sextupole or dipole-octupole) using appropriate generalizations of the usual electron ring formulae. The program will also optimize performance (e.g. maximise luminosity, minimise energy spread, etc.) by calculating appropriate schemes for varying damping partition numbers, coupling, wiggler fields, RF voltage, etc. It also produces the "loofa" diagrams, which provide a gobal picture of the potential performance of a ring.

   Although the program is still under development with a view to integration into the MAD environment, a working version is available.

Publications describing the code:

   J. M. Jowett, "Luminosity and Energy Spread in LEP," CERN-LEP-TH/85-4 gives many examples of output.

   J. M. Jowett, "Description of the WIGWAM Program," CERN internal report LEP Note 521.

Is code documentation available? ☒ Yes ☐ No

How may the code be obtained?
   Contact J. M. Jowett.

Source language: FORTRAN

Computers it runs on: IBM

It is available as: ☒ Source code, ☐ Executable only

Source Media: ☒ Listing, ☒ Tape, ☐ Diskette, ☐ Cards, ☒ Networks
   Tape format:
   Diskette size & format:

Available through: ☐ DECNET, ☐ ARPANET, ☒ BITNET
   ☒ EARNET

Network Address: JOWETT@CERNVM

Date of Latest Version: 1985   Program Name: WOLF

Person to Contact: K. Halbach
    Address: Lawrence Berekeley Laboratory
    1 Cyclotron Road
    Berkeley, CA 94720
    USA

Telephone Number: (415) 486-5868, FTS 451-5868

Classification of Computer Code:
    Component Design:
        ☒ Ion Source, ☐ Magnet, ☐ RF-cavity, ☐
    Accelerator Optimization:
        ☐ Linac, ☐ Cyclotron, ☐ Synchrotron, ☐
    Tracking or Simulation:
        ☐ Linac, ☐ Cyclotron, ☐ Synchrotron, ☐
    Analysis:
        ☐ Stability, ☐ Impedances, ☐
    Other:

Short Description: (Purpose, capabilities, algorithms, special features, etc.)

The WOLF code solves POISSON's equation within a user-defined problem boundary of arbitrary shape. The code is compatible with ANSI FORTRAN and uses a two-dimensional Cartesian coordinate geometry represented on a triangular lattice. The vacuum electric fields and equipotential lines are calculated for the input problem. The user may then introduce a series of emitters from which particles of different charge-to-mass ratios and initial energies can originate. These non-relativistic particles will then be traced by WOLF through the user-defined region. Effects of ion and electron space charge are included in the calculation. A subprogram PISA forms part of this code and enables optimization of various aspects of the problem. The WOLF package also allows detailed graphics analysis of the computed results to be performed.

Publications describing the code:

K. Halbach, "Mathematical Models and Algorithms for the Computer Program WOLF," Lawrence Berkeley Laboratory Internal Report no. LBL-4444 (1979).

D. L. Vogel, "WOLF: A Computer Code Package for the Calculation of Ion Beam Trajectories," LBL Internal Report no. LBL-18871 (1985).

Is code documentation available? ☒ Yes ☐ No

How may the code be obtained?
    Ludmilla Soraka, LBL, (415) 486-5011

Source language: FORTRAN

Computers it runs on: VAX, CDC

It is available as: ☒ Source code, ☐ Executable only

Source Media: ☐ Listing, ☒ Tape, ☐ Diskette, ☐ Cards, ☐ Networks
    Tape format:
    Diskette size & format:

Available through: ☐ DECNET, ☐ ARPANET, ☐ BITNET
    ☐
Network Address:

Date of Latest Version: June 1986   Program Name: ZAP

Person to Contact: Michael S. Zisman
Address: Mail Stop 47/112
Lawrence Berkeley Laboratory
1 Cyclotron Road
Berkeley, CA 94720

Telephone Number: (415) 486-5765, FTS 451-5765

Classification of Computer Code:
Component Design:
☐ Ion Source, ☐ Magnet, ☐ RF–cavity, ☐
Accelerator Optimization:
☐ Linac, ☐ Cyclotron, ☐ Synchrotron, ☐
Tracking or Simulation:
☐ Linac, ☐ Cyclotron, ☐ Synchrotron, ☐
Analysis:
☒ Stability, ☐ Impedances, ☒ Intra-Beam Scattering (IBS) effects;

Lifetimes (Touschek, gas scattering)

Other:

Short Description: (Purpose, capabilities, algorithms, special features, etc.)
ZAP can be used to calculate single bunch instability thresholds and instability growth rates from coupled bunch instabilities in storage rings, bunch lengthening, Touschek and gas scattering lifetimes, equilibrium emittances for electron beams in the presence of radiation damping, quantum fluctuations, and intrabeam scattering. It is an interactive, self-prompting code.

Publications describing the code:
Lawrence Berkely National Laboratory Report no. LBL-21270

Is code documentation available? ☒ Yes ☐ No

How may the code be obtained?
From Michael S. Zisman.

Source language: FORTRAN 77

Computers it runs on: VAX, Ridge

It is available as: ☒ Source code, ☐ Executable only

Source Media: ☐ Listing, ☐ Tape, ☒ Diskette, ☐ Cards, ☒ Networks
Tape format:
Diskette size & format: RX50

Available through: ☒ DECNET, ☒ ARPANET, ☒ BITNET
☐

Network Address: ESGVAX::zisman (node 41.190)

zisman@lbl.arpa

zisman@lbl.bitnet

Date of Latest Version: unknown

Program Name: ZFIELD

Person to Contact: Curry Sawyer
Address: E.G. & G. Energy Measurements, Inc.
Santa Barbara Operations
Goleta, CA 93117
USA

Telephone Number:

Classification of Computer Code:
Component Design:
☐ Ion Source, ☐ Magnet, ☐ RF–cavity, ☐
Accelerator Optimization:
☐ Linac, ☐ Cyclotron, ☐ Synchrotron, ☐
Tracking or Simulation:
☒ Linac, ☐ Cyclotron, ☐ Synchrotron, ☐
Analysis:
☐ Stability, ☐ Impedances, ☒ Space Charge

Other:

Short Description: (Purpose, capabilities, algorithms, special features, etc.)
ZFIELD, a trajectory computer code for a linac beam, has been written as a design aid to complement the TRANSPORT code. It includes space charge, plots the emittance ellipse at axial values, and plots beam radius. Comparing ZFIELD and TRANSPORT in drift regions, significant differences in beam radius predictions are found in the 2-Mev-region for currents above 200 A and above 400 A in the 4-Mev-region. Using ZFIELD, beam envelope growth for beams with different emittance can be compared, and the effect of space charge on emittance growth can be shown graphically.

Publications describing the code:
C. Sawyer and N. Norris, "Estimation of Space Charge and Emittance Growth Effects in a Drift Region," Proceedings of the 1984 Linear Accelerator Conference, Darmstadt-Seeheim, report no. GSI-84-11, pp.349-51.

Is code documentation available? ☐ Yes ☐ No

How may the code be obtained?
unknown

Source language:
Computers it runs on:
It is available as: ☐ Source code, ☐ Executable only
Source Media: ☐ Listing, ☐ Tape, ☐ Diskette, ☐ Cards, ☐ Networks
Tape format:
Diskette size & format:
Available through: ☐ DECNET, ☐ ARPANET, ☐ BITNET
☐
Network Address:

SUPERCONDUCTING MAGNET SYSTEM

Per Dahl

Brookhaven National Laboratory, Upton, NY 11973

TABLE OF CONTENTS

1	Introduction	1329
2	Background for the SSC Magnet Choice	1330
3	Overall Description	1332
4	Magnetic Design	1335
	4.1 Magnet Load Line, Central Field, and Saturation	1337
	4.2 Multipoles: High and Low Field	1340
	4.3 Inductance, Stored Energy, and Leakage Field	1342
	4.4 Superconductor Magnetization	1342
5	Superconductor	1344
	5.1 Critical Current Density and Filament Size	1345
	5.2 SSC Superconducting Cable	1346
	5.3 Operating Current Density	1348
6	Coils	1349
	6.1 Coil Winding	1349
	6.2 Cold Molding and Curing	1349
	6.3 Coil Assembly and Collaring	1351
7	Yoke and Helium Containment	1352
	7.1 Iron Yoke	1354
	7.2 Helium Containment Vessel	1354
	7.3 Heat Generation, Cooling, and Temperature Distribution in the Magnets	1355
8	Beam Tube Assembly	1356
	8.1 Beam Tube	1356
	8.2 Correction Coils	1357
	8.3 The "Multiwire" Technique	1359
9	Quench Protection Features	1360
10	Cryostat	1360
	10.1 General Cryostat Arrangement	1360
	10.2 Cryogenic Piping	1361
	10.3 Suspension System	1361
	10.4 Thermal Shields	1364
	10.5 Insulation	1364
	10.6 Vacuum Vessel	1364
	10.7 Interconnections	1365

TABLE OF CONTENTS (Cont'd)

11	Magnet Testing	1365
	11.1 Cold Measurements	1366
12	Quadrupoles	1367
13	Correction Magnets, Spool Pieces	1371
14	Special Magnets	1373
	14.1 Vertical Bending Magnets	1373
	14.2 IR Quadrupoles	1375
	14.3 Additional Insertion Magnets	1375
15	References	1375

Chapter 5

SUPERCONDUCTING MAGNET SYSTEM

Per Dahl
Brookhaven National Laboratory, Upton, NY 11973

1 INTRODUCTION

The superconducting magnet system for the SSC is unprecedented in technical scope and, manifestly, the costliest machine component as well. A total of 7680 superconducting dipoles are required for bending the two proton beams in the regular arcs of the lattice, 1356 quadrupoles for focusing, and ~1600 "spool pieces" housing correction windings and other instrumentation. In addition, ~650 special magnets are required to bring the two counter-rotating beams into collision in the interaction regions of the experimental areas. Altogether, the magnet system comprises nearly 13,000 superconducting magnets representing an investment of three quarters of a billion dollars.

The most critical magnets, because of their demanding operating specifications and sheer number, are the regular dipoles; consequently, the design effort has concentrated on them. The dipole field strength determines the circumference of the collider rings and thus the cost of the tunnel. Therefore the operating field is chosen to be as high as practical, consistent with a judicious balance between sound magnet design principles and minimum unit cost. The minimum required coil aperture (coil i.d.) and available superconductor current density dictate the overall magnet size and hence its cost. (The superconductor alone accounts for ~30% of the cost.) The "good-field" aperture must be large enough to accommodate the circulating beam, particularly at injection. The uniformity of the magnetic field determines the lifetime of the stored beams. The operational life of the collider facility and fractional "on" time for physics experiments depends on magnet reliability and on a conservative design. (During a projected machine lifetime of 20 years, an individual magnet must tolerate 10^4 acceleration cycles, a substantial number of thermal cycles to and from cryogenic temperatures, numerous quenches to the normal state, and an estimated radiation dose of 10^6 Gys. Finally, the feasibility of producing 8000 cost-effective, high quality dipoles on a tight schedule (10 magnets per day over 4 years) is vitally predicated on industrial involvement in the project from early in its R&D phase. It also assumes a magnet design exceptionally well tailored for mass production practices.

Virtually all hadron accelerators and colliders under construction, planned, or contemplated are based on superconducting magnets and the advantages they offer. Their importance stems basically from two considerations. First, their high current density makes possible very high operating fields produced by compact coil structures: e.g., field strengths typically a factor two or three higher (4 to 6 T) than possible with copper-iron magnets. Ten-Tesla accelerator magnets may

© 1989 American Institute of Physics

be feasible with additional R&D. Second, their relatively high initial capital cost is obviated by drastically lower long-term operating costs.

It goes without saying that the feasibility of large superconducting magnet systems is partly due to concomitant progress in large-scale helium refrigeration technology. To be sure, superconducting accelerator magnets pose many technical challenges.[1] Their cryogenic design must be highly efficient to overcome the low Carnot efficiency at helium temperatures. The cryostat housing the magnet renders accessibility to the magnet, once deployed in the accelerator, well-nigh impossible without a major perturbation of machine operation. Failure of even minor components requires cycling to room temperature of a substantial string of magnets and costly machine down-time. Surveying magnets into position in the tunnel and subsequent adjustments or monitoring are also severely encumbered by the intervening cryostat. Quench protection considerations impact strongly on the magnet and magnet system designs. Finally, in superconducting magnets the field distribution in the aperture is determined primarily by conductor location and its tolerances, not simply by the shape of accurately punched iron yoke laminations as is the case in conventional magnets. A field uniformity of $\Delta B/B_0 \sim 10^{-4}$ requires accuracy in conductor placement of typically 2 mils or 50 μm over magnet lengths 10 m or longer.

2 BACKGROUND FOR THE SSC MAGNET CHOICE

Early SSC magnet design studies were driven by the search for an economic optimum between magnet field strength and tunnel circumference. Low-field dipoles are presumably relatively inexpensive, requiring modest R&D and engineering (or so the argument went); in this case the machine cost is dominated by the civil construction of a correspondingly longer tunnel. Very-high-field dipoles imply a shorter tunnel but should be inherently more costly and technically challenging. The several magnet types seriously suggested are worthy of brief scrutiny, since they illustrate competing design approaches to superconducting accelerator magnets. They are shown schematically in Fig. 1.

In order of increasing field strength, the first or "low-field" solution (Fig. 1C) is commonly known as a "superferric" magnet — a term coined relatively recently.[2] This magnet was a major contender in the final stages of the magnet selection process for the SSC.[3] Both beam tubes and both sets of coils are mounted in a common iron yoke, one above the other but spaced sufficiently apart to be magnetically decoupled, in a common cryostat. The term "superferric" implies that the iron contributes a substantial fraction (about two thirds) of the magnetic field, and the field shape is made to depend mainly on the profile of the iron pole face by (a) limiting the central field to 3 T and (b) exploiting a conductor geometry in the form of simple current sheets delineating the vertical boundaries of an approximately rectangular good-field region. Below 2 T the field is fully determined by the iron. Above 2 T iron saturation gradually sets in, and nonlinear field contributions require substantial corrections. Because of the over/under configuration, there is little flux linkage between the two magnet rings. The SSC ring circumference based on this magnet would be 164 km.

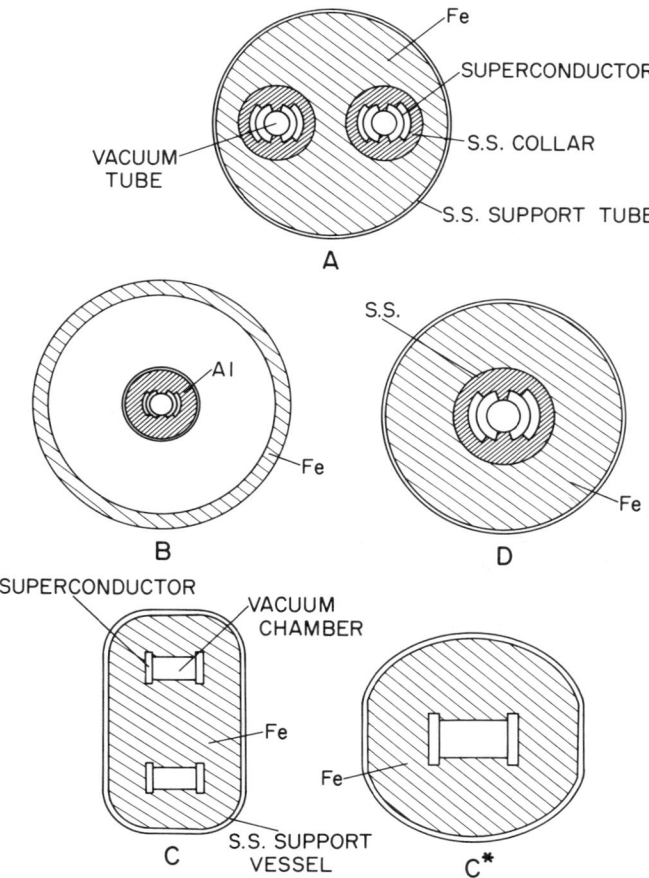

Fig 1. Various magnet configurations considered for the SSC: (A) 2-in-1 high-field cosine theta dipole; (B) medium-field ironless dipole; (C) 2-in-1 low-field ("superferric") dipole; (C*) 1-in-1 version of low-field dipole; (D) 1-in-1 version of high-field dipole.

Major cost savings were envisaged by installing the magnets in very long units, 35 m each, thus minimizing the number of interconnecting cryostat sections which tend to be particularly costly. Magnet C* is a 1-in-1 version of magnet C; i.e., a collider lattice based on C* would incorporate two fully separated magnet rings. It would be a more costly, but flexible, solution.

Another contender was a "medium-field" design (Fig. 1B). It assumes a central field of 5 T with each beam tube and coil in its own cryostat. Its prominent feature is no iron close to the coils; hence, relatively more superconductor is needed to reach the desired field. The circular coils, of the "cosine theta" type, are constrained by aluminum collars. Each cryostat has only enough iron to shield one coil from the field of the other. The "warm" iron also serves as the vacuum vessel of the cryostat. In this design, although it allows great flexibility in the spacing of the magnet rings, forces between the coil package and vacuum vessel require a more robust support structure for the coils (not shown). Designing a strong support with concomitant low heat leak was a major challenge for all the magnet designs considered, but particularly for the medium field "no-iron" dipole.

The upper end of the field scale considered was 6.5 T, represented by two related candidates A and D. Magnet D, a high-field magnet also employing coils of the cosine theta type, was the magnet ultimately chosen, after a rigorous review process, as the basis for the Conceptual Design of the SSC.[4] It is described in some detail in the sections that follow. Note, however, an earlier version of this magnet which figured prominently in the SSC Reference Designs Study[5] of 1984: magnet A, a 2-in-1 version of D. Two-in-one magnets were first introduced (and the term coined) late in the Colliding Beam Accelerator Project at BNL.[6] Only in proton-proton colliders are 2-in-1 magnets feasible, since two apertures are required – one with the field vector pointing up and the other with the vector pointing down. In these magnets the field lines from one aperture are returned, not through the midplane, but across the other magnetic aperture located beside the first. The total weight of iron is about the same as would be required for each of the single magnets. Important drawbacks of the scheme are greater complexity in magnet assembly and loss of flexibility in machine operation.

An important justification for the specific recommendation of the magnet selection panel constituted for the purpose was the extensive, world-wide experience with magnets of type D, culminating in their successful deployment in the Tevatron. A high-field magnet was recommended for machine operational as well as for cost reasons. Before turning to a detailed discussion of the various magnet components, we review briefly the main characteristics of the magnet design and the reasons for particular design choices.

3 OVERALL DESCRIPTION

The dipole[7] is broadly classified as a "cold iron, cold bore" dipole incorporating coils of the so-called cosine theta configuration; this configuration is defined and discussed in a later section. It suffices to note that this dipole, with some variations, has become the workhorse of superconducting accelerator magnet designs, including the magnets for the Tevatron, HERA, and UNK. The coil is clamped firmly with stainless steel collars, another commonly adopted practice, and the collared sub-assembly is held in an iron yoke also at helium temperatures. The 40-mm aperture is adequate to accommodate the circulating beam at the high injection energy of the SSC, 1 TeV. (For comparison, the Tevatron and HERA apertures are about 75 mm, and those of the CBA magnets were ~130 mm!) The magnet length, ~17 m, was the maximum length deemed practical in view of numerous constraints, many of then nonquantifiable, including beam dynamics, thermal contraction, quench protection, field measurements, handling, transportation, and installation. The two rings of magnets are magnetically, electrically, and cryogenically independent of one another for operational reasons and in the interest of machine flexibility. They are arranged in an over/under configuration in the SSC tunnel, which simplifies beam injection and abort and magnet installation and generally results in a more efficient use of tunnel space.

The high operating field of the dipole magnet, 6.6 T at 4.35 K, is made possible by the availability of high-current-density NbTi alloy. This superconductor is the result of a highly successful R&D program carried out in a collaboration between universities, industry, and the national laboratories.[8] The high current density (2700 to 3000 A/mm^2 at 4.2 K, 5 T) of the present "high homogeneity" NbTi alloy is a consequence of steadily improving metallurgy and optimized heat treatment procedures. In addition, incorporation of "barriers" to prevent intermetallic compound formation during heat treatment has promoted filament diameter uniformity during the drawing of the wire. This last advance has made possible the fabrication of high-current-density material with filaments in the few-μm range, resulting in a reduction in superconductor persistent currents (magnetization) at low field levels. The present SSC design assumes 5 μm filament size.

Further improvements in superconductor current density can be expected. They would allow additional cost reductions, either by reducing the amount of NbTi in the magnets while retaining the present operating field, or by increasing the magnetic field with a constant amount of superconductor. The latter possibility, however, is bounded by the need to provide an increasing amount of copper as the current in the superconductor increases.

The coils of the magnets are wound on automated machinery and then formed to a precise size in a molding operation in which the epoxy in the epoxy-impregnated cable wrap is cured, thus forming a rigid, precise structure guaranteeing coil-to-coil reproducibility and lending itself well to industrial fabrication. The random magnet-to-magnet errors, which are largely determined by the variation in coil sizes, can be held within the required tolerances through this technique.

The molded coils, including the beam tube assembly with pre-attached trim coils (also superconducting), are assembled and held in place at the required prestress through the use of nonmagnetic stainless steel collars. The collars, formed from precision stampings, provide the basis for accurate control of the coil positioning and furnish the support for withstanding the large forces present in a high-field magnet, typically 7×10^5 Nt/m (4000 lb per linear inch). Collars were first introduced in the Tevatron magnets. They have another advantage in that they allow testing a magnet at room temperature before proceeding to the next assembly step, insertion into the iron yoke. In this way the electrical integrity of the Kapton insulation and the harmonic content of the field produced by the collared subassembly can be checked at a propitious point in the assembly line.

Final assembly of the magnet "cold mass" involves insertion of the collared coil into the iron yoke laminations and closure of the yoke support vessel by the welding together of two half shells of stainless steel around the iron yoke. (When end plates are subsequently welded to the assembly, the yoke support shell also acts as the outer wall of the helium containment vessel; its inside surface is the stainless steel beam tube.) Through the use of alignment tabs and slots, the position of the coils is accurately fixed relative to the outside of this assembly and can thus be precisely aligned through the cryostat suspension system to the outside of the vacuum enclosure. The need for precision surveying during the assembly of a magnet is thus minimized. The iron yoke, in addition to contributing substantial field enhancement, acts as a magnetic shield and makes possible a low-heat-leak suspension system for the cold mass. There is some magnetic saturation of the iron at high field levels, but this nonlinear perturbation of the dipole field can easily be compensated by the distributed trim coils located interior to the main dipole coils. The need to cool the great mass of iron lengthens somewhat the cooldown time of the machine, but the price is not severe. (The mass can be reduced in the 2-in-1 design, as in magnet A of Fig. 1.)

The cryostat[9] is designed to provide stable yet simple and low-heat-leak support of the cold mass in a way that can be reliably implemented in industrial production. Support posts, constructed as a nested pair of thin-walled fiberglass-epoxy tubes loaded in compression, provide the required support for the vertical and horizontal loads. The support posts are located and fastened to the cold mass by using precision fixturing such that the magnet location, including the required slight sagitta, is determined as the cold mass is assembled into the cryostat and later installed onto the tunnel supports. The folded posts, located at five points along the length of the magnet, minimize the number of penetrations through the insulating shields and barriers. This reduces the chance of introducing heat leaks by thermal radiation through imperfect application of the insulation. Minimizing the number of support posts reduces the danger of introducing unequal stresses into the support system and into the cold mass.

As noted earlier, quench protection is a major consideration for large accelerators based on high-field superconducting magnets. The SSC magnets are quench-protected by electrical heaters built into each magnet. These heaters are activated when a quench is sensed electronically and serve to spread the quench front quickly throughout the entire magnet, thereby avoiding local hot spots or excessively high voltage. This "active" quench protection scheme, with room temperature solid-state diodes whose function is to bypass a half cell of magnets, is similar to that used in the Tevatron and represents a conservative approach. A passive system may in fact be adequate. Such a system would employ one or two diodes across each magnet. When a quench occurs, a large voltage is induced, causing the diodes to become conducting, shunting the main current around the magnet. Such diodes may, however, be vulnerable to the neutron radiation caused by beam-gas interaction during operation. The main virtue of a passive system is its simplicity and reliability.

4 MAGNETIC DESIGN

The coil geometry is a two-layer approximation to the cosine theta configuration, or to the ideal current density distribution giving a uniform vertical dipole field within a circular aperture. (Strictly speaking, a two-layer coil of this type approximates better a variant on the pure cosine theta distribution, namely the constant-current-density or "intersecting-ellipse" solution.) A quadrant of the coil cross section is shown in Fig. 2. The near-rectangular conductors are arranged in blocks with intervening wedges and pole pieces whose azimuthal extent is varied in the field-shaping optimization procedure. With the pole pieces and slightly keystoned conductors, the wedges provide a "Roman arch" type of coil support. They are of copper to aid quench propagation from one block to the next.

The coils are surrounded by a nearly circular, split and laminated iron yoke of low carbon steel, as shown in Fig. 3. They are separated from the yoke by nonmagnetic stainless steel collars that provide the necessary restraint to maintain the coils under a compressive radial stress. The collars are also designed to impart a preload in the azimuthal direction to ensure that the elastic forces exceed the magnetic forces. The minimum preload, about 3.5 kpsi at operating temperature, is that stress just sufficient to prevent separation of the turn adjacent to the pole from the pole piece surface when the magnet is energized. (The room temperature assembly prestress is greater.) Elastic motion of coil components relative to the azimuthal coil ends is still possible but is minimized by a high elastic coil modulus ($E \sim 1 \times 10^6$ psi). The iron yoke contributes ~ 1.4 T, or 25%, to the magnetic field. The relatively thin radial width of the collars (15 mm in the midplane region) minimizes the loss of this contribution. Conversely, a magnetic benefit from the collars is a sharp decrease in the effect of iron saturation on the allowed field harmonics (mainly the sextupole term) and on the transfer function (the ratio of central field to coil current), as discussed in the next section. If necessary, the saturation effects could be made even smaller by shaping the iron (removing some iron) in the region of the poles where saturation starts.

Both coil layers are wound from slightly keystoned, flat superconducting cable. It is based on strands of NbTi alloy embedded in high-purity copper, drawn into wire, twisted, and incorporated in a cable that is compacted to precise dimensions. The cable and its constituent wires are discussed in some detail in Section 5. A consequence of partially keystoned cable (about 1.5°) instead of the full value commensurate with the cable size and coil aperture (4° in the present design), is that the turns are not everywhere aligned radially, as is evident in Fig. 2. (Small departures from "radial stacking" are merely an esthetic eyesore, but more extreme ones can lead to mechanical instability in the assembled coil.) The field in the outer coil is lower than in the inner coil: B_{max}(outer) = 0.844 B_0 vs. B_{max}(inner) = 1.055 B_0, where B_0 is the field on the magnet axis (central field). Therefore, the cable is "graded" in the two sections with the inner layer (16 turns distributed in four blocks in each coil section) wound from a 23-strand cable

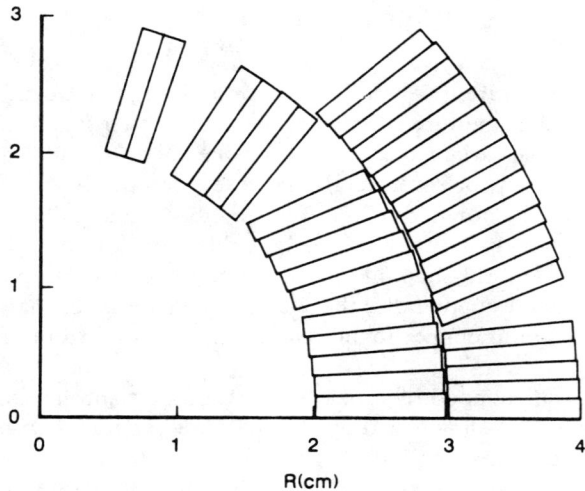

Fig. 2 Quadrant of dipole cross section, showing distribution of slightly keystoned superconduction cable.

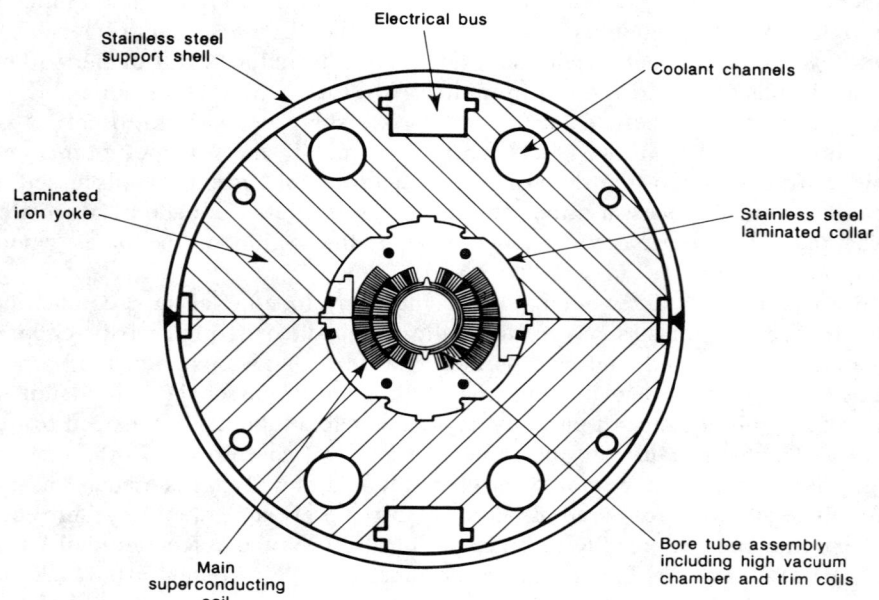

Fig. 3. Overall cross section of dipole cold mass, showing bore tube, coil, collars, iron yoke, and yoke support tube.

with a copper to superconductor ratio of 1.3:1 and the outer layer (20 turns in two blocks) from a 30-strand cable (Cu:SC = 1.8:1). The current density in the superconductor of the outer layer is 1.19 times that in the inner layer. The two layers are powered in series.

The coil end windings are of the usual saddle-shaped configuration. To ensure proper "lay" of the relatively wide cable and to minimize field enhancement as well as field harmonics in the coil ends, the end turns are spaced out in the direction of the magnet axis by interleaving molded spacers and wedge extensions. The iron yoke extends over the coil ends, but with an inside diameter larger than in the magnet "straight section." This further minimizes the field enhancement in the magnet end, yet retains the advantages of providing good local mechanical coil support and minimizes the external fringe field, which cannot be tolerated in a collider. Note that the end-field harmonics are relatively insensitive to the iron end configuration. The accuracy requirements for locating the conductors in the end are also less stringent than those for placing them in the body of the magnet, since the ends contribute less to the total (integrated) magnetic field. The principal magnet design parameters are listed in Table I.

4.1 Magnet Load Line, Central Field, and Saturation

As noted, the two coil layers are powered in series. The operating (central) field B_0 of 6.6 T (based on an assumed temperature of 4.35 K) corresponds to a current of ~6.5 kA, or a transfer function of 1.015 T/kA. The drop in B_0/I at 6.6 T relative to its low-field value, because of iron saturation, is about 2%. At $B_0 =$ 6.6 T, the calculated (two-dimensional) peak field in the inner coil layer is 7.0 T; in the outer layer it is about 5.6 T. The approximate location of the respective peak fields is indicated in Fig. 4, which shows the distribution of the calculated

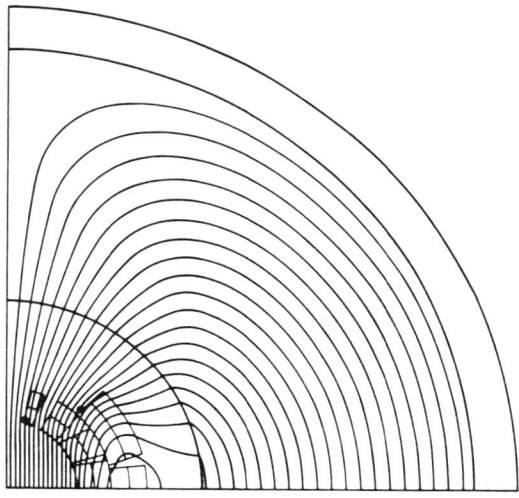

Fig. 4. Quadrant of dipole, showing calculated magnetic field lines. The peak field locations in the inner and outer coils are also indicated.

Table I. Dipole Magnet Parameters

General	
No. of dipoles per ring	3840
Overall length/interface length [m]	17.35/0.8
Magnetic length [m]	16.54
Mass of conductor [kg]	208
Cold mass [kg]	6759
Central field [T]	6.6
Current [A]	6500
Inductance [mH]	53
Stored energy [MJ]	1.12
Bore Tube Assembly	
Material	stainless steel
Inner diameter [cm]	3.23
Outer diameter [cm]	3.45
Copper coating [mm]	0.13
Winding	
Inner layer	
No. of turns (per coil section)	16
Inner diameter [cm]	4.000
Outer diameter [cm]	5.923
Cable length [m] (for 2 coil sections)	1076
Cable mass [kg] (for 2 coil sections)	100
Maximum field [T]	7.0
Outer layer	
No. of turns (per coil section)	20
Inner diameter [cm]	5.974
Outer diameter [cm]	7.986
Cable length [m] (for 2 coil sections)	1341
Cable mass [kg] (for 2 coil sections)	108
Maximum field [T]	5.6
Conductor	
Inner layer	
Cross section, bare [mm]	9.30 × (1.59-1.33)
Keystone [deg.]	1.61
Strand diameter [mm]	0.808
No. of strands	23
Strand twist pitch [per cm]	0.4
Cable twist pitch [per cm]	0.126
Copper-to-superconductor ratio	1.3:1
No. of superconductor filaments	11000
Filament diameter [μm]	5

Table I. Dipole Magnet Parameters (Cont'd)

Outer layer	
Cross section, bare [mm]	9.73 × (1.27-1.07)
Keystone [deg.]	1.21
Strand diameter [mm]	0.648
No. of strands	30
Strand twist pitch [per cm]	0.4
Cable twist pitch [per cm]	0.136
Copper-to-superconductor ratio	1.8:1
No. of superconducting filaments	6000
Filament diameter [μm]	5
Collars	
Material	stainless steel, nitrogen hardened
Lamination thickness [mm]	1.52
Outer diameter [cm]	11.09
Radial thickness, nominal [cm]	1.5
Iron Yoke	
Material	low-carbon steel
Inner diameter [cm]	11.14
Inner diameter, magnet ends [cm]	17.51
Outer diameter [cm]	26.67
Lamination thickness [mm]	1.5
Weight of iron [kg]	5171
Yoke Containment Structure	
Material	stainless steel
Outer diameter [cm]	27.62
Thickness [cm]	0.47
Weight of shell [kg]	540
Cryostat	
Vacuum vessel material	steel
Vacuum vessel outer diameter [cm]	60.96
Wall thickness [cm]	0.635
Heat shield material	aluminum
80 K heat shield, outer diameter [cm]	45.72
20 K heat shield, outer diameter [cm]	40.64
Superinsulation layers, outside 80 K shield	52
Superinsulation layers, outside 20 K shield	13
Cold mass support type	reentrant post
Support materials	FRP unidirectional
Number of supports	5
Support interval [m]	3.5
Load per support [kg]	1352

magnetic field lines in a somewhat simplified model of a dipole quadrant. The maximum field seen by the conductor occurs in the base of the inner turn adjacent to the pole. As observed earlier, the central field end enhancement normally found in saddle coil-wound dipole magnets is largely eliminated by spacing the end turns and enlarging the iron inner radius in the end region. The margin in critical current at operating field, J_c/J_0, where J_c is the minimum critical current density of the strands before cabling, is 1.25, as discussed in Section 5.

4.2 Multipoles: High and Low Field

The required magnetic aperture and dipole tolerances have been discussed in Section 3. The present coil design gives a very uniform magnetic field; expressed in terms of a multipole analysis, the calculated field quality is shown in Table II.

Table II. Calculated Systematic Multipoles at Low and High Field from the Presence of the Iron Yoke (units are $10^{-4}\ B_0\ cm^{-n}$)

Multiple	Low field ($\mu = \infty$)	High field (6.6 T)
Quadrupole (b_1)	not allowed	--
Sextupole (b_2)	0.00	1.18
Octupole (b_3)	not allowed	--
Decapole (b_4)	0.00	−0.05
Dodecapole (b_5)	not allowed	--
14-Pole (b_6)	0.00	0.01
16-Pole (b_7)	not allowed	--
18-Pole (b_8)	0.00	0.02

The multipole coefficients a_n and b_n are defined in terms of the usual expansion:

$$B_y + iB_z = B_0 \sum_{n=0}^{\infty} \left(b_n + ia_n\right)(x + iy)^n \qquad (1)$$

where B_0 is the design bending field, the even b_n are the normal multipole components with dipole symmetry, and the a_n are the skew coefficients. The multipole order follows from the index n as $2(n + 1)$; thus, b_2 is the normal sextupole component, and so forth. The dimensions of the coefficients are (length)$^{-n}$. For dipoles meeting the SSC specifications the order of magnitude of the lower-order coefficients is 10^{-4} cm^{-n} (Table II) which, by convention, is known as 1 "unit." Higher-order coefficients for dipoles of good field quality fall off rapidly with increasing n. Departures from a pure dipole field can be caused by systematic or random contributions. Systematic contributions may come from conductor or coil placement errors, geometric errors (lack of dipole symmetry), iron saturation (Table II), or persistent current magnetization (discussed below). Random contributions stem from conductor placement errors and random variations in superconductor properties. The random errors are typically 10% of the systematic errors; their tolerances used in SSC tracking calculations were established in three independent ways: scaling data from Tevatron magnets, scaling magnet data from the Colliding Beam Accelerator project (the random positional error multipoles scale with the coil diameter d roughly as $d^{-n-1/2}$), and measurements on SSC model dipoles of types A, B, and D. The specifications for the rms random multipole errors for the SSC dipole magnet with inner coil diameter of 4 cm and superconduction filaments of 5 μm diameter (Section 5.2) are summarized in Table III. Also tabulated are rms values of random multipole coefficients measured at BNL in a series of model dipoles of type D.

Table III. Specifications for the rms random multipole errors in the SSC dipole magnet with d_c = 4 cm and 5-μm filament size. Also tabulated are rms values of the random multipole coefficients measured on 4.5-m BNL models. The units of a_n and b_n are 10^{-4} cm^{-n}.

Multipole coefficient	Specified tolerances	Measured random errors (BNL, 6 models)
a_1	0.7	0.75
a_2	0.6	0.23
a_3	0.7	0.46
a_4	0.2	0.12
a_5	0.2	0.11
a_6	0.03	0.03
a_7	0.2	0.01
a_8	0.05	0.01
b_1	0.7	0.78
b_2	2.0	1.51
b_3	0.3	0.16
b_4	0.7	0.28
b_5	0.1	0.03
b_6	0.2	0.05
b_7	0.2	0
b_8	0.1	0.01

The low-field values (typically 1 T) of Table II were calculated for infinite-permeability iron; high-field calculations (6.6 T) include the effect of iron saturation. The only multipole coefficient of note at high field, the sextupole term, is easily canceled with the distributed sextupole correction coil mounted on the beam tube inside the main dipole winding. The main function of the correction coil, however, is to compensate for expected magnetization in the injection field region (0.33 T), not included in the multipole listing of Table II but discussed shortly.

4.3 Inductance, Stored Energy, and Leakage Field

The inductance of the magnet at full field is computed to be 3.2 mH/m, giving a total magnetic stored energy of 1.12 MJ. At 6.4 kA, the mean flux density on the median plane in the iron is computed to be 2.13 T, just equal to the saturation flux density. The fringe field at the iron surface is about 640 gauss and falls off rapidly with radius.

4.4 Superconductor Magnetization

Superconducting accelerator magnets differ in two essential respects from conventional magnets. One has been implicitly discussed in the last several sections: the field distribution in the aperture is determined mainly by the conductor placement; in all but superferric magnets the iron plays a minor role. The second is strictly a low-field effect, associated with the basic properties of superconductors. When exposed to changing magnetic fields, superconductors exhibit an inherent magnetic hysteresis, or magnetization. This is caused by local persistent, diamagnetic current loops induced within the individual filaments of the twisted composite wire of the superconducting cable. These currents produce a uniform field within the cylindrical filament which is equal and opposite to the applied field change. The induced currents constitute elementary dipoles whose strength and orientation depend on the magnetic history of the superconductor, the local magnetic field in the magnet winding, and the superconducting properties. In terms of the "critical state" model[10] (a model in which all regions of the superconductor are carrying either the full critical current density J_c or no current at all), the field-dependent magnetization $M(H)$ can be expressed in terms of J_c and the superconducting filament diameter d as follows:

$$-M(H) = \frac{1}{3\pi} \lambda J_c(H) d , \qquad (2)$$

where λ is the fraction of superconductor in the composite. By summing the contributions of the elementary current dipoles over the coil winding, the magnetization can be formulated in terms of residual, systematic multipole fields whose contributions diminish rapidly with increasing field because of the field dependence of J_c. As noted, these multipoles, primarily a sextupole, dominate the systematic error fields near injection, and are compensated locally by a distributed correction winding, if necessary.

Figure 5 illustrates the typical magnetization behavior exhibited in measured sextupole (and decapole) harmonics as a function of current in several recent 4.5-m-long SSC model dipoles.[11] About 25 units of sextupole are evident at the injection field, or at a current of about 300 A. (The small saturation effect at high field, due to the intervening collars, is also apparent in Fig. 5.) These magnets, however, utilized a conductor with relatively large filaments, ~20 μm in diameter. The linear dependence on filament size has prompted a vigorous R&D effort over the past several years to develop commercially available materials with substantially finer filaments without sacrificing current density. The present dipole design is predicated on a filament diameter of 5 μm, by now routinely available. As a result, the demands on the correction coil system (discussed in a later section) are quite modest.

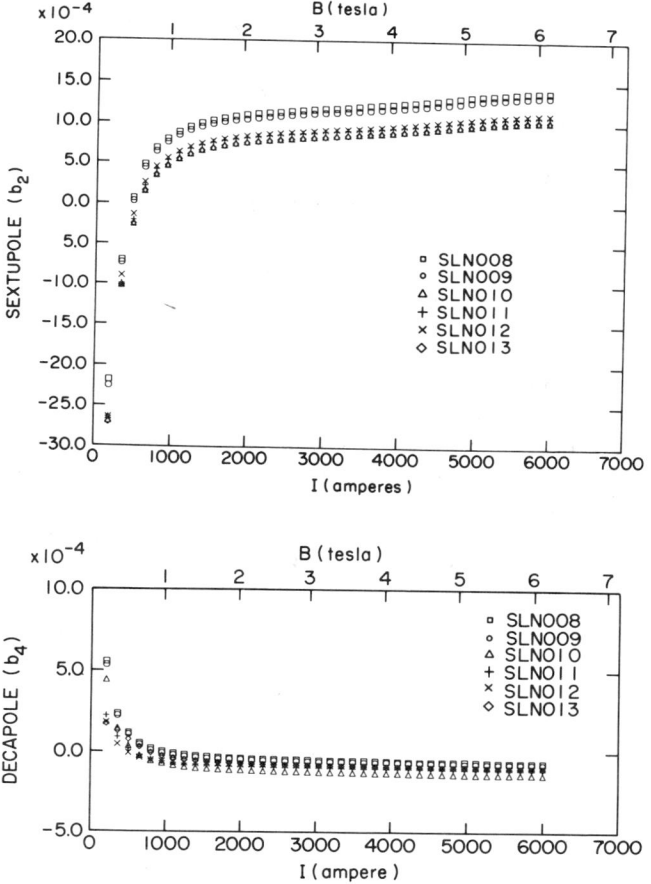

Fig. 5. Variation of measured sextupole (top) and decapole (bottom) harmonics with field in six 4.5-m-long SSC model dipoles.

In addition to the mean sextupole contribution to be expected from superconductor magnetization (about 6 units for 5-μm filaments), its variation must also be considered since, at low fields, this may increase the variation already present from errors in construction. Based on experience with the Tevatron (800 dipoles or 400 billets of conductor), a 10% variation, or about 0.5 units in b_2 at 0.33 T, can be expected. Temperature variation adds another 0.1 unit (through the temperature dependence of J_c). The total is of the same order as the random errors.

5 SUPERCONDUCTOR

The high-current conductors used in pulsed superconducting accelerator magnets combine many, typically 20, composite wires in parallel. Since each wire may contain several thousand to tens of thousands of fine superconducting filaments, as many as 10^5 filaments are connected in parallel. Fine filaments are necessary to "stabilize" the conductor against effects associated with changing fields (today's superconductors are routinely stabilized against "flux jumping," once a major problem plaguing superconducting magnets), reduce ac energy loss effects from pulsed operation, and minimize magnetization currents. A desirable feature of the conductor for accelerator magnets is a high aspect ratio — i.e. a conductor of large radial dimension (height) compared with its azimuthal width in the winding cross section. This minimizes the number of coil layers and thus simplifies construction, and permits a large number of turns in each coil layer, thereby achieving a close approximation to the cosine theta distribution. The superconductor (NbTi roughly 50/50 by weight) is invariably combined with a normal conductor that can carry current for short periods — the simplest form of stabilization and a method of protecting against burnout during a quench. In most magnets the normal conductor material is copper of moderate purity with a ratio of one or two parts by weight of copper to superconductor and a resistance at 4.2 K about 1% that at room temperature. A conductor billet is formed by coating rods of NbTi with copper and stacking a number of these elements in a copper "can." The entire assembly is then extruded at a temperature high enough to ensure a good metallurgical bond between the NbTi and the copper, and then drawn in many passes to final wire size; these wires are subsequently cabled together.

The particular superconductor selected for the main SSC dipoles and quadrupoles is the result of a collaboration, as noted, over the past several years between national laboratories (Brookhaven and Lawrence Berkeley Laboratories), universities (mainly University of Wisconsin), and industry (Intermagnetics General, Supercon, Teledyne Wah Chang, Oxford Superconducting Technology, among others). The goals of this development have been to obtain high-current-density, fine-filament material in commercial quantities. (The SSC will require ~800 tons of NbTi alloy, or 2000 tons of NbTi wire.) The effort has been highly successful, as will be seen below, and further progress is expected. In parallel with the metallurgical effort, collaboration between BNL and LBL has concentrated on incorporating the superconducting strands in cables of acceptable physical parameters; the most important of these are copper-to-superconductor ratio, number and size of strands, keystone angle, amount of compaction, and allowable tolerances in dimensions and mechanical modulus.

5.1 Critical Current Density and Filament Size

Although the superconducting parameter of primary importance is the overall current density at the appropriate operating temperature and field (e.g. 3.5 K and 6.6 T for the SSC), a convenient figure of merit in judging a superconductor is the maximum critical current density J_c (A/mm^2) of the superconductor itself at 4.2 K and in a field of 5 T. The remarkable improvement in J_c in recent years is clear from Fig. 6, where the lower rectangle represents the

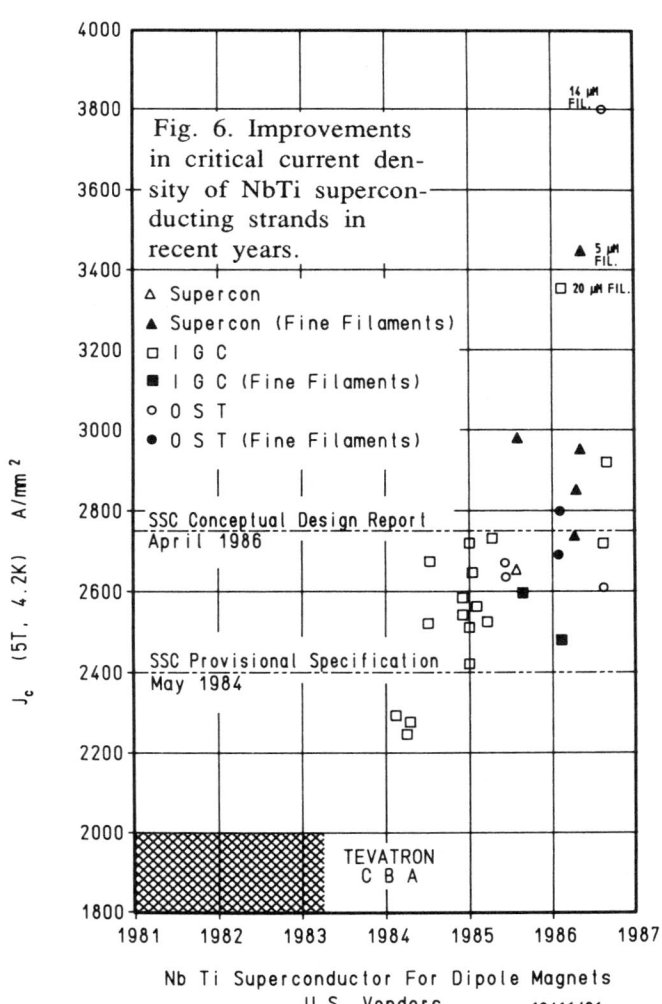

Fig. 6. Improvements in critical current density of NbTi superconducting strands in recent years.

Nb Ti Superconductor For Dipole Magnets
U.S. Vendors   10/16/86

spectrum of J_c values for the strands used in the Tevatron. The improvement has gone hand in hand with steadily decreasing filament size, from typically 20 µm at the time of the Tevatron to diameters now in the 3- to 5-µm range. These improvements are due to several factors, among them more chemically homogeneous NbTi alloy, techniques for preventing formation of titanium-copper intermetallic compounds at the filament-copper interface during extrusion and drawing, and improved heat-treatment and cold-working schedules.

In the vicinity of 2 to 3 µm, magnetization no longer decreases linearly with filament size because of proximity "coupling" of neighboring filaments. In addition, evidence is accumulating that, as the filament diameter becomes very small while attempting to maintain spacing between filaments, the mechanical properties of the composite shift from those of copper to those of the alloy, giving rise to brittle behavior. For these reasons, and since trim coils are needed in any case to control the tune of the machine, the SSC target specification has been set at 5 µm and the metallurgical effort is now focused on optimizing J_c at this filament diameter.

5.2 SSC Superconducting Cable

The final conductor is a flat, high aspect ratio cable produced from a hollow cable of twisted multi-filamentary superconducting strands which is flattened and compacted into a slightly keystoned cross section. Twisting the filaments "decouples" the individual filaments with respect to a uniform changing magnetic field and ensures uniform current sharing between them; twisting the cable decouples the strands. The cable is commonly known as a "Rutherford" or "Rutherford-Fermilab" cable in honor of the laboratories where it was pioneered and first developed on a large scale, respectively. For the present application no electrical insulation is needed between strands of the cable to suppress eddy currents because the cycle time of the SSC is quite long (1000 s).

Since the fields are different in the two coil layers, the conductor is graded with strand and cable parameters adjusted to achieve higher overall current density in the outer layer, as explained earlier, which results in a significant reduction in outer conductor cost. The parameters for the two strand types are given in Table IV, and the corresponding cable parameters in Table V. Cross sections of the two cables are shown in Fig. 7, in which the slight keystone is discernible. The keystone angles are the maximum possible ones for each cable without incurring severe wire distortion that could damage the strands. The number of strands in each cable is chosen to ensure that the inner and outer layers can be energized in series at maximum efficiency. The copper-to-superconductor ratio is chosen to assure approximately equal quench protection behavior in the two layers. Cable insulation consists, first, of a spiral wrap of 0.025-mm Kapton with 50% overlap, or effectively 0.0508 mm (2 mils) thick, followed by a single layer of 0.010-mm (4-mil) fiberglass tape impregnated with B-stage epoxy of nominal 0.0508-mm thickness when the winding is under compression. The fiberglass tape is wrapped with nominal zero spacing between turns. The epoxy serves to hold the coil together during handling.

Table IV. Parameters for NbTi Superconducting Wire

	Inner layer	Outer layer
NbTi composition (by weight)[a]	46.5% Ti	46.5% Ti
Critical current,[b] [A]	613	323
Critical current density (non-Cu) [A/mm^2]	2750	2750
Copper-to-superconductor ratio	1.3:1	1.8:1
Filament diameter [μm]	5	5
Strand diameter [mm/in.]	0.808/0.0318	0.648/0.0255
No. of filaments	11,000	6,000
Strand twist pitch [per cm]	0.4	0.4
Copper residual resistance ratio, R_{295}/R_{10}	>80	>90

[a]High-homogeneity material (see text).
[b]At 5 T, 4.2 K, and a resistivity of 10^{-14} Ω − m.

Table V. Parameters for Superconducting Cable

	Inner layer	Outer layer
Critical current[a] [A]	11,970	8,239
Critical current density[b] (non-Cu) [A/mm^2]	2400	2400
Number of strands	23	30
Strand diameter [mm/in.]	0.808/0.0318	0.648/0.0255
Cable width (bare) [mm/in.]	9.30/0.366	9.73/0.383
Cable thickness (bare)		
Wide side [mm/in.]	1.59/0.0625	1.27/0.050
Narrow side [mm/in.]	1.33/0.0522	1.07/0.042
Keystone angle [deg.]	1.61	1.21
Cable twist pitch [per cm]	0.126	0.136
Residual resistance ratio, R_{295}/R_{10}	>70	>70

[a]Assuming 19% cabling degradation.
[b]At 5 T, 4.2 K, and a resistivity of 10^{-14} Ω − m.

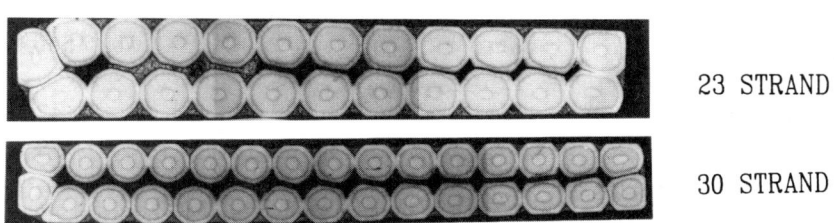

Fig. 7. Cross sections of 23-strand ("inner") and 30-strand ("outer") keystoned cable.

5.3 Operating Current Density

The minimum critical current density of the strands, Table IV, is specified to be 2750 A/mm^2 at 4.2 K, 5 T, and $\rho = 10^{-14}$ ohm-m. (In modern superconducting composites the transition to the normal state with increasing current or field is not instantaneous. By convention in the community of superconducting accelerator magnet builders, the critical current is defined as the current at which the effective resistivity is 10^{-14} ohm-m.) The critical current of NbTi conductors decreases linearly with field; in boiling liquid helium at atmospheric pressure it vanishes at a field of about 10 T. For fields >5 T, the critical current at a given field is linearly dependent on the difference between the bath temperature and the critical temperature. This behavior allows scaling $J_c(T,B)$ with temperature and field to the SSC operating conditions at 4.35 K and 6.6 T by using a relationship of the type

$$J_c(T,B) = a\left[1 - \frac{T - 4.2}{b - B_c}\right](1 + Bd) \qquad (3)$$

where a, b, c, d are empirical constants. Table VI gives the operating parameters thus determined. Here J_0 is the operating current density and J_c the critical current density of the uncabled strand, determined for the maximum calculated field value seen by the cable (corresponding to a central field of 6.6 T) and for the maximum assumed operating temperature. A minimum value of 1.25 has been chosen for J_c/J_0 to ensure an "operating margin." This margin allows for some degradation during cabling, small temperature excursions caused by radiation heating during beam abort, and some additional operating margin. Performance of SSC model dipole magnets has shown that 25% is, indeed, a reasonable margin.

Table VI. Dipole Operating Conditions

	Inner coil	Outer coil
T [K]	4.35	
B_0 [T]	6.595	
I [A]	6485	
B_{max} [T]	6.961	5.566
J_0 [A/mm^2]	1265	1837
J_c [A/mm^2]	1581	2347
J_c/J_0	1.250	1.278

6 COILS

The two-layer coil assembly is manufactured in four sections, from two inner and two outer coil segments. Each coil segment is wound separately on an automatic winding machine, then transferred to a curing press where it is molded to precise dimensions under relatively high temperature and pressure. The finished coil sections are assembled around the bore tube with appropriate electrical interconnecting splices and additional insulation introduced, and then the collaring operation commences. These various fabricating operations are described next.

6.1 Coil Winding

Each coil section of a dipole is wound on a precision-laminated, convex mandrel in a microprocessor-controlled winding machine twice the length of a 17-m dipole (Fig. 8). The cable is fed from a quasi-stationary supply spool onto the mandrel, which acts as a shuttle in weaving parlance. (The more conventional arrangement, a stationary mandrel and moving spool, is impractical because of the length of the SSC dipole.) A turn is initiated by winding the cable around one coil end with the mandrel rocking in unison to ensure a proper "lay." Next, one side of a straight section is wound as the mandrel travels the requisite distance with the stationary reel paying out cable. The supply spool retraces its step, laying cable around the other end of the racetrack-shaped path, and the turn is completed with the mandrel returning shuttle-fashion to its starting position. Wedges and end spacers are introduced between the appropriate number of turns in accordance with the coil design. When all turns have been wound, molded end spacers are mounted; these are used to square off the ends of the coil. Finally, the entire coil-mandrel package is wrapped with half-lapped tensioned Tedlar. In addition to securing the coil firmly to the mandrel for the next operation, the Tedlar will also perform the subsequent role of mold-release agent after the coil has been cured.

6.2 Cold Molding and Curing

A hydraulically operated, oil-heated, coil-curing fixture, or mold, is shown in Fig. 9. It is assembled from 11 module sections, each ~1.5 m long and fabricated from precision-stamped laminations, like the winding mandrel. The mandrel with coil still attached is lowered into the concave body of the fixture, previously prepared with additional mold-release and various tooling components in place. The top cover is placed in position with "stop" shims and clamping rods, and the fixture is hydraulically closed. The coil is cycled through a heating sequence with side and end pressure applied in a particular order. The curing portion of the cycle is done at 150°C for a period of ~100 minutes.

The stop shims, an important feature of the curing process, are shown in the simplified schematic cross section of the fixture in Fig. 10. They provide an adjustable method of closing the molding fixture and thus assuring dimensional uniformity of the coil despite possible variations in cable thickness. (The

Fig. 8. Coil-winding fixture for full-length SSC dipoles. The fixture is designed to accomodate dipole coil segments of arbitrary length up to a maximum of 16.7 m.

Fig. 9. Oil-heated coild-curing fixture, assembled from eleven 1.5-m module sections.

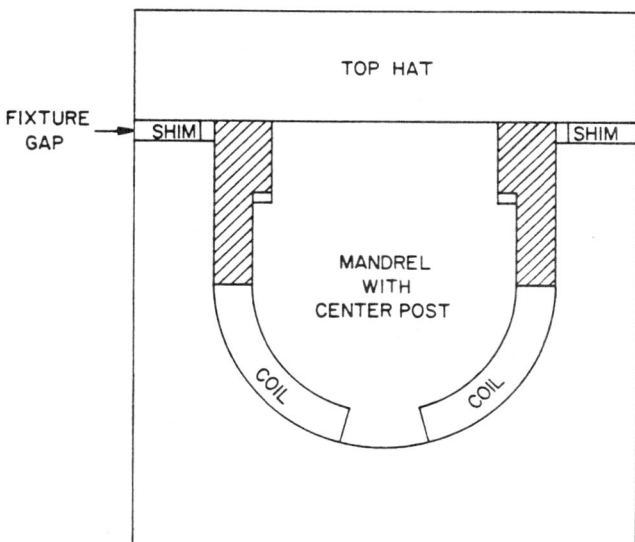

Fig. 10. Schematic cross section of curing fixture.

specified tolerance on cable thickness, 0.5 mil or 0.01 mm, verges on what is industrially feasible.) Essentially two curing steps are involved, a standard shim followed by recure with a new shim selected to yield the proper coil size. Coil thickness is determined from dimensional measurements made on the coil while still in the fixture after the first cure (measurement of the fixture gap, Fig. 10).

6.3 Coil Assembly and Collaring

The inner coil sections are nested in mating outer coil sections, a soldered "ramp" splice is made between them, and the upper and lower coil pairs are assembled around the stainless steel beam tube to which the trim coils have already been attached. (The beam tube assembly is discussed in Section 8.) During the coil assembly additional insulation is introduced: Kapton on the midplane (two layers, each 0.05 mm thick), Kapton (0.05 mm) and Teflon (0.05 mm) between coil layers. The Teflon is a low-friction "slip plane" to guard against frictional heating from any possible wire motion in the coil interface region. Next, the collars, pre-assembled in packs, are mounted over the coil package already insulated on the o.d. with three layers of Kapton (0.127 mm thick each), adequate to withstand 5 kV to ground. Between the first and second of these Kapton layers, one 0.1-mm-thick stainless steel strip heater per quadrant will also have been placed; these heaters are about two thirds the width of a coil quadrant and the full length of the magnet. They are an integral part of the quench protection system.

The interlocking, nonmagnetic collars provide the basis for accurate positioning of the coil sections and the concomitant ability to perform "warm" field measurements early in the magnet production line, before final coil-in-yoke assembly. They must also counteract the substantial Lorentz forces (43,000

Nt/m) generated when the magnet is energized to operating field. To achieve the minimum required preload at operating temperature, 3 to 4 kpsi, a considerably larger preload must be applied during room temperature assembly because of (a) "springback" when the collared coil assembly is removed from the collaring press (the largest source of prestress loss, about 50 to 60%), (b) some additional stress relaxation or creep with time of the plastic insulation (15 to 25%), and (c) differential thermal contraction on cooldown. The coils are fully self-supporting on their inside surface because of the Roman arch geometry noted earlier. The material specified for the collars is Nitronic 40 stainless steel, a nitrogen-hardened product of Armco. It was selected because it meets the strength requirements and also exhibits excellent (low-permeability) properties at cryogenic temperatures.

The shape of the collars, Fig. 11, is similar to those of the Tevatron magnets.[12] Fabrication and assembly of the collar packs, each 15 cm long, is also similar except that pins and keys are basically used instead of welding. The pins hold the collar packs together. The collared assembly is compressed incrementally down the magnet's length with a moving press about 1 m long (Fig. 12), and the square keys are inserted to lock mating collar packs around the compressed coil. Keys, unlike welds, facilitate repair of magnets as well. (Additional spot welding is, in fact, under consideration for further stiffening of the collars and to prevent relative rotation of adjacent collars.) The external key tabs on the midplane and the vertical centerline will accurately register the collared coil in the surrounding iron yoke (to an angular tolerance ≈ 0.5 mrad), and the key slots in the inner coil post register the bore tube relative to the collared coil. The diameters of the collar and yoke are dimensioned so that the collars will not bear against the yoke and are therefore self-supporting. ("Hybrid" support by both yoke and collars is thereby avoided.)

Aluminum collars offer an alternative to stainless steel as, for example, in the HERA dipoles. Aluminum is cheaper, and it shrinks more than stainless steel on cooldown so that lower initial prestress at room temperature is required. It has, however, a lower elastic modulus than stainless steel, requiring wider collars or support by the iron yoke as well. Since, in addition, the apparent cost benefits of aluminum are obviated by the technical modifications needed to reap its advantages, the decision has been made to retain stainless steel collars for the SSC.

7 YOKE AND HELIUM CONTAINMENT

The iron yoke provides the flux return path of the magnetic field, contributes about 25% of the central field, and serves as a magnetic shield. It also ensures rigid alignment of the collared coil assembly, but plays no further mechanical role, since the collared coil package is self-supporting. The yoke is contained within a cylindrical, welded stainless steel support shell which also acts as a stiffening member and serves as the outer wall of the helium containment vessel, — i.e. delineates the boundary of the "cold mass."

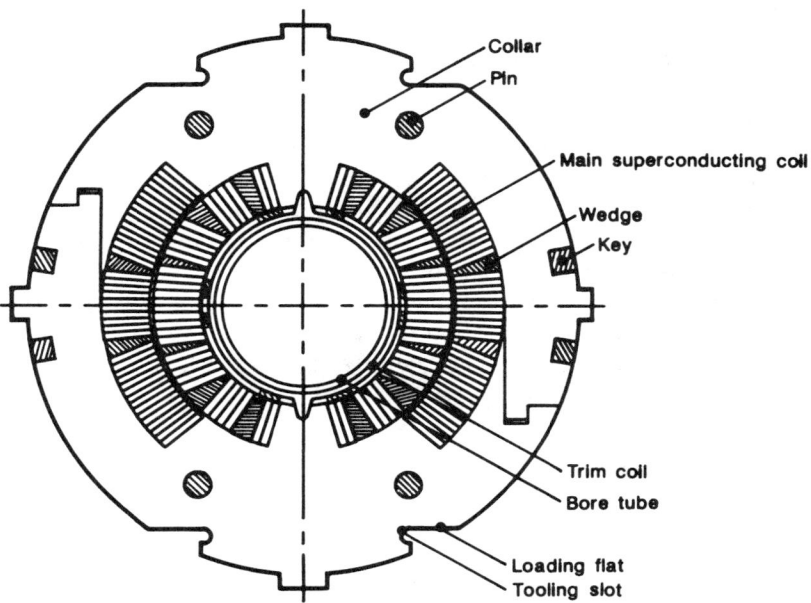

Fig. 11. Cross section of collared coil assembly showing bore tube, main coils, and stainless steel collar.

Fig. 12. Collaring a dipole coil. The incremental press can be seen at the far end.

7.1 Iron Yoke

The magnet yoke consists of stacked cylindrical iron laminations, split on the midplane, 266.7 mm in diameter. They are accurately die-punched from 1.5-mm-thick sheets of low-carbon, cold-rolled electrical steel having magnetic properties typical of those required for conventional accelerators, e.g. low coercive force, high permeability at both low and high induction, and high saturation induction. The two main departures from conventional magnet applications are that the iron is at helium temperatures and that the field in the iron is well above saturation. The magnetic effect of low temperatures is almost entirely a small increase in saturation inductance in accordance with what is known as Bloch's $T^{3/2}$ law. The effect of iron saturation in reducing the transfer function is due to the sharp reduction of permeability at high field, given approximately by the so-called Frolich-Kennely relation

$$\frac{1}{\mu - 1} = a + \frac{H}{M_s} \qquad (4)$$

where a is a constant, H the field in the iron, and M_s the saturation induction. Of greater significance are the changes in the multipole content with saturation, as discussed in Section 4.

Since essentially no gap remains between lamination halves in the assembled dipole, there is no significant magnetic motivation for selecting a horizontal rather than vertical split in the yoke; the choice is primarily a mechanical one. Shrinkage in the welds in the stainless steel containment shell supplies the tension that squeezes the lamination halves together. Because of friction between shell and laminations, compression is greatest if the welds coincide with the split in the laminations. The welds are incompatible, however, with the two rectangular conduits for the electrical bus; for magnetic reasons these conduits are located on the outer periphery of the yoke at the poles, top and bottom. The top conduit houses the main electrical bus which carries current from one magnet to the next in a string; the bottom conduit provides similar space for the correction coil leads. The main helium flow passes through the four large symmetrically placed "bypass" holes.

7.2 Helium Containment Vessel

The helium containment vessel consists of yoke support shell, "bonnets," end plates, and the beam tube described in the next section. All components are of stainless steel. The shell, which fits closely around the yoke, is formed initially as two half shells that are welded with automatic equipment to produce the finished assembly. As noted above, the shell is deliberately stressed during the welding process. The bonnets are short, thicker circular cylinders which are circumferentially welded to the ends of the shell. They provide a way of retaining the end plates that complete the helium vessel or "single phase" assembly, and also present a smooth cylindrical surface to which the interconnection bellows can be welded. The end plates help retain the yoke within the shell and provide a solid end for the coil to bear against during operation.

7.3 Heat Generation, Cooling, and Temperature Distribution in the Magnets

As the magnetic field is raised, heat is generated in the conductors by magnetization of the superconducting material, by induced currents in the conductor cables, and, in the beam tube, by induced currents in the copper used for plating the inside of the tube; finally, heat is generated by synchrotron radiation impinging on the inside of the tube. Near full energy, for the anticipated beam intensity, and for a ramp-up time of 1000 s, the total amount of power to be dissipated increases to 2.8 W/dipole, 80% of which is from the synchrotron radiation. The static heat from the magnet support and radiation and conduction from the inner heat shield, kept at 20 K, amount to 0.3 W. When a steady field is reached, only the static heat and the synchrotron radiation, amounting to 2.3 W/dipole, are left.

The magnets are cooled by supercritical helium flowing longitudinally through the magnets at a rate of 100 g/s. Only 1 g/s of this amount flows through a passage between the correction coil located around the beam tube and the inner surface of the inner main coil. Here the annular gap width is 1.3 mm, partially blocked at the locations of the spacers between trim and main coils. Because of the great length of the magnet and the small pressure drop, the flow is only 1 g/s. The bulk of the helium flows through the four circular bypass holes in the magnet yoke, each 29 mm in diameter.

Some of the heat generated in coils or beam tube is carried off by the helium flowing through the coil passage; however, most of the heat diffuses radially toward the bypass holes. From the inner surface of the beam tube, heat must pass through tube wall, correction coil assembly and insulation, helium in the coil passage, more insulation and main coil assemblies, coil-prestressing collars, a 0.25- to 0.5-mm annular gap filled with helium, and finally through the yoke laminations into the helium flowing in the bypass holes. If the narrow gap of the coil passage were blocked by accumulation of frozen gases or for other reasons, it would be stagnant, but it could still pass the required amount of heat because of the large surface for heat exchange. Since synchrotron radiation preponderantly impinges on the outer side of the curved beam tube, a small temperature gradient is established during machine operation which causes the helium in the coil passage also to circulate around the beam tube by convection, thus promoting heat exchange.

For the described cooling method, a magnet temperature increase of only 0.016 K is expected during field ramp-up. This increase is small because of the large amount of helium contained in magnets and interconnection regions, and, within limits, it does not depend on the helium mass flow.

Table VII shows calculated distributions, during operation, of maximum temperatures at the helium outlet ends of the dipoles for both turbulent flow (1 g/s) and stagnant helium in the coil passage.

Table VII. Calculated Temperature Distributions

	T_p	T_t	T_0	T	T_a
Turbulent helium flow [1 g/s]	4.43	4.38	4.36	4.35	4.30 K
Stagnant helium	4.52	4.44	4.39	4.36	4.30 K

T_p = Temperature at inside of beam tube wall.
T_t = Temperature in trim coil.
T_0 = Helium temperature in coil passage.
T = Average temperature in main coils.
T_a = Helium temperature in bypass.

These temperatures are based on an input helium temperature of 4.3 K. After exiting from the passages, the helium flows mix, giving a resulting temperature increase of 0.089 K per cell, or an average of 0.0074 K per dipole. There is to be one recooler available per cell. Half the temperature difference between T and T_a is due to the collars. The pressure drop per refrigerator loop (21 cells) is calculated to be 0.35 atm.

8 BEAM TUBE ASSEMBLY

The following components comprise the starting point for the magnet assembly, and they should logically come first in describing the assembly. Since, however, they are produced in operations quite separate from those involving the main dipole coils and coil-in-yoke assembly procedures, we have deferred their description.

The beam tube, or bore tube, separates the helium-cooled portion of the magnet structure from the beam high vacuum chamber (which is, however, "cold," not at room temperature) and is thus the inner wall of the cylindrical helium containment vessel. On its outside surface this tube supports the distributed correction coils. The correction coils are described in a following subsection. They are supported and electrically insulated from the main coil by plastic spacers bonded to their outer surface; these spacers also define longitudinal helium cooling passages between the two coils. Figure 13 shows a cross section of the beam tube assembly.

8.1 Beam Tube

The beam tube is fabricated from stainless steel tubing (Nitronic 40, as used for the collar laminations) of 1.0-mm thickness; it is copper-plated on the inside

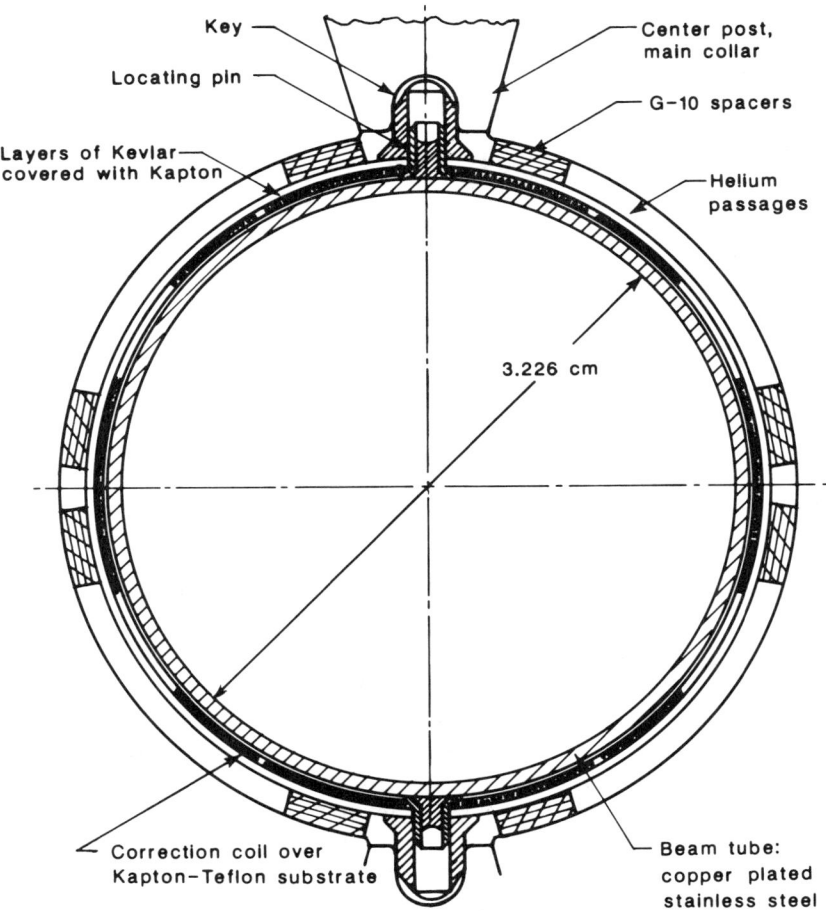

Fig. 13. Cross section of bore tube assembly, including beam tube, superconducting correction coil winding, and spacers bearing against the inner surface of the main dipole coil.

to reduce the electrical resistance of the wall to image currents. The copper coating is 0.127 mm (5 mils) thick. The inside diameter of the beam tube is 32.5 mm. Its outer surface is electrically insulated with 0.025 mm of Kapton and a very thin layer (~0.0025 mm) of Teflon with an infrared sensitive adhesive; 50% overlap. On top of that comes the correction coil layer, described next.

8.2 Correction Coils

The main function of the distributed correction coils (or "trim" coils) located between the main dipole winding and the beam tube is to compensate for the field distortions due to magnetization currents induced in the main winding during field ramping, a predominantly low-field effect discussed in Section 4, and to a lesser extent saturation of the iron yoke near maximum field. (The extensive array of correction elements grouped together at each quadrupole location in the lattice are not as effective for compensating these particular distortions.) Although the field component affected is mainly the sextupole, higher terms are also affected but not as strongly. The SSC Conceptual Design Report of 1986 specified a full-length sextupole as well as a decapole correction winding, one nested on top of the other, but more recent analysis indicates the desirability of including a distributed octupole winding as well. Tentatively, these coils are assumed to be arranged in a single layer as follows: a sextupole coil occupying half of the magnet length (8 m) and asymmetrically placed (starting at either end of the magnet and extending to its center), followed by a 5-m decapole winding and a 3-m octupole winding. The coils will be wound with a 0.2-mm diameter (bare) monofilamentary superconducting wire by the procedure described in the following section. Designed for an operating current of ~5 A, they will provide the following correction at full field: 4 units of sextupole, 0.4 units each of octupole and decapole. Since the critical current at 6.6 T exceeds 20 A, the operating margin exceeds a factor of 4 — substantially more at injection. (The available sextupole correction at injection is ~10 units.)

The exact specifications for the trim coils are somewhat tentative, but Table VIII lists the principal, typical mechanical parameters for the coils. Figure 14 illustrates the winding scheme for the sextupole.

The effects of systematic and random multipole contributions from the sextupole correction coil have been estimated. The first allowed (systematic) term is the 18-pole. Random harmonics will be generated from imperfect wire placement and imperfect trim coil placement with respect to the main dipole coil. The analysis shows that the only multipoles of significance will be the skew sextupole and the skew and normal quadrupole terms, as a result of which the respective random errors are increased by ~5% over those from construction errors in the dipole alone.

Table VIII. Mechanical Parameters for the Trim Coils

	Sextupole	Octupole	Decapole
Overall length [m]	8.04	3.04	5.08
Magnetic length [m]	8.04	3.04	5.08
Mean diameter [mm]	35.15	35.15	35.15
Number of turns/pole	19	14	11
Wire diameter (bare) [mm]	0.203	0.203	0.203
Wire diameter (insulated) [mm]	0.254	0.254	0.254
Copper-to-superconductor ratio	1.6	1.6	1.6

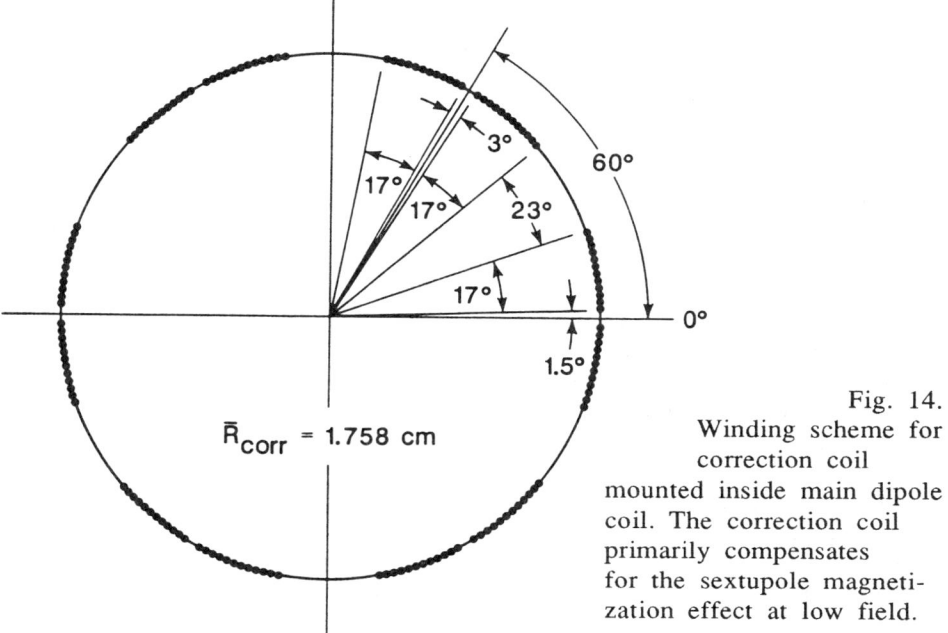

Fig. 14. Winding scheme for correction coil mounted inside main dipole coil. The correction coil primarily compensates for the sextupole magnetization effect at low field.

8.3 The "Multiwire" Technique

The most recent SSC model magnets utilize trim coils manufactured by a promising industrial technique of the Multiwire Corporation, and the parameters of Table VIII are essentially predicated on this technique. The technique is highly automated, akin to printed circuit fabrication, whereby a numerically controlled head ultrasonically embeds wires in a special substrate/adhesive. The device can lay wire at ~10 m/minute and with greater precision (±25 μm) than can be achieved with conventional winding methods. The substrate consists of a film of Kapton and Teflon, on top of which is bonded a proprietary fiberglass-epoxy (FEP) composite. The coil pattern is embedded while the substrate is flat. The substrate is subsequently secured to the bore tube with a FEP-Teflon adhesive and an overwrap of FEP-impregnated Kevlar. The substrate is initially prepared with accurately cut slots; the wire is applied relative to these slots, and then the completed assembly is located over precision guides affixed to the bore tube. The bore tube assembly is keyed to the external dipole collars (center posts of the main dipole winding, Fig. 13) by keys and pins so as to maintain orientation of the multipole fields with respect to the primary vertical dipole field. It is also accurately centered within the dipole winding by longitudinal G-10 spacer strips (1.4 mm thick) which, in addition, define helium cooling passages, as shown in Fig. 13.

An alternative, somewhat simpler version of the Multiwire technique is under study, whereby the coil pattern would be applied directly to the bore tube with the substrate already attached to it.

9 QUENCH PROTECTION FEATURES

Quenches will be initiated in the SSC magnets by a variety of events including beam loss, vacuum failure, and possibly conductor motion associated with training. The magnet quench protection system must reduce the current fast enough to eliminate the possibility of magnet or system damage for any conceivable event. The part of the magnet that initially quenches will reach a slightly higher temperature than the rest of the system because of the extra 20 to 50 ms of heating. The peak temperature will be about 350 K and the average temperature in the quenched region will be <300 K.

Unbalanced voltages can occur during quenching of superconducting magnets because the entire coil volume of the magnet may not revert to the normal state. With the protection system adopted, the outer layer of the dipole will be driven normal by heaters, and the inner layer will follow a short time later. In the extreme case where the inner layer does not revert to the normal state, the peak internal voltage will reach about 1000 V. The insulation in the magnets is designed and tested for >2 kV.

Essential elements of the active protection system are the heater strips located within each dipole. In case of a quench they are activated by discharging a capacitor bank into them, thereby inducing additional quenches and accelerating the normal resistive zone, ensuring that the magnetic energy is distributed rapidly over a large coil volume to prevent local overheating of the conductor. There are four heaters in each dipole, one per magnet quadrant, embedded in the Kapton ground insulation on the outer surface of each outer coil segment. They are fabricated from stainless steel strips, 0.10 mm thick by 30 mm wide, extending over the full length of the magnet and terminating at the end plates of the helium containment vessel. At 12 equidistant locations along the strips, or at ~1.38-m intervals, the strip surface area is reduced by cutouts. These regions constitute the resistive elements proper, and the stainless steel between cutouts serves as the current connection between heater elements. The design current for the heater is 70 A, and the expected cold resistance of the heater is 5.4 Ω. Thus, the design voltage across each heater is 378 V. The speed of the induced quench is weakly dependent on the duration of the heater pulse, but pulse lengths >15 msec do not appreciably reduce the peak heating in the magnet.

10 CRYOSTAT

The cryostat houses the "cold-mass" assembly or magnet proper described thus far. It includes a support system to hold the cold mass in the proper position inside a vacuum tank as well as heat shields and insulation to minimize the heat leak from room temperature into the cryogenic components.

10.1 General Cryostat Arrangement

The general cryostat arrangement is shown in Fig. 15. The major elements are the cryogenic piping, cold-mass assembly, suspension system, thermal shields, insulation, vacuum vessel, and the interconnection region.

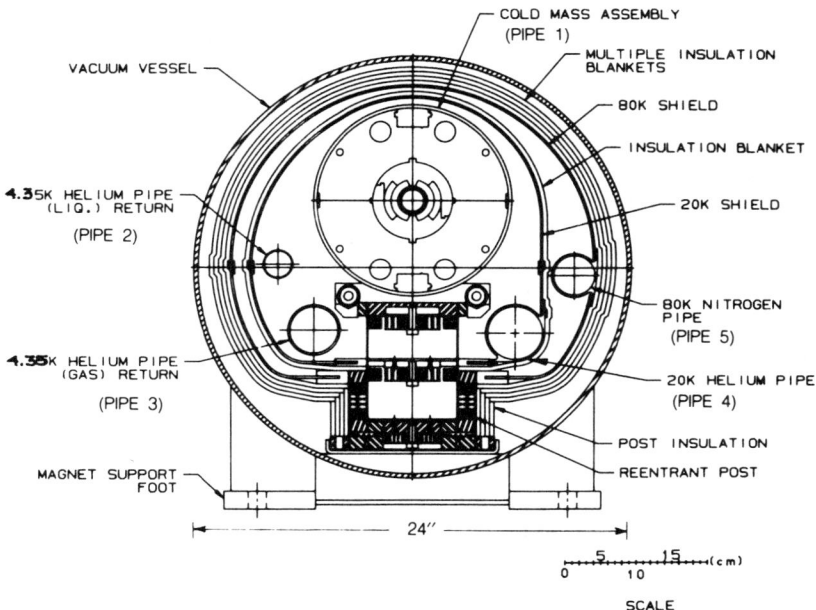

Fig. 15. Cross section of the magnet and cryostat assembly.

10.2 Cryogenic Piping

The cryostat assembly contains all piping that interconnects the magnet refrigeration system throughout the circumference of the ring. A system consisting of five pipes has been selected for cryogenic and magnet safety reasons.

Pipe 1: The complete cold-mass helium-containment assembly that contains the supply of single-phase helium fluid flowing through the magnet's iron and collared coil assembly.

Pipe 2: The 4.35 K fluid helium return and recooler supply pipe.

Pipe 3: The 4.35 K helium gas return pipe. The gas is generated in the helium recooler assemblies spaced around the accelerator ring to regulate the temperature of the single-phase helium fluid.

Pipe 4: The 20 K thermal shield cooling pipe. This pipe connects to the helium relief header during system cooldown, or to the return or supply headers during operation. The 20 K line provides quench buffering.

Pipe 5: The 80 K thermal shield cooling pipe. This pipe connects to the liquid nitrogen return or supply header.

10.3 Suspension System

The cold-mass assembly and thermal shields with their distributed static and dynamic loads are supported relative to the vacuum vessel by the suspension system. The system functions under conditions that include transportation and installation, magnet cooldown and warmup, and magnet steady-state operation and transient conditions. Its requirements include low heat leak, high reliability, dimensional stability, ease of installation and adjustment, and low cost. The cold mass and shields are supported at five points along their length. The number and location of the support points were determined by analysis of the deflection of the cold-mass assembly as a beam, limited to 1 mm between supports, and by the need to minimize the number of support points for ease of magnet fabrication and low heat leak. The support is a cylindrical post type, which has the following significant features:

(a) Load Carrying Versatility – The cylindrical section results in a versatile support member that can carry tension, compression, bending, and torsional loads.

(b) Structural Insensitivity to Thermal Contraction – The post configuration connections to the cold-mass and thermal shield assemblies allow them to move axially relative to the post as they extend or contract, thus eliminating changes in post loading due to thermal transients.

(c) Low Heat Leak – The use of fiber-reinforced plastic (FRP) materials with effective heat intercepts results in predictably low heat leaks. The required number of penetrations through the shields and their insulation is minimal, which reduces the potential thermal radiation heat flux.

(d) Integral Restraint – The hollow central region of the post permits the installation of an integral restraint member that connects the warm and cold ends. The restraint is employed during transportation and handling and is removed prior to operation.

(e) Ease of Installation and Adjustment – The fact that the post involves a single support member at a suspension point simplifies installation and adjustment.

The details of the reentrant post support are indicated in Fig. 16. The insulating sections are of FRP tubing with metallic interconnections and heat intercepts. The junctions between the FRP tubing and the metallic connections, which must be able effectively to transmit tension, compression, bending, and torsional loads, are made by shrink fitting. A post support is subject to internal axial and radial thermal radiation that can significantly affect thermal performance. Control of such radiation is achieved by the use of multilayer insulation. Proper design of the thermal connections between the 20 K and 80 K intercepts of the supports and the thermal shields is essential to minimize the heat leak. A 5 K temperature rise at 80 K and a 1 K temperature rise at 20 K are budgeted. The support post is fixed at its 300 K end and incorporates a slide at the 4.35 K end to accommodate the axial differential contraction between the mid-span anchored cold-mass assembly and the vacuum vessel. Analysis of transient conditions indicates that significant transient bowing of the cold-mass assembly is not expected. Since the 20 K and 80 K shields will undergo transient bowing, the shield-post interfaces are designed to allow relative motions while providing support.

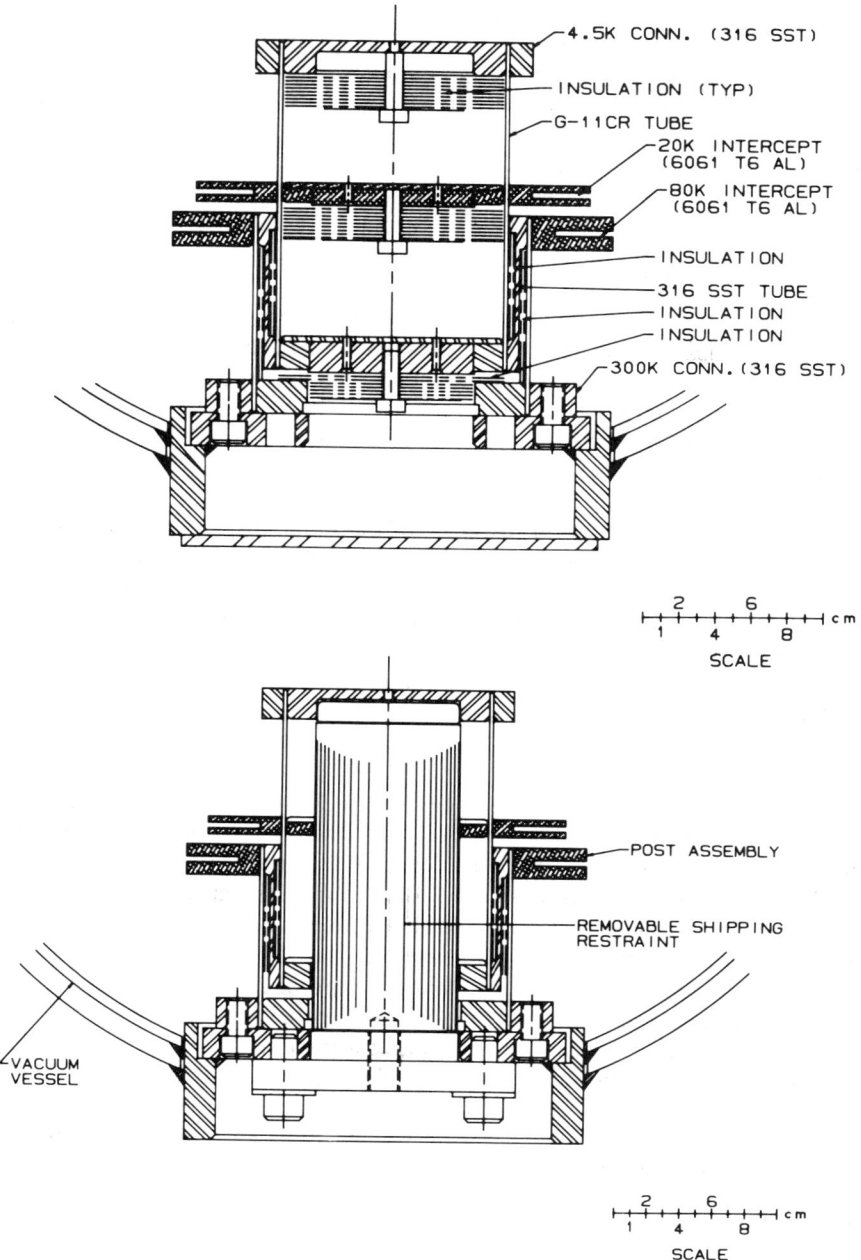

Fig. 16. Cross section: reenetrant support post (top), and reentrant post support assembly with removable shipping restraint.

In order to permit the post support to withstand the lateral handling loads without incurring a severe operational heat leak penalty, the design incorporates an integral, coaxial, removable shipping restraint, as shown in Fig. 16. The restraint is installed at the time of magnet assembly and is removed at the site. The cryostat incorporates a similarly removable axial shipping restraint. This restraint provides a strong structural axial connection between the cold-mass assembly and the vacuum vessel shell. It is installed during magnet assembly and is also removed before cryogenic operation.

10.4 Thermal Shields

Thermal shields, maintained independently at 20 K and 80 K, surround the cold-mass assembly. They absorb the radiant heat flux and provide heat-sink stations for the suspension system intercepts. The shields are aluminum and are fabricated from a combination of extruded flow channels and rolled shapes. The 4.35 K liquid and gaseous helium return pipes are supported from the cold-mass assembly by hangers. The 20 K and 80 K shields are supported by, and are thermally anchored to, the cold-mass assembly post support system.

The thermal and structural response of the inner shield, when subjected to a 100 K temperature differential across the section, was measured and was also modeled and predicted by finite element methods. The observed thermal bowing was within acceptable limits. Based on this analysis of shield behavior, the refrigeration system design limits the transient shield thermal excursions to 100 K. During steady-state operation, the temperature difference across the shield is only 1 K.

10.5 Insulation

Thermal insulation is installed between the 300 K and 80 K surfaces and between the 80 K and 20 K surfaces. The insulation system consists of flat, reflective radiation shields of aluminized Mylar film with randomly oriented fiberglass mat spacers (commonly called superinsulation). The system is prefabricated in blankets of 13 Mylar and 12 fiberglass layers. Four blankets are installed on the 80 K surface and one on the 20 K surface. Prefabricated transition pieces and well-defined installation procedures are utilized to eliminate voids in the insulation system, incurred either during assembly or from differential thermal contraction. The plastic substrate of the insulation system and the fiberglass mat should not suffer performance degradation when subjected to the estimated radiation environment of $<10^6$ Gy over the 20-year machine life. A vacuum pumpout space equal to the thickness of each insulation blanket is provided around one boundary of each insulated assembly.

10.6 Vacuum Vessel

The vacuum vessel, of carbon steel, provides the insulating vacuum space as well as the connection for the support system anchoring the magnet to the tunnel floor. Carbon steel was selected on the basis of cost. The composition of the steel will be a compromise between the material's mechanical properties and fabrication cost.

10.7 Interconnections

Mechanical and electrical interconnections are required at the magnet ends. It is essential that the connections be straightforward to assemble and disassemble, compact, reliable, and economical. The dipole-to-dipole mechanical connections are between beam tube sections, helium containment vessel sections, helium lines, liquid nitrogen shield lines, insulation vacuum vessel sections, thermal radiation shield bridges, and insulation. The dipole-to-dipole electrical connections include magnet current bus bars, quench bypass bus bars, quench protection diodes (if used), instrumentation leads, quench detection voltage taps, correction coil leads, etc. The interconnection design has stressed ease of assembly and disassembly operations in the SSC tunnel. The resulting geometry permits the use of automated welding and cutting equipment that is essential for installation efficiency and interconnection reliability. To reduce costs, straight pipe connections are used between magnets. All vacuum-tight connections are circular. Bellows, needed for axial thermal contractions, are also able to accommodate minor misalignments. For the pipes, two bellows are employed in series better to accommodate offsets. The interconnection region incorporates 20 K and 80 K heat-shield bridges to maintain radiation heat-transfer barriers. The bridges are attached to the magnet shields by riveting, with thermal connections provided by conductive braids. A single blanket of multilayer insulation is installed on the 20 K shield bridge. Four blankets of multilayer insulation surround the 80 K shield bridge.

11 MAGNET TESTING

A thorough program of quality assurance and quality control must be carried out during all stages of design and manufacture. All magnets will be tested in the factory at room temperature to assure that they meet electrical and field quality specifications. The electrical tests cover resistance, inductance, and Q of individual windings and of the complete magnet, as well as the dielectric strength between windings and collars, and between coil and bore tube. The collared coil, as noted earlier, allows magnetic field measurements at room temperature, i.e. of B_0 and the leading multipoles with the current limited to ± 15 A. (The field at 15 A is about 150 G.)

It is anticipated that approximately the first 100 magnets will undergo a complete warm and cold measurement program for full evaluation of field quality and training characteristics of production units. As experience is gained, a sampling method will be employed; a sampling rate of 10% appears reasonable.

11.1 Cold Measurements

The training behavior of a series of magnets of the type described is shown in Fig. 17. Each magnet is seen to undergo several quenches at ever higher fields until finally the magnet is "trained," or able to reach its predicted performance level based on measurements made on sections of the superconductor actually used in the magnet. Also indicated in Fig. 17 are the fields reached at reduced temperature, where the critical current density of the superconductor is higher. These magnets are built with superconductor of critical current ~15% below that expected for the SSC.

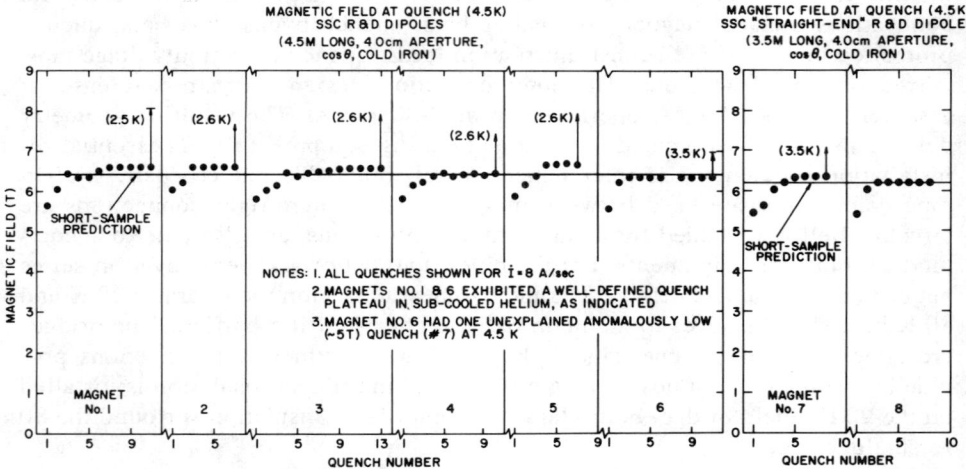

Fig. 17. Training performance of 3.5- and 4.5-m-long model dipoles at 4.5 K, with performance in subcooled helium also indicated.

The reasons why most superconducting magnets require training of the type shown are not well understood. The generally accepted picture of this phenomenon is that conductor motion somewhere in the magnet leads to heat generation, driving a local piece of conductor normal with a subsequent rapid spread of this normal zone throughout the magnet. After several quenches, all elements that can move have done so and are now locked firmly in place. It is known that a lack of coil prestress leads to erratic training behavior, but experimental results indicate that, above some very minimum level, the amount of prestress does not affect the training behavior. It has been observed that magnets built with adequate prestress in all parts of the coil show little or no training if the ratio of copper to superconductor is high enough ($\geq 1.6:1$), but here too the evidence is not conclusive. It is known that epoxy must be kept off the superconductor; the forces at high fields can cause epoxy bonds to break, releasing surprisingly large amounts of energy that drive the local superconductor normal. It is not known whether the number of required training quenches scales with magnet length. Recent experiments at LBL indicate that a magnet can be conditioned not to train below the desired operating field by first subcooling the magnet and then powering it to just above the operating field. This presumably forces all conductor elements to "settle in" but not to quench because of the extra critical current capability of the superconductor at the reduced temperature. The inability to measure the precise point of origin of training quenches has hampered better understanding of the phenomenon.

Magnet field measurements are made with a cylindrical coil with various windings on the surface. The voltages induced in the windings are digitized and Fourier analyzed to give the harmonic content of the magnetic field. These harmonics are a measure of the field uniformity and provide clues on the accuracy of conductor and coil placement in the magnet, and give an indication of various material properties such as superconductor filament size and iron permeability. Even small effects, e.g. slightly magnetic welds in the bore tube, show up readily in this type of analysis. Of course, the beam dynamics is determined largely by the overall field uniformity, obtained by summation over the various harmonics.

12 QUADRUPOLES

There are 678 standard arc quadrupole magnets per ring, or a total of 1356 such magnets. The quadrupoles have the same bore tube diameter and the same inner coil diameter as the dipoles. To maximize the operating gradient, the coil winding has 19 turns distributed in two layers, as shown in Fig. 18(a). The inner layer has 8 turns with a wedge inserted in each octant to maximize uniformity of the gradient. Four such double coil layers are arranged in the usual way to generate the quadrupole field.

The cable is identical to the cable used in the outer coil layer of the dipole: 30 strands of 0.65 mm diameter each, with a copper-to-superconductor ratio of 1.8:1, and partially keystoned. The quadrupoles are designed to operate in series with the dipoles. At $I = 6500$ A, the operating margin is somewhat higher than that of the dipole ($J_c/J_0 = 1.35$ vs. 1.25).

The quadrupole gradient is 212 T/m at 6500 A. (The dipole field is 6.6 T at this current.) An overall 0.5% decrease in gradient per ampere occurs between injection and maximum current because of saturation effects in the iron. This can be compared with a saturation drop of about 2% in B_0/I in the dipole. To maintain proper focusing, the correction package includes trim quadrupoles. Table IX lists the principal quadrupole parameters.

The coils are constructed by techniques developed for the Tevatron and are very similar to those used for dipole fabrication – basically, precision molded after winding, with an identical insulation scheme. Field uniformity of a theoretically perfect winding is shown in Table X, expressed in terms of the multipoles b_n. The $n = 1$ term is the quadrupole, $n = 2$ the sextupole, etc. Symmetry ensures that only the $n = 1, 5, 9, 13$, etc., terms are allowed. Terms above $n = 17$ are $<10^{-7}$. The listed terms do not change significantly as current varies from zero to maximum value. Figure 18(b) shows calculated field lines in a simplified rendering of a magnet octant. Calculated rms random multipole errors for the arc quadrupoles are given in Table XI.

Mechanical coil support and accurate alignment are provided by a system of collars, as in the dipoles. The arrangement is shown in Fig. 19, an overall cross section of the quadrupole. The collaring arrangement is nearly identical to the system used at Fermilab to construct the quadrupoles for the Tevatron. The collared coil assembly is installed in a split iron yoke similar to the dipole yoke, and held together by a welded stainless steel shell that is also the helium containment vessel.

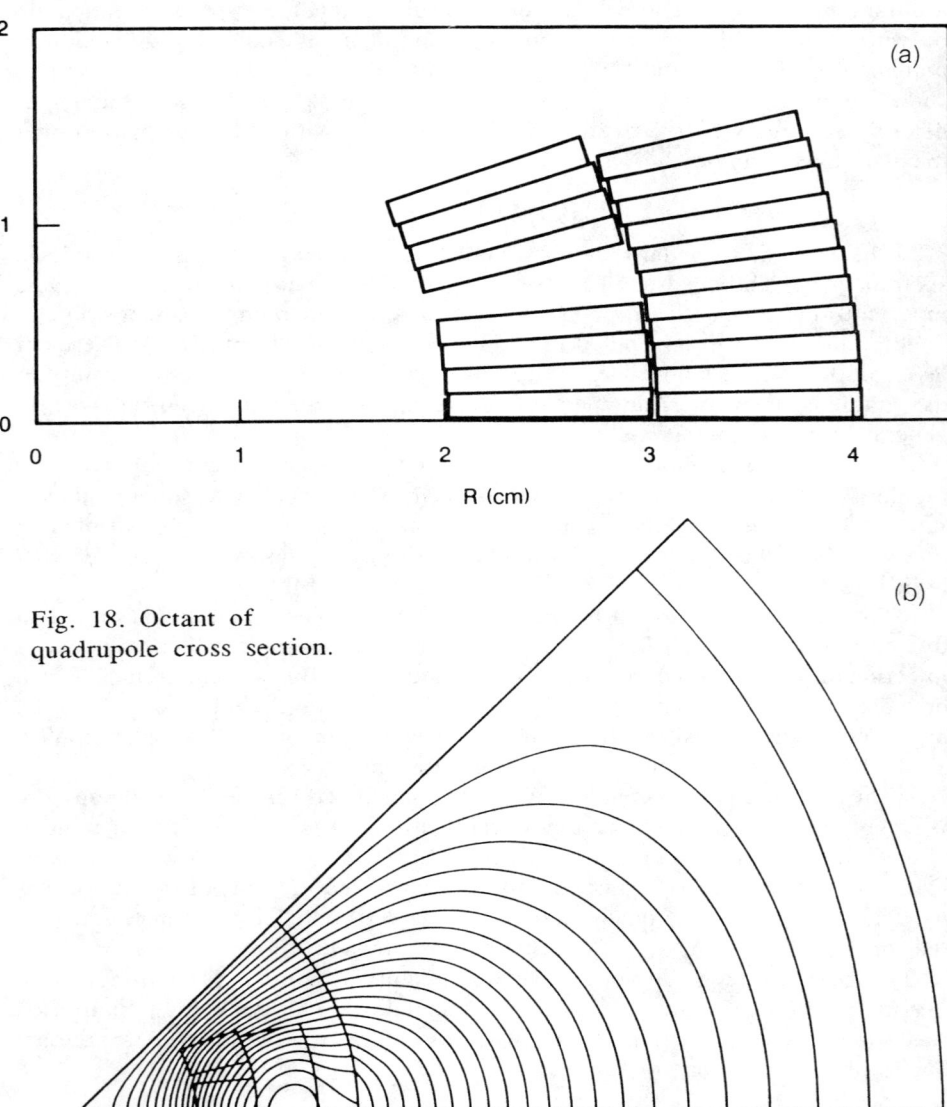

Fig. 18. Octant of quadrupole cross section.

Table IX. Regular Cell Quadrupole Magnet Parameters

General: No. of quadrupoles per ring	678
Overall length [m]	4.32
Magnetic length [m]	3.32
Bore tube inner diameter [cm]	3.226
Mass of conductor [kg]	83
Cold mass [kg]	1114
Central field gradient [T/m]	212
Current [A]	6500
Inductance [mH]	0.57
Stored energy [kJ]	12.0
Winding: Inner layer	
No. of turns (one coil per pole tip)	8
Inner diameter [cm]	4.000
Outer diameter [cm]	5.966
Cable length [m] (for 4 coils)	219
Cable mass [kg] (for 4 coils)	17.5
Maximum field [T]	5.2
Outer Layer:	
No. of turns (one coil per pole tip)	11
Inner diameter [cm]	5.986
Outer diameter [cm]	7.952
Cable length [m] (for 4 coils)	300
Cable mass [kg] (for 4 coils)	24
Maximum field [T]	~4.0
Conductor: Cross section, bare [mm]	9.73 × (1.27-1.06)
Keystone [deg.]	1.21
Strand diameter [mm]	0.648
No. of strands	30
Strand twist pitch [per cm]	0.4
Cable twist pitch [per cm]	0.136
Copper-to-superconductor ratio	1.8:1
No. of superconducting filaments	6000
Filament diameter [μm]	5
Iron Yoke: Material	Low carbon steel
Inner diameter [cm]	10.31
Outer diameter [cm]	22.86
Lamination thickness [mm]	1.5
Weight of iron [kg]	1005
Yoke Containment Structure:	
Material	Stainless steel
Outer diameter [cm]	27.62
Thickness [cm]	0.47
Weight of shell [kg]	109

Table X. Calculated Systematic Multipoles, in units of 10^{-4} cm^{-n}.

n	b_n
1	10^4
5	0.00023
9	0.06345
13	0.076
17	−0.0016

Table XI. Estimated rms random multipole tolerances in the SSC quadrupole magnets with d_c = 4 cm. The units of a_n b_n are 10^{-4} cm^{-n} with the quadrupole field evaluated at 1-cm radius.

n	1	2	3	4	5
a_n	--	2.6	1.2	0.3	0.14
b_n	--	2.9	0.5	0.3	0.56

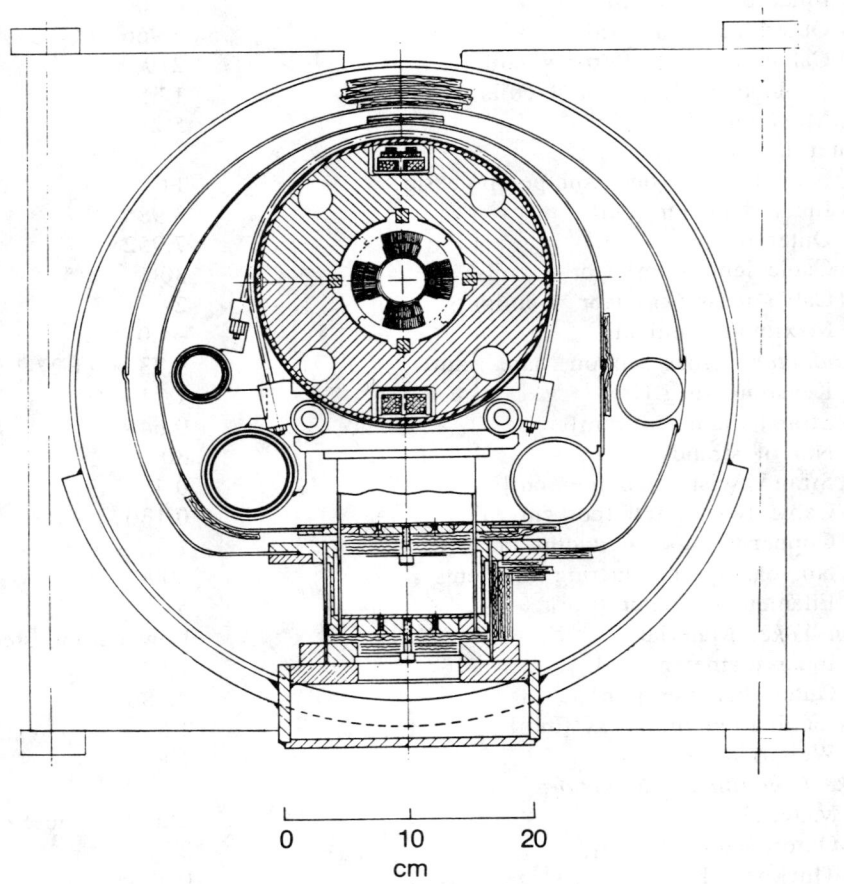

Fig. 19. Overall cross section of quadrupole magnet and cryostat assembly.

The cryostat is also shown in Fig. 19. It is identical in cross section to the dipole, with the minor exception that the two support posts will have a reduced cross-sectional area to minimize heat leak. End connections are nearly identical to those for the dipole.

13 CORRECTION MAGNETS, SPOOL PIECES

As in all large accelerators, the SSC requires correction magnets and a variety of other equipment that cannot be integrated into the dipoles or quadrupoles of the regular lattice. These devices are located in the drift spaces adjacent to every quadrupole in the rings, and are incorporated in a cryogenic enclosure known as a "spool piece." (Coils whose specific function is to correct for effects in the main lattice magnets – e.g. magnetization or saturation – are better distributed within the magnets themselves.) In addition to containing one or more superconducting correction elements, the spool pieces contain the following equipment: current bypass leads to protect the main magnets in case they quench; pumps and associated ports, gauges, and valves for the beam tube and cryostat vacuum systems; barriers that isolate the cryostat vacuum space of each half-cell from its neighbor; electrical leads for the correctors, pulsed quench protection heater strips, and magnet warmup heaters; and pressure relief valves for the supercritical helium circuit. The different types of spool pieces are enumerated in the SSC Conceptual Design Report; here we simply note some typical magnet correction elements. Altogether, the machine design calls for 1656 spool pieces.

The most numerous correction elements, in the so-called primary correction packages, are dipole correctors used to control the central orbit of the machine, quadrupole correctors for adjusting the tune, and sextupoles for controlling the chromaticity (variation of tune with momentum). Each primary correction package consists of a dipole, quadrupole, and sextupole, all 1.5 m long and nested radially on top of each other: the dipole innermost, followed by the sextupole, and the quadrupole outermost. The dipole is strong enough to correct six times the predicted rms closed-orbit error. The quadrupoles are capable of a tune change of ±2 units, and the sextupoles can adjust chromaticity by plus or minus twice the natural chromaticity. A flat, keystoned ribbon conductor is used; the winding geometry is of the cosine theta type.

The secondary correction coils are 2 m long. They consist either of a quadrupole and skew quadrupole nested within the same assembly, or a separate sextupole or octupole; the latter are in separate units to allow the required high fields.

The primary correction package is shown in Fig. 20(a) and the secondary correction coils in Fig. 20(b,c,d).

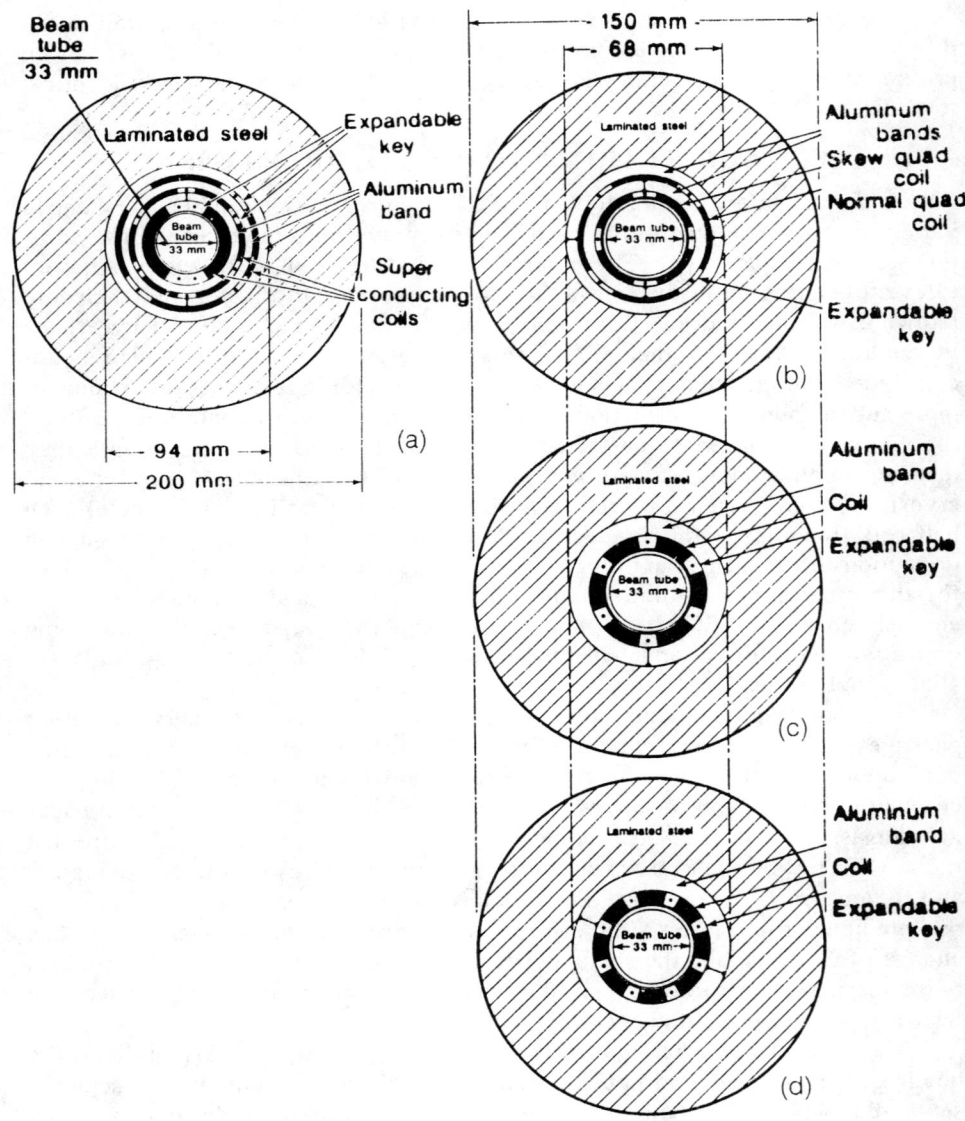

Fig. 20. (a) Primary correction package. (b,c,d) Secondary correction coils.

14 SPECIAL MAGNETS

In addition to the magnets described so far, several types of special superconducting magnets are needed in the interaction regions. Their design, like that of the correction magnets, has not yet been carried to the degree of completion of that of the main arc dipoles and quadrupoles. A set of conceptual designs is outlined in the CDR; only the highlights are presented here.

The special dipoles and quadrupoles differ from the arc magnets in some combination of length, aperture, and field, as demanded by the special requirements of the beam transport in the insertion regions. The list of these special magnets is given in Table 4d of Attachment A, the SSC Parameter List.[13] Though relatively few, these 500 or so special magnets are crucial to the proper functioning of the interaction regions and must at the same time preserve the beam quality to assure long beam storage times. All special dipoles are used to bend the beams in the vertical plane. The special quadrupoles, on the other hand, serve a variety of purposes from dispersion suppression to the final strong focusing at the interaction point in the low-beta IRs.

14.1 Vertical Bending Magnets

There are two types of IRs in the SSC. Near the low-beta regions the beams are fairly large and their separation varies. To accommodate the large beam size and, in the case of some of the special dipoles, two beams, a coil inner diameter of 7.50 cm was selected for the majority of the dipoles. The peak operating field is 5.2 T. Cold iron is used, as for the arc dipoles. Because in certain regions the beams are separated by 35 cm or more, two dipole assemblies are placed in a single cryostat, as in Fig. 21. Moreover, the beams in certain low-beta and medium-beta interaction regions are separated by between 60 and 70 cm; here two dipoles in individual cryostats are stacked on top of each other.

The medium-beta IRs require additional, special dipoles to accommodate the two beams when they are widely separated but converging near the interaction point. A coil aperture of 16 cm is selected for these. Note that 5-T, 16-cm-bore accelerator dipoles would appear to stretch the magnet parameters to the limit. In fact, the CBA magnets were only slightly smaller (13-cm coil i.d.) and reached the 5.28-T operating field with the NbTi conductor of 1980; cf. Fig. 6. The coil configuration chosen for these particular special dipoles is thus a modified two-layer CBA design with modern conductor.

Some dipoles in the medium-beta IRs will be constructed with coil packages similar to those of Fig. 21, but the two dipoles will be incorporated in a single iron yoke as in Fig. 22 – i.e., two-in-one dipoles.

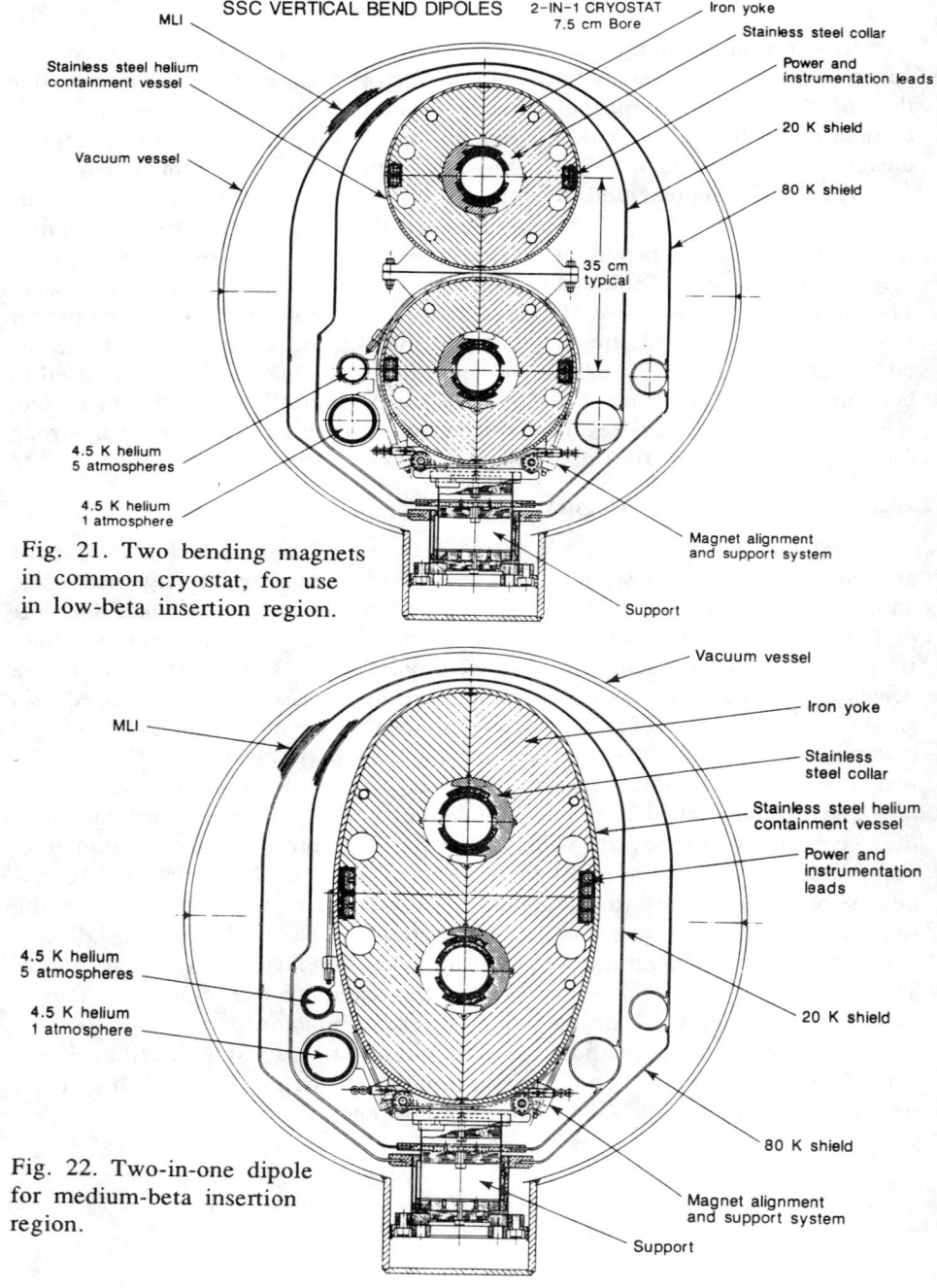

Fig. 21. Two bending magnets in common cryostat, for use in low-beta insertion region.

Fig. 22. Two-in-one dipole for medium-beta insertion region.

14.2 IR Quadrupoles

At each interaction region special quadrupole magnets are required to maximize the collision rate. A list of these is also given in the aforementioned Table 4d.[13] The focal length, and therefore the overall distance required for the IR regions, is directly affected by the achievable gradient. For the low-beta IRs, a gradient of 230 T/m has been assumed, based on a 4-cm magnet aperture and the SSC conductor. The magnetic length of these quads ranges between 13 and 14.5 m. Geometrically, the design closely resembles that of the arc quadrupole, except for the omission of collars to maximize the gradient. (The yoke itself provides pre-compression and force containment.) Iron saturation, nevertheless, has an insignificant effect on the field multipoles, although it does affect the peak gradient. The IR quadrupoles are connected to the main 6500-A bus; because of the saturation, there is a $\pm 10\%$ shunt supply across each IR dipole.

Other IR quadrupoles, all with 4-cm aperture, have the standard gradient of 212 T/m and their cold mass is identical in cross section to the arc quads but with varying lengths up to 15 m. Because of the varying beam separation in the IR regions, some of the quadrupoles will be packaged in 2-in-1 cryostats as well.

14.3 Additional Insertion Magnets

In addition to the aforesaid special dipoles and quadrupoles, the large beam size in the utility insertion regions will require various quadrupoles with larger than standard bore diameter, namely 5 cm. Accordingly, the gradient will be decreased proportionally, to 170 T/m; the focusing power will be maintained by increasing the magnetic length from that of the regular arc quadrupoles to about 11.6 cm. Other quadrupoles in the utility insertion regions can be standard arc quadrupoles, in terms of cross section and gradient, but again with longer effective lengths. There are, finally, a number of horizontal dispersion suppressor quadrupoles in the various insertion regions. These magnets are also identical to the quadrupoles in all respects except for length.

15 REFERENCES

1. M.N.Wilson, *Superconducting Magnets* (Oxford, 1983). R. Palmer and A.V. Tollestrup, "Superconducting Magnet Technology for Accelerators," *Ann. Rev. Nucl. Part. Sci. 34* (1984), 247-284. *Proceedings of ICFA Workshop on Superconducting Magnets and Cryogenics,* P.F. Dahl, Editor, BNL 52006 (Brookhaven National Laboratory, 1986).
2. R. Huson et al, *Proc. DPF Summer Study on Elementary Particle Physics and Future Facilities, Snowmass,* R. Donaldson, R. Gustafson and F. Paige, Editors (Fermilab, 1982), 315; R.R. Wilson, *ibid,* 330.
3. SSC Magnet Selection Advisory Panel Report to the Director of the Central Design Group, September 9, 1985.

4. *Conceptual Design of the Superconducting Super Collider*, J.D. Jackson, Editor, March 1986, SSC-SR-2020; C. Taylor and P. Dahl, "Dipole Magnets: A Brief Description," *Attachment B, SSC Conceptual Design: Magnet Design Details,* SSC-SR-2020B, 1-13.
5. Report of the Reference Designs Study Group on the Superconducting Super Collider, Draft II, May 1984.
6. "2-in-1 Magnets," *CBA Newsletter No. 2*, N.V. Baggett, P.F. Dahl and H.L. McNally, Editors, November 1982.
7. P. Dahl et al, "Construction of Cold Mass Assembly for Full-Length Dipoles for the SSC Accelerator," *Proc. 1986 Applied Superconductivity Conference*, Baltimore, MD (to be published); J. Strait et al, "Full Length Prototype SSC Dipole Test Results," *ibid.*
8. A.F. Greene, D.C. Larbalestier and R. Scanlan, "Superconductor: Conductor Development, Final Specifications," *Attachment B, SSC Conceptual Design: Magnet Design Details,* SSC-SR-2020B, 23-49; R. Scanlan, A.F. Greene and M. Suenaga, "Survey of High Field Superconducting Materials for Accelerator Magnets," *Proc. ICFA Workshop* (Ref. 5-1), 51-55.
9. R.C. Niemann et al, "SSC Dipole Long Magnet Model Cryostat Design and Initial Production Experience," *Proc. ICFA Workshop* (Ref. 5-1), 254-258.
10. C.P. Bean, *Rev. Mod. Phys. 36* (1964), 31.
11. A.K. Ghosh and W.B. Sampson, "Magnetization and Critical Currents of NbTi Wires with Fine Filaments," BNL Informal Report 37044 (1987).
12. *A Report on the Design of the Fermi National Accelerator Laboratory Superconducting Accelerator,* F.T. Cole et al, Editors (Fermilab, May 1979).
13. *Attachment A, SSC Conceptual Design Parameter List,* SSC-SR-2020A.

SUPERCONDUCTOR DEVELOPMENTS

David Larbalestier

University of Wisconsin-Madison

TABLE OF CONTENTS

1 Old-Fashioned Superconductors 1379
2 Niobium-Tin 1384
3 The Materials Science of Superconductors 1385
4 The New Oxide Superconductors 1387
5 The Future of the Oxide Superconductors 1390
6 Closure Note (December 1988) 1392
  Acknowledgment. 1377

Acknowledgment

I would like to thank Peter Lee, Pattie Lee, Mel Month, and Dick Carrigan for editorial help in the preparation of this article.

The work of our group has received long-term support from the Department of Energy, Division of High Energy Physics, Office of Fusion Energy, and Electric Power Research Institute.

Superconductor Developments*

David Larbalestier
University of Wisconsin-Madison

In the space of six months, superconductivity has become a household word. It's been in *Time, Business Week*, the *Wall Street Journal*, and the *New York Times*. Everybody, from children in grade school to chief executive officers, now knows what superconductivity is. There is discussion of the SSC and high-temperature superconductors. I am amazed to hear that the theory of particle physics and the ultimate constituents of matter may be determined by superconductivity theory. We have been through stone ages, iron ages, steel ages, and the semiconductor age. Now we seem to have reached the age of superconductivity. A colleague of mine reminded me of that great piece of advice in the film, The *Graduate*, in the sixties: "The future lies in plastics, young man." The 1987 version of that would certainly be, "The future lies in superconductors," or so the press would have us believe.

I will cover the technology of the new oxide superconductors and how they might relate to the existing superconductors. Fermilab and the high energy physics community in general have had much to do with the development of the "old" superconductors. Bednorz and Müller must be credited with the idea of setting out on the search for the new superconductors. They submitted their paper on April 17, 1986 (see Fig. 1). The work had been going on for a couple of years. Notice that the title is "*Possible* [italics added] High T_c Superconductivity in the Ba-La-Cu-O System." One of my colleagues, a theoretician, was at ETH in Zurich at the time the work was in progress, meeting with Müller every week or so. When he came back to Wisconsin in September he still didn't know about this. Müller was very modest about what was going on.

The history of high-temperature superconductivity since 1973 (when John Gavaler and his group at Westinghouse made the last advance, and a relatively minor advance at that), has been studded with loose ends, fakes, and all sorts of off-beat things. The subject had almost fallen into disrepute. Note the first sentence in the Bednorz-Müller paper: "At the extreme forefront of research in superconductivity is the empirical search for new materials." Müller was one of the few people who actually believed that. I can claim a bit of reflected glory here because that is a quotation from a report of which I am a co-author. It was put together about four or five years ago at Copper Mountain when the National Science Foundation ran a workshop on the science of superconductivity because the scientific excitement behind it was declining. The workshop asked what was needed to stir things up. Four or five of us put together a study group of 40 or 50 people and we wrote this report. The report had no effect that we could see. No

---

*Transcribed version of a talk given to the Fermilab Industrial Affiliates, May 1987 (with some update in Dec. 1988).

Fig. 1. The first page of Bednorz-Müller.

more money came into the field. This may be one of the reasons why people in the condensed-matter community have been attacking the SSC or point out that it would be great if oxide magnets could be put into the SSC. But one also has to say that very few people came up with new ideas.

After the Bednorz-Müller article appeared in September, things started out rather quietly and then the madness began sometime near the beginning of December. Some people claim they have barely eaten or slept or done much else ever since.

1 Old-Fashioned Superconductors

What is one looking for in a superconductor for accelerator magnets? A very high transport critical current density is needed, something like $2 \times 10^5 A/cm^2$ at a field of 4 to 6 T. This can be done with today's metallic superconductors. However, it's not so easy. Partly that's because one needs fine filaments. One would also like good stress-strain tolerance. The basic material must also be reasonable in cost and tolerant of mass production. The material is not the be-all and end-all. The device has to be assembled reasonably and cheaply by an average production-line person. One would like considerably more, but certainly one requires a strength of 20,000 or 30,000 lb/in.^{2} and failure strains > 0.5%. Put differently, something is needed that is not going to break when you look at it. Since the oxides are brittle, the work done over many years on niobium-tin should be of benefit. For example, one knows that a brittle material like glass is a lot more strain tolerant when it is a fine fiber. It may well be that although one does not need fine filaments for electromagnetic reasons as in the case of helium-temperature superconductors, one may go to relatively small dimensions simply to get good mechanical properties. Issues such as positional tolerances and so on are also important.

Figure 2 shows the phase diagram for a type II superconductor. This shows the phase space of temperature, magnetic field, and critical current density. Two of these properties, temperature and field, tend to be determined by details of the immediate atomic environment and not much can be done about them. The critical current density, however, is extremely sensitive to the local defect structure, and in fact can vary over orders of magnitude. The great excitement of the last six months is due to only a factor of four increase in temperature, but one can easily arrange for the critical current density to vary over three or four orders of magnitude.

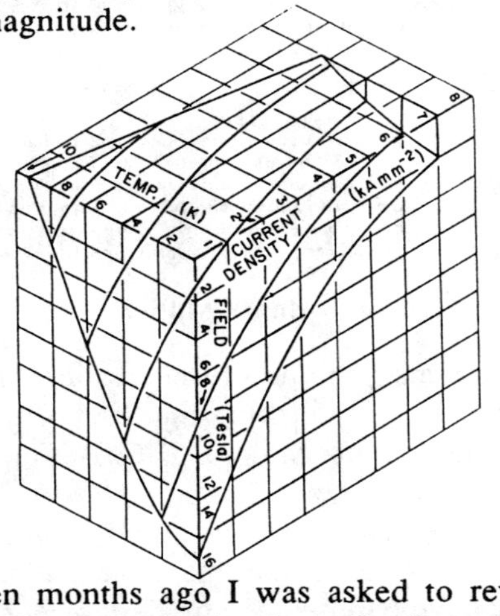

Fig. 2. Phase diagram for Nb-Ti, a typical superconductor.

Eighteen months ago I was asked to review where superconductors were going, for the Magnet Technology Conference — the community of people who will actually use these things and make devices out of them. At that time one looked to the practical superconductors (the ones some people in this audience actually sell), niobium-titanium and niobium-tin (Nb_3Sn) conductors. In 1985 the community was looking ahead at some other materials, such as niobium-aluminum, niobium nitride, and lead-molybdenum sulfide, that might become practical. One of these has an upper critical field of abut 60 tesla, about six times that of niobium-titanium. Some people, particularly in the high energy physics community, say it would be really neat if we could have 20- or 30-tesla magnets. Thus, those materials might have become practical. In the past six months the groups of people working on these materials have been drastically reduced because those people have turned to the oxides.

By 1985 some important recent advances had been made in the materials available, many of them due to advances in measurement and characterization. The oxide explosion has generated tremendous scientific interest in people all across the community — both basic and applied science people, including theorists, condensed-matter experimentalists, chemists, mathematicians, and the engineering community. With 77°K superconductors, there are all sorts of applications of superconductivity never dreamed of before, which we (even those in this audience) clearly are incapable yet of appreciating. There is a real reason for this great excitement, and, I think we will see a big push towards applications at a very early stage.

With that in mind, let's consider a few features of niobium-titanium and see if they might give any hints about what may happen in the oxides. One of the very important things that happened in "old-fashioned" superconductivity was that the TEVATRON established a proper superconducting-materials industry. The industry did not exist before the TEVATRON, except as a large garage-type operation. The necessity of making miles of magnets meant that wire production had to be put on a mass-production basis; 1500 to 2000 quarter-ton composite superconductor extrusions were made into cable and wound into magnets. The critical current density, a very important parameter, was about 180 A/mm^2 at 5 tesla. In the last few years we have been able to raise that to about 2750 A/mm^2 in production. At the same time the filament sizes have gone down to 3 microns through some remarkable developments in industrial fabrication. Now it seems as if the filament size may go down a little more for some other reasons. This has not been discussed much in the United States, but Japan and Europe have developed reasonable programs aimed at 0.1-micron filaments for power frequency applications of superconductors.

The basic understanding of what is going on in these materials is much improved. One of the reasons is a very close coupling between the basic science and the industrial production. A lot of the advances can be credited to precisely that sort of interface.

Let me emphasize something with which our group at Wisconsin has been closely linked. DOE has said: "It's great, we appreciate the fact that you are attempting to achieve a fundamental understanding of these materials, but high energy physics is really a user and a consumer of superconductivity. We want you to get the message out to industry and the users." Thus we were forced (at first a little unwillingly, but in fact it has been a very pleasant experience) to hold a series of workshops on these applications. By now we have had about six of them. Out of this has come a tripartite relationship among basic researchers, the users (essentially the laboratories, including the SSC through its manifestations here at Fermilab, at Brookhaven, and at Lawrence Berkeley), and the producers in industry such as Intermagnetics General Corporation (IGC), Teledyne Wah-Chang, Supercon, Oxford, and Cabot. Groups of 50 people have met together about every nine months. Collectively, there has been a tremendous coupling between the problems on the industrial scale and those on the research scale. The trade-offs between filament size and critical current density have been examined. It has been an example of a situation where people are prepared to get together and talk over problems. One can make tremendous strides in such situations.

Figure 3 illustrates this graphically. The SSC requirements greatly stimulated the recent developments both fundamentally and on an industrial scale. We are seeing a 50 to 60% increase in deliverable production material and we are still pushing. Niobium-titanium production has been described as a 25-year-old technology. That is true, but it is not a technology that has been stagnant. It continues to improve. The properties of the magnets have also continued to improve. It's possible now to buy much better material in very long lengths with very fine filaments. That is not just of use for high energy physics; it is also useful for MRI and all sorts of other things. A Supercon billet made several years ago represented a very considerable advance and contained about 4000 filaments. Now IGC and others have made 40,000-filament composites with rather good properties. This technology is industrially available and can be put into magnets; it can be thought of as a realized engineering technology.

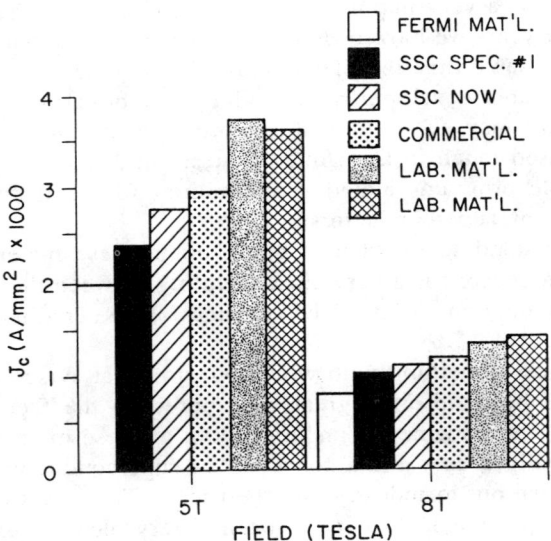

Fig. 3. Progress in raising the J_c of Nb-Ti.

Obviously, critical current density is important, so where do we turn the handle in order to get more critical current density? In anything other than a very weak magnetic field, the superconductors that we are interested in contain flux. The unit of flux in a superconductor is quantized and very small — about 10^{-15} webers. Around the flux is a circulating current vortex, and that current penetrates over some characteristic distance into the superconductor. Between the vertices are superconducting regions, as illustrated in Fig. 4. About 20 years ago Träuble and Essmann in Germany did a very elegant experiment whereby the perfection of this flux structure was revealed with a nice evaporation technique.

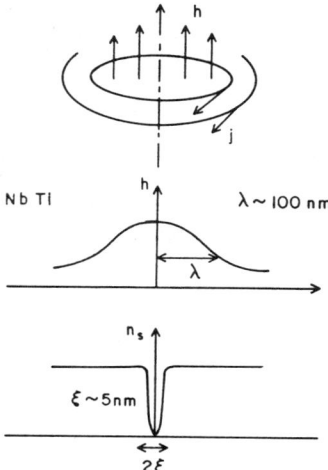

Fig. 4. Structure of one vortex line in a type II superconductor. The field is h, contained within a distance λ, the penetration depth, and screened from the superconductor by the vortex current j; ξ is the coherence length, the length over which the superconductor wave function can vary; n_s is the density of superconductor electrons.

Figure 5 is a view looking down these fluxoids. It shows a very nice triangular array, and this unfortunately means that the flux density is rather uniform and the material can carry only a very small current. In high-current-density materials we perturb this ideal state very severely by putting in may fluxoid-pinning centers. In some of the optimized wires that we developed, the defect structure has a scale of 1 to 2 nanometers, that is, only about 3 to 5 atoms. This extremely localized scale is one major reason why the materials science of superconductivity is such a fascinating challenge.

How high can we raise the critical current density? One way to estimate this is to divide the field within one of these fluxoids by the depth over which the surrounding vortex current circulates. With this phenomenological approach we find that the current density is something like 10^6 A/mm^2. In practice, in niobium-titanium, we get at best 10^4 A/mm^2, so we really have achieved only a fraction of this local supercurrent by pinning the fluxoids. On the other hand, we have established an upper bound. It is nice to know that this upper bound is a very high number because, perhaps in the future as we get more clever, we may get closer to that number.

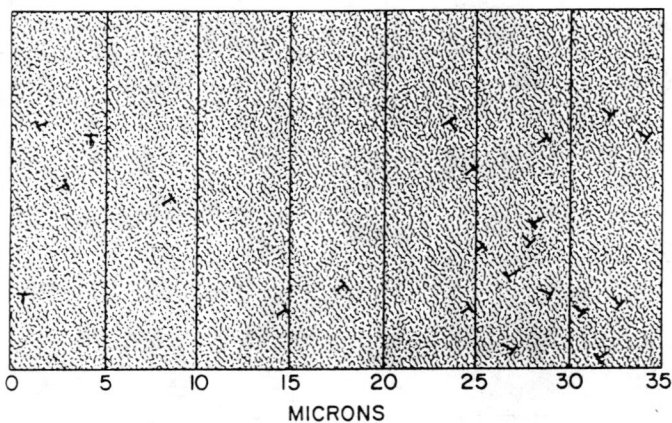

Fig. 5. Individual fluxoids in a type II superconductor.

2 Niobium-Tin

Perhaps 95% of the wire being fabricated is niobium-titanium. The other 5% is niobium-tin. Things have not been standing still in niobium-tin. The basic researchers have found that by adding a bit of titanium to the niobium-tin they get usefully higher critical fields. People like our group, and those who thought it still worthwhile to investigate superconductors, have also been beavering away trying to understand what controlled the critical current density. This is not an easy or short-term problem. Figure 6 gives some idea, from the critical current densities as a function of magnetic field, of what one could do with niobium-tin. Note that this graph starts at 8 tesla. These data are for liquid-helium temperature and go to about 20 tesla. The curve with V's shows how pure niobium conductor behaves.

This type of conductor has been used for very big fusion devices. For example, Westinghouse built a 3-m-diameter coil, which is at Oak Ridge being tested. One can now purchase 16-tesla magnets for $200,000 to $300,000 from one or two companies. These are solenoids and they typically utilize material having properties like those in the lower curves of Fig. 6. In the last few years companies in the United States, Intermagnetics General and Teledyne Wah-Chang in particular, have been developing a process based on using pure tin, rather than a dilute tin alloy, as a source for tin. With this, one can get very much higher critical current densities, although at some sacrifice of other properties. Thus, for making high-field magnets, this does indeed make for a considerable advantage. Nb_3Sn is, unfortunately, a brittle compound. The application of Nb_3Sn for a magnet is considerably more complicated than that of niobium-titanium. This is a point worth returning to when oxides are discussed.

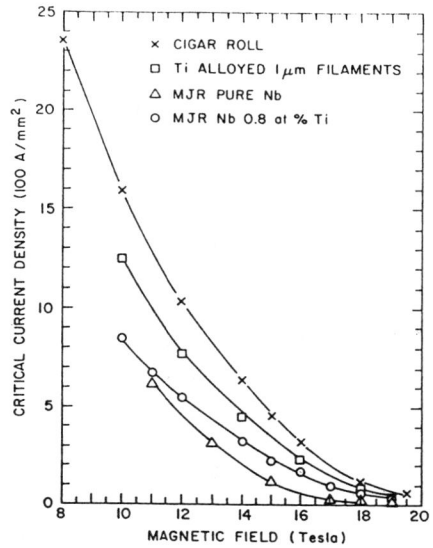

Fig. 6. Critical current density as a function of magnetic field for niobium-tin.

3 The Materials Science of Superconductors

Many times I have asked the Department of Energy to fund what amounts to fundamental materials science. Sometimes I've had to ask them to take it on trust that we have an application in mind. The materials-science community makes a considerable investment in transmission electron microscopes and the like. (The materials-science community also has some very expensive tools!) Typically, we would like $2 or $3 million worth of advanced electron optical devices plus the people to go with them. We are trying to get both structural information about where the atoms are and information about the local chemistry. In particular for Nb-Ti and Nb_3Sn, utilizing these tools has been very useful for the developments mentioned earlier. The next few figures give some of these developments.

Figure 7 shows a conductor made in our own laboratories, where much of the basic processing of niobium-titanium is being worked out on a prototype scale. The figure shows a very small region within these filaments. Note a crucial piece of information: the magnification marker. It is 1/4 micron across. One starts out making materials that are heterogeneous on a scale of about 1/4 micron. Note the particles of virtually pure titanium. These act as very effective flux-pinning centers and thus increase the critical current density. The characteristic separation of the fluxoids is something like 1/10 of this scale; we therefore had to develop methods whereby we can make things on a much, much finer scale. Fortunately, it turns out that just by wire drawing we can turn those relatively round white blobs into the very elegant long ribbons. Figure 8 shows this on a much finer scale. This enables us to match up the fluxoid lattice fairly closely to the defects. This shows how we get these very high critical current densities. We are, in fact, developing nanometer-scale structures 1 to 2 nm or only 4 or 5 atoms across. I expect to see materials science going into fine-scale engineering much more.

Fig. 7. Niobium-titanium micrograph at heat treatment size. TEM transverse cross-section after final heat treatment, 5 × 80/420; diam. = 8.25 mm; 1045 filaments.

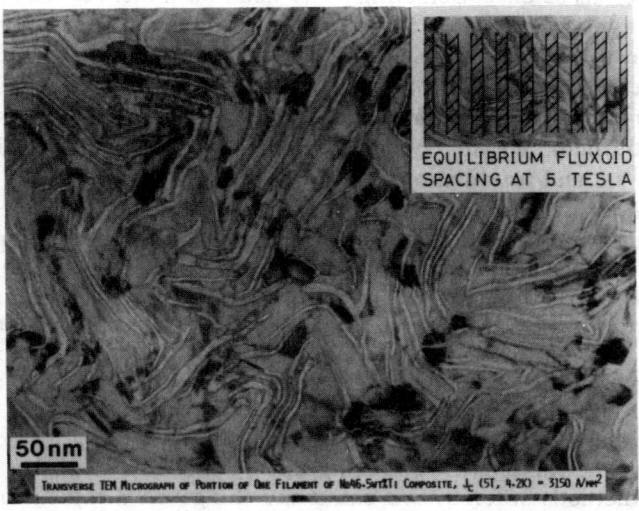

Fig. 8. Optimized Nb-Ti microstructure. The microstructure has been greatly refined by drawing the wire to a finer size, thus greatly raising J_c. Transverse TEM micrograph of portion on one filament of Nb 46.5wt%Ti composite, J_c (5 T, 4.2 K) = 3150 A/mm^2. (Courtesy Peter Lee.)

4 The New Oxide Superconductors

Figure 9 shows a micrograph of the new high-temperature superconductor yttrium-barium-copper oxide. The characteristic size here is a relatively large 15 microns. Note that, because of its anisotropy, the structure shows up very well in polarized light. The characteristic bands that appear in the material are at the level of 10 microns. The grains are very finely self-divided. The features are so-called stacking defects, known as twins.

Fig. 9. Photomicrograph of yttrium-barium-copper oxide. (Courtesy J. Seuntjens.)

Why can't we just simply start to build oxide magnets? The push has clearly been so strong in the press, and in the scientific and the engineering communities, that there are people looking for exactly this challenge who are prepared to do it. But why are some people in the community, while enthusiastic about the oxides, nevertheless saying that some caution is in order if what one really wants are magnetic devices?

The measurement of the critical current density is normally done by passing a transport current through the material. It can also be done by inducing a current in the material and detecting the magnetic moment associated with that circulating current. Knowing something about the dimensions over which the currents flow, we can derive the current density. We call this the magnetization J_c. In any reasonable superconductor that we might make, it would be very unusual for the results of these measurements to differ by more than a factor of two. For anything we were thinking of putting into a device, we would expect them to agree at least to within roughly 10%. A problem in the oxides that is just beginning to be resolved is that the magnetization current density is much greater – orders of magnitude greater – than the current density measured by transport current methods. This raises a question: Why is the transport critical current density so low? I'll return to this point at the end.

The other nasty factor is that weak magnetic fields produce a very big change in the properties. We know that superconductivity exists in these materials to fields of at least 40 to 50 tesla. If our current ideas are true, we have every expectation that it exists to perhaps 100 tesla or even more. So why are weak fields creating so much difficulty? The strong anisotropy of the crystal structure is certainly an important factor. People don't show crystal structures of the old-fashioned superconductors niobium-titanium and niobium-tin. Those structures are rather straightforward; they are cubic. Perhaps wrongly, they are considered to be so simple that they are not worth showing. That is certainly not true of the oxides. Figure 10 shows one formula unit of the 90°K superconductor. What one notices is two chains of copper and oxygen, the key players in the superconductivity, then a plane of barium and oxygen atoms, next a copper-oxygen plane that is somewhat rumpled, and then a very curious feature, just one lonely yttrium atom.* One thing the theorists have said is that there is very little electron

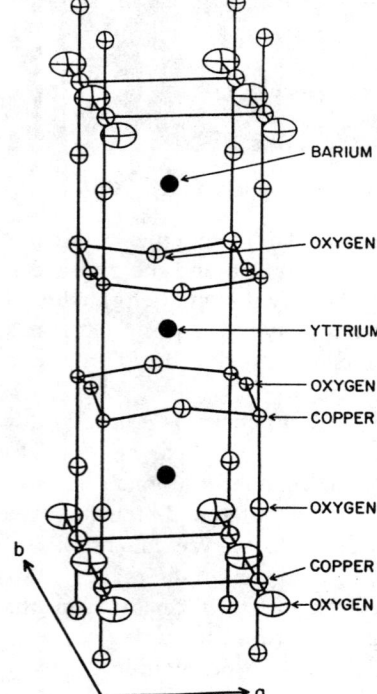

Fig. 10. One formula unit of the new 90°K superconductor.

---

*It is now (1988) generally agreed that the rumpled CuO_2 planes are the determining sites for high T_c properties.

transport in the vertical direction. Looking at this sort of structure we expect properties that are rather anisotropic. In fact, all three axes of the crystal are different.

Right in the early work of Müller and of Chu, it was noticed that the width of the superconducting transition was very broad in a magnetic field. This was particularly the case in polycrystalline, bulk samples. Those interested in the fundamental physics of high T_c didn't worry too much about this to start with, but groups like ours were very concerned, because it limited the transport current density to values of 1 to 10 A/cm^2 in fields of more than a few millitesla. As materials scientists, we asked ourselves if the samples were good ones. The answer is clearly no. Polycrystalline, sintered samples contain cracks, impurity phases, and other unknown defects. the materials-science community is trying to catalog and understand the defects. It's a tremendous job. (It's still going on full speed in 1988 and we still don't have much idea as to why grain boundaries so effectively impede the flow of current.) At present, J_c for bulk samples of sintered oxides is ~100 to 1000 A/cm^2, falling to 10 A/cm^2 or less in fields of 0.1 T or more. (There has been very little improvement in 1988; current can still not easily be transmitted across grain boundaries.) On the other hand, the current density circulating within the grains is very high.

Suenaga at Brookhaven noticed this by grinding a bulk sample up into powder and noting that the magnetic moment stayed about the same. That implies that the currents are not circulating over the whole region of the material but over very restricted local regions. One limit is the grain size, which is abut 10 microns. (Much work in 1988 has shown that barriers also exist within the grains.) If the depth to which the currents are flowing is 10 microns instead of 1 millimeter, it raises these critical current densities by about a factor of 100. Striking confirmation of these ideas was produced recently by Chaudhari's group at IBM. They showed that the transport current density of single-crystal thin films of $YBa_2Cu_3O_7$ was of order 10^5 to 10^6 A/cm^2, thus confirming the conclusions of the earlier magnetization experiments. (Further elegant work by this group during 1988 has shown that the misorientation angle between grains strongly affects the current crossing the boundary. The current diminishes markedly as the misorientation increases.)

A real concern is that although in zero field we start with superconductivity existing all through these materials, very weak fields cause a decoupling of superconducting regions from each other. The most likely explanation at present is that this takes place in grain boundaries. (This has been confirmed by work done in 1988; however, regions within grains can also cause decoupling.) The real question is, is this because of defects in the quality of the material, or is this an intrinsic property of the large anisotropy of the materials? The disappointing news at the moment is that, when we have polycrystal aggregates, we turn on resistance at low J_c values in low magnetic fields.

It is very important from a materials-science standpoint to understand whether this sort of property is intrinsic to the materials or whether it is determined by defects in the materials as we now prepare them. This weak coupling between the grains needs to be defeated. If it is an intrinsic problem, all may not be lost. It may just mean that methods are needed to orient over very long lengths of these conductors so that we can get good conductivity. We need to understand this question so that we can understand where we should put our efforts in the materials-processing business.

(This problem has been worked on extensively during 1988. Many improvements have been made to the processing of polycrystalline bulk high T_c conductors, but these have not yet raised the transport J_c to interesting values in magnetic fields of a tesla or more. Methods of fabrication that emphasize the production of large grains, such as the melt-texturing approach of Jin at Bell Labs, do raise J_c, but they do not appear to have made a solution to the current transfer problem across grain boundaries any more feasible.)

5 The Future of the Oxide Superconductors

How long will it take to develop oxide devices? I think the IBM work establishes that for electronic applications, such as putting superconductors and semiconductors together on the same chip, those basic conditions can be achieved. We are going to see a tremendous effort on the part of IBM, and Bell, and other companies to pursue that type of technology. Of course, the same things will happen in Japan, and Europe, and everywhere else that we compete.

For high-field applications, there is the disconnect problem that I have just described. Although that problem was not encountered with niobium-titanium and niobium-tin, we now have a very strong push for applications because everybody is intrigued by superconductors. One can foresee all sorts of cheap uses of superconductivity at 77°K that were never that attractive or realistic at 4°K.

These oxides are brittle. While it is a favorite dream of many people that we just need to put a bit of basic science and money into making them ductile and then we will be able to do it, I believe that I'm on firm ground in saying that this just ain't so. Glass has been around for 5000 years and it's still brittle. Brittleness and ductility are associated with the inter-atomic forces. We can't muck around with these forces without completely altering the very properties that are crucial for superconductivity. So we will have to live with the brittleness. We can do this for glass and concrete and silicon, but we have to design for the brittleness of the material. One way to do this is to engineer back from the device. That is to say, we need to be device oriented and then go back to the material rather than saying, "Here's an oxide conductor, a wire that looks like niobium-titanium or niobium-tin, and we will treat it just like them." The innovation and the real applications will come when we think in a fresh fashion about the device. Only when we have really worked out the device are we likely to be able to apply this brittle material.

Current density is also needed. Table I gives the history of niobium-tin. Studying niobium-tin is useful because, if we do things the wrong way in the oxides, it may be like niobium-tin. I hope that will not be the case.

Table I.
Development of Nb_3Sn

- Conductor Process Available in 1960 (Bell)

Tape↓

- Nb_3 Tape Drove out the Nb Tube Process (RCA/GE)
- 1964-1974
  Unstable Nb_3Sn Tape Magnets (RCA/GE/IGC)
- 1974-Present
  Relatively Stable Nb_3Sn Tape Magnets, but Restricted to Solenoid Geometry (IGC)

Filamentary↓

- 1970 Bronze Process (AERE, NRIM, BNL)
- 1973 Filamentary Nb_3Sn Magnet Conductors (AERE/Rutherford)
- 1974 Laboratory Solenoids of FM Nb_3Sn (RL)
- 1978-1984 Large Fusion FM Nb_3Sn Magnets (LLL - W)
- No Accelerator-Quality Nb_3Sn Magnets Yet

When niobium-tin came along as a superconductor, in 1960, it came along as a prototype conductor, and the concept electrified the applied scientists thinking of applications for superconductivity. In a couple of years a better process was developed and this drove out the original tube process that was used in 1960. RCA, General Electric, Bell, and other big companies were involved in this. From 1964 to 1974, large numbers of magnets were made. Most of these were not particularly good. One has to credit Intermagnetics General for finally taking hold in about 1974 and learning how to make reliable niobium-tin magnets using tape. Filamentary niobium-tin is much more useful for devices, and the basic science of how to make it was worked out in 1970. The Rutherford Laboratory group that I led was the first to come up with magnet conductors in 1973. Pretty good solenoids were made from this in 1974. Unfortunately, defeating the problems of tape actually set back Nb_3Sn technology for a few years. It has taken a number of years to get laboratory solenoids of filamentary niobium-tin. There has been a large fusion program in niobium-tin. That has indeed produced some big coils, but, because of the great complexity of accelerator magnets, no successful accelerator-quality niobium-tin magnets have yet been built.

If we try to analyze why Nb_3Sn magnets have not been more widely applied, several reasons become apparent. One important reason is that it is much simpler to use a very strong and tough material like Nb-Ti, that winds in almost every way like a strong copper, than to wind a brittle material or one that must be carefully insulated in order to spend one or two weeks at 650 to 700°C to allow the Nb_3Sn phase to form. A second reason follows the first: The added complexity of the reaction process and the brittleness have discouraged the formation of groups of critical size with long-term commitment to Nb_3Sn magnet construction. A sad aspect of this failure is that *all* superconductors with properties better than Nb-Ti are brittle. Magnet builders must face the brittleness problem if they want magnets of >10 to 12 T, whatever superconductor they use.

The optimism that I spoke of a moment ago with respect to oxide magnet construction comes from the certainty that, if oxide conductors having a reasonable J_c at 77°K are produced, then hundreds of groups around the world will start to construct magnets. If attractive motors and other electro-technical devices can be designed with oxide conductors, then the devices will certainly be built, and this should be to the long-term advantage of all high-field superconducting applications.

6 Closure Note (December 1988)

This talk was prepared in May 1987 at a time of great excitement in superconductivity, both for applications using traditional metallic superconductors and for all sorts of potential applications using high T_c superconductors (HTSC). At that time, some believed that helium-temperature superconductors might rapidly be supplanted by HTSC. This does not now seem likely to occur in the near future.

High T_c work in 1988 has been stimulated by many factors: for example, the award of the 1987 Nobel Physics Prize to Bednorz and Müller, the discovery of even higher T_c Bi- and Ti-based cuprates ($T_c \sim 120$ K), and a large federal, industrial, and political interest in HTSC. The fundamental physics (there is still no agreed explanation for the HTSC phenomenon) and basic materials science are still fascinating and in large measure quite uncertain. These materials pose so many challenges to our basic understanding of materials that understanding their science is bound to be of immense value.

Replacement of metallic helium-temperature superconductors, such as Nb-Ti or Nb_3Sn, by HTSC does not look at all feasible at present. The low transport J_c, granular behavior that I described in May 1987 is now widely recognized. Thin films can apparently be better, but much characterization of thin films is needed before we can assess whether the best thin-film characteristics can be transferred to bulk conductors. Disconnected behavior, at least in tesla fields at 77°K, continues to preclude even demonstration magnets. With so much unknown about this exciting class of materials, however, we still have no reason for pessimism for applications in the longer term. Superconductivity has again become a very exciting field and this is stimulating scientists and engineers all over the world to enter the field.

RADIO − FREQUENCY ACCELERATION

Mario Puglisi

Sincrotrone Trieste

TABLE OF CONTENTS

Introduction		1394
Organization of the Lectures		1395
1.	Basic Concepts	1395
2.	The Transit-Time Effect	1397
3.	The Skin Effect and Surface Resistance	1400
	Important Observation	1406
4.	The Resonant Cavity	1407
5.	Physical Considerations	1411
6.	The Accelerating Structure	1413
7.	The Uniform Transmission Lines	1416
8.	Technical Considerations on the Lossless Transmission Lines	1422
9.	The Hybrid Circuits	1426
10.	The Wave Guides	1432
11.	Special Topics on Cavity Resonators	1441
	a. Physics of the Acceleration	1441
	b. The Quality Factor Q	1443
	c. The Shunt Impedance	1446
	d. The Equivalent Scheme	1448
	e. The Coupling to the Cavity	1451
	f. Technical Considerations	1456
	g. Deviations from the Predicted Behaviours	1459
12.	Linear Accelerators	1463
	a. Proton and Ion Linacs	1463
	b. Electron Acceleration	1468
Acknowledgements		1475
Bibliography		1475

RADIO-FREQUENCY ACCELERATION

Mario Puglisi

Sincrotrone Trieste

Introduction

In the early days of cyclic RF accelerators the number of particles orbiting inside the machine was so small that the RF accelerating system was the least expensive and the most simple part of the machine.

Less than twenty years later the situation was very much changed and now the opposite is true.

The intensity of the accelerated current is so high and the required performances are so _high_ that the RF accelerating systems are now a very sophisticated and expensive part of the whole machine.

Moreover, it should be added that the major limitations to the possibility of an orbital machine now only come from the RF system.

The same applies to linear accelerators where the machine itself is (as it was at the beginning) an RF device.

Again at the dawn of the electromagnetic accelerators the linear accelerator was a small RF system with a very limited task.

Now a tremendous effort is being made around the world to increase the fields and the intensity of the accelerated beam and this is generating a family of linear accelerators of increasing cost and complexity.

A significant concluding remark can be the following: many important industries well-rooted in the field of telecommunications have for many years been designing and constructing RF systems both for orbital machines and linear accelerators. Other firms already exist only for the design and the construction of RF systems and/or RF components for accelerators.

© 1989 American Institute of Physics

A significant increase of those activities is easily foreseen for the future.

Organization of the lectures

A volume of more than 500 pages should not be enough to cover a schematic presentation of the whole topic that, on the other hand, has been <u>treated</u> only in three lectures. For this reason and because of the modest limit given to this work our treatment will be based simply on drawings and schematic comments.

Long algebraic manipulations are deliberately omitted because after a first run through this paper the interested reader should study the subjects in specialized textbooks and original articles where any detail is analyzed both from the physical and mathematical points of view.

1. BASIC CONCEPTS

An accelerating gap is the volume between two metallic (good conductors) tubes with the same axis. If the tubes are connected to a dc voltage then the field on the axis can be as indicated in figure 1-1.
A negative particle already inside the "negative" electrode gains energy passing through the gap.

An electrostatic linear accelerator can be realized assembling many gaps in series and giving each gap an appropriate voltage.

This technique has many technical limits that become evident as soon as one realizes that the total energy available at the exit is equal to the charge of the particle times the voltage of the last gap as shown in figure 2-1.

The series of accelerating gaps can be excited with an AC voltage generator and an electromagnetic linac is realized as shown in figure 3-1.

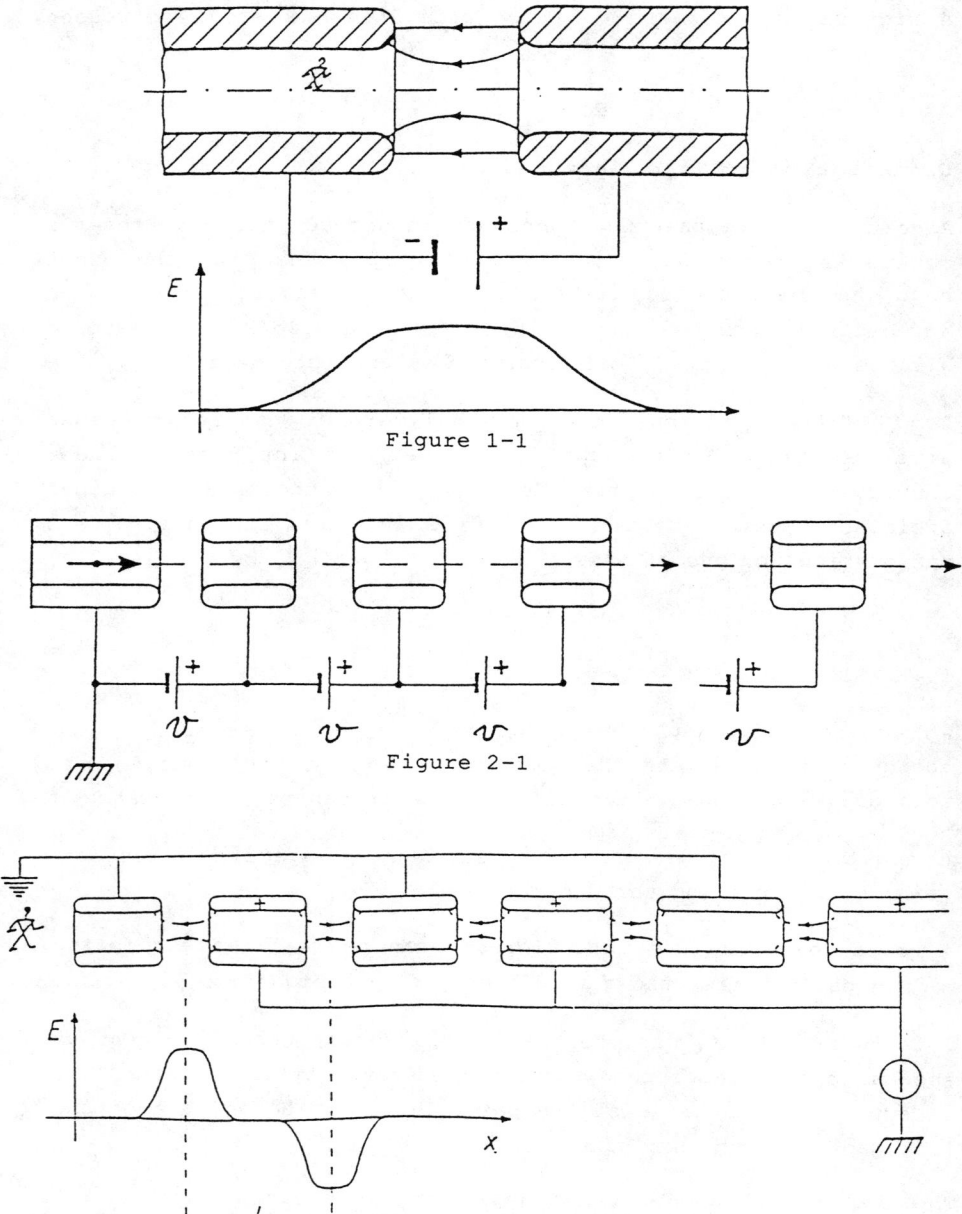

Figure 1-1

Figure 2-1

Figure 3-1

We assume that the accelerating voltage is sinusoidal and consequently the field inside each gap can be

$$E = E_0 \cos \frac{2\pi}{T} t \; ; \; E_0 = E_0(z,r)$$

while the field is negligible inside the electrodes.

If a particle that transits the gaps, with a normalized velocity $\beta = V_p/C$, has to be accelerated in each gap then the length L between the center of two adjacent gaps should be

$$L = V_p \cdot \frac{T}{2} = \frac{V_p}{C} \cdot \frac{TC}{2} = \beta \frac{\lambda}{2}$$

where λ is the free-space wavelength of the applied field. This condition (synchronism) reflects the fact that in the described situation the lines of force of the electric field can only enter or leave each pipe.

With this accelerator the final energy of the particles does not depend upon the voltage of the generator.

Two important considerations:

a) Strictly speaking we need only an alternating field. The sinusoidal form has very many merits and simplifies enormously the engineering of the problem.

b) The speed of the accelerated particles, the length of the machine and many other technical reasons suggest the use of very high frequencies. The frequencies used in the radio-TV communication systems are normally used for the accelerators.

For the above reasons the accelerating voltages (always sinusoidal) used in the electrodynamic machines are referred to as "The Radio Frequency Voltages" or more simply the RF.

2. THE TRANSIT-TIME EFFECT

A charged particle passing through a gap experiences a field that changes with the time and the position. This has an important effect on the energy gain.

Let us examine the effect due to the time-varying field.

Suppose that our reference plane is placed at the center of the gap and that

$$E(t) = E_0 \cos\left(\frac{2\pi}{T} t + \phi\right) \quad ; \quad E_0 = \text{constant}$$

is the field seen by the particle when it crosses the reference plane. (Observe that we have assumed uniform the field along the gap).

If g is the length of the gap then the voltage gain ΔV for a particle crossing the gap is

$$\Delta V = \int_{-g/2}^{+g/2} E(t) \, dz .$$

In general this is a difficult integral, but it becomes very elementary if we can assume $z = V_p t$ where V_p is the average speed of the particle.

This is a reasonable approximation for two reasons:

1. The energy gain is always very small if compared with the energy of the particle.

2. A relativistic particle has a speed always very near to c, where c is the speed of the light.

With the previous hypothesis we have

$$\Delta V = \int_{-g/2}^{+g/2} E_0 \cos\left(\frac{2\pi}{T} \cdot \frac{z}{V_p} + \phi\right) dz = E_0 g \cos\phi \frac{\sin \theta/2}{\theta/2}$$

where

$$\theta = 2\pi \frac{g}{V_p T} = \text{transit angle}.$$

A particle that could pass the gap in a time interval equal to zero (that is impossible) would have a voltage gain $V = E_0 g \cos\phi$.

It follows that

$$\tau = \frac{\Delta V}{V} = \frac{\sin \theta/2}{\theta/2}$$

is the reduction factor that we must introduce in order to take into account the fact that the particle crosses the gap in a finite time ($V = E_0 g \tau$ becomes the so-called useful voltage).

It should be noted that the majority of the authors define the shunt impedance of a gap taking into account the peak voltage that can be measured across the gap (according with the convention of General Electrical Engineering). It follows that

$$W = \frac{(Eg)^2}{2 R_{eq}}$$

is the power wasted for exciting, with the voltage Eg, a gap with physical shunt impedance equal to R_{eq}.

Some people assume that if the same power is wasted for creating the useful voltage $Eg\tau$ then we should define, accordingly, a new shunt impedance of the gap:

$$W = \frac{(Eg\ \tau)^2}{2 R}.$$

Because the power is the same it follows that

$$R = R_{eq}\ \tau^2$$

is the transit-time-corrected shunt impedance.

We note that while for small angles the transit-time factor tends to one for angles near π we have $\tau = \sim 0.636$ and this reduces nearly by a factor 0.4 the transit-time-corrected shunt impedance.(Some people in the linac area define the shunt impedance with reference to the RMS voltage and consequently they omit the factor two). I personally disagree with those semantic adventures and strongly advise the reader to check, each time, the definition of the shunt impedance when he reads the original papers.

3. THE SKIN EFFECT AND SURFACE RESISTANCE

Because of the high value of the speed of light ($\sim 3.10^8$ m/s) the frequency of the accelerating voltages is normally very high and this, in turn, means that the wavelength of the accelerating fields becomes of the same order of magnitude as the mechanical sizes of the circuits used.

Consequently the "RF" circuits to be treated are mainly "distributed constant circuits" for which special mathematical techniques are to be applied.

An important simplification in the distributed circuits' treatment comes from the assumption that the RF currents flow only on the surface of the conductors (wires, planes, corrugated surfaces... etc) without passing through the solid metal as normally happens for the direct current.

The above assumption and its most relevant consequences can be justified by the following considerations.

<u>Assuming sinusoidal steady state</u> the components of the electromagnetic fields can be written as follows:

$$E_x = E_x(x,y,z) \; e^{j\omega t} \qquad \text{etc.}$$

The Maxwell equations for this state are

$$\nabla \cdot \overline{E} = \rho$$
$$\nabla \cdot \overline{B} = 0$$
$$\nabla \times \overline{H} = \overline{J} + j\omega\varepsilon_0 \overline{E}$$
$$\nabla \times \overline{E} = - j\omega\mu_0 \overline{H}$$

where ω is the angular frequency of the fields, ε_0 and μ_0 are the electric and the magnetic permittivity of the vacuum ($\varepsilon_0 = 8.8542 \; 10^{-12}$ Farad/m; $\mu = 4\pi \; 10^{-7}$ Henry/m).

Because we want to study the distribution of the current inside a conductor we assume $\rho=0$ (neutralization of the charges) and $J \neq 0$.

We take the curl of the two sides of the last equation and substitute the $\nabla \times \bar{H}$ from the third. The double curl can be expanded as follows: $\nabla \times \nabla \times \bar{E} = \nabla (\nabla \cdot \bar{E}) - \nabla^2 \bar{E}$. Taking into account that $\rho=0$ we obtain

$$\nabla^2 \bar{E} + \omega^2 \varepsilon \mu \bar{E} = j \omega \mu \bar{J}.$$

Assuming that the Ohm law is valid ($\bar{J} = \sigma \bar{E}$ where σ is the conductivity) we can write

$$\nabla^2 \bar{J} + \omega^2 \varepsilon \mu \left(1 - j \frac{\sigma}{\omega \varepsilon}\right) \bar{J} = 0 \qquad (1-3)$$

where $\omega^2 \varepsilon \mu$ and $\sigma/\omega\varepsilon$ are respectively proportional to the displacement and to the conduction current.

It is very important to note that even at very high frequencies the conduction current remains dominant in a good conductor.

For instance $\sigma = 5.8 \; 10^7$ for copper. This means that, in that case, even for a free-space wavelength of one millimeter the conduction term is six orders of magnitude larger than the one due to the displacement.

We conclude that in the whole RF field we can neglect the displacement current inside a good conductor and the equation (1-3) becomes

$$\nabla^2 \bar{J} = +j \omega \mu \sigma \bar{J}. \qquad (2-3)$$

We assume the ideal case of a conductor limited by a plane surface and filling the whole emispace, as shown in figure 1-3.

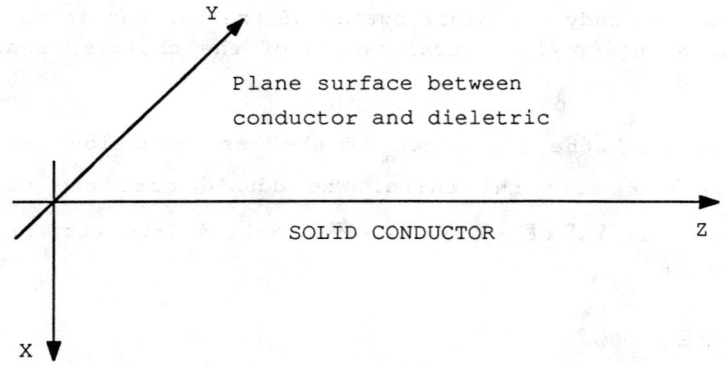

Figure 1-3

If we let $J_x = J_y = 0$ and no variations with z, equation (2-3) becomes

$$\frac{d^2 J_z}{dx^2} - j\omega\mu\sigma J_z = 0 \qquad (3-3)$$

because for x<0 the conductivity is <u>identically zero</u>; then the solution of (3-3) is

$$J_z = J_0 e^{-\alpha x} \; ; \; \alpha = \frac{1+j}{\sqrt{2}} \sqrt{\omega\mu\sigma} .$$

Following the current literature we write

$$\delta = \frac{1}{\sqrt{\pi f \mu \sigma}}$$

where δ is known as the penetration depth.

In fact it is evident that after ~ 3 penetration depths the current becomes negligible.

Integrating along x, and assuming y=1 (one unit of width) we obtain

$$J_z = \int_0^\infty J_0\, e^{-\frac{1+j}{\delta}x}\, dx = \frac{J_0 \delta}{1+j} \qquad (4-3)$$

and we observe that $J_0\delta$ has the dimensions of amp/meter.

From (4-3) it is possible to define the "impedance" of the plane surface as the ratio between the surface field and the total current crossing the infinite surface of width equal to one.

$$Z = \frac{E_{z0}}{J_z} = \frac{\frac{J_0}{\sigma}}{\frac{J_0 \delta}{1+j}} = \frac{1+j}{\delta \sigma} = R_s + j\omega L_s \qquad (5-3)$$

and it is clear that the above impedance is defined for unit length and unit width of a plane surface.

The above considerations are exact only if the previous hypotheses are verified. Nevertheless, the formulae for the distribution of the current and for the surface impedance are applicable every time the mechanical sizes of the conductor are much larger than the penetration depth.

The following table shows the values of δ and R_s for some conductors, technically pure, as function of the frequency.

Table 1-3.

	σ (mhos/m)	δ (m)	R_s (ohm)
Silver	6.17×10^7	$0.064/\sqrt{f}$	$2.52 \times 10^{-7}\sqrt{f}$
Copper	5.80×10^7	$0.066/\sqrt{f}$	$2.61 \times 10^{-7}\sqrt{f}$
Aluminum	3.72×10^7	$0.082/\sqrt{f}$	$3.28 \times 10^{-7}\sqrt{f}$
Solder	0.706×10^7	$0.185/\sqrt{f}$	$7.73 \times 10^{-7}\sqrt{f}$

We see, for instance, that at 16 kHz the δ reduces to half millimeter for copper, and 16 kHz is definitely a very low frequency in the field of accelerators.

The "small value of δ" and the general fact that the currents tend to go near the fields justify the name given to this phenomenon: the "skin effect". In fact due to the skin effect the RF currents flow pratically on the surface of the conductors and the simplifying hypothesis for the treatment of the distributed circuits is completely justified.

Now we ideally isolate an infinitely long cylinder, with square basis ($y_0=z_0=1$), inside the conductor, as shown in figure 2-3.

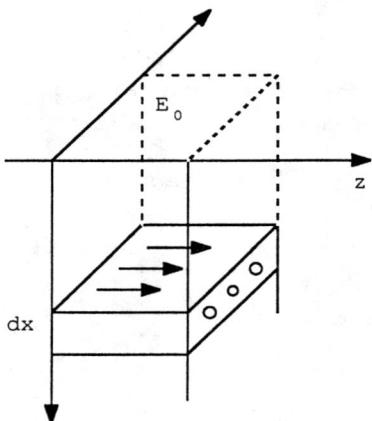

Figure 2-3

The modulus of the infinitesimal current flowing through the infinitesimal rectangle indicated in the cylinder is

$$|dI| = \sigma E_0 \, e^{-\frac{x}{\delta}} \, dx$$

and consequently the power wasted becomes

$$dW = \frac{1}{2} \frac{1}{\sigma\, dx} \left[\sigma E_0 e^{-\frac{x}{\delta}} dx \right]^2 \quad ;$$

simplifying and integrating from zero to infinity we obtain

$$W = \frac{\sigma E_0^2}{2} * \frac{\delta}{2}$$

which is the total power that is wasted inside the infinitely long cylinder if E_0 is the electric field on the reference cross-section.

Now we can think that the total current moved by the electric field induces the same losses passing through some equivalent resistance R_{eq}.

In this case we should have

$$\frac{1}{2} R_{eq} \left| \frac{\sigma \delta E_0}{1+j} \right|^2 = \frac{\sigma E_0^2}{2} \frac{\delta}{2}$$

and we obtain the very important result

$$R_{eq} = \frac{1}{\sigma \delta} = R_s \quad ;$$

the obvious extension is to consider that the cross-section of the cylinder has length equal to L (in the z direction) and width equal to W (in the y direction) and we obtain

$$R_{eq} = R_s \frac{L}{W} .$$

The pratical case involves cylindrical conductors with any cross-section.

In that case the RF equivalent resistance becomes

$$R_{eq} = R_s \frac{\text{LENGTH}}{\text{PERIMETER}}$$

It should be observed that the above formula is valid if:

a) - The current is uniform along the length considered.

b) - The thickness of the conductor is much larger than δ.

c) - The curvature radii of the cross-section are larger than δ.

From this point on we consider a general surface which supports, on top, an RF linear density current

$$J = \sigma E_0(x,y,z) = J(x,y,z) \, ;$$

as a natural extension of the previous consideration we calculate the power losses with the formula

$$W = \frac{1}{2} R_s \int_s J^2(x,y,z) \, ds \, .$$

Important Observation

The distributed constant circuits that we have to treat are normally made of very good conductors and this means that the power wasted is small if compared with that handled by the circuit.

In those cases we consider that the circuit is lossless at least to the first approximation.

This introduces a remarkable simplification in the mathematical treatments.

When the solution is found for the ideal case then the surface currents ($\overline{J} = \overline{n} \times \overline{H}$) are completely determined.

At this point the procedure already described permits the calculation of the power that we must spend for maintaining the fields.

4. THE RESONANT CAVITY

A simple gap or a series of gaps cannot be excited just by connecting the pipes to a source of RF voltage. The losses, the radiation and the complication of the system would render impossible the construction of the accelerator.

Nevertheless the problem can be solved in a very elegant way using the cavity resonator.

While a full treatment of the theory of the cavity resonators is not undertaken (because it is really far beyond the purpose of these lectures) an effort is made to introduce the beginner to the physics of this very important device.

As a first step we will show that an alternating electrical field can be "contained" inside a cylindrical empty volume limited by perfectly conducting walls (figure 1-4).

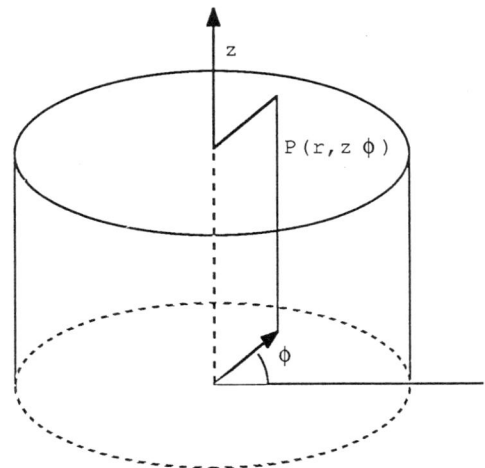

Figure 1-4

Assuming that $\rho = J = 0$ inside the volume, from the Maxwell equation we obtain

$$\nabla^2 \overline{E} + \omega^2 \varepsilon_0 \mu_0 \overline{E} = 0.$$

That is the well-known vector wave equation for the lossless case.

We will solve the above equation with the boundary conditions dictated by the perfectly conducting walls of the pillbox and for a particular case.

In fact we should look for a solution that allows the acceleration of a beam injected along the z axis of the structure.

If the beam is to be accelerated then inside the box an electric field should exist with "at least" one component along the z axis. A transverse electrical field should be avoided, if possible, and a field "constant" with z should be preferred.

Then we look for an "optimal solution" that could be

$$E_r = E_\phi = 0 \; ; \quad E_z \neq 0 \; ; \quad \frac{\partial E_z}{\partial z} = 0$$

where E_z, from now on indicated with E, is a function of r and ϕ.

With the above hypotheses the vector wave equation reduces to the following scalar equation:

$$\frac{\partial^2 E}{\partial r^2} + \frac{1}{r}\frac{\partial E}{\partial r} + \frac{1}{r^2}\frac{\partial^2 E}{\partial \phi^2} + K^2 E = 0$$

where

$$K^2 = \omega^2 \varepsilon_0 \mu_0 = \left(\frac{2\pi}{\lambda}\right)^2$$

depends upon the free wavelength of the field.

Assuming $E = R(r) \cdot y(\phi)$, after substitution and elementary algebraic manipulation we obtain

$$\frac{r^2}{R(r)} \frac{\partial^2 R(r)}{\partial r^2} + \frac{r}{R(r)} \frac{\partial R(r)}{\partial r} + K^2 r^2 = - \frac{1}{y(\phi)} \frac{\partial^2 y(\phi)}{\partial \phi^2}.$$

The left-hand side is only function of r, while the right-hand side is only function of ϕ.

It follows that both sides must be equal to a constant; the same constant.

If we call this constant υ^2 we can write

$$\begin{cases} \dfrac{\partial^2 R(r)}{\partial r^2} + \dfrac{1}{r} \dfrac{\partial R(r)}{\partial r} + \left(K^2 - \dfrac{\upsilon^2}{r^2}\right) R(r) = 0 \\ \dfrac{\partial^2 y(\phi)}{\partial \phi^2} = - \upsilon^2 \phi. \end{cases}$$

The constant υ^2 is not totally arbitrary. With a full rotation around the z axis we should find the same field. This means that υ^2 must be integer.

With the above physical condition the solution is as follows:

$$E = E_0 \, J_\upsilon \, (Kr) \cos \upsilon \phi$$

where J_υ is the Bessel function of the first kind of order υ. The Neuman function cannot be a solution because the z axis (r=0) is part of the domain.

Now E (tangent) should vanish on a conducting surface: consequently the solution will fit the boundary condition, at r=a, if and only if

$$Ka = P_{\upsilon l} \quad \text{or} \quad f = \frac{c}{2\pi a} P_{\upsilon l}$$

where $P_{\upsilon l}$ indicates the lth zero of the Bessel function of the first kind of order υ.

So we have seen that "fortunately" an electromagnetic field of the indicated characteristics can be contained in the ideal volume already seen.

It is important to note that the doubly infinite set of possible values for f (the proper values) define a doubly infinite set of field distributions. Each one is named as a resonant mode or simply as a mode.

If we set $\upsilon=0$; $l=1$ we obtain the so-called "FUNDAMENTAL MODE" and the fields are as follows:

$$\begin{cases} E_z = E_0 \, J_0 \left(\frac{P_{01}}{a} r \right) \\ H_\phi = j \, \frac{E_0}{\eta} \, J_1 \left(\frac{P_{01}}{a} r \right) \end{cases}$$

where $P_{01} = 2.405$ is the first zero of J_0 and $\eta = \sqrt{\frac{\mu_0}{\varepsilon_0}} = 377$ is the impedance of the free space.

The proper value of the frequency for the fundamental mode is consequently

$$f = \frac{3.10^8 * 2.405}{6.28} \cdot \frac{1}{a} = \frac{114.8}{a} \quad \text{in MHz}$$

or $\lambda = 2.61a$ where λ is the free-space wavelength.

It is interesting to note that in this particular situation the proper values of the frequency do not depend upon the z dimension (length) of the box. This was to be expected because we introduced the condition $\partial E_z / \partial z = 0$.

5. PHYSICAL CONSIDERATIONS

The cylindrical volume limited by perfectly conducting walls can contain the electromagnetic field in many different modes. Each mode can exist only if excited at its own frequency. That is known as the resonant frequency of the mode. Removing the previous hypotheses, other modes, described with much more complicated formulae, can be obtained. It is possible to show that for any volume limited by perfectly conducting walls there exist an infinite number of solutions of the wave equation. Each solution (a mode of oscillation) has its own resonant frequency.

For the above reason an empty volume limited by perfectly conducting walls is known as an ideal cavity resonator (other names are: cavity; resonant cavity; resonator; cavity resonator; hollow cavity...). As already said the theory of the resonant cavities is very complicated, but for our purposes it is sufficient to know that many powerful computer programs now exist that allow the determination of both the resonant frequency and the field distribution for the cavities with axial symmetry (super-fish, lala, oscar-2D). For three-dimensional cavities some programs are in fast development.

It is evident that the fields remain trapped into the cavity because the cavity is limited by perfectly conducting walls. This is equivalent to saying that the cavities do not radiate and do not waste the stored energy.

Moreover, it is interesting to note that for each resonant mode while the total energy must remain the same during each period there is a full exchange from magnetic to electric energy [exactly as it happens in a lossless tuned circuit].

At this point it is evident that the argument can be inverted and the resonant frequency that pertains to each mode is the one for which the "electrical energy is equal to the magnetic one".

This concept, useful in many cases, is now applied to a really technical example. Consider the cavities shown in figure 1-5 a and b.

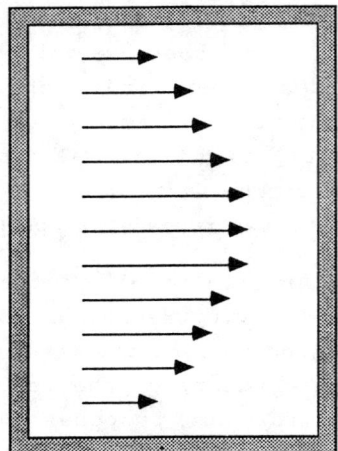 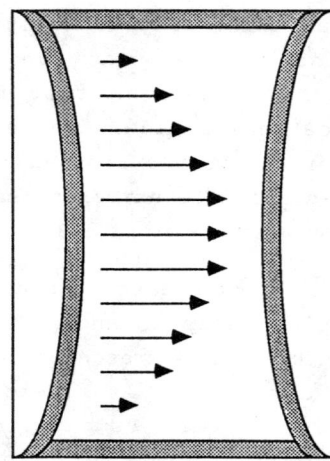

Figure 1-5

The "b" cavity can be thought of as derived from the "a" one with a "small" deformation of the circular faces. In view of the small deformation, the mode (i.e. the shape of the fields) will not be changed but, due to the "capacitive" perturbation, the electrical energy tends to be sligtly increased. The resonant frequency changes and the energy balance is reset for a frequency slightly below the original one.

We pass from "ideal" to "real" cavities, introducing two families of perturbations:

a. The mechanical boundaries of the cavity are realized with non-perfect conductors and non-perfect dielectrics.

b. The continuity of the mechanical boundary of the cavity is interrupted by the holes that are "needed" for coupling the cavity to the outside.

The perturbations introduce energy losses and many other "unwanted" effects such as "radiation of energy", anomalous concentration of fields and currents, terminal effects.

These effects are "small effects" as long as the causes are small perturbations. Consequently we expect that the modes are preserved and that some shifts on the resonant frequencies are introduced.

Much more important is the fact that, due to the losses, we must waste some power for maintaining the fields in the cavity.

Summarizing we can say that a physical cylindrical cavity can contain the required field at the expense of some power.

Moreover, we can deliberately introduce an important perturbation in the geometry just by creating a "gap" inside the cavity.

Apart from the small amount of energy that can be radiated the main effect of the introduction of the gap is to lower the resonant frequency of the fundamental mode of the resonator without changing the mode.

6. THE ACCELERATING STRUCTURE

A sketch of a single-gap accelerating cavity is shown in figure 1-6 a and b.

The "gap" is now fed by the resonant cavity, and the power needed for keeping the cavity excited is fed to the whole device via a coupling loop.

The situation can be repeated and a longer accelerating structure is built up, as shown in figure 2-6, where many gaps are energized, for instance, with a single source of RF power.

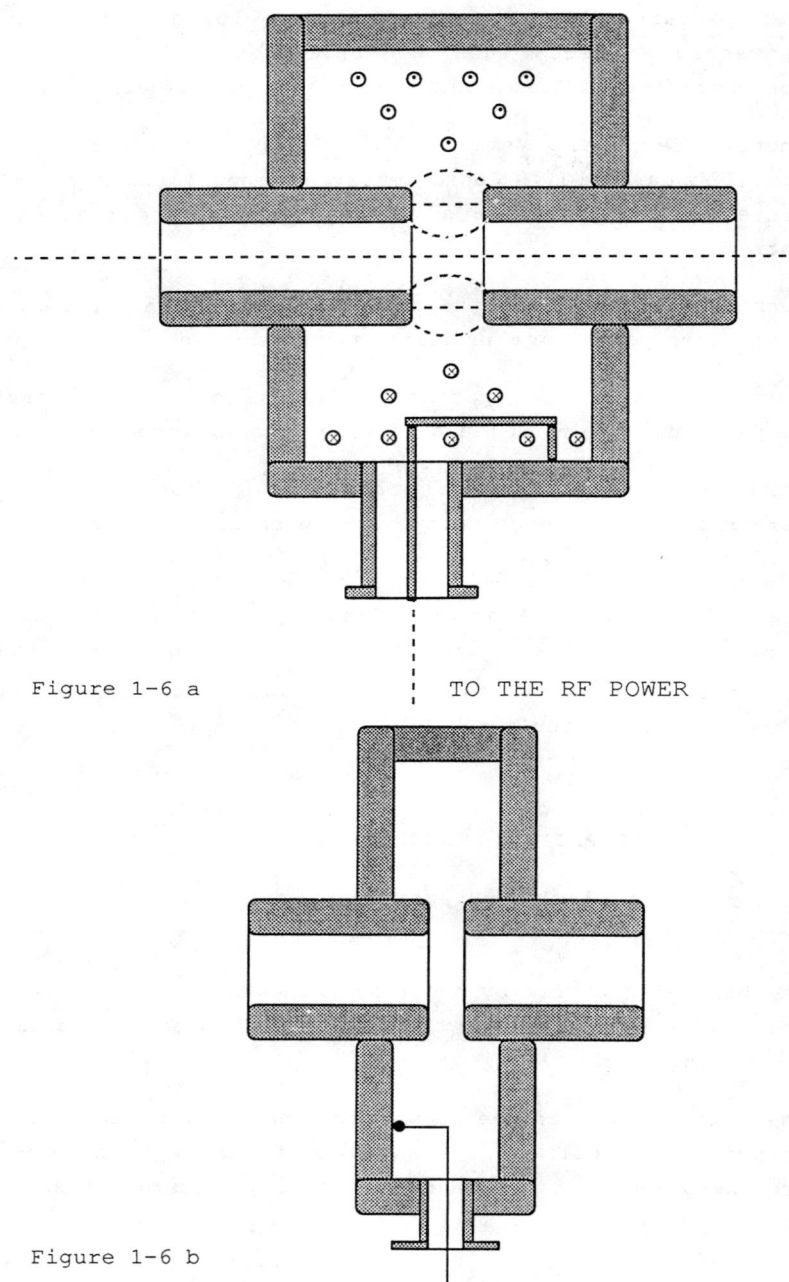

Figure 1-6 a

TO THE RF POWER

Figure 1-6 b

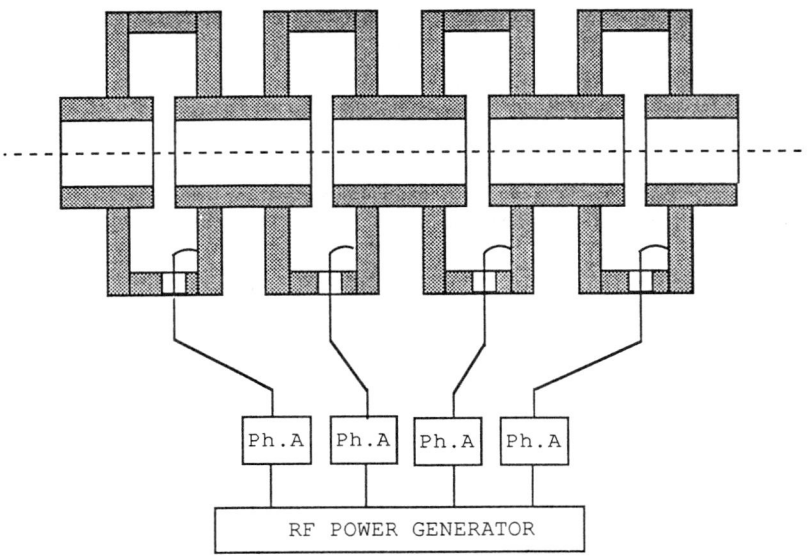

Figure 2-6

If all the cavities, and consequently the gaps, are excited in phase then the distance "L" between the centers of two adjacent gaps should be L=βλ and this "mode" of operation is named the "0" mode in order to signify that the phase shift between adjacent cavities is zero.

If, on the contrary, the phase shift between adjacent gaps is equal to π then the distance L becomes shorter.

In fact if a travelling particle must see the same field at each gap crossing then $L = \frac{\beta\lambda}{2}$ is the condition that should be applied on the lengths, when the operating mode is π, as we have already seen in the case of figure 3-1.

In any case a very accurate phasing of each cavity is needed and for this reason the phase adjusters (Ph.A) are explicitly shown in the drawing as an important part of the whole structure.

From a mechanical point of view the structure shown on figure 2-6 is unnecessarily complicated and many other "more attractive" solutions have been invented, but this topic will follow some more information on the resonant cavities.

7. THE UNIFORM TRANSMISSION LINES

As the cavity resonator the uniform transmission line is a second example of a "distributed constant circuit".

An element of transmission line can be used as:

a. A tool for transmitting "RF" energy.
b. A circuit element capable of handling very high RF power.
c. A cavity resonator.

A uniform transmission line is a transmission line where:

1. The axis is straight.
2. The physical properties of the device are independent of the abscissa.

The cross-section of a uniform transmission line can have very different shapes. In figure 1-7 four fundamental shapes are indicated.

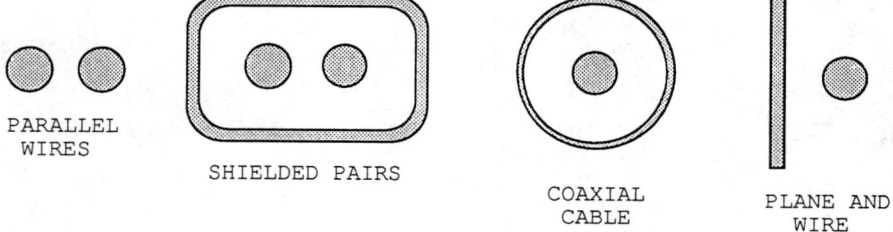

Figure 1-7

The electrical behaviour of a uniform transmission line is completely described by four parameters R, L, C, G which are respectively named RESISTANCE, INDUCTANCE, CAPACITY and CONDUCTANCE per METER.

The above parameters can be defined by energetic considerations. If a piece of line terminated on a load is excited with a voltage whose wavelength is much larger than the physical length of the line then the amplitude of the voltage and of the current can be considered as "constant" along the sample; and then the four quantities---power loss W_R due to surface resistance, power loss W_G due to the conductance between the conductors, & maximum energy stored U_E and U_H due to electric and magnetic field around the conductors---can be calculated or measured.

If l is the length of the sample excited with the voltage V and the current I we have

$$W_R = \frac{1}{2}(lR)I^2 \quad ; \quad W_G = \frac{1}{2}(lG)V^2 \quad ;$$

$$U_E = \frac{1}{2}(lC)V^2 \quad ; \quad U_H = \frac{1}{2}(lL)I^2 .$$

Solving the above equation we obtain the characteristic parameters of the considered line.

The rather intuitive procedure just described should be completed with two important observations:

a. The parameters R, L, C, G depend upon the conductors and the dielectric which form the line and upon the frequency. (For instance the skin effect modifies the distribution of the fields).

b. The parameters R, L, C, G, derived from energy considerations coincide with the corresponding ones that are obtained, in a more rigorous way, using the Maxwell equation.

The problem of the uniform transmission line can be solved, in general, with the formalism of the classical electromagnetic theory. In our case a simpler treatment will be followed because it gives a deeper physical insight of the phenomena.

Consider a piece of line of length dx and the infinitesimal fourpole that can be considered its equivalent scheme (figure 2-7)

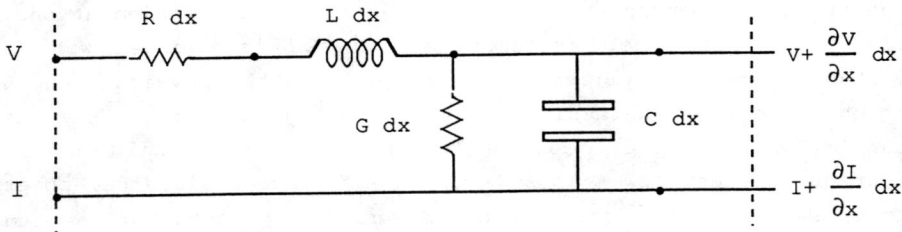

Figure 2-7

where V and I are now functions of the abscissa x and the time t.

Neglecting the second-order terms the equilibrium equations are as follows:

$$\begin{cases} V - (V + \frac{\partial V}{\partial x} dx) = (R + L\frac{\partial}{\partial t}) I \, dx \\ I - (I + \frac{\partial I}{\partial x} dx) = (G + C\frac{\partial}{\partial t}) V \, dx. \end{cases}$$

Simplifying and taking the Laplace transform of the two equations we obtain

$$\begin{cases} -\frac{\partial V}{\partial x} = (R + SL) I \\ -\frac{\partial I}{\partial x} = (G + SC) V. \end{cases} \quad (1.7)$$

As already seen we assume that the line is lossless (R=G=0) and once the solution is found for the ideal case the effect of the losses will be taken into account as a perturbation.

In this case the solution of Eq. (1-7) becomes straightforward and we obtain

$$\begin{cases} V(x) = A_1 e^{-s\sqrt{LC}x} + A_2 e^{s\sqrt{LC}x} \\ I(x) = \dfrac{A_1}{\sqrt{\dfrac{L}{C}}} e^{-s\sqrt{LC}x} - \dfrac{A_2}{\sqrt{\dfrac{L}{C}}} e^{s\sqrt{LC}x} \end{cases} \qquad (2\text{-}7)$$

where:

1. A_1 and A_2 are two integration constants (independent of x) that depend upon the parameter of the line (length, L and C) and upon the boundary conditions (the impedances) at the ends of the line.

2. The functions of x are only exponentials as $EXP[\pm s\sqrt{LC}\ x]$. This indicates propagation without attenuation of waves of voltage and waves of current.

3. Because $\sqrt{L/C}$, the characteristic impedance, is a constant then it follows that the incident wave of voltage has the same shape of the incident wave of current, and the same happens for the reflected waves.

4. Due to the presence of the minus sign in the equation of the current at those points where the total voltage is maximum, there the total current reaches the minimum.

As a very useful application consider, for example, a uniform transmission line inserted between a voltage generator V(s), with output impedance ρ, and an impedance Z (figure 3-7).

Figure 3-7

Let $\alpha=s\sqrt{LC}$; $Z_0=\sqrt{L/C}$ and l the length of the line.

For $x=0$ the voltage across the line should be $V(0)=V(s)-\rho I$, that is

$$A_1+A_2 = V(s)-\rho\left(\frac{A_1}{Z_0} - \frac{A_2}{Z_0}\right); \qquad (3-7)$$

vice versa for $x=l$ the ratio $V(l)/I(l)$ should be equal to Z and we write a second condition

$$\frac{A_1 e^{-\alpha l} + A_2 e^{\alpha l}}{\frac{A_1}{Z_0}e^{-\alpha l} - \frac{A_2}{Z_0}e^{\alpha l}} = Z. \qquad (4-7)$$

Making a system of the two equations (3-7) and (4-7) and solving we obtain the integration constants A_1 and A_2 that should be substituted into equation (2.7).

At this point the reader should be aware of the fact that A_1 and A_2 normally result in fractions where the numerator and the denominator contain exponentials.

With rather simple algebraic manipulations it is <u>always</u> possible to develop the final result in a series (normally infinite) of power of x. Apart from the zero-order terms, each power accounts for the effect of the incident and reflected wave.

The previous example is a very important one and with a proper choice of ρ and Z any transient behaviour of the lossless transmission lines can be calculated.

The calculation of the steady-state behaviour of a lossless transmission line is more simple and with the substitution $s=j\omega$ in equation (2.7) we obtain the considered case

$$\begin{cases} V(x) = A_1 e^{-j\beta x} + A_2 e^{j\beta x} \\ I(x) = \dfrac{A_1}{Z_0} e^{-j\beta x} - \dfrac{A_2}{Z_0} e^{j\beta x} \end{cases} \qquad (5.7)$$

where $\beta = \omega\sqrt{LC} = 2\pi f\sqrt{\varepsilon\mu} = \frac{2\pi}{\lambda}$ is the phase constant and z_0 is the characteristic impedance.

Normally the dielectric around the line is air or a material with very low electrical permittivity. It follows that

$$\frac{1}{\sqrt{LC}} \leq c \quad ; \quad (c = 3 \, 10^8 \text{ m/s}).$$

As an example consider a coaxial transmission line where R_1 is the outer radius of the inner conductor and R_2 is the inner radius of the outer braid (the penetration depth is zero).

Then the parameters are

$$L = \frac{\mu}{2\pi} \ln \frac{R_2}{R_1} \quad ; \quad C = \frac{2\pi\varepsilon}{\ln \frac{R_2}{R_1}}$$

and it follows that

$$\beta = \omega\sqrt{LC} = \omega\sqrt{\varepsilon\mu}; \quad z_0 = \frac{1}{2\pi}\sqrt{\frac{\mu}{\varepsilon}} \ln \frac{R_2}{R_1} \; ; \text{ if } \varepsilon=\varepsilon_0 \text{ and } \mu=\mu_0 \text{ then } \sqrt{\varepsilon_0\mu_0} = \frac{1}{c} \text{ and}$$

$z_0 = 377 \, \Omega$.

Consider the incident wave $V^+ = A_1 e^{-j\beta x}$ where A_1 is a complex constant $A_1 = a_1 e^{j\phi_1}$.

Returning to the time domain we should write

$$V^+(x,t) = \text{Re}\{a_1 e^{j\phi_1} \cdot e^{-j\beta x} \cdot e^{j\omega t}\}$$

$$= a_1 \cos[\omega t + \phi_1 - \beta x]$$

because Re means "Real part of".

The argument of the cosine is $\psi = \omega t - \beta x + \phi_1$; if ψ remains constant the function does not change. Constant ψ means that

$$\frac{\partial \psi}{\partial t} = 0 = \omega - \beta \dot{x} \; ;$$

in this case, the shape of the wave is preserved for an ideal observer moving with a speed $x=V_p=\frac{\omega}{\beta}$, the so-called phase velocity.

The propagation already described is the so-called transverse electromagnetic propagation (TEM) because the fields E and H are both normal to the axis of the line, that is, to the propagation ray.

At this point we will complete the theory with some technical details because the circuits realized with elements of lossless transmission lines play a very important role in all the RF applications. For instance we will see that an important class of cavity resonators comes directly from transmission line elements.

8. TECHNICAL CONSIDERATIONS ON THE LOSSLESS TRANSMISSION LINES

Consider the situation indicated in figure 1-8 and suppose that at the origin x=0 we have the generator while the load Z_L is inserted at x=l, that is, at the end of the line.

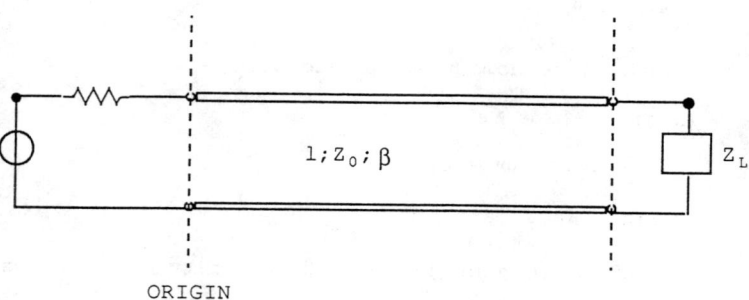

ORIGIN

Figure 1-8

Let V_L and I_L be the voltage and the current at the end of the line (two boundary conditions).

Substituting into the general equations we obtain

$$\begin{cases} V_L = A_1 e^{-j\beta l} + A_2 e^{+j\beta l} \\ Z_0 I_L = A_1 e^{-j\beta l} - A_2 e^{+j\beta l} \end{cases}$$

from which

$$A_1 = \frac{V_L + Z_0 I_L}{2} e^{j\beta l} \; ; \; A_2 = \frac{V_L - Z_0 I_L}{2} e^{-j\beta l} ; \quad (1-8)$$

substituting into (5-7) and rearranging we obtain

$$\begin{cases} V(x) = V_L \cos\beta(l-x) - jZ_0 I_L \sin\beta(l-x) \\ I(x) = I_L \cos\beta(l-x) - j\frac{V_L}{Z_0} \sin\beta(l-x). \end{cases} \quad (2-8)$$

In the practical applications it is convenient to have the load at the origin and the generator at the end of the line. This can be obtained by setting $z = x - l$; then Eq. (2-8) assumes the so-called canonical form

$$\begin{cases} V(z) = V_L \cos\beta z + jZ_0 I_L \sin\beta z \\ I(z) = I_L \cos\beta z + j\frac{V_L}{Z_0} \sin\beta z \end{cases} \quad (3-8)$$

where, as already said, the load is placed at the origin $z=0$ and the generator (or the first discontinuity) at $z=l$.

The ratio $V(l)/I(l)$ indicates the input impedance of an element of line of length l which is terminated on a load Z_1.

A very simple calculation shows

$$Z(1) = Z_0 \frac{Z_L + j Z_0 \tan\beta l}{Z_0 + j Z_L \tan\beta l} \qquad (4-8)$$

A careful examination of (4-8) shows:

a) In general an element of line acts as an impedance transformer.

b) If the load $Z_L = Z_0$ the input impedance is always equal to Z_0. This is the so-called matched condition. The line is matched to the load.

c) The load Z_L is a pure reactance $Z_L = jx$. In this case the input impedance will be, in general, a pure reactance, but for particular frequencies the input impedance can be zero and for others can be infinite.

d) The load is infinite (open end). In that case the input impedance is

$$Z_i = -j \frac{Z_0}{\tan\beta l}$$

and obviously depends upon the frequency. It is important to note that if we choose $\beta l = \pi/2$ (that means $l = \lambda/4$) then the input impedance becomes exactly zero. This property is fully exploited when in a part of the circuit the DC component must be blocked and the AC component must pass undisturbed.

It is evident that we are considering a "resonant phenomenon" that by its very nature is technically verified only inside a narrow band of frequency. Nevertheless, it is a very important one. For instance it is widely used in the construction of the high frequency power amplifier. (A termination with an infinite load is obviously impossible. A small stray capacity is always present and in some cases it is added on purpose. The case c takes into account this possibility).

e) The load is a short circuit. In this very important case the input impedance becomes

$$Z_i = j Z_0 \tan\beta l \qquad (5-8)$$

and we see that for $\beta l = n\frac{\pi}{2}$; n=odd the input impedance becomes infinite and the short-circuited element of line becomes a resonator (the so-called quarter-wavelength resonator). We will come back to this topic.

We now consider the physical phenomenon of the reflection at the ends of a transmission line.

For this purpose we expand the trigonometric functions that appear in the first of equation (3-8) and collecting the phasors we obtain

$$V(z) = \frac{V_L - Z_0 I_L}{2} e^{-j\beta z} + \frac{V_L + Z_0 I_L}{2} e^{+j\beta z}.$$

Again the voltage at each point of the line is given by the sum of two migrating waves. The one going in the positive direction of the z axis [with phasor $\exp(-j\beta z)$] has amplitude $A_R = (V_L - Z_0 I_L)/2$ and it is really the reflected wave because it propagates from the load towards the generator. The second wave, which has amplitude $A_i = (V_L + Z_0 I_L)/2$ is really the incident one because it goes from the generator toward the load.

From the above considerations we define the voltage reflection coefficent

$$\rho = \frac{A_r}{A_i} = \frac{Z_L - Z_0}{Z_L + Z_0} \qquad (6-8)$$

and analogous considerations can be made for the "generator end" of the line. As already seen if the load is equal to Z_0 then the line is matched at the end. This is confirmed by Eq. (6-8) because if $Z_L = Z_0$ then $\rho = 0$.

It is evident that the incident wave interacts with the reflected one and the amplitude of the voltage will vary periodically along the line going through maxima and minima:

$$V_{max} = |A_i| + |A_R| \quad ; \quad V_{min} = |A_i| - |A_R|.$$

Analogous considerations can be made for the current and we obtain

$$I_{max} = \frac{V_{max}}{Z_0} \;;\quad I_{min} = \frac{V_{min}}{Z_0}\;.$$

In general the interaction between two waves going in opposite directions creates standing waves. In particular when $|\rho|=1$ we have purely standing waves on the line and the minimum amplitude of the voltage and of the current is zero (nodal points of the voltage and nodal point of the current).

The voltage standing-wave ratio (WSVR) is defined as follows:

$$S = \frac{V_{max}}{V_{min}} = \frac{1+|\rho|}{1-|\rho|} \;;\quad \rho = \frac{S-1}{S+1}\;;$$

this parameter is very important in the practical applications because, being easily measurable, it permits a continuous monitoring of the match of the line to the load. Besides this information the standing-wave ratio indicates the maximum and the minimum value of the impedance along the line because

$$Z_{max} = S Z_0 \;;\quad Z_{min} = Z_0/S.$$

It should be emphasized that all the previous considerations are exact only for a lossless transmission line. On the other hand ~50 years of experience have proved that the above results are "technically acceptable" in the largest majority of pratical cases.

9. THE HYBRID CIRCUITS

Circuit elements with mechanical sizes smaller than $\frac{1}{50}\lambda$ ($\Delta\varphi \leq 4\pi\ 10^{-2}$) are normally treated as lumped-constant elements while electrical devices with at least one mechanical size larger than $\lambda/10$ ($\Delta\varphi \geq \pi/5$) must be definitely treated as distributed-constant elements.

The most suitable treatment for circuits of intermediate sizes has to be chosen "each time" in view of the particular problem to be solved.

Electrical circuits made of lumped- and distributed-constant elements are known as "hybrid circuits".

A pill-box cavity with an accelerating gap is a good example of hybrid circuit.

Consider the scheme indicated in figure (1-9).

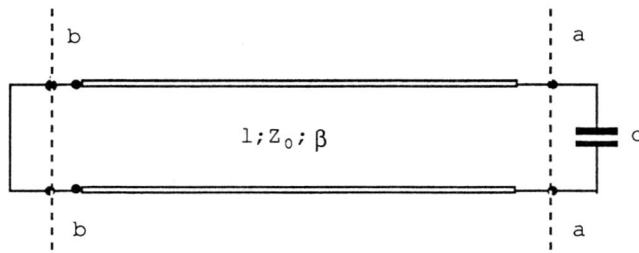

Figure 1-9

An element of uniform lossless transmission line is shorted at one end (terminals bb) and is connected to a capacitor "c" at the other (terminals aa).

The admittance between the terminals aa becomes

$$Y_{aa} = j\omega c + \frac{1}{jZ_0 \tan\frac{\omega l}{V_p}} \qquad (1-9)$$

where V_p is the phase velocity along the line and ω the angular frequency. The admittance becomes zero (parallel resonance) if

$$l = \frac{V_p}{\omega} \text{ATAN}(1/\omega c Z_0). \qquad (2-9)$$

This means that a sinusoidal voltage with frequency $f=\omega/2\pi$ oscillates, forever, between the terminals aa if and only if Eq. (1-9) is verified.

The physical device shown in figure (2-9) is the hybrid circuit whose scheme (for the first TEM mode) is shown in figure (1-9).

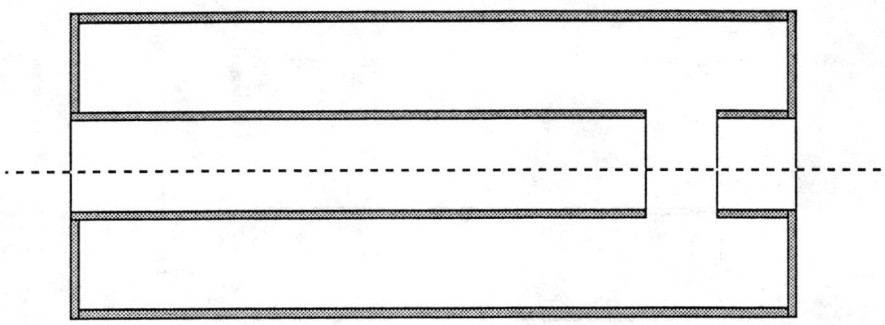

Figure 2-9

We immediately recognize a "new type" of cavity resonator where the lumped capacity of the gap is a fundamental part of the structure.

A large number of variations of the above structure is obviously possible without substantial departures from the hypothesis of coaxial line resonator.

An example is given in figure (3-9), where the gap capacity is made larger.

Sometimes "push-pull" coaxial resonators are used. A very simple example of coaxial push-pull cavity is shown in figure (4-9).

The electrical length of each section of the resonator can be calculated with Eq. (1-9), where twice the gap capacity must be introduced.

This class of resonators, made for a push-pull mode of operation, can oscillate also in the so-called "common mode". In

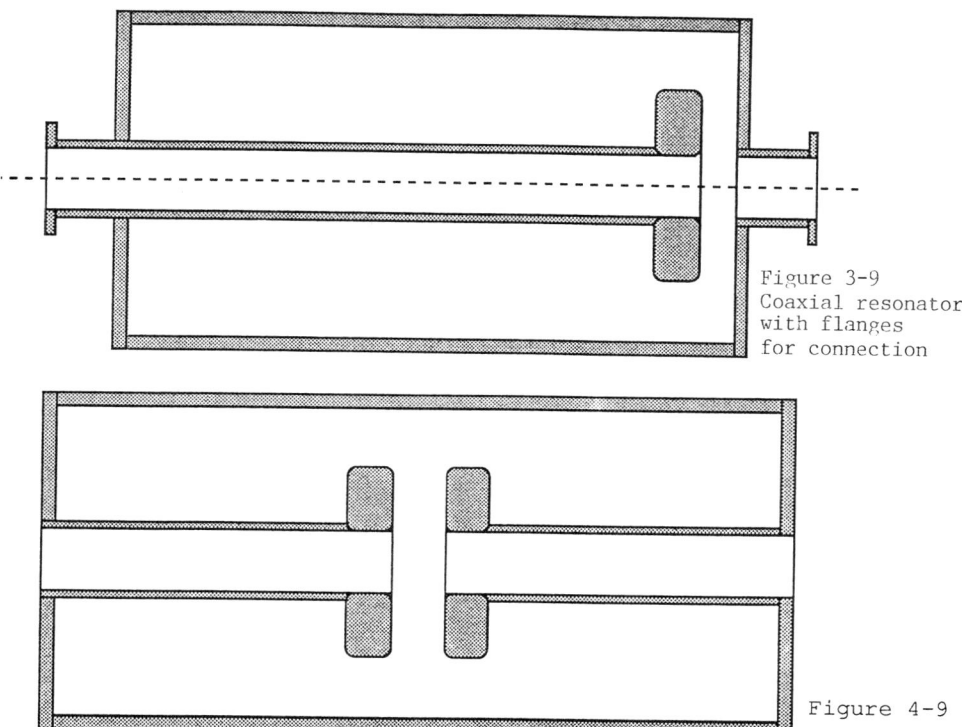

Figure 3-9
Coaxial resonator
with flanges
for connection

Figure 4-9

the push-pull mode (the one which permits the acceleration of the charged particles) the two halves of the gap swing in opposite directions and the displacement current flows through the gap ($I_g = \omega c V_g$).

Vice versa in the "common mode" both the two halves of the gap swing in the same direction, the field across the gap is identically null and the cavity cannot accelerate the charged particles which transit the gap.

It is clear that the two frequencies can be very close especially if the gap capacity, being too small, does not "absorb" a significant portion of the electrical length of the resonator.

From Eq. (1-9) we understand that the electrical length of the coaxial resonator is always less then $\lambda/4$. Nevertheless, those resonators are always called in current laboratory language the "quarter-wavelength resonators" and in many cases this causes misunderstanding.

A continuous variation of the mechanical sizes of the "coaxial line resonator" can lead to extreme shapes where the fundamental hypothesis of "a coaxial transmission line short-circuited at one end and capacitively loaded at the other" is basically violated.

This happens for instance in the "radial line resonator" whose sketch is given in figure 5-9.

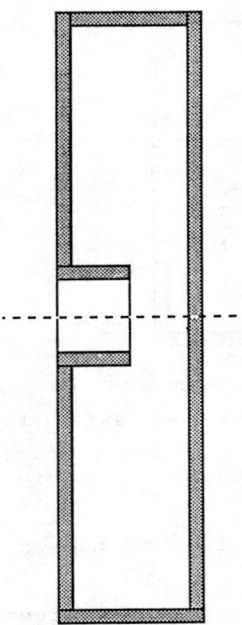

Figure 5-9

RADIAL TRANSMISSION LINE CAVITY

The propagation of the waves is in the radial direction while the E field is parallel to the z axis.

The theory of the radial transmission line can be developed following the same procedure already used for the uniform transmission case with the remarkable difference due to the fact that the characteristic parameters "L" and "C" are now functions of the radius as follows:

$$L = \frac{\mu d}{2\pi r} \quad ; \quad C = \frac{2\pi \varepsilon r}{d}.$$

From this it is immediately recognized that a radial transmission line is not a uniform one.

Another important "limit case" is represented by the "klystron" cavity, often known as "re-entrant" or "foreshortened" coaxial cavity.

Three indicative cross-sections for a re-entrant cavity are shown in figure (6-9) and it should be noted that in a re-entrant cavity

$$r_{max} - r_{min} \equiv 1-g$$

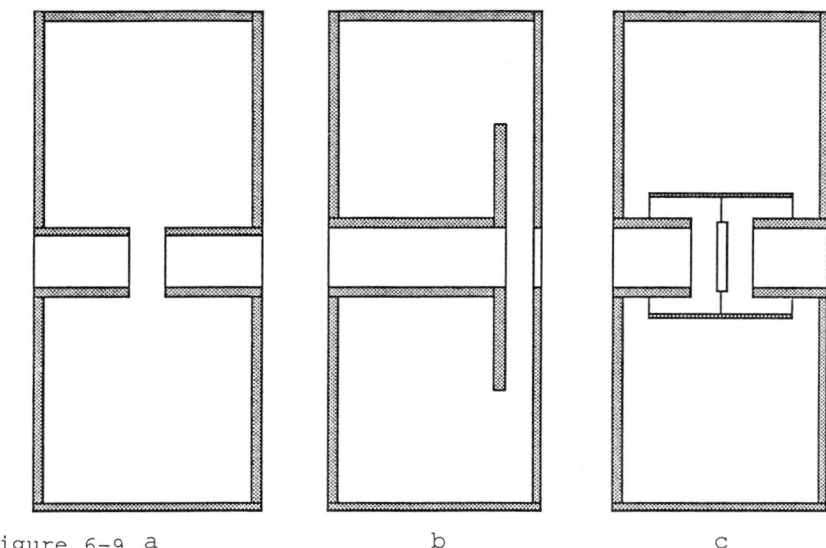

Figure 6-9   a   b   c

where l and g are respectively the lengths of the cavity and of the gap.

If we suppose that across the gap (of equivalent capacity Ceq) a sinusoidal voltage is developed with angular frequency ω, then $I \cong \omega C_{eq} V$ is the amplitude of the current that flows, uniformly, along the inner surface of the toroid that encloses the gap. This current creates a flux than can be calculated, in a very crude approximation, with the law of Biot-Savart:

$$\phi = \int_{r_{min}}^{r_{max}} \frac{\mu I}{2\pi r} l \, dr = \frac{\mu}{2\pi} I l \ln \frac{r_{max}}{r_{min}} \;;$$

from the above formula we obtain

$$L_{eq} = \frac{\mu}{2\pi} l \ln \frac{r_{max}}{r_{min}}$$

where Leq is the equivalent inductance of the cavity.

From this we obtain that the approximate value of the resonant frequency, in the fundamental mode, is given by the formula

$$\omega_r = 1/\sqrt{L_{eq} C_{eq}} \ . \qquad (3-9)$$

Eq. (3-9) should be taken as indicative, and in general a reasonable accuracy is obtained if the calculated resonant wavelength is much larger than the largest mechanical size of the cavity.

More examples of hybrid cavities could be added but the modest limits of this work prevent any further expansion of this topic.

It should be noted that the schemes of analytical calculations outlined above are important whether because they permit a quick estimation of the resonant frequency of a cavity (in the fundamental mode) or because they enhance the physical description of the phenomena.

Moreover, while it is perfectly true that about twenty years ago the analytical calculations were the only available tool (and the sophistication of those calculations was really remarkable) now powerful computer codes solve the problem of the cavity resonators quickly and with a high degree of accuracy.

10. THE WAVE GUIDES

Besides the transmission lines also the waveguides are another important distributed-constant device that can transmit power. For lack of space this important topic will be summarized in order to show how many cavity resonators can be derived from those devices.

A linear time-invariant lossless medium bounded with uniform perfectly conducting cylindrical walls constitutes an ideal waveguide. (In the majority of practical cases this medium is the vacuum or a dielectric gas).

Assuming no charges and no currents in the medium we look for a solution of the Maxwell equations inside this region.

Assume a cylindrical system of reference where $z \equiv s$ is the propagation axis and suppose that the fields inside the guide depend sinusoidally upon the time.

Following a heuristic method we try a solution:

$$\left.\begin{array}{c} E \\ H \end{array}\right\} = F(r,\vartheta) \; \text{EXP}(j\omega t - \gamma s). \qquad (1\text{-}10)$$

Taking the r and ϑ components of the Maxwell curl equations we obtain

$$\begin{cases} \dfrac{1}{r}\dfrac{\partial E_s}{\partial \vartheta} - \dfrac{\partial E_\vartheta}{\partial s} = -\dfrac{\partial B_r}{\partial t} \\[6pt] \dfrac{\partial E_r}{\partial s} - \dfrac{\partial E_\phi}{\partial r} = -\dfrac{\partial B_\vartheta}{\partial t} \\[6pt] \dfrac{1}{r}\dfrac{H_s}{\partial \vartheta} - \dfrac{\partial H_\vartheta}{\partial s} = -\dfrac{\partial D_r}{\partial t} \\[6pt] \dfrac{\partial H_r}{\partial s} - \dfrac{\partial H_s}{\partial r} = -\dfrac{\partial D_\vartheta}{\partial t} \; . \end{cases}$$

Introducing the above tentative solution and rearranging gives

$$\begin{cases} 0 \;+\; \gamma E_\vartheta \;+\; j\omega\mu H_r \;+\; 0 \;=\; -\dfrac{1}{r}\dfrac{\partial E_s}{\partial \vartheta} \\[6pt] -\gamma E_r \;\;\; 0 \;\;\;\;\; 0 \;\;\; - j\omega\mu H_\vartheta \;=\; \dfrac{\partial E_s}{\partial r} \\[6pt] -\gamma\omega\varepsilon E_r \;\;\; 0 \;\;\;\;\; 0 \;\;\; - \gamma H_\vartheta \;=\; -\dfrac{1}{r}\dfrac{\partial H_s}{\partial \vartheta} \\[6pt] 0 \;+\; j\omega\varepsilon E_\vartheta \;-\; \gamma H_r \;+\; 0 \;=\; \dfrac{\partial H_s}{\partial r} \; . \qquad (2\text{-}10) \end{cases}$$

We can solve the system and obtain the four components E_r, E_ϑ, H_r, H_ϑ as functions of the longitudinal components E_s and H_s:

$$E_r = -\frac{1}{K_c^2}\left[\gamma\frac{\partial E_s}{\partial r} + \frac{j\omega\mu}{r}\frac{\partial H_s}{\partial \vartheta}\right]$$

$$E_\vartheta = \frac{1}{K_c^2}\left[-\frac{\gamma}{r}\frac{\partial E_s}{\partial \vartheta} + j\omega\mu\frac{\partial H_s}{\partial r}\right]$$

$$H_r = \frac{1}{K_c^2}\left[\frac{j\omega\varepsilon}{r}\frac{\partial E_s}{\partial \vartheta} - \gamma\frac{\partial H_s}{\partial r}\right]$$

$$H_\vartheta = -\frac{1}{K_c^2}\left[j\omega\varepsilon\frac{\partial E_s}{\partial r} + \frac{\gamma}{r}\frac{\partial H_s}{\partial \vartheta}\right] \qquad (3-10)$$

where $K_c^2 = \omega^2\varepsilon\mu + \gamma^2$. It is immediately evident that the derivatives of the longitudinal fields determine the fields in the cross-section of the guide (the so-called transverse fields).

For $E_s \neq 0$ and $H_s = 0$ we have the TM modes,
for $E_s = 0$ and $H_s \neq 0$ we have the TE modes,
and from (3-10) it is evident that there are no other possibilities.

TM means "transverse magnetic" because the wave has no magnetic field in the direction of propagation. For the same reason when $E_s = 0$ then there is no component of the electric field in the propagation direction. The two modes are equivalent for transmitting energy but have a very different effect on charged particles injected in the guide. The TM modes are known in fact as "accelerating modes" while the TE are known as deflecting due to Eq. (3-10); the problem is now to determine H_s and/or E_s.

From the Maxwell equation we derive the vector wave equation

$$\nabla^2\left\{\frac{\overline{E}}{\overline{H}}\right\} + \omega^2\varepsilon\mu\left\{\frac{\overline{E}}{\overline{H}}\right\} = 0 \; ;$$

then we start with the search of TE modes; that means we start from the H component of the wave and assume that $H_s \neq 0$.

Expanding the wave equation for the H_s component we have

$$\frac{\partial^2 H_s}{\partial r^2} + \frac{1}{r}\frac{\partial H_s}{\partial r} + \frac{1}{r^2}\frac{\partial H_s}{\partial \vartheta^2} + K_c^2 H_s = 0. \qquad (4-10)$$

Attempting to separate the variables we try

$$H(s) = R(r) * \Theta(\vartheta)$$

where R and Θ are two unknown functions respectively of r and of ϑ. Substituting into Eq. (4-10) and manipulating we obtain

$$r^2 \frac{R''}{R} + r \frac{R'}{R} + K_c^2 r^2 = \frac{\Theta''}{\Theta}. \qquad (5-10)$$

Again the two sides of the equation (5-10) must be equal. The L.H.S. is function of r alone and the R.H.S. is function of ϑ alone. Consequently the two sides should be equal to a constant; the same constant.

As we have already seen, we decide that each side must be equal to a constant, for instance v^2, and separating we obtain

$$\begin{cases} R'' + \frac{1}{r} R' + \left(K_c^2 - \frac{v^2}{r^2} \right) = 0 \\ \frac{\Theta''}{\Theta} = -v^2. \end{cases}$$

Solving, we obtain

$$H_s = \left(A\ J_v(K_c r) + B\ N_v(K_c r) \right) \left(C \sin v\vartheta + D \cos v\vartheta \right)$$

where v should be an integer as already seen for the pill-box cavity and J_ϑ N_ϑ are the Bessel and Neuman functions of order ϑ.

Now we look for the appropriate boundary conditions.

At $r=0$, H_s cannot be infinite and consequently $B = 0$.

Selecting the origin for ϑ,

$$H_s = H_0 \, J_\nu(K_c r) \cos \nu \vartheta \,.$$

On the inner surface of the cylindrical wall $E_\vartheta \equiv 0$. Consequently from the second equation of (2-10) we obtain

$$E_\vartheta = \frac{j\omega\mu}{K_c^2} \frac{\partial H_s}{\partial r} = 0 \qquad \text{for} \quad r = a$$

if a is the radius of the guide.

This means that we should have $J'_\nu(K_c a) = 0$ and this condition determines an infinite number of solutions which are known as $TE_{\nu l}$ modes, in fact the value of the Ith zero of the Bessel function of order ν and the value of ω determine the corresponding value of γ.

We now examine the TM modes and start from the wave equation written for the electric field.

Following the same guideline we calculate the field E_s for the TM modes,

$$E_s = E_0 \, J_\nu(K_c r) \cos \nu \vartheta \,,$$

and this time the appropriate boundary condition is $E_s = 0$ on the perfectly conducting cylindrical surface.

Concluding:

TE Modes $\qquad J'_\nu(K_c a) = 0$ ⎫ are the boundary conditions
TM Modes $\qquad J_\nu(K_c a) = 0$ ⎬ that determine the value of γ.

If we call $R_{\nu l}$ the lth root of the appropriate equation we obtain

$$\gamma^2 + \omega^2 \varepsilon \mu = \left(\frac{R_{\nu l}}{a}\right)^2 \,; \qquad (5\text{-}10)$$

if the fields should propagate along the guide then γ must be imaginary: $\gamma = j\beta$, and we obtain

$$\beta = \sqrt{\omega^2 \varepsilon \mu - \left(\frac{R_{\nu 1}}{a}\right)^2} \quad . \qquad (6\text{-}10)$$

To have propagation without attenuation β must be real and this happens if the frequency is larger than the minimum prescribed by Eq. (5-10). Calling, as usual, λ_g the wavelength inside the guide it follows that $\beta = 2\pi/\lambda_g$. In fact, when we pass through a length equal to λ_g the field has to repeat because the argument of exp(jβλ) changes by 2π.

Substituting and rearranging Eq. (5-10), we obtain

$$\left(\frac{2\pi}{\lambda}\right)^2 = \left(\frac{R_{\nu 1}}{a}\right)^2 + \left(\frac{2\pi}{\lambda_g}\right)^2 , \qquad (7\text{-}10)$$

that is, the fundamental dispersion formula for the uniform waveguide where λ is the free-space wavelength of the field.

The dispersion formula is normally rewritten as

$$\omega^2 = c^2 \beta^2 + \text{COST}$$

and can be plotted as the ω-β diagram (the famous "Brillouin" diagram in figure 1-10).

We verify, by inspection, that $\text{TAN}\Phi = \dfrac{\omega}{\beta}$ is the phase velocity.

$\text{TAN}\psi = \dfrac{d\omega}{d\beta}$ is the group velocity.

For β=0 the angle Φ reaches π/2 and the phase velocity becomes infinite while the angle ψ reaches the zero and the group velocity goes to zero. In those conditions the guide resonates. There is no propagation of energy and the mode is at the cut-off.

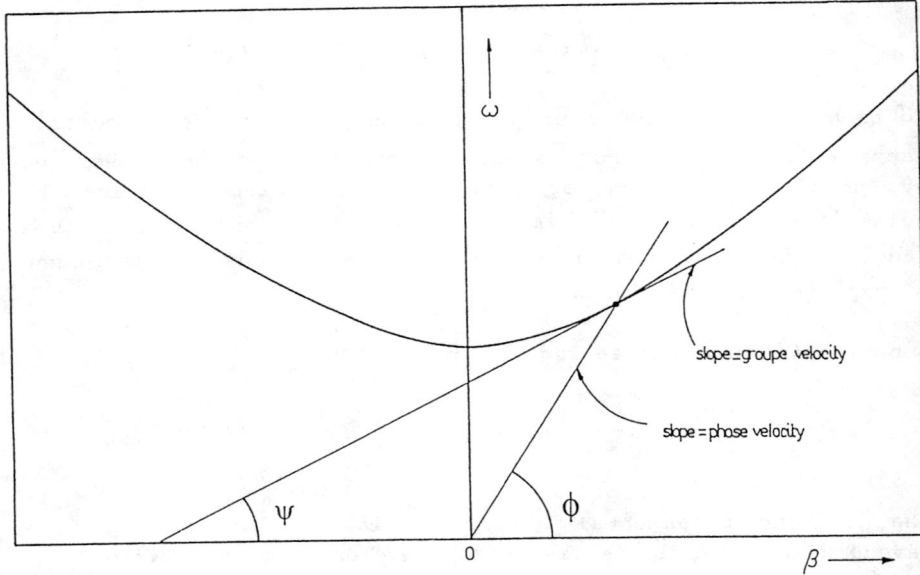

Figure 1-10   ω–β diagram

Wave guides are very important devices for transmitting power at very high frequencies as the calculation of the pointing vector would show immediately.

Moreover, with some important modifications they can support the so-called slow waves, that is TM modes whose phase velocity is less or equal to the speed of the light in the vacuum.

Those modes can accelerate the charged particles and, in fact, the modern linacs are realized with the so-called corrugated waveguides.

Now we can show that many cavity resonators can be derived from the uniform waveguides.

In fact both ends of the cylindrical pipe can be short-circuited with conducting walls normal to the axis (see figure 2-10). The fields specified by the considered mode can remain trapped into this resonator if another condition is fulfilled:

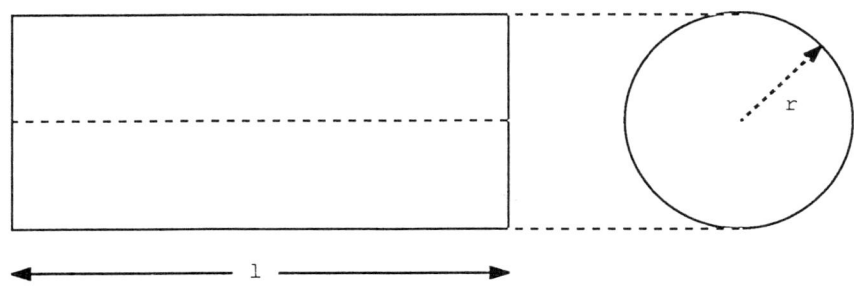

Figure 2-10

"The electric field must be zero or normal on the short-circuiting surfaces".

It is nearly obvious that both for the TE and the TM modes this condition is fulfilled if the distance l is an integer multiple of half wavelength of the field inside the guide:

$$l = p\frac{\lambda_g}{2} \qquad p = 1, 2, \ldots$$

Substituting into the dispersion equation we obtain

$$\frac{1}{\lambda^2} = \left(\frac{R_{\nu 1}}{2\pi a}\right)^2 + \frac{p^2}{(2l)^2} \ .$$

The above formula is normally written as

$$\lambda = \frac{2l}{\sqrt{p^2 + \left(\frac{R_{\nu 1}}{2\pi a} * 2l\right)^2}} \qquad \left.\begin{array}{c}TE\\TM\end{array}\right\} \nu l p$$

and applies both for the TE and TM cavity resonators. (Besides the above conditions a resonant mode is met, for the TM modes, when $\gamma = j\beta = 0$ and, in this case, the resonant wavelength does not depend upon the distance l because the electric lines of force are straight lines and it is exactly the case of the pill-box cavity already examined.

The cavity resonators that we have seen derive from a cylindrical waveguide. In figure 3-10 some examples are given.

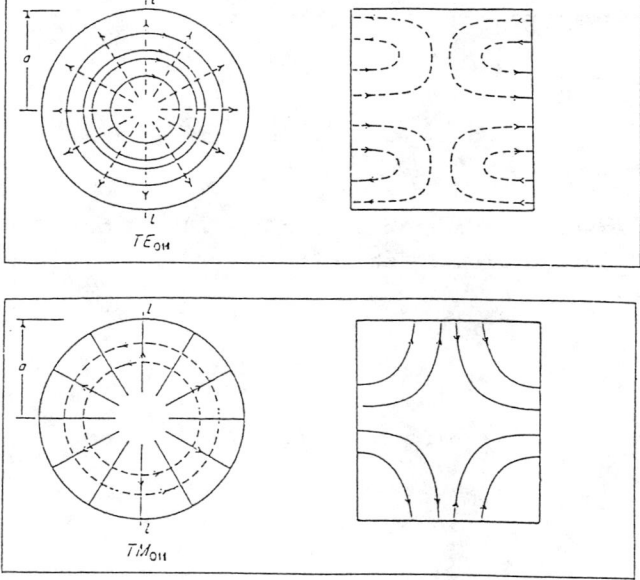

Figure 3-10   Examples of fields in a cylindrical cavity
              Dotted line: H field
              Straight line: E field

Obviously a waveguide can have any cross-section, but apart from the rectangular waveguides (mostly used for transmitting RF power) no practical use has been found for those guides, whose calculation, on the other hand, is extremely difficult. While in any case a cavity resonator can be always derived from a short-circuited waveguide, the reverse is not true.

Many classes of resonators are not derivable from a uniform guide as, for instance, the spherical resonator.

Important note

The rectangular wave guides can be treated following the outlined procedure and using rectangular coordinates. The solutions are obviously given by trigonometric functions.

Besides the important purpose of transmitting RF power the rectangular wave guides and the derived rectangular resonators can be used around the particle accelerator as RF separators, RF beam choppers, RF deflectors and in general for the RF manipulation of beams of charged particles.

11. SPECIAL TOPICS ON CAVITY RESONATORS

a. Physics of the Acceleration

The big merit of the cavity resonator lies in the fact that the cavity "contains" the electromagnetic field so that there is no field outside the device.[*] In other words a cavity resonator creates a region of the space where an electric field of suitable characteristics can be localized.

---

[*] Here we disregard the small amount of field that comes out of the holes of the cavity.

The above property is used for accelerating the charged particles in two different ways as is shown in figure 1-11 (a, b).

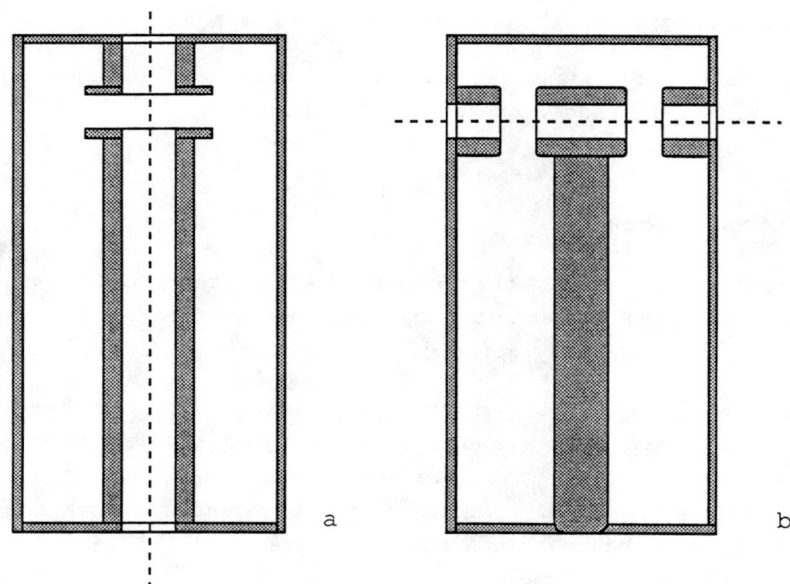

Figure 1-11

In fact both the cavities sketched in a and b are coaxial resonators loaded with the gap capacity and both can be considered as "shortened" quarter-wavelength resonators. In the "a" case the charged particles pass along the axis of the resonator and the corresponding current is coupled to the magnetic flux that corresponds to the gap voltage.

In the "b" case the particle must pass through two gaps where the "E" fields are in opposition (π radians out of phase) and the maximum efficiency is obtainable when the distance "l" between the centers of the two gaps is covered in half of the period of the accelerating voltage and the length of each gap is negligible if compared with l.

For stressing the difference between the two ways of operation consider the case in which the transit-time effect is negligible for the gaps of the case a and b. Then it follows that the net acceleration of a particle is independent of the speed of the particle for the "a" case but is strongly dependent on it for the case b.

Nevertheless, some authors indicate the "b" way as superior because a "synchronous" particle gains twice the energy that any particle could gain in the "a" mode.

This is not the place for going deeper into this topic but the reader should be aware that a technical choice always depends upon many factors and that in the RF field absolute truths are yet unknown.

b. The Quality Factor Q

An ideal cavity is a lossless device and this means that if a mode is excited in the cavity this mode will oscillate forever. This is not the case of a physical cavity, where the losses are always present. In this case a very important parameter is the so-called quality factor of the considered mode. This quality factor Q is defined as

$$Q = 2\pi \frac{\text{Energy stored in the mode}}{\text{Energy lost per cycle}}. \quad (1-11)$$

The quality factor is a fundamental quantity at least for the following reasons:

1. The quality factor for each mode of a cavity is uniquely defined and depends upon both the geometry of the cavity and the properties of the material that limits the cavity.

2. As for all the linear resonant systems, if a cavity, oscillating in a mode, has resonant frequency f_0 and quality factor Q_0 then the time constant τ characteristic of the cavity in that mode is equal to $Q_0/\pi f_0$.

3. As for a parallel resonant circuit, the band pass of the cavity $\Delta f_0/f$, for each mode, is equal to the reciprocal value of the Q_0 corresponding to the mode.

4. The quality factor is a dimensionless quantity that can be measured easily and with high precision.

The quality factor of a cavity (in a given mode) can be very high. In fact for a lossless cavity the quality factor should be infinite.

In the practical cases a quality factor of $10^4 - 2 \times 10^4$ should be considered normal. Quality factors of ~1000 should be considered as very low while above 5×10^4 we are in the region of very high Q.

The superconducting cavities can exhibit quality factors of ~ 10^9.

It should be noted that the theoretical calculation of the quality factors always gives values higher than the measured ones.

This is to be expected because the largest part of power dissipation inside the cavity depends upon the total resistance that the inner wall of the cavity offers to the RF currents.

The penetration depth is always very small (~ 6 microns at 100 MHz for copper!) then it follows that the path of the currents is determined by the status of the inner surface of the cavity walls (this means that the tooling and the finishing of those surfaces is determinant). Moreover, the dielectrics that can be introduced in the cavity together with the unavoidable presence of holes and electrodes also contribute to the total losses.

Speaking of the quality factor of a cavity, the fundamental mode of excitation is normally understood, but it should be pointed out that, as already said, for each mode, a quality factor exists and it is not at all said that the highest value of Q pertains to the fundamental mode.

In the current literature some confusion has been introduced after the definition of the so-called "loaded Q" often indicated with Q_L.

This new definition depends upon the fact that in practical cases the cavity resonator of an accelerator is coupled to the driving amplifier and consequently the output impedance of the amplifier (which always has a real component) loads the cavity.

In fact, let us suppose that the driving (final) amplifier that is on but not excited, is connected to the cavity. If in this condition we try to measure the quality factor of the cavity we necessarily waste energy in the cavity and in the output impedance of the final amplifier. The result of the measurement is Q_L, always lower than the Q of the cavity alone.

On the other hand the loaded Q is not a complete redundant parameter because the knowledge of both Q and Q_L gives some information about the degree of coupling between cavity and final amplifier.

The relation between the unloaded Q and Q_L is given by

$$Q = Q_L (1+\beta)$$

where β is the coupling factor that for complete matching becomes 1 (critical coupling); it follows that for instance if $Q_L = 0.5$ Q then the cavity is matched to the driving amplifier.

The above situation becomes more complicated if the power absorbed by the accelerated beam is not negligible if compared with the power needed for exciting the cavity. In this case the amplifier should provide the power for exciting the cavity and for accelerating the beam, and the coupling between cavity and amplifier must take into account the load due to the beam.

The attempt of defining a new loaded quality factor which takes into account the steady-state beam equivalent admittance gives a parameter that, by definition, is not applicable in a transient situation and consequently the resulting quality factor is nearly spoiled of its original physical meaning. In conclusion, because the question of the loaded Q is not a fundamental one, the student is advised to leave it alone.

c. The Shunt Impedance

This is a very important topic that deserves a lot of attention.

First of all the name "shunt impedance" is an abbreviation for "resonant shunt impedance" and it is always a real quantity.

Normally a real impedance relates the voltage across the impedance with the power needed for creating the voltage, and for lumped-constant circuit there is no ambiguity. Vice versa we meet a fundamental indetermination when we try to apply this concept to a cavity resonator.

In fact if we spend the power W for exciting a resonator then we can invent an infinite number of "integration paths" inside the cavity and for each path the line integral of the electric field could define a voltage due to the power W.

Clearly even the most reasonable choice for the integration path is an "arbitrary choice" and we conclude that for a cavity resonator the shunt impedance is intrinsically undetermined.

At this point we recall that each cavity is constructed for a very definite purpose, and in the case of the particle accelerators each cavity is designed for exciting an accelerating gap.

In that case we could assume that the axis of the gap is the integration path and consequently we define the cavity voltage V_{gap} as

$$V_{gap} = \int E(z)\,dz \qquad r=0\,;\ \varphi=0. \qquad (2\text{-}11)$$

Consequently the cavity shunt impedance "referred to the gap" is

$$R = \frac{V_{gap}^2}{2W} . \qquad (3-11)$$

With the above formula the ambiguity is removed but it should be pointed out that we made an arbitrary assumption (quite reasonable and useful but arbitrary).

In conclusion the shunt impedance of an accelerating cavity is the real quantity that correlates the gap voltage to the power spent for exciting the cavity.

Unfortunately because of the above-mentioned arbitrarity some "variations on the theme" have been introduced and the net result is a little confusion.

In order to avoid confusion here are listed some of the quoted variations:

1. Because the gap voltage is always sinusoidal then its root-mean-square value could be used. In this case the factor two is missing in the formula.

2. Economic arguments could suggest to consider as "useful" only the transit-time-corrected gap voltage (peak or r.m.s. value). In this case a factor α^2 must be introduced, as already seen.

3. Accelerating structures can have more than one accelerating gap. In that case a shunt impedance per meter has been proposed.

4. Sometimes in the power used for exciting the cavity also the power delivered to the beam is considered and the resulting shunt impedance is referred to as the loaded shunt impedance.

Taking into account the endless imagination of the RF people, other new shunt impedances might be invented, published and used even only once. Because too many definitions do not make for clarity, a "grand unification" with a modest return to Eq. (3-11) is highly desirable.

d. The Equivalent Scheme

So far we have concluded that for a given mode a resonant frequency and a quality factor Q are uniquely defined. Moreover, assuming that the integration path is along the axis of the gap we have that for each mode we can calculate or measure the shunt impedance and consequently for each mode we have three characteristic numbers:

$$f_m \; ; \; Q_m \; ; \; R_{shm}$$

where the index m indicates the mth mode. Normally no index is used for the fundamental mode.

The physics of a linear resonant phenomenon is described by a second-order linear differential equation. This consideration suggests that also the behaviour of a resonant cavity, in each mode, can be described by a second-order linear differential equation and this in turn means that a parallel resonant circuit can be a good equivalent circuit of the cavity, for that mode and in a narrow interval around the resonant frequency characteristic of the considered mode.

Assuming that the considered mode is characterized by the three quantities f, Q, R_{sh} then the equivalent circuit should exhibit the same resonant frequency, the same quality factor and the same shunt impedance. It follows that the lumped elements R, L, C of the equivalent circuit are related to the characteristics of the mode by the well-known formulae

$$R = R_{sh} \; ; \; C = \frac{Q}{2\pi f R_{sh}} \; ; \; L = \frac{R_{sh}}{2\pi f Q} \; . \qquad (4-11)$$

For a better understanding of the above representation some statements should be added:

1. The behavior of a resonant cavity (in each mode) depends upon a linear differential equation if and only if the materials that support the electromagnetic field can be considered linear (for the ferrite-loaded cavities the previous model should be implemented with at least a non-linear element).

2. The frequency interval for which the approximation is technically acceptable is roughly equal to f/Q, where f and Q are the resonant frequency and the quality factor of the considered mode. This is valid if and only if the cavity is "clean" inside the same interval (to be clean inside an interval means, in laboratory jargon, that no other modes show, or are predicted, in the same interval).

3. The transit-time coefficient does not affect the value of the elements of the equivalent circuit.
   If a beam excites a mode then it is the equivalent current of the beam that should be multiplied by the transit-time factor. If instead we are considering the impedance offered to the driving amplifier then the transit-time factor should be ignored because the driving amplifier is concerned only with the fields created in the cavity. The transit-time-corrected impedance $R_{eq} = \alpha^2 R_{sh}$ tells us which power should be wasted for "giving" a prescribed voltage to the beam and has nothing to do with the peak voltage that the amplifier should create.

4. Taking into account all the previous considerations, the equivalent scheme of a cavity resonator coupled to the beam is, for each mode, as shown in figure 2-11, where the current generator indicates that the beam current is <u>independent</u> of the gap voltage. (The shape of the beam current obviously depends upon the shape of the bunches that transit the cavity with a transit-time factor equal to α).

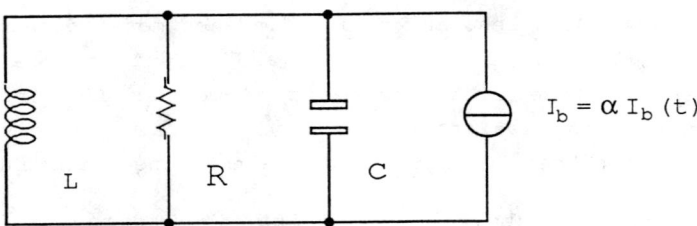

Figure 2-11

5. Besides the parallel tuned circuit other resonant circuits can be used for schematizing the behaviour of a given mode, and the circuit given in figure 3-11 is also frequently used.

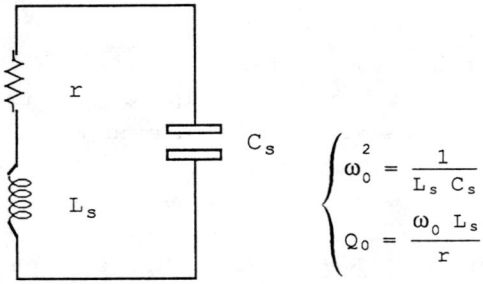

$$\begin{cases} \omega_0^2 = \dfrac{1}{L_s C_s} \\ Q_0 = \dfrac{\omega_0 L_s}{r} \end{cases}$$

The gap voltage is developed across the capacitor.

Figure 3-11

If the quality factor Q is large (greater than ~1000) then a very simple analysis could show that there is no practical difference between the two representations.

Moreover if the parallel RLC circuit must be equivalent to the r, L_s, C_s one then the following equalities must be verified:

$$C_s = C \quad ; \quad L_s = L \quad ; \quad r = \frac{1}{R}\frac{L}{C} \, .$$

6. In dealing with the resonant cavities the parameter R/Q is frequently used.

   The value of this parameter depends upon the selected integration path but its usefulness is related to the fact that its value can be measured with a very simple perturbative technique along the prescribed integration path. Consequently once the quantities ω, Q, R/Q are given (for a mode) then the equivalent circuit is immediately known.

   From the physical point of view the parameter R/Q defines the characteristic impedance of the cavity in the considered mode.

   With reference to a parallel RLC circuit we have

   $$\frac{R}{Q} = \frac{R}{\omega RC} = \frac{1}{\omega C} = \sqrt{L/C}$$

   where ω and Q are the resonant frequency and the quality factor of the mode.

e. The Coupling to the Cavity

A resonant cavity can be coupled to the driving amplifier in many ways. Examples of the most common coupling systems are shown in figure 4-11.

The inductive (loop) coupling is by far the most used and for this reason it will be illustrated with some detail.

In particular we will study the input impedance of this coupling device.

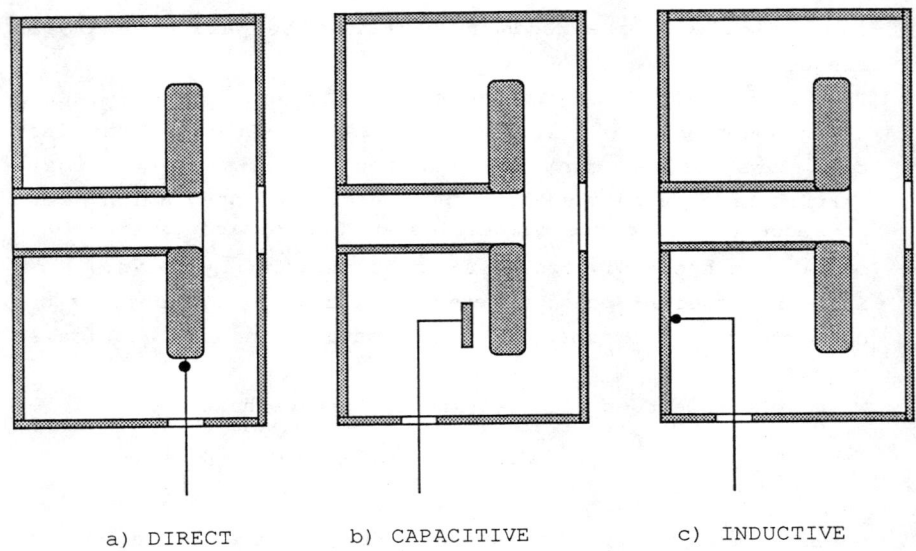

a) DIRECT   b) CAPACITIVE   c) INDUCTIVE

Figure 4-11

Consider the lumped-constant circuit given in figure 5-11,

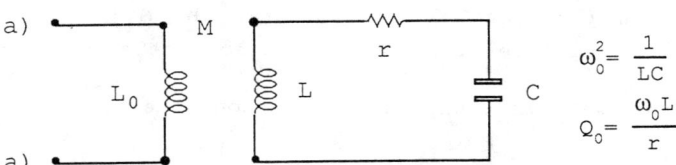

Figure 5-11

$$\omega_0^2 = \frac{1}{LC}$$

$$Q_0 = \frac{\omega_0 L}{r}$$

where ω_0 and Q_0 are respectively the resonant frequency and the quality factor of the series circuit.

From the mesh analysis we calculate the input impedance seen from the terminals as

$$Z_i = j\omega L_0 + \frac{\omega^2 M^2}{j\omega L + \frac{1}{j\omega C} + r} \quad ; \qquad (5\text{-}11)$$

the above formula can be transformed into canonical form:

$$Z_i = j\omega L_0 + j\frac{\omega^3 M^2}{L\left(\omega_0^2 - \omega^2 + j\frac{\omega r}{L}\right)} = j\omega L_0 + j\frac{\omega^3 \frac{M^2}{L}}{\omega_0^2 - \omega^2 + j\frac{\omega \omega_0}{Q_0}} \quad . (6\text{-}11)$$

With simple algebraic manipulation the canonical form can be expanded and we obtain

$$Z_i = j\omega L_0 - j\omega \frac{M^2}{L} + \frac{j\omega \omega_0^2 \frac{M^2}{L}}{\omega_0^2 - \omega^2 + j\frac{\omega \omega_0}{Q_0}} - \frac{1}{Q_0}\frac{\omega^2 \omega_0 \frac{M^2}{L}}{\omega_0^2 - \omega^2 + j\frac{\omega \omega_0}{Q_0}} \qquad (7\text{-}11)$$

as can be checked adding the last three terms.

If the quality factor Q_0 is high then Eq. (7-11) can be simplified and we obtain

$$Z_i \cong j\omega \left(L_0 - \frac{M^2}{L}\right) + \frac{j\omega \omega_0^2 \frac{M^2}{L}}{\omega_0^2 - \omega^2 + j\frac{\omega \omega_0}{Q_0}} \qquad (8\text{-}11)$$

where the second term has exactly the form of the impedance of a parallel resonant circuit. In fact the impedance Z_i offered by a parallel RLC circuit is, as is well-known, equal to

$$Z_i = \frac{1}{\frac{1}{R} + j\omega C + \frac{1}{j\omega}} = \frac{j\omega \omega_0^2 L}{\omega^2 - \omega_0^2 + j\frac{\omega \omega_0}{Q}} \quad . \quad (9-11)$$

Consequently the circuit interpreted by Eq. (8-11) is the one given in figure 6-11,

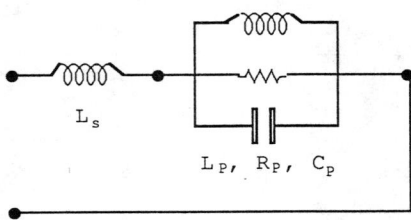

Figure 6-11

provided that:

a) $L_p = \frac{M^2}{L}$; $L_s = L_0 - L_p$ because of the form of (8-11) and (9-11).

b) The resonant frequencies and the quality factors of the two resonant circuits are respectively equal.

Now we apply the previous consideration to a loop-coupled cavity using the equivalent circuit given in figure 5-11.

Assuming that the loop is coupled to the whole set of modes of the cavity (and extrapolating from figure 5-11) the equivalent scheme should be as shown in figure 7-11.

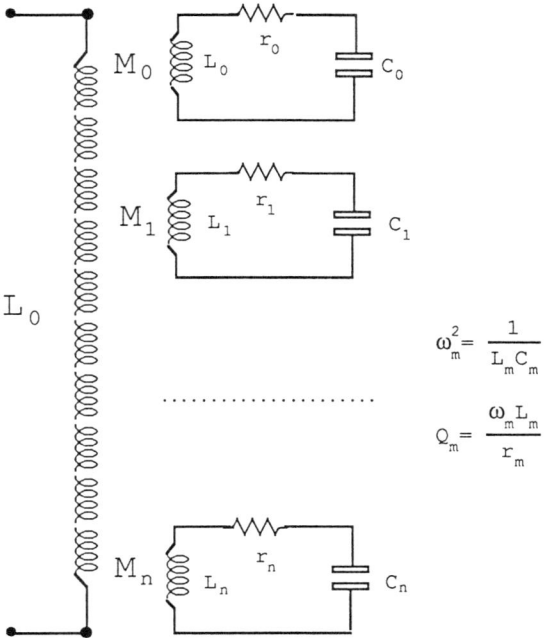

Figure 7-11

$$\omega_m^2 = \frac{1}{L_m C_m}$$

$$Q_m = \frac{\omega_m L_m}{r_m}$$

Following the method described above, the whole circuit can be reduced, as shown in figure 8-11,

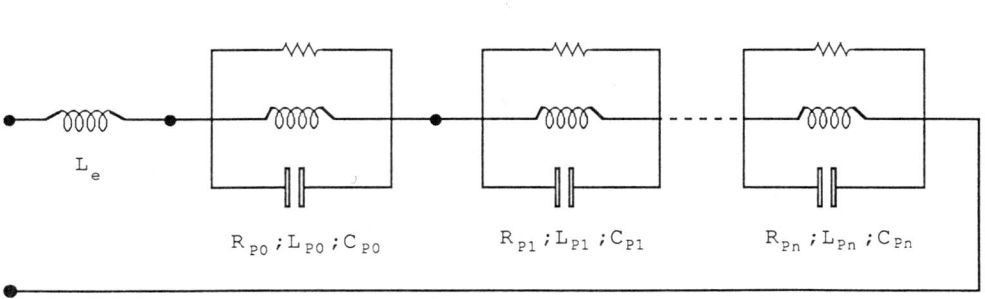

Figure 8-11

where the identification of the parameters is as follows:

$$L_e = L_0 - \frac{M^2}{L} \quad ; \quad C_{pm} = \frac{L}{M} C \quad ;$$

$$L_{pm} = \frac{M^2}{L} \quad ; \quad R_{pm} = \frac{1}{r} \frac{\frac{M^2}{L}}{C} \ . \tag{10-11}$$

Consequently the input impedance that appears at the terminals of the coupling loop becomes

$$Z_i = \sum_{m}^{n} \left(L_0 - \frac{M_m^2}{L_m} \right) + \sum_{m}^{n} j \frac{\omega \omega_0^2 L_{pm}}{\omega_m^2 - \omega^2 + j \frac{\omega \omega_m}{Q_m}} \tag{11-11}$$

and the problem is solved.

The formula (11-11) shows that when the operating frequency approaches the resonant frequency of a mode then the input impedance tends to a peak.

For other coupling systems similar formulations can be found because the physics involved is basically the same.

It should be pointed out that while the scheme given in figure 7-11 is quite general (and can be used as good representation of the physical reality) the one given in figure 8-11 is valid only for calculating the impedance that appears at the terminals of the coupling loop.

f. Technical Considerations

In the majority of cases an accelerating cavity is driven by a power amplifier. This means that the output impedance of the amplifier appears transformed at the gap or, said in other words, the impedance seen by the gap also depends upon the power amplifier.

This is very important because in many applications (and especially in the storage rings) the impedance that the gap offers to the beam cannot be arbitrarily large.

Once the cavity, the amplifier and the coupling system are specified then the total admittance transferred to the gap can be calculated. (Beware! This calculation can be difficult and in many cases the linearization of the active element in the power amplifier becomes imperative).

If Y_a is the admittance transferred to the gap then the equivalent scheme for the whole system (operating in the proximity of a resonant mode of the cavity) is given in figure 9-11,

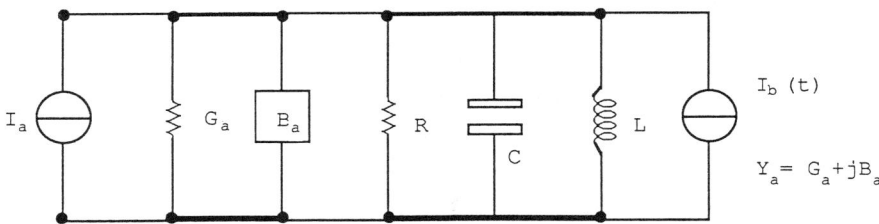

Figure 9-11

where G_a and B_a may be also very complicated functions of the frequency, while I_a is the amplitude of the sinusoidal driving current referred to the gap and $I_b(t)$ is the transit-time-corrected current due to the beam.

As we already know the cavity is operated in an extremely narrow range of frequencies ($\Delta f = \sim f_a/Q$) and this means that the admittance Y_a can be schematized with the one of another resonant circuit (whose resonant frequency does not coincide, generally, with the one of the considered mode).

Consequently the equivalent scheme of the cavity coupling system, seen by the gap, and for a given mode can be as shown in figure 10-11,

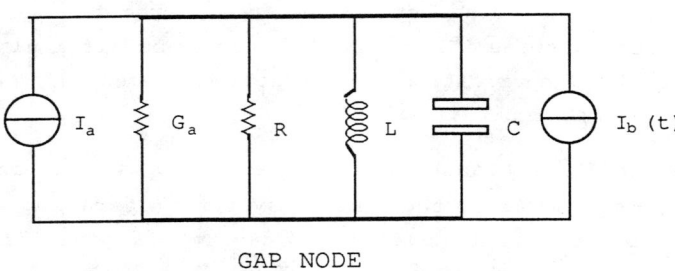

GAP NODE

Figure 10-11

where the elements "L" and "C" now take into account the contribution of the driving amplifier output susceptance. The cavity shunt impedance R and the equivalent shunt impedance $R_a=1/G_a$ due to the driving amplifier are normally indicated with separated elements only for practical purposes.

The above schematization, although very general, is the most used when the driving amplifier is directly coupled to the cavity.

When a transmission line or a waveguide is inserted between the driving amplifier and the coupling circuit then it is mandatory to draw the scheme of the whole system where the coupling device and the cavity are the last elements of the chain.

Beyond ~300 MHz, in the microwave domain, more complicated coupling methods are used. These cannot be analyzed in this context because microwave theory should be used and this is beyond the purpose of this elementary and intuitive presentation. Here we quote only the "hole" coupling where the power flowing along a guide is transmitted to the cavity through a hole made in the wall of the cavity.

The mechanics of the hole coupling lies in the fact that the unperturbed mode in the guide and the unperturbed mode in the cavity should have at least one component oriented in the same way in the plane that contains the hole. When the cavity and the guide are perturbed by the "hole" then the unavoidable distortion of the fields creates the so-called "reactance of the coupling" that can be calculated and introduced in the scheme.

g. Deviations from the predicted Behaviors

In dealing with the cavity resonators we have always ignored the fact that the whole cavity or at least the gap region are under vacuum but, paradoxically, the vacuum is a new element that must be taken into account.

a. If the pressure of the residual gas is not sufficiently low (less than 10^{-5} Tor) then the glow discharge hinders completely the functioning of the cavity and there is also the risk of serious damage of the amplifier if the amplifier is "on" under such conditions.

b. Craters, scraps, sharp edges or unclean parts of the inner walls sooner or later become localized sources of current and the behavior of the cavity becomes "strange".
For some levels of excitation the emission from those points can be important and the cavity behaves in a non-linear way. It is important to recognize that the internal emission affects the shunt impedance, the quality factor and the resonant frequency of the operating mode.
Normally a cavity is a high impedance device and consequently even modest emissions of current from the walls can have a dramatic impact on the cavity impedance.
Moreover a change in the parameters of the cavity greatly reduces the efficiency of the coupling to the driving

amplifier and the two effects compromise the good functioning of the cavity. (Permanent damage to the cavity and to the power amplifier are possible in those conditions).

c. The potential well at the inner surface of the cavity is reduced by the electric field. At some levels of excitation even the most even and clean surface must emit electrons. (Field emission current). This emission is often accompanied by the heating of the surface with localized avalanche effects. The material, the frequency and the tooling of the surface play an important role on the field emission threshold. Unfortunately against this phenomenon there is nothing to do.
The cavity should be designed in such a way as to avoid totally the risk of field emission.

d. The Multipacting.
This is an interesting and complicated phenomenon that can be clarified as follows.
Consider two metallic plane and parallel surfaces under vacuum and suppose that between the two surfaces there exists a sinusoidal uniform electric field. (From the Maxwell equations we know that this E-field must be accompanied by the corresponding magnetic field but for the moment we ignore this H-field).
The cosmic ray and/or other radiation extract electrons from the facing surfaces, the electrons are accelerated by the existing field and may reach the opposing surface.
If the following conditions are simultaneously verified:

1) the energy of the accelerated electrons is sufficient for extracting other electrons at the impact with the wall,

2) the extraction efficiency is larger than "one", no matter how much larger it is,

3) the transit time of the extracted electrons is roughly equal to half period of the electric field,

then an avalanche multiplication of the extracted electrons occurs.

This phenomenon is limited by many factors, nevertheless a cloud of electrons is created that moves back and forth between the plates. It is clear that this oscillating cloud introduces power dissipation and changes the "normal" storage of energy between the plates.

The magnetic field associated with the E-field bends the trajectories of the electrons that, extracted from one wall, may come back on the same wall extracting new electrons and so on. Consequently this avalanche multiplication can take place also on a single wall.

It is now clear--the origin of the name given to this phenomenon. Multipacting comes from "many paths" even if each electron makes only one path.

The above phenomenon, which is really important for its perverse effects on the cavity resonators, has been honoured with many nicknames: multipactoring, multipactor or "resonant electron discharge" for the sake of exactness.

In a cavity resonator the multipacting may occur at different levels of excitation because many different parts of the resonator can be involved, and fortunately it happens more likely at low levels of exitation. Moreover just because the multipacting is an "avalanche" phenomenon it requires "some time" for its build-up: the so-called time constant of the multipacting τ_m. This means that if the rise time of the voltage in the cavity is somewhat smaller than τ_m then the multipacting can be prevented.

Other methods for preventing the multipacting are:

1) To introduce a fixed magnetic field that strongly contributes to rendering the trips there different to the ones back.

2) To introduce, if possible, a fixed electric field that is very efficient in rendering "asymmetric" both the transit times and the trajectories.

3) To treat the inner surface of the cavity in order to reduce the extraction energy. (Artificial glow discharges in an atmosphere of nitrogen, acquadag painting, ...).

4) To design the driving amplifier with the lowest possible output impedance and to realize a strong coupling with the cavity in order to reduce as much as possible the rise time of the RF voltage.

5) To study carefully the shape of the cavity in order to create a geometry not favorable to the multipacting. (Computer programs exist that calculate the "multipacting orbits" as function of the shape of the cavity).

Care should be taken to distinguish the multipacting from the other disturbing phenomena. As a rule in a cavity with "clean" inner surfaces the multipacting occurs at very precise and repetitive levels of excitation. On the other hand if the multipacting is established then any attempt to raise the cavity voltage would result in an increase of the multipacting current.

Multipacting is always associated with the production of x rays and if the cavity walls are not sufficiently thick then strong beams of x rays surround the cavity.

Another clue comes from the fact that the multipacting renders the cavity insensitive to the action of the tuner device. (This is rather obvious because the swinging cloud of electrons introduces a virtual capacity that throws the cavity tune completely out of the range of the tuner).

12. LINEAR ACCELERATORS

a. Proton and Ion Linacs

A natural extension of the "concept" of cavity resonator leads to the idea of linear accelerator.

In fact, as was shown in chapter 10, many cavity resonators aligned on the same axis and properly phased can accelerate a particle along a straight trajectory.

Although it should be perfectly possible to realize a linear accelerator using many resonant cavities, the technique followed exactly the opposite direction: that is, to excite many accelerating gaps using only one cavity resonator.

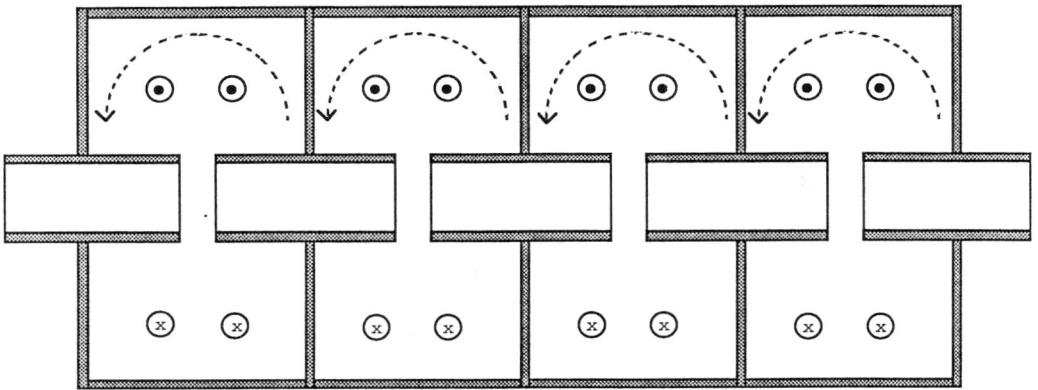

Figure 1-12 Four equal cavities in series and excited in phase

Consider for example the scheme given in figure 1-12, where four cavities, equals, are excited with the same phase and the same power (i.e. the voltages across the gaps are equal).

It follows that the two sides of each septum are run along by opposed currents and, consequently, if the septums are eliminated we obtain a resonant cavity with the same resonant frequency as the single cavities because, as a matter of fact, nothing has been changed. (This is true if the septums are "thin". If the thickness of the septums is not negligible then the total volume available for the electromagnetic field increases and the resonant frequency of the resultant cavity undergoes a small change).

The removal of the septums introduces the problem of supporting the drift-tubes. This can be solved by using metallic posts that, on the other hand, introduce some perturbations on the electromagnetic field.

Dielectric posts should not be used for many important reasons and in particular:

a. Dielectric materials are easily damaged by the intense electric field and by the radiations always present inside the cavity.

b. Dielectrics can sustain strong multipacting currents and the poor thermal conductivity of those materials enhances the creation of hot spots with localized intense emission of electrons.

c. The metallic posts can be realized with tubes of good conductors. This creates a series of shielded ways from the inside towards the outside of the cavity that are normally used for feeding the special focussing coils that must be placed inside the drift tubes.

d. The electrical "stability" of a conductor is always superior to the one offered by a dielectric.

Taking into account the previous consideration, the linear accelerator appears as sketched in figure 2-12. (A structure generally known as the "Alvarez" cavity after the name of its inventor).

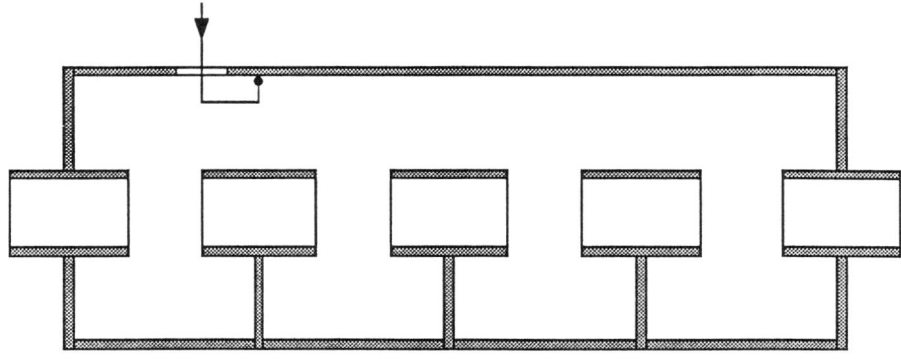

Figure 2-12   Sketch of an "Alvarez" structure

Some observations are in order:

a. A drift tube structure is normally designed for accelerating protons or heavy ions. In fact the mechanical sizes of the drift tubes and of the cavity come out reasonably small only because the β of the accelerated particles has always low values even for very high value of the energy. (For instance the β of a 10-MeV proton is equal to 0.145).

b. Because all the gaps are excited in phase then each drift tube should have a length near βλ (as we already know) and the structure tends to be "long". From the radiofrequency point of view this is a defect because increasing the length of the cavity increases the number of parasitic modes that bunch around the operating frequency.

c. In many structures the energy of the accelerated particles (ions) undergoes a very large change per unit. This means that the length of the drift tubes must vary along the machine. In that case it is convenient to vary accordingly also the outer diameter of the drift tubes in order to maintain constant the resonant frequency of the "virtual" cavities that generate the structure.

d. As shown in figure 1-12 the structure is excited by a coupling loop. When very high power is required many loops are used. This "distributed" excitation helps in suppressing many parasitic modes and from the electrical point of view contributes to the stability of the structure.

Reconsidering the figure 1-12 we could hypothesize that the gaps are excited in opposition. (That is, the fields of adjacent cells have phases π radian apart).

In this case the two sides of each septum are run along by equal currents and the septums cannot be eliminated.

Nevertheless we can arrive at a single resonant cavity, making apertures in each septum that can couple each cavity with its nearest neighbour. With a suitable choice of the shape of the coupling apertures we can obtain the required phase shift per cell (π radians for our purpose).

In this case the linear accelerator becomes much shorter because now the length of each drift tube should be near $\beta\lambda/2$. The couplings through the septums and the feeding with many loops render those structures stable and capable of accelerating very intense beams and, in conclusion, the machine is reduced to a single resonant cavity.

It should be considered that a chain of equal cavities coupled to each other constitute a periodic system than can oscillate in an endless number of modes.

The simplest model that can clarify the mechanism of the phase shift introduced by the coupling is given in figure 3-12. Assuming the sinusoidal steady state, from the mesh analysis we obtain

$$-j\omega MI_{n-1} + \left(j\omega L + \frac{1}{j\omega C}\right) I_n - j\omega MI_{n+1} = 0.$$

Now each cell is indistinguishable from the others so that when the structure, infinitely long, is displaced along its length by one period it cannot be distinguished from its original self.

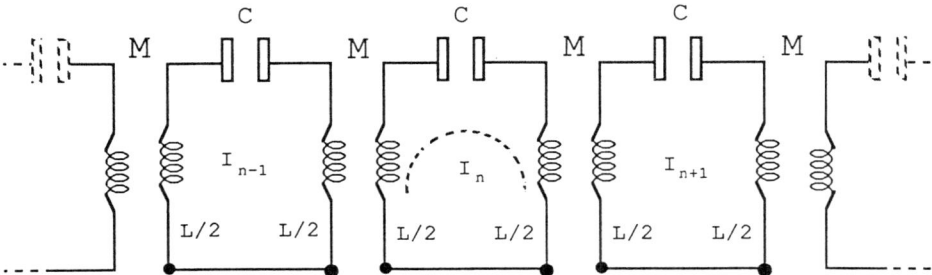

Figure 3-12 Chain of lossless resonators coupled by mutual inductance. The voltage that develops across each capacitor schematizes the gap voltage of the corresponding cavity.

Consequently at a given frequency the current in one mesh differs from those one period away only by a complex constant. (Careful! The above considerations are an intuitive proof of a particular case of the famous Floquet theorem).

If, as in our case, there are no losses then the constant can be only imaginary (propagation) or only real (attenuation).

If imaginary we put

$$I_{n-1} = I_n e^{-j\varphi} \quad ; \quad I_{n+1} = I_n e^{+j\varphi}.$$

Substituting, simplifying, and solving for the radian frequency we obtain

$$\omega = + \frac{1}{\sqrt{LC\left(1 - 2\frac{M}{L}\cos\varphi\right)}}. \qquad (1-12)$$

If the phase advance per cell must be equal to π then $\cos\varphi = -1$ at the radian frequency:

$$\omega_\pi = \frac{1}{\sqrt{LC\left(1 + 2\frac{M}{L}\right)}}.$$

We note, in passing, that the frequency is a periodic even function of the propagation constant as it is shown in (1-12).

Although it was derived in a very particular case the above statement is general for the periodic structures.

b. Electron Acceleration

Electrons and positrons have such a low mass (equal to 0.51 MeV) that those particles becomes ultrarelativistic even for moderate values of the kinetic energy. The well-known formula

$$\beta = \sqrt{1 - \frac{1}{\left(1 + \frac{E}{m_0 c^2}\right)^2}} \qquad (2-12)$$

(where E is the kinetic energy of the particle) and the table calculated for a rest mass equal to 0.51 MeV show that for energies larger than ~10 MeV the velocity of such particles can be considered technically equal to "c" for any practical purpose:

E	1	10	100	1000
β	0.94123	0.99882	0.99998	0.99999

Consequently the large value of the speed of light (2.99768 m/s) would render prohibitive the length of an accelerator equipped with drift tubes and, in fact, a different technique has been used.

A TM electromagnetic wave can accelerate indefinitely a charged particle if the phase velocity of the wave is at any time equal to the speed of the particle. [When the phase velocity of the wave is different from the velocity of the particle then the particle experiences an alternating field and no acceleration can take place on the average. However it is possible to demonstrate that if the phase velocity of the accelerating wave is equal to "c" then for each value of the speed of the particle exists a minimum value of the accelerating field that is sufficient for "capturing" the particle. If this condition is verified then the captured particle travels along the machine with an increasing phase lag on the accelerating wave].

Unfortunately in a uniform waveguide the phase velocity of a mode must be smaller than c, then the K_c of the mode becomes imaginary! In fact

$$K_c^2 = \left(\frac{2\pi}{\lambda}\right)^2 - \left(\frac{2\pi}{\lambda_g}\right)^2 = \left(\frac{2\pi}{cT}\right)^2 - \left(\frac{2\pi}{V_pT}\right)^2$$

becomes negative if the phase velocity V_p is less than c.

This fact has a very important consequence. Consider, for instance, a uniform waveguide with circular cross-section. The accelerating component of the electric field, in the fundamental mode, is

$$E_z = E_0 \, J_0 \, (K_c r)$$

as we already know.

If now we let $K_c = j\tau$ then the Bessel function $J_0(j\tau r)$ becomes the modified Bessel function $I_0(\tau r)$. This function is monotically increasing and has no zeros. This means that $I_0(\tau r)$ cannot be zero for r=a as the uniform boundary condition demands.

What has been said for the fundamental mode can be extended to the others and we conclude than a uniform waveguide cannot support a slow wave (i.e. a wave with $V_p < c$).

In the following we will show that slow waves, in the TM mode, can propagate along the so-called corrugated waveguides.

Consider the structure obtained by loading periodically a circular wave guide with conducting rings having an outer diameter slightly smaller than the one of the guide.

Figure 4-12 shows an axial section of this new device.

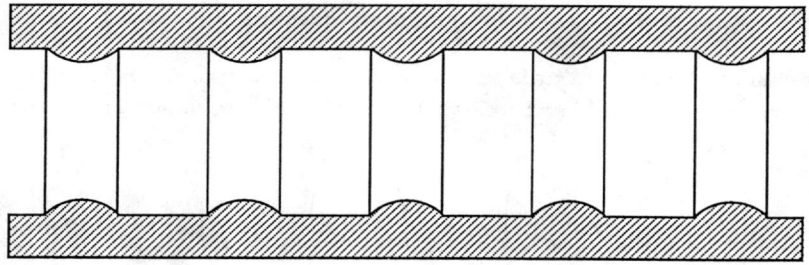

Figure 4-12. Axial section of a corrugated waveguide.

We may ask what happens if a TM_{0n} monochromatic wave is injected into the guide.

It is rather obvious that the periodic small disturbance will not affect the cut-off property of the guide and, consequently, the $\omega-\beta$ diagram of the guide will be unchanged in the neighborhood of the origin.

Increasing the frequency, the unperturbed wavelength λ_g of the electromagnetic wave tends to decrease and we are inclined to think that the propagation along the guide will take place.

This is true for a "small" increase of the frequency, but the reflections due to the periodic obstacles cannot be ignored.

In fact an observer who studies the field inside the guide would notice that by increasing the frequency the reflections from the successive obstacles tend more and more to add in phase. At a frequency for which the phase between obstacles is equal to π the interference is totally constructive and the propagation is stopped.

If we continue to increase the frequency then we will reach another frequency band where the accumulated reflections from an infinite length of guide mostly add up to zero. This is the second pass band.

Further increases of the frequency will create situations where the interference among forward and backward waves will be mainly constructive (stop-bands) or mainly destructive (pass-bands).

We conclude that a corrugated wave guide exhibits both pass and stop bands.

This is very important because it implies that the $\omega-\beta$ diagram of a corrugated waveguide must deviate from the hyperbola, bending, with horizontal tangent, at the start and at the end of each band. [In fact we recall that $d\omega/d\beta$ measures the group velocity, and the group velocity is zero, by definition, at the cut-off frequencies].

The bending of the $\omega-\beta$ diagram implies that the phase velocity of the wave in the corrugated guide can be lower (or much lower) than the one that, at the same frequency, propagates inside the unperturbed guide.

Figure 5-12 shows the $\omega-\beta$ diagram of a simple corrugated guide.

Let us suppose that the fields inside the guide depend upon the time as EXP($j\omega t$) and that the function of the coordinates

$$\overline{E} = \overline{E}\ (r,\theta,z) \qquad (3-12)$$

is periodic in z with period equal to the distance between the corrugations.

Then it easy to prove, by substitution, that the function

$$\overline{E} = \overline{E}\ (r,\theta,z)\ e^{-\gamma z} \qquad (4-12)$$

obeys the Floquet theorem and consequently it can represent, for instance, the longitudinal component of the electric field inside the corrugated guide.

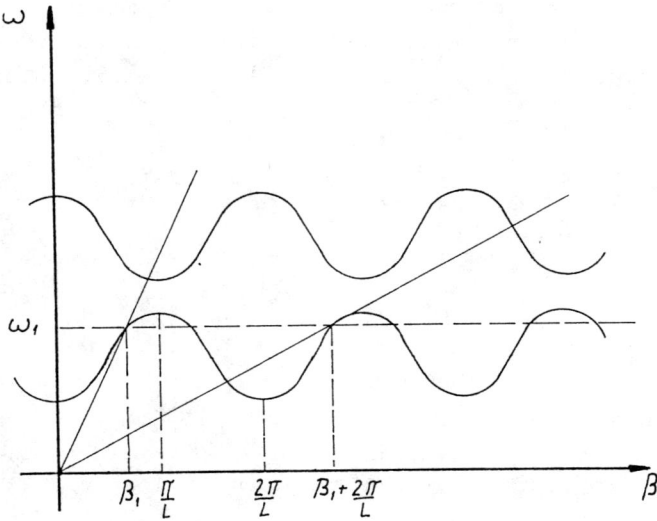

Figure 5-12. Brillouin diagram for a waveguide periodically loaded with washers. β_1 is the first value of the propagation constant that corresponds to the radian frequency ω_1. Other values are shown. The upper solid line indicates the second band-pass.

The function $\overline{E}$ can be expanded in the exponential Fourier series

$$\overline{E} = \sum_{-\infty}^{+\infty} E_n(r,\theta) \; e^{-j\frac{2\pi n}{L} z} \qquad (5-12)$$

where

$$E_n = \frac{1}{L} \int_0^L \overline{E}(r,\theta,z) \; e^{+j\frac{2\pi n}{L} z} \; dz$$

and consequently we can write

$$\overline{E} = \overline{E}(r,\theta,z) \; e^{-\gamma z} = \sum_{-\infty}^{+\infty} E_n(r,\theta) \; e^{-\left(\gamma + j\frac{2\pi n}{L}\right)z} . \qquad (6-12)$$

If there are no losses in the system then γ is either purely real (stop band) or purely imaginary (pass band) and obviously is function of the radian frequency.

If γ is imaginary we write, as usual,

$$\gamma = j\beta_0$$

and we obtain

$$\beta_n = \beta_0 + \frac{2\pi n}{L}.$$

The nth term in the equation (6-12) is known as the nth space harmonic and its phase velocity, V_{gn}, is

$$V_{gn} = \frac{\omega}{\beta_0 + \frac{2\pi n}{L}}. \qquad (7-12)$$

From (7-12) it is immediately clear that at any frequency for a given mode of propagation there is an infinite number of discrete phase velocities characterizing the mode. That is, for each frequency and mode there is no unique phase velocity.

In figure 5-12 we considered the case of small corrugations. Small corrugations means small perturbations of the uniform guide. Consequently we expect that a large change in frequency is required for going through the first pass-band.

Conversely the larger are the corrugations the more the whole structure tends to a series of lightly coupled cavities, and in this case a small change in frequency can produce a large phase shift per cell.

The conclusion is that the larger are the corrugations the smaller are the amplitudes of the pass-bands.

It is obvious that losses, power transmission, attenuation, group and phase velocity depend upon the corrugations. The technique in this field is now very sophisticated and needs special knowledge in the microwave domain; for this reason it cannot be clarified in the modest limit of these lectures.

As a practical consideration we can say that in the current technique the phase velocity of the fundamental component of the accelerating wave is made equal to "c" while the associated group velocity is kept around 1/100 of this value.

At this point a nice conclusion of this part is to demonstrate that if the phase velocity is maintained equal, and not superior, to "c" then the injected particles can be captured and accelerated to any value of the energy.

Let $dl = (c-V_p)dt = c(1-\beta)dt$, the difference in the path between the wave and the particle. This difference can be expressed in terms of phase angle and we obtain

$$d\varphi = \frac{2\pi}{\lambda} dl \quad \text{or} \quad \frac{d\varphi}{dt} = \frac{2\pi c}{\lambda} (1-\beta) . \qquad (8-12)$$

From the relativistic equation of the motion we have

$$\frac{d}{dt}\left(\frac{m_0}{\sqrt{1-\beta^2}} c\beta\right) = q E_0 \sin\varphi \qquad (9-12)$$

where m_0 and q are respectively the rest mass and the charge of the particle.

E_0 is the amplitude of the accelerating field and φ is the phase seen by the particle.

Expanding equation (9-12) and using the identity

$$\frac{d\beta}{dt} = \frac{d\beta}{d\varphi} \frac{d\varphi}{dt}$$

we obtain

$$\frac{1}{(1-\beta^2)\sqrt{1-\beta^2}} \frac{d\beta}{d\varphi} \frac{d\varphi}{dt} = \frac{q E_0}{m_0 c} \sin\varphi . \qquad (10-12)$$

Eliminating the $\frac{d\varphi}{dt}$ with the aid of Eq. (8-12) we obtain

$$\frac{d\beta}{(1+\beta)\sqrt{1-\beta^2}} = \frac{q E_0 \lambda}{2\pi m_0 c^2} \sin\varphi \, d\varphi .$$

With the change of variable $\beta = \cos\alpha$ the integration becomes straightforward and we obtain

$$\left.\text{TAN } \alpha/2\right|_{\alpha_0}^{\alpha} = + \frac{q\, E_0 \lambda}{2\pi m_0 c^2} \cos\varphi \Big|_{\varphi_0}^{\varphi} . \qquad (11\text{-}12)$$

Now we consider that:

1. $\text{TAN } \alpha/2 = \sqrt{\dfrac{1-\cos\alpha}{1+\cos\alpha}} = \sqrt{\dfrac{1-\beta}{1+\beta}}$.

2. The absolute value of the difference between two cosines cannot be larger than two.

3. At the end of the acceleration $\beta = \sim 1$; thus $\sqrt{\dfrac{1-\beta}{1+\beta}} = \sim 0$.

Consequently from (11-12) it follows that

$$E_0 \geq \frac{\pi m_0 c^2}{q\, \lambda} \sqrt{\frac{1-\beta_0}{1+\beta_0}} \qquad (12\text{-}12)$$

where β_0 is the normalized velocity of the particle at injection and E_0 is the minimum value of the amplitude of the field that is required for capturing the particle.

Acknowledgements

At the end of this small piece of work the author would thank Marina Nadalin for having interpreted and typed the manuscript together with the painstaking work of drawing all the figures.

BIBLIOGRAPHY

The treated topics followed a development that went on for at least 50 years. Consequently the articles with the original ideas are now much too old to be easily found. On the other hand the more recent ones are very specialized and difficult for beginners.

Our choice is to suggest, to those interested, only a careful study of one of the many fundamental books in the RF field. The following list of books, ordered in terms of increasing complexity, could be a good guide.

1. L.B. Arguinbau
   "Vacuum Tubes Circuits", McGraw-Hill
   [Excellent introduction to the general problems of radiofrequency circuits].

2. S. Ramo, J. Whinnery, T. Van Duzer
   "Field and Wave in Communication Electronics"
   [It contains a very good exposition of the electromagnetic theory, is simple to study and very complete].

3. A. Massarotti, M. Puglisi
   "Fondamenti di Radiotecnica", Zanichelli, Bologna
   [Elementary, but complete, treatment of radio engineering].

4. J.P. Heyboer, P. Zijlstra
   "Transmitting Valves", Phillips Technical Library
   [Contains the basic information concerning the power termoionic tubes].

5. J.C. Slater
   "Microwave Electronics", Dover, New York
   [A very good treatment of cavity resonators and linac accelerators (microwave tubes are considered). Can be studied only with some background from Ref.2].

6. R.E. Collin
   "Field Theory of Guided Waves", McGraw-Hill
   [Exhaustive treatment of the problem at high level].

7. Ginzton, Edward, L.
   "Microwave Measurements", McGraw-Hill, 1857
   [An invaluable book because, apart from an accurate description of many measurement techniques, it contains a physical picture of the microwave electronics].

PRINCIPLES AND TECHNOLOGY OF RFQs

A. Schempp

Institut für Angewandte Physik, Universität Frankfurt
D 6000 Frankfurt/M, W. Germany

TABLE OF CONTENTS

1.	Introduction	1478
2.	RFQ Design	1479
3.	The 4–Vane RFQ	1483
4.	RF Tuning	1486
5.	The 4–Rod RFQ	1490
6.	Beam Experiments	1492
7.	Applications	1495
8.	References	1498

PRINCIPLES AND TECHNOLOGY OF RFQs

A. Schempp

Institut für Angewandte Physik, Universität Frankfurt
D 6000 Frankfurt/M, W. Germany

1. INTRODUCTION

The RFQ is a linear accelerator structure for low velocity ions which focuses and accelerates with the help of electrical radio-frequency (rf) quadrupole fields. It was first proposed by Kapchinskij[1] and has since been applied in numerous labs mainly as injector for Alvarez accelerators. The RFQ can simultaneously focus, bunch, and accelerate a high current beam directly from the ion source, which can be on a relativly low dc potential. This allows the the use of special ion sources which can be bulky and can have high power consumption like sources for polarized or highly charged ions.

First, general principles of RFQ accelerators and typical limits imposed by particle dynamics are presented. Second, typical problems of rf and mechanical design of this new type of accelerator are discussed, with the RFQ injectors built for DESY as examples to demonstrate typical properties of this type of low energy, high current injector.

For the HERA project at DESY in Hamburg an RFQ has been chosen as injector (20 mA H$^-$, 750 keV) for the 50-MeV Alvarez linac[2,3]. Two different RFQ cavities have been built, tuned, and tested. The first RFQ-resonator is a Four-Vane cavity with resonantly coupled quadrants. The second structure is the Four-Rod RFQ resonator, which was developed in

Frankfurt as an alternative solution for ion injectors. Both cavities operate at a frequency of 202 MHz and fit into the HERA injection system. Properties and special features of both structures are discussed together with results of beam tests and proposed new applications of RFQs.

2. RFQ DESIGN

The concept of spatial homogeneous focusing[1] has closed the low velocity gap of high frequency ion accelerators. The work has ignited numerous activities starting with the thorough work in Los Alamos[4]. While first the aim was the improvement of high energy proton accelerators, possible applications for heavier ions were seen very early and research started early too[5-8]. For protons and light ions the 4-Vane RFQ is the structure mostly used. But frequencies as low as 80-100 MHz, as chosen for FMIT and TALL[9,10], seem to be the lower limit. For low charged heavy ions, frequencies between 10 MHz and 30 MHz must be chosen, which results in a too large diameter for the 4-Vane resonator. Therefore, at GSI, a Split-Coaxial RFQ[5] for heavy ions is being built, in Frankfurt the 4 – Rod RFQ structure[7,8,11,12] has been developed for low frequency applications.

Criteria for RFQ structures are, first, to provide beam dynamics requirements like sufficient acceptance, current limits, small emittance growth, possible high fields, and small quadrupole multipole components. The rf structure must supply the quadrupole voltage on the electrodes. The efficiency, described by the shunt impedance R_p, is as important as a high group velocity, which means strong coupling and good tolerances. A "good design" for an rf structure will give a flat field distribution along the cavity without dipole components, which is assumed in the beam dynamics calculations.

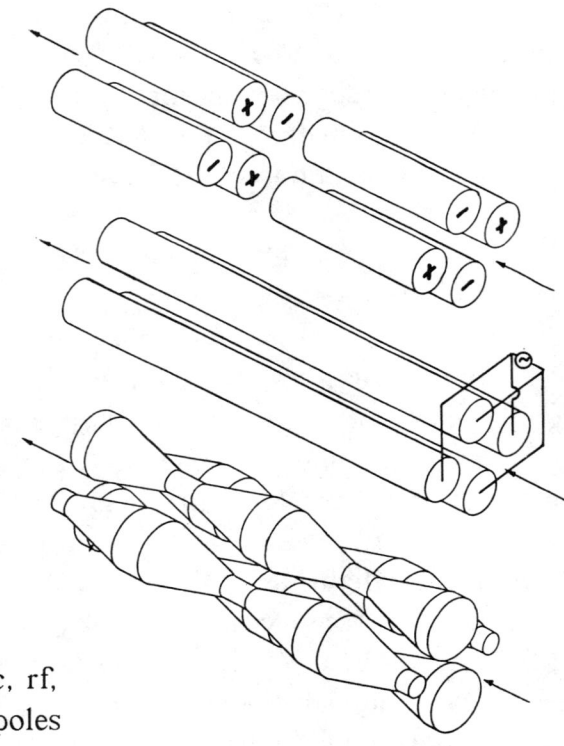

Fig. 1
Schemes of static, rf,
and RFQ quadrupoles

The basic structure of an RFQ consists of four electrodes arranged symmetrically around the beam axis and excited such that adjacent electrodes have the same voltage amplitude but $180°$ phase shift. This corresponds to an array of quadrupole singlets and is a pure focusing channel as first proposed by Paul[13] without accelerating field components. A mechanical modulation of the electrodes as indicated in Figs. 1 and 2 now introduces an axial accelerating field E_z. For a given electrode voltage U the amplitude C of the accelerating field will be proportional to the modulation m of the electrodes while the focusing strength, which is proportional to the radial field E_r, will decrease with increasing modulation.

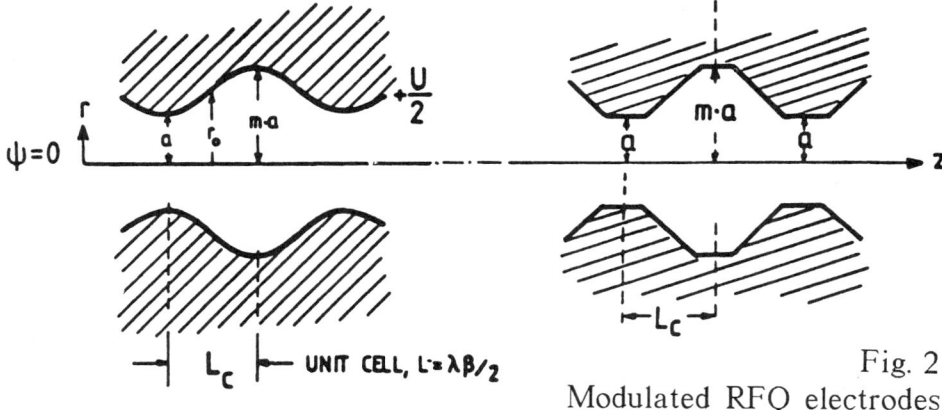

Fig. 2
Modulated RFQ electrodes

The pure rf-quadrupole focusing channel fields are described by
$$E_r = E_\psi = \frac{U}{a^2} r \cos 2\psi \, e^{i\omega t} ; \qquad E_z = 0 .$$
Adding a lowest-order accelerating field with $E_z \neq 0$ as in a normal linac changes also the focusing:
$$E_z = C \cdot I_0(kr) \sin(kz) e^{i\omega t}; \quad C = k A U/2 , \quad k = \omega/v = \beta \, 2\pi/\lambda_0 ;$$
$$E_r = \{ X \frac{U}{a^2} r \cos 2\psi - I_1(kr) \cos(kz) \} e^{i\omega t}$$
with the abreviations $\quad A = \frac{m^2 - 1}{m \, I_0(ka) + I_0(mka)} ; \quad X = 1 - A \, I_0(ka);$

particle velocity v, $\beta = v/c$, rf frequency ω, aperture a.
These fields can be derived from a potential
$$\Phi = \frac{U}{2} \{ X \left(\frac{r}{a}\right)^2 \cos(2\psi) + A \, I_0(kr) \cos(kz) \} e^{i\omega t}$$
which also determines the electrode shape. The focusing fields are weakened by introduction of the acceleration field indicated by the parameters X and A.

Since the parameters are dependant on each other, in the "design" of an RFQ one must choose suitable values for the frequency, the injection and final energies T_i, T_F, the aperture a, the electrode modulation m, and the electrode voltage U. Major constraints are the required beam current to be accelerated, which sets limits to the modulation, and the maximum applicable voltage, which is described by Kilpatrick's [14] criterion. This indicates how technical questions like sparking have to be taken into account early in the particle dynamics design.

The RFQ system provides homogeneous strong electrical AG focusing and allows short accelerating cells and therefore relatively high operational frequency. This makes adiabatic bunching[1] attractive, which efficiently converts a dc beam to a bunched beam by continuously changing the stable phase φ_s and the accelerating field E_z with very small emittance growth.

The beam dynamics design for the ion beam, i.e. the design of the shape of the RFQ electrodes, results in a continuous change of aperture, modulation and cell length L (corresponding to the stable phase φ_s) along the RFQ, as demonstrated in Fig. 3, which gives results for the HERA RFQ design. Optimisation (choice of parameters along the RFQ) with respect, e.g., to beam current, emittance, transmission, and length leads to different "designs" and lengths of the RFQ structure. Beam properties are fixed by the special design because these parameters are "milled" into the electrodes and only the voltage can be adjusted during operation ("one knob structure").

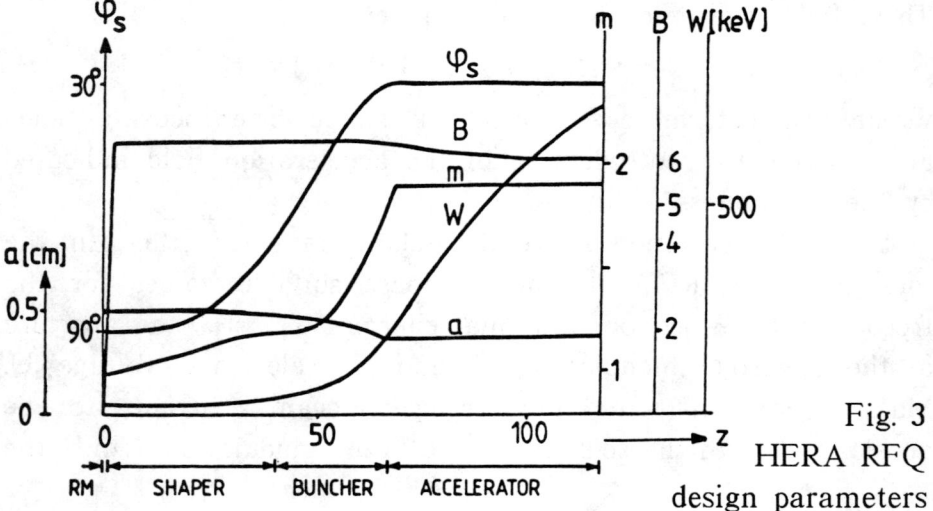

Fig. 3 HERA RFQ design parameters

The injector for the 50 – MeV Alvarez linac for the HERA Project at DESY [2,3] has a low duty cycle, and the H⁻ scheme for injection into the first synchrotron ring (DESY III) allows for a linac design current of only 20 mA. Figure 4 shows a schematic of the Alvarez injector setup at DESY.

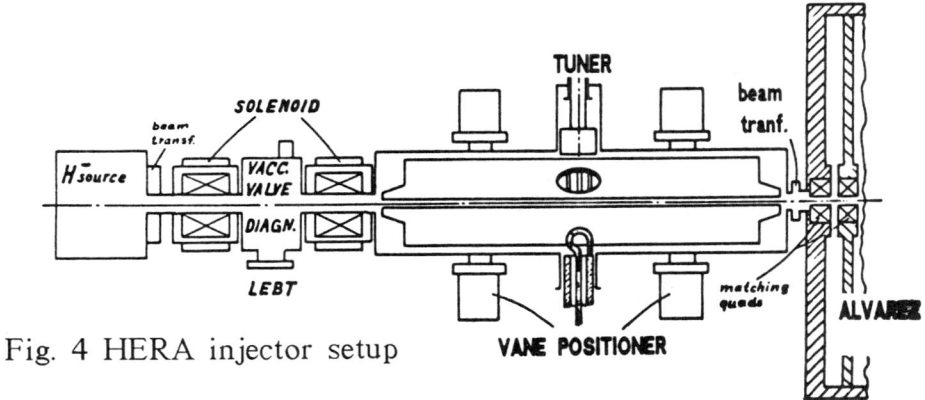

Fig. 4 HERA injector setup

The beam dynamics calculations have shown that, starting with the extraction energy of the FNAL-type Magnetron source of 18 keV, a modest RFQ elecrode voltage of 70 kV can be chosen, resulting in a short RFQ of length 1.2 m to reach the final energy of 750 keV with proper beam properties[15]. The beam dynamic design shown in Fig. 3 was made by using the standard LANL[4] approach with minor modifications and has been tested with PARMTEQ. The results show high transmission and good beam emittance.

3. THE 4-VANE RFQ

It was decided to built a 4-Vane cavity, the RFQ-structure developed by LANL[4,16,17] and succesfully operated at several big labs. Mechanical and rf problems go along with this structure. To make the tuning easier and ensure operational stability we looked for ways to simplify the mechanical design and for an appropriate rf stabilisation.

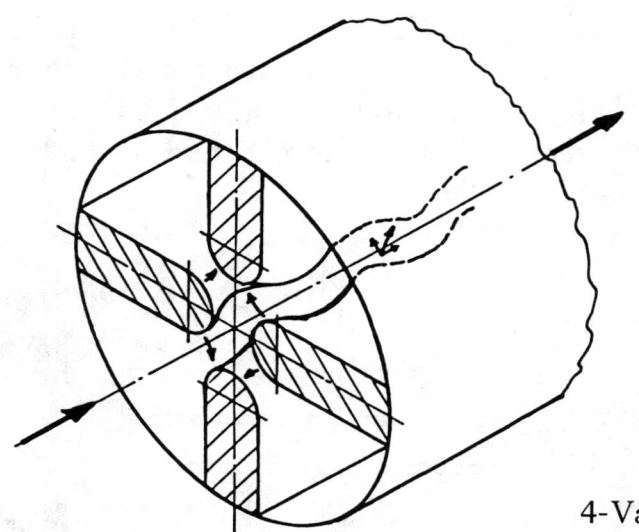

Fig. 5
Scheme of
4-Vane RFQ resonator

The 4-Vane resonator consists of a cylinder in which four vanes with sine-wave-like vane tips are mounted symmetrically as shown in Fig. 5. The resonator is excited in a TE_{210}-like mode which provides a electrical rf-quadrupolefield on the axis. In addition to the precision with which the pole shape must be produced to give proper axial field distribution, the rf properties require a highly symmetric cavity to avoid dipole components in the axial field which deteriorate the beam quality.

The mechanical design differs from other 4-Vane RFQs. We tried to make a separated function structure. Mechanical adjustment, rf contacts, rf tuning, stabilization, and vacuum cooling can be done independently, i.e. changed independently. There are two 3D positioners per Vane for mechanical adjustment and two contact bars on each side of the Vane for rf contacting. The rf tuning and stabilizing is done by modifying the ends of the Vanes ("end cells"). Only the outer cylinder is cooled, and vacuum can be applied from the outside at the very last. Figure 6 shows a cross section of the RFQ.

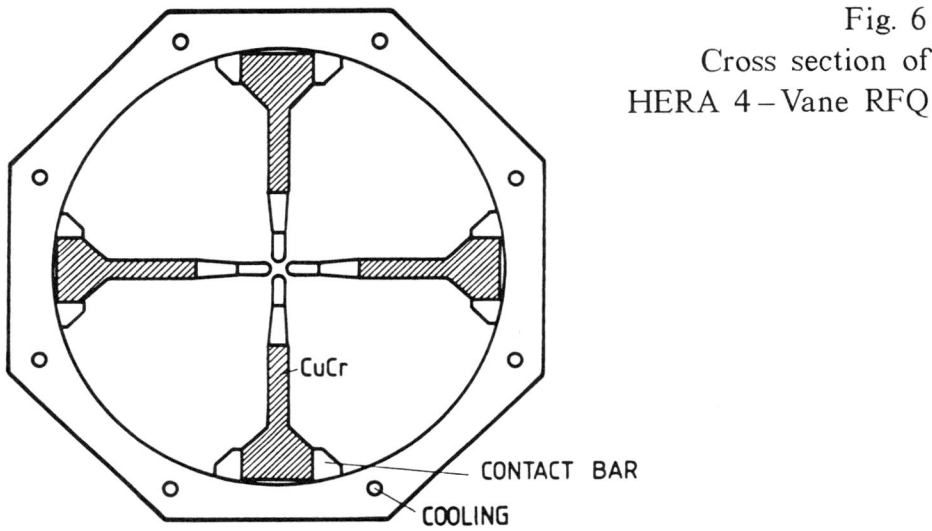

Fig. 6
Cross section of
HERA 4–Vane RFQ

The vanes, as the most critical part, are made of CuCr alloy, which has favourable mechanical properties but still very good electrical and heat conductivity. The Vanes were machined from a solid forged block which had been annealed at high temperature in order to avoid bending by residual internal stresses after milling. Careful machining has been done at Pfeiffer (Balzers) at Asslar. The Vane tip dimensions and the modulation proved to have deviations of < 10 μm from the theoretical values. This was measured independently from the production by Komeg (Zeiss). Measurement of the modulation and the reference edges was done by comparing 2300 points along the vane with theoretical values. Figure 7 shows plots for the mechanical measurements at the RFQ entrance, the middle, and the high energy end of the vane. The deviations are enlarged by a factor of 150. The maximum deviation occurs at the low energy end at the transition from the radial matching section to the shaper. There the theoretical set of values had not been smoothed out as had been done for the milling machine values. The vanes have been installed and adjusted, again with use of the Zeiss 3D machine to get a good basis for rf tuning.

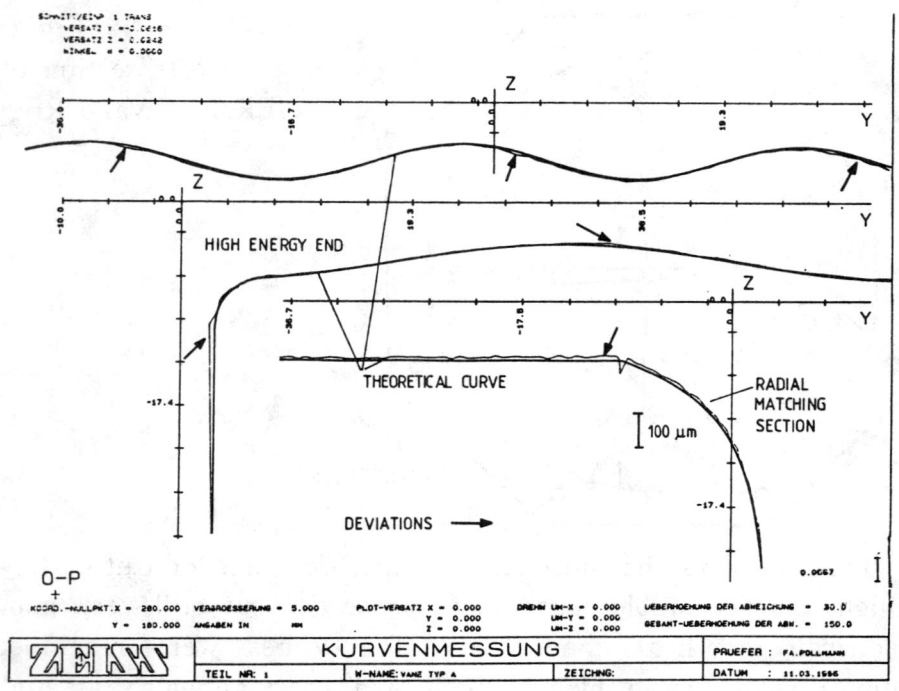

Fig. 7 Measurement of Vane tip modulation

The position of the axis was defined by the cavity, which is round within +- 10 μm. The middle of the first vane was then taken as a reference plane by adjusting the parallel parts of the vane and correcting with the measurements of the individual vanes. Within two weeks all vanes had been adjusted with all critical values being well within the 10-μm margin, with the fabrication deviations of the individual parts taken into account.

4. RF TUNING

After installation of the contact bars the frequency of the quadrupole mode was 198.1 MHz (dipole mode frequencies 198.6 MHz, 198.3 MHz). Frequency coarse tuning had been done by modifying the cut-back areas of the Vanes, as indicated in Fig. 8. Fine tuning was done with perturbation rings screwed to the end plate, which is a simple and symmetric way of tuning.

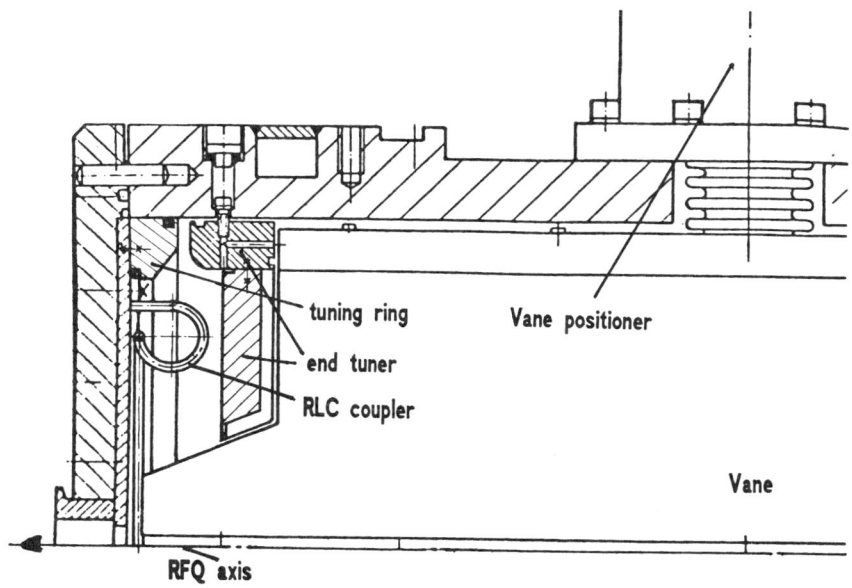

Fig. 8   Vane end cell configuration

It was planned to get an operating frequency slightly above the cut-off frequency, which results in a somewhat concave longitudinal field distribution, as shown in Fig. 9, which shows measurements near the vane tips. The magnetic field B near the outer cylinder in the midplane between the vanes is shown for comparison. The magnetic field shows a stronger variation due to the different distributions at the end of the resonator. The small bumps indicate the position of the four pick-up loops and the central drive loop in this quadrant.

A basic problem for 4-Vane resonators is the balancing of the four quadrants, which is one reason for the required precision of the manufacturing and tuning. The four quadrants are very weakly coupled, with the result that the frequencies of

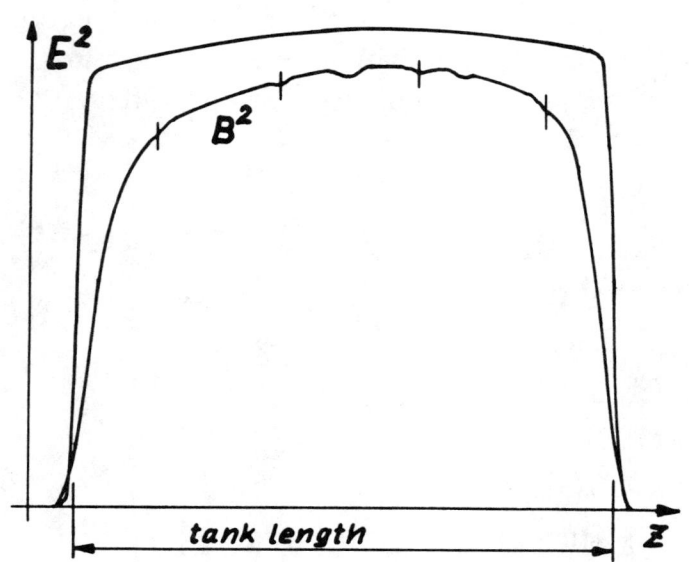

Fig. 9
Measured fields along the RFQ after tuning

two dipole modes, which have opposite polarity of the electrode voltages, differ by < 0.5 MHz from those of the quadrupole modes. This gives rise to mode mixing, which makes the voltage distribution in the resonator "unflat".

After frequency tuning, the fields in the four quadrants varied by up to 40% and were very sensitive to changes of the drive loop and the tuning ball because the closest dipole mode was only 0.11 MHz higher. The azimuthal fields were stabilized with one RLC [18] stabilizing ring at each end of the cavity, which converts the azimuthal π mode to a $\pi/2$ mode. The mechanical design is simple because the RLCs are fixed to the tank end plate (Figs. 8, 10) and act on the fields in the end cells. By tuning the RLCs to a frequency of 201.9 MHz (outside the tank, as shown in Fig. 10) a symmetric shift of the dipole modes could be achieved and a very symmetric (+-2%), very stable field distribution in the quadrants. With the RLC couplers, even after detuning of the cavity asymmetrically with the plunger, placed in the quadrant opposite the feeding loop, by 0.5 MHz, no change of the field distribution could be detected.

Part of the tuning procedure had been the operation of the cavity with rf power up to 1.5 kW and up to 100% duty cycle. Thermal effects of rf heating could be studied, and operation in this typical multipactor level should facilitate rf conditioning under higher power and small duty cycle at DESY. The results of the tuning procedure show that the design goals have been achieved successfully[19]. The field flatness is excellent (+-2% longitudinally and azimuthally) and stable against detuning and temperature changes, and the impedance is very high for this kind of cavity. The R_p value (ratio of electrode voltage U to rf power N ; $R_p = U^2/N$) is as high as 65 kΩ with an unloaded Q value of Q = 12000 (Q for Superfish = 13500). The ratio Q/Q_{SF} = 0.9 is very high; thus we are close to the minimum rf power of N = 66 kW calculated for the design voltage of 70.5 kV (without beam loading).

Fig. 10 View of the RLC coupling ring

5. THE FOUR-ROD RFQ

Generally for accelerator cavities the shunt impedance as well as the tolerances are going directly into costs of the system. Simplicity and reliability mostly come together and are most important criteria for RFQs. A basic problem of the 4-Vane structure is the high symmetry necessary to avoid dipole components of the axial fields even when resonant coupling schemes can be applied. The Split-Coaxial as well as the 4-Rod RFQ don't have these symmetry problems. Precision is only determined by beam dynamics requirements. Thus these structures are simpler and therefore cheaper.

The 4-Rod RFQ applies cylindrical rods with conical varying diameter as electrodes. The influence of higher field harmonics on beam properties has been investigated thoroughly[20] and is proven to be negligible as long as the full aperture is not used. The resonator's basic cell consists of two $\lambda/2$ oscillators excited in the transverse π mode to give the proper quadrupole field distribution between the electrodes. The accelerating structure consists of a chain of these cells operating in longitudinal 0-mode. Figure 11 shows two cells of a linear version of this structure.

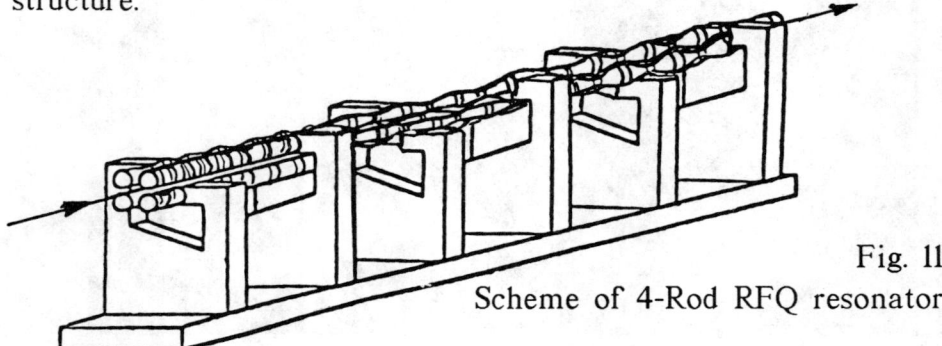

Fig. 11
Scheme of 4-Rod RFQ resonator

Optimizing with respect to shunt impedance results in an equidistant arrangement of the stems[12]. The impedance R_P, the ratio between electrode voltage and rf power consumption, of the 4-Rod RFQ is astonishingly high also for high frequencies.

A 4-Rod RFQ for light ions has been designed and built as a second RFQ injector for HERA to prove the properties of this new type of RFQ resonator[21]. The beam dynamics parameters, (i.e. the electrode modulation along the RFQ) are the same as for the 4-Vane RFQ, allowing good comparison of the structures' characteristics and beam properties.

The rf structure consists of a massive water-cooled ground bar with fourteen specially shaped stems to carry the electrodes. Figure 12 shows a cross section of the resonator. The electrodes were made by taking the HERA RFQ design data and approximating the sinusoidal shape by trapezoidal elements.

Tuning of the RFQ has been done with shorting plates in two cells at each end. Conditioning with rf power up to 25 kW (5% dc) has been done in Frankfurt. High power operation (up to 120 kW, 3×10^{-4} dc) with beam has been done at DESY.

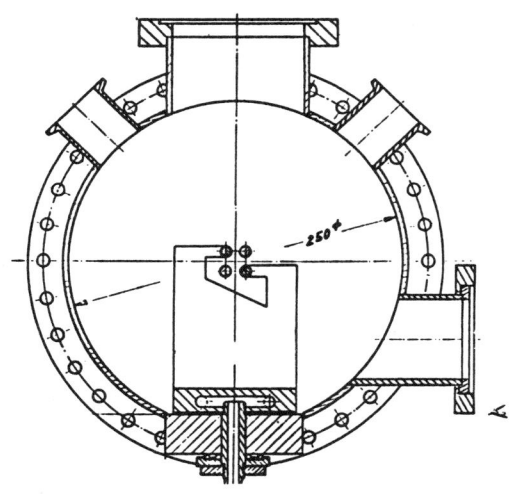

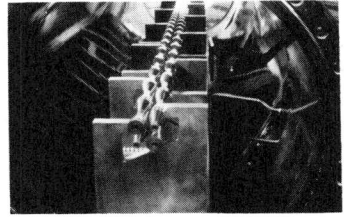

Fig. 12
Cross section
of The 4-Rod RFQ

Fig. 13
View of the
4-Rod RFQ resonator

6. BEAM EXPERIMENTS

Beam tests have been done at DESY at the HERA injector beam line, first with the 4-Rod RFQ that was used as rf and beam dump for the initial tests of the injector equipment.

After carefully conditioning the RFQs to get rf power levels up to 100 kW we had a first beam through the RFQ easily, using approximately injection design solenoid settings. Starting with an injected beam of 25 mA we increased the rf power up to 100 kW and analysed the beam[22]. Figure 14 shows the acceleration effects for different rf powers (electrode voltages). For the determination of the impedance a beam of only 0.2 mA has been used. For the 4-Rod RFQ the minimum power to accelerate ions up to 750 keV was 50 kW, corresponding to an electrode voltage of 61 kV (stable accelerating phase $\varphi_s = 0$ corresponding to $U = U_0 \times 0.866$, no beam loading), resulting in an impedance of the 4-Rod RFQ of $R_P = 90$ kΩm.

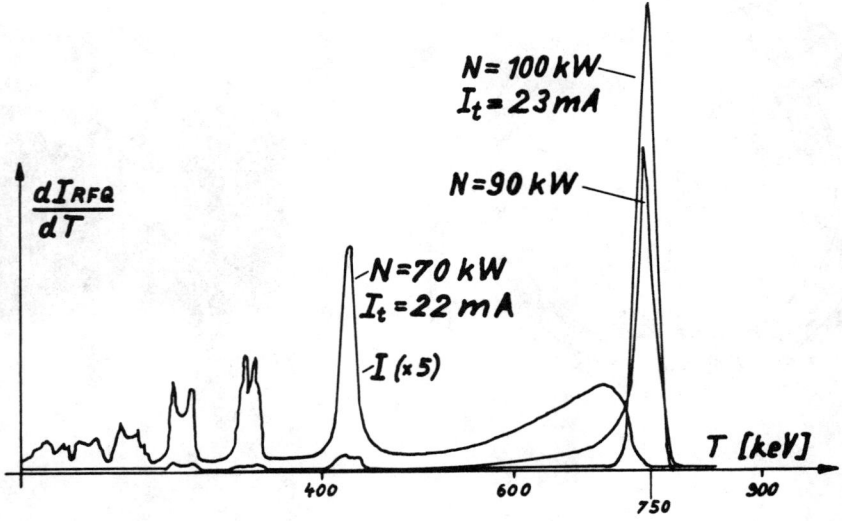

Fig. 14 Beam spectra for different rf powers

This value is ~ 10% higher than for the 4-Vane RFQ, for which 73 kW were needed for the design voltage of 70.5 kV (R_P = 80 kΩm). For the 4-Rod RFQ the maximum value I_m of the accelerated current has been I_m = 36 mA; due to beam loading the rf power then had to be increased to ~ 105 kW. The corresponding value for the 4-Vane was 43 mA with an ion-source beam current of 55 mA. After improvements of the injection system this value could be increased to I_m = 54 mA.

In a next step the microstructure was measured with a fast Faraday cup and the emittance of the accelerated beam pulse was analyzed. A diagnostic box containing a two-plane emittance measurement device with a distance of 16 cm between RFQ exit and the slit (slit-grid 5 cm) was used.

In Fig. 15 emittance ellipses are shown for both x and y plane for N = 90 kW (design power plus beam loading) and 20-mA beam pulses. Figure 16 shows the measured emittances ε_n as a function of the rf power N (constant beam current I) and as a function of the RFQ current I (constant rf power N).

The emittance predicted with the multiparticle code PARMTEQ is 1.0 πmm·mrad at an output current of 20 mA. The current limit is calculated to be 43 mA, which shows how close the performance of both RFQs comes to the theoretical values. Due to space-charge forces there is an expected aditional blow-up of the beam emittance also in the short drift behind the RFQ. This is one reason for coupling the RFQ as closely as possible to the following Alvarez accelerator.

The tests of both RFQs have been successful. The accelerated beam current is clearly higher than the design value, and beam emittances are better than the theoretical predictions, which could be explained e.g. by a better input emittance. The 4-Vane RFQ, which has been produced, aligned, and tuned with highest precision, delivers even a 50-mA H$^-$ beam which has the design emittance (90% of the beam is taken into account).

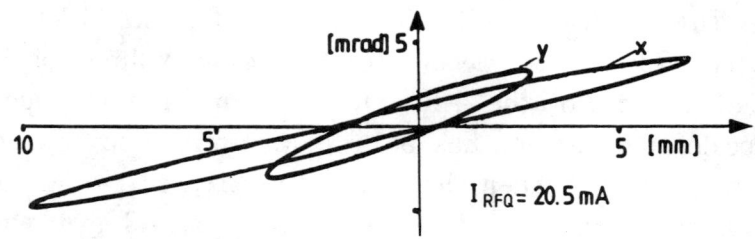

Fig. 15
Radial emittance of the RFQ beam

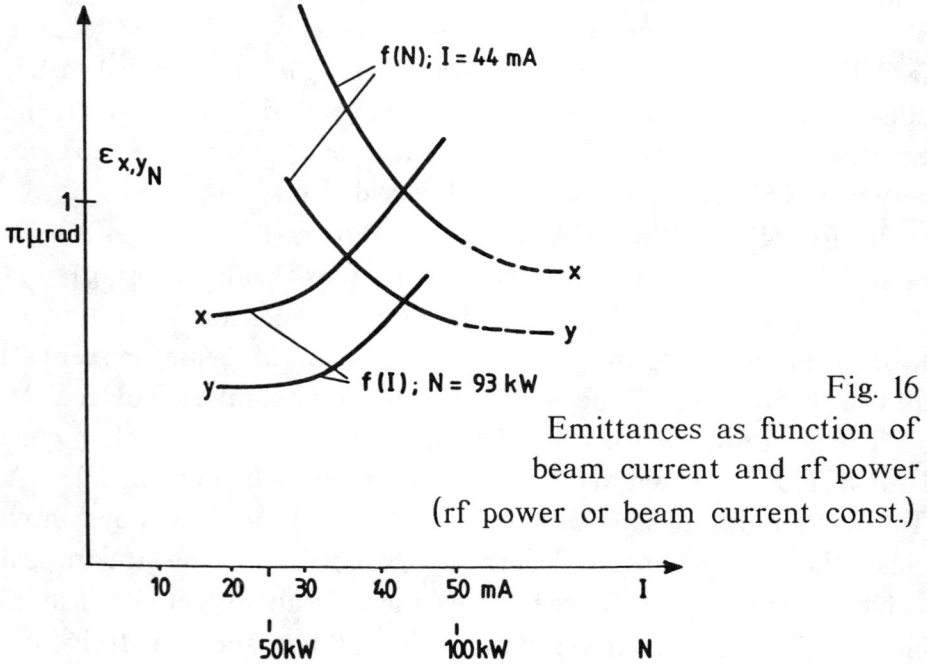

Fig. 16
Emittances as function of beam current and rf power
(rf power or beam current const.)

Although the time for measurement was very short, we were able to demonstrate that the 4-Rod RFQ delivers a nearly identical beam. The small differences can be reconstructed with beam dynamics codes because the first 20 cells with modulation < 0.1 mm were omitted, resulting in a small emittance growth and an ~ 10% smaller transmission.

The resonant coupling worked very well; therefore the 4-Vane RFQ also operated very reliably and stably.

7. APPLICATIONS

Numerous applications for RFQs make use of the high current capability and the good beam emittance essential for injectors, e.g. for high energy machines.

Other applications use the fact that the ion source operates with low potential and can replace big and power-consuming sources like those for polarized or highly charged ions. Particulary for the EBIS and ECR sources the combination with RFQs is attractive, especially the 4-Rod structure with the potential for lower frequencies.

The "HERA" Four-Rod RFQ is a prototype for future light ion injectors. It could be used to accelerate highly ionized ions from an EBIS source[23]. Exchanging the electrodes of the Hera RFQ, for example specific charges e/m = i.e. 0.25 (for protons e/m = 1) could be accelerated to 400 keV/N. More flexibilty can be achieved with lower frequencies. With an operation frequency of 108 MHz specific charges down to e/m = 0.1 could be accelerated efficiently.

We are working on a RFQ accelerator for the CRYRING[24] project in Stockholm, where ions from an EBIS are accelerated to 300 keV/amu and injected into a storage ring, as indicated in Fig. 17. The RFQ will have the following parameters: operating frequency 108.5 MHz, length 1.55 m, electrode voltage 70 kV, aperture 3.5 mm, T_i 10 keV/amu. With an additional debuncher the energy spread will be as low as 0.15% while the transmission is still 90%. The 4-Rod structure is favourably used there because of the efficient acceleration and the good vacuum properties needed for CRYRING and because of the relatively easy manufacturing, good operational stability, and the price, which will be lower by a factor of ~ 5.

The electrode design procedure has been revised and changes have been made which make the structure significantly shorter, and/or give higher ion beam currents without degrading the beam

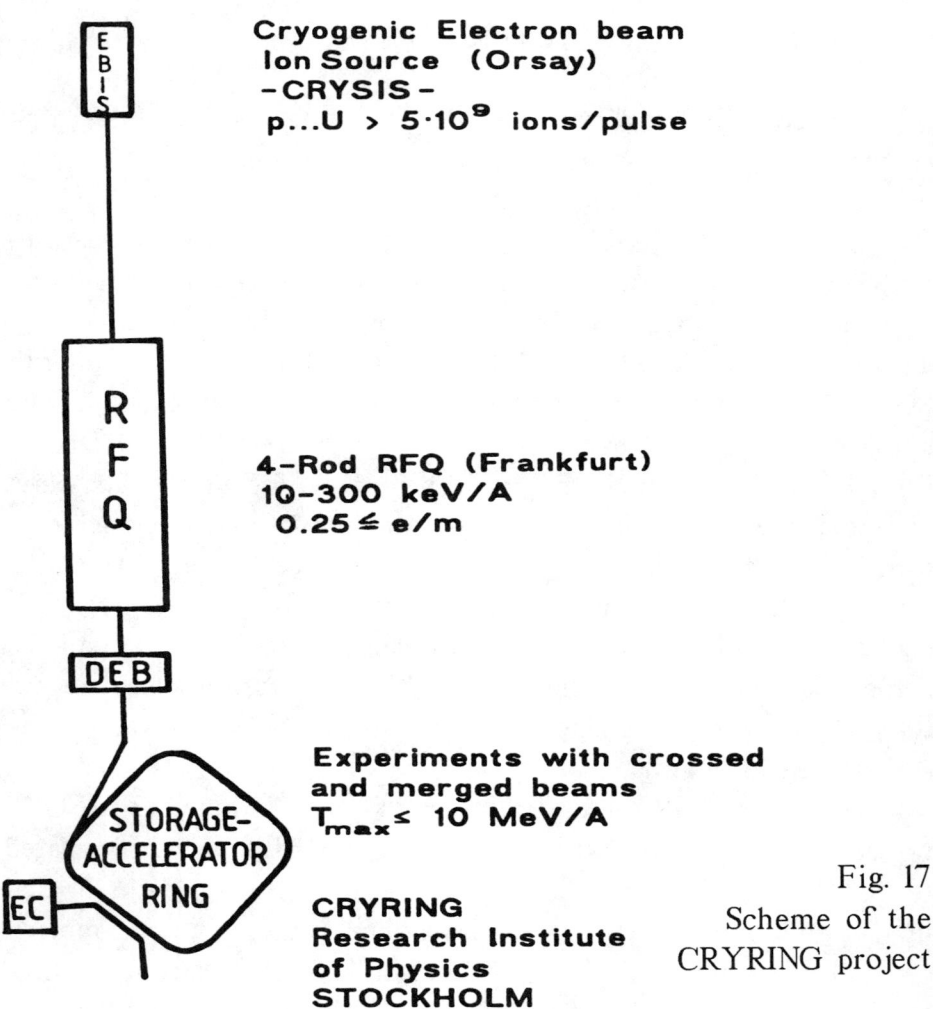

Fig. 17
Scheme of the CRYRING project

quality[25]. A solution similar to that for CRYRING is planned to be used as second injector for the UNILAC heavy ion accelerator at GSI[26]. A high duty cycle (50% d.c.) combination of an ECR ion source, a RFQ, and an IH structure could bypass the 27–MHz Wideroe prestripper part.

For lower charge states and higher currents the operating frequency has to be lower. The High Current Injector project at GSI makes use of a 13-MHz Split Coaxial RFQ (MAXILAC)[27] which is designed for acceleration of U^{2+} (25 mA) from

2.4 keV/amu to 130 kev/amu (this corresponds to an accelerator voltage of 15 MV and a beam power of 370 kW). This beam will be further accelerated in the second Wideroe part and the Alvarez post stripper part of the Unilac for injection into the heavy ion synchrotron SIS now under construction. An alternative 4-Rod RFQ solution for this heavy ion injector would allow a similar output current but higher final energy and better beam emittance, which better suits the "postaccelerator". These heavy ion injector RFQ schemes are shown in Fig. 18.

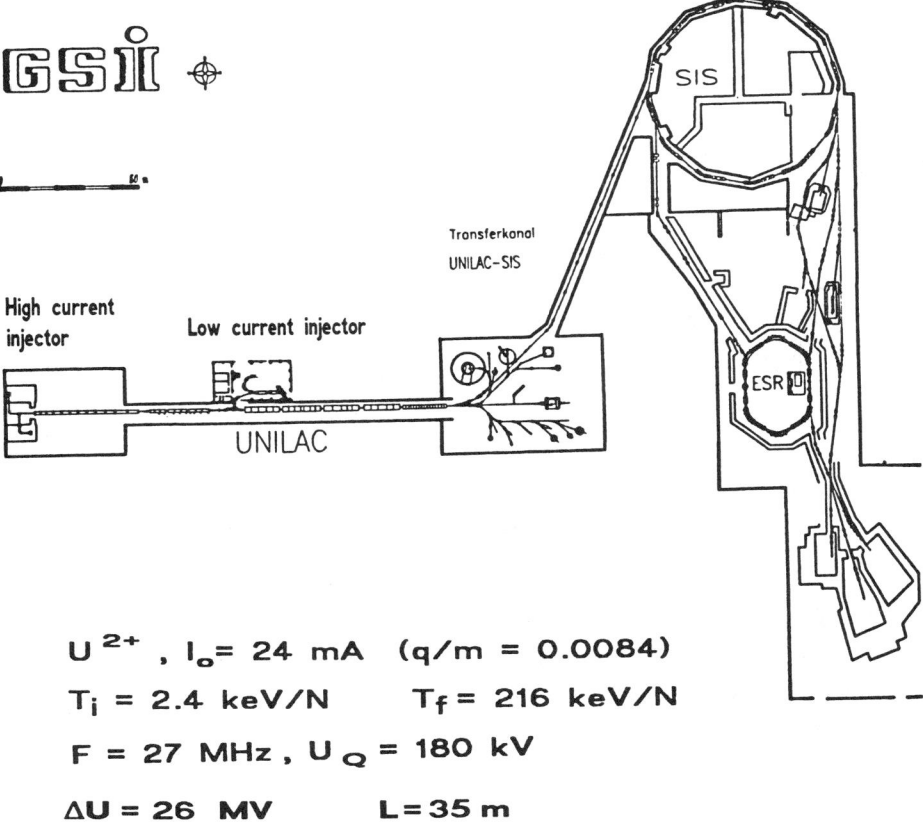

U^{2+}, I_o = 24 mA   (q/m = 0.0084)

T_i = 2.4 keV/N T_f = 216 keV/N

F = 27 MHz, U_Q = 180 kV

ΔU = 26 MV   L = 35 m

Fig. 18 The GSI heavy ion accelerator system

This is thought to solve some of the questions of the injector for a "Heavy Ion Fusion" accelerator[28] as well, where high energy heavy ion beams are planned to ignite D-pellets for power production.

Developments in increasing the high current capability of RFQs and the brightness of the ion beamsare under way.

Numerous low current applications, e.g. in the field of heavy ions, are suited for smaller labs as well. Industrial applications like ion implantation and neutron generators can also be seen. The RFQ structure development will furthermore simplify the mechanical and rf designs. But there could also be an application of superconductivity for special accelerators.

8. REFERENCES

1. I. M. Kapchinskij, V.A. Tepliakov, Prib. Tekh. Eksp., No. 4 (1970), p. 17, p. 19
2. U. Timm, HERA Report 84/12, DESY Hamburg (1984)
3. B.H. Wiik, IEEE NS 32, No. 5 (1985), p. 1587
4. K.R. Crandall, R.H. Stokes, T.P. Wangler, BNL51143 (1980) p. 20
5. R.W. Müller et al., Linac 79, BNL 51143 (1980) p. 148
6. J. Faure et al., Linac 84, GSI 84-11 (1984) p. 103
7. H. Klein et al., HIF 82, GSI 82-7 (1984) p. 150
8. A. Schempp et al., IEEE NS-30, No. 4 (1983) p. 3536
9. W.D. Cornelius, IEEE NS-32, No. 5 (1985) p. 3139
10. N. Ueda et al., IEEE NS-32 No. 5 (1985) p. 3178
11. J. Müller, A. Schempp, IAP Rep. 79-1; LANL LA-TR 82/ 28
12. A. Schempp et al. NIM B10/11 (1985) p. 831
13. W. Paul et al, Z. Physik 140 (1959)
14. W. D. Kilpatrick, UCRL-2321 (1953)

15. A. Schempp et al., IEEE NS-32, No. 5 (1985) p. 3252
16. S.O. Schriber, IEEE NS-32, No. 5 (1985) p. 3134
17. H. Klein, IEEE NS-30, No.4 (1983) p. 3313
18. A. Schempp, Linac 86, SLAC Rep. 303, p. 251
19. A. Schempp et al., IEEE No. 87CH2387-9 (1987) p. 361
20. P. Junior, H. Deitinghoff, IEEE NS-32, No. 5 (1985) p. 3249
21. A. Schempp, M. Ferch, H. Klein, PAC 87, IEEE No. 87CH2387-9 (1987) p. 267
22. M. Ferch, Thesis, Univ. Frankfurt (1987)
23. R. Hamm, Linac 79, BNL 51143 (1980) p. 432
24. C.J. Herrlander et al., IEEE NS-32.No. 5 (1985) p. 2718
25 A. Schempp, IAP Frankfurt, Int. Rep. 87-2, 87-4
26 N. Angert et al. GSI Rep. Dez. 87
27 J. Klabunde et al., Linac 86, SLAC Rep. 303 (1986) p. 299
28. R. Bock, HIIF Workshop 86, AIP Conf. Proc. 152 (1986) p. 23

PHOTOCATHODE RF GUNS*

Richard L. Sheffield
MS H825, Los Alamos National Laboratory
Los Alamos, NM 87545

1	INTRODUCTION	1501
2	PHOTOCATHODE TECHNOLOGY	1502
	2.1 Introduction	1502
	2.2 Metal Photocathodes	1502
	2.3 Semiconductor Photocathodes	1502
	2.4 Other Laser-activated Cathodes	1503
	2.5 Semiconductor Photocathode Fabrication	1504
	2.6 Semiconductor Photocathode Vacuum Requirements	1505
	2.7 Selection of a Suitable Photocathode Source	1505
	2.8 Alkali-Semiconductor Lifetime Experiments	1506
3	INTRINSIC SOURCE BRIGHTNESS	1508
4	PHOTOINJECTOR EXPERIMENTS	1509
	4.1 Los Alamos Experiment	1509
	4.1.1 Experiment Design	1509
	4.1.2 Experiment Results	1511
	4.2 Duke-Rockwell Experiment	1513
5	THEORY OF PHOTOINJECTORS	1515
	5.1 What is Emittance?	1515
	5.2 Space-Charge Effects in Emittance Calculations	1516
	5.2.1 Radial Correlations in ε_{rms} caused by Space Charge	1516
	5.2.1.1 Uniform Continuous Beam	1516
	5.2.1.2 Nonuniform Continuous Beam	1517
	5.2.1.3 Nonuniform Short-pulse Beams	1520
	5.2.2 Longitudinal Correlations in ε_{rms} caused by Space Charge	1520
	5.3 Effects of RF Fields on Emittance	1522
	5.3.1 Growth in ε_{rms} caused by Time-varying RF Fields	1522
	5.3.2 Growth in ε_{rms} caused by Radial RF Fields	1522
	5.4 Integrated Design Approach	1523
6	PRESENT DESIGNS	1524
	6.1 Los Alamos National Laboratory	1524
	6.1.1 Design of a Compact Linac	1524
	6.1.2 Upgrade of Los Alamos FEL Accelerator	1525
	6.2 Brookhaven National Laboratory	1525
	6.3 Lawrence Berkeley Laboratory	1526
	6.4 Bergische Universität-Gesamthochschule Wuppertal	1527
	6.5 LEL-HF in Bruyères-le-Chatel	1527
7	SUMMARY	1529
	REFERENCES	1529

*Work supported by Los Alamos National Laboratory Program Development funds under the auspices of the US Department of Energy.

PHOTOCATHODE RF GUNS
Richard L. Sheffield
MS H825, Los Alamos National Laboratory

1 INTRODUCTION

Free-electron oscillators and amplifiers require electron accelerators capable of delivering pulse trains of electron bunches of high charge density in a wiggler or undulator[1]. A high electron density implies a high peak current (100 A to 2000 A) and a low transverse beam emittance (<80 $\pi\cdot$mm$\cdot$mrad, determined by matching the transverse size of the electron beam to the optical beam in the wiggler). Electron beam collider machines also require high peak currents (>8 nC in picoseconds) with extremely small emittances (<10 $\pi\cdot$mm$\cdot$mrad)[2].

Several approaches have been proposed to attain such performance[3-5]. This article discusses the use of photocathodes in attaining the aforementioned performance requirements. Photocathodes have been used as electron sources in lasertrons[6-8] and for the production of spin-polarized electrons[9,10]. A photocathode is a light-activated electron source that gives unprecedented control over all aspects of the electron distribution: peak current, spatial profile, and temporal profile. This control is possible because the electron distribution is not determined by grids or a cathode, but rather by an incident laser pulse on the photocathode. Lasers have a wide range of variability in pulse format. Pulse lengths can range from femtoseconds to continuous and, for pulses greater than several picoseconds, can have almost any conceivable temporal profile[11-13]. The flexibility in spatial distribution is equally great, with wide latitude in beam diameter and radial profile. By placing the photoemissive material directly in the first rf cavity (Fig. 1) of an accelerator structure, the high gradients available in rf cavities can be used to minimized space-charge effects.

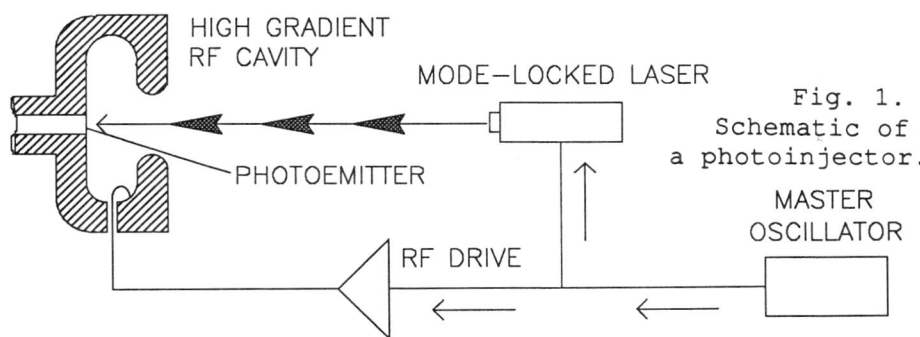

Fig. 1. Schematic of a photoinjector.

2 PHOTOCATHODE TECHNOLOGY

2.1 Introduction

This section contains a brief history of photocathodes and mentions most of the currently studied photoemissive electron sources. Where possible, typically attained values of the quantum efficiency (QE) at 532 nm (wavelength of doubled YAG) are given. The development of photoconductors from 1800 to 1968 and the basic physics of photoconductors are given in a book by Sommer[14]. Further information on alkali-semiconductors may be found in articles by Spicer[15] and Oettinger[16].

2.2 Metal Photocathodes

The discovery of metal photoemitters (non-alkali metals: 10^{-2} QE at 10 eV) occurred in the late 1800s. Two main classes of metal photoemitters exist: cesium-coated (10^{-4} to 10^{-3} QE at 5 eV) and pure metal. The cesium-coated metals require a background pressure of 10^{-6} torr of cesium and thus are not practical for accelerator structures. Because metal photocathodes cannot equal the QE of UHV semi-conductor photoemitters described later, the performance of metal photocathodes at high pressures (10^{-6}) is of interest[17-20]. The QE at 5 eV (excimer laser's wavelength region) varies from 10^{-6} to 10^{-4} depending on the photocathode metal.

2.3 Semiconductor Photocathodes

The first high quantum-efficiency semiconductor photoemitter (1-2% QE) was Ag-O-Cs, discovered in 1929. From 1929 to 1963, a small number of photocathodes were found which had an appreciable quantum efficiency in the visible wavelength region and all used alkali compounds, the most important of which are Cs-Sb (2-4% QE), Bi-Ag-Cs-O (1-2% QE), and K-Cs-Sb-O (4-10% QE). A spectral response curve for Cs_3Sb is shown in Fig. 2.

From 1963 to 1980, the III-IV compounds and silicon semiconductors [which are characterized by a negative electron affinity (NEA) surface] were developed. A thorough study of NEA photoconductors is given in *Semiconductors and Semimetals*[21]. The most significant of the NEA materials is GaAs:CsO, where :CsO indicates a surface dipole layer of cesium and oxygen. GaAs has a very high QE, up to 30%, throughout the visible wavelength region. Another recently reported experiment uses LaB_6 illuminated with a tripled YAG laser light[22]. The QE in this experiment started at 10^{-7} and ended near 10^{-3}; the improvement was attributed to the laser light that cleaned the surface during operation of the experiment.

The experiment at Los Alamos National Laboratory used a $CsK_2Sb:O$ photocathode. During the first experiment Cs_3Sb was chosen for ease of fabrication. However, Cs_3Sb is not thermally stable above room temperature and the measured lifetimes were very short. If a slight background pressure of cesium is present (not practical in accelerator cavities), Cs_3Sb lifetime improves dramatically. When BiAgOCs was used, the lifetimes were longer but the maximum QE is limited to 1-2%. The present photocathode material is $CsK_2Sb:O$, which has a very long lifetime (thermally stable up to 100°C), and has a potential QE of >12% at the double YAG wavelength of 532 nm.

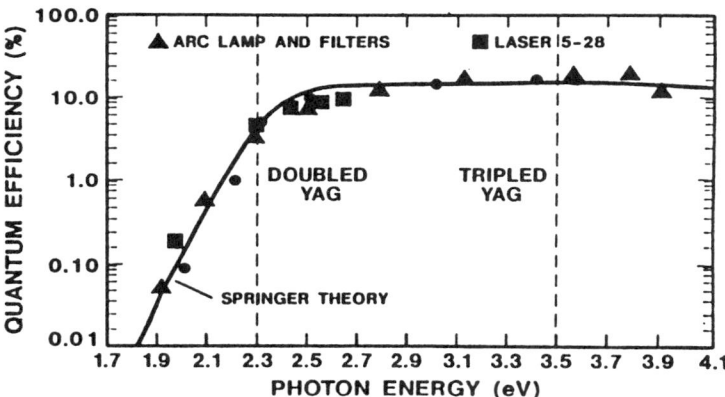

Fig. 2. The measured data, represented by discrete points, and the solid curve, generated using a theory on photocathode response, are the work of Robert Springer, Los Alamos National Laboratory.

2.4 Other Laser-activated Cathodes

The photoemitters discussed so far use only the photoelectric effect for electron production. However, another class of laser-activated photoemitters is possible. This class has two main categories: field-assisted emission and thermal-assisted emission.

Field-assisted emission relies on setting the electric field on a surface to just below the value required for field emission. A laser is then used to free the electrons from the surface. One method for attaining the very high electric field gradients on the surface is to use a fine matrix of needles[12]. However, at present two difficulties exist with this technique: (1) minimizing the needle erosion that occurs because of the very large currents drawn though the needle tips, and (2) producing very bright large-radius beams (where the beam radius is much larger than the needle radius). Although the brightness of an individual beamlet produced by a single needle is very high, the electron trajectories of each of the small beamlets cross downstream of the cathode, giving a much lower overall brightness.

Thermal-assisted emission occurs for thermionic cathodes that run just below their turn-on temperature and, again, a laser is used to free the electrons from the surface. LaB_6 has been successfully operated in this mode with reported results of 10^{-4} to 10^{-3} QE[23]. The major advantage of these sources is their tolerance to poorer vacuum conditions ($>10^{-9}$ torr).

2.5 Semiconductor Photocathode Fabrication

Even though the alkali photocathodes were discovered more than 50 years ago, well-defined fabrication formulas are not available. Several recipes are given in Sommer's book, but procedures depend on several factors: the vacuum conditions present in the preparation chamber, the sources of the photocathode constituents that have varying amounts of impurities, which affect the QE, the exact temperatures of the sources affect deposition rates, etc. Therefore, new users of photocathodes invariably have had to develop their own fabrication techniques. The following brief discussion on Cs_3Sb, $CsK_2Sb:O$, and GaAs:CsO fabrication techniques will provide an appreciation of the procedure.

The photocathode substrate affects the QE response of the photocathode. The 'best' substrate for these photocathodes is unknown, but several substrates (nickel, molybdenum, and gold) have been shown to be acceptable[14]. The fabrication procedures for photocathodes are very similar. Typically, pure sources of the photocathode constituents are present in front of the substrate. First a layer of 0.5 to 1.0 μm of antimony is deposited. Antimony is made from either a compound that releases antimony when heated or from pure beads of antimony placed on a heater coil. Then, depending on the type of photocathode, either potassium and then cesium or cesium interlaced with antimony is deposited on the antimony layer in the presence of oxygen. The amount of potassium and cesium is determined by maximizing low-current photoemission produced by a low-power light source incident on the photocathode. In the case of GaAs, an atomically clean surface on a GaAs crystal is prepared. Then cesium and oxygen are deposited on the surface as described above for the Cs_3Sb photocathode, with the amounts of cesium and oxygen determined by low-current photoemission.

2.6 Semiconductor Photocathode Vacuum Requirements

The sensitivity of photocathodes to contamination depends on the source of the contamination. For instance, pure hydrogen or nitrogen at greater than 10^{-8} torr do not appear to affect cathode QE, but small amounts (10^{-12} torr) of water or carbon dioxide can significantly decrease the QE. The damage caused by contamination can be estimated by the time for monolayer formation on a clean surface in a vacuum system. The time in seconds to form a monolayer is roughly 10^{-6} divided by the pressure in torr. So a monolayer will form in 10^6 s (278 h) at a pressure of 10^{-12}. Because a small fraction (<10%) of a monolayer can severely degrade the QE, gas contaminant pressures of 10^{-12} will degrade QE in a day. As a consequence, only UHV systems that have been baked above 200°C (with a better base vacuum as the bake temperature is increased) can support a photocathode. However, the system cannot then be back-filled with nitrogen without re-baking because even nitrogen pure to one part in a billion at atmospheric pressure has much greater than 10^{-12} torr of water.

2.7 Selection of a Suitable Photocathode Source

The incident laser power (in watts) required to generate a photocurrent of I amperes with light of energy E (in electron volts) is

$$P = E*I/(\text{quantum efficiency}).$$

For example, to produce a current of 200 A from a photocathode with a quantum efficiency of 2% at the double YAG energy of 2.35 eV requires 24 kW of incident laser power.

The choice of photoemitter is determined by the laser required to produce the electron pulses. The electron pulse formats fall into three classes: (1) long trains of high-charge pulses requiring a high average-power laser; (2) long trains of low-charge pulses requiring low average laser power (less than 5 W average); and (3) single or few pulses (independent of charge). The first class of applications that requires high-average currents is limited to the higher QE photocathodes because of the cost of high average power lasers (>20 W average at 532 nm costs more than $1M).

Low average-current beams are limited by present laser systems. The present state of the art in commercial cw YAG lasers at 532 nm is less than 2-W average power. Therefore, high-QE cathodes (>5%) can give an average electron current of 40 mA. The lower QE cathodes

(10^{-4} to 10^{-3}) can give 0.04 to 0.4 mA of average electron current.

The last application group requiring either a single pulse or a few pulses has many photocathode options. Because single-pulse laser power (greater than hundreds of megawatts) is relatively inexpensive, either a very high power YAG laser ($120k), which can be doubled (2.35 eV) or quadrupled (4.7 eV), or an excimer laser ($120k) can be used. The very high powers available mean that all of the cathodes discussed above, including those with a very low QE, can be considered. A mode-locked YAG can be phase locked to the rf source that drives the accelerator with greater than 150 mj at 4.7 eV. However, the overall system is more complicated than an excimer laser. The excimer laser produces higher energy photons directly (3.5 - 6.4 eV) with 100 mJ of energy per pulse. However, the excimer laser has long pulses (10 ns) and cannot be synchronized to a master oscillator. With either laser system, if a small number of pulses are required, a single laser pulse can be easily multiplexed.

Experiments requiring picosecond pulses set another requirement on the photocathode: the ability to produce prompt emission with no temporal tails. Electrons in metal and alkali-semiconductor photocathodes have very short relaxation times (<1 ps) and, consequently, do not have noticeable tails in the picosecond regime. By contrast, the intrinsic emission-time uncertainty of GaAs has been measured in the range of 8 to 71 ps for active layers between 50 nm and 2 µm in thickness[24]. Measurements in the picosecond regime have not been performed on either thermal-assisted or field-assisted emission cathodes. Field-assisted cathodes typically form plasmas on the emission tips and probably have thermal tails that last a fraction of a nanosecond.

2.8 Alkali-Semiconductor Lifetime Experiments

The lifetime of the photocathode electron source has improved markedly since the initial measurements two years ago. The measurements are presented in Table I.

The first major improvement in cathode lifetime (9/87 to 2/88) occurred when we switched from Cs_3Sb to CsK_2Sb photocathodes. After the switch of photocathode materials, we noticed that the photocathode response (current extracted versus laser power input) would be nearly constant for the first hour and then begin to degrade at a much faster rate. From this observation we determined that the electron spectrometer quartz screen (used as the beam dump for the lifetime experiments) was getting excessively hot (see Fig. 3). The temperature rise increased the outgassing rate of the beam dump to the point that the photocathode performance was affected. This problem was eliminated by using a water-cooled copper beam stop (initiated in 8/88). We believe that the final lifetime value (8/88) in Table I is limited primarily by the outer fringe of the beam hitting the downstream beam-transport tubes because of the long drift between the rf gun and beam dump.

Table I. Lifetime Measurements

Photocathode	Rep rate (Hz)	Pulse length (µs)	Average Macropulse Current (A)	1/e life (h)
Cs_3Sb (9/87)	1	5	1	0.5
CsK_2Sb (2/88)	10	10	1	2.0
CsK_2Sb (8/88)	10	10	0.5	17.5

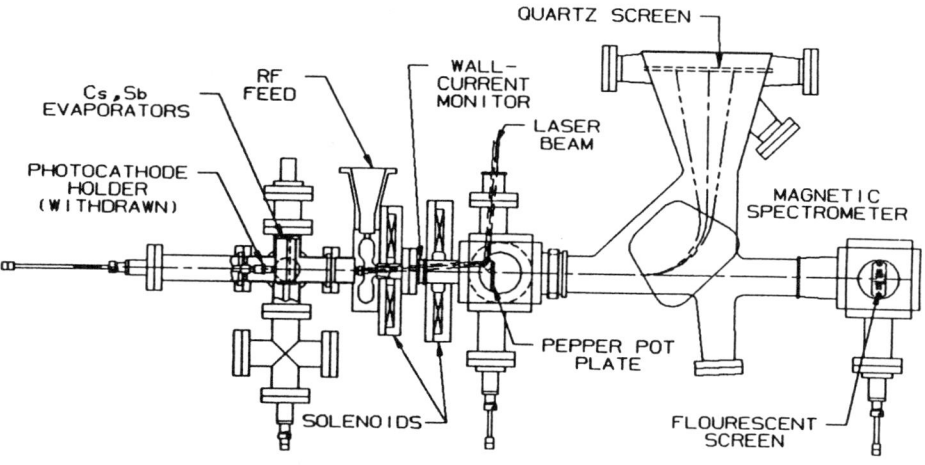

Fig. 3. Single-cavity experimental setup, Los Alamos National Laboratory.

In separate experiments, photocathode lifetimes were measured in low-current dc operation. More than 400°C have been extracted from a CsK_2Sb photocathode during a three-day period. In a third apparatus that was used for shelf-life testing, another photocathode was fabricated with an initial quantum efficiency of 7.5% and did not have any observable degradation over a two-month period.

3 INTRINSIC SOURCE BRIGHTNESS

The normalized peak brightness is defined as

$$B_n = 2I/(\varepsilon_x \varepsilon_y) \quad [\text{units:} \quad A/(m^2 \cdot rad^2)] \quad ,$$

where I is the peak current and ε_x and ε_y are the normalized transverse emittances of the beam[25]. For a thermal distribution or a distribution that does not have recoverable correlations in phase space, it is constructive to use the rms emittance formulation, defined to be the area in phase space, which is

$$\varepsilon_x = 4 \cdot \pi \cdot [<x^2> \cdot <x'^2> - <x \cdot x'>^2]^{\frac{1}{2}} \quad ,$$

where x and x' are the particle's transverse coordinate and angle of divergence from the optic axis, respectively, and <> means an average over the electron distribution $f(x,y,z)$:

$$<x^2> = \frac{\int \int \int f(x,y,z) \cdot x^2 \cdot dxdydz}{\int \int \int f(x,y,z) \cdot dxdydz} \quad .$$

Another common definition of emittance is as the area in phase space divided by π, with the π included in the units.

Using the above formulation, the rms emittance is equal to the total phase-space area for a Kapchinskii-Vladimirskii distribution[26]. The normalized emittance is then

$$\varepsilon_n = \gamma \beta \varepsilon \quad ,$$

where for an azimuthally symmetric beam $\varepsilon = \varepsilon_x = \varepsilon_y$.

The lower limit of the beam's normalized emittance from a thermionic electron source is governed by the emitter size and by the transverse component of the thermal motion of the electrons. The thermal limit of the normalized rms emittance of a beam from a thermionic emitter of radius r_c at a uniform absolute temperature T is

$$\varepsilon_n = 2 \cdot \pi \cdot r_c [k \cdot T/m_o c^2]^{\frac{1}{2}} \quad (\text{units:} \quad m \cdot rad)$$

because $<x \cdot x'> = 0$ at the cathode[27]. For a typical thermionic emitter at 1160 K, the average transverse energy of emitted electrons is 0.1 eV. For a uniform current density J, the total current is $I = \pi \cdot r_c^2 \cdot J$ and the lower limit on the rms emittance is

$\varepsilon_n = 5.0 \times 10^{-6} \pi (I/J)^{\frac{1}{4}} \pi \cdot mm \cdot mrad$, with J in A/cm².

The corresponding normalized peak brightness is limited to

$$B_n = 2 I/\varepsilon_n^2 = 8.2 \times 10^9 J \text{ A}/(m \cdot rad)^2.$$

The current density from a dispenser cathode is typically not more than 20 A/cm²; therefore, the maximum achievable brightness is 1.6×10^{11}.

Semiconductor photoemitters have an effective temperature of 0.2 eV[28]. The electron thermal temperature is not simply the difference between the incident photon energy and the semiconductor bandgap (a difference of 0.7 eV) because of phonon scattering in the semiconductor crystal lattice. Semiconductor cathodes are capable of delivering[29] over 600 A/cm², giving a brightness of 2.5×10^{12} A/(m·rad)².

The brightness of the source normally does not limit the final brightness of the beam. Instead, the acceleration process and transport through a beamline can decrease the beam brightness by orders of magnitude.

4 PHOTOINJECTOR EXPERIMENTS

A photoinjector is a photoemissive electron source placed directly in an rf cavity. The photoinjector design depends on the electron bunch produced from a photocathode being rapidly accelerated to relativistic energies in a single rf cavity, hence eliminating the conventional bunching process entirely. The emittance growth of the electron beam is reduced because electron-beam transport at low energies has been significantly reduced.

4.1 Los Alamos Experiment

4.1.1 Experiment Design

The Los Alamos experiment uses a laser-driven photocathode electron source situated on-axis in the first rf cavity. The electron-pulse shape is easily tailored in both time and space by appropriately shaping the incident laser pulse. The configuration of the experiment is shown in Fig. 4. The linac has two 1300-MHz rf cavities with independent amplitude and phase controls. Both rf cavities have loops to measure the phase and amplitude of the rf fields present in the cavities. Following the second cell are the diagnostics for bunch charge, beam energy, emittance, and temporal profile. The details of the rf cavity design are presented elsewhere[30].

The photocathodes are fabricated in a preparation

chamber vacuum coupled to the rf linac. Following fabrication in the preparation chamber, the photocathode is inserted into the rf cavity. When the quantum efficiency of the photocathode decreases below some arbitrary minimum value, the substrate is pulled back and heat cleaned at 400°C. A new photocathode is then fabricated over the existing substrate without opening the UHV system.

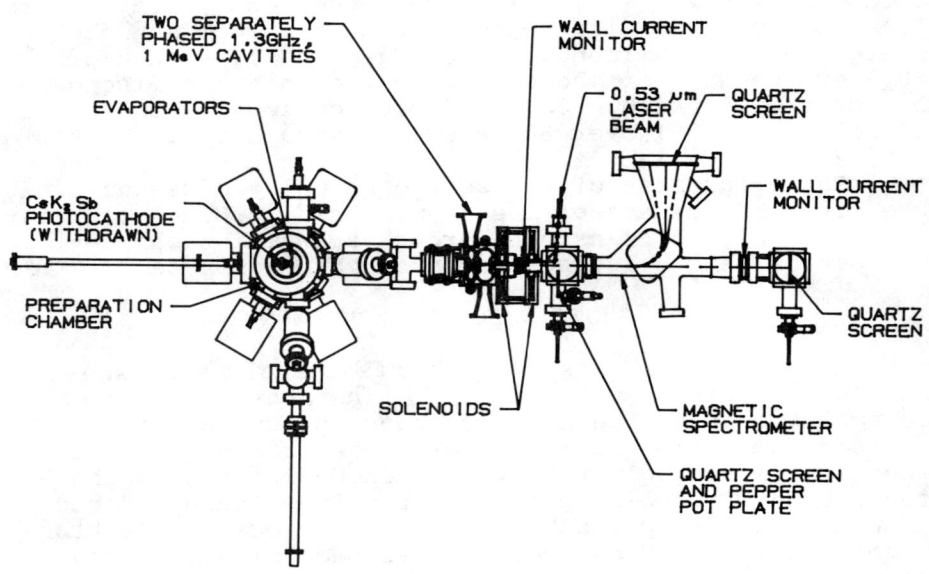

Fig. 4. Two-cavity experiment showing accelerator, beam transport, and diagnostics.

The photocathode is illuminated with a frequency-doubled Nd:YAG laser (Fig. 5). The laser is mode locked at the twelfth subharmonic of 1300 MHz, 108.33 MHz. The mode-locking crystal is driven by the same master oscillator that drives the 1300-MHz rf klystron and is phase locked to the rf. The laser generates 100-ps pulses at 1.06 µs that, after frequency doubling to 532 nm, become 70-ps-long pulses. A Spectra-physics pulse compressor was added to the optical train for generation of 4- to 20-ps pulses. The power available at 532 nm is approximately 250 kW average over 10 µs.

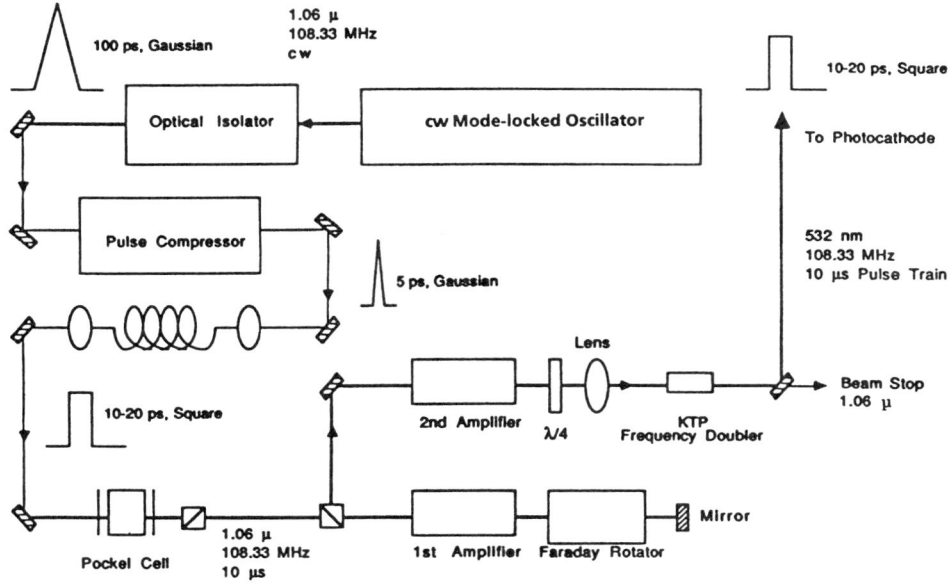

Fig. 5: Block diagram of the photocathode laser consisting of a cw mode-locked Nd:YAG oscillator and pulsed amplifiers.

4.1.2 Experiment Results

The electron energy distribution for a 10-µs train of 28-ps pulses is shown in Fig. 6. The electron energy gain for typical operation was 0.9 MeV in the first cavity and 1.8 MeV in the second cavity. This corresponds to operating both cavities at approximately 2 Kilpatrick (58 MeV/m peak surface field).

The laser pulse length was limited by the gain bandwidth of the Nd:YAG amplifiers to approximately 16 ps. The maximum charge extracted for this pulse was 13.2 nC from 1 cm² of photocathode surface. This gives 820 A/cm² of current density at the cathode. However, PARMELA simulations predict that a 16-ps electron pulse increases to 22 ps on passage through the first cavity, giving a peak current after the first cavity of 600 A.

The beam emittance cannot be directly measured in this experiment because of the space-charge-induced emittance growth in the long drift between the second cavity and the pepperpot plate. However, an emittance measurement of the beam was performed in the previous single-cavity experiment, and this result was compared with PARMELA, MASK [31], and ISIS [32] simulations. The results of the MASK calculations are shown in Fig. 7.

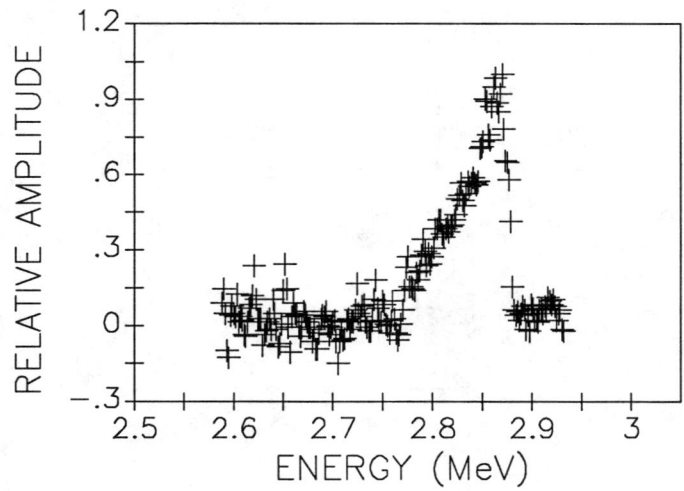

Fig. 6. Electron energy distribution.

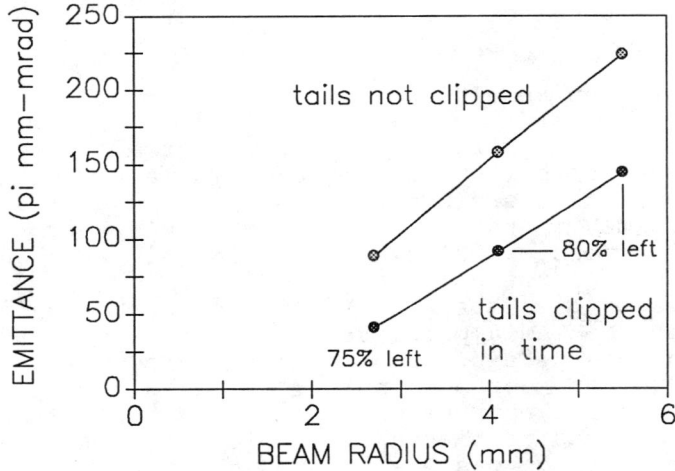

Fig. 7. The beam emittance from Mask simulations (performed by Bill Herrmannsfeldt of SLAC), based on the configuration shown in Fig. 3, are within the experimental error in beam radius if the temporal tails of the Gaussian pulse are not included. The two curves show the difference in emittance gained by excluding a small fraction of the charge at the front and tail of the pulse.

The experimental and simulated electron-beam diameter at the pepperpot and the diameters of the beamlets produced by the pepperpot at the second quartz screen are in close agreement, confirming the accuracy of the simulations. The emittance of the electron beam for that experiment, with 10 nC per bunch, was calculated from the simulations to be 120 $\pi \cdot$mm$\cdot$mrad for 100% of the beam. Simulations show that, if the beam is clipped in time and left with 75% of the original charge, then the emittance of the remaining beam was calculated to be 40 $\pi \cdot$mm$\cdot$mrad [5] in agreement with the experimental results. The large decrease in beam emittance with a small decrease in the charge is due to the temporal tails of the long Gaussian pulse used in the previous experiment. Because the focusing solenoid downstream of the cavities can only be properly matched for one space-charge density, the beam is only matched for the peak of the Gaussian pulse, and the head and tail of the electron bunch are overfocused. The low-intensity tails from all the beamlets overlap on the pepperpot screen; therefore, an individual beamlet's spatial distribution cannot be resolved unambiguously. Hence an experimental emittance value was obtained only for the temporal core of the electron bunch.

The experimental parameters were 11 nc (200-A peak), 70-ps Gaussian temporal width, <0.4-cm beam radius at the cathode (was not accurately measured at the time of the experiment and only the upper bound is known), 1.0-MeV beam energy, and a solenoid field of 1.8 kg. The measured emittance was 40 $\pi \cdot$mm$\cdot$mrad (the long temporal tails could not be measured for the emittance calculation).

Although neglecting the temporal tails of the distribution consequently gives low emittances, most applications of bright electron beams depend upon only the bright central core of the electron bunch. More importantly, the accuracy of the simulation codes have been verified for future linac design.

The theoretical analysis discussed later in this paper shows that for a properly tailored light pulse (square in time and in spatial extent), the whole beam will have an emittance better than the core emittance measurement described above. However, because a drifting electron beam experiences emittance growth over tens of centimeters even at a relativistic gamma factor of 5, the only way to accurately determine the electron-beam emittance from an rf photocathode gun is to measure the beam quality after acceleration greater than 10 MeV.

4.2 Duke-Rockwell Experiment

The construction of the Mark III accelerator has been described in detail elsewhere[3]. The layout of the experiment is shown in Fig. 8. The machine parameters are as follows: macropulse length of 2-5 µs, micropulse length of 2.2 ps, gun energy of 1 MeV, and a magnetic compression of 10 from the alpha magnet. The alpha magnet is a momentum filter and is able to limit the electron energy spread to less than 0.5%.

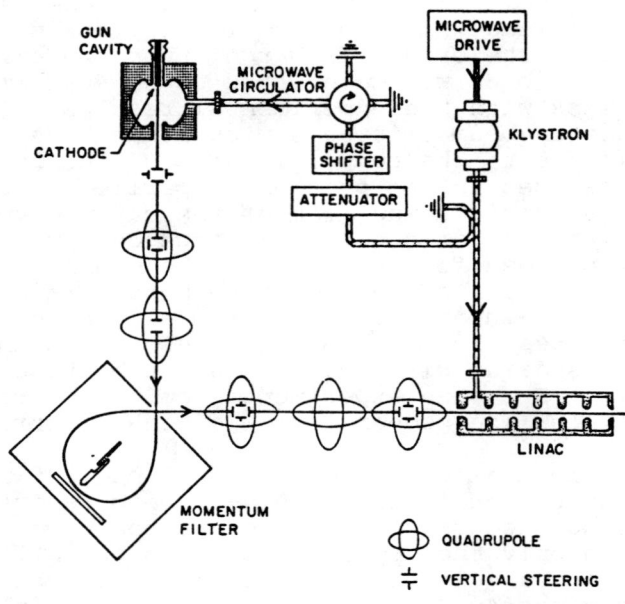

Fig. 8. Schematic of the experiment showing the microwave feed system and the path of the electrons from the laser-switched thermionic gun to the accelerator.

The electron source in the Mark III is a LaB_6 cathode. Originally the cathode produced electrons by pure thermal emission. However, because the electrons are emitted at all phases of the rf, many of the electrons are accelerated at wrong phases for matching into the main linac. The current emission from the cathode is limited by average-power heating; therefore, using the laser to limit the emission to the correct rf phase, higher peak currents can be obtained[33]. In this mode the LaB_6 was operated just below its normal emission temperature, and a laser was used to pulse the cathode. Operation with the laser resulted in an increase in peak current from 33 A to 75 A with no observable loss in beam emittance. Therefore the gun brightness increased by approximately a factor of 3 to near 10^{12} A/(m-rad)^2.

During operation, the gun pressure was about 5×10^9. Not enough operation time has been available to study the cathode lifetime; but based on previous performance, the expected lifetime should be much greater than 1000 hours.

5 THEORY OF PHOTOINJECTORS

Placing the electron source in the first rf cavity has several benefits. Because the photocathode can be switched on and off in picoseconds, no subharmonic bunchers are required to generate short pulses. Short-pulse high-charge beams require a high-voltage gradient, which can be produced in rf cavities, so that space-charge forces do not degrade the electron beam. An added benefit of the high gradient is that the electrons become relativistic in a very short distance (several centimeter) and the electron distribution does have time to thermalize or mix. Therefore, correlations that are present during the initial acceleration phase can be removed by proper beamline design. This section describes the source of these correlations and methods of reducing their effect.

5.1 What is Emittance?

A detailed study of emittance and brightness is given in the paper by LeJeune and Aubert[21]. A brief summary of this work follows. Liouville's theorem states that the volume an ensemble of particles fills in six-dimensional (6-D) phase space is invariant as the ensemble moves through phase space. To analyze beam quality, a series of approximations are made. First a projection of that 6-D space into 4-D is used, a projection that is only valid when the axial and transverse motions are not coupled (which breaks down in the region where the beams are emitted in conventional guns). The hyperemittance ε_4 is then defined to be the hypervolume enclosing all particles in 4-D space and is subject to a number of conditions. In particular, the axial velocity must be constant over any cross section and along the optic axis, and the magnetic field must be entirely transverse.

If the motion in the two transverse directions is independent, then the projection of 4-D into 2-D phase space is allowed. The 2-D phase space plot is the typical x-x' or y-y' plot. If the above assumptions are fulfilled, then the emittance, ε_{actual}, is defined to be the area in 2-D phase space. If ρ is the particle density distribution function, then

$$\varepsilon_{actual} = \int\int \rho \, dx \, dx' \quad (rad \cdot m) \ .$$

Finally, the root-mean-square emittance, ε_{rms}, can be defined. This restrictive definition of the beam quality assumes that an ellipse drawn around the particle distribution in x-x' space characterizes the beam quality. In most accelerator applications, this definition is very useful because the electrons move to reduce the free energy during propagation and consequently fill the bounding ellipse. The definition for the rms emittance is (as given above)

$$\varepsilon_{rms} = 4 \cdot \pi \cdot [<x^2> \cdot <x'^2> - <x \cdot x'>^2]^{\frac{1}{2}}$$

or

$$\varepsilon_{rms} = 2 \cdot \pi \cdot [<r^2> \cdot <r'^2> - <r \cdot r'>^2]^{\frac{1}{2}} .$$

The problem with ε_{rms} is that correlations may exist in the electron distribution in x-x' space and distort the linearity of the distribution but do not increase ε_{actual}. If the correlations can be removed by proper matching, ε_{rms} will overestimate the actual emittance. Because most computer codes calculate ε_{rms}, then if removable correlations are present, the emittance of the beam can first increase and then decrease during propagation.

The photoinjector is unique in beam production because before the electron distribution has a chance to thermalize, the electrons are relativistic. This means that correlations present during the beam formation can be removed to a large degree by careful beamline design. The real difficulty is developing a definition of emittance, ε_{eff}, that will accurately characterize the beam quality at the target but can be calculated upstream.

5.2 Space-Charge Effects in Emittance Calculations

5.2.1 Radial Correlations in ε_{rms} caused by Space Charge

The nonlinear radial expansion of a drifting electron beam causes an increase in ε_{rms} as a result of radial space-charge forces. In general, all beams will exhibit this radial motion; one important exception is a uniform continuous beam. The two examples that follow demonstrate how the ε_{rms} is affected by the electron-beam radial distribution. The two examples are a radially-uniform continuous beam and a continuous beam with a charge profile proportional to the radius. Two assumptions are made to arrive at simple analytic solutions. First, no accelerating cavities are present (drifting beam), and, second, any expansion of the beam is a small fraction of its initial radius.

5.2.1.1 Uniform Continuous Beam

The radial electric field $E(r, \zeta, t)$ for the first case of a continuous, uniform beam is

$$E(r,\zeta,t) = -m \cdot k(t,\zeta) \cdot r/e,$$

where m, e, r are the particle's mass, charge, and radius, respectively; ζ refers to the relative longitudinal position along the beam; and $k(t,\zeta)$ is the component of the electric field that is not a function of the radius multiplied by e/m. The resulting equation of motion for any point in the beam is (for nonrelativistic motion in the radial direction)

$$F = m \cdot r'' = -e \cdot E = m \cdot k(t,\zeta) \cdot r.$$

Let t be small, so that the beam radius has not changed appreciably. Then the above equation has solutions of

$$r = A \cdot \cosh(k(\zeta) \cdot t + \phi)$$

and

$$r' = A \cdot k \cdot \sinh(k(\zeta) \cdot t + \phi),$$

where $A = \sqrt{r_0^2 - (r_0'/k(\zeta))^2}$, $\phi = \tanh^{-1}(r_0'/(k(\zeta) \cdot r_0))$, and r_0' and r_0 are the initial (t=0) radial velocity and position, respectively. If the beam starts off initially as a straight line in r-r' space and k is not a function of ζ, then r_0' is a constant (a) times r_0. The above equations can then be written as

$$A = r_0 \sqrt{|1 - (a/k)^2|} \quad , \quad \phi = \tanh^{-1}(a/k),$$

and

$$r'/r = k \cdot \tanh(k \cdot t + \phi) = f(t).$$

Therefore, r' is linearly proportional for all time to r and independent of the initial position inside the beam. Using the definition given above for the rms emittance,

$$\varepsilon_{rms} = 2 \cdot \pi \cdot [<r^2><r'^2> - <r \cdot r'>^2]^{\frac{1}{2}},$$

and substituting $r' = f(t) \cdot r$, ε_{rms} is zero for all time. The more general case, in which the charge distribution expands radially (k is a function of time) can also be solved. The conclusion of r' being proportional to r is the same because the charge density at any one time is still uniform.

5.2.1.2 Nonuniform Continuous Beam

We next examine the case of a beam with an electric field independent of the beam radius, i.e. a continuous beam with the charge inversely proportional to the radius. The radial dimension of the beam is from r_d, a small radius close to zero, to r_m, the maximum radius.

This case is not chosen to represent an actual beam, but to demonstrate the effect of a nonuniform distribution, which can be easily solved. The electric field is

$$E(r,\zeta,t) = E_o(t).$$

The equation of motion gives

$$F = m \cdot r'' = -e \cdot E_o(t).$$

Let t be small, so that the beam radius has not changed appreciably. Then the above equation has solutions of

$$r' = e \cdot E_o \cdot t/m + r_o'$$

and

$$r = e \cdot E_o \cdot t^2/2m + r_o' \cdot t + r_o.$$

As before, assume that the beam initially, $t_o=0$, is a straight line in r-r' space, then $r_o' = a\, r_o$ and the above equations can be written

$$r' = e \cdot E_o \cdot t/m + a \cdot r_o$$

and

$$r = e \cdot E_o \cdot t^2/2m + a \cdot r_o \cdot t + r_o.$$

In this case, however, r_o cannot be separated to give a solution where r' is proportional to r. The ε_{rms} increases rapidly with time for this beam, although ε_{actual} has only changed slightly. To aid in understanding, a sample case has been generated using the above equations. The beamline for this case consists of a drifting beam starting with a zero divergence and a radius $r_o = 1$. A focusing lens of strength f is positioned at $t = t_1 = 10$. The parameter k equals $e \cdot E_o/m$ (see Fig. 9).

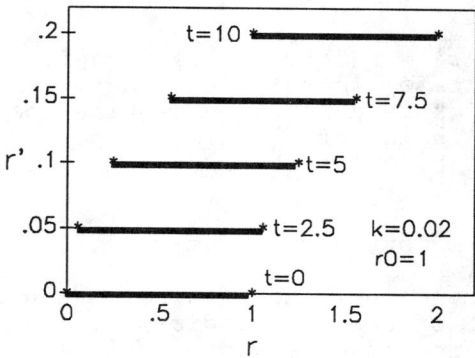

Fig. 9. The r-r' plots for $t < t_1$.

The actual emittance of this beam has not increased with time because the width of the line in r-r' phase-space is the same. However, ε_{rms} has increased quadratically. By putting this beam through a thin lens and reversing r' at location t_1 and then drifting the beam, a minimum in emittance can still be found, as shown in Figs. 10 and 11. The minimum in emittance occurred during the clockwise rotation of the beam phase space distribution, between t = 15 and 20, at the moment it pointed to the origin.

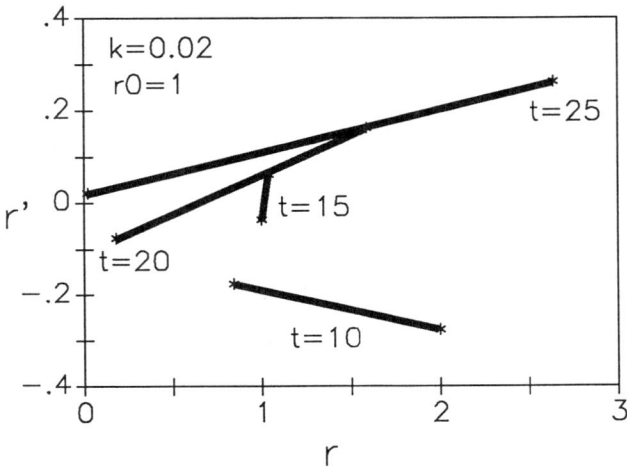

Fig. 10. The r-r' plots for $t > t_1$.

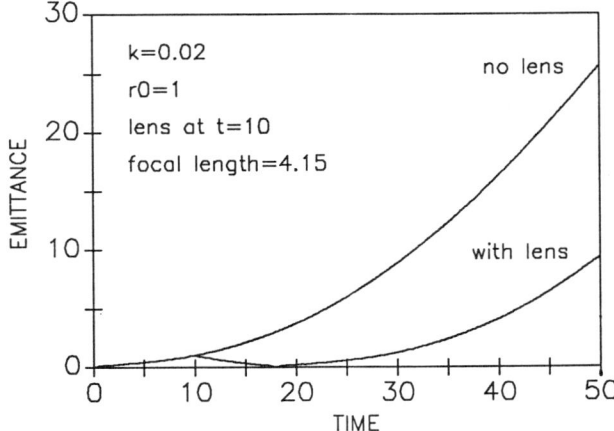

Fig. 11. Plot of the rms emittance versus time.

5.2.1.3 Nonuniform, Short-pulse Beams

The preceding discussion demonstrates that an increase in ε_{rms} does not necessarily correspond to an increase in ε_{actual}. As long as no mixing has occurred, for instance through the action of a radial rf field, the radial correlations can be removed by appropriately focusing the beam. However, the removal of the correlations in general only occurs at one location downstream of the accelerator[34,35,36]. Consequently, a careful design process is required to ensure that the minimum in the emittance occurs where the electron beam's target is located.

5.2.2 Longitudinal Correlations in ε_{rms} caused by Space Charge

The effect of longitudinal variations in density can be understood by using the same equations developed in the last section,

$$r' = e \cdot E_o \cdot t/m + a \cdot r_o$$

and

$$r = e \cdot E_o \cdot t^2/2m + a \cdot r_o \cdot t + r_o$$

and letting E_o be a function of ζ.

As before, the beamline consists of a drifting beam starting with a zero divergence and a radius $r_o = 1$. A focusing lens of strength f is positioned at time $t_1=10$. We define the parameter k to be $e*E_o(\zeta)/m$. Since k is a function of ζ, then the proportionality between r' and r is also a function of ζ. This, in turn, creates an increase in ε_{rms} as a function of t. The plot of r' versus r shows a ``bow-tie'' effect that is caused by the electric field variations. Plots showing the ``bow-tie'' for two time steps before the lens are given in Fig. 12.

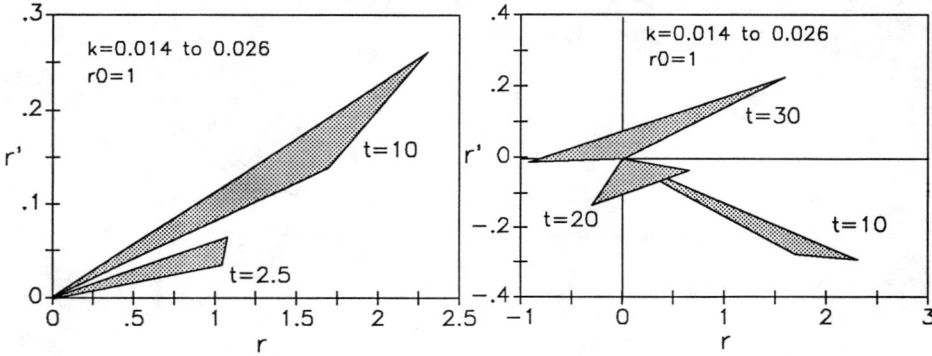

Fig. 12. Bow-tie effect before the lens.

Fig. 13. Bow-tie effect after the lens.

Again the emittance can be reduced by using a thin lens to reverse r' at location t_1 and then drifting the beam. The emittance reduction is possible because the beam in an rf gun is rapidly accelerated and, consequently, very little longitudinal mixing can occur, thus preserving the correlations present. The transformation of the "bow-ties" in phase space is seen in Fig. 13.

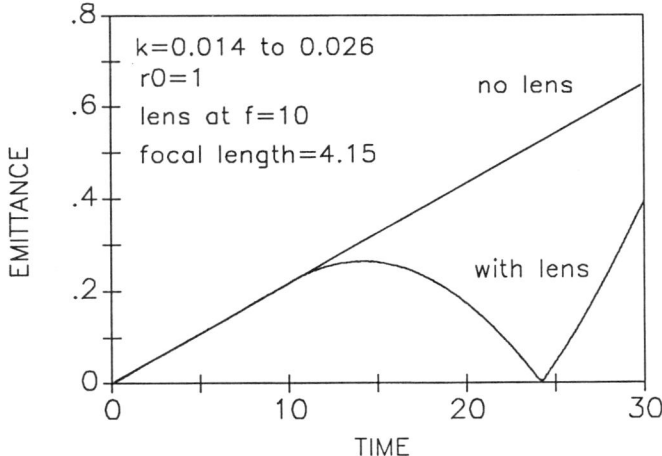

Fig. 14. Plot of rms emittance versus time.

Particles that have gone through a crossover are given negative r values. The emittance minimum occurs as the "bow-tie" of t = 20 is turned inside out to become the "bow-tie" at t = 30. The resulting emittance growth and decrease is shown in Fig. 14.

The above discussion depends on self-similar expansion of the beam. Simplistically, self-similar expansion implies that the electron beam does not distort appreciably during propagation. A more detailed study of self-similar expansion and its effects can be found in a paper by Carlsten[36].

Two effects give rise to an increase in ε_{rms}: radial effects and variations in longitudinal density. Both effects can be eliminated with the appropriate lens placement assuming that no other mixing of the density distribution has occurred. If the beam did not undergo self-similar expansion, then more lens can be used to correct higher-order distortions, but the design process rapidly becomes very complicated.

5.3 Effects of RF Fields on Emittance

5.3.1 Growth in ε_{rms} caused by Time-varying RF Fields

The emittance growth caused by space-charge effects can be reduced by increasing the gradient in the rf gun. The gradient can be increased by raising the frequency of the gun. However, if the frequency is too high, the time-varying fields will exert different forces on the front of the pulse as compared to the end of the pulse and thereby will increase the ε_{rms}. The emittance growth caused by time-varying fields for a uniform distribution is given in a paper by Jones and Peter[32]. Their result (in cgs units) to first order is

$$\varepsilon_{rms} = r_b^2 \cdot \omega \cdot z_0 \cdot \tau \cdot \cos(\phi_0) / (6c\sqrt{3}) ,$$

where r_b, ω, z_0, τ, ϕ_0, c are, respectively, the beam radius, rf frequency in radians, position of rf gun end, pulse length, start phase of pulse, and the speed of light. This effect can be eliminated by appropriate phasing of a downstream rf cavity, if space-charge effects are not present, or by use of a low-frequency rf cavity. For beams where space-charge effects are important, $\omega \cdot \tau$ should be less than 0.1 to minimize this source of rf emittance growth. At an rf frequency of 1.3 GHz, the maximum pulse length subject to the above constraint is 12 ps.

The use of first- and third-harmonic frequencies to produce the equivalent of a dc field[4] during the pulse transit through the cavity can also eliminate time-dependent effects. However, two separate cavities, one at the first and one at the third, are not the equivalent of a single cavity operating both frequencies because the space-charge forces act to expand the pulse at different locations in the beam during the pulse transit between separated cavities. Therefore, the third harmonic does not properly cancel the higher-order temporal terms.

5.3.2 Growth in ε_{rms} caused by Radial RF Fields

Emittance growth can occur in accelerator cavities because of nonlinear components of the radial rf fields. The radial fields in a cavity can be written as

$$E_r(r) = r E_1 + r^3 E_3 + \ldots .$$

As long as the higher-order terms, greater than or equal to E_3, are negligible compared to E_1, then the radial fields will not contribute to the overall emittance growth. If the beam is small in radius compared to the cavity bore, the higher-order terms can be immediately neglected. However, when the beam fills a substantial fraction of the cavity bore, no simple analysis will properly calculate the relative magnitude of the coefficients far off-axis, because these radial components are cavity dependent. Therefore, computer simulations with SUPERFISH on the particular cavity being considered are required. A study of radial fields based on Fourier decomposition is given in a paper by McDonald[37].

5.4 Integrated Design Approach

A design approach based on the above discussion of emittance effects can now be developed. A small high-gradient linac has been designed using the following criteria: minimize the rf effects by keeping the pulse short relative to the rf frequency, minimize the non-linear transverse electric fields in the rf cavities, and correct the correlations that exist in phase space. The rf cavities were designed by Lloyd Young using SUPERFISH. The accelerator was designed by the author and Bruce Carlsten using PARMELA simulations. The accelerator design details are given in the next section of this chapter. To show the effect of removing the correlations in ε_{rms}, a plot of emittance versus position down the accelerator is shown in Fig. 15.

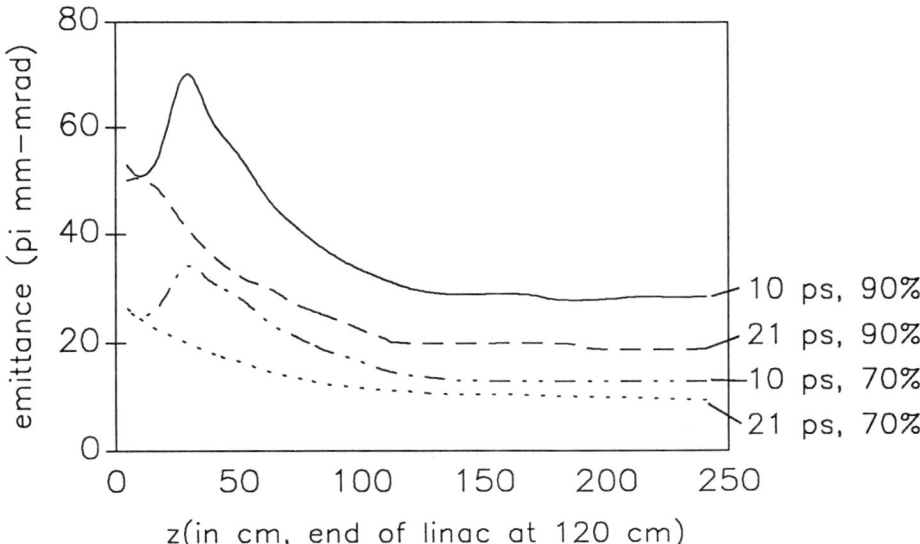

Fig. 15. Emittance plot for compact linac; cathode radius of 0.4 cm, micropulse charge of 5 nC, initial pulse width of 10 ps, which expands to 16 ps at the exit of first linac cavity.

6 PRESENT DESIGNS

6.1 Los Alamos National Laboratory

Two separate initiatives are now underway at Los Alamos based on photoinjector technology.

6.1.1 Design of a Compact Linac

Design of a 20-MeV compact linac (see Fig. 16) based on the photoinjector has been completed. The linac will be approximately 1.2 m long and will be operated with a 10 μs macropulse at up to 15 Hz with a 0.5 A average during the macropulse. The final electron-beam characteristics from PARMELA simulations are a beam emittance of less than 20 $\pi \cdot$mm$\cdot$mrad and peak currents in excess of 350 A. Magnetic compression of the 16-ps electron pulse can increase the peak current to greater than 500 A. The limit in peak current depends on the application. For instance, a free-electron laser (FEL) oscillator is very sensitive to the jitter in the arrival time of the electron bunches in the wiggler. Because variations in the electron bunch charge cause variations in the final electron beam energy, the amplitude stability of the photocathode laser system, which produces the electron bunches, will determine the maximum amount of pulse compression allowed (a change in the electron-beam energy maps into a change in time in the magnetic compressor).

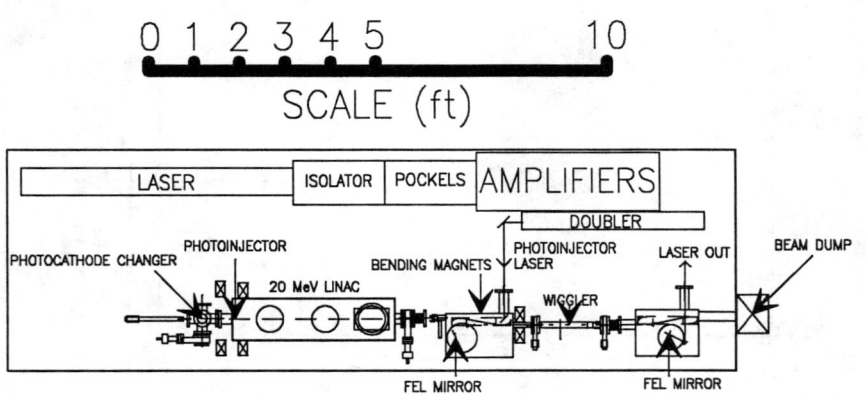

Fig. 16. Compact FEL with photoinjector

6.1.2 Upgrade of Los Alamos FEL Accelerator

The Los Alamos FEL is now being upgraded to provide electron beams of the quality and intensity required by advanced FELs. The improved electron beam is primarily the result of adding a photoinjector to the accelerator. However, the entire device is being modified to demonstrate that the beam quality can be transported to the FEL without degradation. The facility should provide initial data by June 1989. This facility will provide a good benchmark of the computational models used to design advanced FELs because the same models will design the photoinjector, beam transport, oscillator, and amplifier. The design goals of the accelerator are 40 MeV of electron energy, peak currents of 400 A, and a normalized emittance less than 50 $\pi \cdot$mm$\cdot$mrad (90%). An experiment layout is given in Fig. 17.

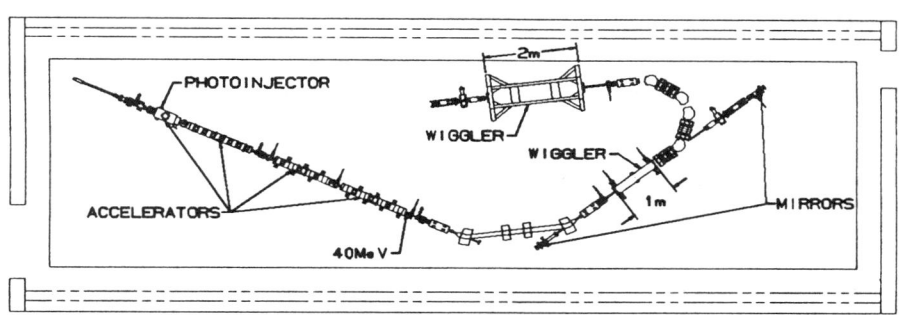

Fig. 17. Upgrade of Los Alamos FEL accelerator with photoinjector.

6.2 Brookhaven National Laboratory

The Accelerator Test Facility at Brookhaven National Laboratory (BNL) is being developed into a research facility for laser acceleration and FELs. The design goal for the accelerator is 50 MeV at an emittance of 30 $\pi \cdot$mm$\cdot$mr. They are building (scheduled for operation in spring of 1989) a 2.856-GHz photoinjector to drive the linac[38]. The S-band, standing wave, disk-loaded structure will operate in the short rf pulse regime (6 μs). The gun is designed for a maximum surface field of 120 MV/m and a pulse repetition rate of 5 Hz. The schematic of the photoinjector is shown in Fig. 18.

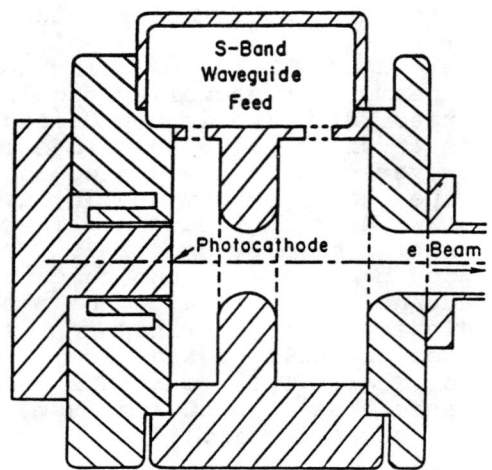

Fig. 18. Schematic of BNL photoinjector.

The surface field at the cathode is 102 MV/m. The energy gain in the 1 1/2 cell structure is 4.9 MeV. The disk-loaded structure was designed to minimize the ratio of the peak surface field to the field at the cathode surface and is not optimized for maximum shunt impedance. To match to the π-mode in the cells, a side-wall coupling scheme is used. In this configuration, the TE_{10} waveguide mode couples strongly to the π-mode and does not, to first order, couple to the zero mode. The π-mode operation was chosen to minimize emittance growth caused by rf defocussing fields in the accelerating gaps[37].

6.3 Lawrence Berkeley Laboratory

A photoinjector design[39] at Lawrence Berkeley Laboratory (LBL) to produce bright beams for linear colliders, compact FELs, propagation of intense bright beams, and coherent x-ray holography has been completed. The rf cavity design is a 1.269-GHz rf cavity consisting of 2 1/2 cells with a peak surface field of 60 MV/m and a cathode field of 30 MV/m. The design goals are to obtain a 3- to 5-ps pulse length and 1 nC of charge at a gun exit energy of 5 MeV. A layout of the facility is shown in Fig. 19.

The photoinjector parameters were obtained by extensive PARMELA simulations[40] and theoretical analysis[41]. The exiting pulse from the gun has a rms length of 6 ps and a 0.6% energy spread. The calculated emittance is 8 to 15 $\pi\cdot$mm$\cdot$mrad.

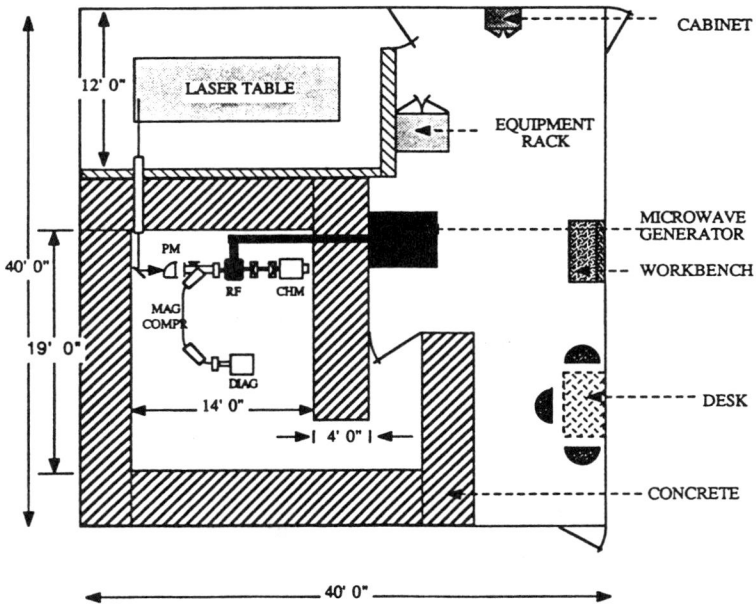

Fig. 19. Layout of LBL photoinjector.

6.4 Bergische Universität-Gesamthochschule Wuppertal

This design for a photoinjector is unique in that the rf gun cavity is superconducting[42]. The design parameters are 1.3 MeV, 5-70 ps, and pulse charge of 0.15-14 nC. The experiment is shown in Fig. 20.

6.5 LEL-HF in Bruyères-le-Chatel

This photoinjector design[43,44] has a much lower cavity frequency, 144 MHz, than the previous designs. A lower frequency can reduce the rf effects because the cavity apertures are larger and the fields approximate dc conditions during the electron transit. The design parameters are a beam with 10-20 nC, a 1 to 1.5-MeV exit energy from the first cavity, bunch lengths of 50-100 ps, an accelerator gap of 7 cm, and a surface field at the cathode of 15-20 MV/m. A schematic of the experiment is shown in Fig. 21.

The design was developed using ATHOS, PARMELA, and OAK. The expected emittance is approximately 20 π·mm·mrad. After initial acceleration to greater than 4 MeV, magnetic compression would be used to increase the peak current.

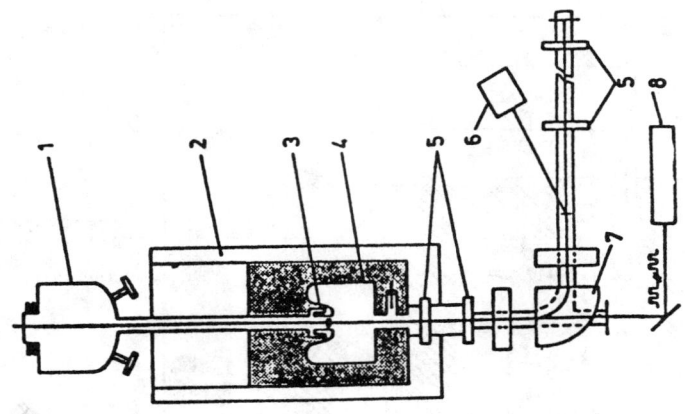

Fig. 20. Experimental setup of the injector designed at Wuppertal: 1. photocathode preparation chamber, 2. bath cryostat, 3. photocathode, 4. reentrant cavity, 5. wire scanner monitor, 6. streak camera, 7. spectrometer, 8. Nd:YAG laser.

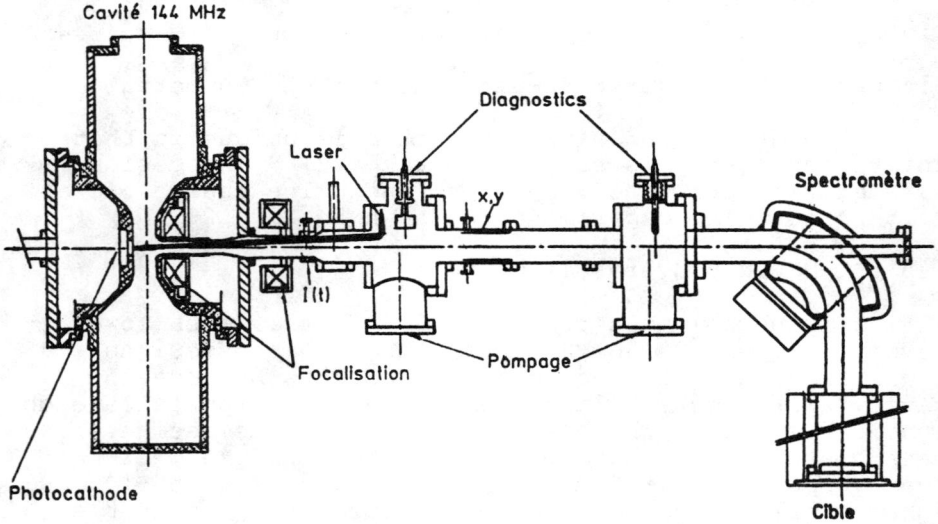

Fig. 21. Photoinjector prototype and electron diagnostics at LEL-HF.

7 SUMMARY

The production of high-current high-brightness electron beams has enjoyed considerable progress over the last several years, mainly because of changes in the requirements imposed by free-electron lasers. The photoinjector shows considerable potential for producing very bright electron beams. The concept of placing a photoemissive source in an accelerating structure has been demonstrated. The basic physics of photoinjectors is understood and the technology is now in the initial engineering phases. Several groups around the world are designing bright beams based on this technology and continued improvement in photoinjector design is expected. The real limit of electron brightness using a photoinjector is still to be determined.

ACKNOWLEDGMENTS

The author thanks John S. Fraser with whom the initial work on the photoinjector program was accomplished. The author is also grateful to Bruce E. Carlsten for the development of the theoretical analysis given in this paper, to Lloyd Young for help in cavity physics and the use of PARMELA, to Bill Herrmannsfeldt, Roger Miller, and Mike Jones for discussion on the theory of photoinjectors, and to Jerry Watson and Stanley Schriber for their continued support and encouragement. The author is indebted to Scott Apgar, Renee Feldman, Paul Giles, Robert Hoeberling, Theodore Gibson, Donald Greenwood, Valerie Loebs, Richard Martinez, Dinh Nguyen, Noel Okay, Louis Rivera, Jake Salazar, Roberta Salazar, Boyd Sherwood, Robert Springer, Robert Stockley, Floyd Sigler, Scott Volz, and Reine Mussett for assistance in the design, construction, and operation of the Los Alamos photoinjector experiment. Finally, I thank all the authors of papers in this field ,whether cited or unmentioned, for laying the groundwork or further developing the photoinjector concept.

REFERENCES

[1] J. M. Watson, IEEE Trans. Nucl. Sci., NS-32, (5) 3363 (1985).
[2] R. B. Palmer, SLAC-PUB-4295, April (1987).
[3] S. V. Benson, J. Schultz, B. A. Hooper, R. Crane, and J. M. J. Madey, Nucl. Instr. and Meth., A272, 22-28 (1988).
[4] T. I. Smith, 1986 Linear Conf. Proc., Stanford Linear Accelerator Center report, SLAC-303, 421-425 (1986).
[5] J. S. Fraser and R. L. Sheffield, IEEE J. Quant. Electron. QE-23 (9), 1489 (1987).

[6] M. Yoshioka, M. Mutuo, Y. Fukushima, T. Kamei, H. Matsumoto, H. Mizuno, S. Noguchi, I. Sato, T. Shidara, T. Shintake, K. Takata, H. Kuroda, N. Nakano, H. Nishimura, K. Soda, M. Miyao, Y. Kato, T. Kanabe, and S. Takeda, Proc. 1984 Linac Conf., Gesellschaft fur Schwerionenforschung, Darmstadt report GSI-84-11,469-471 (1984).
[7] C. K. Sinclair, AIP Proc. on Advanced Accel. Concepts, Madison, WI, 156 (1986).
[8] P. J. Tallerico, R. L. Sheffield, W. D. Cornelius, E. R. Gray, M. T. Wilson, D. C. Nguyen, K. L. Meier, and R. L. Stockley, 1988 Linear Accel. Conf., Williamsburg, VA, Oct. 2-7 (1988), to be published.
[9] D. T. Pierce, R. J. Celotta, G. C.Wang, W. N. Unertl, A. Galens, C. E. Kuyatt, and S. R. Mielczarek, Rev. Sci. Instrum. 51, 478-499 (1980).
[10] C. K. Sinclair and R. H. Miller, IEEE Trans. Nucl. Sci. 28 (3), 2649-2651 (1981).
[11] A. M. Weiner, J. P. Heritage, and R. N. Thurston, Opt. Lett., 11 (3), 153 (1986).
[12] M. Haner and W. S. Warren, Opt. Lett. 12, (6) 398 (1987).
[13] D. C. Nguyen, D. E. Watkins, and M. E. Weber, SPIE Proc., January 11-16, Los Angeles, CA (1988).
[14] *Photoemissive Materials*, A. H. Sommer, (John Wiley & Sons, Inc., New York, 1968).
[15] W. E. Spicer, Phys. Rev. 112, 114-122 (1958).
[16] P. E. Oettinger, R. E Shafer, D. L. Birx, and M. C. Green, Proc. 9th Int. Free Electron Laser Conf., Williamsburg, VA, September 14-18, 264 (1987).
[17] Y. Kawamura, K Toyoda, and M. Kawai, Appl. Phys. Lett. 45, 307 (1984).
[18] S. W. Downey, L. A. Builta, and D. C. Moir, Phys. Lett. 49, 911 (1986).
[19] J. D. Saunders, T. J. Ringler, L. A. Builta, T. J. Kauppila, D. C. Moir, and S. W. Downey, Proc. 1987 Part. Accel. Conf., IEEE Catalog No. 87CH2387-9, 1, 337 (1987).
[20] J. P. Girardeau-Montaut, C. Girardeau-Montaut, R. Dei-Cas, H. Corvet, G. Haouat, J. P. Laget, E. Michaud, M. Renaud, and J. Siguad, Euro. Part. Accel. Conf., Rome, June 7-11, 1988.
[21] *Semiconductors and Semimetals*, Vol. 15, Eds. Willardson and Beer (Academic Press, New York, 1981).
[22] M. Boussoukaya, M. Bergeret, R. Chehab, M. Leblond, and M. Franco, 1987 Proc. Part. Accel. Conf., IEEE Catalog No. 87CH2387-9, 1, 325 (1987).
[23] Private communication from Steve Benson.
[24] C. C. Fllips, A. E. Hughes, and W. Sibbett, Ultrafast Phenomena IV, D. H. Auston and K.W.B. Eisenthal, Eds., (Springer-Verlag, Berlin, 1984), pp. 420-422.
[25] C. LeJeune and J. Aubert, "Emittance and Brightness: Definitions and Measurements", *Applied Charge Particle*

Optics, A. Septier, Ed., Advances in Electronics and Electron Physics, Supp. 13A, 159-259 (1980).
[26] P. Lapostolle, IEEE Trans. Nucl. Sci. $\underline{18}$ (3), 1101-1104 (1971).
[27] J. D. Lawson, *The Physics of Charged Particle Beams*, (Oxford University Press, 1977), p.199.
[28] P. Oettinger, I. Bursuc, R. Shefer, and E. Pugh, Proc. 1987 Part. Accel. Conf.,IEEE Catalog No. 87CH2387-9, $\underline{1}$, 288 (1987).
[29] J. S. Fraser, R. L. Sheffield, E. R. Gray, P. M. Giles, R. W. Springer, and V. A. Loebs, Photocathodes in Accelerator Applications, Proc. 1987 Particle Accelerator Conf., Washington, D.C., March 16-19, $\underline{3}$, 1705 (1987).
[30] E. R. Gray and J. S. Fraser, Proc. 1988 Linear Accel. Conf., Williamsburg, VA, October 3-7, 1987, to be published.
[31] W. Herrmannsfeldt, R. Miller, and H. Hanerfeld, SLAC, PUB 4663, June (1988).
[32] M. E. Jones and W. K. Peter, Proc. 6th Int. Conf. High-Power Particle Beams, Kobe, Japan (1986).
[33] S. V. Benston et al., Lasers 88, to be published (1988).
[34] B. E. Carlsten and R. L. Sheffield, 1988 Linear Accel. Conf., Williamsburg, VA, October 2-7, 1988, to be published.
[35] B. E. Carlsten and R. L. Sheffield, 1989 Part. Accel Conf., Chicago, IL, March 20-23, 1989, to be published.
[36] B. E. Carlsten, 10th Int. FEL Conf., Jerusalem, Israel, August 29-September 2 (1988).
[37] K. T. McDonald, Princeton University Report DOE/ER/3072-43, March (1988).
[38] K. Batchelor, J. Sheehan, and M. Woodle, Brookhaven Internal Report, BNL-41766 (1988).
[39] S. Chattopadhyay, Y. J. Chen, D. Hopkins, K. J. Kim, A. Kung, R. Miller, A. Sessler, and T. Young, 1988 Linear Accel. Conf., Williamsburg, VA, October 2-7, 1988, to be published.
[40] Y.-J. Chen, LBL internal notes, BES 4, ESG Tech Note-74.
[41] K.-J. Kim and Y.-J. Chen, 1988 Linear Accel. Conf., Williamsburg, VA, October 2-7, 1988, to be published.
[42] H. Chaloupka, H. Heinrichs, H. Piel, C. K. Sinclair, F. Ebeling, T. Weiland, U. Klein, and H. P. Vogel, Euro. Part. Accel. Conf., Rome, June 7-11 (1988).
[43] S. Joly, R. Dei-Cas, C. Bonetti, F. Cocu, J. P. De Brion, J. Frehaut, G. Haouat, A. Herscovici, H. Leboutet, and J. Siguad, Euro. Accel. Conf., Rome, June 7-11, 1988, to be published.
[44] R. Dei-Cas et al., 10th Int. FEL Conf., Jerusalem, Israel, August 29-September 2, (1988).

LOW EMITTANCE THERMIONIC ELECTRON GUNS*

W. B. HERRMANNSFELDT

*Stanford Linear Accelerator Center,
Stanford, California 94309*

TABLE OF CONTENTS

1. SELF FIELDS 1533
2. EXTERNAL FIELDS 1535
3. THERMAL-EMITTANCE-LIMITED GUNS 1535
4. GUNS FOR ELECTRON LINACS 1539
   ACKNOWLEDGEMENTS 1541
   REFERENCES 1541

---

*Work supported by the Department of Energy, contract DE–AC03–76SF00515.

LOW EMITTANCE THERMIONIC ELECTRON GUNS*

W. B. HERRMANNSFELDT
Stanford Linear Accelerator Center
Stanford, California 94309

1. SELF FIELDS

The beam emitted from the cathode of a "well-designed" electron gun is born with uniform current density. Almost by definition, this "well-designed" gun is a Pierce[1] design with the focusing electrode carefully matched to the edge of the cathode.

The advantage of the Pierce design is that the focusing fields at the edge of the beam exactly cancel the defocusing fields from space charge. Thus there are no transverse forces on the edge of the beam. This "Pierce condition", as it is sometimes called, cannot be maintained for very long, however. Using, for example, Child's law for space-charge limited flow in a plane parallel diode,

$$j = \frac{4}{9}\sqrt{2e/m}\ \epsilon_0 \frac{V^{3/2}}{x^2}\ ,\qquad(1)$$

it is readily found that, for a given current density, j, the voltage V varies as the 4/3 power of distance x, from the cathode. Because of the limitation on high voltage, the space between cathode and anode is typically between 1 and 3 times the radius of the cathode. Since the Pierce condition can only be maintained as far as the anode, space-charge spreading of the beam will begin as the beam approaches the hole in the anode.

Space-charge spreading does not necessarily cause emittance growth. So long as the beam maintains uniform current density, the transverse electric field, due to space charge, is given by

$$E_r = \frac{r}{2\epsilon_0 c}\left(\frac{j}{\beta}\right)\ ,\qquad(2)$$

where r is the distance from the axis within a uniform beam and βc is the (assumed uniform) axial velocity. Particles which are focussed only by forces that vary linearly with the displacement variable can always be transported without emittance growth.

The above condition on a uniform beam of course does not include the familiar idea of a Gaussian beam profile. Thus, the Gaussian beam will encounter nonlinear transverse fields. The designer of a low emittance gun wants to delay the onset of this condition as long as possible. The greater the kinetic energy of the beam when it encounters nonlinear forces, the less damaging these nonlinearities will be.

© 1989 American Institute of Physics

There is a lower limit to the emittance of an emitted beam, resulting from the temperature of the cathode. Using the Lapostolle[2] definition of rms emittance for a uniform beam

$$\epsilon_n = 4\left[\langle x^2 \rangle \langle x'^2 \rangle - \langle xx' \rangle^2\right]^{1/2} \quad (3)$$

and the condition that $\langle xx' \rangle = 0$ on the cathode, Lawson[3] has noted that for a thermionic emitter of radius r_c at a uniform temperature T

$$\epsilon_n = 2r_c \left(kT/m_0 c^2\right)^{1/2} \text{ m} - \text{rad}. \quad (4)$$

The practical lower limit for emittance as pointed out by Fraser,[4] is found by substituting for r_c in (4) from $I = \pi r_c^2 j$. Noting that kT is 0.1 eV for 1160K, which is a typical cathode temperature, the lower limit on rms emittance is

$$\epsilon_n = 5.0 \times 10^{-6} \left(I/j\right)^{1/2} \text{ m} - \text{rad with } j \text{ in } A/cm^2. \quad (5)$$

A typical cathode area for a linac injector is of order 1 cm². Since the nominal current density from dispenser cathodes is 5–10 A/cm^2, the ratio I/j is of order unity, and the minimum practical emittance is about 5×10^{-6} m-rad.

The rms divergence $\langle \theta \rangle$, implied by this emittance, for a cathode radius of 0.5 cm, is 1 m-rad. Normally such a low divergence does not significantly affect the beam profile; the beam would have to drift for one meter to cause a 1-mm rounding-off of the profile. However, if it is desired to significantly compress a beam transversely, then by the conservation of transverse emittance, the divergence increases inversely to beam radius. A beam thus compressed from a few millimeters diameter to about one millimeter, will almost immediately assume a Gaussian profile, with its attendant nonlinearities.

The above treatment completes our discussion of self-field effects in the gun region. We have deliberately ignored self-magnetic fields because,

(1) They are not significant until the beam achieves semi-relativistic velocities, ($\beta > 0.5$), which usually does not happen within the gun region, and

(2) The self-magnetic fields act only to partially neutralize the space–charge forces. As such, the self-magnetic fields depend on radial uniformity of the beam in exactly the same way as do space charge forces.

The only consideration should be that, if it is desired to avoid the nonlinear self fields altogether, then it is only necessary to maintain a uniform beam profile until the beam has relativistic velocity.

2. EXTERNAL FIELDS

By definition, the first-order radial fields, from external electric and magnetic structures, are all linear. The scale, by which one determines if first-order considerations are adequate, is the ratio of beam radius to electrode or pole piece radius. As a rule of thumb, if this ratio is less than one-half, the significant nonlinear forces have been avoided.

Since the above criterion, i.e., keeping the beam radius small, argues precisely against the conclusions for self fields, it is apparent that there is a preferred dependence of radius as a function of axial position, for the best beam conditions. However, there are usually other criteria for transverse matching of a beam to a subsequent transport system that preempt trying to rigidly apply the above considerations.

There is, in addition, the option of attempting to avoid nonlinear fields in the design of the electrodes and pole pieces. Peter and Jones[5] have devised a formalism for designing electrodes that result only in linear fields in the acceleration cavity. Similarly, one can shape magnetic pole pieces to reduce saturation and nonlinear fields. Both applications have the added advantage of minimizing peak fields. The cost is frequently some increase in size.

In many cases, the current desired in a beam is much greater than a cathode can emit in an area similar to that of the desired beam profile. In these cases, it is usual to use a spherical cathode with a focus electrode inclined to the edge of the beam at the Pierce angle, defined by $4/3\tan^{-1}(y/x) = \pi/2$, i.e., the slope of the electrode at the edge of the beam, relative to the edge of the beam, is 67.5°. A general treatment of the Pierce structure for different beam profiles including hollow beams and curved paths, was given by Sar-El.[6]

3. THERMAL-EMITTANCE-LIMITED GUNS

Most electron gun applications do not allow for a gun designed according to the foregoing discussion. The most frequent additional requirement is for the control of pulse length. In the limit of an injector to an accelerator, pulse length is ultimately determined by a bunching system that will increase the emittance far beyond thermal limits. This subject is treated by T. Smith in another chapter in this volume.

The next level of pulse length control is that from about 1 to 1000 ns, that can be controlled by a pulsed grid. Since such guns are frequently used in accelerator injectors, the subsequent bunching process will dominate over emittance induced by the grid.

There are, however, applications for which the very best possible emittance is required and in which it is possible to avoid grids and rf bunching. Two of these are:

1) A gun for an electron cooling system, and

2) An injector for an electrostatic free electron laser (FEL).

Electron cooling is a process in which a low emittance ("cold") electron beam is made to mix with a circulating beam of ions to improve the emittance of the ion beams. The process was suggested by Budker[7] and has been implemented at several laboratories world wide. The most typical application is to cool a beam of anti-protons in a storage ring. To provide adequate dc current at reasonable power, the gun and collector are both operated at the same high voltage (except for a small bias). The beam power can be, for example, 20A × 120 kV = 2.4 MW with only a few kilowatts of beam energy lost to heat.

The specific problem in the design of an electron gun for electron cooling is to avoid any type of transverse motion. A uniform magnetic field is employed in the interaction area to keep the beam from spreading. There is an unavoidable, but quite small azimuthal motion induced by the $\vec{E} \times \vec{B}$ drift from space-charge forces and this magnetic field. The design problem then is reduced to matching the beam into the magnetic field in the drift region. The method that was devised to do this was used in guns for both CERN[8] and Fermilab.[9] The technique is illustrated in Fig. 1 which is a computer simulation of the flat cathode gun used for the CERN ICE experiment. The gun region is shown in Fig. 1A and the matching section is shown in Fig. 1B. The boundary conditions for the two segments are matched so that 1B is a continuation of 1A.

Counting the gap at the end of Fig. 1A, there are four gaps which can have more or less arbitrary voltages across them. The system is constrained by the voltage between cathode and first anode, which determines the current, and by the overall cathode-to-drift-tube voltage, which determines the total kinetic energy. The kinetic energy is chosen to give the electrons the same velocity as the orbiting proton or antiproton beam. There are thus three free parameters for voltages that can be adjusted to minimize transverse velocity in the beam. An empirical approach to minimizing the transverse energy can be shown to succeed to within about 1 eV. If these voltages are not properly adjusted, or if the "resonant lenses" as these electrodes are called, are not used, the beam is found to continue with quite large scallops representing unacceptable transverse energy. In the Fermilab experiment,[9] a spherical cathode and fully immersed flow (meaning the same total magnetic flux in the beam as through the cathode) was used to obtain higher current density. The same technique of resonant lenses was employed. In both devices, the best residual transverse motion was similar to the transverse thermal velocity.

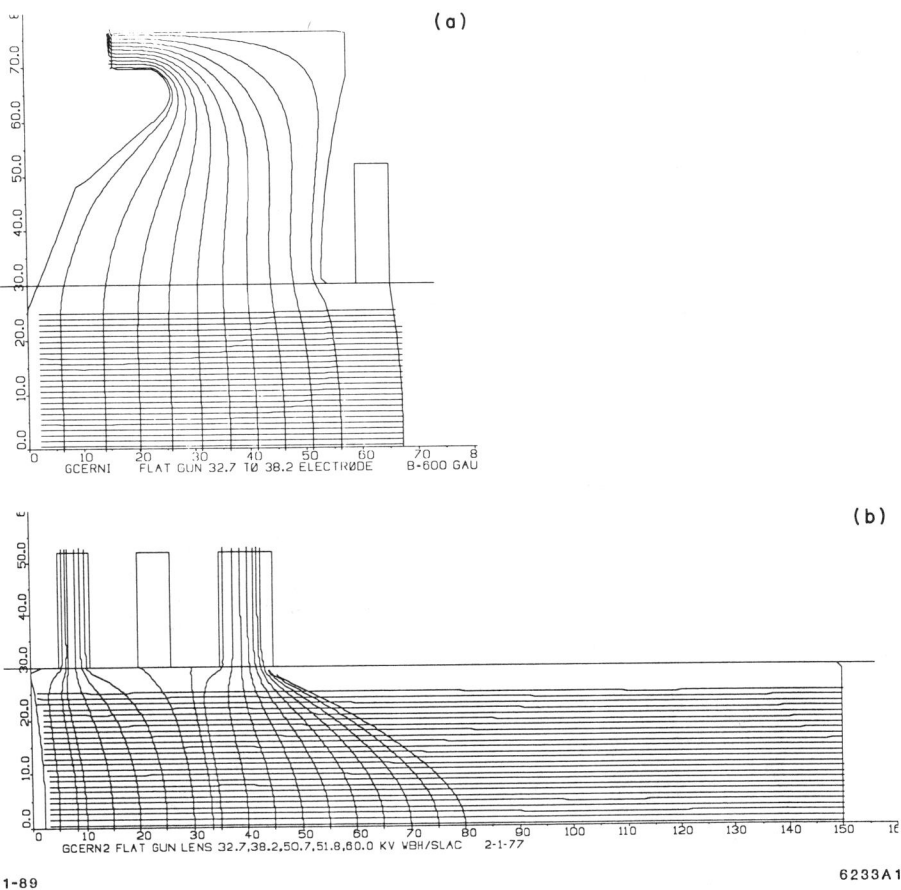

Fig. 1A. Electron gun region modelled for the CERN initial cooling experiment (ICE). Fig. 1B. Continuation of the simulation of Fig. 1A showing the resonant lens configuration.

The SLAC Electron Trajectory Program known as EGUN[10] was used for the design of both of the above electron cooling guns and also for the FEL injector described below. The use of computer programs to design electron guns and beam transport systems was described by Herrmannsfeldt[11] and others at the Beam Optics Codes Workshop.

Free Electron Laser (FEL) injectors must provide a beam with emittance

$$\epsilon = \langle x \rangle \langle \theta \rangle \approx \lambda$$

in order for there to be adequate coupling between the electron and the photon distributions.[12] Note that this is the laboratory emittance, not the "normalized" or invariant emittance $\epsilon_n = \beta\gamma\epsilon$. Thus, ϵ is inversely proportional to the momentum $\beta\gamma$. In most FEL's, an electron accelerator is used to achieve a high γ, typically 200 (for a 100 MeV beam) yielding sufficient "adiabatic" damping to permit the FEL to operate.

In one case, however, using an electrostatic accelerator, the achievable energy is only a few MeV. Then it is necessary to achieve the best possible emittance from the electron gun. The gun designed for the UCSB FEL project was described by Elias and Ramian.[13] The design of this gun is shown in Fig. 2. The first electrode is known as a "mod anode" and is used to gate the beam on and off. At the design voltage, the fields in front and behind the plane of the mod anode still obey the Child's law criterion, so that there is no defocusing of the beam. In order to smoothly maintain this field, the mod anode is made as thin as possible, tapering to a sharp edge at the inner ring.

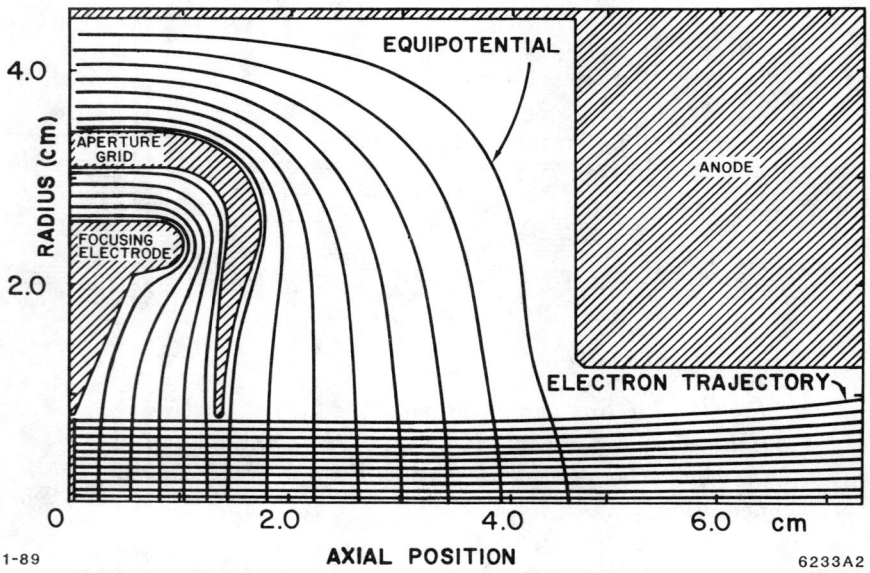

Fig. 2. Design of the electron gun for the UCSB electrostatic accelerator FEL.

A section view of the hardware design of the UCSB gun is shown in Fig. 3. As the beam enters the accelerator column, the field becomes quasi-constant so that there can be some space-charge spreading. However, the velocity is now so great that the residual forces are very weak. There is, in addition, a sort of "weak strong focusing" effect caused by the presence of the system of rings and gaps in the electrostatic accelerator column. Cline et. al.[14] made measurements of this gun, evaluating it for a recirculating electron cooling system. They found the emittance to be very near the thermal limit as defined above.

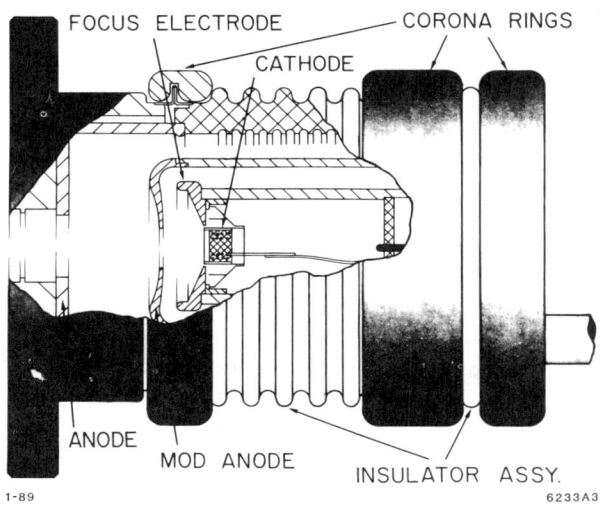

Fig. 3. Mechanical configuration of the UCSB FEL gun.

Note that the configuration for an electron cooling system is essentially the same as for an FEL, i.e., the collector must operate at the same voltage as the gun, usually slightly positive relative to the cathode, but much less than the high voltage of the electrostatic column. In this mode, all of the power that is available to the optical process of the FEL comes from the power supplied to the gun, and not from the electrostatic power system for the accelerator column. The limit to duty cycle usually occurs when the laser process causes some beam current loss which then must be replaced. Thus improving the collection efficiency and eliminating transport losses allow approaching cw operation with a very high efficiency FEL.

4. GUNS FOR ELECTRON LINACS

A criterion known as the "Lawson-Penner Relation" has been used by several authors to compare emittance of different linacs. Although Lawson and Penner[15] have stated that it is not a "law" and better injectors probably can be built, it is still useful to examine the implications. In the form studied by Roberson[16], the Lawson-Penner relation is

$$\epsilon_n(\text{m} - \text{rad}) = S\sqrt{I(A)} \qquad (6)$$

where S is an empirical constant found by comparing existing accelerator systems. Usually this comparison should be made within the same general category of accelerators, i.e., rf bunching or induction linac, etc. Penner found $S = 160 \times 10^{-6}$, in the units used for this paper. Roberson made a better fit to the data he used with $S = 95 \times 10^{-6}$. Both numbers are much greater than the thermionic limit of Eq. 5, thus seemingly confirming the statements earlier that bunching and/or pulse transients dominate the emittance.

If, as these authors have supposed, better injectors can be built, it is worthwhile to speculate what techniques need to be used. If, as seems to be the case, the gun does not limit the emittance, then it follows that it must be the transient fields introduced by the bunching and/or gating devices. Since this discussion quickly leads into the subject areas covered by Sheffield and Smith in their respective chapters in this volume, it may be sufficient to point out that while nonlinear fields may cause emittance growth, other nonlinear fields can cancel the effect. Usually this approach only works if done quickly, as in the case of the electron cooling guns discussed earlier. In the particular case of interest here, Sheffield and Carlsten[17,18] have used the nonlinear space-charge fields themselves to correct the damage done to the emittance by the beam expanding earlier due to those same nonlinear fields. This approach is analagous to the use of properly spaced sextupole and octupole lenses in a beam transport line. Numerical simulations of space-charge spreading and refocusing, with attention to the effects of nonlinear forces, are the subject of work by Hanerfeld et al.[19]

Before leaving the subject of electron guns for linear accelerators, it is important to note the significant number of gun assemblies that have been supplied commercially by Ron Koontz.[20] For short pulse, high current applications, these guns use cathode grid assemblies supplied by the Eimac Corporation. Grid-to-cathode spacings of under 0.1 mm are employed to reduce the grid voltage that must be pulsed. This is especially important for very short pulses.

Although grids are a necessary evil in many guns, it is possible to design the grid so that it is electrically invisible to the beam. There will still be some percentage interception, given by the opacity of the grid, but if a grid is placed on an equipotential line for a diode, and is then pulsed to the potential of that equipotential line, there will be no electric field deflection of the beam particles by the grid.

ACKNOWLEDGMENTS

Thanks to Frank Krienen, CERN, Peter McIntyre and Fred Mills, FERMILAB, for their encouragement during the design work for the electron cooling guns. Thanks to Luis Elias and Gerry Ramias for support for the design of the PEL gun and to Gerry for the illustrations used here. I wish to acknowledge helpful discussions with Ron Koontz, Roger Miller, Rich Sheffield and Todd Smith.

REFERENCES

1. J. R. Pierce, *Theory and Design of Electron Beams* (van Nostrand, 1949).
2. P. Lapostolle, IEEE Trans. Nucl. Sci., NS–18, pp. 1101–1104 (1971).
3. J. D. Lawson, *The Physics of Charged Particle Beams,* Oxford University Press (1977).
4. J. S. Fraser, et al., High–Brightness Photoemitter Injector for Electron Accelerators, IEEE Trans. Nucl. Sci., NS–32, p. 1791 (1985).
5. W. Peter and M. E. Jones, *High-Brightness Electron-Beam Production from Photocathodes,* Particle Accelerators, to be published.
6. H. Sar-El, *Revised Theory of Pierce-Type Electron Guns,* Nucl. Inst. and Methods, **203** 21–33 (1982).
7. G. I. Budker, Atomic Energy **22**, No. 5 (1967).
8. M. Bell, J. Chaney, H. Herr, F. Krienen, P. Moller-Peterson, G. Petrucci, *Electron Cooling in ICE at CERN*, Nuclear Instruments and Methods, **190** pp. 237–255 (1981).
9. W. B. Herrmannsfeldt, W. Kells, P. M. McIntyre, F. Mills, J. Misek and L. Oleksiuk, IEEE Trans. on Nucl. Sci., NS–26, pp. 3237–3239 (1979).
10. W. B. Herrmannsfeldt, EGUN-An Electron Optics and Gun Design Program, SLAC-331, Stanford Linear Accelerator Center, Nov. 1988.
11. W. B. Herrmannsfeldt, *Electron Ray Tracing Programs for Gun Design and Beam Transport*, Proceedings of the Linear Accelerator and Beam Optics Codes Workshop, AIP Conference Proceedings No. 177, San Diego, CA (1988).
12. C. W. Roberson, *Free-Electron Laser Beam Quality*, IEEE Journal of Quantum Electronics, Vol. QE–21, p. 860 (1985).
13. L. R. Elias and G. Ramian, *UCSB FEL Electron Beam System,* in *Free-Electron Generators of Coherent Radiation*, Ed., S. F. Jacobs et. al., Phys. Quant. Elect. **9**, p. 577, Addison-Wesley (1982).
14. D. B. Cline et al., IEEE Trans. on Nucl. Sci., NS–30, pp. 2370–2372 (1983).

15. J. D. Lawson and S. Penner, *Note on the Lawson-Penner Limit*, IEEE Journal of Quantum Electronics, Vol. QE–21, p. 174 (1985).

16. C. W. Roberson, *Bright Electron Beams for Free Electron Lasers*, Proc. Soc. Photo-Opt. Instrum. Eng., pp. 320 (1984).

17. B. E. Carlsten and R. L. Sheffield, *Photoelectric Injector Design Considerations*, Proceedings of the 1988 Linear Accelerator Conference, Williamsburg, Virginia, (Oct. 1988).

18. B. E. Carlsten and R. L. Sheffield, Photoelectric Injector Design Code, Proceedings of the 1989 Particle Accelerator Conference, Chicago, Illinois (March 1989).

19. H. Hanerfeld, W. Herrmannsfeldt and R. H. Miller, High Order Correlations in Computed Particle Distributions, Proc. 1989 Particle Accelerator Conference, Chicago, Illinois (March 1989).

20. R. Koontz, Hermosa Electronics, Menlo Park, CA.

INTENSE, LOW EMITTANCE INJECTORS FOR RF ELECTRON LINACS

Todd I. Smith
High Energy Physics Laboratory
Stanford University, Stanford, California 94305

TABLE OF CONTENTS

1. Introduction... 1544
   1.1 Goals.. 1544
2. Gun.. 1545
   2.1 Transient Guns... 1545
   2.2 Steady State Guns.. 1546
   2.3 Choice of Gun Type... 1547
3. Space Charge... 1547
4. RF Dynamics.. 1548
5. Wake Fields.. 1550
6. Magnet Bunching.. 1551
7. Conclusion... 1551
8. References... 1552

INTENSE, LOW EMITTANCE INJECTORS FOR RF ELECTRON LINACS

Todd I. Smith
High Energy Physics Laboratory
Stanford University, Stanford, California 94305

1. INTRODUCTION

A workshop on low emittance beams was held at Brookhaven National Laboratory from March 20 to March 25, 1987.[1] The workshop was held under the auspices of the International Committee on Future Accelerators (ICFA), and was intended to explore issues relating to the production of electron beams with an unprecedented combination of high charge density and low emittance. The following material is largely a result of discussions held during the workshop, and represents an attempt to estimate the severity of some of the problems which would be encountered in the design of an RF linac injector consistent with the workshop goals.

1.1 Goals

For the purpose of the present discussion, an injector is defined as the low energy part of a linac. It starts with an electron gun and delivers a beam to the rest of the linac at whatever energy is high enough that emittance growth in the linac due to space-charge forces will not be a significant issue. Parameters defined for a target injector which were consistent with (or exceeded) the criteria established for the workshop as a whole are shown in Table I. These parameters will serve as a background for estimates of the seriousness of various issues which might prevent the target injector from performing as desired. Since the workshop's interest was directed towards FEls and colliders, which generally can function with a large separation between bunches, the injector will be imagined to operate in a single-bunch mode. Therefore, multi-bunch effects will not be considered.

Table I

Charge/bunch	10 nC
Normalized transverse emittance (90%)	5 π mm mr
Normalized longitudinal emittance	500 π keV ps
(equivalent to .1% δ E/E at 1 GeV and 2 ps)	
Final energy - whatever required to allow bunching the beam to 2 ps without excessive emittance growth due to space charge.	

Note that the 90% transverse emittance quoted is nearly the same as an rms emittance of 10^{-6} m. If the rule of thumb is applied that the wavelength of an FEl must be less than the absolute value of the emittance, then these injector parameters are compatible with a 1-GeV FEL operating at 50 Å. Finally, the brightness of the beam, defined as the current (5000 Amperes!) divided by the square of the absolute 90% emittance, is 2×10^{13} amperes/m^2 - a very respectable value.

1.2 Issues

The task now is to determine if it it at all reasonable to imagine an injector which can deliver an electron beam with the parameters listed in Table I. To do this, various accelerator physics problems need to be considered. Not only is it necessary to ascertain whether or not an electron gun is capable of initiating such a beam, it is also necessary to examine effects tending to destroy the quality of the beam as it propagates along the length of the injector. The purpose of the present discussion is not to examine all of the possible problems in detail, but it is an attempt to estimate the severity of the most serious set. The issues included in this set are: gun, space charge, RF dynamics, wake fields, and bunching.

2. GUN

It is probable that the most critical issue is that of the electron gun. If the phase space occupied by the beam from the gun has a volume which exceeds that desired at the end of the injector, there isn't much hope that the goals will be met. It is convenient to make a distinction between two types of guns: transient, and steady state. A transient gun is characterized by a short current pulse, while a steady state gun is characterized by a long current pulse. "Short" and "long" are defined with respect to the transit time of electrons through the gun.

2.1 Transient Guns

The accelerating fields in a transient gun are frequently produced by high frequency rf (generally the fundamental frequency of the linac), while the short current pulses seem best produced by a photocathode excited by a mode locked laser. The clear advantages of such a system are: high field gradients minimizing the time and distance required for electrons to become relativistic and thus less susceptible to space-charge effects; very high current densities from the photocathode, minimizing the beam size for a given current or charge and thus maximizing the inherent brightness of the beam; current vs. time(from the gun)tracking the driving laser pulse rather faithfully, thus needing little or no bunching in the rest of the system.

The state-of-the-art example of this type of gun has been described by the group at Los Alamos National Laboratory[2]. Typical parameters of interest regarding their system are: current/pulse = 10 nC; normalized transverse emittance = 20 π mm mr; bunch length = 60 ps; peak surface electric fields = 60 MV/m; cavity frequency = 1.3 GHz; longitudinal emittance = unknown.

These results are close enough to those needed for compatibility with the target injector that they can almost be regarded as an existence proof. The LANL group feels that since the work is in its early stages, substantial improvements can be expected. They feel that the longitudinal phase space, although not yet measured, will not be a serious limitation.

The question of improvements is an interesting one. A list of parameters to be varied in an optimization procedure includes cavity frequency, cavity field strength, cavity shape, cathode size, current pulse length, and current pulse shape. It is clear that the transient gun is complicated and that the interplay among various parameter is presently understood only to a limited extent. For example, high field strengths are generally believed to be good since the beam bunch becomes relativistic more quickly, while high frequencies are generally bad since the time dependence of the fields is more likely to adversely affect the bunch. A conflict arises with the realization that the most promising way to reach higher fields is through the use of higher freuencies!

Charge delivered by a gun in a pulse short compared with propagation times in the gun structure is stored charge. Thus the process of delivering charge by a transient gun can be viewed as similar to the discharge of a capacitor. The maximum charge which can be delivered is determined by the product of the electric field gradient at the cathode, the area of the cathode, and dielectric constant of free space. If reasonable values of 100 kV/cm and 1 cm^2 are used for the gradient and area, the maximum charge which can be delivered turns out to be 10 nC. The LANL gun used a field of 600 kV/cm, and was illuminated with a laser pulse such that 10 nC were produced out of a possible 60 nC. It would appear that large variations in the output energy of electrons delivered during such a pulse are unavoidable. The first electrons emitted will see the full electric field in the gap, while the last to leave the cathode will see a considerably smaller field due to the retarding fields of the electrons preceeding them. If the maximum charge is used, then the last electrons will see no field at all. Assuming the gun geometry to be planar, and thus neglecting effects associated with the edges of the beam, the field seen by an electron depends linearly on the electron's position within the bunch. In this case, even though there might be a large energy variation within the bunch, the variation is highly correlated with linear position, and does not necessarily imply a large longitudinal phase space. On the other hand, if the planar approximation is inappropriate, then it seems likely that the longitudinal phase space could be a significant problem.

It is interesting to note that even with the high electric fields and high current densities used in this class of gun, the transverse phase space of the emitted electrons is not reduced to trivial proportions. From the preceding paragraph, it is clear that for charges in the 10-nC range, cathode areas on the order of 1 cm^2 will be necessary unless surface electric fields can be raised significantly. The precise number to assign to the inherent thermal emittance of the beam from the cathode is somewhat unclear, due to lack of agreement on coefficients, factors of π, whether to use rms or 90% values, what temperature to assume, etc. However, the emittance is certainly proportional to the beam radius and to the square root of the effective cathode temperature. Reasonable assumptions for various parameters and a 1-cm^2 cathode give values for the normalized emittance between 1 and 5 π mm mr.

2.2   Steady State Guns

The accelerating fields in a steady state gun can be produced by a dc source, by a pulsed source, or by a low frequency rf source. The requirement is that time variations of the fields be insignificant during the current pulse. The choice of cathode material is not particularly critical. In principle, photocathodes and thermal cathodes are both adequate. The critical issues in a steady state gun are the electrode design to prevent emittance growth due to non-linear fields, and the technique used to control the beam current. If a thermal cathode is used, control can be provided either with a grid or with a modulating anode. A grid will respond faster than an anode, but its presence can degrade the beam emittance somewhat. A good example of a steady state gun with parameters of interest to the workshop is that developed by Elias at the University of California at Santa Barbara[3]. This gun has a 1-cm^2 cathode area, runs at 50 kV, produces a current of 1 ampere controlled by a modulating anode, and is mounted at the top of the column of a 6-MeV Van de Graaff. The emittance of the beam measured at 6 MeV is 7 π mm mr, consistent with that calculated from the area and temperature of the cathode.

The current pulse from a steady state gun is by definition "long". This is consistent with the philosophy of minimizing charge density and thus space-charge forces during the time that the beam is at relatively low energies. The length of the pulse is limited by the fact that the pulse must be bunched in order to be compatible with the rest of the injector. A single-frequency bunching system has good linearity (and thus preserves longitudinal emittance) over about $\pm 30°$ while a system which incorporates a second harmonic is quite good over $\pm 60°$. If the injector is assumed to operate at 500 MHz and uses a second-harmonic bunching system, then the pulse from the gun can be 666 ps long. If 10 nC is desired, then the gun current must be 15 amperes. If Elias' gun is scaled to work at 40 A/cm^2 and 15 A, the emittance would be expected to be 3 π mm mr. Both photocathodes and thermal cathodes are capable of providing current densities of about 40 A/cm^2 for reasonably long pulses.

2.3   Choice of Gun Type

The choice of one or the other of these types of guns implies a choice of philosophy in the injector design. A transient gun uses a short high current pulse to provide a given charge. Space-charge forces are high, but if the pulse is accelerated rapidly enough emittance growth due to space charge may be kept within bounds. Since the pulse is short, emittance growth due to time variation of the rf fields seen by the beam is automatically minimized. The longitudinal phase space of the beam from a transient gun may present some problems. A steady state gun uses a long, modest current pulse to provide the desired charge. Space-charge forces can be one to two orders of magnitude smaller than those in a transient gun, but the long pulse complicates the acceleration process and care must be taken to assure that time-varying fields do not destroy the initial beam quality. There is no reason why a system using an intermediate length pulse cannot be considered. With proper care and ingenuity, such a compromise system presumably could combine the best features of the others.

The transient gun as embodied by the rf photocathode gun at LANL is clearly promising. The ongoing work, both theoretical and experimental, should provide a better understanding of its limits and potential.

The use of a carefully designed steady state gun is assumed throughout the remainder of this discussion. The long (~1-ns) current pulses can probably be best controlled through the use of a photocathode. The gun voltage will be high to minimize emittance growth during transport. Voltages of 500 keV to 1 MeV may be necessary. This approach can certainly provide the necessary charge/bunch in an acceptable emittance. However, the long bunch must be accelerated and utlimately compressed longitudinally without much growth in either transverse emittance or longitudinal sphase space.

3. SPACE CHARGE

Once a beam of adequate quality has been provided by a gun, it must be transported and accelerated through the remainder of the injector. During this process many effects will have an opportunity to degrade the beam quality. Probably the most difficult of these effects is that of space charge. It is clearly an unavoidable problem, and can be minimized only by attempting to control the charge density in the beam and by accelerating the beam as rapidly as possible so that relativistic effects can help.

Although the mechanism by which space charge causes emittance growth is easy to understand, predictions regarding the magnitude of the effect can be difficult to make. Fortunately, a paper by Jones and Carlsten[4] deals directly with the problem. Numerical studies of the emittance growth of drifting beams of various form factors, shapes, energies, and total charge are presented. From their results, estimates can be made of the normalized emittance of a 15-ampere, 1-cm-diameter, 20-cm-long bunch after drifting for a distance of one meter. (These parameters are the same as those used for beams produced by the steady state gun discussed above: 10 nC and 666 ps.) Assuming that the initial emittance of the beam is zero, the final emittance is approximately 10 π mm mr if the beam energy is 500 keV ($\gamma = 2$), and is 0.1 π mm mr if the beam energy is 10 MeV. ($\gamma = 20$.) It must be emphasized that these numbers are only approximations, as many parameters which affect the problem (such as pulse shape and charge distribution) have not yet been specified.

A great deal of the emittance growth of the beam is a result of the fact that the magnitude of the space-charge forces varies along the length of the beam. This variation causes the orientation of the ellipse in transverse phase space which encloses a longitudinal section of the beam to rotate as the section moves along the beam axis. The area of any section may be small, but the superposition of all of the ellipses necessary to represent the entire bunch results in a large area and thus a large emittance. Three possibilities suggest themselves as helpful. One is to take advantage of the fact that most of the emittance growth due to this effect takes place near the bunch ends. This implies that the ends of the bunch can be used as a guard for the rest of the beam. Accelerate a longer beam than actually desired, and chop off the excess when the beam energy is large enough that the guard is no longer needed. A second possibility is to use time-dependent rf focussing to remove the longitudinal dependence of the phase-space ellipse orientation. This approach would almost certainly require rf fields at harmonics of the fundamental linac frequency in order to adequately approximate the required time dependence of the focussing forces. A third approach, suggested by Carlsten[5], uses a carefully placed lens in an accelerator and drift system to reverse the initial emittance growth.

Emittance growth due to transport of a charged beam can be serious. However, it appears that the target charge of 10 nC in less than 5 π mm mr can be delivered by a steady state gun operating at 0.5 to 1 MeV with a 15-A   beam.

4. RF DYNAMICS

Emittance and longitudinal phase-space growth due to the time dependence of the rf fields seen by a long beam can be devastating. These effects can almost certainly be reduced to acceptable levels through the use of a low fundamental frequency in the linac and through the inclusion of accelerating fields at harmonics of the fundamental[6].

The use of harmonics in the accelerator is clearly related to the use of harmonics to Fourier synthesize any desired waveform. If it were possible to accurately mimic a square wave, then a beam less than one half a cycle long would see no time-dependent fields at all. There would be no time-varying accelerating fields to cause longitudinal phase-space growth, and no time-varying magnetic fields to cause emittance growth. Of course, only an approximation to such an ideal can be attained. However, simulations indicate that merely adding third harmonic fields to the accelerator can have dramatic effects on both the longitudinal and transverse emittance.

Figure 1 illustrates the effects of including fields at the third harmonic. The phase space of an electron beam which is 36 degrees long is shown after being accelerated from 1 MeV to 2 MeV by a single cavity designed to support accelerating fields at f and 3f simultaneously.[7] In (a) only the fundamental fields are used. The large energy spread due to the time variation of the fields is evident, as is the large emittance of the output beam. In (b) the third-harmonic field has been included, with phase and amplitude adjusted to minimize the output energy spread. The energy spread is too small to be seen on the same scale as before. The growth in the output emittance has also been reduced substantially.

SINGLE CAVITY PHASE SPACE
Input Beam Diameter = 1 cm,   No Space Charge

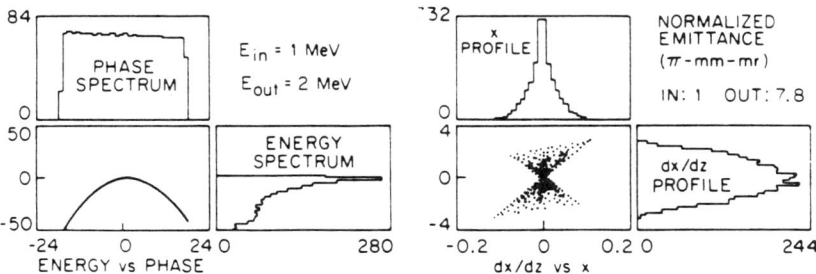

1(a)   Phase space after acceleration by fundamental fields only.

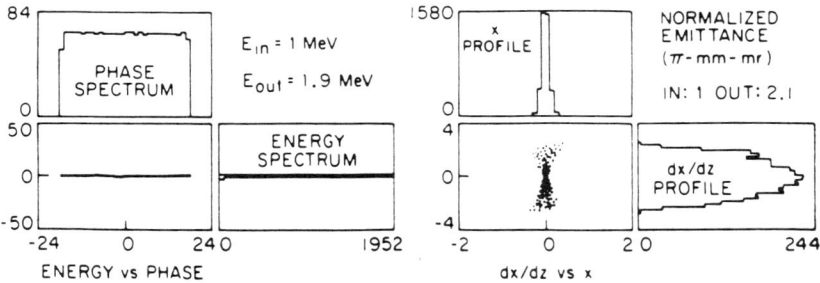

1(b)   Phase space after acceleration by 'flat-topped' fields formed by the fundamental and third harmonic. Note the improvement in the energy spectrum and the transverse phase space relative to (a).

Fig. 1. PARMELA simulation of an electron beam after being accelerated from 1 MeV to 2 MeV by a single cavity. The input beam is 1 cm in diameter and is ± 18 degrees (1 GHz) long. It has a transverse phase space of 1 π mm mr. The units in the figure are cm, mr, keV, and degrees for x, dx/dz, energy, and phase respectively. The simulation used 2000 particles.

5. WAKE FIELDS

Another phase-space diluting effect which needs to be examined is that due to the excitation by the beam of wake fields in the accelerating cavities. Wake-field effects are conventionally divided into two areas: longitudinal and transverse. For a bunch of constant charge but variable length, the magnitude of longitudinal effects tends to decrease as the bunch lengthens while the magnitude of transverse effects tends to increase.

The magnitude of emittance growth which can be expected from the transverse wake may be estimated with the two-particle model as described by Wilson[8]. For the current study it is assumed that the beam drifts a distance z with no focussing (worst case). In this case, the normalized emittance due to wake fields can be written as

$$\varepsilon_n = \pi\gamma \frac{q^2 x_0^2 w_d^2}{8 V_0^2} z^3$$

where x_0 is the initial beam displacement from the axis, q is the bunch charge, w_d is the transverse wake evaluated at 1 (the bunch length), and V_0 is the constant beam energy. Several assumptions were made in order to apply this formula to the present problem. Since the study is concerned with an accelerating beam, the <u>average</u> beam energy is used in place of V_0. The wake field is estimated by scaling the measured wake from the SLAC structure to the frequency and geometry of interest and it is assumed that the strength of the wake field is proportional to the bunch length for cases of interest. The wake field is thus assumed to be given by

$$w_d \sim 2.8 \times 10^{15} \left(\frac{1}{7°}\right)\left(\frac{1.2 \text{ cm}}{\lambda_{2856}}\right)^2 \left(\frac{\lambda}{a}\right)^2 \left(\frac{f}{2.856 \text{ GHz}}\right)^3 \text{ V/C/m}^2$$

where l is the bunch length, 1.2 cm is the size of the SLAC aperture, λ_{2856} is the SLAC wavelength, 2.856 GHz is the SLAC frequency, and λ, f, and a are the wavelength, frequency and aperture of the structure of interest. For the calculation assume $x_0 = 1$ mm; $q = 10$ nC; $l = 36$ deg; $f = 0.5$ GHz; $\lambda/a = 4$; $V_0 = 100/2$ MeV (100 MeV final energy); and $z = 20$ m (gradient = 5 MeV/m). With these values,

$$\varepsilon_n = 2\pi \text{ mm mr.}$$

This is a reasssuringly small number, particularly in view of the fact that the parameters have not been pushed.

A similar estimate can be made for the longitudinal wake, again scaling from the wake field of the SLAC structure. From Ref. 8, including appropriate scaling factors, the longitudinal wake is

$$w_l(t) = 226 \left(\frac{f}{2.856 \text{ GHz}}\right)^2 e^{-\left(\frac{t}{6 \text{ ps}} \frac{f}{2.856 \text{ GHz}}\right)^{.605}} \left(\frac{1.2}{\lambda_{2856}}\right)\left(\frac{\lambda}{a}\right) \text{ V/PC/m.}$$

If this wakefield is convoluted with the 200-ps 15-Ampere current pulse assumed for the beam, the resulting energy drop along the pulse is monotonic, with a value of about 100 keV at the end of the pulse. (The same accelerating structure parameters are assumed as in the transverse wake example above.) The longitudinal emittance arising from the energy variation along the beam calculated using the usual rms definition[9] is about 250 π keV ps. This longitudinal emittance implies that when the beam is compressed to 2 ps, there will be an energy spread of at least 500 keV on the beam. If this value is too large, the longitudinal emittance can be reduced through the addition of harmonic fields to the accelerator.[10] A reduction by at least a factor of five is probably feasible.

6. MAGNETIC BUNCHING

The final energy of the injector will be defined as that necessary to allow the beam to be bunched without undue emittance growth. The bunching would presumably be accomplished through the use of a system of magnets to provide a path length which is a function of energy, coupled with an energy-position correlation within the bunch.

Since the optics of the bunching system can presumably be made as good as necessary, the emittance growth is assumed to arise from space-charge forces. As in section 2, the calculations of the LANL group can be applied. The data and information for the following estimate were provided by Carlsten.[11]

Calculations with PARMELA[12] show an emittance growth from 42 π mm mr to 54 π mm mr when a 6-MeV beam is compressed magnetically from 60 A to 200 A, and a growth from 25 π mm mr to 26 π mm mr when a 12-MeV beam is similarly compressed. If the assumption is made that the emission growth due to space charge in the bunching process is uncorrelated with anything else, then the final emittance is obtained by adding the original emittance in quadrature with the space-charge contribution. This leads to value of 34 π mm mr due to bunching at 6 MeV, and 7 π mm mr at 12 MeV. Since the space-charge forces are only large at high charge densities, the emittance growth due to bunching is largely independent of the density at the beginning of the bunching process. Scaling arguments suggest that the product of $\gamma^2 \times I \times \lambda\varepsilon$ is a constant, where γ is the beam momentum, I is the final bunch current, and $\lambda\varepsilon$ is the emittance growth due to space-charge forces. PARMELA results are consistent with this scaling. This scaling implies that if a bunch current of 5000 A (10 nC at 2 ps) is desired with no more than 3 π mm mr growth then the beam energy must be at least 100 MeV.

7. CONCLUSION

A possible approach to building the target injector begins with the use of a high voltage steady state gun to provide a high quality low density electron beam which is less susceptible to space-charge forces than a tightly bunched beam. The low density beam would be accelerated to a final energy of about 100 MeV using a low frequency (perhaps 500 MHz) incorporating fields at harmonics of the fundamental in order to preserve the small phase space of the beam emitted by the gun. After acceleration, when space-charge forces have been reduced to negligible proportions by relativistic effects, the beam would be compressed by a magnetic bunching system to its final high density value. It could then be injected into a conventional accelerator.

No mechanism has been identified as clearly incompatible with the goals proposed for the target injector. However, it should be clear that only a small set of the possible problems facing such an injector have been considered, and even these have been examined only to the extent of estimating their severity.

8. REFERENCES

1. Proceedings of the ICFA Workshop on Low Emittance e^- - e^+ Beams. Brookhaven National Laboratory, March 20-25, 1987. J. B. Murphy and C. Pellegrini, eds. [BNL 52090.]

2. J. S. Fraser, et al. "Photocathodes in Accelerator Operations," Proceedings of the 1987 IEEE Particle Accelerator Conference, March 1987, p. 1705.

3. L. Elias and G. Ramian, "Status Report of the UCSB FEL Experiment Program," Free Electron Generators of Coherent Radiation, Charles A. Brau, Stephen F. Jacobs, Morton O. Scully, editors; Proceedings SPIE 453, pp. 140-145.

4. M. E. Jones and B. E. Carlsten, "Space-Charge Induced Emittance Growth in the Transport of High-Brightness Electron Beams," Proceedings of the 1987 IEEE Particle Accelerator Conference, March 1987, p. 1319.

5. B. E. Carlsten, "New Photoelectric Injector Design for the Los Alamos National Laboratory XUV FEL Accelerator," to be published in the Proceedings of the Xth International FEL Conference, Jerusalem, Israel, August 29-September 2, 1988.

6. T. I. Smith, "Intense Low Emittance Linac Beams for Free Electron Lasers," 1986 Linear Accelerator Conference Proceedings, Stanford Linear Accelerator Center, June 2-6, 1986, p. 421. [SLAC-PUB-303].

7. C. E. Hess, et al. "Harmonically Resonant Cavities for High Brightness Beams," IEEE Trans. Nucl. Sci. NS-32 (1985) p. 2924.

8. P. B. Wilson, "High Energy Electron Linacs: Application to Storage Ring RF Systems and Linear Colliders," Physics of High Energy Particle Accelerators, (Fermilab Summer School 1981), R. A. Carrigan, F. R. Houson, M. Month, editors, AIP Conf. Proc. # 87, pp. 450-582, [SLAC-PUB-2884].

9. P. Lapostolle, IEEE Trans. Nucl. Sci, NS-18 (1971) pp. 1101-1104.

10. K. C. D. Chan and J. S. Fraser, "Minimum Beam-Energy Spread of a High-Current RF Linac," Proc. of the 1987 IEEE Particle Accelerator Conference, March (1987) p. 1075.

11. B. E. Carlsten (private communication).

12. R. Crandall and L. Young, "PARMELA: Particle Motion in Electron Linear Accelerators," LANL (private communication).

FINAL FOCUS SYSTEMS FOR LINEAR COLLIDERS*

ROGER A. ERICKSON

Stanford Linear Accelerator Center,
Stanford, California 94305

TABLE OF CONTENTS

1. INTRODUCTION 1554

2. ELEMENTARY OPTICAL CONSIDERATIONS 1555
   2.1 Bunch Length 1556
   2.2 Arrival–Time Synchronization 1557
   2.3 Demagnification 1557
   2.4 Telescopes 1561

3. CHROMATIC ABERRATIONS 1562
   3.1 Mathematical Preliminaries 1562
   3.2 Chromatic Correction with Sextupoles 1564
   3.3 Third–Order Aberrations 1567
   3.4 The Optimum Design 1568

4. DESIGN OF THE SLC FINAL FOCUS 1569
   4.1 The Design Process 1569
   4.2 The η-Matching Section 1571
   4.3 The First Telescope and Extraction System . . . 1572
   4.4 The Chromatic Correction Section 1572
   4.5 The Final Bend 1573
   4.6 The Final Telescope 1574
   4.7 Expected Performance 1577

5. TUNING AND DIAGNOSTICS 1579
   5.1 Tuning the SLC 1579
   5.2 Luminosity Monitoring 1581
   5.3 Profile Monitoring Devices 1581
   5.4 Beamstrahlung Radiation 1582

6. STEERING TO COLLISION 1584
   6.1 Beam–Beam Deflections 1584
   6.2 Application to the SLC 1585
   6.3 Maintaining Collisions 1588

   ACKNOWLEDGEMENTS 1589

   REFERENCES 1589

---

*Work supported by the Department of Energy, contract DE–AC03–76SF00515.

FINAL FOCUS SYSTEMS FOR LINEAR COLLIDERS

ROGER A. ERICKSON

Stanford Linear Accelerator Center
Stanford, California 94305

1. INTRODUCTION

The final focus system of a linear collider must perform two primary functions: it must focus the two opposing beams so that their transverse dimensions at the interaction point are small enough to yield acceptable luminosity, and it must steer the beams together to maintain collisions. In addition, the final focus system must transport the outgoing disrupted beams to a location where they can be recycled or safely dumped. As of this writing, the only full-scale collider final focusing system actually constructed is that of the SLAC Linear Collider, better known simply as the SLC.

The SLC is designed to collide 50-GeV beams of electrons and positrons at an initial luminosity of about 10^{29} cm^{-2} sec^{-1} and ultimately to reach a luminosity of 6×10^{30} cm^{-2} sec^{-1} when several planned improvements are completed.[1] While this luminosity is appropriate for the SLC energy range, a linear collider built for a much higher energy will need much greater luminosity. This is because the cross sections for interesting physics, described by point-like fundamental processes, drop with increasing center-of-mass energy as $1/E^2_{cms}$. The cross section for producing new quark-antiquark states, for example, is expected to be roughly

$$\sigma_{Q\overline{Q}} \approx 10^{-37} E^{-2}_{cms} \text{ [TeV}^{-2}\text{] cm}^2 \quad . \tag{1}$$

Thus, the luminosity appropriate for a machine with energy E_{cms} is about

$$\mathcal{L} = 10^{33} E^2_{cms} \text{ [TeV}^2\text{] cm}^{-2} \text{ sec}^{-1} \quad . \tag{2}$$

At this luminosity, roughly 10 $Q\overline{Q}$ states would be produced per day if they exist at all at that energy. The event rate for interesting physics processes is expected to be about the same for all machines satisfying this relationship.[2]

Luminosity is given by the expression

$$\mathcal{L} = \frac{1}{4\pi} \frac{f \, N^+ \, N^-}{\sigma_x \sigma_y} \tag{3}$$

where N^+ and N^- are the numbers of positrons and electrons per bunch, and f is the frequency of collisions. σ_x and σ_y are the RMS radii of the bunches in the horizontal and vertical dimensions,

© 1989 American Institute of Physics

respectively, and are given in lowest order by

$$\sigma = \sqrt{\frac{\epsilon_N \beta^*}{\gamma}} \qquad (4)$$

where ϵ_N is the normalized emittance, β^* is the value of the betatron function at the collision point and $\gamma = E/m_e c^2$. Thus, for identical round beams,

$$\mathcal{L} = \frac{1}{4\pi} \frac{f N^2}{\beta^*} \frac{\gamma}{\epsilon_N} . \qquad (5)$$

Luminosity increases linearly with energy if all other machine characteristics can be held constant, because the effective emittance shrinks as the longitudinal momentum increases.

To reach collision energies much higher than the SLC while maintaining the luminosity implied by Eq. (2) requires some fundamental improvement in the design. This may be an increase in bunch population N or the repetition rate f, either of which increases the average beam power and the electric bill, or a decrease in ϵ_N or β^*. The prospects for reducing ϵ_N below 3×10^{-5} rad m (the SLC design value) are discussed elsewhere.[3]

There are no fundamental limitations to producing very small values of β^*. There are, however, practical limits imposed by the strength, field quality and alignment tolerances of focusing magnets, the space requirements of the experimental detector, and the overall size of the focusing system as limited by the site boundaries. In this paper, the functions of final focus systems in general will be described, along with a discussion of performance limitations arising from some of these practical design considerations. To illustrate the main ideas, the final focus system of the SLC will be presented, and the basis for the key parameter choices will be discussed.

The SLC is not a true linear collider in the sense of having two linacs pointed at each other. Instead, a single linac is used to accelerate bunches of positrons and electrons simultaneously to the desired energy. The bunches are then separated and guided around two arcs of alternating-gradient bending magnets to opposite ends of the final focus system. At the time of this writing, the construction of the SLC has been essentially completed, and initial commissioning tests are in progress. While most of the examples in this paper are drawn from the SLC, the principles and conclusions are generally applicable to other linear collider designs as well.

2. ELEMENTARY OPTICAL CONSIDERATIONS

An arbitrarily small value of β^*, and hence an arbitrarily small focused spot, can be obtained by using sufficiently strong quadrupole magnets close to the interaction point. There are, however, both first-order and higher-order considerations that define limits on β^*, below which further reductions are useless. Among these limiting considerations are the bunch lengths, the synchronization precision and the momentum spread of the beams. Because of these and other considerations discussed below, the SLC will be operated with a β^* of 4 to 10 mm, depending on how the final focus is set up and tuned.

2.1 Bunch Length

The RMS radius $\sigma(z)$ of a beam near the interaction point is given by

$$\sigma(z) = \sigma^* \sqrt{1 + \frac{z^2}{\beta^{*2}}} \qquad (6)$$

where z is the distance along the beam direction from the point where the minimum radius σ^* occurs. The depth-of-focus scale is set by β^*. If the RMS bunch length σ_z is much greater than β^*, only part of the bunch is in focus at any instant during the collision. The resulting loss in luminosity has been computed for Gaussian bunches (Gaussian in all three dimensions) by integrating over the lengths of both bunches as they pass through each other. Figure 1 shows the relative luminosity as a function of σ_z/β^*.

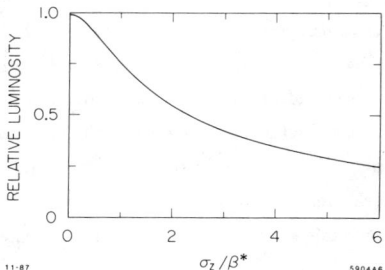

Fig. 1. Luminosity falls with increasing bunch length σ_z. Curve is normalized to β^*.

The RMS bunch length in the SLC is about 1 mm, while the design values of β^* are in the range of 4 to 10 mm. Thus, the bunch-length limit is not yet an issue for the SLC. In a future linear collider, however, this may become a very important consideration. For example, a set of parameters for a 1-TeV center-of-mass linear collider recently suggested by R. Palmer[4] includes $\beta_y^* = 50~\mu$m, which would require a very short σ_z. Palmer suggested a σ_z of 43 μm for this design.

It should be noted that Eqs. (3) through (6) are strictly accurate only when the densities of the opposing bunches are low enough that they do not disrupt each other. When the bunches are very dense, they experience a mutual focusing force due to the electromagnetic attraction between them.[5] This focusing force, which is sensitive to the bunch lengths, reduces the effective beam size near the interaction point and enhances the luminosity.

2.2 Arrival-Time Synchronization

In order to achieve the full potential luminosity with a small β^*, the opposing bunches must arrive at the focal point simultaneously. If the second bunch is late by an amount Δt when the first bunch arrives at the minimum-waist point, the actual collision will occur at a point offset from the desired minimum-waist point by an amount $c\Delta t/2$. Figure 2 shows luminosity loss as a function of this timing error. Controlling the arrival time difference to the necessary precision is easier in the SLC, in which both the e^+ and e^- bunches are accelerated in the same linac, than it might be in a collider built with two separate linacs. In the SLC linac, the e^- bunch trails the e^+ bunch by $168\frac{1}{2}$ S-band buckets, a separation which can be controlled by adjusting the relative r.f. phase of the e^+ and e^- damping rings, which are physically close together and synchronized to the same oscillator. The position of each bunch within its bucket is then determined by monitoring the final energy of each beam. In this way, a reproducible separation can be maintained with a precision of about 1 mm (corresponding to about 3 psec). In the Palmer machine mentioned above $(\beta_y^* = 50~\mu\text{m})$, two independent linacs would have to be synchronized to a precision of about 150 femtoseconds. Maintaining this level of synchronization over a distance of several kilometers would pose an interesting problem.

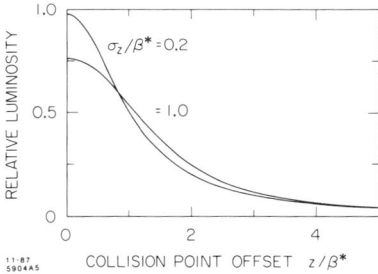

Fig. 2. Luminosity loss as a function of the distance from the collision point to the minimum β point for two bunch lengths. Curves are normalized to β^*.

2.3 Demagnification

A focusing system with a minimum number of components is illustrated schematically in Fig. 3. Here a single thin lens of focal length f is used to form an image at the Interaction Point (IP). The source is a round spot of radius x_1. It can be seen from elementary geometric optics that the condition for bringing an image into focus at the IP is

$$\frac{1}{f} = \frac{1}{L_1} + \frac{1}{L_2} \; . \qquad (7)$$

The image size, x_2 is then

$$x_2 = -\left(\frac{L_2}{L_1}\right) x_1$$

$$= -M\, x_1 \qquad (8)$$

and

$$\theta_2 = -\frac{1}{M}\, \theta_1 \qquad (9)$$

where L_1 and L_2 are the source and image distances from the lens, and their ratio defines the magnification M. For a given total system length, the smallest image is obtained when L_2 is as short as possible and the lens is strong (short f).

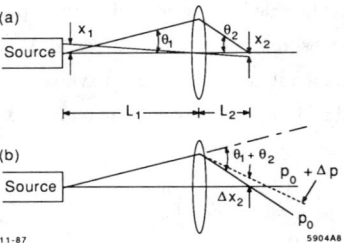

Fig. 3. A thin lens focuser (a). Particles of higher momentum come to a focus downstream of the nominal focal point (b).

The source spot at the input to the SLC final focus (after correction for linear dispersion) has a radius of approximately $x_1 = 50$ μm at a point about 145 m from the IP, with an angular divergence θ_1 of about 6 μrad. Approximating the final set of quadrupole magnets by a thin lens at their common midpoint 5 m from the IP leads to $M = 1/28$, $x_2 = 1.8$ μm and $\theta_2 = 168$ μrad. These results are surprisingly close to the design values for the final spot size of the SLC (cf. Table 1).

The shortcomings of such a simple design become apparent when the finite momentum spread of the beam is considered. A particle near the outer envelope of the beam with momentum higher than the mean beam momentum by a fraction $\Delta p/p$ will be deflected $\Delta \theta/\theta$ less by the lens, as shown in Fig. 3b. Such particles will come to a focus at a point downstream of the nominal IP, resulting in a larger, defocused spot at the point where they collide with the opposing beam.

$$\Delta x_2 \approx L_2\, \Delta\theta$$
$$\approx L_2\left(1 + \frac{1}{M}\right) \theta_1\, \frac{\Delta p}{p} \, . \qquad (10)$$

The actual spot size at the collision point will be the quadratic sum of the "geometric spot size" and this "chromatic spot size."

$$\sigma_z^2 = M^2 x_1^2 + \left[L_2 (1 + \frac{1}{M}) \theta_1 \frac{\Delta p}{p} \right]^2 , \quad (11)$$

where x_1 and θ_1 are now considered the RMS width and angular spread of the input beam. A momentum spread of 0.5% increases the final spot to $\sigma_z = 4.7$ μm in the previous example. This is substantially larger than the pure geometric size and reduces the luminosity by a factor of seven compared to the "monochromatic" calculation.

For fixed L_2 there is an optimum value of M, below which there is no improvement in actual final spot size. Differentiating Eq. (11) with respect to M gives a minimum spot size when

$$\frac{\sqrt{M+1}}{M^2} = \frac{1}{L_2 \frac{\Delta p}{p}} \frac{x_1}{\theta_1} . \quad (12)$$

In the example above, this occurs when $M \approx 1/18$, and results in $\sigma_z = 4.0$ μm.

Substituting

$$\beta^* = M^2 \frac{\sigma_{x_1}}{\sigma_{\theta_1}} \quad (13)$$

in Eq. (12) gives the optimum β^* for the thin-lens system:

$$\beta^* = \sqrt{1+M} \, L_2 \frac{\Delta p}{p} . \quad (14)$$

Since M is small but always positive, and L_2 is greater than or equal to f, it follows that

$$\beta^* \geq f \frac{\Delta p}{p} \quad (15)$$

for any final focus system without chromatic correction features. Unless some exotic[6] new focusing technique is employed, the focal length f is limited by magnet technology and mechanical interference near the experimental detector. The minimum possible spot size is then determined by the momentum spread of the accelerator.

For more complicated examples, it is convenient to use TRANSPORT[7] notation, in which optical elements and drifts are represented by matrices, and a particle trajectory can be represented by a six-element vector[8] $\mathbf{x}_i = (x, \theta, y, \phi, \ell, \delta)$. In this notation, $\theta = x'$, $\phi = y'$, ℓ is the deviation in pathlength from the central trajectory, and $\delta = \frac{\Delta p}{p}$. The one-dimensional, monoenergetic thin lens example can then be represented by:

$$\begin{pmatrix} x_2 \\ \theta_2 \end{pmatrix} = \begin{pmatrix} 1 & L_2 \\ 0 & 1 \end{pmatrix} \begin{pmatrix} 1 & 0 \\ -f^{-1} & 1 \end{pmatrix} \begin{pmatrix} 1 & L_1 \\ 0 & 1 \end{pmatrix} \begin{pmatrix} x_1 \\ \theta_1 \end{pmatrix} \quad (16)$$

which simplifies to

$$\begin{pmatrix} x_2 \\ \theta_2 \end{pmatrix} = R \begin{pmatrix} x_1 \\ \theta_1 \end{pmatrix} \quad (17)$$

where R is the first-order matrix

$$R = \begin{pmatrix} (1 - L_2 \, f^{-1}) & (L_1 + L_2(1 - L_1 \, f^{-1})) \\ -f^{-1} & (1 - L_1 \, f^{-1}) \end{pmatrix} . \quad (18)$$

Focusing implies $R_{12} = 0$, i.e., the position of a particle at the end is independent of its angle at the beginning. Setting this element to zero gives Eq. (7). Then

$$R = \begin{pmatrix} -M & 0 \\ -(\frac{1}{L_1} + \frac{1}{L_2}) & -\frac{1}{M} \end{pmatrix} \quad (19)$$

where again $M = \frac{L_2}{L_1}$.

To see the effect of a finite momentum spread, replace f by $(f + \triangle f) = f(1 + \delta)$ in Eq. (18) above. $\triangle f/f$ is a small fractional change in focal length due to a small fractional change δ in momentum. To simplify the algebra, define $d = \delta/(1 + \delta)$. Then f^{-1} is replaced by $f^{-1}(1 - d)$, and the resulting matrix is designated R'. Note that $d = \delta$ in the limit of $\triangle p \ll p$. With this substitution, Eq. (18) becomes

$$R' = \begin{pmatrix} -M + (M+1)d & L_2(1 + \frac{1}{M})d \\ -f^{-1}(1-d) & (1 - L_1 f^{-1}) + L_1 f^{-1}d \end{pmatrix} . \quad (20)$$

With this substitution, R'_{11} becomes $-M + (M+1)d$, the original part plus a term proportional to momentum spread. In the language of TRANSPORT, the monoenergetic part, $-M$, is R_{11}, and the second part is identified as the second-order term T_{116} times d. Similarly, R'_{12} picks up a non-zero part $L_2(1 + \frac{1}{M})$ which multiplies d and is identified as the T_{126} term [cf. Eq. (10)]. Thus,

$$x_2 \approx R_{11}x_1 + T_{116} \, x_1 \, d + T_{126} \, \theta_1 \, d \quad . \quad (21)$$

In this example, the identification of these second-order coefficients as the T_{116} and T_{126} TRANSPORT terms is an approximation that is strictly correct only in the limit of small $\triangle p$ as d approaches δ.

A more complicated system will have additional terms in this series. The general form of the series expansion for the ith element of the x-vector is:

$$x_i = \sum_j R_{ij} x_j + \sum_j \sum_k T_{ijk} x_j x_k$$
$$+ \sum_j \sum_k \sum_m U_{ijkm} x_j x_k x_m \qquad (22)$$
$$+ \text{ higher-order terms }.$$

The goal in designing any final focus system is to make the magnification terms, R_{11} in the x dimension and R_{33} in the y dimension, small enough to give the desired spot size, while keeping all other terms negligibly small. In this example, the first-order magnification is $R_{11} = -M$, and R_{12} is identically zero at the focal point. The second-order term $T_{116}x\delta$ is negligible, but $T_{126}\theta\delta$ is large and dominates the final spot size. In realistic examples with thick quadrupole lenses, this is still generally true; the T_{126} term dominates the final spot size unless specific features are included to cancel it. It is assumed throughout this paper that any final focus system will be built with mid-plane symmetry, i.e., that all xy coupling terms will be zero.

2.4 Telescopes

The properties of a simple optical telescope can be shown to have direct application to a charged particle transport system. Consider two lenses of focal lengths f_1 and f_2 as shown in Fig. 4. If the spacings are chosen to be $L_1 = f_1$, $L_2 = f_2$ and $L_3 = L_1 + L_2$, then for a monoenergetic beam,

$$R = \begin{pmatrix} -\frac{f_2}{f_1} & 0 \\ 0 & -\frac{f_1}{f_2} \end{pmatrix} = \begin{pmatrix} -M & 0 \\ 0 & -\frac{1}{M} \end{pmatrix} . \qquad (23)$$

Comparing to the single lens case, Eq. (19), it can be seen that the R_{21} term has dropped out, i.e., the angle of a particle leaving the system depends only on its angle at the entrance and not on its position.

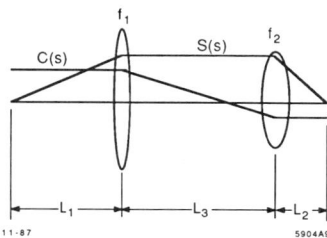

Fig. 4. A telescope constructed with two thin lenses, showing the characteristic sine-like and cosine-like trajectories.

A useful property of telescopes of this kind is that they can "slide" to vary the relative lengths of the leading and trailing drift distances. If the drift length L_1 upstream of the first lens is shortened by an amount $\triangle L_1$, the focal point will move downstream by an amount $\triangle L_2 = -M^2 \triangle L_1$. The first-order R matrix remains unchanged. This property was used to advantage in the design of the final telescope of the SLC, as discussed in Section 4.6 below.

The chromatic properties of a simple telescope can be seen by replacing f_1^{-1} by $f_1^{-1}(1-d)$ and f_2^{-1} by $f_2^{-1}(1-d)$ in the matrix for each lens, as in the single lens example. The R-matrix for the complete telescope then becomes

$$R' = \begin{pmatrix} -\frac{L_2}{L_1} + \left(\frac{L_2}{L_1}\right)d^2 & L_3(d+d^2) \\ \frac{-L_3}{L_1 L_2}(d-d^2) & -\frac{L_1}{L_2} + \frac{L_1}{L_2}d^2 \end{pmatrix} . \quad (24)$$

Comparing to Eq. (20), it can be seen that third-order chromatic terms proportional to d^2 (U_{1166} and U_{2266}) have appeared on the diagonal, but the second-order terms, T_{116} and T_{226}, have vanished. Other third-order contributions (U_{1266} and U_{2166}) have appeared on the off-diagonal terms. T_{126} is unchanged and still the dominant aberration, and the β^* limit in Eq. (15) still applies.

A final focus design based on this simple telescope arrangement was considered briefly for the SLC at an early stage in the design process.[9] A final spot radius of 2 μm could be obtained, even with $\triangle p/p = 0.5\%$, by positioning the nearest face of the innermost quadrupole 20 cm from the IP. In this design, the innermost triplet of quadrupoles required magnetic field gradients of 150 kG/cm. It is conceivable that this gradient could be achieved with a quadrupole constructed from SmCo$_5$ permanent magnets (7.5-kG pole-tip fields) at a radius of 0.5 mm. The field quality requirements, and thus the dimensional and magnetic tolerances on the individual pole pieces, would be severe. While it may be practical in a few years, such a magnet is still beyond present technology. Another practical problem is that such a magnet in the SLC would be destroyed by just a few errant beam pulses. This scheme was rejected in favor of an approach based on active cancellation of chromatic aberrations.

3. CHROMATIC ABERRATIONS

3.1 Mathematical Preliminaries

It is useful at this point to define three functions which in combination can describe any trajectory through a given system. These functions, designated $C(s)$, $S(s)$ and $D(s)$, are sometimes called the "cosine-like trajectory," the "sine-like trajectory," and the dispersion function. $C(s)$ is simply the transverse offset of a particle from the central trajectory as it traverses a given system, when it starts with an offset of 1 unit and zero slope at the input ($C(0) = 1$, $C'(0) = 0$). The value of $C_x(s)$ at any point s along a beam transport system is equal to R_{11} at that point and is simply the magnification. Similarly, a function $C_y(s)$ can be defined for the y-dimension and equals R_{33}. $S(s)$ is the function that starts with zero offset, but a slope of 1, and is equal to R_{12} (in the x-dimension) at each point in the system. At any image point along the way, and in particular at the IP, $R_{12} = 0$. This is a first-order property of any telescopic system.

The dispersion function $D(s)$ relates the position offset to fractional momentum change and is equal at each point to R_{16}. Its value at the focal point can be calculated as an integral over the system:

$$R_{16} = D(s_2) = M \int_{s_1}^{s_2} \frac{S(s)}{\rho} ds \qquad (25)$$

where ρ is the radius of curvature of the central trajectory as it passes through any dipole field. R_{16} will be zero at the IP if the dipoles in the system (if there are any) are arranged in such a way that the integral over the whole system is zero. This will be the case, for example, if dipoles are incorporated into the design only in equal-strength pairs 180° apart in betatron phase, i.e., at points where $S(s)$ has equal amplitude but opposite sign.

Each of the T-matrix terms and higher terms can be expressed as integrals over various combinations of these three functions and their derivatives. These integral expressions are derived and tabulated in Reference 10. The three second-order chromatic aberrations in a point-to-point imaging system are T_{116}, T_{166} and T_{126}. T_{116} for a quadrupole focusing array is given approximately by

$$T_{116} \approx M \int C' S' \, ds \qquad (26)$$

This integral vanishes for a pure telescopic system. For the example in Fig. 4, this is obvious from inspection; one or the other of $C'(s)$ and $S'(s)$ is zero everywhere.

In a real design using sets of thick quadrupoles rather than thin lenses, T_{116} will generally not be identically zero. In the design of the SLC, the pure telescopic conditions were relaxed further to reduce the total length of the final focus system, keeping $R_{11} = 0$, but allowing R_{21} to become significant.[11] With this compromise, T_{116} was still small (≈ 6) but not completely negligible.

The T_{166} term is given by

$$T_{166} \approx -\frac{1}{2}M \int D'^2 S \frac{ds}{\rho} - M \int k_1 S D \, ds + M \int k_2 S D^2 \, ds \qquad (27)$$

where the three integrals represent the contributions from dipoles, quadrupoles and sextupoles, respectively. k_1 and k_2 are the strengths of the quadrupoles and sextupoles. Since each term contains a first power of $S(s)$, each integral can be made zero by designing the system with appropriate symmetries. The contribution from any section of the system with non-zero D or D' can be cancelled by adding another identical section 180° downstream. This is shown schematically in Fig. 5 for a system of two half-wave telescopes separated by dipoles. In the SLC final focus design T_{166} is approximately $-1. \times 10^{-4}$ m, which when multiplied by $(\Delta p/p)^2$ makes an insignificant contribution to the final spot size.

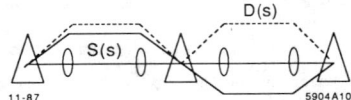

Fig. 5. The T_{166} aberration vanishes in a system of two modules with identical dispersion functions and opposite sine-like trajectories.

The T_{126} term for a point-to-point imaging system is given approximately by

$$T_{126} \approx -M \int S'D'S \frac{ds}{\rho} - M \int S'^2 \, ds + 2M \int k_2 \, S^2 \, D \, ds \qquad (28)$$

where again the three terms correspond to contributions from dipoles, quadrupoles, and sextupoles, respectively. The dipole term can, in principle, be cancelled exactly by appropriate design, but it is typically small in any case. The quadrupole term being quadratic in S' is always positive and increases in value the more focusing elements are in the system. Note that in a typical design, this term comes mostly from the last quadrupoles, which give the particles large angles (S') as they impinge on the IP. In the final telescope of the SLC, T_{126} is approximately 41 m, considerably larger than the size expected from the thin-lens approximation [Eq. (24)]. This aberration would enlarge the final spot by an order of magnitude if left uncorrected. The key to minimizing the T_{126} contribution is to add sextupoles to the system with strengths k_2 chosen to cancel the first two terms.

3.2 Chromatic Corrections with Sextupoles

To see intuitively how this works, think of a sextupole as a quadrupole with a strength that depends on the radial position of the beam passing through its aperture. The y-component of the field seen by the particles of a beam displaced transversely is given by

$$B_y = B_0 \, x^2 = B_0 \, (x_0 + \Delta x)^2 \qquad (29)$$

where x is given by the sum of the position x_0 of the beam centroid and an offset Δx of a particular particle from the centroid. Hence,

$$B_y = (2 \, B_0 \, x_0)\Delta x + B_0 \, x_0^2 + B_0 \, (\Delta x)^2 \quad . \qquad (30)$$

The first term in parentheses corresponds to the position-dependent quadrupole strength. The $B_0 \, x_0^2$ term represents the net dipole deflection felt by the beam as a whole, and $B_0 \, (\Delta x)^2$ is a higher-order term that comes from the fact that the slope of the field is not constant across the geometric width of the beam (as it would be in a true quadrupole field) but increases slightly with increasing $\triangle x$.

Figure 6 shows how the chromatic aberration of a quadrupole can be corrected with a sextupole. A dipole is first used to create linear dispersion. Particles with momentum $p + \triangle p$ are deflected less than those of momentum p, and thus pass through a different part of the sextupole aperture, where they see a stronger focusing field than those on the central trajectory. Of course, the linear dispersion must be removed with one or more dipoles downstream of this point. Sextupoles can be used in this way not only to cancel the chromatic effects of an adjacent quadrupole, but any other quadrupole in the system, either upstream or downstream.

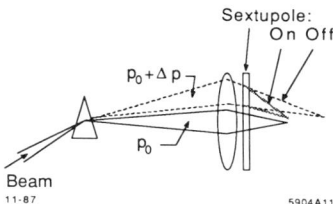

Fig. 6. A dipole and sextupole can be added to a quadrupole focuser to bring off-momentum particles to a focus at the same longitudinal position as the reference beam.

In a quadrupole focusing system without sextupoles, particles with below-nominal momentum come to a focus at a point upstream of the nominal IP; those with higher momentum come to a focus downstream. In a system with sextupoles, the strengths can be chosen to bring a range of momenta to a focus at a single position or even to overcorrect and bring the higher momentum particles to a focus upstream of the nominal image point. This overcorrection can be done in a special Chromatic Correction Section (CCS) upstream of the Final Telescope (FT), compensating in advance for the chromatic aberrations of the final quadrupoles. Specifically, if

$$M \ T_{126}(CCS) = T_{126}(FT) \tag{31}$$

then the T_{126} term for the combined system will vanish. In the final telescope of the SLC, $M = 1/5$ and $T_{126}(\text{CCS}) \approx 41$ m. The chromatic correction section alone, with the sextupoles turned off, has $T_{126}(\text{CCS}) \approx -83$ m. With the sextupoles turned on, $T_{126}(CCS) \approx 205$ m. The net value of T_{126} for the combined system has been calculated to be less than 10^{-4} m.

While T_{126} could be cancelled using only a single sextupole, this would introduce two new problems. These are caused by the second and third terms in Eq. (30): the "even-order" terms, so called because they multiply even powers of $\triangle x$. The $B_0 x_0^2$ term, which is simply the dipole component of the field seen by the off-axis beam in the sextupole, would steer the beam sideways and introduce linear dispersion. The third term, which varies quadratically across the geometric width of the beam, introduces a geometric aberration, i.e., particles with the same momenta would be focused more or less strongly depending on their transverse positions in the beam. Both problems can be solved by using two sextupoles positioned 180° apart, in two identical half-wave telescopes, as shown in Fig. 7. With this arrangement, the effects produced by the even-order terms of the two sextupoles cancel, while the odd terms (needed to cancel the T_{126} aberration) add.

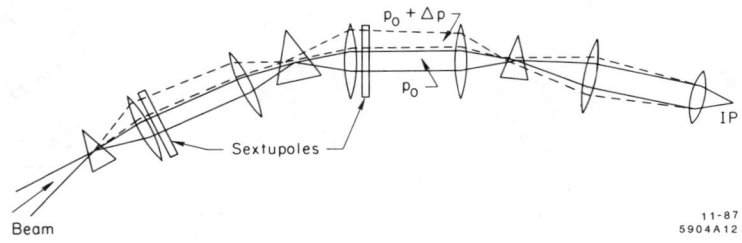

Fig. 7. Two telescopic modules with sextupoles can compensate for the chromatic aberrations of a strong final telescope.

The discussion up to this point has been limited to the x-dimension. The corresponding aberration in the y-dimension, T_{346}, must be cancelled as well. In general, this would require at least one additional pair of sextupoles. One way to do this would be to use two complete chromatic correction sections on each side of the final focus; one pair of half-wave telescopes with sextupoles adjusted to cancel the T_{126} aberration of the complete system, and another pair to take out the T_{346} aberration. In designing the SLC, it was found that both tasks could be done by a single pair of telescopes on each side, but laced with four sextupoles in each telescope, one adjacent to each quadrupole. The odd-numbered sextupoles (counting in the beam direction) are located at points where β_x is larger than β_y and, thus, couple more strongly to the T_{126} term, while the even-numbered sextupoles are located where they couple more strongly to T_{346}. Note that the number of sextupoles in the SLC is twice the minimum number suggested by the discussion above. This came about because the interlacing of the two families of sextupoles introduces higher-order aberrations which would otherwise drop out. By using the additional pairs, it was possible to minimize these other detrimental effects while still cancelling both T_{126} and T_{346} in the space of a single pair of half-wave telescopes.

3.3 Third–Order Aberrations

If a system is chromatically corrected to second order, then the third-order aberrations introduced by the sextupoles become the principal source of the residual optical distortions limiting the system's performance. The ratios of the third-order chromatic aberrations to the third-order geometric aberrations are functions of the strengths of the dipoles used for the second-order chromatic corrections. By adjusting the dipole and sextupole strengths in inverse proportions, the total third-order optical distortion can, in general, be minimized without disrupting the second-order chromatic correction. This minimum value occurs when the angular dispersion introduced by the dipoles is approximately equal to the geometric (monoenergetic) angular spread in the beam. This is expressed by the equation

$$D'_x \frac{\Delta p}{p} = \sqrt{\frac{\epsilon}{\beta_D}} \qquad (32)$$

where D'_x is the angular dispersion introduced by the dipole, and β_D is the value of β_x at the dipole. The final adjustment of the strengths of the dipoles is done empirically with computer simulations, observing the final spot for various combinations of magnet strengths and spacings.

Another factor that must be considered when designing a chromatic correction system is the emittance growth that will occur in the dipoles due to fluctuations in the synchrotron radiation. This emittance growth is related to the energy E, the magnet length L, and the bending radius ρ as follows:

$$\frac{\Delta \epsilon}{\epsilon} \propto \frac{A}{\epsilon} \frac{E^5}{\rho^5} L^4 \qquad (33)$$

where A is a factor that depends on the properties of the beam. The strong dependence on the dipole parameters drives the design to small deflection angles and thus to a long system with strong sextupoles.

As a practical matter, any linear collider must fit within the boundaries of the laboratory. In the case of the SLC, the final focus system was constrained to fit in a section of tunnel approximately 300 m long. This led to a design in which the sextupoles were chosen to be as strong as could be achieved with a conservatively large aperture and conventional iron construction. The total length of each chromatic correction section is about 60 m. While these constraints did not significantly affect the luminosity in the overall system, they did lead to inconvenient mechanical designs in a few places where beamline components were packed closely together. The total loss of luminosity due to emittance growth in these dipoles is about 15%.[11]

In addition to the third-order effects due to the sextupoles, there are third-order effects inherent in the telescopic modules. It was shown above that the effective β^* cannot be less than $f \frac{\Delta p}{p}$ [Eq. (15)] unless some mechanism for cancelling the second order chromatic aberrations is included in the design. This analysis can be taken a step further to get a rough estimate of the limiting third-order aberration. In the case of the simple telescope (Section 2.4), the largest third-order contribution to the final spot size is U_{1266} [Eq. (24)]. From this it follows that

$$\beta^* \geq f\left(\frac{\Delta p}{p}\right)^2 \ . \tag{34}$$

This sets the lower limit on β^* for telescopic systems with perfect second-order corrections. It is instructive to consider, as an example, the final telescope of the SLC. With $\frac{\Delta p}{p} = 0.5\%$ and $f \approx 4$ m, this particular third-order effect would become dominant for $\beta^* \approx 100$ μm. It is interesting to note that this particular example sets a limit on β^* that is far below the chosen β^* for the SLC but is twice the β^* suggested in recent design studies for a 1-TeV machine.[4]

In any high-energy final focus design, as elements are added to cancel aberrations of any given order, they will generally introduce other aberrations of higher orders. In the SLC design, the dominant second-order aberrations, T_{116}, T_{166} and T_{126} (and their counterparts in the y dimension) can be reduced to very small levels, leaving third-order aberrations that ultimately limit the performance of the system. In the case of the SLC, the dominant chromatic aberrations are U_{1266} and U_{3466} which cause about a 20% loss in the ultimate luminosity. The dominant geometric aberrations are U_{1222} and U_{3442} which cause about a 10% loss in luminosity.[11] In principle, aberrations of any order can be cancelled by introducing appropriate nonlinear elements in the system; however, no practical scheme has been invented for cancelling aberrations above second order without introducing larger higher-order effects. Ideas that have been considered for cancelling third-order effects typically require excessive magnet strengths or unfeasible alignment tolerances.

3.4 The Optimum Design

There is no unique final focus design that is well-suited for all possible linear colliders. The optimum architecture, as well as the detailed parameters of the design, will depend on the energy, energy spread, emittance and bunch lengths of the beams. A design that is optimized for one set of beam parameters cannot generally be scaled to a much higher energy, even if the other parameters are the same, because of technical limitations. Magnetic field strengths in particular are limited by the maximum attainable current densities in electrical conductors and the magnetic properties of materials and cannot be scaled up to arbitrarily high values.

The use of flat, rather than round, beams is an interesting option for large TeV-class linear colliders. Flat beams are usually mentioned in the context of beamstrahlung radiation, as a way of reducing the spread of collision energies for a given luminosity by reducing the amount of energy lost as electromagnetic radiation.[12] Flat beams also have interesting implications for the optical design of a final focus system. If the system is designed to cancel only the aberrations that enlarge the vertical beam size, while ignoring horizontal effects, a simpler system with fewer elements is possible, and some aberrations that would otherwise arise from the interplay of these various elements will never appear. Furthermore, quadrupole fields, which are intrinsically asymmetric in the x and y dimensions, can be used to best advantage in arrays that do not require round beams. For fixed field strengths, a smaller, sharper focus can be achieved in one dimension than is possible for a round beam, but at the expense of the other dimension. Whether or not this is a sensible approach for a particular collider depends on the emittance and energy spread of the beams and the details of the final focus system. An interesting example of a 1-TeV flat beam design is given in Reference 13.

4. DESIGN OF THE SLC FINAL FOCUS

4.1 The Design Process

The optical design of the SLC final focus system started with an overall architecture suggested by Karl Brown. The magnet strengths and positions were initially chosen based on a combination of analytic calculations and reasonable guesses. The final design evolved over a period of many months as the implications of various parameter choices were studied and impractical options were discarded. Most of this work was done with the aid of the computer programs TRANSPORT[7] and TURTLE.[14]

TRANSPORT is a second-order matrix multiplication program for designing the first and second-order optics of a proposed system. It has a fitting feature which is used to determine the magnet strengths required to achieve the desired optical and geometric properties of the system. TURTLE is a ray-tracing program that uses the second-order matrices from TRANSPORT to represent each element of the system. Individual rays are traced through each element separately so that higher-order cross-coupling effects between elements are included. This program was especially useful for evaluating the severity of the third and fourth-order cross-coupling effects introduced by the interleaved sextupole families in the SLC. The results are displayed in the form of histograms which show the density distribution of rays as a function of any of the phase-space coordinates.

At an early stage in the SLC project, it was recognized that the final focus system should be designed to accommodate variations in the last telescope. It was anticipated that shorter-focal-length quadrupole lenses would eventually become available, and that the final focus system should be able to exploit any such developments with minimal changes to existing hardware. Ideally, such changes in the final telescope should not require any changes to the other optical modules. Two designs for the final telescope have been developed. Conventional iron quadrupoles have been built for the commissioning and initial operation because of their simplicity and relatively low cost. In parallel with this effort, a design based on superconducting quadrupoles has been developed.[15]

There are two reasons for upgrading the SLC final focus with superconducting quadrupoles. The first is that with higher gradient quadrupoles, a luminosity improvement of roughly a factor of two is possible over the initial machine configuration. The second reason is that superconducting quadrupoles can be operated inside the experimental detector, immersed in a 6-kG solenoid field. This becomes especially important when the SLD,[16] a large second-generation detector, is brought into the SLC.

The final optical design is illustrated in Fig. 8. The β and η functions are plotted in Fig. 9. Each side consists of five half-wave modules: an azimuthal bending section that immediately follows the arc and cancels the off-energy function, η, a telescopic β-matching section that accommodates the extraction system for the outgoing beam, two modules which together correct

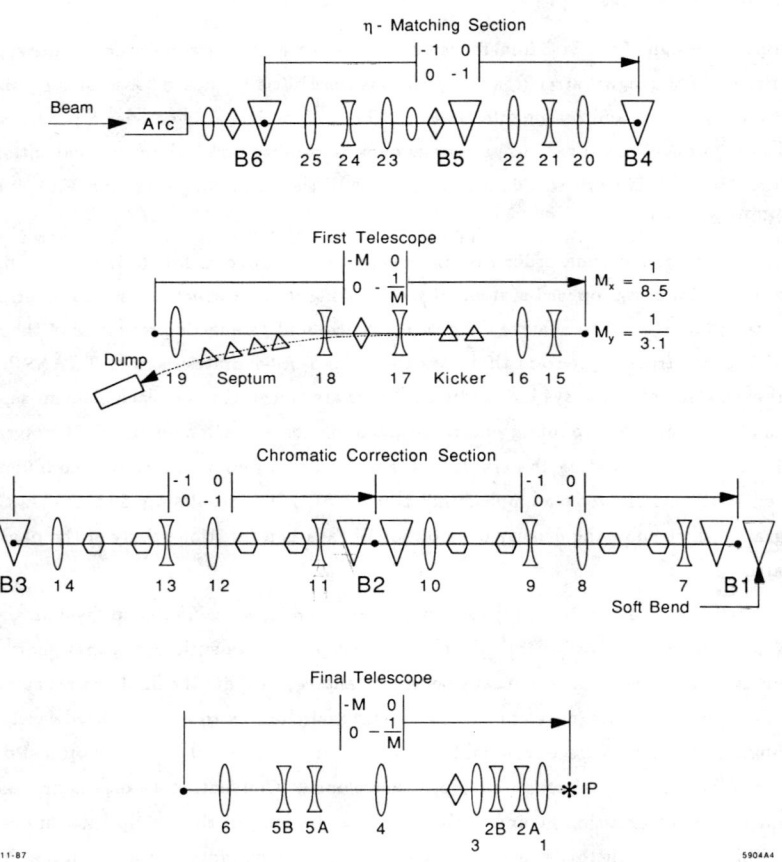

Fig. 8. The optical design of the SLC final focus system.

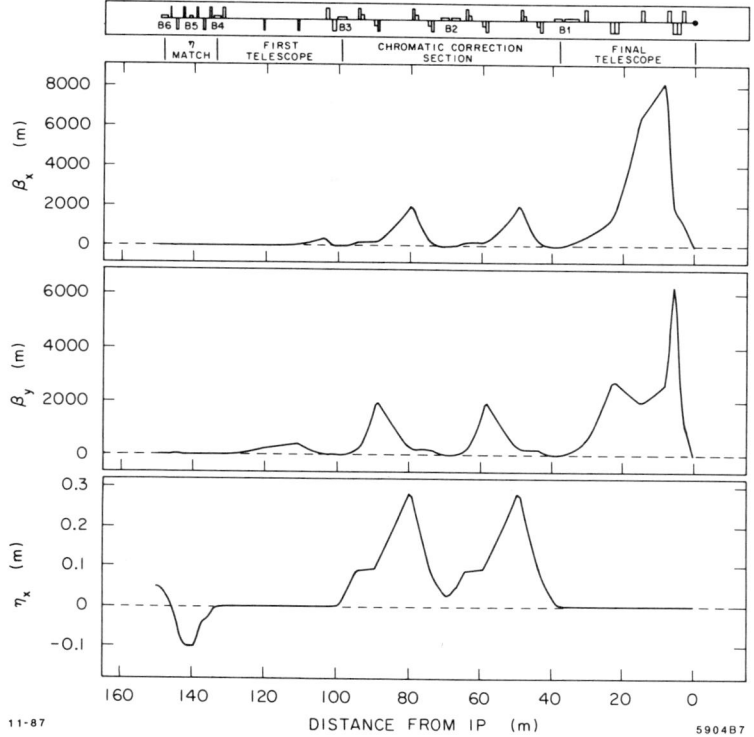

Fig. 9. β and η functions of the SLC final focus. The η function in the y-dimension (not shown) is zero everywhere.

the second-order chromatic distortion of the system, and finally a telescope which demagnifies the beam to the final spot size and provides for fast precise steering.

4.2 The η-Matching Section

Beginning at the end of the arc, the first half-wave module consists of three dipoles separated by quadrupole triplets. The quadrupoles were made strong (0.50 inch full aperture, 8.22-kG pole-tip fields) to minimize the total length of this section. The strength of the middle dipole, B5, was chosen so that its momentum dispersion cancels the residual η from the arcs (about 47 mm) by zeroing the η' at the quarter-wave point. The first and third dipoles, B6 and B4, are identical and separated by a half wavelength. Thus, the additional dispersion introduced by B6 is exactly cancelled by B4. Because of the terrain-following feature of the arcs, the beam enters the final focus system pitched up about 0.10°, and with the beam ellipse rolled "inward" a few degrees, i.e., with the y-axis rolled toward the center of the arcs. The net bend of this module, which is oriented in the plane of the last arc magnet, has a vertical component that brings the beam into the horizontal plane of the IP and a horizontal component that aligns the dispersion-free beam with the first telescopic transformer.

4.3 The First Telescope and Extraction System

Following the η-matching section is a telescopic module, consisting of quadrupoles Q15 through Q19, with demagnifications 1/M of 8.51 and 3.09 in the x and y planes, respectively. These values were chosen to transform the elliptical beam at the end of the arc to a round beam with $\sigma_x = \sigma_y \approx 6.0$ μm at the beginning of the chromatic correction section. Because the envelope of the incoming beam is a rotated ellipse in the xy plane, the magnets in this section must be mounted with a slight roll (approximately 4.1° on the north side, 5.9° on the south). Since the beam leaving this section is circular, all downstream elements are oriented with a vertical y axis.

This module was designed with drift spaces long enough to accommodate the extraction magnets for the outgoing beam. The space between quadrupoles Q16 and Q17 ($\approx$ 7.3 m) accommodates the pulsed kicker. Located 107 m from the IP, the kicker must turn on after the incoming beam has passed and reach full field before the outgoing beam arrives 700 nsec later. The kickers deflect the outgoing beams approximately 1.2 mrad, sending them off toward water-cooled aluminum dumps. The space between quadrupoles Q18 and Q19 ($\approx$ 11 m) accommodates the dc septum magnet. Quadrupoles Q17 and Q18, through which the outgoing beam must pass off-axis, are both horizontally defocusing and thus assist in extracting the beam rapidly.

4.4 The Chromatic Correction Section

The next two half-wave modules, used for the chromatic correction system, each have unity magnifications in the x and y planes. These modules each consist of an array of four identical quadrupoles with alternating polarities interspersed with sextupoles in a sequential symmetry arrangement. Dipoles were introduced at the junctions between the modules to produce a symmetric momentum dispersion about the center of this section. The first and third dipoles are of the same strength; the second dipole, located at the center, has twice the strength of the others. The sextupoles are connected in pairs at corresponding locations in the two modules, such that the momentum dispersion is the same at each member of the pair. Similarly, the R_{12} and the R_{34} matrix elements, the sine-like functions, have the same magnitude but opposite sign at each member of a given sextupole pair. This arrangement results in a natural cancellation of certain second-order aberrations, as discussed in Section 3.2 above. The strengths of the sextupole pairs were adjusted to minimize the dominant second-order chromatic aberrations, T_{126} and T_{346} of the combined CCS and final telescope system.

4.5 The Final Bend

The last dipole of the chromatic correction section, which provides the final bend before the IP, was actually built as two separate magnets, designated the "hard bend" and "soft bend" magnets, mounted end-to-end and powered in series. The hard bend, with a field of 12.5 kG, bends the beam through an angle of 17.3 mrad. The soft bend, 4.0 m long but with a field of only 417 gauss, bends the beam through the last 1.00 mrad. This arrangement was included to reduce the critical energy of the synchrotron radiation that reaches the detector. The critical energy is given by

$$k_c = \frac{3}{2} \frac{\gamma^3}{\alpha} \frac{r_e}{\rho} m_e c^2$$
$$\approx 0.666 \text{ (MeV)} \left[\frac{E}{100 \text{ GeV}}\right]^2 B(\text{kG}) \tag{35}$$

where ρ is the radius of the bend and E is the beam energy in GeV. The hard-bend synchrotron radiation, with a critical energy of 2.08 MeV for a 50-GeV beam, is 29 mm from the beam at the entrance to the final triplet and can thus be masked off with a simple collimator. The soft-bend radiation which unavoidably passes into the detector, has a critical energy of about 69 keV.

A novel feature of the SLC final focus not found in conventional storage rings is the S-bend geometry formed by the antisymmetric bends of the north and south chromatic correction sections. Specifically, the final dipoles on the north side bend the electron beam to the west, while the final dipoles on the south side bend the positron beam to the east. This is illustrated in Fig. 10. This feature has been incorporated in the hope of further reducing the synchrotron radiation background in the experimental detector.

In a conventional storage ring, most of the dipole synchrotron radiation from both directions can be blocked by a pair of circular masks, symmetrically arranged on each side of the detector. Unavoidably, however, some of the radiation passing through the center hole of the first mask will strike the inside (detector side) surface of the second mask and back-scatter into the detector. With the beam energies and intensities planned for the SLC, these hot spots on the inside surfaces of the masks could be troublesome sources of hard x-ray background.

This problem can, in principle, be curtailed by introducing the S-bend. With this arrangement, the synchrotron radiation from the electron beam (approaching from the north) is all in a thin horizontal sheet to the east of the beam itself. The radiation from the positron beam is all in a horizontal sheet to the west. Masks with open slots on one side can then be arranged anti-symmetrically to block the incoming radiation but pass the outgoing radiation from the other beam. Whether or not this feature of the S-bend geometry will make a significant improvement in the overall background problem has yet to be verified.

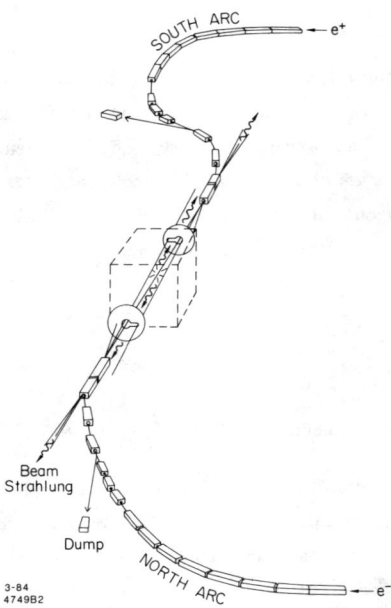

Fig. 10. The final bend configuration in the SLC. All bend angles are exaggerated in this figure.

4.6 The Final Telescope

The function of the final telescopic module is to focus the beam to the smallest possible spot size at the interaction point with equal demagnification factors in the x and y planes. In general, the higher the field gradient in the final quadrupole triplet, Q1, Q2, Q3, and the shorter the distance, ℓ^*, from the IP to the face of the first quadrupole, the smaller is the achievable beam spot and the higher the luminosity. As discussed in Section 4.1, two designs for the final telescope have been developed: one based on conventional iron quadrupoles, the other on high-gradient superconducting quadrupoles.

During the design of the final telescope, the first parameter to be frozen was its overall length, the distance between the interaction point and the waist in the beam envelope one-half betatron wavelength upstream. This distance was chosen somewhat arbitrarily to be approximately 38 m, based on general optical considerations and site boundary constraints. Studies demonstrated that a wide variety of telescope designs were possible using conventional, superconducting or permanent magnet[17] quadrupoles, all constrained to this total length.

Choosing the drift distance ℓ^* from the face of the innermost quadrupole to the interaction point required a compromise between the ultimate luminosity and the space requirements of the detector. For the design using conventional quadrupoles, this distance was chosen to be 2.82 m, the minimum consistent with avoiding any adverse saturation effects of the Mark II solenoid field on the field quality of the nearest quadrupole. At this distance, the stray solenoid field decreases to about 10 gauss. The use of the conventional triplet at this ℓ^* is not compatible with the SLD detector, however, because of mechanical interference and because of the strong solenoid field extending out to about 4 m. The design using superconducting final triplets is based on an ℓ^* of 2.21 m. This was deemed to be the shortest ℓ^* that still allows room for beam position monitors, synchrotron radiation masks, and a precision vertex detector. The superconducting quadrupoles, cryostat, and SLD detector are outlined in Fig. 11. The conventional quadrupoles and the Mark II detector are indicated by dashed lines.

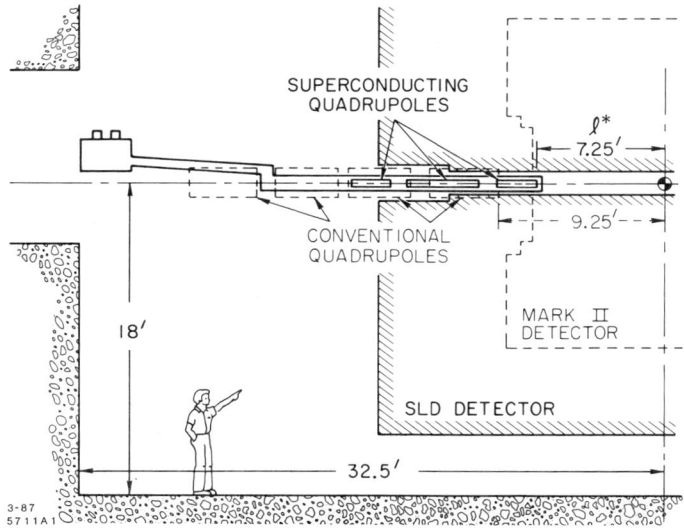

Fig. 11. Elevation view of the SLC interaction region showing the conventional quadrupoles and Mark II detector (dashed lines) with the superconducting quadrupoles and SLD detector superimposed.

The next step was to select the highest practical magnetic field gradient for the final quadrupole triplet. For the conventional iron quadrupoles, the pole-tip field was conservatively chosen to be 10 kG (at 50 GeV) to ensure against any magnetic saturation effects detrimental to the quality of the pure quadrupole field. For a given pole-tip field, the maximum gradient that can be achieved in a quadrupole depends on its aperture. In this triplet, the minimum aperture required is determined primarily by background considerations. Off-axis electrons originating from beam-gas interactions and slit scattering upstream follow trajectories that reach their maximum excursions in the final triplet. Computer simulations indicated that a clear aperture of about 40 mm is necessary in this section to avoid an intolerable flux of secondary particles scattering into the experimental detector. Studies also showed that when the SLC reaches its full performance specifications, beam-beam disruption effects will generate an intense beacon of electromagnetic radiation out to an angle of about 2 mrad. (See Section 5.4 below.) An unobstructed path for this radiation requires a minimum aperture of about 35 mm. Subject to this constraint, a pole-tip gap of 1.625 inches and a gradient of 4.85 kG/cm were chosen for the conventional quadrupoles at 50 GeV. This gradient can be increased for beam energies up to at least 55 GeV before saturation effects become significant.

A gradient of 12 kG/cm has been chosen for the superconducting quadrupoles at 50 GeV. The aperture, measured to the innermost surface of the conductor, was chosen to be 2 inches, a practical lower limit imposed by the geometry of the keystone-shaped superconducting cable.[18] Allowing space for an inner helium passage and for the beam pipe, the clear aperture is about 40 mm, a value consistent with the other requirements discussed above.

Prototype magnets with these characteristics have been tested at Fermilab to gradients of over 18 kG/cm at 4.2° K. When installed in the SLC, the maximum attainable gradient will be lower than the test value, due to the external 6-kG detector solenoid field and to a higher and more economical operating temperature of 4.6° K (each causing about a 6% reduction in the maximum current). When tuned for 55 GeV, these quadrupoles will be running at about 80% of the quench limit.

Each quadrupole triplet is powered as a series string by a single power supply. Thus, the field gradient is the same in each member of each triplet. The quadrupoles of the conventional final triplet have separate trim windings with individual power supplies. These trims correct for fabrication errors and allow separate adjustment of the focal lengths in the horizontal and vertical planes. The superconducting quadrupoles do not have separate trim windings. Instead, the trim functions will be performed by bucking or boosting the triplet current with small supplies connected across each individual quadrupole.

The desired quarter-wave focusing properties of each triplet were obtained by adjusting the length and spacing of the magnets and the length of the middle magnet relative to the first and third (which were constrained to be identical to preserve the symmetric triplet properties). With ℓ^*, the total length, and the gradients of the final triplet quadrupoles fixed, and with the additional constraint that the system as a whole must preserve half-wave telescopic properties in both dimensions, the lengths and spacings of the final triplet quadrupoles and the overall strength and spacings of the upstream triplet were varied to obtain the desired demagnification. Note that in the conventional configuration, Q2 is actually two magnets mounted closely together, as evident in Figs. 8 and 11, rather than a single long magnet. Splitting Q2 in this way was done to simplify fabrication and improve mechanical stability.

The "sliding telescope" property discussed in Section 2.4 was used to advantage in this part of the design to increase ℓ^* and at the same time to shorten the overall length of the telescope. In the superconducting design, for example, ℓ^* at 2.21 m is about 20 cm longer than the effective focal length of the final triplet alone, while the distance between the soft-bend dipole and Q6, and thus the total length of the telescope, is about 5 m shorter than it would have been with the simplified configuration of Fig. 4.

For a fixed total demagnification, the magnitudes of the sextupole fields in the chromatic correction section and, by implication, the severity of the uncorrected aberrations, grow as the final triplet strength is reduced. Demagnification factors (1/M) of 4 and 5 were chosen for the conventional and superconducting configurations, respectively. These magnifications are "diminishing returns" points in each case. Attempts to reduce the spot sizes by choosing stronger demagnifications were offset by increasing third-order aberrations.

The optimum configurations using conventional and superconducting final triplets do not have identical upstream triplets. However, a pair of solutions was found in which the same upstream magnets, Q4, 5, 6 can be used by adjusting their locations and field strengths.

An unusual feature of the SLC is the use of an anti-symmetric optical arrangement across the IP.[19] This was adopted to improve the masking of synchrotron radiation background originating in the final triplet quadrupoles. This is the quadrupole analog of the S-bend scheme described above for masking dipole radiation. The synchrotron radiation from the quadrupoles is generated in an elliptical pattern. By choosing the polarities to give the triplet on the electron side a focus-defocus-focus configuration while the positron triplet has a defocus-focus-defocus configuration, the major axes of the synchrotron light beams will be oriented horizontally on one side and vertically on the other. Elliptical masks with the same aspect ratio but rotated 90° with respect to the incoming radiation pattern will block much of the incoming radiation, while allowing outgoing radiation (that which passed the mask on the incoming side) to pass out of the detector without back-scattering.

4.7 Expected Performance

Figure 12 shows horizontal and vertical betatron functions (β_x and β_y) for the final telescope. The solid lines show these functions for the superconducting quadrupoles. The dashed lines show these functions for the conventional triplet configuration.

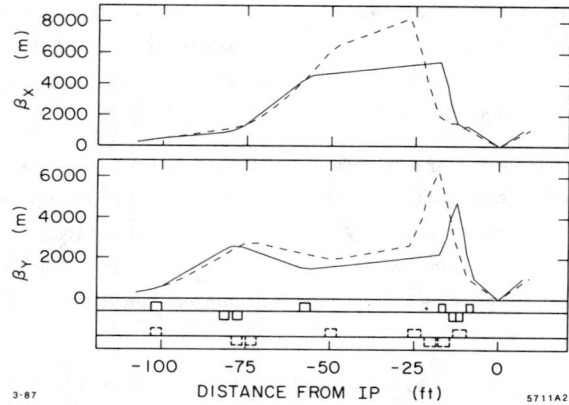

Fig. 12. Horizontal and vertical beta functions in the final telescope. The solid lines indicate the superconducting configuration; the dashed lines indicate the conventional configuration. The positions of the quadrupoles are shown along the bottom of the figure.

The main differences between the conventional and superconducting configurations are summarized in Table 1. The spot size listed is the rms radius (σ) of the beam at the interaction point,

Table 1. Parameters for the two final focus configurations at 50 GeV.

	Initial	Upgraded	
Quadrupoles	Conventional	Superconducting	
Clear Aperture	3.95	4.13	cm
Gradient	4.85	12.0	kG/cm
Drift Length (ℓ^*)	9.25	7.25	ft
Demagnification	× 4	× 5	
β^*	0.75	0.50	cm
First-Order Focus	1.5	1.2	μm
Spot Size (σ^*)	2.07	1.65	μm
Repetition Rate	120	180	sec^{-1}
Particles/Bunch	5×10^{10}	7.2×10^{10}	$N^\pm$/bunch
Emittance at IP	4.2×10^{-10}	4.2×10^{-10}	$\varepsilon_{x,y}$ (rad m)
Disruption Parameter	0.34	0.76	D
Pinch Factor	1.14	2.2	H
Luminosity	6.4×10^{29}	6.0×10^{30}	cm^{-2} sec^{-1}

including residual aberrations. For low beam currents, the luminosity is inversely proportional to σ^2; thus, the superconducting configuration is expected to yield at least 1.6 times greater luminosity than the conventional. Note that the "Upgraded" column also shows planned improvements in the repetition rate and in the number of particles per bunch, which contribute factors of 1.5 and 2, respectively. In addition, as the number of particles per beam pulse is increased, this relative luminosity gain is expected to increase further, multiplied by a "pinch factor" due to the beam-beam disruption effect. This would enhance the luminosity by an additional factor of 2 for the full design current.[1]

5. TUNING AND DIAGNOSTICS

The tolerable errors in the positions and strengths of the magnetic elements of a final focus system may be very small, depending on the amount of demagnification, the absolute size of the beam, and other factors. The alignment tolerance on the last focusing element, for example, must be less than σ^*, and the tolerance on the strength must be small enough that $\Delta f \ll \beta^*$, if the resulting loss of luminosity is to be negligible.

In the SLC, the position of the final quadrupole triplet must be held stable to about 1 μm, and the field strengths must be controlled to the level of 1 part in 10^4. Calibrating the absolute strength of an iron-core quadrupole to an accuracy of 1 part in 10^4 is a formidable task requiring careful attention to end-field effects, hysteresis, core temperature and the measurement of the excitation current. Achieving the necessary accuracy for all components of a large system directly by careful construction is impractical for any plausible collider design.

In practice, a final focus system must be instrumented to monitor the position and size of the beams at critical points, and procedures must be available to use this information to diagnose the problems that arise from any deviations from design specifications. Controls must then be provided to correct for steering and optical errors. These may be implemented with trim windings on the main magnets, special "corrector" dipoles and quadrupoles, or remotely controlled mechanisms for moving critical focusing elements.

5.1 Tuning the SLC

Tuning the final focus of the SLC is done in several steps.[20] The first step is simply to steer the beams through the system to their respective dumps. This is done with pairs of small vertical and horizontal dipoles distributed throughout the final focus system, at least one pair per quarter-wavelength of betatron phase. This steering is guided by strip-line Beam Position Monitors (BPMs), one adjacent to each sextupole in the chromatic correction section, and one or two per triplet in the other sections. In the SLC, this steering can be done to an accuracy of about 100 μm at each BPM.

In a simple optical telescope like the one in Fig. 4, there is a one-to-one correlation between the location of the last lens and the image point. This is also true for the final telescope of the SLC. Moving the final triplet 1 μm transversely moves the focused beam spot 1 μm. At this level, temperature variations in support structures and floor movements due to ground settling become important.

Movements of a few microns can be compensated with small steering dipoles immediately upstream of Q3. Larger dipole corrections are undesirable at this point, however, because the resulting large betatron oscillations that result from an off-axis trajectory through the triplet generate linear dispersion, increased synchrotron radiation, and possible geometric aberrations from imperfections in the quadrupole fields. In the SLC, these problems are avoided with a remotely-controlled motorized support system that allows the final triplet to be moved over a range of about ± 2 mm with a step size of about 10 μm.

After the beam trajectories have been centered in all the quadrupole apertures, the residual dispersion is corrected. This dispersion is measured by observing beam centroid shifts in each of the BPMs as the incoming beam energy is varied and comparing to the nominal design values. Corrections are made using four small quadrupoles near B5 and B6 in the η-matching section (Fig. 8), two upright quads which couple mainly to η_x (near B5) and η'_x (near B6), and two skew quads which couple mainly to η_y and η'_y. An automated procedure has been developed to measure and correct the residual dispersion in a few seconds.

The amplitude and phase of the betatron oscillations in the chromatic correction section and final telescope can be adjusted with the quadrupoles in the upper transformer, each of which has a separate power supply to facilitate these adjustments. When the betatron phase is properly adjusted to center the waist in B3, quadrupoles Q16 and Q17 can be used as a zoom lens to adjust the magnification.

Because of the roll variations introduced by the terrain-following feature of the SLC, small alignment and magnet setting errors tend to introduce coupling between the x and y components of the beam parameters. This coupling typically enlarges the beam size along the diagonal direction. Skew quadrupoles were introduced near Q3 and Q17 to untangle this coupling. They are symbolized by diamonds in Fig. 8. The effect of the skew quad near Q17 can be seen on a profile monitor in the high-β region near Q4. This skew quad is adjusted to rotate the elliptical beam spot upright. The skew quad near Q3 is used to remove any remaining xy coupling, including the coupling introduced by the solenoid field of the experimental detector. This skew quad can only be adjusted by observing the final spot size at the IP, or some direct manifestation of the small spot size, as discussed below.

The final focusing of the beam at the IP is done with small adjustments to the strengths of Q2B and Q3. These two magnets, adjusted in combination, move the x-waist and y-waist along the z-dimension to bring them together at the collision point. Like the last skew quad adjustment, this can only be done by observing the properties of the beam at the IP. Ultimately, some observable quantity related to luminosity, a figure of merit for the performance of the machine, must be available to guide in tuning the machine.

5.2 Luminosity Monitoring

The e^+e^- elastic scattering ("Bhabha" scattering) rate has been a useful figure of merit for conventional storage rings. The Bhabha rate can be calculated from fundamental principles with high accuracy and thus gives an absolute measure of luminosity. The scattering rate into an angular region bounded by $\theta_{\min}$ and $\theta_{\max}$ is $R = \mathcal{L}\triangle\sigma$, where

$$\triangle\sigma = \frac{4\pi\alpha^2}{E^2}\left(\frac{1}{\theta_{\min}^2} - \frac{1}{\theta_{\max}^2}\right) \quad (36)$$

is the Bhabha cross section, E is the energy of each beam, and α is the fine structure constant. It is interesting to note that the Bhabha scattering rate will be the same for any e^+e^- collider having the luminosity and energy relationship of Eq. (2).

The small-angle-monitor ("SAM") built for the Mark II detector at the SLC is an array of detector elements designed to identify back-to-back e^+e^- pairs scattered into an angular range from 50 to 150 mrad of the beam direction. At a luminosity of 10^{30} cm^{-2} sec^{-1} and 50 GeV per beam, a detectable Bhabha rate of 0.014 events/sec is expected. Mini-SAM, a device that covers the region from 15 to 25 mrad to supplement the SAM measurements, will raise the total detected Bhabha rate to 0.32 events/sec. The usefulness of the mini-SAM will depend on its ability to distinguish Bhabha scatters from off-axis background particles. Assuming this can be done successfully, a 10% measurement of luminosity will require 100 counts and will take about five minutes at this rate with the combined data of SAM plus mini-SAM. With the SAM alone, this measurement will take about two hours. Starting in an untuned condition with a luminosity of, say, 10^{27} cm^{-2} sec^{-1}, a single 10% measurement with the combined SAM/mini-SAM system will take 3.6 days.

The conclusion to be drawn from this is that Bhabha scattering may be useful for charting the integrated luminosity and normalizing large blocks of experimental data but is far too slow to be of any value in guiding any beam-tuning procedures.

5.3 Profile Monitoring Devices

Profile monitors based on a fluorescent screen in the path of the beam, viewed by a TV camera, have proven to be immensely valuable at many places in the SLC including the interaction point. They have been used for diagnosing hardware problems, verifying the proper operation of magnets, measuring emittance, and tuning out cross-plane coupling. Resolutions of 50 μm are routinely achieved, and the prospect of reaching 10 μm resolution with this technique seems likely. Beyond this point, the diffraction limits of camera lenses and the graininess of fluorescent screens become important.

Another approach to monitoring small spot sizes is based on detecting the secondary emission signal generated when the beam is scanned across a thin carbon fiber. A device using fibers as small as 4 μm in diameter has been in use at the interaction point during the commissioning tests of the SLC. Figure 13 shows a 5.2-μm beam profile measured with a carbon fiber. This particular measurement was significant because it marked an official milestone in the SLC commissioning: it was the first beam spot smaller than 6 μm.

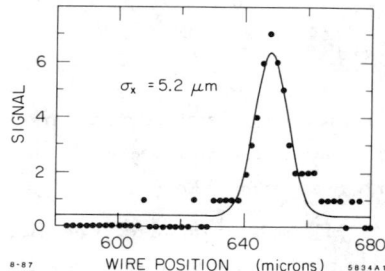

Fig. 13. Electron beam profile measured at the SLC interaction point using a 7μm diameter carbon fiber.

An important limitation on the usefulness of the wire scanning technique is the survivability of the fiber when blasted by the charged particle beam. Fibers of various sizes up to 25 μm diameter have been used successfully for many hours in the SLC with no evidence of beam-induced failure, but only at a low bunch population ($\leq 5 \times 10^9$ particles/bunch) and low repetition rate (5 pps). The fibers are not expected to survive with SLC design-spec beams.[21] It will be necessary to focus the machine at reduced beam current and then increase the current only after the fibers are moved safely to the side, clear of the beam. In future colliders with smaller spot sizes and higher bunch densities, wire scanners may prove useful in performing the functions now performed by fluorescent screens in the SLC, but for final adjustments at the interaction point a different technique will be needed.

5.4 Beamstrahlung Radiation

Beamstrahlung is the name given to the radiation emitted by each bunch as it passes through the electromagnetic field of an opposing bunch. The radiation emitted by a particular particle depends on the net force it feels in this passage. The maximum radiation is emitted by particles offset about 1.5 σ from the center. In contrast, the maximum contribution to the luminosity comes from particles at the center of the bunch, where they are most likely to collide with particles of the opposing bunch.

When two Gaussian bunches collide, the fraction of one bunch's energy radiated away is given approximately by the classical synchrotron radiation formulation:

$$\delta_c = \frac{r_e^3 N^2 \gamma}{3\sqrt{3}\, \sigma_z\, \sigma_r^2} \qquad (37)$$

where σ_z and σ_r are the longitudinal and radial sizes of the bunches, N is the number of particles in the "target" bunch, r_e is the classical radius of the electron, and $\gamma = E/m_e c^2$. For equal size opposing beams, the number of photons emitted is proportional to $1/\sigma_r$, and the critical energy of the radiation spectrum is also proportional to $1/\sigma_r$. The radiation is peaked strongly forward in the direction of the beam motion.

Beamstrahlung radiation is a macroscopic manifestation of beam-beam interactions that can, in principle, be detected and quantified for each collision. While it is not strictly proportional to luminosity, the beamstrahlung flux above a given threshold is an observable quantity that generally increases with luminosity as the beams are brought into focus. In the SLC design, this radiation is intercepted by special detectors[22] located along a line-of-sight from the IP, just upstream of the last large dipole bending magnet on each side as shown in Fig. 10. The expected performance of the SLC beamstrahlung system is shown in Fig. 14.

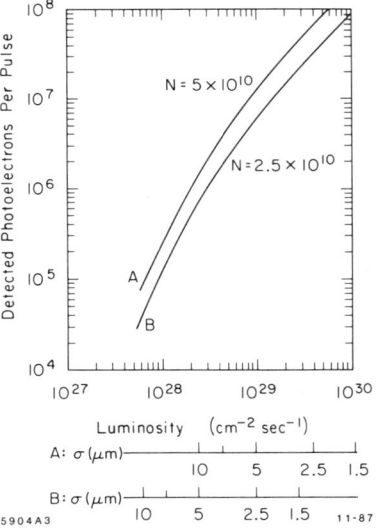

Fig. 14. Predicted signal strength in the SLC beamstrahlung monitor increases monotonically with luminosity for identical round beams.

The small spot and high luminosity parameters of the TeV-class collider studies lead to beamstrahlung fluxes that carry off a substantial fraction of the total beam energy. These machines enter the quantum beamstrahlung regime, where the classical synchrotron radiation formula is no longer adequate. Himel and Siegrist have shown[23] that in this regime, the fraction of the beam energy radiated is proportional to $\sigma_r^{-2/3}$. As in the classical regime, tuning for maximum beamstrahlung will generally maximize the luminosity.

6. STEERING TO COLLISION

In an e^+e^- storage ring with a purely magnetic guide field, the counter-rotating beams follow exactly the same central trajectory and thus head-on collisions are unavoidable. There is no *a priori* reason why this should be true in linear colliders, however. In any linear collider, including the SLC, the opposing beams must be actively steered into collision, guided by some observable that is sensitive to the impact parameter. Using state-of-the-art strip-line BPMs, it may be possible to direct the two beams independently to the intended interaction point with an accuracy of perhaps 100 μm. In order to achieve useful luminosity, the beams must be steered to within about one beam radius of each other. This corresponds to about 2 μm in the SLC and is likely to be much smaller in a TeV-class linear collider. It is in this regime, far below the resolution limits of single-beam diagnostic devices, that the beam-beam deflection is strongest.

6.1 Beam–Beam Deflections

The electromagnetic force acting between two intense colliding beams of oppositely charged particles will cause them to be deflected in passing by an angle that depends on the offset between the bunches and the distribution of charge within the bunches. This is illustrated in Fig. 15. This deflection, measurable with nondestructive techniques, is expected to be the key to the final steering of the e^+e^- beams in the SLC.[24] Furthermore, this deflection is expected to be a measurable manifestation of the interactions between even the extremely small beams envisioned for a large future linear collider.

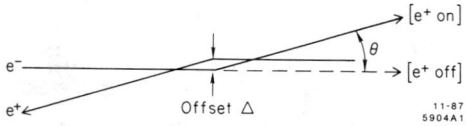

Fig. 15. The trajectory of each beam is deflected by the opposing beam passing at an offset $\triangle$. The dashed line shows the trajectory followed by the e^- beam when the e^+ beam is suppressed.

The deflection of a single particle of charge e, passing at an offset $\triangle$ from the centroid of an oppositely charged Gaussian distribution, is given by

$$\theta(\triangle) = \frac{-2r_e N_T}{\gamma} \frac{1 - \exp[-\triangle^2/2\sigma^2]}{\triangle} , \qquad (38)$$

where r_e is the classical radius of the electron, N_T the number of particles in the target bunch and σ the RMS transverse size of the Gaussian distribution.

When two beams pass with offsets large compared to their transverse sizes, they see each other as point charges and Eq. (38) is a good approximation for their mutual deflection. When colliding with a small offset, the finite sizes of the beam distributions must be taken into account. This can be done by convoluting Eq. (38) with the distribution of the opposing beam. The result of such a calculation, carried out in the limit of small $\triangle$, is expressed in terms of a form factor[25] which reduces the average deflection:

$$F(R) = \frac{\ell n(1 + R^2)}{R^2} . \qquad (39)$$

Here R is the ratio of the transverse sizes of the two beams.

6.2 Application to the SLC

Deflection versus offset is plotted in Fig. 16 for 50-GeV beams consisting of 5×10^{10} particles, with transverse spot sizes σ of 2, 5 and 10 μm. 10 μm is the typical size of the beams at the SLC interaction point after beam steering and dispersion corrections have been made but before optical corrections are completed. Magnet setting errors and misalignments contribute to this estimate.

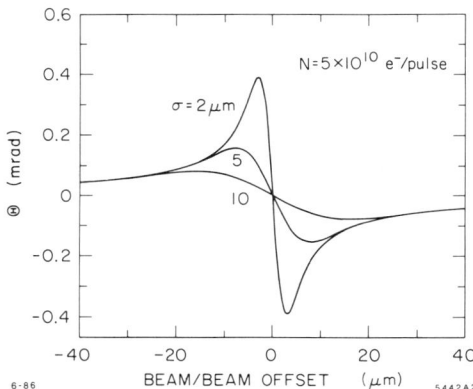

Fig. 16. The deflection angle θ as a function of offset $\triangle$, plotted for three spot sizes.

By adjusting the final focus corrector magnets and quadrupole trims, σ can be reduced to about 2 μm. The form factor described above has been incorporated in the curves as a multiplying reduction factor, assuming in each case $R = 1$.

Several methods have been studied for detecting and measuring the beam-beam deflections. The most obvious is to use a pair of BPMs straddling the interaction point. If the drift length "lever arm" is long enough, a deflection at the IP will result in a measurable position shift at the BPM. The power of this method can be greatly enhanced by suppressing the opposing beam on some pulses and watching the measured beam jump back to its undeflected position. To make this possible, a pair of pulsed magnets, the "single-beam dumpers," have been provided to kick either beam out of the transport system on command. These are shown in Fig. 17.

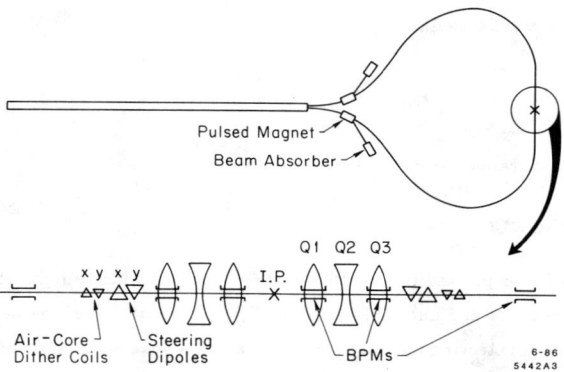

Fig. 17. Schematic of beamline components relevant to the deflection technique.

Another approach is based on detecting beamstrahlung radiation. The angular distribution of the beamstrahlung photons, strongly peaked forward in the direction of the outgoing beam and undeflected by intervening magnets, can be measured with a segmented detector with the arrangement described in Section 5.4 above.

Yet another approach is to observe the sizes and positions of the outgoing beams after they have been extracted from the main transport system but before they reach the dumps. The long lever arm from the IP to the dump on each side of the SLC makes the beam position at this point sensitive to the angle at the IP. Furthermore, destructive monitoring devices, such as fluorescent screens, can be left in place at the dumps without interfering with the incoming beams.

The procedures described here are based on relative measurements of the outgoing beam position at locations where the angular deflection produced in the collision leads to a transverse position shift. There are several BPMs in the outgoing transport system of the SLC that are suitable for this purpose. The best locations, however, are in the final telescope quadrupoles where the β-functions reach their largest values, thereby magnifying the deflections the most, and where dispersion is negligible (which minimizes confusion with energy variations). These locations are indicated in Fig. 17.

A three-step tuning procedure is envisioned:

1. Initial beam finding: One beam — designated the "target" in this case — is momentarily suppressed with a single-beam dumper while position measurements are made on the "probe" beam. In this way, the shift induced by the target beam can be determined. When the offset between the beams is large, the magnitude of the shift is inversely proportional to the offset, and its sign tells in which direction to steer. This can be seen by taking the limit of Eq. (38) for large $\triangle$:

$$\theta(\triangle) \simeq \frac{2r_e N_T}{\gamma} \frac{1}{\triangle} . \qquad (40)$$

The useful range of this technique, i.e., the maximum offset that still gives a measurable deflection, is limited only by the ability of the BPMs to resolve beam centroid movements. For example, assume the BPM near the innermost quadrupole in the SLC can resolve the centroid position of a single bunch of 5×10^9 particles to a level of 20 μm. It will then be possible to detect relative beam-beam offsets up to a maximum of

$$\triangle(\mu m) \simeq 40 \frac{N_T}{5 \times 10^9} . \qquad (41)$$

If the initial offset between the beams is too large to give a measurable deflection, the wire-scanning technique described in Section 5.3 can be used to bring both beams within a fiber-width of each other. Notice, however, that this "capture range" drops as $1/E$. For a larger collider with ten times the energy but the same N_T and BPM arrangement, the beams will have to be steered to within 4 μm of each other before the deflection can be detected. If this can not be done with a wire scanner or other direct means, an automated "raster scan" could be used to search a small area, scanning one beam while looking for deflections of the other. For larger beam currents, it may be possible to do better than the limit indicated in Eq. (41), because the BPM resolution also improves with increasing current. By chopping one beam off and on and averaging over many pulses, the resolution can be improved further.

2. Beam centering: Scanning the target across the probe and recording a plot similar to Fig. 15 for the probe will facilitate optimal steering of the two beams. The zero-deflection symmetry point is reached when the beams are perfectly centered.

3. Spot size tuning: Taking the limit of Eq. (38) for small $\triangle$ and multiplying by the form factor (39) gives

$$\theta(\triangle) \simeq \frac{-r_e\, N_T}{\gamma} \frac{\triangle}{\sigma^2} F(R) \quad . \tag{42}$$

The slope of the deflection of the probe beam near the zero-deflection symmetry point is inversely proportional to the cross-sectional area of the target. By differentiating Eq. (38), it can be seen that the deflection is maximum for offsets of about 1.6 standard deviations of the target distribution, and that the maximum deflection scales as the inverse of the transverse spot size:

$$\theta_{\max} = 0.451 \frac{2 r_e\, N_T}{\gamma} \frac{1}{\sigma} \quad . \tag{43}$$

A relative measure of spot size can thus be obtained by scanning one beam across the other as in Step 2 above. Guided by these measurements, an operator can adjust optical elements of the transport system to minimize this final size.

6.3 Maintaining Collisions

It is expected that even when the static crossing errors have been corrected as described above, the two beams will not remain centered on each other without an active feedback system. Many sources of drift and jitter that could cause the beams to wander at the IP have been identified at the SLC. In most cases, these effects can be minimized with careful attention to hardware designs. Magnet power supplies, for example, must be well-regulated, and support structures must be rigid. Natural ambient ground vibrations at frequencies above 1 Hz have been shown[26] to be negligible at the level important for the SLC, although some local man-made vibration sources, such as reciprocating pumps, require special isolation. On a slower time scale, thermal effects will cause mechanical support structures to expand and power supplies to drift enough to adversely affect the luminosity unless steering corrections are made. Studies of feedback schemes for the SLC have focused on simple and relatively slow algorithms, although the BPM electronics, control system and other key components have been built to allow pulse-by-pulse feedback to accommodate faster or more complex schemes.

A simple feedback algorithm for correcting relatively slow drifts is based on automatically suppressing one beam periodically using a single-beam dumper, while observing the motion of the other beam. Of course, luminosity would be sacrificed on these occasional pulses, but they would enable a steering correction to be computed from the measured position shifts of the outgoing beam. Because each measured deflection can correspond to two possible offsets, the operation has to be carried out frequently enough to ensure that the actual offset does not drift outside the domain of the IP, bounded by the deflection maxima, in the time between updates. This technique is adequate to track the thermal expansion of support structures and other mechanical effects in the SLC.

An approach that does not require sacrificing any beam pulses would be to excite small "dither coil" dipoles (Fig. 17) in a preprogrammed way to induce small periodic offsets at the IP with an amplitude of a fraction of a standard deviation. In this way, one beam can be made to trace out a pattern such as a small circle at the IP. The deflections of the opposing beam will then project the same pattern at the BPM. When the offset between the beams corresponds to a point on a steeply rising positive slope in Fig. 16 (beyond the 1.6 σ peak on either side), the projection is a magnified image of the dither pattern. When the offset is less than 1.6 σ, the projection is an inverted image of the dither pattern. Synchronous position measurement would then allow a determination of whether the beams were colliding within or beyond 1.6 standard deviations of each other. If necessary, a correction could be applied to bring them back to within 1 σ. The sign of the deflection would indicate the direction in which to steer. In both these algorithms, corrections are applied using steering correctors immediately upstream of the final triplet.

ACKNOWLEDGEMENTS

It is a pleasure to acknowledge the help of T. Fieguth and J. Murray, who worked out the innumerable details of the SLC optical design, and Karl Brown, chief architect of the underlying final focus theory, who first encouraged me to prepare this paper. I would also like to thank P. Bambade, A. Chao, H. DeStaebler, C. Field, A. Hutton, W. Kozanecki, R. Palmer, N. Phinney, J. Rees, J. Seeman, R. Servranckx and the other members of the SLC commissioning project for many interesting and provocative discussions. Finally, I would like to thank Mary Lou Arnold for transcribing my scrawls and scratch-outs into crisp, neat text and Mel Month for his immeasurable patience and encouragement.

REFERENCES

1. *SLC Design Handbook* (December 1984).

2. See, for example, B. Richter, *Very High Energy Colliders,* SLAC–PUB–3669 (May 1985) and *Proceedings of the 1985 Particle Accelerator Conference,* IEEE Trans. Nucl. Sci. NS–32 No. 5 (October 1985), 3828.

3. See, for example, the section on *Emittance Production and Preservation,* of the *Proceedings of the Symposium on Advanced Accelerator Concepts,* Madison, WI, August 1986, AIP Conference Proceedings 156 (1987).

4. R. Palmer, *Interdependence of Parameters for a 1-TeV Collider*, SLAC AAS Note–31 (May 1987).

5. R. Hollebeek, *The Linear Collider Beam–Beam Problem*, these proceedings.

6. See, for example, P. Chen, J. J. Su, T. Katsouleas, S. Wilks and J. M. Dawson, *Plasma Focusing for High Energy Beams*, IEEE Trans. Plasma Sci. PS–15 (1987) 218; and P. Chen, *A Possible Final Focusing Mechanism for Linear Colliders*, Particle Accelerators 20 (1987) 171.

7. K. L. Brown, F. Rothacker, D. C. Carey and Ch. Iselin, *TRANSPORT, A Computer Program for Designing Charged Particle Beam Transport Systems*, SLAC–91 (May 1977).

8. The subscripts 1 and 2 on x and θ are used in this paper to indicate the input and output, respectively, of a final focus system. Beware of possible confusion with the subscripts 1 through 6 of the x-vector and R matrix, which indicate the variables, e.g., $x_6 = \delta$.

9. *Proceedings of the SLC Workshop on Experimental Use of the SLAC Linear Collider*, SLAC–Report–247 (March 1982).

10. K. L. Brown, *A First- and Second-Order Matrix Theory for the Design of Beam Transport Systems and Charged Particle Spectrometers*, SLAC–Report–75 (June 1982).

11. J. J. Murray, K. L. Brown and T. Fieguth, *The Completed Design of the SLC Final Focus System*, SLAC–PUB–4219 (February 1987).

12. R. Hollebeek and A. Minten, *Disruption and Luminosity of Flat Beams*, SLAC CN–302 (1985).

13. A. Chao, J. Hagel, F. Ruggiero and B. Zotter, *A Flat Beam Final Focus*, CERN–LEP–TH/87-48 (September 1987).

14. D. C. Carey, K. L. Brown and Ch. Iselin, *DECAY TURTLE, A Computer Program for Simulating Charged Particle Beam Transport Systems, Including Decay Calculations*, SLAC–246 (March 1982).

15. R. Erickson, T. Fieguth and J. J. Murray, *Superconducting Quadrupoles for the SLC Final Focus*, SLAC–PUB–4199 (January 1987).

16. *SLD Design Report*, SLAC–273, (May 1984).

17. J. Spencer, *Some Optics Alternatives for the FFS*, SLAC CN–264 (February 1984).

18. R. A. Lundy, B. C. Brown, J. A. Carson, H. E. Fisk, R. H. Hanft, P. M. Mantsch, A. D. McInturff, R. H. Remsbottom and R. Erickson, *High Gradient Superconducting Quadrupoles*, IEEE Trans. on Nucl. Sci. NS–32, No. 5 (1985), 3707.

19. The anti-symmetrical optical scheme using elliptical masks was suggested by T. Himel.

20. P. S. Bambade, *Beam Dynamics in the SLC Final Focus System*, SLAC-PUB-4227 (June 1987).

21. C. Field, private communication.

22. G. Bonvicini, C. Field and A. Minten, *Beamstrahlung Monitor for the SLC Final Focus Using Gamma Ray Energies*, SLAC-PUB-3980 (May 1986), and *Proceedings of the 1986 Linear Accelerator Conference*, SLAC-Report-303 (September 1986).

23. T. Himel and J. Siegrist, SLAC-PUB-3572 (February 1985).

24. P. Bambade and R. Erickson, *Beam-Beam Deflections as an Interaction Point Diagnostic for the SLC*, SLAC-PUB-3979 (May 1986), and *Proceedings of the 1986 Linear Accelerator Conference*, SLAC-Report-303 (September 1986).

25. This parametrization was suggested by P. Bambade.

26. G. Bowden, *Mechanical Vibrations of the Final Focus*, SLAC CN-314 (1985); G. E. Fischer, *Ground Motion and Its Effects in Accelerator Design*, in Physics of Particle Accelerators, AIP Conference Proceedings 153, Vol. 2 (1987).

POSITRONS FOR LINEAR COLLIDERS[*]

STAN ECKLUND

Stanford Linear Accelerator Center
Stanford, California 94305

TABLE OF CONTENTS

1.	REQUIREMENTS	1593
2.	METHODS OF POSITRON PRODUCTION	1593
3.	ELECTROMAGNETIC CASCADE SHOWERS	1594
4.	CROSS SECTION FOR PAIR, COMPTON	1594
5.	THE EGS PROGRAM	1595
6.	GENERAL OVERVIEW OF POSITRON DISTRIBUTIONS	1595
7.	PHOTONS FROM SYNCHROTRON RADIATION	1599
8.	PHOTONS FROM CHANNELING	1599
9.	POSITRON COLLECTION	1602
10.	TRANSVERSE FOCUSING TECHNIQUES	1603
11.	LONGITUDINAL CAPTURE	1604
12.	COMPUTER RAY TRACING	1605
13.	SPACE-CHARGE EFFECTS	1606
14.	THERMAL HEATING AND STRESS DUE TO SHOWER	1606
	REFERENCES	1614

---

[*]Work supported by the Department of Energy, contract DE-AC03-76SF00515.

POSITRONS FOR LINEAR COLLIDERS

STAN ECKLUND

Stanford Linear Accelerator Center
Stanford, California 94305

1. REQUIREMENTS

In general, the requirements of a positron source for a linear collider are similar to those for the electron source. This arises because of intensity limits caused by wake fields in the linac structure and the desire to maximize the luminosity of the machine. For example, if the luminosity limiting factor is wake field disruption of the beams, one would push to that limit with both positrons and electrons. In the usual case where the method of production uses an electron beam to initiate the positron production process, a system with a yield or efficiency greater than or equal to one is needed.

Ultimately, a low emittance beam is needed for small size beams at the interaction region. Most positron sources give adequate numbers of particles into extremely large emittances; hence the need for emittance reduction in one or several damping rings. However, the positron beam emittance must be suitable for injection into a reasonable damping ring. We must consider the full six–dimensional phase volume (coordinates: x, x', y, y', dP/P, dz). Both the transverse (x, x') or (y, y') and the longitudinal $(dP/P, dz)$ emittances present challenges to the design and realization of an adequate positron system.

Another important factor in obtaining large luminosity for a linear collider is to produce a high instantaneous current (small dz spread). This requires a positron source with the particular time structure desired or a storage device to collect and periodically release the charge.

2. METHODS OF POSITRON PRODUCTION

Positrons are not stable in matter because of their annihilation with atomic electrons. As a source for accelerators, positrons are normally produced by electromagnetic interactions, namely by pair production from a photon in the field of the nucleus. Positrons also are readily available by nuclear beta decay via the weak interactions. Relatively hot sources would be required to achieve intensities required for linear colliders. Also, because of its random time structure, beta decay is most useful as a source of positrons when a continuous beam is desired. The pair production process has the advantage of being essentially instantaneous, allowing production of beams with a time structure or bunching required for the linear collider. For sources which use pair production, the varieties of design depend on the method of obtaining photons.

© 1989 American Institute of Physics

3. ELECTROMAGNETIC CASCADE SHOWERS

The most common method to produce the required photons is by an electromagnetic cascade shower. Usually an electron beam is used to initiate the shower. The multiplication process involving Compton scattering, bremsstrahlung, and pair production is well-known[1] and is calculated in excellent detail for particular geometries by a Monte Carlo computer program.[2] The particle multiplication ratio can be in the hundreds for incident beams of tens of GeV, which provides an easy way to obtain large numbers of positrons, albeit in rather large phase-space volumes.

4. CROSS SECTION FOR PAIR, COMPTON

Cross sections for the electromagnetic cascade shower processes of positron-electron pair production and Compton scattering are shown in Fig. 1.

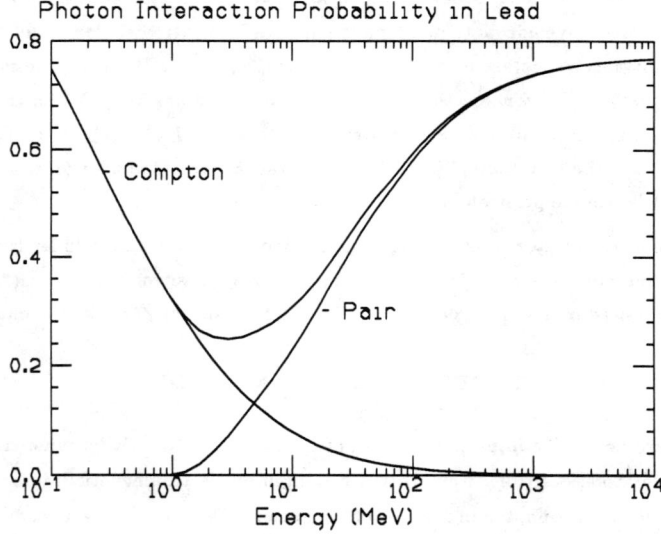

Fig. 1. Probability per radiation length of e^+e^- pair production and Compton scattering as a function of photon energy. These data come from the EGS program.

Note that pair production, which is the supplier of positrons, is of a good magnitude above 50 to 100 MeV. This means that, practically, incident beam energies should be at least that high or the efficiency of the system will be low.

5. THE EGS PROGRAM

The electromagnetic cascade is sufficiently complicated so that a computer Monte Carlo program is the efficient way to obtain meaningful results for a particular geometry. A program called EGS is commonly used for this purpose. Actually two separate programs are used; PEGS, which calculates cross sections and makes appropriate tables, and the main EGS code, which does the Monte Carlo particle tracking for the geometry of interest. PEGS is specific to a particular element, compound, or mixture of compounds and an energy range. PEGS is usually run only once at the beginning of solving a problem (or not at all if one of the standard files can be used), and EGS numerous times for the various geometries, energies, *etc.*, of interest. The majority of the EGS program is "standard code," but the user must write a portion of the code which is descriptive of the geometry of the problem to be solved, the incident beam conditions, and the output desired from the program. For many cases and those reported here, a simple cylindrically symmetrical geometry with infinite slabs, of thickness one radiation length, perpendicular to the beam axis will be used. Positron yields will be a function of positron momentum or energy (three components), and position, as well as the initial conditions.

6. GENERAL OVERVIEW OF POSITRON DISTRIBUTIONS

Outputs from a variety of EGS computer runs are given here to provide a characterization of the positron beam emerging from a thick production target. In the following, E is the energy of the positron in MeV, x, y, z are the position of the positron in cm, where the incident beam is parallel to the z axis, and P_x, P_y, P_z are components of the momentum of the positron in units of MeV/c. Figures 2 to 7 are calculated from EGS output for an incident electron beam of energy 33 GeV, incident upon a tungsten target.

Several properties of the positron distributions are worth noting since they relate to the number and phase-space volume of positrons. First note that yields substantially greater than unity are realizable for an energy as large as 33 GeV. The energy spread, however, is large and characteristic of a $1/E$ distribution. The transverse size of the positrons in a dense target is small, of order 1 mm, which is good from a phase-space point of view but bad for target stress and pulse heating. The transverse momentum appears dominated by multiple scattering, almost independent of z, but dependent on the high energy cut of the sample considered. Correlations in transverse phase-space (*i.e.*, x, P_x) are present at about the 20% level, which may be significant but not terribly useful. Note that this correlation simply reflects that the effective source z location is a finite distance up-beam of the target down-beam face where the positron trajectories are tabulated.

The yield dependence on atomic number is especially striking and is due to a lower critical energy for high–Z materials. Effectively the positron or electron energy loss per radiation length

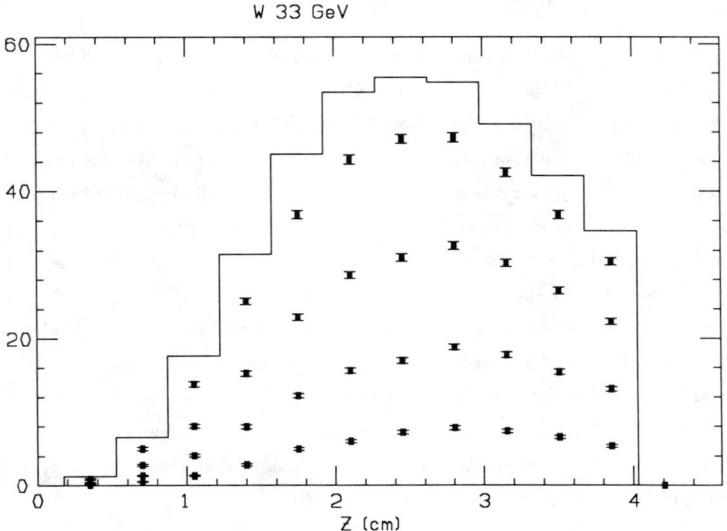

Fig. 2. Positron flux in tungsten per incident electron vs z for incident energy of 33 GeV. The different curves are for successively bigger cutoffs in maximum positron energy of 5, 10, 20, 50, and 100 MeV. The minimum energy cutoff is 2 MeV. The z bins are one radiation length. Note the shower maximum is around seven radiation lengths for this energy. The calculation covers the first eleven radiation lengths.

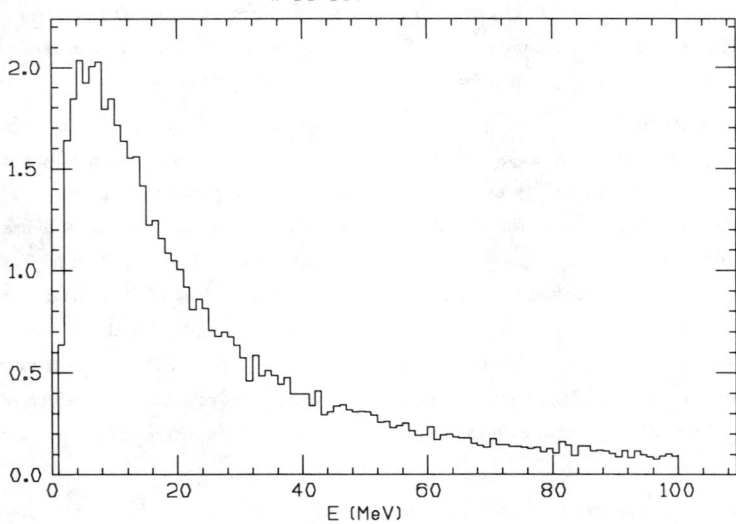

Fig. 3. Yield per 1-MeV energy (E) bin versus E at $z = 6$ radiation lengths.

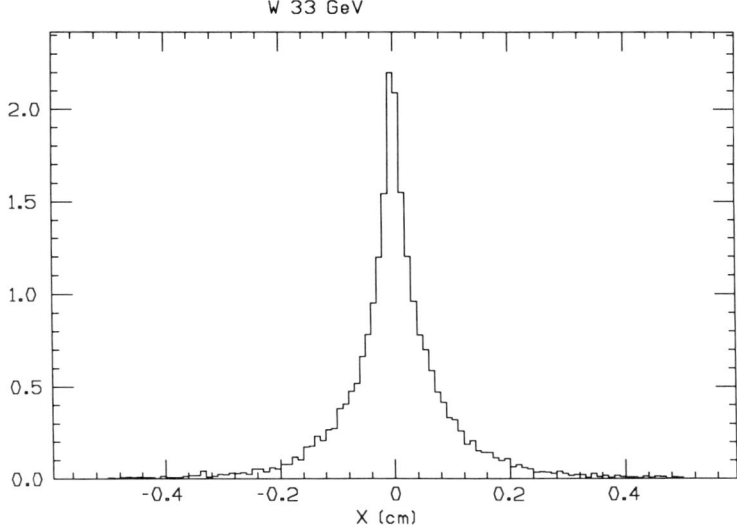

Fig. 4. Yield per 0.01–cm bin versus x at $z = 6$ radiation lengths.

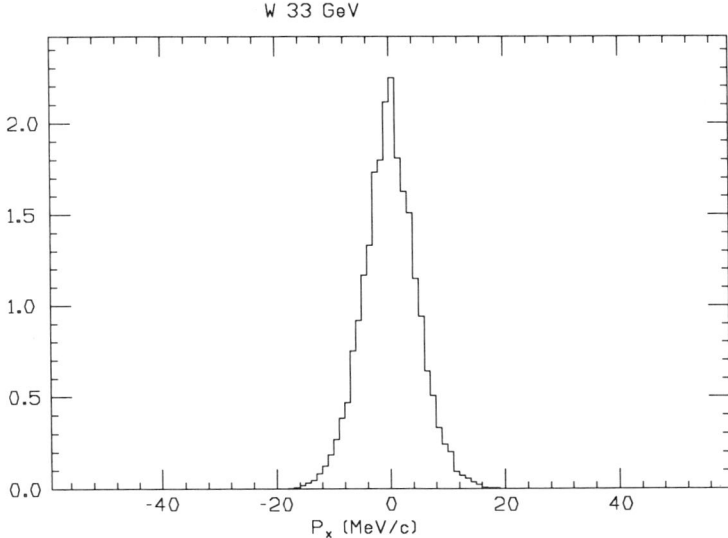

Fig. 5. Yield per 1–MeV bin versus P_x at $z = 6$ radiation lengths.

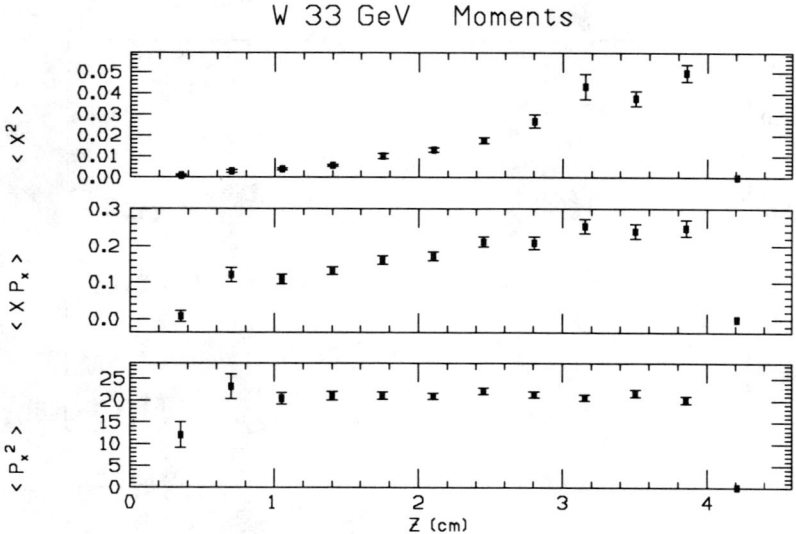

Fig. 6. Moments in x, P_x versus z for $5 \leq E \leq 20$ MeV.

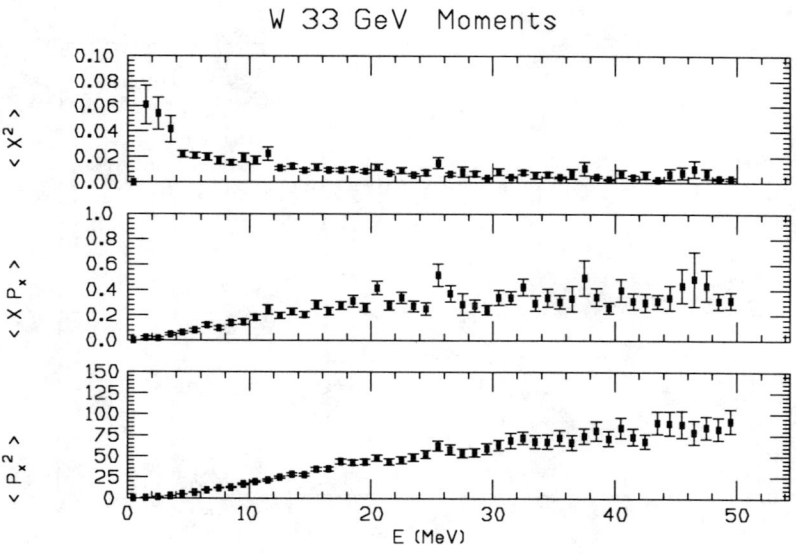

Fig. 7. Moments in x, P_x versus E for $z = 6$ radiation lengths.

is less in high–Z materials. Table I shows the data from an EGS run which illustrates this effect for copper and tungsten at 17– and 50–GeV incident beam energies. Also note that the higher–Z material results in a desirable smaller phase-space of positrons due to a smaller beam size.

Table I. Positron Yield Properties from Copper and Tungsten

Incident Energy:	17 GeV		50 GeV	
Material	Cu	W	Cu	W
Radiation Length	14.3	3.5	14.3	3.5
Yield at $z = 6$ r.l. for $2 < E < 5$	2.8	5.8	7.4	13.4
σ_x (mm)	2.1	1.1	1.8	1.1
σ_{P_x} (MeV/c)	3.0	3.3	3.0	3.5

Of interest and possibly relevant for high energy linear colliders are the yields with higher energy electron beams. The relevancy will of course depend on the practicality of using a spent beam after collision or accelerating extra scavenger electron bunches to initiate an electromagnetic cascade shower in a thick target. Figures 8 to 12 summarize an EGS calculation with 1–TeV electrons incident on a thick tungsten target. Results are shown at six-radiation-lengths depth into the shower and at shower maximum, which occurs at ten radiation lengths. The yields obtainable are large because of the high multiplication of the shower. Note, however, that the properties of the positrons are quite similar to the properties at 33 GeV.

7. PHOTONS FROM SYNCHROTRON RADIATION

In the case of a high energy collider, an attractive method to obtain photons is to pass the spent electron or positron beam (after collision) through an alternating magnetic field called a wiggler or undulator.[3] The photons produced by synchrotron radiation are then impinged on a thin target where pair production provides the positrons. It is required, of course, to produce a large quantity of photons of sufficient energy to be above threshold for this production process. An important benefit from this method is that, by using a helical undulator, circular polarized photons are obtained and from them polarized positrons. To obtain an adequate number of positrons (yield ratio of unity or better) via synchrotron radiation, a beam energy of at least 100 GeV and a long (300–meter) wiggler magnet with high field (1 Tesla) is needed.[4] Probably this method will be useful for colliders with beam energies above 200 GeV.

8. PHOTONS FROM CHANNELING

Another method to obtain an enhanced photon flux is via channeling of electrons in a crystal. Enhancements of order 10 to 100 over ordinary bremsstrahlung have been calculated.[5]

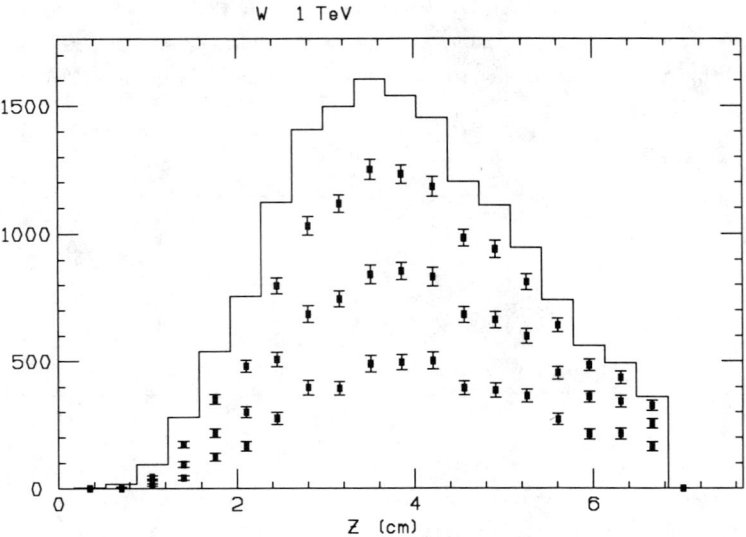

Fig. 8. Total positron yield versus z for positron energies, $E < 10$, 20, 50, and 1000 MeV.

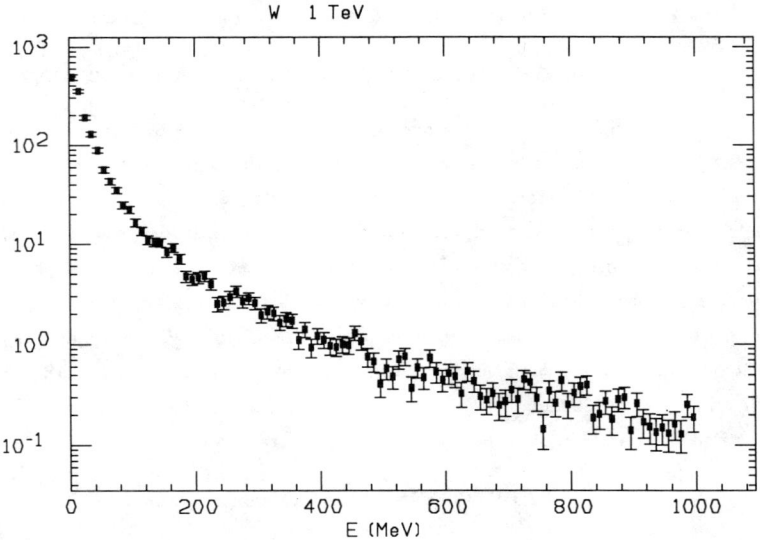

Fig. 9. Positron yield per 10–MeV bin versus E at $z = 10$ radiation lengths.

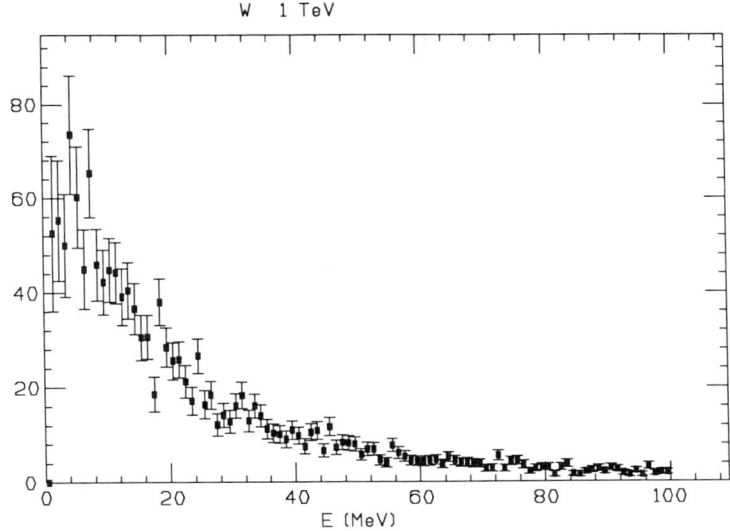

Fig. 10. Positron yield per 1–MeV bin versus E at $z = 10$ radiation lengths.

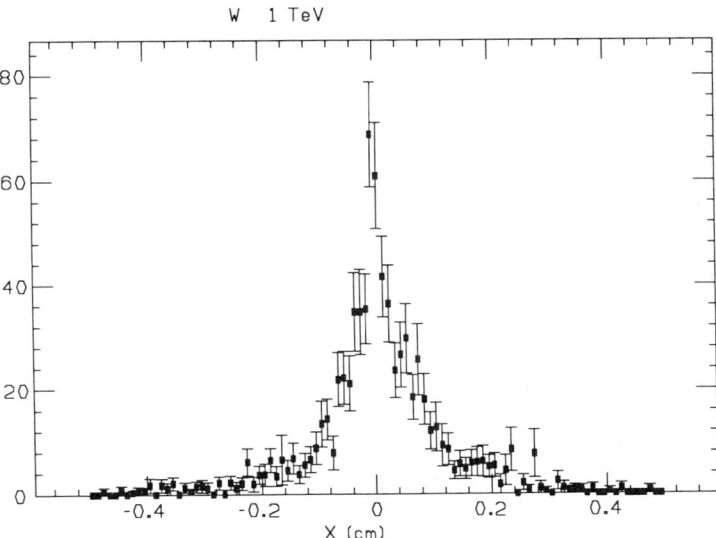

Fig. 11. Yield per 0.01–cm bin versus x at $z = 10$ radiation lengths.

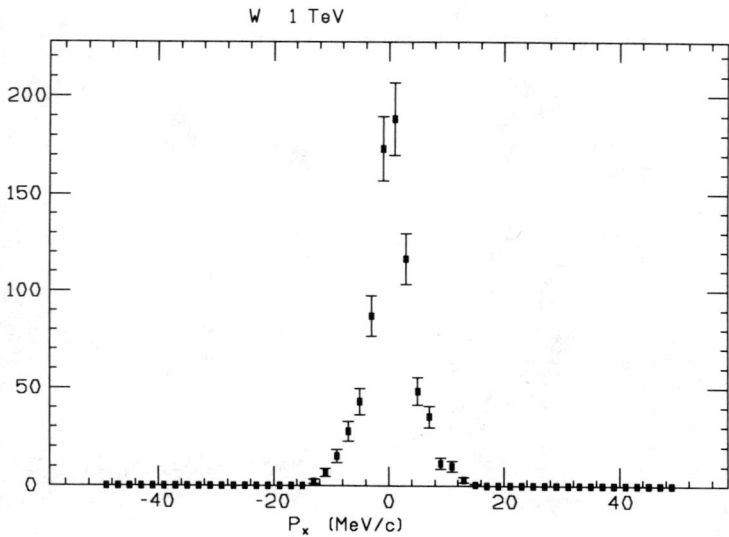

Fig. 12. Positron yield per 2–MeV/c bin versus P_x at $z = 10$ radiation lengths.

This method can be thought of as using an atomic wiggler, which provides strong electric fields in crystal planes. Channeling may prove to be useful for positron production, but a number of practical problems would need to be solved such as radiation damage to the crystal.

9. POSITRON COLLECTION

A major task in a positron source design is efficient collection of the positrons coming out of the target. The phase-space distributions shown in the figures reveal large energy ranges like 2 to 20 MeV and large angular divergences. The large energy spread leads to the need for wide-band transverse focusing. That is, the focusing must work for a large range of positron energies. Solenoidal fields, which deflect particles in proportion to their transverse momentum, are appropriate when adiabatic or slowly varying. The EGS calculations give beam sizes of order 1–mm radius and transverse momentum up to the energy of the positron, which implies angles of order one radian. If one wanted optical matching of the positron beam coming from a target, it would imply a beta-function of order one millimeter. To achieve such a beta-function value would require extremely strong magnets.

Positrons ultimately must be accelerated and stored in a damping ring in order to reduce their large emittance. The storage or damping ring must operate at an energy that provides adequate damping, typically about 1 GeV. The acceleration mechanism must efficiently capture the bunch from the target. Given the low energy, not all the positrons are relativistic. It takes some finite acceleration to obtain a fully relativistic beam, during which time the lower energy positrons are falling behind the higher. This leads to an increase in the longitudinal (z) size of the beam and hence the longitudinal emittance. Subsequent acceleration even after the beam is relativistic will usually lead to additional energy spread due to the sine wave shape of the RF accelerating field. This effect, along with the damping ring energy acceptance, then limits the useful number of positrons. Acceleration with high gradient immediately after the target is desirable in most instances. The positron longitudinal capture and damping ring should be designed and optimized as a whole system. Often the longitudinal phase-space can be rotated and matched by use of RF compressors and nonisochronous beam transports to obtain the best overall performance.

10. TRANSVERSE FOCUSING TECHNIQUES

Several types of focusing devices have been proposed and used for collecting positrons. Some provide solenoidal magnetic fields (in the z direction along the beam direction) and others use transverse fields. The simplest configuration is an adiabatically changing solenoidal field. It can be shown[6] that this transforms the phase-space radii by the square root of the magnetic field ratio, *i.e.*, Ba^2 is a constant proportional to beam emittance, where B is the magnetic field and a the beam radius. The final field value, away from the target, must be large enough to keep the beam inside the accelerator aperture. The initial value then needs to be high to achieve a useful transformation of the phase-space aspect ratio. The adiabatic solenoid is probably the best solution except for the limit of the maximum field obtainable, which is usually in the 1– to 2–Tesla range.

Another approach is to use a 1/4–wave transformer. One uses a solenoidal field of strength and length just right to rotate phase space 90° or 1/4 of a cyclotron wave length. This has the advantage of transforming phase-space by the ratio of initial to final magnetic fields, which is a better ratio than for the adiabatic case. The disadvantage is that 1/4–wave is correct only for a unique energy, and a relatively narrow energy band or set of bands is collected. Devices that are compromises between 1/4–wave and adiabatic may offer some kind of an optimum.

A device commonly used to provide a high magnet field is a pulsed flux concentrator.[7] This type of magnet is basically an eddy current transformer that steps up current. Its name comes from the concentration of magnetic flux lines from a large area of a primary coil to a smaller area single-turn secondary. It is made without the aid of magnetic materials and relies on the inability of AC fields to penetrate beyond a skin depth into a metal. As such its efficiency is strongly dependent upon primary to secondary coupling or conductor spacing. The single-turn secondary is typically a cylinder of copper with a radial slot cut to a central conical hole where the field is concentrated. Neglecting leakage flux, it follows that the field varies inversely with the cross-sectional area of the open conical hole. The small end of the cone is located near the target where the strongest field is desired.

A flux concentrator has been used at SLAC in the SLC positron source to provide 4-Tesla peak fields. Thin 0.2-mm sheets of mica are used for insulation. In time we found in some units an abrasion of the mica, presumably due to mechanical motion of the coil. This leads to voltage breakdown at the higher operating points. Because of this, a new single coil was designed which does not need an insulating material but instead relies on vacuum insulation. It is basically a simple coil with the inside dimensions conical as in the mica insulated design. This results in the same boundary conditions, and therefore the same magnetic field, except for any leakage flux between the turns. The coil is manufactured by wire electronic discharge machining, which results in about 0.25-mm spaces between conductors turns. Even with such small spaces for leakage flux, this device is about 50% less efficient than the design utilizing mica insulation. Its chief advantage is expected to be its durability.

Several types of focusing devices provide azimuthal fields by running a pulsed current in the beam direction. In particular use, both for positron collection and antiproton collection, are lithium lenses. Lithium is chosen as the current-carrying conductor to minimize multiple scattering of the beam. If the current is uniform in the conductor, the magnetic field will increase linearly with radius, providing a linear focusing element. The current is pulsed to reduce the average power, but the pulse length is chosen to obtain good current penetration and a nearly uniform distribution. Currents of order 100 kA are used to achieve fields of 10 Tesla in lenses 1 or 2 cm long. Another variation on focusing with an azimuthal field is the parabolic lens. A current in a wire will produce a field that falls off inversely with radius or distance from the wire. To produce a linear focusing force a parabolic horn-shaped conductor is used which provides a path in the field which increases as the square of the radius.

11. LONGITUDINAL CAPTURE

The longitudinal motion of positrons in the accelerator is relatively straightforward to calculate in the case of negligible motion transverse to the accelerator axis. A constant of the motion can be obtained relating energy and phase of the particle relative to the RF accelerating field.

Let

G = accelerator gradient of RF ,

λ = RF wavelength ,

β_p = phase velocity of RF ,

ϕ = phase of particle relative to RF ,

E = energy of particle ;

then

$$d\phi = \frac{2\pi}{\lambda} \left(\frac{1}{\beta_p} - \frac{1}{\beta}\right) dz ,$$

$$\frac{dE}{dz} = -G \sin \phi$$

give the equations of motion. Combining the two above equations to eliminate z and integrating gives

$$\frac{2\pi}{\lambda} \left(\frac{E}{\beta_p} - \sqrt{E^2 - m^2}\right) = G \left(\cos \phi - \cos \phi_\infty\right)$$

where $\cos \phi_\infty$ is a constant of integration chosen to correspond to the phase at infinite energy or time. This relation gives the directed path of particle motion in $E\phi$ space under the accelerating field. The parameter for this family of paths is the constant of integration $\cos \phi_\infty$. From an initial bunch of positrons the final asymptotic phase distribution is easily calculated by solving for ϕ_∞. This formulation is convenient, but the approximation of small transverse motion is usually not valid and the full space and time equations must be solved. This usually requires computer ray tracing.

12. COMPUTER RAY TRACING

Depending on the devices used in collection of positrons, the calculation of net yield can be of varying degrees of difficulty. It is convenient to have analytic calculations which lend themselves to an understanding of the quantitative performance of the system. Often, however, in pushing for highest positron yield, one is driven to nonideal devices which may be nonlinear and not adiabatic. Also the positron shower properties need four variables (r, P_r, P_θ, and P_z) to be properly described. The net effect is to make analytic calculations difficult except in idealized cases. Computer ray tracing provides a way to calculate and optimize the performance of the collection system. At SLAC a program named ETRANS[8] has been used to integrate the positrons' trajectories in three-dimensional space and time through a flux concentrator field, slowly varying solenoidal fields, and the acceleration to 200 MeV. The space dependence of fields is taken to be cylindrically symmetrical. Input trajectories are taken from EGS Monte Carlo positron output trajectories. The output from the program is in the form of histograms of accepted and rejected positrons as a function of input or output space and momentum coordinates. Output trajectories were also used as input to other ray tracing programs such as TURTLE.[9] Space-charge effects were ignored in this program. Computer runs could be made for a variety of collection parameters in order to optimize the positron yield.

13. SPACE–CHARGE EFFECTS

The effects of space charge or wake fields are more difficult to calculate. A program called MASK[10] has been used at SLAC and at several other laboratories to solve for the induced electromagnetic fields and particle transport. Calculating the beam trajectory for as short a time as 40 picoseconds is a significant job for existing computers. In a limited number of runs of the SLC geometry, noticeable space-charge effects were observable only 12 mm from the target for intensities 6×10^{12} positrons. Note that both positrons and electrons are present in nearly equal numbers and must be included in any space-charge calculation. It is fortuitous that the two charges will induce opposite fields which will cancel. Once the positrons and electrons enter the accelerator, however, the positrons and electrons separate and large wake fields will be present.

14. THERMAL HEATING AND STRESS DUE TO SHOWER

A major concern in using a target for production of positrons is whether the target material can withstand the thermal shock caused by the incident beam. Energy loss by ionization locally heats the target, causing an area of stress that propagates as a shock wave in the material. The EGS program can sum the energy deposition in various regions of the target. A convenient quantity is the fractional energy density

$$\frac{1}{E}\frac{dE}{dV} ,$$

where E is the total incident beam energy and dE is the energy accumulated in a volume dV. Note that this quantity is independent of the incident beam intensity or power.

The temperature rise in a localized region is given by

$$\Delta T_p = \frac{N E_0}{\rho C_p}\left(\frac{1}{E}\frac{dE}{dV}\right) ,$$

where

N = number of particles in beam pulse ,
E_0 = beam energy ,
C_p = material heat capacity ,
ρ = density .

Examples of EGS runs for 50–GeV electrons incident on Al, Cu, and W target materials are given in Figures 13–24.[11] Note the logarithmic scaling of the radial bins (cylindrical symmetry is assumed about the beam axis). The temperature rises are calculated for $N = 5 \times 10^{10}$ incident electrons. Incident beam sizes of 0.05, 0.10, 0.5, and 1.0 mm are shown by the respective symbols ×, ◇, +, and □. Note that temperature rises from a single beam pulse can be dangerously large in the case of small beams hitting high-Z materials. A common technique used to reduce the maximum temperature is to use a low-Z material as a spoiler up-beam of the main target material. This spoiler then increases the beam size at the target by multiple scattering but because of its lower density does not itself have too high an energy deposition density.

Although it is easy to find target materials that can stand high temperatures, the problem of excessive stress to the target is often difficult to solve. Solving the stress equation in three dimensions with time dependence of the shock wave after a short beam pulse is seemingly beyond our present computational capabilities. Often the problem is simplified by calculating the maximum stress occurring just after the beam hits the target and assuming the z (beam direction) dependence to be slowly varying. One then can use the stress equations for a long cylinder with axis parallel to the beam direction. For a long cylinder of radius b, the stress equations are[12]

$$\sigma_r = \frac{\alpha E}{1-\nu} \left(\frac{1}{b^2} \int_0^b T(r)\, r\, dr - \frac{1}{r^2} \int_0^r T(r)\, r\, dr \right) ,$$

$$\sigma_\theta = \frac{\alpha E}{1-\nu} \left(\frac{1}{b^2} \int_0^b T(r)\, r\, dr + \frac{1}{r^2} \int_0^r T(r)\, r\, dr - T(r) \right) ,$$

$$\sigma_z = \frac{\alpha E}{1-\nu} \left(\frac{2\nu}{b^2} \int_0^b T(r)\, r\, dr - T(r) \right) ,$$

α = coefficient of expansion ,

E = modules of elasticity ,

ν = Poisson's ratio (typically 0.25–0.3) .

Note that the maximum stress is of order

$$\sigma_{max} = \frac{\alpha E T_{max}}{2(1-\nu)} .$$

This quantity in comparison to the ultimate tensile strength is a figure of merit for selection of materials and judgment of the survivability of a target.

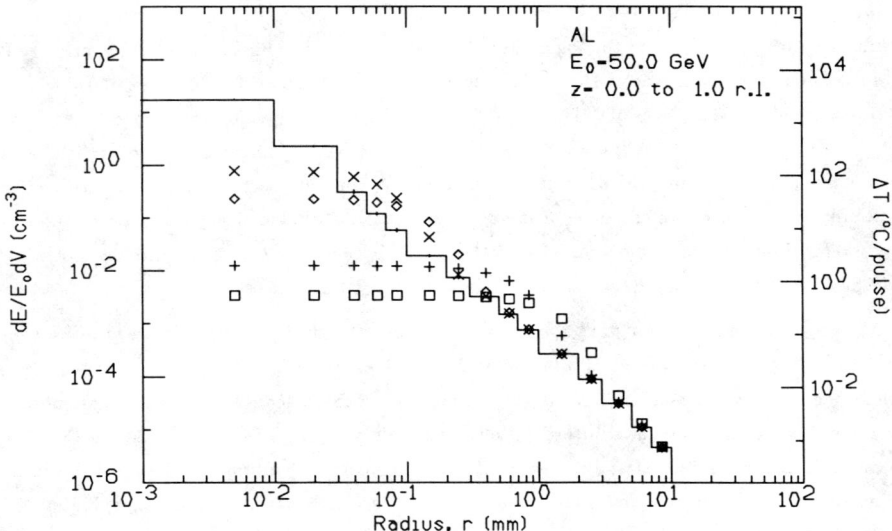

Fig. 13. E-deposition and temperature rise for electron beams of various sizes for aluminum at $0 < z < 1$ radiation length.

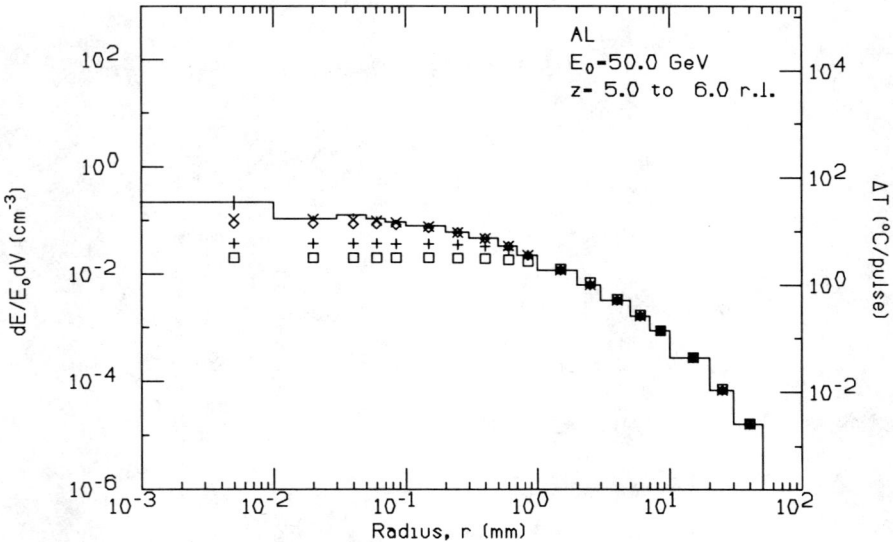

Fig. 14. E-deposition and temperature rise for electron beams of various sizes for aluminum at $5 < z < 6$ radiation lengths.

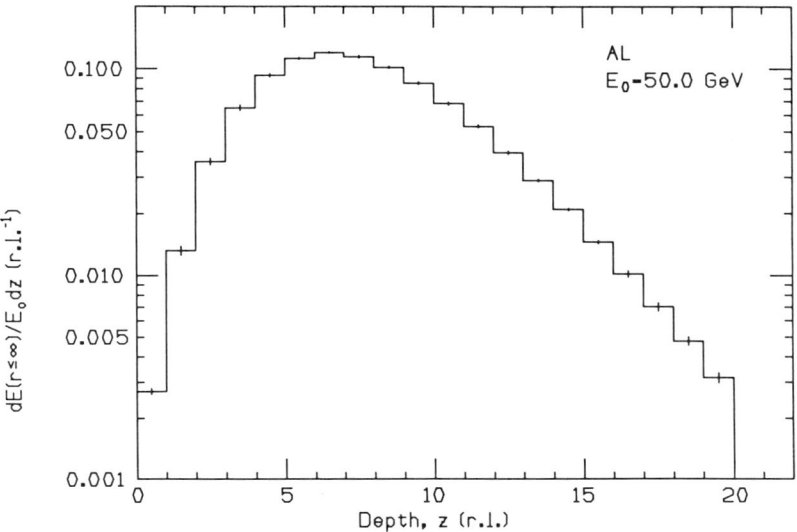

Fig. 15. Longitudinal E-deposition for aluminum.

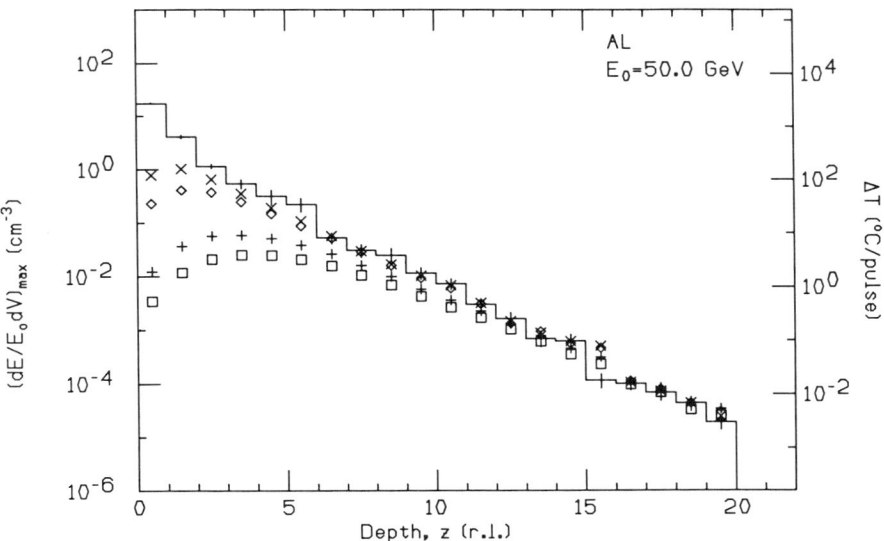

Fig. 16. Maximum E-deposition and temperature rise for electron beams of various sizes for aluminum.

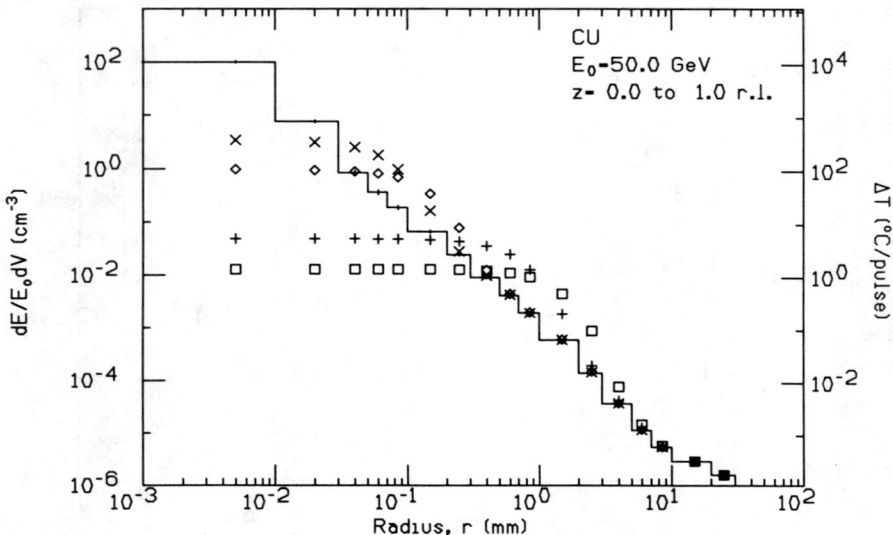

Fig. 17. E-deposition and temperature rise for electron beams of various sizes for copper at $0 < z < 1$ radiation length.

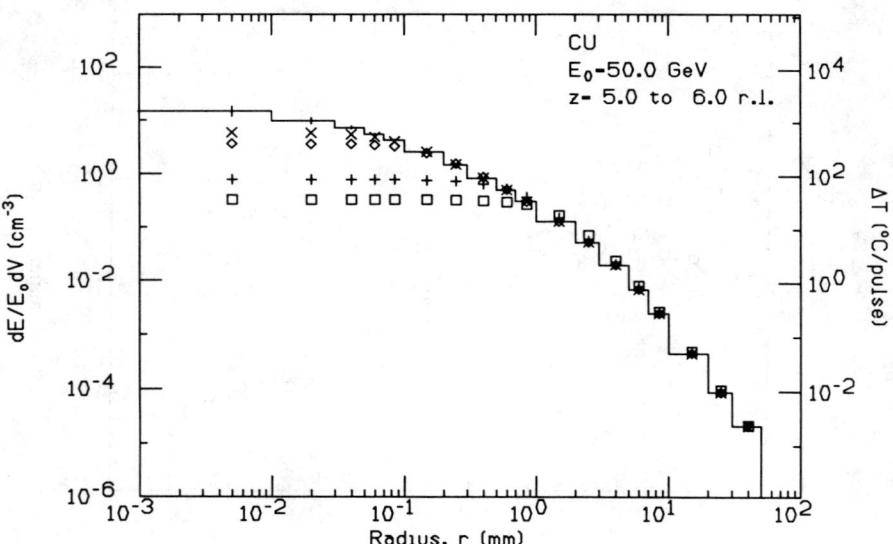

Fig. 18. E-deposition and temperature rise for electron beams of various sizes for copper at $5 < z < 6$ radiation lengths.

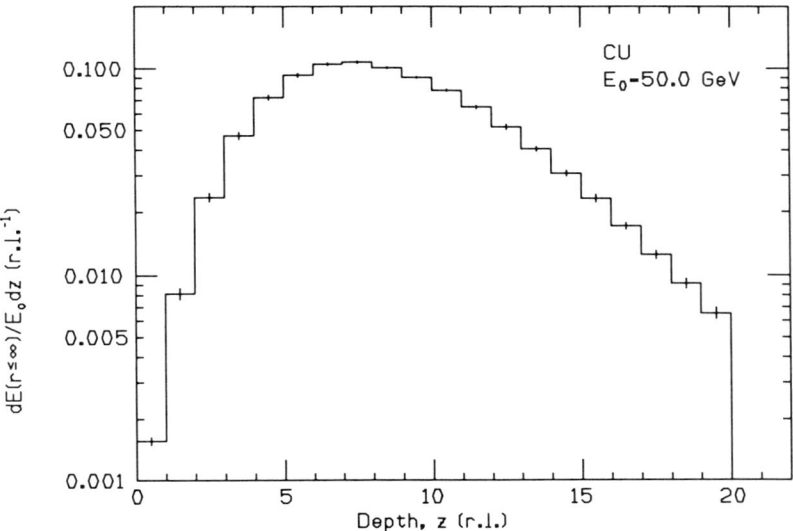

Fig. 19. Longitudinal E-deposition for copper.

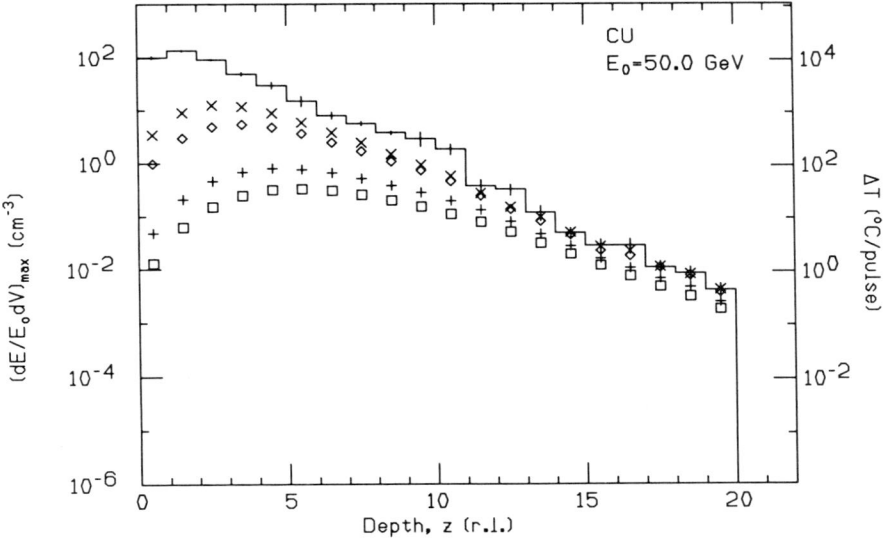

Fig. 20. Maximum E-deposition and temperature rise for electron beams of various sizes for copper.

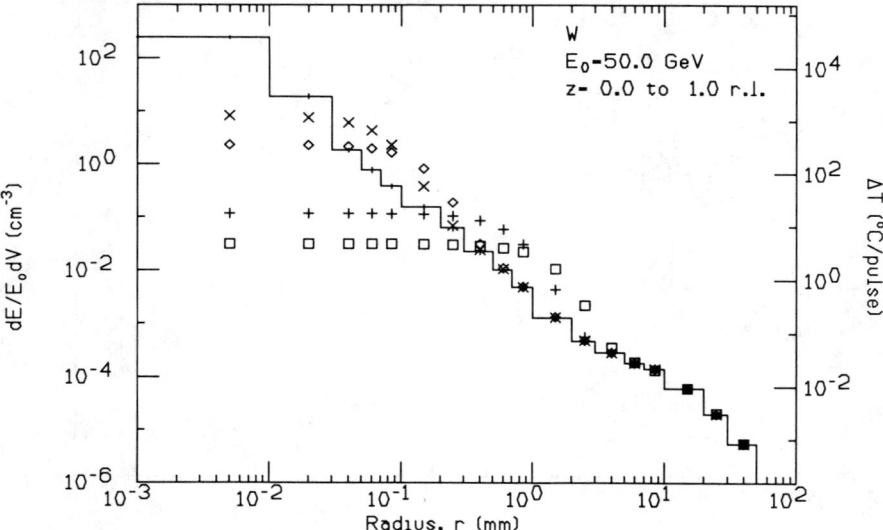

Fig. 21. *E*-deposition and temperature rise for electron beams of various sizes for tungsten at $0 < z < 1$ radiation length.

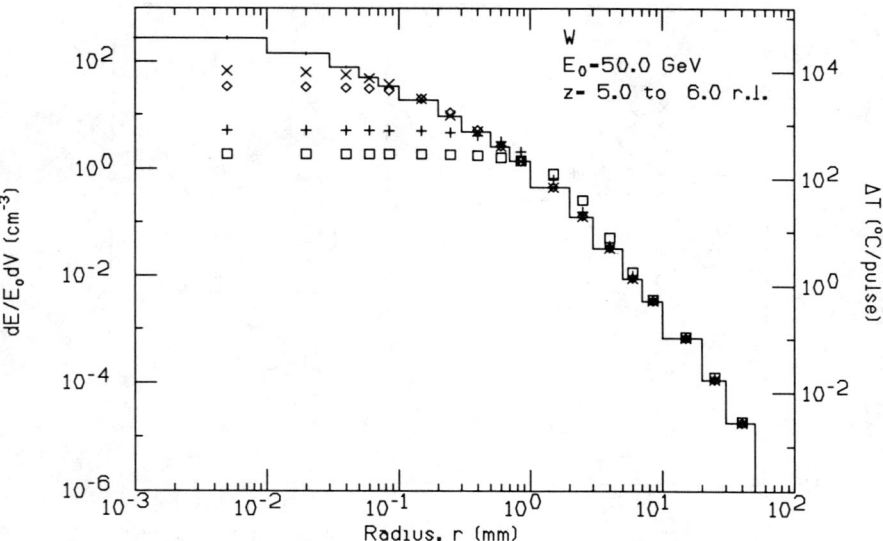

Fig. 22. *E*-deposition and temperature rise for electron beams of various sizes for tungsten at $5 < z < 6$ radiation lengths.

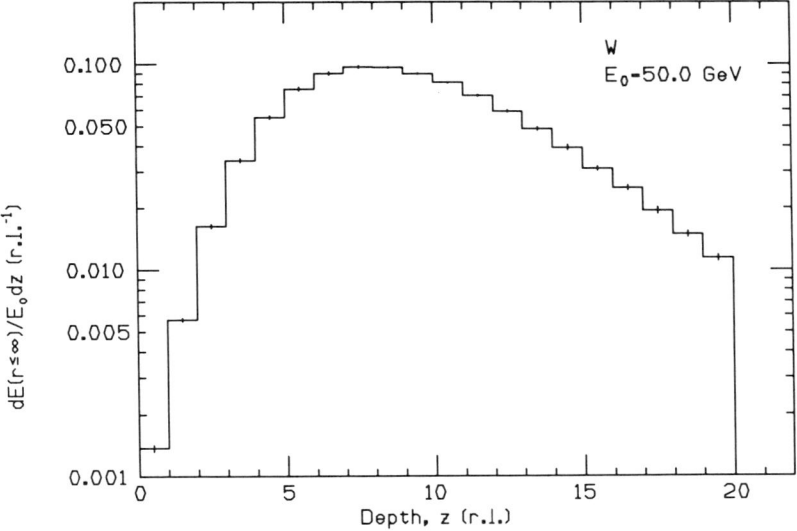

Fig. 23. Longitudinal E-deposition for tungsten.

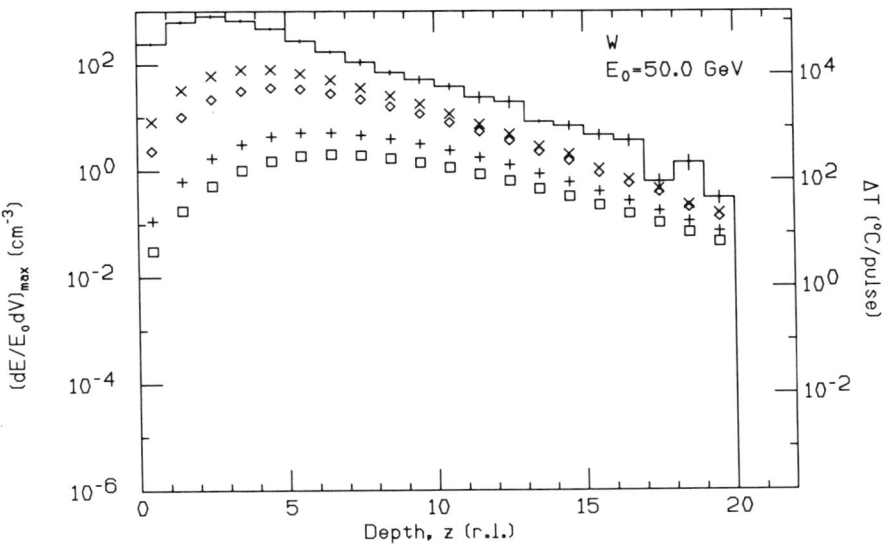

Fig. 24. Maximum E-deposition and temperature rise for electron beams of various sizes for tungsten.

Also important are the fatigue properties of the target material. Since most colliders have repetition rates of over 100 Hz, a target must withstand over 10^7 pulses per day of operation. This is essentially an infinite number of cycles when considering a fatigue curve (allowable stress versus cycles). In addition, some safety factor should be applied, *e.g.*, two or three times less stress than ultimate tensile strength, to account for imperfect materials, radiation damage, and other seen or unforeseen circumstances. Even with all precautions a target might structurally fail; hence its environment should be arranged to allow minimal damage to adjacent equipment and facilitate easy replacement or repair.

REFERENCES

1. Bruno Rossi, *High-Energy Particles,* Prentice–Hall, 1952.

2. W. R. Nelson, H. Hirayama and D. W. O. Rogers, *The EGS4 Code System,* SLAC–Report–265 (1985).

3. G. Brown, H. Winick, J. Harris and K. Halbach, Wiggler and Undulator Magnets, Nucl. Instrum. Methods **208**, 65–77 (1983).

4. U. Amaldi, *Proc. ICFA Workshop on Possibilities and Limitations of Accelerators and Detectors,* Les Diablerets, Switzerland, 1979; V. E. Balakin and A. A. Mikhailichenko, The Conversion System for Obtaining High Polarized Electrons and Positrons, Institute of Nuclear Physics, Novosibirsk, USSR Preprint 79–85 (1979).

5. J. C. Kimball and N. Cue, Synchrotron Radiation and Channeling of Ultrarelativistic Particles, Phys. Rev. Lett. **52**, 1747 (1984).

6. R. B. Neal, *The Stanford Two Mile Accelerator,* Benjamin, 1968; R. H. Helm and R. Miller, in *Particle Dynamics, Linear Accelerators,* p. 115–145, P. M. LaPostolle, A. L. Septier, Ed., North–Holland, Amsterdam, 1970.

7. Y. B. Kim and E. D. Platner, Rev. Sci. Instrum. **30**, 524 (1959); M. N. Wilson and K. D. Srivastana, *ibid.* **36**, 1096 (1965).

8. H. Lynch, SLAC, Private communication, 1984.

9. D. C. Carey, K. L. Brown and Ch. Iselin, *Decay Turtle,* SLAC–246 (1982).

10. W. B. Herrmannsfeldt and A. Drobot, Private communication, 1984.

11. W. R. Nelson and S. Ecklund, *Energy Deposition and Thermal Heating in Materials Due to Low Emittance Beams,* SLAC–CN–135 (1981).

12. Stephan Timoshenko and J. N. Goddier, *Theory of Elasticity,* McGraw–Hill, 1951.

FREE-ELECTRON LASERS
Charles A. Brau
Vanderbilt University, Nashville, TN 37235

CONTENTS

1. SURVEY 1616
   1.1. Historical overview 1616
      1.1.1. Basic ideas 1616
      1.1.2. Early work 1617
      1.1.3. Current status 1618
   1.2. Basic Principles 1619
      1.2.1. Fundamental theory 1619
      1.2.2. Comparison with conventional lasers 1627
   1.3. Laser Experiments 1628
      1.3.1. Technology vs. wavelength 1629
      1.3.2. Electrostatic accelerators 1631
      1.3.3. Induction linacs 1631
      1.3.4. rf linacs and microtrons 1633
      1.3.5. Storage rings 1635
      1.3.6. Technology directions 1637
   1.4. Applications 1637
      1.4.1. Research 1638
      1.4.2. Medicine 1639
      1.4.3. Industrial processing 1643
      1.4.4. Other applications 1645
   1.5. Conclusion 1646
2. BASIC THEORY 1647
   2.1. Basic Physics 1647
      2.1.1. Quantum effects 1648
      2.1.2. Collective effects 1651
   2.2. Electron Dynamics 1654
      2.2.1. Theoretical approach 1655
      2.2.2. The Hamilton equations 1655
   2.3. Optical Dynamics 1662
      2.3.1. The Maxwell equations 1662
      2.3.2. Conservation, gain, and refraction 1664
   2.4 Dimensionless Equations of Motion 1666
      2.4.1. Dimensionless variables 1666
      2.4.2. Dimensionless equations 1667
3. ONE-DIMENSIONAL BEHAVIOR 1670
   3.1. Analytical Results 1670
      3.1.1. Small-signal gain 1670
      3.1.2. The Madey theorem 1676
      3.1.3. Saturation 1680
   3.2. Numerical Simulations 1686
      3.2.1. Short pulses 1686
      3.2.2. Synchrotron instabilities 1695
      3.2.3. Long pulses 1700
   REFERENCES AND NOTES 1703

This article is based in part upon the first three chapters of the book <u>Free Electron Lasers</u>, by Charles A. Brau (Supplementary volume to <u>Advances in Electronics and Electron Physics</u>, edited by P. W. Hawkes) (copyright © 1989 by Academic Press, Inc.) adapted and reprinted by permission of the publisher.

FREE-ELECTRON LASERS
Charles A. Brau
Vanderbilt University, Nashville, TN 37235

1. SURVEY

Free-electron lasers represent an altogether new and exciting class of coherent optical sources. Making use of a simple and elegant gain medium - an electron beam in a magnetic field - they have already demonstrated broad tunability and excellent optical-beam quality. For the future they offer the possibility of generating the greatest focused power ever achieved by a laser. Even before this is achieved, the unique advantages of free-electron lasers, especially their tunability, will make them useful for a variety of important applications in science and medicine.

The purpose of this chapter is to introduce the broad field of free-electron lasers before launching into the details of their operation and the technologies which support them. The chapter begins with a brief history of free-electron lasers, and then continues with a simple discussion of the basic principles of their operation. This is followed by a survey of the experiments which have been accomplished to date, and finally a discussion of some of the applications for which free-electron lasers seem particularly well suited.

1.1. Historical Overview

Like so many inventions, free-electron lasers have been discovered and rediscovered several times, with different names and from different backgrounds. It is interesting and useful to trace the origins of the ideas now embodied in free-electron lasers, and to understand the relationship between the various branches of the field which have developed.

1.1.1. Basic ideas

"Free" electrons, on which these devices are based, are distinguished from electrons which are "bound" in atoms or molecules. The electrons in a free-electron laser have the form of an electron beam in a vacuum, much like the beam in the picture tube of a television set except that the electrons have much higher energy and intensity. Electrons bound in atoms and molecules vibrate only at specific frequencies. Thus, the laser light from conventional lasers, which make use of bound electrons, appears only at these specific frequencies. On the other hand, the electrons in free-electron lasers are forced to vibrate by their passage through an alternating magnetic field. Thus, the vibration frequency can be adjusted by altering the construction of the magnetic field or by changing the speed of the electrons passing through the magnetic field. The broad tunability of free-electron lasers from the far infrared to the visible and beyond was the origin of the great interest in these lasers. More recently it has been recognized that

© 1989 American Institute of Physics

free-electron lasers have unique advantages for operation at high average power levels, and this has made them attractive for military applications.

Because they depend on an electron beam in a vacuum magnetic field, free-electron lasers have as much in common with microwave devices as with lasers. For this reason, they can be regarded as an extrapolation of microwave technology to optical wavelengths. In fact, the first free-electron laser was operated in the microwave portion of the spectrum. It was not known as a free-electron laser because at that time lasers had not yet been invented.

Free-electron lasers as we know them today actually developed independently, as an outgrowth of synchrotron-radiation research. Synchrotron radiation is the very short wavelength radiation which is given off by electrons in synchrotrons and storage rings. This radiation can be enhanced by adding magnets to a storage ring to wiggle the electrons, with the magnets arranged in the same configuration now used for free-electron lasers. The synchrotron radiation from such wigglers (or undulators) is identical to the incoherent, spontaneous radiation observed from free-electron lasers before they begin to lase.

1.1.2. Early work

The earliest work on wiggler or undulator radiation dates back to 1951, when Hans Motz[1] proposed the wiggler magnet configuration now used in free-electron lasers. Subsequently, he demonstrated incoherent radiation from such devices in both the millimeter and optical regimes[2].

Following this work, Robert M. Phillips, then at General Electric, developed a device he called the "ubitron," which would now be called a low-voltage free-electron laser[3]. The ubitron uses the same configuration of electron beam and magnetic field as proposed by Motz, but at a high enough electron density that space-charge waves are excited in the electron beam[4]. High power (>1 MW) and high efficiency (>10%) were obtained at wavelengths from 10 cm to 5 mm, but other devices such as the travelling-wave tube offered higher gain and other advantages, and the ubitron was not actively pursued.

Then, in 1970, John M. J. Madey, of Stanford University, independently proposed what he called the free-electron laser[5]. Influenced by research on synchrotron-radiation sources, Madey conceived his device for the optical region, using a beam of highly relativistic electrons. In his original proposal, Madey actually described his device with a quantum-mechanical theory. This made it immediately clear that the new invention was a laser, that is, a quantum-mechanical device that is based on stimulated emission from a population inversion between quantum states of the electron beam. Subsequently it was discovered that the device could be accurately described using classical mechanics[6,7]. This made it possible to develop much more powerful descriptions of free-electron laser performance.

In 1976, Madey and his co-workers at Stanford succeeded in demonstrating gain with a free-electron laser, using a 24-MeV electron beam and a 5-m long wiggler to amplify the beam from a CO_2 laser[8]. A year later, they added mirrors to the system and operated the accelerator at 43 MeV to demonstrate laser oscillation at a wavelength of 3.5 µm, in the near-infrared part of the optical spectrum[9]. Although the power (300 mW) and efficiency (0.01 %) were small, there could be no doubt that the device had lased. Because of the obvious potential of the device for high power and broad tunability, these results immediately attracted a great deal of interest. Theoretical work on free-electron lasers expanded rapidly, and experimental work began at several laboratories which already had suitable accelerators.

Unfortunately, none of the electron-beam sources available at that time had enough electron-beam current and satisfactory electron-beam quality to make lasing easy, and it was not until six years laser, in 1983, that the second free-electron laser was operated in the optical part of the spectrum. In that year three devices began to lase. The first was at Orsay, France, where the electron beam in the storage ring ACO was used to achieve lasing in the visible[10]. The second was at Stanford, where a team from TRW, Inc. used the superconducting accelerator previously used by Madey to achieve lasing in the near infrared[11]. The third was at Los Alamos, where a newly constructed electron accelerator was used to achieve lasing in the mid-infrared[12].

During this same period, development of ubitron-type devices began again at several laboratories. Because the threshold electron-beam current at which space-charge waves can be excited increases as the third power of the electron energy, these devices are limited to low electron energy (no more than a few MeV), and long wavelength. Nevertheless, Marshall and his co-workers at Columbia and Naval Research Laboratory achieved lasing at 400 µm with an electron beam having an energy of 1.2 MeV and a peak current of 25 kA[13]. Because these devices are limited to wavelengths in the submillimeter region and beyond, where the optical radiation is transmitted through a waveguide, and because the physics in this regime involves collective oscillations in the electron beam, which are not important at shorter wavelengths, we do not address ubitron-type devices in this book. The interested reader is referred to the excellent text by Marshall[14].

1.1.3. Current status

Up to the present time, about ten free-electron lasers have been successfully operated in the short-wavelength region, in addition to a number of ubitron-type devices which have been operated in the microwave region. Several types of accelerators have been used, including radio-frequency linear accelerators (rf linacs), storage rings, microtrons, induction linacs, and electrostatic accelerators. Since Madey's original demonstration, the performance of free-electron lasers has expanded in all directions. The shortest wavelength (460 nm) was recently achieved with a free-electron laser operating on a storage ring at the Laboratoire pour l'Utilisation du Rayonment Electromagnetique (LURE), in Orsay, France[15]. The highest power (1 GW) and gain (40 dB) have been achieved with the induction linac ETA at Livermore, at a very long wavelength (9 mm)[16]. In between lie a variety of devices of intermediate power and wavelength. In addition, about a dozen more devices are currently under construction.

It should be pointed out that of the free-electron lasers which have been constructed thus far, considering only those built to operate in the optical regime, only about half have worked. The reason, in most cases, has been that the available electron accelerator was not satisfactory for free-electron laser experiments and could not, within time and budget restrictions, be suitably modified. One or two of these lasers may yet be brought into operation, but free-electron lasers remain subtle, expensive devices.

1.2. Basic Principles

Conceptually, free-electron lasers are rather simple, consisting of an electron beam in a magnetic field. Because of this essential simplicity, the theoretical description of free-electron lasers has often, especially in the early days, preceded the experimental development of these devices. This is a significant departure from the mode of development of conventional lasers, and is quite fortunate in view of the expense of building and operating high-energy electron accelerators for free-electron laser experiments.

1.2.1. Fundamental theory

The configuration of the electron beam and wiggler magnets in a free-electron laser is shown in Fig. 1.1[17]. The magnets are arranged with their poles alternating so that the magnetic field reverses every few centimeters. The overall length of the wiggler is typically a few meters, which corresponds to about 100 periods. The electron beam is injected into the end of the wiggler, and travels down its length.

As the electrons proceed down the wiggler they are deflected alternately left and right by the magnetic field and follow a wiggly path. The motions are simple forced oscillations; no subtle resonant

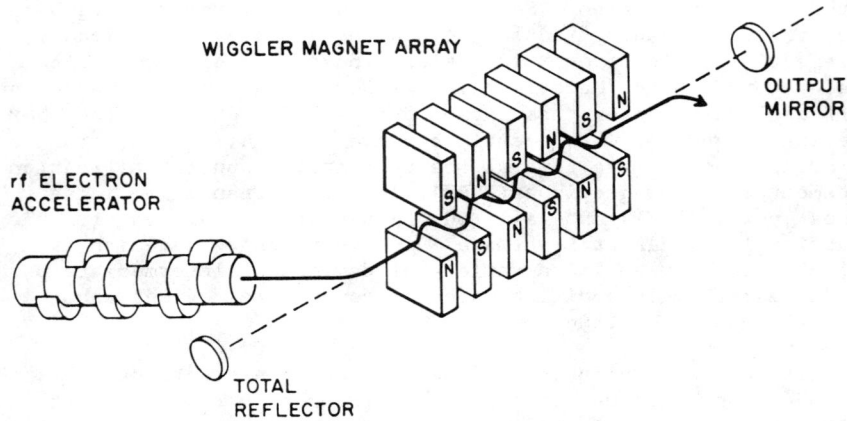

Fig. 1.1. The radiation from free-electron lasers is created by electrons which are forced to execute a wiggly motion as they pass througha series of magnets which form an alternating magnetic field called the wiggler.

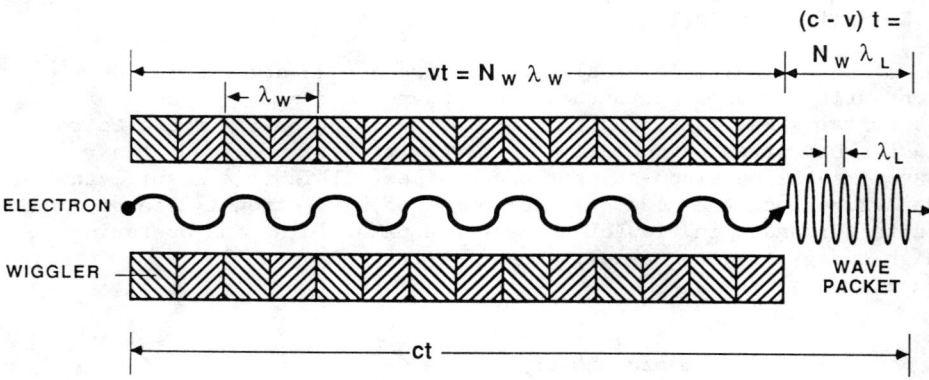

Fig. 1.2. As an electron moves down the length of the wiggler it emits a wave packet whose length is (c-v)t, where c is the velocity of light, v is the electron velocity, and t is the time for the electron to move the length of the wiggler. If the wiggler contains N_W periods, then the wave packet contains N_W periods as well.

effects are involved. If we place ourselves in a frame of reference moving down the wiggler at the mean velocity of the electrons, we observe the electrons to oscillate back and forth in a straight line perpendicular to the wiggler axis[18]. The situation is similar to that of an electric current running up and down the antenna on a police car, and like the police antenna the electrons radiate energy at the frequency with which they are oscillating.

Observed in the moving frame, the radiation from the electrons goes in all directions, like the radiation from the antenna. However, in the stationary frame of the laboratory the electrons are moving at nearly the speed of light, and radiation directed toward the sides cannot move very far from the wiggler axis before the electron and its radiation field have moved down the wiggler. As a result, the radiation appears to be moving almost entirely in the forward direction, parallel to the electron beam. This phenomenon is well known in high-energy particle physics, where Bremmsstrahlung (the radiation from the electrons in the wiggler corresponds to magnetic Bremmsstrahlung) is observed to be confined in a small cone around the direction of motion of the electron. To an observer standing in the laboratory, then, looking at the radiation produced by a source moving toward him, the frequency is Doppler shifted to a higher frequency, which corresponds to a shorter wavelength.

To calculate this wavelength, we examine the radiation from a single electron travelling down the wiggler with a velocity $v = \beta c$, where c is the velocity of light. As shown in Fig. 1.2, the electron executes N_W wiggles as it passes through the wiggler in a time $t = L_W/v$, where $L_W = N_W \lambda_W$ is the length of the wiggler and λ_W is the length of one period of the wiggler (the distance over which the magnetic field undergoes one complete alternation and returns to its original value). By the time the electron has reached the end of the wiggler, the front of the wave packet, which was emitted by the electron at the beginning of the wiggler, has moved a distance ct. But the back of the wave packet is just at the end of the wiggler, so the total length of the wave packet is (c-v)t. Since the wave packet, like the wiggler, contains N_W oscillations, the laser wavelength λ_L is given by the expression

$$\lambda_L = \frac{(c-v)t}{N_W} = \frac{\lambda_W(1-\beta)}{\beta}. \tag{1.1}$$

We may simplify this expression by recognizing that for relativistic electrons the velocity is nearly that of light, $\beta \approx 1$, so that $(1-\beta)/\beta \approx (1-\beta)(1+\beta)/2 = (1-\beta^2)/2$. To evaluate this we may use Einstein's formula for the energy of a relativistic electron,

$$\gamma = \frac{1}{(1-\beta^2)^{1/2}}, \qquad (1.2)$$

where γ is just the energy of the electron expressed in units of its rest energy. The rest energy of an electron is $mc^2 = 0.511$ MeV, where m is the electron rest mass. The wavelength is then given by the formula

$$\lambda_L = \frac{\lambda_W}{2\gamma^2}. \qquad (1.3)$$

Actually, we have ignored the effect of the wiggle motions on the velocity of the electron through the wiggler. For moderate to strong magnetic fields, the wiggles can significantly increase the path length of the electron trajectory through the wiggler. This slows the average velocity of the electrons through the wiggler and increases the wavelength of the light. When this effect is properly accounted for, we obtain the formula

$$\lambda_L = \frac{\lambda_W}{2\gamma^2}\left[1 + \left(\frac{eB_W\lambda_W}{2\pi mc}\right)^2\right], \qquad (1.4)$$

where e is the electron charge and B_W is the rms average magnetic induction. This formula is quite accurate for all cases of interest. It implies that in a frame of reference moving with the mean motion of the electrons the frequency of oscillation of the electrons is the same as that of the optical field. Consequently, Eq. (1.4) is referred to as the resonance condition for free-electron lasers.

Typically, the magnetic field is made as large as possible to maximize the wiggle motion and electron radiation, subject to the constraint that the wavelength not become too long. This means that the magnetic field term in Eq. (1.4), $eB_W\lambda_W/2\pi mc$, is generally of the order of unity. Using available permanent magnet material, the average magnetic field strength is typically of the order of 0.5 T for a wiggler period of the order of 2 cm. For an electron energy of 100 MeV, corresponding to $\gamma = 200$, the laser wavelength is $\lambda_L \approx 0.5$ μm, in the green portion of the visible spectrum. Thus, for convenient values of the wiggler period and the electron energy, we obtain a very useful wavelength.

Since the wave packet emitted by each electron contains only a finite number, N_W, of oscillations, the frequency is imperfectly defined. The spectrum of the radiation corresponds to the Fourier transform of the wave packet radiated by the electron. Since the electron radiates uniformly from one end of the wiggler to the other, the wave packet has a square envelope, as shown in Fig. 1.2. The intensity spectrum is therefore given by the expression

$$I(\lambda_L) \propto \left| \int_0^{N_W \lambda_L/c} \exp[-i(\omega-\omega_L)t] \, dt \right|^2$$

$$\propto \left\{ \frac{\sin[2\pi N_W(\omega-\omega_L)/2\omega]}{2\pi N_W(\omega-\omega_L)/2\omega} \right\}^2 , \qquad (1.5)$$

where the laser frequency is $\omega_L = 2\pi c/\lambda_L$. The shape of the spectrum is shown in Fig. 1.3, where we see that the linewidth of the radiation is of the order of $1/N_W$. Practical wigglers which have been built so far typically contain of the order of 100 periods, so that the radiation has a linewidth of about 1 percent.

The radiation we have just discussed is identical to what is called spontaneous emission in conventional lasers. It is also the incoherent radiation which is called wiggler or undulator radiation in synchrotron radiation sources. Because the electrons passing through the wiggler are randomly positioned in the electron beam, the waves they emit have random phases with respect to one another, as illustrated in Fig. 1.4. The field amplitude corresponding to the sum of these randomly phased waves, like the displacement from a series of steps in a random walk, is proportional to the square root of the number of electrons in the beam. The optical intensity, or power, which is proportional to the square of the field amplitude, is therefore proportional to the number of electrons in the beam, that is, to the electron-beam current. This is what one would might have expected intuitively.

When this radiation is used to amplify an incident optical beam, forming a laser, the radiation becomes coherent and the linewidth becomes much narrower, comparable to the linewidth of conventional lasers. The radiation also becomes much more powerful.

We can see how a free-electron laser develops coherence by examining the behavior of a single electron in the wiggler when a coherent optical beam is present. If the optical beam is propagating parallel to the electron, then the electric field is transverse to the motion of the electron. In a coordinate system moving with the mean motion of the electron, as before, we see the electron oscillating transverse to the axis of the wiggler. Therefore, the motions are parallel or antiparallel to the electric field of the optical beam. Near resonance, given by Eq. (1.4), the optical field has the same frequency as the the electron in the moving coordinate system. Thus, if the electron and the electric field are in phase, the electric field always points in the same direction as the electron motion and the electron is accelerated. If, on the other hand, we consider an electron which is half a wavelength ahead of or behind the first electron, it is out of phase with the electric field and it is decelerated. After a short time, those electrons that are accelerated catch up to those that are decelerated, and the electron beam which initially consisted of randomly distributed electrons soon

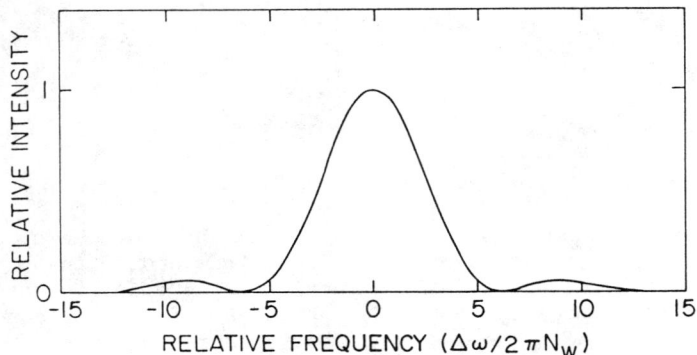

Fig. 1.3. The spectrum of the spontaneous emission from electrons passing through a wiggler magnet has a characteristic shape whose relative width is $1/N_W$.

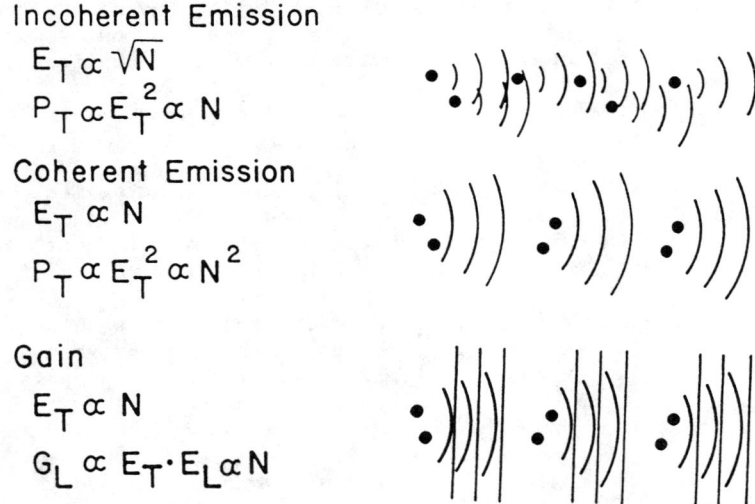

Fig. 1.4. Bunching of the electrons by their interaction with the laser beam causes the electron emission to become coherent and powerful.

consists of bunches of electrons spaced at the optical wavelength, as illustrated in Fig. 1.4. The waves radiated by the initially random electrons then add in phase with one another, and the amplitude of the sum is proportional to the number of electrons. The intensity of the radiation, which is proportional to the square of the field amplitude, is then proportional to the square of the number of electrons. Since the number of electrons is very large, typically of the order of 10^8 in the length of the wave packet from a single electron (see Fig. 1.2), the coherent emission is much more powerful than the incoherent emission. Surprisingly, the intensity of coherent emission is proportional to the square of the electron-beam current.

This unexpected behavior has been observed experimentally, as shown in Fig. 1.5[19]. The radiation required to achieve the bunching was provided by a CO_2 laser. To see the emission from the bunched electrons it was necessary to remove the CO_2 laser radiation from the emission coming out of the wiggler. To do this, the CO_2 laser was polarized at an angle with respect to the wiggler plane of symmetry, and a crossed polarizer used to discriminate against the CO_2 radiation. In addition, a cell containing CO_2 was used to absorb the narrow-band radiation from the CO_2 laser, while passing the relatively broad-band radiation from the picosecond pulses of electrons used in the experiment. For coherent radiation, the linewidth is given not by Eq. (1.5), but by the Fourier transform of the envelope of the electron-beam pulse, which is much narrower. As shown in Fig. 1.5, the radiation observed in this way is quadratic in the electron-beam current, confirming that the radiation is from coherently bunched electrons.

Not only are the electrons bunched, but, perhaps more importantly, the bunches are in phase with the incident optical field. Thus, the emission from the bunched electrons adds coherently to the incident field and amplifies it. This is illustrated in Fig. 1.4. The gain, at least in the small-gain regime, is proportional to the electron-beam current. We may see this in the following way. If **E** is a phasor representing the electric field of the incident laser, and **e** is a phasor representing the emission of the bunched electrons, where e/E ≪ 1, then the field at the wiggler exit is **E**+**e**, and the intensity is

$$(E+e)^2 = E^2 + 2E \cdot e + e^2 \approx E^2 + 2E \cdot e. \qquad (1.6)$$

The intensity gain is therefore given by the expression

$$g = \frac{E^2 + 2E \cdot e}{E^2} - 1 = 2e\cos(\phi)/E, \qquad (1.7)$$

where ϕ is the phase between the electron emission and the incident field. Since e is proportional to the electron-beam current, the gain is also. The phase around which the electrons bunch in the

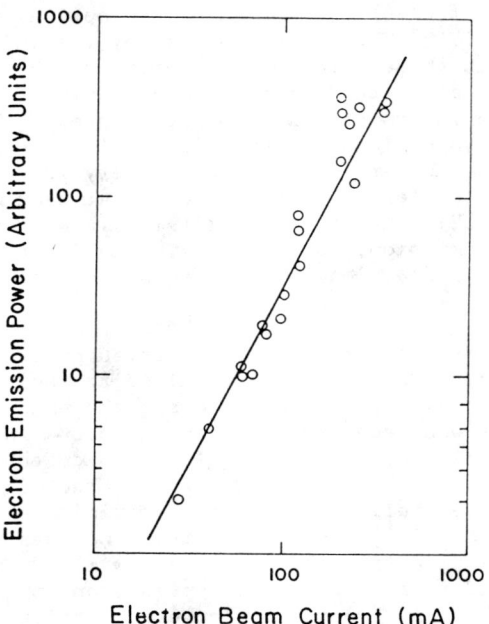

Fig. 1.5. When electron bunching makes the electron emission coherent, the emission intensity becomes quadratic in the electron-beam current.

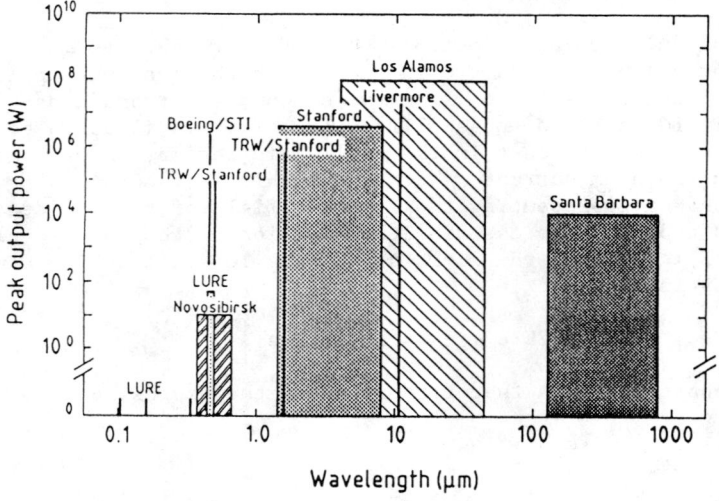

Fig. 1.6. Free-electron lasers have been operated at a variety of wavelengths from the visible to the millimeter region of the spectrum. Several devices have been tuned over broad wavelength ranges.

incident optical field depends on the energy of the electrons. For electrons initially at precisely the resonant energy, the phase ϕ is $\pi/2$, and the gain vanishes. This is because half the electrons have been accelerated and half have been decelerated by the interaction with the field, so that there is no net change of energy. Clearly, the amplification of the optical field must come at the expense of the electron energy. If, however, the electrons enter the wiggler with slightly more than the resonant energy, then they bunch around a phase such that ϕ is less than $\pi/2$, so that the gain is positive; the electrons, on the average, lose a little energy to the field. When the electrons lose enough energy so that the average electron energy is the resonant energy, no more energy can be converted into light and the laser saturates. Conversely, if the electrons enter the wiggler with less than the resonant energy, then they bunch around a phase such that ϕ is greater than $\pi/2$, and the gain is negative. We see in Chapter 2 that the central problem in computing the laser gain is finding the phase of the electrons.

The source of the optical beam incident on the free-electron laser discussed above can be either external, from a separate laser, or regenerated from the output of the free-electron laser itself. If the source is external then the free-electron laser is referred to as an amplifier, and the source is called the master oscillator. Clearly, the wavelength of the output beam is the same as that of the master oscillator, and the electron energy and magnetic field must be adjusted to satisfy Eq. (1.4). If the master oscillator is a conventional laser, then the wavelength is restricted to the specific values achievable with that laser.

Alternatively, the input optical beam may be obtained by taking part of the output beam from previous electrons and returning it to the input end of the free-electron laser. This is illustrated in Fig. 1.1, where a partially reflecting mirror is used to split the beam coming out of the wiggler into the output beam and the return, or feedback beam, which is reflected back through the wiggler. In this configuration the free-electron laser is referred to as an oscillator. In this case the wavelength can have any value, and corresponds, in general, to the wavelength for which the laser gain, or amplification, is a maximum. This wavelength is given by Eq. (1.4)[20].

1.2.2. Comparison with conventional lasers

Compared with conventional lasers, free-electron lasers offer a number of important advantages. The first and most obvious is the degree to which free-electron lasers can be tuned. Wavelength tuning can be accomplished initially by the design of the wiggler, and various devices have already been operated at wavelengths from the far-infrared to the visible portions of the optical spectrum, as shown in Fig. 1.6. In addition, it is possible to tune a given device over a large range by varying the electron energy. Most accelerators can be operated over a range of electron energy exceeding a factor of two, which corresponds to more than a factor of four variation of the optical wavelength. The wavelength ranges achieved in the experiments at Los Alamos and Santa Barbara, shown in Fig. 1.6, represent the broadest tuning range ever achieved by any laser of any type[21,22].

An equally important advantage of free-electron lasers is the high power to which their output can be scaled. There are two reasons for this scalability. In the first place, free-electron lasers are able to reject enormous amounts of waste heat. In all lasers, most of the input energy is converted into waste heat. This must be removed by cooling the laser medium or, in the case of gas lasers, which are currently the most powerful lasers, by flowing the hot laser gas out of the laser at very high speed. In the free-electron laser, the waste heat is in the electron beam, which is moving at nearly the speed of light, about a million times faster than a high-speed flowing gas. The electron beam exits the laser in a few billionths of a second, carrying the waste heat with it. Moreover, recent experiments at Los Alamos have shown that the spent electron beam from the laser can be decelerated to recover as much as 70% of the leftover energy, thereby increasing the overall efficiency[23]. The second reason why free-electron lasers can be scaled to high power is the availability of high-power acclerators. Although research must be continued to develop electron beams with the very high quality which is required for free-electron lasers, high-power beams already exist. For example, the large electron accelerator at the Stanford Linear Accelerator Center (SLAC) has a 200-kW electron beam operating around the clock, and the large proton accelerator at the Los Alamos Meson Physics Facility (LAMPF) has an 800-kW proton beam which likewise operates 24 hours per day.

These advantages don't come without disadvantages, however. The single, most important disadvantage of free-electron lasers is their high cost. Electron accelerators are expensive, costing several million dollars even for a small one, so free-electron lasers don't lend themselves to small devices. On the other hand, in very large sizes the unit cost of laser power becomes quite competitive, and it may be possible, some day, to build free-electron lasers of 100 kW and larger at unit costs of less than $500 per Watt. In the mean time, applications will be limited to those specialized circumstances which can bear the high cost. Fortunately, there are some very important applications which can do this.

1.3. Laser Experiments

Although the early developments in free-electron lasers were dominated by theory, in recent years a growing number of experiments have been carried out which demonstrate the potential of these devices. In general, the experiments have vindicated the theoretical predictions, but to accomplish this has required some exciting and important technological innovations. The maturity of the experiments has now reached the point where the experiments often lead the theory, and many of the unique advantages of free-electron lasers, especially their tunability, have been demonstrated with hardware.

1.3.1. Technology vs. wavelength

The broad wavelength range spanned by the various free-electron lasers shown in Fig. 1.6 corresponds not only to a broad range of electron energy but to a variety of accelerator technologies as well. Since almost all wigglers have a period of the order of a few centimeters long, the wavelength of a free-electron laser is largely determined by the electron energy. As a result, since any given type of accelerator is most useful over a certain energy range, it is possible to correlate each type of accelerator with a wavelength range over which it is most useful. This correlation is shown in Fig. 1.7, where the spectral regions of interest to a number of applications are also indicated. Of course, the relation between energy and wavelength is approximate, since the wiggler period and magnetic field are somewhat variable, and the spectral regions of interest for the various applications are fuzzy and in some cases overlap. Most importantly, the operating regions for the various types of accelerator technology should not be regarded as well-defined. In particular, the boundaries keep on expanding as the technology improves.

It must also be borne in mind that the electron energy is not the only factor determining the useful wavelength range for each type of accelerator. Both induction linacs and rf linacs can be operated at electron energies sufficient to achieve much shorter wavelengths than indicated in Fig. 1.7. However, as the wavelength becomes shorter, it is necessary to focus the electron beam more carefully inside the laser beam, and this requires a better electron-beam quality. Specifically, the laws of diffraction show that the product of the radius of a coherent optical beam and the angle with which it diverges is just the optical wavelength, which is preserved through any optical system. Similarly, the product of the radius and divergence of an electron beam represents a measure of the electron-beam quality called the emittance. This quantity is preserved through a good electron transport system. As a rule of thumb applicable to most free-electron lasers, if the emittance is very much larger than the wavelength, then the electron beam cannot be focused inside the laser beam. The emittance is determined largely by the electron source, called the electron gun or injector, and the injector is different for each accelerator technology. Thus, the short-wavelength limit of each technology is set by the injector technology. An exception to this rule is provided by electrostatic accelerators. These devices have excellent emittance, which would

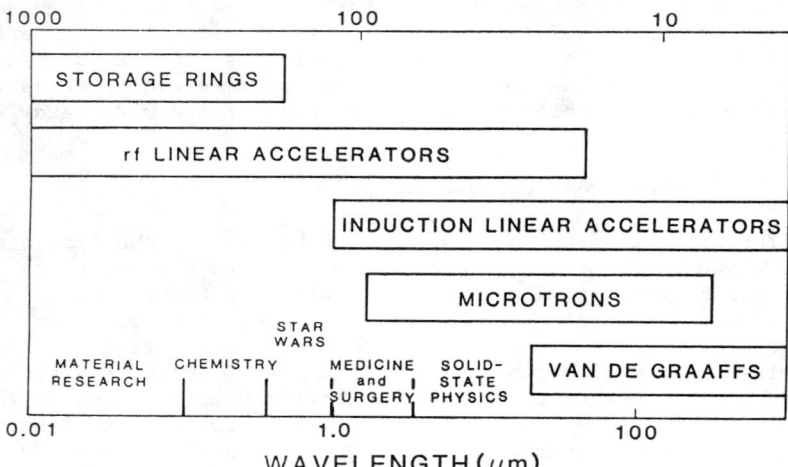

Fig. 1.7. Various types of accelerator technology are useful for different electron-energy ranges and, accordingly, different laser wavelength regions.

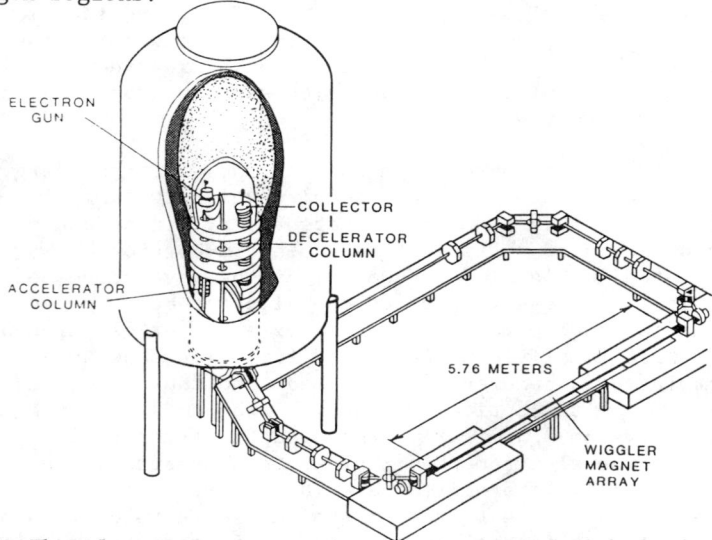

Fig. 1.8. The electrostatic accelerator at Santa Barbara produces electrons with energy up to 6 MeV with extremely good electron-beam quality. These are transported through the wiggler and then back up into the high-voltage terminal on the left.

make them suitable for operation at very short wavelengths, but they become very large and expensive at voltages above 20 MV.

1.3.2. Electrostatic accelerators

Because of their simplicity and reliability, electrostatic accelerators are ideal for the far-infrared spectrum. The first such device is now operating at the University of California at Santa Barbara, where it is already in use not only for free-electron laser research, but also for research in solid-state physics and biophysics[24]. Graduate students routinely operate the free-electron laser in support of the various experiments. A schematic diagram of the device is shown in Fig. 1.8. The operating range of the accelerator is approximately 2.5 to 6 MeV, which corresponds to wavelengths in the range from 120 to 800 μm. The laser operates with an extremely narrow linewidth, estimated to be better than one part in 10^8[25]. Since electrostatic accelerators are limited to very low average current, typically less than a milliampere, the electron beam leaving the wiggler is transported back to the high-voltage terminal to keep it charged up. Nearly 90% of the current can be recovered in this fashion, allowing the accelerator to run for 10 to 30 μs before the current loss discharges the terminal and the accelerating voltage drops appreciably.

To overcome the wavelength limitations imposed by the low voltage of electrostatic accelerators, an interesting set of experiments is about to begin at Santa Barbara in which a high-power microwave beam will be used to act as a very short wavelength wiggler[26]. Such a wiggler field is quite weak, and the gain is expected to be small. But if the experiment is successful, it will point the way to very much shorter wavelengths, in the visible and ultraviolet, and perhaps even X-ray regions[27].

1.3.3. Induction linacs

Induction linacs are another type of accelerator which has been used for free-electron laser experiments. The outstanding feature of induction linacs is the very large current which they can acclerate. Typically, these devices can accelerate as much as 10 kA in pulses lasting about 50 ns. However, the large current is usually difficult to focus, making these devices most useful at long wavelengths. The ETA at Lawrence Livermore National Laboratory is shown in Fig. 1.9, and its wiggler is pictured in Fig. 1.10. This device operates at 3.5 MeV, which corresponds to a wavelength of about 9 mm, in the microwave part of the spectrum[28]. Because of the large current, the gain, or amplification, is very large, and very high peak power and efficiency have been achieved. The laser uses a wiggler specially designed to preserve the electron resonance even when the electrons lose energy as they amplify the optical field. As much as

Fig. 1.9. ETA, shown in this photograph, is a powerful induction linac used for free-electron laser experiments at Lawrence Livermore National Laboratory. In produces an intense electron beam with energies up to 3.5 MeV, and is used in microwave free-electron laser experiments.

Fig. 1.10. This photograph shows a section of the pulsed electromagnetic wiggler used in the experiments at Lawrence Livermore National Laboratory.

40% of the electron energy has been converted into microwave radiation, which corresponds to a peak output power of 1 GW[29].

Because the electron-beam pulse from induction linacs is so short, these devices do not have time to feed the optical signal back into the wiggler and operate as an oscillator. Thus, induction-linac free-electron lasers operate as amplifiers for conventional lasers. At the present time, a more powerful, 50-MeV accelerator is being prepared at Livermore to extend the performance achieved in the microwave region to the mid-infrared spectrum, using a CO_2 laser as the master oscillator. Extension to shorter wavelengths will depend on the development of electron injectors with improved emittance.

1.3.4. rf linacs and microtrons

Since the original free-electron laser experiment of Madey and his co-workers at Stanford, several free-electron lasers have been operated with rf (radio-frequency) accelerators. rf accelerators are characterized by high energy and low current, but improvements in injector technology have increased the current and electron-beam quality for free-electron laser applications. The rf linac used in the experiments at Los Alamos, pictured in Fig. 1.11, is typical of this type of accelerator. The accelerator has an active length of 3 m, and accelerates electrons to 21 MeV with a peak current of more than 100 A. The 5-m long wiggler developed by TRW for free-electron laser experiments at Stanford is shown in Fig. 1.12. The shortest wavelengths achieved with an rf-linac free-electron laser were obtained recently by a collaboration between teams from Boeing and Spectra Technology[30]. Using a 114-MeV electron beam from a newly built rf accelerator, the groups achieved laser oscillation at 500 nm, in the green part of the visible spectrum. Free-electron lasers of this type also produce extremely good optical-beam quality. In experiments at Los Alamos the collimation of the laser beam was observed to be within 4% of the physical limit of diffraction, in spite of deliberate misalignment of the optical system and electron beam[31].

Microtrons are similar to rf linacs except that instead of running the electron beam through many accelerator sections to achieve high energy, microtrons pass the beam through the same accelerator section many times. In principle, this can greatly reduce the cost and size of the accelerator: a 20-MeV accelerator can fit into a few square meters of space. Unfortunately, microtrons are generally characterized by low current and poor electron-beam quality, and experiments to use them have not yet been successful[32,33,34]. Improvements in these compact devices may make much smaller and cheaper free-electron lasers possible in the future.

In the next few years it should be possible to extend the operation of free-electron lasers using rf accelerators to much shorter wavelengths. The key to this development will be improved

Fig. 1.11. The rf linac used in the Los Alamos free-electron laser experiments is of conventional design and typifies accelerators of this type used in free-electron laser experiments at several laboratories.

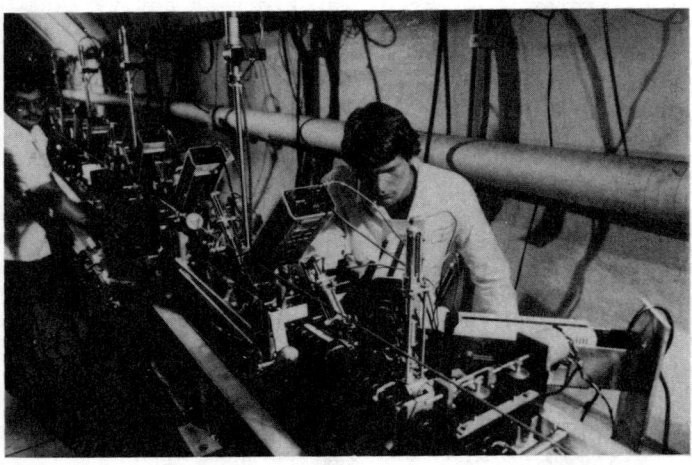

Fig. 1.12. The wiggler used in the TRW/Stanford experiments uses a permanent magnet structure which is 5-m long overall. The television cameras are used to align the electron beam on the wiggler centerline.

electron injectors. New concepts are being developed in which the cathode is placed inside a high-power accelerator cavity, rather than in a dc electric field[35,36]. This should produce much higher current and improved emittance, making it possible to extend free-electron laser operation to the vacuum-ultraviolet and even soft X-ray regions[37].

1.3.5. Storage rings

Storage rings offer the most elegant approach to free-electron laser technology. In operation, storage rings circulate a "stored beam" of electrons around a ring which is typically several meters in diameter. In modern storage rings the electron beam may be stored for many hours. As the electrons circulate around the ring, they lose a small amount of energy to radiation, which is called synchrotron radiation. This energy must be replaced at each turn by a small accelerator placed in one section of the ring. In addition to being useful for a variety of physics experiments, this radiation has the effect of damping and cooling the electrons stored in the ring to a very low temperature. In this way, the electrons achieve a very small energy spread and become very well collimated. Because they work best at high electron energy, over 100 MeV, and have such good electron-beam quality, storage rings are well suited to free-electron laser operation at very short wavelengths.

The first such laser is in operation at the Laboratoire pour l'Utilisation du Rayonment Electromagnetique (LURE), in Orsay, France. A photograph of the storage ring is shown in Fig. 1.13, and the wiggler is shown in Fig. 1.14. Operating at an electron energy of 195 to 233 MeV, the laser produces radiation tunable over the range from 486 to 463 nm, in the blue portion of the spectrum[38]. This is the shortest wavelength which has been achieved by any free-electron laser. Because of the low current stored in the ring, which is called ACO, the gain is quite small. Successful operation was made possible by the use of low-loss mirrors for the optical resonator and by the development of a very clever wiggler, called an "optical klystron." Although the laser beam is of rather low average power, about 60 mW, it continues for several hours at a time, which corresponds to the storage time of the electron beam in the ring.

At the present time, several new experiments are being undertaken to extend storage-ring free-electron lasers to the ultraviolet portion of the spectrum. Recently, the first vacuum-ultraviolet radiation was obtained in the form of coherent harmonics of a wiggler operating at a fundamental wavelength of 530 nm[39]. Because of the odd harmonics of the wiggler field, spontaneous harmonic radiation occurs naturally. To increase the harmonic power and make it coherent, the electrons were bunched at the fundamental wavelength by saturating the laser at that wavelength with a doubled Nd:YAG laser. Due to the nonlinearity of the bunching process, density modulation occurs at the harmonics as well. The radiation at

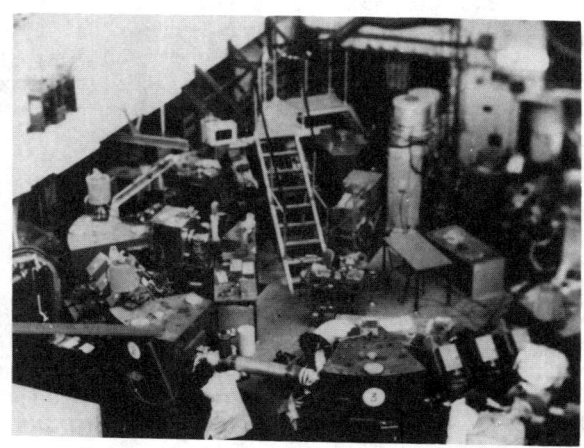

Fig. 1.13. The free-electron laser at LURE operates on the electron beam circulating in the electron storage ring ACO. The wiggler is shown in the near section of the ring.

Fig. 1.14. The wiggler used at LURE, shown in this photograph, consists of magnets above and below the beam line which can be moved away from the beam line while the electrons are being stored in the ring.

177 nm was enhanced a factor of about 350 above the spontaneous emission, while that at 106 nm was enhanced about threefold.

Because the synchrotron radiation, upon which storage rings depend for electron cooling, exceeds the laser radiation, storage rings are not very efficient, and are not suitable for very high power. Nevertheless, they will continue to be interesting for operation at short wavelengths.

1.3.6. Technology directions

In the future, free-electron lasers will be extended to much higher power and shorter wavelengths. Higher power will depend in part on the development of higher-current accelerators, but even more important will be the development of optics capable of operating at high power without damage. Damage to optical coatings has already been observed in free-electron laser experiments at several laboratories, owing to the fact that the optical beam is narrowly confined in the optical resonator and becomes quite intense. However, new optical resonator configurations using grazing-incidence mirrors are being developed, and with improved coatings and higher laser gain it should be possible in the future to achieve power levels exceeding those of any other lasers.

Shorter-wavelength lasers will depend not only on the development of improved injectors, as discussed above, as well as better mirrors. In parts of the extreme ultraviolet and soft X-ray regions (roughly 10 to 100 nm), mirrors have been developed with reflectivity exceeding 50%, but in other spectral regions no mirrors exist. For those regions where mirrors exist, this reflectivity is good enough to lase if the gain is more than a factor of four per pass. This appears quite possible, but better mirrors will help. Alternatively, it may be possible to build a very high gain wiggler which can amplify the spontaneous radiation produced at the beginning of the wiggler and produce a coherent optical beam without any mirrors at all[40].

1.4. Applications

As with all developments in science, the importance of free-electron lasers is not measured by the effort - or even success - of their development, but rather by the usefulness of the results for other human endeavors. Thus far, free-electron lasers have been limited in their applications because of the small number of devices available, and because of the difficulty of operating these new, experimental devices reliably. Now, however, reliability and even "user friendliness" are being considered in the development and refinement of free-electron lasers, and several institutions have or are constructing user facilities dedicated primarily to applications, with development oriented to the needs of the user community. The paradigms for these user facilities are the high-energy accelerator centers such as the Stanford Linear Accesserator Center, and the synchrotron radiation sources such as the National Synchrotron Light Source at Brookhaven, although the free-electron laser facilities will, at least initially, be much smaller in scale. A variety of potential applications are being addressed by these facilities.

1.4.1. Research

As has been the case with many new technological advances born in research laboratories, the first applications of free-electron lasers will be to research itself, in a variety of fields. Compared with alternative light sources, free-electron lasers offer several advantages for research applications. Foremost among these are their tunability and their high peak power and intensity. Among lasers, free-electron lasers are uniquely able to tune throughout previously inaccessable regions of the spectrum, especially in the far infrared, beyond about 20 μm, and (someday) in the far ultraviolet, beyond about 200 nm. Although conventional lasers already exist in the far-infrared and far-ultraviolet spectral regions, none are continuously tunable and only a few have very high peak power. Besides lasers, other sources of light are also available in these spectral regions. Some, like synchrotrons in the far ultraviolet and thermal sources in the infrared, offer continuously tunable radiation, but none offer the coherence and high intensity (focused power) of free-electron lasers (or other lasers, for that matter).

High intensity is necessary for opening up a whole new class of phenomena - called nonlinear phenomena - which have become accessible only since the development of the laser. Nonlinear phenomena are those in which the effect caused by the laser is not strictly proportional to the power of the laser. For example, when light is directed through a filter, the power output from the filter is ordinarily proportional to the power input. However, when the extreme intensity of a laser is directed through a filter, the filter can become temporarily "bleached", that is, transparent to the light, or the filter can become dark in regions where it was previously transparent. This darkening can occur, for example, if the filter transmits a single photon at a time but absorbs two photons if they appear at the same time. Such nonlinear phenomena tell us a great deal about the properties of matter. More important, however, they also form the basis for many new and exciting technologies from laser medicine to optical computers.

In the field of biology, for example, a laser has recently been used to remove a selected segment as small as a quarter of a micrometer in size from a chromosome. This is accomplished by tuning the laser to a wavelength at which the segment absorbs of the laser light[41]. It is believed that under certain circumstances the process takes place by the simultaneous absorption of two photons. This nonlinear phenomenon is one of the first illustrations of a "laser scalpel," with which it may someday be possible to make local surgical alterations in large molecules. In other experiments, tunable lasers may be used to unravel the spectroscopy of the genetic material DNA in the far-infrared spectral region[42]. With sufficient laser intensity it may someday be possible to effect conformational changes in the DNA or accomplish other operations. Working at the opposite end of the spectrum, in or near the X-ray region, it may someday become possible with free-electron lasers to create the ultimate image: a hologram of a large biological macromolecule.

Free-electron lasers offer exciting opportunities in chemistry reserch, too. No other tunable sources exist for precision spectroscopy - such as Doppler-free spectroscopy - and nonlinear spectroscopy in the far-infrared and far-ultraviolet regions. Moreover, the extremely short pulses from certain types of free-electron lasers will make it possible to examine the dynamics of chemical reactions on time scales as short as picoseconds. Such fast times are particularly important in the chemistry of liquids. In the emerging field of surface chemistry, tunable free-electron lasers in the far-infrared spectrum will be able to analyze chemical processes occurring on time scales shorter than a microsecond.

Solid-state physics is another research area rich in applications for free-electron lasers. Fig. 1.15 illustrates the variety of phenomena which occur in the far-infrared region of the spectrum[43]. The first free-electron laser built specifically for research in this spectral region is now in active use at Santa Barbara for research in solid-state physics and other applications[44]. The first experiments at Santa Barbara used only the tunability of the free-electron laser for linear spectroscopy. Future experiments will include nonlinear spectroscopy such as excitation of coherent phonons and transient spectroscopy of electron-phonon and electron-electron coupling in semiconductors on a nanosecond time scale.

1.4.2. Medicine

Medicine also offers a large number of opportunities for free-electron lasers. The interaction of laser radiation with mammalian tissue produces a wide variety of effects,[45] which can be used for a broad spectrum of medical and surgical applications[46]. The nature of the tissue effect depends on both the wavelength and the pulse length of the laser radiation. Fig. 1.16 shows how the absorption of the laser beam depends on the wavelength of the laser: because of the properties of water, which is the principle constituent of most tissue, both infrared and ultraviolet light are absorbed in a very short distance, while light in the visible and near-infrared portions of the spectrum is absorbed more weakly. At a

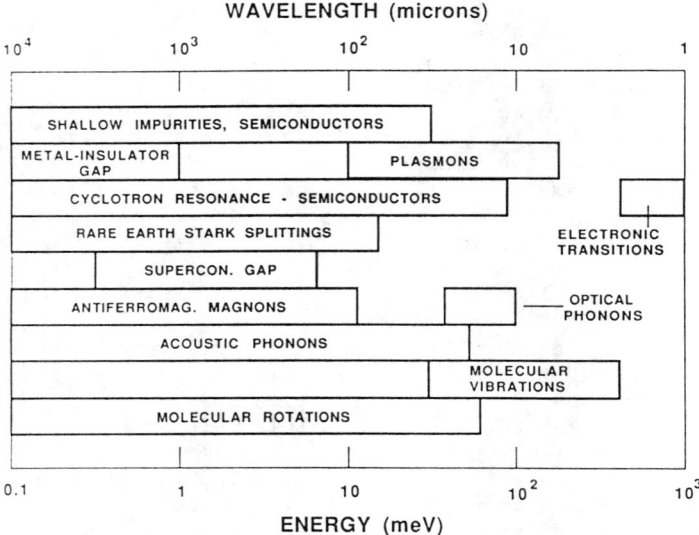

Fig. 1.15. A variety of solid-state physics phenomena are observable in the infrared part of the spectrum[65].

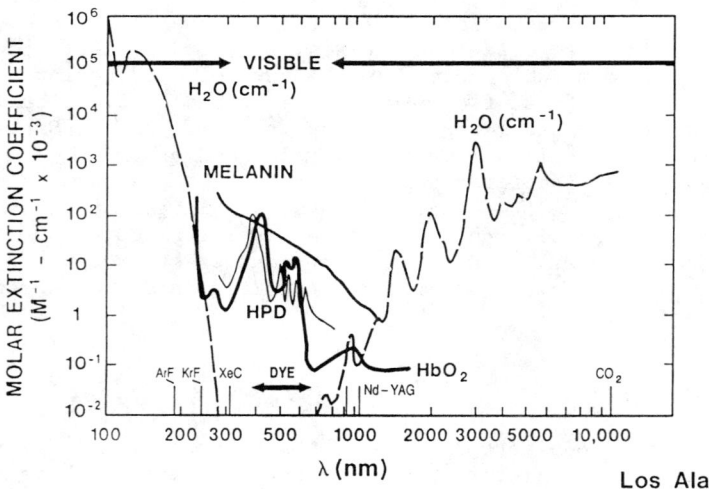

Fig. 1.16. The absorption of radiation by mammalian tissue is dominated by the absorption properties of water[66].

wavelength of 3 µm the light is absorbed in less than 10 µm, and the effects of the radiation are extremely localized near the surface. On the other hand, at a wavelength of 1 µm the penetration of light is limited principally by scattering, rather than absorption, and the effects of the radiation spread out over a larger volume. Pulse length is also important in determining the nature of the tissue interaction, as shown in Fig. 1.17, since thermal conduction can diffuse the heat away from the laser spot. For example, for pulses longer than a second the heat can diffuse more than a millimeter, whereas the heat from microsecond pulses is localized to dimensions of the order of a micrometer. Diffuse heating of tissue is observed to produce burning, cauterization, coagulation, and, in special cases, chemical reactions. On the other hand, intense local heating of very small volumes leads to instantaneous vaporization of the tissue or even to plasma formation and shock-wave effects.

These various types of tissue interaction can be used to effect a variety of surgical procedures[47]. For example, continuous or long-pulse Nd:YAG lasers operating at 1 µm have been used to coagulate blood and to actually "weld" arteries together long enough for healing to occur[48]. Continuous-beam argon lasers operating at a wavelength of 0.5 µm are conveniently transmitted through optical fibers but strongly absorbed by thrombus and arterial plaque. These properties have made them useful for removing plaque which is obstructing blood flow through arteries[49]. Pulsed CO_2 lasers, which deposit their energy in very small, well-defined volumes, are used to remove tumors in the central nervous system by vaporizing them[50]. Very intense pulses from Nd:YAG lasers have been shown to fracture kidney stones by means of the shock produced at the surface of the stone, so that the fragments can be passed from the body. Helium-neon lasers, operating at a visible wavelength which is transmitted through the eye, have long been used to reattach retinas by a thermal process. Recently, however, excimer lasers operating in the ultraviolet part of the spectrum, to which the eye is opaque, have been used to reshape the cornea in a process which is believed to be chemical in nature[51].

Given the diversity of tissue effects and procedures which are possible, the advantages of a free-electron laser become clear. Not only are free-electron lasers able to provide wavelengths and pulse lengths unavailable from any other laser - or even combination of lasers - but they can be adapted to the requirements flexibly and in real time. The flexibility makes it possible to get combinations of laser parameters which not only achieve the desired result but which adapt to the limitations of the surgical environment, such as the need to propagate the radiation through an optical fiber to the site of the operation. The real-time adaptability can be used to vary the laser parameters while the surgical procedure is in progress, even using feedback from the operation to control the laser by means of an expert system. For example, the fluorescence from the laser-produced vapor and plasma might be used to control a laser operating on plaque near the arterial wall. If the arterial wall is punctured, the laser

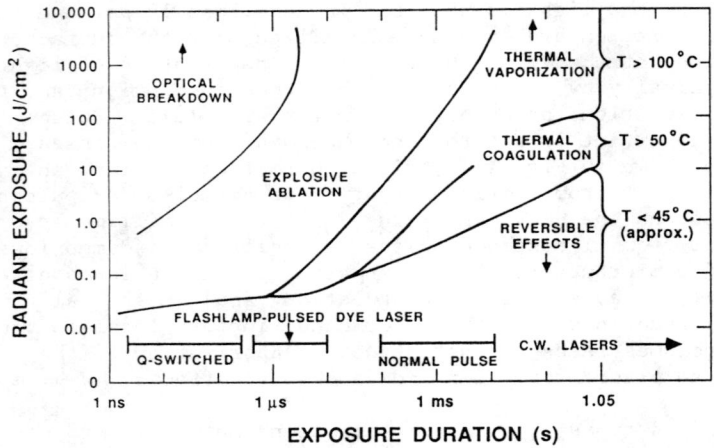

Fig. 1.17. The nature of the interaction of laser light with mammalian tissue depends on both the laser energy and the pulse length[67].

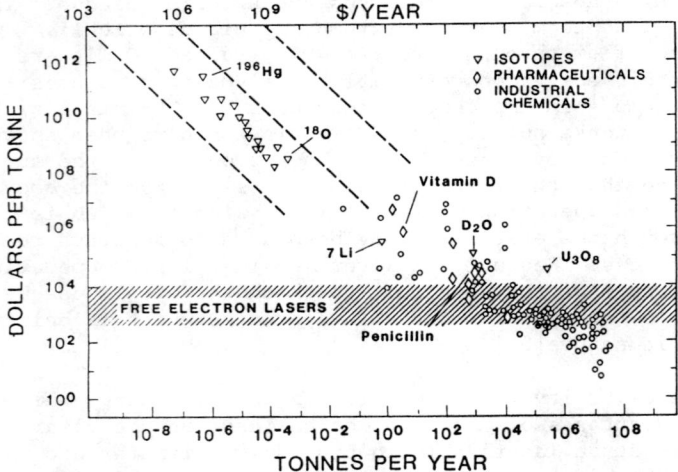

Fig. 1.18. A large number of specialty chemicals are sufficiently expensive to justify laser processing with a free-electron laser. However, large-scale chemical processing is economical only when the quantum yield (product molecules per laser photon) is large enough to bring down the laser costs.

wavelength and pulse length can be adjusted to cauterize the wound and seal the perforation.

Photodynamic therapy presents a new and different approach to cancer treatment[52]. It is observed that certain dyes, such as hematoporphyrin derivative (HpD) are more persistently absorbed by tumors than by normal tissue. About 48 hours after the dye is administered, the tumors are selectively stained. When irradiated by a laser wavelength near 630 nm, the dye releases chemically active singlet oxygen which kills the cells stained by the dye, that is, the tumor cells. This procedure works best, of course, on tumors located in the skin and other accessible parts of the body, such as the trachea. In the future, it may be possible to find more effective dyes, such as dyes which can be activated by nonlinear, two-photon processes at longer wavelengths. This would make the dyes insensitive to low-intensity light, such as sunlight, which is troublesome for patients treated with HpD.

1.4.3. Industrial processing

Many opportunities for free-electron lasers exist in the industrial sector as well. One of the most promising is microfabrication of semiconductor circuits. In the conventional process, the circuit is imaged from a mask onto the surface of a silicon chip covered with a thin layer of SiO_2 and a layer of photoresist. When the photoresist is exposed to light it is either hardened of softened by the photochemical effect of the light and can be selectively removed by a chemical bath. Subsequently, the SiO_2 can be etched from the exposed areas, where the photoresist has been removed, to form the image of the microcircuit on the silicon chip. At the present time, shorter wavelengths are being exploited to extend this technology to circuit features with dimensions smaller than than one micrometer, and synchrotron radiation sources are being built for this purpose. In the future, however, these processes are likely to be replaced by new technologies which are just now emerging. For example, it is possible to use a scanning laser to print the pattern on the photoresist by direct writing, the way a laser prints the pattern in a computer laser printer. With sufficient laser intensity, the resist can be removed directly by the laser without the need for chemically developing the image[53]. In fact, with sufficiently high laser intensity, the SiO_2 layer can be removed directly by the laser. It is even possible to use an ultraviolet laser to initiate chemical reactions in a gas or liquid over the surface of the silicon chip to deposit the microcircuit on the chip[54]. By tuning the laser wavelength and/or varying the composition of the gas or liquid, it should be possible to construct various features of the microcircuit. The potential advantages of free-electron lasers in terms of high intensity and tunability at very short wavelengths for applications of this type are manifest.

In addition to microfabrication of semiconductor integrated circuits, lasers may be used for the microfabrication of microminiature thermionic vacuum-tube circuits. Almost as small as semiconductor integrated circuits, the thermionic circuits operate in environments too harsh for semiconductors. In addition to electrical circuits, a variety of miniature mechanical devices, such as gas chromatographs, can be microfabricated from silicon.

The power and wavelength tunability of free-electron lasers makes them intrinsically well suited to large-scale chemical processing. The issue is generally one of the cost of a laser photochemical process compared with the conventional process[55]. This is illustrated in Fig. 1.18, where the cost of commercially available chemicals is compared with the cost of laser photons[56]. To calculate the cost per ton of laser photons, the following assumptions have been used:

(1) The free-electron laser device has a capital cost of $40/W to $400/W, amortized over 5 years (1.6×10^8 s), which corresponds to 2.5×10^{-7}/J to 2.5×10^{-6}/J for photons. Electricity and operating costs are ignored.

(2) Each molecule of product requires one 4-eV (300-nm) photon for its formation, and has an atomic weight of 100, which corresponds to 6.02×10^{27} molecules per metric ton.

It is clear from Fig. 1.18 that a wide variety of specialty chemicals and isotopes are, at least in principle, economically accessible to photochemical production with free-electron lasers. However, to address the interesting chemicals in the lower right-hand corner of the chart, the process must be highly leveraged. That is, each (expensive) photon must produce a large number of product molecules. In other words, the quantum yield (product molecules per photon) must be large, generally greater than 1000. This is especially true when it is recognized that for large-scale chemical processes the cost of the raw materials typically represents 60% to 80% of the cost of the final product. Thus, the laser cost must be held to a small fraction of the value of the final product unless laser processing increases the yield and reduces the cost of raw materials, which has been observed.

Two important processes have been identified in which such a large quantum yield is possible. In the first process, a laser is used to purify chemical compounds by removing a few impurity molecules from a large number of desired molecules. A good example is the purification of silane (SiH_4)[57]. High-purity silane is the starting material for the production of certain types of semiconductors and optical materials. Experiments with an excimer laser at 190 nm have demonstrated that this method produces extremely pure product material with high quantum yield.

A second class of photochemical processes having high quantum yield is that of laser-initiated chain reactions. In this case, a single photon is used to initiate a chain reaction which may yield thousands of molecules of product before the chain is broken. A good example of such a process is offered by the synthesis of vinyl chloride[58]. Vinyl chloride is the starting material for the production of polyvinyl chloride (PVC). In experiments with ultraviolet lasers the quantum yield has been observed to be as high as 20,000, and pilot plants are now under consideration in several countries.

1.4.4. Other applications

At the very highest power levels, other applications become possible. Two that come immediately to mind are nuclear fusion and strategic defense. Fusion as a source of commercial power is being pursued along two lines, magnetic-confinement fusion and inertial-confinement fusion. In magnetic-confinement fusion, the reacting plasma is contained in a magnetic field. To initiate the thermonuclear reaction, a microwave or far-infrared radiation source is used to heat the plasma. Experiments are now underway at Lawrence Livermore National Laboratory to develop a free-electron laser for this purpose[59]. In inertial-confinement fusion, extremely high power, short pulses of laser light are used to compress small targets of deuterium and tritium to densities and temperatures at which thermonuclear reactions occur. Free-electron lasers may also be used for this application[60].

In the end, military applications may present the most demanding and challenging opportunities of all. The ability of lasers to transport energy over long distances in almost no time makes them attractive for strategic defense against missiles. Using lasers, the missiles can be attacked when they are most vulnerable, in their boost phase. To do this, the laser may be sited on the ground, with mirrors in orbit to redirect the beam over the horizon to the target, or the laser itself may be sited in orbit. Tremendous amounts of laser power - estimates are as high as 10 GW - will be required to defend against a large number of simultaneously launched missiles[61]. The unique ability of free-electron lasers to operate at high power and short wavelength has made them a leading candidate for this application. Nevertheless, it will ba a long time before even free-electron lasers achieve the power levels required to defend against a massive attack[62]. However, it may not be necessary to defend against large numbers of simultaneously launched missiles. As part of a broad disarmament program, relatively small defensive systems might make it safe to negotiate the complete elimination of nuclear missiles by protecting each side against the possibility of being held hostage by a few missiles kept hidden by the other side.

1.5. Conclusion

Free-electron lasers represent a complete departure from conventional lasers, and they offer performance not available from other sources. By varying the electron energy it is possible to tune the wavelength over a broad range, and to operate at wavelengths where no other lasers exist. Equally important, by avoiding the cooling problems associated with conventional lasers and using high-power accelerator technology, it is possible to operate free-electron lasers at extreme power levels. Clearly, there are manifold applications for lasers with these qualifications, and two new free-electron lasers for the near-infrared and visible regions of the spectrum are being developed to support research in medicine and materials science at Vanderbilt University[63] and the National Bureau of Standards[64]. These will supplement the facilities at Santa Barbara and Stanford. Unfortunately, free-electron lasers are expensive, especially in small sizes, and until cheaper free-electron lasers are developed only a limited subset of all the possible applications can be addressed. Nevertheless, it is clear that free-electron lasers are developing rapidly. As they do so, they will play an increasingly important role in a broad variety of fields.

2. BASIC THEORY

In this chapter we develop the basic theory of free-electron lasers. We assume that, apart from the wiggles, the electrons move parallel to the wiggler axis, and in developing the Maxwell equations we ignore the diffraction of the laser beam transverse to the wiggler axis. This theory is more useful than it might seem, and in any event forms the basis for more elaborate theories.

We begin this Chapter with a discussion of the basic physics of the interaction of the electrons with the wiggler and optical fields, and with one another. We then develop the one-dimensional theory for the dynamics of the electrons and the optical field in a free-electron laser.

2.1. Basic Physics

In Chapter 1 we discussed the basic theory of free-electron lasers in terms of the classical motion of individual electrons. Of course, like all things in nature, free-electron lasers are quantum mechanical devices and under certain circumstances the quantum effects may not be ignored. Similarly, under certain circumstances the electrons do not behave individually, but display collective effects. Both quantum and collective effects may generally be neglected for free-electron lasers in the optical regime, which is the focus of this book. However, it is worthwhile examining the limits of validity of these approximations and the nature of the effects which appear when they are violated.

To see the nature of the quantum and collective effects most clearly, it is convenient to look at the electron-wiggler interaction in the frame of reference moving with the mean motion of the electron. In this reference frame the electron motion is nonrelativistic, at least as long as the magnetic field is not strong enough to deflect the electron transversely with a relativistic velocity. This corresponds to the limit when the magnetic field of the wiggler satisfies the restriction

$$\frac{eB_W \lambda_W}{2\pi mc} \ll 1. \tag{2.1}$$

This approximation is not necessary, and does not change the conclusions, but it simplifies the algebra. In many free-electron lasers the quantity $eB_W \lambda_W / 2\pi mc$ is of the order of unity. In the electron reference frame, the wiggler magnetic field is coming toward the electron with the velocity

$$v = \beta_z c = c/(1-\gamma_z^{-2})^{1/2}, \qquad (2.2)$$

and appears in the electron reference frame as an oscillating magnetic field. To satisfy the Maxwell equations, this field has associated with it an oscillating electric field, so the wiggler field appears to the electron to be an oscillating transverse electromagnetic field, that is, an optical beam. This is known as the Weizsacker-Williams approximation[68]. Because of the relativistic velocity of the electrons in the z direction, the wiggler period is shortened in the electron frame by the Lorentz contraction, so the electromagnetic field has the wavelength

$$\lambda = \lambda_W/\gamma_z. \qquad (2.3)$$

In this reference frame, the electrons are observed to oscillate transversely in the oncoming electromagnetic field of the wiggler. Actually, if the wiggler field is strong enough, the magnetic field of the oncoming electromagnetic wave interacts with the transverse motion so that the electrons are accelerated slightly in the longitudinal direction. Because of this they execute a "figure-eight" motion as indicated in Fig. 2.1. In the electron reference frame, the radiation emitted by the electrons may be thought of as backward scattering of the oncoming wiggler "radiation."

2.1.1. Quantum effects

Quantum phenomena appear in the theory of free-electron lasers both in the quantization of the electromagnetic field and in the wave nature of the electrons. The quantization of the electromagnetic field manifests itself in the Compton effect[69], which is observed in the scattering of X-rays from electrons. Compton observed that the scattered X-rays appear at a wavelength longer than the incident X-rays by an amount which depends on the angle at which the X-rays are scattered. This is explained by treating the X-rays as photons with momentum and energy given by the expressions

$$p_X = (1/2\pi)hk, \qquad (2.4)$$

$$E_X = (1/2\pi)h\omega, \qquad (2.5)$$

where h is Planck's constant, $k = 2\pi q/\lambda$ the wave vector for the X-ray, with q a unit vector in the direction of propagation of the X-ray, and ω is the frequency of the X-ray. When the X-ray is scattered from the electron, the total energy and momentum of the X-ray photon and the electron are conserved. The wavelength shift observed by Compton is due to the momentum absorbed by the electron as it recoils from the incident photon. In the nonrelativistic case, the velocity of recoil is given by the expression

$$v_R \approx 2p_X/m. \qquad (2.6)$$

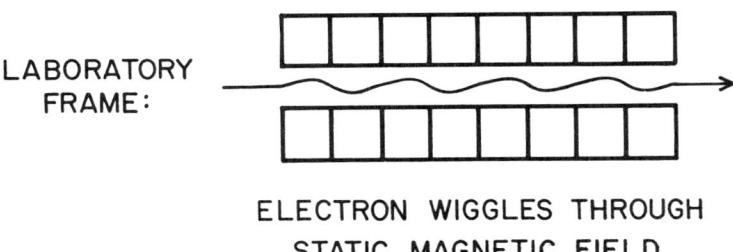

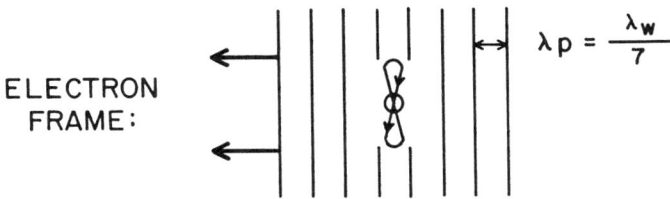

Fig. 2.1. Viewed in the electron frame of reference, the wiggler appears to be an oncoming electromagnetic field with a wavelength λ_w/γ_z, and the electrons oscillate vertically with a slight "figure-eight" motion.

If the electron recoils a distance comparable to or greater than the wavelength of the oncoming wave in the time it takes for the wave to move past the electron,

$$t = N_W \lambda/c, \qquad (2.7)$$

then quantum effects cannot be ignored. If, specifically, we require that $2\pi v_R t/\lambda$ be small compared with unity in the classical limit, then the classical limit is valid so long as

$$\lambda \gg 4\pi \lambda_C N_W, \qquad (2.8)$$

where the Compton wavelength is defined by the expression

$$\lambda_C = h/mc = 2.42 \times 10^{-12} \text{ m}. \qquad (2.9)$$

We may use the resonance condition derived in Chapter 1,

$$\lambda_W = 2\gamma_z^2 \lambda_L, \qquad (2.10)$$

to express Eq. (2.8) in the form

$$\lambda_L \gg 2\pi N_W \lambda_C/\gamma_z = 1.52 \times 10^{-11} \, N_W/\gamma_z \text{ (m)}. \qquad (2.11)$$

Because the Compton wavelength is so small, quantum effects are not observable even for wavelengths as short as the X-ray region ($\lambda \approx 10^{-10}$ m) unless N_W is very large and γ_z is small. This can occur, for example, when an intense laser beam is used for the wiggler, with a length of 10^4 periods, and the electrons have a low energy[70]. Otherwise, quantum effects can be ignored.

To be totally correct, a theoretical description of the free-electron laser must also treat the electron, as well as the field, quantum mechanically, that is, as a wave packet. Due to the uncertainty relations for the electron position and momentum, the electron wave packet spreads with time. When the width of the wave packet becomes comparable to or larger than the wiggler wavelength, the electron can no longer be treated as a classical particle[71]. To see when this is important, we consider a wave packet initially localized to the width Δz_0. According to the Heisenberg relations, the corresponding uncertainty in the longitudinal momentum is given by the formula

$$\Delta p_z \Delta z \geq h/2\pi. \qquad (2.12)$$

For a Gaussian wave packet, the width after a time t is given by the equation[72]

$$\Delta z^2(t) = \Delta z_0^2 + \left(\frac{\Delta p_z t}{m}\right)^2 = \Delta z_0^2 + \left(\frac{ht}{2\pi m \Delta z_0}\right)^2. \qquad (2.13)$$

Differentiating with respect to the initial width of the wave packet, we find that the minimum width of the wave packet at time t is given by the expression

$$\Delta z^2(t) = ht/\pi m. \qquad (2.14)$$

If, as a condition for the validity of the classical approximation, we require that $2\pi\Delta z/\lambda$ be small at the end of the wiggler, then we obtain the restriction

$$\lambda \gg 4\pi\lambda_c N_W. \qquad (2.15)$$

This is identical to the result (2.8) which we obtained by requiring that the effects of field quantization be small.

These arguments in favor of the classical approximation show that the mean values of all observables are correctly computed by classical mechanics. Somewhat more complicated arguments are required to show that the fluctuations about the mean are likewise dominated by classical effects, such as shot noise[73]. This is important because the startup of laser oscillators and certain regimes of laser operation depend on noise. It may also be shown that noise limits the practical gain of free-electron laser amplifiers[74]. Since noise can be treated classically in free-electron lasers, these regimes of operation can be described within the classical approximation. On the other hand, because the fluctuations are dominated by classical shot effects, the photon fluctuations of free-electron lasers do not obey Poisson statistics, as they do in conventional lasers. However, this is generally of little consequence.

In the remainder of this book we ignore quantum effects and address the classical regime. This is appropriate for all free-electron lasers which have been built so far, and will be satisfactory for all free-electron lasers in the future with the exception, perhaps, of those operating at X-ray wavelengths using low-energy electron beams with laser beams as wigglers.

2.1.2. Collective effects

When the electron density in the electron beam is sufficiently high, the electrons interact with one another and execute collective motions. In the free-electron laser these motions take the form of space-charge waves, that is, longitudinal waves oscillating at the plasma frequency. In this case it is necessary to solve the Vlasov equation together with the Maxwell equations, using methods which are conventional in plasma physics[75]. The results of these calculations are generally expressed in the form of dispersion relations, which relate the real (wavelength) and imaginary (gain) parts of the propagation constant for the waves to the optical frequency. It is found that when collective effects are important, the gain is enhanced by the generation of space-charge waves in the electron beam, but that the wavelength is shifted to longer values by the effect of the space-charge oscillations. Because of the wavelength shift, the collective regime is often called (somewhat confusingly) the Raman regime.

Intuitively, it is clear that collective oscillations are not important if the number of space-charge oscillations executed by the electrons travelling down the wiggler is small compared with unity. To calculate the plasma frequency in the rest frame of the electron mean motion, it is necessary to allow for the Lorentz contraction of the moving electron beam when it is observed in the laboratory reference frame. In the electron rest frame, therefore, the electron density is reduced by the factor γ_z. The plasma frequency in the electron frame is then given by the formula[76]

$$\omega_p^2 = \frac{n_e e^2}{\epsilon_0 \gamma_z m}, \tag{2.16}$$

where n_e is the electron density in the laboratory frame, e the electron charge, m the electron mass, and ϵ_0 the permittivity of free space. The number of collective oscillations completed by the electrons as the wiggler passes by is therefore given by the expression

$$N = \frac{\omega_p t}{2\pi} = \left[\frac{n_e e^2}{\epsilon_0 \gamma_z m}\right]^{1/2} \frac{\lambda N_W}{2\pi c} = \left[\frac{n_e e^2}{\epsilon_0 \gamma_z m}\right]^{1/2} \frac{\lambda_W N_W}{2\pi \gamma_z c}, \tag{2.17}$$

where we have used Eq.(2.3) to express the result in terms of λ_W. If we wish to ignore collective effects, then it is necessary that N be small compared with unity. This leads immediately to the restriction

$$J_e = n_e ec \ll \frac{2\pi \epsilon_0 \gamma_z^3 mc^3}{eL_W^2} = 8500 \frac{\gamma_z^3}{L_W^2} \text{ (A/m}^2\text{)} \tag{2.18}$$

which must be satisfied by the current density J_e, where $L_W = N_W \lambda_W$ is the length of the wiggler in the laboratory frame. This relation may be cast in a more useful form by recognizing that for a Gaussian optical beam whose Rayleigh length is about half the wiggler length, the area of the beam is given approximately by the formula

$$A \approx L_W \lambda_L / 4. \tag{2.19}$$

Eq. (2.18) may therefore be expressed as the restriction

$$I = \frac{n_e ec \lambda_L L_W}{4} \ll \frac{\pi \epsilon_0 mc^3 \gamma_z}{4 e N_W} = 1000 \frac{\gamma_z}{N_W} \text{ (A)}, \tag{2.20}$$

where I is the total electron-beam current inside the laser beam. For a typical free-electron laser operating with a wiggler period of 25 mm, an electron energy of 20 MeV ($\gamma_z = 40$), and a wiggler length of 1 m, the limiting current is about 1000 A. Clearly, collective effects diminish rapidly in importance as the electron energy increases for a given wiggler length. Care must be exercised, however, since the wiggler length of practical devices generally increases sharply with increasing electron energy. When an intense laser beam is used as the wiggler, Eq. (2.20) is generally not satisfied. For example, for a laser operating at a wavelength of 10 nm with a 10-mm long wiggler having a period of 1 μm ($\gamma_z \approx 5$), the critical current is about 500 mA[77].

However, the fact that Eq. (2.20) is violated does not insure that collective oscillations occur. It is well known in plasma physics that when the the wavelength of a space-charge wave is comparable to or less than the Debye length, electron thermal motions damp the oscillations. For longitudinal oscillations, the Debye length in the electron reference frame is given by the formula[78]

$$\lambda_D = \left(\frac{\epsilon_0 \gamma_z KT_e}{n_e e^2}\right)^{1/2}, \qquad (2.21)$$

where n_e is the electron density in the laboratory frame, K is Boltzmann's constant, and the longitudinal temperature is defined by the expression

$$KT_e = m\langle u_z'^2 \rangle, \qquad (2.22)$$

in which $\langle u_z'^2 \rangle$ is the mean square longitudinal velocity in the electron rest frame. To see how this temperature is related to the energy spread of the electron beam in the laboratory frame, we use the relativistic velocity addition law[79]

$$u_z = \frac{u_z' + v}{1 + vu_z'/c^2}, \qquad (2.23)$$

where u_z is the velocity in the laboratory frame and v is the velocity of the electron frame, together with the Einstein formula

$$\gamma_z^2 = 1/(1-\beta_z^2), \qquad (2.24)$$

to show that for $\gamma_z \gg 1$ and $\delta\gamma/\gamma_z \ll 1$

$$KT_e \approx mc^2 \frac{\langle \delta\gamma^2 \rangle}{\gamma_z^2}. \qquad (2.25)$$

Provided that the plasma frequency is not too large, the period of the space-charge modulation is approximately that of the wiggler in the electron rest frame. Combining these results we obtain the restriction

$$\lambda \ll \left[\frac{\epsilon_0 mc^2 \langle \delta\gamma^2 \rangle}{n_e e^2 \gamma_z}\right]^{1/2}, \qquad (2.26)$$

which must be satisfied if collective oscillations are to be damped. Using the resonance condition (2.10) and the laser beam area (2.19), we can recast this equation in the more convenient form

$$I \ll \frac{\epsilon_0 mc^3}{8e} \gamma_z N_W \frac{\langle \delta\gamma^2 \rangle}{\gamma_z^2} = 170 \; \gamma_z N_W \frac{\langle \delta\gamma^2 \rangle}{\gamma_z^2}. \qquad (2.27)$$

For the 20-MeV laser discussed above, if we assume an energy spread $\delta\gamma/\gamma \approx 10^{-2}$, the limiting current is about 30 A, so collective oscillations are only marginally damped in most practical devices. For the 10-nm laser discussed above, if we assume an energy spread of 10^{-4}, the limiting current is about 100 mA, so Eq. (2.27) is not satisfied. In this case, however, the mean distance between the electrons at the limiting current $((n_e/\gamma_z)^{-1/3})$ is of the order of 0.2 μm in the electron reference frame. This is larger than the wavelength $(2\gamma_z \lambda_L \approx 0.1$ μm$)$ in that reference frame, and therefore a higher current density is required to support collective oscillations.

When space-charge waves are important, the first-order effect of the space-charge fields is to decrease the performance of the free-electron laser by resisting the bunching of the electrons[80]. For most free-electron lasers, however, space-charge forces have only a small effect on the performance. Therefore, in the remainder of this book we ignore the effects of plasma oscillations, and focus our attention on the single-particle regime, which is sometimes called the Compton regime. For a thorough discussion of the collective regime, the interested reader is referred to the excellent text by Marshall[81].

2.2. Electron Dynamics

As in all of classical dynamics, the interaction of electrons and radiation fields is divided into two problems, the first being the motion of the electrons in the radiation field and the second being the evolution of the radiation field of the electrons. As the electrons move, the fields are updated to include the effects of the electron radiation so that the electrons move in the corrected fields. In this way, the energy lost by the electrons is properly accounted for in the evolution of the fields. In carrying out the solution of the two problems, it is arbitrary whether one begins with the electron motions or the evolution of the optical field. Since it seems, somehow, more intuitive to discuss the motions of the

electrons in a specified electromagnetic field, we begin with the
problem of the electron motions.

2.2.1. Theoretical approach

Within the classical approximation, two theoretical approaches
can be used to describe the electron dynamics of free-electron
lasers. The first, pioneered by Scully and his co-workers, describes
the electrons in terms of a velocity distribution function which
depends on the spatial coordinates and time[82]. To find the
distribution function, it is necessary to solve the relativistic
Boltzmann equation, together with the Maxwell equations. This
approach is necessary when collective interactions are important.
When such interactions are not important, the electrons can be
treated individually. Their interaction with each other takes place
only through the optical field, which must be computed self-
consistently. This latter approach, which was pioneered by Colson
and his co-workers, is conceptually simpler, and has proved to be
more powerful for computing the saturated behavior of free-electron
lasers both analytically [83] and with computers[84]. Therefore, it
enjoys broader currency for free-electron lasers operating at short
wavelengths, where collective effects are not important, and we adopt
it here. The effects of space charge can also be included in this
formalism[85].

In describing the motions of the electrons, it is possible, in
principle, to follow the individual wiggles of the electrons as they
travel down the wiggler. In fact, powerful computer codes developed
originally for plasma physics problems have been used for just this
purpose[86]. However, this approach becomes intractable for even the
most powerful computers when the electron energy exceeds 10 MeV,
owing to the great disparity between the length scales of the wiggler
and the optical field, which are related by γ^2. Fortunately, it is
not necessary to solve the equations of motion in such detail.
Because the motion of the electrons in the wiggler consists, in
general, of many wiggles, and neither the electron energy nor the
optical field change very much over the length of a few wiggles, it
is possible to find equations which average over the individual
wiggles of the electrons but describe quite accurately the secular
development of the electron mean energy and phase and the optical
field. This approach is described in the following sections.

2.2.2. The Hamilton equations

Since we ignore the effects of space charge, it is convenient
to describe the fields in terms of a vector potential **A**, and use the
Coulomb gauge so that the scalar potential Φ is equal to zero. In
the single-particle picture of the free-electron laser, the motion of
an electron is described by the relativistic Hamiltonian

$$H = \gamma mc^2 = [c^2(P-eA)^2 + m^2c^4]^{1/2}, \quad (2.28)$$

where γ is the total energy of the electron. The canonical momentum is given by the expression

$$P = p + eA, \quad (2.29)$$

in which p is the electron momentum and

$$A = A_W + A_L \quad (2.30)$$

represents the combined vector potential of the wiggler and laser fields.

To simplify the solution, we restrict ourselves to the one-dimensional case and limit the fields to transverse components only. For a magnetostatic wiggler the fields are assumed to have the form

$$A_W = A_W(z), \qquad A_W \cdot z = 0, \quad (2.31)$$

$$A_L = A_L(z,t), \qquad A_L \cdot z = 0, \quad (2.32)$$

where z is a unit vector in the z direction. Clearly, the wiggler field described by Eq. (2.31) cannot satisfy the Maxwell equations (specifically, the Laplacian of A_W cannot vanish as required) unless A_W is identically constant. Near the axis Eq. (2.31) is satisfactory, however. We explore the consequences of the off-axis variation of A_W in Chapter 3.

To simplify the solution further, we assume that the vector potential of the laser field is small compared with that of the wiggler field:

$$A_L/A_W \ll 1. \quad (2.33)$$

This is not equivalent to assuming that the magnetic field of the laser is small compared with that of the optical field, which is not necessarily true in practical cases. The vector potential of a periodic field is of the order of

$$A = B\lambda, \quad (2.34)$$

where λ is the period of the field. Since the period of the laser field is smaller than that of the wiggler field by a factor of the order of γ^2, according to the resonance condition, the magnetic vector potential of the laser field is correspondingly smaller than that of the wiggler field.

The electron energy is found from the Hamilton equation:

$$\frac{dH}{dt} = mc^2 \frac{d\gamma}{dt} = \frac{\partial H}{\partial t} = -\frac{e}{\gamma m}(P-eA) \cdot \frac{\partial A}{\partial t}. \tag{2.35}$$

Since $A_L/A_W \ll 1$, but $\partial A_W/\partial t = 0$, this simplifies to the equation

$$\frac{d\gamma}{dt} = \frac{e}{\gamma m^2 c^2}(P-eA_W) \cdot E_L, \tag{2.36}$$

where the electric field of the laser is given by the equation

$$E_L = -\partial A_L/\partial t. \tag{2.37}$$

Since the electric field of the laser is purely transverse, only the transverse part of the canonical momentum survives the dot product in Eq. (2.36). However, the Hamiltonian is independent of the transverse coordinates, so the transverse canonical momenta are conserved. Therefore, the energy equation assumes the simple form

$$\frac{d\gamma}{dt} = \frac{e}{\gamma m^2 c^2}(P_0-eA_W) \cdot E_L, \tag{2.38}$$

where P_0 is the initial value of the electron canonical momentum. To solve Eq. (2.38) it is necessary only to know the longitudinal position of the electron at time t so that the values of A_W and E_L can be determined.

The longitudinal position is found from the Hamilton equation

$$\frac{dz}{dt} = \frac{\partial H}{\partial P_z} = \frac{P_z}{\gamma m}. \tag{2.39}$$

Since the transverse components of the canonical momentum are conserved, we may find the longitudinal component of the electron momentum by rearranging the Hamiltonian in the form

$$P_z = [(\gamma^2-1)m^2c^2 - P_{0x}^2 - P_{0y}^2 + 2eP_0 \cdot A_W - e^2 A_W^2]^{1/2}, \tag{2.40}$$

where we have neglected the laser field compared with the wiggler field. For highly relativistic electrons, the right-hand side of Eq. (2.40) is dominated by the first term. We may therefore expand the square root about this term and substitute into the Hamilton equation to obtain the result

$$\frac{dz}{dt} = c\left[1 - \frac{1}{2\gamma^2}\left(1 + \frac{P_{0x}^2}{m^2c^2} + \frac{P_{0y}^2}{m^2c^2} + \frac{e^2 A_W^2}{m^2c^2} - 2e\frac{P_0 \cdot A_W}{m^2c^2}\right)\right]. \tag{2.41}$$

The first four terms represent motion at the speed corresponding to the energy γ, at the angle corresponding to the initial momentum P_0. The last two terms represent deflections due to the wiggler field, and one or both may oscillate with the wiggler period.

To proceed further, it is necessary to define the wiggler field explicitly. We therefore choose a plane-polarized wiggler with the magnetic field oriented in the y direction, so that the vector potential is given by the equation

$$A_W(z) = 2^{1/2} \underline{A}_W \mathbf{x} \sin(k_W z), \qquad (2.42)$$

where $\underline{A}_W$ is the rms wiggler vector potential, $\mathbf{x}$ is a unit vector in the x-direction, and

$$k_W = 2\pi/\lambda_W \qquad (2.43)$$

is the wiggler wave number. The magnetic field is then given by the equation

$$B_W(z) = 2^{1/2} \underline{B}_W \mathbf{y} \cos(k_W z) = 2^{1/2} \underline{A}_W k_W \mathbf{y} \cos(k_W z), \qquad (2.44)$$

where $\mathbf{y}$ is a unit vector in the y-direction. The optical field is assumed to be a plane wave with the same polarization as the electron motions. The frequency spectrum of the optical field is assumed to be confined to the region near ω_R, so that the field can be described by a wave at the frequency ω_R with smoothly varying phase departures and a smoothly varying envelope. The electric field of the laser is therefore given by the expression

$$E_L(z,t) = 2^{1/2} \underline{E}_L \mathbf{x} \cos(k_R z - \omega_R t + \phi_L), \qquad (2.45)$$

in which

$$k_R = 2\pi/\lambda_R = \omega_R/c \qquad (2.46)$$

is the resonant wave number, and the rms electric field $\underline{E}_L(z,t)$ and phase $\phi_L(z,t)$ are slowly varying functions of z and t. Slow variations, in this context, take place over distances large compared with a wavelength. To simplify the equations still further, we assume that the electrons are initially travelling parallel to the z axis, so that

$$\mathbf{P}_0 = p_0 = p_0 \mathbf{z}. \qquad (2.47)$$

The effects of off-axis motions of the electrons are discussed in Chapter 3.

The longitudinal position then satisfies the equation

$$\frac{dz}{dt} = c\left[1 - \frac{1}{2\gamma^2}(1+a_W^2)\right] + \frac{ca_W^2}{2\gamma^2}\cos(2k_W z), \qquad (2.48)$$

where the dimensionless vector potential is defined by the expression

$$a_W = \frac{e\underline{A}_W}{mc} = \frac{e\underline{B}_W \lambda_W}{2\pi mc}. \qquad (2.49)$$

For λ_W small compared with distances on which the energy changes significantly, the motion described by Eq. (2.48) consists of a mean motion plus a longitudinal oscillation. We may therefore express the motion in the form

$$z = z_0 + z', \qquad (2.50)$$

where the mean motion satisfies the equation

$$\frac{dz_0}{dt} = c\left[1 - \frac{1}{2\gamma^2}(1+a_W^2)\right]. \qquad (2.51)$$

The oscillations satisfy the equation

$$\frac{dz'}{dt} = \frac{ca_W^2}{2\gamma^2}\cos[2k_W(z_0+z')]. \qquad (2.52)$$

A solution to this equation may be obtained in the form of a power series in the quantity $1/\gamma^2$. To first order in $1/\gamma^2$ the oscillations are described by the formula

$$z' = \frac{a_W^2}{4\gamma^2 k_W}\sin(2k_W z_0). \qquad (2.53)$$

For the plane-polarized fields described by Eqs. (2.42) and (2.45), with the initial conditions of Eq. (2.47), the energy equation (2.38) assumes the form

$$\frac{d\gamma}{dt} = -\frac{e^2}{\gamma m^2 c^2}\underline{A}_W \underline{E}_L [\sin(\psi+\phi_L) - \sin(\psi+\phi_L - 2k_W z)], \qquad (2.54)$$

where we have expanded the product of the sine and cosine, and the phase angle between the wiggler and laser fields is defined by the expression

$$\psi = k_W z + k_R z - \omega_R t. \qquad (2.55)$$

If we substitute Eq. (2.50) into this equation, we find that for $k_W \ll k_R$ the phase may be expressed in the form

$$\psi = \psi_0 + \psi', \tag{2.56}$$

where for $k_W \ll k_R$ the mean phase satisfies the equation

$$\frac{d\psi_0}{dt} = ck_W - ck_R \frac{1+a_W^2}{2\gamma^2} \tag{2.57}$$

and the phase oscillations are given by the formula

$$\psi' = \frac{k_R}{k_W} \frac{a_W^2}{4\gamma^2} \sin(2k_W z_0). \tag{2.58}$$

If we define resonance by the criterion that the phase be stationary, then the resonant energy is given by the equation

$$\gamma_R^2 = \frac{1+a_W^2}{2} \frac{k_R}{k_W} = \frac{1+a_W^2}{2} \frac{\lambda_W}{\lambda_R}, \tag{2.59}$$

which agrees with the resonance condition derived in Chapter 1. The mean phase then satisfies the equation

$$\frac{d\psi_0}{dt} = ck_W \left[1 - \frac{\gamma_R^2}{\gamma^2}\right]. \tag{2.60}$$

Substituting Eq. (2.56) into the energy equation, Eq. (2.54), and expanding the sines, we find, to lowest order in $1/\gamma^2$, that

$$\frac{d\gamma}{dt} = -\frac{e^2}{\gamma m^2 c^2} A_W E_L \{\sin(\psi_0+\phi_L) \cos[\xi \sin(2k_W z_0)]$$
$$+ \cos(\psi_0+\phi_L) \sin[\xi \sin(2k_W z_0)]$$
$$- \sin(\psi_0+\phi_L) \cos[\xi \sin(2k_W z_0) - 2k_W z_0)]$$
$$- \cos(\psi_0+\phi_L) \sin[\xi \sin(2k_W z_0) - 2k_W z_0)]\}, \tag{2.61}$$

where

$$\xi = \frac{1}{2} \frac{a_W^2}{1+a_W^2} \frac{\gamma_R^2}{\gamma^2}. \tag{2.62}$$

For long wigglers, in which

$$N_W = L_W/\lambda_W \gg 1, \qquad (2.63)$$

where L_W is the length of the wiggler, γ and $\underline{E}_L$ change only slightly over a wiggler period. Therefore, we may average the oscillating terms over a wiggler period. When this is done, the second and fourth terms vanish by symmetry. To evaluate the averages of the first and third terms we may use the integral representation of the Bessel functions[87] given by the expression

$$J_n(z) = \frac{1}{\pi}\int_0^\pi \cos[z\sin(\theta) - n\theta]\, d\theta. \qquad (2.64)$$

We then obtain the equation

$$\frac{d\gamma}{dt} = -\frac{e^2}{\gamma m^2 c^2}\, \underline{A}_W \underline{E}_L\, [J_0(\xi) - J_1(\xi)]\, \sin(\psi_0 + \phi_L). \qquad (2.65)$$

Because of the slowly varying phase and amplitude approximation for $\underline{A}_W$ and $\underline{E}_L$, we may evaluate these functions using the mean position z_0.

In this same limit of long wigglers, a further simplification is possible. As shown in Chapter 1, the width of the emission spectrum is of the order of $1/N_W$. This is also the width of the resonance in frequency space. The width of the resonance in electron energy space is found by differentiating the resonance equation, Eq. (2.59):

$$\frac{\delta\gamma_R}{\gamma} = \frac{1}{2}\frac{\delta k_R}{k_R} = O(1/2N_W) \ll 1. \qquad (2.66)$$

For electrons close enough to resonance to interact with the optical field, we may use the approximation

$$\frac{\gamma - \gamma_R}{\gamma_R} = O(1/2N_W) \ll 1. \qquad (2.67)$$

To lowest order, then, the energy equation may be expressed in the form

$$\frac{d\gamma}{dt} = -\frac{ea_W}{\gamma_R mc}\, E_L\, [J_0(\xi) - J_1(\xi)]\, \sin(\psi_0 + \phi_L), \qquad (2.68)$$

where

$$\xi = -\frac{1}{2}\frac{a_W^2}{1+a_W^2}. \tag{2.69}$$

Similarly, the position and phase equations simplify to the forms

$$\frac{dz_0}{dt} = c\left[1 - \frac{1+a_W^2}{2\gamma_R^2}\right] \tag{2.70}$$

and

$$\frac{d\psi_0}{dt} = 2ck_W \frac{\gamma-\gamma_R}{\gamma_R}. \tag{2.71}$$

2.3. Optical Dynamics

To complete the theory of a free-electron laser, it is necessary to describe the optical fields generated by the electrons. These are described by the Maxwell equations, using the microscopic currents of the electrons.

2.3.1. The Maxwell equations

In the single-particle regime, as discussed earlier, it is possible to ignore the effects of the space charge of the electron beam. In this case, it is convenient to describe the electromagnetic fields using the magnetic vector potential in the Coulomb gauge. In this case the Maxwell equations may be expressed in the form[88]

$$\nabla^2 \mathbf{A}_L - \frac{1}{c^2}\frac{\partial^2 \mathbf{A}_L}{\partial t^2} = -\mu_0 \mathbf{J}_T, \tag{2.72}$$

where $\mathbf{J}_T$ is the transverse current. As before, the optical field is assumed to be a plane wave polarized in the direction of the electron motions, with a frequency near the resonant frequency ω_R, and a slowly varying amplitude and phase:

$$\mathbf{A}_L = 2^{1/2}\underline{A}_L \mathbf{x}\, \sin(k_R z - \omega_R t + \varphi_L), \tag{2.73}$$

where $\underline{A}_L$ and ϕ_L are slowly varying functions of z and t. In the slowly varying amplitude and phase approximation, the electric field is given by the expression

$$\mathbf{E}_L = -\frac{\partial \mathbf{A}_L}{\partial t} \approx 2^{1/2}\omega_R \underline{A}_L \mathbf{x}\, \cos(k_R z - \omega_R t + \phi_L), \tag{2.74}$$

and we may identify ϕ_L with the previously defined phase and

$$\underline{E}_L = \omega_R \underline{A}_L \qquad (2.75)$$

with the previously defined electric field. Substituting Eq. (2.73) into the Maxwell equation, and ignoring second derivatives and squares of derivatives of the slowly varying functions $\underline{A}_L$ and ϕ_L, we obtain the equation

$$\frac{D\underline{E}_L}{Dt}\cos(k_R z - \omega_R t + \phi_L) - \underline{E}_L \frac{D\phi_L}{Dt}\sin(k_R z - \omega_R t + \phi_L)$$

$$= -\frac{1}{2^{3/2}\epsilon_0} J_T \cdot x, \qquad (2.76)$$

where the total, or convective, derivative is defined by the expression

$$\frac{D}{Dt} = c\frac{\partial}{\partial z} + \frac{\partial}{\partial t}. \qquad (2.77)$$

The transverse current is given by the sum over all electrons of the point currents:

$$J_T \cdot x = \sum_n ev_x \, \delta(x-x_n) \, \delta(y-y_n) \, \delta(z-z_n), \qquad (2.78)$$

where the transverse velocity is found from the Hamiltonian (Eq. (2.28)) by means of the equation

$$v_x = \frac{\partial H}{\partial P_x} = \frac{P_x - eA_x}{\gamma m} \approx -2^{1/2} \frac{e\underline{A}_W}{\gamma_R m} \sin(k_W z) \qquad (2.79)$$

for near-resonant electrons entering the wiggler parallel to the axis ($P_x = P_{0x} = 0$) when the laser vector potential is small compared with the wiggler field.

The current in Eq. (2.78) is clearly neither smooth nor one-dimensional in its microscopic form. However, the macroscopic currents, which are driven by the macroscopic fields, are both smooth and one-dimensional. To calculate these macroscopic fields, we average the microscopic currents over a volume V which is large compared with the wavelength and contains a large number of electrons, but is small compared with the distance over which the field varies significantly. To do this, we substitute the expression for the microscopic current into Eq. (2.76). If we then multiply by the factor $\cos(k_R z - \omega_R t + \phi_L)$ and integrate over V, the average of the

$\cos^2$ on the left is 1/2, so the first term survives, while the average of the sin cos is 0, so the second term vanishes. We therefore obtain the equation

$$\frac{DE_L}{Dt} = \frac{a_w J_e}{\epsilon_0 \gamma_R} \langle \sin(k_w z) \cos(k_R z - \omega_R t + \phi_L) \rangle, \qquad (2.80)$$

where the electron-beam current is

$$J_e = \frac{ecN_V}{V}, \qquad (2.81)$$

in which N_V is the number of electrons in the Volume V. The brackets $\langle \rangle$ indicate the average over all the electrons in the volume V, defined by the expression

$$\langle \sin(k_w z) \cos(k_R z - \omega_R t + \phi_L) \rangle = \frac{1}{N_V} \Sigma_V \sin(k_w z_n) \cos(k_R z_n - \omega_R t + \phi_L)$$

$$= \frac{1}{2N_V} \Sigma_V [\sin(\psi_n + \phi_L) - \sin(\psi_n + \phi_L - 2k_w z_n)], \qquad (2.82)$$

in which Σ_V is the sum over all the electrons in the volume V and ψ_n is to be evaluated at z_n. The phase ϕ_n is slowly varying and is the same for all electrons in the volume V. As discussed above, the phase ψ consists of a mean phase plus a small oscillation. Because of the slow variation of the electric field, this small oscillation may be averaged over a wiggler period. When this is done as before (see Eqs. (2.54)-(2.65)), we obtain the result

$$\frac{DE_L}{Dt} = \frac{a_w J_e}{2\epsilon_0 \gamma_R} [J_0(\xi) - J_1(\xi)] \langle \sin(\psi_0 + \phi_L) \rangle. \qquad (2.83)$$

In the same way, if we multiply Eq. (2.76) by $\sin(k_R z - \omega_R t + \phi_L)$, integrate over the volume V, and then average over a wiggler period we obtain the equation

$$E_L \frac{D\phi_L}{Dt} = \frac{a_w J_e}{2\epsilon_0 \gamma_R} [J_0(\xi) - J_1(\xi)] \langle \cos(\psi_0 + \phi_L) \rangle. \qquad (2.84)$$

2.3.2. Conservation, gain, and refraction

The conservation of energy follows from Eqs. (2.68) and (2.83). If we examine the rate of change of the total energy in a small volume moving along the wiggler at the speed of light, and treat the electrons as moving along the wiggler at nearly the speed of light, we obtain the equation

$$\frac{D}{Dt}(n_e mc^2 \langle\gamma\rangle + \epsilon_0 \underline{E}_L^2) = -\frac{a_W J_e}{\gamma_R} \underline{E}_L [J_0(\xi) - J_1(\xi)] \langle\sin(\psi_0 + \phi_L)\rangle$$
$$+ \frac{a_W J_e}{\gamma_R} \underline{E}_L [J_0(\xi) - J_1(\xi)] \langle\sin(\psi_0 + \phi_L)\rangle = 0. \quad (2.85)$$

In other words, the energy lost by the electrons appears directly in the optical field. There is no exchange of energy with the wiggler field, which must be true since charged particles cannot be accelerated or decelerated by a static magnetic field.

If we define the differential gain by the expression

$$\frac{DG}{Dt} = \frac{1}{\underline{E}_L^2} \frac{d\underline{E}_L^2}{dt} = \frac{a_W J_e}{\epsilon_0 \underline{E}_L \gamma_R} [J_0(\xi) - J_1(\xi)] \langle\sin(\psi_0 + \phi_L)\rangle, \quad (2.86)$$

we see that the gain is due to the average of the sine of the phase. The average of the cosine of the phase is responsible for the index of refraction of the laser medium. To see this, we observe that the phase velocity of the wave described by Eq. (2.45) is to be found from the equation

$$\left[v_{ph} \frac{\partial}{\partial z} + \frac{\partial}{\partial t}\right](k_R z - \omega_R t + \phi_L)$$

$$= (v_{ph} - c)k_R + (v_{ph} - c)\frac{\partial \phi_L}{\partial z} + \frac{D\phi_L}{Dt} = 0. \quad (2.87)$$

If, using the slowly varying phase approximation, we ignore $\partial\phi_L/\partial z$ compared with k_R, we find that the index of refraction is given by the formula

$$n - 1 \approx 1 - \frac{v_{ph}}{c} = \frac{1}{\omega_R} \frac{D\phi_L}{Dt}$$

$$= \frac{a_W J_e}{2\omega_R \epsilon_0 \underline{E}_L \gamma_R} [J_0(\xi) - J_1(\xi)] \langle\cos(\psi_0 + \phi_L)\rangle, \quad (2.88)$$

for $n-1 \ll 1$. Comparing this with Eq. (2.86), we see that the index of refraction is related to the gain by the formula

$$n-1 = \frac{1}{2\omega_R} \frac{\langle\cos(\psi_0+\phi_L)\rangle}{\langle\sin(\psi_0+\phi_L)\rangle} \frac{DG}{Dt}. \qquad (2.89)$$

The factor of 2 appears because G is the intensity gain rather than the amplitude gain. As shown by Eq. (2.88), the sign of n−1 depends on $\langle\cos(\psi_0+\phi_L)\rangle$. We shall see that in general, at least for wigglers with a uniform field strength and period, the electrons tend to bunch around the phase $\psi_0+\phi_L \approx 0$, so that n−1 > 0. This tends to focus the optical field toward the axis of the electron beam, and can even guide the optical beam in the manner of a fiber-optic waveguide.

2.4. Dimensionless Equations of Motion

2.4.1. Dimensionless variables

To simplify the notation, it is useful at this point to change to a set of dimensionless variables[89]. Properly chosen dimensionless variables have several additional advantages. In the first place, they illustrate the importance of various terms in the equations, and therefore, the importance of various physical effects. In the second place, they generally have magnitudes of the order of unity, so that they do not lead to overflow in computer simulations. The dimensionless variables are defined by the expressions

$$\tau = ct/L_W, \qquad (2.90)$$

$$\zeta = (z_0-ct)/N_W\lambda_R, \qquad (2.91)$$

$$\mu = 4\pi N_W(\gamma-\gamma_R)/\gamma_R, \qquad (2.92)$$

$$\nu = N_W\lambda_R(k-k_R), \qquad (2.93)$$

$$\epsilon_L = 4\pi \frac{a_W e L_W N_W}{\gamma_R^2 mc^2} [J_0(\xi)-J_1(\xi)] \underline{E}_L. \qquad (2.94)$$

$$j_e = 2\pi \frac{a_W^2 e L_W^2 N_W}{\epsilon_0 \gamma_R^3 mc^3} [J_0(\xi)-J_1(\xi)]^2 J_e. \qquad (2.95)$$

The significance of this normalization is as follows. The time is normalized to the time L_W/c for the optical field to traverse the wiggler. The distance is measured relative to a point in the moving optical field and normalized to the slip length $N_W\lambda_L$. As discussed in Chapter 1, this length is the length of a wave packet emitted by an electron as it traverses the wiggler, and is also the distance by which a resonant electron slips behind the optical field as they both traverse the wiggler. The electron energy is measured relative to the resonant value and normalized by the energy bandwidth of the laser. The wavenumber, introduced for completeness, at this point, is measured relative to the resonant value and normalized to the slip length. The significance of the normalization of the electric field and electron-beam current will become apparent in later sections.

It is worth digressing for a moment to discuss the formal separation of the variables ζ and ψ, and the meaning of the transformation to τ and ζ. Strictly speaking, the phase is determined by the position, that is, ψ is a function of z. However, over a small interval of distance, too small to affect the macroscopic amplitude E_L and phase ϕ_L of the laser, the phase ψ of the electrons can change by a large amount. Thus, in an interval around the position ζ which is small enough that the amplitude and phase are constant, electrons can be found with all phases. Thus, since it is useful to do so, it is perfectly logical, within the slowly varying phase and amplitude approximation, to characterize each electron by the independent variables z and ψ or, equivalently, ζ and ψ.

In the same vein, the transformation from z and t to ζ and τ is not an orthogonal transformation. Thus, when it is assumed that the wiggler begins at $\tau = 0$ this corresponds to the same time for all electrons when, in fact the electrons toward the back of the pulse enter the wiggler later, at the time $\tau = -(k_W/k_L)\zeta$. However, so long as $k_W/k_L \ll 1$, this introduces little error. The non-orthogonality of the transformation also becomes apparent when the spectrum (that is, the Fourier transform with respect to time) of the optical field is calculated.

2.4.2. Dimensionless equations

In terms of these variables, the equations of motion of the electrons assume the form

$$\frac{d\mu}{d\tau} = -\epsilon_L \sin(\psi+\phi_L), \qquad (2.96)$$

$$\frac{d\psi}{d\tau} = \mu, \qquad (2.97)$$

$$\frac{d\zeta}{d\tau} = -1, \qquad (2.98)$$

where we have dropped the subscript 0 from ψ and ζ since there is no longer any confusion. Similarly, the Maxwell equations assume the form

$$\frac{\partial \epsilon_L}{\partial \tau} = j_e \langle \sin(\psi+\phi_L) \rangle, \qquad (2.99)$$

$$\epsilon_L \frac{\partial \phi_L}{\partial \tau} = j_e \langle \cos(\psi+\phi_L) \rangle. \qquad (2.100)$$

It is often more convenient, especially in numerical simulations, to deal with the more symmetric field variables

$$\epsilon_R = \epsilon_L \cos(\phi_L), \qquad (2.101)$$

$$\epsilon_I = \epsilon_L \sin(\phi_L). \qquad (2.102)$$

As shown below, these can be used as the real and imaginary parts of a complex electric field. In terms of these variables, the electron equations of motion may be expressed in the form

$$\frac{d\mu}{d\tau} = - \epsilon_R \sin(\psi) - \epsilon_I \cos(\psi), \qquad (2.103)$$

$$\frac{d\psi}{d\tau} = \mu, \qquad (2.104)$$

$$\frac{d\zeta}{ds} = -1, \qquad (2.105)$$

Expanding the sine and cosine in Eqs. (2.99) and (2.100), and recognizing that in the averages indicated there the phase ϕ_L is constant, we see that the Maxwell equations may be expressed

$$\frac{\partial \epsilon_R}{\partial \tau} = j_e \langle \sin(\psi) \rangle, \qquad (2.106)$$

$$\frac{\partial \epsilon_I}{\partial \tau} = j_e \langle \cos(\psi) \rangle. \qquad (2.107)$$

In the simple case when the optical field is uniform in space and time and resonant with the electron energy (ϕ_L = constant = 0), the first two equations become the equations for the motion of a simple pendulum, as shown in Fig. 2.2. In the case of the pendulum, the angle with respect to the vertical is ψ_0, and the angular velocity is μ. Eqs. (2.96) and (2.97) may be combined to give the second-order differential equation

$$\frac{d^2\psi}{d\tau^2} = -\epsilon_L \sin(\psi). \tag{2.108}$$

For small motions of the pendulum around the vertical, the frequency is just $\epsilon_L^{1/2}$. As we shall see, the analogy between the motions of the electrons in a free-electron laser and those of a pendulum is quite useful for understanding the behavior of a free-electron laser in the saturated regime.

Finally, it is sometimes convenient to write the free-electron laser equations in the more compact notation of complex variables. To do this, we introduce the complex electric field

$$\varepsilon_L = \epsilon_R + i\epsilon_I. \tag{2.109}$$

The Maxwell equations then assume the form

$$\frac{\partial \varepsilon_L}{\partial \tau} = ij_e \langle \exp(-i\psi) \rangle, \tag{2.110}$$

and the energy equation is expressed

$$\frac{d\mu}{d\tau} = -\mathrm{Im}[\varepsilon_L \exp(i\psi)]. \tag{2.111}$$

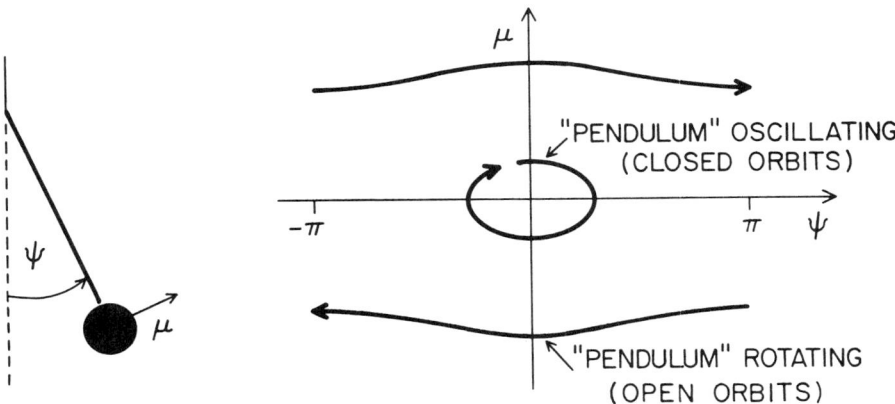

Fig. 2.2. The phase-plane motions of an electron in the ponderomotive potential of the wiggler are similar to the phase-plane motions of a simple pendulum.

3. ONE-DIMENSIONAL BEHAVIOR

Although free-electron lasers are three-dimensional devices, the one-dimensional theory includes most of the important phenomena encountered in these devices. Moreover, even with high-speed computers it remains impossible to compute the time-dependent behavior of free-electron lasers in three dimensions except in some limited circumstances. Fortunately, the one-dimensional theory provides a good description of the experimentally observed performance of free-electron lasers for a wide variety of devices.

In this Chapter, we explore the solutions of the equations derived in Chapter 2. We begin with a discussion of some of the analytic results which have been obtained in both the small-signal and saturated regimes of laser operation. This is followed by an introduction to the numerical solution of the free-electron laser equations, and a discussion of some of the results which have been obtained from simulations.

3.1. Analytic Results

While numerical simulations now provide the most accurate and general description of free-electron laser performance, they provide relatively less insight into what is going on with the electrons in the laser. In addition, useful formulas can be derived for the small-signal gain and some other parameters of interest. We examine first the small-signal regime, where quantatively useful results can be obtained. Results for the saturated regime are more qualitative in nature, but provide a great deal of insight into the behavior of free-electron lasers. We examine the results for the saturated regime last.

3.1.1. Small-signal gain

The most important single parameter of any laser is the small-signal gain, for it is this parameter which determines if the laser will reach threshold and lase. The behavior of free-electron (and other) lasers in the small-signal regime is linear in the intensity ϵ_L^2, and therefore tractable analytically[90]. To simplify the calculations, we further assume that the gain is small, so that the optical field may be regarded as uniform through the wiggler. We therefore examine the electron motion in the uniform optical field given by the expression

$$E_L = 2^{1/2} \underline{E}_L x \sin[k(z-ct) + \phi_L], \qquad (3.1)$$

where k is the laser wavenumber. The dimensionless field parameters are therefore

$$\epsilon_L = \text{constant} \ll 1, \tag{3.2}$$

$$\phi_L = \omega\zeta, \tag{3.3}$$

where

$$\omega = N_W \lambda_R (k-k_R). \tag{3.4}$$

The equations of motion are given by the expressions

$$\frac{d\mu}{d\tau} = -\epsilon_L \sin(\psi+\omega\zeta), \tag{3.5}$$

$$\frac{d\psi}{d\tau} = \mu, \tag{3.6}$$

$$\frac{d\zeta}{d\tau} = -1. \tag{3.7}$$

For the case when ϵ_L is small, we look for a solution in the form of a power series

$$\mu = \mu_0 + \epsilon_L \mu_1 + \epsilon_L^2 \mu_2 + \ldots, \tag{3.8}$$

$$\psi = \psi_0 + \epsilon_L \psi_1 + \epsilon_L^2 \psi_2 + \ldots, \tag{3.9}$$

$$\zeta = \zeta_0 + \epsilon_L \zeta_1 + \epsilon_L^2 \zeta_2 + \ldots. \tag{3.10}$$

If we substitute these series into the equations of motion, expand the sines and cosines, and collect all the terms on one side of each equation, we obtain a set of equations involving powers of ϵ_L on the one side and zero on the other. For these equations to be valid for all values of ϵ_L, the coefficient of each power of ϵ_L must vanish identically. We therefore obtain the heirarchy of equations

$$\frac{d\mu_0}{d\tau} = 0, \tag{3.11}$$

$$\frac{d\mu_1}{d\tau} = -\sin(\psi_0+\omega\zeta_0), \tag{3.12}$$

$$\frac{d\mu_2}{d\tau} = -(\psi_1+\omega\zeta_1)\cos(\psi_0+\omega\zeta_0), \tag{3.13}$$

and so on from Eq. (3.5),

$$\frac{d\psi_0}{d\tau} = \mu_0, \tag{3.14}$$

$$\frac{d\psi_1}{d\tau} = \mu_1, \tag{3.15}$$

$$\frac{d\psi_2}{d\tau} = \mu_2, \tag{3.16}$$

and so on from Eq. (3.6), and

$$\frac{d\zeta_0}{d\tau} = -1, \tag{3.17}$$

$$\frac{d\zeta_1}{d\tau} = 0, \tag{3.18}$$

$$\frac{d\zeta_2}{d\tau} = 0, \tag{3.19}$$

and so on from Eq. (3.7). The initial conditions are given by the equations

$$\mu_0 = \omega+\alpha, \qquad \mu_{n>0} = 0, \tag{3.20}$$

$$\psi_0 = \phi_0, \qquad \psi_{n>0} = 0, \tag{3.21}$$

$$\zeta_0 = 0, \qquad \zeta_{n>0} = 0, \tag{3.22}$$

where

$$\alpha = 2\pi N_W \left[2\frac{\gamma-\gamma_R}{\gamma_R} - \frac{k-k_R}{k_R} \right] = 2\pi N_W \left[2\frac{\gamma-\gamma_R}{\gamma_R} + \frac{\lambda-\lambda_R}{\lambda_R} \right]. \tag{3.23}$$

The solution through zeroth order is

$$\mu_0 = \omega+\alpha, \tag{3.24}$$

$$\psi_0 = (\omega+\alpha)\tau + \phi_0, \qquad (3.25)$$

$$\zeta_0 = -\tau, \qquad (3.26)$$

and the solution through first order is

$$\mu_1 = \frac{1}{\alpha}[\cos(\alpha\tau+\phi_0) - \cos(\phi_0)], \qquad (3.27)$$

$$\psi_1 = \frac{1}{\alpha^2}[\sin(\alpha\tau+\phi_0) - \sin(\phi_0) - \alpha\tau\cos(\phi_0)], \qquad (3.28)$$

$$\zeta_1 = 0. \qquad (3.29)$$

In second order we need only the energy change, which is given by the expression

$$\mu_2 = -\frac{1}{\alpha^3}\Big\{1 - \cos(\alpha\tau) - \alpha\tau\sin(\alpha\tau)\cos^2(\phi_0)$$

$$- \alpha\tau\cos(\alpha\tau)\sin(\phi_0)\cos(\phi_0)$$

$$+ \frac{1}{2}[\sin^2(\alpha\tau+\phi_0) - \sin^2(\phi_0)]\Big\}. \qquad (3.30)$$

Because the electrons can enter the wiggler at random times, the average energy change for all the electrons in a beam is found by averaging over all possible values of the optical phase ϕ_0. When this is done, it is found that $\langle\Delta\mu_0\rangle = \langle\Delta\mu_1\rangle = 0$, and that the final energy change of the electrons at the end of the wiggler ($\tau = 1$) is given by the formula

$$\langle\Delta\mu\rangle = \epsilon_L^2\langle\Delta\mu_2\rangle = -\epsilon_L^2 g(\alpha), \qquad (3.31)$$

where the function $g(\alpha)$ is given by the expression

$$g(\alpha) = -\frac{1}{\alpha^3}\left[1 - \cos(\alpha) - \frac{\alpha}{2}\sin(\alpha)\right] \qquad (3.32)$$

$$= -\frac{d}{d\alpha}\left[\frac{\sin(\alpha/2)}{\alpha}\right]^2. \qquad (3.33)$$

To find the optical gain, it is sufficient to recognize that since the electrons cannot exchange energy with the static magnetic field of the wiggler, the energy which they lose in the wiggler appears in the optical beam. That is, energy is conserved as shown in Chapter 2. The actual power change of the beam, per unit area, is given by the formula

$$\Delta\Phi_B = \frac{mc^2}{e} J_e \langle\Delta\gamma\rangle = \frac{\gamma_R mc^2}{e} J_e \frac{\langle\Delta\mu\rangle}{4\pi N_W}, \quad (3.34)$$

where J_e is the electron-beam current density. The power change of the optical beam, per unit area, is given by the formula

$$\Delta\Phi_L = (G_0-1)\Phi_L = (G_0-1)\epsilon_0 c \underline{E}_L^2 = -\Delta\Phi_B \quad (3.35)$$

where G_0 is the ratio of the laser power out of the wiggler to the laser power in. Combining these equations, we obtain for the gain coefficient the expression

$$G_0 - 1 = \frac{\mu_0 e^3 B^2 \lambda_W L_W^3}{\pi \gamma_R^3 m^3 c^3} [J_0(\xi) - J_1(\xi)]^2 J_e g(\alpha) \quad (3.36)$$

$$= 2 j_e g(\alpha), \quad (3.37)$$

where j_e is a dimensionless current density defined by Eq. (2.95), which may also be expressed in the form

$$j_e = \frac{\mu_0 e^3 B^2 \lambda_W L_W^3}{2\pi \gamma_R^3 m^3 c^3} [J_0(\xi) - J_1(\xi)]^2 J_e. \quad (3.38)$$

It is interesting to note that because the average electron motion vanishes to first order in ϵ_L, the optical power increase is proportional to the laser intensity (ϵ_L^2), rather than the laser amplitude, and the small-signal gain coefficient G_0 is independent of the intensity. This is expected from the behavior of conventional lasers, of course.

The gain function $g(\alpha)$ is shown in Fig. 3.1. Unlike the gain profile of conventional lasers, $g(\alpha)$ is antisymmetric, and vanishes at resonance. One of the early triumphs of the theory of free-electron lasers was the experimental demonstration of the gain profile $g(\alpha)$, as shown in Fig. 3.2[91]. The gain is largest for positive values of the resonance parameter α, that is, for $\gamma > \gamma_R$ and $\lambda > \lambda_R$. The peak value of the gain function is

$$g_{MAX} = 6.75 \times 10^{-2}, \quad (3.39)$$

which corresponds to a peak gain

$$G_0 - 1 = 0.135 j_e. \quad (3.40)$$

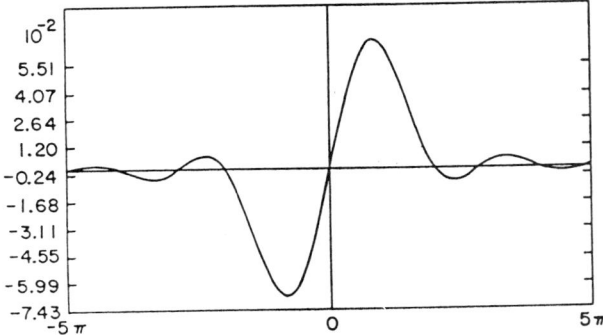

Fig. 3.1. Small-signal gain function for free-electron lasers in the small-gain approximation.

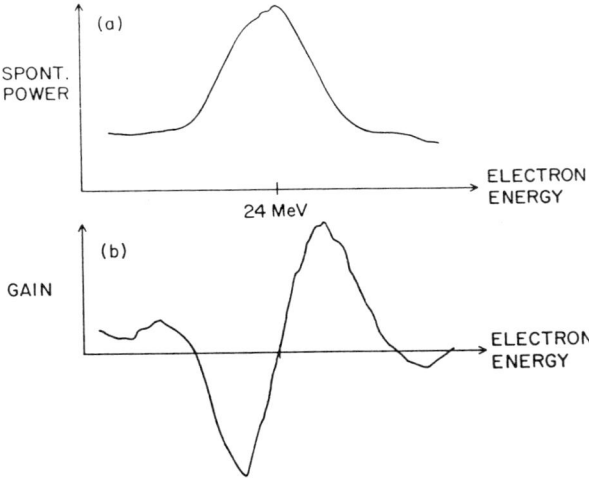

Fig. 3.2. Spontaneous-emission and gain profiles observed in the experiments of Madey and co-workers.

This occurs at the value

$$\alpha = 2.6 \tag{3.41}$$

of the resonance parameter. The width of the largest positive peak of the gain curve is

$$\Delta\alpha = 2\pi, \tag{3.42}$$

which corresponds to a wavelength width

$$\frac{\Delta\lambda}{\lambda_R} = \frac{1}{N_W}, \tag{3.43}$$

and an energy width

$$\frac{\Delta\gamma}{\gamma_R} = \frac{1}{2N_W}. \tag{3.44}$$

The fact that the width in energy space is half the width in wavelength space follows from the fact that the wavelength is quadratically related to the energy, as indicated by the resonance condition, Eq. (2.59). If the energy spread is comparable to or larger than $1/2N_W$, then the gain must be averaged over the electron energy distribution function.

3.1.2. The Madey theorem

Two things are worth noting in the discussion of the small-signal gain given above. The first is the tedious nature of the derivation of the gain formula, which owes to the fact that the mean energy change of the electrons vanishes in first order and must therefore be calculated to second order. The second is the relation of the gain profile to the spontaneous emission profile. Comparing the spontaneous emission and gain in Fig. 3.2, we see that the gain profile is proportional to the derivative of the spontaneous emission profile[92]. This relationship, called the Madey gain-spread theorem, is quite general, and applies to complex wiggler designs in which the wiggler period and field are not uniform[93]. By means of this theorem it is possible to compute the gain from the first-order motions of the electrons, which is an important simplification for complex wigglers. The theorem also has important consequences for the operation of free-electron lasers in storage rings.

The Madey theorem consists of two parts. In the first, the relationship

$$\langle\Delta\gamma\rangle = \frac{1}{2} \frac{d}{d\gamma} \langle\Delta\gamma^2\rangle \qquad (3.45)$$

between the first and second moments of the electron energy change is shown to hold in the limit of small signal and small gain. In the second part, the relationship

$$\langle\Delta\gamma^2\rangle = \frac{16\pi^3 \epsilon_0 \underline{E}_L^2}{m^2 c \omega^2} \frac{d^2W}{d\Omega d\omega}, \qquad (3.46)$$

between the second moment of the energy fluctuations and the spontaneous emission is shown to hold in the same limit, where $d^2W/d\Omega d\omega$ is the energy spontaneously radiated per unit solid angle Ω around the wiggler axis, per unit frequency ω, by an electron as it passes through the wiggler. Using the conservation of energy, Eqs. (3.34) and (3.35), we obtain from Eqs. (3.45) and (3.46) the result

$$G_0 - 1 = -\frac{8\pi^3 J_e}{m e \omega^2} \frac{d}{d\gamma} \frac{d^2W}{d\Omega d\omega}, \qquad (3.47)$$

where it is assumed that ω is nearly constant over the width of the spontaneous emission spectrum.

To simplify the calculations, we prove the theorem only for the case of a uniform wiggler. To prove the more general case, the procedure is the same except that the integrals cannot be explicitly evaluated and must be manipulated formally to show that the required expressions are equal.

To prove the first part of the theorem, we evaluate the energy change at the end of the wiggler ($\tau=1$) using Eqs. (3.27) and (3.30). To lowest nonvanishing order we see that

$$\langle\Delta\mu\rangle = \epsilon_L^2 \langle\Delta\mu_2\rangle = \epsilon_L^2 \frac{d}{d\alpha}\left[\frac{\sin(\alpha/2)}{\alpha}\right]^2, \qquad (3.48)$$

$$\langle\Delta\mu^2\rangle = \epsilon_L^2 \langle\Delta\mu_1^2\rangle = \frac{\epsilon_L^2}{\alpha^2} \langle[\cos(\alpha+\phi_0) - \cos(\phi_0)]^2\rangle$$

$$= \frac{\epsilon_L^2}{\alpha^2}[1 - \cos(\alpha)] = 2\epsilon_L^2\left[\frac{\sin(\alpha/2)}{\alpha}\right]^2, \qquad (3.49)$$

after averaging over ϕ_0. Comparing these results, we see that

$$\langle\Delta\mu\rangle = -\frac{1}{2}\frac{d}{d\alpha}\langle\Delta\mu^2\rangle. \qquad (3.50)$$

Since, from Eqs. (2.92) and (3.23), we see that

$$\Delta\mu = 4\pi N_W \frac{\Delta\gamma}{\gamma_R} \qquad (3.51)$$

and

$$d\alpha = 4\pi N_W \frac{d\gamma}{\gamma_R}, \qquad (3.52)$$

the first part of the theorem is proved.

To demonstrate the second part of the Madey theorem, we calculate the spontaneous emission from the formula[94]

$$\frac{d^2W}{d\Omega d\omega} = \frac{e^2\omega^2}{16\pi^3\epsilon_0 c^3}\left|\int n\times(n\times v)\exp[ik(n\cdot r - ct)]\,dt\right|^2, \qquad (3.53)$$

where n is a unit vector from the origin to the point of observation, v is the electron velocity and $r(t)$ is the position of the electron. For an electron initially moving parallel to the z axis, we may use the fact that the transverse canonical momentum is constant to write

$$p = P_0 - eA_W = -eA_W. \qquad (3.54)$$

For an observation point on the z axis, then, the velocity factor is given by the expression

$$n\times(n\times v) = -\frac{p_x}{\gamma m}x = -\frac{eA_W}{\gamma m}x, \qquad (3.55)$$

for a plane-polarized wiggler. Thus, the on-axis radiation from an electron passing through the wiggler is

$$\frac{d^2W}{d\Omega d\omega} = \frac{e^4\omega^2}{16\pi^3\epsilon_0\gamma^2 m^2 c^3}\left|\int A_W\exp[ik(z-ct)]\,dt\right|^2. \qquad (3.56)$$

But, for an electron moving in the plane-polarized optical field of Eq. (2.45), which may be expressed in the equivalent form

$$E_L = \frac{E_L}{2^{1/2}}x\,\exp[ik_R(z-ct)+i\phi_L] + \text{c.c.}, \qquad (3.57)$$

we see from Eq. (2.38) that the electron energy change is given by the expression

$$\Delta\gamma = -\frac{e^2 E_L}{2^{1/2}\gamma m^2 c^2} \exp(i\phi_L) \int A_W \exp[ik_R (z-ct)] + \text{c.c.} \quad (3.58)$$

Squaring and averaging over all ϕ_L, we find that

$$\langle\Delta\gamma^2\rangle = \frac{e^4 E_L^2}{\gamma^2 m^4 c^4} \left|\int A_W \exp[ik_R (z-ct)] dt\right|^2. \quad (3.59)$$

Comparing this with Eq. (3.56), we see that the second part of the theorem, Eq. (3.46), is proved.

Although we have proved the Madey theorem only for the case of a uniform, plane-polarized wiggler, it holds for arbitrary wigglers of any elliptical polarization. In fact, we proved the second part of the theorem for an arbitrary, plane-polarized wiggler. The theorem is valid only in the small-signal, small-gain limit, in the one-dimensional approximation, and only for the gain on the axis of the wiggler and the electron beam, which must be parallel[95].

The importance of the theorem is two-fold. In the first place, it allows one to compute the small-signal gain of a free-electron laser from the quantity $\langle\Delta\gamma^2\rangle$. Since this quantity is nonvanishing in first order, this approach can simplify the calculation of the gain for complex wigglers[96]. In fact, it is even possible to calculate the gain from the experimentally measured spontaneous emission. On the other hand, it is also possible to show that the gain can be calculated from the first-order motions of the electrons using the Maxwell equations.

The second reason for the importance of the Madey theorem is the fact that the gain is related directly to $\langle\Delta\gamma\rangle$, as we have seen, whereas the energy spread is described by $\langle\Delta\gamma^2\rangle$. For this reason, the theorem relating these two quantities is also called the gain-spread theorem. It tells us that there can be no gain without energy spread, and that the small-signal energy loss and energy spread are generally of the same order of magnitude. This has implications for the operation of free-electron lasers in storage rings, where the energy spread of the electrons after one pass through the wiggler reduces the gain during the next pass. Wiggler designs have been proposed to reduce the energy spread for storage-ring operation, at least in the saturated regime[97]. However, they are generally very long, since they depend on adiabatic motions of the electrons in the phase plane, and have low gain.

Having calculated the small-signal gain straightforwardly, as described above, we may turn Eq. (3.47) around to find the spontaneous emission on axis. Using Eq. (2.10), we find that

$$\frac{d}{d\gamma}\frac{d^2W}{d\Omega d\omega} = \frac{\mu_0 e^4 B^2 L_W^2}{8\pi^3 \gamma_R m^2 c} [J_0(\xi) - J_0(\xi)]^2 \frac{d}{d\gamma}\left[\frac{\sin(\alpha/2)}{\alpha}\right]^2. \qquad (3.60)$$

Integrating with respect to γ, we obtain the result

$$\frac{d^2W}{d\Omega d\omega} = \frac{\mu_0 e^4 B^2 L_W^2}{8\pi^3 \gamma_R m^2 c} [J_0(\xi) - J_0(\xi)]^2 \left[\frac{\sin(\alpha/2)}{\alpha}\right]^2. \qquad (3.61)$$

where

$$\alpha = -2\pi N_W \frac{\omega - \omega_R}{\omega_R} \qquad (3.62)$$

for resonant electrons.

3.1.3. Saturation

Once the laser gain exceeds threshold, the optical intensity builds up until the electron motion becomes nonlinear and the laser saturates. To understand the motions of the electrons at saturation, we examine the behavior of the electrons in a uniform, resonant, optical field, so that ϵ_L = constant, and ϕ_L = 0. As described by Eqs. (2.96) and (2.97), or, equivalently, by Eq. (2.108), the electrons in the wiggler then behave as though they were particles in a sinusoidal potential given by the expression

$$V_P(\psi) = -\epsilon_L \cos(\psi). \qquad (3.63)$$

The potential described by Eq. (3.63) is called the ponderomotive potential. The corresponding motions, shown in Fig. 3.3, can be divided into trapped oscillations in the troughs, or "buckets," of the pondermotive potential, and open motions above the oscillating potential.

To understand these motions in more detail, it is useful to consider the phase plane (μ vs. ψ), shown in Fig. 3.4. Motions of the electrons, [$\mu(\tau)$, $\psi(\tau)$], correspond to curves, or trajectories, in the phase plane. From Eq. (3.63) we see that the origin of the phase plane corresponds to a minimum of the ponderomotive potential, and the electron trajectories in the phase plane are periodic with a period 2π.

The motions in the phase plane are found from Eqs. (2.96) and (2.97). From Eq. (2.96), the energy equation, we see that the motions are generally upward for phases between $-\pi$ and 0, and downward for phases between 0 and π, as indicated in Fig. 3.4.

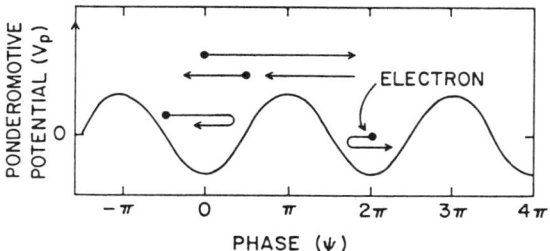

Fig. 3.3. Electron motions in the periodic "ponderomotive" potential formed by the fields of the wiggler and laser.

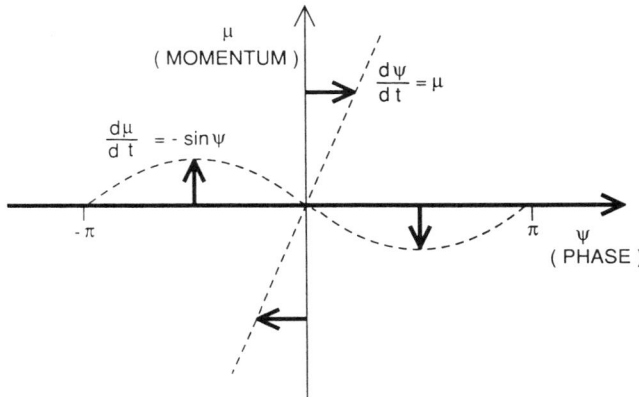

Fig. 3.4. Phase plane for describing the energy and phase motions of the electrons in a wiggler.

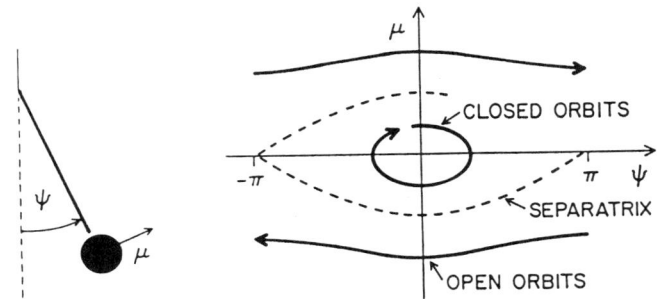

Fig. 3.5. Trajectories of the electrons in the phase plane.

Similarly, from Eq. (2.96) we see that the motions are to the right in the upper half plane, and to the left in the lower half plane. Thus, the electrons generally describe a circulation clockwise around the origin, as indicated in Fig. 3.5. The "closed" trajectories near the origin correspond to oscillations of the pendulum, while the open trajectories above and below the region of closed trajectories correspond to rotation of the pendulum completely around its fulcrum.

To compute the trajectories, we multiply Eq. (2.96) by μ and Eq. (2.97) by $\epsilon_L \sin(\psi)$. Adding the results, we obtain the equation

$$\mu \frac{d\mu}{d\tau} + \epsilon_L \sin(\psi) = \frac{d}{d\tau}\left[\frac{\mu^2}{2} - \epsilon_L \cos(\psi)\right] = 0. \tag{3.64}$$

This equation is, of course, just the conservation of energy of the pendulum, and could have been written down from Eq. (3.63). Integrating this, we see that the trajectories correspond to the curves

$$\frac{\mu^2}{2} - \epsilon_L \cos(\psi) = U, \tag{3.65}$$

where U is the total "energy" of the pendulum.

The curve separating the closed and open motions, called the separatrix, corresponds to the trajectory passing through the points $(0, \pm\pi)$, for which $U = \epsilon_L$. This is shown by the dotted curve in Fig. 3.5, and the area it encloses is called the "bucket" in the phase plane. Since the boundary of the bucket is a trajectory, electrons initially inside the bucket remain trapped inside it and electrons initially outside the bucket remain outside. The height of the bucket, which may be compared with the initial distribution of electron energy in the electron beam, is given by the expression

$$\Delta\mu_{MAX} = 2 (2\epsilon_L)^{1/2}, \tag{3.66}$$

which in physical variables corresponds to the expression

$$\frac{\Delta\gamma}{\gamma_R} = \left(\frac{2a_W e L_W E_L}{\pi \gamma_R^2 mc^2 N_W}\right)^{1/2}. \tag{3.67}$$

For a free-electron laser operating at a wavelength of 1 μm, with a wiggler 4 m long having 200 periods, the laser beam is typically focused to a spot of the order of 10^{-6} m^2. At a laser power of 1 MW, which corresponds to an electric field of the order of 2×10^7 V/m, the bucket height is of the order of 0.5 percent. This is broader than the small-signal gain bandwidth, $1/4\pi N_W \approx 0.04$ percent, and illustrates the general situation in which the saturated behavior is more tolerant of a broad energy spread than is the small-signal gain.

Electrons near the origin of the phase plane undergo small oscillations around the origin. These phase oscillations are called synchrotron oscillations because of their similarity to the phase oscillations originally studied in in synchrotrons, but subsequently observed in all electron accelerators. For small motions $\psi \ll 1$, and we may describe them using the approximation (see Eq. (2.107))

$$\frac{d^2\psi}{d\tau} = -\epsilon_L \psi = -\Omega_S^2 \psi. \qquad (3.68)$$

From this we see that the synchrotron frequency for small oscillations is given by the expression

$$\Omega_S = \epsilon_L^{1/2}. \qquad (3.69)$$

In physical variables, the synchrotron period is given by the formula

$$\lambda_S = 2\pi \frac{L_W}{\Omega_S} = 2\pi \frac{L_W}{\epsilon_L^{1/2}} = \frac{L_W}{N_W} \left[\frac{\pi}{2} \frac{1+a_W^2}{a_W} \frac{v_0}{E_L \lambda_L} \right]^{1/2}, \qquad (3.70)$$

and depends on the fourth root of the laser intensity. For the case of a free-electron laser operating at 1 μm with a wiggler 4 m long having 200 periods, and a 1-MW laser beam focused to an area of the order of 10^{-6} m^2, as discussed above, the synchrotron period is of the order of 6 m.

For large oscillations, the synchrotron frequency decreases. In fact, near the points $\pm\pi$ the motion in the phase plane vanishes, as shown by Eqs. (2.96) and (2.97). Thus, a distribution of electrons in the phase plane does not rotate as a solid body inside the separatrix. Instead, electrons near the nodes lag behind.

The motions of a set of electrons initially near resonance but distributed randomly in phase is shown in Fig. 3.6. The electrons near the origin wrap around the origin, but because the electrons are initially near resonance, the final distribution is symmetric about the ψ axis and the average electron energy remains near resonance. In this case, there is no net gain or loss of electron energy, so there is no laser gain.

To get gain in the saturated case, as in the small-signal case, the electrons must enter the wiggler at an energy slightly above resonance. The motions of the electrons in this case are illustrated in Fig. 3.7. After about one-half synchrotron period, the electron distribution has moved to a lower average energy. Thus, there is net laser gain.

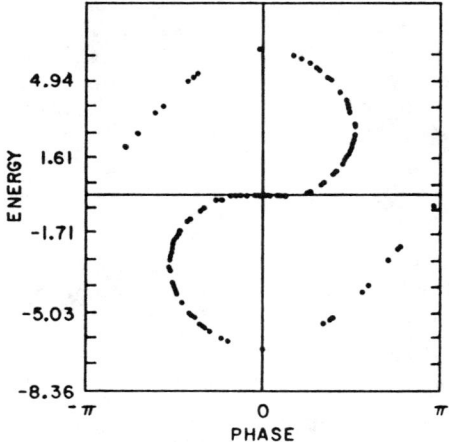

Fig. 3.6. Phase-plane motions of electrons initially distributed with random phase at the resonant energy.

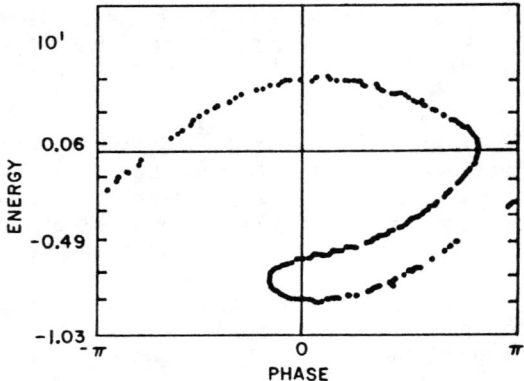

Fig. 3.7. Phase-plane motions of electrons initially distributed with random phase at an energy above the resonant value.

It is of interest to note that in both cases discussed above (Figs. 3.6 and 3.7), the energy spread induced by the interaction in the wiggler is of the order of the bucket height, and is larger than the energy extraction. This is generally the case. However, proposals have been put forward for wiggler designs which induce a much smaller energy spread by moving the electrons adiabatically in a wiggler whose period increases along the length of the wiggler[98]. Unfortunately, the wiggler must be very long, which presents other difficulties. Thus, "heating" of the electron beam by the laser interaction is a serious problem in storage rings, where the electron beam must be repeatedly re-used[99].

If the optical intensity increases so that the distribution in Fig. 3.7 moves farther around the origin, the electrons near the "nose" of the distribution will begin to move upward, increasing the average electron energy. This means reduced energy extraction from the electrons and, therefore, reduced net gain, and corresponds to saturation of the laser. Saturation therefore occurs at an intensity corresponding to about one-half synchrotron rotation, which is equivalent to a laser field

$$\epsilon_L(\text{saturation}) = \Omega_S^2 = O(\pi^2). \qquad (3.71)$$

We may estimate the saturated efficiency in the following way. The average electron energy change in the small-signal regime is given by Eqs. (3.31) and (3.39):

$$\langle\Delta\mu\rangle = -\epsilon_L^2 g(\alpha) = -6.75\times10^{-2} \epsilon_L^2. \qquad (3.72)$$

Extrapolating this to saturation, we see that the extraction efficiency (the fraction of the electron energy converted to laser radiation) is given by the approximate expression

$$\eta_X = \frac{\Delta\gamma}{\gamma_R} = -\frac{\langle\Delta\mu\rangle}{4pN_W} = O\left(\frac{\pi^4 g(\alpha)}{4\pi N_W}\right) = O(1/2N_W). \qquad (3.73)$$

Actually, when more careful calculations are made it is found that Eq. (3.73) is a bit of an overestimate of the saturated efficiency. Another way of estimating the saturated efficiency is to argue that the saturated energy change is of the order of half the bucket height (actually, this is also a bit of an overestimate). Then, from Eqs. (3.66) and (3.71) we find that the saturated extraction efficiency is of the order of $1/2^{3/2}N_W$.

Clearly, in the strongly saturated regime the optimum initial electron energy for maximum energy extraction is a fraction of the bucket height, and therefore depends on the laser intensity. Equivalently, if the incident electron energy cannot change as the laser saturates, the laser wavelength must shift, or "chirp." This is generally observed to occur in simulations of laser performance. In general it is difficult to observe in experiments because it is important only at the beginning of the laser pulse, and is obscured by other effects such as transients in the start-up of the accelerator and by line-broadening instabilities in the free-electron laser. These instabilities, which are called synchrotron instabilities because they are caused by the synchrotron oscillations of the electrons, are discussed at the end of this chapter.

3.2. Numerical Simulations

Subject to the approximations described in Chapter 2, the performance of free-electron lasers in one dimension is described by Eqs. (2.103) to (2.107). Unfortunately, it is not possible to solve these equations analytically except in the limit of a small, time-independent optical field. Thus, even in the one-dimensional case, the general time-dependent, saturated behavior of the laser can be described only by numerically integrating the equations of motion of the electrons and fields on a computer.

In this section, we first set up the equations to be integrated for a general, time-dependent free-electron laser pulse. We then look at the saturated behavior of a typical free-electron laser and examine the so-called synchrotron instabilities which occur in saturated lasers. Finally, we look briefly at simplifying assumptions which may be used when the electron-beam and laser pulses are so long that available computer time and memory are exceeded.

3.2.1. Short pulses

To illustrate the problem, we consider the important case of an rf-linac free-electron laser. The electron beam from an rf linac appears in short pulses (called micropulses) which are typically 3 to 10 ps long, and which have a repetition frequency of the order of 20 MHz to 3 GHz. The optical beam has the same pulse format, and is called "mode-locked" in a conventional laser. This pulse format is clearly difficult to model analytically, but detailed numerical simulations have provided an excellent description of the performance of these devices.

Because the electron-beam current in rf linacs is limited, the gain is generally not sufficient for useful operation of the free-electron laser as a single-pass amplifier. However, by placing mirrors around the wiggler as shown in Fig. 3.8, the optical pulses can be repeatedly amplified, and the free-electron laser operated as a laser oscillator. However, because there is no optical gain in the absence of electrons, the optical pulse from a previous pass must arrive at the beginning of the wiggler coincident with the arrival of an electron pulse from the accelerator. This places strict constraints on the length of the optical resonator. One important

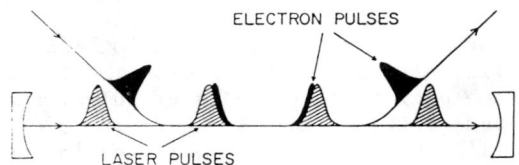

Fig. 3.8. Schematic diagram of a free-electron laser oscillator operating on the pulsed electron beam from an rf accelerator.

test of any simulation is the ability to predict the effect of detuning the length of the resonator from exact synchronism. Clearly, the effect of desynchronism depends on the shape of the electron pulse. We assume, in the following, that the electron current density is given by the expression

$$j_e = j_{max}\left[1 - 4\left(\frac{\zeta+\tau}{\Delta\zeta_p}\right)^2\right], \qquad \frac{|\zeta+\tau|}{\Delta\zeta_p} < \frac{1}{2},$$

$$= 0, \qquad \text{otherwise.} \qquad (3.74)$$

We set up the problem by computing the trajectories of N_E electrons, where N_E is typically of the order of 10^3 to 10^4. This is much smaller than the actual number of electrons in a micropulse, which is typically of the order of 10^{10}. The problems introduced by inadequate statistics, which appear in the form of excess shot noise, or spontaneous emission, are discussed below. The electrons are described by the coordinates

$$\gamma_n(\tau), \ \psi_n(\tau), \ \zeta_n(\tau), \qquad n = 1,2,\ldots N_E. \qquad (3.75)$$

The optical field of a micropulse is described at a set of points which are generally spaced uniformly across the "window" in which the field is to be computed. Thus, the field is described by the coordinates

$$\epsilon_{R,m}(\tau), \ \epsilon_{I,m}(\tau), \qquad m = 1,2,\ldots M. \qquad (3.76)$$

Between these points, the field can be found by interpolation. For complex optical pulses up to 1000 (or more) points may be necessary.

Initially, the electrons must be distributed over the range of energy and space occupied by the electron-beam pulse. Various methods have been used to distribute the electrons, but in the code OKPulse, for example, which runs on an IBM PC (R) microcomputer and has been used to generate the results shown below, the electrons are distributed randomly in (μ, ζ) space, with a probability in each dimension which is parabolic. The phase ψ is assigned randomly, with uniform probability over the range from $-\pi$ to π.

The initial field can be set to zero, but if the actual startup from zero is not of interest, it is more economical to initialize the field at some value. In the code OKPulse, the field is initialized at the wavelength of maximum small-signal gain, so that

$$\epsilon_{R,m}(0) = \epsilon_L(0) \cos(\nu_{max}\zeta_m), \qquad (3.77)$$

$$\epsilon_{I,m}(0) = \epsilon_L(0) \sin(\nu_{max}\zeta_m), \qquad (3.78)$$

where ζ_m is the mth field point and $\nu_{max} = -2.6$ is the frequency of maximum gain.

For numerical integration, the equations of motion are written in the form of difference equations, specifically

$$\mu_n(\tau+\Delta\tau) = \mu_n(\tau)$$
$$- [\epsilon_R(\zeta_n) \sin(\psi_n) + \epsilon_I(\zeta_n) \cos(\psi_n)] \Delta\tau, \quad (3.79)$$

$$\psi_n(\tau+\Delta\tau) = \psi_n(\tau) + \mu_n \Delta\tau, \quad (3.80)$$

$$\zeta_n(\tau+\Delta\tau) = \zeta_n(\tau) - \Delta\tau, \quad (3.81)$$

for the electrons, and

$$\epsilon_{R,m}(\tau+\Delta\tau) = \epsilon_{R,m}(\tau) + \frac{\Sigma_m \sin(\psi_n)}{\Sigma_m 1} j_e \Delta\tau, \quad (3.82)$$

$$\epsilon_{I,m}(\tau+\Delta\tau) = \epsilon_{I,m}(\tau) + \frac{\Sigma_m \cos(\psi_n)}{\Sigma_m 1} j_e \Delta\tau, \quad (3.83)$$

for the fields, where the space interval is

$$\Delta\zeta = \zeta_{m+1} - \zeta_m, \quad (3.84)$$

in which ζ_m is the coordinate of the mth optical field point, the time interval is

$$\Delta\tau = 1/N_S, \quad (3.85)$$

in which N_S is the number of steps used for the integration, and Σ_m means the sum over all electrons in the interval centered around the field point m.

These equations may be integrated straightforwardly by iterating Eqs. (3.79) to (3.83) N_S times. More sophisticated integration techniques, such as Runge-Cutta or Adams-Bashforth-Moulton algorithms are difficult to implement because of the memory required to store the intermediate results. In general, the number of time steps required increases with the degree of saturation of the laser, since in saturated lasers the electrons describe complex orbits in the phase plane. Typically, 50 time steps are sufficient to achieve better than 10% accuracy.

When the laser is operated as an oscillator, the output from one pass reflects off the resonator mirrors and becomes the input for the next pass. In this case the amplitude is reduced by the reflectance of the mirrors:

$$\epsilon_{R,m}(\text{input}) = \epsilon_{R,m}(\text{output})\, (R_1 R_2)^{1/2}, \qquad (3.86)$$

$$\epsilon_{I,m}(\text{input}) = \epsilon_{I,m}(\text{output})\, (R_1 R_2)^{1/2}, \qquad (3.87)$$

where R_1 and R_2 are the reflectances of the resonator mirrors. The square root is appropriate because each mirror reduces the laser intensity (which is proportional to ϵ^2) in direct proportion to the reflectance.

In the case of a free-electron laser oscillator, the optical pulse makes repeated passes through the laser, and is amplified by successive pulses of the electron beam, as shown in Fig. 3.8. If the round-trip time of the optical pulse in the resonator is not exactly the same as the time between electron pulses, then the overlap of the two pulses is not the same from pass to pass. To account for this, the coordinate of each field point is shifted by the amount

$$\Delta \zeta = -\, 2 \Delta L_C / N_W \lambda_L, \qquad (3.88)$$

where ΔL_C is the change of the cavity length from the synchronous length, and the factor of two is necessary to account for the change in the round-trip length. When a field point shifts out of the window, it is discarded and a new point is introduced at the other end of the window.

In computing the field from Eqs. (3.82) and (3.83), the averages of the sine and cosine of the phase are required. When the laser is saturated, the electrons are strongly bunched at certain phases and the averages are easily computed. However, when the laser is not saturated, then the electrons are randomly distributed and the averages depend on the statistics of the random distribution. In general, the sum of a large number of samples of a random distribution is proportional to the square root of the number of samples. The random-walk problem, in which the final displacement is proportional to the square root of the number of steps, is an example of this principle. As a result, the average of a large number of random samples is proportional to the reciprocal of the square root of the number of samples. Since the number of electrons for which the trajectories are computed is very small compared to the number of electrons in the actual electron beam, the statistical, or "shot noise" contribution to the field is much larger in the computations than in the actual laser. This leads to important errors when the laser is starting up, and when it is lasing weakly, such as when the cavity length is tuned far from synchronism, as discussed above.

In most cases, the shot noise is a small effect, and it is sufficient to eliminate the shot noise from the computations altogether. This is accomplished in the following way. Initially, the electrons are distributed with random phases. To eliminate the shot noise, the electrons are distributed in pairs such that both electrons have the same initial energy $\mu(0)$ and position $\zeta(0)$, and the first has a random phase $\psi(0)$ while the second has the same phase advanced by π, that is, $\psi(0) + \pi$. This exactly cancels the averages in Eqs. (3.82) and (3.83) as long as the electrons retain their initial random position and energy, but allows the electrons to bunch coherently as described earlier and radiate as they should. Both the small-signal gain and the saturated performance are then correctly predicted, except when spontaneous emission is a dominant effect.

In some cases, the noise, which is actually spontaneous emission, is important[100,101]. For example, when the laser is starting up with no initial field, noise is dominant. Also, when the cavity length is out of synchronism, the optical pulse slips away from the electron beam after many successive passes and must be replaced by a field generated from noise. In such cases, the noise may be approximated by re-introducing a small amount of randomness to the initial distribution. To do this, the initial phases of the electrons in the pair are given by the equations

$$\psi_1(0) = 2\pi R_0, \qquad (3.89)$$

$$\psi_2(0) = 2\pi R_0 + \pi + 2\pi K R_1, \qquad (3.90)$$

where R_0 and R_1 are random numbers between 0 and 1, and K is a constant. This approximation is found to give satisfactory results for the laser performance when spontaneous emission is important. The computed noise spectrum is shown in Fig. 3.9.

Strictly speaking, the constant K cannot be evaluated from first principles within the one-dimensional approximation. Electrons radiate spontaneously into all angles, and how much of this may be regarded as forward directed must be determined by calculating the radiation into the three-dimensional modes of the optical resonator. Alternatively, it is possible to compare the results of simulations with experimental data to determine the degree of randomness required. Fortunately, the results are not very sensitive to the value of K. A value of the order of 10^{-3} works for a variety of typical cases, and is used in the results shown below.

Some typical simulation results are shown in Figs. 3.10 to 3.13. The parameters used for the computations correspond to the conditions of the Los Alamos free-electron laser[102]. In Fig. 3.10 we see the increase of the laser extraction efficiency (the fraction of the electron energy converted into light)as the laser grows from noise to saturation, and the corresponding decrease of the gain at saturation.

The laser pulse itself is shown in Fig. 3.11 for times which correspond to the period during which the laser grows from noise to saturation and to the period during which the laser is fully saturated. During the period of growth to saturation, the pulse is smooth. When the laser reaches saturation, the pulse becomes strongly modulated due to the growth of so-called synchrotron

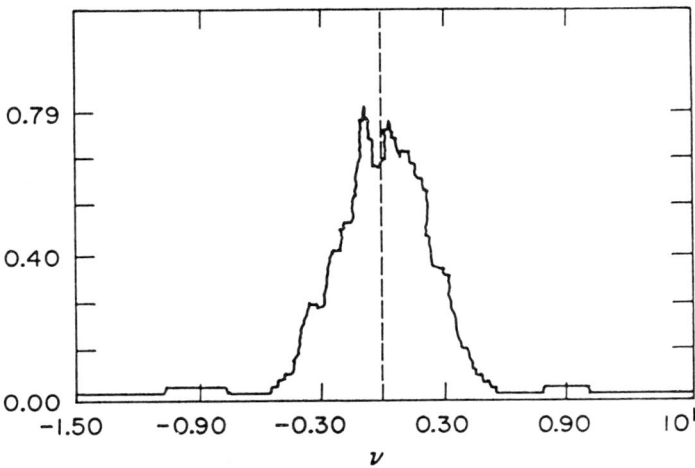

Fig. 3.9. Computed noise (or spontaneous-emission) spectrum.

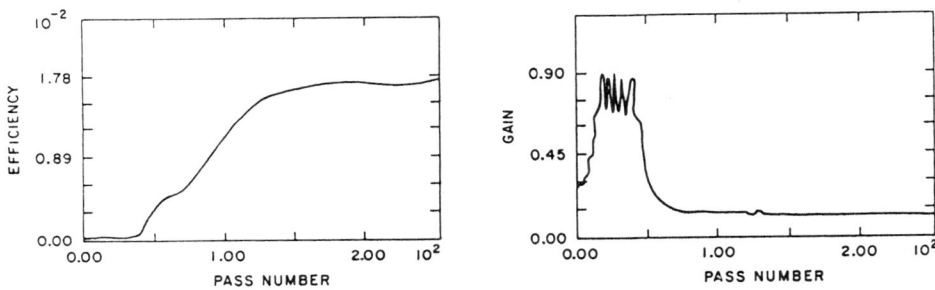

Fig. 3.10. Growth of the laser efficiency from noise to saturation, and saturation of the laser gain.

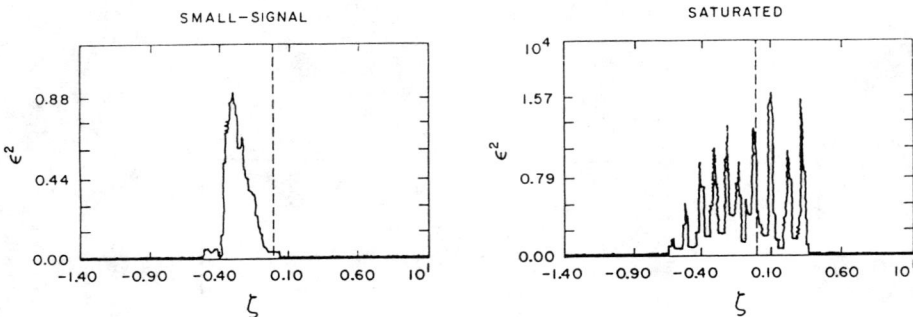

Fig. 3.11. Laser optical pulse in the small-signal and saturated regimes.

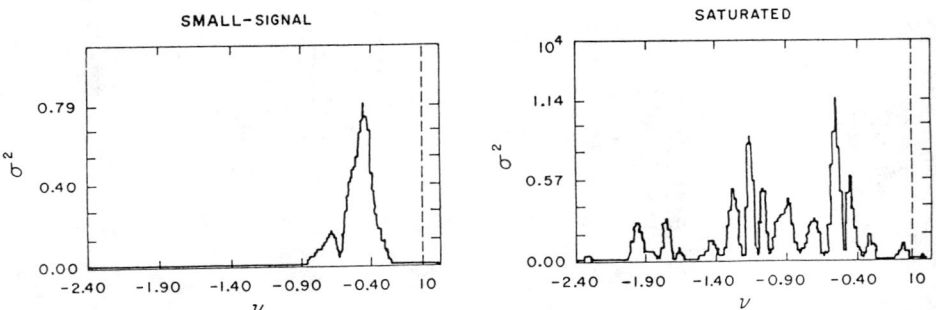

Fig. 3.12. Laser optical spectrum in the small-signal and saturated regimes.

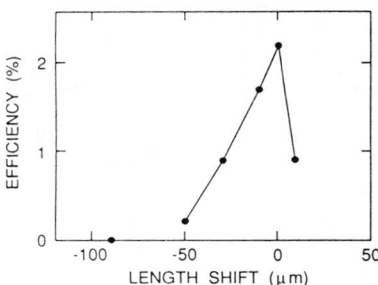

Fig. 3.13. Computed resonator-length desynchronism curve.

instabilities. These are discussed in detail in the following section.

The spectrum is shown in Fig. 3.12. As shown there, the spectrum is narrow during the growth phase, and is centered on the frequency $\nu = -2.6$, which is the frequency for maximum small-signal gain. When the laser becomes fully saturated, the spectrum spreads to lower frequencies (longer wavelengths) due to the synchrotron instabilities.

The effect of the length of the laser cavity on the synchronism of the laser and electron-beam pulses is illustrated in Fig. 3.13. The laser power and efficiency are maximum when the pulses overlap at each pass, that is when the round-trip time of the optical resonator is exactly the spacing of the electron-beam pulses.

The variation of the laser power with the cavity length observed in experiments at Los Alamos is shown in Fig. 3.14[103]. The qualitative agreement with the theoretical computations is rather good. Clearly, the width of the synchronism curve is quite narrow, amounting to only 0.25 mm. The asymmetry of the curve is also interesting, and is related to the effect called laser lethargy.[104] Laser lethargy describes the finite time required for a gain medium suddenly immersed in a laser field to establish a coherent polarization. In a free-electron laser, the electrons begin to wiggle as soon as they enter the wiggler, but the synchrotron motions which modulate the electron density and establish coherence between the electron radiation and the laser field require some time to occur. Thus, the electrons radiate more strongly at the downstream end of the wiggler than they do at the upstream end. Since the electrons are travelling at less than the speed of light, they slip backward through the optical pulse as the two pulses proceed down the length of the wiggler, as shown in Fig. 3.8. Thus, toward the end of the wiggler, when the electrons are radiating most strongly, they have slipped toward the back of the optical pulse. As a result, the back of the optical pulse is more strongly amplified, and the centroid of the pulse appears to be retarded. In fact, when the electrons slip out of the laser pulse to the left, they continue to radiate coherently due to the density modulation they have built up. This stretches the optical pulse to later times. To compensate for this, the free-electron laser may be operated with the cavity slightly shortened. This advances the position of the optical pulse at the beginning of each pass through the wiggler. On the other hand, if the optical cavity is lengthened, then the optical pulse is shifted to the left at the beginning of each pass. Since the electrons at the front of the pulse have no polarization other than the random distribution of the electrons in the beam, they radiate almost nothing at the front of the pulse to make up for the displacement of the optical field out of this region. Thus, after several passes the optical pulse slips completely behind the electron pulse and the laser turns off. For this reason, the synchronism curve is truncated almost immediately on the long-cavity side of

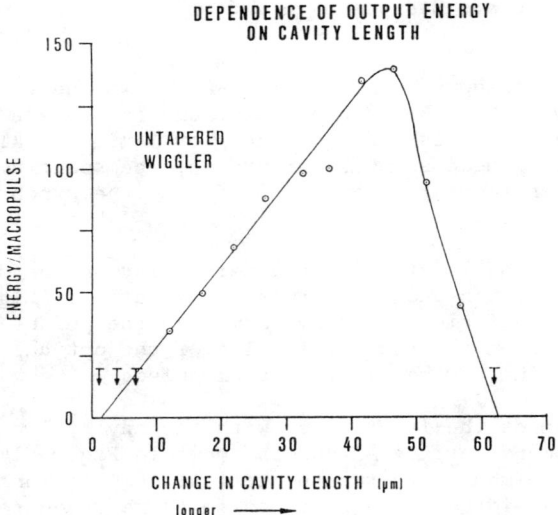

Fig. 3.14. Resonator-length desynchronism curve observed in the experiments at Los Alamos.

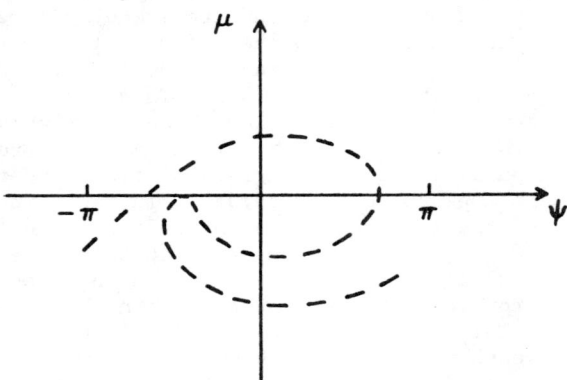

Fig. 3.15. Phase-plane motion of electrons when the laser intensity is high (saturated regime).

exact synchronism. The truncation would be absolute if not for the
contribution of spontaneous emission.

3.2.2. Synchrotron instabilities

The complex structure evident in the saturated optical pulse of
Fig. 3.11 is the result of synchrotron instabilities. These
instabilities may be understood by analogy to the Raman effect in
molecular spectroscopy. In the Raman effect, molecular vibrations
cause the polarizability of the molecule to oscillate at the
molecular vibrational frequency. As a result, light scattered by the
molecule is modulated at the vibrational frequency of the molecule.
The scattered radiation then contains not only the original
frequency, but also sidebands shifted from the original frequency by
the vibrational frequency. The lower (long-wavelength) sideband is
called the Stokes line and the upper sideband is called the anti-
Stokes line. In coherent Raman scattering, which occurs when the
incident laser is very intense, gain is observed at the sideband
frequencies, and radiation at the sideband frequency is amplified.
Generally the long-wavelength line has the highest gain, and since
the outgoing photon has less energy than the incoming one, the
molecule is left in a vibrationally excited state. In free-electron
lasers the electrons are oscillating in the buckets of the
ponderomotive potential with the synchrotron frequency, as discussed
previously. These oscillations can mix with the emission and
absorption processes to create radiation at frequencies shifted from
the fundamental by the synchrotron frequency, given by Eq. (3.69).
At sufficiently high intensity, gain is observed on the long-
wavelength sideband. Because of the analogy with the molecular Raman
effect, the synchrotron instability is sometimes referred to as the
Raman instability. This is unfortunate because the parameter region
in which collective oscillations are excited is also referred to as
the Raman region, as discussed in Chapter 2. In addition, the
synchrotron frequency depends on the laser intensity, as indicated by
Eq. (3.69), and is lower for electrons in orbits near the separatrix
in phase space. Thus, there are important differences between the
effect observed in free-electron lasers and the molecular Raman
effect.

The mechanism by which the synchrotron oscillations modulate
the optical field may be understood as follows. At sufficiently high
intensity, the electrons describe trajectories in the phase plane
which circulate around the origin. For the first half of this
circulation, electrons which begin with energies above the resonant
value lose energy as they move to the lower half of the phase plane,
as illustrated in Fig. 3.15. During this phase of the electron
motion, the local optical field experiences gain. Once the electrons
complete half of a synchrotron oscillation, they begin to increase in
energy at the expense of the optical field, as indicated in Fig.
3.15. The local optical field then experiences absorption. Since
the electrons have slipped relative to the optical beam during the

absorption phase, the absorption is shifted in space relative to the gain, so that a modulation of the optical beam, as shown in Fig. 3.16, results.

When the optical beam and electron beam are uniform, electrons with different phases average against one another so that the optical beam sees a uniform net gain. However, when the optical beam is modulated, the modulations affect the electron motions in such a way that the peaks are amplified and the troughs are attenuated, and random fluctuations grow in an unstable fashion. This is illustrated in Fig. 3.16 for the case of a strongly modulated, intense optical field, and may be understood as follows. If we consider first an electron which enters the wiggler coincident with the leading edge of a modulation peak, we see that the electron describes the first half of a synchrotron oscillation inside the peak, loses energy, and amplifies the peak. When it reaches the trough, the electron moves only slowly in the phase plane, due to the weak field in the trough, and causes weak absorption. These processes increases the strength of the modulation. Similarly, if we examine the behavior of an electron which enters the wiggler coincident with the leading edge of the trough, we see that as the electron slips through the trough it moves only slowly along the beginning of its trajectory in the phase plane, slightly amplifying the field in the trough. However, when the electron reaches the next peak, it accelerates its motion in the phase plane and completes the first half of a synchrotron period, strongly amplifying the peak. In this way, the synchrotron motions become coherent with respect to the modulation and amplify it.

Since the modulation is caused by the slip of the electrons, the period of the modulation is the slip length. This periodicity is evident in Fig. 3.11, and is discussed in detail in the next section. As is well known in the theory of radio transmission, amplitude modulation of a carrier wave introduces sidebands into the frequency spectrum. To see this, we expand the signal shown in Fig. 3.16, consisting of a carrier wave at the resonant wavelength modulated by the amount α, to show that

$$[1 + \alpha \cos(\zeta)] \cos(2\pi N_W \zeta) = \cos(2\pi N_W \zeta)$$
$$+ \frac{\alpha}{2} \{\cos[(2\pi N_W + 1)\zeta] + \cos[(2\pi N_W - 1)\zeta]\}. \qquad (3.91)$$

A phasor diagram of the components of the modulated wave is shown in Fig. 3.17. Experimentally and in simulations, as indicated in Fig. 3.13, it is observed that the lower sideband predominates. This may be understood in the following way[105]. For small values of α, we see from Fig. 3.17 that the carrier with only the lower sideband is given approximately by the expression

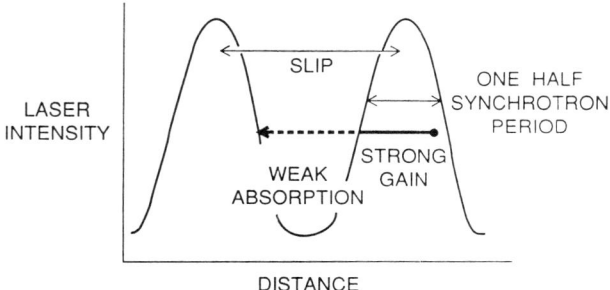

Fig. 3.16. Growth of strong modulation of the optical beam by the synchrotron motion and slip of the electrons.

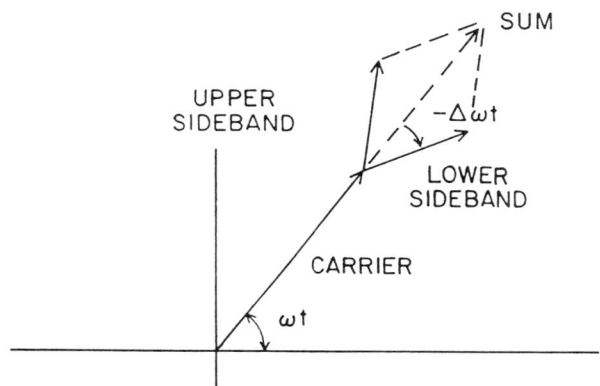

Fig. 3.17. Phasor diagram of a carrier wave modulated by the presence of sidebands.

$$\cos(2\pi N_W) + \frac{\alpha}{2} \cos[(2\pi N_W - 1)\zeta]$$

$$\approx \left[1 + \frac{\alpha}{2}\cos(\zeta)\right] \cos\left[2\pi N_W \zeta - \frac{\alpha}{2}\sin(\zeta)\right], \tag{3.92}$$

which shows that the signal is both amplitude and frequency modulated. The instantaneous shift of the frequency from resonance is given by the expression

$$\nu = -\frac{\alpha}{2}\cos(\zeta). \tag{3.93}$$

The effect of this frequency modulation is illustrated in Fig. 3.18. As shown there, the frequency is lower in the peaks than in the troughs. Thus, an electron is farther above resonance in the peaks than in the troughs. Since the energy loss is smaller for an electron near (or below) resonance, we see that the gain in the peaks is enhanced by the frequency modulation. The opposite is true for the upper sideband, which accounts for the predominance of the lower sideband.

Various theories have been advanced to calculate the gain of the sidebands[106,107]. A simple theory would be extremely useful for predicting the threshold for the synchrotron instabilities, and the attenuation required to suppress them. However, the theories all depend upon the details of the distribution of the electrons in the phase plane, and are therefore difficult to compare with each other and with experiment. For this reason, simulations of the type discussed above and shown in Figs. 3.11 and 3.12 must be used. It is clear, however, from both experiments and simulations, that the instabilities appear only when the intensity is high. This is evident in Fig. 3.19. It is to be expected that the synchrotron instabilities will be strong when the synchrotron period is comparable with the wiggler length, although other effects such as resonator-length desynchronism also play a role[108]. Since saturation also occurs when the synchrotron period is comparable to the wiggler length, as indicated in Eq. (3.71), synchrotron instabilities are generally observed in strongly saturated lasers.

When the instability exceeds threshold, the peaks grow higher and narrower until they become isolated spikes, as shown in Fig. 3.11. The width is controlled by the fact that as the peak increases, the synchrotron period becomes shorter and the electrons enter the absorption phase before they slip completely through the peak. As the spikes become stronger and narrower, the spectrum becomes correspondingly broader. It is also found that when the simulations include a random spread of electron energy, the peaks are

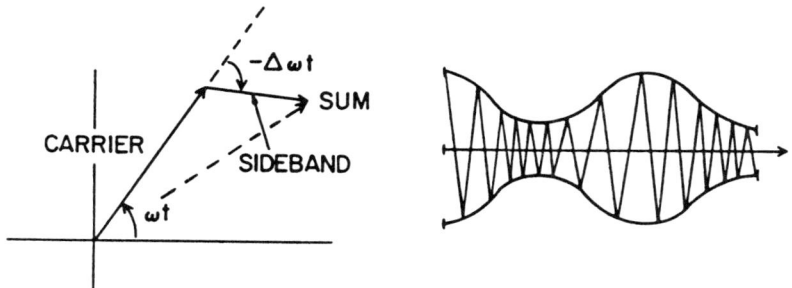

Fig. 3.18. Diagram showing the frequency modulation of a carrier wave modulated by only a single sideband (the lower sideband).

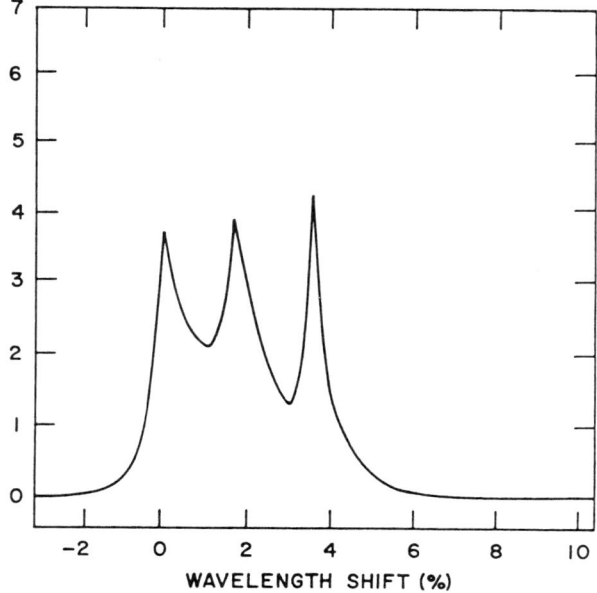

Fig. 3.19. Experimentally observed spectra showing modulation due to synchrotron instabilities at high laser power.

not as regularly spaced as indicated in Fig. 3.11, and the spectrum likewise becomes more complex than that shown in Fig. 3.12.

Although the instabilities degrade the spectrum, they can be useful by increasing the laser efficiency severalfold[109]. This happens because the electrons interact with the ponderomotive potential only when they are within the spikes. Since the width of the spikes is only a fraction of the spacing between them, the electrons interact only during a correspondingly short section of the wiggler. Since the saturated efficiency is generally less than $1/2N_W$, as found in Eq. (3.73), an effectively shorter wiggler will have higher efficiency. This effect is evident in Fig. 3.10, where we see that the efficiency jumps when, after beginning to saturate, the laser becomes unstable with respect to synchrotron modulations. The highest observed efficiency for a uniform wiggler is about 1.3 percent[110].

When desired, the synchrotron instabilities may (at least conceptually) be suppressed in one of several ways. The most straightforward is to place a filter in the optical resonator to introduce absorption at the wavelength of the synchrotron sideband. This has been demonstrated theoretically[111], and experimentally, using a narrow-bandpass mirror as both the optical resonator and the filter[112]. Recently, a grating in the cavity has been used as a filter for the same purpose[113], and a dispersive grating pair which desynchronizes the cavity length for the sidebands has also been proposed[114]. It has also been found that under certain circumstances the sidebands can be suppressed by detuning the cavity length[115].

3.2.3. Long pulses

For electron pulses which are many slip lengths long, a large number of electron trajectories and a large number of optical field points must be computed to accurately simulate the laser performance. This can become prohibitively expensive, in terms of computer time, and can exceed available computer memory. However, within such long pulses, one section of the pulse looks much like another, and evolves more or less independently. During a single pass of the optical pulse through the wiggler, the influence of each electron does not extend beyond one slip length. For this reason, near the center of a long pulse the solution becomes roughly periodic, with a period equal to one slip length, allowing for small differences due to statistical fluctuations arising from the random distribution of the electrons in the pulse. This behavior is evident in Fig. 3.11, and suggests that the behavior near the central part of the pulse can be described by looking at a window in the pulse which is an integral number of slip lengths long. This is the basis of the periodic-boundary-condition approximation[116].

In this approximation, the window is defined by the boundaries

$$\zeta_- = -n_-, \qquad \zeta_+ = n_+, \tag{3.94}$$

where n_- and n_+ are small integers. The trajectories and optical fields are computed as before, except that when an electron slips out of the window to the left it "wraps around" and reappears at the right edge with the same energy and phase:

$$\mu_n \to \mu_n, \tag{3.95}$$

$$\psi_n \to \psi_n, \tag{3.96}$$

$$\zeta_n \to \zeta_n + n_+ - n_-. \tag{3.97}$$

These are sometimes referred to as periodic boundary conditions. Within the window the electrons are initially distributed with a uniform, random distribution, and the field is initialized as before.

Some results obtained with the code OKPulse are shown in Figs. 3.20 to 3.22. The parameters are the same as those used for Figs. 3.10 to 3.12. Comparing these figures, we see that the periodic model reproduces quite well the behavior of the center of the pulse. Even the spectra are quite comparable except, of course, that the spectrum in the periodic approximation becomes a series of lines because the periodic field can be represented by a Fourier series rather than a Fourier integral. The efficiency calculated with the periodic model is slightly higher since the average efficiency of the entire pulse is reduced by the low efficiency at the ends of the pulse where the intensity is lower.

Of course, it is not possible to compute all the properties of the complete pulse using the periodic approximation. In particular, all effects due to cavity-length synchronization are completely outside the scope of the model, and effects caused by variations in the electron energy distribution through the micropulse can be approximated only by repeating the computation for various positions in the pulse, using different electron-energy distributions. Nevertheless, the great savings of computer time and memory which are possible with the periodic approximation make it a useful tool even in the one-dimensional approximation. When three-dimensional effects are included, the computer time and memory required for complete pulse computations with sufficient time resolution to describe synchrotron instabilities exceed the limits of even the largest supercomputers. In this case, the periodic-boundary-condition approximation becomes essential.

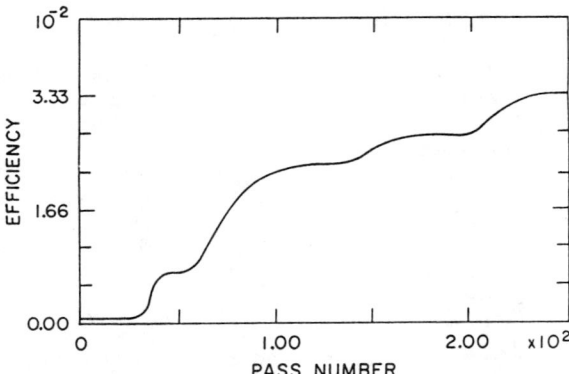

Fig. 3.20. Growth of the laser power from noise to saturation, and saturation of the laser gain computed using the periodic-boundary-condition approximation.

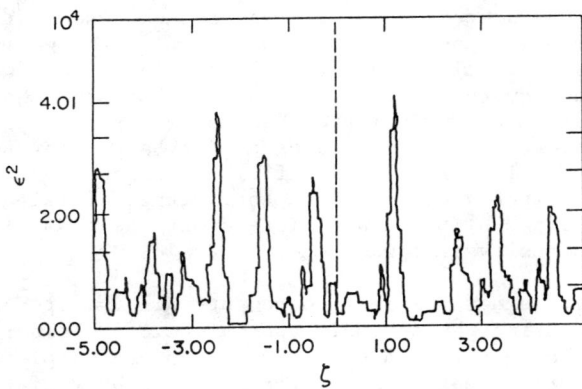

Fig. 3.21. Laser optical pulse in the saturated regime computed using the periodic-boundary-condition approximation.

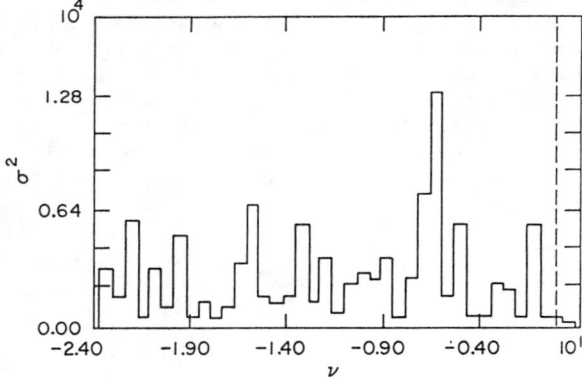

Fig. 3.22. Laser optical spectrum in the saturated regime computed using the periodic-boundary-condition approximation.

REFERENCES AND NOTES

1. H. Motz, J. Appl. Phys. 22, 527 (1951).
2. H. Motz, W. Thon, and R. N. Whitehorst, J. Appl. Phys. 24, 826 (1953).
3. R. M. Phillips, IRE Trans. Elec. Dev. 7, 231 (1960).
4. N. M. Kroll and W. A. McMullin, Phys. Rev. A17, 300 (1978).
5. John M. J. Madey, J. Appl. Phys. 42, 1906 (1971).
6. F. A. Hopf, et al., Opt. Comm. 18, 4 (1976).
7. W. B. Colson, Phys. Lett. 59A, 187 (1976).
8. L. R. Elias, et al., Phys. Rev. Lett. 36, 717 (1976).
9. D. A. G. Deacon, et al., Phys. Rev. Lett. 38, 892 (1977).
10. M. Billardon, et al., Proc. SPIE 453, 137 (1984).
11. J. A. Edighoffer, et al., Phys. Rev. Lett. 52, 344 (1984).
12. R. W. Warren, et al., " "First Operation of the Los Alamos Free-Electron Laser," Proceedings of the International Conference on Lasers '83, R. C. Powell, ed. (STS Press, McLean, VA, 1983).
13. D. B. McDermott, et al., Phys. Rev. Lett. 41, 1368 (1978).
14. T. C. Marshall, *Free-Electron Lasers* (MacMillan, New York, 1985).
15. M. Billardon, et al., Europhys. Lett. 3, 689 (1987).
16. T. J. Orzechowski, et al., Phys. Rev. Letters 54, 889 (1985).
17. In synchrotron-radiation sources, the magnetic field arrangement shown in Fig. 1.1 is usually referred to as an "undulator." "Wigglers" are similar, but have only a few, large, periods and much stronger magnetic fields. However, in free-electron lasers the term wiggler is more commonly used.
18. Actually, if we look closely we observe the electrons to oscillate back and forth with a slight figure-eight motion, but this is of no consequence for the present discussion.
19. B. E. Newnam, et al., Proc. SPIE 453, 137 (1984).
20. Actually, the wavelength is slightly longer than that given by Eq. (1.4) since the electrons must be at an energy slightly greater than resonance to provide positive gain. This difference is of the order of $1/N_W \ll 1$, and can ordinarily be ignored.
21. B. E. Newnam, et al., IEEE J. Quant. Electron. QE-21, 867 (1985).
22. L. R. Elias, IEEE J. Quant. Electron. QE-23, 1470 (1987).
23. D. W. Feldman, et al., IEEE J. Quant. Electron. QE-23, 1476 (1987).
24. L. R. Elias and G. J. Ramian, Proc. SPIE 453, 137 (1984).
25. L. R. Elias, et al., Phys. Rev. Lett. 57, 424 (1986).
26. L. R. Elias, et al., Phys. Rev. Lett. 42, 977 (1979).
27. J. Gea-Banacloche, et al, J. Quant. Electron. QE-23, 1558 (1987).
28. T. J. Orzechowski, et al., Phys. Rev. Letters 54, 889 (1985).
29. T. J. Orzechowski, et al., Phys. Rev. Lett., 57, 2172 (1986).
30. J. Adamski, et al, "Visible-Wavelength Oscillator," Proceedings of the Ninth International Free-Electron Laser Conference, Williamsburg, VA, September 14-18, 1987.
31. B. E. Newnam, et al., IEEE J. Quant. Electron. QE-21, 867 (1985).
32. A. Ts. Amatuni, et al., "Generation of Coherent Induced Undulator Radiation," Yerevan Physics Institute Report EÒN-727 (42)-84 (1984).
33. E. D. Shaw, et al., Nucl. Inst. Meth. A250, 44 (1987).

34. F. Ciocci, et al., "Status of the ENEA-FEL in the Medium Infrared," Proceedings of the Ninth International Free-Electron Laser Conference, Williamsburg, VA, September 24-28, 1987.
35. S. V. Benson, et al., Nucl. Inst. and Meth. **A250**, 39 (1986).
36. J. S. Fraser and R. L. Sheffield, IEEE J. Quant. Electron. **QE-23**, 1489 (1987).
37. B. E. Newnam, Proc. SPIE **738**, 155 (1987).
38. M. Billardon, et al., Europhys. Lett. **3**, 689 (1987).
39. R. Prazeres, et al., "First Production of Vacuum-Ultraviolet Coherent Light by Frequency Multiplication in a Relativistic Electron Beam," Proceedings of the Ninth International Free-Electron Laser, Williamsburg, VA, September 24-28, 1987.
40. J. Murphy and C. Pelligrini, J. Opt. Soc. Am. **1B**, 530 (1984).
41. M. W. Berns, et al., Science **213**, 505 (1981).
42. A. Wittin, et al., Phys. Rev. **A34**, 493 (1986).
43. "The Free-Electron Laser," C. K. N. Patel, et al., report of the Free-Electron Laser Subcommittee of the Solid-State Physics Committee of the National Academy of Science (National Academy Press, Washington, DC, 1982).
44. W. M. Yen, et al., J. Phys. Colloq. **C7**, 413 (1982).
45. D. H. Sliney, Clin. Chest Med. **6**, 2 (1985).
46. J. A. Parrish and T. F. Deutsch, J. Quant. Elec. **QE-20**, 1386 (1984).
47. J. A. Dixon, "Lasers in Surgery," Current Problems in Surgery **XXI**, No.9 (Yearbook Medical Publishers, Chicago, 1984).
48. K.K. Jain, IEEE J. Quant. Electron. **QE-20**, 1401 (1984).
49. D. S. J. Choy, IEEE J. Quant. Electron. **QE-20**, 1420 (1984).
50. K.K. Jain, IEEE J. Quant. Electron. **QE-20**, 1401 (1984).
51. E. J. Dehm, et al., Arch. Opthamology **104**, 1364 (1986).
52. J. A. Dixon, "Lasers in Surgery," Current Problems in Surgery **XXI**, No.9 (Yearbook Medical Publishers, Chicago, 1984).
53. R. Srinivasan, Science **234**, 559 (1986).
54. D. J. Erlich, Nucl. Inst. and Meth. **A239**, 397 (1985).
55. R. B. Hall, Laser Focus **18**, 57 (1982).
56. "Laser Induced Chemistry," Jason Report JSR 78-11 (1979).
57. A. Hartford, et al., J. Appl. Phys. **51**, 4471 (1980).
58. J. Wolfrum, et al., W. German Patent #293,853 (1981).
59. D. Prosnitz and A. M. Sessler, in *Free-Electron Generators of Coherent Radiation*, Phys. Quant. Electron. **9** (Addison-Wesley, Reading, MA, 1982).
60. D. Prosnitz and L. Schlitt, Proc. SPIE **270**, 102 (1981).
61. Aviation Week and Space Technology, August 18, 1986, p40.
62. Report to the American Physical Society of the Study Group on Science and Technology of Directed Energy Weapons (American Physical Society, New York, 1987).
63. G. Edwards, et al., Proceedings of the Ninth International Free-Electron Laser, Williamsburg, VA, September 24-28, 1987.
64. C.-M. Tang, et al., Nucl. Inst. Meth. **A250**, 278 (1986).
65. "The Free-Electron Laser," C. K. N. Patel, et al., report of the Free-Electron Laser Subcommittee of the Solid-State Physics Committee of the National Academy of Science (National Academy Press, Washington, DC, 1982).

66. R. S. Straight, private communication.
67. D. H. Sliney, Clin. Chest Med. **6**, 2 (1985).
68. J. M. J. Madey, J. Appl. Phys. **42**, 1906 (1971).
69. D. Bohm, *Quantum Theory* (Prentice-Hall, Englewood Cliffs, NJ, 1951), page 33.
70. J. Gea-Banacloche, *et al.*, IEEE J. Quant. Electron. **QE-23**, 1558 (1987).
71. A. Renieri, "Free-Electron Generation of Extreme Ultraviolet Coherent Radiation: Theoretical Aspects," in *Free-Electron Generation of Extreme Ultraviolet Coherent Radiation*, AIP Conference Proceedings 118-4, edited by J. M. J. Madey and C. Pelligrini (American Institute of Physics, New York, NY, 1984), page 1.
72. D. Bohm, *Quantum Theory* (Prentice-Hall, Englewood Cliffs, NJ, 1951), page 65.
73. S. Benson and J. M. J. Madey, Nucl. Inst. Meth. **A237**, 55 (1985).
74. H. A. Haus, IEEE J. Quant. Electron. **QE-17**, 1427 (1981).
75. N. M. Kroll and W. A. McMullin, Phys. Rev. **A17**, 300 (1978).
76. J. D. Jackson, *Classical Electrodynamics, Second Edition* (Wiley, New York, NY, 1975), page 490.
77. J. Gea-Banacloche, *et al.*, IEEE J. Quant. Electron. **QE-23**, 1558 (1987).
78. J. D. Jackson, *Classical Electrodynamics, Second Edition* (Wiley, New York, NY, 1975), page 497.
79. J. D. Jackson, *Classical Electrodynamics, Second Edition* (Wiley, New York, NY, 1975), page 524.
80. C.-C. Shi and A. Yariv, IEEE J.Quant. Electron. **QE-17**, 1387 (1981).
81. T. C. Marshall, *Free-Electron Lasers* (McMillan, New York, 1985).
82. F. A. Hopf, *et al.*, Opt. Comm **18**, 413 (1976).
83. N. M. Kroll, *et al.*, IEEE J. Quant. Electron. **QE-17**, 1436 (1981).
84. W. B. Colson, Phys. Lett. **59A**, 187 (1976).
85. C.-C. Shih and A. Yariv, IEEE J. Quant. Electron. **QE-17**, 1387 (1981).
86. T. J. T. Kwan, "An Investigation of Efficiency Optimization in Free-Electron Lasers," in *Free-Electron Generators of Coherent Radiation*, Physics of Quantum Electronics, Volume 7, edited by S. F. Jacobs, *et al.* (Addison-Wesley, Reading, MA, 1980), page 491.
87. M. Abramowitz and I. A. Stegun, *Handbook of Mathematical Functions, Tenth Printing* (U. S. Government Printing Office, Washington, DC, 1972), page 360.
88. J. D. Jackson, *Classical Electrodynamics, Second Edition* (Wiley, New York, NY, 1975), page 222.
89. W. B. Colson, "Optical Pulse Evolution in the Stanford Free-Electron Laser and in a Tapered Wiggler," in *Free-Electron Generators of Coherent Radiation*, Physics of Quantum Electronics, Volume 8, edited by S. F. Jacobs, *et al.* (Addison-Wesley, Reading, MA, 1982), page 447.
90. C. A. Brau, IEEE J. Quant. Electron. **QE-16**, 335 (1980).
91. L. R. Elias, *et al.*, Phys. Rev. Lett. **38**, 892 (1977).
92. J. M. J. Madey, Nuovo Cim. **50B**, 64 (1979).
93. N. M. Kroll, *et al.*, IEEE J. Quant. Electron. **QE-17**, 1436 (1981).

94. J. D. Jackson, *Classical Electrodynamics*, Second Edition (Wiley, New York, NY, 1975), page 671.
95. W. B. Colson, et al., Phys. Rev. A **31**, 828 (1985).
96. P. Elleaume, J. Physique 44, C1-333 (1983).
97. N. M. Kroll, et al., "Variable-Parameter Free-Electron Laser," in *Free-Electron Generators of Coherent Radiation*, Physics of Quantum Electronics, Volume 7, edited by S. F. Jacobs, et al. (Addison-Wesley, Reading, MA, 1980), page 89.
98. N. M. Kroll, et al., "Variable-Parameter Free-Electron Laser," in *Free-Electron Generators of Coherent Radiation*, Physics of Quantum Electronics, Volume 7, edited by S. F. Jacobs, et al. (Addison-Wesley, Reading, MA, 1980), page 89.
99. A. Renieri, IEEE Trans. Nucl. Sci. **NS-26**, 3827 (1979).
100. H. Haus, IEEE J. Quant. Electron. **QE-17**, 1427 (1981).
101. J. M. J. Madey, et al., Proc. SPIE **453**, 315 (1984).
102. B. E. Newnam, et al., Nucl. Inst. Meth. **A237**, 187 (1985).
103. B. E. Newnam, et al., Nucl. Inst. Meth. **A237**, 187 (1985).
104. H. Al-Abawi, et al., Opt. Commun. 30, 235 (1979).
105. R. W. Warren, et al., Nucl. Inst. Meth. **A250**, 19 (1986).
106. N. M. Kroll, et al., IEEE J. Quant. Electron. **QE-17**, 1436 (1981).
107. S. Riyopoulos and C.-M. Tang, Nucl. Inst. Meth. **A259**, 226 (1987).
108. D. C. Quimby, Proc. SPIE **738**, 103 (1987).
109. R. W. Warren, et al., Nucl. Inst. Meth. **A250**, 19 (1986).
110. D. W. Feldman, et al., IEEE J. Quant. Electron. **QE-23**, 1476 (1987).
111. R. W. Warren and J. C. Goldstein, "The Generation and Suppression of Synchrotron Sidebands", Proceedings of the Ninth International Free-Electron Laser Conference, Williamsburg, VA, September 14-18, 1987.
112. J. C. Goldstein, et al., "Sideband Suppression by an Intracavity Optical Filter in the Los Alamos Free-Electron Oscillator," Proceedings of the Ninth International Free-Electron Laser Conference, Williamsburg, VA, September 14-18, 1987.
113. J. E. Sollid, private communication.
114. R. L. Tokar, J. Quant. Electron. **QE-24** (1988), to be published.
115. R. W. Warren and J. C. Goldstein, "The Generation and Suppression of Synchrotron Sidebands," Proceedings of the Ninth International Free-Electron Laser Conference, Williamsburg, VA, September 14-18, 1987.
116. W. B. Colson, Nucl. Inst. Meth. **A250**, 168 (1986).

FREE-ELECTRON LASERS

R.H. Pantell
Electrical Engineering Department
Stanford University
Stanford, CA 94305

TABLE OF CONTENTS

1	Introduction...	1708
2	Electron Beam Sources for FELs.........................	1714
	2.1 RF Linac..	1714
	2.2 Superconducting Accelerator (SCA)...............	1715
	2.3 Microtron.......................................	1715
	2.4 Van de Graaf Accelerator........................	1716
	2.5 Induction Accelerator...........................	1716
	2.6 Pulse-Line Accelerator..........................	1717
	2.7 Storage Ring....................................	1718
3	Operating Characteristics of the FEL...................	1718
	3.1 Gain..	1718
	3.2 Optical Guiding.................................	1722
	3.3 Cavity Design...................................	1722
	3.4 Beam Quality Requirements.......................	1723
	3.5 Optical Klystron................................	1725
4	Summary and Conclusion.................................	1727
	References..	1728

FREE-ELECTRON LASERS

R.H. Pantell
Electrical Engineering Department
Stanford University
Stanford, CA 94305

1 INTRODUCTION

The purpose of the free-electron laser[1-6] (FEL) is to convert the kinetic energy of an electron beam to electromagnetic energy. There are two primary components to the FEL:
  (1) A device for providing energetic particles.
  (2) A device for converting particle kinetic energy to radiation.
Thus far, there has been a greater variety of approaches in the first category than in the second.

To achieve energy transfer between a particle beam and a wave over an extended distance requires a synchronism condition between the electron velocity and the phase velocity of the wave. This presents a problem since, in vacuum, a plane wave has a phase velocity equal to c, the velocity of light, and the electrons have a velocity v such that v < c. In addition, since the electrons enter the system uniformly distributed along the direction of motion, bunching must occur so that a majority are in a phase of the field for which the wave extracts energy from the particles.

At microwave frequencies the synchronism problem may be solved by using a periodic structure, as in a linac or TWT, such that a component of the wave that is established to match the periodic boundary propagates with a phase velocity equal to or less than c. This is not a good solution for infrared and visible wavelengths because:
  1. The structure becomes difficult to fabricate because structure dimensions are a fraction of a wavelength.
  2. The synchronous fields decay exponentially from the conducting surface with the e^{-1} distance = $\gamma\lambda/2\pi$, where γ is the ratio of particle energy to rest energy and λ is the wavelength. At short wavelengths this distance is small, thereby limiting the interaction volume.

The solution provided by the FEL is shown in Figure 1, wherein a periodic magnet produces an undulating beam trajectory. A wave absorbs energy from the transverse motion and, for an oscillator, buildup occurs between mirrors of an optical oscillator. For the "hybrid wiggler" design, as illustrated in the lower portion of Figure 1, permeable pole faces are used to concentrate the flux in the plane normal to the direction of relativistic particle motion.

The rate of energy change of an electron beam in a wave is given by

$$mc^2 \frac{d\gamma}{dt} = e\, \mathbf{v} \cdot \mathbf{\mathcal{E}} \qquad (1)$$

where   m = rest mass
        e = electron charge
        v = velocity
        $\mathcal{E}$ = electric field.

To extract energy from the electron it is necessary that $v\cdot\mathcal{E} > 0$ averaged over an extended interval.

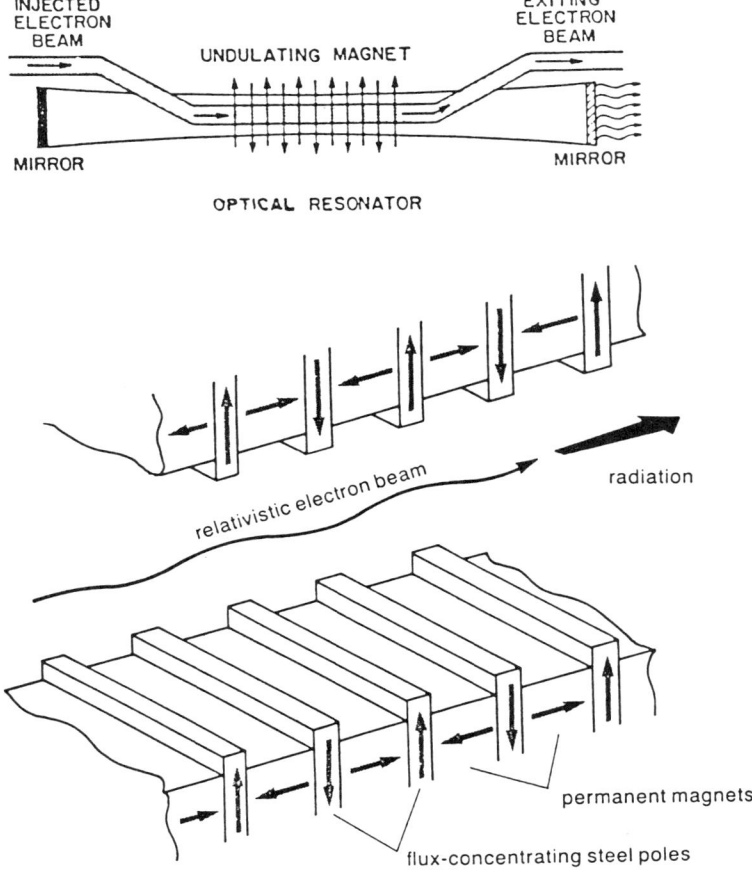

Figure 1: Cumulative energy transfer from an electron beam to a wave is accomplished in the FEL by introducing an undulating beam trajectory.

Figure 2 illustrates how this is accomplished by means of a beam wiggling in the x-z plane. The periodicity of the motion is λ_w, the wiggler period, and an electron is shown at two locations on its trajectory: one where the transverse velocity is in the positive x-direction and the other with a transverse velocity in the negative x-direction. The electric field vector, which is in the x-direction, is drawn such that at both locations $v\cdot\mathcal{E} > 0$. To accomplish this the electron slips half an optical wavelength in traversing half the wiggler period, giving the synchronism condition

$$\frac{\lambda}{\lambda_w} = \frac{c}{\bar{v}_z} - 1 \qquad (2)$$

where $\bar{v}_z$ is the z-component of electron velocity averaged over $\lambda_w/2$.
For $\gamma \gg 1$,

$$\frac{\lambda}{\lambda_w} = \frac{1 + a_w^2}{2\gamma^2} \qquad (3)$$

where a_w, the wiggler parameter, is given by

$$a_w = \frac{|e|\lambda_w B}{2\pi mc} = 0.093\, B(kG)\, \lambda_w(cm) \qquad (4)$$

and B is the r.m.s. value of the wiggler field. Typical values for a_w lie in the range 0.5 to 1.0. For example, for $\lambda_w = 2.5$ cm, $a_w = 1.0$ and $\gamma = 100$, from Eq.(3) we find that the wavelength is 2.5 μm.

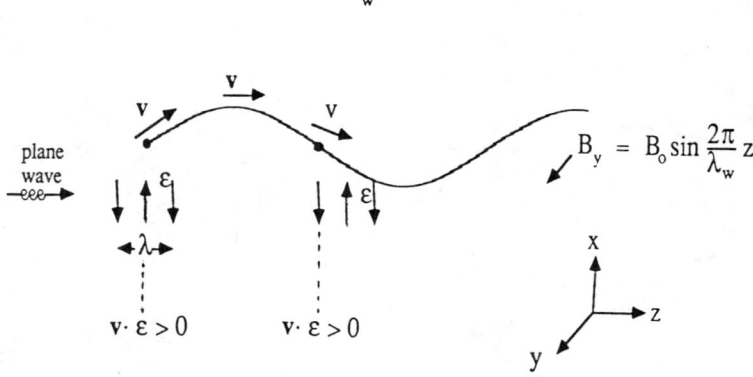

Figure 2: The dot product of electron velocity and electric field is positive on both half cycles of the wiggling motion.

Two methods for changing the wavelength of an FEL can be seen from Eq.(3): the beam energy can be altered or a_w can be changed by varying the strength of the wiggler field. For a relativistic beam it is a rather slow process to obtain broadband tuning by changing γ because this requires readjusting magnets in the beam transport system. Magnetic field strength can be altered by changing the gap spacing in the wiggler magnet, but this results in an expensive wiggler.

An alternate derivation for the synchronism condition is obtained from the relationship between the wiggler period and the radiation wavelength that results from a Lorentz contraction and a Doppler shift. In the electron rest frame the periodicity of the motion is reduced to λ_w/γ, which results in radiation in the laboratory frame that is Doppler shifted to the value given by Eq.(3) in the direction of relativistic motion.

Some of the novel features of the FEL are:
1. It can operate over a wide range of frequencies since no material resonance is involved as in the usual laser. Wavelength is selected by specifying γ, λ_w and a_w.
2. The FEL has the capability of delivering high peak and average powers because of the high power capacity of a relativistic electron beam. i.e., there is not a problem of material damage except for the interferometer mirrors.

Figure 3 shows, as dots, the peak power obtained from FELs, along with the power from conventional lasers and microwave sources. From this figure it is seen that for wavelengths between 10 and 100 μm and below 1000Å there are essentially no alternative sources, so these are the wavelength intervals for which FELs are likely to be important.

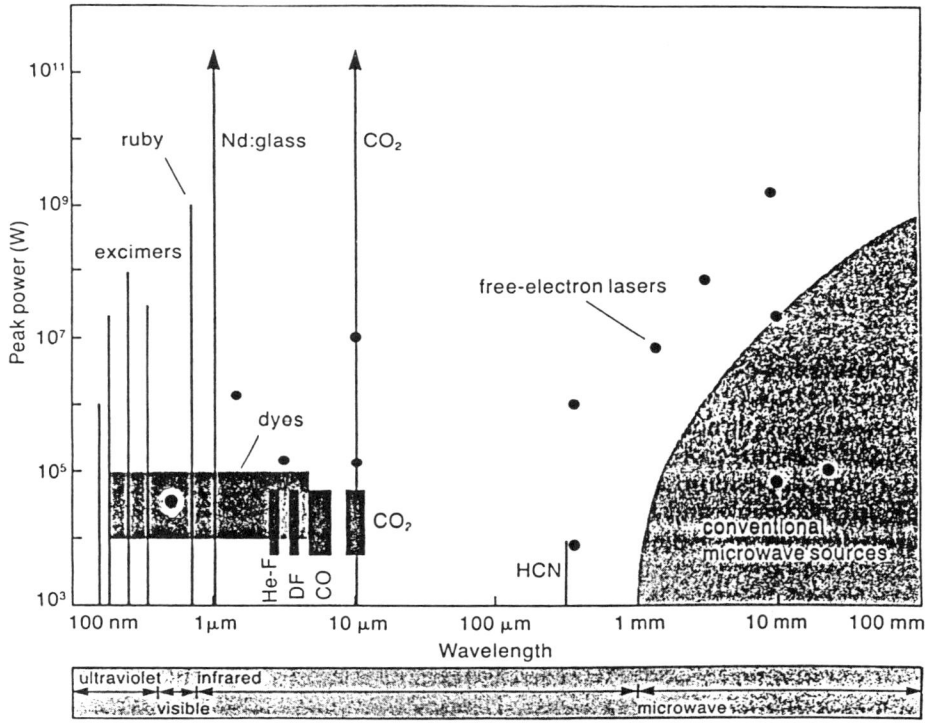

Figure 3: Conventional laser and free-electron laser sources.

Figure 4 shows the types of accelerators that have been used or might be used for FELs, the wavelength intervals appropriate for each accelerator, and the peak FEL power derivable from each type of machine. The Raman regime applies to high current density, low energy electron beams where space-charge forces affect the interaction, and the Compton regime is where space-charge forces are small (i.e., the beam plasma wavelength >> the interaction length). Inductions linacs have been used for FELs at the Lawrence Livermore National Laboratory (LLNL); r f linacs have been used at the Los Alamos National Laboratory (LANL), at Boeing Aircraft Company and at Stanford University; and a storage ring was used at Orsay to provide the first visible wavelength FEL.

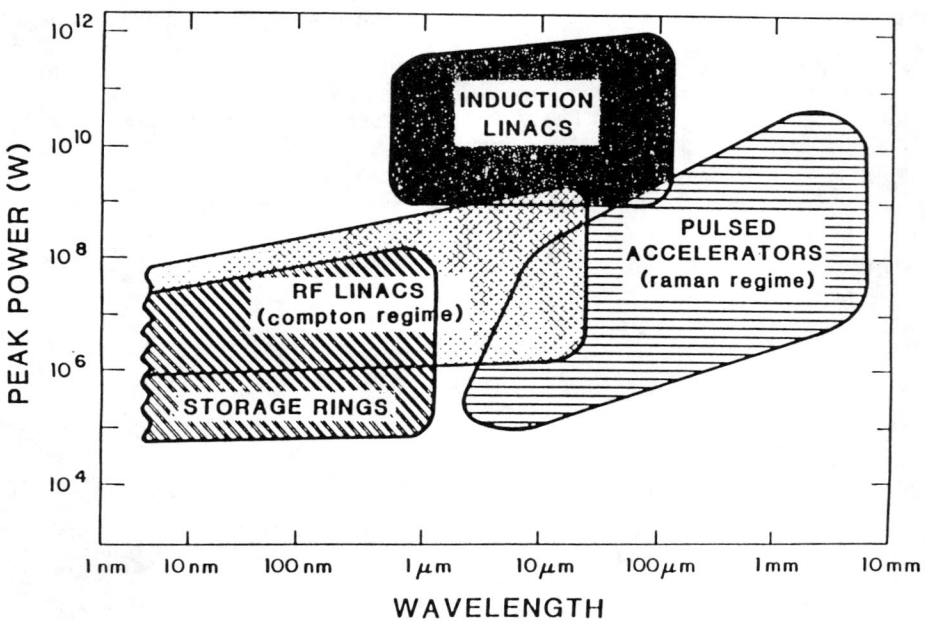

Figure 4: Accelerator technologies for FELs for wavelength intervals. (Courtesy of Dr. Brian Newnam, LANL)

The peak powers and wavelengths for various FELs operating at wavelengths below one millimeter are illustrated in Figure 5. A significant portion of the far IR spectrum has been covered by the Santa Barbara machine using a Van de Graaff accelerator. At the University of Paris the ACO storage ring provides the relativistic particles, and the remaining FELs in Figure 5 employ RF linacs.

A brief history of FELs is listed in Table I, divided into three time periods. John Pierce's analysis for the traveling-wave tube was non-relativistic and applied to longitudinal rather than transverse interactions. However, he presented all the essential elements for the FEL calculations. Hans Motz built a magnetic undulator and observed spontaneous emission at submillimeter wavelengths. Phillips, at General Electric Co., constructed the first FEL, then called a Ubitron, operating at 150 keV beam voltage and providing over 100 kW of power at S-band wavelengths. John Madey and coworkers achieved an infrared realization of the FEL, observing amplification at 10 μm wavelength.

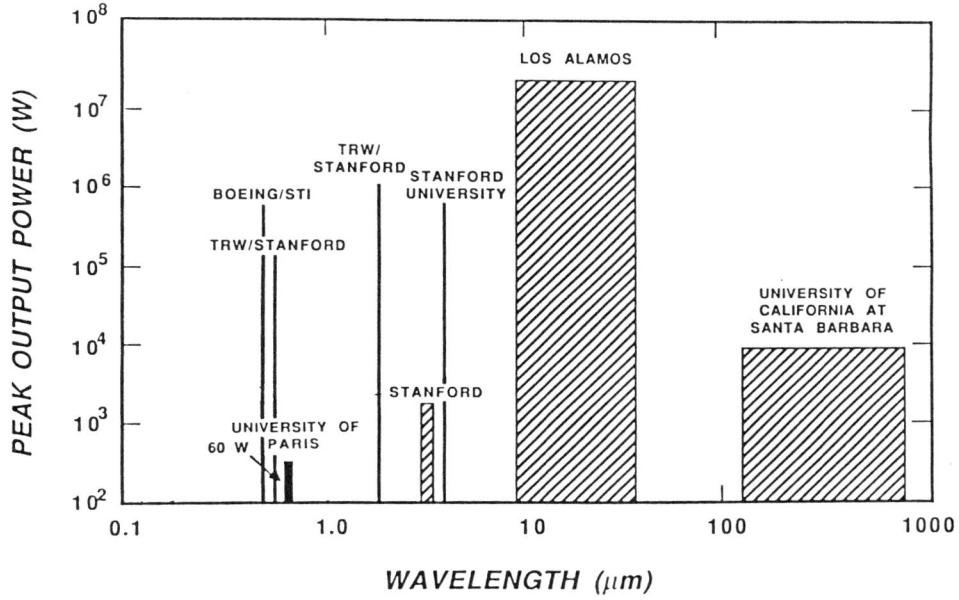

Figure 5: Free-electron laser oscillator wavelengths.
(Courtesy of Dr. Brian Newnam, LANL)

During the first half of the 1980s, several FELs came into operation. An induction linac was used at the LLNL to obtain amplification at $\lambda = 8.6$ mm, with over 30 db power gain and an efficiency greater than 40%. (Efficiency is defined as the ratio of the optical power to the electron beam power.) The Santa Barbara Van de Graaff FEL emitted in the far IR; the LANL facility operated at $\lambda = 10$ µm; and with gains on the order of 0.1%, the Orsay storage ring produced visible light.

In the second half of the 1980s a tunable FEL was assembled at Stanford University by John Madey and coworkers, designed to operate from 2.5 to 10 microns, and using a 40-MeV electron beam from an RF linac. Half the wavelength range was derived from a variation in the gap spacing of the wiggler magnet, so that only one change in beam energy was required for full tuning. At the Los Alamos National Laboratory the efficiency of the 10-µm-wavelength FEL was approximately doubled, from one to two percent, using a tapered wiggler. A Cherenkov laser was successfully operated at submillimeter wavelengths by passing a beam through a hollow dielectric cylinder. Third and fifth harmonics of the FEL oscillation wavelength were observed at Orsay, giving XUV radiation. Finally, an IR FEL operating at $\lambda = 4$ µm was shifted 0.7 µm towards the visible by introducing hydrogen gas into the wiggler, thereby reducing the phase velocity of the wave.

Table I: FEL History

Early History of the FEL (before 1980)			
1950	John Pierce	B.T.L.	FEL (non-relativistic) analysis: J.R. Pierce, *Traveling Wave Tubes*, D. Van Nostrand Co., N.Y. (1950)
1951	Hans Motz	Stanford	Spontaneous emission from a magnetic undulator ($\lambda < 500$ μm)
1960	R.M. Phillips	G.E.	Ubitron: the first FEL (V = 150 kV, $\eta \approx$ 10%, $\lambda \approx 10$ cm, $P \geq 100$ kW)
1975	J.M.J. Madey and coworkers	Stanford	Infrared realization of the FEL (7% amplification at $\lambda = 10.6$ μm)
Middle History (1980-1985)			
	ETA LLNL	Induction Linac	5 MeV, 850A, $\eta \geq 40\%$, $\lambda = 8.6$ mm (Amplifier in a waveguide)
	UCSB	Van de Graaff	2.5 MeV, 1.2A, $\lambda \approx 400$ μm
	LANL	RF Linac	20 MeV, 100A, $\lambda \approx 10$ μm
	Orsay	Storage Ring	160 MeV, 100 mA, $\lambda \approx 6500$ Å, $G \approx 10^{-3}$
Modern History (after 1985)			
	Stanford	Linac	$\lambda \sim 3$ μm (tunable)
	LANL	Linac	$\lambda \sim 10$ μm (tapered wiggler, $\eta \sim 2\%$)
	Dartmouth/U of Paris		$\lambda \sim 500$ μm (cherenkov)
	Stanford	SCA	$\lambda = 0.52$ μm (superconducting)
	U of Paris, Orsay	Storage Ring	$\lambda = 0.532$ μm = 0.177 μm (3rd harmonic) = 0.106 μm (5th harmonic)
	Stanford	Linac	H_2 gas in the wiggler $\lambda \approx 4$ μm ≈ 0.7 μm tuning

2. ELECTRON BEAM SOURCES FOR FELs

As noted previously, most of the diversity in FEL design is associated with the device for providing energetic particles, rather than in the device for converting particle kinetic energy to radiation.

2.1 RF Linac

The RF linac utilized a microwave field, propagating with a phase velocity about equal to the velocity of light, to accelerate electrons. Such accelerators have been used for FELs at Stanford, Los Alamos, Osaka, Boeing, and in the United Kingdom.

The linac at Stanford is powered by a 30-MW, S-band ($\nu = 2.856$ GHz) klystron. A typical acceleration gradient is 15 MV/m, so that in a 3-meter length the beam energy is 45 MV. (Recent experimental linacs have provided acceleration gradients of 350 MV/m for a 50-nsec pulse, yielding much higher energies in a given length.) The electron beam has the time structure illustrated in Figure 6, with a 2-psec micropulse and a 3-μsec macropulse. Time between micropulses, 350 psec, equals the period of the S-band microwave signal. The repetition rate is adjustable between 1.5 and 30 Hz, with usual operation at 15 Hz, giving 67 msec between macropulses. The beam power during the micropulse is $\approx$ 1.2 GW, $\approx$ 7 MW over the macropulse, and $\approx$ 300 W for the time average. Fractional energy spread on the electron beam is $\approx$ 0.5% and the unnormalized emittance is $\approx$ 0.5 mm-mrad. (The unnormalized emittance is the product of the beam radius and half-width, half-maximum divergence angle.)

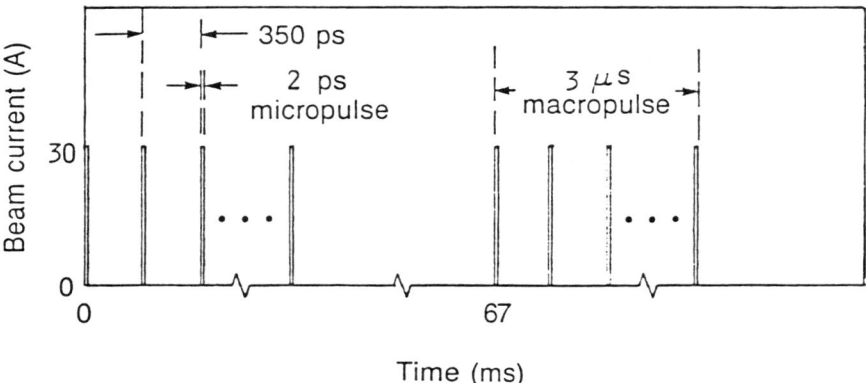

Figure 6: Time structure of the current from the Stanford Mark III RF linac.

2.2 Superconducting Accelerator (SCA)

A subclass of the RF linac is the SCA, wherein the circuit for the microwave acceleration field is superconducting. For the Stanford SCA, niobium is used at 1.9K, giving a cavity Q for the circuit in excess of 10^8, whereas the typical room temperature Q is less than 10^4. This allows for a long macropulse duration, 2-5 msec, with a high duty cycle and efficient linac acceleration. The long macropulse duration permits optical power buildup to saturation with low gain per pass.

For typical operation the micropulse current is 2.4 A, the micropulse duration is 3.2 psec, and the beam energy is 55 MV. By recirculating the beam through the accelerator several times, the beam energy can be as high as 150 MV. The microwave frequency is 1.3 GHz, the repetition rate is 5-10 Hz, and the time between micropulses is 84.6 nsec.

The beam quality is high, with a fractional energy spread of about 0.1% and unnormalized emittance < 0.07 mm-mrad. This linac was used for the first FEL experiments, and has since been used for FELs operating in the wavelength interval from 0.5 to 10 µm.

2.3 Microtron

The microtron, sketched in Figure 7, consists of a single accelerating cavity through which the beam makes repeated passes. This is accomplished by having the particles orbit in a magnetic field which returns the particles to the cavity. The parameters are adjusted such that each succeeding orbit is an integral number of microwave periods longer than the previous orbit, so as to have the particles enter the cavity at a phase of the field to produce acceleration on every pass.

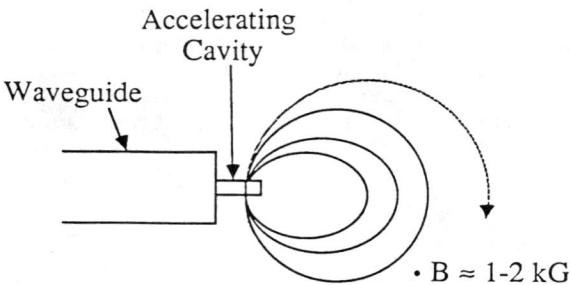

Figure 7: The microtron accelerator.

Since only a single cavity is used, in contrast with the RF linac, the cost and size of the accelerator are appreciably less for the microtron. However, the beam quality is not as good, so that it has been difficult to achieve appreciable FEL gain. Microtrons have been used for infrared FELs at Frascati, Bell Laboratories, and in the Soviet Union.

The electron beam parameters for the Frascati microtron are as follows: particle energy = 20 MV, micropulse duration = 20 psec, micropulse current = 6.5A, macropulse duration = 12 μsec, macropulse current = 0,35 A, microwave acceleration frequency = 3 GHz, repetion rate = 150 Hz, fractional energy spread = 0.12%, and unnormalized emittance = 5 mm-mrad.

2.4   Van de Graaff Accelerator

At the University of California in Santa Barbara, Dr. Luis Elias has used a Van de Graaff accelerator to provide relativistic electrons for a far IR FEL, with wavelengths in the 100-800 μm regime. A current slightly greater than one ampere is obtained over a 50-μsec pulse duration, with beam energies of 2-6 MV. The unnormalized emittance is 0.5 mm-mrad; height of the Van de Graaff machine is ten feet; and by decelerating the electron beam after it passes through the wiggler approximately 95% of the beam energy can be recovered.

2.5   Induction Accelerator

Figure 8 is a drawing of the induction accelerator, which is capable of producing kiloamperes of peak current at MV energies. The beam is subjected to a series of accelerating voltages, each derived from a changing magnetic flux. At the Lawrence Livermore National Laboratory (LLNL), 5-MV (the ETA) and 50-MV (the ATA) machines have provided 10 kA of current for pulse durations of 30 nsec and 50 nsec, respectively. The ETA has a length of 10 m, and the ATA is 53 m long.

Using the ETA, and allowing 850A into the wiggler with an unnormalized emittance slightly under 10^3 mm-mrad, laser emission was obtained at 8.6 mm wavelength, achieving a conversion efficiency from beam energy to optical energy in excess of 40%. The ATA has provided a beam for an FEL operating at 10 μm wavelength.

These FELs are single-pass amplifiers, rather than oscillators, and, at $\lambda = 8.6$ mm, 40 db gain has been achieved with a peak output power of one GW.

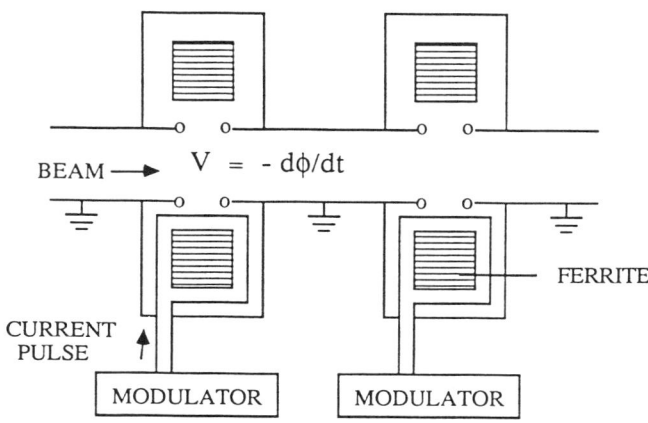

Figure 8: Schematic of the induction linear accelerator.

2.6   Pulse-Line Accelerator

At several MV energy, a kiloampere electron beam can be obtained from a pulse-line accelerator, wherein a capacitor bank is discharged through a coaxial line to provide a high voltage pulse between a cathode and anode. Because of the relative low voltage available, the pulse-line accelerator would be used for the microwave to far infrared portion of the spectrum. A typical pulse duration is 100 nsec with a fractional energy spread of several percent.

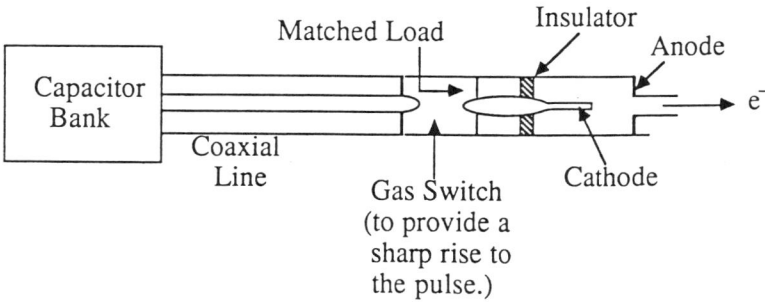

Figure 9: Sketch of a pulse-line accelerator.

2.7 Storage Ring

A high current density, low emittance beam can be obtained in the straight section of a storage ring, and therefore storage rings are proposed for the shorter wavelength (visible to x-ray) FELs. (The emittance requirement for FEL interaction scales linearly with wavelength.) Indeed, the first visible FEL was operated at $\lambda \approx 6500\text{Å}$ on the ACO storage ring in France with a gain of $\approx 0.1\%$. The parameters of the ACO ring are as follows: beam energy = 160 MV, peak current = 0.1A, and pulse duration = 1 nsec.

A disadvantage of the storage ring is that there is a maximum energy spread allowable to maintain stable orbits in the ring, which limits the energy spread (and therefore the output power) introduced by the FEL interaction. Energy spread caused by the FEL is reduced by synchrotron radiation cooling, leading to a relationship between maximum FEL power and synchrotron power:

$$\text{FEL power} = \frac{1}{2N_w} (\text{Synchrotron Power}) \qquad (5)$$

where N_w = number of wiggler sections.

3 OPERATING CHARACTERISTICS OF THE FEL

3.1 Gain

There are a variety of procedures for calculating FEL gain, but for all approaches the necessary input relationships are the force equation

$$\frac{d\mathbf{p}}{dt} = e\left[\mathcal{E} + \mathbf{v} \times \mathbf{B}\right] \qquad (6)$$

relating particle momentum **p** to electric $\mathcal{E}$ and magnetic **B** fields; Maxwell's equation (MKS units)

$$\nabla \times (\nabla \times \mathcal{E}) + \frac{1}{c^2}\frac{\partial^2 \mathcal{E}}{\partial t^2} = -\mu \frac{\partial \mathbf{J}}{\partial t} \qquad (7)$$

where μ is permeability and **J** is current density; and a continuity relationship

$$\nabla \cdot \mathbf{J} = \nabla \cdot (\rho \mathbf{v}) = -\frac{\partial \rho}{\partial t} \qquad (8)$$

where ρ is the induced charge density.

Defining gain G as the ratio of the increment in optical power to the input power, for $G \ll 1$ (which applies to most infrared and visible FELs), the simultaneous solution of the above equations gives

$$G = \frac{j}{4} \frac{d}{d\theta}\left[\frac{\sin\theta}{\theta}\right]^2 \qquad (9)$$

where

$$j = \left[\frac{\pi Z e}{mc^2}\right] \frac{I L^3 a_w^2(1+a_w^2)}{A \lambda \gamma^5} \qquad (10)$$

(for the helical wiggler)

$$= \left[\frac{\pi Z e}{mc^2}\right] \frac{I L^3 a_w^2(1+a_w^2)}{A \lambda \gamma^5} \left[J_0(\xi) - J_1(\xi)\right]^2 \qquad (11)$$

(for the planar wiggler)

Z = impedance of free space

$\xi = \dfrac{a_w^2}{2(1+a_w^2)}$

L = interaction length
I = beam current
A = beam area plus optical mode area
J_0, J_1 are Bessel functions of the first kind.

The parameter θ equals one-half the phase slippage from synchronism in a distance L. At synchronism, an electron slips one optical period per wiggler period, so that

$$\omega t - k z - 2\pi = 0 \qquad (12)$$

where $t = \lambda_w/v_s$, $z = \lambda_w$, and v_s is the electron velocity for which synchronism occurs. If the average value of the z-component of velocity $\overline{v}_z$ is not equal to v_s, then the phase slippage in one wiggler period is $\omega t - kz - 2\pi$, where $t = \lambda_w/\overline{v}_z$ and $z = \lambda_w$. For an interaction length L, the phase slip is

$$\frac{L}{\lambda_w}\left[\omega\left(\frac{\lambda_w}{\overline{v}_z}\right) - k(\lambda_w) - 2\pi\right]$$

$$= L\left[k\left(\frac{1}{\overline{\beta}_z} - 1\right) - k_w\right] \equiv 2\theta \qquad (13)$$

where $\overline{\beta}_z = \overline{v}_z/c$.

Figure 10 is a plot of FEL gain, as given by Eq.(9), as a function of θ. In the left-half plane the electron velocity exceeds the synchronous velocity, and in the right-half plane electron velocity is less than the synchronous velocity. At exact synchronism the gain is zero, and maximum gain occurs at θ = −1.3 where G = 0.135j.

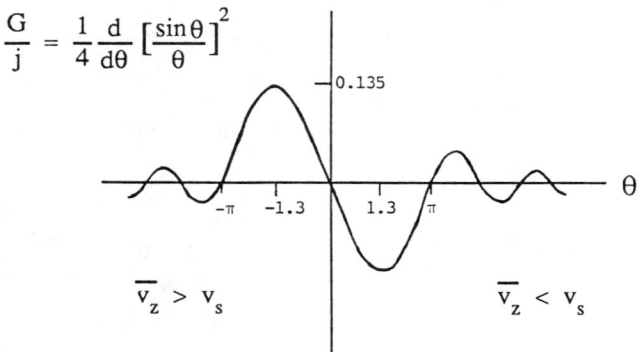

Figure 10: FEL gain as a function of phase slippage.

Figure 11 is a sketch explaining why the electron velocity must exceed the synchronous velocity for gain to occur. At the wiggler entrance, where z = 0, the electrons are uniformly distributed in the phase of the optical field. Electron bunching occurs at a distance z = L as a consequence of some particles gaining energy and others losing energy. However, at synchronous velocity bunching occurs around the zero field point so that no net energy transfer occurs. As illustrated in the last diagram, if the electron velocity is slightly greater than synchronous velocity, most electrons can bunch in a phase of the field for which there will be net energy extraction from the beam.

As an example of FEL gain, consider the following parameters: I = 10 A, L = 1 m, a_w = 0.7, λ = 1 μm, γ = 100, A = 2 x $10^{-6}m^2$. This gives j = 0.84 and a maximum gain of 11%.

When the gain per unit length becomes large, the electromagnetic field grows exponentially

$$\mathcal{E} \approx \frac{\mathcal{E}_0}{3} \exp\left\{ \frac{\sqrt{3}}{2} \left(\frac{j}{2}\right)^{1/3} + \frac{i}{2} \left(\frac{j}{2}\right)^{1/3} + ikL - i\omega t \right\} \qquad (14)$$

where $\mathcal{E}_0$ is the field for z = L = 0. The corresponding expression for power P is

$$P \approx \frac{P_0}{9} \exp\left\{ \sqrt{3} \left(\frac{j}{2}\right)^{1/3} \right\} \qquad (15)$$

where P_0 is the power at z = L = 0. Since j varies as L^3, from Eq.(14) it is seen that the field has an exponential dependence upon L. Figure 12 is a plot of P/P_0 as a function of j, with the exact solution drawn as a solid curve and the small and large-signal approximations shown as dashed curves.

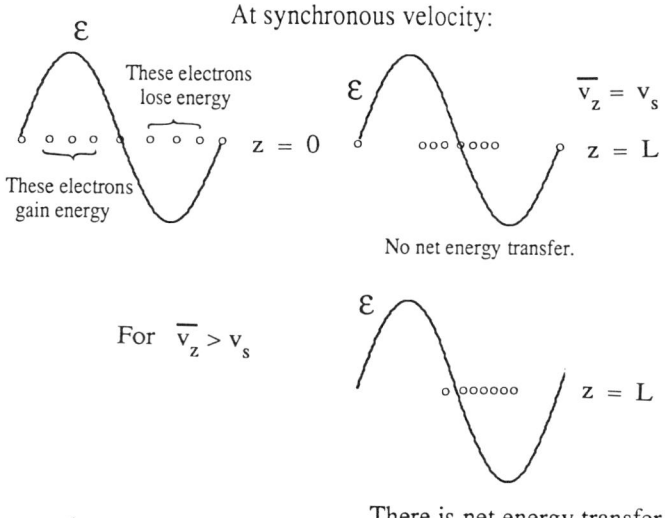

Figure 11: Electron velocity must exceed the synchronous velocity to obtain energy transfer from the beam to the wave.

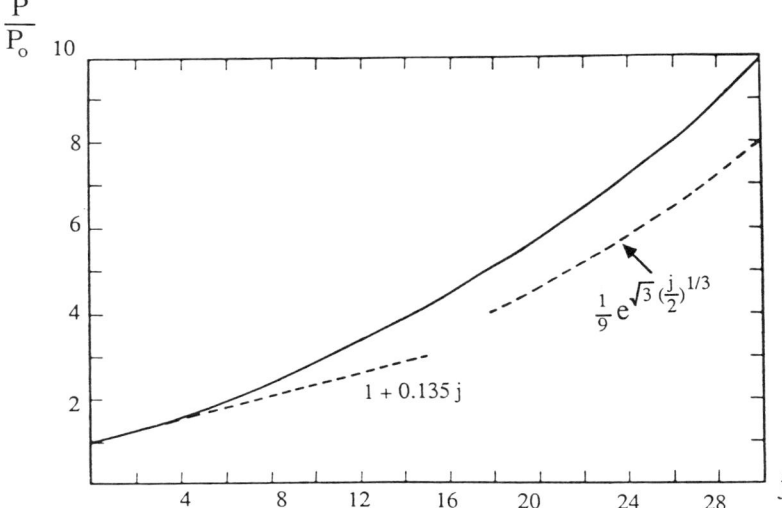

Figure 12: Power as a function of the parameter j is shown as a solid curve, and the small-signal and large-signal solutions are indicated as dashed lines.

3.2 Optical Guiding

In the exponential gain regime, the phase velocity v_p of the wave is less than the velocity of light c. A refractive index n of a medium is the ratio v_p/c, so that from Eq.(14) we have that

$$n = 1 + \frac{\lambda}{4\pi L}\left(\frac{j}{2}\right)^{1/3}. \quad (16)$$

Thus, the electron beam under high gain conditions reduces the phase velocity of the wave, introducing a refractive index greater than unity. Analogous to an optical fiber, this means that the wave is evanescent outside the beam and optical guiding occurs. Figure 13 is a sketch of this guiding effect, where

$$a = e^{-1} \text{ radius for the field} \approx \frac{\lambda}{2^{1.5}\pi(n-1)^{0.5}} \quad (17)$$

For the parameter values $I/A = 5 \times 10^4$ A/cm^2, $a_w = 1.0$, $\lambda = 10$ μm and $\gamma = 100$, then $n - 1 = 2.3 \times 10^{-6}$ and $a = 0.7$ mm.

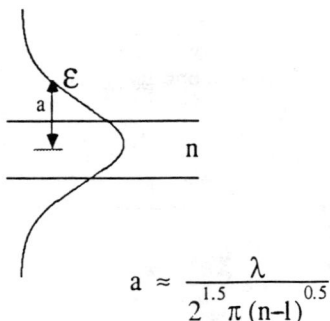

Figure 13: Optical guiding.

3.3 Cavity Design

Typically, two spherical mirrors are used in a confocal arrangement. For an RF linac beam, where the electron bunches are only a few millimeters long (corresponding to picosecond time durations), the cavity length L_c must be set at a value for which the round-trip transit time of the wave in the cavity is an integral number times the time between micropulses. The latter condition results from the requirement that the reflected light pulse from the downstream mirror must overlap a new incoming pulse of electrons for growth of the wave to continue. If λ_m is the distance between micropulses, then

$$\frac{2L_c}{c} = p\frac{\lambda_m}{c} \quad (18)$$

where p is an integer. Figure 14 is a sketch of the cavity showing the electron beam pulses and the wave pulses reflected from the downstream mirror M_2.

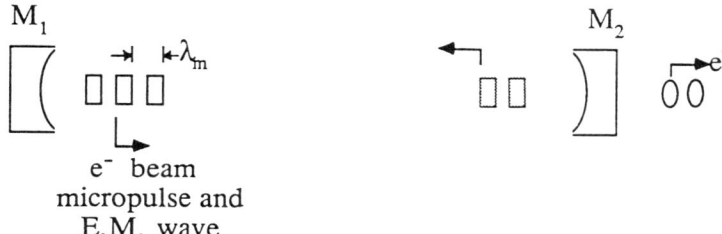

Figure 14: Electron beam micropulses and reflected wave pulses in an interferometer cavity.

The buildup time for the electromagnetic wave in the cavity can be determined from the relationship

$$P = \text{cavity power} = P_0 (1 + G_{NET})^N \qquad (19)$$

where P_0 is the initial power in the cavity, G_{NET} is the net gain (i.e. electronic gain minus cavity loss) per round trip in the cavity, and N is the number of round trips. Since N is expressible in terms of buildup time τ_b as

$$N = \tau_b \left(\frac{2 L_c}{c}\right)^{-1}, \qquad (20)$$

from Eqs. (19) and (20) we have that

$$\tau_b = \frac{2 L_c}{c} \frac{\log P/P_0}{\log (1 + G_{NET})}. \qquad (21)$$

To reach saturation, $P/P_0 \approx 10^{10}$, so that for $G_{NET} = 0.2$ and $L_c = 2m$, the buildup time is 1.7 μsec.

3.4 Beam Quality Requirements

From Figure 10 we see that the FEL exhibits gain for a range of phase slip θ given by

$$0 > \theta > -\pi \qquad (22)$$

where, from Eq. (13), $\theta = \frac{L}{2} \left[k \left(\frac{1}{\beta_z} - 1\right) - k_w \right]$. Equation (22) imposes a limit on the energy spread of the beam since a change in θ can be related to a change in γ. Using Eq. (13) and the fact that

$$(\bar{\beta}_z)^{-1} \approx 1 + \frac{1 + a_w^2}{2\gamma^2} \tag{23}$$

we obtain

$$\Delta\theta = -\frac{kL}{2} \frac{1 + a_w^2}{\gamma^3} \Delta\gamma \tag{24}$$

where $k = 2\pi/\lambda$.
From the synchronism condition, Eq. (3), and limit on θ from Eq. (22), Eq. (24) leads to the constraint

$$\left| kL \frac{\lambda}{\lambda_w} \frac{\Delta\gamma}{\gamma} \right| \leq \pi \tag{25}$$

or

$$\left| \frac{\Delta\gamma}{\gamma} \right| \leq \frac{\lambda_w}{2L} = \frac{1}{2N_w} \tag{26}$$

where N_w is the number of wiggler periods and $\Delta\gamma$ is the full width of the energy spread on the beam.

Similarly, there is a limit on the angular divergence of the beam. If an electron is traveling at an angle φ to the z axis, then $\bar{\beta}_z$ is changed by an amount

$$\Delta\bar{\beta}_z = -\frac{\varphi^2}{2} ; \tag{27}$$

Eq. (22), in conjunction with Eq. (13), then leads to a limit on the half-angle divergence φ of the beam given by

$$\varphi \leq \left[\frac{\lambda}{2L}\right]^{1/2} . \tag{28}$$

For example, with $L = 2m$ and $\lambda = 1\ \mu m$, then $\varphi \leq 0.5$ mrad.
The unnormalized beam emittance ε is defined as

$$\varepsilon = w\varphi \tag{29}$$

where w is the spot size of the beam. To obtain maximum energy from the beam, w should not exceed the spot size of the optical mode so that

$$w \leq \left(\frac{\lambda R}{\pi}\right)^{1/2} \tag{30}$$

where R is the Rayleigh length (i.e. the distance in which the area of a diffracted wave doubles) for the optical mode. For a typical FEL, the Rayleigh length is chosen to be one-half the interaction length L, so that from Eqs. (28) - (30) we obtain

$$\varepsilon \leq \frac{\lambda}{2\sqrt{\pi}}. \tag{31}$$

Thus, as the wavelength of the FEL decreases it is necessary that the emittance of the beam decrease in direct proportion.

3.5 Optical Klystron

The optical klystron (OK) is a device for converting a change in the energy of an election to a change in relative position of the electron. Figure 15 illustrates the manner in which the klystron functions. Electrons enter from the left into a magnetic field B for a distance $L_D/4$; then pass through a region in which the field B is reversed for a distance $L_D/2$; and then exit through a field B in the original direction for a distance $L_D/4$. Electron (1) enters first and has a lower γ then Electron (2), as reflected by their positions in the electric field $\mathcal{E}$. Since Electron (1) follows the path shown by the darker line and Electron (2) follows the path of the lighter line, Electron (1) takes longer to pass through the klystron and so the two electrons are closer together upon exit. (The radius of the particle trajectory in the magnetic field is the Larmor radius given by $\gamma mc/eB$.)

Enhancing the bunching by means of an OK can increase FEL gain without increasing the length of interaction. This is an important consideration if the available interaction length is restricted or if the beam divergence is a limiting factor. From Eq. (28) it is seen that a long length necessitates low beam divergence.

Two conditions imposed on the OK are

$$\int_0^{L_D} B \, dz = 0 \quad \text{so that the particle exits in the same direction it enters,}$$

$$\int_0^{L_D} dz \int_0^z B \, du = 0 \quad \text{so that there is no transverse displacement of the particle.}$$

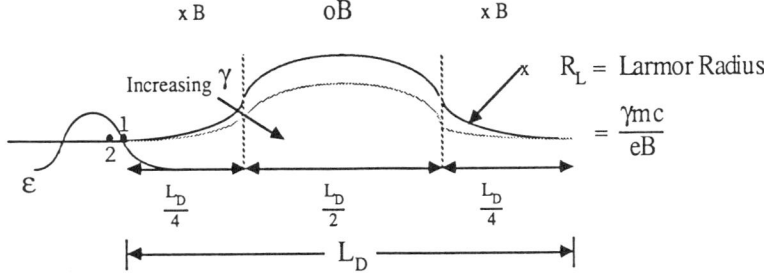

Figure 15: The optical klystron converts a change in γ to a change in position, thereby enhancing the bunching.

The magnetic field arrangement shown in Figure 15 is referred to as a "dispersive magnet," converting a change in γ to a change in position. Figure 16 shows how an OK can be incorporated into an FEL to increase gain. Electrons enter from the left and pass through a modulation section to introduce a change in γ. The dispersive magnet bunches the electrons, and energy is extracted from the bunched beam in the output section.

If it is assumed that the modulation and output sections are identical and that all the bunching occurs in the dispersive magnet, then the gain of the optical klystron G_{OK} is given by (MKS units)

$$G_{OK} \approx 16.5 \, K \, L^2 L_D^3 B^2 \qquad (32)$$

where L = length of modulation and output sections
L_D = length of the dispersive magnet
B = magnet field in the dispersive section

$$K = \frac{I \, a_w^2}{A \, \lambda \gamma^5} (J_0 - J_1)^2 \, .$$

The maximum FEL without the OK as determined from Eq. (9) is

$$G_{FEL} = 3.12 \times 10^{-4} \, K \, (1 + a_w^2) \, L_{FEL}^3 \qquad (33)$$

where L_{FEL} = length of the FEL.

For the parameter values L_D = 0.2 m, L = 0.4 m, B = 0.3 T, a_w = 1.0, L_{FEL} = 1.0 m, we find that G_{OK} = 3 G_{FEL}. It should be noted that the length of the FEL has been chosen to equal the total length of the OK.

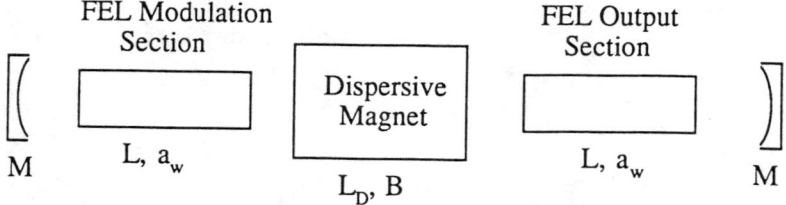

Figure 16: Incorporation of an optical klystron dispersive magnet into an FEL oscillator.

A disadvantage of the OK is that the allowed energy spread is less than that for an FEL. From Eq. (27) we have that the energy acceptance for the FEL is given by

$$\left(\frac{\Delta \gamma}{\gamma}\right)_{FEL} = \frac{1}{2 N_W} \qquad (34)$$

and for the OK (MKS units)

$$\left(\frac{\Delta\gamma}{\gamma}\right)_{OK} = 0.7 \times 10^{-4} \frac{\lambda\gamma^2}{L_D^3 B^2} \quad . \tag{35}$$

The assumptions used for the derivation of Eq. (35) are the same as those used for the gain calculation. A rather simple relationship exists for the product of gain and energy acceptance,

$$\left(G\frac{\Delta\gamma}{\gamma}\right)_{OK} = \left(\frac{2L}{L_{FEL}}\right)^2 \left(G\frac{\Delta\gamma}{\gamma}\right)_{FEL} \quad . \tag{36}$$

For the example considered previously, in which $\frac{2L}{L_{FEL}} = 0.8$ and $G_{OK} = 3\, G_{FEL}$, the energy acceptance for the klystron is only obout one-fifth the energy acceptance for the FEL.

4 Summary and Conclusion

Free-electron lasers can provide high peak-power radiation from the microwave to the visible portion of the spectrum. Perhaps the most interesting region for application is from 10 to 1000 μm wavelength, where there are few alternative sources of coherent emission. A reason for considering the far infrared in preference to shorter wavelengths is that the lower beam quality and beam energy requirements reduce the cost of the FEL and increase the likelihood of achieving successful operation.

Figure 17 illustrates some of the areas in solid state physics amenable to investigation at infrared wavelengths.

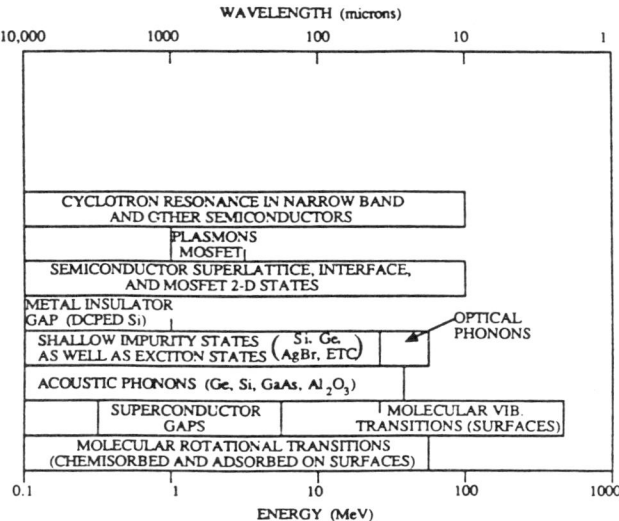

Figure 17: Solid state studies in the infrared.

Another portion of the spectrum where few sources are available is the ultraviolet shorter wavelength regime. FELs can probably operate here, but the devices will be expensive and will require high quality beams. A number of interesting biological problems can be explored in the 2000 - 4000 Å range utilizing the tunability, high peak power, and picosecond pulse duration available from certain types of FELs. These include the study of electronic and vibrational excitation of DNA,[7] the investigation of the oxygenation of hemoglobin,[8] and the role of rhodopsin in the retina.[9]

References

1. A.M. Sessler and D. Vaughan, "Free-electron lasers," Am. Scientist 75, 34-42 (Jan.- Feb. 1987).
2. C.A. Brau, "Recent developments in free-electron lasers," Laser Focus, 40-46 (Feb. 1987).
3. B.G. Levi, "Free-electron lasers take small steps toward distant goal," Physics Today, 17-19 (June 1987).
4. W.B. Colson, "Tutorial on classical free-electron laser theory," Nuclear Instruments and Methods A237, 1-9 (1985).
5. T.C. Marshall, Free-Electron Lasers, Macmillan Pub. Co.,New York (1985).
6. IEEE Transactions on Quantum Electronics, Special Issue on Free Electron Lasers, ed. by C. Brau and B. Newnam (Sept. 1987).
7. D.A. Angelov, G.G. Gruzadyan, P.G. Kryukov, V.S. Letokhov, D.N. Nikogosyan, and A.A. Oraevsky, "High-power UV ultrashort laser action on DNA and its components," in Picosecond Phenomena II, ed. by R.M. Hochstrasser, W. Kaiser and C.V. Shank, 331-335, Springer-Verlag, N.Y. (1980).
8. D. Houde, J.W. Petrich, O.L. Rojas, C. Poyart, A. Antonetti, and J.L. Martin, "Reactivity and dynamics of hemeproteins in the femtosecond and picosecond time domains," Ultrafast Phenomena V, ed. by G.R. Fleming and A.E. Siegman, 419-422, Springer-Verlag, N.Y. (1986).
9. T. Kobayashi, H. Ohtani, and M. Tsuda, "Primary process of vision: hypsorhodopsin," Ultrafast Phenomena V, ed. by G.R. Fleming and A.E. Siegman, 416-418, Springer-Verlag, N.Y. (1986).

FUNDAMENTALS OF INTENSE RELATIVISTIC ELECTRON BEAMS

R. B. Miller

TITAN Technologies, Albuquerque, New Mexico

TABLE OF CONTENTS

1	INTRODUCTION	1730
2	BEAM ENVELOPE EQUATIONS	1733
	2.1 Solenoidal Field Transport in Vacuum	1734
	2.2 Charge and Current Neutralization by a Background Plasma	1736
3	INTENSE BEAM DIODES	1737
	3.1 One-D Planar Diode	1737
	3.2 High-Current Electron Beam Diodes	1738
4	ELECTRON BEAM PROPAGATION IN VACUUM	1740
	4.1 Space-Charge Limiting Current for a Thin Annular Beam	1740
	4.2 Virtual Cathode Formation	1742
	4.3 Single-Particle Orbits	1743
	4.4 Laminar Flow Equilibria	1744
	4.5 Electron Beam Normal Modes	1745
5	BEAM PROPAGATION IN PLASMA	1746
	5.1 Current Neutralization	1747
	5.2 Macroscopic Beam-Plasma Equilibria	1749
	5.3 Macroscopic Beam-Plasma Instabilities	1750
	5.4 Microscopic Beam-Plasma Instabilities	1752
6	BEAM PROPAGATION IN NEUTRAL GAS	1753
	6.1 Beam Current Less Than the Space-Charge Limit	1754
	6.2 $I_1 < I < I_A$	1755
	6.3 $I > I_A$	1757
	6.4 Summary	1757

ABSTRACT

An intense charged particle beam can be characterized as an organized charged particle flow for which the effects of beam self-fields are of major importance in describing the evolution of the flow. Research employing such beams is an important field with applications ranging from the development of high power sources of coherent radiation to inertial confinement fusion. In this series of lectures the primary goal is to provide a basic description of intense beam transport in a variety of situations of practical importance. We first present a brief summary of the development of the pulse power technology which is often used to generate these beams, followed by an elementary description of intense beam behavior based on single-particle beam envelope equations. We then describe in some detail the behavior of intense beams in high-voltage diodes, and the important processes involved in beam transport in vacuum, in plasma, and in neutral gas. Much of this material is taken from my book, <u>An Introduction to the Physics of Intense Charged Particle Beams</u> (Plenum, 1982), which may be consulted for more information and references.

FUNDAMENTALS OF INTENSE RELATIVISTIC ELECTRON BEAMS

R. B. Miller
TITAN Technologies, Albuquerque, New Mexico

1 INTRODUCTION

The field of intense beam physics has its origins in the development of high-voltage pulse power systems in the early 1960s. J. C. Martin at the Atomic Weapons Research Establishment in England was able to produce short (10 to 100 ns), high-voltage pulses (several hundred kilovolts) by pulse-charging a high speed transmission line using an existing high-voltage Marx generator technology. Applying these high-voltage pulses to a simple vacuum diode caused the generation of very high current electron beams (10 to 100 kiloamperes). With a high atomic number material used for the anode, the resulting intense bremsstrahlung x-ray burst could be used to take flash radiographs (x-ray pictures) of artillery shells and the implosion systems used in nuclear weapons.

Since that time progress in pulse power technology has been very rapid, with development work being performed at a number of major laboratories throughout the world. In the United States these include Physics International Co., Ion Physics Corp., Maxwell Laboratories, Cornell University, the Naval Research Laboratory, the Air Force Weapons Laboratory, and Sandia National Laboratories, to name a few. To illustrate this development, I will briefly review the evolution and applications of this technology at Sandia National Laboratories, where I was employed for several years.

Sandia National Laboratories has the broad responsibility of nuclear ordnance engineering for the United States. An important part of this job is ensuring that critical electronic components (such as the fuse and fire sets) can survive the harsh radiation environment associated with a nuclear explosion. As a result, it has been necessary to develop devices capable of partially simulating the effects of nuclear explosions, including neutron, gamma, and x-ray dose and dose rate effects. Soon after Martin's pulse power research became known in this country, it was recognized that this technology could be used to simulate the hard gamma spectrum of a nuclear explosion. With assistance from J. C. Martin, Sandia's Tom Martin and Ken Prestwich subsequently developed the HERMES I and HERMES II accelerators in the late 60s. HERMES II is still being used extensively for gamma dose and dose rate effects testing.

A schematic drawing of this machine is shown in Figure 1.1. A large Marx generator is used to pulse-charge a Blumlein pulse-forming line. The dielectric energy storage medium is transformer oil. A solid dielectric insulator in the form of a stacked, voltage-graded, ring assembly is used to separate the liquid-insulated transmission line section from the evacuated beam generation region of the diode. A high-voltage pulse is produced at the diode by a self-breaking oil switch in the Blumlein. Typical HERMES II peak beam parameters are 10 MeV and 100 kA, with a FWHM pulsewidth of about 70 ns. Using a tantalum insert in the anode, this machine generates an intense bremsstrahlung x-ray pulse, with a total dose exceeding 1 krad at one meter. The gamma-ray dose rate exceeds 10^{11} rads/sec.

© 1989 American Institute of Physics

In order to simulate the softer x-ray spectrum of a nuclear weapon, it is necessary to lower the electron kinetic energy. However, since the bremsstrahlung production efficiency scales as almost the cube of the electron energy, much higher electron currents are needed to generate the same x-ray photon flux. The impedance (voltage/current) of a pulse power device is generally established by the pulse-forming-line (PFL) geometry and the dielectric constant of the insulating medium. For low-impedance applications, the high dielectric constant of deionized water (~ 80, compared to ~ 2.4 for transformer oil) makes it the more suitable medium. As a result, the next generation of pulse power machines at Sandia were the HYDRA water-line pulsers. These devices typically operated at voltages of 1 to 2 MV, and generated beam currents of several hundred kiloamperes.

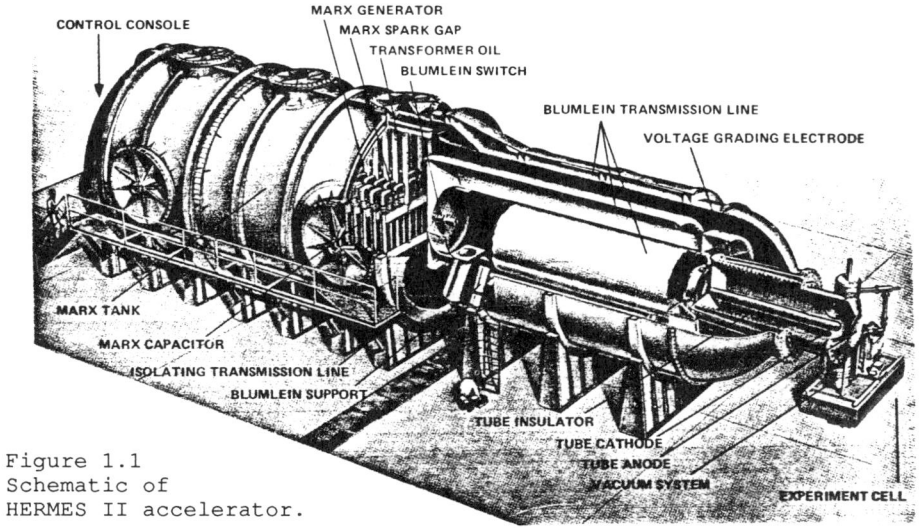

Figure 1.1
Schematic of
HERMES II accelerator.

Arriving at Sandia in 1972, Gerry Yonas began to explore the possibility of using these pulse power generators for inertial fusion applications. The basic idea was to focus a very high current electron beam symmetrically onto a target containing a mixture of deuterium and tritium, thus creating the conditions for rapid heating, compression, and significant D-T burning prior to pellet disassembly. In comparison with laser fusion, particle-beam fusion drivers have much higher efficiencies, but there are major problems associated with beam transport and focusing onto the target. In order to study these problems several new generators were designed and constructed, beginning with PROTO I and II, and culminating in the PBFA I and II accelerators. The typical layout of these machines is indicated by the PROTO II accelerator shown in Figure 1.2. Multiple pulse power modules are now arranged at large radii. When they are simultaneously switched in parallel, they deliver their electrical energy to a common, central, diode load region, thus generating very high beam currents (many megamperes) at relatively low voltages (few megavolts). Research has shown that ions have more favorable energy deposition characteristics, as well as no bremsstrahlung pre-heat problem. As a result, particle-beam fusion has now become synonymous with ion-beam fusion.

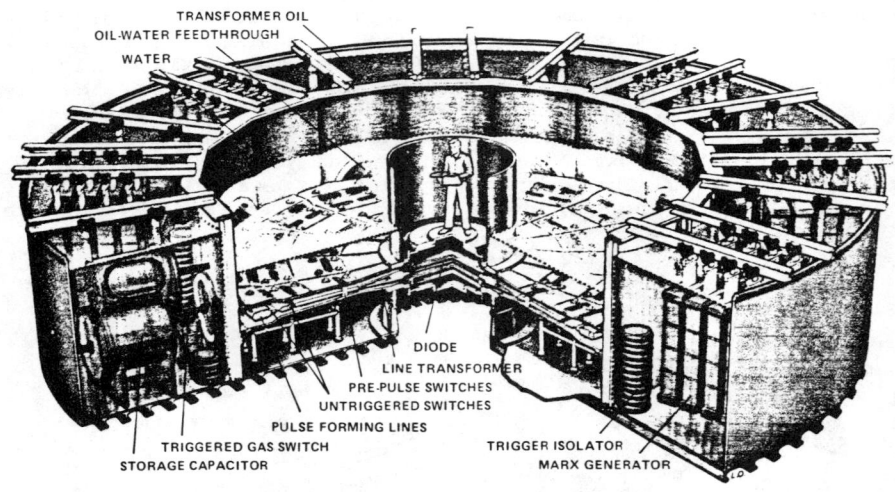

Figure 1.2 Schematic of PROTO II accelerator.

Yonas also realized that it might be possible to arrange this modular pulse power technology in an effective series connection to create very high electron kinetic energies (tens and hundreds of MeV) at modestly high beam currents (many tens of kiloamperes). The RADLAC I and II accelerators (developed by Prestwich and Miller) are thus induction linacs which use essentially the same modular pulse power technology as employed by the PROTO and PBFA accelerators. The RADLAC devices (see Figure 1.3 for an example) have been used to explore a number of beam transport and propagation phenomena associated with a variety of directed energy weapon concepts.

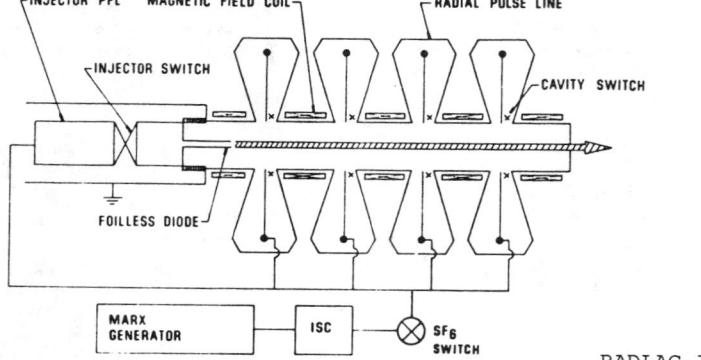

Figure 1.3
Schematic of
RADLAC I accelerator.

As a final point, it is important to note that the PBFA I accelerator has been modified, and is now called SATURN. It is being used for x-ray effects testing. In addition, induction linac technology has provided the essential building blocks for the development of the HERMES III accelerator (20 MeV, 800 kA), which is being used for gamma-ray effects testing. Thus, the research and development associated with Sandia's inertial fusion and directed energy programs have served to enhance Sandia's capability for performing its primary mission.

In the course of this brief historical summary of Sandia's pulse power program, many important developments (including applications involving excimer laser excitation, high power microwave generation, recirculating linac development, etc.) have necessarily been neglected. In addition, there are now several potentially important commercial applications of intense beam/pulse power technology, including biological waste processing, food preservation, medical instrument sterilization, explosives detection, etc. For almost all these applications, a good basic understanding of the physics of intense beams is essential. This will be the primary goal of the remainder of these lectures.

2   BEAM ENVELOPE EQUATIONS

It is often useful to be able to predict quickly the qualitative motion of an intense charged particle beam under various projected operating conditions. In this regard, beam envelope equations, which neglect the detailed internal state of the beam, serve a very useful function. In this section we present a simple derivation of an envelope equation and apply it to several situations of practical interest. Other derivations can be found in most books on electron optics.

The motion of a single particle of mass γm and charge e is described by the Lorentz force law

$$(d\mathbf{p}/dt) = e[\mathbf{E} + (\mathbf{v} \times \mathbf{B})/c)] \tag{1}$$

where $\mathbf{p} = \gamma m \mathbf{v}$ is the relativistic momentum, with $\gamma = (1 - \beta^2)^{-1/2}$, $\beta = |\mathbf{v}|/c$, where $\mathbf{v}$ is the particle velocity vector, and c is the speed of light. $\mathbf{E}$ and $\mathbf{B}$ represent the macroscopic electric and magnetic fields. Taking the dot product of $\mathbf{v}$ with Eq. (1) indicates that the relativistic particle energy changes according to

$$d(\gamma m c^2)/dt = e\ (\mathbf{v} \cdot \mathbf{E})\ , \tag{2}$$

i.e. only electric fields in the direction of the particle motion will change the particle's energy.

We now assume that the beam is cylindrically symmetric about the z-axis, as shown in Figure 2.1. The velocity is decomposed as

$$\mathbf{v} = v_r \mathbf{e}_r + v_\theta \mathbf{e}_\theta + v_z \mathbf{e}_z\ . \tag{3}$$

It is further assumed that $|v_r|, |v_\theta| \ll v_z$, in which case $\gamma \cong (1 - \beta_z^2)^{-1/2}$. The variables v_z and γ are thus regarded as known functions of z and t.

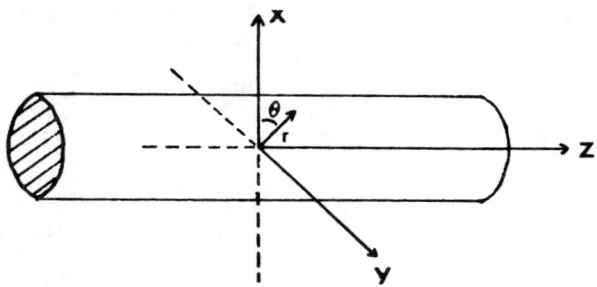

Figure 2.1
Charged particle beam propagating along the z-axis.

With the assumed symmetry the important field components are E_r, B_θ, and B_z. We further assume that the longitudinal magnetic field is generated by external coils only, and is constant and uniform across the beam radial profile; i.e. $B_z = B_0$. In this case the radial component of Eq. (1) becomes

$$\ddot{r} + \dot{\gamma}\dot{r}/\gamma - v_\theta^2/r = (e/\gamma m)[E_r + (v_\theta B_0 - v_z B_\theta)/c] \qquad (4)$$

while the azimuthal component becomes a statement of conservation of canonical angular momentum,

$$\gamma r v_\theta + \Omega r^2/2 = P_\theta/m = \text{const} \qquad (5)$$

where $\Omega = eB_0/(mc)$ is the cyclotron frequency. Solving Eq.(5) for v_θ, substituting this quantity into Eq. (4), and restricting attention to a single particle at the edge of the beam gives an equation for the radial motion of the beam envelope,

$$\ddot{r}_b + \dot{r}_b\dot{\gamma}/\gamma + (\Omega/2\gamma)^2 r_b = (e/\gamma m)(E_r - v_z B_\theta/c) + (P_\theta/m)^2/\gamma^2 r_b^3 \; . \qquad (6)$$

We will now examine the use of Eq. (6) for a few situations of practical importance.

2.1 Solenoidal Field Transport in Vacuum

Consider the case of an intense electron beam propagating in vacuum with constant axial velocity v_0 along a longitudinal magnetic field B_0. The beam kinetic energy is also assumed to remain constant. Eq. (6) then becomes

$$\ddot{r}_b + (\Omega/2\gamma)^2 r_b = -(e/\gamma_0 m)(E_r - \beta_0 B_\theta) + (P_\theta/m)^2/\gamma_0^2 r_b^3 \; . \qquad (7)$$

The field components E_r and B_θ at the edge of the beam are calculated using Gauss' and Ampere's laws

$$\int \mathbf{E}\cdot d\mathbf{A} = -4\pi e \int n \, dV, \qquad \int \mathbf{B}\cdot d\mathbf{l} = -(4\pi e/c) \int n\mathbf{v}\cdot d\mathbf{A}. \qquad (8)$$

For a beam of constant density ($n = n_0$) out to the radius r_b, we have

$$E_r = -2\pi e n_o r_b, \qquad B_\theta = -2\pi e n_o r_b \beta_o = \beta_o E_r . \qquad (9)$$

Substituting these quantities into Eq. (7) yields

$$\ddot{r}_b + (\Omega/2\gamma_o)^2 r_b - \omega_e^2 r_b/2\gamma_o^3 = (P_\theta/m)^2/\gamma_o^2 r_b^3 \qquad (10)$$

where $\omega_e = (4\pi n_o e^2/m)^{1/2}$ is the beam plasma frequency. Hence, the unneutralized beam self-fields cause the beam to expand, while the B_o supplies a restoring force. The equilibrium radius r_{bo} ($\ddot{r}_b = 0$) is given by

$$r_{bo}^2 = (2P_\theta/m)(\Omega^2 - 2\omega_e^2/\gamma_o)^{-1/2} \qquad (11)$$

and there can be no equilibrium unless $\Omega^2 \geq 2\omega_e^2/\gamma_o$. A quick substitution of numerical values indicates that solenoidal magnetic fields can easily satisfy this criterion for vacuum transport of intense electron beams, but substantial charge neutralization is generally required for intense ion beam transport.
Eliminating the beam plasma frequency in favor of the beam current ($I = -\pi r_b^2 n_o e \beta_o c$), because I does not vary as the beam expands or contracts, Eq. (10) becomes

$$\ddot{r}_b + (\Omega/2\gamma_o)^2 r_b - 2Ie/(\gamma_o^3 m \beta_o c r_b) = (P_\theta/m)^2/\gamma_o^2 r_b^3 \qquad (12)$$

The general solution is oscillatory. Assuming small perturbations about the equilibrium radius, $r_b = r_{bo} + \delta r$, the linearization of Eq. (12) yields

$$\ddot{\delta r} + [(\Omega/\gamma_o)^2 - 4eI/(\gamma_o^3 m \beta_o c r_{bo}^2)] = 0 \qquad (13)$$

which has sinusoidal solutions with frequency

$$\omega = [(\Omega/\gamma_o)^2 - 4eI/(\gamma_o^3 m \beta_o c r_{bo}^2)]^{1/2} \qquad (14)$$

and spatial wavelength $\lambda = 2\pi \beta_o c/\omega$.

I first became aware of the potentially severe consequences of such radial oscillations of the beam envelope while conducting beam transport experiments in the RADLAC I accelerator. When injecting a 40-kA beam through the first accelerating gap it was relatively easy to get 40 kA of beam current out. When the same beam was injected through two accelerating gaps, however, only 20 kA was successfully transported. Use of radiochromic films (which are sensitive to dose, and can thus be used to monitor the radial profile of the beam) soon made it apparent that radial force imbalances arising from the large beam self-fields and the accelerating fields in the gap region were exciting radial oscillations of the beam envelope. Depending on the spatial phase of the oscillations in the second gap, the oscillations could further grow, until substantial beam current was being lost to the drift-tube wall. The problem was eventually solved by carefully tailoring the gap shapes and magnetic field profiles in the gap region, such that radial force balance was ensured.

2.2 Charge and Current Neutralization by a Background Plasma

When an intense electron beam is injected into a plasma, the plasma charges will tend to move in such a fashion as to neutralize the beam self-fields. These phenomena can be described qualitatively by introducing the charge and current neutralization fractions f_e and f_m into Eq. (9) as

$$E_r = -2\pi e n_o r_b (1 - f_e), \qquad B_\theta = -2\pi e n_o r_b \beta_o (1 - f_m). \qquad (15)$$

In the absence of an external magnetic field and assuming $P_\theta = 0$, Eq. (10) becomes

$$\ddot{r}_b - [2eI/(\gamma_o m \beta_o c r_b)][1 - f_e - \beta_o^2(1 - f_m)] = 0 \qquad (16)$$

and the equilibrium condition is identified as $f_e = \gamma_o^{-2} + \beta_o^2 f_m$. If the charge neutralization fraction exceeds the equilibrium criterion, the beam will constrict, or "self-pinch." In the opposite limit the beam will expand.

All these possibilities can be observed when an electron beam is injected into neutral gas at various pressures. At very low pressures the beam-generated ionization is small ($f_e = f_m = 0$), and the beam expands radially. At somewhat higher pressures, $f_e > \gamma^{-2}$, but the conductivity remains too low to provide significant current neutralization ($f_m = 0$), and pinched beam propagation is observed. As the pressure is further increased both the charge and neutralization fractions can increase to unity, and the beam streams freely in a force-neutral fashion with beam expansion depending on the perpendicular temperature (emittance). Finally, at still higher pressures, the frequency of collisions between plasma electrons and neutrals increases, and the conductivity decreases ($f_m < 1$). Since the beam space charge is still neutralized ($f_e = 1$), pinched beam propagation is again possible.

These various conditions can also be created by other means. For example, Lawrence Livermore scientists have successfully transported the 10-kA electron beam through the 50-MeV accelerator structure of the Advanced Test Accelerator utilizing a small-radius plasma channel created by laser photoionization of a low density benzene background. The subsequent injection of the ATA beam into the channel caused expulsion of the plasma electrons from the channel, and the residual ion core then focused and guided the beam. Thus, $f_e \leq 1$ and $f_m = 0$. It has also been suggested that a laser ionization channel could successfully transport an electron beam across the earth's magnetic field in the tenuous upper atmosphere of a few hundred kilometers altitude.

In addition, Sandia scientists have used a low energy electron beam guided by a small magnetic field to produce a 90° curved plasma channel. As for the ATA, the plasma channel was able to guide a high current, relativistic electron beam around the bend without significant losses. This result has become the basis for a high-current recirculating electron linac concept at Sandia.

3 INTENSE BEAM DIODES

The application of very high-voltage pulses to a simple vacuum diode can produce very high-current electron and ion beams. The process is termed explosive field emission. Microscopic examination of the surface of almost any material will reveal small protrusions (whiskers) that are typically 10^{-4} cm in height with a base radius of $<10^{-5}$ cm and an even smaller tip radius. When high voltage is applied, the electric field at a whisker tip can be geometrically enhanced by a factor of several hundred, leading to intense electron field emission from the tip of each cathode whisker. The resulting current flow resistively heats the whisker. If the current flow is sufficiently high (the electric field exceeds a critical value), the whisker can explosively vaporize, and the expansion and merger of the resulting plasma blobs can dramatically increase the effective electron emission area. Since the cathode plasma can be considered as a metal surface whose work function is effectively zero, the cathode current supply is essentially unlimited. In this case, the space charge of the electron flow itself provides the current limitation mechanism.

To illustrate this self-regulating emission mechanism, assume that a plasma completely covers the cathode surface at the instant that high voltage is impressed on the diode. The evolution of the potential variation between the electrodes is described in Figure 3.1. Initially, there is no electron charge in the gap and the potential variation is simply a linear function of the distance from the cathode to the anode (I). As electrons are drawn into the gap, the potential at all positions becomes lower (II). As more and more electrons enter the gap, a potential minimum can form outside the cathode surface (III), and only those electrons having an initial energy greater than the barrier potential (resulting from thermal spread in the cathode plasma) can escape from the cathode and reach the anode. Eventually, the condition is reached when most of the electrons are reflected by the potential barrier (IV). The decrease of space charge in the gap causes the barrier to collapse (V), and the process can repeat.

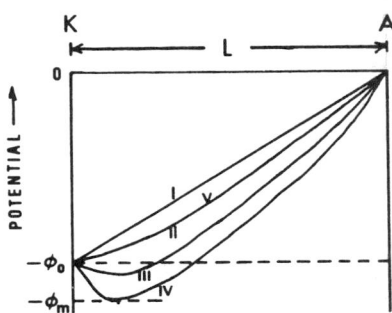

Figure 3.1
The space-charge-limiting emission mechanism.

3.1 One-D Planar Diode

For high-voltage diodes the magnitude of the potential minimum is negligible compared with the high applied potential, and the location of the potential minimum is practically coincident with the cathode surface. Hence, it may be assumed that the electric field vanishes at the cathode surface and that electrons are emitted from this surface with zero velocity. Thus, the electron velocity at any position between the electrodes is determined from conservation of energy. For the planar geometry, Poisson's equation is

$$(d^2/dz^2)\phi = 4\pi e n_e \tag{17}$$

where z is the distance from the cathode and n_e is the electron number density. In the steady state, and for the one-dimensional geometry, the continuity equation requires that the current density, $j = e n_e v$, be constant. Solving for the electron density, substituting into Eq. (17), and using energy conservation (in the non-relativistic limit) to express the electron velocity v in terms of the electric potential yield

$$(d^2/dz^2)\phi = 4\pi j (m/2e\phi)^{1/2} . \tag{18}$$

Integrating Eq. (18) twice and setting z = d (the anode-cathode gap separation) and $\phi = \phi_0$, the applied potential, we obtain the familiar Langmuir-Child emission current density as

$$j = (9\pi)^{-1} (2e/m)^{1/2} \phi_0^{3/2}/d^2 . \tag{19}$$

Although this expression has been derived for the case of an infinite planar diode, it is also approximately correct for large aspect ratio ($r_c/d \gg 1$), cylindrically symmetric diodes, provided the electron current flow is not too large. For small aspect ratio diodes undergoing space-charge-limited flow, the dominant electron motion is still along the electrostatic lines of force (which may have a large radial component), but the geometrical factor is different. Also, an approximate expression appropriate for relativistic voltages is

$$j = (3^{1/2}/6\pi)(mc^3/e)(\gamma_0^{2/3}-1)^{3/2}/d^2 . \tag{20}$$

3.2 High-Current Electron Beam Diodes

When the effects of the reduction of the diode gap separation resulting from the expanding cathode plasma are included ($d = d_0 - v_c t$), the previous discussion provides a good description of the flow. However, in the limit of low diode impedances and high electron currents the effect of the azimuthal self-field on the electron trajectories (self-constriction or "pinching") cannot be neglected. A rough transition criterion is that diode pinching will occur when the relativistic gyroradius, $r_b = \gamma mc^2/eB$, of an electron emitted at the cathode radius is equal to the anode-cathode gap separation. Assuming $B = 2I/cr_c$, then the diode critical current is approximately given by

$$I_c = (\gamma mc^3/e)(r_c/2d) . \tag{21}$$

When the diode current exceeds the critical current no exact analytical solution for the electron flow is known. However, assuming that the dominant feature of the flow is self-consistent E x

B drifting along equipotential surfaces yields an estimate for the total diode current given by

$$I_p = (mc^3/2e)(r_c/d)\gamma_0 \ln[\gamma_0 + (\gamma_0^2 - 1)^{1/2}] \ . \tag{22}$$

This "parapotential" model was found to give good agreement with experimental measurements of high-current diode impedance, and diode designs based on its scaling law predictions operated in the correct parameter regimes. Nevertheless, the model suffers from serious deficiencies. First, it is unable to treat the details of electron flow near the cathode or anode, i.e. how the electrons get on and off the equipotential contours. Second, in analogy to the external magnetic field required for nonrelativistic Brillouin flow, the parapotential model requires a non-physical bias current flowing on the diode axis. As a result, a newer, more physically correct, theory has been developed which describes the time-dependent evolution of the flow in terms of four distinct phases: (1) pure Child-Langmuir flow at low voltages, (2) weak pinching when the beam current first exceeds the diode critical current, (3) a collapsing pinch due to time-dependent <u>ion emission</u> from an anode plasma, and (4) steady-state pinched electron flow and laminar ion flow (Figure 3.2).

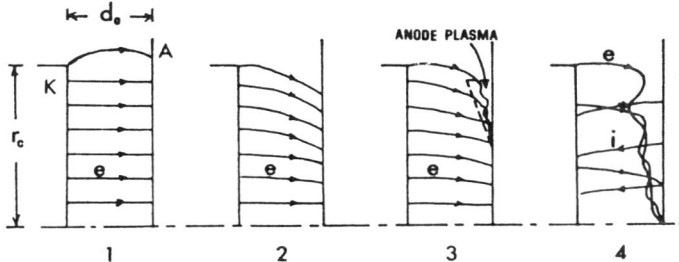

Figure 3.2 Evolution of the flow in a high-current electron diode.

The essential new feature is the formation of an anode plasma. This can occur during either the low-voltage phase or the weak pinch phase depending on the generator characteristics (voltage risetime) and the anode material. If the anode plasma forms during phase 1, a non-relativistic Child-Langmuir bipolar solution with space-charge-limited electron and ion emission is a good approximate description of the flow. If the plasma formation occurs during phase 2, however, the plasma will initially form at large radii because the grazing incidence of the electron orbits permits more efficient heating of the anode material. This plasma then provides a space-charge-limited source of ions to be accelerated across the gap. The space charge of the ion flow neutralizes the electron space charge, while the ion current adds to the electron current resulting in strongly pinched electron flow. When these electrons enter the anode plasma at grazing incidence there is no electric field to counteract the $v_r B_\theta$ magnetic force, and they are reflected back toward the cathode while continuing to drift radially inward until they encounter an anode region where plasma has not yet formed. This process is then repeated until eventually the condition is reached in which the electrons follow complicated orbits drifting inward with multiple reflections thereby resulting in a tight electron pinch on the diode axis.

Assuming that a steady-state condition is eventually reached which is characterized by laminar ion flow and pinched electron flow, one can estimate the magnitude of both the electron and ion current contributions. Since the electric field vanishes in both the cathode and anode plasmas, the total space charge between the two electrodes must be nearly zero. Thus, the ratio of the ion current to the electron current is inversely proportional to the ratio of the average diode crossing times for each species. For the pinched electron flow, the electrons move radially inward and their characteristic crossing length scale is the diode radius. For the ions on the other hand, their larger mass implies that their trajectories will be nearly laminar, and the length scale remains the diode gap separation. Hence, the ratio of the crossing time is enhanced by the geometrical factor r_c/d, yielding

$$I_i/I_e \sim (r_c/2d)(\beta_i/\beta_e) . \tag{23}$$

Depending on the aspect ratio (r_c/d), a substantial ion current, in excess of the electron current, may characterize the output diode current. Recognizing that the total diode current is well approximated by the parapotential current, Eq. (22), we have

$$I_i = I_p/[1 + 2(d/r_c)(\beta_e/\beta_i)] \tag{23}$$

and there is no upper limit for the diode ion current since $I_i \sim r_c/d$.

Thus, these pinched electron diodes can be considered as being self-magnetically insulated. Simply increasing the diode aspect ratio in an attempt to increase the electron current will increase the total diode current, but this increase will be due primarily to an increased ion beam current. This was an important consideration in the shift from electron beam fusion to ion beam fusion at Sandia in the late 1970s.

4   ELECTRON BEAM PROPAGATION IN VACUUM

From the discussion concerning Eq. (11) we observed that for an electron beam propagating in vacuum it was relatively easy to ensure an equilibrium using a solenoidal magnetic field. However, even if the strength of the magnetic field is infinite there is always a limit to the amount of current (space charge) that can propagate in an evacuated drift space. If the magnitude of the electrostatic potential associated with the beam space charge exceeds the beam kinetic energy, then an intense beam equilibrium with total transmission of the beam current is clearly not possible. The physics of this situation is very similar to that associated with the space-charge-limited Child-Langmuir emission current of the previous section. The principal difference is the initial condition describing the electron velocity. We next calculate the space-charge-limiting current for a particular geometry to illustrate the procedure.

4.1 Space-Charge-Limiting Current for a Thin Annular Beam

Consider an annular beam, whose thickness is much smaller than both the beam radius r_b and the distance between the beam and the chamber wall, propagating through a cylindrically symmetric drift space of radius R. The electrostatic potential is essentially constant across the thin beam, and is described by the solution of Laplace's equation

$$(1/r)(d/dr)r(d/dr)\phi = 0 \tag{24}$$

subject to the boundary conditions $\phi(R) = 0$, $E_r(r_b) = -2I/r_b v_b$, where I is the beam current, and v_b is the longitudinal electron velocity (uniform across the thin beam). The solution of Eq. (24) is

$$\phi(r) = (-2I/v_b)\ln(R/r) \tag{25}$$

where $v_b = c[1 - (\gamma_0 + e\phi_b/mc^2)^{-2}]^{1/2}$, and $\phi_b = \phi(r_b)$. Hence, at the beam radius we have the following result:

$$F(\phi_b) = (-e\phi_b/mc^2)[1 - (\gamma_0 + e\phi_b/mc^2)^{-2}]^{1/2} = (2eI/mc^3)\ln(R/r_b). \tag{26}$$

Note that the left side of Eq. (26) is a function of ϕ_b only, while the right side involves only the beam current and the geometry. Thus, for a fixed geometry, there will be a maximum beam current that can flow (in steady state) that will be determined by maximizing $F(\phi_b)$. Performing this calculation yields

$$F_{max}(\phi_b) = (\gamma_0^{2/3} - 1)^{3/2} \tag{27}$$

and the maximum current (the space-charge-limiting current) is evaluated as

$$I_l = (mc^3/2e)(\gamma_0^{2/3} - 1)^{3/2}[\ln(R/rb)]^{-1}. \tag{28}$$

From Eq. (28) it is apparent that the space-charge limit increases as the beam radius approaches the wall radius (because the electrostatic potential depression at the position of the beam is decreased).

For more complicated geometries it is generally very difficult to analytically derive a limiting current expression. However, there is a very simple procedure for getting a good approximation. The key steps involved (assuming cylindrically symmetric systems) are

1. Use Gauss' law to calculate the radial electric field E_r as a function of radius.
2. Integrate E_r to find the radial variation of the potential.
3. Evaluate the potential on axis, and set $|e\phi(0)| = (\gamma_0 - 1)mc^2$.
4. Replace the factor $\beta(\gamma_0 - 1)$ by $(\gamma_0^{2/3} - 1)^{3/2}$, Eq. (27).

An easy exercise for the reader is to use this procedure to show that the space-charge-limiting current for a solid electron beam of radius r_b propagating in a drift tube of radius R is given by

$$I_l = (\gamma_0^{2/3} - 1)^{3/2}(mc^3/e)[1 + 2\ln(R/r_b)]^{-1}. \tag{29}$$

4.2 Virtual Cathode Formation

Returning to the example of the previous section, suppose the injected current were raised above the limiting current value of Eq. (28). Since a steady flow of current in excess of the space-charge limit cannot exist, a potential well (virtual cathode) must develop that reflects some of the electrons backward into the injector region. Although it is possible to obtain a mathematical solution to the static problem of virtual cathode formation for simple geometries, such solutions are not physically realizable, as demonstrated by numerous experiments and computer simulations. Again, the physics of this time-dependent behavior is very similar to that described in Section 3, except that the injected electron velocities are now much larger than any thermal spread, and the dominant feature is total disruption of the flow. As an introduction to this time-dependent phenomenon, the motion of a single charge sheet in a one-dimensional drift cavity will be analyzed.

Consider an infinitely thin sheet of charge density $(-\rho)$ that is positioned between two infinite, parallel, grounded conducting walls (Figure 4.1). The sum of the charges induced in the walls is equal to the total charge on the sheet. Resolving this charge into two parts $-\rho_1$ and $-\rho_2$, which represent the portions of the charge on which lines of force from the left and right boundaries terminate, from Gauss' law the electric fields in the two regions are given by

$$E_1 = 4\pi\rho_1 , \qquad E_2 = -4\pi\rho_2 . \tag{30}$$

Since the fields are uniform, the electrostatic potential decreases linearly to a potential minimum ϕ_{min} at the sheet. Solving for ϕ_{min} yields

$$\phi_{min} = -4\pi\rho(L - z_1)z_1/L \tag{31}$$

and the electric fields can be expressed as

$$E_1 = 4\pi\rho(L - z_1)/L , \qquad E_2 = -4\pi\rho(z_1/L) . \tag{32}$$

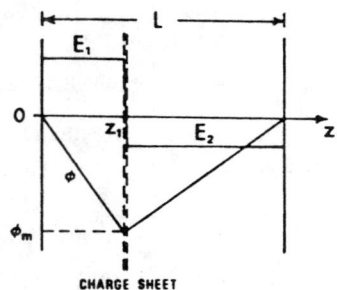

Figure 4.1
A thin charge sheet positioned between conducting walls.

It is easily shown that the net electric field acting to move the sheet is just the average field $E_a = 4\pi\rho(0.5 - z_1/L)$. With the mass per unit area of the sheet designated as M, the (non-relativistic) equation of motion of the sheet is

$$\ddot{z}_1 = (4\pi\rho^2/ML)(z_1 - L/2). \tag{33}$$

With initial conditions $z_1 = 0$, $\dot{z}_1 = v_0$ at $t = 0$, the solution of Eq. (33) is

$$z_1 = (L/2)[1 - \cosh(\omega t)] + (v_0/\omega)\sinh(\omega t) \tag{34}$$

where ω is given by $\omega^2 = 4\pi\rho^2/ML$. As $\omega \to 0$ (the charge decreases to zero), Eq. (34) reduces to $z_1 = v_0 t$, and the sheet has a transit time given by $t_0 = L/v_0$. As the charge density increases, the sheet can slow considerably in the center of the gap. In fact, when the charge density reaches a limiting value determined by $\omega = 2/t_0$, Eq. (34) becomes

$$z_1 = (L/2)[1 - e^{-2t/t_0}]. \tag{35}$$

As $t \to \infty$, $z_1 \to L/2$, and the charge sheet is trapped in the center of the gap. For still larger charge densities, the sheet is reflected back to the injection plane by its own self-electric field.

Although the single-charge sheet model appears to be a drastic oversimplification, it does form the basis of numerical simulation codes that follow the motion of many such sheets (in one dimension). Furthermore, this simple model demonstrates many correct features of the detailed simulations, including the existence of a limiting charge density (or current), and oscillation of the negative sheet charge with its multiple positive images when the limiting density is exceeded.

4.3 Single-Particle Orbits

We now assume that the beam current is less than the space-charge limit, and that a radial force balance equilibrium is enforced with a strong, externally applied, longitudinal magnetic field. We wish to examine the detailed motion of individual electrons. In Cartesian coordinates the beam self-fields are given by

$$E_x = 2Ix/(\beta_z c r_b^2), \quad E_y = 2Iy/(\beta_z c r_b^2),$$
$$B_x = -2Iy/cr_b^2, \quad B_y = 2Ix/cr_b^2 \tag{36}$$

where $I = -\pi r_b^2 e n_0 \beta_z c$ is the beam current. Hence, the Cartesian components of the equation of motion become

$$\ddot{x} = (-\Omega^2/\gamma^3)x - (\Omega_0/\gamma)\dot{y}, \qquad \ddot{y} = (-\Omega^2/\gamma^3)y + (\Omega_0/\gamma)\dot{x}$$

where $\Omega_0 = eB/mc$ and $\Omega^2 = 2eI/m\beta_z c r_b^2$. Defining the parameter $s = x + iy = re^{i\theta}$, the x and y components can be combined into a single equation as

$$\ddot{s} = (-\Omega^2/\gamma^3)s + i(\Omega_0/\gamma)\dot{s} \tag{37}$$

which has the formal solution $s(t) = Ae^{i\omega_+ t} + Be^{i\omega_- t}$, where

$$\omega_\pm = (\Omega_o/2\gamma)[1 \pm (1 + 4\Omega^2/\gamma\Omega_o^2)^{1/2}] \ .$$

With the initial conditions $s(0) = s_o$, $\dot{s}(0) = 0$, in the limit of large Ω_o the solution becomes

$$s(t) \sim s_o[e^{i\omega_- t} + (\Omega^2/\gamma\Omega_o^2)e^{i\omega_+ t}] \ . \tag{38}$$

Hence, the single-particle motion can be considered as the sum of two rotations: (1) a large amplitude rotation at the slow E × B rotation frequency; and (2) a small amplitude rotation at the fast cyclotron frequency. If $\Omega^2/\gamma\Omega_o^2 \ll 1$, then the detailed fine structure of the particle orbits can be ignored and fluid equations can be used to described the equilibrium.

4.4 Laminar Flow Equilibria

With the assumption of the previous section, the basic equations for examining the equilibrium are the fluid and Maxwell equations. For the electron beam these are

$$(\partial/\partial t)n + \nabla\cdot(n\mathbf{v}) = 0 \ , \quad (\partial/\partial t)\mathbf{p} + \mathbf{v}\cdot\nabla\mathbf{p} = -e[\mathbf{E} + (\mathbf{v}\times\mathbf{B})/c];$$
$$\nabla \times \mathbf{E} = -c^{-1}(\partial/\partial t)\mathbf{B} \ , \quad \nabla \times \mathbf{B} = -(4\pi/c)en\mathbf{v} + c^{-1}(\partial/\partial t)\mathbf{E} \ ;$$
$$\nabla\cdot\mathbf{E} = -4\pi en \ , \quad \nabla\cdot\mathbf{B} = 0 \ .$$

In these equations, n, **v**, and **p** represent the macroscopic equilibrium electron fluid density, velocity, and momentum, and **E** and **B** represent the self-consistent equilibrium electric and magnetic fields. Assuming the steady state yields the radial force balance given by

$$m\gamma(r)v_\theta^2(r)/r - e\{E_r + [v_\theta(B_o + B_z^s) - v_z B_\theta]/c\} = 0 \tag{39}$$

where $\gamma = [1 - (v_\theta/c)^2 - (v_z/c)^2]^{-1/2}$. The equilibrium radial electric field is determined from Poisson's equation, and the equilibrium magnetic field is expressed as

$$\mathbf{B}(r) = (B_o + B_z^s)\mathbf{e}_z + B_\theta\mathbf{e}_\theta \tag{40}$$

where B_o is the externally applied uniform field. The axial, diamagnetic self-field, B_z^s, is given by

$$B_z^s(r) = (-4\pi e/c) \int_r^R dr' n(r') v_\theta(r') + B_c \ . \tag{41}$$

The constant B_c represents the uniform field due to the azimuthal image current in the drift-tube wall; it is easily computed from flux conservation. The first term in Eq. (39) is the outward centrifugal force, and the second term is the outward force of the electric field due to the beam space charge. The third and fourth terms represent the constraining forces of the magnetic fields. The third term also represents the diamagnetic character of the beam, i.e. the reduction of the applied magnetic field strength interior to the beam.

Assuming $v_\theta \ll v_z$, and introducing the dimensionless variables $\beta_\theta = v_\theta/c$ and $\beta_z = v_z/c$, Eq. (39) becomes

$$\gamma(r)\omega_\theta^2(r) + r^{-2} \int_0^r dr' r' \omega_p^2(r') - \omega_\theta \Omega_o$$

$$- [\beta_z(r)/r^2] \int_0^r dr' r' \beta_z(r') \omega_p^2(r') = 0 \qquad (42)$$

where $\Omega_o = eB_o/mc$, $\omega_p^2 = 4\pi e^2 n/m$, and $\omega_\theta(r) = \beta_\theta(r)c/r$.

Note that Eq. (42) has three unknowns, $\gamma(r)$, $\omega_p^2(r)$, and $\omega_\theta(r)$. This arbitrariness may be removed by using conservation of energy and conservation of canonical angular momentum, and some specific solutions have been obtained for simple cases with some difficulty. For our purposes we will simply assume that the beam density and velocity are uniform across the beam and obtain the radial profile of the angular velocity.

For a solid beam of radius r_b, the solution is

$$\omega_\theta(r) = \omega_\theta^\pm = (\Omega_o/\gamma_o)[1 \pm (1 - 2\omega_{po}^2/\gamma_o\Omega_o^2)^{1/2}] . \qquad (43)$$

In this case, the rotation rate is constant across the beam, and this case is termed a rigid rotor equilibrium. Note also that the equilibrium condition is $\Omega_o^2 > 2\omega_{po}^2/\gamma_o$, which is identical to that obtained with the beam envelope equation. In addition, there are fast and slow rotation modes which correspond to the fast and slow rotations of the single-particle orbits.

As a second example consider a uniform hollow beam of inner radius r_o. In this case the rotation rate is a function of radius, as described by

$$\omega_\theta(r) = \omega_\theta^\pm(r) = (\Omega_o/2\gamma_o)\{1 \pm [1 - (2\omega_{po}^2/\gamma_o\Omega_o^2)(1 - r_o^2/r^2)]^{1/2}\} . \qquad (44)$$

The free energy associated with the angular velocity shear can drive the growth of a filamentation instability called the diocotron instability.

4.5 Electron Beam Normal Modes

Having developed various equilibria, we next analyze the stability of these configurations to small harmonic perturbations of the form $\delta\psi = \delta\psi(r)e^{-i\omega t}e^{im\theta}e^{ik_z z}$. General instability analyses are quite complicated and beyond the scope of these lectures. Nevertheless, we can illustrate the important concept of normal modes of the electron beam by considering the simple rigid rotor equilibrium. For this configuration it can be shown that the linearized Poisson's equation for the perturbed potential $\delta\phi$ becomes

$$(1/r)(\partial/\partial r)[r(\partial/\partial r)\delta\phi] - (m/r)^2 \delta\phi + k_\perp^2 \delta\phi = 0 \qquad (45)$$

where
$$k_\perp^2 = -k_z^2\{1 - (\omega_p^2/\gamma^2)/(\omega - k_z v_z - m\omega_\theta)^2\}/[1 - \omega_p^2/\gamma v^2]$$
and
$$v^2 = (\omega - k_z v_z - m\omega_\theta)^2 - (-\Omega/\gamma + 2\omega_\theta)^2.$$

Equation (45) is recognized as a form of Bessel's equation. The solutions that remain finite at the origin are $\delta\phi \sim J_m(k_\perp r)$. Also, if there is a conducting wall at $r_b = R$, then $J_m(k_\perp r_b) = 0$. This latter condition implies that $(k_\perp r_b)^2 = \rho_{mn}^2$, where ρ_{mn} is the nth zero of J_m. Thus, we can write

$$1 - (k_z/k)^2 (\omega_p^2/\gamma^3)/(\omega - k_z v_z - m\omega_\theta)^2$$
$$- (k_\perp/k)^2 (\omega_p^2/\gamma)/[(\omega - k_z v_z - m\omega_\theta)^2 - \omega_v^2] = 0 \quad (46)$$

where $k^2 = k_z^2 + k_\perp^2$, and ω_v is the vortex frequency defined by $\omega_v = -(-\Omega/\gamma + 2\omega_\theta)$. Equation (46) relates the frequency ω of electrostatic waves to the wavenumber k. If we solve for ω as a function of k in the limit $(2\omega_p^2/\gamma\Omega^2) < 1$ (such that the equilibrium condition is strongly satisfied), it can be shown that the imaginary part of the frequency, Im(ω), vanishes, which implies undamped, stable oscillations. Further, for axisymmetric perturbations (m = 0), the dispersion relation factors approximately as

$$\omega \sim k_z v_z \pm (k_z^2 \omega_p^2/k^2)^{1/2} \gamma^{-3/2}, \quad (47)$$

$$\omega \sim k_z v_z \pm (\Omega^2/\gamma^2)^{1/2}. \quad (48)$$

Thus, the normal modes can be identified as two Doppler-shifted plasma modes corresponding to longitudinal bunching of the beam space charge [Eq. (47)], and two Doppler-shifted cyclotron modes, which correspond to transverse constrictions of the beam. For several applications (such as microwave generation, collective ion acceleration, etc.), these modes are excited and caused to grow in a more or less controlled fashion.

5 BEAM PROPAGATION IN PLASMA

In this section we will generally be interested in the resulting motion of the plasma species following the injection of a high-current electron beam into a plasma, and the resulting influence on the behavior of the electron beam. Since most high-current electron beam pulses are relatively short (< 100 ns), we will neglect the motion of the plasma ions to simplify the analysis. It should be understood that the self-fields surrounding an intense electron beam are large. For example, a simple Gauss' law calculation of the radial electric field at the edge of an unneutralized 1-cm-radius, 100-kA beam gives about 6×10^6 volts/cm! Hence, it is to be expected that when such a beam is injected into a plasma with mobile electrons, the beam space charge will be rapidly neutralized by the expulsion of the plasma electrons.

To estimate this charge neutralization time, assume that a spherically symmetric clump of excess positive charge is instantaneously deposited in a plasma characterized by a conductivity σ. We then expect that a radial plasma current, j_r, will flow as a result of the radial electric field of the charge clump, and that the electric field (and plasma current) will then decrease with time in a self-consistent manner. The relevant Maxwell equation is

$$\nabla \times \mathbf{B} = 4\pi \mathbf{j}/c + (\partial/\partial t)\mathbf{E}/c . \tag{49}$$

Because of the spherical symmetry, derivatives with respect to the angular coordinates vanish, and $(\nabla \times \mathbf{B})_r = 0$. Hence,

$$|4\pi j_r/c| \sim (\partial/\partial t)E_r/c \sim E_r/c\tau_n \tag{50}$$

where τ_n is the phenomenological charge neutralization time. Introducing Ohm's law for the plasma as $j_r = \sigma E_r$ then yields

$$\tau_n \sim (4\pi\sigma)^{-1} . \tag{51}$$

Thus, the excess charge is neutralized in a characteristic time τ_n. As an order-of-magnitude scaling, $\sigma \sim 10^9$ sec^{-1} is easy to achieve, and it is reasonable to expect that the excess charge of an intense electron beam will be neutralized in < 0.1 ns.

Assuming that the radial electric field has been reduced to zero by expelling the plasma electrons, we can qualitatively estimate the resulting motion of the electron beam by using the beam envelope equation. Assuming that the kinetic energy is constant and that there is no external magnetic field, then

$$\ddot{r}_b = -(e\beta/\gamma m)(2I/cr_b) \tag{52}$$

and the beam will radially pinch. A first integration yields

$$(d/dt)r_b = -[(4eI\beta/\gamma mc)\ln(r_{bo}/r_b)]^{1/2} . \tag{53}$$

Clearly, if the radial pinch velocity becomes equal to the initial axial injection velocity the beam can not propagate. Setting $|(d/dt)r_b| = \beta c$ at $r_b = r_{bo}/2$ and solving for the current I yields

$$I_l = (\beta\gamma mc^3/e)/(4 \ln 2) . \tag{54}$$

The numerical factor 4 ln2 is of order unity.

5.1 Current Neutralization

From Eq. (54), electron beam currents in excess of the <u>Alfven</u> limit, $I_A = \beta\gamma mc^3/e$, should not be able to propagate. Since $mc^3/e = 17$ kA, for a 2-MeV electron beam ($\gamma \sim 5$), severe pinching should prevent the flow of currents exceeding about 85 kA. With the development of the high-voltage pulse power technology it became possible to produce sufficiently high beam currents to test this limit. In fact, when beams exceeding the Alfven limit were injected into a dense plasma they were found to propagate! In retrospect it seems obvious from Lenz' law that the injection of beams into a plasma should generate an induction electric field that would tend to cause plasma electrons to flow in opposition to the beam. This process, called current neutralization, will now be examined qualitatively.

Starting with the appropriate Maxwell equation

$$\nabla \times \mathbf{E} = -(1/c)(\partial/\partial t)\mathbf{B},$$

using $|B| = 2I/cr$, and assuming a linear current rise to a maximum I_o in a time t_r, we find the correct scaling to be

$$E_z/r \sim (-1/c)(2I_o/crt_r) \tag{55}$$

and estimate the magnitude of the inductive electric field as

$$E_z \sim -2I_o/c^2 t_r.$$

As a numerical example, for a beam rising to a peak current of 50 kA in 10 ns the electric field has a magnitude of about 10^6 volts/meter.

Introducing the plasma Ohm's law and defining the plasma current as $I_p = \pi r_b^2 \sigma E_z$, we use Eq. (55) to obtain

$$|I_p/I_o| \sim 2\pi r_b^2 \sigma/c^2 t_r \sim t_m/2t_r \tag{56}$$

where $t_m = 4\pi r_b^2 \sigma/c^2$ is recognized as the magnetic diffusion time of the plasma. Thus, if the beam risetime is shorter than, or comparable to, the plasma magnetic diffusion time, then a substantial plasma return current can be expected to flow.

To include the plasma current in a self-consistent fashion we must use $B = 2I_n/cr$, where $I_n = I + I_p$ is the net current. In this case Eq. (55) is modified as

$$E_z \sim (-2/c^2)(\partial/\partial t)[I(t) + I_p]$$

or

$$2I_p/t_m = -(\partial/\partial t)[I(t) + I_p]. \tag{57}$$

For scaling purposes we again assume that $I(t) = I_o t/t_r$. Performing the integration then yields

$$I_p/I_o = -t_m/2t_r[1 - \exp(-2t/t_m)]. \tag{58}$$

Thus, if the beam risetime is short compared to the plasma diffusion time, then the return current closely follows the beam current and the net current will be very small.

Note also that these same equations can be used to examine the situation at the end of the beam pulse. In this case, Lenz's law causes the plasma current to flow in the same direction as the beam current. The exponential decay of the return current can then be used to estimate the effective conductivity of the plasma.

If the plasma were contained in a conducting cylinder then, depending on the plasma conductivity, it is possible that some of the return current can (inductively) flow through the conducting walls. We can qualitatively analyze this situation by referring to the simple circuit diagram of Figure 5.1. In this case the beam current can be thought of as dividing between the current I_p flowing through the resistance R of the plasma, or the wall current, $I_w = I_n$ returning through the inductance L of the wall. The circuit equation then becomes

$$(d/dt)I_p = (d/dt)I - RI_p/L \tag{59}$$

which has the solution

$$I_p = I(L/Rt_r)[1 - \exp(-tR/L)] . \tag{60}$$

Note that the functional form of this equation is identical to that of Eq. (58). If the risetime of the pulse is short compared to the L/R time of the drift tube, then the return current will be carried entirely in the plasma.

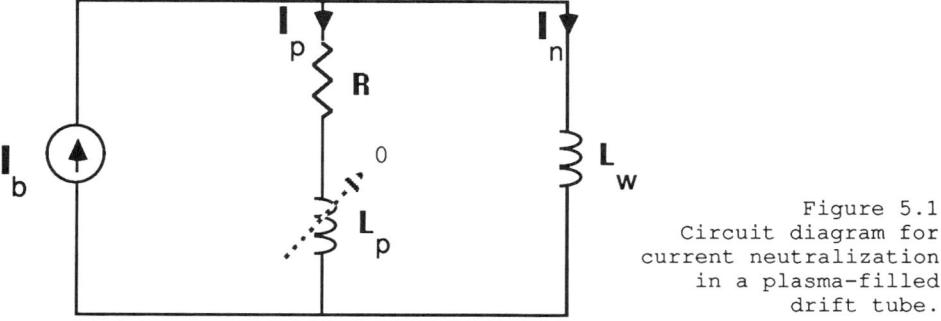

Figure 5.1
Circuit diagram for current neutralization in a plasma-filled drift tube.

As a final point, we can still use the concept of an Alfven current limit by writing $I_A^* = I_A(1 - f_m)^{-1}$, where f_m, the current neutralization fraction, is defined by $f_m = |I_p/I|$ (f_m is the quantity that we previously used in the envelope equation analysis). If $f_m \to 1$, then $I_A^* \to \infty$, and all of the beam current can be transported.

5.2   Macroscopic Beam-Plasma Equilibria

In the previous discussion we were interested primarily in the response of the plasma to the injection of a beam, i.e. time-dependent charge and current neutralization phenomena. We now wish to couple the beam and plasma fluid components in order to develop and examine various equilibrium configurations. We will generalize the fluid - Maxwell equations to include the effects of finite, non-zero, beam and plasma temperatures, by adding a pressure-dependent term, $\nabla \cdot \mathbf{P}/m$, to the equation of motion. $\mathbf{P}$ is the the fluid pressure tensor.

If there is no external magnetic field, and beam and plasma centrifugal effects are negligible, the equations of motion for the beam and plasma electrons reduce to

$$(kT_b/n_b)(d/dr)n_b = -e(E_r - v_{bz}B_\theta/c) ,\qquad(61)$$

$$(kT_e/n_e)(d/dr)n_e = -e(E_r - v_{ez}B_\theta/c) .\qquad(62)$$

In writing these equations we have further assumed that components of the pressure tensors perpendicular to the z-axis are given by $p_\alpha = n_\alpha k T_\alpha$, where k is the Boltzmann constant, and T_e and T_b are the isotropic plasma and beam-electron kinetic temperatures.

We now suppose that the beam has been injected into an underdense plasma, i.e. the plasma density is less than the beam density. In this case, the beam space charge will rapidly expel all plasma electrons from the region of the beam, leaving only the (assumed) stationary ions. Using the charge neutralization fraction $f_e = n_i/n_b$, and assuming that the beam axial velocity is a constant, Eq. (61) becomes

$$(kT_b/n_b)(d/dr)n_b = (4\pi e^2/r)(1 - f_e - \beta^2)\int_0^r rn_b dr .\qquad(63)$$

It is easy to show that if n_b is a monotonically decreasing function of radius and $f_e > \gamma^{-2}$, then an equilibrium exists in which the outward forces of the beam fluid pressure and space-charge self-repulsion are balanced by the magnetic self-pinch. Differentiation of Eq. (63) with respect to r yields

$$(d^2/dr^2)n_b + r^{-1}(d/dr)n_b - n_b^{-1}(d/dr)^2 n_b + Kn_b^2 = 0 \qquad(64)$$

where $K = (4\pi e^2/kT_b)[\beta^2 - (1 - f_e)]$. This expression and its solution were first derived by Bennett in 1934; its solution, termed the Bennett profile, is given by

$$n_b = n_0[1 + (r/a)^2]^{-2} \qquad(65)$$

where n_0 is the on-axis beam density, and $a^2 = 8/Kn_0 = 8\lambda_d^2/[\beta^2 - (1 - f_e)]$ is the Bennett radius. (λ_d is the Debye length, defined by $\lambda_d^2 = kT_b/4\pi n_0 e^2$, at r = 0.) Note that a^2 is a minimum for $f_e \sim 1$, and becomes large if $f_e \sim \gamma^{-2}$.

Generalized Bennett relations can also be obtained for the case of space-charge-neutral equilibria ($E_r = 0$), which are assumed by setting $n_e + n_b = n_i$ = const. However, physically acceptable solutions are obtained only if the beam temperature exceeds the plasma electron temperature. In this case the degree of current neutralization is easily shown to be $f_m = T_e/T_b$.

5.3 Macroscopic Beam-Plasma Instabilities

Given the existence of various beam-plasma equilibria, we must next investigate the stability of these systems. There are several possible instability mechanisms, which are generally grouped as macroscopic, in which the typical scale length is the beam radius, and microscopic, in which the length scale is much smaller than the beam radius. As an example of a macroscopic instability we will examine the resistive hose instability, in which the beam is hydromagnetically unstable to the growth of transverse perturbations because of the finite plasma conductivity.

A simple picture which illustrates the important features of this instability can be developed by referring to Figure 5.2. The first important feature is that if a beam is perturbed, it will tend to perform sinusoidal oscillations about its equilibrium position. The axial length scale of the oscillations is the betatron wavelength $\lambda_\beta = 2\pi r_b (17\beta\gamma/2I_b)^{1/2}$, and the betatron radian frequency is $\omega_\beta = 2\pi\beta c/\lambda_\beta$. The second important feature is that the plasma conductivity causes the displacement of the magnetic axis to lag behind the displacement of the beam axis by the magnetic diffusion time, $\tau_m = 4\pi\sigma(r_b/c)^2$. Hence, there is a restoring force, due to the interaction between the longitudinal beam current and the perpendicular component of the residual magnetic field, that tends to push the beam back to its original position. If the plasma is a perfect conductor ($\tau_m \rightarrow \infty$), then the magnetic field of the beam is "frozen" into the plasma and beam displacements simply result in stable betatron oscillations. If σ is finite, however, the beam displacements and the restoring forces can become phased such that the displacements grow.

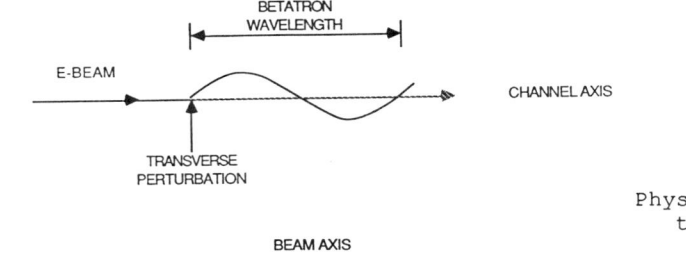

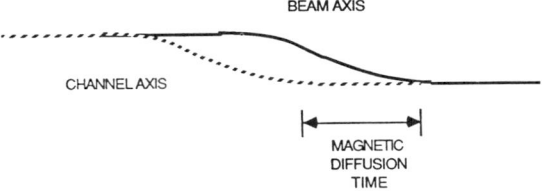

Figure 5.2
Physical concepts for the resistive hose instability.

Let Y and D represent the transverse displacements of the beam centroid and the magnetic field centroid from the nominal beam propagation axis, and assume that their variation is harmonic

$$Y, D \sim e^{-i(\omega t - kz)} .$$

We know that separation of the beam and magnetic field axis will result in betatron oscillations. Since the beam is propagating, we must use the convective derivative. Thus, the phenomenological equation that describes the transverse motion of the beam is

$$[(\partial/\partial t)Y + v(\partial/\partial z)Y]2 + \omega_\beta^2 (Y - D) = 0 . \tag{66}$$

However, the time-dependent displacement of the magnetic field axis depends only on the separation between the beam and magnetic field axis,

$$(\partial/\partial t)D = (k/\tau_m)(Y - D) . \tag{67}$$

Substituting for the harmonic dependence, and defining a Doppler-shifted frequency, $\Omega = \omega - kv$, gives the dispersion relation as

$$-i\omega\tau_m/k = \Omega^2/(\omega_\beta^2 - \Omega^2) . \tag{68}$$

Instability ($-i\omega>0$) thus corresponds to $\Omega^2 < \omega_\beta^2$.

For low frequency perturbations this simple phenomenological picture gives surprisingly good results. For higher frequencies, however, the intrinsic assumption of a rigid beam and a rigid magnetic field breaks down, and phase mixing of the betatron orbits causes the instability to become convective in character.

5.4 Microscopic Beam-Plasma Instabilities

In contrast to the macro-instabilities, the microscopic beam-plasma instabilities are usually associated with the streaming motion of one species through another. As we have seen, there are normal modes of an electron beam associated with longitudinal clumping of the beam space charge (space-charge modes) and transverse oscillations in the presence of a magnetic field (cyclotron modes). In addition, there are also natural oscillation modes of the plasma. Thus, it is reasonable to expect an instability to develop when the phase velocity of a particular plasma wave is very close to the phase velocity of a particular beam wave. In this circumstance, the interaction time between the waves and particles is very much longer, and significant particle acceleration and deceleration can occur. This, of course, leads to wave growth, which leads to more bunching, etc.

From Section 4.5 we have seen that the dispersion relation for the normal modes of the beam can be described as

$$\omega = kv - \omega_b/\gamma^{3/2} \quad \text{(space charge)} ,$$

$$\omega = kv - \Omega/\gamma \quad \text{(cyclotron)} .$$

The normal modes of the plasma correspond to space-charge oscillations at the plasma frequency, ω_e. Hence, we may draw a dispersion diagram, such as that in Figure 5.3. The instability associated with the intersection of the beam space-charge line and the plasma frequency is called two-stream (or Cerenkov) instability, and the intersection of the beam cyclotron wave is called cyclotron instability.

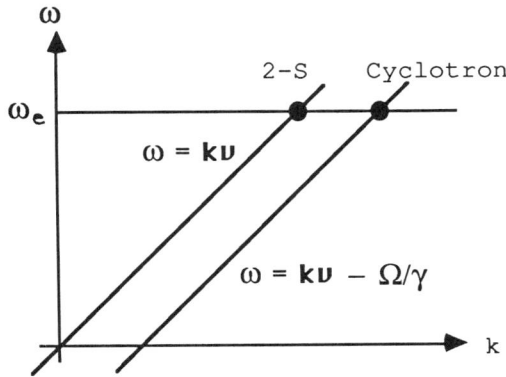

Figure 5.3
Simple dispersion diagram
for beam-plasma instabilities.

In order to calculate instability growth rates it is necessary to have the full beam-plasma dispersion relation. This is generally a tedious procedure involving the linearization of the fluid - Maxwell equations. An example of this type of analysis was in fact given in determining the beam normal modes. Rather than carry out a new calculation, we will simply illustrate the use of a dispersion relation to obtain an instability growth rate. Our example will be the so-called Buneman instability, which is associated with the streaming of a very tenuous electron beam through a stationary ion background. The dispersion relation for this case can be given as

$$k_z^2[1 - (\omega_b^2/\gamma^3)(\omega - k_zv)^2 - (\omega_i/\omega)^2] + k_\perp^2[1 - (\omega_i/\omega)^2] = 0 \qquad (69)$$

and we expect instability when $\omega \sim k_zv$. Writing $\omega = k_zv + \delta$, and substituting into Eq. (69), we obtain

$$\delta^2 = (k_z\omega_b/k_\perp)^2\gamma^{-3}[1 - (\omega_i/k_zv)^2]^{-1} . \qquad (70)$$

In obtaining Eq. (69) it was assumed that the time variation of all perturbed quantities was $e^{i\omega t}$. Hence, the negative root of Eq. (70) corresponds to unstable growth if $\omega_i > k_zv$, and $|\delta|$ is the growth rate.

In many instances efficient transportation of intense electron beams is desirable. For these cases the growth of instabilities is not desired. In many other cases, however, the growth of a particular unstable mode is the desired effect. Important examples include high power microwave generation, plasma heating with relativistic electron beams, and various collective ion acceleration techniques.

6 BEAM PROPAGATION IN NEUTRAL GAS

When an intense relativistic electron beam is injected into neutral gas, the beam transport can have the characteristics of either propagation in a vacuum or propagation in a plasma, depending on how rapidly the beam can ionize the gas. Because of the several parameters that determine the ionization rate, a quantitative analysis of beam transport through neutral gas in the general case is very difficult.

In descriptive terms it is helpful to characterize the various situations by the ratio of the beam current to the space-charge limit, I/I_l, and the ratio of the beam current to the Alfven limit, I/I_A. Writing the space-charge-limiting current as $I_l(t) = I_{lo}[1 - f_e(t)]^{-1}$, where f_e is the charge neutralization fraction, if the peak injected beam current I_o never exceeds I_{lo}, then a virtual cathode can never form, and propagation is possible regardless of the gas pressure. However, if $I_o > I_{lo}$ then the propagation phenomenology depends critically on the ionization rate, and thus the gas pressure.

Assuming that $f_e \leq 1$, then efficient propagation can occur if the beam current does not exceed the Alfven current limit. However, if $I_o > I_A$, then propagation of the total beam current can occur only if there is some degree of current neutralization, $f_m > 0$. Since current neutralization is largely determined by the plasma conductivity, and the conductivity depends in turn on the plasma-electron - neutral-atom collision frequency, current neutralization is also a sensitive function of the background gas pressure.

Assuming that the conditions for beam propagation are satisfied, the resulting behavior can be qualitatively analyzed in terms of the beam envelope equation

$$\gamma m \ddot{r} = -eE_r [1 - f_e - \beta^2 (1 - f_m)] . \tag{71}$$

Depending on the beam parameters, the background gas ionization, and the plasma conductivity, a number of situations are possible. To facilitate the analysis, we will further order the different phenomena according to the background gas pressure.

6.1 Beam Current Less than the Space Charge Limit ($I < I_l$)

In this case a virtual cathode cannot form, and the beam electrons are not slowed appreciably in the axial direction. If the background gas pressure is low enough that the charge neutralization time t_n exceeds the beam pulse time t_p, then the gas ionization is volumetric and due entirely to electron impact ionization. For the case of hydrogen, this criterion can be stated approximately as

$$p_N(\text{Torr}) < 5/t_p \text{ (ns)} .$$

In this limit the beam behavior can be analyzed in terms of the radial force equation

$$\gamma m \ddot{r} = -eE_r(\gamma^{-2} - f_e) .$$

For regions in which $f_e < \gamma^{-2}$, the beam electrons experience a net repulsive force and the beam expands. If $f_e > \gamma^{-2}$, however, the beam electrons will undergo betatron oscillations in r and z, maintaining an envelope radius on the order of the initial beam radius.

For somewhat higher pressures, such that $t_n \leq t_r$, where t_r is the beam risetime, $f_e \sim 1$ early in the pulse, and electron avalanche can be an important ionization process. For a weakly ionized gas the conductivity is given by $\sigma \sim e^2 n_e / m \nu_c$, where ν_c is the momentum transfer collision frequency. Typically $\nu_c(\text{sec}^{-1}) \sim 10^9 P$ (Torr). Hence, the exponentiation of the secondary plasma electron density due to avalanche can lead to the formation of a reverse plasma current according to $j_p = \sigma E_z$. For beam currents of about 10 kA with risetimes of typically 10 ns, these processes imply that $f_m \sim 0$ for background pressures in the range of 1 Torr. Under these conditions the beam self-fields are essentially cancelled by the plasma and the beam freely streams.

For much higher background pressures electrostatic breakdown and the formation of a plasma channel occur very rapidly ($f_e \sim 1$). However, ν_c is sufficiently high that the rise in plasma conductivity is limited, and $f_m < 1$. Thus, the possibility exists for pinched beam transport. In addition, the plasma is sufficiently resistive that hose instability is likely to occur (see Figure 6.1).

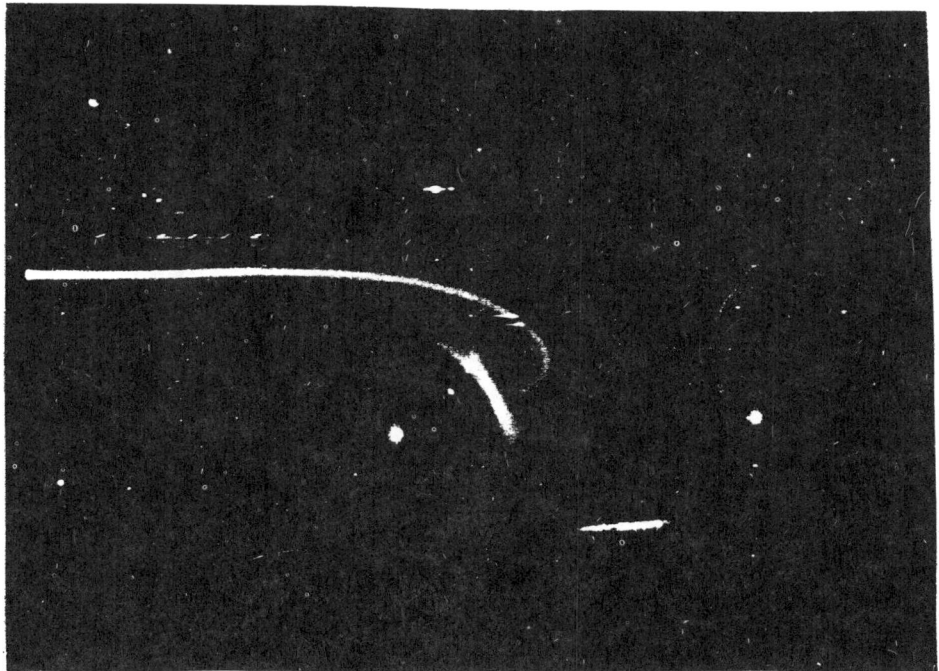

Figure 6.1 Hose instability of the 10-MeV, 100-kA beam in air.

6.2 $I_l < I < I_A$

When the electron beam current exceeds the space-charge limit, beam propagation is prevented by the formation of a virtual cathode. This potential barrier will persist until ionization of the background gas (and expulsion of the plasma electrons) produces an ion density sufficient to charge-neutralize the beam space charge almost completely. Depending on the specific current and voltage waveforms, and the background gas pressure, there are several possibilities, as indicated in Figure 6.2. Assuming the voltage rise is zero, I_1 can be written as $I_1 = I_{10}[1 - f_e(t)]^{-1}$, where I_{10} depends only on the beam kinetic energy and the geometry. If the peak injected current I_0 exceeds I_{10}, then the beam behavior will depend critically on the background pressure. During the time that the rising portion of the beam current does not exceed the limiting current, the beam will propagate. Meanwhile, the background gas will be ionized by electron impact ionization, as described in the previous section. When the beam current exceeds the space-charge limit, a virtual cathode forms in the vicinity of the injection plane, and beam propagation ceases. Ions previously generated can now also contribute to the ion density through ion impact processes. For hydrogen, the effective ion avalanche time is $\tau_i \sim 0.33[P(\text{Torr})]^{-1}$ ns.

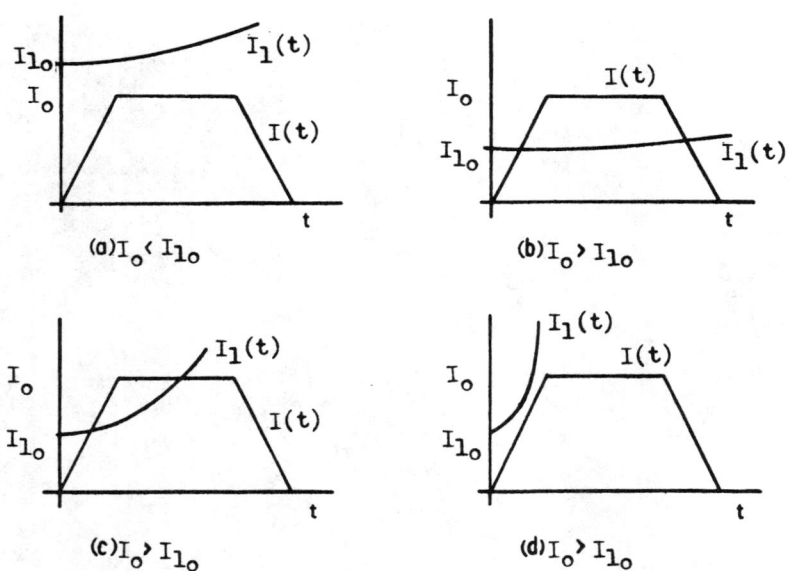

Figure 6.2 Idealized beam current and space-charge-limit vs time.

In the low pressure regime (typically <10 mTorr) described by Figure 6.2(b), the gas density is insufficient for these processes to rapidly neutralize the beam space charge, and most of the beam will not propagate. For somewhat higher pressures, Figure 6.2(c), ionization is fast enough that the virtual cathode collapses early in the pulse and the beam can propagate, although the processes that control the beam front velocity can be quite complicated. For still higher pressures (≥ 1 Torr), the situation is described by Figure 6.2(d). Ionization of the background gas due solely to electron impact ionization is rapid enough that the beam current never exceeds the space-charge limit. A virtual cathode never forms, and the beam propagates immediately away from the injection plane.

This "runaway" pressure is also the effective lower limit of the pressure regime in which current neutralization effects become important. (It should be recognized, however, that a lack of current neutralization does not prevent propagation, since $I < I_A$.) The experimentally observed net current behavior has been adequately explained in terms of a simple model in which the plasma current is assumed to obey the simple Ohm's law formula. The conductivity is assumed to be zero until the onset of charge neutralization and exponentiation of the secondary electron density by electron avalanche ionization. After this gas "breakdown" time, the conductivity is assumed to be constant. According to this theory, if gas breakdown occurs early in the pulse and the breakdown conductivity is high, then substantial current neutralization can occur. For air this situation occurs roughly in the 1 to 10-Torr pressure regime. At somewhat lower pressures, two-stream instability turbulence decreases the effective current neutralization. At somewhat higher pressures, the increased momentum transfer collision rate increases, and the conductivity again decreases. As a result the net current increases and the resistive hose instability can develop.

6.3 $I > I_A$

When the injected beam current exceeds the Alfven limit, the efficient transport of such beams can occur only if there is a substantial degree of both charge and current neutralization, i.e. the 1 to 10-Torr regime for air. For very low pressures propagation is first limited by virtual cathode formation. Even after the beam has become space-charge neutralized, the lack of current neutralization can prevent efficient propagation. In the high pressure regime the space charge is effectively neutralized, but the plasma conductivity is too low to permit a high degree of current neutralization. Effectively, all of the beam kinetic energy is consumed in establishing the magnetic field of the beam current.

6.4 Summary

In view of the previous discussion the general features of beam transport in neutral gas can be summarized as indicated in Table I. It should be realized that this summary provides only general guidelines, however. In particular, the possible development of beam-plasma instabilities can markedly alter this simple picture. Once beam propagation has been established, the subsequent beam behavior can be qualitatively analyzed in terms of the radial force (envelope) equation, including the effects of both charge and current neutralization. Finally, for pinched beams that are transported over several betatron wavelengths it is necessary to include the effects of beam scattering and various energy loss processes (collisional ionization and bremsstrahlung).

Table I Summary of beam transport characteristics as a function of beam current and background gas pressure.

	Low pressure $f_e \sim f_m \sim 0$ ($\lesssim 0.5$ Torr)		Intermediate pressure $f_e \sim f_m \sim 1$ (1 – 10 Torr)		High pressure $f_e \sim 1, f_m < 1$ ($\gtrsim 10$ Torr)
$I < I_l$	Propagation with radial expansion	Two-stream instability	Stable beam propagation: Beam expansion controlled by transverse kinetic energy	Hose Instability	Pinched beam Propagation: Nordsieck Expansion
$I_l < I < I_A$	Virtual cathode formation				
$I > I_A$					Magnetic power balance limit

APPLICATION OF PULSE POWER TECHNOLOGY TO ULTRA HIGH ENERGY ELECTRON ACCELERATORS

John A. Nation
Laboratory of Plasma Studies and School of Electrical Engineering
Cornell University, Ithaca, N Y 14853, U.S.A.

TABLE OF CONTENTS

1 Summary 1760
2 Introduction 1760
3 Externally Driven Sources 1762
   3.1 RF Coupled Devices 1762
   3.2 Non-Resonant Accelerator Structures 1765
4 Direct Coupled Accelerators 1767
   4.1 RF Sources 1767
   4.2 Non-Resonant Accelerators 1770
5 Induction Linacs 1771
6 Concluding Comments 1773
7 Acknowledgements 1774
8 References 1774

APPLICATION OF PULSE POWER TECHNOLOGY TO ULTRA HIGH ENERGY ELECTRON ACCELERATORS

John A. Nation
Laboratory of Plasma Studies and School of Electrical Engineering
Cornell University, Ithaca, N Y 14853, U.S.A.

1 SUMMARY

We present in this paper a review of the application of pulse power technology to the development of high gradient electron accelerators. The technology demands are relatively modest compared to the ultra high power technology used for inertial confinement fusion drivers. With the advent of magnetic switching intense electron beams can be generated with a sufficiently high repetition rate to be of interest for high energy electron accelerator driver applications. Most of the techniques considered rely on the excitation of large amplitude waves on the beams. Within this framework there are two broad categories of accelerator, those in which the waves are directly excited in and supported by the medium and, secondly, those where the waves are used to generate radiofrequency signals which are then coupled via structures to the beam being accelerated. In what follows we shall consider both approaches. Present-day pulse power technology limits pulse durations to about 100 nsec. Consequently, if we are to use these sources we will need to use high group velocity structures to avoid the need for short accelerator module lengths. An advantage of the short pulse duration is that the available acceleration voltage gradient increases compared to that obtained using conventional rf drivers.

2 INTRODUCTION

Single-pulse high-power electron beam generators were first developed at the 30-GW level in 1962[1]. Since that time the technology has been developed to the point where reliable 1-kHz repetition rate operation has been achieved using magnetic switching to control the pulse development[2]. The next generation of high energy particle accelerators will require new primary driver technology. There appears to be a match between these requirements and the present-day capabilities of pulse power sources.

We summarize in this paper various approaches, based on the use of pulse power technology, for the development of a high gradient, ultra high energy electron accelerator. The discussion is

© 1989 American Institute of Physics

restricted to electron accelerators. Since the peak energy available in conventional accelerators has for many years followed the development of rf sources, much of the work has been centered on the development of intense high frequency rf sources which may then be used as the driver for the accelerator. The rf may be coupled to the accelerated beam in two ways: firstly, via structures following traditional accelerator technology or, secondly, by a direct coupling to the high frequency waves on the low energy, high current electron beam. The latter approach, a collective accelerator, requires excitation of a suitable eigenmode of the primary beam with the correct phase velocity.

To be useful the driver must have a repetition rate capability, and be capable of generating a low emittance, high current, weakly relativistic electron beam which can then be used to generate the rf fields used to accelerate the main electron beam. Electric field gradients well in excess of 100 MV/m are desired for the application. The former requirement can be met by use of a Linear Induction Accelerator as the primary driver. Perhaps the most important recent development is the use of Magnetic Pulse Compression techniques for switching. With the use of this technology the Lawrence Livermore group, for example, report the construction of the ETA II accelerator with design parameters of 7 MeV, 3 kA, in 50-ns pulses with a 5-kHz repetition rate over at least a 30-sec duration[3]. In the same laboratory a 1-MeV, 1-kA electron beam runs over extended periods at a high repetition rate.

The approach to the design of high field gradient accelerators from pulse power, as opposed to the use of conventional rf source drivers, raises a number of issues. Pulse power drivers normally have only relatively short pulse durations (~50 ns), although a 2.0-microsec, 800-kV induction linac pulser was fabricated almost 20 years ago[4]. The short pulse durations imply that we will either have to use low Q or high group velocity structures if the filling time $(2Q/\omega)$ is to be made short compared to the pulse duration. For cavity Q's of order 10^4 the filling time requirement implies operating frequencies in the 30-GHz range so it might also be necessary to spoil the Q in order to operate at lower frequencies, e.g. at X band, where the requirements on structure alignment are less demanding. In traveling-wave systems typical wave group velocities used in today's accelerators are about 0.03c and accelerator modules would need to be located every 50 cm if the pulse duration were dropped to 50 ns. It would be desirable to increase the pulse duration by up to an order of magnitude and at the same time increase the group velocity of the waves in the accelerator significantly. Recent designs suggest the use of group velocities approaching 0.1c. An alternate approach currently under consideration is to pulse-compress existing rf sources to increase the available power and to roughly double or treble the design group velocity.

A second important factor in the design of suitable structures arises from the desire to keep the r f field from the walls so the accelerated particles sample the maximum field region whereas the wall fields, which can lead to breakdown, are held as low as possible. This criterion is met in collective beam accelerators, where the accelerated particles travel in the region of maximum field and the wall fields are small. A similar class of structures exists for externally driven devices, however the particles must then sample the fields across the radius of the guide.

In the following sections we describe concepts being investigated for electron accelerators. The work falls into four categories as shown in figure 1. Broadly speaking these may be classified as internally and externally driven devices which are either harmonically or non-harmonically excited. As stated earlier

Coupling / Method	Structures	Direct
RF Drive	Two-Beam Accelerator New RF Devices	Ultralac (Upper Hybrid Mode Coupling)
Non-Resonant Systems	Wake-Field Accelerator Switched Power Linac	Auto Accelerators Collective Implosion

Figure 1 Classification of Accelerator Schemes

we shall emphasize the work on rf driven accelerators and only briefly indicate work on other approaches.

3 EXTERNALLY DRIVEN SOURCES

3.1 RF Coupled Devices

Sessler and his colleagues have presented two different scenarios for a 'two-beam accelerator' in which an induction linac powers a low energy, high current beam (~5 MeV, 1 kA) used for rf generation[5,6]. The rf is then coupled to a slow wave structure where the second, lower current beam is accelerated. The device is shown schematically in figure 2. It differs from a conventional accelerator in the sense that there is a single drive beam rather than a large number of primary 'beams' driving discreet rf sources, e.g. klystrons. A key feature in the proposal is the idea that the energy lost to the rf is replenished by post-acceleration of the primary beam and the extraction process is then repeated. This increases the efficiency of the rf generation, provided that the beam quality is still sufficiently good to allow rf generation, at the cost of developing stable low energy beam transport over long

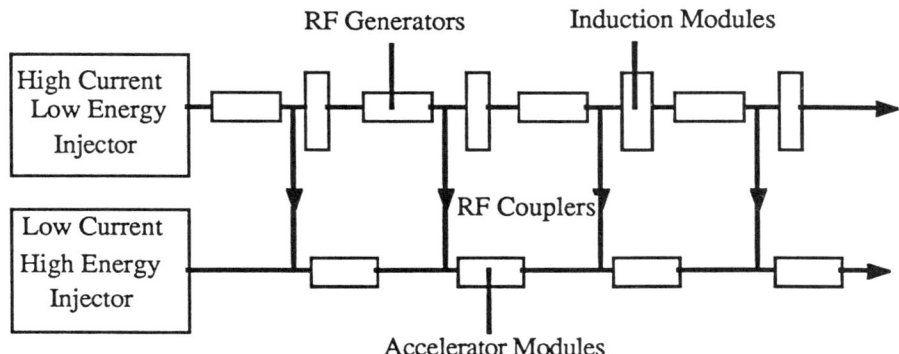

Fig 2 Two-Beam Accelerator Schematic

distances. Beam transverse instabilities, especially those associated with traversing acceleration cavities, must be adequately controlled.

Two scenarios have been proposed. In the former an FEL interaction was explored and in the latter a relativistic klystron was proposed as the rf source. In the FEL work peak powers of 1.5 GW were obtained at a frequency of 35 GHz using the ETA accelerator with a tapered transverse wiggler to enhance the interaction efficiency (~30%). The relatively high efficiency was obtained through use of a tapered wiggler to keep the electron momentum at a value satisfying the phase slip condition for interaction in the wiggler field. No post-acceleration of the electrons after rf extraction was attempted. Concerns with this approach include the control of the rf phase, power extraction, and the suppression of side-band instability.

More recently work has centered on the development of an 11.4-GHz relativistic klystron. In the initial investigations of this approach a six-cavity klystron is driven by a 1.2-MeV, 1-kA, 50-ns thermionically emitted electron beam. The beam is compressed in a converging guide field from 12.5 cm to 0.9 cm in diameter. Peak output powers of 80 MW with 60 dB gains have been obtained with full beam pulse duration and 200 MW has been generated in a ten-ns pulse[7]. The shorter pulse duration is believed to be associated with secondary electron emission and multipactoring.

Preliminary measurements on a prototype relativistic klystron have also been carried out at Cornell using a two-cavity system operating at 9 GHz. The cavities are run in the TM020 mode and are powered by a 400-kV, 300-A, 50-ns electron beam.

The input to the first cavity is about 2 kW from a magnetron, and gains of up to 30 dB have been observed[8]. To date only the two-cavity klystron has been studied and no power extraction has been attempted. The interaction of the beam with the first cavity has been explored using magnetic field probes in the drift region between cavities. Unlike a conventional klystron, in which the beam is ballistically bunched, we observe a bunching which is apparently due to the excitation of slow and fast space-charge waves on the beam. The process is similar to the collective interaction observed by Friedman and Serlin[9] at 1 GHz in beam modulation experiments. Similar interactions have been reported by Anselmo and Nation[10] using travelling-wave tube structures. The scale length for ballistic bunching is large compared to the experimentally observed bunching length. In this experiment the large beam loading in the first cavities eliminates the need for artificially reducing the cavity Q's as was done in the SLAC/LLNL/LBL experiments.

In other rf generation experiments for particle accelerators Granatstein[11] at the University of Maryland is developing a relativistic gyroklystron at 10 GHz. The projected output power is 30 MW for the prototype and 300 MW for a practical device. The beam is driven from an artificial transmission line with a step-up transformer to produce a 500-keV output with a pulse length of order 1 microsecond. The gyroklystron has four cavities and is driven from a thermionic cathode. The efficiency is limited to about 50% as a result of the need to provide free energy from beam rotation to couple to the TE modes in the system. The 30-MW device is expected to come on line shortly.

At Cornell work is in progress on the development of travelling-wave tube amplifiers using an 850-keV, 1.0-kA, 100-ns electron beam[12]. A 6-mm-diameter field emission source is used as the cathode for the electron beam. Two amplifiers have been built, both using a 150-kW input source. Gains of 29 dB and 17 dB have been achieved at 8.76 GHz and a peak output power of 100 MW has been achieved. In these devices the input power is coupled in to the amplifier from the side walls of the tube with the rear face of the anode serving as a shorting plate a half wavelength behind the feed arms. The configuration is sketched in figure 3. All interactions take place in the TM01 mode of the system to reduce the possibility of competing modes. The interaction point is chosen to be close to the $2\pi/3$ point so that the coupling is guaranteed to be to a forward wave on the structure. This eliminates the possibility of an absolute instability leading to oscillation such as that found in backward-wave devices. The coupling to higher-order modes is smaller than that to the TM01 mode and no evidence is found of oscillation in higher modes. Heterodyning of the amplified signal with the output from a local oscillator confirms that the amplified wave has the same frequency as the input signal and shows that the bandwidth of the output is only about 11.5 MHz and comparable to

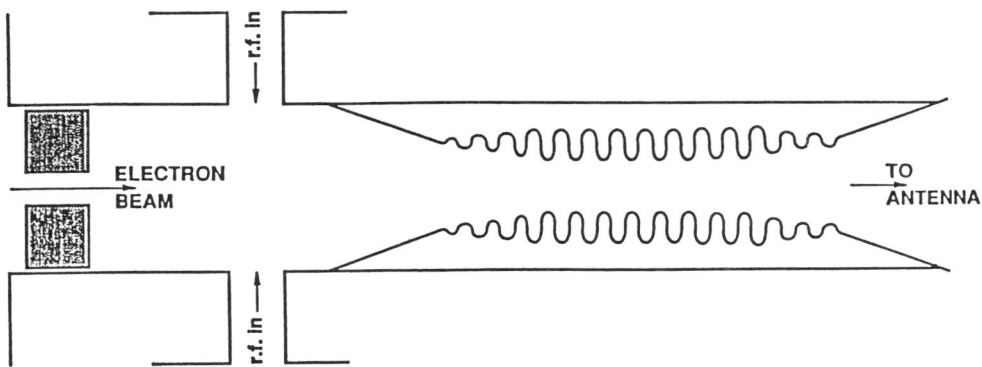

Fig. 3 Schematic showing high power traveling-wave amplifier experiment

the natural bandwidth expected for the 100-nsec pulse duration.
Unlike recent results from high power BWO's, where the rf pulse duration is much shorter than that of the driver beam, the pulse duration is equal to the duration of the high voltage pulse. The output power is also independent of the strength of the applied guide magnetic field. This feature also differs from that in the BWO work, where the output power varies dramatically with the guide field strength.

In all of the above cases work still needs to be done to establish that the frequency and the phase stability of the rf output meet the stringent requirements for accelerator applications.

In other laboratories pulse power facilities have been used to develop rf sources using magnetrons, FEL's, BWO's, reditrons, vircators and various other devices[13]. The phase stability requirements of an accelerator suggest that amplifier configurations such as the relativistic klystron and the TWT amplifier may be the preferred configurations. Recent measurements on an S-band magnetron by the group at Physics International show frequency locking and phase stable output of tightly coupled magnetron oscillators. The phase stability of the two oscillators was measured at better than 10⁰ for output powers of about 1 GW. The coupling between the two oscillators was accomplished by directly coupling the outputs from one vane of each magnetron. With oscillators, even with strong coupling, the locking takes many cycles of the rf wave.

3.2 Non-Resonant Accelerator Structures

In the preceding section we have described various rf accelerator configurations. We now describe briefly some proposals for non-resonantly driven accelerators, although these do not fit directly within the framework of pulse power technology as used in this paper. An interesting technique for particle acceleration is the Wake-Field Accelerator originally proposed by Voss and Weiland[14]. In their proposal the accelerator uses an annular relativistic electron beam injected into a structure similar to that shown in figure 4. By appropriate shaping of the wall a unidirectional axial electric field is developed as the head of the beam passes the structure. The field is aligned in the sense required to extract

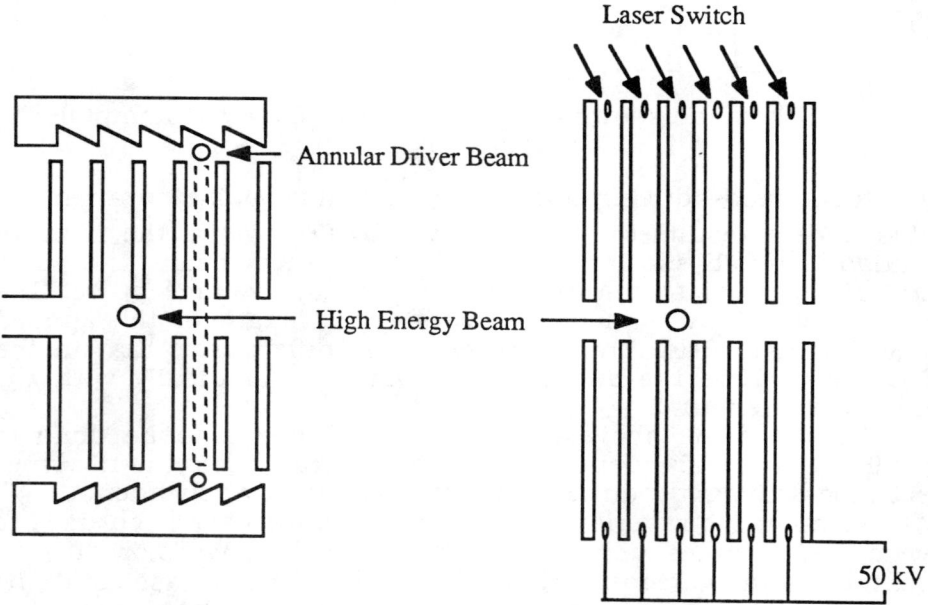

Fig.4 Schematics showing the Wake-Field Accelerator and the Switched Power Linac

energy from the annular beam. An electromagnetic wave propagates towards the axis with the electric field being enhanced as the wave propagates inward in the radial pulse line. Field enhancements of up to about 20 are achievable. Since the electron velocity in the annular beam is approximately equal to the speed of light, the accelerating field on axis is automatically synchronized with the electrons being accelerated. In the DESY facility a proof of principle experiment is in progress. It uses an 8-MeV, 1-cm-long electron beam having a charge of 100 nC. Axial fields of 8 MV/m have been achieved. Parameters for more useful devices require scaled-up systems having a charge of ~1 microcoulomb in a 10-cm-diameter annular beam of minor radius 2 mm and a beam length of about 2 mm. The central hole, which is used for the accelerated beam, has a diameter of about 1 mm and should yield an axial voltage gradient in excess of 150 MeV/m.

As a result of the difficulties in control and generation of the driver beam, the Switched Power Linac has been proposed[15]. Figure 4 also shows schematically the proposed technique. The switched high voltage wire replaces the driver beam. Work is in progress in several laboratories to study the switching problem and to identify the azimuthal switching uniformity needed to prevent dipole modes being excited in the acceleration region. Prototype devices have been proposed. Typical device parameters have an outer radius of 12 cm, an inner radius of 1 mm, and disk separations of about 2 mm. Pulse lengths are of order 10 psec, and power levels of 6 GW give gap voltage gradients of about 600 MeV/m with field gains of 12 and charge voltages of 80 kV.

Friedman and Serlin[16] at NRL have suggested a 'resonant' version of the switched power linac.

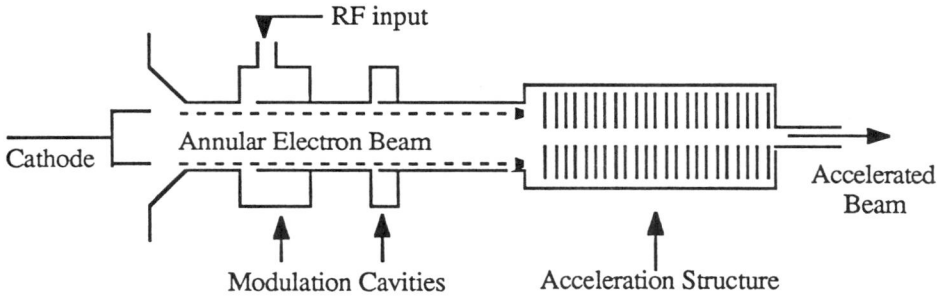

Fig 5 Schematic of Resonant Radial Line Accelerator

This device strictly belongs in the first category described but, for convenience, has been deferred until after we have presented the switched power linac concept. In this configuration an annular beam is modulated at high frequency. The modulation frequency is chosen to match the resonant frequency of a series of disks in a cylindrical cavity as shown in figure 5. After the cavity filling time, the second beam, which is to be accelerated, is injected into the cavity. A 1-GW beam is predicted to yield voltage gradients in excess of 100 MeV/m. To date almost complete modulation of the beam at a frequency of about 1.5 GHz has been achieved, and work is in progress to increase this frequency into the X band.

4 DIRECT COUPLED ACCELERATORS

4.1 RF sources

If the accelerating and source beams coexist in the same pipe and suitable rf wave sources are identified, there may be substantial advantages for accelerator design. For example, this eliminates all of the coupling and slow wave structures and maximizes the rf field strength on the drift tube axis rather than at the tube or structure walls, while eliminating or at least drastically reducing the effects of wake-fields due to the structures degrading the beam emittance.

A recent proposal by Anselmo and Nation[17] addresses a possible configuration for an electron accelerator of this type. The use of the fast upper hybrid wave on a mildly relativistic electron beam propagating in a guide has been suggested. The Doppler-shifted fast upper hybrid wave has the characteristic of being the only non-radiative mode of a beam in a pipe which cuts the light line and hence is suitable for electron acceleration to high energies.

Two ways have been suggested for the excitation of this mode. In the first, parametric excitation was suggested to couple the slow and fast hybrid waves. No external source of free energy is required to drive the wave growth since the slow wave has negative energy and the fast wave is a positive energy mode and both can grow simultaneously. In the second approach the free energy is provided by propagating the beam through a wiggler field having a periodicity such that $k_w v = \Omega_c$, where k_w is the bifilar helix wiggler wavenumber, v the electron velocity, and Ω_c the relativistic cyclotron frequency of the uniform applied guide magnetic field. In this field the electrons oscillate resonantly with the wiggler and their drift motion in the z direction is converted into transverse motion. The transverse energy then provides the free energy to drive the interaction between the beam and an electromagnetic mode in a cavity as shown in figure 6. The output taper on the cavity cuts off the electromagnetic mode and leaves the upper hybrid mode body wave on the beam. The cavity is excited in a low-order $TE01_n$ mode using a 150-kW magnetron. In the cavity section a TE01 mode couples to the fast upper hybrid wave and has a maximum growth rate when the cyclotron frequency is approximately equal to the cut-off frequency. Beyond the cavity the radiative mode is cut off but the beam upper hybrid mode can continue to propagate. This process has been observed with large signal levels (> 30 dB signal to noise with no external cavity excitation) for the 50-cm length of the experiment. The signal level maximizes at approximately the correct value of the magnetic field. In addition the same mode has been excited at larger amplitude by feeding a high power signal from a magnetron in the low Q cavity. The detected signal amplitude after the taper follows the input power.

From the accelerator viewpoint the excited kinetic mode is a hybrid TE/TM-like mode with a strong axial electric field in the region of parameter space at large values of the wavenumber. We

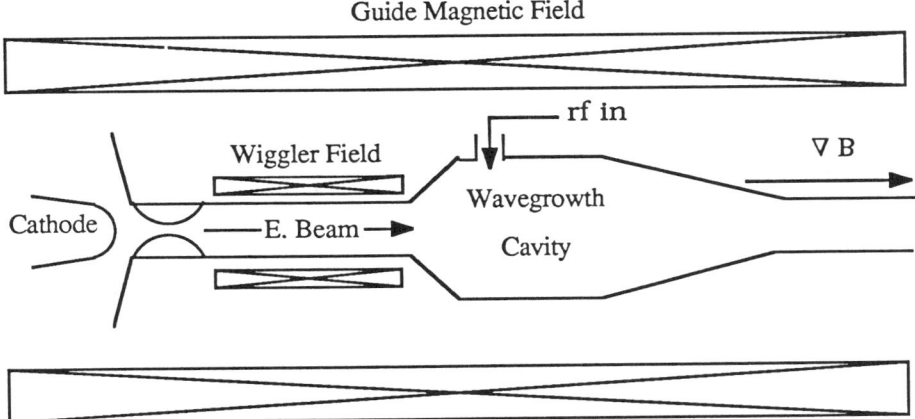

Fig 6 Schematic showing the wave growth configuration for excitation of the upper hybrid mode on an electron beam

plan to adiabatically change the value of the cyclotron frequency to move towards this region and specifically to a value of wavenumber such that the wave phase velocity is equal to the speed of light and the system is matched for acceleration of electrons. In addition the group velocity of this mode at the operating condition of interest is close to the electron velocity so that long staging lengths are possible. Table 1 shows characteristic values of some of the salient parameters of the proposed accelerator.

Table 1

Electron Injection Energy (keV)	500	900
Beam Current (kA)	1.9	5.0
Guide Magnetic Field (kG)	5.4	4.3
Wave Frequency (GHz)	30	30
Maximum Acceleration Field (MV/m)	225	430

In the table the electric field quoted is based on limits set by electron self-trapping in the slow hybrid wave obtained when parametric excitation is used for the wave excitation. Using the technique suggested in figure 6 the field strength obtained is limited by the free energy available.

Recent attempts to excite the waves using axisymmetric magnetic field wigglers have not succeeded yet. The use of the second technique as described above has been used to successfully excite a wave which, based on the evidence available to date, seems to be the correct mode.

For the sake of completeness we mention at this point that it is possible to achieve an externally driven r f accelerator configuration where the fields are a minimum at the walls of the guide if one employs a phase slip device[18]. In this device a TM wave propagates through an undulating guide so the accelerating beam, which travels in a straight line, samples the fields across the cross section of the guide. In general the acceleration of the particles will be zero, since the wave phase velocity is greater than, and the electron velocity less than, the speed of light so that no trapping can occur. Acceleration can occur however when the electrons slip one cycle of the rf wave every period of the undulation in the guide. This criterion and concept is a variation on the inverse free electron laser, in which the particles travel in straight lines and the waves propagate along a curved path, in comparison with conventional FEL geometries, in which the electron trajectory is along a curved path and the wave motion is rectilinear.

4.2 Non-Resonant Accelerators

Another new and interesting idea for an electron accelerator has been dubbed the collective implosion accelerator. The idea, first reported by Briggs[19], is a result of work on the ion column focussing technique needed for transport of an intense electron beam through the ATA accelerator. The geometry is shown in figure 7. In the proposed accelerator a KrF laser beam is used to ionize a

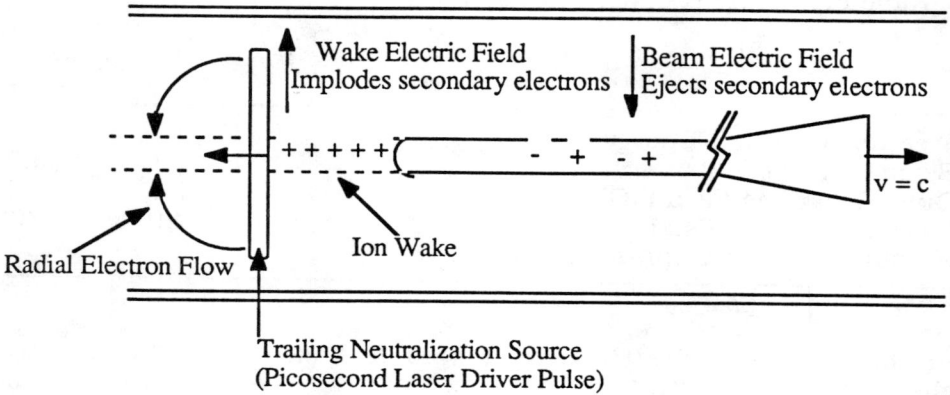

Figure 7   Collective Implosion Electron Accelerator

beam channel in a low pressure benzene background. The radial electric field of the electron beam causes the secondary (plasma) electrons to be ejected from the channel. A sharp end to the beam leaves behind an unneutralized ion channel. The positive-ion-rich region is then space-charge neutralized by electrons generated by photoionization from a picosecond laser pulse propagating through the surrounding gas. The radial current flow caused by the imploding electrons causes an axial electric field which results in the electron acceleration.

The scheme has a high intrinsic efficiency (a wall-plug efficiency of 5% has been estimated) with a long single-stage length, about 200-400 m. Use of the ATA accelerator as the driver would give an accelerated beam parameters of about 300 MeV/m, a bunch length of 1-2 mm, and a particle loading of 3×10^{10} per bunch. Each stage would be capable of generating peak energy increments of about 100 GeV. The sharp turn-off of the primary electron beam in about 1 nsec, of order of one plasma period, is essential to the accelerator. This can be achieved by use of a laser ionization 'kicker' before injection of the beam into the accelerating channel. Obviously this interesting proposal needs much additional work but it has nonetheless much intrinsic appeal

Before leaving the topic of non-resonant accelerator concepts we should point out that autoaccelerator configurations have been proposed which yield interesting voltage gradients. The main feature of these proposals is the shaping of the primary driver beam pulse to give high transformer ratios. This approach is not described in this presentation.

5 INDUCTION LINACS

As we have indicated earlier, induction linacs will play an important role in the development of pulse power devices for any repetition rate system. For high energy electron accelerators this is the case if the high luminosities required are to be reached. In addition, if the two-beam accelerator concepts come to fruition, the ability to increment the beam energy will be essential.

In an induction linac the pulse power is inductively coupled to the electron beam being accelerated. The coupling arrangement is shown schematically in figure 8. The ferrite core is looped by the output conductor from the pulse line and is driven from remnant magnetization in one sense to saturation in the opposite sense. The flux swing achievable is limited by the magnetic properties of the material and its cross-sectional area. The rate of change of flux in the loop formed by the conductors encircling the core drives the acceleration of the beam electrons. An alternate way of viewing the induction linac core is as an inductor which isolates the two different potential sides of successive accelerating gaps. The key feature of the induction linac which distinguishes it from a regular pulse power diode is that the acceleration electric field is derived from the rate of change of the magnetic vector potential and not from the gradient of a scalar potential. Hence it is possible to stack

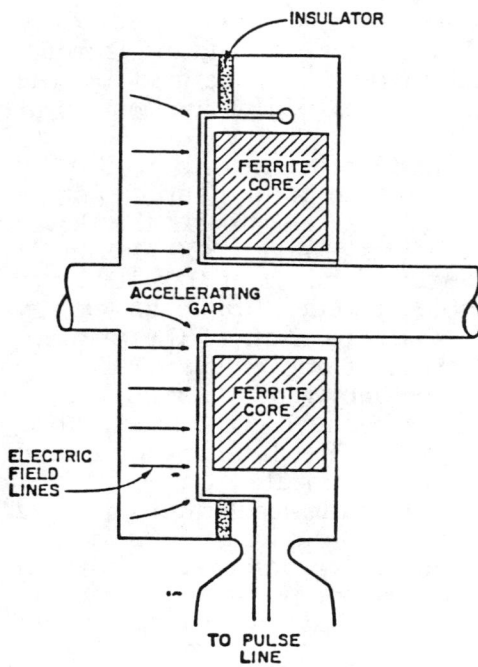

Fig. 8 Coupling arrangement used in Linear Induction Accelerators

n induction modules in series to generate a beam with an output energy of neV, where V is the voltage across a single diode gap, and to never have any component at a voltage greater than V with respect to ground. This feature allows one to increment the beam energy as required for example in the two-beam accelerator concept.

The practical realization of the repetition rate induction linac depends on the revitalization of an old technology, magnetic switching, using new materials. The basic principle is illustrated in figure 9. In magnetic switching a pulse generator feeds a series of cascaded LC circuits. The inductors are saturable reactors using thin (~1-mil) Metglas, an amorphous magnetic glass, rolled into tightly wound coils.

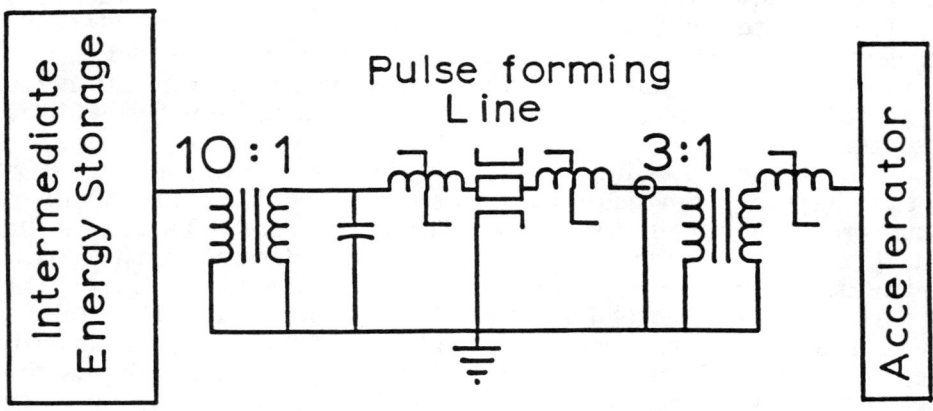

Fig. 9 Simplified schematic illustrating the use of saturable reactors for pulse forming networks

Prior to use the cores are set to their remnant magnetization state in the reverse sense of that to be used in the pulse generation. For the sake of simplicity consider all of the capacitors in the ladder network to be of equal capacitance. If the time taken T_{sat} to charge capacitor C_n is equal to that taken for the reactor L_n to saturate we obtain $T_{sat} = [\Delta(B\,A)]N/V$ where N is the number of times the core is linked by the winding, V is the average value of the charging voltage across L_n, and $\Delta(B\,A)$ is the flux swing between remnant magnetization and the core saturation. As L_n saturates and C_n is fully charged C_{n+1} starts to charge. Pulse compression (switching) is achieved by making the charging times for successive capacitors, and the corresponding times for inductor saturation, smaller. The time scales as the square root of the ratio of the saturated inductances of successive saturable inductors. By this process one can shape and compress a pulse fed into the network. Typical compressions of about 25 are obtained in a three- to four-stage network with the last stage a pulse line or Blumlein line. The final stage may also include a step-up transformer to increase the output diode voltage. The shortest pulses obtainable are determined by the properties of the magnetic material used in the final inductor and are typically tens of nanoseconds. Half-mil Metglas has a slower response time than ferrite cores using TDK P11 and P14 ferrite, but has a much greater flux swing. For example, Metglass has a flux swing of about 25 kG whereas P14 ferrite has only 6.3 kG. Both of these figures are less than those obtainable in a good steel but steels are not available in thin enough sheets to have the response times required for these applications. Magnetic switching is really a technique used for pulse compression. The typical induction linac module is powered using a capacitor discharge in about 10 microseconds to initiate the pulse forming cycle. The capacitor is resonantly charged though a transformer and the output stage, typically a Blumlein line, is fed from a step-up transformer so that output gap voltages of a few hundred kilovolts are obtained for pulse durations of order 50 ns from a single module. In the more conventional pulse power systems switching is achieved through the use of pressurized spark gaps which cannot be cycled rapidly without a huge investment in blowers to sweep the remnant plasma from the switch prior to the arrival of the next pulse. Output pulse energies per module of order 1 kJ can be obtained at 80% efficiency and with repetition rates of at least a few kHz.

6 CONCLUDING COMMENTS

In the above material we have presented a brief overview of the potential uses of pulse power technology for the development of new high energy electron accelerators. The recent developments in production of ultra high power coherent rf sources is encouraging and could well find immediate application in new high gradient accelerators. Pulse power technology, especially magnetic switching, provides a very good match to the driver requirements, although longer duration pulse power sources would be useful.

We have also briefly reviewed other technologies including wake-field, laser switched power linacs, and two novel collective electron accelerators. No effort has been made to provide an account of wake-field accelerators or autoaccelerators. Success with these devices will depend on technology developments in the next few years.

7 ACKNOWLEDGEMENTS

This work was supported in part by the United States Department of Energy, by the AFOSR, and by the SDI, Office of Innovative Science and Technology.

8 REFERENCES

1. J. C. Martin et al., Unpublished (1962).
2. D. Birx, S. Hawkins, S. Poor, L. Reginato, M. Smith, IEEE Trans. Nucl. Sci. NS-32, 2743 (1985).
3. D.S. Prono et al., Proc. 7th Int. Conf. on High Power Particle Beams, Karlsruhe, 1988.
4. J. E. Leiss, IEEE Trans. Nucl. Sci. NS-26, 3, 3870-3876, 1979.
5. T.J. Orzechowski, B.Anderson, W.M. Fawley, D. Proznitz, E.T.Scharlemann, S. Yarema, D. Hopkins, A.C. Paul, A. M. Sessler, and J.S. Wurtele, Phys. Rev. Lett. 54, 889 (1985).
6. A. M.Sessler and S. S. Yu, Phys. Rev Letts. 58, 23, 2439-2442 (1987).
7. M.A. Allen, R.S. Callin, H. Deruyter, K.R. Eppley, W.R. Fowkes, W.B. Herrmannsfeldt, T. Higo, M.A. Hoag, T.L. Lavine, T.G. Lee, G.A. Loew, R.H. Miller, P.L. Morton, R. B. Palmer, J. M. Patterson, R.D. Ruth, H.D. Schwarz, Y. Takeuchi, A.E. Vlieks, J.W. Wang, P.B. Wilson, D.B. Hopkins, A.M. Sessler, W. A. Barletta, D. L. Birx, J.K. Boyd, T. Houck, G.A. Westenskow, and S.S. Yu, Relativistic Klystron Research for High Gradient Accelerators, European Particle Accelerator Conference, Rome, Italy, June 1988, To be Published.
8. E. Chojnacki et al., Proc. 7th Int. Conf. on High Power Particle Beams, Karlsuhe, 1988.
9. M. Friedman et al., To be published. See also M. Friedman and V. Serlin, Phys. Rev. Lett. 55, 26, 2860-2862 (1985).
10. A. Anselmo, G. Kerslick, J.A. Nation, and G. Providakes, Phys. Fluids 28, 358 (1985).

11. K. R. Chu, V.L. Granatstein, P.E. Latham, W. Lawson, and C. D. Striffler, I.E.E.E. Trans. Plasma Science PS-13,6, 424-434 (1985).
12. J.A. Nation and D. Shiffler, Proc. SPIE Conf. 873,78-83, 1988.
13. See for example, High Power Microwave Sources, Artech House, Norwood, Mass., Eds. V.L. Granatstein and I. Alexeff, 1987.
14. G. Voss and T.Weiland, DESY report M-82-10, 1982, and T. Weiland, IEEE Trans. Nucl. Sci. NS-32, 3471(1985).
15. See, for example, S. Aronson, Proc. New Developments in Particle Acceleration Techniques, ECFA-CAS/ CERN-IN2P3-IRF/ CEA-EPS Workshop, 48-57, Orsay, France, 1987.
16. M. Friedman and V. Serlin, Appl. Phys. Lett. 49, 10, 596-598 (1986).
17. A. Anselmo and J. A. Nation, Proc. IEEE 1987 Particle Accelerator Conf. 1, 145-148, 1987.
18. A. Anselmo, S. Greenwald, and J.A. Nation, Proc. AIP Conference on Advanced Accelerator Concepts, Madison Wisc., Ed. F. E. Mills, Vol. 156, 46 (1986).
19. R. J. Briggs et al., IEEE Trans. Nucl. Sci. NS-28 (3), 3360-3364, 1988.

COHERENCE AND STATISTICAL PROPERTIES OF PHOTON BEAMS WITH APPLICATION TO THE FREE–ELECTRON LASER

A. Bhattacharjee and I. Gjaja

Department of Applied Physics, Columbia University, New York, New York 10027

Table of Contents

1. Introduction 1777
2. A thought experiment 1778
3. Coherence 1778
4. The quasi-probability distribution and the characteristic function 1785
5. Application to the free-electron laser 1790
   Acknowledgments 1796
   References 1797

COHERENCE AND STATISTICAL PROPERTIES OF PHOTON BEAMS WITH APPLICATION TO THE FREE-ELECTRON LASER

A. Bhattacharjee and I. Gjaja

Department of Applied Physics, Columbia University, New York, New York 10027

1. INTRODUCTION

The subject of quantum optics has undergone considerable development in the last twenty-five years. Spurred by the invention of the laser, the theoretical understanding of the coherence and statistical properties of optical beams has now attained a stage of maturity where it is possible to treat different kinds of light sources. Many excellent review papers and texts, instances of which are references 1 and 2, dealing extensively with different ramifications of this subject, have appeared.

Over approximately the latter half of this period of development in quantum optics, the free-electron laser (FEL) has been developed to the point where it is now regarded as a tunable light source of considerable versatility, with prospects of producing radiation over a wide range of wavelengths at large levels of power. (There are many excellent review papers in this rapidly growing field, and the first text /3/ has appeared recently!) Since the basic mechanism for radiation gain in an FEL can be understood in classical terms, much of the literature on the subject does not need to venture outside the domain of classical physics. However, it is natural that the coherence and statistical properties of photon beams produced by FEL's should be investigated as they are in conventional lasers, and that a quantum theory of the FEL should be necessary for this purpose.

In this paper, we review the theory of coherence and statistics of photon beams in general, with application to the FEL in particular. As far as the general theory is concerned, this paper may be regarded as a poor man's version of existing wisdom in quantum optics, described in much greater detail in excellent references such as 1 and 2. The attempt here has been to distill the existing principles in a compact, self-contained way that will enable one to use them for a light source such as the FEL, which is the application of interest in this paper. If the paper succeeds in generating some interest in experts on quantum optics in FEL's, and in FEL enthusiasts for quantum optics, it will have served its purpose.

© 1989 American Institute of Physics

2. A THOUGHT EXPERIMENT

We imagine an idealized variant of Young's double-split experiment, which is a textbook paradigm for studies of interference phenomena (Fig. 1). The light source is imagined to be a point with no spatial extent. Rays (or photons) emitted by the source are converted to a parallel, paraxial beam by a lens; the beam then falls on two slits centered on $\vec{r} = \vec{r}_1$ and $\vec{r} = \vec{r}_2$, respectively. Interference effects are investigated by placing a screen (or photodetectors) to the right of the two slits. This thought experiment will provide the framework for the concepts to be discussed in this paper.

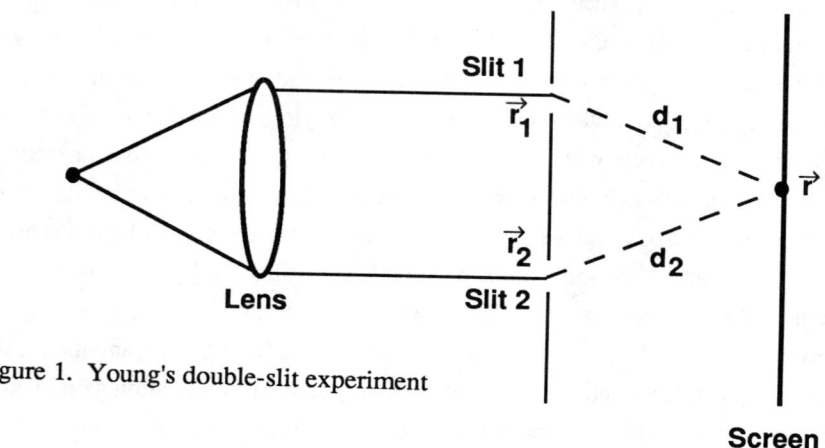

Figure 1. Young's double-slit experiment

3. COHERENCE

Let us consider at first the classical description of the interference pattern observed on the screen. The intensity of light at a point $\vec{r}$ on the screen at time t can be determined by the linear superposition of the electric fields at the two slits located at $\vec{r}_1$ and $\vec{r}_2$ respectively at times $t_1, t_2 < t$. For the present discussion, we will treat the electric fields as if they were scalars. In the context of our thought experiment, the screen is assumed to be fitted with polarizers which allow only one component of the electric field to pass through them. The total electric field $E(\vec{r}, t)$ at a point $\vec{r}$ observed at time t can be written as

$$E(\vec{r}, t) = C_1 E(\vec{r}_1, t_1) + C_2 E(\vec{r}_2, t_2) \, , \tag{1}$$

where $t_1 = t - d_1/c$, $t_2 = t - d_2/c$, c is the speed of light, and the constants C_1 and C_2 are determined from purely geometric considerations, such as the size and shape of the slits, and the distance of the point $\vec{r}$ from each of the slits at $\vec{r}_1$ and $\vec{r}_2$. We then obtain the intensity

$$I(\vec{r}, t) = \frac{1}{2} |E(\vec{r}, t)|^2 \ . \tag{2}$$

A measurement of the interference pattern over a large time-interval T provides information on the average intensity

$$<I(\vec{r}, t)> \equiv \lim_{T \to \infty} \frac{1}{T} \int_0^T d\tau \ I(\vec{r}, t+\tau) \ ,$$

$$= \frac{1}{2}[|C_1|^2 <|E(1)|^2> + |C_2|^2 <|E(2)|^2> + 2\text{Re}C_1^*C_2 <E^*(1)E(2)>] \ , \tag{3}$$

where we denote $1 \equiv \vec{r}_1, t_1$ and $2 \equiv \vec{r}_2, t_2$ and define

$$<E^*(1)E(2)> = \lim_{T \to \infty} \frac{1}{2} \int_0^T d\tau \ E^*(\vec{r}_1, t_1+\tau)E(\vec{r}_2, t_2+\tau) \ ,$$

$$\equiv G^{(1)}(1,2) \ , \tag{4}$$

which is known as the <u>first-order temporal coherence</u> of the light source. The <u>degree of first-order coherence</u> is defined to be

$$g^{(1)}(1,2) = \frac{<E^*(1)E(2)>}{[<|E(1)|^2> <|E(2)|^2>]^{1/2}} \ . \tag{5}$$

If $E(\vec{r}, t)$ is defined statistically, then $<\cdot>$ should include an average with respect to an <u>ensemble</u>. The function $G^{(1)}$ obeys the Schwarz inequality

$$G^{(1)}(1,1) G^{(1)}(2,2) \geq |G^{(1)}(1,2)|^2 \ , \tag{6}$$

which guarantees positivity of the average intensity $<I(\vec{r}, t)>$, and implies that

$$0 \leq |g^{(1)}(1,2)| \leq 1 \ . \tag{7}$$

When $g^{(1)}(1, 2) = 0$, the light source is temporally incoherent, and no interference fringes will be observed. When $g^{(1)}(1, 2) = 1$, the light of 1 and 2 has perfect first-order, temporal coherence, and the fringe will display maximal contrast. In between these extreme values of $g^{(1)}$ the light is said to have partial, first-order temporal coherence.

In analogy with Equation (5), we can define the degree of nth-order coherence as

$$g^{(n)}(\vec{r}_1, t_1, \ldots \vec{r}_n, t_n; \vec{r}_{n+1}, t_{n+1}, \ldots \vec{r}_{2n}, t_{2n})$$

$$= \frac{|\langle E^*(\vec{r}_1,t_1)\ldots E^*(\vec{r}_n,t_n) E(\vec{r}_{n+1},t_{n+1})\ldots E(\vec{r}_{2n},t_{2n})\rangle|}{[\langle |E(\vec{r}_1,t_1)|^2\rangle \ldots \langle |E(\vec{r}_n,t_n)|^2\rangle \langle |E(\vec{r}_{n+1},t_{n+1})|^2\rangle \ldots \langle |E(\vec{r}_{2n},t_{2n})|^2\rangle]^{1/2}}. \quad (8)$$

It should be pointed out that the upper bound of unity, which holds for $g^{(1)}$, does not hold, in general, for the higher-order functions $g^{(n)}$ ($n \geq 2$).

We now consider the quantum-mechanical analog of the concepts introduced here. The field E is regarded, not as a complex number, but as an operator. Instead of E and its complex conjugate E^*, we introduce, equivalently, the positive and negative frequency parts

$$E = E^+ + E^-, \quad (9)$$

where the hermiticity of E requires that $E^- = (E^+)^\dagger$. The degree of first-order coherence is then defined as

$$g^{(1)}(1, 2) = \frac{|\langle E^-(1) E^+(2)\rangle|}{[\langle E^-(1)E^+(1)\rangle \langle E^-(2)E^+(2)\rangle]^{1/2}}, \quad (10)$$

where

$$\langle E^-(1) E^+(2)\rangle = \text{Trace}\,[\rho E^-(1) E^+(2)],$$

$$\equiv G^{(1)}(1, 2), \quad (11)$$

and ρ is the density operator given by

$$\rho = \sum_i p_i |i\rangle\langle i| , \qquad (12)$$

where p_i is a normalized probability distribution over possible states $|i\rangle$, and obeys the relation $\sum_i p_i = 1$. A pure state is one for which all but one of the p_i's are zero, in which case $\rho = |i\rangle\langle i|$ for the particular pure state $|i\rangle$. Pure states obey the condition $\rho^2 = \rho$, but mixed states do not.

Analogous to the classical definition, the quantum-mechanical degree of nth-order coherence is defined to be

$$g^{(n)}(\vec{r}_1,t_1,...,\vec{r}_n,t_n;\vec{r}_{n+1},t_{n+1},...,\vec{r}_{2n},t_{2n})$$

$$= \frac{|\langle E^-(\vec{r}_1,t_1)....E^-(\vec{r}_n,t_n) E^+(\vec{r}_{n+1},t_{n+1})....E^+(\vec{r}_{2n},t_{2n})\rangle|}{[\langle E^-(\vec{r}_1,t_1) E^+(\vec{r}_1,t_1)\rangle....\langle E^-(\vec{r}_{2n},t_{2n})E^+(\vec{r}_{2n},t_{2n})\rangle]^{1/2}} . \qquad (13)$$

The positive-frequency part E^+ can be written as

$$E^+(\vec{r},t) = i\sum_k (\hbar\omega_k/2V)^{1/2} a_k u_k(\vec{r}) \exp(-i\omega_k t) , \qquad (14)$$

where $\omega_k > 0$ are a sequence of real, positive frequencies which correspond to the spatial eigenmodes $u_k(\vec{r})$ in a volume V. a_k is the amplitude operator, known otherwise as the annihilation operator. Similarly,

$$E^-(\vec{r},t) = -i\sum (\hbar\omega_k/2V)^{1/2} a_k^\dagger u_k^*(\vec{r}) \exp(i\omega_k t) , \qquad (15)$$

where $a_k^\dagger$ is the creation operator. The operators $a_k, a_k^\dagger$ obey the commutation relations

$$[a_k, a_{k'}^\dagger] = \delta_{kk'} , \quad [a_k, a_{k'}] = [a_k^\dagger, a_{k'}^\dagger] = 0 . \qquad (16)$$

In relation (13), the operator E^+ containing the annihilation operator a_k always appears to the right of E^- containing the creation operator $a_k^\dagger$. This ordering of terms is natural from the point of view of the detection process in which existing photons are absorbed by a detector.

For the classical plane wave, $g^{(1)} = 1$ for all n. Thus, the classical plane wave is coherent at all orders. The quantum-mechanical analog is the <u>coherent state</u>, which is an eigenstate $|\{\alpha_k\}\rangle$ of E^+, i.e.

$$E^+(\vec{r},t) | \{\alpha_k\}\rangle = i \sum_k \{(\hbar\omega_k/2V)^{1/2} \alpha_k u_k(\vec{r}) | \{\alpha_k\}\rangle , \qquad (17)$$

where α_k constitute a set of complex numbers (c-numbers). The kth mode, represented by the state $|\alpha_k\rangle$, obeys the relation

$$a_k |\alpha_k\rangle = \alpha_k |\alpha_k\rangle , \qquad (18)$$

which represents the fact that $|\alpha_k\rangle$ is an eigenstate of the annihilation operator a_k. (We note that $|\alpha_k\rangle$ is <u>not</u> an eigenstate of the creation operator $a_k^\dagger$.) For the pure state the nth-order quantum-mechanical correlation function is exactly factorizable, and we obtain $g^{(n)}(\vec{r}_1,t_1,...\vec{r}_n,t_n;\vec{r}_{n+1},t_{n+1},...,\vec{r}_{2n},t_{2n}) = 1$ for all n. In our thought experiment, the observation of the interference pattern and correlation functions for the coherent state is identical to that for the classical plane wave. However, the uncertainty principle determines the uncertainties in, for instance, the amplitude and phase of the coherent state $|\alpha_k\rangle$. The coherent states belong to a family of states with the special property that they reduce the uncertainty product to the minimum value allowed by quantum mechanics. It is in that sense that they are as close to classical states as quantum states can be.

The number of photons in an electromagnetic mode of wavevector $\vec{k}$ is determined by eigenstates $|n_k\rangle$ of the operator $\hat{n}_k = a_k^\dagger a_k$, which has possible eigenvalues $n_k = 0,1,2...$. For specificity, we consider a single mode $\vec{k}$ and drop the subscript k. From the commutation relations (16) and the definition of $\hat{n}$, it follows that

$$a^+ | n \rangle = \sqrt{n+1} | n+1 \rangle , \qquad (19a)$$

$$\langle n | a = \sqrt{n+1} \langle n+1 | . \qquad (19b)$$

Using Equation (19b), we get

$$\langle n | a | \alpha \rangle = \sqrt{n+1} \langle n+1 | \alpha \rangle = \alpha \langle n | \alpha \rangle , \qquad (20)$$

which in turn implies that

$$\langle n|\alpha\rangle = (\alpha/\sqrt{n})\langle n-1|\alpha\rangle \ . \tag{21}$$

From Equation (21), we obtain, by recursion,

$$\langle n|\alpha\rangle = (\alpha^n/\sqrt{n!})\langle 0|\alpha\rangle \ . \tag{22}$$

Exploiting the arbitrariness in the definition of the quantum-mechanical phase, we choose $\langle 0|\alpha\rangle = \exp(-|\alpha|^2/2)$. We then obtain

$$|\alpha\rangle = \sum_n |n\rangle\langle n|\alpha\rangle = \exp(-|\alpha|^2/2)\sum_n \frac{\alpha^n}{\sqrt{n!}}|n\rangle \ . \tag{23}$$

Since coherent states can be constructed for a continuum of values of the c-numbers α, each of which has a real and an imaginary part, it is clear from Equation (23) that the $|\alpha\rangle$-states are more numerous than the $|n\rangle$-states, i.e., the $|\alpha\rangle$-states are overcomplete. It is therefore not surprising that $|\alpha\rangle$-states are not orthogonal, as can be seen by the relation

$$\langle\alpha|\beta\rangle = \exp[-|\alpha|^2/2 - |\beta|^2/2 + \alpha^*\beta] \ . \tag{24}$$

However, $\langle\alpha|\alpha\rangle = 1$, which shows that the $|\alpha\rangle$-states are normalizable. It is easy to check by direct substitution that $|\alpha\rangle$, represented by the expansion (23), satisfies the defining relation (18).

The probability that a coherent state $|\alpha\rangle$ contains n photons is given by

$$|\langle n|\alpha\rangle|^2 = \frac{|\alpha|^{2n}}{n!}\exp(-|\alpha|^2) \ , \tag{25}$$

which is a Poisson distribution, with a mean

$$\langle n\rangle = \langle\alpha|\hat{n}|\alpha\rangle = |\alpha|^2 \ . \tag{26}$$

For any mode $\vec{k}$, the classical ("strong-field") limit is given by $|\alpha|^2 \gg 1$. In the context of our thought experiment, we have noted that the interference pattern observed on the screen is the same independent of whether we choose to describe the pattern as a superposition of coherent states or classical plane waves. However, we should bear in mind that the photodetection counter is an intrinsically quantum-mechanical device, and that the number of photon counts registered by such a device is strictly without meaning in classical physics. This point can be dramatized in the weak-field limit $|\alpha|^2 \ll 1$, which corresponds to a weak light source in the thought experiment. We imagine a light source so weak that only one photon passes at a time through the apparatus. If we carry out observations over a sufficiently large time T, the interference pattern will be exactly the same as that for a strong light source. Hence, the interference pattern must be essentially a single-photon effect, and is not the result of the interaction of photons with one another, which on first sight would appear to be at odds with the classical picture. The two pictures are, of course, reconciled when we ascribe probability amplitudes to the photon for each of the paths passing through slits 1 and 2, and calculate the intensity at the point $\vec{r}$ on the screen by taking the sum of the amplitudes and the modulus square of the resultant amplitude. Any attempt to determine which slit a single photon actually passes through will at the same time disturb the interference pattern, which is consistent with the role of the detection process in quantum mechanics. The appropriate description of the interference pattern is therefore in terms of the probability amplitudes ascribed to a single photon. A point which emerges from the discussion given above is that whereas some aspects of the coherence and statistical description of a given light source have simple correspondences in classical theory, there are other aspects which require a quantum theory for a proper interpretation.

We recall that the coherent states $|\alpha>$ of a single mode constitute an overcomplete state. Overcompleteness subsumes completeness, and a closure relation for $|\alpha>$-states is given by the identity

$$(1/\pi)\int d\alpha_R d\alpha_I |\alpha><\alpha| = 1 , \qquad (27)$$

where $\alpha = \alpha_R + i\alpha_I$, with real α_R and α_I. We remark that Equation (27) is not the only possible representation of unity in terms of the overcomplete set $|\alpha>$, but its use will give us a prescription for representing an arbitrary state by means of an expansion in $|\alpha>$-states. We first represent an arbitrary state in terms of the number states $|n>$ which constitute a complete, orthogonal set,

$$| > = \sum_n f_n |n> = \sum_n \frac{f_n (a^\dagger)^n}{n!} |0> , \qquad (28)$$

where the f_n's are c-numbers and $\sum_n (f_n)^2 = 1$. We can define an analytic function $f(z)$ of the complex variable z,

$$f(z) = \sum_n f_n z^n/\sqrt{n!} \quad , \tag{29}$$

in terms of which Equation (28) can be written as

$$|> = f(a^\dagger)|0> \quad . \tag{30}$$

We now write

$$|> = (1/\pi)\int d\alpha_R d\alpha_I |\alpha><\alpha|> \quad , \tag{31}$$

in which, using Equation (30), we get

$$<\alpha|> = <\alpha|f(a^\dagger)|0> = f(\alpha^*)\exp(-|\alpha|^2/2) \quad , \tag{32}$$

which is an analytic function. Equations (31) and (32) specify the representation of an arbitrary state $|>$ in terms of the coherent states $|\alpha>$. A corollary of this result is the representation of a coherent state $|\beta>$ in terms of the other members of the overcomplete set $|\alpha>$; we get

$$|\beta> = (1/\pi)\int d\alpha_R d\alpha_I \exp[\alpha^*\beta - |\alpha|^2/2 - |\beta|^2/2]|\alpha> \quad . \tag{33}$$

4. THE QUASI-PROBABILITY DISTRIBUTION AND THE CHARACTERISTIC FUNCTION

In any representation of a pure state of the field, the exact linear combination of the basic states, such as the number states $|n>$, is known. In most practical situations, this is not the case — what may be known is the probability that the field is in one of the basic states. The field is then said to be in a mixed state, for which the density operator is given by Equation (12). Using identity (27), we get

$$\rho = (1/\pi^2)\int d\alpha_R d\alpha_I d\beta_R d\beta_I |\alpha><\alpha|\rho|\beta><\beta| \quad . \tag{34}$$

In order to compute the weight function $<\alpha|\rho|\beta>$, we first consider the representation of ρ in number states $|n>$,

$$\rho = \sum_{n,m} |n> \rho_{nm} <m| , \qquad (35)$$

from which by straightforward manipulation, we obtain

$$<\alpha|\rho|\beta> = \sum_{n,m} \rho_{nm} \frac{\alpha^{*n}\beta^m}{\sqrt{n!\,m!}} \exp(-|\alpha|^2/2 - |\beta|^2/2) . \qquad (36)$$

The advantage of the representation (34) is that it enables us to reduce the problem of computing quantities such as the degree of nth-order coherence $g^{(n)}$, given by Equation (13), to one of evaluating integrals of complex functions. We note that a corollary of the representation (34) is the identity

$$1 = (1/\pi^2) \int d\alpha_R d\alpha_I d\beta_R d\beta_I |\alpha><\alpha|\beta><\beta| , \qquad (37)$$

the "diagonal-form" of which is identity (27). We are thus motivated to ask if a "diagonal-form" exists for the density operator ρ. If such a form exists, ρ may be represented as

$$\rho = \int d\alpha_R d\alpha_I P(\alpha) |\alpha><\alpha| , \qquad (38)$$

where $P(\alpha)$ is known as the <u>quasi-probability distribution</u>, introduced by R. J. Glauber and E. C. G. Sudarshan in 1963. Since Trace $\rho = 1$, we must have

$$\int d\alpha_R d\alpha_I P(\alpha) = 1 . \qquad (39)$$

Moreover, since ρ is hermitian, $P(\alpha)$ must be a real function. However, $P(\alpha)$ is not a true probability density function — it can have negative values and is, in general, not even a function but a distribution. However, this presents no difficulties in so far as the calculation of the averages of normally ordered operators such as $a^{\dagger n}a^m$ is concerned, and we can write

$$<a^{\dagger n}a^m> = \text{Trace}\{\rho a^{\dagger n}a^m\} = \int d\alpha_R d\alpha_I P(\alpha) \alpha^{*n}\alpha^m , \qquad (40)$$

which resembles formally a classical ensemble average. Thus the quasi-probability distribution enables us to establish a formal equivalence between the classical and quantum descriptions of coherence.

The relation between $<\alpha|\rho|\beta>$, given by Equation (36), and $P(\alpha)$ is obtained by writing

$$<\alpha|\rho|\beta> = \int d\gamma_R d\gamma_I <\alpha|\gamma><\gamma|\beta> P(\gamma)$$

$$= \exp(-|\alpha|^2/2 - |\beta|^2/2)\int d\gamma_R d\gamma_I P(\gamma)\exp(-|\gamma|^2 + \alpha^*\gamma + \beta\gamma^*), \qquad (41)$$

which, for the special case $\alpha = \beta$, reduces to

$$<\alpha|\rho|\alpha> = \int d\gamma_R d\gamma_I P(\gamma) \exp(-|\alpha-\gamma|^2)$$

$$= \pi P_A(\alpha) . \qquad (42)$$

The function $P_A(\alpha)$ is known as the <u>anti-normal quasi-probability distribution</u>, and plays the same role for anti-normally ordered operators such as $a^m a^{\dagger n}$ as $P(\alpha)$ does for normally ordered operators. Thus

$$<a^m a^{\dagger n}> = \int d\alpha_R d\alpha_I \alpha^{*n} \alpha^m P_A(\alpha) . \qquad (43)$$

In the theory of probability, a well-known device used for the computation of statistical averages is the so-called characteristic function, which is the Fourier transform of the probability density. We can define analogously a quantum-mechanical <u>characteristic function</u>

$$\chi(\mu) = \text{Trace } \{\rho \exp[\mu a^\dagger - \mu^* a]\} , \qquad (44)$$

where the exponential operator is unitary, which implies that

$$|\chi(\mu)| \leq 1 . \qquad (45)$$

A related set of characteristic functions are

$$\chi_N(\mu) = \text{Trace } \{\rho \exp[\mu a^\dagger] \exp[-\mu^* a]\} ,$$

$$= \int d\alpha_R d\alpha_I \, P(\alpha) \exp(\alpha^*\mu - \alpha\mu^*) \; , \tag{46}$$

and

$$\chi_A(\mu) = \text{Trace} \{\rho \, \exp[-\mu^* a] \exp[\mu a^\dagger]\} \; ,$$

$$= \int d\alpha_R d\alpha_I \, P_A(\alpha) \exp(\alpha^*\mu - \alpha\mu^*) \; , \tag{47}$$

where the subscripts N and A designate respectively normal and anti-normal characteristic functions. Using the identity

$$\exp(A) \exp(B) = \exp\{A + B + [A,B]/2\} \; , \tag{48}$$

where A and B are operators with the commutator [A,B], we get

$$\chi(\mu) = \chi_N(\mu) \exp[-|\mu|^2/2] = \chi_A(\mu) \exp[|\mu|^2/2] \; . \tag{49}$$

From inequality (45), it follows that $\chi_A(\mu)$ must decay at least as $\exp[-|\mu|^2/2]$ for $|\mu| \to \infty$. We note that the averages of the ordered operators can be computed from the characteristic functions by differentiation. We have

$$<a^{\dagger m} a^n> = \frac{\partial^m}{\partial \mu^m} \frac{\partial^n}{\partial(-\mu^{*n})} \chi_N(\mu) \bigg|_{\mu=0} \; , \tag{50a}$$

$$<a^m a^{\dagger n}> = \frac{\partial^m}{\partial \mu^m} \frac{\partial^n}{\partial(-\mu^*)^n} \chi_A(\mu) \bigg|_{\mu=0} \; . \tag{50b}$$

It is clear by inspection of Equation (46) that when χ_N is square integrable, so is P. However, since $\chi_N(\mu)$ can diverge as rapidly as $\exp(|\mu|^2/2)$ as $|\mu| \to \infty$, P can be a highly singular function. We now consider two examples. The first one is a light source in a coherent state $|\beta>$, for which

$$P(\alpha) = \delta(\alpha - \beta) \; , \tag{51}$$

where δ is a two-dimensional Dirac delta function, and consequently,

$$\chi_N(\mu) = \exp(\mu\beta^* - \mu^*\beta) \; . \tag{52}$$

We note that even in this simple special case, it is more advantageous to work with $\chi_N(\mu)$ than $P(\alpha)$, which is more singular. The second example is a chaotic light source, with maximal disorder (entropy), for which the density operator is

$$\rho = \frac{1}{1+<n>} \exp\left\{a^\dagger a \log\left[\frac{<n>}{1+<n>}\right]\right\} , \qquad (53)$$

where $<n> = \text{Trace}(\rho a^\dagger a)$. It is straightforward to show that

$$P(\alpha) = \frac{1}{\pi <n>} \exp\left\{-\frac{|\alpha|^2}{<n>}\right\} , \qquad (54)$$

and

$$\chi_N(\mu) = \exp[-<n>\mu^2] . \qquad (55)$$

For the chaotic light source, $P(\alpha)$ is not singular, in contrast to that for the coherent state.

Another distribution of considerable interest is the photon number distribution $p(n)$, defined as

$$p(n) = <n|\rho|n> , \qquad (56)$$

which represents the probability that there are n photons in the normalization volume V. Using Equation (38), we can obtain a relation which enables us to calculate $p(n)$ from $P(\alpha)$:

$$p(n) = \int d\alpha_R d\alpha_I \, P(\alpha) \, |<n|\alpha>|^2$$

$$= \int d\alpha_R d\alpha_I \, P(\alpha) \, \frac{|\alpha|^{2n}}{n!} \exp(-|\alpha|^2) . \qquad (57)$$

For a coherent state, for which $P(\alpha)$ is given by Equation (51), it follows, upon using Equation (56), that

$$p(n) = \frac{<n>^n}{n!} \exp(-<n>) , \qquad (58)$$

which is the Poisson distribution. For a chaotic state, with $P(\alpha)$ given by Equation (54), we obtain, similarly,

$$\rho(n) = \frac{<n>^n}{(1 + <n>)^{1+n}} \quad . \tag{59}$$

It should be pointed out that the reverse calculation, from $\rho(n)$ to $P(\alpha)$, is not possible, in general, since $\rho(n)$ is not a complete description of the density matrix. (The off-diagonal elements are not specified.)

Thus far, we have collected the basic tools necessary for the study of coherence and statistical properties of any light source, given its density operator ρ. We have made the case that a knowledge of the quasi-probability distribution, or more conveniently, the characteristic function enables us to compute the degree of coherence or other correlation functions. We now turn to the computation of these entities for the FEL.

5. APPLICATION TO THE FREE-ELECTRON LASER

The FEL is a device in which energy is transfered from a beam of relativistic electrons to the radiation field. The electrons pass along the axis of a static periodic magnetic field (wiggler), which causes them to oscillate in the direction perpendicular to their motion. The radiation produced is peaked in the forward direction and centered at the frequency given by /3/

$$\omega = \frac{2\gamma^2 k_w c}{(1 + e^2 A_w^2/m^2 c^4)} \quad ; \tag{60}$$

here $k_w = 2\pi/\lambda_w$, λ_w is the period of the wiggler magnet, A_w is the amplitude of the vector potential of the wiggler, and γ is the electron energy in units of its rest mass. The relative width in frequency, $\Delta\omega/\omega$, is approximately $1/N$ where N is the number of wiggler periods /3/.

In a quantum mechanical description, the FEL interaction can be conveniently described by the Hamiltonian /4/

$$H = \sum_j \left(E_j + \frac{e^2 \vec{A}_w \cdot \vec{A}_s}{E_j} \right) ; \qquad (61)$$

here $E_j = \sqrt{p_j^2 + M^2}$, $M^2 = m^2 + e^2 A_w^2$, where m is the electron rest mass, and we have set $\hbar = c = 1$. The vector potentials of the wiggler and the radiation fields are taken to be circularly polarized, i.e.

$$\vec{A}_w = \frac{iA_w}{\sqrt{2}} (\hat{\varepsilon} e^{-ik_w z} - \hat{\varepsilon}^* e^{ik_w z}) , \qquad (62)$$

and

$$\vec{A}_s = \sqrt{\frac{2\pi}{V\omega}} (a(t) e^{ikz - \omega t} \hat{\varepsilon} + h.c.) , \qquad (63)$$

where h.c. denotes hermitian conjugate, $\hat{\varepsilon} = (\hat{x} + i\hat{y})/\sqrt{2}$, and V is the quantization volume. a(t) is the annihilation operator for the mode of frequency $\omega(= k)$ with the time dependence $\exp(-i\omega t)$ factored out, and $[a(t), a^\dagger(t)] = 1$. Considering only one mode of the radiation field in Equation (63) (no sum over k) is the usual approximation in FEL calculations. In linear theory to be presented below, it is rigorously justified since the different modes are not coupled, and thus the many-mode problem is equivalent to the single-mode one. In deriving the Hamiltonian above, we have neglected the electron spin /5/, the effect of space charge /6,7/, and the overlap of electron wavepackets /8/ (i.e. there are no exchange terms in the Hamiltonian). We have also assumed that $|\vec{A}_s| \ll |\vec{A}_w|$. Under these assumptions, Equation (61) can be obtained by a rigorous reduction of the Dirac Hamiltonian /4/.

In the Heisenberg picture, the equations of motion for the electron and field operators are given by

$$\frac{d\psi_j}{dt} = -\omega_\kappa + \frac{(k+k_w)p_j}{E_j} \left[1 + \frac{ie^2 A_w R}{E_j^2} (a e^{i\psi_j} - h.c.) \right] , \qquad (64a)$$

$$\frac{dp_j}{dt} = -\frac{Re^2 A_w(k + k_w)}{E_j} (a e^{i\psi_j} + h.c.) , \qquad (64b)$$

$$\frac{da}{dt} = R \sum_j \frac{e^2 A_w}{E_j} e^{-i\psi_j} , \qquad (64c)$$

where $\psi_j = (k + k_w) z_j - \omega t$ is the relative phase between the electrons and the radiation field, and $R \equiv \sqrt{\pi/V\omega}$. In order to examine how the photon statistics of an assumed initial radiation field changes as a result of the FEL interaction, we need to solve the equation of motion for $a(t)$. Since the equation for $a(t)$ is coupled to the equations for the electron operators, it is necessary to solve all Equations (64a-c). This, however, cannot be accomplished analytically for the full equations. We therefore resort to an approximation which is frequently used in the study of FEL's /4,6,9–11/: we linearize Equations (64a-c) about their equilibrium values and then solve for $a(t)$. A limitation of this method is that it is strictly valid only for small departures from equilibrium and is not valid at saturation. As an alternative, one can use the perturbation expansion in which $a(t)$ is expanded in a power series in the coupling parameter $(e^2 A_w / \sqrt{p_0^2 + M^2})$ /12-14/. Here p_j in the denominators of Equations (64a-c) is replaced by a constant p_0, and then made part of the coupling constant. In yet another approximate method a perturbation series is developed in $p_j - p_0$ /15–17/. While these perturbation theories can trace the time evolution of the system for arbitrary initial conditions, the perturbation expansions computed to a small number of terms are valid only for short times, or equivalently, for low electron densities. In this review we do not treat these approximations, but refer the reader to the existing literature.

Equations (64a-c) admit a stationary solution which corresponds to zero radiation field and to a classical monoenergetic beam (i.e. precisely determined position and momentum of each electron) with a continuous, uniform distribution of electrons. We take small departures from equilibrium and obtain the appropriate linearized equations; details are given in Refs. 4, 9, and 10. The linearized equations can be derived directly from a Hamiltonian /4, 10/, which guarantees that the commutation relations for the linearized operators will be preserved in time. If we assume a solution of the form $\exp(i\lambda t)$, we obtain from the linear analysis the standard cubic equation which permits three regimes of FEL operation: the small gain, the exponential gain and the inverse gain regimes /7/. Upon solving the linearized equations we obtain

$$a(t) = F_1(t) x(0) + F_2(t) y(0) + F_3(t) a(0) , \qquad (65)$$

where F_1, F_2, F_3 contain exponential dependence on t, and x(0) and y(0) are electron collective operators describing the departures of phase and momentum, respectively, from their equilibrium values /4,9,10/. We can also introduce the creation and annihilation operators for electron transitions /10/, $b_1 = (1/\sqrt{2})(x+iy)$, $b_2 = (1/\sqrt{2})(x^\dagger + iy^\dagger)$, which satisfy $[b_i, b_j^\dagger] = \delta_{ij}$, $[b_i, b_j] = 0$, where i,j = 1,2. The solution for a(t) is now

$$a(t) = G_1(t)b_1(0) + G_2(t)b_2^\dagger(0) + F_3(t)a(0) , \qquad (66)$$

where $G_{1,2} = (1/\sqrt{2})(F_1 \mp iF_2)$. The reader is referred to Ref. 10 for explicit expressions for $F_{1,2,3}(t)$. The quantity $|F_3(t)|^2$ is usually referred to as the gain; for $|F_3(t)|^2 < 1$, the gain is designated as negative (initial radiation attenuated), and for $|F_3(t)|^2 > 1$, positive (initial radiation amplified).

We now examine the normal ordered characteristic function for the radiation field produced by an FEL. We assume that the field is initially in a coherent state with an amplitude α_0 (vacuum, or an initially present weak laser radiation). The density matrix is assumed to be of the form $\rho = \rho_{field} \otimes \rho_{electrons}$. Into the general expression for $\chi_N(t)$, given by (46), we substitute the expression for a(t) from Equation (66), and use the identity exp A exp B = exp (A + B) exp (1/2) [A,B], which is valid provided A and B commute with their commutator. Then

$$\chi_N(t) = e^{\mu F_3^* \alpha_0^* - \mu^* F_3 \alpha_0} \langle e^{\mu G_1^* b_1^\dagger - \mu^* G_1 b_1} e^{\mu G_2^* b_2 - \mu^* G_2 b_2^\dagger} \rangle$$

$$\times e^{(1/2)|\mu|^2(|G_1|^2 - |G_2|^2)} . \qquad (67)$$

Here, for brevity, we have suppressed the arguments of G_1, G_2, F_3 which are evaluated at the time t and of b_1 and b_2, evaluated at t = 0.

From the commutator $[a(t), a^\dagger(t)] = 1$, we obtain the relation $|G_1(t)|^2 - |G_2(t)|^2 + |F_3(t)|^2 = 1$. Since the operator in angular brackets is unitary, and hence its expectation value is less than or equal to 1, we see that for positive gain $\chi_N(t)$ drops off at least as fast as $e^{-(1/2)|\lambda|^2(|F_3|^2 - 1)}$ as $|\mu| \to \infty$. Hence $P(\alpha)$ will be less singular than the delta-function, and the resulting state of the radiation field will be more chaotic than the coherent state. This is true regardless of the initial state of the electrons.

It is possible to show /18/, in fact, that under assumptions reasonable for most electron beams, even in the region of negative gain, the initial electron fluctuations destroy the coherence of the output radiation. The loss of coherence with respect to the initial state is due to quantum-mechanical fluctuations, which can be traced to the finite width of electron wavepackets in position and momentum. In addition, statistical fluctuations in the initial distribution of electrons in phase cause a further loss of coherence; in what follows, we do not consider this effect.

We return to the explicit computation of $\chi_N(t)$, and choose the initial electron state in which the electron fluctuations are at a minimum. This is the ground state of oscillators 1 and 2, and represents a monoenergetic electron beam, with a uniform distribution of electrons in phase, where each electron is described by a minimum uncertainty wavepacket symmetric in $\delta\phi$ and $\delta p/p_0$. Then

$$\chi_N(t) = e^{\mu F_3^* \alpha_0^* - \mu^* F_3 \alpha_0} e^{-|\mu|^2 |G_2|^2}. \tag{68}$$

The quasiprobability distribution function is

$$P(\alpha) = (\pi |G_2|^2)^{-1} e^{-|\alpha - F_3 \alpha_0|^2 / |G_2|^2}. \tag{69}$$

A comparison of Equation (69) with Equations (51) and (54) will suggest that the field is a superposition of a coherent state and a chaotic state (noise). We notice that the width of the distribution is independent of the amplitude of the initial coherent field; rather it is a function only of the "spontaneous emission" /10/. Thus it seems that the radiation produced by an FEL cannot be made less chaotic than that given by Equation (69). We also point out that the calculation given in Ref. 17 indicates that the photon statistics of spontaneous emission is thermal with $\langle n^2 \rangle = 2\langle n \rangle^2 + \langle n \rangle$, or $p(n) = \langle n \rangle^n/(1 + \langle n \rangle)^{1+n}$. This is in agreement with the result obtained above (setting $\alpha_0 = 0$), except that Equation (69) describes amplification of the spontaneous emission as well as other higher-order effects which also have a thermal distribution.

We now take a brief look at photon statistics of an FEL oscillator. In such a set-up a light pulse makes many round trips inside a cavity and is amplified in each pass through the interaction with an electron beam. To calculate the evolution of the radiation field in the cavity, we can use Equations (64a-c) with the addition of a term that describes losses. For the start-up from noise and the evolution of the radiation field prior to saturation, the method of choice for solving Equations (64a-c) has been perturbation theory /14,19,20/. Here we do not present the calculations and results obtained using this method; instead, we refer the reader to the literature. We remark, however, that the single-mode calculation given in Ref. 14 states that the photon statistics of spontaneous emission is thermal, which is the same result as that obtained above in linear and in perturbation theory for a single pass.

In the steady-state operation of the oscillator, perturbation theory is not adequate to obtain photon statistics /14/. Instead, if we assume small gain per pass, and restrict ourselves to the single-mode problem, we can follow the approach given in Ref. 21. Equations (64a-c) support the conservation of momentum, which can then be used to obtain the number of photons emitted in one round trip,

$$\delta n = -(k+k_w)^{-1} \sum_j (p_j(T) - p_j(0)) , \qquad (70)$$

where $T (T \simeq L/c)$ is the interaction time, and L is the length of the cavity. We can now make the small-gain approximation: the radiation field inside the cavity changes little in one pass, so that Equations (64a-c) need not be solved consistently to obtain the electron motion. Instead, we can take the radiation field to be approximately constant during one pass, and then calculate $p_j(T)$ using Equations (64a-b). If we also assume that the initial beam is monoenergetic (which simplifies calculations but is not an essential assumption), then $p_j(T)$ will only depend on the initial phase ψ_{j0} of the jth electron. Finally, if we let the photon number distribution after the $(k+1)^{th}$ pass depend only on that after the kth pass, we can write

$$\rho_{k+1}(n) = \int dn' \rho_k(n') \delta(n - n' - \delta n(n') + fn') , \qquad (71)$$

where f represents the fractional loss per round trip. Expanding this relation to second order in $\delta n(n') - fn'$, and performing the integrations, we obtain

$$\rho_{k+1}(n) - \rho_k(n) = \frac{d}{dn}(A\rho_k) + 1/2 \frac{d^2}{dn^2} [(A^2 + B^2)\rho_k] , \qquad (72)$$

where $A \equiv N(k+k_w)^{-1}(\bar{p} - p_0) + fn$, $B \equiv (k+k_w)^{-1}[N(\overline{p^2} - \bar{p}^2)]^{1/2}$ and $\bar{p}$ is the average over all electrons, $\bar{p} = 1/N \sum_j p_j(T, \psi_{j0})$, which is equivalent to the average over initial phases. In steady state, $\rho_{k+1} = \rho_k$, and it can be shown from Equation (72) that ρ_k is peaked near the value $n = n_{ss}$ for which $A(n_{ss}) \equiv 0$, i.e. where the gain is equal to the loss. In the vicinity of n_{ss}, ρ_k is approximately a Gaussian of variance

$\sigma = B(n_{ss})/(2A'(n_{ss}))^{1/2}$. Therefore, in steady state the photon statistics is not that of the coherent state. The photon number fluctuation is

$$\langle <n^2> - <n>^2 \rangle^{1/2} = \sigma = (k+k_w)^{-1} \left(\frac{(N/2)(\overline{p^2} - \overline{p}^2)}{f + \frac{N}{k+k_w} \frac{d}{dn}(\overline{p}-p_0)} \right)^{1/2}. \quad (73)$$

Equation (73) shows that the photon number fluctuation is related to the spread in the momentum of the electrons introduced by the interaction. In Ref. 21, σ is calculated numerically for some typical FEL experimental values, and compared to the photon number fluctuations that would result from a coherent state. It is found that the radiation generated by an FEL in steady state has larger photon number (intensity) fluctuations than the coherent state, and that unlike the ordinary laser, it does not approach the coherent state as the FEL is brought above threshold.

We close this review with two remarks: first, the theory presented above is a many-electron theory, and the predicted photon statistics is completely different from the one obtained from a single-electron theory (e.g., a single-electron theory predicts that the photon number distribution for spontaneous emission is Poissonian /15/). Thus, when dealing with statistical properties of FEL radiation we cannot use single-electron results and scale them with N ; it is essential that many-particle effects be incorporated from the beginning. Second, the possibility of producing squeezed states of radiation from an FEL has recently received some amount of attention, even though a cursory inspection of Equation (68) may suggest that such a task would be very difficult /18,22/ .

ACKNOWLEDGMENTS

We thank Dr. M. Month for the kind invitation to write this review paper, and Prof. T. C. Marshall for his encouragement and interest.

This work is supported by Award No. BNL 27 4067-S from the Brookhaven National Laboratory.

REFERENCES

1. R. J. Glauber in <u>Laser Handbook</u>, Vol. 1, F. T. Arecchi and E. O. Schulz-Dubois, eds. (North-Holland, Amsterdam, 1972), pp 1-43
2. R. Loudon, The Quantum Theory of Light, second edition (Clarendon, Oxford, 1981)
3. T. C. Marshall, <u>Free-electron Lasers</u> (Macmillan, 1985)
4. I. Gjaja and A. Bhattacharjee, Phys. Rev. A $\underline{37}$, 1009 (1988)
5. For an estimate of its importance, see, for instance, W. Becker and H. Mitter, Z. Physik B $\underline{35}$, 399 (1979)
6. J. B. Murphy, C. Pellegrini and R. Bonifacio, Opt. Comm. $\underline{53}$, 197 (1985)
7. I. Gjaja and A. Bhattacharjee, Opt. Comm. $\underline{62}$, 39 (1987)
8. A. Renieri in <u>Free Electron Generation of Extreme Ultraviolet Coherent Radiation</u>, J. M. J. Madey and C. Pellegrini, eds. (APS, New York 1984), pp. 1-11
9. R. Bonifacio, L. Narducci and C. Pellegrini, Opt. Comm. $\underline{50}$, 373 (1984)
10. R. Bonifacio and F. Casagrande, Opt. Comm. $\underline{50}$, 251 (1984)
11. S. Krinsky and L. H. Yu, Phys. Rev. A $\underline{35}$, 3406 (1987)
12. S. T. Stenholm and A. Bambini, IEEE J. Quant. Electron. $\underline{17}$, 1363 (1981)
13. P. Bosco, W. B. Colson and R. A. Freedman, IEEE J. Quant. Electron. $\underline{19}$, 272 (1988)
14. J. Gea-Banacloche, Phys. Rev. A $\underline{31}$, 1607 (1985)
15. W. Becker and M. S. Zubairy, Phys. Rev. A $\underline{25}$, 2200 (1982)
16. W. Becker and J. K. McIver, Phys. Rev. A $\underline{27}$, 1030 (1983)
17. W. Becker and J. K. McIver, Phys. Rev. A $\underline{28}$, 1838 (1983)
18. I. Gjaja and A. Bhattacharjee, Phys. Rev. A $\underline{36}$, 5486 (1987)
19. A. T. Georges in <u>Free-Electron Generators of Coherent Radiation</u>, C. Brau, S. F. Jacobs and M. O. Scully, eds. (SPIE, Washington, 1984), pp 297-305
20. J. Gea-Banacloche and M. O. Scully, Nuclear Inst. and Methods A $\underline{237}$, 100 (1985)
21. J. Gea-Banacloche, Phys. Rev. A $\underline{33}$, 1448 (1986)
22. W. Becker, M. O. Scully and M. S. Zubairy, Phys. Rev. Lett. $\underline{48}$, 475 (1982)

PLASMA ACCELERATION OF PARTICLE BEAMS

T. Katsouleas and J.M. Dawson
University of California, Los Angeles, CA 90024

TABLE OF CONTENTS

    Introduction 1799
I.   Plasma Wave Excitation by Beating Lasers 1801
    A. Theoretical Models 1801
    B. Computer Modelling 1805
    C. Experimental Results 1809
II. Acceleration Mechanisms 1811
    A. Injection/Trapping Thresholds 1812
    B. Simple Beat Wave Acceleration 1813
    C. The Surfatron Acceleration Mechanism 1815
III. Plasma Wave Excitation by Particle Beams
    (The Plasma Wakefield Transformer) 1818
IV. Plasma Focusing of Particle Beams (the Plasma Lens) 1823
V.   Summary 1825
    References 1826

PLASMA ACCELERATION OF PARTICLE BEAMS

T. Katsouleas and J.M. Dawson
University of California, Los Angeles, CA 90024

ABSTRACT

Plasmas, being fully ionized gases, are immune from electrical breakdown and so can support ultra-high accelerating fields (order GeV/cm) in the form of relativistic plasma waves. Several schemes to excite these waves and use them for particle acceleration are reviewed in this article. These include the beat wave accelerator (laser driven) and the plasma wakefield transformer (particle beam driven). In addition, the possible use of plasmas to provide strong final focusing of beams (the plasma lens) is described.

INTRODUCTION

The first question one might ask when considering a plasma as a medium for supporting large amplitude electric fields for particle acceleration is "Why use a plasma?" In light of their history of unexpected instabilities and complexity, plasmas may seem an unlikely path toward the goal of high-gradient particle accelerators. In this paper, the results of investigations of plasma as a medium for high-gradient acceleration are presented. Recent work has demonstrated that plasma properties are somewhat desirable and often unavoidable when contemplating advanced accelerator concepts. Some of these attractive properties may make plasmas useful in roles other than linear acceleration. For example, they can provide strong focusing of particle beams in a scheme known as the plasma lens and can provide a strong (electric) wiggler for generating synchrotron radiation from particle beams.[1]

As a first step toward addressing the question of why accelerate in plasma, let us briefly examine some alternatives. Conventional accelerators can be expected to produce accelerating gradients of about 20 to 100 MeV/m. At field strengths not too much higher, the problem of breakdown in the accelerator walls arises. Similarly, many of the near-field laser accelerator schemes[2] will unintentionally become plasma schemes if the laser intensity is too high. This suggests the first advantage of a plasma accelerator: it is already fully ionized and cannot be destroyed any more than it already is.

Next, one might look to high powered lasers with fields as high as 5 GeV/cm to directly accelerate particles. However, the laser fields are transversely polarized and coupling to these fields (the far-field schemes) necessarily leads to transverse particle acceleration and radiation losses. Some direct acceleration in the longitudinal direction is possible from the radiation pressure of the light.

© 1989 American Institute of Physics

However, for a relativistic particle this force decreases as one over the Lorentz factor γ of the particle.[3]

On the other hand, the accelerating force of a longitudinal electric field is invariant under Lorentz transformations to any frame moving along the field direction. Hence, the second advantage of a plasma accelerator: plasmas are capable of supporting large, longitudinal electric field waves (i.e., space-charge waves for which the wavenumber k is parallel to the electric field E).

We can estimate just how large the plasma wave fields can be from Poisson's equation:

$$\nabla \cdot \vec{E} = 4\pi e \delta n_e \qquad (1)$$

where δn_e is the perturbed electron density of the plasma and the ion background is assumed uniform and immobile. Now the largest density compression or rarefaction that can occur roughly is when <u>all</u> of the plasma electrons are removed. In that case $\delta n_e \sim n_o$, the equilibrium density. Assuming a plasma wave with phase velocity near c and frequency near the plasma frequency $\omega_p = (4\pi n_o e^2/m_e)^{1/2}$ so that $k \approx \omega_p/c$ and approximating $\nabla \cdot E \sim ikE$, we obtain from (1) the maximum electric field amplitude[4]:

$$eE_{MAX} \sim \frac{4\pi n_o e^2}{(\omega_p/c)} = m_e c \omega_p \approx .97 \sqrt{n_o} \text{ eV/cm} . \qquad (2)$$

This is sometimes referred to as the cold wavebreaking field because a more rigorous treatment shows that it is the amplitude at which a cold plasma wave steepens and becomes double valued so that the crest begins to fall into the trough.[5] This field can be quite large. For example, in a plasma of density $n_o = 10^{18}$ cm^{-3} the cold wavebreaking field is of the order 1 GeV/cm, or about one thousand times the gradient of conventional linacs.

One might be concerned that a medium such as a dense plasma would interfere with the acceleration process via Coulomb scattering or Cerenkov radiation. The dominant energy loss mechanism for electrons above a few MeV is due to multiple scattering from plasma nuclei. Since the radiation length $(-U/[\partial U/\partial x])$ in a hydrogen plasma of density 10^{20} cm^{-3} is roughly 2 km,[6] and the scattering cross section decreases as energy squared, this should not pose a problem. Furthermore, collisional damping of the laser scales as one over laser intensity to the three halves and is minimal for intense lasers.

Having illustrated two principal advantages of plasma accelerators—immunity from breakdown and potential for large longitudinal electric fields—we turn to describing the progress that has been made toward realizing this potential.

In Section I we review the most developed of the plasma wave excitation schemes: plasma wave generation by the beating of two lasers.[7,8] In Section II we examine two mechanisms for particle acceleration in the fields of high phase velocity plasma waves: the simple beat wave acceleration mechanism and the surfatron mechanism (using an imposed DC magnetic field). In Section III we explore another plasma wave excitation scheme: the plasma wakefield accelerator.[9] The plasma wakefield scheme employs the electron (or proton) bunches of a conventional accelerator rather than lasers as the free energy source to drive the plasma waves. In Section IV we describe the use of plasmas to focus particle beams to unprecedented spot sizes in the scheme known as the plasma lens.

I. PLASMA WAVE EXCITATION BY BEATING LASERS

   A. THEORETICAL MODELS

   The basic mechanism for excitation for a longitudinal plasma wave by laser light is illustrated in Fig. 1.

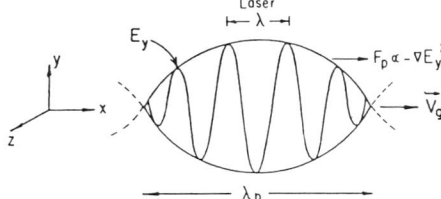

Fig. 1 The radiation pressure or pondermotive force of a packet of light drives longitudinal oscillations of plasma electrons.

Consider a packet of incident electromagnetic radiation which might be created in one of the following two ways: (1) by a very short pulse from a single laser (wake plasmon excitation) or (2) by the beat envelope of two lasers of slightly different frequency (beat wave excitation). The packet of radiation exerts a force per unit area on the plasma which is given by the gradient of the stress tensor[10]:

$$F_p \approx \nabla \cdot \ddot{T}_p - \nabla \cdot \ddot{T}_v \approx \frac{\varepsilon - 1}{8\pi} \frac{\partial}{\partial x} E_y^2 \qquad (3)$$

where $\ddot{T}_p$, $\ddot{T}_v$ refer to the stress tensor in plasma and vacuum, respectively, ε is the dielectric function of the plasma ($\varepsilon = 1 - \omega_p^2/\omega^2$, ω is laser frequency and $\omega \gg \omega_p$), $\partial/\partial x$ describes the gradient of the envelope of the packet and E_y is the laser electric field. Since we want just the force on the plasma, we subtracted $\ddot{T}_v$, the component of $\ddot{T}$ which would exist even if there were no plasma. This so-called pondermotive force displaces plasma electrons and creates a temporary charge imbalance in the plasma (the plasma ions are too massive to respond with the electrons). After the packet passes, the space-charge force acts to restore the plasma electrons leading to oscillations in the $\hat{x}$ direction and and hence a longitudinal electric field wave.

In the first scenario of a very short pulse, known as the wake plasmon excitation scheme, there is no resonant interaction with the plasma, and the plasma essentially receives only a single kick. As a result, the electric force of the plasma wave can never exceed the ponderomotive force of the laser. Furthermore, all the energy required for a full stage of the accelerator must be contained in a single pulse shorter than a plasma period. For a 10^{16}-cm^{-3} density plasma, the pulse must be 1 picosecond or less. Nevertheless, a short pulse laser with normalized electric field amplitude $V_{osc}/c \equiv eE/m\omega_p c$ of order one has sufficient ponderomotive force to create a plasma wave near the cold wavebreaking amplitude (Eq. 2).

In the second scenario, involving the beat of two lasers, the plasma wave can be built up resonantly over many plasma periods if the beat frequency of the lasers coincides with the natural frequency at which the plasma electrons oscillate (i.e., $\omega_1 - \omega_2 \approx \omega_p$). In this way, even relatively small laser fields can lead to very large longitudinal plasma wave fields.

Consider the resonant beat wave excitation scheme. By energy and momentum conservation of the laser photons and the plasmon, we have that

$$\omega_1 - \omega_2 = \omega_p ,$$
$$k_1 - k_2 = k_p . \tag{4}$$

If the plasma is very underdense, such that $\omega_1 \approx \omega_2 \gg \omega_p$, we find that the phase velocity of the plama waves is

$$V_{ph} \equiv \frac{\omega_p}{k_p} = \frac{\omega_1 - \omega_2}{k_1 - k_2} = \frac{\Delta\omega}{\Delta k} \approx \frac{\partial\omega}{\partial k} = V_g^{light} = c(1 - \omega_p^2/\omega^2)^{1/2} . \tag{5}$$

Thus, the phase velocity of the plasma waves equals the group velocity of the light. The light pulse and the wake of excited plasma waves move as a unit into undisturbed plasma, as depicted in Fig. 2.

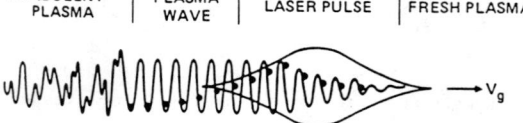

Fig. 2 The laser pulse, the wake of plama waves and the accelerating particles move together at $V_g \approx c$ into fresh plasma in the beat wave excitation scheme.

If the laser frequencies are much higher than ω_p (i.e., a very underdense plasma), then V_{ph} is very close to c and injected relativistic particles may stay in phase with the electric field of the plasma wave for a sufficient distance to be accelerated to high energy.

As illustrated in the Fig. 2, the plasma waves well behind the light pulse become turbulent. This can be due to competing instabilities, such as parametric decay (the decay of an electron plasma wave into a secondary plasma wave and an ion acoustic wave) or, as we will discuss shortly, due to relativistic effects which detune the plasma wave frequency from the laser beat frequency. The onset of turbulence does not affect the acceleration of the particles trapped in plasma waves nearer to the light pulse since these continually move into fresh plasma. The onset of turbulence does limit the number of plasma waves that can be used for particle acceleration. For example, if the first onset of turbulence is due to ion instabilities which grow on a time scale characterized by the ion plasma period [$\tau_i = 2\pi/\omega_{pi}$, $\omega_{pi} = (4\pi n e^2/M_i)^{1/2}$, where M_i is the ion mass], then plasma waves can be created for a time $2\pi/\omega_{pi}$ or a distance $2\pi c/\omega_{pi}$ behind the laser pulse. For a hydrogen plasma this corresponds to 43 (= $\sqrt{M_i/m_e}$) electron plasma wave periods. Plasma waves further than 43 plasma wavelengths behind the laser will begin to become degraded and will probably not be suitable for acceleration. This does not pose a problem since most of the plasma wave energy can be extracted by particles loaded in a few or even one accelerating bucket (see the end of Section IIB).

The growth of the plasma waves behind the laser pulse can be quantified by considering the oscillation of a single background plasma electron driven by the ponderomotive force F_p [from Eq. (3) divided by n_0 to get the force on a single electron] of the beating lasers:

$$\frac{d}{dt}(\gamma V_x) + \omega_p^2 x = \frac{e^2}{2m\omega_{1,2}^2} \Delta k E_1 E_2 \sin(\Delta\omega t - \Delta k x) \qquad (6)$$

where $E_{1,2}$ are the electric field amplitudes of each laser, $\Delta\omega$ and Δk are the beat frequency and beat wave number of the lasers (i.e., $\Delta\omega = \omega_1 - \omega_2 \approx \omega_p$), $\gamma = (1 - V_x^2/c^2)^{-1/2}$, and x is the displacement of the electron from its equilibrium position (i.e., the Lagrangian coordinate). Neglecting the factor γ, this is a simple harmonic oscillator equation with a resonant driver if the beat frequency $\Delta\omega = \omega_p$.

Now the amplitude of the electron's oscillation is proportional to the amplitude E_p of the plasma wave field that its motion supports (the coefficients of proportionality are[11] $\Delta k x = eE_p/m\omega_p c \equiv \varepsilon$). Thus, renormalizing the equation, applying the chain rule to γV and expanding γ for small V/c, we obtain the equation for the beat-driven plasma wave field normalized to the wavebreaking plasma wave amplitude:

$$\ddot{\varepsilon} + \omega_p^2(1 - \frac{3}{2}\dot{\varepsilon}^2)\varepsilon \approx \frac{\alpha_1\alpha_2}{2} \sin \Delta\omega t . \qquad (7)$$

Here $\alpha_{1,2} \equiv eE_{1,2}/m\omega_{1,2}c$. This is the model of Rosenbluth and Liu[12] which includes a correction to the plasma frequency due to the relativistic mass increase of the oscillating plasma electrons. Initially, if $\Delta\omega = \omega p$, the plasma wave amplitude exhibits secular growth

$$\varepsilon(t) = \frac{\alpha_1 \alpha_2}{4} \omega_p t \tag{8}$$

where ε is now understood to represent the amplitude of the plasma wave. As ε grows, the effective plasma frequency becomes[12] $\omega_{ef} \approx \omega_p(1 - 3\varepsilon^2/16)$. At a time t such that $\int^t (\Delta\omega - \omega_{ef}) dt' = \pi/2 \approx \omega_p \int 3/16\ \varepsilon^2(t') dt'$, the driver has become $\pi/2$ out of phase with the oscillator and actually begins to drive the oscillation back down. Solving for the value of ε from (8) at this time gives $\varepsilon_{max} = (2\pi\alpha_1\alpha_2)^{1/3}$. A more careful solution of the model equation (7) gives the maximum value of ε as[12]

$$\varepsilon_{max} = (\frac{16}{3} \alpha_1 \alpha_2)^{1/3} . \tag{9}$$

In principal, Eq. (9) is valid only for $\alpha, \varepsilon \ll 1$, but simulations indicate that it is a good approximation for values of ε as high as 0.8.

In order to model the rise time of the lasers, we modify the Rosenbluth and Liu model (7) slightly by allowing the normalized laser amplitudes $\alpha_{1,2}$ to be functions of time. This leads to modified expressions for growth and saturation:

$$\varepsilon(t) = \int_0^t \frac{\alpha_1(t)\ \alpha_2(t)}{4} d(\omega_p t) , \tag{8b}$$

$$\varepsilon_{max} = (\frac{16}{3} \alpha_1(\tau)\alpha_2(\tau))^{1/3} , \tag{9b}$$

where τ is the value of t at which (8b) and (9b) become equal. Fig. 3 shows a numerical solution of the modified model equation (7) with the analytic expressions

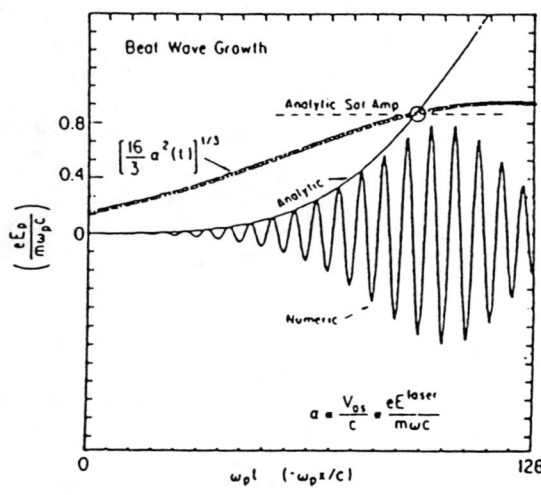

Fig. 3
Numerical solution of the modified plasma beat wave model Eq. (7) and the corresponding analytic growth and saturation for (8b) and (9b).

corresponding to Eqs. (8b) and (9b) plotted. Given the details of the laser pulse, Eqs. (8b) and (9b) enable predictions for the time for growth and the saturated amplitude of the resulting plasma space-charge wave.

In order to minimize the effect of the detuning between the driving frequency $\Delta\omega$ and the effective plasma frequency ω_{ef}, C.M. Tang et al.[13] have considered altering the beat frequency. They find that the beat wave growth is optimal for

$$\Delta\omega = \omega_p \left[1 - \frac{1}{2}(9\alpha_1\alpha_2/8)^{2/3} \right] \qquad (10)$$

with corresponding maximum amplitude

$$\varepsilon_{max} \approx 4 \, (\alpha_1\alpha_2/3)^{1/3} , \qquad (11)$$

an increase of about 50%. For lasers of $V_{osc}/c \equiv a = .1$, Eq. (10) corresponds to a 2% shift in the laser beat frequency below the natural plasma frequency.

The results of Tang et al. also provide a guideline for plasma homogeneity requirements. Eq. (10) suggests not only the optimal detuning, but also a scaling law for the fractional density change ($\delta n/n \sim \delta\omega^2_p/\omega^2_p \sim 2\delta\Delta\omega/\omega_p$) that will not have a significant effect on beat wave growth. We might expect to have to maintain the plasma homogeneity to within

$$\frac{\delta n}{n} \sim (\alpha_1\alpha_2)^{2/3} \qquad (12)$$

or about 5% for lasers of $V_{osc}/c = .1$. The effect of noise in the plasma density is considered by Horton and Tajima,[14] who give conditions for the resonant noise components moving with the beat wave to limit growth of the plasma wave. An in-depth description of beat wave excitation that includes not only rise time and detuning effects but also damping, harmonics, transverse dimensions and drifts, by W.B. Mori, can be found in Ref. 15.

B. COMPUTER MODELLING

There are a multitude of mechanisms which might cause the behavior of a real plasma to deviate from the simple model of Eq. (7). Some of these are nonlinear wave steepening, thermal corrections, and ion effects. Fully self-consistent particle simulations have been invaluable in separating out the most important effects. In Fig. 4, we show the results of a 1-D simulation in which beating lasers were injected from the right.[16] In (a) the beat pattern of the rising lasers is visible, and in (b) we see the characteristic growth and saturation of the plasma space-charge wave. Remarkably, Fig. (4b) agrees almost identically with the numerical solution to the model Eq. (7) shown below it. This is true despite the fact that the saturation amplitude is more than 70% of the cold wavebreaking maximum. Many of the nonlinear effects associated with particle trapping normally expected at such amplitudes simply do not appear because the plasma wave phase velocity is too high to allow trapping of the background particles. Only particles with velocity V such that $V/c \gtrsim 1 - \varepsilon - 1/\gamma_{ph}$ can be trapped (see Eq. 19).

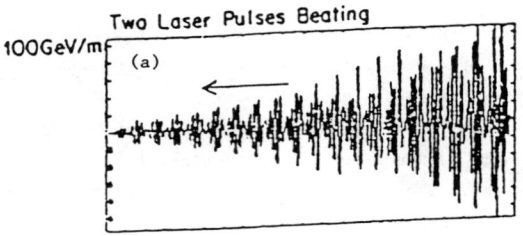
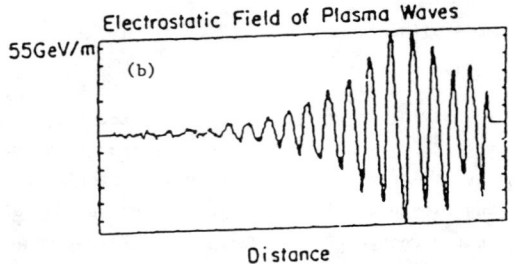
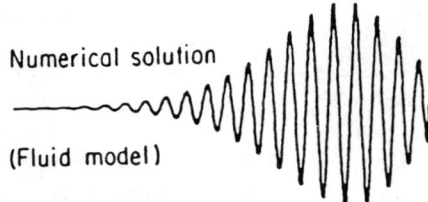

Fig. 4 1-D simulation showing (a) the laser beat pattern, (b) the space-charge plasma wave. The numerical solution of the modified fluid model is shown below. $\omega_1 = 4\omega_p$, $\omega_2 = 5\omega_p$, $\alpha_1 = \alpha_2$.

Recently, the computer models have been extended to two dimensions[17,18], and a sampling of these results is shown in Figs. 5-7. The 2-D simulations are consistent with the 1-D results and provide insight into transverse properties, such as filamentation and self-focusing.

The 2-D contour plot in Fig. 5 shows the generation of coherent plane wave fronts (moving left to right). A slice down the axis, Fig. 5b, shows the growth of the beat wave to be in nice agreement with both the model of section A and the 1-D simulations.

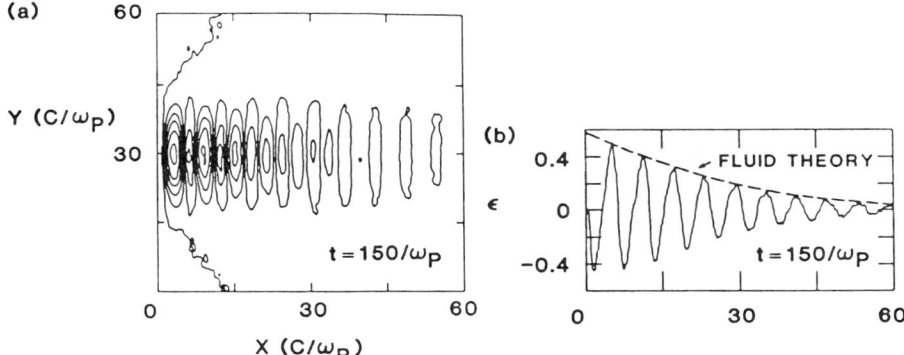

Fig. 5 2-D simulation (a) contour plot of the beat wave and (b) a slice down the y-30 axis.

Laser filamentation and self-focusing are phenomena familiar in laser fusion and can be seen clearly in Figs. 6 and 7. In Fig. 6, intense lasers ($V_{osc}/c \approx .5$) of width 20 c/ω_p incident from the left at first begin to filament into two narrower beams, then the strong radial gradient provides a ponderomotive force which blows plasma out of the laser beam channels.[19,20] Ponderomotive blowout is thought to be one of the major factors in the lack of accelerated electrons from a recent Los Alamos experiment. For very intense lasers, blowout can take place on a time scale as fast as the ion plasma period, necessitating the use of laser pulses shorter than this.

Laser self-focusing is a potentially beneficial phenomenon which enables propagation of the lasers through the plasma over distances much longer than a Rayleigh length. In addition, as in Fig. 7, it may compensate for the depletion of pump intensity on axis (see Section II for a discussion of pump depletion). The mechanism for self-focusing can be understood quite simply by noting that the refractive index of a plasma is $n = (1 - \omega_p^2/\omega^2)^{1/2}$. In a region of lower ω_p^2, the index of refraction is higher. Inside the plasma channel where the laser exists, ω_p^2 ($= 4\pi n_0 e^2/m_e$) is lower so the channel acts like an optical fiber to confine the light. Inside the channel ω_p^2 is lower for two reasons. First, the plasma density can be slightly lower because of ponderomotive blowout, and second, the oscillatory motion of the background electrons in the intense laser fields gives them a relativistic mass increase. The latter effect is probably more important because it may occur faster than an ion time scale. The time scale for self-focusing can be estimated from the inverse of the growth rate ν obtained by C. E. Max and J. Arons[21]:

$$\nu = \frac{\omega_p^2}{8\omega} \left(\frac{V_{osc}}{c}\right)^2 . \tag{13}$$

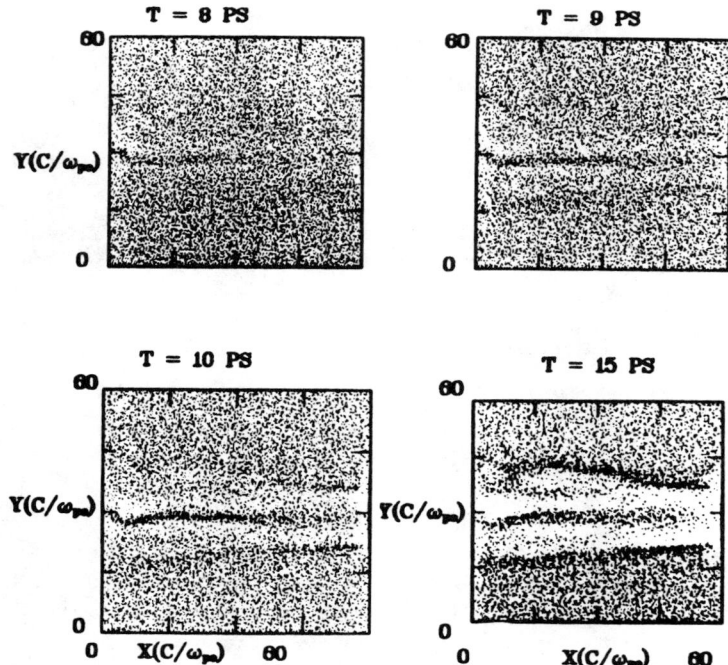

Fig. 6 X-Y particle plot of ions, showing evolution of filamentation.

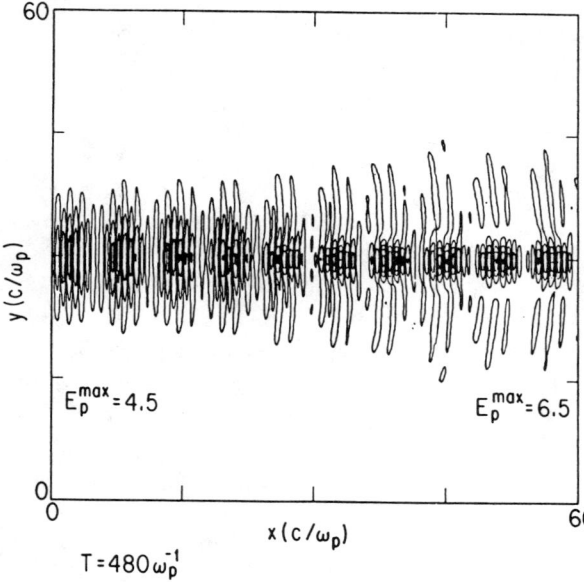

Fig. 7
2-D simulation contour plot of laser fields showing self-focusing (l to r) and increasing laser intensity on axis.

The power threshold for self-focusing to overcome Rayleigh diffraction is approximately[22]

$$P \geq \frac{9}{2} \left(\frac{\omega}{\omega_p}\right)^2 \times 10^9 \text{ watts} .\tag{14}$$

Some question remains as to whether the final beam radius will asymptote or oscillate about a value of order c/ω_p.

C. EXPERIMENTAL RESULTS

A number of experimental programs have started on the beat wave accelerator in the U.S. and in Canada, England, France and Japan. These are all in their early stages; although the results that have been obtained so far are encouraging, they must still be considered preliminary.

The first experimental verification of the theoretical and simulation models was performed at UCLA by C. Joshi and his group. The 9.6 and 10.6-micron lines of a CO_2 laser ($\alpha_{10.6} \approx .03$, $\alpha_{9.6} \approx .015$) were used to resonantly drive the beat wave in a 10^{17}-cm^{-3} density plasma. Modelling the laser's rise to be linear over 1 nsec, the model equations (8b) and (9b) predict a maximum normalized wave amplitude $\varepsilon = .08$ (or $E_L = 2.8$ GeV/m) at a time of roughly 500 picoseconds.

In order to conclusively diagnose the beat wave, one needs to know the phase velocity and amplitude of the wave. The phase velocity was obtained by measuring the frequency shift of the light from a ruby laser which is scattered by the density modulations of the beat wave. A sample of the frequency spectrum of the scattered light is shown in Fig. 8. The frequency shift of the scattered light shows that the plasma wave frequency is centered at the beat frequency of 9.6 and 10.6-µ light. By varying the angle of the Thomson scattering, the k-spectrum of the plasma wave was obtained (Fig. 9). Figs. 8 and 9 together confirm that the phase velocity of the plasma wave was near c as predicted by Eq. (5).

The time-averaged plasma wave amplitude was obtained by measuring the intensity of the scattered light and calculating (assuming Bragg scattering from a moving grating) the corresponding n_1/n_0 (= ε) of the plasma wave. The resulting beat wave amplitude was $n_1/n_0 = 1$ to 9% or $E_p = 300$ Mev/m to 3 GeV/m. This was the time-averaged field; the peak field may have been larger.

An experiment quite similar to that carried out at UCLA by Joshi has been carried out by F. Martin and his group at INRS-Energy, Université du Québec, Canada. They had a more powerful laser and were able to use part of its energy to create some preaccelerated electrons by the following method. A portion of the laser pulse was focused on a solid target producing a laser plasma. At high intensities such plasmas produce many energetic electrons. By passing the electrons through an energy analyzer which also focused them into the region of the beat wave, they were able to inject electrons with energies of over 0.5 MeV. These were fast enough so

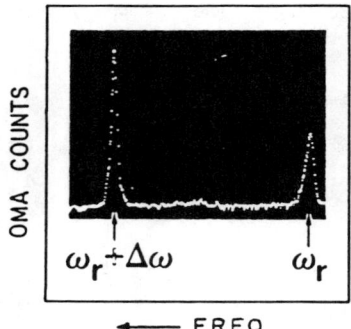

Fig. 8 Experimental small-angle Thomson scattering indicating a plasma wave at the beat frequency ($\Delta\omega$) of the driving lasers (ω_r is the frequency of the diagnostic ruby laser).

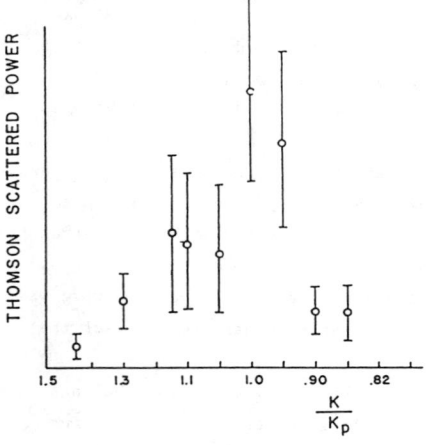

Fig. 9
K-spectrum of the beat wave normalized to the theoretical beat wave wavenumber.

that they could be accelerated by the beat wave generated. They analyzed the energy of the electrons coming from the beat wave region and found electrons with energies up to 2.0 MeV. Since the size of the beat wave region was only 1.5 mm in length, accelerating fields of about 10^9 volts/meter can be inferred.

A third experiment on the beat wave accelerator is a collaboration between a group under A.E. Dangor from Imperial College and Roger Evans and his group at the Rutherford Appleton Laboratory in England. This experiment uses a neodymium glass laser with beams at roughly 1 and 1.1-µ wavelength. Only preliminary experiments have been carried out and as yet no results on the generation of the beat wave have been reported. However, they have obtained one very important auxiliary result. Dangor and Evans have shown that, by using their laser to give multiphoton ionization of a gas, a reproducible and very uniform plasma (the uniformity was perfect within the limits of their measurements) can be created. In order to make a successful beat wave accelerator for high energy physics a uniform quiescent plasma several meters long will have to be produced. Multiphoton ionization appears to be a promising means; it ionizes virtually every atom quickly so that the plasma density and hence plasma frequency is determined by the initial gas density. It also produces a low temperature plasma at rest with no turbulence.

Recently programs on the beat wave accelerator have been started in France at Ecole Polytechnique and in Japan at the Institute of Laser Engineering of Osaka University. The Japanese program builds on the strong program in laser fusion which they have. It will use CO_2 lasers built for that program and it appears that it will yield results in the near future.

II. ACCELERATION MECHANISMS

We now treat the acceleration of charged particles in the wave fields described in the previous section. The treatment is not rigorous; we neglect the damping and distortion of the plasma waves by the accelerated particles. This is a fair approximation for moderate beams in which the number of beam particles per trapping potential is much less than the number of background plasma electrons in a plasma wavelength. Furthermore, we neglect the effect of the lasers on the accelerated particles. This is partially justified by the fact that, in the beat wave frame, the laser electric field E_0 is reduced by the Lorentz transformation to E_0/γ_{ph} and the magnetic field transforms away completely. The longitudinal plasma wave field on the other hand is invariant to transformations along the direction of particle acceleration, so we consider it alone in our treatment of charged particle motion. We rely on simulations to validate these approximations and to identify the important self-consistent effects.

A particle in the plasma cannot be picked up or trapped by the accelerating wave unless it exceeds a minimum velocity in the direction of the wave, much as a surfer must paddle to catch an ocean wave.

For this reason, the accelerated particles either must be injected externally or picked up out of the high energy tail of the background plasma distribution. Before proceeding to the beat wave and surfatron acceleration mechanisms, we first calculate the minimum velocity or injection threshold for particle trapping. This calculation will also provide an estimate of the maximum plasma wave amplitude ε before trapping of background particles becomes significant (catastrophically damping the wave).

A. INJECTION/TRAPPING THRESHOLD[24]

Consider a particle of velocity V and momentum γmV in the lab frame moving in the direction of a wave of high phase velocity, $V_{ph} \lesssim c$. Its energy γ' in the wave frame is given by the Lorentz transformation to be

$$\gamma' = \gamma_{ph}\left(\gamma - \beta_{ph}\frac{P}{mc}\right) \approx \gamma_{ph}\gamma(1 - \beta_{ph} V/c) \tag{15}$$

where $\beta_{ph} = V_{ph}/c$ and $\gamma_{ph} = (1 - \beta^2_{ph})^{-1/2}$. A particle is just trapped by the wave when the wave's potential (in the wave frame) can overcome the particle's kinetic energy in the wave frame. That is, when

$$e\phi \geq (\gamma' - 1)mc^2 \tag{16}$$

the wave potential in the lab frame is simply

$$\phi = \phi'/\gamma_{ph} \tag{17}$$

since $E = k\phi$ is an invariant ($k'\phi' = k\phi$) and $k = \gamma_{ph}k'$. Combining Eqs. (15) to (17) gives the trapping potential

$$e\phi = mc^2 [\gamma_{ph}\gamma(1 - \beta_{ph}V/c) - 1]/\gamma_{ph} . \tag{18}$$

For $V = V_{ph}$, the minimum ϕ is zero as expected. Eq. (18) is valid for all particle and wave velocities. For $V/c \ll 1$ and $\gamma_{ph} \gg 1$, we find that

$$\phi/\phi_{cold} \approx 1 - V/c - 1/\gamma_{ph} \tag{19}$$

where ϕ_{cold} is the cold trapping threshold $\approx mc^2/e$ (from [18] with $V = 0$).

Interestingly, for electrons in the beat-wave example, cold trapping and cold wavebreaking (Eq. 2) have approximately the same threshold, as can be verified from the above expression for ϕ_{cold} and $k = \omega_p/c$:

$$E_{\substack{cold \\ trapping}} \approx k\phi_{cold} = kmc^2/e \approx mc\omega_p/e = E_{\substack{cold \\ wavebreaking}}$$

Thus, to the extent that $|E| \approx |k\phi|$ for the plasma wave, $\phi/\phi_{cold} \approx \epsilon \approx E/E_{max}$ (for $\gamma_{ph} \gg 1$). We comment that, for ions, the cold trapping threshold would be larger than the cold wavebreaking threshold by the ion to electron mass ratio. Thus, trapping of ions will be much more difficult and will require the ions' velocity to be essentially V_{ph}.

We expect to trap few background particles if (19) is satisfied for V = a few (e.g., 3) times the thermal velocity (V_t). Thus, one should take ϵ less than $1 - 3V_t/c$ ($V_t \ll c$) to insure a minimum of background plasma trapping.

Expression (18) may be used to estimate what injection velocity or injection energy is required for injected particles to be picked up by a wave of given amplitude. Inverting (18) and solving for the injection energy $(\gamma - 1)$ in terms of $\phi/\phi_{cold} \equiv e\phi/mc^2 \approx \epsilon$ yields

$$\gamma - 1 \approx \gamma^2_{ph} \{\varepsilon + 1/\gamma_{ph} - \beta_{ph} [(\varepsilon + 2/\gamma_{ph})\varepsilon]^{1/2}\} - 1 \ . \tag{20}$$

Thus, for example if $\phi/\phi_{cold} = .1$ and $\gamma_{ph} = 10$, then the injection energy must be about 900 keV.

The injection energies versus normalized wave amplitude ε from Eqs. (18) or (20) are plotted for several values of γ_{ph} in Fig. 10.

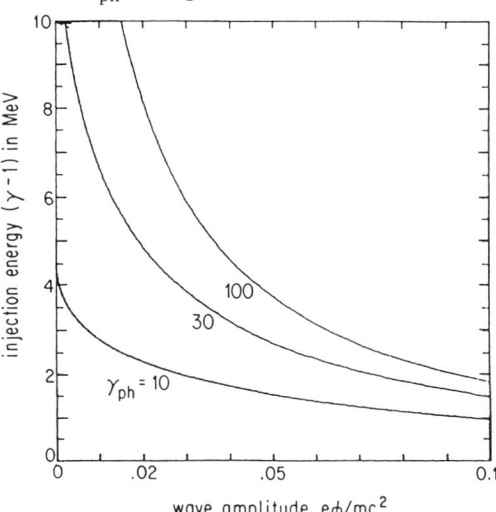

Fig. 10 Injection energy thresholds vs. normalized wave amplitude for various wave phase velocities (from Eq. 20).

B. SIMPLE BEAT WAVE ACCELERATION

In the original beat wave accelerator (BWA) concept, a particle accelerates by simply riding from the top to the bottom of the potential well of a plasma wave (i.e., a half of the plasma wavelength λ_p). Since the plasma wave moves at $V_{ph} = c(1 - \omega_p^2/\omega^2)^{1/2} \approx c(1 - \omega_p^2/2\omega^2)$ [see Eq. (5)] and the particle moves at nearly c,[25] the particle outruns the wave (reaches the bottom of the potential well) in a distance

$$\ell = \frac{\lambda_p}{2} \frac{c}{c - V_{ph}} \approx \frac{\omega^2}{\omega_p^2} \lambda_p \ . \tag{21}$$

The maximum energy gain of the particles is of order $eE_p \cdot \ell$ or

$$W_{max} = 2\varepsilon \frac{\omega^2}{\omega_p^2} mc^2 \tag{22}$$

where we have substituted $\lambda_p = 2\pi c/\omega_p$, ε is the plasma wave amplitude normalized to wavebreaking and we have taken the average value of E_p to be its amplitude over π in order to give a result consistent with other derivations.[25]

The energy gain can be quite large. For example, for 1-μ laser radiation incident on a 10^{17}-cm^{-3} plasma, $\omega/\omega_p = 100$ and the maximum energy gain is 20 GeV in a distance less than a meter (for ε near 1).

We comment that the actual region of the plasma wave usable for acceleration is less than $\lambda_p/2$. Radial fields due to the finite width of the plasma wave are defocusing to particles over half of the accelerating portion of the plasma wave.[26] If the width of the plasma wave is given by the laser beam width and this in turn self-focuses to a few times c/ω_p as discussed in Section I, then the radial fields will be quite large and it will be necessary to avoid the regions of defocusing.

1-D simulations performed prior to the first laser accelerator workshop verified electron acceleration consistent with Eq. (22) (see D.J. Sullivan and B.B. Godfrey in Ref. 8). 2-D simulations[18] have substantiated the 1-D results and added information about the transverse particle dynamics. The phase-space plots of Fig. 11 illustrate both electron acceleration in the plasma wave troughs and the focusing of the accelerated particles in the transverse direction. The peak γ of particles in Fig. 11 is around 25, consistent with Eq. (22) and the simulation parameters ($\epsilon \sim .5$, $\omega_1 = 4\omega_p$, $\omega_2 = 5\omega_p$).

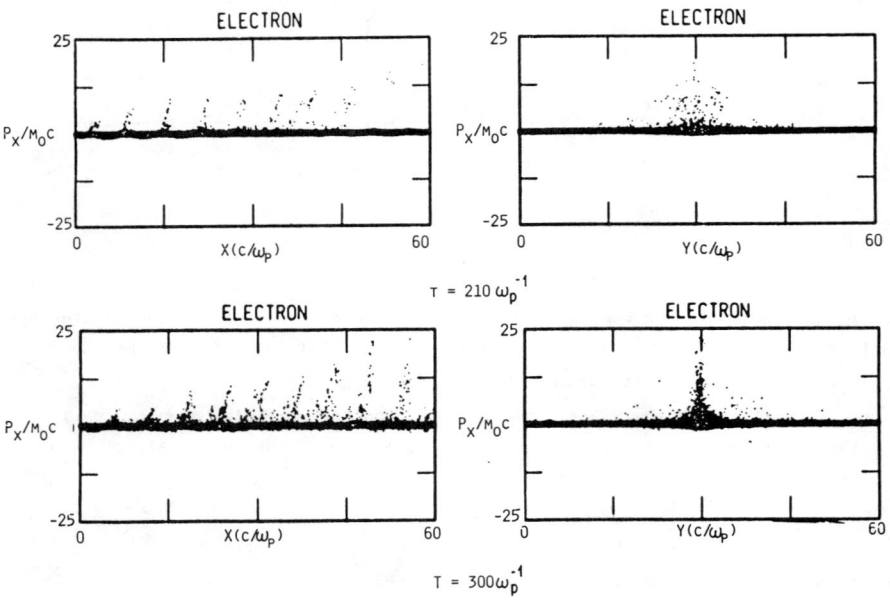

Fig. 11 2-D simulations showing acceleration in plasma wave troughs (left) and particle focusing (right) at two different times.

Particle focusing is important to the emittance quality of the accelerated beam. The particle focusing in Fig. 11 is caused by a combination of the radial plasma wave fields and the self-generated magnetic fields of the accelerated electron bunches. These fields are quite strong in a plasma; the magnetic field near each bunch would correspond roughly to 2 megagauss for CO_2 laser parameters (see Fig. 12).

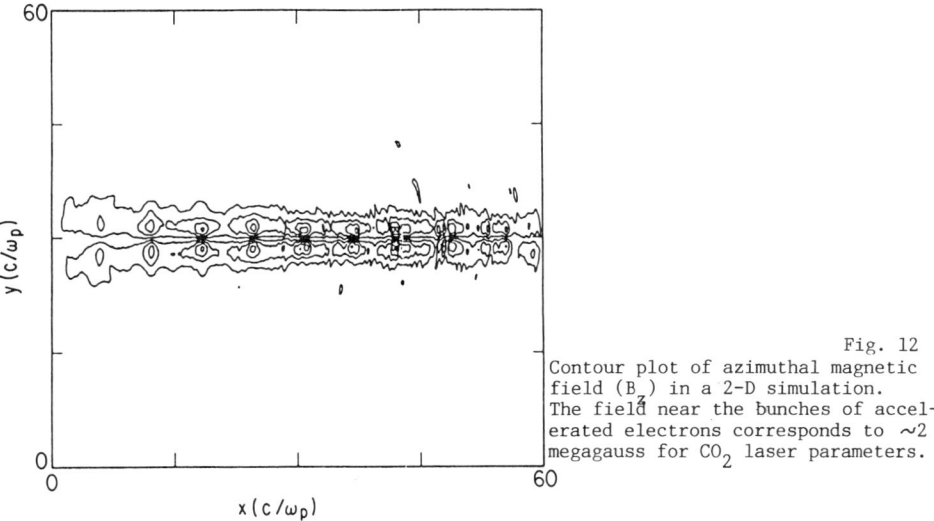

Fig. 12
Contour plot of azimuthal magnetic field (B_z) in a 2-D simulation. The field near the bunches of accelerated electrons corresponds to ~2 megagauss for CO_2 laser parameters.

While the acceleration gradient and focusing properties of a plasma wave accelerator are attractive, it is also important that it be able to accelerate a large number of particles. Recently, this beam loading issue has been addressed.[27,28] The total number of particles (of charge ±e) that can be accelerated (per shot) can be summarized by

$$N = \frac{n_0}{k_p} \cdot \varepsilon \cdot A_{eff} \approx 5 \times 10^5 \sqrt{n_0} \cdot \varepsilon \cdot A_{eff}$$

where again n_0 is in cm^{-3}, ε is the normalized plasma wave amplitude (<1) and A_{eff} is the effective cross-sectional area of the plasma wave in cm^2. A_{eff} is roughly the larger of the area of the driving source (e.g., the lasers) and $\pi c^2/\omega_p^2$. The above expression is an upper limit based on total absorption of the wave energy by the beam-loaded particles. By reducing the beam number by a factor of about two it is possible by shaping the beam distribution and phase to obtain high beam-loading efficiency ($\gtrsim 75\%$ in 1-D, $\gtrsim 20\%$ for very narrow beams) with small energy spread ($\lesssim 10\%$) and emittance growth.[27] The absorption of wave energy by a narrow beam load is illustrated in Fig. 13. As an example, for $n_0 = 10^{16}$ cm^{-3} and $A_{eff} \approx 1$ mm^2, N is on the order of 10^{11} particles.

C. THE SURFATRON ACCELERATION MECHANISM

Expression (22) for the maximum energy gain of the BWA suggests that the lower is ω_p^2 (i.e., the lower the plasma density), the higher is the final energy that can be obtained. Then why not eliminate the plasma so that the maximum energy can become infinite? The answer of course is contained in Eq. (2) which shows that the acceleration gradient ($\Delta W/\Delta x$) goes to zero:

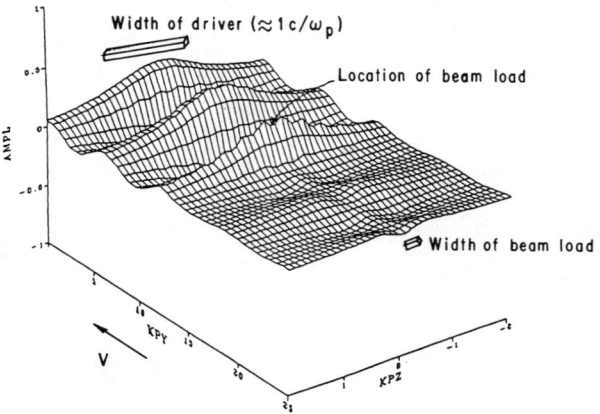

Fig. 13 Numerical solution of 2-D plasma wave electric fields (longitudinal above, transverse below) illustrating the absorption of wave energy by a narrow beam load. In this case the wave is excited by the wakefield mechanism described in Section III.

$$W_{MAX} \sim \frac{1}{n_o}, \qquad \Delta W/\Delta x \sim \sqrt{n_o}.$$

The higher the final energy required of the beat wave accelerator, the slower the acceleration gradient.

The surfatron[26] is proposed as a mechanism to overcome this limitation of the BWA by phase locking the particles in the plasma wave, thereby allowing them to gain energy for as far as the plasma wave can be maintained. The basic idea, illustrated in Fig. 14, is to impose a DC magnetic field (B_z) perpendicular to the beat wave. The $-e\vec{V}_{ph} \times \vec{B}/c$ force on a trapped particle deflects the particle across the wave fronts in a way similar to the way a surfer cuts across the face of an ocean wave. Once the particle acquires y-velocity the $-e\vec{V}_y \times \vec{B}/c$ force presses the particle up against the side of the wave, where it remains phase locked.

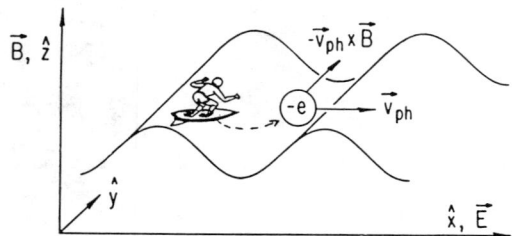

Fig. 14
A DC magnetic field deflects particles across the waves in the surfatron, preventing them from outrunning the waves.

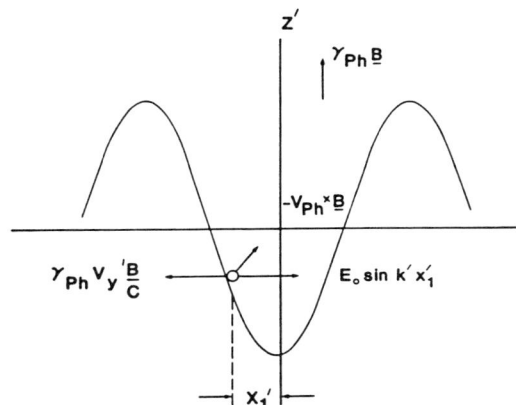

Fig. 15 In the wave frame, a surfatron particle comes to equilibrium against the side of the potential well.

In the wave frame (Fig. 15), the accelerated particle comes to a phase-stable point in the potential well where the electric force ($E_0 \sin k'x_1'$, primes denote wave frame quantities) of the plasma wave balances the magnetic force ($V_y'B_z'/c \approx \gamma_{ph}B_z$) due to motion across the wave.

Since $\sin k'x_1' \leq 1$, we must have

$$E_0 > \gamma_{ph}B_z \tag{23}$$

in order to have a phase-stable point. Too large a magnetic field will cause the particle to detrap and exhibit cyclotron motion.

If the wave phase velocity is near c, the particle travels a nearly linear trajectory at a small angle (θ) to the direction of the plasma wave. This can easily be seen from the fact that for a trapped relativistic particle

$$V_x \approx V_{ph}, \quad V_x^2 + V_y^2 \approx c^2 . \tag{24}$$

Thus

$$\tan\theta = V_y/V_x \approx 1/\gamma_{ph} . \tag{25}$$

The rate of energy gain for a surfatron particle is easily obtained from the $\hat{y}$ equation of motion in the lab frame:

$$\frac{d}{dt}(\gamma V_y) = \omega_c V_x \tag{26}$$

where $\omega_c = eB_z/mc$, the non-relativistic cyclotron frequency. Substituting for V_x and V_y from (24) and integrating once gives the energy gain versus time or distance:

$$\gamma = \gamma_{ph}(V_{ph}/c)\omega_c t = \gamma_{ph}\omega_c x/c . \tag{27}$$

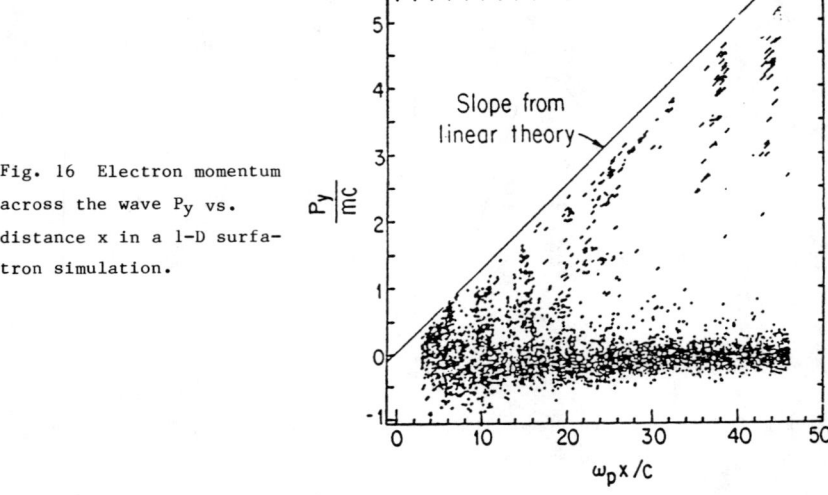

Fig. 16 Electron momentum across the wave P_y vs. distance x in a 1-D surfatron simulation.

In Fig. 16, the phase-space plot from a 1-D simulation shows surfatron motion in agreement with Eq. (26).

Unfortunately, simulations of surfatron acceleration failed to show energy more than about 30% higher than corresponding BWA runs. Energy gain in the surfatron scheme is not limited by particle dynamics; it is limited instead by wave dynamics. That is, the laser pulse/plasma wake package of Fig. 2 cannot be propagated indefinitely through the plasma. The laser pulse is continually feeding energy into plasma waves. Since the plasma wave energy convects at the plasma wave group velocity ($\sqrt{3}\ V^2_{th}/V_{ph}$), which is effectively zero, all of the plasma wave energy is left behind. When the energy left in plasma waves becomes comparable to the energy originally in the laser pulse, the laser pulse will have become completely depleted and no more plasma waves can be produced. An estimate of the pump depletion length L_d from this simple energy balance argument indicates that L_d is only slightly higher than the dephasing length for the beat wave accelerator (Eq. 22), consistent with the simulations.

III. PLASMA WAVE EXCITATION BY PARTICLE BEAMS (THE PLASMA WAKEFIELD TRANSFORMER)

The second method that has been proposed to generate the accelerating plasma wave is the so-called plasma wakefield method.[29-33] In this method a bunch of moderate energy electrons from, say, a linear accelerator is made to pass through the plasma. These electrons generate a plasma wave wake by pushing the plasma electrons aside much as a motor boat generates a wake of water waves by pushing the water aside. The phase velocity of the wake must match the velocity of the driving

bunch, which, if it is relativistic, is very close to the velocity of light. A
second bunch of electrons suitably placed in the wake so as to be accelerated can
be accelerated to a higher energy than that of the driving bunch.

The whole process is illustrated in Fig. 17. This figure shows the driving
bunch (the long bunch) and the accelerated bunch (the short bunch). The upper
surface shows the acceleration an electron would feel; it is low in the region of
the driver and has large peaks and valleys in the wake. The lower surface shows the
radial acceleration an electron feels. There are places where electrons are
accelerated and focused towards the axis.

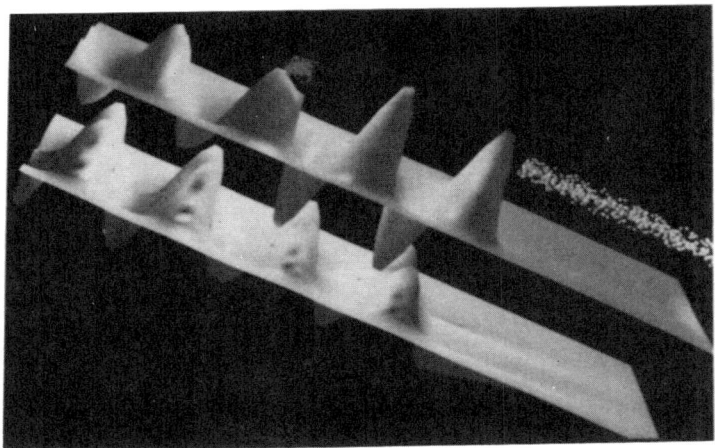

Fig. 17 Numerical solutions of longitudinal (upper) and transverse (lower)
wakefields excited by a long driving beam and loaded with a short
accelerating beam.

At first it may seem paradoxical that we are using one bunch of energetic
electrons to accelerate another; why not simply accelerate the first bunch to the
desired energy in the first place? The method makes sense only if we can use a
bunch of modest energy electrons containing a large number of electrons to
accelerate a bunch containing fewer electrons to a much higher energy. This, in
fact, is the case and the large bunch of modest energy electrons (1 GeV, say)
appears relatively straight-forward and cheap to obtain. The key to doing this is
the achievement of a much larger electric field in the wake than the retarding field
slowing down the driving bunch; the ratio of these fields is called the transformer
ratio and is analogous to the voltage step-up of a transformer. The trick to
obtaining a good transformer ratio is to make the density of the driving bunch rise
slowly and then cut it off rapidly, as illustrated in Fig. 18. As the bunch moves

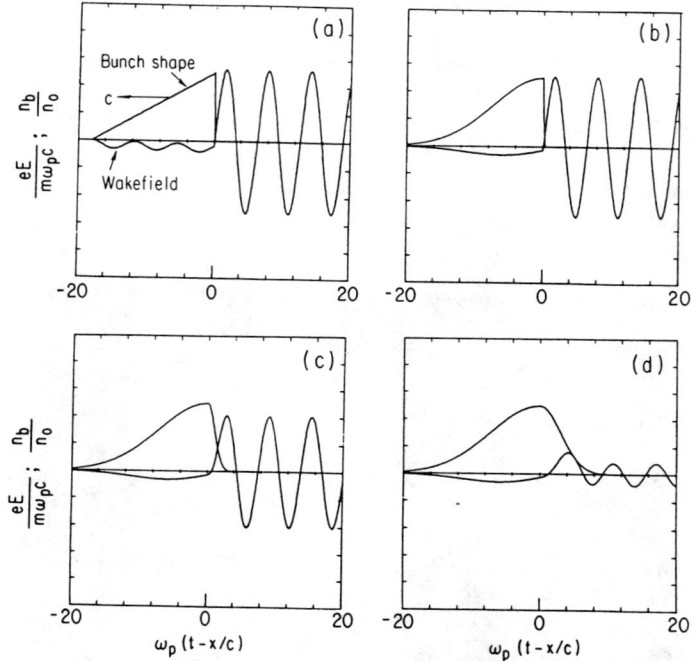

Fig. 18 Numerical solutions of plasma wakefields excited by the various bunch shapes shown (from Ref. 31).

into a region of plasma, the plasma electrons move out of the way so that the plasma plus beam is almost neutral. The rising current of the beam also produces a rising magnetic field which by Faraday's law of induction produces an electric field just like the time-varying magnetic field in a transformer. This electric field accelerates plasma electrons in the direction opposite to the driving beam's motion and almost, but not quite, cancels out the beam current. Thus in the beam there is very little electric or magnetic field; they are just strong enough to extract the energy required to generate the wake. Now when the bunch moves out of a region the plasma suddenly finds itself in a nonneutral state with respect to both charge and current, and an intense oscillation is left behind.

The equations which describe the acceleration gradient and energy gain (i.e., the transformer ratio) in the plasma wakefield scheme are[31]

$$eE = \varepsilon \sqrt{n_o} \text{ eV/cm} , \qquad \Delta\gamma \lesssim N\pi\gamma_b \frac{\varepsilon\gamma_b}{N + \varepsilon\gamma_b} ,$$

where $\varepsilon = n_b/n_0$, n_b is the peak driving beam density, N is the length of the driving bunch divided by λ_p (= $2\pi c/\omega_p$). Thus for large γ_b, the transformer ratio is approximately $N\pi$. Using a one-dimensional computer model for this process J.J. Su has shown that a 75-MeV driving bunch can be used to accelerate electrons from 75 MeV to 1 GeV (Fig. 19).

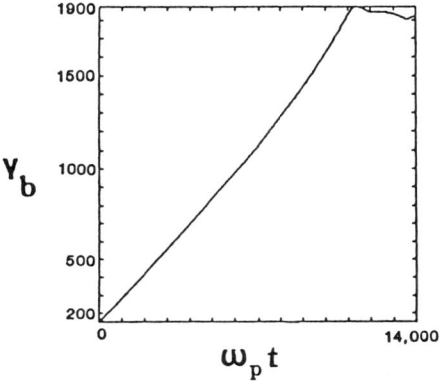

Fig. 19
Test particle acceleration from 75 MeV to ~1 GeV in a PIC simulation of the plasma wakefield accelerator.

When streams of electrons pass through plasma they are subject to a variety of instabilities; if the wakefield approach is to be successful both the driving and the accelerated bunches must be stabilized.[34] Since the accelerated bunch has fewer electrons and is shorter, and the relativistic mass of its electrons is greater, it is far less subject to such instabilities than the driving bunch and so we may concentrate our effort on the latter.

There are two critical types of instabilities to consider. The first is known as the two-stream instability, which in some ways is analogous to waves generated by wind blowing over water; in this case, however, the wind corresponds to the moving driving bunch and the water to the background plasma and, rather than being separated by a surface, they are interpenetrating. Because the bunch is moving, disturbances produced in the plasma are left behind, which greatly helps stability. Disturbances created on the beam itself are however carried along with it and can grow as it goes along. However, for high energy drivers the mass of the driver electrons is increased and they become stiff. One-dimensional simulation studies of Su indicate that something like 80% of the driver energy can be extracted before breakup of the driver due to the two-stream instability.

A more serious instability is the filamentation or so-called Weibel instability. This instability results from the tendency of parallel currents to attract each other. If the driving bunch contains filaments of slightly enhanced charge density, then because of their motion these constitute current filaments and they tend to contract and compress further; the tendency of the filaments' charge density to oppose this contraction is neutralized by the expulsion of the background plasma electrons from the filaments. Since the filaments move along with the driving bunch, this instability can be quite serious. Fig. 20 shows results of a two-dimensional simulation showing filamentation. Fortunately, this instability appears to be controllable by giving the driver electrons a little random motion perpendicular to their direction of propagation. The random motion tends to disperse any filamentary concentration that occurs; if the time to disperse is less than the contraction time, stabilization is achieved. Recent work[34] indicates a transverse energy of about 50 keV is required for stabilization independent of the beam energy. For driving bunch energies of 1 GeV this is one part in 20,000 and is totally negligible. Fig. 20b shows how the filamentation appearing in Fig. 20a has been suppressed by introducing this random motion.

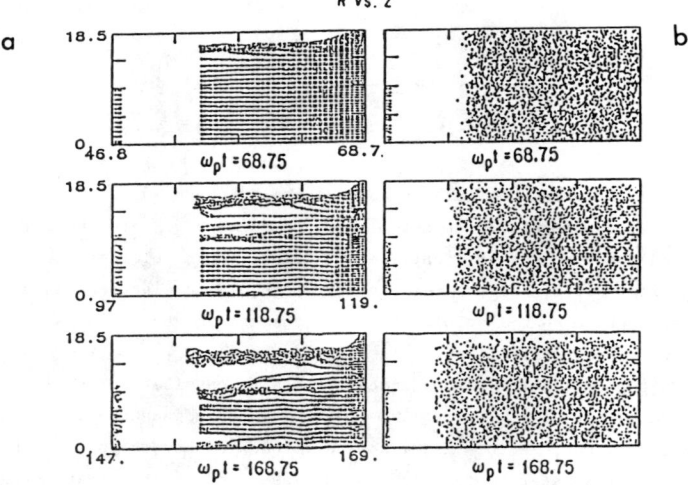

Fig. 20 2-D PIC simulations of the plasma wakefield driving beam in real space ($k_p r$ vs. $k_p z$) at various times, showing filamentation instability of a cold beam (a) and no instability in a warm beam (b).

So far there is only one experiment planned on the plasma wakefield accelerator. This was started by David Cline as a collaborative effort between the University of Wisconsin and Argonne National Laboratory; UCLA will be joining this effort by carrying out laser scattering diagnostics of the plasma waves in its later stages and by carrying out computer modeling for it. The plasma source, a hollow-cathode-arc magnetic-mirror confined plasma was built by J. Rosenweig of the University of Wisconsin.

Argonne National Lab has a 20-MeV linear accelerator that they have adapted to this experiment; under the direction of J. Simpson it provides bunches of electrons for producing the wakefield. A small fraction of a bunch is slowed down to 15 MeV by passing through a thin foil. This is deflected from the main beam magnetically and passes through a trombone leg. Fig. 21 shows a picture of the apparatus. The 15-MeV bunch enters the plasma behind the driving bunch and acts as a witness pulse for the waves generated. By varying the length of the trombone leg the position of the witness pulse behind the driver can be varied so that different regions of the wake can be probed. Computer modeling of this experiment[35] by Rhon Keinigs and M. Jones of Los Alamos and J. J. Su of UCLA indicate that the electrons in the witness pulse can be accelerated from 15 to 25 MeV by the wake.

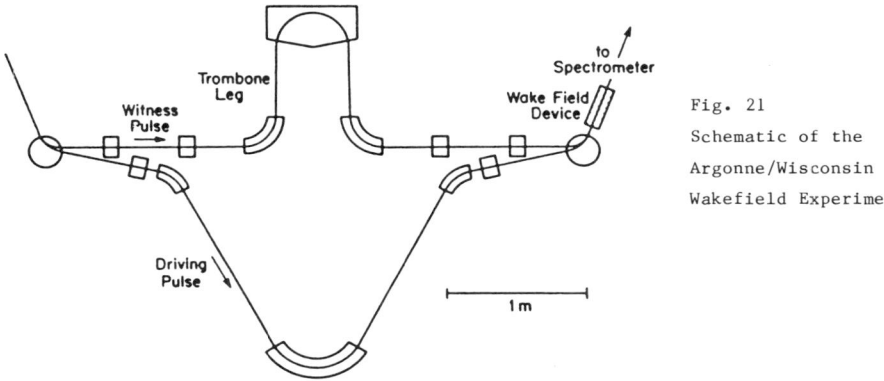

Fig. 21 Schematic of the Argonne/Wisconsin Wakefield Experiment

IV. PLASMA FOCUSING OF PARTICLE BEAMS (THE PLASMA LENS)

One of the challenges for future e^+e^- high energy linear colliders is to increase the luminosity according to the square of the center-of-mass energy in order to keep the event rate constant. For a fixed repetition rate and number of particles this can be accomplished only by reducing the spot size at the interaction point. Here again plasmas may play a role. When a dense particle beam traverses a plasma it experiences a strong radial focusing force due to the transverse wakefields excited by the beam itself. If the beam density is appropriately shaped and/or a precursor pulse (laser or particle beam) is injected ahead of the beam, this focusing force can be made uniform and linear (in r) over the length of the beam.[36,37] In this case a thin slab of plasma placed in front of the interaction point becomes a powerful final focusing lens, pinching the beam and enhancing the luminosity.

A formal description of the plasma lens in terms of the longitudinal and transverse wakefield response of the plasma to the beam (or beam and precursor) is given in Ref. 36. Here we give a physical description of the focusing mechanism and use this physical argument to estimate the focusing strength of a plasma lens. The physical mechanism is as follows: When an electron beam enters the plasma, the plasma electrons respond to the excess charge by shifting away from the beam particles. The remaining plasma ions neutralize the space-charge force within the beam. For positron beams the charge neutralization is equivalent but is due to the plasma electrons shifting in the opposite direction. While the plasma is very effective at shielding the beam's space charge, it is less effective at shielding its current. For beams narrower than c/ω_p (the plasma skin depth) most of the plasma return current flows on the outside of the particle beam. Thus the beam experiences almost the full effect of its self-generated azimuthal magnetic field. From Ampere's law this is $B_\theta \approx 2\pi n_b e r \beta$ for a uniform beam density n_b, where $\beta = v/c \approx 1$. This gives a radial Lorentz force

$$F_r \approx 2\pi n_b e^2 r .$$

The focusing gradient F_r/r can be expressed conveniently as

$$F_r/r \approx 2\pi n_b e^2 \approx 3\times 10^{-9} n_b \text{ gauss/cm}$$

for n_b in cm^{-3}. For example, a beam similar to that required for a future 5-TeV collider[36] might have 4×10^8 particles, 3-μ radius, and be 100 μ long, so that $n_b \approx 1.4\times 10^{17}$ and $F_r/r \approx$ megagauss/cm. This exceeds by four orders of magnitude the equivalent focusing strength of conventional quadrupole magnets. Neglecting limitations due to aberrations, the beam radius at the interaction point (a*) is inversely proportional to the focusing strength of the final lens (for fixed lens thickness and beam emittance), and the luminosity is proportional to a*$^{-2}$. Thus the luminosity enhancement from a plasma lens may be considerable.

The simple physical argument given above based on plasma shielding of space charges neglects some important effects. First is electron inertia: When the plasma electrons are displaced by the particle beam they tend to overshoot and oscillate. This causes the focusing force to oscillate sinusoidally (about the value we estimated above) over the length of the bunch. Two solutions to this problem have been proposed. One is to exactly cancel the oscillating portion of the focusing field by superimposing an appropriately phased wave field from a precursor bunch (or precursor beat wave). For a square beam profile (n_b = constant over length b and zero elsewhere), a possible precursor arrangement would be[36] a short ($\ll c/\omega_p$) bunch placed $\pi/2$ c/ω_p in front of the beam with total charge $Q_{pre} = Q_b/k_p b$, where Q_b is the charge in the main beam and $k_p = \omega_p/c$.[38]

A second, perhaps simpler, solution to the problem of oscillations in the focusing force is to allow the beam density to increase slowly at the head compared to c/ω_p. In this way the plasma electrons respond adiabatically without oscillating appreciably.

The difficulties associated with realization of a plasma lens center around
beam shaping and alignment. Without proper shaping of the radial and longitudinal
beam density profiles, the plasma lens will be limited by aberrations. In this case
the final spot size will be limited to

$$a^* = \frac{\Delta F}{F} a$$

where a is the spot size at the entrance to the lens and $\Delta F/F$ is the fractional
variation in the focusing force due to aberrations. Alignment difficulties can
arise because the particle beam focusses to its own axis of symmetry rather than the
axis of an externally controllable lens. With attention to these issues a plasma
lens for significant (order of magnitude) luminosity enhancement appears very
feasible. Some other effects of the plasma such as thermal noise have been
considered[39] and shown not to degrade the final spot size. Finally, the plasma lens
has the advantage of providing simultaneous focusing in both transverse directions.

V. SUMMARY

The experiments on acceleration of particles using plasma waves are still in
their very early stages. It is too early to know whether or not they can produce
beams of particles of energies, intensities and quality required to do present-day
high energy physics experiments. The high energy accelerators in use today have
developed over the last 50 years and are remarkable machines. It is not easy to
challenge them, but history has given us abundant proof that advances often come
from looking at radically new technologies. Besides the original goal of
acceleration these studies have led to the realization that there are other
potential applications of plasmas in high energy physics. One such is the use of
the powerful focusing forces mentioned in this article to focus high energy
particles in linear colliders to unprecedented small spot sizes.

In addition to the possibilities for use in high energy accelerators, research
on generating intense plasma waves opens new areas in plasma physics and related
science. For example, a beat excited or wakefield excited plasma wave provides an
ideal initial condition for studying the complex couplings between various plasma
modes[23] and possibly for understanding the path to turbulence. Another area of
interest is the radiation generated when relativistic electrons are made to
propagate perpendicular (or even antiparallel)[1] to the direction of the beat wave.
The transverse acceleration of these electrons by the intense longitudinal fields
(or transverse fields in the anti-parallel case) causes them to radiate. Their
relativistic motion Doppler-shifts the frequency into the x-ray or even γ-ray range.
Such sources of light could extend that available from synchrotron sources.

Finally, such investigations also may lead to an understanding of particle acceleration processes occurring in nature. Although nothing as organized as the beat wave or wakefield accelerators described in this article is likely to occur naturally, effects like the surfing effect used in the surfatron could be important. It is known that around pulsars particles can be accelerated to extremely high energies (as high as 10^{17} eV for Cygnus X-3 [Scientific American, Nov. 1985]) in relatively short distances. Such acceleration almost certainly takes place in the plasma surrounding the pulsar.

This work was supported by NSF Contract Nos. PHY 85-12390 and PHY 86-11235, DOE Contract No. DE-AS03-ER40120 and ONR Contract No. N00014-86-K-0585.

REFERENCES

1. C. Joshi, T. Katsouleas, J.M. Dawson, Y.T. Yan and F.F. Chen, Proc. 1987 Particle Accelerator Conf., Washington, DC; C. Joshi, T. Katsouleas, J.M. Dawson, Y.T. Yan and J.M. Slater, IEEE J. Qtm.-Electron QE-23, 1571 (1987).

2. Laser Acceleration of Particles (Malibu, 1985). C. Joshi and T. Katsouleas, Eds., AIP Conf. Proc. 130 (AIP, New York, 1985).

3. The radiation or pondermotive force on a single electron at rest in a light wave $\vec{E} = \hat{y} E_y \sin(kx - \omega t)$ is $F_p' = k'e^2 E_y'^2 / 2\omega'^2$, where primes denote quantities in the particle's rest frame. If the particle were moving relativistically along $\hat{x}$, then $E_y' = \gamma(E_y - \beta B_z) \approx E_y/2\gamma$, $\omega' \sim \omega/2\gamma$, $k' \sim k/2\gamma$, $F_p = \frac{dp}{dt} = 2\frac{dp'}{dt'} = 2F_p'$, so that $F_p = ke^2 E_y^2 / 2\gamma\omega^2 \propto 1/\gamma$. This derivation was suggested to me by G. Schmidt; a Hamiltonian-Jacobi treatment is in L.D. Landau and E.M. Lifshitz, Classical Theory of Fields, Section 47 (Addison-Wesley, Reading, MA, 1971), 3rd ed.

4. An exact nonlinear solution to the relativistic cold fluid equations [A.I. Akhiezer and R.V. Polovin, Sov. Phys. JETP 3, 696 (1956)] shows that the maximum electric field of a highly relativistic wave can exceed this value. Their result is $eE \approx m\omega_{pc}(\gamma_{ph} - 1)^{1/2}$. However, a waterbag model of a warm plasma gives a value on the order of Eq. (2) for temperatures of 1 to 500 eV; see T. Katsouleas and W.B. Mori, UCLA PPG-1040 (1987).

5. J.M. Dawson, Phys. Rev. 113, 383 (1959).

6. Berkeley Particle Data Group, "Review of Particle Properties," Phys. Lett. 111B, 39-41 (1982).

7. T. Tajima and J.M. Dawson, Phys. Rev. Lett. 43, 267 (1979).

8. Laser Acceleration of Particles (Los Alamos, 1982), AIP Conf. Proc. 91, P. Channel, Ed. (New York, 1982).

9. P. Chen, R.W. Huff and J.M. Dawson, UCLA PPG-802 (1984); P. Chen, J.M. Dawson, R.W. Huff and T. Katsouleas, Phys. Rev. Lett. 54, 693 (1985).
10. L.D. Landau and E.M. Lifshitz, Electrodynamics of Continous Media, Section 15, (Pergamon Press, Oxford, 1960).
11. This can be shown from a 1-D plasma model in which the electrons are treated as a series of parallel negatively charged plates in a uniform background of positive charge density $n_0 e$. In displacing an electron (or plate) by an amount x, it passes over an amount of positive charge $n_0 ex$ giving an electric field $E = 4\pi n_0 ex$ at the plate. But from Poisson's equation, $\nabla \cdot E = 4\pi e n_1 \sim k \cdot E$, from which $kx \approx n_1/n_0$. From the considerations in the introduction, $n_1/n_0 \approx eE/m\omega_p c$ for a plasma wave at c, so $kx \approx eE/m\omega_p c$.
12. M. Rosenbluth and C.S. Liu, Phys. Rev. Lett. 29, 701 (1972); to obtain the frequency shift $3/16 \; \varepsilon^2 \omega_p$ from Eq. (7), one must note that if $\varepsilon \sim |\varepsilon|\cos\phi$, then $\dot{\varepsilon}^2 \varepsilon \sim \sin^2\phi\cos\phi = 1/4\cos\phi$ + higher harmonics of ϕ.
13. C.M. Tang, P. Sprangle and R. Sudan, Appl. Phys. Lett. 45 (1984).
14. W. Horton and T. Tajima, Phys. Rev. A 31, 3937 (1985).
15. IEEE Trans. Plasma Sci., Special Issue on Plasma-Based High Energy Accelerators, PS-15 (2), 1987.
16. W.B. Mori, C. Joshi and J.M. Dawson, IEEE Trans. Nucl. Sci. NS-30, 3244 (1983).
17. C. Joshi, W.B. Mori, T. Katsouleas, J.M. Dawson, J.M. Kindel and D.W. Forslund, Nature 311, 525 (1984).
18. D.W. Forslund, J.M. Kindel, W. Mori, C. Joshi and J.M. Dawson, Phys. Rev. Lett. 54, 558 (1985).
19. W.B. Mori, et al., Private communication.
20. C. Joshi, C.E. Clayton and F.F. Chen, Phys. Rev. Lett. 48, 874 (1982).
21. C.E. Max, J. Arons and A.B. Langdon, Phys. Rev. Lett. 33, 209 (1974).
22. G. Schmidt and W. Horton in Ref. 2; P. Sprangle, C.M. Tang and E. Esarey in Ref. 15; G.Z. Sun, E. Ott, Y.C. Lee and P. Guzdar, Phys. Fluids 30, 526 (1987); W.B. Mori, C. Joshi, J.M. Dawson, D. Forslund and J. Kindel, submitted to Phys. Rev. Lett. (1987).
23. C. Clayton, C. Joshi, C. Darrow and D. Umstadter, Phys. Rev. Lett. 55, 1652 (1985); C. Joshi, et al. in Ref. 2; C.B. Darrow, et al., Phys. Rev. Lett. 56, 2629 (1986); C. B. Darrow, et al. in Ref. 15.
24. T. Katsouleas, Ph.D. Thesis, Appendix C, UCLA PPG-769 (1984).
25. The derivation of maximum energy gain given in Ref. 7 does not make the simplifying assumption that the particle velocity is roughly c. Also, the factor 2 in Eq. (22) is obtained by assuming that the particle gains energy by the amount of the wave's potential amplitude $e\phi'$ in the wave frame. Since the particle can actually gain $2e\phi'$ in going from top to bottom of the potential

well, the result (22) should actually be $4\varepsilon(\omega^2/\omega_p^2)mc^2$. The latter expression is closer to the maximum particle energy observed in simulations. See also the discussion of energy gain in the review by C.J. McKinstrie, submitted to **Applications of Laser Plasmas**, L.J. Radziemski and D.A. Cremers, Eds. (Dekker, NY, 1987).

26. J. Lawson, J. Allen, R. Bingham, J. Butteworth, F. Close, R. Evans, G. Rees and R. Ruth, Rutherford Appleton Laboratory Report RL 83057 (1983); R. Fedele, U. de Angelis and T. Katsouleas, Phys. Rev. A 33, 4412 (1986).
27. T. Katsouleas, S. Wilks, P. Chen, J.M. Dawson and J.J. Su, Particle Accelerators 22, 81 (1987); S. Wilks, et al. in Ref. 15.
28. S. Van der Meer, CLIC Note No. 3, CERN/PS/85-65(AA)(1985); R.D. Ruth and P. Chen, Proc. 13th SLAC Summer Institute on Particle Physics, July 1985.
29. P. Chen, J.M. Dawson, R.W. Huff and T. Katsouleas, Phys. Rev. Lett. 54, 693 (1985); P. Chen and J.M. Dawson in Ref. 2; P. Chen, J.J. Su, J.M. Dawson, P.B. Wilson and K.L. Bane, Phys. Rev. Lett. 56, 1252 (1986).
30. R.D. Ruth, A.W. Chao, P.L. Morton and P.B. Wilson, Particle Accelerators 17, 171 (1985).
31. T. Katsouleas, Phys. Rev. A 33, 2056 (1986).
32. R. Keinigs and M.E. Jones, Phys. Fluids 30, 252 (1987).
33. J. Rosenzweig in Ref. 15.
34. J.J. Su, T. Katsouleas, J.M. Dawson, P. Chen, M. Jones and R. Keinigs in Ref. 15.
35. R. Keinigs, M.E. Jones and J.J. Su in Ref. 15.
36. P. Chen, Particle Accelerators 20, 171 (1986); P. Chen, J.J. Su, T. Katsouleas, S. Wilks and J.M. Dawson in Ref. 15.
37. J. Rosenzweig, B. Cole, D.J. Larson and D.B. Cline, WISC-EX-85-0000 (July 1987), submitted to Particle Accelerators.
38. The use of a precursor can also facilitate strong focusing of dilute beams which are shorter than c/ω_p; in this case all the focusing is provided by the precursor's wakefield and none by the beam itself. Another plasma lens scheme employs a z-pinch plasma current to focus dilute beams (see B. Autin, et al. in Ref. 15). Yet another plasma-type lens scheme using a pure ion column is described by S. Wilks, et al. in UCLA PPG-1064 (1987).
39. R.G. Evans, Proc. Orsay Workshop on New Developments in Particle Acceleration Techniques, Orsay, France, June 29-July 4, 1987.

THE TEVATRON

INTRODUCTION

D. A. Edwards
Fermi National Accelerator Laboratory
P. O. Box 500, Batavia, IL 60510

TABLE OF CONTENTS

1 Evolution of the Tevatron Design............................ 1831
   1.1 Context and Initial Decisions......................... 1831
   1.2 Development of the Fixed-Target Design................ 1832
   1.3 Transformation to a Collider.......................... 1833
2 Components of the Tevatron................................. 1834
   2.1 Superconducting Magnets............................... 1835
   2.2 Cryogenics 1837
   2.3 Power System and Quench Protection.................... 1838
   2.4 Vacuum System... 1839
   2.5 Conventional Systems.................................. 1840
3 Performance and Outlook.................................... 1841
   3.1 The Fixed-Target Program.............................. 1841
   3.2 Colliding Beams....................................... 1842
   3.3 Outlook... 1843

INTRODUCTION

D. A. Edwards
Fermi National Accelerator Laboratory
P. O. Box 500, Batavia, IL 60510

1 EVOLUTION OF THE TEVATRON DESIGN

1.1 Context and Initial Decisions

When the Fermi National Accelerator Laboratory and its accelerators were designed in the late 1960's, it was too early to give serious consideration to the use of superconducting magnets in the main proton synchrotron, for the necessary technology was in its infancy. However, space was reserved in the Main Ring enclosure and in the service buildings for the eventual addition of a superconducting synchrotron. Because the "supermagnets" might be expected to reach twice the field of their room-temperature counterparts, the superconducting ring was called the Energy Doubler at that time.

After high energy physics experiments were underway in 1972, it became possible to devote some attention to the Doubler idea. A major factor in the design context was already settled: the superconducting synchrotron would occupy the same tunnel as the Main Ring. A number of other design principles were established at the very outset, a few of which deserve mention here. The decision to use a cold beam tube was controversial at the time; today it is the natural choice. A warm iron magnet design was chosen, but the debate over the relative virtues of cold or warm iron still continues and successful magnets have been constructed with both approaches. The decision to use NbTi was a recognition of the state of materials technology.

The superconducting magnet designs of that time were quite unsuited to the needs of a large synchrotron. Nor was there an established production base for filamentary NbTi in the volume or quality required. A substantial research and development program was necessary, focused almost exclusively on the magnets and their cryogenic system. The evolution and verification of a successful magnet design for a mass production environment led to the constuction of over 200 full scale prototypes.

In 1978, after six years of magnet development, it was possible to expand the Doubler effort to include the design of the entire accelerator-collider synchrotron. As a result of this research and development program, a design report was published in 1979 that served as the basis for the superconducting accelerator construction project. A parallel effort on an antiproton source followed close behind, with the intent that colliding-beam physics commence soon after the beginning of fixed-target physics in the new energy region.

© 1989 American Institute of Physics

The "Tevatron" projects included the construction of three new synchrotrons, but common usage has attached that name to the superconducting ring itself, and we will follow that convention here.

1.2 Development of the Fixed-Target Design

Among questions of overall accelerator design, one concern dominated all others. Could beam extraction be performed so efficiently that beam particles striking the magnets would not cause a superconducting-to-normal transition?

For use in high energy physics experiments, the beam is removed from the synchrotron by a process called resonant extraction. After the beam of over 10^{13} protons has been accelerated, the particles are gradually spilled from the ring at a rate of about 10^7 per turn. The delicate control needed to coax such a small fraction of the protons out on each orbit is provided by nonlinear magnetic fields that excite a resonance in their motion.

Of course, other operational situations involve beam loss, and so the possibility of quenches. But in most cases, the risks attendant on beam loss can be eliminated or reduced by component or systems design. With resonant extraction the situation is different. In any version of the process devised to date, the particles leaving the circulating beam encounter the entrance to an extraction channel, and the particles that will depart are separated from those that will remain within the ring by a material boundary. Inevitably, some particles strike this boundary, or "septum." Protons will scatter or interact with the nuclei in the septum, and the secondary particles from these processes can deposit their energy in superconducting magnets downstream. Experimental studies confirmed the anticipated sensitivity of accelerator-style superconducting magnets to beam loss.

The fact that the Tevatron would occupy the same enclosure as the Main Ring implied that the magnet disposition of the Tevatron would be a near-replica of the resident synchrotron. An exact reproduction of the Main Ring optics would lead to energy deposition exceeding allowances by a factor of fifty to a hundred. Fortunately, it was not necessary that the superconducting ring duplicate its predecessor exactly bend-for-bend and lens-for-lens. By introducing conventional bending magnets in the neighborhood of the extraction septa, local orbit modifications were made that reduced losses from inelastic interactions by an order of magnitude. Similarly, a modification in the focusing order of the quadrupole lenses spread the beam out as it approached the extraction septa and so improved the extraction efficiency. Collimation was introduced within the arcs to intercept particles that had undergone elastic scattering in the primary septum.

Finally, the lattice was modified to permit the introduction of an efficient beam abort system. If beam losses in the magnets during extraction should exceed an operationally determined level, the remainder of the beam would be kicked out of the ring into an external beam dump. Again, use was made of conventional magnets to the degree possible in the critical region near the beam exit channel. A longitudinal gap was left in the beam so that no particles would be deflected during the abort kicker risetime.

Thus, the ring as constructed was a hybrid: mainly composed of superconducting magnets, but with a vital admixture of traditional steel and copper hardware.

1.3 Transformation to a Collider

In the mid-1970's, the growing appreciation of the potential of hadron colliders added an additional function to the requirements for the superconducting ring. It should perform as a storage ring as well. Though there was a period during which proton-proton collisions involving the Main Ring and the Tevatron was contemplated, the decision finally fell on the side of proton-antiproton collisions in the Tevatron.

Interestingly enough, the potentially vexing question of magnet aperture had been settled by the slow extraction requirement. The 7.5-cm inner diameter of the coil was fixed in 1975 following an analysis of the aperture and field quality needed for resonant extraction. It was judged that magnets of the resulting design would be adequate for single-particle stability during beam storage.

Rather, the main issue raised by the additional application was a good deal less subtle; namely, where yet another set of functions was to be put within the existing tunnel. A major colliding-beam experiment needs space surrounding the beam tube for a detector some 20 meters in length, and on either side there must be strong quadrupole lenses to reduce the beam area at the interaction point. In short, at least one of the six long straight sections permitted by the tunnel geometry would have to be reserved exclusively for colliding-beam physics. A plan was devised which compressed the accelerator and fixed-target physics systems into the other five regions in both the Main Ring and the Tevatron.

Use of a second straight section for collisions requires the removal of some function used in the fixed-target physics program. Fortunately, this can be done relatively easily, for the primary extraction septum and the nearby conventional magnets that protect the superconducting magnets from beam loss occupy an entire straight section. Thus, that straight section can be cleared by moving a limited number of components. To be sure, the installation of the focusing optics and the detector is a task of somewhat greater magnitude, but still of reasonable scale.

Production of the principal missing ingredient - the $\bar{p}$'s - was the mission of the antiproton source. The design called for targetting of 120-GeV protons from the Main Ring, followed by capture of the resulting $\bar{p}$'s at 8.9 GeV/c in a debuncher synchrotron. Radiofrequency system gymnastics in the Main Ring deliver short duration but broad momentum-spread bunches to the target. Since the momentum distribution of the $\bar{p}$'s is inherently wide, this procedure provides the optimum use of the longitudinal acceptance of the system. After momentum-spread reduction by debunching, the $\bar{p}$'s transfer to an accumulator synchrotron, where stochastic cooling and stacking take place along the lines of the approach pioneered by CERN.

The layout of the Tevatron facility is shown schematically in Fig. 1, and a short summary of the principle design parameters is given in Table I. The ratio between peak beam energy and injection energy may look surprisingly low, but this is simply a consequence of the Tevatron sharing the tunnel with the Main Ring and so having a high energy injector available.

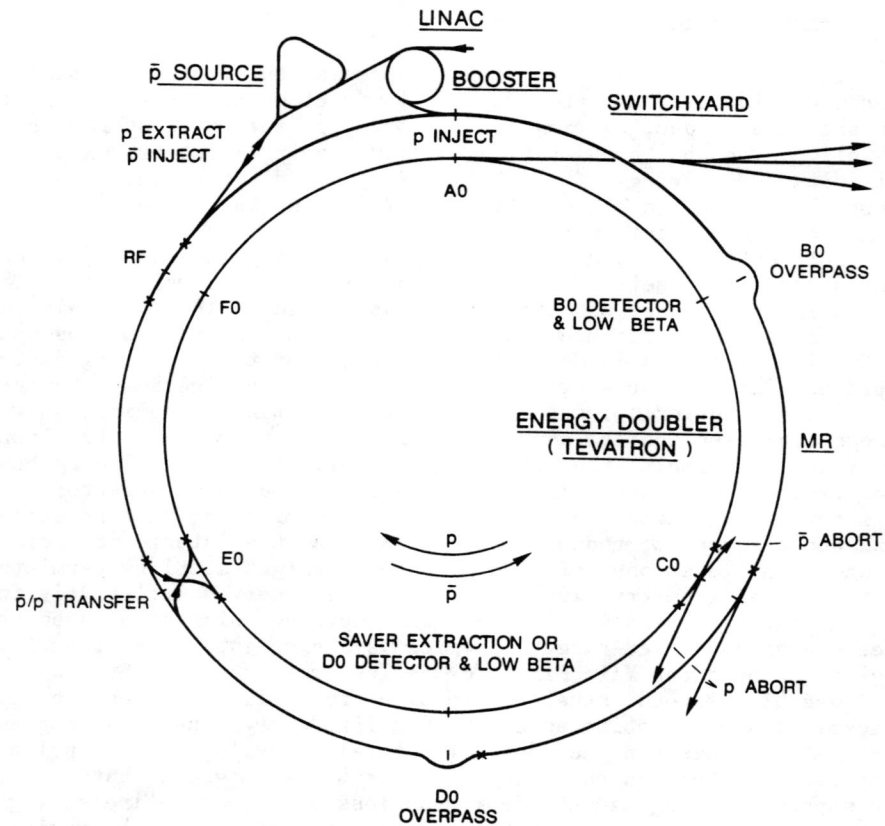

Figure 1 Fermilab accelerator complex including the Antiproton Source rings and the Tevatron. The straight sections are labeled A0 through F0. A0 and D0 are needed for extraction, with D0 as a colliding beam area as well. B0 is the other dedicated colliding beam area. Injection occurs at E0; rf acceleration at F0. Beam abort is at C0.

2 COMPONENTS OF THE TEVATRON

The main special design features of a superconducting accelerator are, of course, the magnets themselves and the cryogenic system to cool them. But other systems are required to assume new roles in the superconducting ring. For example, the magnet power

Table I Tevatron Design Parameters

General	
Accelerator radius	1 km
Peak beam energy	800–1000 GeV
Injection energy	150 GeV
Bend magnetic field at 1000 GeV	4.4 Tesla at 4400 Amperes
Beam emittance ϵ_N (95%)	24π mm mrad
Fixed target	
Intensity	$\sim 2\times 10^{13}$ protons/cycle
Acceleration rate	50 GeV sec^{-1}
Cycle time	60 s
Slow spill duration	20 s
Fast spill	5 pulses at 2×10^{12} protons
Collider	
Intensity per bunch	6×10^{10} expected
Number of bunches	3p, 3p̄
Luminosity	10^{30} cm^{-2} sec^{-1}
Storage time between fills	~ 4 hr
Amplitude function β^*	1 meter (x,y)

system must take on the primary burden of protecting the magnets if they quench. The vacuum system must provide thermal insulation for the cryostats in addition to establishing the empty space in which the beam circulates. Finally, a higher standard of reliability is demanded of the conventional accelerator systems, in order to compensate for the inevitable maintenance demands of the components at the frontier of technology.

2.1 Superconducting Magnets

The potential benefits of superconductivity for synchrotron magnets are high field in a compact package with low power consumption. As the Tevatron design effort began, high field meant 4.4 T, a figure twice that of the Main Ring and one that should be achievable with the materials of the day.

The superconducting magnet complement of the Tevatron includes 772 bending magnets (dipoles), 224 quadrupoles, and 720 small correction and adjustment elements. The main bending magnets occupy 75% of the perimeter of the accelerator - they are the dominant magnetic element, and the discussion here will be limited to them.

The standard dipole is 6.4 m long and 38 cm by 25 cm in cross section. The only exceptions are two half-length dipoles installed in the vicinity of the beam abort. The evacuated beam pipe runs the length of the magnet through its center, and the vertically oriented magnetic field deflects the protons and antiprotons by an angle of 8.1 mrad.

A transverse section of the magnet is shown in Fig. 2. Just outside of the square beam tube, there is space for the liquid helium which cools the inner edge of the coil. Each of the many small rectangles in the coil represents the cross section of the cable. The coil is clamped by stainless steel collars in a highly reproducible, accurate configuration that does not distort during magnet excitation. There are additional spaces for liquid helium flow between the outer surfaces of the collared coil assembly and the enclosing tube. The next annular region contains two-phase (liquid and gas) helium flowing in the direction opposite to that of the single-phase fluid. The two-phase helium is at the lower temperature and so extracts heat from the liquid, which in turn extracts heat from the coil. Outside of the helium container is an insulating vacuum space and then two concentric pipes. The narrow space between these pipes contains liquid nitrogen, which intercepts heat flow inward from room temperature. The insulating vacuum region between the nitrogen shield and the room temperature outer cryostat tube contains superinsulation (aluminized Mylar) as an additional radiation shield. The whole magnet-cryostat assembly is vacuum tight. It is held in a laminated iron yoke that contributes some 18% to the total magnetic field. The cryostat is precisely adjusted relative to center with suspension blocks of epoxy-fiberglass laminate and preloaded suspension cartidges that allow for contraction and expansion during the thermal cycle.

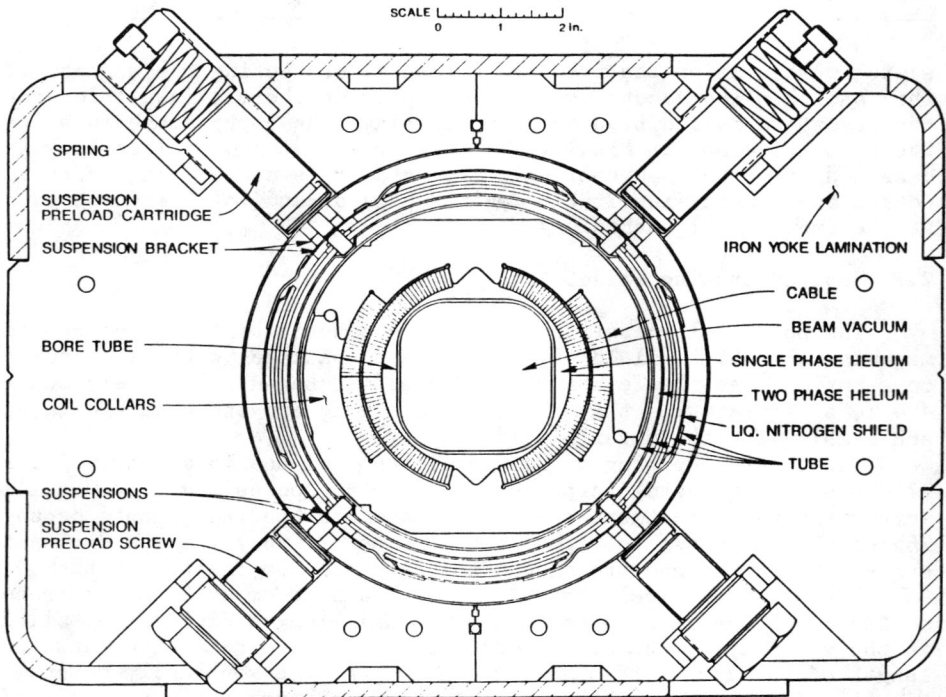

Figure 2 Cross section of the Tevatron dipole magnet showing the collared coil assembly, the cryostat, and the warm iron yoke.

Measurements on superconducting cable for the Tevatron magnets gave an average critical current density in the NbTi of 1800 A/mm^2 at 5 T and 4.2°K. Taking into account the magnet geometry, field, and operating temperature of 4.6°K, one finds that this critical current density should permit the magnets to achieve 4.6 T. All magnets prior to their installation in the ring were measured under two different excitation conditions. In the first test, magnets were ramped at 200 A/s until a quench occurred. In the second test, repetitive ramps approximating the accelerator cycle were used and the peak current was gradually increased until the quench limit was found. The results indicate that excitation of the ring in the 900-950 GeV range should be possible with only a few magnet replacements, whereas to reach 1 TeV will probably require either very slow ramps or cryogenic modifications.

2.2 Cryogenics

The helium refrigeration system is the world's largest. It consists of a large helium liquefier, a nitrogen reliquefier, a distribution system for the cryogens around the ring, and 24 satellite refrigerators spaced around the ring. These components can provide a total of 24 kW of cooling at 4.7°K for the magnets as well as liquid helium for power lead cooling and liquid nitrogen for the cryostat heat shields.

The satellite refrigerators are located directly above the tunnel and feed helium and nitrogen directly into the magnet string below. The flow is split and goes upstream and downstream (with respect to the proton beam) through typically 16 dipoles and associated components in each direction. Throughout this outward flow, the helium is in a single subcooled liquid phase; it is this helium that is in direct contact with the magnet coil. At the end of the string, the helium passes through an expansion valve which lowers its temperature and pressure. These new conditions are adjusted so that the helium is a boiling liquid at a temperature of about 4.5°K, one or two tenths of a degree lower than the temperature of the single-phase fluid. The two-phase fluid is directed back through the string of magnets and absorbs heat from the outgoing stream. Thus each magnet is a counterflow heat exchanger. The ratio of gas to liquid in the two-phase path increases as the distance to the refrigerator feed point decreases, but the temperature remains nearly constant. Energy generated in the coil is thereby removed efficiently by the single-phase liquid and absorbed as heat of vaporization in the two-phase region. Nitrogen for the magnet shields makes a single pass to the ends of each string, where it is discharged into a nitrogen header as 92°K gas.

The central plant plus satellites arrangement offers a wide variety of operating conditions, and provides the redundancy necessary to the continuous operation of the synchrotron. In "satellite mode," the central plant supplies large amounts of cold helium to the magnet strings and thus to the return side of the satellite heat exchangers. The excess flow in the satellite heat exchangers results in 1 kW of refrigeration from the satellites without use of their gas expansion engines. At the other end of the spectrum is the "stand-alone mode." Here, without the availability of helium from the central plant, each satellite is able to deliver 450-500 W of refrigeration plus 25 liters per hour of liquid helium. This capability is adequate to compensate the static heat load of the magnets. A variety of intermediate cases are possible depending on the availability of helium from the central plant.

2.3 Power System and Quench Protection

The magnet power system plays the dual roles of powering the main bend and quadrupole magnet string and protecting these same magnets from the stored energy in the magnetic field should any fraction of the superconductor in the whole ring become normal (resistive) for any reason. Because this system requires quick detection of quenches and consequent action of electrical components in order to save the magnets from self-destruction, it is called "active" as opposed to one that might require little or no external action, that is, a "passive" system.

The main bend and quadrupole magnets form a single series circuit, in which 12 power supplies are uniformly distributed. Each power supply is capable of ramping to 4500 A at 1 kV. Since the resistance of the circuit is small (but not zero, for there are conventional magnets in the circuit), a single well-regulated supply is able to supply during particle injection, flattop, or storage conditions.

Our main interest here lies in the quench protection aspect of the system. Consider what happens in the cable of a magnet coil when a "normal zone" appears and current transfers from the NbTi to the copper in which the superconductor filaments are embedded. The copper, which now conducts most of the current, has too high a resistivity to prevent further heating, and the cable will melt unless some means is found to remove the current expeditiously. The rate at which the cable temperature rises is difficult to calculate because of the nonlinear behavior of the parameters (specific heat, resistivity, thermal conductivity, etc) that describe the cable consituents at low temperature. The system parameters were set as a result of measurements made on the rate of temperature rise in quenched cable. At the maximum operating current, there is less than one-half second available for removing the magnet current to prevent permanent damage.

The magnets and their interconnections are continuously monitored for a resistive voltage component. Once the onset of a quench is detected, the power supplies are turned off, and the current is shunted through dump resistors at the supply locations. The resulting exponential current decay has a 12- second time constant that is too slow to protect the normal zone in the magnet that has quenched, so locally the current must be reduced much more quickly. To do this, the circuit is divided into 24 quench protection units. A "safety lead" connects the superconducting bus to a room temperature bypass circuit at the ends of each unit. This

lead cannot carry steady state operating currents. If it were
designed to do so, the refrigeration load presented by the leads
would be unreasonably high. But the leads can convey the decaying
magnet current around the quench protection unit that contains the
quenching magnet. Current is switched into the safety leads by the
closing of thyrister switches.

The fate of the cable depends on the outcome of a race between
the cable temperature and the decay of the current in the quench
protection unit; the latter depends on the total resistance of the
normal zone. To insure that the race ends in favor of the cable,
heaters are energized in the dipoles of the protection unit to
quench a large quantity of superconductor. The resulting rapid
resistance growth drives the current down with a time constant
appropriate to an in-bounds temperature rise.

2.4 Vacuum System

The vacuum system consists of three separate subsystems with
different characteristics and requirements. The cryostat insulating
vacuum system is the most complex and is completely isolated from
the high-vacuum cold beam tube system inside the magnets. The
straight sections and other noncryogenic regions have warm beam
tube, bakeable, conventional vacuum systems.

All in all there are about 1300 cryogenic interfaces between
magnets or between magnets and other components. A magnet-to-magnet
interface includes a beam tube seal, two liquid helium connections,
one liquid nitrogen connection, and a large external room
temperature insulating vacuum seal. Each of the cryogenic seals
must be able to be verified at room temperature with sufficient
sensitivity to assure that it will not leak liquid helium. By far
the most time-consuming aspect of installation is the interface
connection and leak checking. During initial installation each
interface took on the average one man-week; subsequent work has been
done in about half this time.

The static heat leak of the cryostat-magnet system is due to
thermal radiation and heat conduction through magnet supports and
other structural elements as well as through residual gas in the
insulating space of the cryostat. For the Tevatron geometry, the
static heat load doubles at a pressure of 2×10^{-5} torr (He). An
upper limit of 10^{-5} torr is set for operation, which corresponds to
a reading of 3×10^{-6} on nitrogen-calibrated cold cathode gauges. In
operation, readings are typically at the 10^{-7} level. The insulating
vacuum is pumped with turbo molecular pumps.

The pressure in the cold beam tube is very low if helium leaks
are absent. Pressures of 5×10^{-11} torr cold (5×10^{-10} torr as
measured warm) are normal. The cold beam tube provides an
economical way of obtaining the high vacuum required for beam
storage over the major fraction of the ring circumference. The main
concern in adopting the cold bore approach was the potential of
helium leakage to the beam tube; such leaks are extremely difficult
to detect or pump without warming the system up. Because of this
worry, the cryostat was designed so that the beam tube seam weld is
the only weld between the helium spaces and the bore tube vacuum.

The vacuum systems, with their many flanges, seals, pumps, valves, and gauges, have been remarkably trouble free and reliable. This must, to a large extent, be due to the cryo-pumping ability of the refrigerated surfaces. The fact that most of the circumference of the ring (93%) is cold means that the pressure in the warm regions need not be particularly low. With 5×10^{-11} torr where cold and 10^{-8} in the warm regions, the reduction in luminosity during storage due to interactions in the residual gas is expected to be 23% after 20 hours.

2.5 Conventional Systems

Under this heading, we comment briefly on other systems of the Tevatron, which, though impacted by the superconducting design, are basically required in any accelerator. Correction magnets, beam diagnostics, and the radiofrequency acceleration system are examples.

Correction magnets are used to compensate for field imperfections or alignment errors of the main magnets, and to tune the optics of the ring to desired operating conditions. With a few exceptions, these are superconducting magnets configured in circuits suitable for specific functions. For instance, there are two circuits each containing 90 trim quadrupoles that are used to make fine adjustments in the overall focusing characteristics of the synchrotron. The correction magnet power supplies can be programmed to produce virtually any waveform throughout the accelerator cycle. The strengths of the superconducting elements are sufficient for use at full excitation, a design feature that has proved particularly valuable for steering corrections.

Under beam diagnostics, we will mention only the position and loss monitoring systems; they are essential to the operation of the synchrotron. The fact that recovery from a quench can take an hour or more makes it imperative that the reasons for erratic beam behavior be sorted out with as few beam pulses as possible. The position monitoring system can measure the deviation of the beam in-or-out or up-or-down from center at 200 locations around the ring. The system has a wide dynamic range to permit start-up of operations with a beam so low in intensity that a quench is unlikely. A large amount of information is stored in the system memory, so that an operator can ask for recall of position at every location of the injected beam, position at specific locations for 1000 turns, orbit deviation from center at arbitrary times throughout the accelerator cycle, and detailed position profiles prior to an abort. A similar number of loss monitors are distributed around the ring. These are radiation detectors placed outside of but close to the magnets. The electronics for each detector is designed so that the output signal is related to the probability of quenching the magnets. Outputs of the loss monitors are continuously checked and used to abort the beam automatically (within 200 microseconds) if the signal is larger than tolerances derived from experience.

The radiofrequency accelerating system consists of eight 53-MHz resonant cavities, each of which can produce 1/3 MV. The frequency must be changed by only 2 kHz from 150 to 1000 to compensate for the change in the protons' speed, since the proton is already moving at 99.998% of the speed of light at injection. Because the required modulation is small, the frequency program is a completely dead-reckoned digitally generated function derived from the main magnet excitation program. All eight cavities are used for acceleration of protons during fixed-target operation. The cavities have been positioned relative to one another so that, by appropriate phasing of their radiofrequency excitation, they will function as one set of four for acceleration of protons and the second set of four for independent acceleration of antiprotons. The colliding point of the two beams can be moved circumferentially around the accelerator and frozen at a particular point by frequency and phase adjustment of the two sets.

3 PERFORMANCE AND OUTLOOK

3.1 The Fixed-Target Program

The Tevatron was commissioned in the Summer of 1983. On July 3, protons were accelerated to 512 GeV and the Tevatron became the highest energy accelerator in the world. Operation for fixed-target physics began in October of that year at an energy of 400 GeV; the following February the energy was raised to 800 GeV and the goal of doubling the energy of the Main Ring had been reached.

To date, there have been four fixed-target runs. Generally speaking, the beam dynamics behavior of the Tevatron has been excellent. Of course, a small beam size at injection and an extensive correction magnet system are a big help. But, more to the point, some of the "ghosts" that had been attributed to the new magnet technology failed to materialize. For instance, there had been concern that the coils of magnets would gradually move as many ramps accumulated or suddenly move as a result of quenches. Such has not been the case; operating conditions are more stable and reproducible from day to day than the older synchrotrons in the accelerator chain.

In the fixed-target mode, the peak energy has been limited to 800 GeV, primarily due to energy deposition in the superconducting magnets during the resonant extraction process. Typically, about 1.5×10^{13} protons are accelerated in each one-minute cycle, and extracted throughout a twenty-second interval at peak energy. Intensities as high as 1.8×10^{13} have been reached. Though the intensity that can be achieved varies from day to day, it usually represents a balance between total protons accelerated versus a tolerable level of beam aborts or quenches. During a week, 1000 aborts and 20 quenches would not be unusual figures.

Cyclic operation provides a severe test of the new superconducting magnets. In the 20 months of fixed-target physics, the ring has been put through about three-quarters of a million magnetic cycles, most of them to 800 GeV. During the first run, a production error was uncovered that required repair of half of the magnets in the Summer of 1984. The coil leads at one end of these magnets had not been tied to prevent motion due to their mutual magnetic repulsion; as a result, strands of the superconducting cable began to break. Now, three years later, the present run has been plagued by a succession of magnet failures that are likely due in part to more subtle motions of the magnet leads. Another round of repairs is in the offing.

Reliability is the big issue in the accelerator facility. There are now four accelerators in the chain for fixed-target physics (for colliding-beam physics, there are six). Actual uptime for high energy physics is about 70% of scheduled time, as opposed to 80% for Main Ring operation prior to the Tevatron era. It is obvious that the long-term success of the Tevatron rests on the attainment of high operational reliability.

3.2 Colliding Beams

At the end of the fixed-target run in 1985, four weeks of intensive effort concluded with the observation of a dozen p̄p events in the detector facility, marking the beginning of colliding-beam physics at Fermilab. A long shutdown for a variety of construction projects followed, after which a four-month "engineering" run for collider operation took place in early 1987. During this latter period, a peak luminosity of 10^{29} cm^{-2} sec^{-1} and an integrated luminosity of 70 nb^{-1} were achieved.

The collider mode is much more congenial to the superconducting ring. The number of particles in the accelerator is almost two orders of magnitude less, there is no slow extraction, and the burden of frequent ramping is absent. The reduced beam loss in the magnets permitted an increase in the energy to 900 GeV, only about 30 GeV lower than the quench limit without beam.

The last major dynamical "ghost" associated with superconducting magnets vanished with the observation of single-beam lifetime in excess of 100 hours. Thus far, the luminosity lifetime is in the 5 to 10 hour range, and is determined by emittance growth of unknown origin. A major goal of present collider studies is the identification of the noise mechanism that produces the emittance growth of both particle species.

The peak luminosity in the 1987 run was 10% of the design value. Roughly speaking, beam loss in the various transfer and acceleration stages, and emittance dilution can be blamed in equal proportion. Approximately a factor of two dilution in emittance takes place in the transfer from the Main Ring to the Tevatron as a result of the vertical dispersion mismatch between the two rings and the relatively large momentum spread of the proton or antiproton bunch. That there is a vertical dispersion is a consequence of the undulations added to the Main Ring to move the two accelerators apart at the collision points. This source of dilution will be removed by a further modification to the Main Ring in early 1988.

During acceleration and turn-on of the collision optics in the Tevatron, the p̄'s suffer a further factor of two increase in emittance. The p̄ bunches are about an order of magnitude less intense than the proton bunches, so the beam-beam tune spread differs for the two species. It is likely that inadequate tune control throughout the many steps between the injection optics and the low-beta optics is the source of much of this dilution; if so, improvement will come with further study.

The non-cyclic character of collider operation transformed a hitherto innocuous characteristic of the superconducting magnets into a major irritant. The magnets exhibit a broad spectrum of eddy current time constants, from the second to many hour time scale. In fixed target operation, multipole moments arising from these currents could be lumped together with the persistent current multipoles. But during the lengthy setup for a transfer in collider mode, the variation in, for example, the chromaticity must be compensated. An adequate model of the magnet does not yet exist to account for these effects. In retrospect, it is again fortunate that it was possible to inject at relatively high energy; the advantage of high injection energy for persistent currents was recognized, but the additional advantage for eddy current phenomena was not.

3.3 Outlook

Over four years after its commissioning, the Tevatron is still very much a prototype accelerator, as much a research instrument in its own right as it is a high energy physics tool. The superconducting magnets are not as robust as the better examples of their conventional counterparts, and it is entirely possible that extensive repairs and improvements will be necessary in the near future to achieve adequate reliability of operation in the fixed-target mode.

For the collider run scheduled to begin in April of 1988, the goal is to achieve a peak luminosity of 3×10^{29} cm^{-2} sec^{-1}. Steps have been taken to reduce the beam loss at some of the points in the transfer and acceleration process, and as noted earlier, a major source of emittance dilution will be corrected. In reaching the peak luminosity in the 1987 run, β^* at the interaction point was reduced to 0.7 m in contrast to the design value of 1 m listed in Table I. For 1988, β^* will be further reduced to 0.5 m. Operation will be upgraded to 6 bunches of both particle species.

It is likely that the performance levels of Table I can be reached with time. But to surpass those levels to a significant degree will probably require some modification of the accelerator facility. Both the fixed-target and collider programs would benefit from a reduction in the proton beam emittance. The emittance growth in the first few milliseconds of the Booster synchrotron cycle is attributed to space charge; raising the energy of the Linac injector would ameliorate that situation. The Main Ring lifetime at injection is much less than that which would be predicted from gas scattering; simulations suggest that there is a dynamic aperture limitation. At 20 GeV however, beam lifetime is consistent with gas scattering, and a Post-Booster synchrotron to inject at this level would also eliminate the necessity to cross transition in the Main Ring.

An order of magnitude increase in peak luminosity beyond the present design may be possible. It implies multi-bunch operation with separated beams in the Tevatron. Many bunches are needed in order to achieve the luminosity, and separation is necessary except at the collision points in order to limit the beam-beam tune spread. The total number of $\bar{p}$'s required is almost an order of magnitude larger than the number that can be collected and stored in the present Accumulator, so another ring, that might be called the Depository, is implied. The Depository would accept $\bar{p}$'s from the Accumulator, or receive and recool diluted $\bar{p}$'s from the Tevatron.

Many variations of the foregoing scenario are obviously possible. But the main point is that some major modification of the existing Fermilab facility will be needed to keep the high energy physics program abreast of demands until the SSC becomes available in the middle of the next decade.

ANTIPROTON SOURCE

J. Peoples
Fermi National Accelerator Laboratory
P. O. Box 500, Batavia, IL 60510

TABLE OF CONTENTS

1	Introduction..	1846
2	Physics Program of the Tevatron Collider....................	1846
3	Plan of the Tevatron Collider...............................	1849
4	Antiproton Source Operation.................................	1850
5	The Debuncher...	1859
6	The Accumulator...	1864
7	1985 Commissioning Performance..............................	1874
8	References..	1875

Chapter 3

ANTIPROTON SOURCE

J. Peoples
Fermi National Accelerator Laboratory
P. O. Box 500, Batavia, IL 60510

1 INTRODUCTION

In the past 18 months the Antiproton Source and the Collider have evolved from designs under construction into real machines.[1] On October 11, 1985, a few days before the start of a year-long shutdown to build the D0 collision hall, the p̄ Source stored 10^{10} antiprotons. On October 12 and 13 these antiprotons were accelerated to 800 GeV in the Tevatron and brought into collision with 800-GeV protons. In late 1986 the p̄ Source and the Tevatron Collider were in their early stages of use, but their promise should be fulfilled in 1987.

The Tevatron I project is Fermilab's entry into the colliding beam sweepstakes. It is an extraordinary one, largely because it will be the highest energy collider, by nearly a factor of three, until the early 1990's. It has already reached 1.6 TeV and is expected to approach 2 TeV within the next two years. Collisions take place in the Tevatron, the superconducting proton synchrotron that the Accelerator Division Staff built and commissioned as the Energy Saver over the past five years. This ring was transformed into a collider as part of the Tevatron I project, although strictly speaking only a few additions were needed to make this transformation. Since the Tevatron Collider has been described in other chapters in this book, I will discuss only the performance specifications and their implications for physics.

2 PHYSICS PROGRAM OF THE TEVATRON COLLIDER

A center-of-mass energy of nearly 2 TeV and a luminosity of 10^{30} cm^{-2} sec^{-1} will make it possible to observe collisions of quarks and gluons, the constituents of the proton and antiproton, which have a center-of-mass energy >500 GeV. Most frequently encountered will be gluon-gluon, quark-quark, and quark-gluon collisions caused by the strong forces. When a pair of these particles scatters through a large center-of-mass angle, two or more jets of hadrons are produced. Each jet is the hadronization of one of the final-state quarks or gluons. The cross section for jet production in these collisions has a very simple form when the quarks behave as point-like particles. By measuring the two- and three-jet cross section it will be possible to establish whether quarks are composite at the distance scale of 3×10^{-17} cm. The quarks and antiquarks will in certain instances interact electromagnetically or weakly in an observable manner. The formation of the standard model W through the fusion of an up quark and an anti-down quark is an example of one such distinctive interaction.[2] Should there be a massive W boson with the same electroweak couplings to quarks and leptons as the standard model W, it should be possible to observe its production and subsequent decay into an electron (positron) and an antineutrino (neutrino) if its mass is <350 GeV/c^2.[3] The decay of the standard model W may itself be a source of undiscovered particles. If the top quark exists and if its mass is less than the mass of the W minus the mass of the b quark, then the W$^+$ can decay into a top quark and an anti-b quark. Since W production will be copious at 2 TeV in the center-of-mass, the prospects for discovering it at the Tevatron are good. If the top is more massive than the W then the production process is through gluon fusion. The cross section for this process is large enough at the Tevatron to permit a discovery, while at the Sp$\bar{p}$S it is not feasible. The greater rate for top quark production at the Tevatron should improve the prospects for its discovery, if the top has not already been discovered at the Sp$\bar{p}$S. These are just three examples of the ability of the Tevatron to test the standard model.

The 1-TeV beams in the Tevatron can be thought of as variable momentum beams of the fundamental constituents: quarks, antiquarks, and gluons. While the probability that a single constituent will have all of the momentum of the proton or antiproton is vanishingly small, the average momentum of each constituent is about 160 GeV/c, roughly one-sixth of the beam energy. There is a small but reasonable probability for a single constituent to have a momentum >250 GeV/c. These are very, very large momenta. I emphasize the size of the momenta that the constituents can have with reasonable probability because this provides a way to compare a hadron collider with an electron-positron collider. LEP, the highest energy electron-positron collider, will be able to reach 85 GeV/c per beam only after a great deal of superconducting rf is added. Collisions of elementary constituents with momenta of this magnitude have already been observed at the Sp$\bar{p}$S. The Tevatron with its higher beam energy will be able to reach the equivalent of 250 GeV per constituent. By increasing the luminosity of the Tevatron to 10^{31} cm^{-2} sec^{-1} it will be possible to extend the effective center-of-mass energy at which collisions can be observed by another factor of two.

The Tevatron luminosity should reach 10^{30} cm^{-2} sec^{-1} by 1989, two years after the first run, which will begin early in 1987. Once this goal has been achieved, it should be possible to increase the luminosity by a factor of ten over a period of another four years. It may even be possible to reach a luminosity several times 10^{31} cm^{-2} sec^{-1}.

The plan of the Tevatron and its straight sections is shown in Fig. 1. The experimental program will use four of the six collision points. At B0, a large experimental hall has been built for the CDF detector. The construction of a second large experimental hall for the D0 detector, begun in the fall of 1985, will be completed in early 1988. Nevertheless, it will be sufficiently advanced by mid-1986 to permit the resumption of accelerator operations and the start of construction of the D0 detector. These two detectors are large general purpose detectors capable of observing the very rare

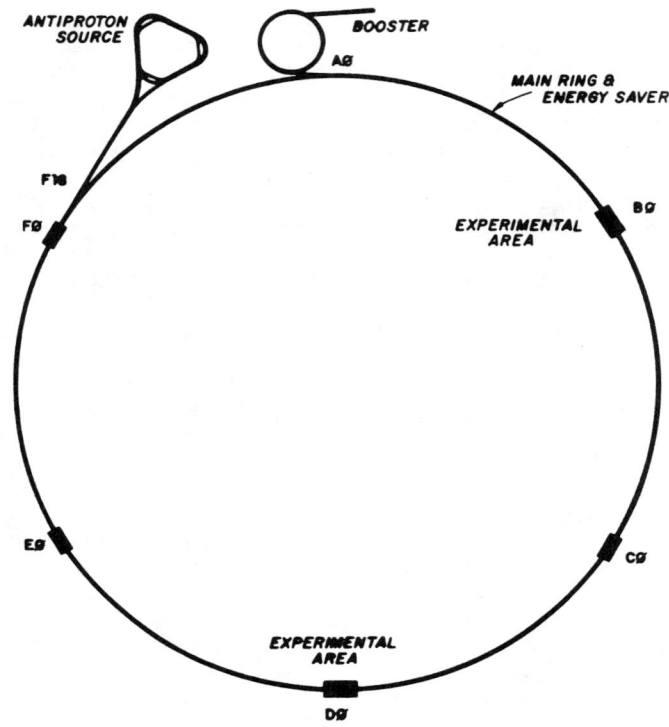

Figure 1   Tevatron I Fermilab Accelerator Complex.

but energetic collision products that characterize the decay of massive particles. A luminosity of 10^{30} cm^{-2} sec^{-1} is needed for these observations. The calorimetry of both detectors is considerably more sophisticated than that of the current (1984) UA1 and UA2 detectors, which, through the discovery of the W and Z, demonstrated the ability of a large acceptance, finely segmented calorimeter to resolve the significant details of hadron collisions from the background.[2,4] In addition to these large detectors several smaller experiments are planned. An elastic scattering experiment, E-710, will be located at E0. At C0, experiment E-735, a search for evidence of phase transitions in the quark-gluon plasma, is designed to observe changes in the particle multiplicity in events with large multiplicities as a function of transverse momentum. Initially, while the D0 detector is being built, the D0 straight section will be used for a monopole search, E-713. It is worth noting that when the luminosity at B0 is 10^{30} cm^{-2} sec^{-1}, the luminosity at C0, E0, and initially at D0 will be about 1.4×10^{28} cm^{-2}. This reflects the fact that C0 and E0 will not have low-β insertions and that the D0 low-β insertion will not be finished until 1989.

3 PLAN OF THE TEVATRON COLLIDER

The way the various parts of the accelerator complex work together to produce colliding beams can be seen from Fig. 1. The long straight section at A0 is used to inject 8-GeV protons into the Main Ring and it also contains the Tevatron extraction channel for the fixed-target program. The long straight section at B0 is used solely for CDF. Most of the remaining long straight sections are used for more than one purpose. C0 is used for the beam abort systems as well as for E-735. Until recently D0 contained just the slow extraction electrostatic septa. In the future, during colliding-beam operation, it will contain the D0 detector. One of the major challenges for the D0 experiment is to configure the extraction system and the detector so that the two can change places in less than one week, since both types of equipment cannot be present simultaneously. E0 contains the proton transfer line between the Main Ring and the Tevatron, the antiproton transfer line between the Main Ring and Tevatron, and Experiment E-710. Although the detectors for E-710 will be deep inside the Tevatron, some counters will be located in the E0 straight section. The Main Ring and Tevatron rf are located in F0.

Each machine has six medium straight sections. The one at F17 in the Main Ring deserves special mention since it contains the extraction channel which deflects 120-GeV protons out of the Main Ring toward the p̄ Source. The small triangular rings represent the Debuncher and Accumulator of the Antiproton Source.

Table I lists the Tevatron I Design parameters. The center-of-mass energy should ultimately reach 2 TeV. The peak luminosity of 10^{30} cm^{-2} sec^{-1} will be achieved when there are three bunches of protons and antiprotons and each proton and antiproton bunch has 6×10^{10} particles. Squeezing 6×10^{10} antiprotons into a phase-space volume corresponding to an invariant emittance in both planes of $<24\pi$ mm-mrad in a few hours is the major technological trick which the Antiproton Source must turn.

Table I Tevatron I Design Parameters

Center-of-mass energy, $\sqrt{s}$	2 TeV
Peak luminosity	10^{30} cm^{-2} sec^{-1}
Number of proton bunches	3
Number of antiproton bunches	3
Number of particles per bunch	6×10^{10}
Invariant emittance of both beams in both planes	24π mm-mrad
Proton longitudinal emittance per bunch	3.0 eV-sec
Antiproton longitudinal emittance per bunch	3.0 eV-sec
β^* at the interaction point (B0 and D0)	1 meter
Luminosity lifetime	12 hours

When the longitudinal emittances are as given in Table I, the rms bunch length is expected to be 40 cm. The insertion at B0 has been commissioned, and β^* was reduced to one meter during the 1985 run. The performance of the insertion has met all expectations.[5]

When the bunches have the properties given in Table I, calculations show that the initial rate at which the luminosity decreases will be dominated by intrabeam scattering.[6,7] When intrabeam scattering is the dominant cause of emittance growth, the growth rate will be inversely proportional to $(\epsilon_x \epsilon_y)^a$, where a is about 1/2.[8] As the emittance grows, intrabeam scattering decreases and the rate at which the luminosity decreases slows. Because the aperture of the Tevatron relative to the beam size is very large at 1 TeV, particles will not be lost until the emittance has increased many times over. The estimated decay rates of the luminosity, τ_L^{-1}, as calculated for the Tevaton I Design Report,[1] are given in Table II at the moment the Tevatron has been filled. The estimate of 12 hr for the luminosity lifetime contains an ad hoc assumption of 70 hr for the contribution due to beam-beam interactions. With somewhat comparable conditions the luminosity lifetime of the Sp$\bar{\text{p}}$S collider was measured to be 30 hr during the 1984 Sp$\bar{\text{p}}$S run, with the most significant contribution to the decay of the luminosity lifetime being due to intrabeam scattering.[9]

Table II Estimated Contributions to the Luminosity Decay Rates (sec^{-1})

Proton-antiproton scattering	5.0×10^{-6}
Beam gas scattering	3.4×10^{-6}
Intra-beam scattering	11.1×10^{-6}
Total decay rate	19.5×10^{-6}

In order to achieve a luminosity of 10^{30} cm^{-2} sec^{-1} with the bunch parameters of Table I, 2×10^{11} $\bar{\text{p}}$'s will need to be injected into the Tevatron every 12 hours. Since the $\bar{\text{p}}$ Source has been designed to accumulate 10^{11} per hour it should be possible to refill the Tevatron twice per day.

4 ANTIPROTON SOURCE OPERATION

The accelerator operations that make it possible to produce and accumulate the required number of antiprotons are as follows. Antiprotons are produced when 120-GeV protons are extracted from the Main Ring at F17 and directed onto a tungsten target in the $\bar{\text{p}}$ target station. Negatively charged particles which emerge from the target with a momentum of 8.89 GeV/c are collected and transported to the Debuncher. After circulating in the Debuncher for a millisecond, only antiprotons survive. The Debuncher reduces the fractional momentum spread of the antiprotons from 3.5% to 0.25% and reduces the transverse emittances of the beam by a factor of three by stochastic cooling. After 2 sec the antiprotons are transferred to the Accumulator, and another batch of antiprotons is injected into the Debuncher. Between Accumulator injection cycles the antiprotons are stochastically cooled and gradually a dense stack of antiprotons is built up. The layout of the extraction line, the target station, the Debuncher, and the Accumulator is shown in Fig. 2.

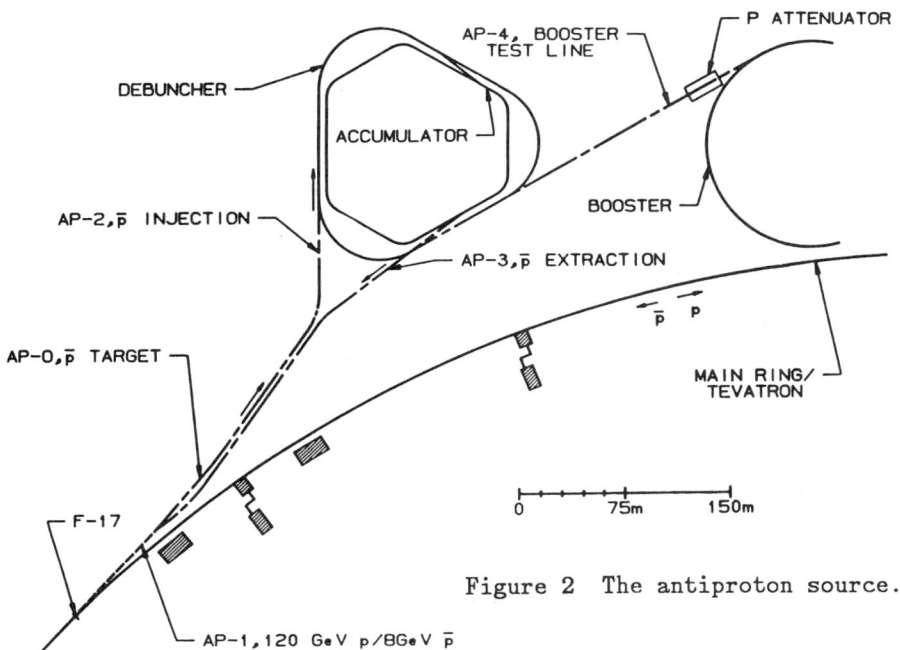

Figure 2 The antiproton source.

A more detailed discussion of this sequence, which follows, explains how the fractional momentum spread of the p̄ is reduced. Every 2 seconds one Booster batch of protons, consisting of 82 bunches spaced by 18.8 ns, is accelerated to a kinetic energy of 8 GeV in the Booster. The ensemble of bunches is accelerated to 120 GeV and transferred to the Main Ring. Just prior to extraction the bunch length of each bunch is reduced to <5 cm rms. This is done by the following steps: The rf voltage is quickly reduced from its nominal value of 4 MV to roughly 300 kV, causing the bunches to rotate in the mismatched buckets, as shown in Fig. 3. After one-quarter of a synchrotron oscillation period, the rf voltage is restored to 4 MV. One-quarter of a synchrotron oscillation period later, the distribution has rotated to its smallest projection on the phase (time) axis, as shown in Fig. 4. Since the time required to increase or reduce the rf voltage is very short compared to the synchrotron oscillation period, the changes in voltage are essentially instantaneous. Although the rms bunch length has been reduced to the desired value at the expense of increasing the fractional momentum spread to about 0.4%, it can be contained within the Main Ring and extracted at F17 efficiently. This rather small energy spread has no effect on antiproton production. The specific values of rf voltage and the bunch lengths achieved during the October 1985 test run are given in the article describing the Main Ring.[10]

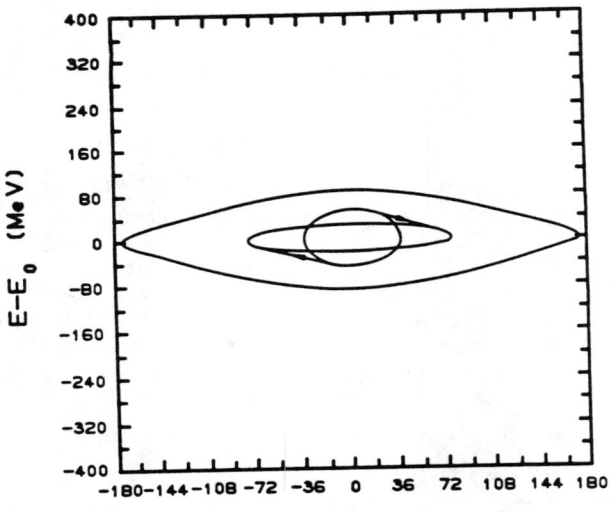

Figure 3 The rf voltage is quickly reduced to a very low value, typically 100 kV, for 2.9 msec, causing the bunches to rotate in the mismatched buckets.

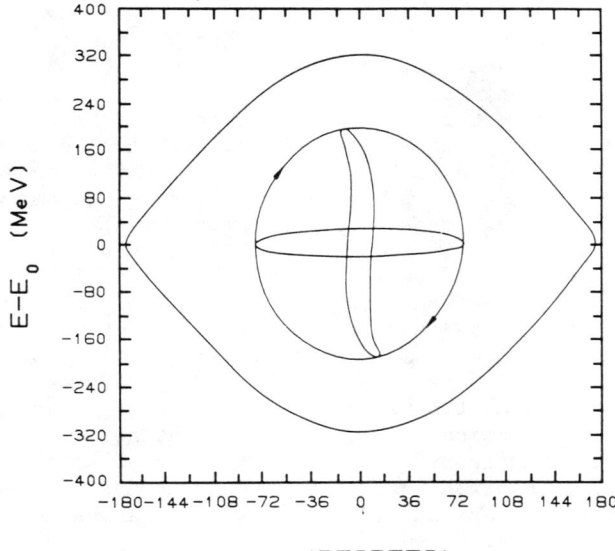

Figure 4 Rotation of Main Ring rf bunches to reduce the phase spread to a minimum prior to extraction.

Why do we carry out such a complicated operation when the protons are just going to hit a piece of tungsten after leaving the Main Ring? The density of antiprotons in the six-dimensional phase space is much too small at production to be used directly in a collider. Cooling is the means by which the density is increased to a useful value. Since cooling is time consuming, we want to make both the density of antiprotons in the six-dimensional phase space and the number of antiprotons as large as possible at the moment of production. Since the acceptance of the beam transport system and the Debuncher define a subspace in this phase space, the two requirements are essentially the same. The antiprotons have the same time structure as the protons, and they also have the same beam diameter as the protons. These quantities correspond to three of the coordinates of the six-dimensional phase space in which the antiprotons are created.

The larger the density of antiprotons in phase space, the smaller the demand that must be placed on the cooling system. It follows that the density will be as large as possible, if the bunch length and beam diameter are as small as possible. The other three coordinates in the phase space are the horizontal and vertical angles of each antiproton and the deviation of its energy from the mean energy. Since the number of antiprotons that can be collected is proportional to the product of the solid angle and the energy spread, nothing is gained by reducing the range of these coordinates. Since the objective is to collect as many antiprotons as possible, the spreads in these coordinates are made as large as technology will permit. Ultimately the number of antiprotons collected is limited by the apertures of the Debuncher stochastic cooling electrode arrays and the Debuncher extraction kicker.

When 2×10^{12} protons traverse the 5-cm tungsten target, we expect that 7×10^7 8.9-GeV/c $\bar{p}$'s will be produced within the phase space of the Debuncher.[11] Since the acceptances of the lithium lens and the AP-2 injection line are greater than that of the Debuncher, these systems do not limit the initial stage of collection. The fractional momentum acceptance of the Debuncher was designed to be slightly greater than 3.5% and the design transverse acceptance is 22π mm-mrad in both planes.

The idea for the lithium lens came from INP at Novosibirsk. A number of lenses were built as positron collectors, and later the design was adapted to a proton focusing lens for the original antiproton source. A lens designed to focus protons was built at INP and brought to Fermilab in 1981.[12] When we changed the $\bar{p}$ Source design from electron cooling to stochastic cooling in July 1981, we decided to use a 2-cm-diameter lithium lens as the first collection lens downstream of the target. The parameters of this lens represented a significant extrapolation from those of the lens we received from Novosibirsk. With considerable consultation from Novosibirsk and development work, we successfully constructed such a lens, although not without difficulty.[13]

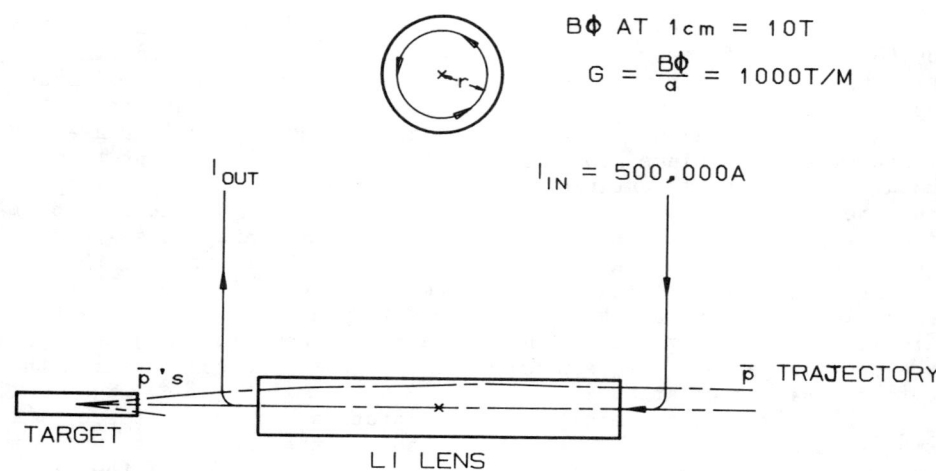

Figure 5 A lithium lens.

The concept of a lithium lens, as shown in Fig. 5, is quite simple. It consists of a cylinder of lithium through which a uniformly distributed current of 0.5 mA flows at the instant the proton beam strikes the target. While the cylinder is only 15 cm long, its effective focal length is 15 cm. The very short focal length is a consequence of the very strong gradient. The field at the edge of the 1-cm radius is 10 T when the antiprotons traverse the Li. At production the antiproton beam has a radius of roughly 0.4 mm rms, and its angular divergence is limited to 50 milliradians by the lithium lens. The lens efficiently matches the rapidly diverging p̄ beam into the AP-2 line. Although the latter can transport a beam that is typically 3 to 5 cm in diameter, it can transport an angular divergence of only a few milliradians.

Two essential complications make the lens very difficult to construct. First a 0.5-MA current pulse will heat the lithium to the melting point if its duration is more than a few hundred μsec. Second, this time is just barely long enough for the current to penetrate into the center of the lithium lens. The uniform current density is obtained by pulsing the lens to a peak current of 0.65 MA starting 220 μsec before the arrival of beam. In the time that follows, the current and field penetrate into the center of the lens while the current and field at the edge of the lens decrease. The pulse is a half sine wave of 360-μsec duration. As the beam passes through the lens, the current density is nearly uniform. Fig. 6 shows the current pulse obtained from the diagnostic readback.

The considerable amount of heat generated by the current cannot be extracted by having water pass directly over the lithium, since Li reacts with water; therefore, the Li is contained in an air- and water-tight jacket cooled by water passing over the jacket. Ideally the jacket should be very thin so that the heat can diffuse through the jacket into the water in 2 sec. The large currents and fields create very strong forces which in turn impose very large stresses

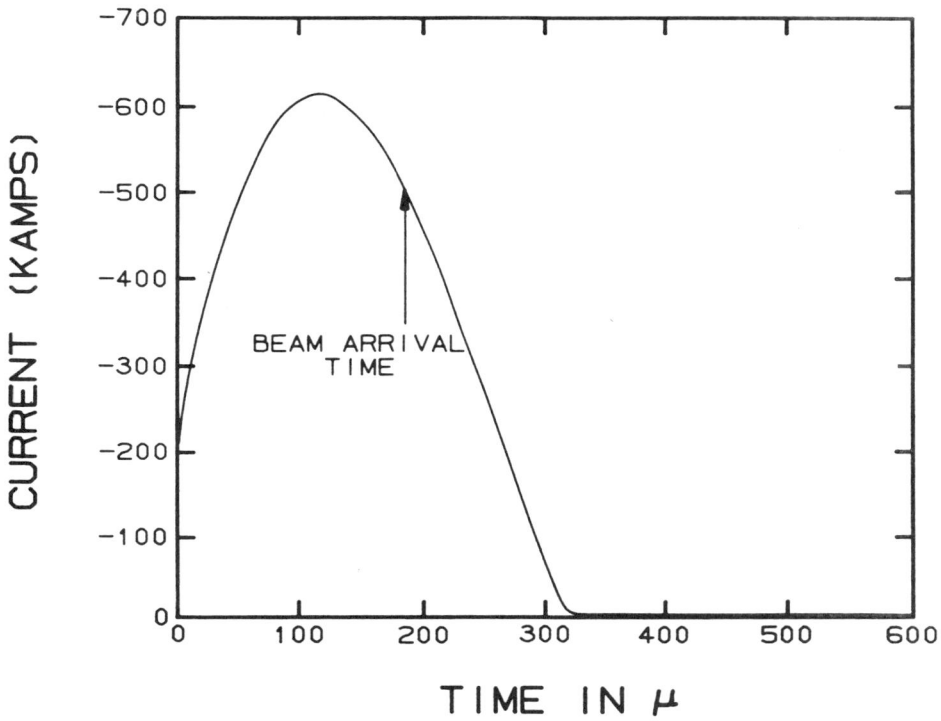

Figure 6   Li lens current waveform.

on the containment jacket and water passages. If the containment jacket is very thin it will not survive the continuous pulsing. A successful compromise between these conflicting requirements is a thin titanium jacket.

Once the antiprotons have been captured in the Debuncher, they undergo a bunch rotation that transforms the bunches with a narrow time spread and large energy spread into bunches with a wide time spread and a small energy spread. The sequence of operations that achieves this proceeds as follows: The antiproton bunches are injected into mismatched 53-MHz rf buckets. After somewhat less than a quarter of a synchrotron oscillation, the rf voltage is rapidly decreased from 5 MV/turn to 0.15 MV/turn to match the bucket to the rotated bunch. The rf voltage is then adiabatically decreased to a few kilovolts. The bunch rotation and adiabatic debunching were designed to reduce the momentum spread from 3.5% to <0.25%.[1,14] The final momentum spread will be greater than this value if the initial spread of the p̄ bunches exceeds 1.0 ns.

The small energy spread is needed for two reasons: (1) the effective momentum aperture of the Accumulator injection orbit with the injection shutter closed is only 0.4%, and (2) the Accumulator stack-tail system cannot cool 7×10^7 antiprotons into the stack in 2 sec if the momentum spread is >0.25%. During the October 1985 colliding-beam test the debunching system achieved a final momentum spread of 0.25% when the initial spread was 3.5%.

The sum of the times required to accelerate a batch of protons in the Main Ring from 8 to 120 GeV and to reset the Main Ring guide field to accept the next batch of protons was 4 sec. Ultimately the Main Ring cycle time will be reduced to a value close to 2.2 sec. Since debunching requires only 12 ms and antiprotons remain in the Debuncher for at least 2 sec before another batch is injected, there is time for stochastic cooling of the transverse emittances of the antiprotons. The horizontal and vertical betatron cooling systems were designed to reduce the emittances of a beam of 7×10^7 particles from 20π mm-mrad to about 5π mm-mrad in 2 sec.[1,15] The longer cycle time made it possible to reduce the emittances to somewhat less than 3π mm-mrad in both planes.

Just before injection of a new batch of antiprotons into the Debuncher the cooled antiprotons are extracted and transferred to the Accumulator. The next sequence of steps constitutes the crucial process of accumulation. The debunched beam of antiprotons circulating on the injection orbit is adiabatically captured with a harmonic 84, 52.8-MHz rf system and then decelerated by 60 MeV to the stacking orbit. As the stacking orbit is approached, the rf voltage is gradually reduced to zero so that an unbunched beam is deposited on the stacking orbit.[16] The stack-tail cooling system is then turned on, and it decelerates most of the fresh batch away from the stacking orbit and toward the core of the stack. It also causes some of the freshly injected p̄'s and some of the already cooled p̄'s to diffuse back toward the stacking orbit. These p̄'s are lost in subsequent stacking cycles.[17] After 2 sec the stacking orbit has been cleared of particles and the Accumulator is ready to accept another batch. Fig. 7 shows the expected energy distribution of antiprotons after four hours of cooling. In that time the peak core density grows to about 10^{11} antiprotons per MeV and the total number of p̄'s in the stack reaches 4×10^{11}. Throughout this time, the Accumulator betatron cooling systems reduce the horizontal and vertical emittances to 2π mm-mrad. Once the density and number of antiprotons reach these values, the dense part of the stack can be extracted and transmitted to the Tevatron by way of the Main Ring.

The energy distribution, shown in Fig. 7, can also be thought of as a horizontal position distribution of the antiprotons as they traverse one of the three high dispersion straight sections. At these locations a decrease of 10 MeV of the energy of a particle causes the particle to move radially inward by 10 mm. Typical cross sections and horizontal positions of the beam in a high dispersion straight section are shown in Fig. 8. The overall increase in phase-space density of the antiprotons in the core relative to the density at production is 1.4×10^6.

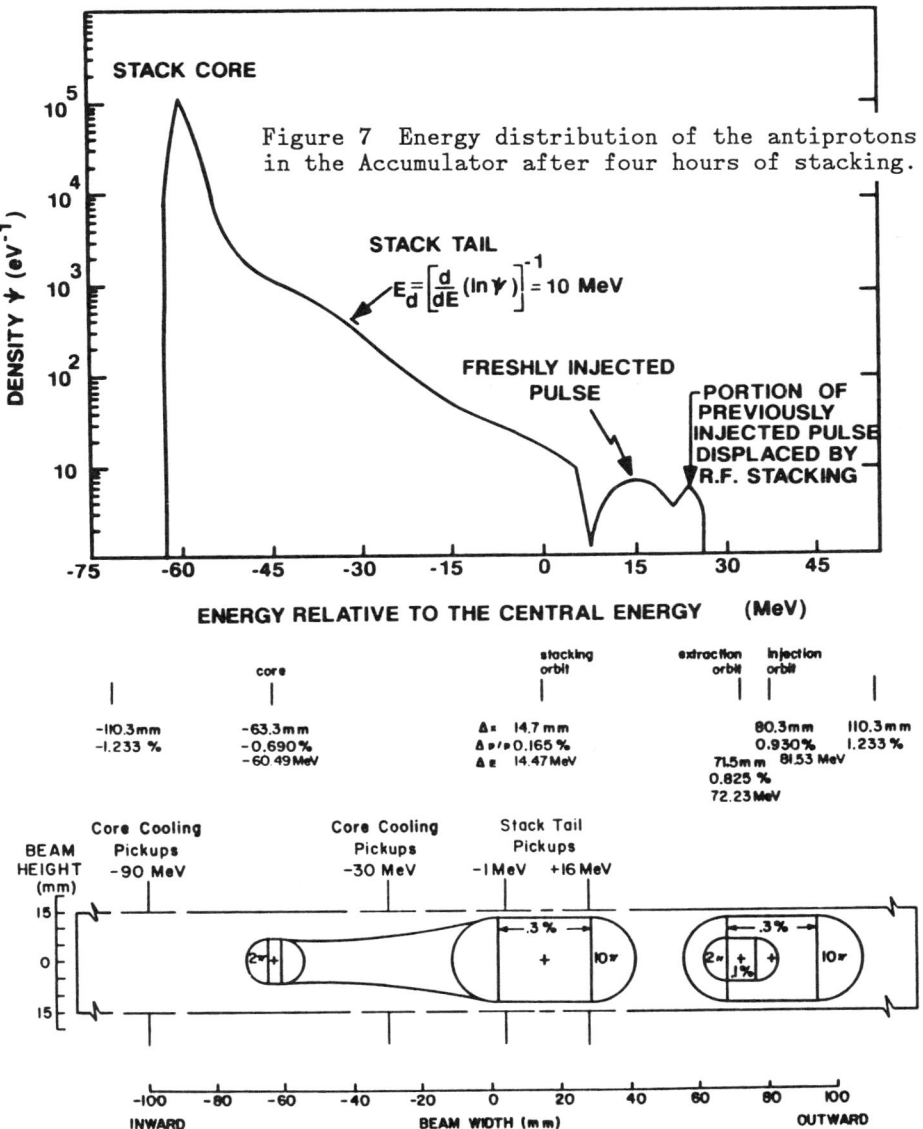

Figure 7 Energy distribution of the antiprotons in the Accumulator after four hours of stacking.

Figure 8 Accumulator beam position at the end of A20 high dispersion straight section. The antiprotons are injected at energies 75 MeV greater than the energy of the central orbit. The harmonic 84 rf system decelerates them to an energy 15 MeV greater than the energy of the central orbit. After the batch is deposited, the stack-tail cooling system is turned on and it pushes the antiprotons, particle by particle, toward the core at -60 MeV. Eventually, after four hours or so, the antiprotons have reached the core and the **longitudinal** density has increased by roughly a factor of 1.5×10^4.

When the Main Ring is ready to receive antiprotons, antiproton production is terminated. The stack-tail system is turned off after all the p̄'s are within the influence of the core cooling system. The core cooling systems are left on until the extraction process starts, since they are needed to counter the effects of intrabeam scattering and beam gas scattering. The dominant term is intrabeam scattering.[9]

The next sequence of steps on the way to achieving colliding beams consists of unpacking the antiprotons from the core and transferring them to the Tevatron by way of the Main Ring. This sequence is very complicated and will be described only in a schematic way.

The Accumulator extraction orbit is very near the 8-GeV (kinetic energy) injection orbit, as noted in Fig. 8. The extraction orbit length is designed to be equal to the 8-GeV extraction orbit length of the Fermilab Booster. Both machines have harmonic number 84, 52.8-MHz, rf systems. The Main Ring rf frequency is also 52.8 MHz at 8 GeV. Each of these three rf systems can be phase-locked to either of the others, and relative phase shifts are adjusted so that the bunch-to-bucket beam transfer can be done from Booster to Main Ring, Main Ring to Accumulator (backward down the Accumulator extraction channel), or Accumulator to Main Ring with minimal synchrotron oscillation, provided magnetic guide fields (orbit lengths) are properly adjusted.

A Main Ring ramp is set up to receive the antiprotons and accelerate them to the Tevatron injection energy, 150 GeV. Before transfer of antiprotons, the transfer between the Accumulator and the Main Ring is tested by injecting 8-GeV protons from the Main Ring into the Accumulator in the reverse direction through AP-3. This is done by injecting a partial batch, consisting of 10 or fewer bunches, from the Booster into the Main Ring and subsequently into the Accumulator by extraction at F17. During this operation the Main Ring rf is phase-locked to an injection frequency reference oscillator. Both the Booster and the Accumulator rf systems are phase-locked to the Main Ring system. The protons are transferred to matched stationary buckets in the Accumulator. Components in the AP-3 line and the relative magnitude of the guide fields in the Main Ring and the Accumulator are adjusted to minimize coherent synchrotron oscillations and betatron oscillations in the Accumulator. When this operation is completed the Main Ring is ready to accept 8-GeV antiprotons from the Accumulator and accelerate them to 150 GeV.

A single bunch of antiprotons of the desired intensity, nominally 7×10^{10} particles, is captured in an rf bucket at the cooled core momentum and accelerated to the extraction orbit, which is 132 MeV more energetic than the core orbit, as shown in Fig. 8. If this were to be done with an rf system operating at the rotation frequency, h = 1, only a few volts of rf would be required, and it would be difficult to create the precisely correct bucket area to obtain the required number of particles. Consequently this unstacking is done with a single "isolated" h = 2 rf bucket.[18] When the bunch reaches the extraction orbit an additional h = 2 fixed frequency rf system capable of generating 600 volts is slowly turned on and the bunch is narrowed to about 150 ns. This rf system generates a normal h = 2 rf wave but only one of the two rf buckets contains antiprotons.

The longitudinal emittance of such an antiproton bunch containing 7×10^{10} particles with the maximum core density is about 1 eV-sec. If this bunch were compressed in time duration to <18.8 ns, the period of a 52.8-MHz bucket, the momentum spread would be too large to be injected into the Main Ring at F17. Furthermore, such a bunch would be badly matched to available Main Ring buckets, and acceleration of such a bunch through transition in the Main Ring would result in a very large loss of beam. These difficulties are avoided by bunching the beam into nine or ten h = 84 buckets, each of which has a small enough momentum spread for the bunch to be transmitted efficiently to the Main Ring and subsequently accelerated through transition. This is done by slowly turning on the h = 84, 52.8-MHz Accumulator rf system while the bunch remains narrowed to 150 ns in the h = 2 bucket. This causes the antiprotons to split into nine adjacent bunches with longitudinal emittances between 0.1 and 0.15 eV-sec. During the adiabatic capture the 52.8-MHz Accumulator rf frequency and the 1.26-MHz h = 2 frequency are both derived from the same source. The Main Ring rf system is phase-locked to the Accumulator h = 84 frequency and the amplitude is set for bunch-to-bucket transfer of the nine antiproton bunches to matched buckets in the Main Ring. When this step is completed the transfer to the Main Ring is initiated. The antiprotons are then accelerated from 8 GeV to 150 GeV. At 150 GeV the nine bunches are coalesced into a single (h = 1113) bunch which is then transferred to the Tevatron.

The process is repeated three times in order to place three separate antiproton bunches in the Tevatron. The same number of proton bunches of similar intensity are prepared in the Main Ring and injected into the Tevatron. A description of the bunch coalescing system can be found in the Tevatron I Design Report[1] and a summary of its performance can be found elsewhere in these proceedings.[10]

When all six bunches are in the Tevatron, the whole ensemble is accelerated to the peak energy. The strengths of the low-beta quads are increased, thereby squeezing the beams down to an rms radius of 0.06 mm. The conditions are then appropriate for an experiment like CDF.

5 THE DEBUNCHER

Having provided a description of the sequence of operation, it is appropriate to provide more details about the Debuncher and Accumulator.[19,20] The Debuncher, the outer of the two rings, is shown in Fig. 9. The pickups of the Debuncher stochastic cooling system are located in the D10 straight section. There are four electrode arrays for both vertical and horizontal cooling systems. Each array consists of 32 pairs of loop couplers. Since an array of

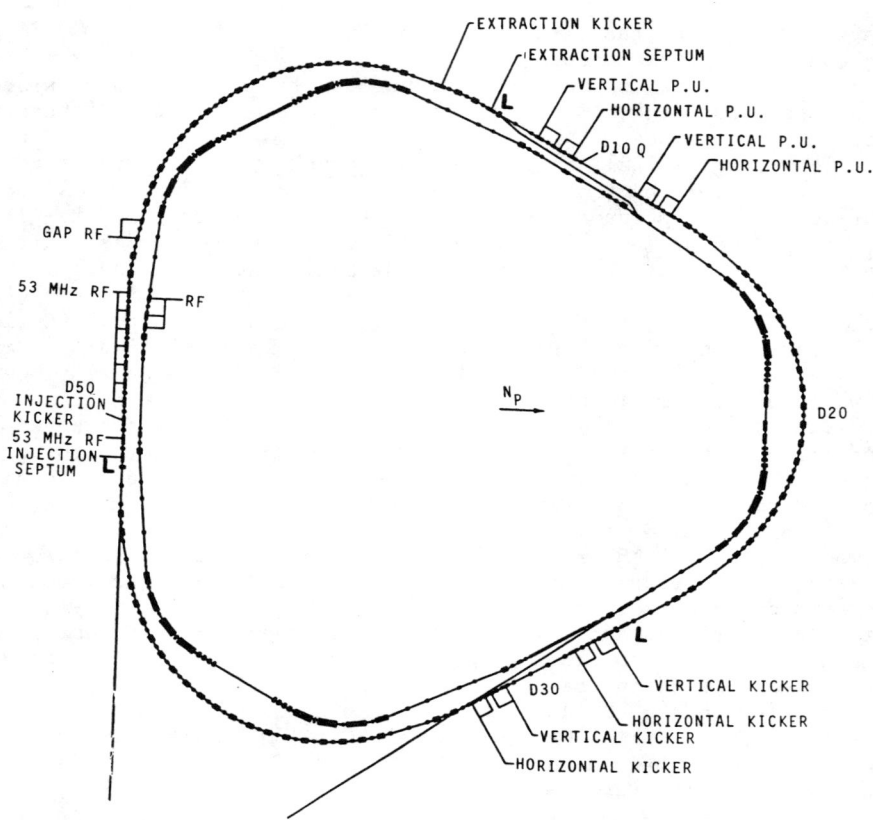

Figure 9 Layout of Debuncher magnets.

128 pairs cannot fit between the quadrupoles, the arrays are distributed between pairs of adjacent quadrupoles. The signals from each pair of loops in a 32-loop array are first added in phase and then the signals from the four arrays are combined in phase.[21] After the signals are added in phase, the sum and difference signals of the loop pairs are formed and transmitted across the ring on coaxial cables located in an underground communication duct. The signal must traverse a chord that is shorter than the beam path if it is to arrive at the kickers simultaneously with the particles that created the signal. There are two reasons for this: (1) the speed of the signal on the cable is only 0.9 times the velocity of light, whereas the speed of the antiprotons is 0.995 that of light; (2) it takes about 75 ns to amplify the signal by 140 dB.

The Debuncher lattice was deliberately kept regular throughout the whole machine in order to fit the 20π mm-mrad beam into the roughly 30-mm opening between the pairs of loops in each pickup array. A 2-GHz bandwidth was chosen for the Debuncher betatron cooling systems in order to reduce the emittance by a factor of 3 in 2 sec. This choice of a 30-mm separation of the loops provides efficient coupling of the beam signal to the loops. Given a beam emittance of 20π mm-mrad, it follows that the maximum value of β in the pickup array has to be <11 meters. After examining the sextupole strength required to correct the large tune variation brought on by the 3.5% fractional momentum spread, we chose a phase advance per cell of 60°.[19] The lattice functions for the Debuncher are shown in Fig. 10 and the layout of a cell is shown in Fig. 11. A typical operating point in the tune diagram during the 1985 run was 9.782 horizontal and 9.778 vertical. Table III lists some of the more important properties of the Debuncher lattice.

During the October 1985 run these parameters were achieved with two exceptions: η was 0.0054 and the betatron acceptance was 12π mm-mrad in the vertical plane and 10π mm-mrad in the horizontal plane. The cause of the reduced acceptance has since been traced to misalignment of a number of devices.

The value of 0.006 for η represents a compromise between two conflicting demands. For a cooling system with optimal gain, the cooling rate can be written as

$$\frac{1}{\epsilon}\left(\frac{d\epsilon}{dt}\right) = \frac{W}{NM}, \quad M = (W\Lambda T|\eta|\Delta E/\beta^2\bar{E}) \text{ and } W\Lambda T|\eta|\Delta E/\beta^2\bar{E} < 1, \quad (1)$$

while the rf voltage, V, needed to contain the energy spread, ΔE, of the $\bar{p}$'s before debunching is

$$V = \frac{\pi h|\eta| (\Delta E)^2}{2\sin^2(\phi/2) \beta^2\bar{E}} \quad (2)$$

where ϕ is the bunch spread.

A value of η of 0.03 would have been ideal for stochastic cooling, while a value <0.006 would have been better for the rf system. Since it was possible to obtain an rf voltage of 5 MV/turn, the value required to rotate the $\bar{p}$ bunches without significant filamentation when η is 0.006, η was chosen to be 0.006. This meant that the cooling system bandwidth had to be at least 1 GHz. The bandwidth was chosen to be 2 GHz, since that seemed to be the limit for practical construction at the time. As noted earlier, the rf system and the cooling system performed as designed.

Each set of pickup arrays, one horizontal and one vertical, is subdivided into four sub-arrays each containing 32 loop pairs. Two of the vertical arrays are placed on either side of a horizontally focusing quadrupole, roughly the point where the vertical betatron function is a minimum. The betatron phase advance from the furthest upstream loop to the furthest downstream loop is 25°. Two horizontal arrays are placed next to the vertically focusing quad, which is just downstream of the above-mentioned horizontally focusing quad. The remaining vertical arrays are placed next to the horizontally focusing quad which is displaced by three full cells or

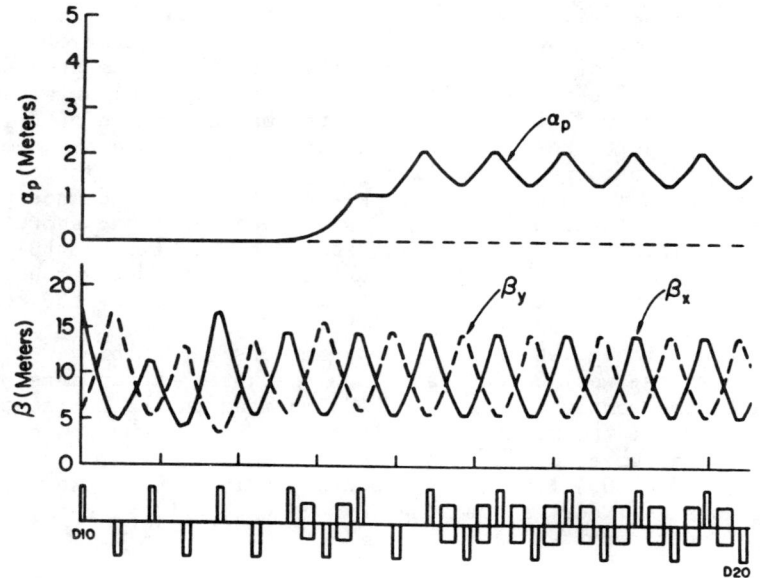

Figure 10 Debuncher lattice function for one sextant.

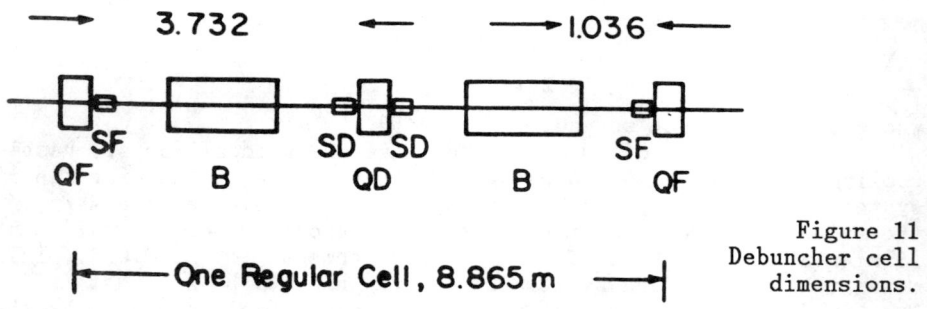

Figure 11
Debuncher cell dimensions.

Table III Some Properties of the Debuncher Lattice

γ_t, transition gamma	7.65
η at 8 GeV	0.006
Circumference	505 m
Momentum aperture, $\Delta p/p$	3.5%
Betatron acceptance, H and V	20π mm-mrad
Betatron tunes, H and V	9.782, 9.778
Natural chromaticity, ξ_H and ξ_V	-10.4, -10.6
Phase advance per cell, H and V	60°

180° in betatron phase advance from the first quad. The signals from the two pairs of horizontal arrays are added in phase, by inverting the signal from one of the array pairs. The remaining horizontal arrays are placed next to the next most downstream quadrupole, a vertically focusing quadrupole, and the signals are combined in the same manner as the signals from the vertical arrays.

The kicker arrays are located in the 30 straight sections, following the pattern described for the pickup arrays. The locations of the kicker arrays were chosen so that the betatron phase advance between kickers and pickups was $(6.5)\pi$ radians.

Although this system employs the principles of stochastic cooling of betatron oscillations proposed by van der Meer to reduce the transverse size of circulating beams at the ISR[22] it broke new ground in stochastic cooling because it was the first high power multigigahertz bandwidth system to be used for fast betatron cooling. A number of difficult problems had to be solved to adapt this type of microwave equipment to our use. The sheer number of loop couplers was a significant engineering problem. The design of loops that provided reasonable transfer impedance and sensitivity was difficult but was successfully done.[23] The terminating resistors on the pickups and the preamplifiers are cooled by liquid nitrogen to suppress the thermal noise generated by random motion of the electrons.[24] A combiner board that combines the signals from 128 loop pairs without excessive loss and phase shifts had to be developed.[20] Table IV lists some of the more important properties of these systems.

Table IV Debuncher Betatron Cooling System

Frequency band	2-4 GHz
Pickup characteristic impedance, Z_o	75 Ω
Sensitivity (coupler factor)	1.73
Number of pickup pairs per system, N_p	128
Terminating resistor temperature, θ_B	80°K
Amplifier equivalent noise temperature, θ_A	60°K
Separation of pickup plates	30 mm
Amplifier gain	138 db
Maximum output power	500 W
Initial emittance, H and V	20π mm-mrad (100%)
Final emittance, H and V	5π mm-mrad (95%)

In the Debuncher the antiprotons are simply cooled and then transferred to the Accumulator. The design of Accumulator cooling systems is more complicated because successive batches of antiprotons have to be merged together in the presence of a very dense core.

6 THE ACCUMULATOR

The Accumulator is designed to accommodate the rf stacking and stochastic cooling systems.[20] To explain how this is done it is useful to begin with the standard description of the horizontal orbit of a single particle as given in the following formula:

$$x_i(s) = [\sqrt{\epsilon_i \beta(s)} \cos(2\pi\nu_x f_o t + \phi_i)] + \left[\alpha_p \left(\frac{E_i - \bar{E}}{\beta^2 \bar{E}}\right)\right] . \qquad (3)$$

The first term in square brackets describes the betatron motion as a function of time as witnessed by an observer at a fixed position in the ring. The second term accounts for the displacement caused by the deviation of the antiproton energy from the energy of the central orbit. In the absence of time-dependent fields created by rf systems and the cooling systems, ϵ_i, ϕ_i, and E_i are constants of the motion. The first term changes rapidly every turn since the phase change per turn, $2\pi\nu_x$, is about 2400°. RF stacking takes place on a time scale characterized by the synchrotron phase oscillation period, a few thousand turns, while cooling takes place on a time scale characterized by a few hundred thousand to a few billion turns. One can approximate the betatron term by its average value, zero, when studying momentum cooling. By exploiting the momentum dependence of the position, it is possible to inject, extract, rf stack, and cool without letting the various operations interfere with one another. This is done either by changing the momentum of each antiproton (stochastic cooling) or of an ensemble of antiprotons (rf stacking or unstacking), thereby moving the antiproton or the ensemble horizontally within the vacuum chamber. Of course α_p must be different from zero in order to carry out these operations. The high dispersion straight sections in the Accumulator are designed so that α_p is 8.7 meters. Thus if the energy of an antiproton is decreased by 10 MeV it will move radially inward by 10 mm in the high dispersion straight sections.

The equation for x(s) can also be used to illustrate why it is not desirable to decelerate or accelerate the beam in a region of zero dispersion. When the antiproton energy is abruptly changed in a high dispersion location, its position cannot change instantaneously. Since the energy-dependent term changes, there must be a compensating change in the betatron term. Unless the betatron phase advance between pickup and kicker can be carefully selected, these changes in the betatron term will on average lead to an increase in the betatron amplitude. Since stochastic stacking is achieved by making small accelerations and decelerations of samples of the beam, the betatron amplitudes will generally increase if these small accelerations are made in a region where α_p is large. In the Tevatron I p̄ Source α_p is designed to be zero. The betatron cooling systems compensate for any heating caused by small values of dispersion. The layout of equipment in the Accumulator is shown in Fig. 12. The kickers of the several momentum cooling systems are located in sector A-30, a zero dispersion region. The rf systems

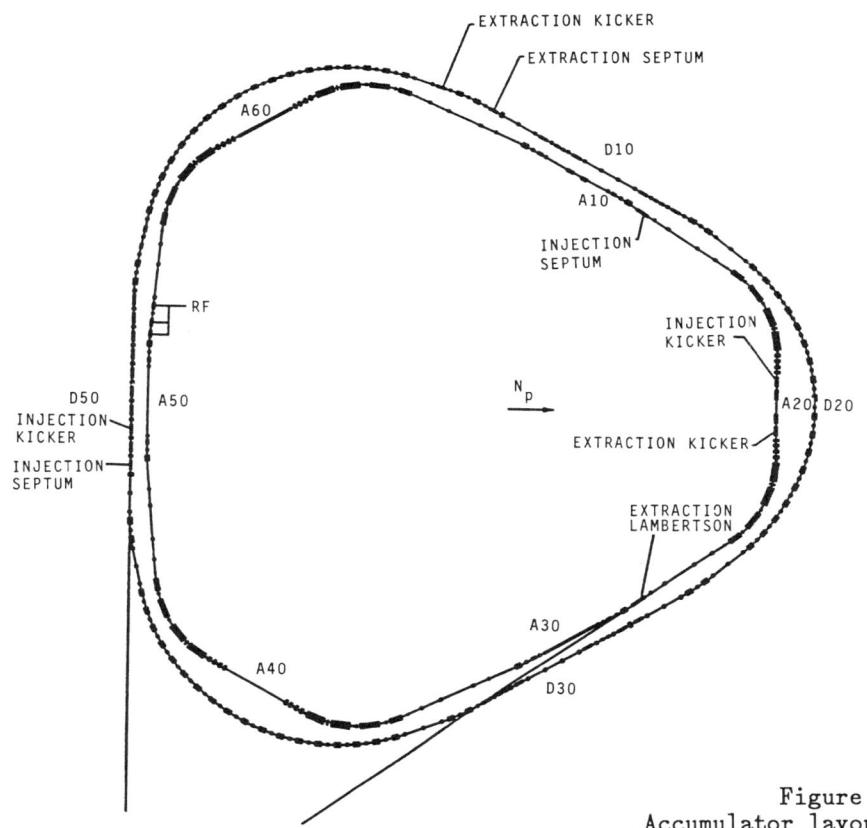

Figure 12
Accumulator layout.

are located near sector A-50, also a zero dispersion region. The stack-tail momentum and core momentum pickups are located in A-60, a high dispersion region. The locations of the injection and extraction kickers and of the pickups of the remaining cooling systems will be discussed after the Accumulator lattice is described.

The lattice of one sixth of the Accumulator is shown in Fig. 13, which shows that nearly all of the bending is concentrated just ahead of the four quadrupoles that comprise the triplet in the high dispersion straight section. The small 5° bend adjacent to the quadrupole triplet in the zero dispersion region was placed there rather than at S7 in order to reduce η from 0.034 to 0.022 without changing α_p very much. The value of 0.022 is nearly optimum for the 1- to 2-GHz stack-tail momentum system. Ideally we would like to restrict the high dispersion regions to just the straight sections since the magnet apertures have to be very large to accommodate the highly dispersed beam. This practical choice leads to a lattice that has mirror symmetry about the high dispersion straight section.

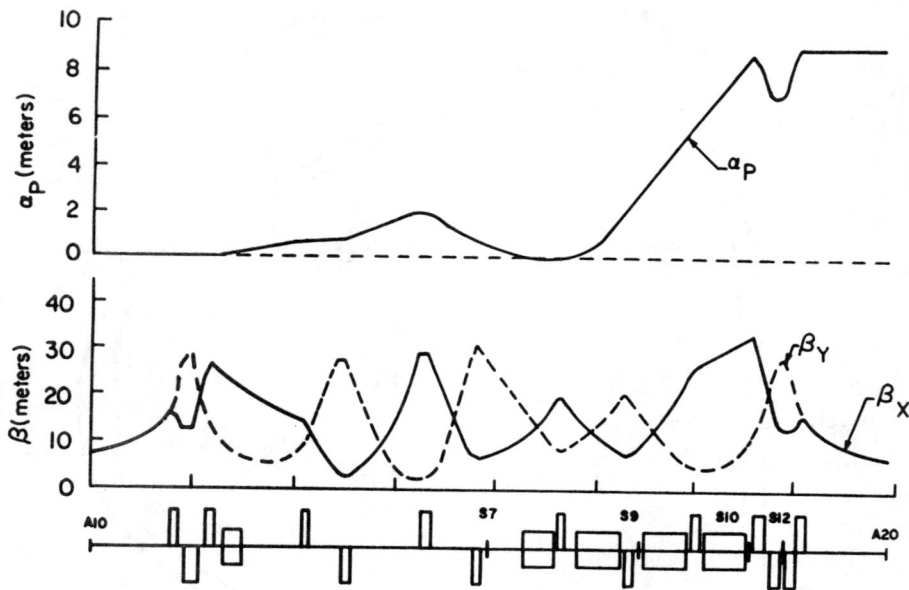

Figure 13 Accumulator lattice functions for one sextant.

The mirror symmetric piece reduces the dispersion back to a small value. Thus high dispersion and zero dispersion straight sections alternate in the lattice. We chose a lattice with three high dispersion straight sections in order to accommodate the immediate needs and to provide a pair of straight sections that could be used for the development of cooling systems going beyond the parameters of the Tevatron I Design. A-10 contains the core cooling betatron pickups and a considerable amount of diagnostics. A-20 contains the injection and extraction kickers and the stack-tail betatron pickups. A-30 contains all of the stochastic cooling kickers. A-40 and A-50 are set aside for future cooling systems. A-60 contains the pickups for the momentum stack-tail system, the momentum core cooling system, and the betatron cooling systems. While A-50 also contains some diagnostics, it is the most likely location for kickers for any new cooling system. Note that stochastic cooling pickups and kickers were not placed in the same straight section or even adjacent ones because it is highly desirable if not essential for the stochastic cooling pickups and kickers to be in separate straight sections. The extreme sensitivity of the pickups and the very high gain of the cooling systems make them very susceptible to unwanted feedback paths. By separating the pickups and kickers so that they are not within the line of sight of one another, one of these feedback paths was eliminated. Given all the equipment that had to be installed in the straight sections, a lattice with three high dispersion straight sections and three zero dispersion straight sections was almost a necessity. The triangular shape was a natural outcome of this choice.

As seen in Fig. 13, the high dispersion triplet actually consists of four magnets. The central magnet of the triplet was split in two in order to provide room for the sextupole which corrects the vertical chromatic aberrations. It had to go in a position where the beam is large vertically and has a large dispersion. Because everything is packed so tightly in this region, there were no places that satisfied this requirement until the quad was cut in two. Of course the cutting was done in the design stage. Table V lists the lattice parameters on the injection, stacking, and core orbits.

Table V Accumulator Lattice Parameters Corrected for Chromaticity

	Injection orbit	Stacking orbit	Core orbit
Kinetic energy (GeV)	8.03	7.96	7.89
ν_x	6.616	6.611	6.614
ν_y	8.611	8.611	8.611
ξ_x	2.05	1.13	-0.22
ξ_y	0.21	0.32	0.33
γ_T	5.37	5.42	5.50
η	0.023	0.023	0.022

The location of various systems in the Accumulator having been described, the injection and extraction process can be given in more detail. The beam is injected at an energy that is 82 MeV higher than the reference energy. As seen in Fig. 8, the injection orbit is 80 mm radially outward from the central orbit in a high dispersion straight section. While on the injection orbit, the antiprotons traverse the aperture of the injection and extraction kickers. During the injection cycle the stainless steel shutter of the injection kicker is moved into place at a point +50 mm from the central orbit. It forms an eddy current shield for the kicker magnet so that when the kicker is pulsed the field of the kicker is contained within the magnet and the stack is undisturbed. After injection is complete the shutter is lowered and the freshly injected antiprotons are moved to the stacking orbit. This is done by adiabatically capturing the antiprotons on the injection orbit with the harmonic 84 rf system and then decelerating the captured beam to the stacking orbit. As the beam approaches the stacking orbit, the rf voltage is gradually reduced. Any particles left on the stacking orbit during a previous stacking cycle are phase-displaced back toward the injection orbit as the bucket passes through them. Eventually these particles will strike the shutter or be deflected into the aperture walls when the injection kicker is fired.[17] Once the antiprotons are deposited on the stacking orbit, the cooling systems are turned on.

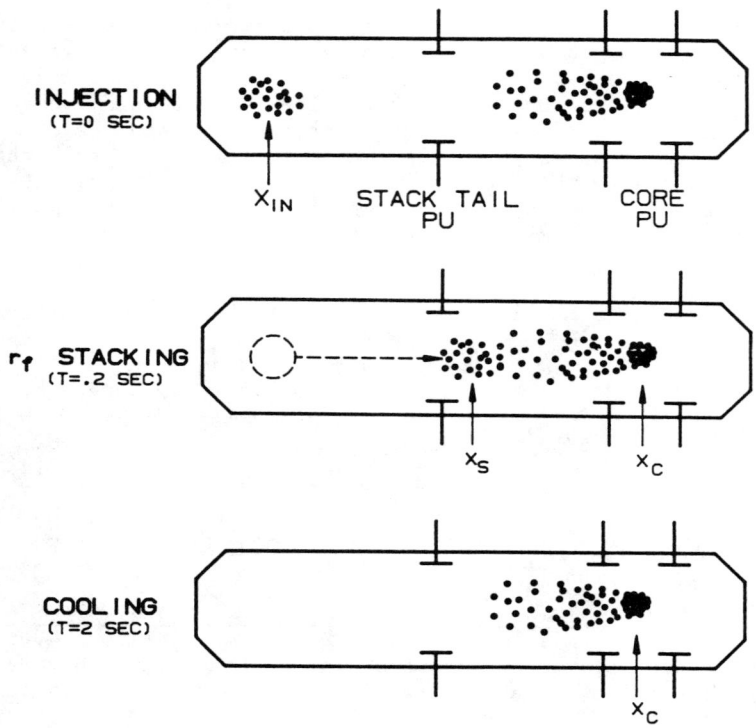

Figure 14 Schematic arrangement of the stochastic cooling pickups in the A-60 straight section.

Figure 14 shows a schematic arrangement of the stochastic cooling pickups in the A-60 straight section. Since the stacking orbit is radially inward from the stack-tail pickups, the sum of the signals, Vi, induced on the upper and lower pickups by a single antiproton decreases exponentially as the antiproton moves away from the pickup. This can be expressed by

$$V_i = ke-(\pi x_i/h) \qquad (4)$$

where V_i is the sum signal, x_i is the particle location, and h is the vertical separation of the pickup loops. If x_i is replaced by its average value from the equation for x(s), the voltage induced on the pickups by the n particles sampled by the pickups is

$$V = [\sum_i^n ke-(\pi \alpha \tfrac{\Delta E}{p} / h\beta^2 \bar{E})] + \text{electric noise} . \qquad (5)$$

The closer the antiprotons are to the stack-tail pickups, the greater their contribution to the signal.

Consider the Fourier components of this signal. If ΔE represents the energy difference between the high energy extreme of the stacking orbit and the low energy extreme of the core orbit, there will be a one to one correspondence between frequency and energy within the n_hth Schottky band if $|\eta n_h \Delta E/(\beta^2 \bar{E})|<1$. Since $|\eta|$ and the maximum n_h are chosen so that this condition holds, the frequency spectrum of a particle on the stack orbit does not overlap that of a particle on the core orbit. This is important because the stack-tail correction signal is amplified and applied to the stack-tail kickers in sector A-30, which has zero dispersion. All particles passing through the kickers are affected by this signal whether or not they are on the stack orbit or on the core orbit. Because a particle behaves like a lossless resonator, it responds only to signals that have the same frequencies as the revolution harmonics of the particle. As a result the signals from the core do not interfere with the stack tail and vice versa.

Consider the case in which only one particle is injected on each pulse. During the first cycle the particle is injected and moved over to make room for the particle injected on the second cycle. After the second particle is injected, it makes a much greater contribution to the signal than the first. It follows that the cooling system moves the second particle over to make room for the particle injected on the third cycle. On any given traversal of the kickers the first particle responds to the pickup signal from itself and the much larger signal from the second particle. Since the Fourier spectrum of the signal from the second particle does not overlap the spectrum of the first particle, on average the kicks due to the second particle neither accelerate nor decelerate the first particle. Since the Fourier spectrum of the kicks due to the amplified signal from the first particle overlaps its own revolution harmonics, this particle continues to be moved toward the core even in the presence of the Schottky noise of the second particle. On each successive injection cycle the particles that were put in earlier move a little less, but the amount that a particle on the injection orbit moves is the same. Gradually a stack of particles will build up and the density will rise exponentially.

When a large number of particles are injected in the presence of a dense stack, the cooling system moves the injected particles over in the same way. Over a period of several hundred milliseconds, the particles in the stack respond only to the Fourier components of the amplified signal that are very close to a multiple of their own revolution frequency. Thus each particle is selectively pushed away from the stack tail pickups by itself. Particles having the same revolution frequency cause heating while particles having different revolution frequencies make no net contribution. The exponentially declining sensitivity causes the density of particles gradually to build up as the particles move away from the core. Note that the Schottky noise, which produces heating, is also weighted by this exponentially decreasing sensitivity. Thus in the vicinity of the stack-tail pickups, where the response is strong, there are, relatively speaking, very few particles with nearly the same frequency as the stack tail, and the system gain can be large. Near the core there are many more particles, but the signal is attenuated by the exponentially decreasing gain, and the Fourier components of their signal do not overlap the revolution harmonics of the particles in the stack tail. As a result, the signal from the high density part of the stack does not heat the core. This process, stochastic stacking, was developed by van der Meer and his colleagues for the Antiproton Accumulator at CERN.[25]

Figure 14 shows another four sets of pickups in the low energy side of the aperture. These pickups are part of the core cooling system, an independent system that operates in the frequency range 2 to 4 GHz. The high energy pair of the four pickups is used to push particles toward the low energy pickups, while the pair of low energy pickups is used to push particles toward the high energy pickups. Particles midway between the pickups will not be accelerated or decelerated. In this manner a well is created to contain the core so that it does not diffuse into the vacuum chamber wall. The stack-tail system is designed to sweep the 7×10^7 particles injected every two seconds away from the stacking orbit and into the stack, to make ready for the next batch. The evolution of the number of p̄'s per unit of energy as a function of time during the stacking process is shown in Fig. 15. Since the stack-tail system does not sweep all of the beam out of the way, the part left behind will be phase-displaced back towards the injection orbit on the next cycle. This is shown in Fig. 15 as a growing

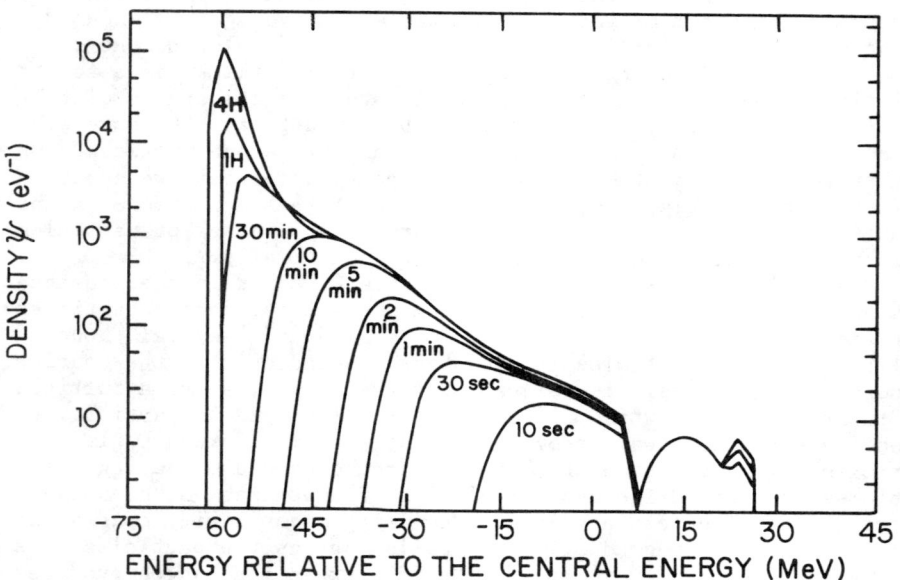

Figure 15 Evolution of the density of antiprotons per unit energy as a function of time.

protuberance on the high energy side of the stacking orbit near 25 MeV. We calculate that <5% of the antiprotons will be lost because of phase displacement and heating.[17] To allow for what is inevitably less than perfect performance, we have assumed that up to 15% of the antiprotons will be lost between Debuncher extraction and containment in the core.

Figure 15 also shows how the stack should look at different times. Note that the density starts to increase rapidly near -45 MeV. This happens because the core cooling system, which has a greater bandwidth than the stack-tail system, cools the beam much more rapidly once the beam is under its influence. The final density in the core will be limited by intrabeam scattering, the Coulomb scattering of particles off one another.[8]

An array of pickups is shown in Fig. 16. The electrode array is enclosed in a large circular vacuum tank. Each array contains a set of loops that are little flat plates, one end of each being connected to a printed circuit transmission line deposited on a Teflon board and the other end resistively terminated by a 50-Ω resistor. The transmission lines are combined by using quarter wave lines printed on the same circuit board.[21] The lines are laid out so that the signals are combined in phase at midband (1.5 GHz) inside the vacuum tank. When the beam passes through the beam chamber, it induces a tiny signal on each plate. The signals on the upper and lower plates are separately combined. Each array has either 8 or 32 pairs of pickup loops. Groups of eight pickups are added together with transmission lines inside the vacuum tank if needed. The combined signals are brought into the preamp box, where the signals from upper and lower plates are added or subtracted depending on the function of the array.

Figure 17 shows the stack-tail system. It shows the notch filters, a very essential feature of all momentum cooling systems, which increase the density by more than two orders of magnitude. In order to cool the antiprotons into the stack in two seconds the signal is amplified by 170 dB. Wide-band noise caused by the motion of electrons in the terminating resistor and the preamps is also amplified by 170 dB. Since the noise is wide-band, the frequency spectrum will overlap the frequency spectrum of the harmonics of the core. This would blow up the core unless something were done. The exponential decrease of the pickup sensitivity does not help since the noise is injected at the beginning of the amplifier chain after the pickup. The slightest amount of thermal noise is amplified and applied directly to the kicker. The purpose of the notch filters is to reduce the amplifier gain by 30 or 40 dB at the revolution harmonics of the core.

It was noted earlier that η and ΔE are chosen so that there is a unique correspondence between the energy and frequency of each particle at all revolution harmonics within the system bandwidth. The uniqueness exists as long as $|\eta|(\Delta E/\beta^2 \overline{E})n_h$ is <1. Since the energy spread is roughly 1%, the frequency spread corresponding to the first harmonic is compressed into 100 Hz. At the 1600th harmonic, which corresponds to a frequency of 1 GHz, the

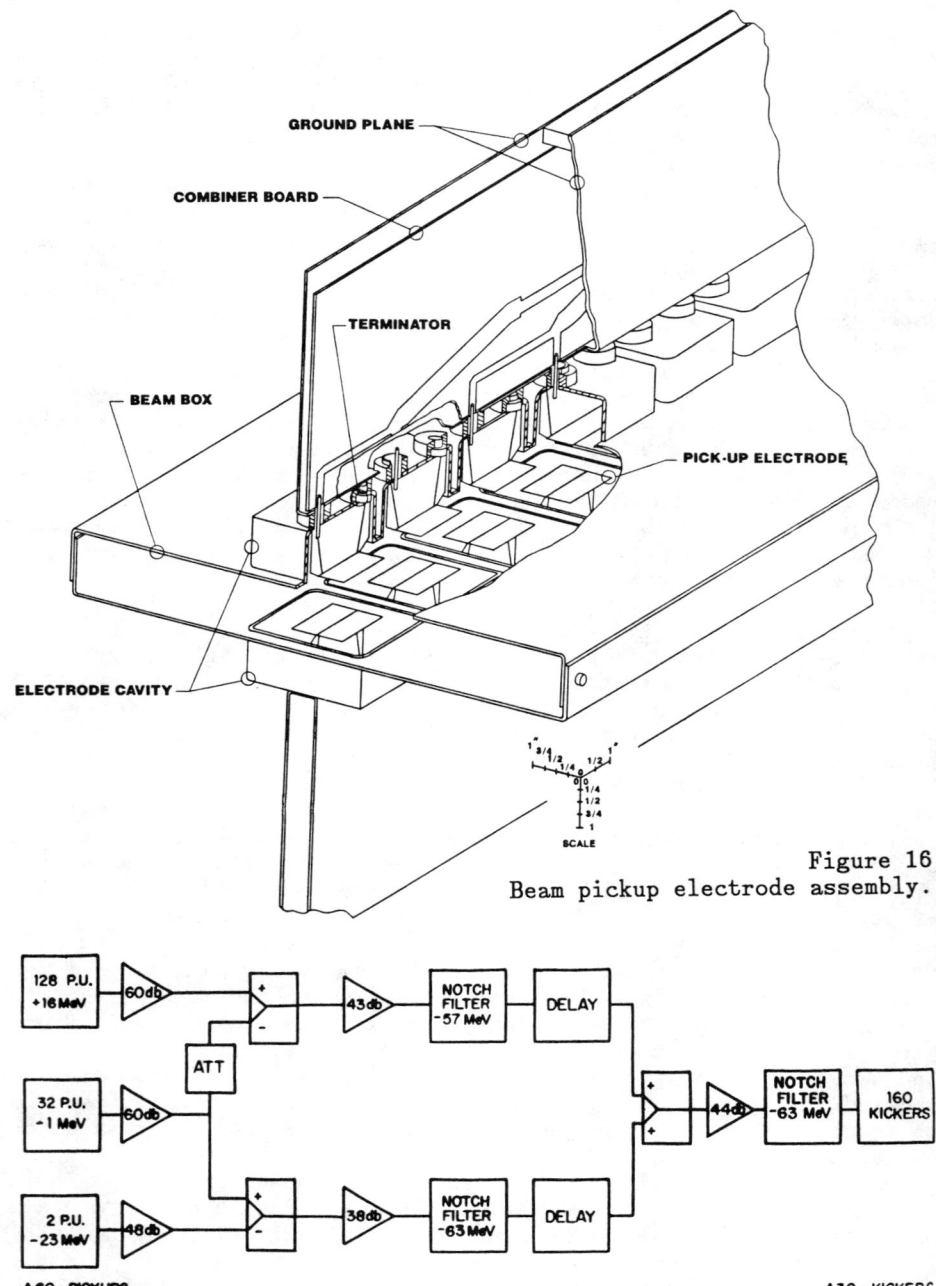

Figure 16 Beam pickup electrode assembly.

Figure 17 Block diagram of the stack-tail system.

distribution has a width 200 kHz which fills about one-third of the Schottky band, and at the 3200th harmonic the distribution is broader and fills two-thirds of the Schottky band. A filter that can reduce the gain by 30 to 40 dB at each of these frequencies without affecting the gain at the revolution harmonics of the stack orbit will reduce the heating due to thermal noise to an acceptable level without reducing the gain at the frequencies of the stacking orbit.

Such a filter is shown in Fig. 18. The filters that we have been using work as follows: A partially amplified signal is split in two, and half of it goes down a long cable that has a delay equal to the revolution period of a particle in the core. The signals are combined out of phase. The resulting transfer function is given by $G(\omega)$:

$$G(\omega) = \frac{1}{2}\{(1 - \cos \omega T_c) + (j \sin \omega T_c)\} . \qquad (6)$$

Ideally the gain of the filter is zero for every harmonic of the core frequency, $1/T_c$. Since real cables have imperfections that cause small reflections, multiple reflections fill the notch in.

The cable has a diameter of about 4 mm and is about 350 m long. Anyone working with microwaves knows that if one tries to transmit a 2-GHz signal down a standard coaxial cable that is 4 mm in diameter and 350 meters long, very little signal will come out the other end: it would not be a very good device. The cables for these notch filters are made of lead-plated copper with a Teflon insulator. By cooling the cable to liquid helium temperature the losses have been very substantially reduced. We measure the loss to be about 1 dB/1000 m.[26] This makes an excellent filter, and the performance of our system is dependent on these notch filters. Such cables were originally developed by the Furukawa Electric Company for the Nippon Telephone Company of Japan.[27] The advent of single mode optical fibers and the solid state laser made this approach to mass communication unattractive, but this type of cable is the best device developed so far for microwave notch filters. Fortunately

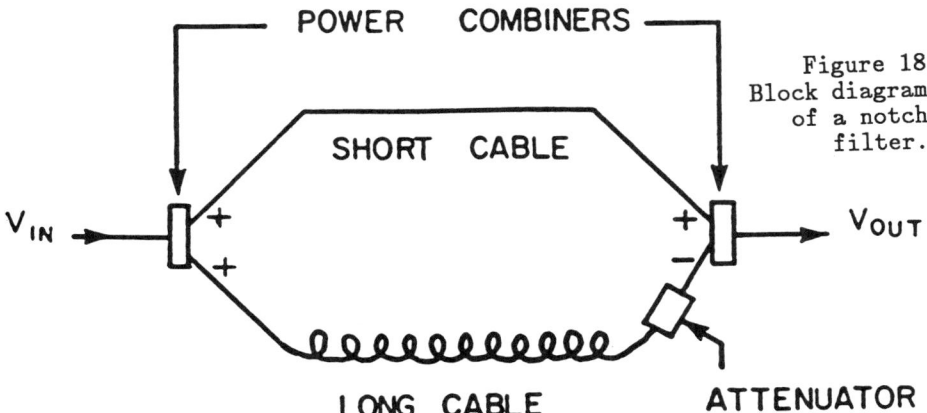

Figure 18
Block diagram of a notch filter.

for us, the development was successful before it was realized that other technology was superior for phone calls. The impressive property of the cable is the absence of imperfections that lead to multiple reflections. The filters have achieved a reduction of 30 dB or more at the core. In this way the contribution of thermal noise to heating the core is adequately suppressed.

7 1985 COMMISSIONING PERFORMANCE

Commissioning of the Antiproton Source began on January 17, 1985, when 120-GeV protons were first extracted from the Main Ring at F17. The next major achievement in the commissioning occurred in April, when 8-GeV protons were stored for the first time in the Debuncher. In August, 8-GeV protons were stored and accelerated from the injection orbit to the core orbit in the Accumulator. Commissioning of the cooling systems for each ring began shortly after a beam was stored in that ring. Antiproton accumulation was successfully carried out on September 6 for the first time. By October 11, it was possible to accumulate 10^9 antiprotons per hour. On October 11 a stack of 10^{10} antiprotons was accumulated, and this proved sufficient to achieve colliding beams early in the morning of October 13.

Although the best performance of the source is a factor of 100 below the design, we believe the causes for the poor performance are well understood and will be eliminated in the next few years. The contributors to this factor of 100 can be broken down as follows:

1. Main Ring intensity was typically 9×10^{11}/pulse, a factor of 2 below design.
2. The Main Ring repetition rate was only 1 pulse/4 sec, a factor of 2 below design.
3. The lithium lens could be operated only at 350 kamp, a factor of 2 below the design current. This reduced the yield of antiprotons per incident proton transported to IQ728 in AP-2 by a factor of 1.5.
4. The acceptance of the remainder of the AP-2 and the Debuncher was such that the flux of antiprotons was reduced by another factor of 6.7. If the aperture of AP-2 and the Debuncher had met the design specifications and if the transfer between AP-2 and the Debuncher had been matched, there should have been no further reduction.
5. The debunching process worked very well since the 3.5% fractional momentum spread was reduced to a 0.25% fractional momentum spread.
6. Since there was typically 3.5 sec for cooling the beam emittance instead of 2 sec and since the number of antiprotons per pulse was down by a factor of 20, the Debuncher betatron cooling systems were not tested anywhere near the limits of their expected performance. At the end of 3.5 sec of cooling the beam emittance was $<3\pi$ mm-mrad.
7. The loss during the transfer between the Debuncher and the Accumulator was a factor of 1.2. This was due to the Accumulator aperture being severely reduced by component misalignment.

8. The transfer of antiprotons from the injection orbit to the stacking orbit reduced the antiproton flux by another factor of 1.8.
9. The cooling of antiprotons into the core reduced the flux by another factor of 1.35.

Since the design allowed for a cumulative reduction in steps 7, 8, and 9 of 1.15, the product of all factors corresponds to a cumulative reduction relative to the design of a factor of 100. Improvements can be made in a straightforward manner in item 2, the Main Ring repetition rate, item 4, the AP-2 and Debuncher Aperture, and item 7, the transfer efficiency. The Main Ring should be capable of reaching a 2.2-sec repetition rate in 1987, because of a number of improvements that have beem made to its power supply since October 1985. Items 4 and 7 will require careful and painstaking searches for the causes of the aperture restrictions. Typically one expects to find many small limitations, none of which would be a cause for concern by itself. The low current in the Li lens, item 3, was due to failure of a transformer housing which is now understood and is being corrected.

It may be difficult to raise the Main Ring intensity to 1.8×10^{12}/pulse while preserving the longitudinal emittance. A major improvement program was initiated to remove smooth sharp discontinuities in the Main Ring vacuum chamber in order to reduce the Main Ring impedance. At the same time a major program is underway to improve the Booster beam quality. It is hoped that these improvements will make it possible to accelerate and bunch-rotate close to 1.5×10^{12} protons/pulse without encountering longitudinal emittance growth caused by microwave instabilities.

On the basis of the improvements in progress, the flux of antiprotons injected into the Debuncher is expected to increase by a factor of 20. Although this will put a substantially heavier burden on the cooling system, the $\bar{p}$ flux that must be cooled is still a factor of 5 below design. On the basis of small improvements being made to the cooling system, we expect it to handle this flux without significant loss.

The time required for the rest of the Tevatron to reach its performance goals is likely to be longer, since improved Antiproton Source performance is a prerequisite for improving the efficiency of the steps that take the antiprotons from the Accumulator to the Tevatron. Nevertheless it is hoped that a luminosity of 10^{29} can be reached in the spring of 1987.

8 REFERENCES

1. J. Peoples, The Fermilab Antiproton Source, IEEE Trans. Nucl. Sci. NS-30, 1970 (1983); Design Report, Tevatron I Project, Sept. 1984 (Fermilab).
2. C. Rubbia, Experimental Observation of the Intermediate Vector Bosons $W^+ W^-$, and Z^0, Rev. Mod. Phys. 57, 699 (1985).
3. E. Eichten, I. Hinchliffe, K. Lane, and C. Quigg, Rev. Mod. Phys. 56, 579 (1984) and Errata, Fermilab-PUB86/75-T (1986).
4. M. Banner et al. (UA-2 collaboration), Phys. Lett. 118B, 203 (1982); G. Arneson et al. (UA-1 collaboration), Phys. Lett. 123B, 115 (1983).

5. D. E. Johnson, The B0 Low-Beta Insertion Design for the Tevatron, IEEE Trans. Nucl. Sci. NS-32, 1672 (1985); K. Koepke, E. Fisk, G. Mulholland, and H. Pfeffer, The Tevatron B0 Low-Beta System, ibid., 1675; D. A. Finley, R. P. Johnson, and F. Willeke, Control and Initial Operation of the Fermilab B0 Low-β Insertion, ibid., 1678.
6. J. Bjorken and S. Mtingwa, Fermilab-Pub82/47-THY, July 1982. This reference contains results appropriate for the Tevatron.
7. A. Piwinski, Proc. 9th Int. Conf. on High Energy Accelerators, 1974, p. 405.
8. L. R. Evans and J. Gareye, Beam-Beam and Single Beam Effects in the SPS Proton-Antiproton Collider, IEEE Trans. Nucl. Sci. NS-30, 2397 (1983).
9. L. R. Evans and J. Gareyte, Performance Limitations of the CERN SPS Collider, IEEE Trans. Nucl. Sci. NS-32, 2234 (1985).
10. K-Y. Ng, these proceedings.
11. C. Hojvat and A. J. Van Ginneken, Calculation of Antiproton Yields for the Fermilab Antiproton Source, Nucl. Instrum. Methods 206, 67 (1983).
12. B. F. Bayonov, J. N. Petrov, G. I. Silvestrov, J. A. MacLachlan and G. L. Nicholls, A Lithium Lens for Axially Symmetric Focusing of High Energy Particle Beams, Nucl. Instrum. Method, 190, 9 (1981).
13. G. Dugan, C. Hojvat, A. J. Lennox, G. Biallas, F. Cilyo, M. Leninger, J. McCarthy, W. Sax, and S. Snowdon, Mechanical and Electrical Design of the Fermilab Li Lens, IEEE Trans. Nucl. Sci. NS-30, 3660 (1983).
14. J. Griffin, J. MacLachlan, A. G. Ruggiero, and K. Takayama, Tune and Momentum Exchange for Production and Collection of Intense Antiproton Beams at Fermilab, IEEE Trans. Nucl. Sci. NS-30, 2630 (1983).
15. B. Autin, J. Marriner, A. Ruggiero, and K. Takayama, Fast Betatron Cooling in the Debuncher Ring in the Fermilab Tevatron I Project, IEEE Trans. Nucl. Sci. NS-30, 2593 (1983).
16. Design Report, Tevatron I Project, Sept. 1984, pp. 5-8 (Fermilab).
17. A. Ando and K. Takayama, Effects of rf Stacking on Cooling in the Fermilab Antiproton Accumulator, IEEE Trans. Nucl. Sci. NS-30, 2601 (1983).
18. J. E. Griffin, C. Ankenbrandt, J. A. MacLachlan, and A. Moretti, Isolated Bucket RF Systems in the Fermilab Antiproton Facility, IEEE Trans. Nucl. Sci. NS-30, 3502 (1983).
19. A. Ruggiero, The Fermilab Tevatron I Debuncher Ring, IEEE Trans. Nucl. Sci. NS-30, 2478 (1983).
20. A. Ando, T. L. Collins, and D. E. Johnson, Design of an 8 GeV Accumulator Ring for the Fermilab Tevatron I Project, IEEE Trans. Nucl. Sci. NS-30, 2031 (1983).
21. J. K. Johnson and R. Nemetz, Power Combines/Dividers for Loop Pickup and Kicker Arrays for FNAL Stochastic Cooling Rings, IEEE Trans. Nucl. Sci. NS-32, 2171 (1985).
22. S. van der Meer, Stochastic Damping of Betatron Oscillations in the ISR, CERN/ISR-PO/72-31, 1972. More complete descriptions

of the principles of stochastic cooling are given in D. Mohl, G. Petrucci, L. Thorndahl, and S. van der Meer, Physics and Techniques of Stochastic Cooling, Phys. Rep. C58, 73-119 (1980).

23. F. A. Voelker, T. Henderson, and J. Johnson, An Array of 1 to 2 GHz Electrodes for Stochastic Cooling, IEEE Trans. Nucl. Sci. NS-30, 2262 (1983); D. A. Goldberg, G. R. Lambertson, F. Voelker, and L. Shalz, Measurement of Frequency Response of LBL Stochastic Cooling Arrays for Tev I Storage Rings, ibid., NS-32, 2168 (1985).

24. C. C. Lo, Stochastic Beam Cooling Amplifier System Front End Components Performance Characteristics, IEEE Trans. Nucl. Sci. NS-32, 2174 (1985).

25. S. van der Meer, Stochastic Stacking in the Antiproton Accumulator, CERN/PS/AA/78-22, Oct. 1978 (CERN).

26. M. Kuchnir, J. D. McCarthy, and R. J. Pasquinelli, Superconducting Delay Line for Stochastic Cooling Filters, IEEE Trans. Nucl. Sci. NS-30, 3360 (1983).

27. Y. Hoshiko, Proc. Fifth Int. Cryogenic Engineering Conf., (ICEC5), p. 282, Tokyo (1975).

PRIMER ON BEAM DYNAMICS IN SYNCHROTRONS

L. Teng
Fermi National Accelerator Laboratory
P. O. Box 500, Batavia, IL 60510

TABLE OF CONTENTS

1	Intensity-Independent Dynamics...............................	1879
1.1	Linear Transverse Motion.............................	1879
1.2	Momentum Effects.....................................	1880
1.3	Nonlinear Transverse Motion..........................	1881
1.4	Longitudinal Motion..................................	1884
2	Intensity-Dependent Dynamics.................................	1885
2.1	Emittance..	1885
2.2	Static Beam-Field Effects............................	1886
2.3	Coherent Instabilities...............................	1889
2.4	Transition Problems..................................	1892
2.5	Incoherent High-Intensity Effect: Intrabeam Scattering.................................	1893
2.6	Luminosity Issues....................................	1893
Appendix A	Space-Charge Tune Shift............................	1896
A.1	Circular Beams.......................................	1896
A.2	Elliptical Beams.....................................	1898
Appendix B	Formulas Related to Longitudinal Oscillation.......	1900
Appendix C	Formulas Related to Transition Crossing............	1903
C.1	Microwave Instability................................	1906
C.2	Space-Charge Mismatch................................	1907

Chapter 4

PRIMER ON BEAM DYNAMICS IN SYNCHROTRONS

L. Teng
Fermi National Accelerator Laboratory
P. O. Box 500, Batavia, IL 60510

1 INTENSITY-INDEPENDENT DYNAMICS

We discuss first the factors controlling the beam dynamics when the beam intensity, hence the self-field generated by the beam, is negligible. In this case the motions of the particles in the beam are independent and we have the so-called single-particle dynamics. The single-particle dynamics is clearly controlled only by external electromagnetic fields and external physical barriers such as beam collimators and vacuum chamber walls. The transverse motion is controlled principally by the magnetic field, and the longitudinal motion is controlled by the radiofrequency electric field. These motions are discussed in more detail below.

1.1 Linear Transverse Motion

The closed orbit, hence the overall geometry of the beam, is determined by the dipole field on the orbit. The closed orbit is generally unique for a given particle momentum. The dipole field is usually designed to be vertical on the closed orbit and to have a high sector-periodicity. This gives a planar closed orbit with a sector-periodic geometry. Since particles travel in a narrow beam the effects of the magnetic field are most simply discussed by expanding the field in powers of the transverse coordinates x(horizontal) and y(vertical) measured from the closed orbit. The coefficients defining these multipole fields are given by

$$B_y + iB_x = B_o \Sigma_n (b_n + ia_n)(x + iy)^n \qquad (1)$$

where all quantities are functions of the distance s along the closed orbit, B_o is the vertical field on orbit (x = y = 0), and b_n and a_n are respectively the normal and skew 2(n + 1)-pole coefficients. So defined $b_o = 1$ and $a_o = 0$ for error-free field. The quadrupole fields specified by b_1 and a_1 produce linear focusing actions. The particles are guided by the quadrupole field to oscillate stably (betatron oscillations) about the closed orbit. Generally the quadrupole field is designed such that the skew component $a_1 = 0$ and the normal component b_1 has the same high orbital sector-periodicity. The horizontal and vertical betatron oscillations are then uncoupled, each given by

$$x(\text{or } y) \propto \sqrt{\beta} \; e^{\pm i \int ds/\beta} \qquad (2)$$

where $\beta = \beta(s)$ having the sector-periodicity is called the amplitude function. The number of oscillations per revolution or the tune, ν, is defined by

$$\nu = \frac{1}{2\pi} \int_0^{2\pi R} ds/\beta \qquad (3)$$

where $2\pi R$ is the circumference of the closed orbit.

Errors in the dipole field cause closed orbit distortions. Closed orbit distortions are undesirable because they reduce the effective aperture of the ring. Vertical dipole field errors are produced by construction errors of the dipole magnets and horizontal misalignments of the quadrupole magnets; horizontal dipole field errors are produced by roll errors of the dipole magnets and vertical misalignments of the quadrupole magnets. Closed orbit distortions can be corrected by realigning the magnets or by using steering dipoles.

Errors in the quadrupole field cause distortions in the amplitude function β. Amplitude distortions also reduce the effective aperture of the ring but their magnitude is generally smaller than that due to closed orbit distortions and correction is seldom necessary.

Both the closed orbit distortion and the amplitude distortion blow up on resonances. It is easy to see that if the tune is an integer, m, orbits with oscillation are also closed, and hence the closed orbit is no longer unique. Any dipole error with harmonic m will drive an arbitrarily large closed orbit distortion. It is less transparent but equally suggestive that if the oscillation is closed in two revolutions (2ν = integer = m) quadrupole errors with harmonic m will drive an arbitrarily large amplitude. Since driving error fields are unavoidable, in either the case of integer or half-integer resonance the motion becomes unstable. The half-integer resonance has, in fact, a finite width $\Delta\nu$ within which the motion is always unstable. This "stopband" width is given by

$$\Delta\nu = \text{mth harmonic amplitude of } \left[\frac{\beta}{4}\frac{b_1}{\rho}\right]$$

$$= c_2 \text{ defined by Eq. (11) below} \qquad (4)$$

where $\rho = \rho(s)$ is the radius of the closed orbit. The coupled linear resonances $\nu_x \pm \nu_y$ = integer = m are driven by the mth harmonic of the skew quadrupole field given by a_1. The sum resonance $\nu_x + \nu_y$ = m also has a finite width stopband.

1.2 Momentum Effects

A ring magnet lattice is capable of confining particles over a limited range of momentum, each particle traveling about the closed orbit corresponding to its own momentum. For planar orbits the orbits of different momenta are separated horizontally. The orbit displacement per relative momentum increment, $\Delta p/p$, is called the dispersion function D, which has, of course, also the sector-periodicity and is given by

$$D \equiv \frac{\Delta x}{\Delta p/p} = \frac{1}{2\nu \sin \pi \nu} \int_s^{s+2\pi R} \frac{1}{\rho} \cos[\nu \pi + \phi(s) - \phi(r)] dr \qquad (5)$$

where $\phi(s) = \int ds/\beta$ is the betatron oscillation phase. The relative circumference increment per $\Delta p/p$ is called the momentum compaction factor α and is given by

$$\alpha = \frac{\Delta R/R}{\Delta p/p} = \frac{1}{2\pi R} \int_0^{2\pi R} \frac{D}{\rho} ds . \qquad (6)$$

The amplitude function is also dependent on momentum. The relative change in the amplitude function β per $\Delta p/p$ is given by

$$\frac{\Delta \beta/\beta}{\Delta p/p} = \frac{1}{2\sin 2\pi \nu} \int_s^{s+2\pi R} \frac{b_1}{\rho} \beta \cos 2[\nu \pi + \phi(s) - \phi(r)] dr . \qquad (7)$$

Integrating around the closed orbit gives the momentum dependence of the tune. The tune change per $\Delta p/p$ is called the chromaticity, ξ, and is given by

$$\xi \equiv \frac{\Delta \nu}{\Delta p/p} = -\frac{1}{2\pi} \int_0^{2\pi R} \frac{\Delta \beta/\beta}{\Delta p/p} \frac{ds}{\beta} = -\frac{1}{4\pi} \int_0^{2\pi R} \frac{b_1}{\rho} \beta ds . \qquad (8)$$

When the momentum deviation $\Delta p/p$ is too large, the dispersion may cause the particle to strike the horizontal aperture, or the chromaticity may run the tune onto resonance values. Thus both the dispersion and the chromaticity act to define the momentum aperture of the ring.

1.3 Nonlinear Transverse Motion

Nonlinear fields are introduced either deliberately or inadvertently through errors and beam-beam interactions in colliders. Sextupole field is introduced at places where the dispersion function D is large to modify or compensate for the natural chromaticity. Dispersion puts orbits with different momenta at different horizontal positions in the sextupole field, hence under the actions of different quadrupole fields. This introduces an additional momentum dependence of the tune which may be adjusted to compensate for the natural chromaticity. To modify the chromaticities independently in both transverse planes we need two sets of sextupoles placed at locations with greatly different ratios β_y/β_x and hence having very different effects on the horizontal and the vertical chromaticities.

Octupoles must be introduced when half-integer is employed for slow extraction.

In any case, the dynamics of the beam particles is, in fact, always nonlinear. With nonlinearities, the tunes ν_x and ν_y are amplitude dependent. As the amplitudes grow the tunes will encounter a succession of resonances. The 4-dimensional phase space is thus crisscrossed with intersecting surfaces in the shape of tori on which the motion has resonant tune values. All resonances, linear and nonlinear, may be summed up in the formula

$$j\nu_x \pm k\nu_y = m \qquad (9)$$

where j, k, and m are positive integers or zero. Each resonance is excited by the mth harmonic of the 2n-pole field where $n \equiv j + k$ is the "order" of the resonance. The linear integer and half-integer resonances thus have orders 1 and 2 respectively. Nonlinear resonances are of orders ≥ 3. Those with $j = n, n - 2, n - 4, \cdots$ are excited by the normal field b_{n-1} and those with $j = n - 1, n - 3, \cdots$ are excited by the skew field a_{n-1}. For example, those excited by norm $\nu_x \pm 2\nu_y = m$, and $2\nu_x \pm \nu_y = m$, $3\nu_y = m$.

To understand or visualize the features of nonlinear motions it is best to start with a single degree of freedom. The proper normalized coordinates in the 2-dimensional phase space are $\sqrt{\beta}\, x$ and $\sqrt{\beta}\, x' - (\beta'/2)(x/\sqrt{\beta})$ (prime means d/ds). Consider the turn-to-turn mapping curve of an oscillation with small amplitude tune ν_0 not on a resonance. In the normalized coordinates the mapping curve of a small linear oscillation is just a circle. For large oscillation, as the tune approaches a resonance, say $n\nu = m$, the mapping curve takes on the shape of a regular n-sided polygon with rounded corners. All stable phase points are contained in the central stable region, which is an area bounded by an n-cornered figure formed by the separatrices of the resonance. The corners are the unstable fixed points. To first approximation the radial distance of the unstable fixed points from the origin is given by

$$r_n = \left(\frac{\left| \nu_0 - \frac{m}{n} \right|}{C_n} \right)^{\frac{1}{n-2}} \qquad (10)$$

where C_n is the resonance driving harmonic amplitude given by

$$C_n \cos(m \frac{s}{R} - \text{"phase"}) = \frac{\beta_x^{\frac{n}{2}}}{2^n} \frac{b_{n-1}}{\rho} . \qquad (11)$$

To this lowest order of approximation the separatrices can also be expressed in simple analytical forms.

This stable region defines the dynamic aperture. Outside this region the motion is at least locally unstable. The stable area is somewhat smaller than $\pi(r_n)^2$ and goes to zero as ν_0 approaches the resonant value m/n. The tune deviation $|\nu_0 - m/n|$ for which the stable area is just enough to contain the beam, namely just equal to the emittance of the beam (see Section 2 below) is defined as the half-width of the resonance. In practical cases this first-order single degree-of-freedom picture always gives resonance widths much smaller than measured values, indicating that more precise two degrees-of-freedom computations are needed. Clearly the dynamic aperture has to be larger than the beam, but, depending on the magnitude of the driving term and the separation of the small amplitude tune from resonance, the dynamic aperture may be smaller than the physical aperture. In this case the stable particle motion is limited by the dynamic aperture.

Nonlinear motions in two coupled degrees of freedom have the same behavior but are more complicated and more difficult to visualize. In the 4-dimensional phase space the mapping points of oscillations not on resonances lie on closed 2-dimensional tori. The projections of these mapping points onto 2-dimensional phase planes corresponding to each degree of freedom cover broad bands which encircle the origins. The projection points seem to scatter all over the bands. The scatter makes the emittance appear larger and have a fuzzier boundary. This makes the beam loss versus tune curves show broad valleys at resonances instead of narrow gulches.

The shapes of the tori and of the stability boundaries, and hence the dynamic aperture, can all be derived analytically to any arbitrary order. To the lowest order the effect of the nonlinear term is to introduce distortion functions on the linear amplitude and phase. However, the algebra involved is rather complicated especially when many high-order nonlinear terms are present. It is easier to use a straightforward tracking program to compute the dynamic aperture numerically. This is the favored approach at present.

The dynamic aperture was investigated long ago and understood in connection with sector-focusing cyclotrons and fixed-field alternating-gradient accelerators, in which the fields are extremely nonlinear and the physical apertures (at least the horizontal aperture) are essentially nonexistent.

Another complication of the nonlinear dynamics is the existence of the stochastic regime of solutions. These solutions generally appear in stochastic layers near separatrices and unstable fixed points, where many high-order resonances overlap. In the projections of motion near the stability limit, sometimes higher-order resonance loops do show up. These stochastic layers make the boundaries of the central stable region fuzzy, but they are fairly narrow and do not sensibly affect the definition of the dynamic aperture. The stochastic regime motion plays a more major role in determining the limitations for the beam-beam interaction in colliders. This will be discussed in Section 2 below.

Distinct from the dynamic aperture is the physical aperture defined by beam collimators or beam pipe walls. The physical aperture is, of course, much more definite, easier to understand, and simpler to calculate.

1.4 Longitudinal Motion

The longitudinal motion of the particles is controlled mainly by the radio-frequency electric field. The motion is intrinsically nonlinear and can be approximated as linear only when the amplitude is small. The nonlinearity is generally that of a sinusoidal electric field. The coordinate is the longitudinal displacement, z, from the synchronous position. The stable region in the (z, z' ≡ dz/ds) phase plane is bounded by a single separatrix passing through a single unstable fixed point. The stable region is shaped like a tear drop and is called the rf bucket. Similar to the transverse motion, the small, approximately linear oscillation can be written as

$$z \propto \sqrt{\beta_z} \, e^{\pm i \int ds/\beta_z} \tag{12}$$

where

$$\beta_z = \sqrt{\frac{2\pi}{h} \frac{E}{eV_o} \frac{\beta^2 R^2}{\eta \cos\phi_s}} \;,$$

$E = mc^2\gamma$ = total energy,

h = harmonic number,

V_o = peak rf voltage per turn,

$\eta = \frac{d\omega/\omega}{dp/p}$ = revolution frequency dispersion,

ϕ_s = synchronous phase angle,

except now β_z is an adiabatic constant and the motion is sinusoidal. The number of oscillations per turn, namely the longitudinal tune ν_z is given by

$$\nu_z = \frac{1}{2\pi} \int_0^{2\pi R} \frac{ds}{\beta_z} = \frac{R}{\beta_z} \;. \tag{13}$$

The longitudinal focusing is generally rather weak, and ν_z is very small, approaching 0.1 only for high energy, high repetition rate synchrotrons. Thus, there are no pure longitudinal resonances. The longitudinal oscillation (also called synchrotron oscillation or phase oscillation) does, however, contribute to transverse resonances through coupling to the horizontal oscillation by the orbit dispersion. The lowest-order coupling term in the Hamiltonian is proportional to

$$(xD' + x'D)z, \quad \text{prime} = d/ds \;. \tag{14}$$

The longitudinal oscillation therefore contributes to the side-bands

$$\nu_x \pm \ell \nu_z, \quad \ell = \text{integer} \;, \tag{15}$$

of the horizontal oscillation and creates the overall resonant conditions

$$j\nu_x \pm k\nu_y \pm \ell\nu_z = m . \qquad (16)$$

This so-called synchro-betatron coupling, and hence the resonance strengths, become progressively weaker at higher ℓ values. The coupling can be eliminated altogether either by placing the rf cavities in zero dispersion (D = D' = 0) straight sections or by judiciously distributing the cavities around the ring lattice such that all the coupling terms [Eq. (14)] add up to zero.

The longitudinal oscillation remains fairly linear within the central half of the area of the bucket. The motion becomes strongly nonlinear only when it gets close to the wall of the bucket (the separatrix). As for all nonlinear motions, the longitudinal motion becomes stochastic within a stochastic layer next to the separatrix and the unstable fixed point, but the stochastic layer is usually very thin.

2 INTENSITY-DEPENDENT DYNAMICS

2.1 Emittance

Particles in a beam bunch oscillate about the synchronous closed orbit (plotted as the origin of the phase space) and populate a central volume of the phase space. The coordinate variables of the 6-dimensional phase space are simply x, y, and z as defined in Section 1 above, and the commonly used momentum variables are listed in Table I, where p_S is the synchronous momentum.

Table I

Coordinate variables	Momentum variables		
	I	II	III
x	p_x	$x' \equiv p_x/p_S$	$\beta\gamma x' = p_x/mc$
y	p_z	$y' \equiv p_y/p_S$	$\beta\gamma y' = p_z/mc$
z	$p_z (=p-p_S)$	$z' \equiv \eta p_z/p_S$	$\beta\gamma z' = \eta p_z/mc$

The independent variable is either time t or the distance s along the closed orbit. Set I gives the proper conjugate momentum variables. With these variables the 6-dimensional volume of the phase space has the unit (eV sec)^3 and is an invariant of the motion. With variables of set II the 6-dimensional phase volume has the simpler unit m^3 but shrinks as p_S^{-3}. The set III variables provide both an invariant phase volume and the simple unit.

The 2-dimensional area formed by projecting the 6-dimensional phase volume, which is populated by particle phase points, on the phase plane of one specific degree of freedom is called the emittance in that degree of freedom. This is not a well-defined concept, because the density distribution in the populated phase volume varies from beam to beam, and the boundary of the volume is generally fuzzy. Two extreme density distributions in the 2-dimensional projection are usually considered, uniform and bi-Gaussian.

If the density is uniform inside an area with sharp boundaries, the emittance is simply the bounded area.

For a linear lattice the "closed" boundary shape (namely one that has the sector periodicity) is an ellipse. The extent of the ellipse along the coordinate variable x, say, is just the x-width of the beam. Denoting the half-width by a_x, we can write the area of the x phase-ellipse or the x emittance as

$$\epsilon_x = \begin{cases} \dfrac{\pi a_x^2}{\beta_x} & \text{(un-normalized in x, x' plane)} \\[2ex] \beta\gamma \dfrac{\pi a_x^2}{\beta_x} & \text{(normalized in x, } \beta\gamma\text{x' plane)} \end{cases} \qquad (17)$$

If the density is bi-Gaussian in x and x', and the rms beam half-width is σ_x, the emittance is usually defined as

$$\epsilon_x = \begin{cases} \dfrac{6\pi\sigma_x^2}{\beta_x} & \text{(un-normalized)} \\[2ex] \beta\gamma \dfrac{6\pi\sigma_x^2}{\beta_x} & \text{(normalized)} \end{cases} \qquad (18)$$

which contains 95% of the beam.

The density function of a real beam is never this simple, especially with nonlinear fields present. One has the choice of using either an iso-density curve or an ellipse that contains, say, 95% of the beam, to define the emittance. The latter is more practical because in all likelihood the beam will be further transported in a linear periodic lattice, and the phase points inside the iso-density curve will be smeared out to fill an ellipse.

The choice of 95% is arbitrary. The CERN convention is to use $4\pi\sigma^2/\beta$ for the Gaussian distribution. Such an emittance contains 86.5% of the beam.

2.2 Static Beam-Field Effects

Assuming the beam is stable and the density distribution in the beam bunch is in a steady state, one can calculate the effect of the electromagnetic field produced by the beam (beam-field or self-field) on the motion of individual particles in the beam. In the transverse plane the effect is a detuning of the betatron oscillations. It is convenient to consider the effect as resulting from two different contributions, the space-charge contribution and the image-charge contribution.

The transverse space-charge (and current) force has an energy dependence of $1 - \beta^2 = 1/\gamma^2$, the result of the cancellation between the electric defocusing force (factor 1) and the magnetic focusing force (factor β^2). The tune shift $\delta\nu$ contains, in addition, the energy factor $1/\beta^2\gamma$, where β^2 arises from the tune being expressed in terms of the angular velocity, and γ arises from the relativistic mass increase. The dependence of the force on the transverse coordinates is related crucially to the particle density distribution in the beam. If the density distribution is uniform inside an elliptical beam, the transverse force is linear up to the edge of the beam, and the tune depression is independent of amplitude for oscillations inside the beam. The x and y tune depressions are given by

$$\begin{cases} \delta\nu_x = -\dfrac{r_p}{\beta^2\gamma^3}\dfrac{1}{\epsilon_x}\int\dfrac{\lambda ds}{1 + \dfrac{a_y}{a_x}}, & \epsilon_x \equiv \dfrac{\pi a_x^2}{\beta_x} \\ \\ \delta\nu_y = -\dfrac{r_p}{\beta^2\gamma^3}\dfrac{1}{\epsilon_y}\int\dfrac{\lambda ds}{1 + \dfrac{a_x}{a_y}}, & \epsilon_y \equiv \dfrac{\pi a_y^2}{\beta_y} \end{cases} \quad (19)$$

where $r_p \equiv e^2/mc^2 = 1.535 \times 10^{-18}$, m = classical radius of proton, $a_x(s)$ and $a_y(s)$ are the semi-axes of the beam cross-section ellipse, and $\lambda = \lambda(s)$ is the local linear particle density.

As expected, the tune shift is larger in the direction of the minor axis of the ellipse. This simple but rather unrealistic distribution is called the Kapchinsky-Vladimirsky distribution. In the 4-dimensional transverse phase space (x, x', y, y') this corresponds to a δ-function distribution on a 4-dimensional ellipsoidal shell.

If the density distribution in the elliptical beam is bi-Gaussian in x and y, we get a spread in the tune depressions. The depressions are greatest for the smallest amplitude oscillations in the dense core of the beam and are

$$\begin{cases} \delta\nu_{x_{max}} = -\dfrac{3r_p}{\beta^2\gamma^3}\dfrac{1}{\epsilon_x}\int\dfrac{\lambda ds}{1 + \dfrac{\sigma_y}{\sigma_x}}, & \epsilon_x \equiv \dfrac{6\pi\sigma_x^2}{\beta_x} \\ \\ \delta\nu_{y_{max}} = -\dfrac{3r_p}{\beta^2\gamma^3}\dfrac{1}{\epsilon_y}\int\dfrac{\lambda ds}{1 + \dfrac{\sigma_x}{\sigma_y}}, & \epsilon_y \equiv \dfrac{6\pi\sigma_y^2}{\beta_y} \end{cases} \quad (20)$$

where σ_x and σ_y are the standard deviations of the Gaussian distributions or the rms half-widths of the beam. With the usual definition of $\epsilon = 6\pi\sigma^2/\beta$ for the Gaussian distribution, the maximum tune shift is 3 times that of the uniform distribution.

Neither of these distributions is realistic, but this discussion shows convincingly that the realistic space-charge tune depression has a spread from a value approaching zero for the largest amplitude oscillations to a value approaching but likely no greater than that given by Eq. (20) for the smallest amplitude oscillations.

The image charge (and current) force does not contain the electric-magnetic cancellation factor γ^{-2} and therefore tends to dominate at high energies. To first order it depends only on the linear density of the beam and the cross-sectional dimensions of the imaging beam pipe and magnet poles, and not on the cross-section of the beam. Blowing up the transverse dimensions of the beam reduces the space-charge tune depression but not the image-charge term. To reduce the image-charge term, one has to enlarge the beam pipe. Also, since the image charge (and current) is external to the beam, the effects of its field on the beam are opposite in the two transverse planes as necessitated by the Laplace equation.

The electric image tune shifts are

$$\begin{cases} \delta\nu_x = c_1 \dfrac{r_p}{\beta^2\gamma} \dfrac{2\pi R\lambda}{\langle\pi h^2/\beta_x\rangle} \\[2ex] \delta\nu_y = -c_1 \dfrac{r_p}{\beta^2\gamma} \dfrac{2\pi R\lambda}{\langle\pi h^2/\beta_y\rangle} \end{cases} \quad (21)$$

where $2h$ = vertical separation of the assumed electric imaging surfaces, c_1 = numerical factor depending on the shape of the imaging surface (= $\pi^2/48$ for parallel planes), and $\langle\rangle$ = averaging around the ring.

The magnetic image tune shifts are

$$\begin{cases} \delta\nu_x = -c_1 \dfrac{r_p}{\gamma} \dfrac{2\pi R\lambda - N}{\langle\pi h^2/\beta_x\rangle} + c_2 \dfrac{r_p}{\gamma} \dfrac{N}{\langle\pi g^2/\beta_x\rangle} \\[2ex] \delta\nu_y = c_1 \dfrac{r_p}{\gamma} \dfrac{2\pi R\lambda - N}{\langle\pi h^2/\beta_y\rangle} - c_2 \dfrac{r_p}{\gamma} \dfrac{N}{\langle\pi g^2/\beta_y\rangle} \end{cases} \quad (22)$$

where $2g$ = vertical separation of the assumed dc magnetic imaging surfaces, $N = \int\lambda ds$ = number of particles in ring, and c_2 = numerical factor depending on the shape of the surface (= $\pi^2/24$ for parallel planes).

In these expressions the first terms are the ac magnetic image terms and the second terms are the dc magnetic terms. The shape factors c_1 and c_2 have been computed for other than the parallel plane geometry.

We have assumed that the image forces are instantaneously in phase with the beam bunch. Since the bunch is moving and since the imaging vacuum pipe, magnet poles, etc., are all electromagnetically active elements, the image force can have out-of-phase components and can therefore induce oscillations in the beam. This will be discussed further below.

For the longitudinal beam-field force, as a simple approximation, we assume the beam bunch to be a line charge with linear density $\lambda(z)$ along the centerline of a beam-pipe that has a capacitance per unit length C. For a circular conducting beampipe of radius b and a circular beam of radius a, $C = [1 + 2\ln(b/a)]^{-1}$ is a reasonable approximation. The charge distribution then produces a voltage distribution $e\lambda/C$ and a longitudinal field $E_z = -(e/C)d\lambda/dz$. If λ is parabolic, say

$$\lambda(z) = \frac{3}{4} \frac{N}{a_z} \left(1 - \frac{z^2}{a_z^2}\right) \qquad (23)$$

with $2a_z$ = bunch length and N = number of particles in bunch, we have

$$E_z = \frac{3}{2} \frac{eN}{Ca_z^3} z , \qquad (24)$$

namely a linear force directed away from the midpoint of the beam bunch (z = 0). The response of the particles is defocusing below transition ($\eta > 0$) and focusing above transition ($\eta < 0$). Thus, the particle behaves as though it has a negative longitudinal mass above transition. The consequences of the longitudinal beam-field force and the negative mass effect on transition crossing will be discussed in further detail later.

In reality the linear density distribution is likely to be more complicated than simple parabolic, and the varying transverse size of the beam will make the longitudinal self-field force dependent also on x and y. The above descriptions of both the transverse and the longitudinal effects are oversimplifications that help create a physical understanding of the basic processes involved and nevertheless give quantitatively reasonable and approximate estimates.

2.3 Coherent Instabilities

The particle beam traveling in an accelerator is surrounded by and coupled to a great number of electromagnetically active elements each of which can be represented electromagnetically by a complex impedance. These include, e.g., the resistive beampipe wall, discontinuities or structures formed in the beampipe such as bellows and rf cavities, apparatus inserted inside the pipe such as kicker magnets, beam position monitors, etc. The bunched beam current I is rich in harmonic content and induces a voltage per turn IZ where Z is the total impedance of all the electromagnetic elements in the ring which are coupled to the beam. This voltage acts back on the beam particles with a force $U + iV \equiv eIZ$. If the action on the particles is a positive feedback, and if the motions of the particles stay coherent for a long enough time that the positive feedback can be considered as acting on the beam as a whole, a coherent instability in the beam will result and the beam may be lost. Depending on the length of the decay time of this wakefield, a beam bunch may feel its own wakefield on the next turn around and become unstable; this is called the self-excited or turn-to-turn instability. Or the wakefield of one beam bunch may be felt by the succeeding bunches and induce the coupled-bunch or bunch-to-bunch instability. Rather complete analyses have been made of the behaviors of the different modes and the onset thresholds of these instabilities. Fortunately, in most practical cases the wakefield of one beam bunch is effectively attenuated before the next bunch arrives, and these instabilities are not excited. Even in the case when they are excited, these stabilities are easily cured or damped either by electronic feedback or by Landau damping ("decoherencing" motions of individual bunches).

We are therefore left with only the intra-bunch or single-bunch instabilities. The frequencies of these single-bunch instabilities are too high for damping by available electronics, and the effect of Landau damping is limited in magnitude. Together with some incoherent effects discussed below, the single-bunch coherent instability usually imposes ultimate limitations on the beam current. To increase the beam current that can be accelerated we must either reduce the impedance or reduce the coherence time (i.e. increase the Landau damping).

Wakefields with long decay times are generally induced by the high-Q parasitic modes of the accelerating rf cavities and can usually be eliminated by damping out these modes in the cavities. Or we can "decoherence" the motions of the different bunches. Longitudinally we can make the synchrotron oscillation frequencies of the bunches different by adding a cavity operating, e.g., at the harmonic number h + 1. Transversely the betatron tunes of different bunches can be made different by using a radiofrequency quadrupole that imposes different quadrupole fields on different bunches. For proton synchrotrons it is generally sufficient just to damp out the harmful parasitic modes in the cavities.

The longitudinal single-bunch instability, commonly known as the microwave instability, induces very short longitudinal lumping of the particles at microwave frequencies within a beam bunch. This instability is stabilized by Landau damping through a spread in the revolution frequency due to the momentum spread. The threshold of the instability expressed as the maximum allowed longitudinal impedance Z_ℓ for given beam current and momentum spread is

$$\frac{|Z_\ell|}{n} \le F_\ell \frac{E}{eI} \beta^2 |\eta| \left(\frac{\Delta p}{p}\right)^2 \qquad (25)$$

where n = (instability frequency)/(revolution frequency) = mode number, I = peak current in bunch, $\Delta p/p$ = peak momentum spread (FWHM) in bunch, and F_ℓ = form factor of order unity.

The attainable value of $|Z_\ell|/n$ has a practical lower limit, but with proper care a value of ~ 1 Ω can easily be attained.

The transverse single-bunch instability is also known as the high-mode head-tail instability. The primary excitation mechanism is the following. The field generated by the transverse oscillation of the head of the beam bunch acts on particles in the tail. Because of the phase difference between the head and the tail produced through a non-zero chromaticity by the momentum swing during synchrotron oscillation, this excitation force has the necessary out-of-phase component to induce instability. The particles in the head and the tail are continually interchanged by synchrotron oscillation. Thus, the instability of the whole beam bunch is self-regenerative. This instability can be "cured" mainly by Landau damping arising from a spread in betatron tune. In principle, it can also be damped by a spread in the synchrotron oscillation frequency which produces a mixing of particles along the length of the bunch. But in practice it is difficult to attain sufficiently rapid longitudinal mixing. The threshold of the instability expressed as an upper limit for the transverse impedance per unit length, Z_t, is

$$|Z_t| \le \pi F_t \frac{E}{eI} \frac{\beta^3 \nu}{R} \Delta \nu \qquad (26)$$

where F_t is another form factor of order unity and $\Delta \nu$ is the betatron tune spread.

Both the momentum spread $\Delta p/p$ in Eq. (25) and the tune spread $\Delta \nu$ in Eq. (26) are limited by resonance. The limitation is stronger for colliders in which, because of the long storage time requirement, much higher resonances have to be avoided. Thus, we must reduce the impedances to the minimum.

Again, the contributions to the impedances Z_ℓ and Z_t have two sources, that from the space charge/current and that from the image charge/current on the beampipe wall. The image contributions to Z_ℓ and Z_t are related to each other. For a circular beampipe of radius b the simple approximate relation is

$$Z_t \simeq \frac{2R}{\beta b^2} \frac{Z_\ell}{n} . \qquad (27)$$

As stated above, a practical lower limit for $|Z_\ell|/n$ is about 1 Ω, a value more or less independent of the size of the ring. This relation then shows that Z_t is larger for higher energy machines, for which R is larger and b is smaller. Therefore, one expects transverse instability to be more troublesome for higher energy machines. A great deal of effort has been devoted to computing the impedances for special geometries of the beam and the chamber wall. But the eventual conclusion must be based on measurements.

2.4 Transition Problems

Several special problems are caused by longitudinal beam self-field forces in crossing transition. Below transition the revolution frequency dispersion

$$\eta \equiv \frac{d\omega/\omega}{dp/p} = 1/\gamma^2 - 1/\gamma_t^2$$

(where γ_t is the transition energy in units of mc^2), is positive (velocity increasing faster with momentum than orbit length) and a particle responds to longitudinal force as though it has a positive mass, i.e. accelerates in the direction of the force. Above transition $\eta < 0$ and a particle responds as though it has a negative mass, i.e. accelerates in the direction opposite to the force. This reversal of response is usually taken care of by making a phase jump in the accelerating rf field from a positive slope (converging force) to a negative slope (diverging force). Together with the change in sign of the "mass," this phase jump keeps the effect always focusing.

When the longitudinal self-field force becomes comparable to the force due to the external rf field, the following problems arise: (1) Unlike the rf force, the self-field force cannot be reversed in sign at transition and remains diverging both before and after; thus it subtracts from the rf force below transition and adds to it above transition, causing mismatch in the force constant and hence a blowup in the longitudinal emittance. (2) Near transition η is sensibly zero, and there can be no Landau damping to stabilize the beam against the longitudinal microwave instability.

Both problems can be resolved or at least alleviated by employing the transition jump scheme. In this scheme fast pulsed quadrupoles are installed in the ring to jump the orbit length dispersion, $(\gamma_t)^{-2}$, at transition crossing so that η is changed abruptly from a non-zero positive value to an appropriate non-zero negative value. With proper adjustments, matching can be reestablished even in the presence of the beam-field force, and, since $|\eta|$ is never zero or even small, Landau damping is always present to damp the microwave instability. In addition to taking care of these longitudinal problems, one must remember to switch the chromaticity ξ from a negative value below transition to a positive value above transition to keep the transverse head-tail instability under control.

All these transition problems, if not properly resolved, although they may not cause direct beam loss, will invariably blow up the longitudinal emittance and perhaps also the transverse emittance if the head-tail instability is not appropriately damped.

2.5 Incoherent High Intensity Effect: Intrabeam Scattering

In the rest frame of a beam bunch the particles are confined in a potential well in all three degrees of freedom. In addition, the particles interact via Coulomb scattering. This intrabeam scattering can be expected to cause noticeable growth in the 6-dimensional emittance if the particle density is sufficiently high. The total growth rate is given by

$$\frac{1}{\tau} = \pi^2 \frac{cr_p^2 N(\log)}{\gamma \Gamma} \langle H(\lambda_1,\lambda_2,\lambda_3)\rangle \tag{28}$$

where $r_p = e^2/mc^2 = 1.535 \times 10^{-18}$ m = classical proton radius, N = number of protons in bunch, log = Coulomb logarithm $\simeq$ 20, $\Gamma \equiv (2\pi\beta\gamma)^3(\sigma_x^2/\beta_x)(\sigma_y^2/\beta_y)(\sigma_z^2/\beta_z)$ = invariant 6-dimensional phase volume occupied by beam in Gaussian distributions, and the expression $\langle H(\lambda_1, \lambda_2, \lambda_3)\rangle$ is a dimensionless and homogeneous "momentum shape factor." The quantities $(\lambda_1)^{-1/2}$, $(\lambda_2)^{-1/2}$, and $(\lambda_3)^{-1/2}$ measure the principal axes of the momentum ellipsoid of the beam bunch, and $\langle\rangle$ denotes averaging around the ring. The function H equals zero if $\lambda_1 = \lambda_2 = \lambda_3$, namely if the momentum spread is isotropic. This shows that the effect of the intrabeam scattering is to equipartition the momentum spread among all three degrees of freedom. On the other hand, in an alternating-gradient lattice the λ's cannot be equal everywhere. Hence the emittance will always grow.

In the general case, the function H can be expressed in terms of elliptic integrals, but in the special case when $\lambda_1 > \lambda_2 = \lambda_3$, namely when the momentum distribution is an oblate circular ellipsoid with the short axis along the 1 direction, one can write

$$H = \frac{2(\lambda_1 + 2\lambda_2)}{\sqrt{\lambda_2(\lambda_1-\lambda_2)}} \sin^{-1}\sqrt{\frac{\lambda_1-\lambda_2}{\lambda_1}} - 6 . \tag{29}$$

Formulas, slightly more complicated, exist also for $1/\tau_x$, $1/\tau_y$, and $1/\tau_z$, namely the individual growth rates for the emittances in each degree of freedom.

For the Fermilab accelerators operating at the present intensity, intrabeam scattering has not been much of a problem.

2.6 Luminosity Issues

Colliding beams give the possibility of reaching high center-of-mass energies. On the other hand the luminosity is naturally lower than that attainable with a single beam on a fixed target. To maximize the luminosity, the beams are focused hard to a tiny spot at the point of collision. This unfortunately increases the electromagnetic forces between the beams. These forces are extremely nonlinear and act to disrupt or at least blow up the beam so that the beam lifetime is reduced.

For the head-on collision of two circular beam bunches, each of radius a and having N particles, the integrated luminosity is just

$$L_b = \frac{N^2}{\pi a^2} = \beta\gamma \frac{N^2}{\beta^* \epsilon_n} \qquad (30)$$

where β^* = amplitude function β at point of collision, and $\epsilon_n = \beta\gamma \, \pi a^2/\beta^*$ = normalized emittance.

For given energy and particle number, to maximize L we must minimize ϵ_n (low emittance beam) and β^* (low β^* obtained by an insertion of strong quadrupoles). If the time interval between bunches in each beam is τ_b, the luminosity is

$$L = \frac{L_b}{\tau_b} = \beta\gamma \frac{N^2}{\beta^* \epsilon_n \tau_b} \,. \qquad (31)$$

The disruptive effects of the nonlinear beam-beam forces are difficult to calculate analytically. But for a given density distribution in the beam, i.e. a given mix of nonlinear forces, the effects in each transverse degree of freedom can be measured in terms of only one parameter, the linear tune-shift. We demonstrate this for the case of one transverse degree of freedom, say x. To begin with, neglecting the effect of the beam pipe, the electric potential of the second beam as seen by a particle in the first beam can be written as

$$V(x,s) = eNf\left(\frac{x^2}{a^2}, s\right) \qquad (32)$$

where N is the number of particles in the bunch and f is a properly normalized function. It is clear that f is an even function of x and that x should be scaled by the half-width a of the beam. The effect of the beam-beam force alone on a particle in the first beam is given by

$$m\gamma \frac{d^2x}{dt^2} = -(1+\beta^2)e\frac{\partial V}{\partial x} = -(1+\beta^2)e^2 N \frac{\partial f}{\partial x}$$

where the factor $(1 + \beta^2)$ arises from the reinforcement (because the beams are going in opposite directions) of the electric and the magnetic forces. With s as the independent variable this equation becomes

$$\frac{d^2x}{ds^2} = -\frac{1+\beta^2}{\beta^2 \gamma} r_p N \frac{\partial f}{\partial x} \,.$$

This shows that the total Hamiltonian of the motion could be written as

$$H(x,p;s) = \frac{1}{2}(p^2 + Kx^2) + \frac{1+\beta^2}{\beta^2\gamma} r_p Nf \,. \qquad (33)$$

We make the usual canonical transformation to the angle-action variables ϕ and J by

$$\begin{cases} x = \sqrt{2\beta_x J} \cos\phi \\ p = -\sqrt{\dfrac{2J}{\beta_x}} \left(\sin\phi - \dfrac{\beta'_x}{2} \cos\phi\right) \end{cases} \quad (34)$$

and obtain the new Hamiltonian

$$k(\phi,J;s) = \nu J + \frac{1+\beta^2}{\beta^2 \gamma} r_p N f\left(\frac{2\beta_x J}{a^2} \cos^2\phi, s\right). \quad (35)$$

We then define a scaled action variable

$$I \equiv \frac{2\beta_x}{a^2} J.$$

Keeping in mind that $a^2 \propto \beta_x$, and hence β_x/a^2 is independent of s, we can write the canonical equations of K as follows:

$$\begin{cases} \dfrac{d\phi}{ds} = \dfrac{\partial K}{\partial J} = \nu - 4\pi(\delta\nu_b) \dfrac{\partial f}{\partial I} \\ \dfrac{dI}{ds} = \dfrac{2\beta_x}{a^2} \dfrac{dJ}{ds} = -\dfrac{2\beta_x}{a^2} \dfrac{\partial K}{\partial \phi} = 4\pi(\delta\nu_b) \dfrac{\partial f}{\partial \phi} \end{cases} \quad (36)$$

where the beam-beam tune shift

$$\delta\nu_b \equiv \frac{1}{2\pi} \frac{1+\beta^2}{\beta^2 \gamma} \frac{r_p N \beta_x}{a^2} = -\frac{1}{2} \frac{1+\beta^2}{\beta} \frac{r_p N}{\epsilon_n} \quad (37)$$

has the form of usual space-charge tune shift except that the factor $1-\beta^2$ is changed to $1+\beta^2$. In Eq. (36) $f = f(I\cos^2\phi, s)$ and the nonlinearity in $I\cos^2\phi$ is derived from the original nonlinearity in x^2/a^2. Equation (36) shows that the motion is characterized only by the linear tune ν and the beam-beam tune shift $\delta\nu_b$.

We can now write a phenomenological expression for the beam-beam blowup rate,

$$\frac{1}{T} \propto \frac{1}{\sqrt{\tau_c}} G\left(\frac{\delta\nu_b}{\delta\nu_c}\right) \quad (38)$$

where τ_c = time interval between collisions of a beam bunch with others, and $\delta\nu_c$ = critical tune shift.

The blowup effects from collisions with different bunches are not coherent, hence the dependence on τ_c is taken to be $\sqrt{\tau_c}$. G is a function that rises sharply for $\delta\nu_b > \delta\nu_c$. Experience at the S$\bar{\text{p}}$pS indicates a critical tune shift value of $\delta\nu_c \sim 0.003$.

To summarize, the disruptive effects of the nonlinear beam-beam forces can be measured by the linear beam-beam tune shift $\delta\nu_b$, and experience shows that, to obtain reasonable lifetime for the colliding beams, $\delta\nu_b$ should not be >0.003.

In addition to the nonlinear effects, the simple linear tune shift per revolution is, as before, limited by resonances. To avoid resonances up to, say, the 7th order, we must have

$$N_c \delta\nu_b \stackrel{\sim}{<} 0.02 \tag{39}$$

where N_c is, for a given bunch, the number of collisions with other bunches in one revolution. If $\delta\nu_b \sim 0.002$, say, this imposes an upper limit on N_c of about 10. For pp colliders, or $\bar{p}p$ colliders with beam separators, the beam bunches collide only at the interaction points (IP's) for experiments, and N_c is equal to the number of IP's. A value of 10 for N_c is quite acceptable. But for $\bar{p}p$ colliders without beam separators, all bunches of beam 1 collide with all bunches of beam 2. A value of $N_c = 10$ will severely limit the number of bunches per beam.

An entirely different limitation on the integrated luminosity per bunch crossing, L_b, is imposed by the resolving power of the detector for the events produced. The total cross-sections for pp or $\bar{p}p$ collisions are of the order of 100 mbarn = 10^{-25} cm^2 at TeV energies. It is difficult to resolve more than, say, 2 events during the collision of two beam bunches. Thus L_b should be of the order of 10^{25} cm^{-2} and no greater, and to obtain a luminosity of $L = 10^{33}$ cm^{-2} sec^{-1} we need $\tau_b \sim 10^{-8}$ sec or ~ 10 nsec. This imposes rather stringent demands on the geometry of the interaction point. The bunches must collide more or less head-on (otherwise the luminosity will be reduced) at the IP, and 5 nsec (1.5 m) away on either side the bunches must be separated, at least by as much as their width. These demands can indeed be met, but only with difficulty.

The strategy of getting $L_b \sim 10^{25}$ cm^{-2} with the lowest particle number N is also clear. For this we want to use the lowest possible emittance ϵ_n and the lowest possible β^*. We should then check that Eq. (37) gives a beam-beam tune shift $\delta\nu_b$ smaller than $\delta\nu_c \sim 0.003$.

APPENDIX A SPACE-CHARGE TUNE SHIFT

A.1 Circular Beams

<u>Case I</u> Uniform density distribution

The electric field is radial and is a function only of the radial coordinate r. As long as one stays inside the beam ($r < a$) the field is given by

$$\frac{1}{r} \frac{d(rE_r)}{dr} = -4\pi e\rho = -\frac{4e\lambda}{a^2} \qquad (A-1)$$

or

$$E_r = -\frac{2e\lambda}{a^2} r \qquad (r < a) \qquad (A-2)$$

where a is the radius and $\lambda(s)$ is the linear density of the beam. The force f_r is then

$$f_r = \frac{eE_r}{\gamma^2} = -\frac{2e^2\lambda}{\gamma^2 a^2} r \qquad (A-3)$$

where the γ^2 factor arises from the cancellation between the electric and the magnetic forces. Thus, the tune shift is

$$\delta\nu_r = \frac{1}{4\pi} \int \frac{\beta_r}{pc\beta} \frac{df_r}{dr} ds = -\frac{r_p}{2} \frac{1}{\beta^2\gamma^3} \frac{N}{\epsilon_r} \qquad (A-4)$$

where $r_p = \frac{e^2}{mc^2} = 1.535 \times 10^{-18}$ m = classical proton radius,

$\epsilon_r \equiv \frac{\pi a^2}{\beta_r}$ = emittance,

$N = \int \lambda ds$ = total number of particles,

and r denotes either x or y.

<u>Case II</u> Gaussian density distribution

In this case we write

$$\rho = \frac{\lambda}{2\pi\sigma_r^2} e^{-\frac{r^2}{2\sigma_r^2}} \qquad (A-5)$$

and obtain

$$\delta\nu_r = -3r_p \frac{1}{\beta^2\gamma^3} \frac{1}{\epsilon_r} \int \lambda \left\{ e^{-\frac{r^2}{2\sigma_r^2}} - \frac{\sigma_r^2}{r^2} \left[1 - e^{-\frac{r^2}{2\sigma_r^2}} \right] \right\} ds \qquad (A-6)$$

where, again, r denotes either x or y. The maximum tune shift is obtained for oscillations with vanishing amplitude or as $r \to 0$. This gives

$$\delta\nu_{r\ max} = -\frac{3r_p}{2}\frac{1}{\beta^2\gamma^3}\frac{N}{\epsilon_r} \qquad (A-7)$$

where
$$\epsilon_r \equiv \frac{6\pi\sigma_r^2}{\beta_r} = 95\% \text{ emittance} .$$

A.2 Elliptical Beams

Case I Uniform density distribution

With uniform density and an elliptical beam boundary with semi-axes a and b, it is easy to see that a potential function

$$V(x,y) = \frac{2\lambda}{a+b}\left[\frac{x^2}{a} + \frac{y^2}{b}\right] \qquad (A-8)$$

satisfies the Poisson equation

$$\nabla^2 V = \frac{\partial^2 V}{\partial x^2} + \frac{\partial^2 V}{\partial y^2} = 4\pi\frac{\lambda}{\pi ab} = 4\pi\rho \qquad (A-9)$$

and the boundary condition that the elliptical beam boundary should be an equipotential. Thus the electric fields are

$$\begin{cases} E_x = -\dfrac{\partial V}{\partial x} = -\dfrac{4\lambda}{a(a+b)} x \\[2mm] E_y = -\dfrac{\partial V}{\partial y} = -\dfrac{4\lambda}{b(a+b)} y \end{cases} \qquad (A-10)$$

and the tune shifts are

$$\begin{cases} \delta\nu_x = -\dfrac{r_p}{\beta^2\gamma^3}\dfrac{1}{\epsilon_x}\displaystyle\int\dfrac{\lambda ds}{1+\dfrac{b}{a}}, & \epsilon_x \equiv \dfrac{\pi a^2}{\beta_x} \\[4mm] \delta\nu_y = -\dfrac{r_p}{\beta^2\gamma^3}\dfrac{1}{\epsilon_y}\displaystyle\int\dfrac{\lambda ds}{1+\dfrac{a}{b}}, & \epsilon_y \equiv \dfrac{\pi b^2}{\beta_y} \end{cases} \qquad (A-11)$$

Case II Bi-Gaussian density distribution

The density distribution is

$$\rho(x,y) = \frac{\lambda}{2\pi ab}\exp\left[-\frac{x^2}{2a^2} - \frac{y^2}{2b^2}\right] \qquad (A-12)$$

where a and b are, now, the standard deviations. The potential function for such a density distribution is

$$V(x,y) = e\lambda \int_0^\infty dt \; \frac{1 - \exp\left[-\frac{x^2}{2(a^2 + t)} - \frac{y^2}{2(b^2 + t)}\right]}{\sqrt{(a^2 + t)(b^2 + t)}} . \tag{A-13}$$

To get the x tune shift, e.g., we have

$$E_x = -\frac{\partial V}{\partial x} = -e\lambda x \int_0^\infty dt \; \frac{\exp\left[-\frac{x^2}{2(a^2 + t)} - \frac{y^2}{2(b^2 + t)}\right]}{(a^2 + t)\sqrt{(a^2 + t)(b^2 + t)}} . \tag{A-14}$$

The x tune shift is then

$$\delta\nu_x = \frac{1}{4\pi} \frac{r_p}{\beta^2 \gamma^3} \int \frac{1}{e} \beta_x \frac{\partial E_x}{\partial x} ds$$

$$= -\frac{1}{4\pi} \frac{r_p}{\beta^2 \gamma^3} \int \lambda \beta_x ds \int_0^\infty dt \; \frac{\left[1 - \frac{x^2}{a^2 + t}\right] \exp\left[-\frac{x^2}{2(a^2 + t)} - \frac{y^2}{2(b^2 + t)}\right]}{(a^2 + t)\sqrt{(a^2 + t)(b^2 + t)}} . \tag{A-15}$$

The maximum tune shift is for vanishingly small oscillations corresponding to x, y → 0. This gives

$$\delta\nu_{x\;max} = -\frac{1}{4\pi} \frac{r_p}{\beta^2 \gamma^3} \int \lambda \beta_x ds \int_0^\infty \frac{dt}{(a^2 + t)\sqrt{(a^2 + t)(b^2 + t)}}$$

$$= -\frac{3 r_p}{\beta^2 \gamma^3} \frac{1}{\epsilon_x} \int \frac{\lambda ds}{1 + \frac{b}{a}} , \qquad \epsilon_x \equiv \frac{6\pi a^2}{\beta_x} . \tag{A-16}$$

Similarly we have

$$\delta\nu_y = -\frac{1}{4\pi} \frac{r_p}{\beta^2 \gamma^3} \int \lambda \beta_y ds \int_0^\infty dt \; \frac{\left[1 - \frac{y^2}{b^2 + t}\right] \exp\left[-\frac{x^2}{2(a^2 + t)} - \frac{y^2}{2(b^2 + t)}\right]}{(b^2 + t)\sqrt{(a^2 + t)(b^2 + t)}} \tag{A-17}$$

and

$$\delta\nu_{y\,max} = -\frac{3r_p}{\beta^2\gamma^3}\frac{1}{\epsilon_y}\int\frac{\lambda ds}{1+\frac{a}{b}}, \qquad \epsilon_y \equiv \frac{6\pi b^2}{\beta_y}. \qquad (A-18)$$

APPENDIX B FORMULAS RELATED TO LONGITUDINAL OSCILLATION (SYNCHROTRON OSCILLATION, PHASE OSCILLATION)

In Section 2, for clarity we used a unified notation for all three degrees of freedom. Here we are under no such constraint. The simplest and most directly obvious starting equations are

$$\begin{cases} \dfrac{d\phi}{dt} = \omega_o - h\omega \\[6pt] \dfrac{dp}{dt} = \dfrac{eV_o}{2\pi R}\sin\phi \end{cases} \qquad (B-1)$$

where ϕ = rf phase as seen by the particle,
$\omega = c\beta/R$ = particle revolution frequency,
h = harmonic number,
p = particle momentum, and
ω_o, V_o = frequency and peak voltage of rf.

To put the equations in canonical form we define the variable W, remembering that $d\phi = (h/R)dz$, by

$$dW \equiv \frac{R}{h}dp = \frac{dE}{h\omega}, \qquad E = mc^2\gamma = \text{total energy}$$

and write the equations as

$$\begin{cases} \dfrac{d\phi}{dt} = \omega_o - h\omega = \dfrac{\partial H}{\partial W} \\[6pt] \dfrac{dW}{dt} = \dfrac{eV_o}{2\pi h}\sin\phi = -\dfrac{\partial H}{\partial \phi} \end{cases} \qquad (B-2)$$

with the Hamiltonian

$$H(\phi,W;t) = \omega_o W - E(W) + \frac{eV_o}{2\pi h}\cos\phi.$$

We define the synchronous values (subscript s) by

$$p_s \equiv (e/c)B\rho \quad \text{(synchronous momentum)}$$

where $B = B(t)$ = guide magnetic field and ρ = constant bending radius, are all given parameters. This then defines the synchronous phase ϕ_s by

$$\frac{dp_s}{dt} = \frac{eV_o}{2\pi R}\sin\phi_s.$$

Expanding the first of Eq. (B-2) to the first-order term in $w \equiv W - W_s$ gives

$$\begin{cases} \dfrac{d\phi}{dt} = (\omega_o - h\omega_s) - \dfrac{h^2 \omega_s}{R p_s} \eta w = - \dfrac{h^2 \eta}{mR^2 \gamma_s} w = \dfrac{\partial K}{\partial w} \\ \dfrac{dw}{dt} = \dfrac{eV_o}{2\pi h} (\sin\phi - \sin\phi_s) = - \dfrac{\partial K}{\partial \phi} \end{cases} \quad (B-3)$$

where

$$\eta \equiv \frac{d\omega/\omega}{dp/p} = \gamma^{-2} - \gamma_t^{-2} = \text{revolution frequency dispersion,}$$

$$\gamma_t^{-2} \equiv \frac{dR/R}{dp/p} = \text{orbit length dispersion, and } \gamma_t = \text{transition-gamma,}$$

and where the rf is tuned such that $\omega_o = h\omega_s$. The new Hamiltonian is

$$K(\phi,w;t) = - \frac{h^2 \eta}{mR^2 \gamma_s} \frac{w^2}{2} + \frac{eV_o}{2\pi h} (\cos\phi + \phi\sin\phi_s) \ .$$

The adiabatic phase trajectories are given by K = constant. The separatrix or the "bucket boundary" is the trajectory that passes through the single unstable fixed point at

$$w = 0 \ , \qquad \phi = \pi - \phi_s \equiv \phi_2$$

and is given by the equation

$$w = \pm \frac{R}{hc} \sqrt{\frac{1}{\pi h} \frac{eV_o E_s}{\eta}} \left[\cos\phi + \cos\phi_s + \phi\sin\phi_s - (\pi - \phi)\sin\phi_s \right]^{1/2} \ . \quad (B-4)$$

This gives the following dimensions for the rf bucket:
<u>Horizontal extent</u>: ϕ_1 to ϕ_2. Both are given by the solutions of

$$\cos\phi + \cos\phi_s + \phi\sin\phi_s - (\pi - \phi_s)\sin\phi_s = 0 \ . \quad (B-5)$$

Then $\phi_2 = \pi - \phi_s$ is an obvious solution. The other solution ϕ_1 is on the opposite side of ϕ_s as ϕ_2 and must be obtained numerically.
<u>Vertical extent</u>: $2w_{max}$. It is easy to see that w is maximum at $\phi = \phi_s$. Thus, we get

$$w_{max} = \frac{RE_s}{hc} \sqrt{\frac{2}{\pi h} \frac{eV_o}{E_s} \frac{1}{\eta}} \ \beta(\phi_s) \quad (B-6)$$

with

$$\beta(\phi_s) \equiv \sqrt{\cos\phi_s + (\phi_s - \frac{\pi}{2})\sin\phi_s} \quad (B-7)$$

so defined that $\beta(\phi_s = 0) = 1$. The other more physical momentum variables introduced in section 2 are related to w by

$$\begin{cases} p_{z\,max} = \dfrac{h}{R} w_{max} = \dfrac{E_s}{c} \sqrt{\dfrac{2}{\pi h} \dfrac{eV_o}{E_s} \dfrac{1}{\eta}} \; \beta(\phi_s) \\[2ex] z'_{max} = \dfrac{\eta h}{p_s R} w_{max} = \dfrac{1}{\beta_s} \sqrt{\dfrac{2}{\pi h} \dfrac{eV_o}{E_s} \eta} \; \beta(\phi_s) \\[2ex] \beta\gamma z'_{max} = \dfrac{\eta h}{mcR} w_{max} = \gamma_s \sqrt{\dfrac{2}{\pi h} \dfrac{eV_o}{E_s} \eta} \; \beta(\phi_s) \end{cases} \qquad (B\text{-}8)$$

<u>Area of bucket</u>: This is given in the basic units of (ϕ, w) by

$$A(\phi, w) = 8 \dfrac{RE_s}{hc} \sqrt{\dfrac{2}{\pi h} \dfrac{eV_o}{E_s} \dfrac{1}{\eta}} \; a(\phi_s) \qquad (B\text{-}9)$$

where

$$a(\phi_s) \equiv \dfrac{1}{4\sqrt{2}} \int_{\phi_1}^{\phi_2} \left[\cos\phi + \cos\phi_s + \phi\sin\phi_s - (\pi - \phi_s)\sin\phi_s\right]^{1/2} d\phi \qquad (B\text{-}10)$$

so defined that $a(\phi_s = 0) = 1$. In the other more physical variables the bucket area is given by

$$\begin{cases} A(z, p_z) = 8 \dfrac{RE_s}{hc} \sqrt{\dfrac{2}{\pi h} \dfrac{eV_o}{E_s} \dfrac{1}{\eta}} \; a(\phi_s) = 8 \dfrac{R}{h} p_{z\,max} \dfrac{a(\phi_s)}{\beta(\phi_s)} \\[2ex] A(z, z') = 8 \dfrac{R}{h\beta_s} \sqrt{\dfrac{2}{\pi h} \dfrac{eV_o}{E_s} \eta} \; a(\phi_s) = 8 \dfrac{R}{h} z'_{max} \dfrac{a(\phi_s)}{\beta(\phi_s)} \\[2ex] A(z, \beta\gamma z') = 8 \dfrac{R\gamma_s}{h} \sqrt{\dfrac{2}{\pi h} \dfrac{eV_o}{E_s} \eta} \; a(\phi_s) = 8 \dfrac{R}{h} (\beta\gamma z')_{max} \dfrac{a(\phi_s)}{\beta(\phi_s)} \end{cases} \qquad (B\text{-}11)$$

The functions $\phi_1(\phi_s)$, $a(\phi_s)$, and $\beta(\phi_s)$ are all given in tabulated form in CERN Report CERN/MPS-SI/Int.DL/70/4.

APPENDIX C FORMULAS RELATED TO TRANSITION CROSSING

We start with Eqs. (B-3) of Appendix B. For small oscillations we expand to linear terms in $\psi \equiv \phi - \phi_s$ (ϕ_s = constant). We also assume a linear increase of energy in time, namely

$$\dot{\gamma} = \frac{c\beta_t}{2\pi R}\frac{eV_o \sin\phi_s}{mc^2} = \text{constant} \qquad (C-1)$$

where $c\beta_t$ is the particle velocity at transition. This gives

$$-\eta = \frac{1}{\gamma_t^2} - \frac{1}{\gamma^2} \simeq \frac{2\dot{\gamma}}{\gamma_t^3} t \qquad \text{(transition at } t = 0\text{)}$$

and

$$\begin{cases} \dfrac{d\psi}{dt} = -\dfrac{h^2 \eta}{mR^2 \gamma_s} w \simeq \left(\dfrac{hc}{R}\right)^2 \left(\dfrac{2\dot{\gamma}}{\gamma_t^4}\right)\dfrac{t}{E_o} w \equiv at\, w \\[2ex] \dfrac{dw}{dt} = \dfrac{eV_o \cos\phi_s}{2\pi h}\, \psi \equiv b\cot\phi_s\, \psi \end{cases} \qquad (C-2)$$

where we have defined

$$a = 2\,\frac{\omega_\infty^2}{E_o}\,\frac{\dot{\gamma}}{\gamma_t^4} > 0 , \qquad E_o = mc^2 = \text{rest energy},$$

$$\omega_\infty = \frac{hc}{R} = \text{rf frequency at } \infty \text{ energy, and}$$

$$b = \frac{c\beta_t}{2\pi R}\frac{eV_o \sin\phi_s}{mc^2}\frac{R}{hc}\frac{mc^2}{\beta_t} = \frac{E_o}{\omega_\infty}\frac{\dot{\gamma}}{\beta_t} > 0 .$$

We also define a scaled time variable $x \equiv t/T$ where T is related to the synchrotron oscillation frequency Ω by

$$\Omega^2 = -abt\cot\phi_s \equiv \frac{|t|}{T^3} \qquad \begin{cases} t < 0, & \phi_s < \frac{\pi}{2} \\ t > 0, & \phi_s > \frac{\pi}{2} \end{cases}$$

or

$$T^{-3} = ab|\cot\phi_s| \quad \text{and} \quad (\Omega T)^2 = |t|/T . \qquad (C-3)$$

So defined T is the time away from transition when the synchrotron oscillation phase has advanced one radian and is hence a measure of the "width" of the transition.

To solve Eq. (C-2) we make the transformations
$$y = \frac{2}{3} x^{3/2} = \frac{2}{3} \left(\frac{t}{T}\right)^{3/2} \quad \text{and} \quad \psi = x\Phi ,$$
and obtain for Φ the equation
$$\frac{d^2\Phi}{dy^2} + \frac{1}{y}\frac{d\Phi}{dy} - \left[1 + \frac{(2/3)^2}{y^2}\right]\Phi = 0 . \tag{C-4}$$
This is the Bessel equation giving the solution
$$\begin{cases} \psi = Ax\left(\cos\chi\, J_{2/3} + \sin\chi\, N_{2/3}\right) \\ aT^2 x^{1/2} w - x^{-3/2}\psi = Ax\left(\cos\chi\, J'_{2/3} + \sin\chi\, N'_{2/3}\right) \end{cases} \tag{C-5}$$
where A and χ are the "amplitude" and "phase" constants. The phase plane trajectory or equivalently the boundary of the phase area covered by the beam is obtained by eliminating χ in Eqs. (C-5). This gives
$$\gamma_\psi \psi^2 + 2\alpha_\psi \psi w + \beta_\psi w^2 = \frac{\epsilon_\psi}{\pi} \tag{C-6}$$
where
$$\begin{cases} \gamma_\psi \equiv \frac{1}{aT^2 x^{1/2}} \left(J^2_{-1/3} + N^2_{-1/3}\right)\left(J_{2/3}N_{-1/3} - N_{2/3}J_{-1/3}\right)^{-1} \\ \alpha_\psi \equiv -\left(J_{2/3}J_{-1/3} + N_{2/3}N_{-1/3}\right)\left(J_{2/3}N_{-1/3} - N_{2/3}J_{-1/3}\right)^{-1} \\ \beta_\psi \equiv aT^2 x^{1/2} \left(J^2_{2/3} + N^2_{2/3}\right)\left(J_{2/3}N_{-1/3} - N_{2/3}J_{-1/3}\right)^{-1} \end{cases} \tag{C-7}$$
and the "amplitude" A is related to the phase-space area or longitudinal emittance ϵ_ψ by
$$A^2 = \left(\frac{\epsilon_\psi}{\pi}\right)\left[\frac{aT^2}{x^{3/2}}\right]\left(J_{2/3}N_{-1/3} - N_{2/3}J_{-1/3}\right)^{-1} \tag{C-8}$$
It is interesting that:
<u>At large y</u> ($y \to \infty$, away from transition)
$$\begin{cases} J_\nu \to \sqrt{\frac{2}{\pi y}} \cos\left(y - \frac{\pi\nu}{2} - \frac{\pi}{4}\right) \\ N_\nu \to \sqrt{\frac{2}{\pi y}} \sin\left(y - \frac{\pi\nu}{2} - \frac{\pi}{4}\right) \end{cases}$$

and Eq. (C-6) becomes

$$\left(aT^2x^{1/2}\right)^{-1}\psi^2 + (aT^2x^{1/2})w^2 = \frac{\epsilon_\psi}{\pi} \tag{C-9}$$

which is an upright ellipse with semi-axes

$$\begin{cases} \hat{\psi} = \sqrt{\frac{\epsilon_\psi}{\pi}} \sqrt{aT^2}\, x^{1/4} \equiv \theta_o x^{1/4} \equiv \theta \\ \\ \hat{w} = \sqrt{\frac{\epsilon_\psi}{\pi}} \frac{1}{\sqrt{aT^2}} x^{-1/4} = \frac{\epsilon_\psi}{\pi} \frac{1}{\theta} \end{cases} \tag{C-10}$$

and Eq. (C-8) becomes

$$A^2 = \frac{\epsilon_\psi}{3} aT^2 . \tag{C-11}$$

<u>At small y</u> (y → 0, at transition)

$$J_{2/3} \to \frac{3}{2\,\Gamma(2/3)}\left(\frac{y}{2}\right)^{2/3}, \qquad J_{-1/3} \to \frac{1}{\Gamma(2/3)}\left(\frac{y}{2}\right)^{-1/3}$$

$$N_{2/3} \to -\frac{2}{\sqrt{3}\,\Gamma(1/3)}\left(\frac{y}{2}\right)^{-2/3}, \qquad N_{-1/3} \to -\frac{1}{\sqrt{3}\,\Gamma(2/3)}\left(\frac{y}{2}\right)^{-1/3}$$

and Eq. (C-6) becomes

$$\frac{4/3}{kaT^2}\psi^2 + 2\frac{1}{\sqrt{3}}\psi w + kaT^2 w^2 = \frac{\epsilon_\psi}{\pi} . \tag{C-12}$$

The ellipse is slightly tilted. The maximum extents of the ellipse are given by the usual relations to be

$$\begin{cases} \hat{\psi}_t = \sqrt{k}\sqrt{\frac{\epsilon_\psi}{\pi}}\sqrt{aT^2} \equiv \sqrt{k}\,\theta_o \equiv \theta_t \\ \\ \hat{w}_t = \frac{2}{\sqrt{3k}}\sqrt{\frac{\epsilon_\psi}{\pi}}\frac{1}{\sqrt{aT^2}} = \frac{2}{\sqrt{3}}\frac{\epsilon_\psi}{\pi}\frac{1}{\theta_t} \end{cases} \tag{C-13}$$

where
$$k \equiv \frac{3^{1/3}}{\pi} [\Gamma(2/3)]^2 = 0.842$$

and
$$\theta_o \equiv \sqrt{\frac{\epsilon_\psi}{\pi}} \sqrt{aT^2} \tag{C-14}$$

will be used for scaling below. Eq. (C-8) again becomes Eq. (C-11). The envelope equation of $\hat{\psi}$ derived from Eq. (C-2) is

$$\frac{d}{dx}\left(\frac{1}{x}\frac{dY}{dx}\right) - \delta Y = \frac{x}{Y^3} \tag{C-15}$$

where
$$x = \frac{t}{T} = \text{scaled time},$$

$$\delta \equiv \frac{\cot\phi_s}{|\cot\phi_s|} = \begin{cases} +1 & \phi_s < \frac{\pi}{2} \\ -1 & \phi_s > \frac{\pi}{2} \end{cases}$$

and
$$Y = \frac{\theta}{\theta_o} = \text{scaled envelope of } \hat{\psi}$$

$$(\theta = \hat{\psi} = \text{envelope of } \psi).$$

We are now ready to investigate the two unpleasant features of transition crossing.

C.1 Microwave Instability

The stability condition (with $F_\ell = 1$) is given by Eq. (25):

$$\frac{|Z_\ell|}{n} < \frac{E}{eI}\beta^2 |\eta| \left(\frac{\Delta p}{p}\right)^2. \tag{C-16}$$

Near transition, we have approximately

$$|\eta| \simeq \frac{2\dot{\gamma}}{\gamma_t^3} |t|$$

and

$$\frac{\Delta p}{p} \simeq \frac{2\frac{h}{R}\hat{w}_t}{mc\beta_t\gamma_t} = \sqrt{\frac{8}{3\pi k}} \sqrt{\frac{\epsilon_\psi}{E_o\dot{\gamma}T^2}} \frac{\gamma_t}{\beta_t}$$

which, when substituted in Eq. (C-16), give

$$|t| > \frac{3\pi k}{16} \frac{eI|Z_\ell|/n}{\epsilon_\psi} T^2 \equiv t_o. \tag{C-17}$$

Thus the beam is unstable against microwave instability from
$t = (-t_o$ to $t_o)$ or for

$$\Delta\gamma \equiv \gamma - \gamma_t = (-\dot{\gamma}t_o \text{ to } \dot{\gamma}t_o) . \qquad (C-18)$$

One can calculate the blowup factor e^S where S is an integral from
$-t_o$ to t_o. The blowup can be avoided by making a γ_t jump of $2\dot{\gamma}t_o$ as
shown below.

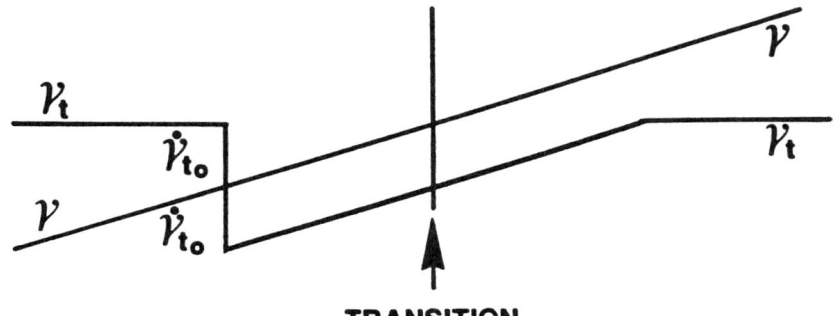

TRANSITION

C.2 Space-Charge Mismatch

We assume a parabolic longitudinal distribution with bunch
half-length

$$\hat{z} = \frac{R}{h}\hat{\psi} = \frac{R}{h}\hat{\theta}$$

as follows:

$$\lambda(z) = \frac{3}{4}\frac{N}{\hat{z}}\left(1 - \frac{z^2}{\hat{z}^2}\right)$$

where λ is the linear density and N is the number of particles in
the bunch. Then the longitudinal electric field is

$$E_z = -eg\frac{d\lambda}{dz} = \frac{3}{2}\frac{egN}{\hat{z}^3}z \qquad (C-20)$$

with
$$g \simeq 1 + 2 \ln\left(\frac{\text{beam pipe radius}}{\text{beam radius}}\right)$$

and

$$\left(\frac{dw}{dt}\right)_{s.c.} = \frac{3}{2}\frac{r_p}{c}\frac{E_o\omega_\infty}{\gamma^2}\frac{gN}{\hat{\theta}^3}\psi \equiv \frac{G}{\gamma^2}\frac{\psi}{\hat{\theta}^3} \qquad (C-21)$$

where, as defined,

$$G \equiv \frac{3}{2}\frac{r_p}{c}E_o\omega_\infty gN$$

and $r_p \equiv \frac{e^2}{mc^2} = 1.535 \times 10^{-18}$ m = classical proton radius.

The synchrotron oscillation Eqs. (C-2) now become

$$\begin{cases} \frac{d\psi}{dt} = a t\, w \\ \frac{dw}{dt} = \left[b \cot\phi_s + \frac{G}{\gamma^2 \theta^3} \right] \psi \end{cases} \quad (C-22)$$

The mismatch arises from the fact that the sign of $\cot\phi_s$ is switched from positive at $t < 0$ (before transition) to negative at $t > 0$ (after transition) by shifting the synchronous phase from ϕ_s to $\pi - \phi_s$, but there is no way to switch the sign of the space-charge term $G/(\gamma^2\theta^3)$. Hence the synchrotron oscillation frequency is effectively shifted from

$$\Omega^2 = a\left[b \cot\phi_s + \frac{G}{\gamma^2\theta^3} \right](-t) \quad \text{before transition, } t < 0$$

to

$$\Omega^2 = a\left[b |\cot\phi_s| - \frac{G}{\gamma^2\theta^3} \right](t) \quad \text{after transition, } t > 0 .$$

The degree of mismatch is exhibited by the Sørenssen parameter η_{oo}, which is defined by

$$\eta_{oo} = \frac{\frac{G}{\gamma_t^2 \theta_o^3}}{b|\cot\phi_s|} = \frac{aG}{\gamma_t^2}\left(\frac{T}{\theta_o}\right)^3 \equiv \frac{K}{\gamma_t^2} . \quad (C-23)$$

(Sørenssen used $\theta_t = \sqrt{K}\, \theta_o = 0.02\theta_o$ instead of θ_o. This of course makes little difference in the discussion.) If $\eta_{oo} \ll 1$ the space-charge mismatch is negligible. Otherwise matching can be reestablished by making γ_t and/or ϕ_s appropriate functions of time.

With space charge, the scaled envelope Eq. (C-15) becomes

$$\frac{d}{dx}\left(\frac{1}{x}\frac{dY}{dx}\right) - \left(\delta + \frac{K}{\gamma^2}\frac{1}{Y^3}\right)Y = \frac{x}{Y^3} \quad (C-24)$$

where, as defined in Eq. (C-23),

$$K \equiv aG\left(\frac{T}{\theta_o}\right)^3 .$$

Many different γ_t or ϕ_s jump schemes are possible. One such γ_t jump that will reestablish matching is as follows:

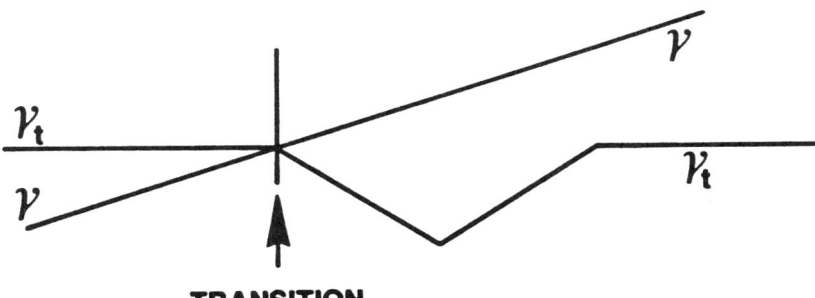

Some ϕ_s jump schemes also do very well in rematching. But ϕ_s jump cannot avoid blowup due to microwave instability.

If one wants both to reestablish matching and to cure microwave instability one must employ γ_t jump or both γ_t and ϕ_s jump. The optimal design of the jump(s) is best done by using multi-particle numerical simulation.

EMITTANCES THROUGH THE FERMILAB ACCELERATOR CHAIN

D. Finley
Fermi National Accelerator Laboratory
P. O. Box 500, Batavia, IL 60510

TABLE OF CONTENTS

1	Emittance Parameterization...................................	1911
2	Emittance Measurement Techniques.............................	1912
3	Sources of Emittance Increases...............................	1913
4	Summary of Measurements......................................	1915
5	Linac Measurements...	1916
6	Booster Measurements...	1917
7	Measurements Between the Booster and Main Ring...............	1917
8	Main Ring Measurements.......................................	1922
9	Tevatron Measurements..	1926
10	Comments...	1927

Chapter 5
EMITTANCE THROUGH THE FERMILAB ACCELERATOR CHAIN

D. Finley
Fermi National Accelerator Laboratory
P. O. Box 500, Batavia, IL 60510

In this chapter we describe the development of the proton beam emittance through the Fermilab accelerator chain from the 750-keV pre-accelerator to the Tevatron. The basic relationships used to describe emittances are contained in Appendix A of Teng's paper in this volume, and the student may wish to refer to it at this time. The measurements given here were taken during late 1984 and early 1986. They indicated where there were some sources of emittance increase. Since that time, some improvements have been made and others are planned.

1 EMITTANCE PARAMETERIZATION

The transverse emittance represents an area in phase space: (x,x') for the horizontal and (y,y') for the vertical. It is usually quoted in units of π mm-mrad. At Fermilab, the measurement technique used upstream of the second tank of the linac directly measures the transverse phase-space area occupied by the beam, but the measurement destroys the beam. Most of the other measurements downstream of this are nondestructive, but all are based on beam profiles. The width of a profile is characterized by σ, the second moment of the profile distribution. From the profile measurements, the emittance is defined to be

$$\epsilon = 6(p/m)(\sigma^2/\beta)\pi . \qquad (1)$$

An emittance normalized to p/m in this way would be a constant as the beam is accelerated if the only effect on the beam were adiabatic shrinking. As we shall see, the measured values ranged from 1π mm-mrad to $>50\pi$ mm-mrad.

In the definition Eq. (1), the lattice amplitude function β enters as an overall scale factor. If the wrong value for β were used, then the absolute value of ϵ would be wrong, but the fractional change $\Delta\epsilon/\epsilon$ would still be correct. The factor 6 is also an overall scale factor. If the beam were distributed in phase space as two Gaussians, then ~95% of the beam would be included in the phase-space area ϵ as defined here. Therefore, the emittance defined here is referred to as "the 95% normalized emittance." (At CERN a scale factor of 4 rather than 6 is usually used; in that case ~87% of the beam would be included in the phase-space area if the beam were distributed as two Gaussians.) Of course, if the real beam happens to be distributed in some other way, then ϵ may not contain 95% of the beam.

At Fermilab, the horizontal profile monitors are usually at locations in the lattice where the dispersion function (η) is not zero. One cannot calculate the transverse emittance from a single profile measurement alone if the monitor is at such a location. In this case one assumes the observed σ to be a combination of a width purely due to the transverse emittance (σ_T) and a width due to the momentum spread ($\eta\sigma_p/p$):

$$\sigma^2 = \sigma_T^2 + (\eta\sigma_p/p)^2 . \qquad (2)$$

This relationship assumes that the momentum spread distribution is symmetric and that the transverse and longitudinal emittances are uncorrelated. If one has only one profile, one must measure the momentum spread by some other technique in order to calculate the transverse emittance. On the other hand, as pointed out in Teng's Appendix A, one can combine the measurements from profile monitors at two different locations to measure both ϵ and σ_p/p.

The longitudinal emittance (ϵ_L) is defined as an area in ($\Delta E, \Delta t$) space and is usually quoted in units of eV-sec. For a stationary bucket, the longitudinal emittance is defined to be

$$\epsilon_L \simeq (\text{bucket area}) \times (\pi/64)\Delta^2(1 - 5\Delta^2/384) . \qquad (3)$$

Here Δ is the measured total bunch length in radians and is usually taken to include all the beam. The approximation is adequate for $\Delta < 4$ radians. The bucket area is given by

$$\text{bucket area} = 16(R/hc)[E_s \text{ (eV)}/(2\pi h|\eta|)]^{1/2} , \qquad (4)$$

with $|\eta| = |1/\gamma_t^2 - 1/\gamma^2|$, where γ_t is the transition gamma. As with the transverse, the longitudinal emittance would be a constant in an ideal chain of accelerators. As we shall see, the measured values range from 0.02 eV-sec to 3.2 eV-sec, where part of the increase is deliberate.

2 EMITTANCE MEASUREMENT TECHNIQUES

As mentioned above, the measurement technique used upstream of the second tank of the linac directly measures the transverse phase-space area occupied by the beam. This is done by stepping a 0.075-mm slot across the beam and recording the width of the image formed on a pickup plate 10 cm away. The pickup plate is divided into 20 segments on 0.2-mm centers. Each segment thus corresponds to 2 mrad of angular divergence. The stepping motion of the slit and the recording and analyzing of the data are done by computer.

The transverse emittances exiting the linac are measured by directing the beam to the linac dump rather than to the booster. Three wire scanners in this line are used to measure the profiles as the beam passes through this region, which is free of quadrupole magnets. The data are analyzed by the main control room computers.

The transverse emittances of the beam passing through the 8-GeV beam-transfer line between the Booster and the Main Ring are measured by a set of mulitwire grids. Additional multiwire grids in the Main Ring prevent the beam from circulating, but they can be used to measure profiles on the first turn. The multiwire data are analyzed by the main control room computers.

The transverse emittance of a beam circulating in the Booster, Main Ring, or Tevatron is measured by passing a thin wire rapidly through the beam. The 50-micron-diameter beryllium wire is moved at a constant speed whose value is changed from 2 to 12 m/sec depending on the measurement desired. As the beam passes through the wire, a tiny fraction of the beam undergoes nuclear interactions. Some of the debris from these interactions passes through a can of liquid scintillator in which a phototube is immersed. The phototube monitors the interaction rate, which is proportional to the beam density at the position of the wire. The phototube response and the wire position are recorded locally by a microprocessor. The data from these flying wires are analyzed by the main control room computers.

Unlike the data for the transverse emittances, the data for the longitudinal emittance are not taken by computers. Instead, a photograph is taken of the signal from a detector that is sensitive to the instantaneous beam current passing through it. The bunch length is then measured by hand from the photograph. This manual measurement, combined with the value of the rf voltage, yields the longitudinal emittance.

3 SOURCES OF EMITTANCE INCREASES

At Fermilab, the proton beam is transferred from one accelerator to the next by four beam-transfer lines: the 750-keV line from the pre-accelerator to the linac, the 200-MeV line from the linac to the Booster, the 8-GeV line from the Booster to the Main Ring, and the E0 line from the Main Ring to the Tevatron. For beam transfers into the three circular accelerators, the transverse emittance of the circulating beam in the receiving accelerator will be larger than that in the delivering accelerator if the beam-transfer line is not set up properly. In addition, the longitudinal emittance will be larger if the individual bunches of protons are not injected into the buckets properly. This last statement applies to transfers from the Booster to the Main Ring, and from the Main Ring to the Tevatron.

Mis-steering in a beam transfer line is one cause of transverse emittance increases. The machine into which the beam is to be injected has a particular horizontal and vertical closed orbit. If the beam is not injected onto the closed orbit with the proper position and angle, the circulating beam will undergo betatron oscillations. Eventually, field nonlinearities will cause a net increase in the transverse emittance. The Main Ring and Tevatron have injection dampers that decrease reasonable betatron oscillations in less time than the time in which significant increases in transverse emittances can occur. The situation with longitudinal emittance increases is similar: if the beam is not placed correctly into the bucket with the proper energy and rf phase, the nonlinearity of the bucket causes a net increase of the longitudinal emittance of the circulating beam.

Another type of injection mis-steering that can cause transverse emittance increases is related to the momentum spread of the beam. The previous paragraph referred to the closed orbit. Since the Fermilab circular accelerators necessarily have horizontal dispersion, there is a different closed orbit for each momentum. Since the Main Ring has vertical overpasses around B0 and D0 for the colliding detectors in the Tevatron, it also has vertical dispersion. The closed orbit referred to in the previous paragraph is the one appropriate to the design momentum. For proper injection of a beam having a momentum spread, the dispersion function of the beam transfer line must agree with that of the lattice at the point of injection. If the magnitudes or slopes of the dispersion functions do not agree, the particles with momenta different from the design momentum will undergo betatron oscillations since they are not being injected onto their closed orbit. Again, this leads to a net increase in the transverse emittance of the circulating beam. The injection dampers are not expected to help in this situation as they act on the centroid of the positions of the particles in a bunch. Since the momentum distribution is typically symmetric, the centroid does not change and thus the dampers have nothing to act on.

Mismatching is another cause of emittance increases. The machine into which the beam is to be injected has a particular set of lattice parameters (β, α) at the point of injection. If the beam-transfer line does not match these, the field nonlinearities will cause a net increase of the transverse emittance of the circulating beam. The matching condition for the longitudinal can be expressed for stationary buckets for transfer from machine 1 to machine 2 as

$$|V/(h\eta)|_1 = |V/(h\eta)|_2 , \qquad (5)$$

where V is the rf voltage and h is the harmonic number. If this condition is not met, the longitudinal emittance of the beam circulating in machine 2 will be larger than that in machine 1.

In addition to improper machine-to-machine transfer, the emittances can also increase during acceleration. As we shall see, this is particularly apparent in the longitudinal emittance as the beam is accelerated through the transition energy in the Booster and Main Ring.

Space charge is another cause of transverse emittance increase which is particularly evident in the very low energy beam upstream of the second tank of the linac. (At this stage of acceleration, the beam is H^- ions; the electron is stripped upon injection into the Booster and then the beam converts to protons.) In principle, space charge is simply the force that an individual charged particle experiences in the presence of other particles. In a real beam, this force is very nonlinear and tends to increase the distance between the particles, and thus the emittance increases. One way to decrease this force is to introduce particles of opposite charge over the volume of the beam. In practice this is done by controlling the vacuum pressure in the 750-keV beam-transfer line so that molecules that are initially neutral become ionized by the passage of the H^- beam itself. The electrons leave much more quickly than the heavier positively charged ions, and these ions dilute the space-charge forces acting among the H^- beam particles.

4 SUMMARY OF MEASUREMENTS

Table I summarizes the normalized emittances as the beam passes through the Fermilab accelerator chain. They are listed in order from the low energy (200-MeV linac) to the highest energy (800-GeV Tevatron collider). These data were taken under a variety of conditions, and in nearly all cases they are not taken on the same machine cycle. This fact alone leads to a spread in the data, but more importantly, the intensity (number of particles per bunch) ranged from <1 to >3x10^{10}. As we shall see, higher intensities are associated with larger emittances. The blanks in the table are mostly indicative of the difficulty in making reliable measurements in a real accelerator environment.

Table I Summary of Emittances

These transverse emittances are calculated from the beam profiles as $\epsilon = 6(p/m)(\sigma^2/\beta)\pi$. The longitudinal emittance represents all the beam.

	ϵ Transverse π mm-mrad		ϵ Longitudinal eV-sec
	Horizontal	Vertical	
Linac out	6	6	
Booster in	15		0.02
Booster out	15		0.1
8-GeV line	10-15	10	
8-GeV Main Ring	10-15	8-12	0.1
Main Ring accelerated	17		0.1-0.4
Main Ring coalesced	25		4-5
Tevatron collider	30-60	20-30	

In an ideal chain of accelerators, the normalized emittance does not change. Even a cursory glance at Table I reveals large changes as one goes down the columns. The only place where we make a deliberate change is with the longitudinal emittance during coalescing in the Main Ring. Coalescing is the process by which the protons from several (~9) buckets are gathered into a single bucket; thus, the act of coalescing is expected to make the longitudinal emittance ~9 times larger. All the other changes are telling us something about the Fermilab accelerators or beam-transfer lines.

The first line in Table I shows that the transverse emittance leaving the linac is 6π mm-mrad. The horizontal emittance has more than doubled by the next time we can measure it, which is after it has been circulating in the Booster for ~1 msec. The reason for this increase is thought to be the space-charge tune shift limit of the Booster rather than mis-steering or mismatching in the 200-MeV line. The transverse emittance does not grow very much during acceleration in the Booster.

Upon transfer to the Main Ring an expected increase of a factor of ~1.3 is due to the mismatch of (β,α) between the Main Ring and the 8-GeV line. The 8-GeV line has since been replaced, and the replacement should allow elimination of emittance increases associated with the previous beam-transfer line. The Main Ring accelerates the beam to 150 GeV with no large increase in the transverse emittance.

A large increase in the transverse emittance occurs, however, after the beam has been coalesced in the Main Ring and transferred into the Tevatron. Additional measurements are needed to separate clearly how much of the increase occurs in the Main Ring after coalescing and how much is due to the EO beam-transfer line.

The longitudinal emittance was increasing by a factor of ~5 during acceleration in the Booster when these data were taken. Much of the increase was associated with passing through transition. There are several ways to improve this, and some are being implemented. Some increase of the longitudinal emittance during acceleration in the Main Ring is again associated with transition. The factor of ~10 increase after coalescing is expected, as indicated in the discussion above.

5 LINAC MEASUREMENTS

The transverse emittance passing through the linac is summarized in Table II. The emittance at the exit of the 750-keV pre-accelerator is 1.0π mm-mrad horizontal and 1.5π mm-mrad vertical. After passing down the 750-keV beam-transfer line, the emittance has doubled in the horizontal and nearly doubled in the vertical before it enters the first tank of the linac. This increase is thought to be due to space-charge forces. After passing through the first tank of the linac the emittances have again doubled, and this is also thought to be due to space-charge forces. Thus, the transverse emittance increases by a factor of ~3.5 to 5 between the pre-accelerator and the end of the first linac tank. However, as the beam passes through the remaining eight tanks of the linac no further increase in the emittances is observed.

Table II Linac Emittances

These transverse emittances are calculated from direct measurements of phase space and represent 90% of the beam.

	Energy	$\beta\gamma$	Current mamp	ϵ Transverse π mm-mrad	
				Horizontal	Vertical
Pre acc out	750 keV	0.04	~50	1.0	1.5
Linac in	750 keV	0.04		2.1	2.6
Tank 1 out	10 MeV	0.15		5.0	5.3
Linac out	200 MeV	0.684	~35	5.0	5.2

6 BOOSTER MEASUREMENTS

Figure 1 shows measurements of the horizontal transverse emittance shortly after injection into the Booster. It demonstrates that the transverse emittance did not depend much on the intensity at this point in the accelerator chain. It also demonstrates, however, that, even for the lowest intensities, the transverse emittance was at least twice as large as that leaving the linac.

Figure 2 shows measurements that demonstrate the development of the longitudinal emittance throughout the 1/30-sec Booster acceleration cycle. The longitudinal emittance not only increases at transition but continues to grow to near the end of the acceleration cycle. The net increase is about a factor of 5 throughout the Booster acceleration cycle.

Figure 3 demonstrates that the longitudinal emittance increases faster than linearly as the bunch intensity is increased.

7 MEASUREMENTS BETWEEN THE BOOSTER AND MAIN RING

Figure 4 shows the transverse emittance measurements, which compare multiwire data and flying-wire data. Previous emittance measurements from 1981 are summarized by the curves. The multiwires were used to measure the profiles of the injected beam as it passed through the 8-GeV beam-transfer line and passed once through part of the Main Ring. After the multiwires were removed, a flying wire was used to measure the profile of the 8-GeV circulating beam. The horizontal emittance measured by the multiwires was ~9π mm-mrad. From the multiwire measurements, one can predict the emittance expected for the circulating beam after dilution due to the mismatch of (β,α) between the Main Ring and the 8-GeV beam-transfer line; this is ~13π mm-mrad. Because only one flying-wire profile was taken, it had to be corrected for the momentum spread. After this correction, the emittance value from the flying-wire data is consistent with the diluted multiwire value, as it should be.

Figure 5 shows the results of contemporaneous horizontal flying-wire measurements in the Booster and the Main Ring. The Booster profiles were taken on the circulating beam just before extraction, and the Main Ring profiles were taken just after injection on the circulating beam. These measurements show a dramatic difference between the transverse and longitudinal: the transverse emittance does not change nearly as much as the longitudinal emittance as the intensity increases. In addition, the longitudinal increases faster than linearly.

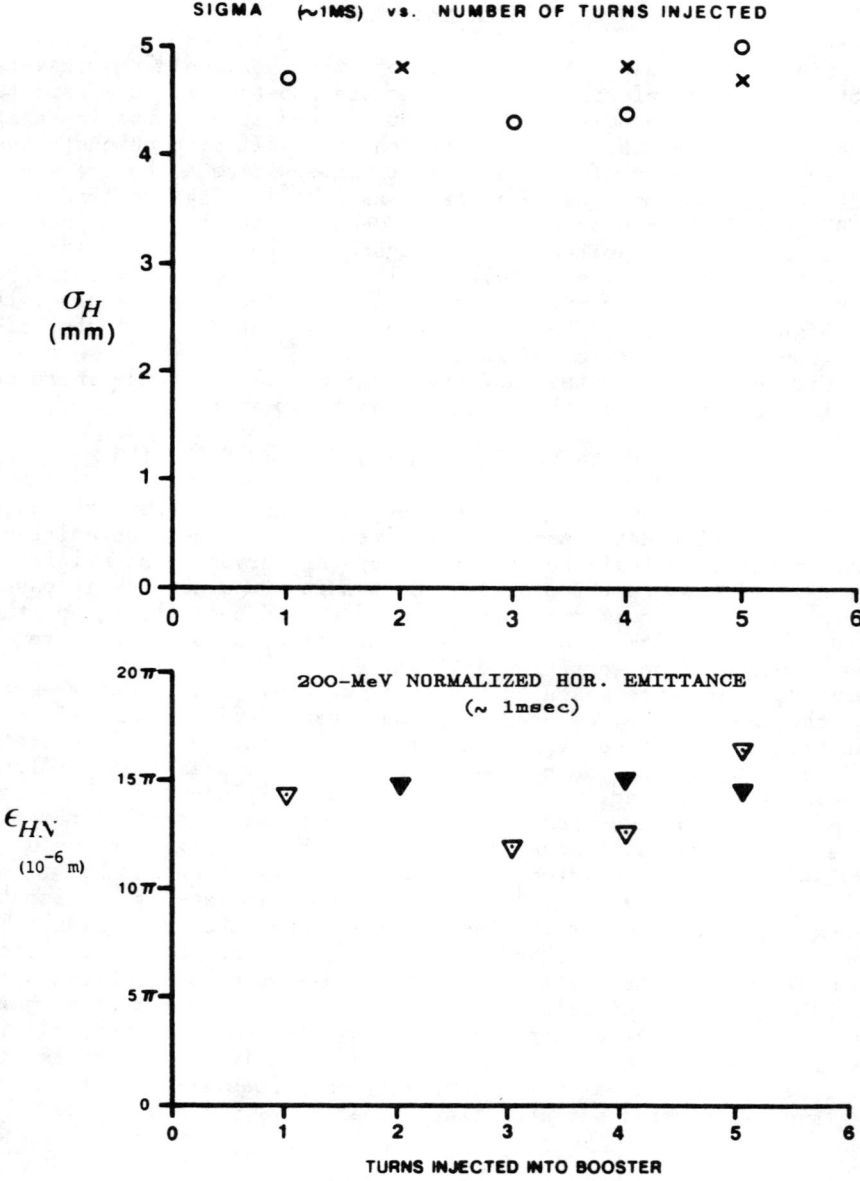

Figure 1 Booster horizontal flying-wire data near injection

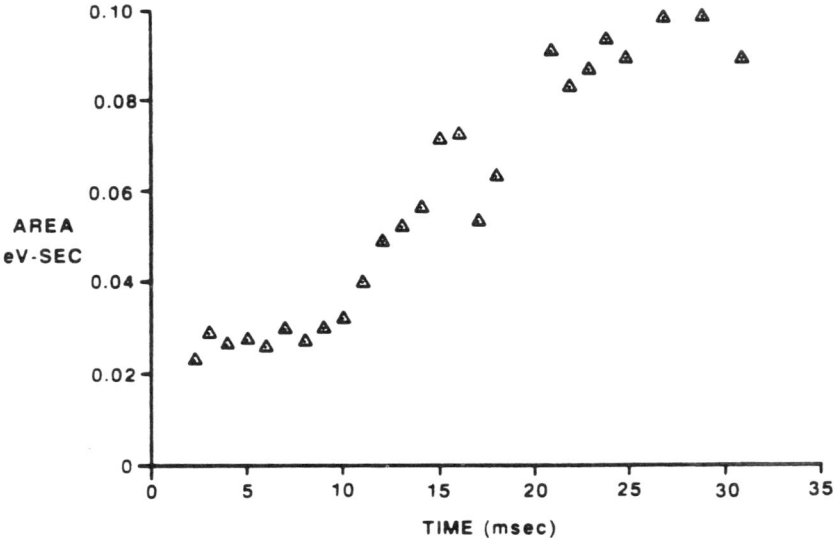

Figure 2 Booster longitudinal phase space

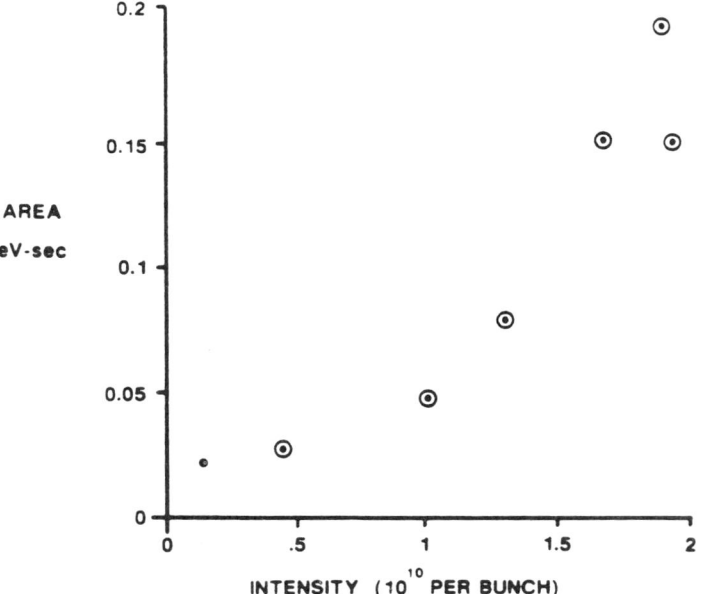

Figure 3 Booster longitudinal phase space

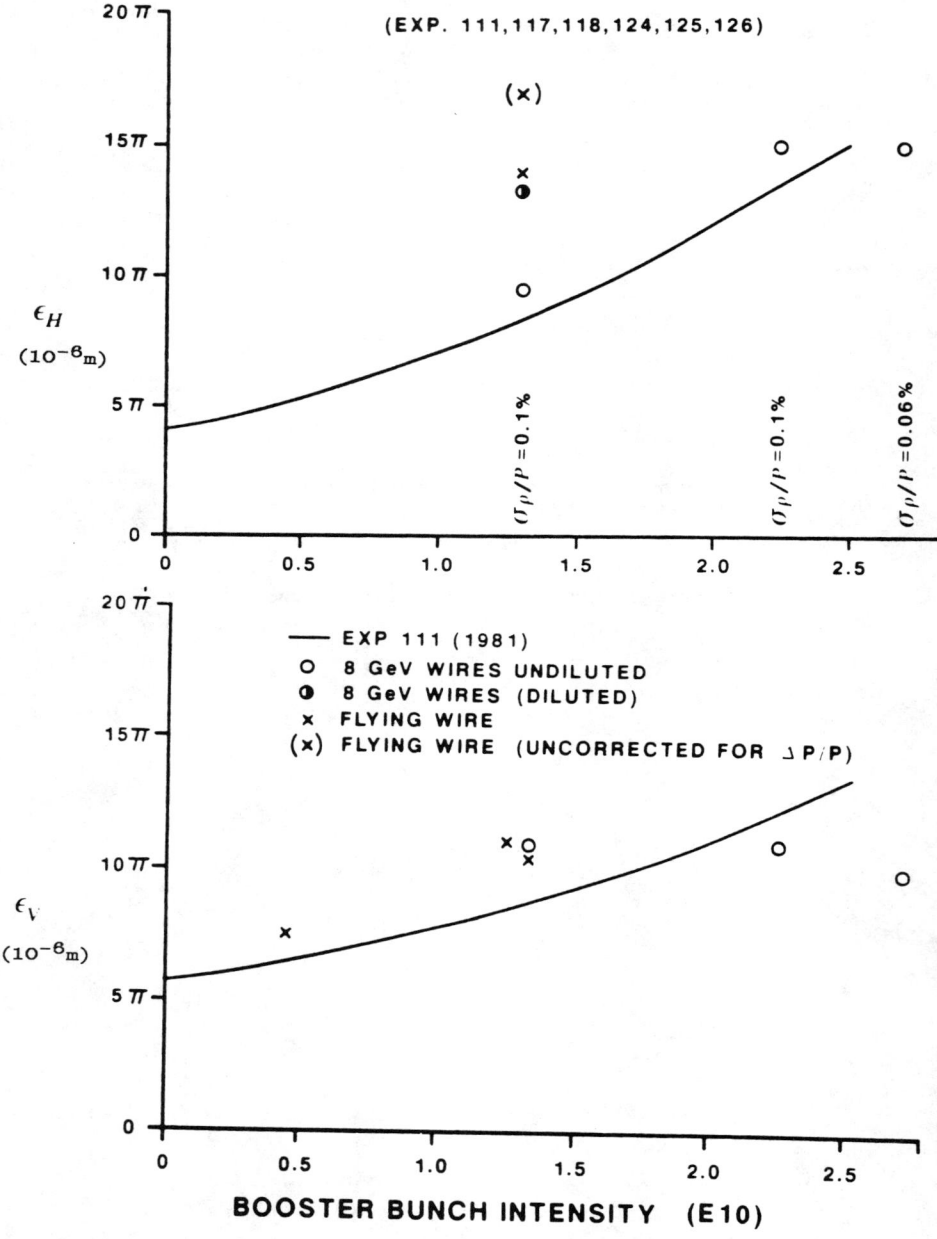

Figure 4 Main Ring 8-GeV emittance

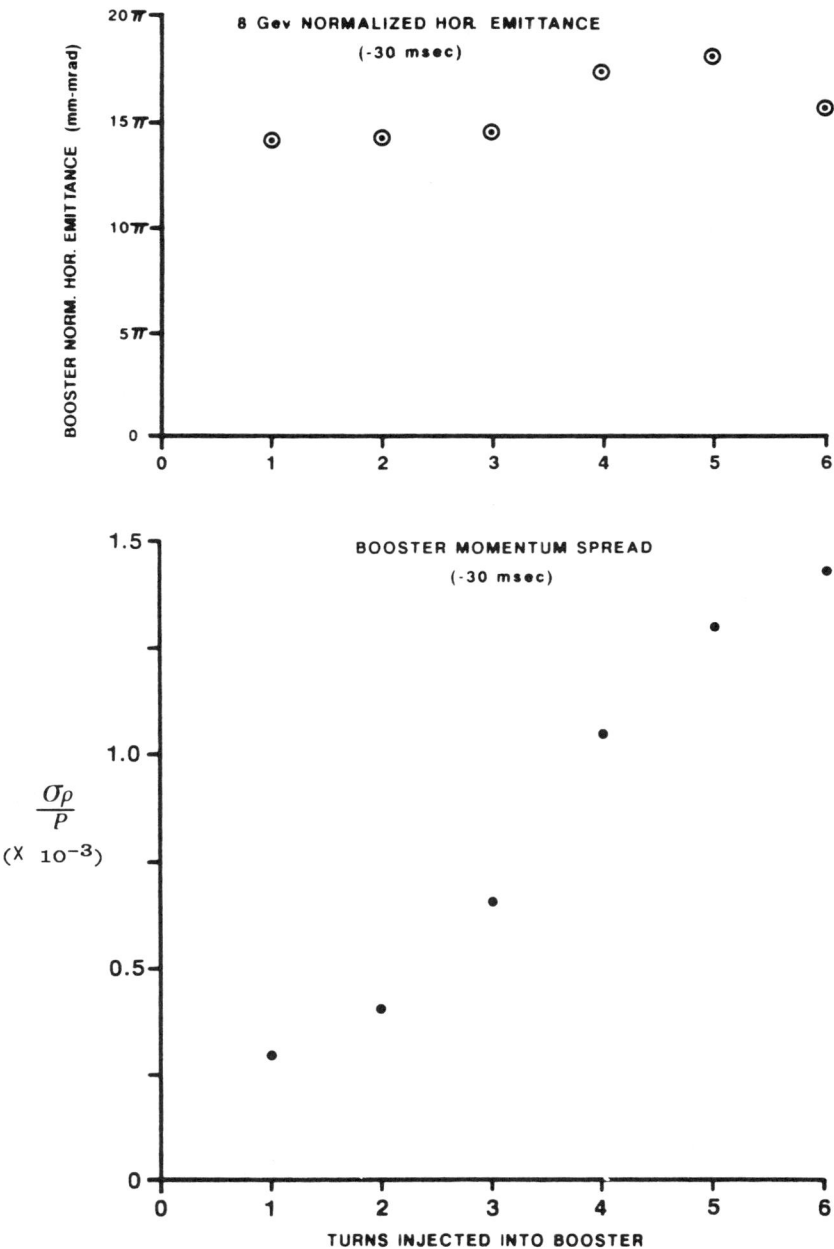

Figure 5 Booster and Main Ring correlated horizontal flying-wire data

8 MAIN RING MEASUREMENTS

Figure 6 shows flying-wire measurements of the horizontal emittance taken on September 5, 1985. Eleven bunches, with ~0.9x10^{10} protons per bunch were injected at 8 GeV into the Main Ring. About half this intensity reaches 150 GeV. On the 150-GeV flat-top, the bunches were coalesced into one. The top part of the figure shows the measured beam sizes. The middle part shows the calculated emittance, and the bottom shows the calculated σ_p/p. The profiles at A17 and F48 were taken simultaneously, but the data from different times in the 4-sec cycle were taken on different machine cycles.

The lattice amplitude and dispersion functions (β,η) at the flying-wire locations are (101 m, 2.3 m) at F48 and (89 m, 5.1 m) at A17. Since the dispersion function is much larger at A17, the beam size there is expected to be more sensitive to the momentum spread, and this is demonstrated in the top plot of Figure 6.

The middle plot of Figure 6 demonstrates that the emittance showed no significant increase as the Main Ring accelerated the beam. The points just after transition show a dip which is probably not real; from these data alone, one cannot tell for sure what is going on here.

Figure 6 shows four measurements taken after the beam had been coalesced. The most obvious change after coalescing is that σ_p/p increased by a factor of ~4, and it actually returned to about the same level it was at 8 GeV. This is reflected in the the top plot by the fact that the higher dispersion A17 beam size increased much more than the F48 beam size after coalescing.

In Figure 6 the transverse emittance measurements after coalescing are not as clearcut as the longitudinal. The data show an increase from ~15π mm-mrad to >25π mm-mrad. A change in the beam size at F48 by <200 microns would be large enough to produce this change. Since the profile monitor system measures beam sizes to ~22 microns, one must conclude that this represents a change in the beam. However, from these data alone one cannot tell whether the emittance increased after coalescing or whether these last two points were associated with beam pulses with unusually large emittance before the coalescing began. We are currently improving the emittance measurement technique so that this type of quandary can be sorted out.

Figure 7 shows measurements of the longitudinal emittance made during the same time period as the transverse emittance measurements shown in Fig. 6. The most striking features of these data are two clear increases in the longitudinal emittance. The first is a factor of ~3 increase as the beam is accelerated through transition. The second is the expected factor of ~9 increase associated with the coalescing process.

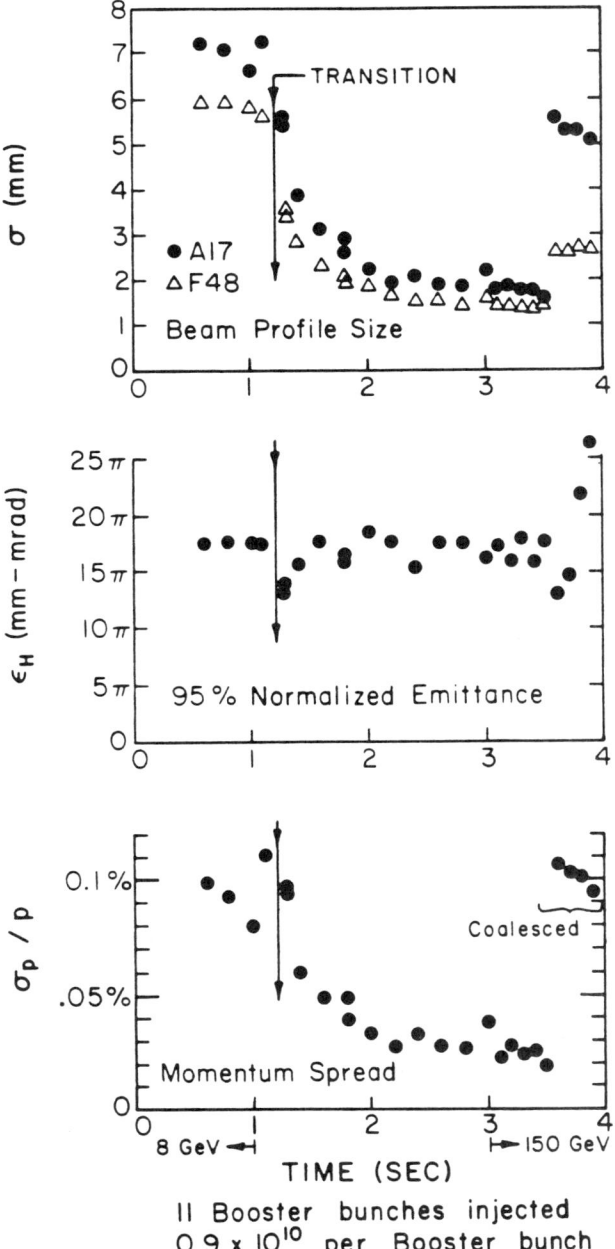

Figure 6 Main Ring flying-wire data, September 5, 1985, with coalescing

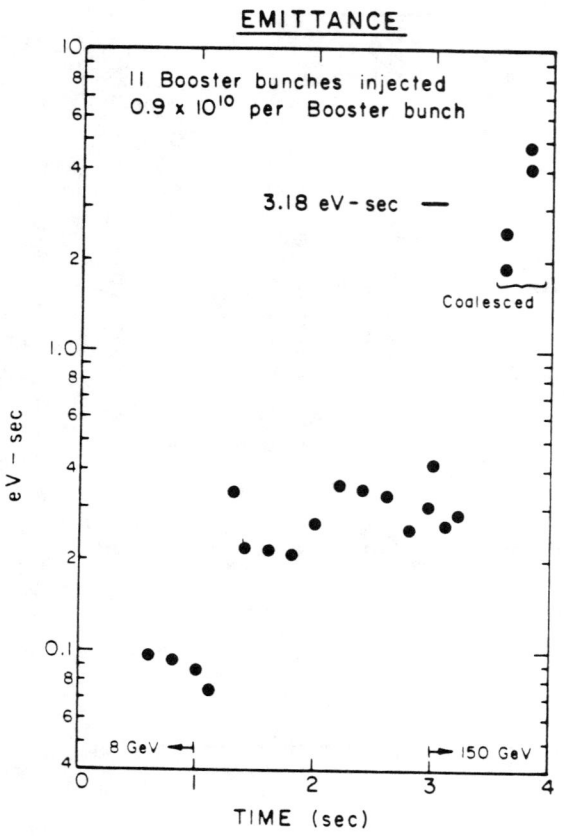

Figure 7
Main Ring bunch length data, September 5, 1985, with coalescing

Figure 8 shows both the horizontal and vertical transverse emittance measurements made on September 8, 1985. These data were taken with the Main Ring at its 8-GeV injection level, and the bunch intensity from the Booster was varied from 0.5 to 2.2×10^{10}. The horizontal emittance increased from ~9 to ~16π mm-mrad, whereas the vertical increased from ~9 to ~13π mm-mrad as the intensity was increased. Since two horizontal profiles were used, the momentum spread σ_p/p was also measured, and it shows an increase from ~0.02 to ~0.06%. From Fig. 8, one would expect a smaller emittance and momentum spread at 8 GeV than those actually observed in Fig. 6; therefore, one must conclude that there were noticeable day-to-day variations in the beam emittance.

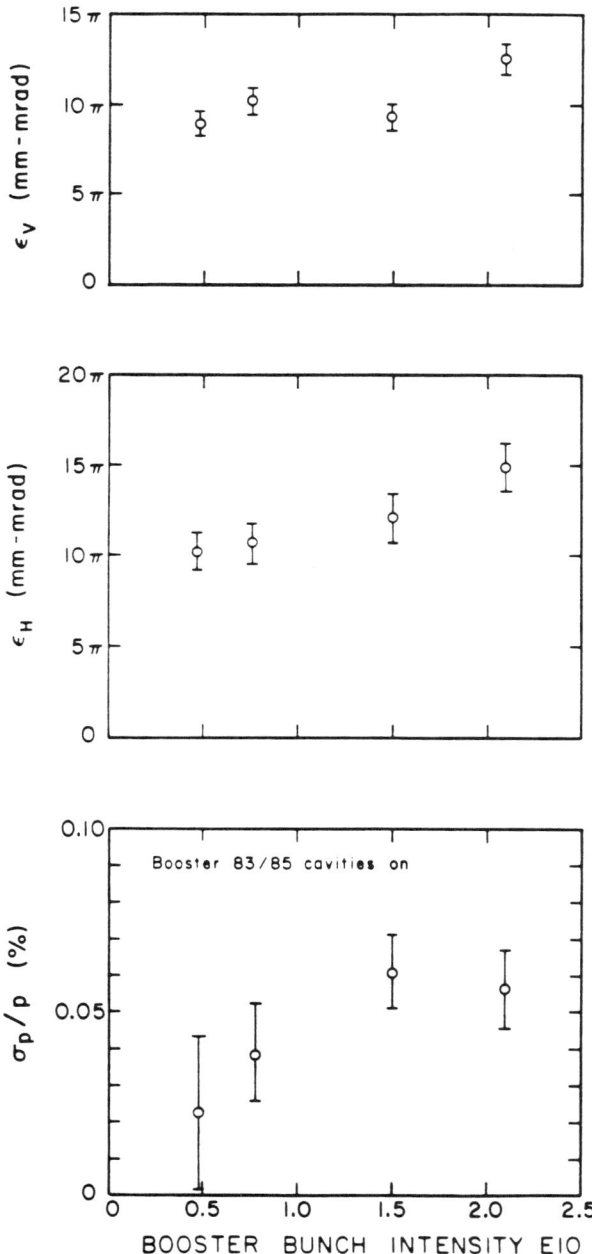

Figure 8 Flying-wire data at 8-GeV injection, Main Ring, September 8, 1985

9 TEVATRON MEASUREMENTS

Figure 9 shows transverse emittances measured in the Tevatron during the initial collider commissioning of September and October 1985. In all cases coalesced beam was injected into the Tevatron. The 150-GeV and 800-GeV points were taken before the low-β quads were used to squeeze the beam at the B0 interaction point where the CDF detector is located. Several things are very striking (and perhaps puzzling) about these data.

The <u>lowest</u> values for the 150- and 800-GeV points are ~20 to 30π mm-mrad, and the typical points are more than double this. These are clearly larger than the values recorded in the Main Ring. This may be due to a combination of (1) the vertical dispersion mismatch between the Main Ring, the E0 beam-transfer line, and the Tevatron, and (2) the large momentum spread associated with the coalesced beam. If it is, additional measurements can verify this speculation and provide a basis for deciding how to ameliorate this problem.

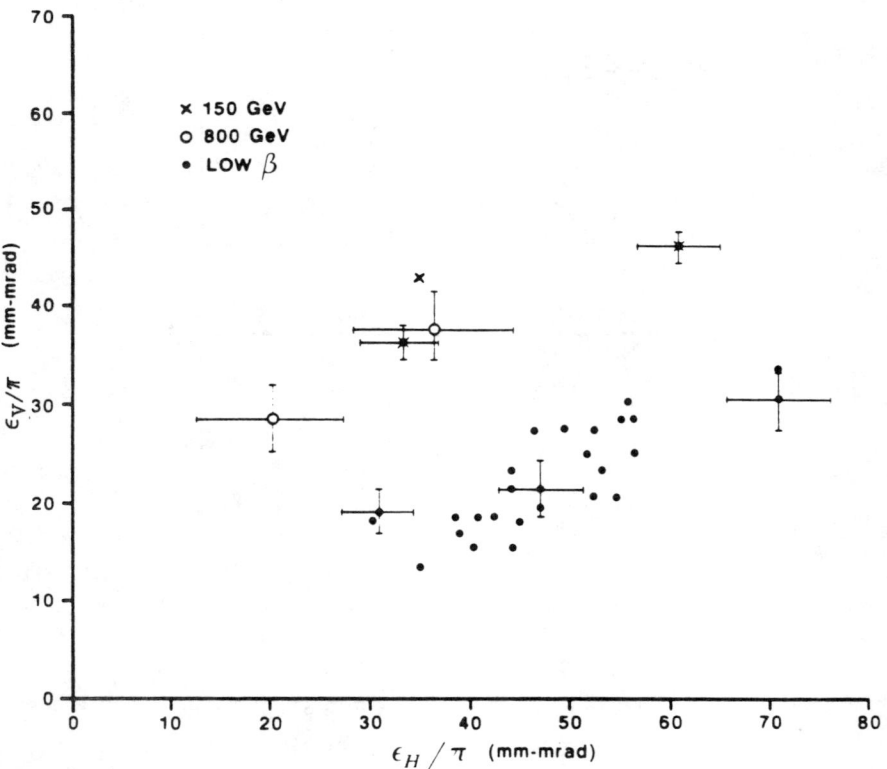

Figure 9 Tevatron normalized transverse emittance (September-October 1985)

The spread in the data is very large. It could be due to cycle-to-cycle variations in the emittance coming into the Tevatron from the Main Ring (or into the Main Ring from the Booster). On the other hand it may simply be due to our not using the Tevatron injection dampers during the initial collider commissioning. The planned improvements in the emittance measuring technique will allow us to make coordinated measurements from the Booster to the Tevatron on the same pulse of protons, and such measurements should help clarify this.

The points with the low-β energized form a group that is clearly disjointed from the others. We now take this as evidence that we do not understand the β-functions in the lattice with the low-β energized as well as we would like. (Recall that β enters as an overall scale factor in the emittance calculation.) One way to have a real decrease in the emittance as one turns on the low-β would be to scrape off part of the beam; however, we do not observe an appreciable loss of beam intensity during the squeeze.

10 COMMENTS

We hope that this discussion of emittances measured at Fermilab has been instructive. Emittance measurements at Fermilab are becoming easier to perform as better techniques and newer technology are applied. The reasons for making the measurements, however, remain the same: to understand how beams behave and how to control them in a real accelerator environment.

DETECTION AND COMPENSATION OF TRANSVERSE RESONANCES

Rod Gerig
Fermi National Accelerator Laboratory
P. O. Box 500, Batavia, IL 60510

TABLE OF CONTENTS

1 Introduction.. 1929
2 The Operational Problem Described........................... 1929
3 Source of Magnetic Errors Driving Resonances................ 1930
4 Techniques for Machine Study and Driving Term Correction.... 1931
5 Analytical Evaluation.. 1933
6 Design of a Correction System............................... 1934
7 Comparison of Main Ring Experience with Tracking Results.... 1937
8 Summary.. 1939
9 References... 1939

Chapter 6
DETECTION AND COMPENSATION OF TRANSVERSE RESONANCES

Rod Gerig
Fermi National Accelerator Laboratory
P. O. Box 500, Batavia, IL 60510

1 INTRODUCTION

This chapter discusses the operational implications of the problem of nonlinear dynamics in synchrotrons.[1] Although accelerator designers try to avoid these problems in the design of their machines it is impossible to avoid the problems completely, and as a result it is important to be able to diagnose and provide compensation for the resulting transverse resonances. The author draws from his experience in working with the original 400-GeV Fermilab Main Ring, where these effects have been routinely observed at the injection energy of 8 GeV.

The article first presents the problem and how it relates to machine performance. Techniques for measuring and compensating the resonance lines are then discussed, and the design of the correction system used to provide compensation is described. The last section discusses efforts to simulate the measurement and compensation process using a tracking program.

2 THE OPERATIONAL PROBLEM DESCRIBED

Early in the operation of Main Ring it was understood that the 8-Gev lifetime was limited by the presence of strong nonlinear driving terms causing loss of particles of tunes near half- and third-integer resonance lines. Although the normal tunes of the Main Ring are set at $\nu_x = 19.42$ and $\nu_y = 19.38$, a variety of factors can cause the tune of individual particles to be far from these normal values. Among these factors are chromatic effects and space-charge tune shifts. Both these factors produce a spread in the tune of the particles (not merely a shift), and therefore it is not possible to find a new value of the normal tunes which prevents individual particles from nearing resonance lines.[2] Because of these resonances, particles can become unstable at amplitudes less than the physical aperture, leading to ever larger oscillations and subsequent beam loss. It is therefore desirable first to determine how much of the observed beam loss is due to these nonlinear effects. Then the individual resonance lines are investigated and compensation is provided so that the beam loss caused by each is minimized.

The Main Ring has correction element families comprised of quadrupoles and sextupoles to provide correction to the driving terms for the following resonance lines:

© 1989 American Institute of Physics

$3\nu_x = 58$ — Sextupole
$2\nu_x + \nu_y = 58$ — Skew sextupole
$\nu_x + 2\nu_y = 58$ — Sextupole
$3\nu_y = 58$ — Skew sextupole
$2\nu_x = 39$ — Quadrupole
$2\nu_y = 39$ — Quadrupole
$\nu_x + \nu_y = 39$ — Skew quadrupole
$\nu_x - \nu_y = 0$ — Skew quadrupole

3 SOURCE OF MAGNETIC ERRORS DRIVING RESONANCES

The derivation of the driving terms has been presented in many summer school lectures.[3] A vector notation will be used in this chapter, with a sine and a cosine component derived according to the following expressions:

$$A_k = \frac{\beta_o}{(B\rho)} B_o \Sigma_i \left(\frac{\beta_{x_i}^m \beta_{y_i}^n}{\beta_o^{m+n}}\right)^{\frac{1}{2}} l c_{m+n-1_i} \cos\left(\frac{k}{m+n}(m\phi_{x_i} + n\phi_{y_i})\right) \quad (1)$$

$$B_k = \frac{\beta_o}{(B\rho)} B_o \Sigma_i \left(\frac{\beta_{x_i}^m \beta_{y_i}^n}{\beta_o^{m+n}}\right)^{\frac{1}{2}} l c_{m+n-1_i} \sin\left(\frac{k}{m+n}(m\phi_{x_i} + n\phi_{y_i})\right) \quad (2)$$

where the summation is over the nonlinear elements, l is the length of the element, the c's are the multipole coefficients (b's for normal and a's for skew), the β's and ϕ's are taken at the nonlinear elements (ϕ is the reduced phase which goes from 0 to 2π for one turn around the ring), and β_o is the beta-function at a reference point in the ring. The parameter k defines the resonance line:

$$m\nu_x + n\nu_y = k. \quad (3)$$

Distributions of field errors will contribute to these driving terms. For example the main dipoles in the Main Ring are known to have a large remanent sextupole. If this sextupole component were the same in all the dipoles, the terms would sum to zero because of the argument of the trig functions. There would still be a need to correct the chromatic effects from this sytematic sextupole field, but it would not be a source of third-integer driving terms. This is not the case; the remanent sextupole varies from magnet to magnet producing non-zero driving terms and therefore we need to compensate with correction elements. The driving term equations can be evaluated to derive a statistical estimate of what can be expected given a sigma of the distribution of sextupole. The expression is

$$\langle A_k \rangle = \frac{\beta_o}{(B\rho)} B_o N^{\frac{1}{2}} l \langle c_k \rangle \overline{\left(\frac{\beta_x^m \beta_y^n}{\beta_o^{m+n}}\right)^{\frac{1}{2}}} \quad (4)$$

where N is the number of elements being considered (in this case the number of Main Ring dipoles, 774) and $\langle c_k \rangle$ is the sigma of the distribution of the sextupole.

Evaluation of this expression for the distribution of sextupole normally considered to be present in the Main Ring yields driving terms on the order of 10 m^{-1}. From this (it will be shown later in this chapter) the aperture is reduced from the physical size of 50 mm to under 25 mm for particles with a tune different by 0.01 from the third-integer resonance lines.

Other sources of driving terms are variation in the gradient of the main quadrupoles which can drive the half-integer resonances; roll alignment errors of the quadrupoles which can produce coupling; special injection or extraction devices which may have some nonlinear fields, and so on. There are also intentional nonlinearities introduced such as chromaticity correcting sextupoles, and, depending on their strength and placement, they can become sources of driving terms.

4 TECHNIQUES FOR MACHINE STUDY AND DRIVING TERM CORRECTION

The first step in the correction of the driving terms is the measurement of their effect on the beam. This is accomplished by measuring and plotting the 8-Gev efficiency as the tunes are varied in such a way that all the resonance lines are individually crossed. Figure 1 is a diagram of tune space showing the second- and third-order resonance lines. Higher-order resonance lines are not shown for the sake of clarity. The line along the diagonal is the path taken through tune space as the efficiency is measured. Figure 2 is the plot which results as the measurements are made. Care must be taken (when the beam is used as a probe in this manner) to be certain that the chromaticity is zero, and that the other factors which can cause tune shift and tune spread are minimized. Thus these studies are done at low intensity.

After the severity of the driving terms is measured, a program to compensate each is undertaken. This is done by adjusting the main quadrupole currents so that the tunes lie on one of the resonance lines but are reasonably far from other resonance lines. Figure 1 indicates the location in tune space where the compensations are done. The console computers have provision for hooking various devices together with different multipliers so that a set of devices can be modified in a prescibed ratio. This feature is used with sets of sextupoles and quadrupoles to provide a "knob" that produces driving terms. The method for defining these ratios is described in Section 6. These knobs are adjusted while the tunes lie on top of the corresponding resonance lines in such a way as to maximize the efficiency.

Attempts have been made to standardize this procedure since very different numbers for 8-GeV efficiency can be obtained depending on the techniques used to derive them. The prescription generally used is to inject two batches of protons from the Fermilab Booster spaced 67 msec apart, measuring the intensity of each at the downstream end of the injection line into the Main Ring. The beam

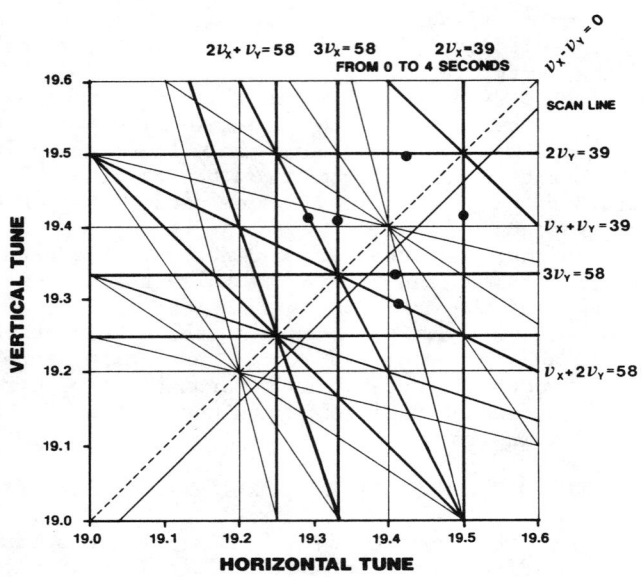

Figure 1 Tune diagram. The current in the main quadrupoles is varied so that the tunes are adjusted along the solid line running along the diagonal.

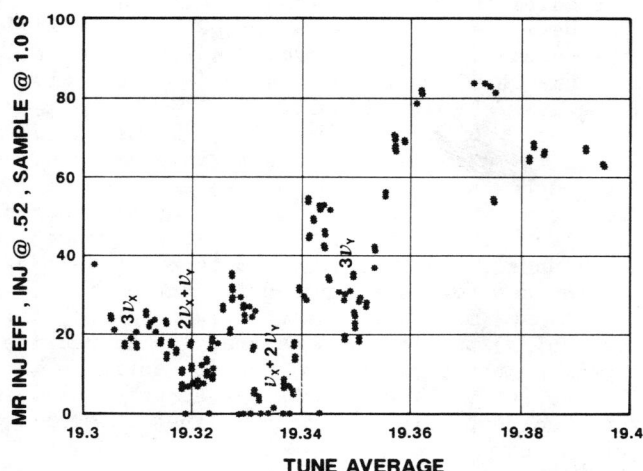

Figure 2 The abscissa is the average of the horizontal and vertical tunes. The ordinate is the efficiency of the beam at 8 GeV. Beam loss due to the third-integer resonances is observed. This scan was done before corrections were made. A scan taken after the corrections were applied shows that the dips in efficiency due to the resonance lines are reduced, but still visible.

is allowed to coast with the rf off, and a final intensity measurement is made in the ring 800 msec after injection of the first booster batch. The efficiency is the ratio of these measurements. The apparent success of the resonance corrections will depend on the length of the coast before the final intensity measurement. This should be on the order of the length of time the beam coasts during machine operation. Since in the Main Ring, for fixed-target physics, the first booster batch coasts for nearly 800 msec, that number has been chosen. With an 800-msec coast time, the resonance lines are still evident after the corrections are made, indicating that complete correction of the driving terms has not been realized thus far in the Main Ring. If the coast time is shortened significantly, the resonances may not be visible.

5 ANALYTICAL EVALUATION

In order to predict the strengths of the driving terms, one must know the distribution of higher-order multipoles around the accelerator. In the Main Ring these data have not been available. A project of measuring the multipoles of some of the magnets from the Main Ring is underway, but measurement of all the dipoles and quadrupoles is not possible. Therefore two approaches have been used to estimate the strengths of the driving terms. The first is a statistical approach. Enough measurements of the remanent sextupole in the main dipoles were made to demonstrate that the variation in remanent sextupole was about half the average value.[4] From this, the expected strength of the correction elements needed to compensate can be determined. Using Eq. (4), with a random distribution equal to half the mean value, leads to a driving term for the $3\nu_x = 58$ and $\nu_x + 2\nu_y = 58$ resonance lines of about 6.0 m^{-1}. From these considerations, appropriate elements could be designed and built. The second way of estimating the severity of the Main Ring's inherent driving terms is to use the results of the studies in which these driving terms were compensated. The assumption can be made that the settings of the correction elements represent driving term vectors equal and opposite to those found in the ring. Then the magnitude of the driving term from just the correction magnets will be the magnitude of that found in the ring. Many of these correction element setting files still exist, and these have been analyzed to extract the magnitude and direction of the driving term vector. The resulting vectors range from 3 m^{-1} up to 22 m^{-1}, indicating either that the random distributions are greater than expected or that there is an additional source of driving term not yet understood. The magnitude will vary depending on the flat-top energy of the Main Ring (the multipole distribution at the 8-GeV injection energy depends on the nature of the remanent fields, which in turn depends on the previous ramp history) but this does not account for the variation from 3 to 22 mentioned above. We also find driving term compensation for the resonances driven by skew sextupole, but these are smaller, in the range of 2 to 6 m^{-1}.

Table I shows the output of the program that calculates the driving terms by evaluating Eqs. (1) and (2) for a set of correction element settings obtained during a compensation study. In this case the only multipoles included in the calculation were the strengths of the 72 resonance correction quadrupoles and sextupoles.

Table I All harmonic correction elements

Resonance line	Amplitude	Sine	Cosine
3 NU X	13.334822	-3.455760	-12.879254
NU X + 2 NU Y	10.723385	-7.799346	-7.359429
2 NU X + NU Y	2.972437	2.777905	1.057650
3 NU Y	2.578212	1.400485	2.164675
2 NU X	0.244122	0.179678	0.165262
2 NU Y	0.411328	0.056660	0.407407
NU X + NU Y	0.028446	0.003452	-0.028236
NU X - NU Y	0.315241	0.076826	0.305736

These numbers can be related to dynamic aperture through a simplified expression.[5] If the particle amplitude is greater than

$$\frac{8\pi\delta}{A} \quad (5)$$

where δ is the distance from the resonance line, and A is the magnitude of a third-integer driving term, then it will be lost. For example if an accelerator has an uncorrected $3\nu_x$ resonance line with a driving term of 10 m^{-1}, then a particle with a fractional tune of 0.35 will become unstable if its amplitude exceeds 42 mm.

6 DESIGN OF A CORRECTION SYSTEM

Synchrotron designers normally include some multipole magnets in the lattice of the accelerator to provide compensation for unwanted driving terms and to reduce coupling. For each driving term these can be as few as two elements, or they can be a family of elements driven in a particular ratio.

This section indicates how such a system is put together using the Main Ring harmonic correction system as an example. The expressions for the driving term vectors [Eqs. (1) and (2) of Section 3] serve as a starting point. These expressions are general and apply to all the different resonance lines for which compensation is provided. As an example, however, the expressions will be worked out for the $3\nu_x = 58$ resonance line. In this case the driving term from a single sextupole at some location i becomes

$$A_{58} = \beta_o \left(\frac{\beta_{x_i}^3}{\beta_o^3}\right)^{\frac{1}{2}} \left(\frac{1}{(B\rho)} \frac{b_{2_i}}{}\right) \cos\left(58\phi_{x_i}\right), \quad (6)$$

$$B_{58} = \beta_o \left(\frac{\beta_{x_i}^3}{\beta_o^3}\right)^{\frac{1}{2}} \left[\frac{1}{(B\rho)} \frac{b_{2_i}}{}\right] \sin\left(58\phi_{x_i}\right) . \tag{7}$$

The second bracketed quantity now contains the deflection given to a particle by the sextupole as a function of position squared.

It would be possible with just two sextupoles to provide independent sine and cosine driving terms. We would merely need to find two locations in the accelerator where the sine and cosine terms vanish and locate the sextupoles there. However, we would have then introduced a zeroth-order sextupole effect, i.e. each time these sextupoles were varied to modify driving terms, we would also change the chromaticity in both transverse planes. Furthermore it is desirable to have distributed multipoles rather than one large one, as will be shown in a later section. For this reason a family of sextupoles is used, with the initial constraint that they do not affect the chromaticity.

The equations above indicate that if these sextupoles are to be very effective they should be at locations at which β_x is large; this is the second constraint. The two remaining variables in setting up the $3\nu_x$ system are where to locate them in ϕ_x and how strongly to power them. The Main Ring was initially built with six-fold symmetry. It is composed of six sectors, each having the same phase advance, identical β-functions, and so on. If the sextupoles are located at a symmetrical point in each sector, then the argument of the trig functions will advance by $2\pi/6$ radians between sextupoles. This also indicates that the β-function and the dispersion function will be the same at each sextupole, which reduces the constraint that the chromaticity not change, to the constraint that the sum of the sextupole strengths equal zero.

We now have six sextupoles located at symmetrical points in the ring. Their placement within the sector does not matter so long as the β-function is large. In fact the location within the sector is normally determined by available beam pipe, available power supplies, etc. The problem is reduced to determining the ratio of changing the strengths of these six sextupoles so that a pure sine term is produced, or a pure cosine term is produced, while preserving chromaticity.

Assume that the point of observation is taken at one of the sextupoles (i.e., $\phi_x = 0$), and that it is given unit strength. The driving term from this sextupole is plotted as vector 1 in Fig. 3. The angle ϕ_x for the next sextupole will be $\pi/3$, and an evaluation of the trig functions indicates that the driving term from this second sextupole will point in the direction of vector 2. The driving term for the third sextupole will point in the direction of vector 3. The vectors for sextupoles 4, 5, and 6 will lie on top of vectors 1, 2, and 3. This combination with all sextupoles at positive unit strength yields no net driving term, and changes chromaticity. However, if sextupoles 2, 3, 5, and 6 are given the opposite sign and half the strength, a pure cosine driving term results, yet the sum of the sextupole strengths is zero, resulting in no chromaticity change.

Table II summarizes the ratios in which the sextupoles are changed to produce a pure cosine driving term for the $3\nu_x = 58$ resonance line, or a pure sine driving term.

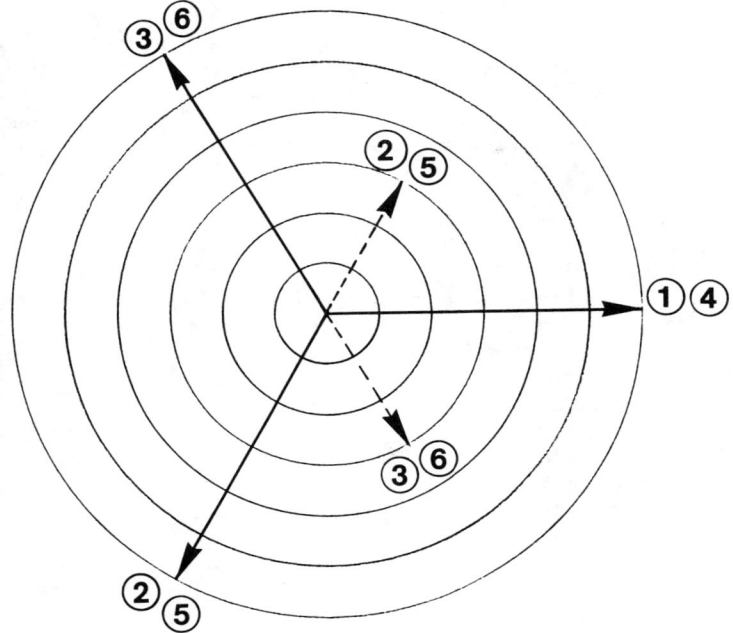

Figure 3 The individual driving terms for sextupoles 1 through 6 are plotted as solid lines. The dotted lines shows how sextupoles 2, 3, 5, and 6 are turned on when ganged together with 1 and 4 to produce a pure cosine driving term with no chromatic contribution.

Table II

	Pure Cosine	Pure Sine
Sextupole 1	times 1.0	times 0.0
Sextupole 2	times -0.5	times -1.0
Sextupole 3	times -0.5	times 1.0
Sextupole 4	times 1.0	times 0.0
Sextupole 5	times -0.5	times -1.0
Sextupole 6	times -0.5	times 1.0

The actual ratios used on the Fermilab control system are not these, because the $\phi_x = 0$ point is not taken at the first sextupole but at the beginning of the first sector. The resulting ratios provide pure sine and cosine terms when evaluated from that point.

If these ratios are used in analyzing the driving term equations for the $\nu_x + 2\nu_y = 58$ resonance, the other resonance driven by normal sextupole, the vectors will be found to be non-zero. In other words in attempting to minimize the driving terms for $3\nu_x = 58$, the driving terms for $\nu_x + 2\nu_y = 58$ are also modified. This is not desirable. To compensate for this effect the proper ratios of the $\nu_x + 2\nu_y = 58$ driving terms (another set of sextupoles in the Main Ring) are added to the above ratios so that only the $3\nu_x$ driving terms are varied.

The console program which provides this control allows manipulation of the angle and magnitude of the driving term vector since it is easier to conceptualize the vector in polar coordinates. A display of the driving term vector is provided on the console graphics device along with a plot of the efficiency of the beam so that the tuning can be done while looking at a single display.

7 COMPARISON OF MAIN RING EXPERIENCE WITH TRACKING RESULTS

Recently attempts have been made to understand the observed Main Ring resonance behavior by tracking studies. Such a project has a number of limitations which are important to keep in mind. First, only one particle is tracked with one momentum and unique tune values. Second, the tracking is typically done for no more than 1000 turns (21 msec real accelerator time vs. 800 msec for the harmonic correction studies done in the Main Ring); note, however, that the stability of the single particle is almost always known by 1000 turns. Third, in order to be accurately represented, the nonlinear fields must be known, and as mentioned previously this is not possible. Nevertheless it is possible to specify the Main Ring lattice and provide a random distribution of higher-order multipoles. Because most of the Main Ring's nonlinear problems are thought to involve the third-order resonance lines, most of the work done so far has concentrated on sextupoles. In particular we have addressed the question of why we continue to be bothered by the third-integer resonances if the driving terms can be compensated. This question is dealt with by introducing large driving terms either through random sextupole distributions in the main dipoles (a distributed driving term) or through one large sextupole error at one location in the ring. The "correction elements" in the program, located as they are in the Main Ring, are then set to compensate, and a driving term calculation is done to verify that the driving terms have vanished. The dynamic aperture is then determined via tracking at tunes that lie along the scan line used in the actual tune scans in the Main Ring.

The resulting simulated resonance scan (see Fig. 4) indicates that, for the distribution of fields used, the resonances are indeed removed. In the case of the single large sextupole error, however, the aperture is reduced at all tunes. Phase-space plots made under these conditions (see Fig. 5) show that the large single sextupole error produces strong coupling, and energy goes from the horizontal

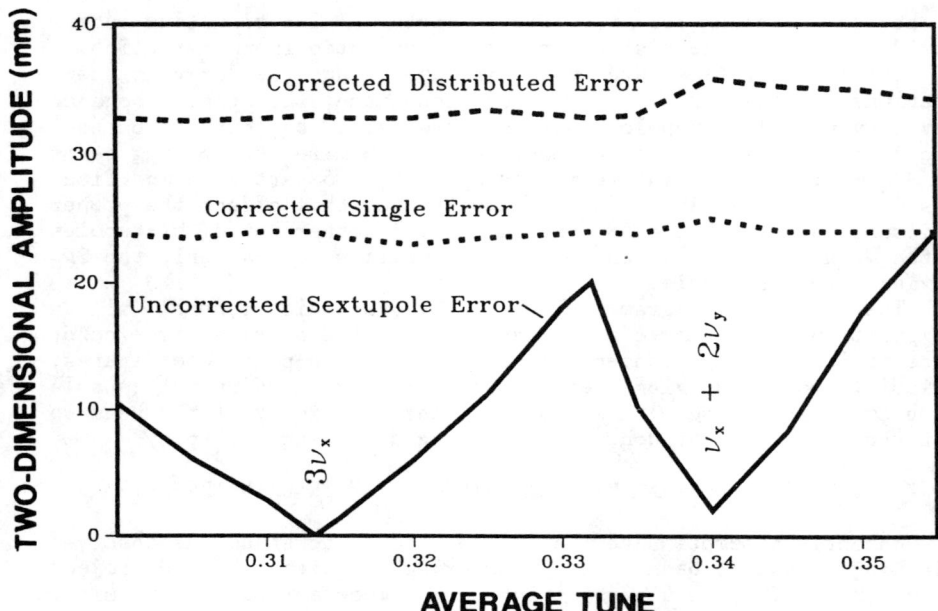

Figure 4 A simulated tune scan, to be compared to Fig. 2. The ordinate in this plot is not efficiency since only one particle is tracked, but is related to beam loss, as explained in Ref. 6.

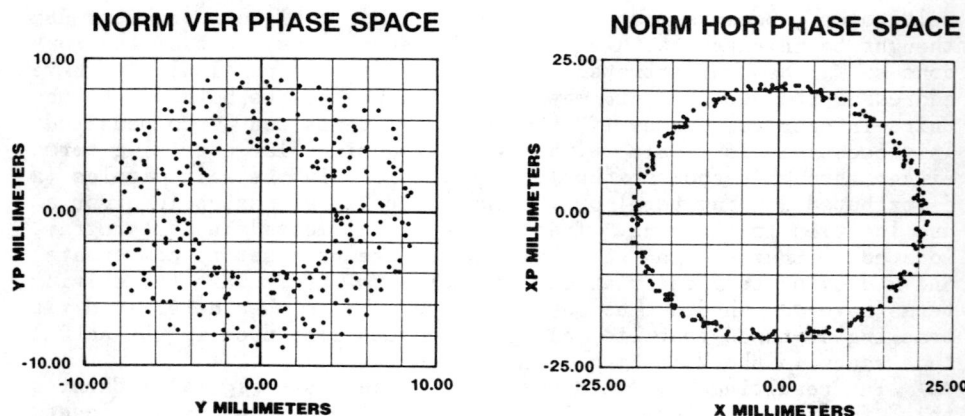

Figure 5 Phase-space plots corresponding to the line in Fig. 4 labeled "Corrected Single Error" showing how the coupling, due to the single large sextupole error, leads to beam loss.

plane to the vertical plane, which has a smaller limiting aperture, thus accounting for the reduction in "efficiency" with a single large sextupole.

Recent magnet measurements have shown that the Main Ring dipoles also have a large distribution in the decapole component of the field at 8 GeV. While this multipole is normally associated with fifth-integer resonance lines, a distribution of decapoles will produce third-integer driving terms as well. A correction family of sextupoles will not be able to compensate for these driving terms at all particle amplitudes. Simulations have been done showing that decapoles can produce more serious beam loss problems at third-integer tunes than at fifth-integer tunes.[6] At present the Main Ring has no decapole correction elements.

In order to study these problems more thoroughly, tracking programs are being modified to handle the more realistic cases, in which the tune of individual particles is modulated by the combinations of synchrotron frequency, chromaticity, and space-charge tune shifts. It may be that the conditions for instability for a particular particle will exist only after many milliseconds.

8 SUMMARY

This article has considered the techniques used to identify and compensate nonlinear errors in the magnets of a circular accelerator. The design of a correction system and the way in which it is used have also been described. Finally, a simulation was described in which the emphasis is on reproducing the results that are observed while measuring and compensating resonance lines. We have found that correction of the resonance lines in the simulation is successful whereas compensation in the real machine in only partially successful. Attempts are being made to enhance the model in order to better simulate the behavior of the Main Ring.

9 REFERENCES

1. Many of the articles from the proceedings of these High Energy Particle Accelerator Summer Schools discuss the problem of nonlinear dynamics, transverse resonances, and the tools to deal with them. In particular the reader is referred to an article by Dave Douglas, "Dynamic Aperture Calculations for Circular Accelerators and Storage Rings," in Physics of Particle Accelerators (Fermilab Summer School, 1984; A.I.P. Conf. Proc. 153). Much of the language used in this article follows from what is introduced in this reference.

2. See article by W.T. Weng, "Space Charge Effects - Tune Shifts and Resonances," Physics of Particle Accelerators (Fermilab Summer School, 1984; A.I.P. Conf. Proc. 153).

3. D.A. Edwards, "An Introduction to Circular Accelerators," Physics of High Energy Particle Accelerators (BNL/SUNY Summer School 1983, A.I.P. Conf. Proc. 127), p. 45-49.

4. D.A. Edwards and T.L. Collins, Fermilab Publication TM-614 (Sept.1975), p. 15.

5. This expression is for a single-dimensional resonance, and the magnitude guarantees an instability (particles of different phase-space coordinates may be unstable at smaller amplitudes). The means of translating driving terms to particle stability is shown on page 48 of Ref. 3.

6. R. Gerig, S. Pruss, and F. Turkot, "Simulations of the Fermilab Main Ring," presented at the 1987 Particle Accelerator Conference in Washington, proceedings to be published.

LONGITUDINAL PHASE SPACE IN CIRCULAR ACCELERATORS

Philip S. Martin
Fermi National Accelerator Laboratory
Batavia, IL 60510

and

Sho Ohnuma
University of Houston
Houston, TX 77004

TABLE OF CONTENTS

1 Introduction.. 1942
2 Beam and Bucket Parameters..................................... 1947
   2.1 Stationary Bucket.. 1947
   2.2 Moving Bucket.. 1948
3 The Case of Two Frequencies.................................... 1952
4 Bunch Coalescing... 1954
   4.1 Debunching Before Rotation................................ 1954
   4.2 Rotation in h=53/h=106 System............................. 1959
   4.3 Recapture in h=1113....................................... 1963
   4.4 Conclusions... 1966
5 References... 1966
Appendix A Bucket Parameters..................................... 1967

LONGITUDINAL PHASE SPACE IN CIRCULAR ACCELERATORS

Philip S. Martin
Fermi National Accelerator Laboratory
Batavia, IL 60510

and

Sho Ohnuma
University of Houston
Houston, TX 77004

1 INTRODUCTION

Circular accelerators use time-varying electric fields to accelerate particles as they cross the fields on each successive revolution around the accelerator. In general, these fields are simply sinusoidally varying, at some harmonic number h times the revolution frequency. Under that condition, there is a reference, or synchronous, particle, with energy E_s, which arrives back at the accelerating gap at the proper phase on the next turn. Thus, if the revolution frequency is $\omega_0 = 2\pi/T_0$, the accelerating, or rf, frequency $\omega_{rf} = h\omega_0$. The energy gain on a given turn is then $dE = eV \sin \phi$, where ϕ is the phase of the voltage at the time the particle passes the accelerating gap. The phase is taken to be zero when the voltage is zero and increasing. The phase for the synchronous particle is denoted ϕ_s; $\sin \phi_s$ is denoted Γ. A particle with a momentum slightly different from the synchronous particle will have a different revolution period, $T = C/v$, where C is the circumference and v is the velocity of the particle. The change in circumference for a particle with a change in momentum is defined as the momentum compaction α, $\delta C/C = \alpha\, \delta p/p$. From relativistic kinematics, $\delta v/v = 1/\gamma^2\, \delta p/p$, and hence

$$\delta T/T = (\alpha - 1/\gamma^2)\delta p/p = \eta \delta p/p \; .$$

This is the definition of η, which is referred to as the frequency slip factor; it describes the change in revolution period with momentum. It vanishes when $\alpha = 1/\gamma^2$, which we refer to as the transition energy γ_t. Below the transition energy, η is negative, and the increase in speed of the particle dominates the increased circumference. Above transition, the opposite is true. In order to preserve phase stability, the rf phase is jumped by an amount $\Delta\phi = \pi - 2\phi_s$ when crossing transition. Thus, the quantity $\eta \cos \phi_s$ is always negative. In the discussion presented here, we will generally assume we are below transition and the stable phase angle $\phi_s < \pi/2$.

The variables we will use for the description of longitudinal motion are

$$q = \phi - \phi_s \quad \text{and} \quad y = (E - E_s)/\omega_{rf}$$

where q is dimensionless and y has the units eV-sec. The parameters $\Delta t = (\phi - \phi_s)/\omega_{rf}$, measured in nsec, and $\Delta E = E - E_s$, measured in MeV, will also be used, since they are properties of a distribution of particles which may be measured directly. Frequently ϕ will be used in place of q for simplicity. The quantity $\Delta p/p$ is also used in place of y or ΔE, particularly to facilitate calculating transverse properties of the beam.

In terms of these variables, the system is described by the Hamiltonian

$$H(q,y) = \frac{1}{2} \omega_{rf}^2 \frac{\eta}{\beta^2 E_s} y^2 + \frac{eV}{2\pi h} [\Gamma q + \cos(q + \phi_s)]. \quad (1)$$

Setting $A = \omega_{rf}^2 \dfrac{\eta}{\beta^2 E_s}$ and $B = \dfrac{eV}{2\pi h}$

we have

$$\frac{dq}{dt} = Ay \quad \text{and} \quad \frac{dy}{dt} = B[\sin(q + \phi_s) - \sin \phi_s]. \quad (2)$$

This system is characterized by two regions of phase space. In one region, the motion is unbounded; in the other, the motion is bounded, and particles undergo oscillations around the synchronous phase angle. The two regions are separated by a trajectory called the separatrix. The separatrix is defined by the equation

$$A \frac{y^2}{2} + B\{\Gamma\phi + \cos \phi - \Gamma\phi_1 - \cos \phi_1\} = 0.$$

where ϕ_1 is defined below. We identify two fixed points on the y = 0 axis; the stable fixed point corresponds to q = 0, i.e. $\phi = \phi_s$. The unstable fixed point, at $\phi_1 = \pi - \phi_s$, marks one point at which the separatrix intersects the y = 0 axis; the other intersection of the separatrix is at the ϕ_2 which is a solution of the equation $\Gamma\phi_2 + \cos \phi_2 = \Gamma\phi_1 + \cos \phi_1$.

The area in phase space bounded by the separatrix is called the bucket area S. The bucket height Y_b is the maximum y value on the separatrix; it occurs at q = 0, i.e. at $\phi = \phi_s$. When switching to ΔE coordinates, the bucket height is denoted ΔE_b. The bucket width is simply $\phi_1 - \phi_2$. The bucket is called a stationary bucket when $\Gamma = 0$, i.e. $\phi_s = 0$ or $\phi_s = \pi$. There is no acceleration of the beam. When $\Gamma \neq 0$, the bucket is called a moving bucket.

The beam distribution occupies an area in phase space which we call the longitudinal emittance ε_ℓ. The beam is usually captured within the bucket, although there are exceptions such as phase-displacement acceleration and the capture of unbunched beam. Unlike the transverse distribution, which is typically characterized by a Gaussian profile, the longitudinal distribution of proton beams, when contained in a bucket, is limited in width and height. Each particle in the distribution traces out a trajectory in phase space determined by the Hamiltonian for that particle (see Fig. 1), and each particle has an emittance corresponding to the area enclosed by its trajectory. The beam emittance is usually taken to be the largest emittance of the particles in the beam, although there are times when one may wish to include only some fraction such as 95% of the particles.

One common parametrization of the beam distribution is the elliptic distribution

$$\rho^2(q,y) \propto y_B^2(q) - y^2$$

where

$$y_B^2(q) = Y^2[1 - \frac{q}{Q^2}]$$

defines the boundary of the bunch. Normalizing to unity,

$$\rho(q,y) = \frac{3}{2\epsilon_\ell} [1 - \frac{q^2}{Q^2} - \frac{y^2}{Y^2}]^{1/2} \tag{3}$$

where Q is the half-width of the beam and Y is the half-height. The longitudinal emittance ϵ_ℓ is the area of the ellipse defined by y_B, πQY. The current distribution is

$$I(q) = e \int_{-y_B}^{y_B} \rho(q,y) dy \propto y_B^2(q)$$

which has the form of a parabola. The term parabolic distribution is often used to describe an elliptic distribution, the difference in terminology arising from ambiguity over whether one is discussing the distribution in phase space or the projection on the space axis. Some of the properties of the elliptic distribution are shown in Fig. 2.

The elliptic distribution is a reasonable choice in many cases. The elliptical boundary can be calculated from the Hamiltonian in the limit of small q, and the emittance is therefore accurately reflected. But in many instances the particle density is not so well-behaved.

The above discussion has been intended to clarify the terminology and notation used here. Other works[1,2] present a fuller mathematical treatment. In the next section, we present the formulae that allow us to calculate bunch and bucket parameters. Then, using these formulae, we will analyze one of the rf manipulations used for the Tevatron I collider operation, and proposed for use in the SSC injector.

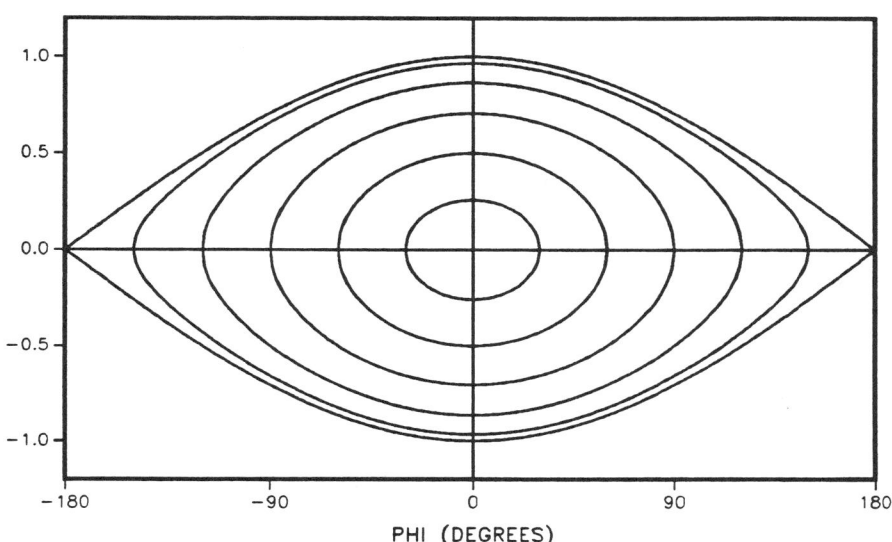
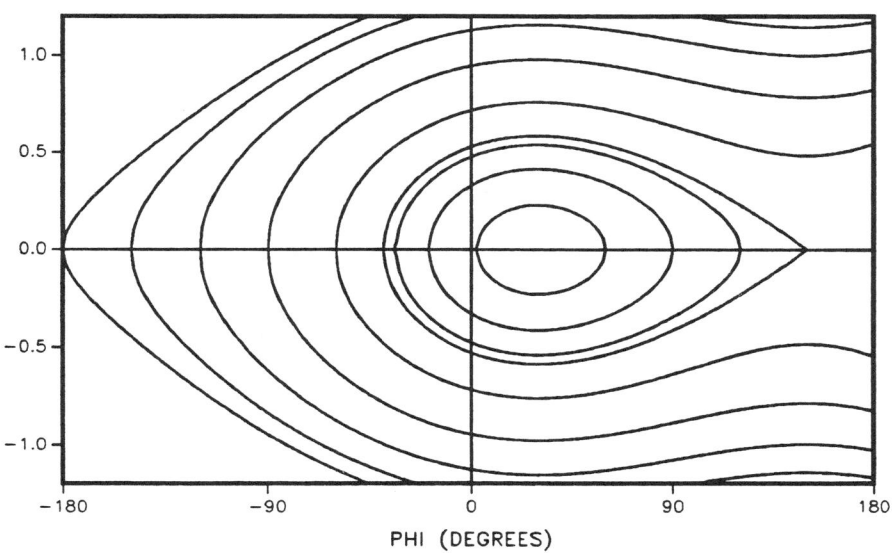

Figure 1. Phase-space trajectories.

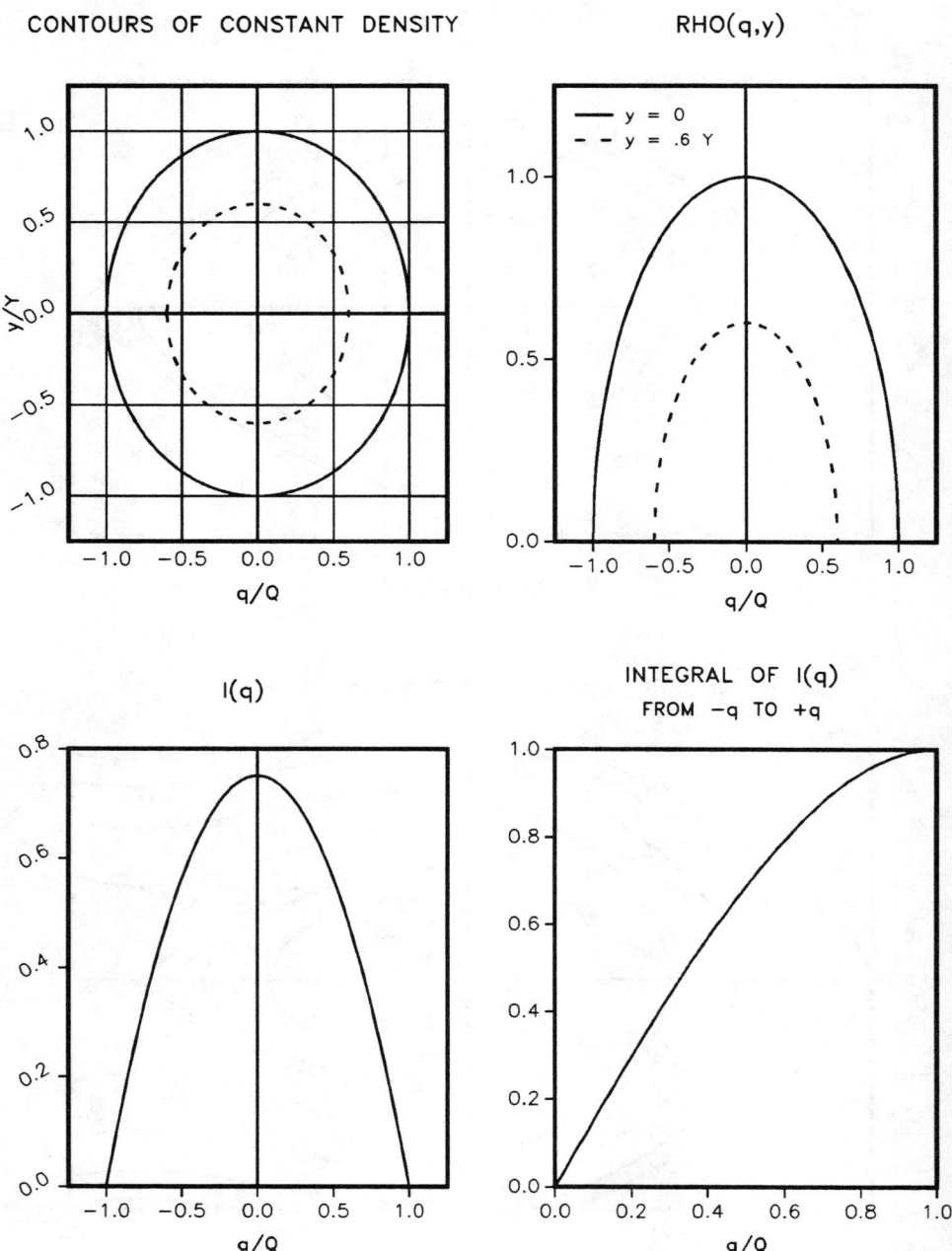

Figure 2. Properties of the elliptic distribution.

2 BEAM AND BUCKET PARAMETERS

2.1 Stationary Bucket

The bucket area is

$$S = 16(B/|A|)^{1/2} = 8(2E_s V/\pi h\eta)^{1/2} \text{ eV-sec} \qquad (4)$$

The bucket height is

$$Y_b = 2(B/|A|)^{1/2} = (2E_s V/\pi h\eta)^{1/2} \text{ eV-sec} . \qquad (5)$$

These follow directly from the equation for the separatrix, which in the case of $\Gamma = 0$ can be readily integrated or solved.

The beam height $Y(Q)$ is the bucket height Y_b times $\sin(Q/2)$. This follows from the Hamiltonian, and is correct for all values of Q.

The beam emittance ε_ℓ is given by solving the Hamiltonian for y and integrating:

$$\varepsilon_\ell = 4\sqrt{2} \; (B/|A|)^{1/2} \int_0^Q [\cos q - \cos Q]^{1/2} dq . \qquad (6)$$

For small Q, we can formulate approximate expressions for the emittance that are easier to use. The simplest is derived from our earlier expression of the emittance as the area of the ellipse $\pi Q Y$. With Y as given above, and the factor of 8 between bucket height and bucket area, the emittance can be written as

$$\varepsilon_\ell = S \frac{\pi}{8} Q \sin \frac{Q}{2} \simeq S \frac{\pi}{16} Q^2 (1 - \frac{1}{24} Q^2) .$$

In the last step, we have expanded $\sin(Q/2)$ keeping the first two terms. This expression was based on the assumption that the beam was defined to be an ellipse. The integrand above can also be expanded as a power series, and then integrated. This yields an expression for the emittance that does not rely on the assumption of an ellipse:

$$\varepsilon_\ell \simeq S \frac{\pi}{16} Q^2 (1 - \frac{5}{96} Q^2) .$$

The latter expression is good to 6% for beam filling the bucket, slightly underestimating the emittance. The expression derived for the ellipse overestimates the emittance, and is somewhat worse. Either is sufficient for most circumstances. Generally the error in measuring Q will exceed the error in the approximations in the calculation.

The particles at $q \neq 0$ undergo synchrotron oscillations at a frequency

$$f_s = \frac{\omega_s}{2\pi} = \frac{1}{2\pi} (|A|B)^{1/2} \frac{\pi}{2K(\sin(q^2))} \qquad (7)$$

where $K(k)$ is the complete elliptic integral of the first kind

$$K(k) = \int_0^{\pi/2} (1 - k^2 \sin^2\theta)^{-1/2} d\theta \ .$$

For small k,

$$K(k) \simeq (\pi/2)(1 + k^2/4),$$

so that

$$f_s \simeq \frac{1}{2\pi} (|A|B)^{1/2} (1 - \frac{1}{16} q^2) \ .$$

This is quite good up to $q \sim 2.5$. Over that range, the synchrotron frequency has changed by 40%. The variation of the synchrotron frequency with q is referred to as the synchrotron tune spread.

2.2 Moving Bucket

The bucket area when $\Gamma \neq 0$ is parametrized by introducing a term $\alpha(\Gamma)$ where the bucket area is the stationary bucket area times $\alpha(\Gamma)$:

$$S = 16(B/|A|)^{1/2} \alpha(\Gamma) \text{ eV-sec} \ .$$

Similarly the bucket height is the stationary bucket height times the function $\beta(\Gamma)$:

$$Y_b = 2(B/|A|)^{1/2} \beta(\Gamma) \ .$$

The functions $\alpha(\Gamma)$ and $\beta(\Gamma)$, which can be calculated from the equation for the separatrix, are plotted in Fig. 3, along with the bucket width, as a function of ϕ_s. See Appendix A for numerical values of these functions.

The beam parameters within the bucket can be calculated either with reference to the stationary bucket parameters or to the moving bucket parameters. In the small angle limit, the Hamiltonian for a moving bucket takes on the form

$$q^2 + \frac{|A|}{B\cos\phi_s} y^2 = Q^2 \ .$$

In this limit, then, we expect the beam height $Y(Q)$, the emittance ε_ℓ, and the synchrotron frequency f_s to behave as in the stationary bucket case, with an extra multiplicative term $\sqrt{\cos\phi_s}$. For larger

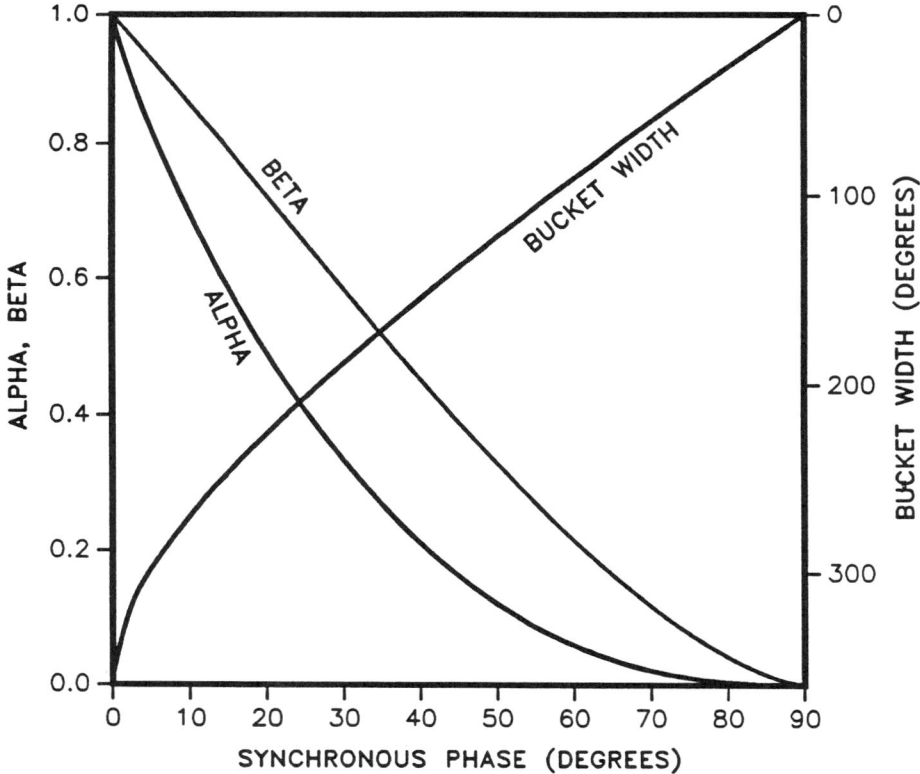

Figure 3. Alpha, beta, and bucket width.

amplitudes, we specify the relationship between these parameters and the bucket parameters through functions K_A, K_H, and K_S:

$$\varepsilon_\ell = S\ K_A(\Gamma,\Delta).$$

$$Y = Y_b\ K_H(\Gamma,\Delta),$$

$$f_s = \frac{1}{2\pi}\ (|A|B\cos\phi_s)^{1/2}\ K_s(\Gamma,\Delta)\ .$$

The quantity Δ is the bunch width divided by the bucket width, where the bucket width, the bucket area S, and the bucket height Y_b all pertain to the moving bucket. The functions K_A, K_H, and K_S are shown in Figs. 4, 5, and 6. Note that in Fig. 6 the ordinate is $1 - \Delta$, and that the term $\sqrt{\cos\phi_s}$, which describes the zero-amplitude synchrotron frequency, is not included in the function K_S.

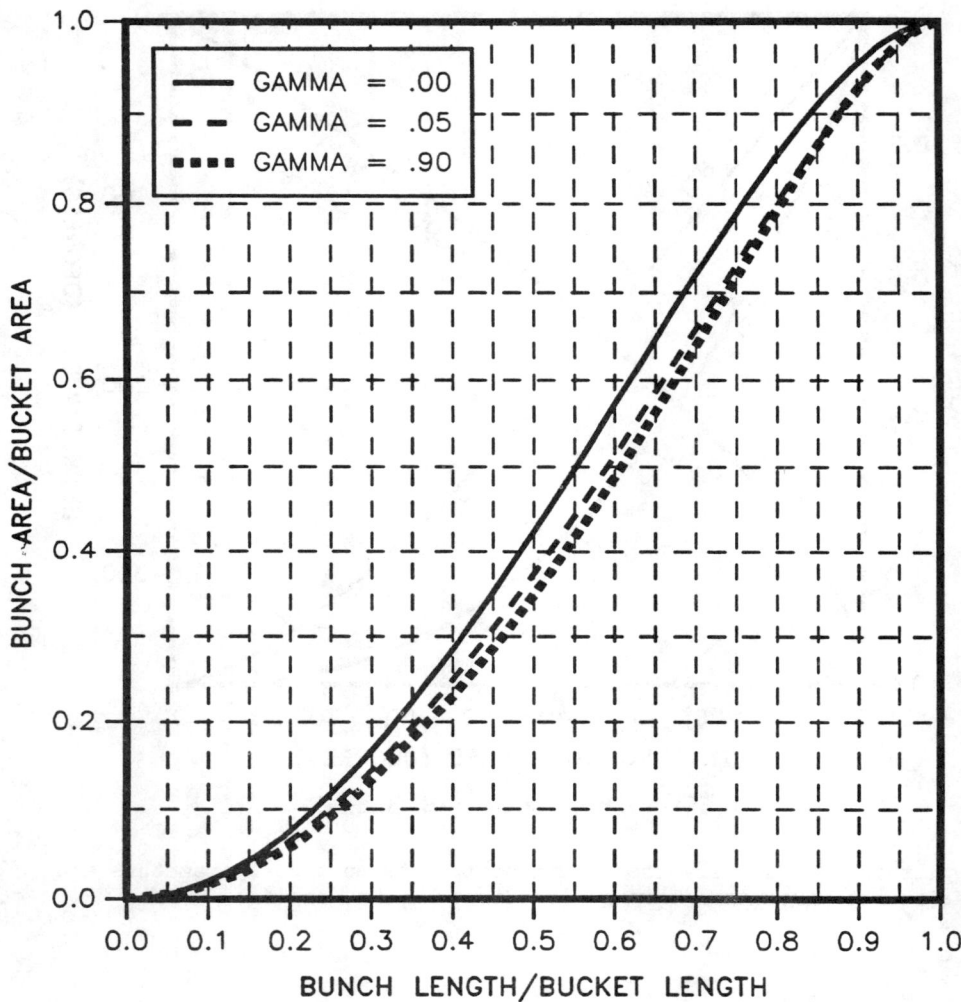

Figure 4. Bunch area vs. bunch length.

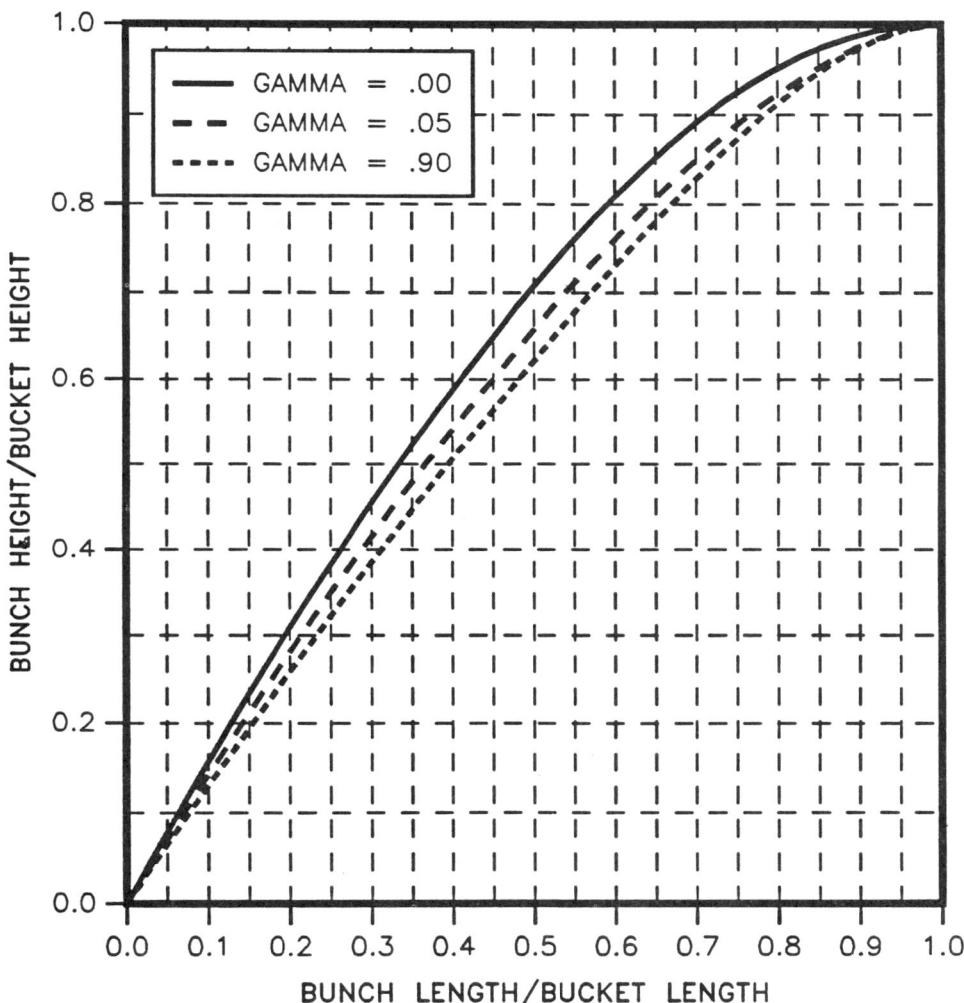

Figure 5. Bunch height vs. bunch length.

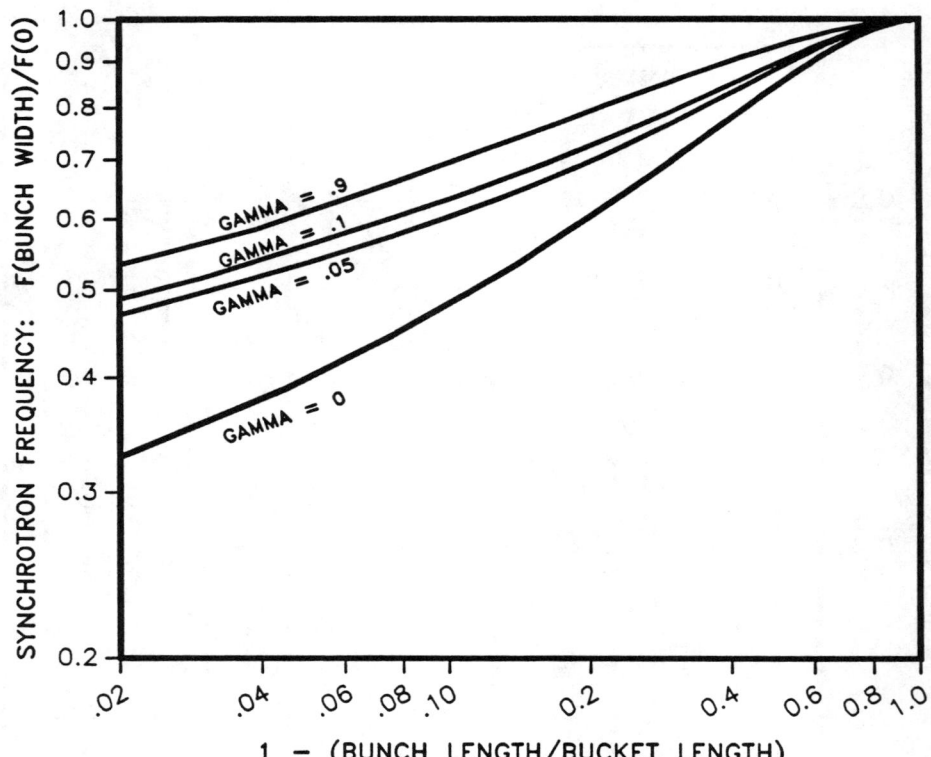

Figure 6. Synchrotron period vs. bunch length.

3 THE CASE OF TWO FREQUENCIES

The above discussion treated the case of a single rf, the most common situation, and the simplest mathematically. Any periodic waveform can of course be expanded in a Fourier series:

$$g(\phi) = \Sigma\ (a_n \sin n\phi + b_n \cos n\phi),$$

$$G(\phi) = -\int g(\phi)\ d\phi = \Sigma\ \frac{1}{n}\ (a_n \cos n\phi - b_n \sin n\phi).$$

The formalism developed in the first section can be replaced, beginning with the Hamiltonian

$$H = \frac{1}{2}\ Ay^2 + B\{\Gamma q + G(\phi)\}$$

where now $\Gamma = g(\phi_s)$. Calculations of parameters such as bucket area and bucket height, previously straightforward, become intractable in the general case. The simplified case of an rf waveform comprised of only the first and second harmonic is often used and will be treated below.

More complicated cases, for example where the higher harmonic is a larger multiple of the lower, are best treated through computer codes that solve the Hamiltonian and track particles. An example of such a code is the program ESME written[3] by MacLachlan of Fermilab. This program has been very useful in simulating rf manipulations required for operation of the Tevatron collider, and produced some of the figures in this paper. Some features of in this program are the following: bunch distributions that are Gaussian, elliptic, or uniform (or combinations of uniform in one projection and Gaussian or elliptic in the other); specification of two frequencies with variable phase, and voltages varying in time either linearly or isoadiabatically; calculation of bunch parameters (rms values of q and ΔE); and particle tracking that stops when the bunch width or height is a minimum, if desired. Space-charge effects may also be included.

The case of an rf waveform

$$g(\phi) = \sin m\phi + f \sin n\phi,$$

(i.e. the Fourier expansion in which $a_m = 1$, $a_n = f$, and all other coefficients = 0), is a circumstance that is not uncommon in accelerators. (Here, the frequency $m\omega_o$ is the principle rf harmonic of the revolution frequency.) The higher (or lower) harmonic frequency $n\omega_o$ may be used to modify the bucket boundary and hence the synchrotron tune spread within a bunch. For instance, it may be desirable to perform a bunch rotation, in which case the synchrotron period should have little variation with phase. An example of this will be discussed in the next section. Or, a large variation in synchrotron tune may be desired to inhibit instabilities through Landau damping; the harmonic cavity operating at frequency $n\omega_o$, some small multiple of the principle frequency $m\omega_o$, is called a Landau cavity, and it usually operates at a voltage of m/n times the lower harmonic voltage. Similarly, coupled bunch instabilities can be cured by using a small amount of a lower harmonic to modulate the synchrotron tune from bunch to bunch in a dominant higher harmonic. An example of this is seen in the Fermilab Booster, in which h=84 is the primary rf harmonic. One cavity can be run at h=77, which results in the desired synchrotron tune modulation and avoidance of higher-mode resonances between bunches.[4]

Let us consider the case where $n = 2m$; then $g(q) = \sin q + f \sin 2q$, and $G(q) = \cos q + (f/2)\cos 2q$. Let us further consider only stationary buckets, $\Gamma = 0$. For $|f| < 1/2$, there is only one stable fixed point,

$$\Gamma = \sin q + f \sin 2q = \sin q + 2f \sin q \cos q = 0$$

at $\sin q = 0$. If $|f| > 1/2$, however, other fixed points will be present at $q = \cos^{-1}(-1/2f)$. The point $q = 0$ will be unstable for f < -1/2. Figure 7 shows the variation in synchrotron period for different values of f. Both positive and negative values of f can be used to increase the synchrotron tune spread, although the effect is small until f approaches -1/2. However, small negative values of f, as shown in Fig. 8, can also be used to decrease the tune spread up to about $\phi = 1.8$. The degree of uniformity achievable depends on how far out in ϕ one wishes to go. For small ϕ, smaller values of $|f|$ (around 0.125) are best while larger values of $|f|$ are required for larger ϕ; however, the degree of uniformity decreases as ϕ increases.

4 BUNCH COALESCING

One of the important requirements of colliding-beam operation is the availability of intense, isolated bunches of particles, where "isolated" means that neighboring buckets should not be populated. The coalescing process merges a number of bunches into a single intense bunch by a bunch rotation process using a lower harmonic rf system. Figure 9 shows various stages of the process in an ESME simulation of bunch coalescing; Fig. 10 is a photograph of the rotation and recapture. The goal is to take between 7 and 13 h=1113 (53 MHz) bunches and produce a particle distribution which may be recaptured in a single h=1113 bucket. The height of the bucket is limited since the bunch must be transferred to and accelerated in the Tevatron. In the Fermilab Main Ring, the lower harmonic rf is h=53 (2.5 MHz) together with h=106 for linearizing the rotation. The time spread of the rotated ensemble of particles is broadened by (i) non-linearities in the rotation, and (ii) insufficient debunching of the initial h=1113 bunches, since the rotation in the h=53/h=106 system exchanges time and energy spread. In addition, increasing (decreasing) the available h=53/h=106 voltage has the effect of narrowing (broadening) the ensemble, at the expense of a larger (smaller) energy spread. The purpose of this discussion is to analyze the coalescing process by separating out the different effects that contribute to the final bunch width, with the aim of identifying the most crucial aspects. The question of longitudinal emittance of the initial bunches will be largely ignored; the bunches will be assumed to have an initial emittance of 0.20 eV-sec, typical of that observed in the Main Ring at 150 GeV.

4.1 Debunching Before Rotation

The coalescing process begins with bunches at 150 GeV in the Main Ring, and an h=1113 voltage of 800 kV. The proton bunch intensity is about 10^{10}; for antiprotons, the intensity may be as much as a factor of ten lower. Sixteen of the 18 rf cavities are paraphased so that their voltages cancel, and then turned off. The remaining two stations are then paraphased, reducing the voltage from 90 kV to some low value. This two-stage reduction permits finer control of the voltage towards the end of the process.

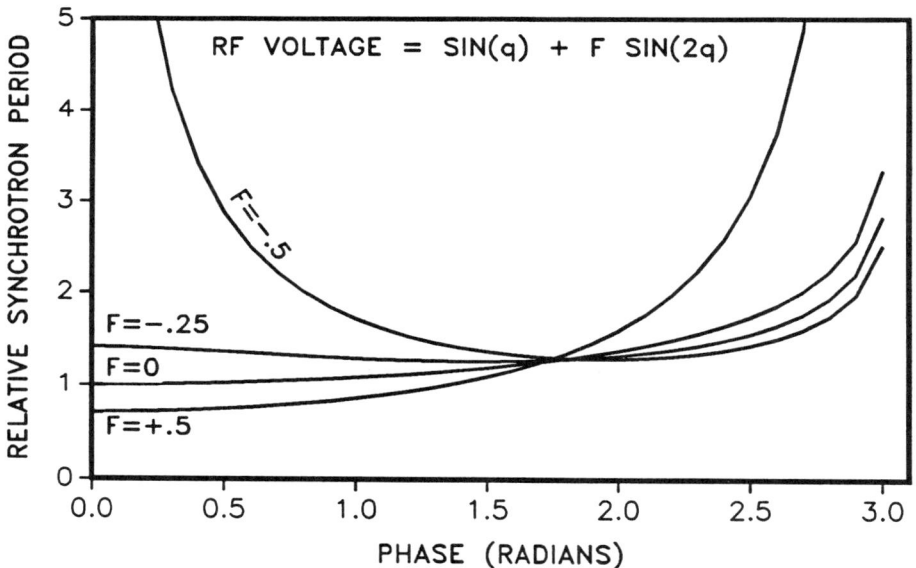

Figure 7. Synchrotron period vs. amplitude.

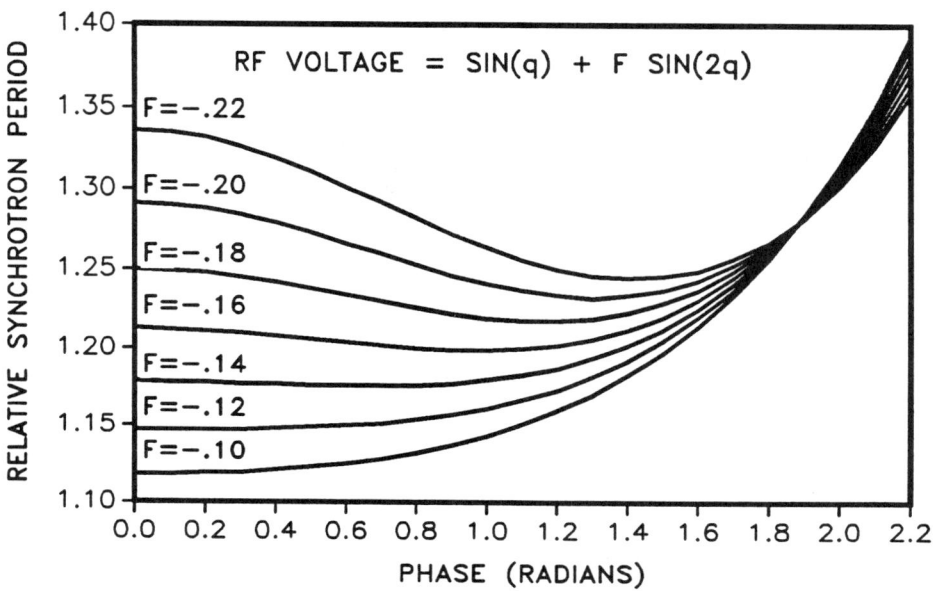

Figure 8. Synchrotron period vs. amplitude.

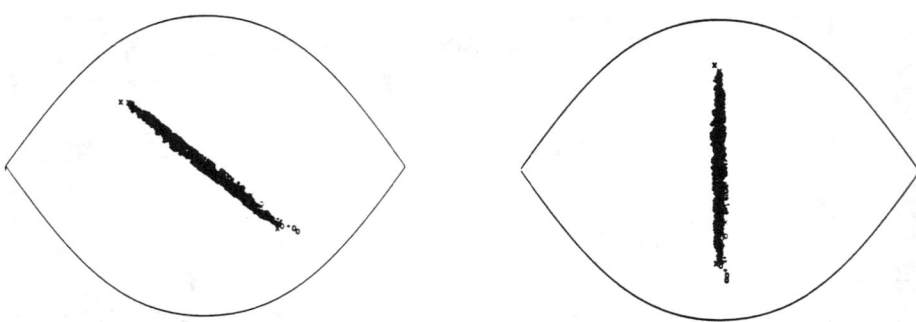

Figure 9. Rotation in bucket formed by 33 kV of h=53 plus 5.94 kV of h=106. Bucket height = 146 MEV.

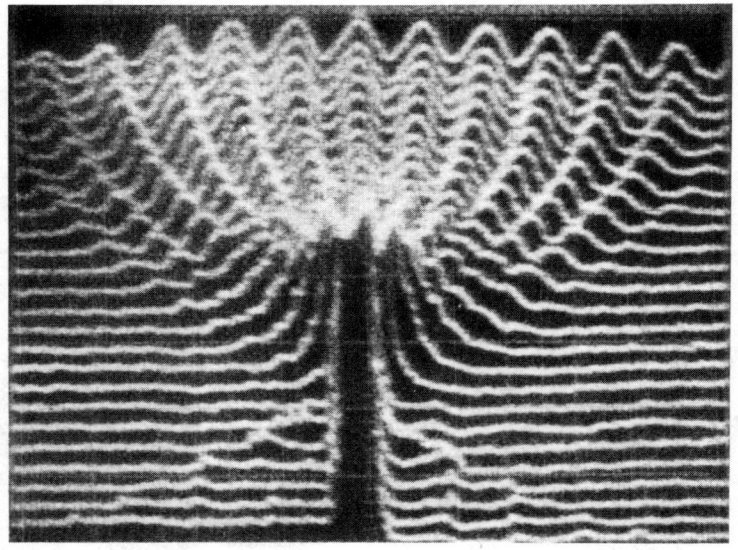

Figure 10. Bunch coalescing operation, sweep speed 20 nsec/div. The earliest trace, at the top, occurs at the end of debunching. Successive traces occur at 6.8-msec intervals during the rotation in H=53 + H=106 and the recapture in H=1113.

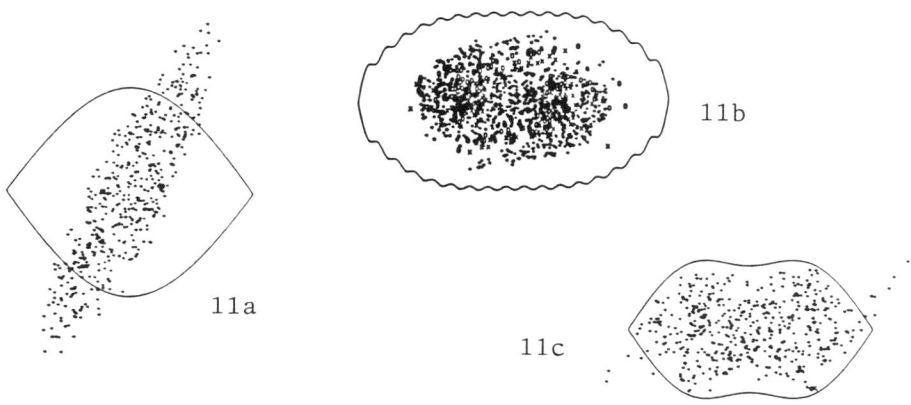

Figure 11 a. Debunching too rapidly in h=1113 bucket.
b. Debunching into h=53 bucket.
c. Debunching into h=1113 + h=2226 bucket.

Simulations (Fig. 11a) demonstrate the disastrous effects of reducing the voltage too quickly, since particles escape the bucket too early (at a large energy spread). The reduction of voltage must be sufficiently slow that particles do not leave the bucket until the bucket area is reduced below the bunch area. The rate of change of an rf bucket is characterized by a parameter a_c, called the adiabaticity coefficient, which relates the change in bucket area to the synchrotron period:

$$\frac{dS}{S} = a_c \frac{dt}{T_s} .$$

The adiabaticity coefficient should be small (compared with unity) to avoid emittance blowup. Thus, an "adiabatic" process is one that is done on a time scale of many synchrotron periods. During the 1987 Collider run, the debunching was being done too quickly, a result of ramp limitations on the time available for the coalescing process.

A bucket formed by 2270 V of h=1113 has an area of 0.20 eV-sec and a bucket height of ΔE_b = 8.3 MeV (η in the Main Ring at 150 GeV is 0.0023). If the beam could be totally debunched into a rectangle of length 18.83 nsec (the length of an h=1113 bucket) and uniform height, the height (again for 0.20 eV-sec) would be 5.3 MeV; the ratio, as one can easily calculate, is $\pi/2$. Thus, if one does nothing beyond reducing the voltage until the beam fills the bucket, one should expect a blowup in emittance of ~1.6. The transformation to a rectangular distribution is difficult; however, distributions are attainable which have a lower momentum spread than that produced by the full bucket. For example, one can allow the beam to squeeze out as the voltage is reduced further. Such a process depends on controlling the rf voltages (including those induced by the beam itself) to some precision. The smallest final energy spread attained in any simulation of this process, with an initial emittance of 0.2 eV-sec, was ΔE = 6.07 MeV, which is 14% higher than the ideal 5.31 MeV. (In this discussion, bunch heights and widths are taken as the values that include 95% of the beam. The rms values generated by the ESME simulations have been multiplied by $\sqrt{6}$ to produce the values presented here.)

The ideal distribution for rotating in the h=53 system is an ensemble that is matched to a very low h=53 (or h=53 + h=106) voltage. Thus, instead of simply lowering the h=1113 voltage to some small value, one applies a few hundred volts of h=53. The voltage is chosen to provide a trajectory that contains all the bunches in the desired phase extent and with an area equal to that of the bunches. Such a process can in principle be done with virtually no emittance blowup (Fig. 11b) but requires many synchrotron periods to accomplish.

Another possibility for lowering the momentum spread during debunching would be to augment the h=1113 with an h=2226 rf system. An rf bucket formed by 1.34 kV of h=1113, plus 1.53 kV of h=2226 (180° out of phase), has an area of 0.2 eV-sec and a height of 7.0 MeV, or ~18% lower than the height of a 0.2-eV-sec bucket formed by h=1113 only. The ratio of -1.14 for the h=2226/h=1113 voltages, which gives the minimum bucket height to bucket area ratio, was determined computationally and is a fairly shallow minimum. As the ratio is changed, both the bucket height and bucket area change slowly; changing the amount of h=2226 to -1.00 or -1.28, for example, changes both the height and area by <3%. Figure 11c shows beam being debunched by lowering the h=1113 voltage adiabatically to 1.34 kV while maintaining a constant h=2226 voltage of 1.53 kV. This procedure results in a beam energy spread of about ΔE = 6.5 MeV. This bunch height should be compared with those attained by debunching to 2.27 kV of h=1113 alone (bunch height ΔE = 8.2 MeV), which still contains all the beam, or by debunching to 1.34 kV of h=1113, with no h=2226 (ΔE = 7.1 MeV). In the latter case, the beam is no longer completely within the bucket. Using the h=2226 voltage along with the h=1113 allows one to get to lower energy spreads while still capturing all the beam. It would not preclude trying to lower the voltages further to achieve additional energy spread reduction, if so desired. A simulation in which both the h=1113 and h=2226 voltages were reduced by a factor of four from their initial values of 1.34 kV and 1.53 kV over 5000 turns resulted in a final energy spread of about ΔE = 5.6 MeV, only 6% larger than the ideal 5.31 MeV. This process is much faster, and hence less susceptible to instabilities, than the process of debunching into a lower harmonic, since the synchrotron frequency varies as $\sqrt{h}$. The present limitation in the debunching is the inability to keep the beam captured in buckets that have a larger (by a factor of 2 to 4) bucket area than the beam emittance. The cause of this problem may be power supply noise or noise in the rf, or it may be an instability. The synchrotron period vs. amplitude for the bucket formed from h=1113 plus h=2226, as shown in Fig. 12, varies appreciably across the bucket. This could reduce beam instability problems. (The bucket produced by this combination has a stable fixed point at $\phi \simeq 1.12$, and unstable fixed points at $\phi = 0$ and $\phi = \pi$. The trajectory that passes through $\phi = 0$ also passes through $y = 0$ at $\phi \simeq 1.7$.)

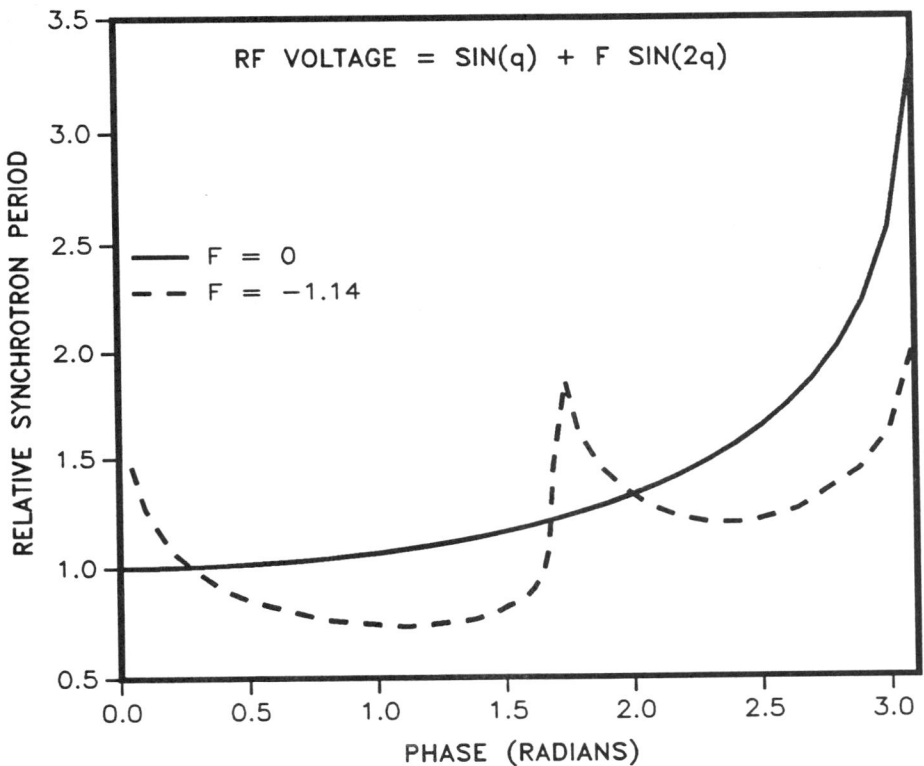

Figure 12. Synchrotron period vs. amplitude.

4.2 Rotation in h=53/h=106 System

In order to understand the bunch rotation process more thoroughly, a number of ESME simulations were done. One goal was to understand how the individual bunches rotated, so that the broadening effects of non-linearities could then be estimated. Since the ideal waveform, i.e. one with no synchrotron tune spread, is linear, $V(\phi) = a\phi$, the task is to determine what value of f produces the best results. Separate simulations were done for individual bunches displaced by n h=1113 buckets from the center of the h=53 bucket, and the resulting bunch widths and energy offsets determined. The coalescing process of an entire ensemble involves the 2n+1 bunches from -n to +n. In all cases, the particle distribution used as input was 0.2 eV-sec, and it was debunched into the bucket shown in Figure 11. The maximum available h=53 voltage is in the range of 25 to 30 kV. Simulations were done with the value of the h=53 voltage fixed at 25.6 kV, for no h=106, and for h=106 voltages of -0.18, -0.20, and -0.27 times the h=53 voltage. The multipliers were chosen for the following reason: the value -0.18 gives the most linear voltage waveform if one restricts oneself to n = 4 ($\phi \sim 1.4$). As seen in Fig. 8, the synchrotron period is uniform to a couple of percent out to $\phi \sim 1.7$. The choice f = -0.27 gives the best fit to a linear voltage over the range $\phi < 1.9$, the center of bunch number 6 ($\phi \simeq 2.1$ is its outer edge.) As might be expected from Fig. 8, however, the variation in synchrotron period is too large. Thus, -0.20 is the choice, based on synchrotron period rather than linear voltage, for trying to coalesce 13 bunches. The rotation of the individual bunches at bucket numbers ±n was carried out until the rms phase spread of the pair of bunches was a minimum. The rotation time was therefore allowed to vary with n.

The energy offsets after rotation are shown in Fig. 13; for no h=106, the curve follows the expected $\Delta E_b \sin(\phi/2)$ (ΔE_b = 129 MeV). The rms phase spreads after rotation are shown in Fig. 14, plotted in units of angle, where 2π equals the ring circumference. The right axis provides the conversion to 95% full width. The large growth in the width of the bunches with increasing bunch number when there is no h=106 is a consequence of phase-space conservation together with the $\sin(\phi/2)$ factor above. As the ΔE points become closer and closer, the phase spread must increase to maintain constant emittance. The addition of the h=106 reduces this effect significantly.

Next, ensembles of 7, 9, 11, and 13 bunches were rotated in h=53/h=106 buckets. The results are shown in Fig. 15. For 7 bunches, one does as well with no h=106; for more than 7, the lower value of the h=106 (-0.18) is clearly beneficial. The bunches at ±6 are lost, however. If one increases the h=106 to -0.27 to try to capture them, the non-linearities indeed increase and the resulting bunch is broader. The f = -0.20 produces a slightly broader bunch when coalescing 11 bunches or less, but does better for 13 bunches than does f = -0.18.

To evaluate the bunch-broadening effects due to the voltage waveform, which is not exactly linear, one can add in quadrature the phase spreads of the individual bunches, which were plotted in Fig. 14. Including only 11 bunches yields the values in Table I.

Table I Coalescing 11 Bunches (Bunch widths in nsec)

Fraction of h=106:	0	0.18	0.20	0.27
Sum of individual bunches	17.0	15.9	15.9	16.5
All bunches simultaneously	32.4	15.9	16.5	24.4

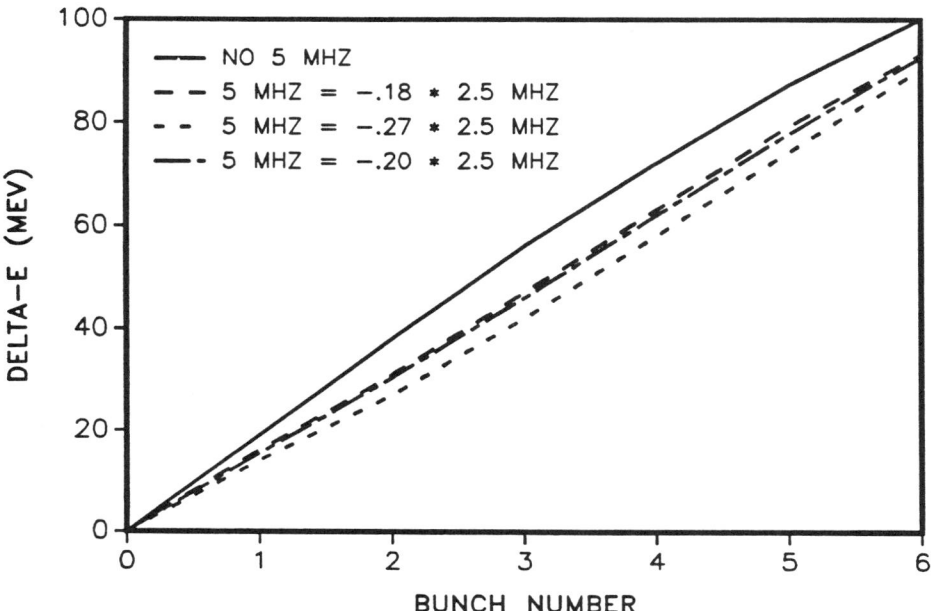

Figure 13. Energy offset after rotation as a function of initial bunch position.

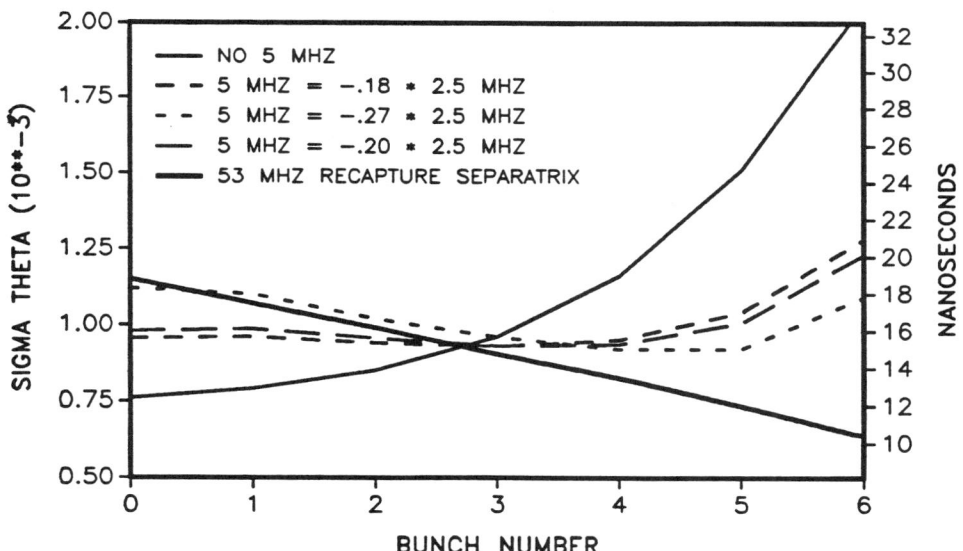

Figure 14. Phase spread after rotation as a function of initial bunch position.

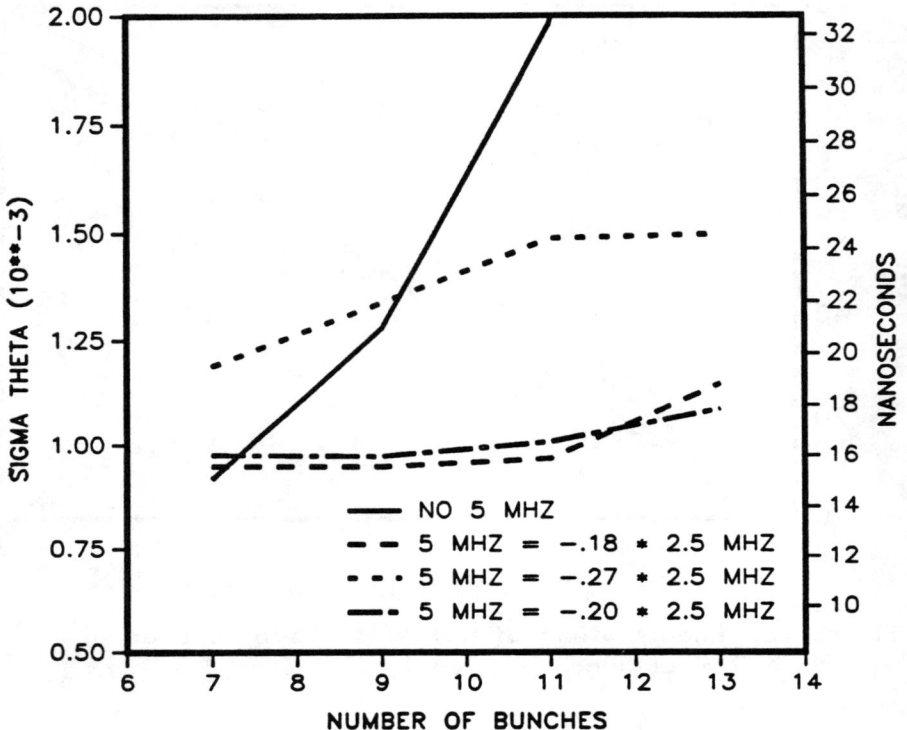

Figure 15 Coalesced bunch width as a function of number of bunches.

The second line in Table I contains the data plotted in Fig. 15 for 11 bunches. It is clear that the non-linearities in the case of no 5 MHz, or of 5 MHz = 0.27, seriously broaden the bunch width after the rotation. On the other hand, for the lower values of 5-MHz voltage, the rotation is sufficiently linearized that there is little or no difference between the individual bunches and the ensemble. The data for only these values of f, but including 13 bunches, are given in Table II.

Table II Coalescing 13 Bunches (Bunch widths in nsec)

	0.18	0.20
Sum of individual bunches	18.5	16.5
All bunches simultaneously	18.8	17.8

The bunches at bunch number ±6 are substantially broader, as seen in Fig. 14, and also lag substantially behind the bunches just inside. A third harmonic would be required to coalesce more than 11 bunches, but up to 11 can be done with reasonable efficiency.

4.3 Recapture in h=1113

Following the rotation by the h=53/h=106 system, the beam must be recaptured in an h=1113 bucket. What is the proper voltage for capturing 11 bunches? The beam had an initial longitudinal emittance of 11 times 0.2, or 2.2 eV-sec. The debunching process in the simulations led to a blowup of 22% to 2.69 eV-sec. The rotation is assumed to lead to no further increase, with a resulting bunch width of 16 nsec. Thus, ΔE for the rotated distribution is 84 MeV. (This is slightly larger than the corresponding point in Fig. 13, which referred to the center of the bunches. Here it is the extremum which is of concern.) To capture all of the beam requires that the bucket height at 8 nsec from the center is this same 84 MeV; the bucket that has this property is produced by 4.2 MV of h=1113, and has a height of 360 MeV and an area of 8.6 eV-sec. This capture, a blowup by a factor of 3, would result in a bunch too large to accelerate in the Tevatron.

Instead of trying to capture all of the beam, what happens with a bucket height of $\sqrt{2}$ times 84 MeV, or 119 MeV? This corresponds to a voltage of 462 kV, and has an area of 2.85 eV-sec. This bucket would have a height of 28 MeV at 8 nsec from the center, and a half-width of 4.7 nsec at a height of 84 MeV (Fig. 16). It would exclude that fraction of the beam in a roughly triangular shaped region between 28 and 84 MeV, and between 4.7 and 8 nsec from the center of the bucket; the four such regions have a total area of 0.37 eV-sec. This capture would enclose 86% of the emittance of the beam, and the captured beam would be blown up from 2.32 to 2.85 eV-sec, or by 23%. The same calculations can be done for 7 and 9 bunches; the results are shown in Table III.

The particle distribution, although no longer a well-defined function, is less dense at the edges, and the part of phase space that is not captured has a lower than average density. Assuming the particle distribution is parabolic in phase and uniform in ΔE, the fraction of particles lost is about one-fourth the fraction of phase space lost, i.e. the particle capture efficiency will be about 96%. For an arbitrary recapture bucket height, the efficiency can be determined by a numerical integration over the energy axis. An example is shown in Fig. 17, which is based on the 11-bunch, 5 MHz = -0.18 rotation, for which the phase spread is 16 nsec and ΔE = 84 MeV. The fraction of particles and the fraction of the beam emittance which are captured by various bucket heights are plotted. It differs somewhat from the rough calculation, indicating a 95% particle capture efficiency and an 88% emittance capture efficiency for a capture bucket height of ΔE = 119 MeV. The expected ratio of one-fourth in the naive calculation is approached as the capture bucket height increases. Also plotted on Fig. 17 is the ratio of bucket area to captured emittance as a function of bucket height. Since the captured beam extends to the separatrix, it will eventually fill the bucket; thus, this ratio is the expected emittance growth during the capture process.

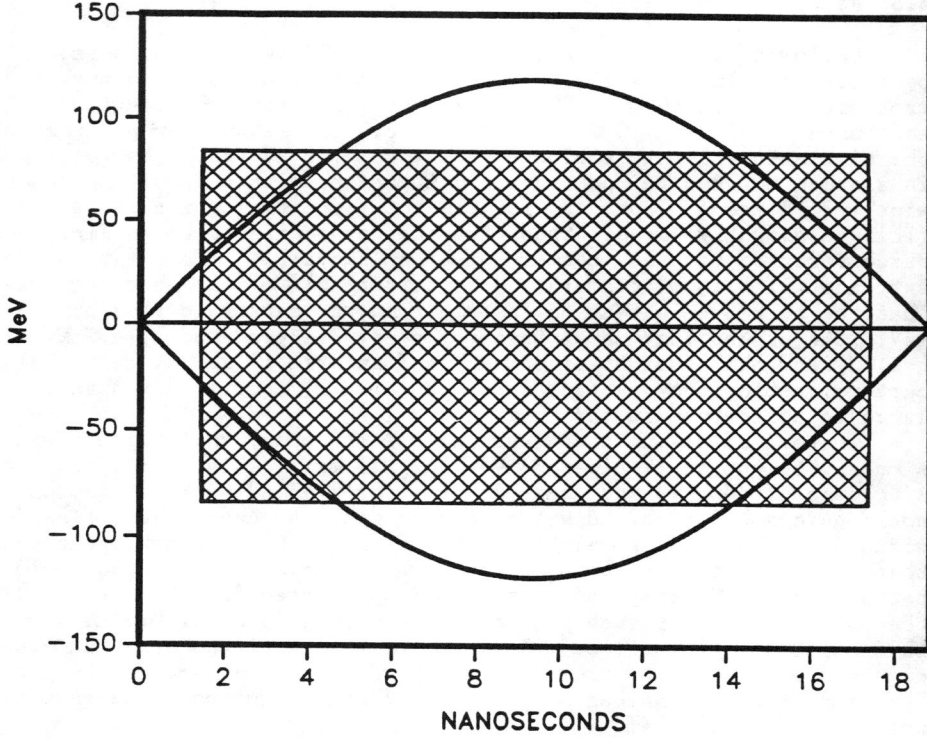

Figure 16. Recapture in 53 MHz.

Table III Capture Voltages for Coalescing

	7 Bunches	9 Bunches	11 Bunches
ΔE	53.4 MeV	68.7 MeV	84 MeV
Emittance	1.71 eV-sec	2.20 eV-sec	2.69 eV-sec
Voltage - 100% capture	1.7 MV	2.8 MV	4.2 MV
Bucket height	229 MeV	294 MeV	360 MeV
Bucket area	5.5 eV-sec	7.1 eV-sec	8.6 eV-sec
Emittance blowup	3.2	3.2	3.2
Voltage - $\sqrt{2}\Delta E$	186 kV	308 kV	460 kV
Bucket height	75.5 MeV	97.1 MeV	119 MeV
Bucket area	1.81 eV-sec	2.33 eV-sec	2.85 eV-sec
Capture efficiency	86%	86%	86%
Emittance blowup	1.23	1.23	1.23

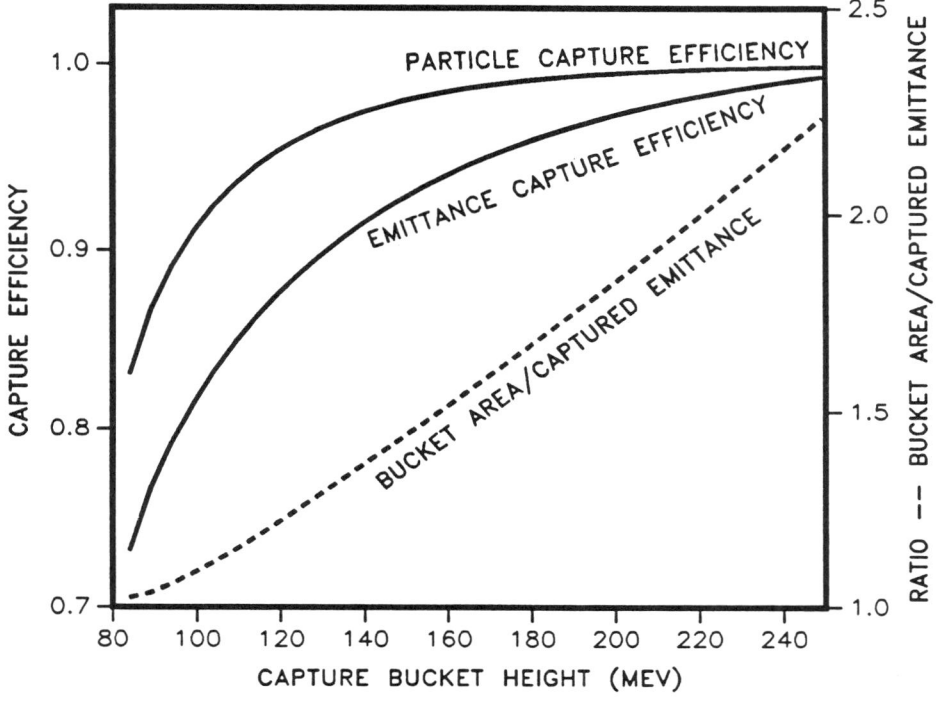

Figure 17. Capture efficiency and emittance blowup assuming elliptic distribution projection on phase axis.

The question posed earlier -- what is the proper voltage for recapturing? -- has no unambiguous answer. It depends on the relative priorities of capture efficiency and emittance blowup, but, based on Fig. 17, the case discussed above of a voltage that produces a bucket height of $\sqrt{2}$ times the bunch height appears close to optimum. The capture efficiency increases slowly for larger bucket heights, while the emittance increases linearly. The analysis will change if the bunch width shrinks. If the longitudinal emittance of the bunches before coalescing is decreased, and the noise/instability problems are solved, the rotated distribution will be narrower and it may be advantageous to change the recapture voltage. As the bunch width decreases, the capture efficiencies, both particle and emittance, increase for a given bucket height. But since the emittance of the bunch is smaller, the blowup has increased. Then it may be advantageous to capture with a higher voltage that produces a trajectory which captures the beam, rather than the separatrix. If the problems in fully debunching the beam can be solved, then clearly one should also try debunching into a low voltage h=53 bucket, which, after rotation, will match an h=1113 trajectory. This avoids the problem of a square peg in a round hole, or in this case a sinusoidal hole, which is really the source of the emittance increase.

4.4 Conclusions

Simulations using the program ESME have shown that the debunching was being done non-adiabatically. Subsequent operation in which the rate of debunching was slowed down have shown considerable improvement in the coalescing, increasing the efficiency from about 70% to 90% or better. The present limitation is the inability to debunch the beam to the point where it fills the bucket. A proposal is made for using an h=2226 cavity which might reduce the momentum spread before rotation further (and faster) than is achievable with only an h=1113 system. Such a system might also reduce the sensitivity to instabilities which may be limiting the performance. The h=53/h=106 system has been shown to be suitable for coalescing up to 11 bunches. Thirteen bunches cannot be coalesced as efficiently. The recapture cannot be made 100% efficient without serious emittance blowup, but efficiencies >95% should be achievable.

5 REFERENCES

1. G. Dôme, "Theory of RF Acceleration and RF Noise," CERN Accelerator School Antiprotons for Colliding Beams Facilities, CERN 84-15 (1984).

2. F. T. Cole, "Longitudinal Motion in Circular Accelerators," Physics of Particle Accelerators, AIP Conf. Proc. 153 (1987).

3. J. MacLachlan, "Particle Tracking in E-ϕ Space as a Design Tool for Cyclic Accelerators," 1987 IEEE Particle Accelerator Conf., p. 1087 (1987).

4. S. A. Bogacz and S. Stahl, "Coupled Bunch Instability in the Fermilab Booster - Longitudinal Phase-Space Simulation," European Particle Accelerator Conf., Rome (1988), to be published.

APPENDIX A   BUCKET PARAMETERS
(all angles in degrees)

ϕ_s	Γ	$\beta(\Gamma)$	$a(\Gamma)$	ϕ_2	ϕ_1	$\phi_2-\phi_1$
0.0	0.0000	1.0000	1.0000	-180.0	180.0	360.0
1.0	0.0175	0.9863	0.9541	-154.0	179.0	333.0
2.0	0.0349	0.9725	0.9175	-143.5	178.0	321.5
3.0	0.0523	0.9587	0.8845	-135.5	177.0	312.5
4.0	0.0698	0.9449	0.8537	-128.8	176.0	304.8
5.0	0.0872	0.9311	0.8246	-122.9	175.0	297.9
6.0	0.1045	0.9172	0.7970	-117.6	174.0	291.6
7.0	0.1219	0.9033	0.7704	-112.6	173.0	285.6
8.0	0.1392	0.8894	0.7449	-108.1	172.0	280.1
9.0	0.1564	0.8755	0.7202	-103.7	171.0	274.7
10.0	0.1736	0.8616	0.6964	-99.6	170.0	269.6
11.0	0.1908	0.8477	0.6733	-95.7	169.0	264.7
12.0	0.2079	0.8337	0.6508	-92.0	168.0	260.0
13.0	0.2250	0.8198	0.6290	-88.4	167.0	255.4
14.0	0.2419	0.8059	0.6078	-84.9	166.0	250.9
15.0	0.2588	0.7919	0.5872	-81.5	165.0	246.5
16.0	0.2756	0.7780	0.5671	-78.2	164.0	242.2
17.0	0.2924	0.7641	0.5476	-75.0	163.0	238.0
18.0	0.3090	0.7502	0.5285	-71.9	162.0	233.9
19.0	0.3256	0.7363	0.5099	-68.9	161.0	229.9
20.0	0.3420	0.7224	0.4918	-65.9	160.0	225.9
21.0	0.3584	0.7085	0.4741	-63.0	159.0	222.0
22.0	0.3746	0.6947	0.4568	-60.1	158.0	218.1
23.0	0.3907	0.6809	0.4400	-57.3	157.0	214.3
24.0	0.4067	0.6671	0.4236	-54.5	156.0	210.5
25.0	0.4226	0.6533	0.4076	-51.8	155.0	206.8
26.0	0.4384	0.6396	0.3920	-49.1	154.0	203.1
27.0	0.4540	0.6260	0.3768	-46.4	153.0	199.4
28.0	0.4695	0.6123	0.3619	-43.8	152.0	195.8
29.0	0.4848	0.5987	0.3475	-41.2	151.0	195.2
30.0	0.5000	0.5852	0.3334	-38.7	150.0	188.7
31.0	0.5150	0.5717	0.3196	-36.2	149.0	185.2
32.0	0.5299	0.5582	0.3062	-33.7	148.0	181.7
33.0	0.5446	0.5448	0.2932	-31.2	147.0	178.2
34.0	0.5592	0.5315	0.2805	-28.8	146.0	174.8
35.0	0.5736	0.5182	0.2682	-26.3	145.0	171.3
36.0	0.5878	0.5050	0.2562	-23.9	144.0	176.9
37.0	0.6018	0.4919	0.2445	-21.6	143.0	164.6
38.0	0.6157	0.4788	0.2332	-19.2	142.0	161.2
39.0	0.6293	0.4658	0.2222	-16.9	141.0	157.9
40.0	0.6428	0.4529	0.2115	-14.6	140.0	154.6
41.0	0.6561	0.4400	0.2011	-12.3	139.0	151.3
42.0	0.6691	0.4273	0.1910	-10.0	138.0	148.0
43.0	0.6820	0.4146	0.1813	-7.7	137.0	144.7
44.0	0.6947	0.4020	0.1718	-5.4	136.0	141.4

ϕ_s	Γ	$\beta(\Gamma)$	$\alpha(\Gamma)$	ϕ_2	ϕ_1	$\phi_2-\phi_1$
45.0	0.7071	0.3895	0.1627	-3.2	135.0	138.2
46.0	0.7193	0.3772	0.1538	-1.0	134.0	135.0
47.0	0.7314	0.3649	0.1453	1.2	133.0	131.8
48.0	0.7431	0.3527	0.1370	3.5	132.0	128.5
49.0	0.7547	0.3406	0.1290	5.6	131.0	125.4
50.0	0.7660	0.3286	0.1213	7.8	130.0	122.2
51.0	0.7771	0.3168	0.1139	10.0	129.0	119.0
52.0	0.7880	0.3050	0.1068	12.2	128.0	115.8
53.0	0.7986	0.2934	0.0999	14.3	127.0	112.7
54.0	0.8090	0.2819	0.0933	16.5	126.0	109.5
55.0	0.8192	0.2705	0.0870	18.6	125.0	106.4
56.0	0.8290	0.2593	0.0809	20.7	124.0	103.3
57.0	0.8387	0.2482	0.0751	22.8	123.0	100.2
58.0	0.8480	0.2372	0.0696	24.9	122.0	97.1
59.0	0.8572	0.2264	0.0643	27.0	121.0	94.0
60.0	0.8660	0.2158	0.0593	29.1	120.0	90.9
61.0	0.8746	0.2052	0.0544	31.2	119.0	87.8
62.0	0.8829	0.1949	0.0499	33.3	118.0	84.7
63.0	0.8910	0.1847	0.0456	35.4	117.0	81.6
64.0	0.8988	0.1747	0.0415	37.4	116.0	78.6
65.0	0.9063	0.1648	0.0376	39.5	115.0	75.5
66.0	0.9135	0.1552	0.0340	41.6	114.0	72.4
67.0	0.9205	0.1457	0.0305	43.6	113.0	69.4
68.0	0.9272	0.1364	0.0273	45.7	112.0	66.3
69.0	0.9336	0.1273	0.0243	47.7	111.0	63.3
70.0	0.9397	0.1183	0.0215	49.8	110.0	60.2
71.0	0.9455	0.1096	0.0190	51.8	109.0	57.2
72.0	0.9511	0.1012	0.0166	53.8	108.0	54.2
73.0	0.9563	0.0929	0.0144	55.9	107.0	51.1
74.0	0.9613	0.0849	0.0123	57.9	106.0	48.1
75.0	0.9659	0.0771	0.0105	59.9	105.0	45.1
76.0	0.9703	0.0695	0.0088	61.9	104.0	42.1
77.0	0.9744	0.0622	0.0073	63.9	103.0	39.1
78.0	0.9781	0.0552	0.0060	65.9	102.0	36.1
79.0	0.9816	0.0485	0.0048	68.0	101.0	33.0
80.0	0.9848	0.0420	0.0038	70.0	100.0	30.0
81.0	0.9877	0.0359	0.0029	72.0	99.0	27.0
82.0	0.9903	0.0301	0.0022	74.0	98.0	24.0
83.0	0.9925	0.0246	0.0016	76.0	97.0	21.0
84.0	0.9945	0.0196	0.0011	78.0	96.0	18.0
85.0	0.9962	0.0149	0.0007	80.0	95.0	15.0
86.0	0.9976	0.0106	0.0004	82.0	94.0	12.0
87.0	0.9986	0.0069	0.0002	84.0	93.0	9.0
88.0	0.9994	0.0038	0.0001	86.0	92.0	6.0
89.0	0.9998	0.0013	0.0000	88.0	91.0	3.0
90.0	1.0000	0.0000	0.0000	90.0	90.0	0.0

LONGITUDINAL INSTABILITIES AND STABILITY CRITERIA

King-Yuen Ng
Fermi National Accelerator Laboratory
P. O. Box 500, Batavia, IL 60510

TABLE OF CONTENTS

1	Criteria for Longitudinal Microwave Instability..............	1970
2	Stability Limits..	1974
	2.1 Across Transition......................................	1974
	2.2 Acceleration...	1976
	2.3 Bunch Coalescence......................................	1977
	2.4 Preparation of p Bunches for p̄ Production.............	1978
3	Longitudinal Impedance Estimation...........................	1981
	3.1 Resistive Wall...	1981
	3.2 Bellows..	1982
	3.3 Beam Position Monitors.................................	1987
	3.4 Kickers..	1990
	3.5 Lambertson Magnets.....................................	1991
	3.6 Summary..	1993
4	Wire Measurement Experiment.................................	1995
	4.1 Experimental Setup.....................................	1995
	4.2 Results of Measurement.................................	1998
	4.3 Conclusions..	2000
5	Measurement of Impedance by Debunching......................	2000
	5.1 The Experiment...	2002
	5.2 Theory	2002
	5.3 Analysis of Results	2004
	5.3.1 Bunch Shape.....................................	2004
	5.3.2 Starting Time of Growth.........................	2005
	5.3.3 Overlapped Bunches..............................	2005
	5.3.4 Determination of Driving Impedance..............	2007
6	References..	2008

Chapter 8
LONGITUDINAL INSTABILITIES AND STABILITY CRITERIA

King-Yuen Ng
Fermi National Accelerator Laboratory
P. O. Box 500, Batavia, IL 60510

1 CRITERIA FOR LONGITUDINAL MICROWAVE INSTABILITY

The electromagnetic fields emitted by a particle beam will interact with the beam itself directly or indirectly through bouncing back from the walls of the vacuum chamber. For a beam with uniform longitudinal density, there is no longitudinal field and no force acting on the particles. However, if a small bump is introduced, it will produce a longitudinal space-charge force $eE = -e\partial\lambda/\partial\theta$ acting in the direction away from the bump, where λ is the local linear charge density and θ is the azimuthal angle around the accelerator ring (Fig. 1). This force will increase the energy at the front of the bump and decrease the energy at the back. However, above transition, this implies decreasing the revolution frequency at the front and increasing it at the back; the bump will increase in magnitude. In other words, any small local perturbation in the uniform density of the beam will grow, resulting in a nonuniform particle distribution.

The growth rate can be easily derived for any longitudinal force even if it is not due to space charge.[1] We start from a beam with uniform linear density λ_0 and fixed energy E. Let us assume a

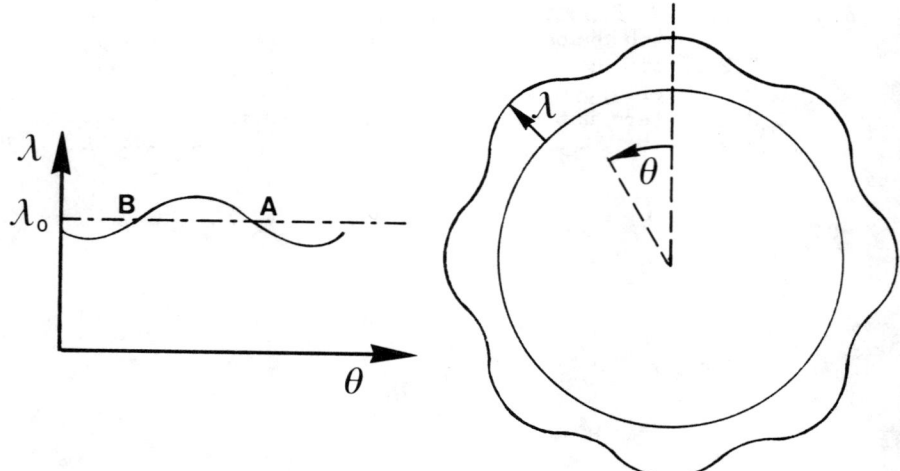

Figure 1   A perturbing longitudinal density wave. Space-charge force will increase the energy at the front of the wave crest A and decrease the energy at the back B.

small perturbation λ_1 of harmonic n and collective frequency $\Omega/2\pi$. We write the density as

$$\lambda(\theta,t) = \lambda_o + \lambda_1 e^{-j(n\theta-\Omega t)} . \quad (1)$$

Because of the continuity equation

$$\frac{\partial \lambda}{\partial t} + \frac{\partial}{\partial \theta}(\lambda w) = 0 , \quad (2)$$

the revolution frequency $w(\theta,t)/2\pi$ must be of the form

$$w(\theta,t) = w_o + w_1 e^{-j(n\theta-\Omega t)} , \quad (3)$$

where the deviation w_1 from the unperturbed value w_o is given by

$$w_1 = (\Omega/n - w_o)\lambda_1/\lambda_o . \quad (4)$$

The current is given by $I(\theta,t) = \lambda(\theta,t)w(\theta,t)$ or

$$I(\theta,t) = I_o + I_1 e^{-j(n\theta-\Omega t)} , \quad (5)$$

with $I_1 = (\Omega/n)\lambda_1$. In the above, only first-order quantities have been retained.

Riding on a particle with angular velocity w_o, we should not see any change in revolution frequency except for that due to a longitudinal electric field from the beam particles bouncing from the walls. Thus, dw/dt is

$$\partial w/\partial t + (\partial w/\partial \theta)(d\theta/dt) = (\partial w/\partial E)(dE/dt) , \quad (6)$$

where $\partial w/\partial E = -\eta w_o/\beta^2 E$ with $\eta = \gamma_t^{-2} - \gamma^{-2}$ denoting the frequency dispersion parameter, β is the particle velocity relative to c (the speed of light), $\gamma^2 = 1/(1 - \beta^2)$, and γ_t is gamma at transition. The rate of increase in energy is $dE/dt = -eI_1 Z_L w_o/2\pi$, where Z_L, the longitudinal impedance per turn around the accelerator ring at harmonic n, is defined as the potential drop around the ring per unit current of harmonic n. We therefore have

$$j(\Omega - nw_o)w_1 = \eta eI_1 Z_L w_o^2/2\pi\beta^2 E . \quad 7)$$

Equation (7) has a physical meaning. As a result of the longitudinal impedance, we can imagine n buckets created, each with peak voltage $-I_1 Z_L$. The particles trapped inside a bucket will execute synchrotron oscillation with

$$w_s^2 = -j\eta neI_1 Z_L w_o^2/2\pi\beta^2 E . \quad (8)$$

In the above, a factor j has been inserted to take into account that, to produce buckets encircling the ripples or small bunches in the beam current, the voltage has to be 90° out of phase. Riding on a particle with angular velocity w_o, we see the particles inside one of those buckets moving, on the average, an azithumal angle of π/n

across the bucket in time π/ω_s (half an oscillation); i.e. they have an angular velocity of $\omega \sim (\omega_s/n)\exp(j\omega_s t)$. Thus, the rate of change of angular velocity is

$$d\omega/dt = j\omega_s^2/n , \quad (9)$$

which is the same as Eq. (7).

Substituting Eq. (4) for ω_1, noting that $I_o = \lambda_o \omega_o$, and letting $\Omega/n \sim \omega_o$, we get for the collective frequency

$$(\Omega - n\omega_o)^2 = -jn\eta e I_o Z_L \omega_o^2 / 2\pi\beta^2 E . \quad (10)$$

This gives the shift in revolution frequency; the imaginary part of Ω, if negative, gives the rate of growth for the wave at harmonic n. Note that I_o appears in the formula although it is I_1 that excites the longitudinal electric field. This is because this electric field interacts with every particle in the beam.

If, however, the beam particles have an energy spread ΔE corresponding to a revolution frequency spread $\Delta\omega = -\eta\omega_o\Delta E/\beta^2 E$, the collective effect cannot take place or the coherence of the perturbing wave will be destroyed when the growth time $-1/\text{Im}\Omega$ is $>2/n\Delta\omega$, the time for one wavelength to go out of phase by 180°. In other words, the beam is stable when

$$4|\eta e I_o Z_L \omega_o^2 / 2\pi n E| \leq (\eta\omega_o\Delta E/\beta^2 E)^2 . \quad (11)$$

This is the famous Keil-Schnell criterion for microwave stability for a coasting beam.[2]

For a single bunch, if the bunch length is much longer than the wavelength of the perturbation and the synchrotron oscillation period is much longer than the growth time, the bunch, at least its central portion, resembles a coasting beam during the unstable growth. Therefore, the above Keil-Schnell criterion applies if we replace the average total current I_o by the local peak current I_p; i.e.

$$\left|\frac{Z_L}{n}\right| \leq \frac{2\pi|\eta|E/e}{\beta^2 I_p}\left(\frac{\sigma_E}{E}\right)^2 . \quad (12)$$

In the above, we have written the rms energy spread σ_E instead of $\Delta E/2$, because this is the exact criterion for a Gaussian bunch when a careful derivation is performed using the linearized Vlasov equation.[3] The peak current is $I_p = 1/\sqrt{2\pi}\sigma_T$, where σ_T is the rms time spread of the bunch. The Vlasov equation leads to a dispersion relation which can be plotted as growth-rate contours in the $(\Delta\Omega_o/n)^2$-plane, where $\Delta\Omega_o = \Omega - n\omega_o$, given by Eq. (10), is the frequency shift without Landau damping. Such a plot is given in Fig. 2, where only the stability contour is shown. A given Z_L/n corresponds to a point in the $(\Delta\Omega_o/n)^2$-plane. If the point lies outside the stability contour, there exists at least one growing solution and the bunch will be unstable. If the point lies inside

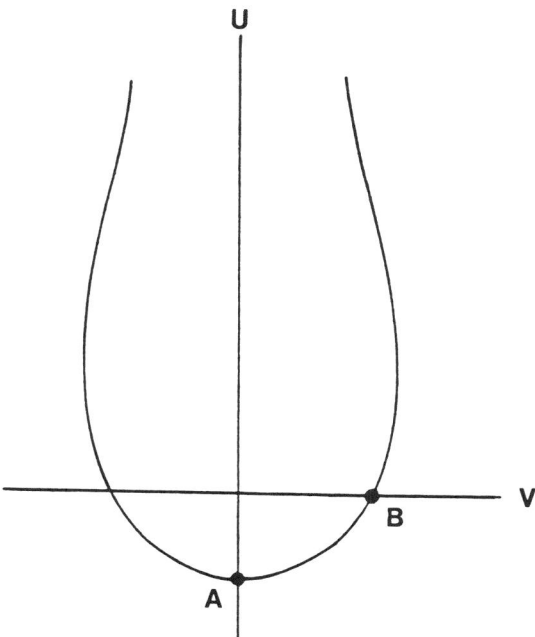

Figure 2 Stability curve of a Gaussian bunch in the $(\Lambda\Omega_o/n)^2$-plane. V and U are the real and imaginary part of $(\Lambda\Omega_o/n)^2$.

the stability contour, all solutions are stable. Noting that $(\Delta\Omega_o/n)^2$ is proportional to $-j\eta Z_L/n$, above transition when $\eta > 0$, the criterion of Eq. (12) corresponds to the point A, where the impedance is capacitive. An inductive Z_L/n lies on the positive vertical axis and is therefore always stable. A real Z_L/n corresponds to the point B which is ~1.434 times farther away from the origin than the point A (for a Gaussian bunch only). Therefore, when the impedance is real, the stability criterion of Eq. (12) can be relaxed by inserting the factor 1.434 on the right side. Below transition when $\eta < 0$, a capacitive impedance, corresponding to a point on the positive vertical axis, will always be stable, while an inductive one, corresponding to a point on the negative vertical axis, can be unstable if it is too big.

As stated above, the longitudinal electric field produced by the impedance interacts on every particle of the bunch. The impedance Z_L in the above criterion is therefore the convolution of the impedance $Z(\omega)$ of the ring and the frequency spectrum $\rho(\omega)$ of the bunch

$$Z_L = \int Z(n\omega_o + \omega)\rho(\omega)d\omega , \qquad (13)$$

$\rho(\omega)$ being normalized to unity. When the impedance is a broad band much broader than σ_τ^{-1}, the frequency spread of the bunch, the convolution depends only on the peak of the broad-band impedance, and therefore Z_L in Eq. (12) represents the peak impedance of a broad band. However, with cavities and pipe-joining housings along the vacuum chamber, the impedance may exhibit sharp resonances with frequency much narrower than that of the bunch. For a narrow resonance of shunt impedance Z_{sh}, quality factor Q, and resonance frequency $\omega_R \sim n\omega_o$, the convolution gives

$$Z_L = [\sigma_T/\sqrt{2\pi}][(\pi\omega_R/2)(Z_{sh}/Q)] , \qquad (14)$$

where the first term is the peak value of the bunch spectrum ρ and the second the area under the narrow resonance. This is just the effective impedance seen by the bunch, which, because of its finite frequency spread, is not able to resolve the narrow peak. Substituting this into Eq. (12), we obtain a criterion for narrow resonances,[4]

$$\frac{Z_{sh}}{Q} \leq \frac{4|\eta|E/e}{\beta^2 I_{AV}}\left(\frac{\sigma_E}{E}\right)^2 , \qquad (15)$$

where $I_{AV} = eN\omega_o/2\pi$ is the average current of the bunch containing N particles.

2 STABILITY LIMITS

In this section we try to estimate the limits[5] of Z_L/n or Z_{sh}/Q for the Main Ring in order to ensure longitudinal instability during its different performances.

The present duties of the Main Ring are a follows:

(1) To accelerate proton (or antiproton) bunches up to 150 GeV and coalesce 7 to 13 of them to form an intense bunch of $\sim 10^{11}$ to be injected into the Energy Saver for further acceleration and eventual collision with an antiproton (proton) bunch. In order to achieve a high luminosity, the initial bunches must be kept at small bunch areas $\sim$0.2 eV-sec before coalescence.

(2) To accelerate intense proton bunches up to 120 GeV for antiproton production. To maximize the production efficiency, these bunches must be of high intensity $\sim 2.4 \times 10^{10}$ p/bunch and small bunch area $\sim$0.2 eV-sec.

We see that, in each performance, the bunch area has to be small, unlike the operation in the fixed-target mode, where the bunch area was deliberately increased by the bunch spreader in order to achieve a better duty cycle.

2.1 Across Transition

When the bunches are accelerated through transition, which corresponds to γ_t = 18.75, the frequency dispersion parameter η changes from negative to positive through zero. Equation (10) tells us that there must be a portion of time when the bunch is unstable. The total growth across transition[5] is $G = \exp(S_a + S_b)$ where

$$S_{b,a} = \int \mathrm{Im}\,\Omega\, dt \tag{16}$$

represents the integration of the positive growth rate $\mathrm{Im}\,\Omega$ before and after transition. The rate of growth is dependent on the number of particles in the bunch N and also the driving impedance Z_L/n. The time t_o from loss of stability to transition or from transition to regain of stability is also dependent on NZ_L/n. In fact, it turns out that $S_b + S_a$ scales as $(Z_L/n)^2 N^2 A^{-9/4}$, where $A = 6\pi\sigma_\tau \sigma_E$ is the bunch area. If space-charge is neglected, we have

$$\frac{S_b}{n} = \frac{S_a}{n} = \frac{F_1[eN(Z_L/n)\gamma_t]^2 (E_o/e)^2 \sigma_\tau}{V_{rf} \sin\phi_o (A/e)^3}, \tag{17}$$

where γ_t is gamma at transition, V_{rf} is the rf voltage at transition, ϕ_o is the synchronized phase at transition, E_o is the rest energy of the particle concerned, σ_τ is the rms time spread of the bunch at the moment when stability is lost (before transition) or when stability is regained (after transition), and $F_1 = 8.735$ is a numerical constant. The rms time spread σ_τ in the above can be evaluated from

$$\frac{\sigma_\tau}{T} = \frac{2}{3^{1/3}\Gamma(\frac{1}{3})}\left[\frac{AeV_{rf}\omega_o \sin\phi_o}{6\pi E_o^2 \beta^2 \gamma_t^4}\right]^{1/2} \left[1 + 0.686\left|\frac{t_o}{T}\right|\right] \tag{18}$$

for $|t_o/T| \ll 1$, and

$$\frac{\sigma_\tau}{T} = \left[\frac{AeV_{rf}\omega_o \sin\phi_o}{6\pi^2 E_o^2 \beta^2 \gamma_t^4}\right]^{1/2} \left|\frac{t_o}{T}\right|^{1/4} \tag{19}$$

for $|t_o/T| \gtrsim 1$. In the above, ω_o is the angular revolution frequency of the bunch and β its velocity in units of c (the velocity of light). The characteristic time

$$T = \left[\frac{4\pi^2 (E_o/e)^2 \beta^2 \gamma_t^4}{h\omega_o^3 V_{rf}^2 \sin 2\phi_o}\right]^{1/3}, \tag{20}$$

where h is the rf harmonic, is a measure of the time for which the bucket change is not adiabatic. The time t_o is given by

$$|t_o| = \frac{F_2 eN(Z_L/n)\gamma_t^4 (E_o/e)^2 \sigma_\tau}{\omega_o V_{rf} \sin\phi_o (A/e)^2}, \tag{21}$$

where $F_2 = 49.42$ is a constant.

For $N = 1.2 \times 10^{10}$, $Z_L/n = 10$ ohms, $A = 0.15$ eV-sec, rf voltage $V_{rf} = 2.5$ MV, and a synchronous phase $\phi_o = 50°$, we get $2t_o = 5.2$ ms and $(S_b + S_a)/n = 3.6 \times 10^{-5}$. Exact numerical solution[6] of the dispersion relation including space-charge effects gives 5.5 ms and 3.86×10^{-5}. The characteristic time T is 2.86 ms. Microwave signals at 1.65 GHz have been observed (see Section 5 below) corresponding to a harmonic of $n = 3.46 \times 10^4$. Thus, the growth across transition is $G \sim 3.86$ times. However, this is the growth of only the microwave amplitude, the perturbing one that we discussed in Section 1. It will increase the energy spread and dilute the bunch area through second-order effects. The relation between the growth of the bunch area and that of the microwave signal is complicated and model-dependent and will be presented elsewhere. But, in some sense, we can consider $G = \exp(S_b + S_a)$ as a measure of the growth of the bunch area. The final size of the microwave amplitude depends on the the size of that amplitude before transition. In our case, the rms bunch length σ_T near transition is ~0.3 ns or the rms frequency spread is $\sigma_\omega \sim 3.3$ GHz. Thus, the amplitude at $\omega/2\pi = 1.65$ GHz is only $\exp(-\omega^2/2\sigma_\omega^2)$ or 0.72% of the amplitude at low frequencies. Thus, a growth of 3.86 times may not be very much. However, if the bunch is ill-behaved and does not fit the bucket well at the beginning, its amplitude in the microwave region can be comparable with the amplitude at low frequencies, and a growth of 3.86 times may be very significant. Growth of bunch area across transition in the Main Ring has been reported to be 20% to ~4 times. We believe that this variation is due mainly to the behavior of the bunch before transition.

When the Main Ring accelerates proton bunches for $\bar{p}$ production, the number of protons per bunch $N \sim 2.4 \times 10^{10}$ is bigger, and the bunch area $A \sim 0.1$ eV-sec is smaller. According to the scaling law, $S_b + S_a$ increases by four times, and the growth of the microwave amplitude becomes $G = 146$. Thus, the growth may become too big to tolerate. But so far we have included linear effects. Generally, nonlinear effects will come in, causing overshoots, and the growth will end.

2.2 Acceleration

After passing through transition, if we assume that the rf voltage V_{rf} and the synchronous phase ϕ_o are changed adiabatically and the bunch area $A = 6\pi\sigma_T\sigma_E$ is small compared with the bucket area, from Eq. (10) the Z_L/n limit becomes

$$\left|\frac{Z_L}{n}\right| = F\left(\frac{h^{1/4}}{I_o\beta}\right)\left(\frac{Ac}{R}\right)^{3/2}\left(\frac{\eta}{E}\right)^{3/4}\left(V_{rf}\cos\phi_o\right)^{1/4}, \qquad (22)$$

where $F = 3^{-3/2}(2\pi)^{-5/4}$, $h = 1113$, the harmonic number of the rf, $R = 1$ km is the average radius of the main ring, I_o is the average current of the bunch, and c is the velocity of light. In deriving Eq. (17), use has been made of the approximate relation between the rms time spread σ_T and the rms energy spread σ_E of the bunch; i.e.

$$\sigma_\tau = (R/h\beta c)(|\eta|h/\nu_s)(\sigma_E/\beta^2 E) \ , \qquad (23)$$

where $\nu_s = |h\eta V_{rf}\cos\phi_o/2\pi\beta^2 nE|^{1/2}$ is the synchrotron tune. After transition, η becomes bigger and bigger, and $|V_{rf}\cos\phi_o|$ still increases; thus the bunch becomes more and more stable. Later, for higher energies, η approaches a constant, and $|V_{rf}\cos\phi_o|$ is constant or decreasing; thus the $|Z_L/n|$ limit becomes smaller as the energy E increases. Figure 3 shows a typical acceleration cycle. We see that the limit is most stringent at the highest energy of 120 GeV or 150 GeV. Therefore we need to discuss stability at these top energies only.

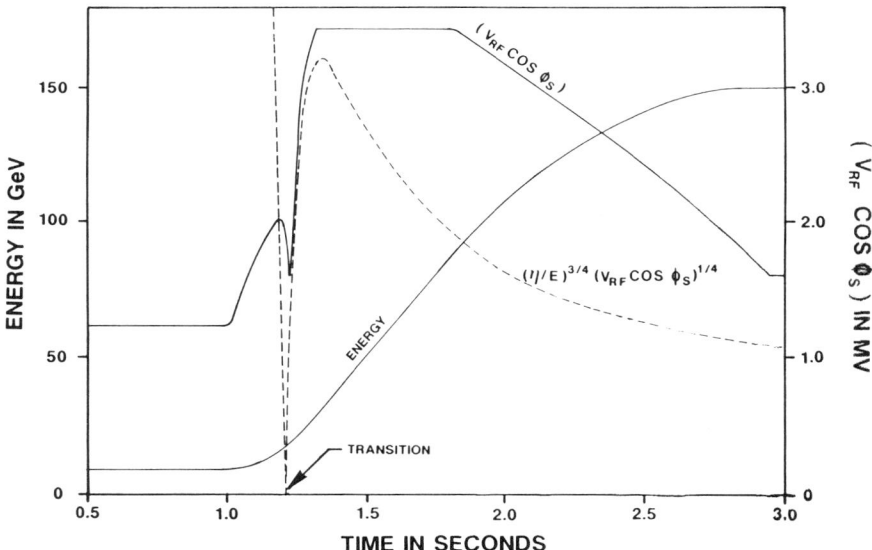

Figure 3 A typical acceleration cycle in the Main Ring. The scale of the dashed curve is arbitrary.

2.3 Bunch Coalescence

Seven proton bunches each of intensity $N = 1.2 \times 10^{10}$ and area $A = 0.20$ eV-sec (we assume that the bunch area increases to 0.20 eV-sec across transition) at 150 GeV are allowed to coalesce into one bunch of intensity $\sim 8 \times 10^{10}$. The procedure consists of lowering the rf voltage adiabatically from ~ 1.5 MV to ~ 2.30 kV so that the bunch will just fill the bucket. This $h = 1113$ rf is then turned off while a $h = 53$ rf of 22 kV is turned on. The seven bunches will lie inside the $h = 53$ rf bucket and are allowed to rotate for 90°. This rf is then replaced by a $h = 1113$ rf of 0.46 MV to capture the seven bunches into a single bucket. The rf is then increased slowly to 1 MV so that the coalesced high-intensity bunch is matched to the bucket. The $|Z_L/n|$ limit of each stage during the coalescence is listed in Table I. We see that the lowest limit is $|Z_L/n| = 6.84$ Ω for a broad band and $Z_{sh}/Q = 9.45$ kΩ for a narrow resonance. This occurs when the rf voltage is lowered adiabatically to match the bunch. With the desire to obtain higher luminosity, sometimes up to 11 or 13 bunches are coalesced. However, under such conditions, a h = 106 rf is added alongside the h = 53 rf so as to linearize the 90° rotation and ensure more efficient capture.

The antiprotons are coalesced at 150 GeV similarly. The only difference is that 13 adjacent bunches each of intensity $\sim 8\times 10^9$ are involved. Because of the lower intensity, the higher limit $|Z_L/n| = 10.3$ Ω or $Z_{sh}/Q = 14.0$ kΩ is obtained. This and other data are listed in Table II.

2.4 Preparation of p Bunches for p̄ Production

For fixed p̄ momentum spread, the bunch area of the p̄ is minimized by making the time spread of the bombarding proton bunches as narrow as possible. At the 120-GeV flat-top, the rf voltage is maintained at 4 MV, and the bunches, about 0.2 eV-sec and intensity 2.4×10^{10}, are matched to the large bucket ($\Delta E \sim 300$ MeV). The rf is reduced to ~300 kV within two turns. The now mismatched bunches begin phase oscillation, which results in a maximum time spread of $\sqrt{6}\sigma_\tau \sim 4$ ns, about half the bucket length. The rf is next increased quickly again to its maximum value of 4 MV, whereupon the mismatched bunches rotate in one-quarter phase oscillation to a large energy spread and narrow time spread configuration ($\sqrt{6}\sigma_\tau \sim 0.3$ ns) and are extracted. The various bunch parameters and $|Z_L/n|$ limits are summarized in Table III. We see that $|Z_L/n|$ reaches 6.61 Ω when the time spread of the proton bunch is biggest. For a sharp resonance narrower than the bunch spectrum, this corresponds to $Z_{sh}/Q = 21.4$ kΩ. Note that $Z_L/n \propto A^{3/2} V_{rf}^{1/4}$ and $Z_{sh}/Q \propto A V_{rf}^{1/2}$, where A is the bunch area. Thus, if the bunch area is reduced to 0.1 eV-sec, these limits will become 2.33 Ω and 10.7 kΩ respectively. Sometimes three batches are accelerated, but each batch must be extracted only every one or two seconds to avoid overheating the lithium lens. In this situation, during the extraction of the first batch, the 4-MV rf is turned on for about 2.5 ms, or half a synchrotron period only, so that the bunches that are not extracted will rotate to the position of maximum time spread. The process is then reversed by lowering the rf to 300 kV to allow the bunches to rotate for a quarter synchrotron oscillation and increasing it abruptly to 4 MV again. The bunches now match the 4-MV buckets. The whole process is repeated again when the next batch is ready for extraction. In this way, the bunches will stay in the most dangerous position of maximum time spread for a quarter of a synchrotron oscillation only. If there is a microwave instability, the microwave amplitude will not have enough time to grow appreciably.

Table I Coalescence of 7 proton bunches at 150 GeV

	Start	V_{rf} lowered to match bunch	7 bunches in H = 53 bucket	End of 90° rotation	Recapture into h = 1113 bucket	V_{rf} increases before extraction
Longitudinal emittance $6\pi\sigma_T\sigma_E$, eV-sec	0.20	0.20	1.40	2.20	2.20	2.20
$\sqrt{6}\sigma_T$, ns	1.34	9.42	65.9	8.89	8.89	4.90
$\sqrt{6}\sigma_E$, MeV ($\sqrt{6}\sigma_E/E$)	47.6 (3.15×10^{-4})	8.31 (5.51×10^{-5})	8.31 (5.51×10^{-5})	61.5 (4.10×10^{-4})	61.5 (4.08×10^{-4})	143 (9.46×10^{-4})
V_{rf}	1.50 MV	2.30 kV	22 kV	22 kV	0.12 MV	1.0 MV
h	1113	1113	53	53	1113	1113
ν_s	2.24×10^{-3}	8.72×10^{-5}	5.91×10^{-5}	5.91×10^{-5}	1.24×10^{-3}	1.83×10^{-3}
Number per bunch	1.2×10^{10}	1.2×10^{10}	8.4×10^{10}	8.4×10^{10}	8.4×10^{10}	8.4×10^{10}
Microwave stability limit Z_L/n, Ω (broad band)	31.8	6.84	6.84	50.5	50.5	150
Z_{sh}/Q, kΩ (narrow)	310	9.45	9.45	74.0	74.0	399

Table II Coalescence of 13 antiproton bunches at 150 GeV

	Start	V_{rf} lowered to match bunch	13 bunches in h = 53 bucket	End of 90° rotation	Recapture into h = 1113 bucket	V_{rf} increases before extraction
Longitudinal emittance $6\pi\sigma_T\sigma_E$, eV-sec	0.20	0.20	2.60	4.08	4.08	4.08
$\sqrt{6}\sigma_T$, ns	1.34	9.42	122	8.89	8.89	6.68
$\sqrt{6}\sigma_E$, MeV ($\sqrt{6}\sigma_E/E$)	47.6 (3.15×10⁻⁴)	8.31 (5.54×10⁻⁵)	8.31 (5.54×10⁻⁵)	114 (7.61×10⁻⁴)	114 (7.61×10⁻⁴)	195 (1.29×10⁻³)
V_{rf}	1.50 MV	2.30 kV	22 kV	22 kV	0.46 MV	1.0 MV
h	1113	1113	53	53	1113	1113
ν_s	2.34×10⁻³	8.72×10⁻⁵	5.91×10⁻⁵	5.91×10⁻⁵	1.24×10⁻³	1.83×10⁻³
Number per bunch	8×10⁹	8×10⁹	1.04×10¹¹	1.04×10¹¹	1×10¹¹	1×10¹¹
Microwave stability limit Z_L/n, Ω (broad band)	47.8	10.3	10.3	141	141	318
Z_{sh}/Q, kΩ (narrow)	323	14.0	14.0	204	240	620

Table III RF maneuvering of proton bunches for p̄ production

	120-GeV flat-top	V_{rf} reduced and bunch rotates to max time spread	V_{rf} raised to 4 MV; bunch rotates for 90°
Longitudinal emittance $6\pi\sigma_T\sigma_E$, eV-sec	0.2	0.2	0.2
$\sqrt{6}\sigma_T$, ns	1.11	4.02	0.31
$\sqrt{6}\sigma_E$, MeV ($\sqrt{6}\sigma_E/E$)	57.4 (4.74×10^{-4})	15.8 ($1.31 10^{-4}$)	208 (1.72×10^{-3})
V_{rf}	4 MV	300 kV	4 MV
Bucket height, MeV	314	85.4	314
ν_s	0.00405	0.00112	0.00405
Number per bunch	2.4×10^{10}	2.4×10^{10}	2.4×10^{10}
Microwave stability limit Z_L/n, Ω (broad band)	23.9	6.61	86.7
Z_{sh}/Q (narrow resonance)	281 kΩ	21.4 kΩ	3.69 MΩ

3 LONGITUDINAL IMPEDANCE ESTIMATION

Single-bunch microwave instabilities are driven by impedances at high frequencies; single-bunch mode-coupling instabilities are driven by impedances at low frequencies; and coupled-bunch instabilities are driven by resonances with high quality factors. Thus we need to estimate impedances of the Main Ring at both low and high frequencies.[7]

3.1 Resistive Wall

For a circular beampipe of length L, radius b, and conductivity σ, the electromagnetic field of the beam will penetrate into the wall of the pipe by a skin depth $\delta = (2/\mu\omega\sigma)^{1/2}$, where μ is the permeability of the pipe material and $\omega/2\pi$ the frequency of the electromagnetic wave. Thus, the beam will see an impedance due to a strip of pipe material of length L, width $2\pi b$, and thickness δ. In other words,

$$Z_L = (1 + j)L/(2\pi b \delta \sigma) . \qquad (24)$$

The inductive part comes about because the electromagnetic wave flowing into the surface of the beampipe has to attenuate by a factor of 1/e after penetrating a distance δ, and therefore the wave number contains an imaginary part.

For a rectangular beampipe of width w and height h, the return current of the beam does not distribute uniformly on the walls. Thus, the longitudinal wall impedance cannot be derived as simply as for the circular beampipe. However, a detailed derivation gives an expression[8] very similar to Eq. (7),

$$Z_L = (1 + j)F_L L/(\pi h \delta \sigma) . \qquad (25)$$

Here h/2 takes the place of the radius b, and a form factor F_L takes into account that the pipe is rectangular in cross section.

The vacuum chamber of the Main Ring consists of beampipes of different cross sections: 8136 ft of 1.5"×5" and 8496 ft of 2"×4" rectangular beampipes inside the dipoles and parts of the straight sections, 1536 ft of rhombic pipes (approximate circular radius 3.8 cm) for the quads, 624 ft of 6" circular pipes in part of the straight sections, and 188.75 ft of 5" circular pipe in the rf region. The pipe in the rf region is of copper and the rest is of stainless steel. The form factors F_L for the 1.5"×5" and 2"×4" pipes are taken as unity[8] with an error <2.5%. The rhombic pipes are approximated by circular pipes of radius 3.8 cm. Assuming a conductivity of σ = 0.14×10^7 mho-m^{-1} for stainless steel and 5.8×10^7 mho-m^{-1} for copper, the total contribution to the wall impedance is

$$Z_L/n = (1 + j)14.5n^{-1/2} \; \Omega \; , \qquad (26)$$

where n is the revolution harmonic of the frequency of the electromagnetic wave under consideration. Comparing with the limits obtained in the previous section, we see that the wall impedance is too small to drive a fast instability in the microwave region, which is usually taken as n = R/$\bar{b}$, where R is the mean radius of the Main Ring and $\bar{b}$ is the mean radius of the vacuum chamber.

3.2 Bellows

The Main Ring bellows housings serve the purpose of joining beampipes together and provide some leeway in both the horizontal and vertical directions. A typical bellows consists of a pill-box cavity of length g ~ 15.85 cm and radius d ~ 7.16 cm with only four to five ripples at one end and a porthole connected to the vacuum pump. Because of the smallness of the ripple width and depth compared with the dimensions of the pill box, the electromagnetic fields hardly go into the ripples. In other words, the main contribution to the bellows impedance comes from the pill box itself: the steps between the pill box and the side-pipes at low frequencies and the resonances of the pill box at high frequencies.

When a particle passes through the pill box, electromagnetic fields are left and trapped inside the pill box. However, at low frequencies, only magnetic flux can be trapped because electric field cannot satisfy the boundary conditions (Fig. 4). These trapped magnetic fluxes ϕ_t will induce a back e.m.f.

$$\int E_z dz = -j\omega\phi_t \qquad (27)$$

on the beam itself. If the length of the cavity g is less than or about the same as the steps $2(d - b)$, the amount of trapped flux is

$$\phi_o = g(\mu_o I_\omega/2\pi)\ln(d/b) , \qquad (28)$$

where b = 3.61 cm is the radius of the side-pipes, I_ω are the components of the longitudinal current wave with frequency ω, and μ_o is the permeability of free space. The longitudinal impedance per

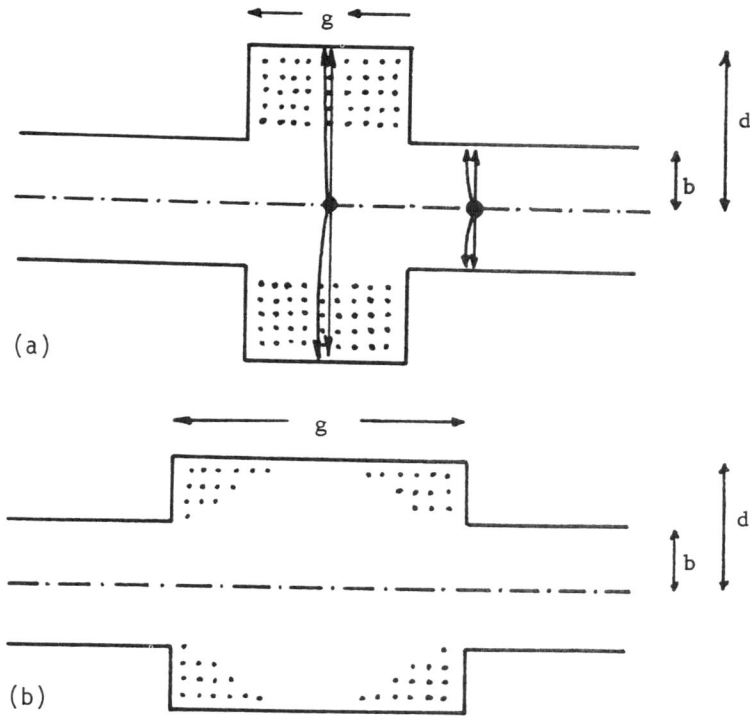

Figure 4 (a) Low frequency magnetic flux (dots) is trapped in the pill box after the passage of a particle whose electric fields are shown as solid arrows. (b) When $g \gg 2(d - b)$, only the flux near the corners is trapped; the flux in the middle leaks away.

harmonic seen at the beam at frequency ω is therefore

$$Z_L = jZ_o(\beta g/2\pi R)\ln(d/b) , \qquad (29)$$

where $Z_o = 377\,\Omega$, βc is the phase velocity of the current wave, and $R = 1$ km is the mean radius of the Main Ring. Eq. (29) was first derived by Keil and Zotter by solving some complicated matrix equations.[9]

When $g \gg 2(d - b)$, which is our case, only the flux near the corners will be trapped and that in the middle will leak away (Fig. 4). Thus, g in Eqs. (28) and (29) should be replaced by $\sim 2(d - b)$. We therefore obtain the impedance per harmonic $Z_L/n \sim j2.9\,\Omega$ for ~ 1000 bellows housings at low frequencies ($n \ll g/2\pi R$). A numerical computation using the code[9] TBCI yields $Z_L/n = j2.4\,\Omega$.

At high frequencies, in order to study the resonances of the pill box, we make the approximation that the pill box is closed (no side-pipes) and neglect the ripples and the porthole. The impedance can then be computed exactly. The first few modes that contribute to the longitudinal impedance are the TM_{01p} modes. The longitudinal electric fields are proportional to

$$\epsilon_p(r,z,t) = J_o(x_1 r/d)\cos(p\pi z/g)\exp(j\omega_p t) , \qquad (30)$$

where J_o is the Bessel function of order zero and $x_1 = 2.405$ is its first zero. The resonance frequency $\omega_p/2\pi$ is given by

$$(\omega_p/c)^2 = (x_1/d)^2 + (p\pi/g)^2 . \qquad (31)$$

The four lowest modes therefore have frequencies 1.60, 1.92, 2.48, and 3.26 GHz. The higher modes need not be considered because they are above the cutoff frequency, ~ 3.2 GHz, of the beampipes. The fields of these modes will no longer be trapped inside the pill box. They will propagate away along the beampipes and therefore will not contribute to the impedance significantly.

The resonance modes are excited by a beam particle passing through the pill box along the central axis. The amount of excitation for a certain mode will therefore be proportional to

$$\int_o^g \epsilon_p^*(0,z,t)\,dz , \qquad (32)$$

where ϵ_p is given by Eq. (30). In the above integration, $t = (z - z_o)/\beta c$ is considered a function of z. It denotes the time when the particle travelling at velocity βc is at a certain location inside the pill box; for example, $t = 0$ when $z = z_o$. The total retarding force seen by the particle traversing the pill box is proportional to ϕ_p^*. The shunt impedance per quality factor of the resonance Z_{sh}/Q is given by

$$(Z_{sh}/Q)_p = Z_o|\phi|^2 C_p/(\omega_p/c) , \qquad (33)$$

where $Z_o = 377$ Ω, and C_p is a normalization constant defined as

$$C_p \int |\vec{e}_p|^2 2\pi r dr dz = 1 \ . \tag{34}$$

Equation (33) then becomes

$$(Z_{sh}/Q)_p = Z_o(c/\omega_p d)(g/d)|\phi/g|^2 a_p/[\pi J_1^2(x_1)] , \tag{35}$$

where J_1 is the Bessel function of order 1 and $a_p = 1.00, 1.48, 0.84, 0.48$ for $p = 1, 2, 3, 4$.

The results for the first four modes are listed in Table IV and the electric field configurations are shown in Fig. 5. We see that Z_{sh}/Q for the first mode is very much smaller than the rest. This is because the longitudinal electric field is constant along the pill-box axis. But at a resonance frequency of 1.60 GHz, this field changes sign as the particle reaches the middle of the pill box. Therefore, the total retarding force seen by the particle is nearly zero. On the other hand, the longitudinal electric field for the second mode behaves like $\cos(\pi z/g)$, or it is of opposite signs for the two halves of the pill box, as shown in Fig. 5. At a resonance frequency of 1.92 GHz, the electric field also changes sign as the particle reaches the middle of the box. But in this case the particle will be seeing retarding force of one sign only, and the impedance becomes enhanced. This effect is known as the transit-time effect, and the information is contained in the transit-time factor $T = |\phi/g|^2$, which is also listed in Table IV.

Table IV The four lowest resonance modes of a Main Ring bellows housing approximated as a closed pill box. The last column shows Z_{sh}/Q without consideration of transit-time effect. Results of numerical computations using TBCI and URMEL are in parentheses.

Mode	Resonance frequency (GHz)	Transit factor T	$(Z_{sh}/Q)/T$ (Ω)	Z_{sh}/Q (Ω)
$p = 0$	1.603 (1.57)	0.030	250	6.16 (2.38)
$p = 1$	1.919 (1.84)	0.202	254	51.3 (37.0)
$p = 2$	2.481 (2.42)	0.232	111	25.6 (24.3)
$p = 3$	3.261 (3.14)	0.242	48.7	11.8 (28.9)

For an open-end pill box, numerical computation is required. We try to run the code[9] TBCI by assuming that the beampipes on both sides are circular and are of radius ~3.61 cm, which corresponds to a cutoff frequency of ~3.2 GHz. The Fourier transform of the wakefield shows four resonances at roughly the positions predicted above (Fig. 6). The addition of the bellows ripples does not change the result appreciably. We next analyze the pill box with closed side-pipes in the frequency domain using the code[10] URMEL, from

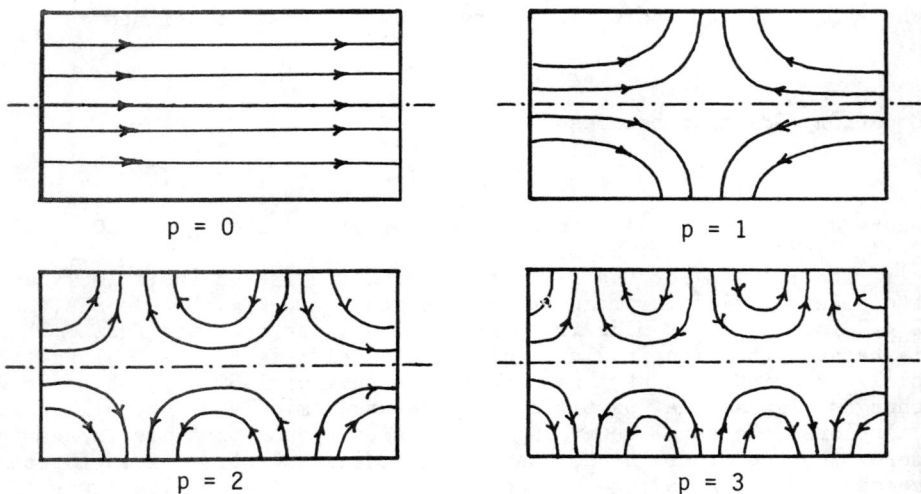

Figure 5 Electric field configurations inside a closed pill box (an approximation of the bellows housing) for the four lowest modes.

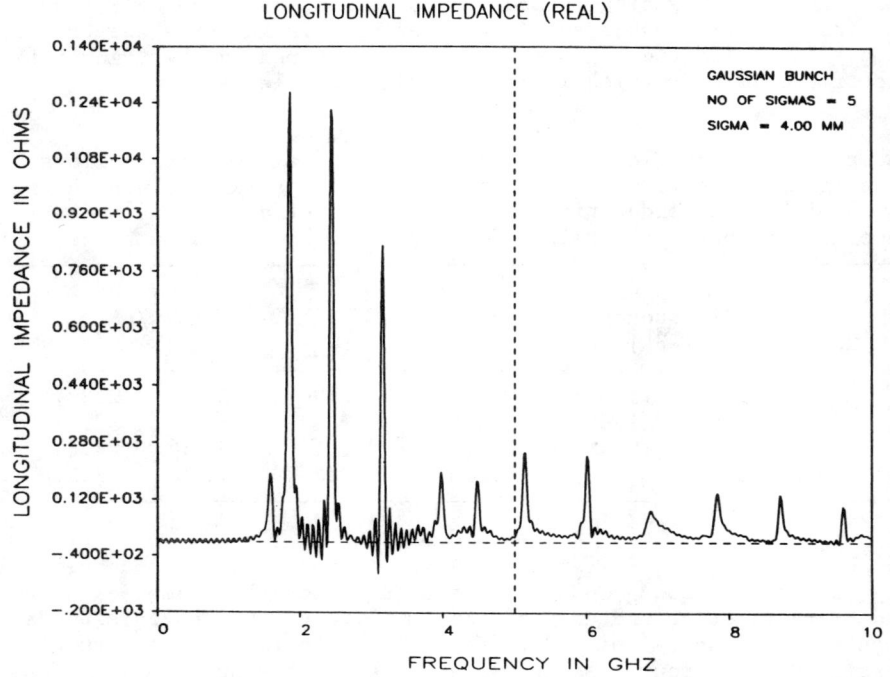

Figure 6 The first four resonances of a bellows system computed through TBCI.

which the quality factors Q and Z_{sh}/Q for the first four modes are obtained. These numerical results are also listed in Table IV for comparison. We note that these results do not differ from those of the closed pill box by very much except for the Z_{sh}/Q of the first mode. However, the impedance of the first mode is very sensitive to the transit-time effect because of the near cancellation. Thus, it is not unexpected that the impedance will change considerably when the side-pipes are added.

The Main Ring has about 1000 such bellows systems; the radii and lengths have standard deviations $\Delta d/d = 0.003$ and $\Delta g/g = 0.034$. Thus, the frequency of resonance will have a spread and the resonance will be broadened. However, the quality factor of one bellows pill box is around $Q \sim 2V/S\delta \sim 4500$, where V and S are the volume and the internal surface area of the pill box and δ is the skin depth. The broadened resonances are still very much narrower than the width of the spectrum of a bunch. As a result, Z_{sh}/Q should be quoted instead of Z_L/n. Using the numerical results, for 1000 bellows systems, $Z_{sh}/Q = $ 2.4, 37.0, 24.3, and 28.9 kΩ for the four resonances. From Table I, the stability limit for proton-bunch coalescence is only $Z_{sh}/Q = 9.5$ kΩ. Therefore these resonances will definitely drive microwave instabilities during bunch coalescence. During the shutdown of the Main Ring from 1985 to 1986, sleeve pieces with tapered ends were installed inside the bellows pill boxes so that a bunch will see a smooth transition from one beam pipe to another. As verified by the wire measurement experiment discussed in Section 4, the four resonances will be completely shielded off and the bellows systems will no longer contribute significantly to the impedance of the machine.

3.3 Beam Position Monitors

About M = 216 beam position monitors are located in the Main Ring. They are rectangular in shape, d = 11.75 cm in length, slant-cut diagonally, and terminated at the center by a resistance of 25 ohms. A typical horizontal monitor is shown in Fig. 7a. We try to approximate it by two half-cylindrical strip plates each with an open angle $\phi_o = \pi$. Each plate can then be viewed as a transmission line with respect to the beampipe (Fig. 7b). The characteristic impedance Z_c is taken as 25 ohms also. At very low longitudinal frequency ω, the input impedance Z_i seen at one end of the plate is nearly Z_c; therefore, there will not be any reflection. If we keep the lowest reactive term,

$$Z_i = Z_c(1 - j\omega d/2c) . \qquad (36)$$

For a beam current of the form $I(t) = I_\omega e^{j\omega t}$ where ω is small or more precisely $\omega d/c \ll 1$, the image current sees a voltage V_u at the upstream end and a voltage V_d at the downstream end:

$$V_u = Z_i(\phi_o/2\pi)I(t) , \qquad (37)$$

$$V_d = -Z_i(\phi_o/2\pi)I(t - d/\beta c) , \qquad (38)$$

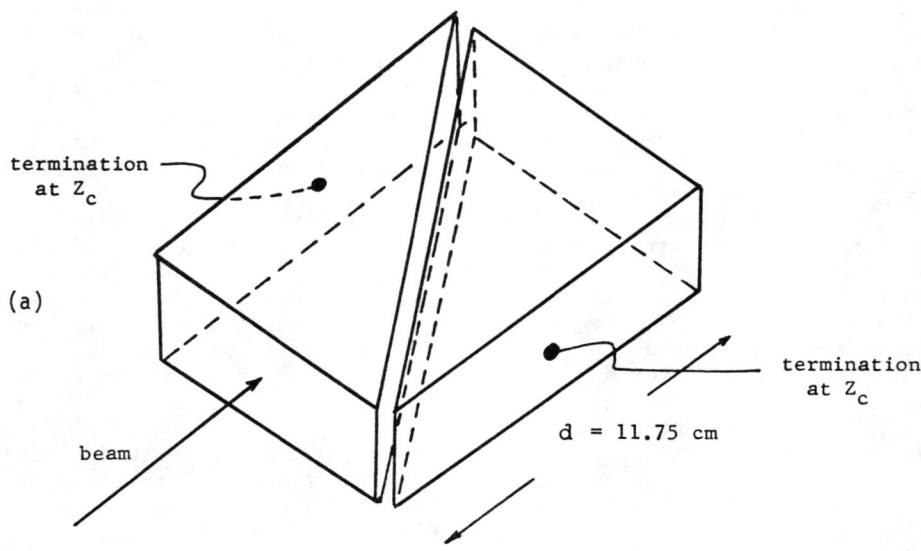

Figure 7 (a) A horizontal beam monitor. (b) A monitor plate is terminated at middle by Z_c. A beam sees a voltage V_u and one V_d at the upstream and downstream ends.

where βc is the phase velocity of the beam current. The voltage seen by the beam is therefore

$$V = (\phi_o/2\pi)(V_u - V_d) \ , \qquad (39)$$

or

$$V = Z_i(\phi_o/2\pi)^2 [I(t) - I(t - d/\beta c)] \ . \qquad (40)$$

Here only a fraction $\phi_o/2\pi$ of the image current passes through the monitor plate and sees the voltage $V_u - V_d$ while the rest just passes by without any resistance. Thus the average voltage V seen by the beam should be multiplied by the extra factor $\phi_o/2\pi$ as in the above equations. Putting in Z_i and $I(t)$, we get

$$V = Z_c I_w (\phi_o/2\pi)^2 (1 - j\omega d/c)[1 - e^{-j\omega d/\beta c}] \ . \qquad (41)$$

Thus, for one plate, the impedance seen by the beam at low frequencies is $Z_L = V/I_w$, or

$$Z_L = Z_c(\phi_o/2\pi)^2(\omega d/\beta c)(j + \omega d/c) \ . \qquad (42)$$

The total impedance for 216 monitors or 432 plates is therefore

$$Z_L/n = (j0.317 + 3.73 \times 10^{-5} n) \text{ ohms} . \tag{43}$$

At high frequencies, the plate can accept resonances with the standing waves having a node at the middle because the termination impedance there does not absorb any power and the resonances will not be disturbed. In other words, the possible resonating wavelength is

$$\lambda_m = md/2 , \tag{44}$$

where $m = 1, 3, 5, \ldots$. This relationship can also be obtained by equating the input impedance

$$Z_i = jZ_c \cot(\omega d/c) \tag{45}$$

of the open-ended transmission line to infinity. To compute the shunt impedance, we include a small surface impedance per unit length

$$R = 1/(\phi_0 b \delta \sigma) , \tag{46}$$

where b is the radius of the cylindrical monitor plate, σ the conductivity, and δ the skin depth. Then, in Eq. (45), $1/c$, which is equal to $(LC)^{1/2}$ where L and C are the inductance and capacitance per unit length of the plate, should be replaced by

$$1/c = \{[L + (1 + j)R/j\omega]C\}^{1/2} . \tag{47}$$

Assuming that $Rd \ll Z_c = (L/C)^{1/2}$, Eq. (45) is expanded in the vicinity of the resonance frequency $\omega_m = m\pi c/d$ to become

$$Z_i \sim \frac{2Z_c^2/R}{1 + j(2m\pi Z_c/Rd)(\omega/\omega_m - 1)} . \tag{48}$$

We can then read off the shunt impedance Z_{sh} and the quality factor Q,

$$Z_{sh} = 2Z_c^2/R , \quad Q = m\pi Z_c/R , \tag{49}$$

or

$$Z_{sh}/Q = 2Z_c/m\pi . \tag{50}$$

It is interesting to point out that Z_{sh}/Q, as given in Eq. (50), depends only on the characteristic impedance of the plate and is independent of its actual shape provided the plate is terminated at the middle by an impedance equal to Z_c. The results for the first four modes are listed in Table V. The contribution of the first mode is the largest; $Z_{sh}/Q = 6.9$ kΩ for 216 monitor systems or 432 monitor plates. Although the most stringent limit for the Main Ring is $Z_{sh}/Q = 9.5$ kΩ, the crudeness of our estimation implies that this mode should be considered dangerous.

Table V The lowest resonance modes of 216 Main Ring
beam monitors estimated as cylindrical strip plates

Mode	Resonance frequency (GHz)	Quality factor	Z_{sh}/Q for 216 monitors (kΩ)
m = 1	1.28	1260	6.9
m = 3	3.84	2190	2.3
m = 5	6.40	2830	1.4
m = 7	9.00	3340	1.0

3.4 Kickers

The Main Ring contains extraction kickers for transferring protons and antiprotons to the Energy Saver and protons for antiproton production; it also contains an abort kicker for protons and injection kickers for transferring protons from the Booster and antiprotons from the Accumulator. These kickers are in the form of window-frame magnets either of size 2" x 6" and length d = 1 m, or 1 1/2" x 3 3/8" and d = 1.9 m. If the beam passes through the center of the magnet, the magnetic flux it links inside the magnet is just equal to the return flux it links outside the magnet; i.e. no voltage will be induced on the beam by the magnet. If the beam deviates from the central axis by x_o, the flux it links (Fig. 8) is

$$\phi = \mu_o I_k (x_o d/h) \ , \tag{51}$$

where h is the height of the magnet whose width is w, I_k the current in the magnet windings, and μ_o the permeability of free space. Therefore, there is a mutual inductance of $M = \mu_o x_o d/h$ between the horizontally deviated beam and the kicker current.

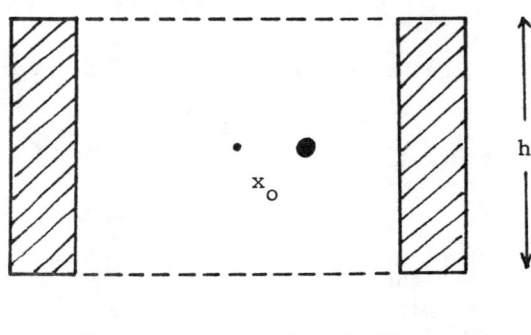

Figure 8 The cross section of a window magnet showing the beam deviating from the center by a distance x_o. The magnet current element is cross-hatched.

Thus, the beam current I_b induces a voltage $V_k = j\omega M$ in the magnet windings, which in turn induce a back e.m.f. $j\omega M V_k/Z_k$ on the beam. The impedance seen by the beam is therefore $Z_L = -\omega_0^2 M^2/Z_k$. Here, $Z_k = j\omega L + Z_g$ is the impedance of the magnetic circuit; the first term, with inductance equal to $\mu_0 wd/h$, is the impedance due to the magnetic fields it produces, the second term is the generator impedance, including cables. If one neglects Z_g one gets[11]

$$Z_L/n = Z_0 x_0^2 d/RA_m , \qquad (52)$$

where $Z_0 = 377\ \Omega$, $A_m = w \times h$ is the cross-sectional area of the window-frame magnet, and R is the mean radius of the Main Ring. Taking the horizontal dispersion function $\eta \sim 5$ m and $\sqrt{6}\sigma_p/p \sim 10^{-3}$, the deviation of the beam is $x_0 \sim 1$ mm. However, at injection the beam has a 95% betatron emittance of 2.5 mm-mr or a transverse spread of ~1.58 cm. Taking the latter as x_0, we obtain for the Main Ring kickers at low frequencies $Z_L/n = 0.03\ \Omega$, which will be too small for instability.

3.5 Lambertson Magnets

Lambertson magnets are required for the injection and extraction of the particle beam. The cross section of a typical Lambertson is shown in Fig. 9a. The beam will see laminations of thickness $\tau \sim 0.953$ cm separated by gaps or cracks $\Delta \sim 28.6$ μm. There are two 120" Lambertsons at E0, two 204" Lambertsons at F0, one 90" Lambertson at A0, and two 181" Lambertsons at C0. In total, the Lambertsons occupy a length of L = 1016"; or there are roughly $N \sim 2.71 \times 10^4$ laminations. The laminations are approximated as annular rings of inner and outer radii b = 1" and b + d = 2" respectively and are shorted at the outer end as shown in Fig. 9b. The relative magnetic permeability and the conductivity of the lamination are taken as $\mu_L \sim 100$ and $\sigma_L \sim 0.5 \times 10^7$ mho/m respectively. The dielectric constant and conductivity of the material filling the cracks are taken as $\epsilon_1 \sim 6$ and $\sigma_1 \sim 0.01$ mho/m respectively.

Each crack can be viewed as a transmission line. The impedance seen by the beam is contributed by the image current flowing around each lamination and also field penetrating into the cracks. At frequencies $f \ll \mu_L c/(\pi\sigma\Delta^2 Z_0)$ or ~6 GHz, the former contribution dominates. Similar to the resistive wall derivation, the longitudinal impedance is

$$Z_L = (1 + j)N\ln(1 + d/b)/(\pi\delta\sigma) , \qquad (53)$$

where δ is the skin depth in a lamination. This amounts to $Z_L/n = (1 + j)11.6 n^{-1/2}\ \Omega$.

If we further approximate the crack to be a parallel plate transmission line of width $2\pi b$, the first resonance can be estimated. The impedance per unit length along the lamination is $z = (1 + j)/(\pi\delta\sigma)$, and the admittance per unit length is $y = j\omega\epsilon_0\epsilon_1\Delta/2\pi b$ where ϵ_0 is the permittivity of free space and the

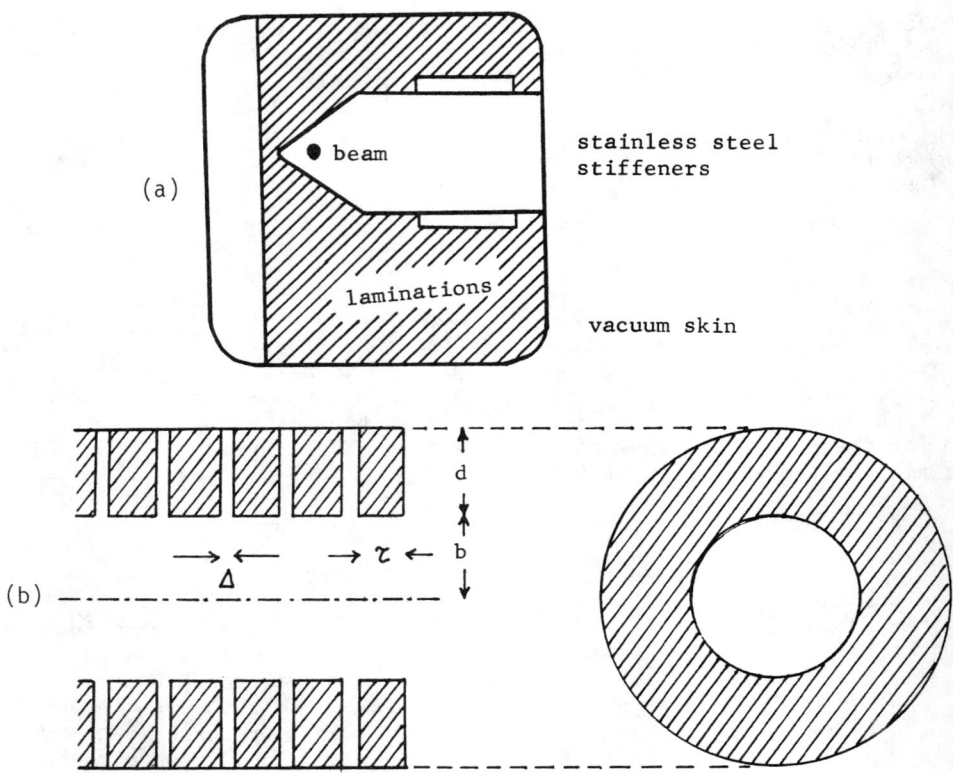

Figure 9 (a) The cross section of a Lambertson magnet. (b) Approximation of the cross section of a Lambertson by an annular ring shorted at the outer circumference.

conductivity of the medium has been neglected. The wave going into the crack has the propagating constant $\gamma = (yz)^{1/2}$. The first resonance is roughly given by $\text{Im}(2\gamma d) = \pi$. Putting in the data, we obtain $\gamma d = (1.11 + j2.69)f^{3/4}$, where the frequency $f = \omega/2\pi$ is in GHz. The characteristic impedance is $Z_c = (z/y)^{1/2} = 0.0615 e^{j3\pi/8} f^{1/4}$. Thus, the first resonance is at 0.488 GHz where $|Z_c| = 0.0736$ Ω. The impedance seen by the beam is $Z_i = Z_c \tanh(\gamma d)$. An estimate of $\tanh(\gamma d)$ at resonance is $(1 + e^{-2a})/(1 - e^{-2a})$ or 1.74, where $a = \text{Re}(\gamma d)$. Therefore, for all $N \sim 2.71 \times 10^4$ laminations, the shunt impedance is $Z_{sh} = 3470$ Ω. The quality factor is $Q \sim \Delta/2\delta$ or ~14. An actual numerical computation gives the first resonance at 0.47 GHz with shunt impedance 3460 Ω. The wall impedance increases with frequency; therefore, we expect the higher resonances to be damped heavily and to appear as ripples only. This is verified by the numerically computed impedance shown in Fig. 10.

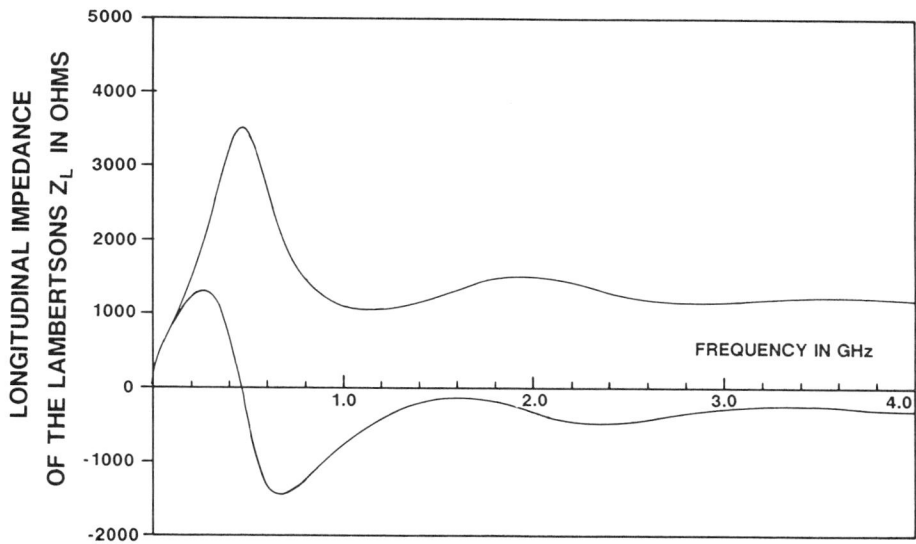

Figure 10 Estimated longitudinal impedance of the Lambertsons as a function of frequency. The upper curve is the real part and the lower curve the imaginary part.

3.6 Summary

The contributions to the longitudinal impedance estimated in the above are plotted in Fig. 11 and listed in Table VI. We see that near cutoff frequency, the most important contributions come from the bellows housings and the beam monitors. They are big enough to drive microwave instabilities. If sleeves are installed inside the bellows to smooth out the steps and shield off the pill-box structure, all the bellows contribution will vanish. In that case, the beam monitors will become the most dangerous threat to stability.

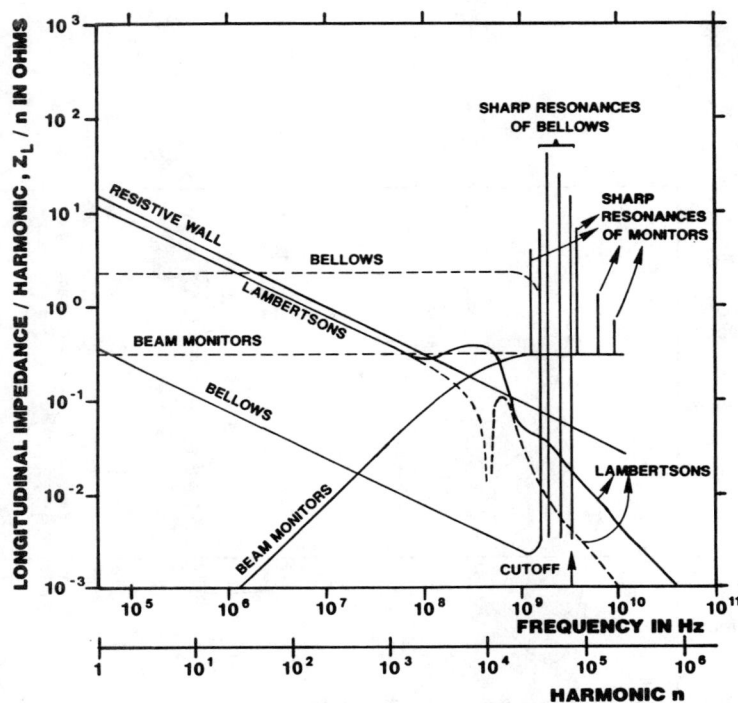

Figure 11 Various contributions to the longitudinal impedance/harmonic Z_L/n of the Main Ring. Solid curves represent the real parts and dashed curves represent the imaginary parts.

Table VI Various contributions to the longitudinal impedance of the Main Ring. $|\bar{Z}_L/n|$ is an average of Z_L/n over the bunch spectrum. This quality will drive the single-bunch coupled-mode instabilities. The high frequency contributions will drive the microwave instabilities. The limit Z_L/n occurs during preparation for p̄ production, and the limit Z_{sh}/Q occurs during p̄ bunch coalescence. See Section 3 for details.

	Low frequency $|\bar{Z}_L/n|$, Ω	High frequency Z_L/n or Z_{sh}/Q
Resistive wall	1.10	
Bellows systems	2.46	$Z_{sh}/Q = 37$ kΩ
Beam monitors	0.32	$Z_{sh}/Q = 6.9$ kΩ
Kickers	0.03	
Lambertsons	0.81	$Z_L/n = 0.35$ Ω
Total	4.72	
Estimated limits	7.2	$Z_L/n = 6.6$ Ω
		$Z_{sh}/Q = 9.5$ kΩ

4 WIRE MEASUREMENT EXPERIMENT

A wire carrying a current is placed inside a beampipe to simulate the effect due to a particle beam. The potential difference across the wire will give a measure of the impedance due to the discontinuities of the beampipe. Such an experiment had been carried out by Reid.[12]

4.1 Experimental Setup

Four sections of the Main Ring beampipe about 26" long with flanges were configured to simulate different devices. Pipe A is the standard 2"×4" rectangular beampipe with flanges at each end and was used as the reference (Figs. 12a and 12b). Pipe B is a Main Ring bellows system inside the same beampipe as A (Fig. 12c), and Pipe C consists of exactly the same bellows as B but with sleeve pieces inside connecting the side-pipes and shielding the bellows ripples and pill-box structure. Pipe D contains a Main Ring bellows system which includes a beam monitor (Fig. 12d); the detector can be vertical or horizontal and the system can be assembled with or without sleeve pieces.

A 1/16" diameter wire was strung down the center of the standard pipe. The characteristic impedance of this wire in the rectangular beampipe is approximately $Z_c \sim 216$ Ω. To match this characteristic impedance to the 50-Ω Times Cables No. SF-142B on each side, a minimum loss L-pad matching network was installed on each side as in Fig. 13a. In order that signals coming out of the beampipe see 216 Ω, we need

$$1/(1/Z_2 + 1/50) + Z_1 = 216 . \tag{54}$$

In order that signals from the Times Cable entering the beampipe see 50 Ω, we need

$$1/(Z_1 + 216) + 1/Z_2 = 1/50 . \tag{55}$$

These give $Z_1 = 189.357$ Ω and $Z_2 = 57.035$ Ω. The actual best match was accomplished by choosing $Z_1 = 191$ Ω and $Z_2 = 57$ Ω. This match is extremely important because we do not want any reflection from each connection between the beam pipe and the RG57 cable, so that the whole system with the standard pipe A will not introduce any voltage drop except that at the L-pads, which can be computed. At the input end, the attenuation is

$$a_i = 216/(Z_1 + 216) , \tag{56}$$

which is 0.4693 or -5.502 db. At the output end, the attenuation is

$$a_o = Z_3/(Z_1 + Z_3) , \tag{57}$$

where Z_3 is Z_2 and 50 Ω in parallel. Therefore, $a_o = 0.1224$ or -18.178 db. Thus, a signal passing through the system will show an

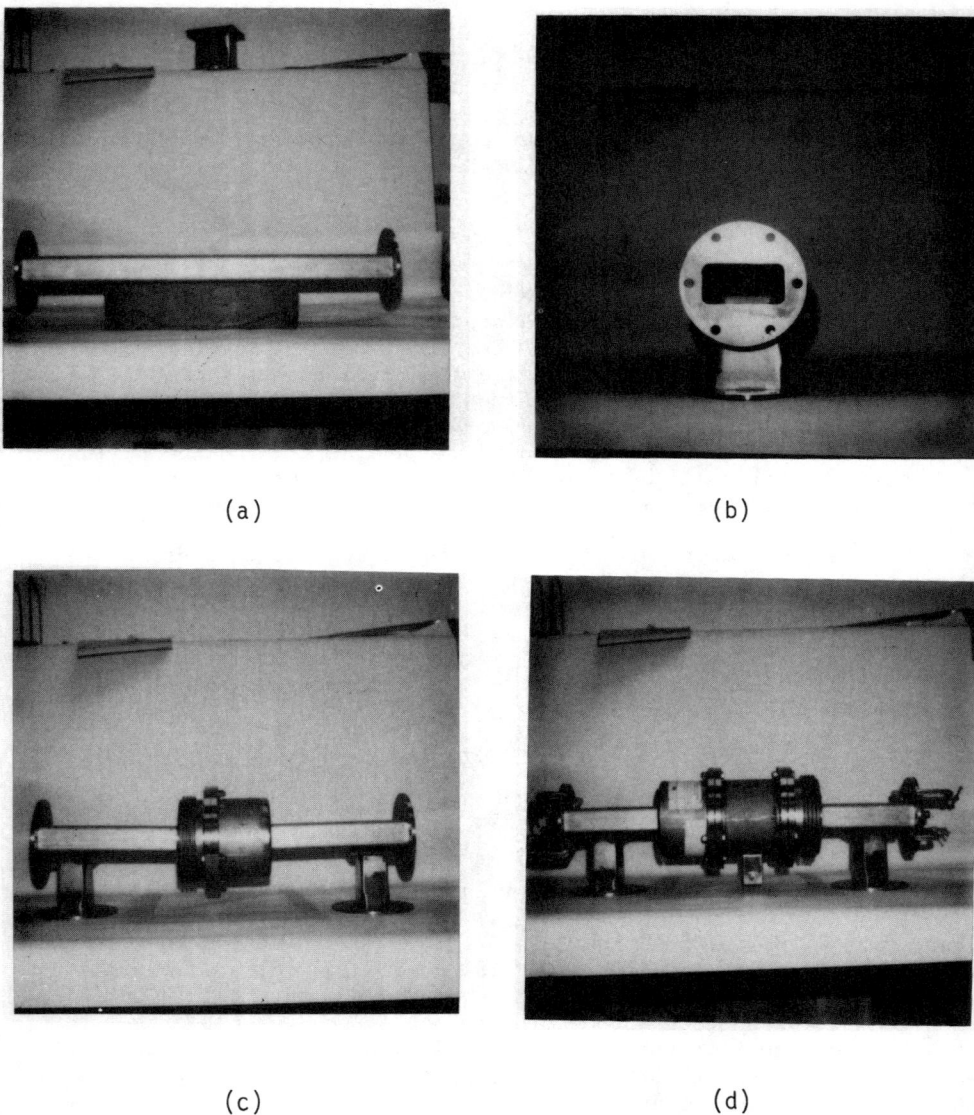

Figure 12 Beampipe sections to be tested. (a) Pipe A: a section of the standard beampipe to be used as a reference. (b) An end view of Pipe A. (c) Pipe B (or C): the standard beampipe containing a typical bellows housing without (or with) sleeve pieces. (d) Pipe D: the standard beampipe containing a beam monitor system with or without sleeve pieces.

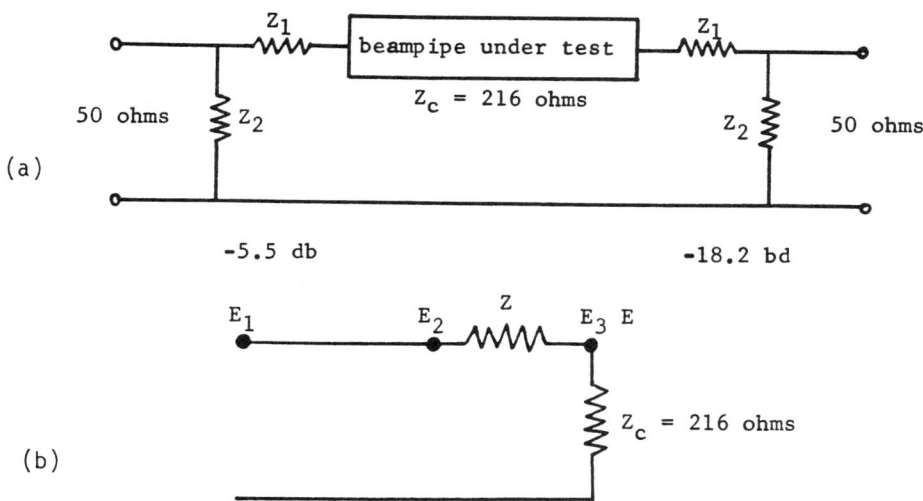

Figure 13 (a) Circuit diagram matching network. (b) Voltage attenuation along the beampipe due to a discontinuity impedance Z. The voltage E_2 before the discontinuity is equal to the incident voltage E_1 plus a reflected voltage ρE_1, where ρ is the reflection coefficient.

attenuation of -23.68 db. If there is a discontinuity in the beampipe, for example a bellows housing or a beam monitor, contributing an impedance Z, a signal with a voltage E_1 entering the beampipe will be reflected at the discontinuity with a reflection coefficient

$$\rho = \frac{(Z_c + Z) - Z_c}{(Z_c + Z) + Z_c} , \qquad (58)$$

so that the voltage at the beginning of the discontinuity is $E_2 = (1 + \rho)E_1$. Then, the voltage at the end of the discontinuity is, from Fig. 13b, $E_3 = E_2 Z_c/(Z_c + Z)$. The additional attenuation is defined as $a = E_3/E_1$. Therefore,

$$a = 2Z_c/(2Z_c + Z) , \qquad (59)$$

or the impedance of the discontinuity is

$$Z = 2Z_c(1 - a)/a . \qquad (60)$$

Two different network analyzers were employed. An HP model 8754A was used for frequencies up to 1.36 GHz, and to frequencies up to 3.56 GHz, a modified version of the HP 8410 automatic network analyzer with HP accuracy enhancement pac was used.

The results of the test pipe calibration (Fig. 14) show how good the impedance matching was. At high frequencies, the calibration curve exhibits some variations. This is in fact

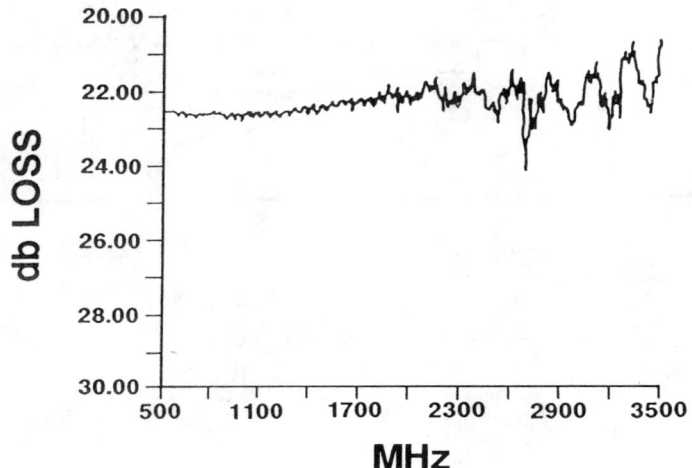

Figure 14 The attenuation as a function of frequency for the reference Pipe A. The nearly frequency-independent curve indicates the excellent matching of the reference pipe to the 50-ohm cables. Note that the attenuation is very close to the designed value of -23.7 db.

expected. The resistors used in the L-pads are of the 1/4-watt Allen Bradley type. At high frequencies, they are not just pure resistors anymore and therefore the matching will not be exact. Figure 14 also defines the resolution of the setup, which was ~1.5 dB, corresponding to ~81 Ω. Since the purpose of this measurement was to search for large impedances, such a resolution was acceptable.

4.2 Results of Measurement

The results with the reference pipe replaced by Pipe B, containing the Main Ring bellows housing (Fig. 15a and Table VII) agree with the estimates in Section 3, which are listed for comparison. The only resonance that does not agree is the fourth one; however, the experimental setup was not designed to give consistent large impedance magnitudes in the upper frequency range. This is due to the ill behavior of the Allen Bradley resistors and also partly to the mechanical placement of the wire and associated pieces of the beampipe.

We should point out that a wire inside a beampipe can support TEM mode, which has no cutoff frequency, i.e., any resonance fields inside the bellows housing can leak away along the beampipe although the frequency is below the TM-mode cutoff frequency of the beampipe. As a result, the quality factors measured here will be very small and have nothing to do with the quality factors of the resonances excited by an actual particle beam, which does not support any TEM

(a)
(b)

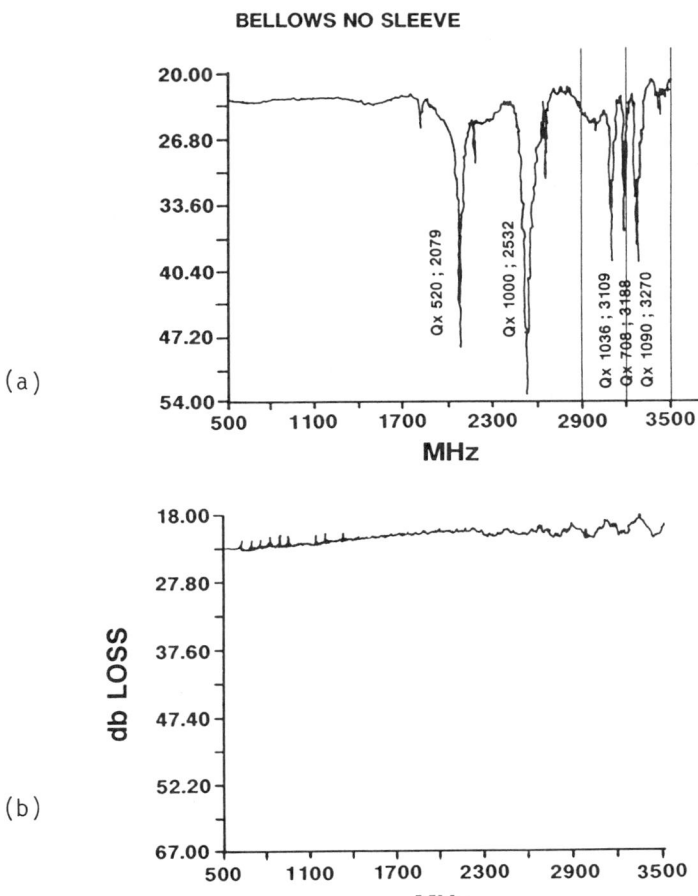

Figure 15 Wire measurements of one bellows system (a) without sleeve pieces and (b) with sleeve pieces.

Table VII Wire measurement results of one bellows housing without sleeve pieces. The estimated values of Section 3 are in parentheses.

Frequency (MHz)		Q	Z_{sh} (KΩ)	Z_{sh}/Q (Ω)	
1820	(1570)	183	0.269	1.47	(2.4)
2050	(1900)	1020	25.0	24.4	(37.0)
2536	(2400)	1000	29.7	29.7	(24.3)
3109	(3100)	1036	2.20	2.1	(28.9)
3188		708	1.35	1.9	
3270		1090	2.2	2.0	

mode. As a result, the measured shunt impedances will be very much smaller also. The only meaningful quantities may be Z_{sh}/Q.

The results for Pipe C, with sleeve pieces shielding the bellows ripples and the pill-box discontinuity (Fig. 15b) show no measurable modes up to 3.5 GHz.

Pipe D, containing the Main Ring horizontal beam detector assembly, was first measured with no sleeve (Fig. 16a and Table VIII). The results show that the lowest resonance mode is at 794 MHz, very much lower than the estimated 1.2 GHz. This may be because the estimation was made assuming two straight cylindrical strips of length 11.75 cm whereas the actual beam monitor is diagonally cut (Fig. 7a) with a cut length of ~15.53 cm. Together with some end effects, this may be able to support a mode of much lower frequency than 1.2 GHz. Thus, the former estimates may have been too crude for comparison here. On the other hand, the lowering of the resonance frequency may be due to the geometry of the experimental setup and the presence of the wire instead of a beam. From Table VIII, we see that Z_{sh}/Q decreases roughly as $1/m$ for $m = 1, 3, 5, \ldots$. In any case, for 216 monitors, a $Z_{sh}/Q = 4.5$ kΩ or 9.7 kΩ is definitely too large for the stability of the beam.

With sleeve pieces installed, the measurements of the horizontal beam monitor are shown in Fig. 16b and Table VIII. The lowest and most dangerous mode at ~800 MHz appears to have been damped, but the second mode is not and is therefore still a threat to stability. The mode at 2.1 GHz disappears.

4.3 Conclusions

The effect of adding sleeves inside the Main Ring bellows is dramatic. It eliminates all measurable modes up to 3.5 GHz. The effect of sleeves in the beam monitor is less dramatic. Although the ~800-MHz mode is suppressed and the 2.1-GHz mode is eliminated, the 1.6-GHz mode is still a threat to stability.

We must also keep in mind that a wire inside a beam pipe is different from an actual particle beam. For example, with the wire there is a continuum of TEM modes which are certainly not present when the wire is replaced by a particle beam. Therefore the results of a wire measurement experiment can be considered only as a guide or reference.

5 MEASUREMENT OF IMPEDANCE BY DEBUNCHING

When the rf voltage is turned off suddenly, the once bucket-matched bunch will start to shear because particles with different energies travel with different angular velocities. The bunch becomes longer and longer while the energy spread becomes smaller and smaller, with the bunch area unchanged. According to the microwave stability criteria, Eqs. (10) and (13), the critical $Z_L/n \propto \sigma_E$ if the driving impedance is a broad band and the critical $Z_{sh}/Q \propto \sigma_E^2$ if the driving impedance is a narrow resonance. Thus, a time will come when the bunch becomes unstable. From the microwave signals measured during the growth, we can determine the longitudinal impedance in the ring that drives the instability.[13] Such an experiment was carried out by Crisp[14] for the Main Ring.

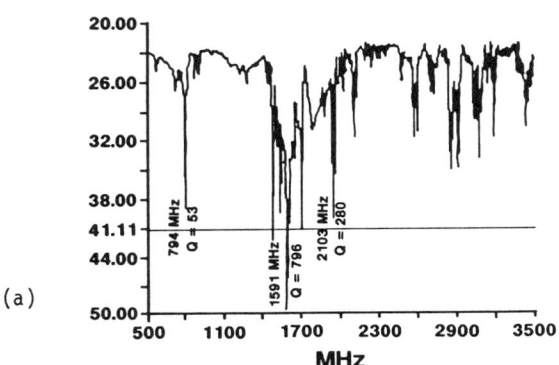

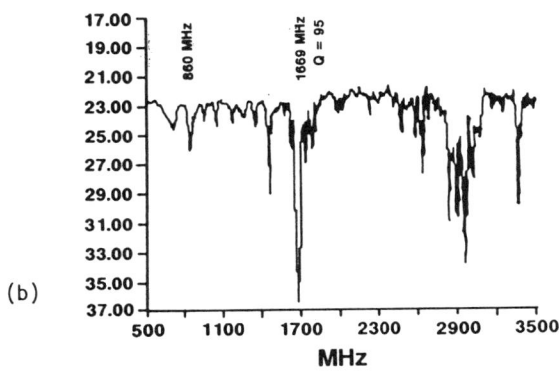

Figure 16 Wire measurement results of one beam monitor system (a) without sleeve pieces and (b) with sleeve pieces.

Table VIII Wire measurement results of one beam monitor system

Frequency (MHz)	Q	Z_{sh} (kΩ)	Z_{sh}/Q (Ω)
(a) Without sleeve pieces			
794	53	2.4	45.0
1591	796	8.5	10.7
2103	280	2.7	9.8
(b) With sleeve pieces			
845	70	0.4	5.7
1685	337	3.9	11.6

5.1 The Experiment

Nine bunches, each of intensity $N = 0.636 \times 10^{10}$ protons, were accelerated to 150 GeV and were left in buckets with rf voltage $V_{rf} = 1.079$ MV. The rf voltage was turned off suddenly and the bunches tended to shear. When the energy spread was small enough, Landau damping failed and microwave signals started to grow.

A coaxial directional coupler designed by Griffin[15] was used to pick up the microwave signals. The detector consists of two concentric pipes, 1/4 wavelength long, having a characteristic impedance of 12.5 Ω. Each end has four symmetrically spaced 50-Ω ports. The signals were transported from the enclosure via a 177-nsec-long 7/8" heliax cable. A spectrum analyzer in the zero-span mode was used to monitor the amplitude of the 31st harmonic of the rf ($f_{MW} = 1.646$ GHz). Zero-span mode basically plots the amplitude of the signal that passes through an equivalent filter with the center frequency and bandwidth specified. The peak detector was not used because it has the disadvantage that the signal must overcome the diode forward voltage drop before it can be detected, whereas the spectrum analyzer has a linear response to small signals and a logarithmic scale. The displays of the spectrum analyzer and the peak detector for a typical run are compared in Fig. 17. The diode detector indicates a time of 60 msec for the onset of the microwave growth, but the analyzer shows a clear minimum at about 30 msec.

5.2 Theory

The original bunch has rms time spread and energy spread $\hat{\sigma}_T$ and $\hat{\sigma}_E$ respectively. After time t, they become σ_T and σ_E respectively. Any particle with an energy ΔE higher than that of the synchronous particle will be ahead of it by an extra time

$$\Delta \tau \sim \eta t (\Delta E/E) , \qquad (61)$$

where η is the frequency dispersion (or frequency slip) parameter. A particle at the rms bunch width σ_T will come from the center of the original bunch with an excess energy equal to $\hat{\sigma}_E$, or

$$\sigma_T \sim \eta t \hat{\sigma}_E / E . \qquad (62)$$

Remembering that $\sigma_T \sigma_E$, being proportional to the bunch area, is equal to $\hat{\sigma}_T \hat{\sigma}_E$, the stability criteria of Eqs. (10) and (13) become

$$\left(\frac{Z_L}{n}\right) t_o = \frac{(2\pi)^{3/2} \hat{\sigma}_T^2 (\hat{\sigma}_E/e)}{eN} , \qquad (63)$$

and

$$\left(\frac{Z_{sh}}{Q}\right) t_o^2 = \frac{4 \hat{\sigma}_T^2 (E/e)}{\eta I_{AV}} , \qquad (64)$$

where N is the number of particles in a bunch with average current I_{AV} and t_o is the time when instability starts. Thus, knowing t_o, one can compute the magnitude of the driving impedance.

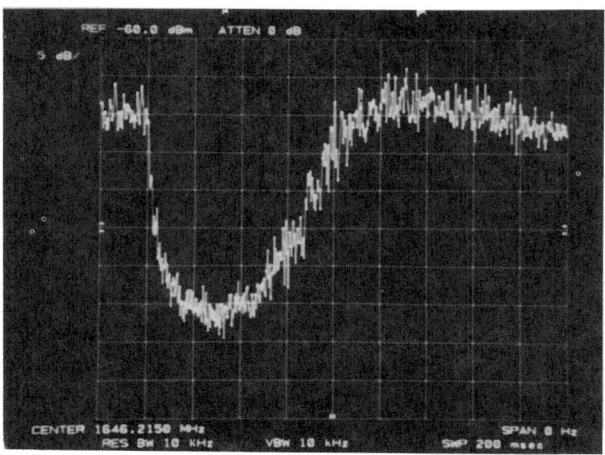

(a) Zero-span mode 10-MHz bandwidth analyzer.
Scales are 5 db/div and 20 ms/div.

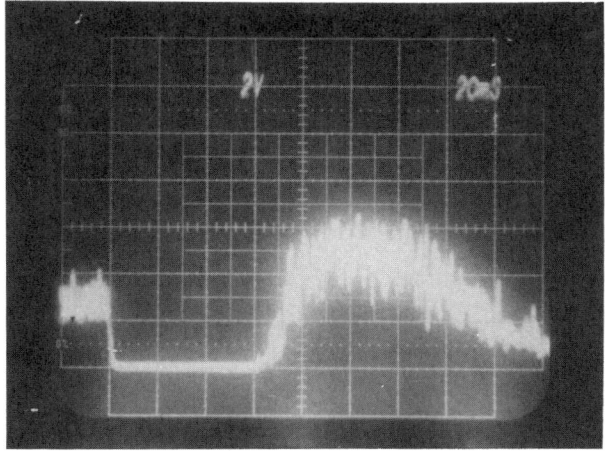

(b) Diode detector. Scales are 2 V/div and 20 ms/div.

Figure 17 Comparison of the starting growth times
by zero-span analyzer and diode detector.

5.3 Analysis of Results

5.3.1 Bunch Shape

Since the stability criteria used in the above are for a Gaussian bunch only, we should check whether our bunches are Gaussian or not. A typical bunch shape is selected and is fitted by a Gaussian curve with rms time spread $\hat{\sigma}_T$ and also a parabolic curve with maximum half spread $\hat{\tau}$ (Fig. 18). Since the normalized sum of deviation of the Gaussian fit is about ten times less than that of the parabolic fit, we accept the bunch to be Gaussian. The best rms time spread obtained is $\hat{\sigma}_T = 0.635$ cm. The rf voltage was 1.074 MV. This leads to a rms energy spread of $\hat{\sigma}_E = 0.0192$ GeV and a bunch area of $6\pi\hat{\sigma}_T\hat{\sigma}_E = 0.230$ eV-sec.

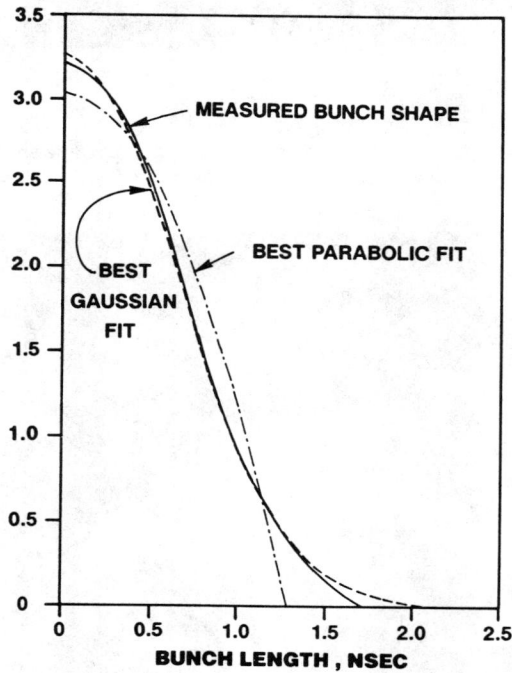

Figure 18 Bunch shape fittings. Solid curve is the measured bunch shape. Dot-dashed curve is the best parabolic fit with half bunch length = 1.278 nsec. Dashed curve is the best Gaussian fit with rms time spread $\hat{\sigma}_t = 0.635$ nsec.

5.3.2 Starting Time of Growth

As seen in the stability criteria of Eqs. (63) and (64), the starting time of growth t_o is very crucial to the determination of the driving impedance. From Fig. 17 it is clear that the bunch is unstable at 30 msec, but the actual growth may have started before that.

To determine t_o more accurately, the growth rate is computed as a function of time from a dispersion relation.[15] At time $t = t_o(1 + x)$, the logarithmic power growth is proportional to the function $F(x)$, which is plotted in Fig. 19. By comparing the maximum slope at $x = 2.25$ and the change in slope from $x = 0$ to 2.25 with the actual logarithmic growth measured in Fig. 17, we conclude that the starting time of growth is $t_o \sim 20$ msec.

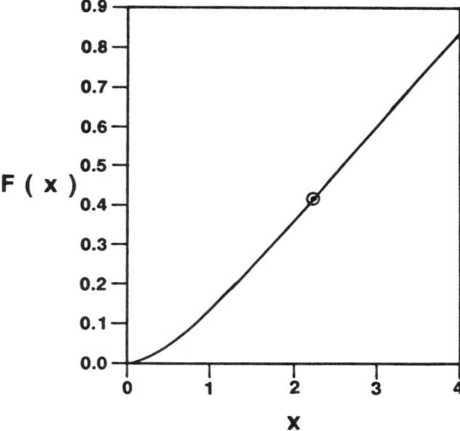

Figure 19 The normalized integrated growth rate $F(x)$ of Ref. 15. The turning point is at $x = 2.25$ where the slope is 0.2194.

5.3.3 Overlapped Bunches

Two adjacent bunches start to touch each other when the half bunch length $\sqrt{6}\sigma_\tau$ is equal to the half length of a bucket $\pi/h\omega_o$, where h is the rf harmonic and $\omega_o/2\pi$ is the revolution frequency. Putting this into the left side of Eq. (61) and letting ΔE equal the half energy spread of the bunch $\sqrt{6}\hat{\sigma}_E$, the time when two adjacent bunches touch each other is

$$t = \pi/[\eta h\omega_o(\sqrt{6}\hat{\sigma}_E/E)] \; , \qquad (65)$$

which gives 9.3 msec. Thus, when instability starts at ~20 msec, two adjacent bunches overlap completely. We are faced with two questions. Which energy spread and which current should we use in the stability criteria? If we use the actual energy difference between the two bunches other than the energy spread σ_E of one

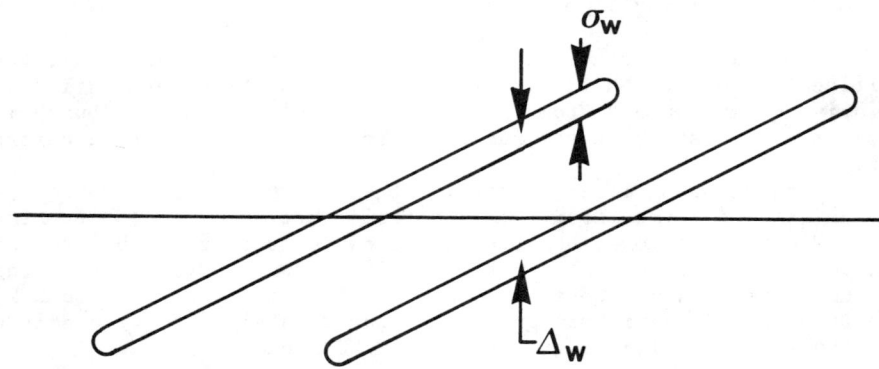

Figure 20 Two adjacent bunches overlap each other during debunching. Note that the frequency spread of each bunch σ_w is very much less than the mean revolution frequency difference Δw for particles in the two bunches at some azimuthal position.

bunch, the bunch will be much more stable (Fig. 20). If we take the sum of the currents of the two bunches as the current in the criteria, then the bunch will be less stable.

This problem can be solved[16] by examining two coasting beams, one with revolution angular frequency w_1 and the other with w_2 and each having a rms angular frequency spread of σ_w. We also assume that $\sigma_w \ll \Delta w$ where $\Delta w = |w_1 - w_2|$. Imagine a small perturbing longitudinal current wave of the form $I_1 e^{j(\Omega t - n\theta)}$ where θ is the azimuthal angle around the accelerator ring. If the coherent frequency $\Omega \sim nw_1$, it will set the particles in the first beam to oscillating with harmonic n and will eventually lead to growth if the frequency spread σ_w is not big enough to destroy the coherency. Since $\sigma_w \ll \Delta w$, the particles in the second beam will not be affected. On the other hand, if the coherent frequency of the perturbation is $\Omega \sim nw_2$, it can drive growth only of harmonic n in the second beam while the first beam is not affected. Thus, the stability criterion is for one beam only; i.e., it involves σ_w or σ_E instead of Δw or ΔE, and the current is the local current of one bunch only. In other words, the stability criterion remains unchanged; it is exactly that for one bunch. An actual solution of the Vlasov equation for two overlapped bunches gives a stability curve that wraps around the origin of the $(\Delta\Omega/n)^2$-plane two times, as in Fig. 21, where the stability curve for one bunch is also shown for comparison. Take the situation when the driving impedance is capacitive and above transition. The stability point for one bunch is point A. For two bunches, there are two such points, A_1 and A_2, where the stability curve crosses the negative imaginary $(\Delta\Omega/n)^2$-axis twice. In fact, these two points coincide with each other and are very close to point A if $\sigma_w \ll \Delta w$. Note that points A_1 and A_2 correspond to very different coherent frequencies Ω, the difference being $n\Delta w$.

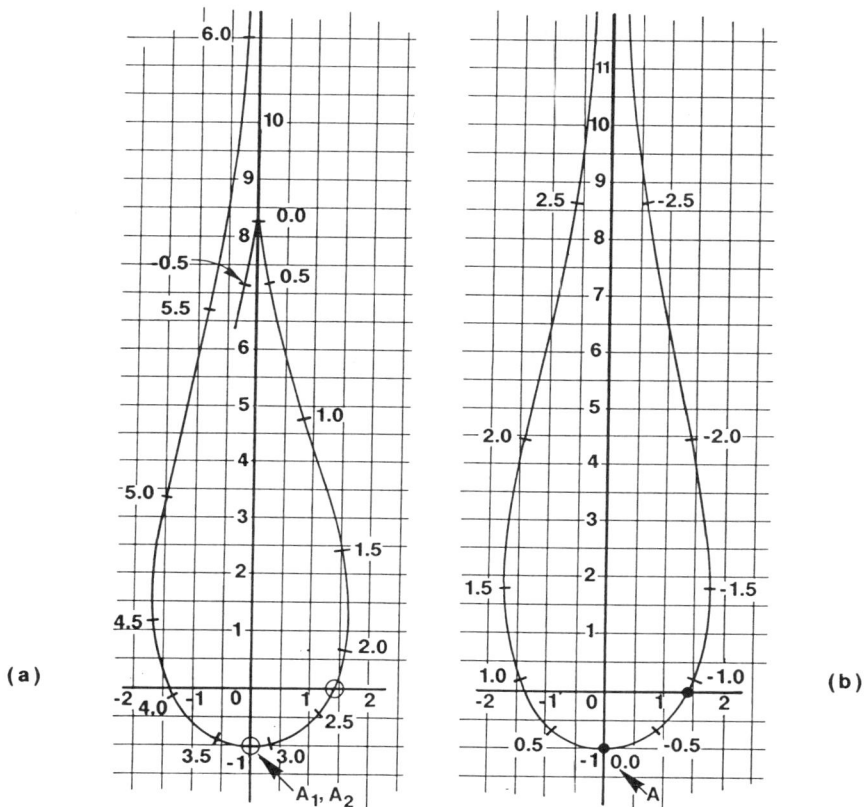

Figure 21 Threshold curves for (a) two overlapped bunches and (b) a single bunch. In each case the abscissa and ordinate are the real and imaginary parts of $(\Delta\Omega_o/n\sigma_w)^2$ respectively. The real coherent frequency shift $\text{Re}(\Delta\Omega/\sqrt{2}n\sigma_w)$ is marked along the curves. In (a), for clarity, only half the curve is plotted.

5.3.4 Determination of Driving Impedance

Having the correct stability criterion and knowing the starting time of growth, we can now compute the driving impedance. Here we assume that the driving impedance is the peak of a broad band or a narrow resonance; or it is real. Therefore, it corresponds to points B_1 or B_2 on the stability curve of Fig. 21 which are about 1.434 times (for Gaussian beam only) farther away from the origin than points A_1 or A_2, for which the stability criteria of Eqs. (63) and (64) are derived. When this factor of 1.434 is included, we obtain at 1.64 GHz, $Z_L/n = 8.6$ Ω for a broad-band and $Z_{sh}/Q = 6.4$ kΩ for a narrow resonance. In the impedance estimation of Section 3, there is no broad-band impedance around 1.64 GHz. Therefore, the driving force may come from sharp resonances. Possible candidates are the first resonance of the beam monitors with Z_{sh}/Q = 6.9 kΩ at ~1.28 GHz and the second bellows housing resonance with Z_{sh}/Q = 2.4 kΩ at ~1.6 GHz.

We should point out that, since the starting time of growth t_o is not evident from Fig. 17, its theoretical determination in the above is not without controversy. Any change in t_o will affect the final results for the driving impedance. For this reason, the computed results may include some big inaccuracy.

6 REFERENCES

1. A. Hofmann, in Proc. First Course of Int. School of Particle Accelerators of the Ettore Majorana Center for Scientific Culture, Erice, 1977, p. 139.
2. E. Keil and W. Schnell, CERN ISR-TH-RF/69-48, 1969.
3. S. Krinsky and J. M. Wang, Particle Accelerators 17, 109 (1985).
4. If we include only the real part of the resonance and let Q tend to infinity, this criterion can be obtained easily. However, the neglect of the imaginary part which carries a long tail is a violation of causality.
5. K. Y. Ng, Fermilab Report TM-1383, 1986.
6. S. Y. Lee and J. M. Wang, IEEE Trans. Nucl. Sci. NS 32, 2323 (1985).
7. K. Y. Ng, Fermilab Report TM-1388, 1986.
8. K. Y. Ng, Particle Accelerators 16, 63 (1984).
9. T. Weiland, DESY 82-015, 1982.
10. T. Weiland, DESY M-82-24, 1982.
11. G. Nassibian and F. Sacherer, Nucl. Instrum. Meth. 159, 21 (1979).
12. J. Reid, Fermilab Internal Report, April 1986.
13. K. Y. Ng, Fermilab Report TM-1389, 1986.
14. J. Crisp, Fermilab Internal Report, April 1986.
15. B. A. Prichard, J. E. Griffin, R. F. Stiening, and E. Wilson, Fermilab EXP. 74, 1975.
16. K. Y. Ng, in Proc. 2nd Conf. on the Intersection between Particle and Nuclear Physics, Lake Louise, May 26-31, 1986, p. 401.

RESONANT EXTRACTION AT THE TEVATRON

M. Harrison
Fermi National Accelerator Laboratory
P. O. Box 500, Batavia, IL 60510

TABLE OF CONTENTS

1 Introduction... 2010
2 Nonlinear Resonant Behavior................................. 2011
3 Machine Aperture Considerations............................. 2017
4 Tevatron Extraction Elements................................ 2019
5 Operational Considerations.................................. 2022
   5.1 Slow Spill... 2022
   5.2 Fast Spill... 2026
6 Fast-Spill Simulation....................................... 2027
7 Energy Deposition... 2030

Chapter 9   RESONANT EXTRACTION AT THE TEVATRON

M. Harrison
Fermi National Accelerator Laboratory
P. O. Box 500, Batavia, IL 60510

1 INTRODUCTION

At the most basic level, extraction is the process of persuading a circulating beam to leave the machine in a controlled fashion. This is accomplished by deflecting a certain proportion of the circulating beam away from the central machine orbit and into the start of the extraction channel, from which point it is then shaped and split before entering the experimental areas. Deflecting the beam into the extraction channel is achieved by moving the beam across a thin electrostatic septum. Several potentially applicable techniques are available for doing this. For a variety of reasons the method we chose to implement at the Tevatron was that of horizontal half-integer resonant extraction. Under this scenario the beam is slowly moved into a nonlinear resonance where the incoherent betatron amplitudes grow in a controlled way from turn to turn until the amplitude is large enough for deflection into the extraction channel.

Another extraction-related effect we shall consider is the not so obvious one of beam losses associated with the extraction process. In conventional accelerators, beam losses pose no operational problems beyond that of residual radioactivity. In the Tevatron this is manifestly not the case. Appreciable levels of energy deposition in the superconducting magnet coils can be caused by relatively small fractions (<0.1%) of the circulating beam impinging on the magnets. This effect is most severe during fast extraction, where the instantaneous rate of energy deposition is several orders of magnitude greater than during slow spill. This energy can heat the superconductor above the transition temperature, causing the superconductor to go normal and the element to quench. The resultant deposition of the internal stored energy causes a rapid rise in temperature, and it can take from 20 minutes to several hours to cool the magnet again. Magnet quenches are thus very disruptive to the machine operation, and an accurate understanding of the beam loss mechanism is vital in avoiding them.

In this chapter we start by examining the resonant process. We then show the effect of the finite machine aperture on extraction efficiency. The specific layout of the extraction elements in the Tevatron is given, together with how they are used. We then consider the operational characteristics of the system. Results from a Monte Carlo simulation of fast extraction are presented. We conclude with a brief discussion of measurements and calculations of energy deposition in the superconducting magnets resulting from the beam losses incurred during the fast spill.

© 1989 American Institute of Physics

2 NONLINEAR RESONANT BEHAVIOR

Bringing the beam into a resonant state can be easily accomplished by raising the machine tune from the nominal operating point of $\nu_x = 19.42$ towards the half-integer stopband ($\nu_x = 19.5$). This, however, is a linear resonance since it is produced by quadrupoles, and thus the beam is either all stable or all unstable. To produce a controlled beam spill, it is necessary to separate the phase space into stable and unstable regions, with the unstable phase-space trajectories following definite paths. This condition is satisfied by the introduction of octopoles into the machine lattice. These elements (of strength k) produce a magnetic field B_y, given by

$$B_y = kx^3 ,$$

and hence a tune change of

$$\Delta\nu_x \propto \frac{dB_y}{dx} \propto x^2 .$$

The particle tune is now a function of betatron amplitude, and, in the situation where the tune is close to 19.5, the small amplitude particles are stable and the large amplitude ones are not.

To present in more detail what this all means, we first define in a crude way what we mean by the half-integer stopband and then consider the effect of introducing octopoles.

Starting with a linear machine, we introduce two quadrupoles of opposite polarity on opposite sides of the ring (see Fig. 1).

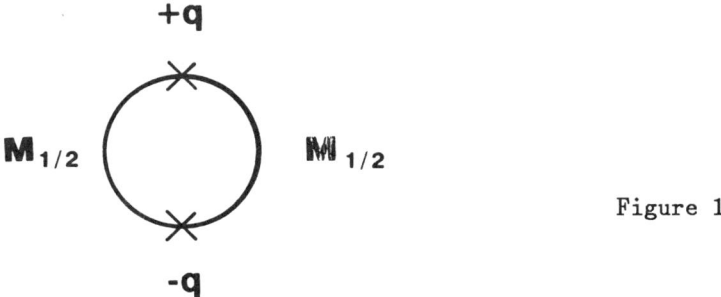

Figure 1

The one-turn transfer matrix M (starting from the center of the quadrupole $a = 0$) is now given by

$$M = M_{+-}M_{-+} = \begin{pmatrix} 1 & 0 \\ q/2 & 1 \end{pmatrix} M_{1/2} \begin{pmatrix} 1 & 0 \\ -q/2 & 1 \end{pmatrix} \begin{pmatrix} 1 & 0 \\ -q/2 & 1 \end{pmatrix} M_{1/2} \begin{pmatrix} 1 & 0 \\ q/2 & 1 \end{pmatrix}$$

where $M_{1/2}$ is the unperturbed half-turn transfer matrix given by

$$M_{1/2} = \begin{pmatrix} \cos\pi\nu_0 & \beta_0 \sin\pi\nu_0 \\ \dfrac{-\sin\pi\nu_0}{\beta_0} & \cos\pi\nu_0 \end{pmatrix}$$

then

$$M_{+-} = \begin{pmatrix} \cos\pi\nu_0 - \dfrac{q}{2}\beta_0 \sin\pi\nu_0 & \beta_0 \sin\pi\nu_0 \\ \dfrac{-\sin\pi\nu_0}{\beta_0} - \dfrac{q^2}{2}\beta_0 \sin\pi\nu_0 & \cos\pi\nu_0 + \dfrac{q}{2}\beta_0 \sin\pi\nu_0 \end{pmatrix}.$$

Hence $M = M_{+-}M_{-+}$ is given to lowest order in q by

$$M = \begin{pmatrix} \cos 2\pi\nu & \beta \sin 2\pi\nu \\ \dfrac{-\sin 2\pi\nu}{\beta} & \cos 2\pi\nu \end{pmatrix}$$

$$= \begin{pmatrix} \cos 2\pi\nu_0 - \dfrac{q^2}{2}\beta_0^2 \sin^2\pi\nu_0 & \beta_0 \sin 2\pi\nu_0 - q\beta_0^2 \sin^2\pi\nu_0 \\ \dfrac{-\sin 2\pi\nu_0}{\beta_0} + \text{many terms} & \cos 2\pi\nu_0 + \dfrac{q^2}{2}\beta_0^2 \sin^2\pi\nu_0 \end{pmatrix}.$$

We can now define the new tune of the machine by equating the (1,1) matrix elements

$$\cos 2\pi\nu = \cos 2\pi\nu_0 - \dfrac{q^2}{2}\beta_0^2 \sin\pi\nu_0 \qquad (1)$$

and close to the half integer say

$$\nu = \dfrac{1}{2} + \mu, \quad \nu_0 = \dfrac{1}{2} + \mu_0 \, ;$$

then

$$\sin\pi\nu_0 \simeq \sin\dfrac{\pi}{2} = 1$$

and

$$\cos(2\pi\nu) = \cos(2\pi\dfrac{1}{2} + 2\pi\mu) = -1 + \dfrac{1}{2}(2\pi\mu)^2 \, .$$

From Eq. (1)

$$-1 + \dfrac{1}{2}(2\pi\mu)^2 = -1 + \dfrac{1}{2}(2\pi\mu_0)^2 + \dfrac{1}{2}q\beta_0^2 \, ,$$

$$\mu^2 = \mu_0^2 - \left(\frac{q\,\beta_0}{2\pi}\right)^2$$

i.e. $\quad\quad\quad\quad\quad\quad \mu \to 0 \quad \text{as} \quad \left|\frac{q\,\beta_0}{2\pi}\right| \to \mu_0 \;.$

Thus we have the situation that the new tune is pulled onto the half integer ($\mu \to 0$) by these quadrupoles as their strength becomes comparable with μ_0, the unperturbed tune separation from the half integer. To lowest order then we can say that the quadrupoles have produced a half-integer stopband of width $\pm |q\beta_0/2\pi|$, and particles with a tune inside this stopband are unstable.

We can demonstrate that the lattice is unstable in this regime by equating the (1,2) matrix elements

$$\beta \sin 2\pi\nu = \beta_0 \sin 2\pi\nu_0 - \beta_0^2 q \sin 2\pi\nu$$

and making the same substitutions as before for the situation close to half integer; then

$$\frac{\beta}{\beta_0} = \left[\frac{\mu + \frac{q\beta_0}{2\pi}}{\mu_0 - \frac{q\beta_0}{2\pi}}\right]^{1/2},$$

hence, as $q\beta_0/2\pi \to \mu_0$, the β-values approach infinity.

At this point the alert student will undoubtedly remark that this is all well and good; however, real accelerators are not built with opposite polarity quadrupoles facing each other across the machine, but half-integer stopbands do exist without them--why?

The answer, of course, is that magnets are not perfect and small quadrupole errors exist all around the lattice. The error "wave" in the lattice due to a quadrupole propragates at twice the tune. In the Tevatron, which operates close to the tune of 19.5, these error waves have an effective "tune" of 39. To calculate the effect of the machine imperfections close to the half integer, we must sum up the quadrupole error terms lying on this 39th harmonic. The effective quadrupole components that contribute to the natural half-integer stopband have two phases and are given by

$$q_c = \frac{1}{B\rho} \oint \beta(z)\cos(39\Psi[z])\,\partial z$$

and

$$q_s = \frac{1}{B\rho} \oint \beta(z)\sin(39\Psi[z])\,\partial z \;.$$

As we shall see later, these are measurable quantities. The illustration we have just used with two quadrupoles is the simplest physical representation of these two ring-wide integrals.

The effect of a zeroth harmonic octopole field on the beam dynamics can be illustrated in a way similar to that used for the quadrupoles. Consider the case of a single octopole of strength k (see Fig. 2).

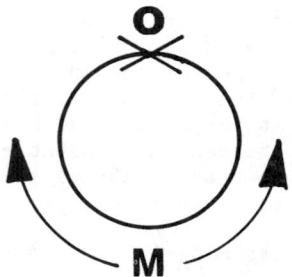

Figure 2

The effect of this element can be represented by a kick on each turn where $\Delta x' = kx^3$. The two-turn behavior of a particle is given by

$$\begin{pmatrix} x_2 \\ x_2' \end{pmatrix} = M^2 \begin{pmatrix} x_0 \\ x_0' \end{pmatrix} + M^2 \begin{pmatrix} 0 \\ kx_0^3 \end{pmatrix} + M \begin{pmatrix} 0 \\ kx_1^3 \end{pmatrix} \quad (2)$$

where M is the single-turn transfer matrix given by

$$\begin{pmatrix} \cos 2\pi\nu & \beta \sin 2\pi\nu \\ \dfrac{-\sin 2\pi\nu}{\beta} & \cos 2\pi\nu \end{pmatrix}$$

where again for algebraic simplicity we have set $\alpha = 0$ (by assuming we start and finish the mapping in the center of a quadrupole). Putting $\nu = 1/2 + \mu$ gives

$$M = \begin{pmatrix} -1 & -\beta\mu \\ \dfrac{\mu}{\beta} & -1 \end{pmatrix} .$$

Substituting in Eq. (2) and neglecting terms in k^2, we get

$$x_2 = x_0 + 2\beta\mu(x_0' + kx_0^3) - \beta\mu k x_1^3$$

$$x_2' = -\dfrac{2\mu x_0}{\beta} + x_0' + kx_0^3 - kx_1^3 .$$

Eliminating x_1 and keeping only first-order terms, we get

$$x_2 = x_0 + 2\beta\mu x_0' , \qquad x_2' = -\frac{2\mu x_0}{\beta} + x_0' + 2kx_0^3 .$$

Transforming to "normalized" coordinates gives

$$x = \frac{x}{\sqrt{\beta}} , \qquad x' = \sqrt{\beta} \; x' ;$$

then

$$x_2 = x_0 + 2\mu x_0' , \qquad x_2' = x_0' - 2\mu x_0 + 2k\beta^2 x_0^3 \qquad (\text{set } 2k\beta^2 = \lambda) .$$

The phase-space equations of motion (for two turns) are given by

$$\partial x = x_2 - x_0 = 2\mu x' , \qquad \partial x' = x_2' - x_0' = -2\mu x + \lambda x^3 ,$$

valid close to the half integer, where ∂x, $\partial x'$ are small.

Let us now consider the function

$$H = \frac{\mu}{2} (x^2 + x'^2) - \frac{\lambda}{8} x^4 .$$

This is a "Hamiltonian" of the system in the sense that, by construction, it is a constant of the motion, and it reproduces the equations of motion:

$$\partial x = \frac{dx}{dt} \times \Delta t = \frac{\partial H}{\partial x'} \times \Delta t = 2\mu x' \quad (\text{note } \Delta t = 2 \text{ for two turns})$$

$$\partial x' = \frac{dx'}{dt} \times \Delta t = -\frac{\partial H}{\partial x} \times \Delta t = -2\mu x + \lambda x^3 .$$

From this Hamiltonian we can construct the phase-space trajectories (curves of constant H), as shown in Fig. 3. The large amplitude trajectories are unstable, the small amplitude ones stable. The stable region is bounded by the separatrix and has two fixed points defined by

$$\partial x = \partial x' = 0$$

which is

$$x' = 0 , \qquad x = \pm \sqrt{\frac{2\mu}{\lambda}} .$$

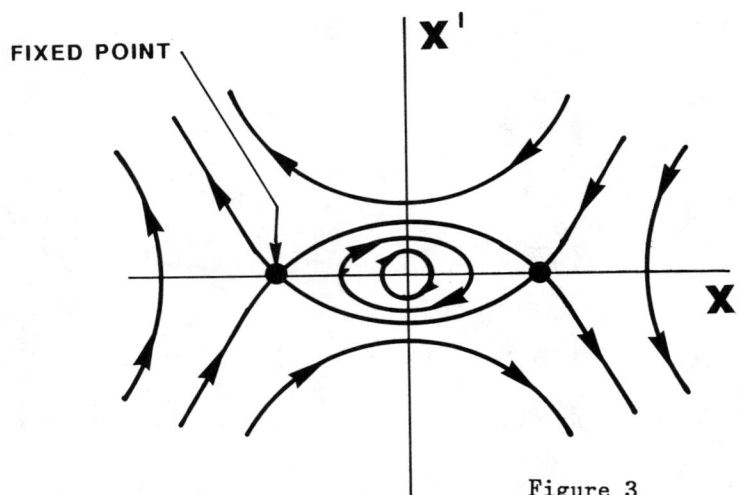

Figure 3

The position of the fixed points (i.e. the area of the stable region) is thus defined by the proximity to the half integer (μ) and the strength of the octopole term (λ). The equations of motion for the separatrix at small amplitudes are dominated by the quadratic terms in the Hamiltonian and are given by the intersection of two circles.

The equations of motion in the presence of both octopoles and quadrupoles can be calculated in a fashion identical to that already employed, although the algebra is tedious. The resulting motion in phase space is given by

$$\partial x = 2\mu x' + 2qx' , \qquad \partial x' = -2\mu x + 2qx + \lambda x^3 .$$

In this case the fixed points are

$$x' = 0 , \quad x = \pm \sqrt{\frac{2(\mu - q)}{\lambda}} .$$

Instead of the stable phase-space region shrinking to zero as $\mu \to 0$, we have $x \to 0$ as $|\mu - q| \to 0$, i.e. a finite stopband width. The phase-space trajectories at small amplitudes are still given by the intersection of two circles although the expressions are modified by the 39th harmonic quadrupole term. The particle oscillation amplitude is increasing monotonically from turn to turn in the unstable region, and the two-turn change in amplitude is given by

$$\sqrt{\partial x^2 + \partial x'^2} ,$$

which for small amplitudes (quadrupole dominated) is linear.

The physical situation we have produced in this example is that of a central stable phase space which is reduced in size by either raising the machine tune close to the half integer or increasing the stopband width. When this stable region is too small to contain the beam (i.e. smaller than the beam emittance) particles are driven into the unstable phase-space region along well-defined trajectories. Here the particles oscillate from side to side with increasing amplitude. The particle density distribution along these trajectories is decreasing inversely with amplitude as the rate of change of the particle amplitude increases.

3 MACHINE APERTURE CONSIDERATIONS

In the preceding section we finished with the unstable particles executing betatron oscillations with an ever-increasing amplitude vector. At some point in this process the amplitude of these oscillations becomes large enough for the particles to jump across (or hit) the electrostatic septum, whereupon they are deflected into the extraction channel. During this process the particle phase-space trajectory must lie on that defined by the quadrupoes and octopoles. In order for this to be the case, the rest of the machine lattice needs to be linear out to the maximum amplitude. In the Tevatron the linear region of the magnetic field is defined by the coil diameter of 75 mm in the dipoles and corresponds to an amplitude of ±20 mm in the standard cells (β_{cell} = 100 m).

The cryogenic nature of the machine means that the extraction septa and the extraction channel must be placed in the warm regions of the machine, i.e. the straight sections. The electrostatic septum is located on the side of the machine (D0) opposite the extraction channel (A0) with a fractional phase advance of $3\pi/2$ between the two. The phase-space picture of the beam at the electrostatic septum looks like that shown in Fig. 4, where x_s is the septum offset from the closed orbit and Δx is the maximum gain in position in two turns, i.e. $x_s + \Delta x$ represents the maximum

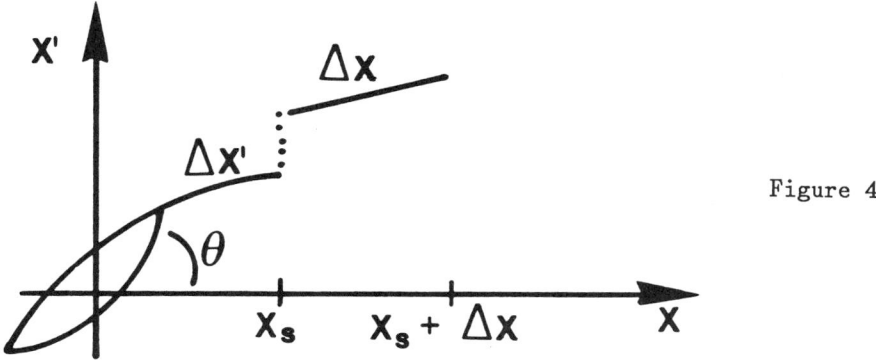

Figure 4

amplitude particle. The kick from the electrostatic septum $\Delta x'$ is a shear in phase space. The relationship between the maximum amplitude in the standard cell (x_{max}) and at the septum is given by

$$x_{max} = \frac{x_s + \Delta x}{\cos\theta} \left(\frac{\beta_{cell}}{\beta_{septum}}\right)^{1/2} .$$

In the standard straight section in the Tevatron, then, β_{septum} = 50 m and θ = 35°, which, together with x_{max} = 20 mm and β_{cell} = 100 m, gives a value of 11.5 mm for $x_s + \Delta x$ (the effective machine aperture at the septum).

The number of particles that hit the septum, the extraction beam losses, are related to this 11.5 mm of aperture in this region, as can be demonstrated by considering the (steady-state) particle density distribution as a function of amplitude. The situation looks something like that shown in Fig. 5.

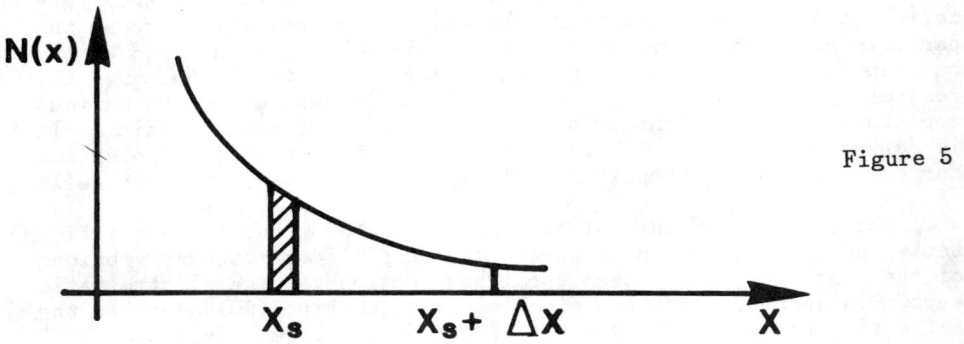

Figure 5

The number of particles that hit the septum is just given by the product of the effective septum width w and the density distribution at the septum $N(x_s)$. It is conventional to express these losses as a percentage of the total beam, i.e.

$$\xi = \frac{w \times N(x_s)}{\Delta x \times \int_{x_s}^{x_s + \Delta x} N(x)\partial x} ,$$

which is just the ratio of those that hit the septum to those that don't.

If we assume a linear distribution for $N(x)$, which corresponds to quadrupole domination at small amplitudes, then

$$\int_{x_s}^{x_s + \Delta x} N(x)\partial x = \frac{N(x_s) + N(x_s + \Delta x)}{2} .$$

The minimum value for ξ is given when $x_S = \Delta x$ and hence

$$\xi_{min} = \frac{2w}{\Delta x} \ .$$

A well constructed electrostatic septum has an effective thickness of 0.1 mm; then, for $x_S = \Delta x = 5.25$ mm, $\xi = 3.8\%$. The effect of a limited machine aperture thus creates the condition of high extraction losses.

It became apparent at an early stage in the Tevatron design that magnet quenching due to extracted beam loss was potentially a major problem and 3.8% was too high a value for the losses. Increasing the effective linear aperture of the machine to accomodate the extraction system was solved neatly by a modification to the machine lattice that raised the value of β_{septum}. This was the so-called high-beta straight section. By changing the focussing strength and polarities of the quadrupoles around the straight section, a local lattice modification that raised β_{septum} from 50 m to 225 m was found.

The effect of this change can be found by substituting $\beta_{septum} = 225$ m in the equations. Under these conditions, $x_S + \Delta x = 24.5$ mm and the extraction losses become 1.6%. This is equivalent to increasing the linear machine aperture to ±35 mm.

4 TEVATRON EXTRACTION ELEMENTS

The main elements of the Tevatron extraction are as follows: (a) the tune quadrupoles, (b) the 39th harmonic quadrupoles, (c) the zeroth harmonic octopoles, (d) the slow feedback quadrupoles, (e) the fast feedback quadrupoles, (f) the splitting septum, and (g) the extraction channel.

The tune quadrupoles are part of the correction elements built into the spool pieces. The machine tune can be varied over a range of ±1.0 units. This range of tune adjustment is determined by the requirements of the collider program rather than the fixed-target operation.

The 39th harmonic quadrupoles are split into two circuits (cosine and sine). Their functions are to correct the natural harmonics of the machine and also to provide part of the driving term for the extraction process. As shown above, the unwanted dipole field harmonics give rise to a random half-integer stopband. This natural stopband can be compensated by these two circuits. This was indeed done as part of the commissioning of the extraction system. The data from the stopband measurement are summarized in Fig. 6.

Using the two harmonic circuits (Q1 and Q2) independently, a low intensity beam was brought into resonance (no extraction, only beam loss) from a fixed tune offset from the half integer. Both positive and negative values for both circuits were measured. In the absence of any nonlinearities in the beam dynamics, it is easy

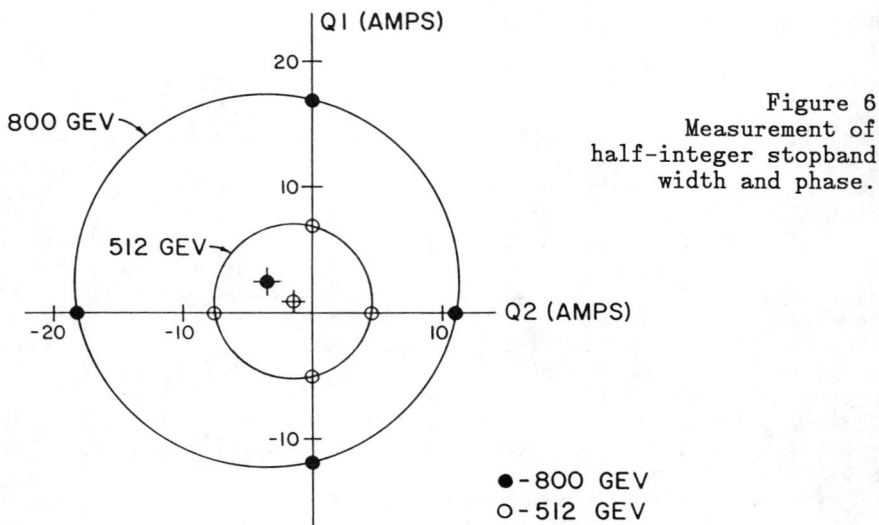

Figure 6
Measurement of half-integer stopband width and phase.

to see that these four points ought to lie in a circle. The origin of this circle defines the width and phase of the natural stopband and hence the quadrupole strengths needed to correct them. The data shown in Fig. 6 were taken at two energies (512 and 800 Gev); the change in the origin of the circle reflects the fact that the dipole field harmonics change with excitation. Each circuit has eight quadrupoles with a strength of 60 kg-in. at 1 in. at full current (50 A).

The zeroth harmonic octopole circuit consists of 36 individual spool piece elements around the ring. The individual octopole strength is 30 kg-in. at 1 in. at 50 A.

The slow feedback quadrupoles are air-core devices at four locations in the ring arranged on the 39th harmonic. The output waveform for these elements is generated by the extraction microprocessor (QXR) and is used to control the overall features of the extraction cycle. The operation cycle during the 1985 800-Gev fixed-target run looked like Fig. 7.

The 20-s slow spill was interspersed with three 3-ms pulses of fast beam (~2×10^{12} ppp) spaced 9.5 s apart. At the start of the flat-top the computer samples the circulating beam intensity via a current toroid in the ring. From this number and the desired extraction parameters (slow-spill length, slow-spill intensity, fast-spill intensity, number of fast-spill pulses, etc.) the computer calculates an ideal beam signal, i.e. the circulating beam intensity signal for a perfect spill. During the extraction process the circulating beam intensity signal is continuously sampled at 720 Hz, and the difference between the ideal beam signal and the measured one is used to generate an error signal that is applied to

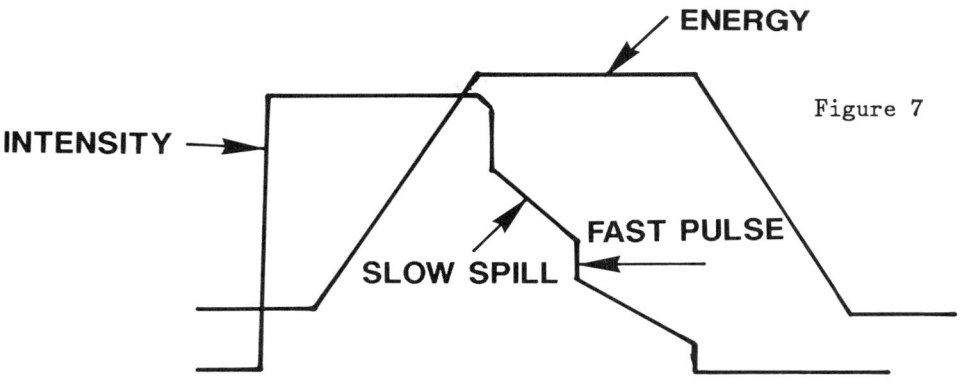

Figure 7

the output waveform of the extraction quadrupoles. This output waveform, which has a theoretical bandwidth of 360 Hz, is heavily filtered to reduce potential noise problems so that the effective bandwidth of this real-time feedback circuit is reduced to 30 Hz. Besides this real-time signal, a cycle-to-cycle learning process is also implemented. At the end of each machine cycle the output waveform of the quadrupoles is still resident in the computer memory. This waveform is digitally filtered to smooth any noise spikes and then combined with the results from earlier cycles to give the projected waveform for the next cycle. In this way any repetitive features in the machine behavior are incorporated into the quadrupole waveforms ("learned"), and under stable operating conditions the real-time error signal can be quite small. The spill feedback system is shown schematically in Fig. 8.

LEARNING INCLUDES
1) I CYCLE DELAY
2) PHASE SHIFT
3) NEIGHBOR GAIN
4) LEARNING RATE

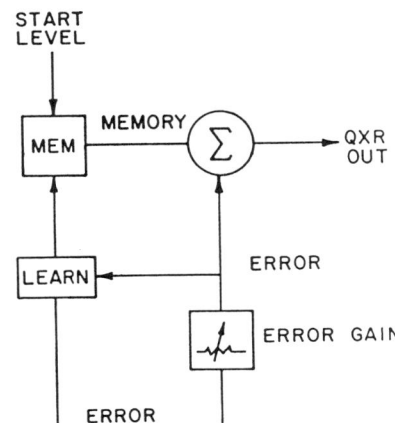

ERROR = $I_{BEAM} - I_{IDEAL}$

Figure 8
Spill feedback system.

The fast feedback quadrupoles are air-core elements located next to the slow quadrupoles in the machine lattice. The main difference between the two systems is that the strength of these devices is an order of magnitude less than that of the slow system, and the bandwidth is much higher (~3 kHz). The output waveforms are generated in a way similar to that in the slow system (cycle-to-cycle learning plus a real-time error signal), but the role of this system is to attempt to linearize the slow spill. This aspect will be examined in more detail in the next section.

The splitting septum consists of two modules located in the upstream end of the D0 straight section, as shown in Fig. 9. Also located in this region are some small bump dipoles for orbit control, a set of motorized collimators, and four Main Ring dipoles that produce a 50-mm orbit bump within the straight section. A cross-section of the "business" end of the septum is shown in Fig 10. The anode consists of 0.05-mm-diameter tungsten-rhenium wires, spring loaded, with a spacing of 2.5 mm. The wire plane is 4.66 m long and is straight to within 0.05 mm. The cathode is made of titanium. The high voltage gap is 14 mm and is designed to operate up to a maximum field of 7.5 kV/mm. The alignment between the circulating beam closed orbit and the septum wires is adjusted by means of motorized supports at each end of each module in the horizontal plane. The septum alignment is determined empirically by minimizing the amount of beam that strikes the wires.

The extraction channel (Fig. 11) is located in the A0 straight section. The first five elements in the channel are Lambertson magnets (magnetic septa). The circulating beam is in the field-free region of the magnet. The splitting septum deflects the extracted beam into the dipole region, where it is bent vertically upwards. Downstream of the magnetic septa the beam separation is sufficient (175 mm) to allow the extracted beam to enter a string of three superconducting dipoles. Rotated by 19° from the vertical plane, these magnets deflect the beam down and out. At this point the extracted beam enters the switchyard, where it is split and sent to the various experimental areas.

5 OPERATIONAL CONSIDERATIONS

The realization of an operational extraction system is never quite as straightforward as one would prefer. In the preceding sections we examined the general features of the half-integer extraction scheme. As a concession to reality we indicate here some of the problems encountered during the implementation and commissioning of the system.

5.1 Slow Spill

The aim of the slow extraction system is to maintain a smooth constant rate of extracted beam throughout the cycle. This is provided by moving the beam smoothly through the resonance, or in this case the resonance through the beam. One can represent the particle distribution in tune space as in Fig. 12.

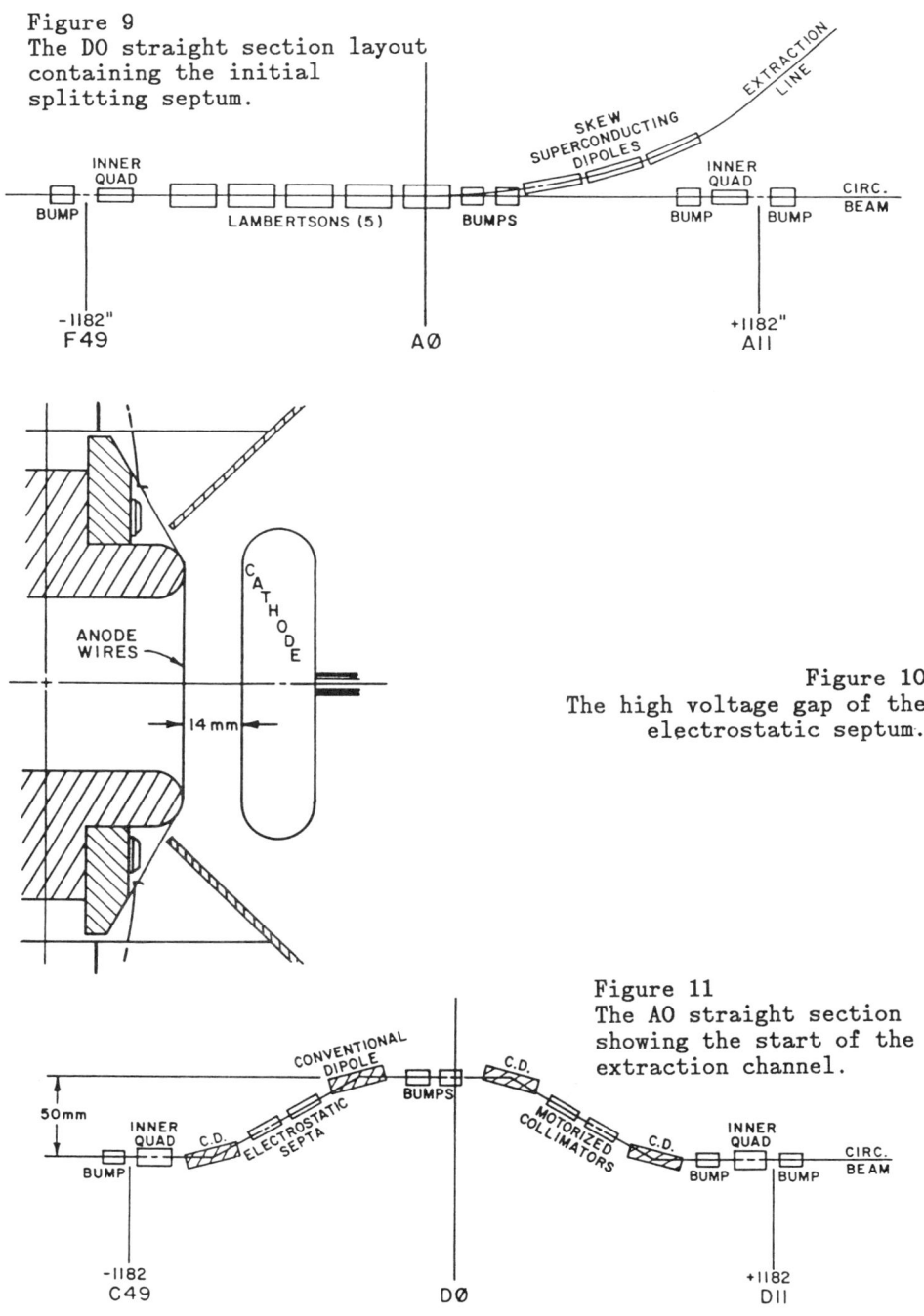

Figure 9
The D0 straight section layout containing the initial splitting septum.

Figure 10
The high voltage gap of the electrostatic septum.

Figure 11
The A0 straight section showing the start of the extraction channel.

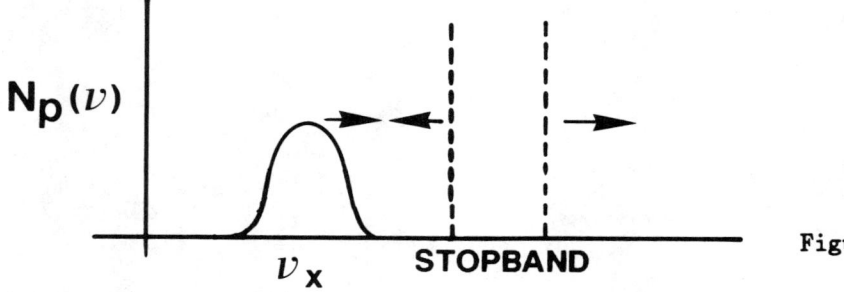

Figure 12

The tune spread inside the beam is generally assumed to be Gaussian and arises from the finite energy spread of the beam (chromatic effects), the space charge within the bunch (intensity effects), and effects due to guide-field nonlinearities that produce tune variations as a function of betatron amplitude (closed orbit and emittance effects). The spill rate is given by the number of extracted particles per unit time:

$$\frac{\partial N}{\partial t} = \frac{\partial N}{\partial \nu} \times \frac{\partial \nu}{\partial t} = \frac{\partial N}{\partial \nu} \times \dot{\nu} .$$

The rate of change of the tune $\dot{\nu}$ can be expressed as $\dot{\nu} = (\dot{\nu}_0 + \dot{\nu}_n)$ where $\dot{\nu}_0$ is the constant term from the extraction system, and $\dot{\nu}_n$ is the unwanted tune modulation inside the machine. The spill rate can thus be expressed by

$$\frac{\partial N}{\partial t} = \frac{\partial N}{\partial \nu} \times \dot{\nu}_0 (1 + \frac{\dot{\nu}_n}{\dot{\nu}_0}) \equiv K(1 + \lambda)$$

where

$$K = \frac{\partial N}{\partial \nu} \times \dot{\nu}_0$$

(assumed to be constant for the sake of simplicity), and

$$\lambda = \frac{\dot{\nu}_n}{\dot{\nu}_0} .$$

It is conventional to talk about the spill duty factor (SDF) where

$$SDF = \frac{\left(\sum_{i=1}^{n} c_i\right)^2}{n \times \sum_{i=1}^{n} c_i^2} = \frac{\langle C \rangle^2}{\langle C \rangle^2} \qquad \langle \rangle \equiv \text{expectation value}$$

where C_i is the number of counts in the spill monitor per unit time [$= K(1 + \lambda)$], and n such samples are taken during the spill. In this situation

$$\langle C \rangle = K, \qquad \langle C^2 \rangle = K^2(1 + 2\langle\lambda\rangle + \langle\lambda^2\rangle), \qquad \langle\lambda\rangle = 0.$$

Then

$$\text{SDF} = \frac{K^2}{K^2(1 + \langle\lambda\rangle^2)} = \frac{1}{1 + \langle\lambda^2\rangle}.$$

In the absence of any noise, $\lambda = 0$ and SDF = 1. In the situation with a large amount of noise ($\dot{\nu}_n$ big) or with $\dot{\nu}_0$ small (long spill times), the SDF will be reduced from the ideal value. During the 1985 fixed-target operation the spill length was 20 s and, sampled at a 5760-Hz rate, produced an SDF of 0.85 to 0.95 during normal operation (10^{13} ppp).

This SDF was achieved by using the fast feedback system and also by identifying and fixing, where possible, the source of the tune modulation. Three major categories of problems were identified during the commissioning: (a) power supply ripple and regulation, (b) various rf phenomena, and (c) mechanical vibration. By injecting a well-defined noise signal into the power supplies, it is possible to get a quantitative measure of a power supply's contribution to the spill. This was done for every major power supply in the Tevatron, revealing two elements that needed attention. Modifications were also made to the system to reduce common-mode ripple. Two spill modulation components were identified in the rf system. One was caused by two oscillators in the low level rf beating against each other. This was eliminated by only using one oscillator during extraction. The other was due to coherent synchrotron oscillations inside the bucket and was alleviated by using bunch spreaders to make the density inside the bucket more uniform. Mechanical vibration was found to be responsible for 4.62-Hz and 19.8-Hz modulations. The 4.62-Hz signal was caused by the central helium liquifier's 4000-HP, 277-rpm motor shaking the bedrock some 200 m from the accelerator tunnel, causing the elements to vibrate with ~20-μm amplitude. By adjusting the accelerator cycle time to be a multiple of 13/15 s these oscillations were made repeatable on a cycle-to-cycle basis, and the learning algorithm was able to reduce the effect substantially. The 19.8-Hz component was traced to vacuum roughing pumps mounted on quadrupoles in places of slow vacuum leaks. The signal was asynchronous. Changing the support stands reduced but did not completely remove the problem.

At this point the remaining spill modulation was dominated by the remaining line frequency components in the power supplies (60, 120, 180, 360, etc.). Since the machine cycle is line locked, it was hoped that the learning system would remove these effects. This

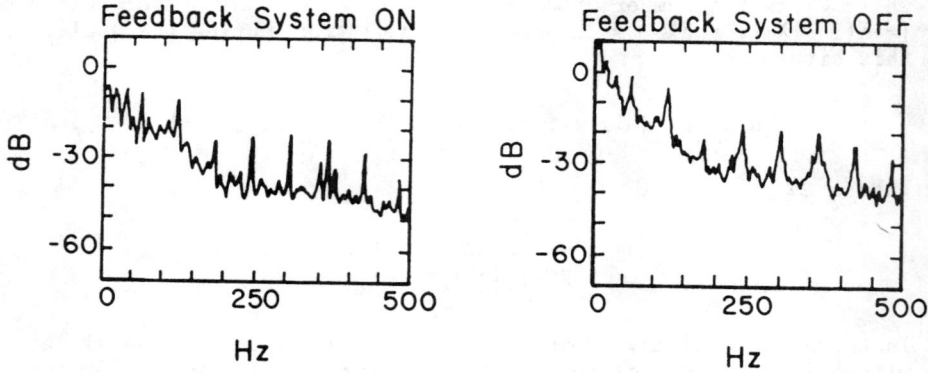

Figure 13 Beam spill frequency spectrum
(averaged over 20-sec spill).

proved to be the case for the lower frequency components (60- and 120-Hz) but not the higher ones. This was due to the transfer function of the machine being dispersive, and to the presence of phase shifts >180° at the higher frequencies, which reduce the effectiveness of learning. The beam spill frequency structure is shown in Fig. 13 with and without the feedback system. The noise reduction corresponds to an increase of ~7% in the SDF.

5.2 Fast Spill

The neutrino experimental program places a requirement on the extraction system for short duration, high intensity pulses of beam. These are provided by an interrupt-driven sequence of events programmed into the extraction-controlling microprocessor and can occur at any point during the slow spill with any reasonable spacing (>1 s). The only requirement is that slow spill be occurring, since the feedback quadrupole current at the time the interrupt is received is assumed to define the resonant condition. The actions taken by the system are as follows:

(1) The slow spill is halted by decrementing the quadrupole currents by a known amount.
(2) A pause (30 ms) is initiated to allow steering elements in the switchyard to direct the beam towards the neutrino area.
(3) Two of the (four) slow feedback quadrupoles receive a fast (5-ms) high current (~300-A) pulse to extract the beam.
(4) There is another pause (250 ms) while the switchyard is reestablished and the loss monitors in the ring recover from the high instantaneous losses (losses of this magnitude would cause an abort condition at any other time in the cycle; the interrupt that initiates fast spill also changes the abort settings on the loss monitors).
(5) Slow spill is turned back on.

The operational problems with this system, besides beam losses, are pulse-to-pulse intensity variations and restarting the slow spill. The beam intensity in a pulse is very sensitive to the amplitude of the current pulse in the quadrupoles since we are removing only a small percentage of the total beam. Since the beam is extracted quickly enough that real-time feedback cannot be used, the current pulse is calculated from the preceding cycles. The inherent error in this technique leads to a 10% intensity variation during stable operation. The total amount of fast beam extracted in a given cycle is held constant, however, by adjusting the last pulse, which takes all the remaining beam out of the machine at the end of the cycle to compensate for pulse intensity variations within that cycle.

Reestablishing slow spill smoothly after a fast pulse has also proved difficult. Non-adiabatically shrinking and then enlarging the stable phase area appears to result in a phase-space distribution that is more diffuse than before, and large spikes on the slow spill would result while the feedback systems stabilized. The problem proved difficult to solve completely, but was reduced considerably by redefining the start of the slow spill to have a short (100-ms) linear rise, which is equivalent to asking for a parabolic waveform from the quadrupoles.

6 FAST-SPILL SIMULATION

The time scale on which the fast-spill extraction takes place (500 turns) is small enough that a computer simulation of the complete process without any approximations is feasible in terms of computer time. The code used to do the simulation is a kick code in which the particle ensemble (~500) is propagated from element to element. The effect of the dipole field harmonics is calculated at the center of each magnet and an appropriate kick given to the particle. Subdividing the dipoles into smaller segments produced no discernable difference in the results, as did the inclusion of the quadrupole harmonics and changing from single to double precision. The correction elements in the spool pieces were faithfully reproduced to provide the same correction circuits as those available on the accelerator, and were used to manipulate the tune and chromaticity as well as the extraction mechanism. The skew quadrupole circuit was used to minimize the orbit coupling. The dipole harmonics were generated randomly with Gaussian distributions that corresponded (mean and sigma) to the measurement data. No shuffling scheme was applied, but the magnet selection procedure was approximated by truncating the distributions at 3σ. Different random number seeds required slightly different settings of the correction elements, but once these corrections had been established the beam dymanics remained essentially constant. The electrostatic septum is included in the program. The tracking of the particles ceases when they reach the entrance to the extraction channel.

The flow of the program is as follows:
(a) Generate initial Gaussian phase-space distributions (x,x', y,y', D_p) corresponding to the measured machine parameters.

(b) Track the particle for ~100 turns to establish initial conditions.
(c) Turn on the slow-spill extraction elements and bring the beam close to resonance; track for another 100 turns.
(d) Pulse the extraction quadrupoles to initiate fast spill.
(e) See what happens.

The results of the simulation are illustrated in Figs. 14 and 15. Figure 14 shows the beam spill produced by a 3-ms half-sine-wave pulse on the extraction quadrupoles. All of the beam is extracted from the machine in 1.4 ms. The beam losses are calculated by observing the number of particles that hit a 1-mm septum (to increase statistics) and then dividing by 10 to give an effective 0.1-mm width.

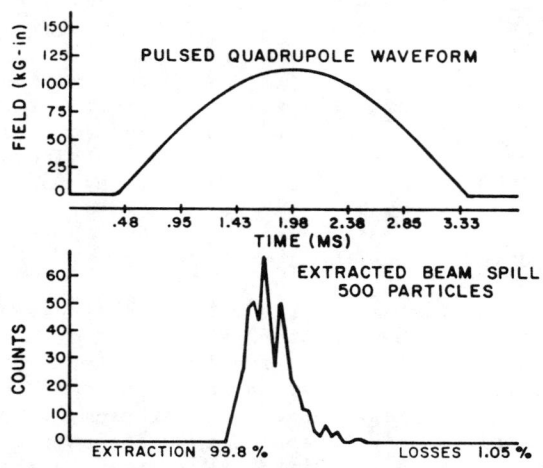

Figure 14
Simulated fast extraction beam spill.

The phase-space evolution during this pulse is shown in Fig. 15. A sample is taken at four different times during the cycle. The initially circular phase space is progressively deformed until all the beam is extracted (or in the process thereof). The pictures are taken at the entrance to the extraction channel, and the ~5-mm split between the circulating and deflected beams is sufficient to allow the two to pass on opposite sides of the string of magnetic septa with a position and alignment tolerance of ~1 mm.

It is of obvious interest to attempt to verify the results of this simulation. Since direct measurement of the phase-space development is not possible, indirect methods must be used. The most straightforward of these uses the fact that the program predicts the phase space of the extracted beam. This phase-space distribution can then be projected in the external beam lines to give a predicted beam profile at the locations of the beam monitors that measure the profile. A comparison between the measured and predicted horizontal beam profiles at the upstream end of the extraction line is shown in Fig. 16. The overall beam width and the asymmetric profile agree well with the measurements.

Less direct comparisons involving internal machine apertures and element strengths also give good agreement with the predictions.

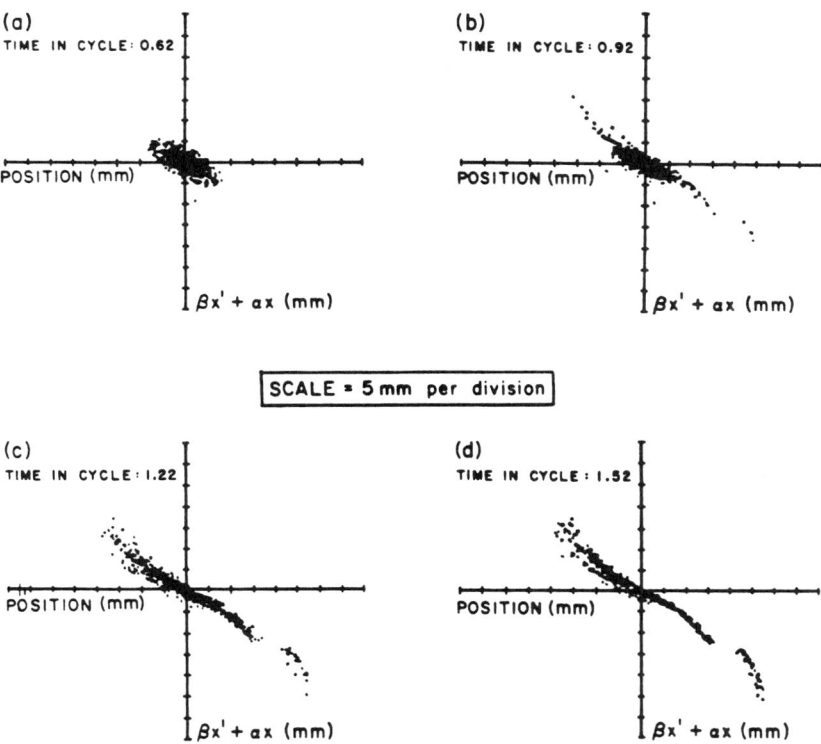

Figure 15 Simulated fast extraction phase-space evolution.

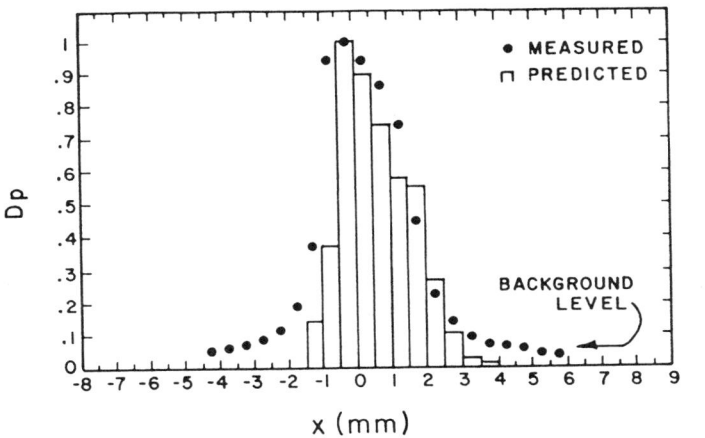

Figure 16
Extracted
beam profiles,
800 GeV.

7 ENERGY DEPOSITION

It was recognized at an early stage in the Tevatron design that energy deposition in the superconducting magnets arising from the fast-spill beam interacting with the electrostatic septum could seriously compromise fixed-target operation. This observation prompted a series of calculations undertaken in order to understand and rectify the problem. The highly detailed nature of this work, which involves an admixture of particle tracking in the machine structure and hadronic and electromagnetic shower production in the elements of the machine, is beyond the scope of this report. The design and operation of the extraction system was strongly influenced by this work, however, and since it represents one of the unique aspects of the Tevatron we shall indicate the more general features of the calculation and the results. The major part of the analysis, involving the energy deposition, was done using the 3-D cascade program CASIM, developed over several years at Fermilab by A. VanGinneken. This energy deposition in the magnets raises the local temperature in the superconducting coils and can cause the magnet to quench. This process is sensitive to beam energy, magnetic field, and the rate at which the heat load is applied.

The calculation proceeds as follows:

(1) The initial phase-space distribution of the beam that strikes the septum is obtained from the output of the fast extraction Monte Carlo program.

(2) These particles are allowed to interact with a homogenized electrostatic septum (i.e. the individual wires are replaced with a uniform density distribution). The resulting spray of particles from the septum is separated into the inelastics (20%) and the elastics.

(3) The inelastics are distributed locally in the accelerator over the straight section and the first two half-cells in the lattice. Individual particles are tracked until they interact in some element of the machine, and the resulting shower is propagated until the complete energy flow is determined. The machine layout is faithfully reproduced inside the computer, including the mechanical details of the region and the magnetic fields in the bodies of the magnets.

(4) The small-angle elastically scattered particles typically stay inside the accelerator for a significant distance. A ray-tracing program is used to propagate each one around the machine lattice. Aperture checks are performed at the front and back of each element in the ring. Particles failing the aperture check are retraced to determine their interaction point, and then the energy deposition program is used. The responses of the nearest beam-loss monitors in the ring are also calculated as part of this sequence. The particles are tracked until they either leave the machine via the extraction channel or interact with some element of the accelerator.

Of the particles that interact in the septum, the global loss distributions are as follows:
   20% inelastics are lost in or around the D0 straight section,
   10% are lost in the first half-turn between D0 and A0,
   40% are extracted on the first turn,
   5% interact in the extraction channel magnets,
   25% are extracted on the third turn.
The particles in the first two categories are the ones that produce operational problems.

The superconducting elements immediately downstream of the septum are protected from the large losses in this region, a series of four conventional dipoles that create a closed orbit deformation internal to the straight section, as shown in Fig. 9. The effect of the magnets is to sweep out the negatives and low energy positives. The neutrals are, of course, unaffected by the field, but they are directed by this closed orbit bump into the return yoke of one of these conventional magnets. Only 7% of the inelastics actually reach the superconducting magnets with this design, with the heat load peaking in the third and fourth dipoles in the first cell. This proved to be sufficient rejection that quenching in this region has never occurred.

A ring-wide loss monitor profile during fast extraction is shown in Fig. 17. The high losses in the extraction straight

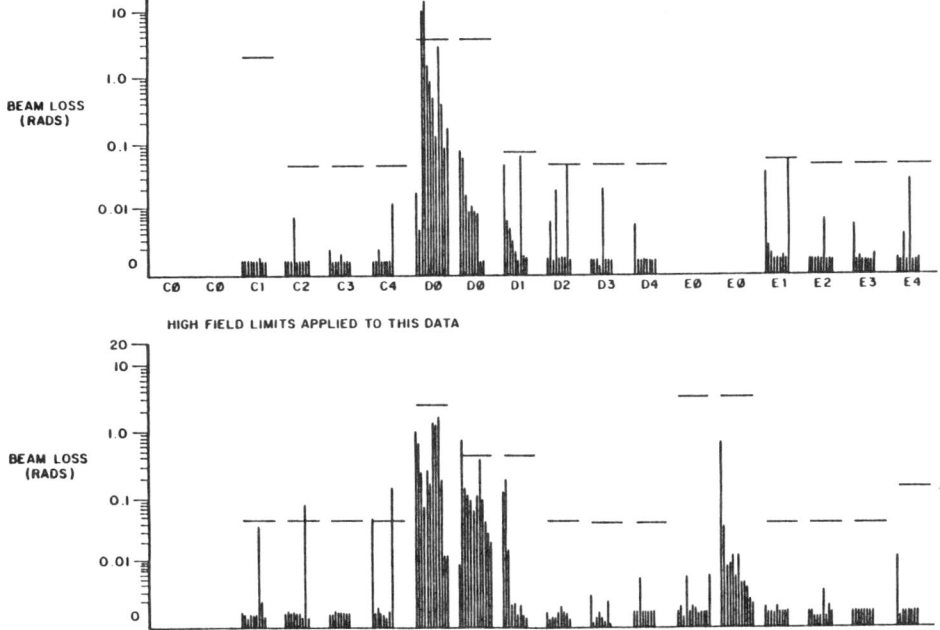

Figure 17 Ring-wide fast extraction losses.

sections are apparent. Besides these areas, well-defined (and reproducible) loss points occur in sectors D, E, and F. It was in these regions that operational problems were encountered. The abort trip levels on the loss monitors are also shown in Fig. 17; the data are taken from a fast extraction quench in the last cell of the F4 house, where these levels were deliberately exceeded. The calculations show that the beam should be lost in about fifteen locations, in excellent agreement with the data. The magnitude of the losses proved more difficult to calculate. The predictions at any given location would vary by up to a factor of 10 for 1-mm changes in the closed orbit. Since we know the relative alignment of the dipoles with respect to the position monitors to about 1 mm, there seems to be a fundamental limit to the accuracy of the calculations, arising from the mechanical tolerances of the machine. Another, rather obvious, point is that the magnitude of the losses in one location affects the readings around the rest of the ring. The losses at any given point can be reduced by moving the beam in the appropriate direction with the consequence of raising the losses elsewhere. The data in Fig. 17 had the closed orbit adjusted in an attempt to equalize the loss points around the three sectors.

The results of the calculations based on three controlled beam-induced quenches predict that 4 to 15 mJ per g of energy deposited locally in the magnet coils at 800 GeV will produce a magnet quench. These results are consistent with measurements made on prototype Tevatron magnets in an external beam, which produced estimates of ~5 mJ per g at this energy.

The breakdown of energy deposition in a typical magnet is as follows:
- 12% - beam pipe,
- 40% - superconducting coils,
- 18% - stainless steel coil collars,
- 4% - cryostat,
- 26% - warm iron.

BEAM LOSS

A. VanGinneken, D. Edwards, and M. Harrison
Fermi National Accelerator Laboratory
P. O. Box 500, Batavia, IL 60510

TABLE OF CONTENTS

1 Introduction.. 2034
2 Accelerator Geometry..................................... 2036
   2.1 Lattice.. 2036
   2.2 Component Description.............................. 2036
3 The Extraction Process................................... 2038
4 Beam on Septum... 2042
   4.1 Beam Phase Space................................... 2042
   4.2 Septum Model....................................... 2043
   4.3 Interactions of Beam in Septum..................... 2043
   4.4 Results of Septum Calculations..................... 2044
5 Particle Tracking.. 2048
6 Energy Deposition in Magnets............................. 2048
   6.1 Elastic Transport Through Magnet................... 2048
   6.2 Energy Deposition.................................. 2049
7 Beam Loss Monitor Response............................... 2050
   7.1 Response Function.................................. 2050
   7.2 Calculation of BLM Response........................ 2051
8 Results.. 2052
   8.1 Inelastics... 2052
   8.2 Elastics... 2061
9 Conclusions.. 2071
10 References... 2072

BEAM LOSS

A. VanGinneken, D. Edwards, and M. Harrison
Fermi National Accelerator Laboratory
P. O. Box 500, Batavia, IL 60510

1 INTRODUCTION

A basic problem in the operation of a superconducting accelerator is that of beam loss. Appreciable levels of energy deposition in the superconducting magnet coils result from relatively small fractions of the total circulating beam impinging on the magnets. This energy can heat the superconducting coil above transition temperature, causing it to go normal and the element to quench. The resultant deposition of internally stored magnetic energy causes a rapid temperature rise. Recovery from a magnet quench varies from about 20 minutes to several hours and is thus very disruptive to the accelerator operation.

The temperature change in the superconducting coil resulting from this energy deposition is related to the time structure of the beam loss. An instantaneous loss results in a temperature change determined by the specific heat of the conductor. A slow (>100 ms) uniform loss, on the other hand, results in an equilibrium condition between the superconducting coil and refrigeration system, with heat transfer taking place from the conductor through the cable insulation to the liquid helium. Between these two extremes one expects an intermediate loss condition where high heat transfers exist for time periods of the order of milliseconds.

The maximum tolerable temperature increase of the magnet coils (so as not to destroy the superconducting state) is related to the product of current in the conductor and magnetic field. For zero current this corresponds to about 6°K (in NbTi) whereas at maximum excitation it is only a few tenths of a degree. All data and most calculations presented here pertain to Tevatron runs at 800 GeV/c, i.e. at 80% of maximum excitation. Under these conditions the allowable temperature rise is about 1°K.

A complete simulation of a radiation-induced quench would proceed in three steps. From a full description of the beam loss (i) calculate the energy deposition as a function of location in the magnet, (ii) calculate the resulting temperature distribution (as a function of time) in the presence of cryogenic cooling as well as the other structural elements of the magnet, (iii) from this information determine if (and when) a quench occurs. While various models address (ii) and (iii), a detailed study of this is not undertaken. For the cases of interest here, at least part of the rather complicated set of calculations this entails may be bypassed by a simple empirical procedure. This consists in comparing the calculated maximum energy density with a design limit, based on experiments wherein similar magnets are made to quench under controlled beam loss conditions. These experiments are interpreted in terms of energy deposition via the same simulation programs used to interpret quench data under operating conditions. The energy deposition design limits arrived at for the Tevatron are 8 mW/g for slow losses and 1 mJ/g for a fast loss.[1] Numerical values of these limits may be made more precise in the light of operational experience, e.g. of the type described in Section 8, but it appears reasonable that for a magnet of given design the maximum energy deposition determines whether or not a quench occurs, and that this limiting value is not overly sensitive to small design changes. It is not clear a priori that a complete three-part simulation would do better than this empirical shortcut.

Beam loss inside an accelerator can be categorized as either accidental or inherent. In principle, accidental beam loss can occur in a large variety of ways. None of these are expected to happen frequently since the sophisticated Tevatron beam abort systems provide effective protection against such occurrences. Therefore a detailed investigation of a single loss mode would be of little value in terms of operational consequences. Inherent beam loss occurs in the Tevatron fixed-target running cycle when beam is resonantly extracted. The standard operating mode calls for 20 seconds of slow spill interspersed with fast (1 to 2-ms) pulses of beam for neutrino experiments, with approximately equal intensity between fast and slow beams. The instantaneous extraction losses are directly proportional to the rate at which beam is extracted, and hence, fast beam pulses with loss rates several orders of magnitude higher than during slow spill are the critical processes in the Tevatron in terms of energy deposition.

In the initial stages of the Tevatron project concerns arose regarding the ability of the superconducting magnets to function in an operational environment where significant beam losses are present. Energy deposition experiments were performed in the Fermilab external beam areas, which then provided the impetus for developing the necessary computer codes. This early work[2] was insufficient to reach unambiguous conclusions on machine performance but was influential in the design of the region around the electrostatic septa.

The next section describes the accelerator geometry. The rest of the paper discusses a detailed attempt to simulate a fast extraction cycle, essentially in chronological order. Beginning with an unperturbed beam, the simulation generates proton phase-space distributions incident on the electrostatic septum. These interact either elastically or inelastically with the septum wires, and the products of these interactions are traced through the machine. Where these particles leave the accelerator, energy deposition levels in the magnets are calculated together with the projected response of the beam-loss monitors in this region. Finally, results of the calculation are compared with experimental data.

The computer codes used in this work are based on a hadronic cascade simulation program (CASIM),[3] upgraded to accomodate the higher energy provided by the Tevatron, and supplemented with a magnet-by-magnet tracking code (see Section 5).

2 ACCELERATOR GEOMETRY

2.1 Lattice

The Tevatron layout consists of six bending arcs broken by symmetrically disposed straight sections (Fig. 1). Each arc consists of 15 normal focusing cells, containing two quadrupoles, eight bending magnets (with an exception to be noted below), and correction magnets located in the spool pieces near each quadrupole location (Fig. 2a). All these are superconducting magnetic elements. The third cell, counting in the direction of the proton beam, is exceptional in that two of the bending magnets are omitted from the lattice to provide space for other components (Fig. 2b). The straight sections each provide 50 m of drift space in which the various machine functions (injection, abort, etc.) are accommodated. The two regions of special interest here are the ones associated with the extraction system. Their detailed layout is given in the next sections.

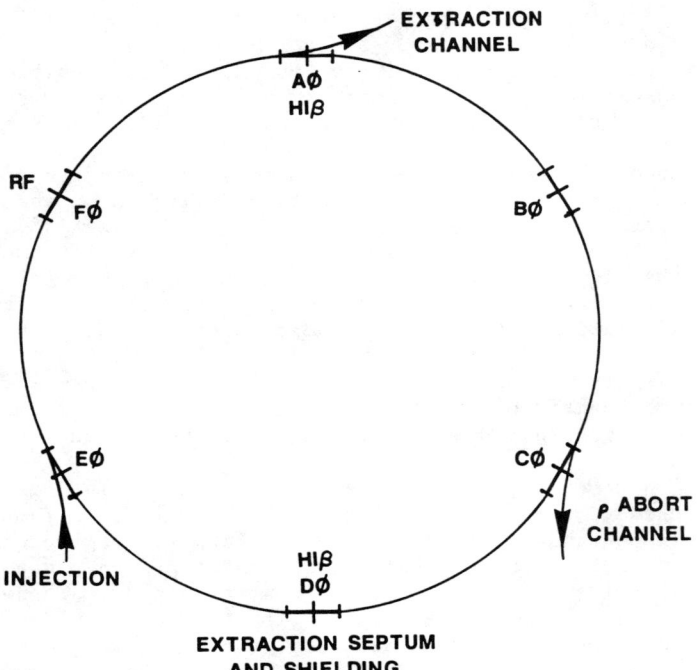

Figure 1 Overall layout of Tevatron and the extraction system.

2.2 Component Description

The superconducting magnets used in the Tevatron are described elsewhere.[1] Simulation of their magnetic properties is discussed in Section 5. Their geometry, for use in radiation transport calculations, is a reasonable facsimile of the main mass of the magnets. Consider, for example, the main bending magnet. A

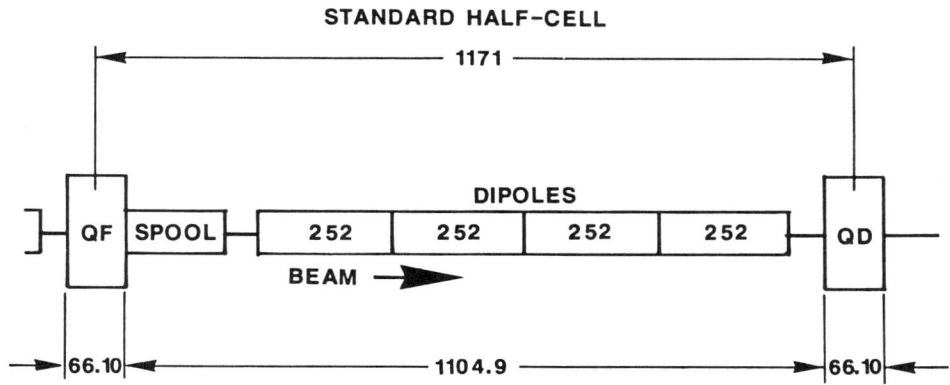

Fig. 2a

Figure 2a. Location of elements in standard cell.
All dimensions in inches.

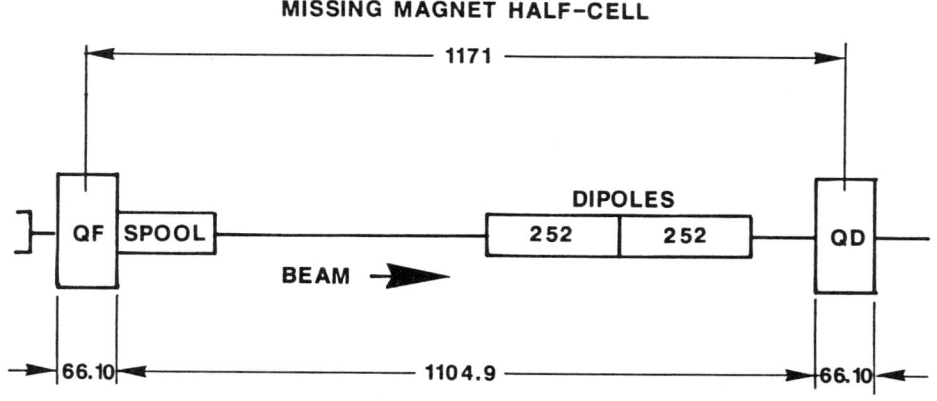

Fig. 2b

Figure 2b. Location of elements in medium straight section.
All dimensions in inches.

technical drawing of the cross section of this magnet is shown in Fig. 3a. From inside to outside, the principal dense components are (i) the stainless steel beam tube, (ii) superconducting coil, (iii) stainless steel collars confining the coil, (iv) several concentric shells of the cryostat, and (v) the iron yoke. Its representation in the simulation is shown in Fig. 3b. The main quadrupoles are treated in analogous fashion.

Spoolpieces are modeled in less detail. The massive elements of interest here are the superconducting correction and adjustment magnets, and they are described in the model by a beampipe, coil, and yoke. The typical excitation of these adjustment magnets is quite low (<20 A), and under these conditions quenching these devices is improbable.

Particle losses in the accelerator structure are measured via the Beam Loss Monitors (BLMs). The BLM system is a network of argon-filled ionization chambers at 1 atm. Nickel electrodes in a sealed glass envelope provide a detector capable of monitoring instantaneous doses in excess of 10 rads without saturating, with excellent uniformity and stability. BLMs are placed at each quadrupole location, and at selected locations in the straight sections. Ring-wide loss profiles are taken automatically in the case of a beam abort and are also available at any preselected times in the cycle.

3 THE EXTRACTION PROCESS

Horizontal half-integer resonant extraction is the process used at the Tevatron. By exciting a mixture of quadrupole and octopole fields, the stable phase-space area available to the circulating beam is gradually reduced in size until it equals the beam emittance. At this point any further reduction in stable area causes a fraction of the beam to become unstable. Under these conditions the particles execute progressively larger amplitude betatron oscillations on each successive turn. These oscillations lie on a well-defined trajectory in phase space. At some point on this trajectory, defined by the relative offset of the septum with respect to the closed orbit, the particles are deflected into the extraction channel by an electrostatic septum. The amount of beam that strikes the septum is the source of the extraction losses and is determined by the particle density distribution in phase space at the septum position and by the septum geometry.

The first step in these loss calculations, therefore, is to determine the transverse phase-space distributions of the beam striking the septum. This is done by a Monte Carlo simulation of the fast extraction cycle. This approach is well suited to this problem since it involves a sufficiently small number of turns so that particle tracking from element to element with full field harmonics is feasible. A typical phase-space evolution, taken at the start of the extraction channel during the fast extraction cycle, is shown in Fig. 4. As the beam is brought into resonance, the circular distribution becomes more elliptical and the particles become unstable. The initial conditions pertaining to the beam striking the septum are directly obtained from these calculations.

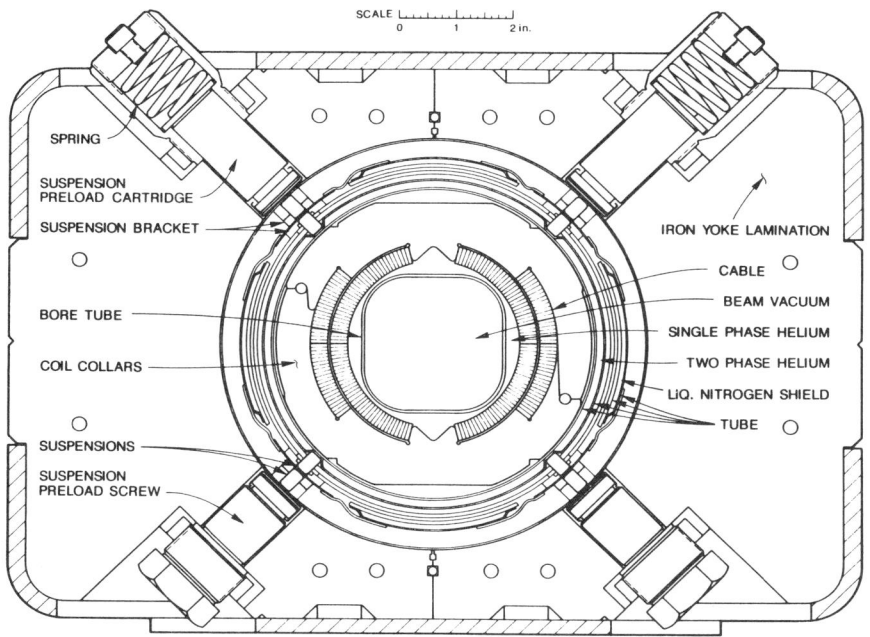

Figure 3a Cross-section of Tevatron dipole. Engineering drawing.

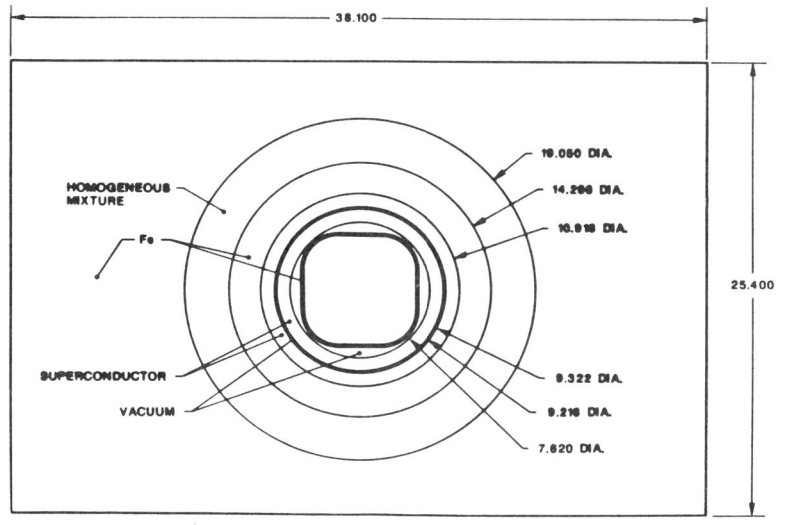

Figure 3b Cross-section of Tevatron dipole as represented in the MC simulation.

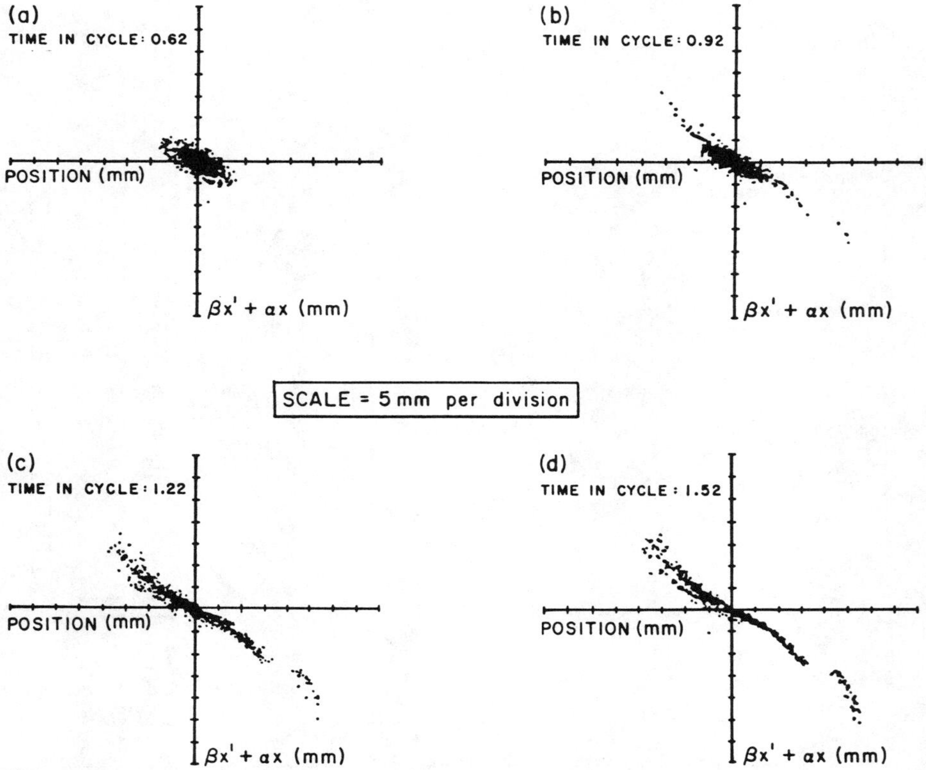

(a) TIME IN CYCLE: 0.62
POSITION (mm)
$\beta x' + \alpha x$ (mm)

(b) TIME IN CYCLE: 0.92
POSITION (mm)
$\beta x' + \alpha x$ (mm)

SCALE = 5 mm per division

(c) TIME IN CYCLE: 1.22
POSITION (mm)
$\beta x' + \alpha x$ (mm)

(d) TIME IN CYCLE: 1.52
POSITION (mm)
$\beta x' + \alpha x$ (mm)

Figure 4 Phase-space evolution during fast extraction at the start of the extraction channel.

The layout of the extraction elements is as follows. The initial splitting septum is located in the D0 straight section (Fig. 5) halfway around the ring from the start of the extraction

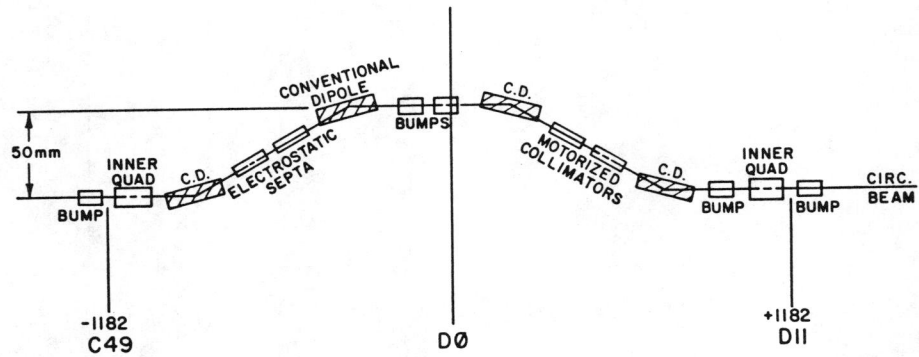

Figure 5. Schematic layout of electrostatic septum and conventional magnet 4-bump in D0 straight section.

channel at A0. The septum area at D0 is designed to protect the downstream superconducting magnets from particles produced in the septum. Internal to the straight section, situated horizontally outside, is a closed 4-bump made up of conventional (model B2) Main Ring dipole magnets. Interspersed with these bending elements are 40-in. bump dipoles, which provide orbit control during the fast extraction cycle. At the downstream end of the long straight are two independently motorized, stainless steel, L-shaped collimators, each 120 in. long with accurately milled flat surfaces. The collimators are oriented in opposite directions and can be moved to the point of closing the machine aperture.

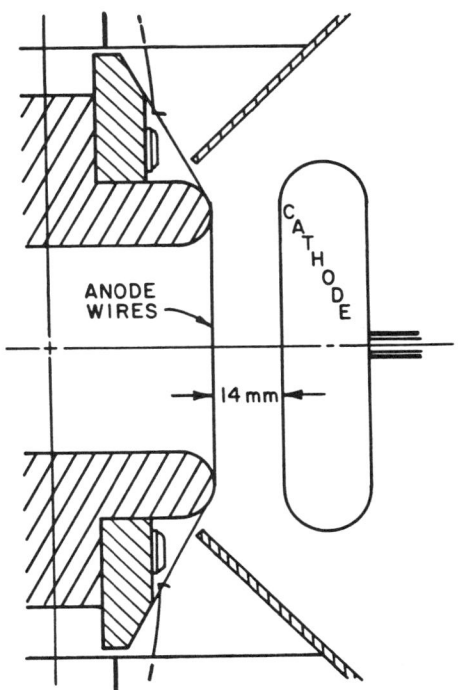

Figure 6
High voltage gap region in the electrostatic septum.

The electrostatic septum, which consist of two independent modules, is located between the first two Main Ring dipoles. Figure 6 shows the high voltage gap of the septa. Each module is 144 in. long and capable of operating up to fields of 75 kV/cm across the gap. The wire plane is made up of 0.002-in. tungsten-rhenium wires spaced every 0.1 in. It has been determined to be straight to within ±0.001 in. Each septum module is motorized at each end and is aligned relative to the beam by minimizing the extraction losses. The angle of deflection produced by the septum is 36 μrad.

The extraction channel (Fig. 7) is located in the A0 straight section. In this area the design problem was to provide sufficient

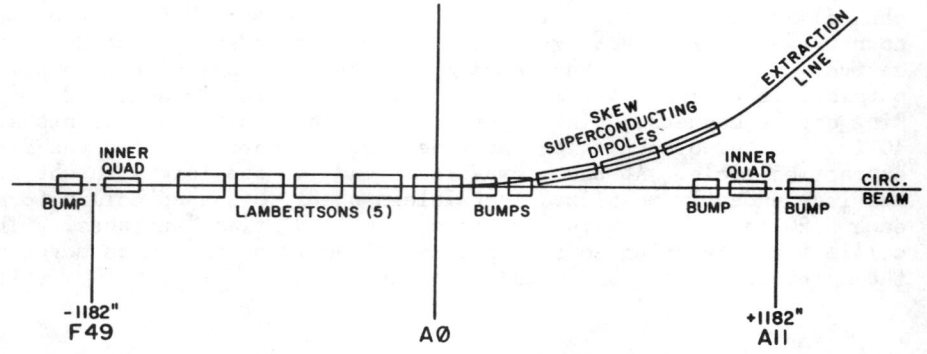

Figure 7 Schematic layout of magnetic extraction septum and extraction channel in A0 straight section.

bending of the extracted beam in the available space; no special measures were taken to reduce beam losses. The extraction channel starts with a string of Lambertson magnets. These magnets provide a total vertical bend of 10.5 mrad and create a 7-in. vertical separation between circulating and extracted beams. This allows the extracted beam to enter a string of three standard superconducting Tevatron dipoles azimuthally rotated by 19° from the horizontal plane to provide both an outward and downward bend. Downstream from these magnets the extracted beam exits the accelerator tunnel and enters the switchyard. Particle tracking ceases at this point in the simulations.

4 BEAM ON SEPTUM

4.1 Beam Phase Space

The phase space of the beam striking the septum varies with the working point chosen for the extraction system. Depending on these conditions one can identify a realistic phase-space prevalent under normal conditions, and a "worst case" distribution. The latter attempts to represent a beam which, while still functional, approaches the worst conditions from a beam loss point of view. For both cases the transverse phase-space distribution of the beam is assumed to factorize into a product of four truncated Gaussians in x, x', y, and y'. Since σ_x is much larger than the septum width, its precise value is immaterial. For a reference momentum of 1000 GeV/c the other σ for the realistic PS distribution are σ_x' = 2.3 μrad (6.5), σ_y = 0.29 mm (0.8), and σ_y' = 71 μrad (165), with the maximum (absolute) value of the variables shown in parentheses in the same units. Similarly for the worst case phase-space σ_x' = 8 μrad, σ_y = 0.7 mm, and σ_y' = 20 μrad. In this case truncation is performed for all at $\pm 3\sigma$. The phase space of the truncated Gaussians is rotated in y, y' space by an angle of 0.223 rad. For momenta other than 1000 GeV/c, σ_y and σ_y' as well as the y and y' truncation limits are multiplied by $(1000/p)^{1/2}$. A

Gaussian beam momentum distribution with $\sigma_p = 8.8 \times 10^{-5}$ is assumed for both cases.

Both phase-space distributions are used simultaneously as input in the beam-on-septum simulation program. The x', y, and y' of the incident particle are chosen from a distribution intermediate between the two cases (but extending to the full range of the "worst case"). The particle carries two weights, w_r and w_w, but $w_r = 0$ whenever one of the phase-space variables exceeds the truncation limits of the realistic case. This saves computer time and provides a complete correlation between the two distributions in the simulations, which facilitates comparisons.

4.2 Septum Model

The details of the septum geometry (Section 3) are faithfully reproduced in the calculation with one exception: for convenience the wire density is homogenized in the beam direction and taken to be equal to $\rho_s = (2/L)(r_0^2 - x^2)^{1/2} \rho_w$ for $r < r_0$ and $\rho_s = 0$ elsewhere. For a given particle trajectory through the wires the approximation improves with the number of wires a trajectory crosses. On average, it should be excellent for incident particles and elastically scattered particles and quite satisfactory for high energy inelastics, which have angles typically much smaller than 10 mrad ($\simeq r_0/L$). The electrostatic field is assumed to increase linearly from zero to full strength across the wire diameter.

To represent effects of mechanical tolerances in septum construction which manifest themselves as a deviation of the wires from a straight wire plane, the width of the septum is increased by a factor of two while the density is decreased by the same factor.

4.3 Interactions of Beam in Septum

The usual distinction between elastic and inelastic interactions serves well here, with each component going its separate way. Products of inelastic interactions of the beam with tungsten nuclei in the septum wires are quickly lost from the aperture. Almost all are deposited within the 4-bump of conventional magnets, which is there for precisely this reason. The exception is positively charged particles sufficiently energetic to survive the magnetic analysis, which are almost exclusively leading particle protons. They will be gradually swept onto the inner wall of the superconducting magnets following the D0 straight section.

The elastically scattered particles are typically transported over long distances in the accelerator. Indeed, the majority of all particles with a trajectory intersecting the septum will be extracted. Those that are not, typically leave the aperture in one of a number of hot spots on the first half-turn or at the Lambertson septa at A0.

The production of inelastics in the septum follows CASIM, i.e. the Hagedorn-Ranft model[4] plus a high p_t component plus low energy nucleons. The Hagedorn-Ranft model includes leading particles explicitly, and the many parameters of that model are adjusted[5] to fit p-nucleus data at 19.2 GeV/c.[6] At low p_t good agreement is found between model predictions and experiment at Fermilab energies.[7] Likewise there is good agreement between predictions of energy deposition and experiment for both small and large targets.[8]

Because of its relative importance in this problem, elastic scattering is treated more carefully than previously in CASIM. Briefly, the present model considers four components: (i) multiple Coulomb scattering, which treats all single scatters below some judiciously chosen cutoff angle via the Gaussian approximation, (ii) Coulomb plus coherent nuclear scattering and their interference, (iii) nuclear incoherent scattering, and (iv) diffractive low-mass target excitation. The last three are treated on an event-by-event basis.

The energy loss of the particles in the wires is estimated as in CASIM. This includes effects of fluctuations for each of the energy loss mechanisms involved. The energy lost in elastic scattering is calculated by using p-nucleus (for coherent) or p-p (for incoherent) kinematics. For low-mass target excitation the mass of excited nucleon target is employed in the kinematics. The elastic scattering program will be documented separately.[9]

4.4 Results of Septum Calculations

The septum calculations create a set of files each corresponding to different initial conditions: beam energy, phase-space distribution, septum alignment, etc. Each file contains the characteristics of ~10^5 particles emerging from the septum, i.e. x, y, x', y', p, w_r, w_w (= weights) and (for inelastics) particle type (p, n, π^+, π^-, γ, e^+, e^-). These files form a new starting point for further calculations towards the main goal, but a look at some of these intermediate results seems worthwhile. All pertain to 800-GeV/c protons.

Figure 8 shows the longitudinal distribution along the septum of elastically scattered particles escaping the wire region. The distribution shows a linear rise over the first 100 cm and then drops exponentially as more and more particles are scattered out of the wires. The momentum loss associated with these particles is given in Fig. 9. This is essentially the convolution of Fig. 8 with the energy losses per unit length, including fluctuations. The distribution of inelastic collisions along the length of the septum is shown in Fig. 10. In the absence of any out-scattering, the interaction rate would fall exponentially with distance along the septum.

Figures 11 and 12 show the transverse phase-space distributions of protons elastically scattered in the septum. The vertical phase-space distributions show a width characterized by the scattering processes. The horizontal ones show a double-peaked structure caused by the electric field and septum shadowing (out-scattering). Note that the distribution does not fall to zero in the valley but is populated by particles passing through one or both sections of the septum. These particles strike the magnetic septum at the start of the extraction channel which is positioned in this notch.

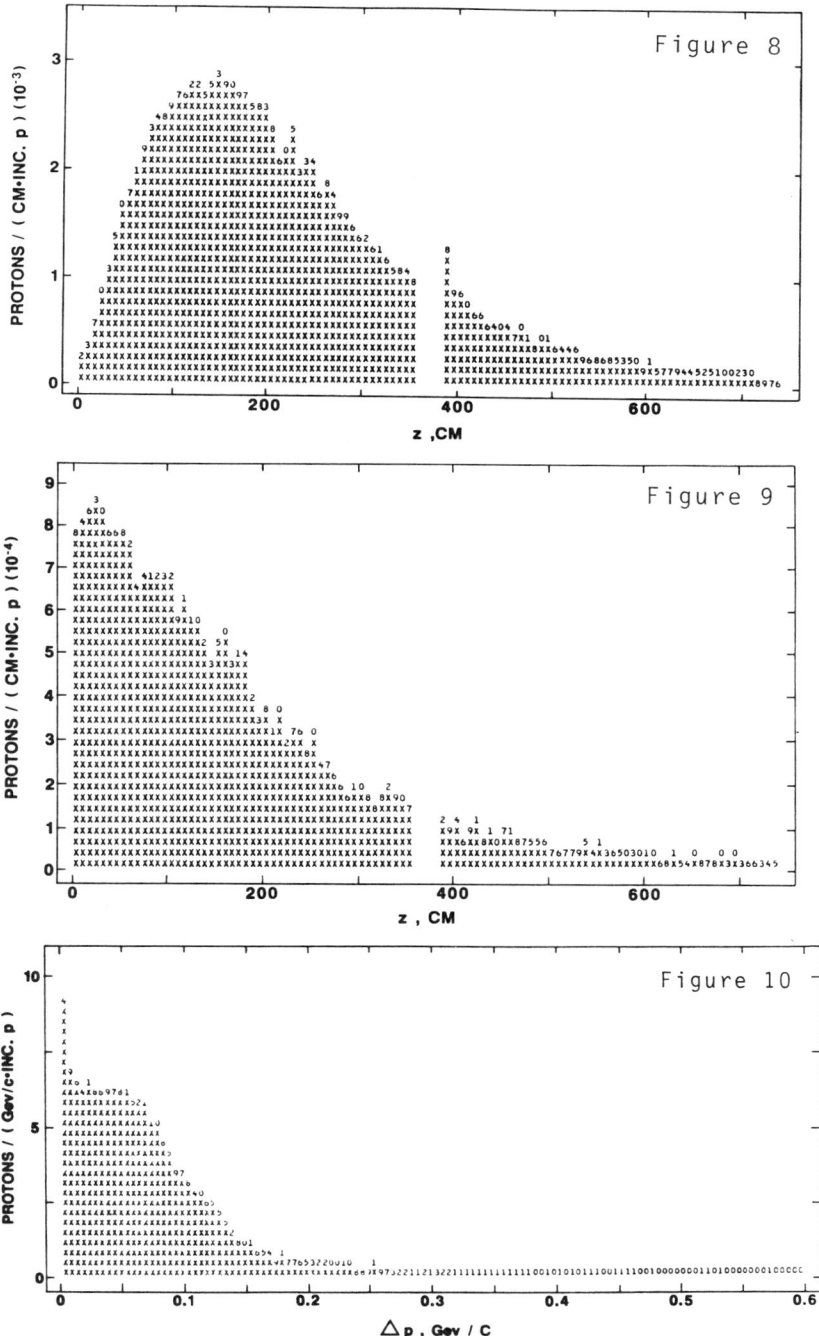

Figure 8

Figure 9

Figure 10

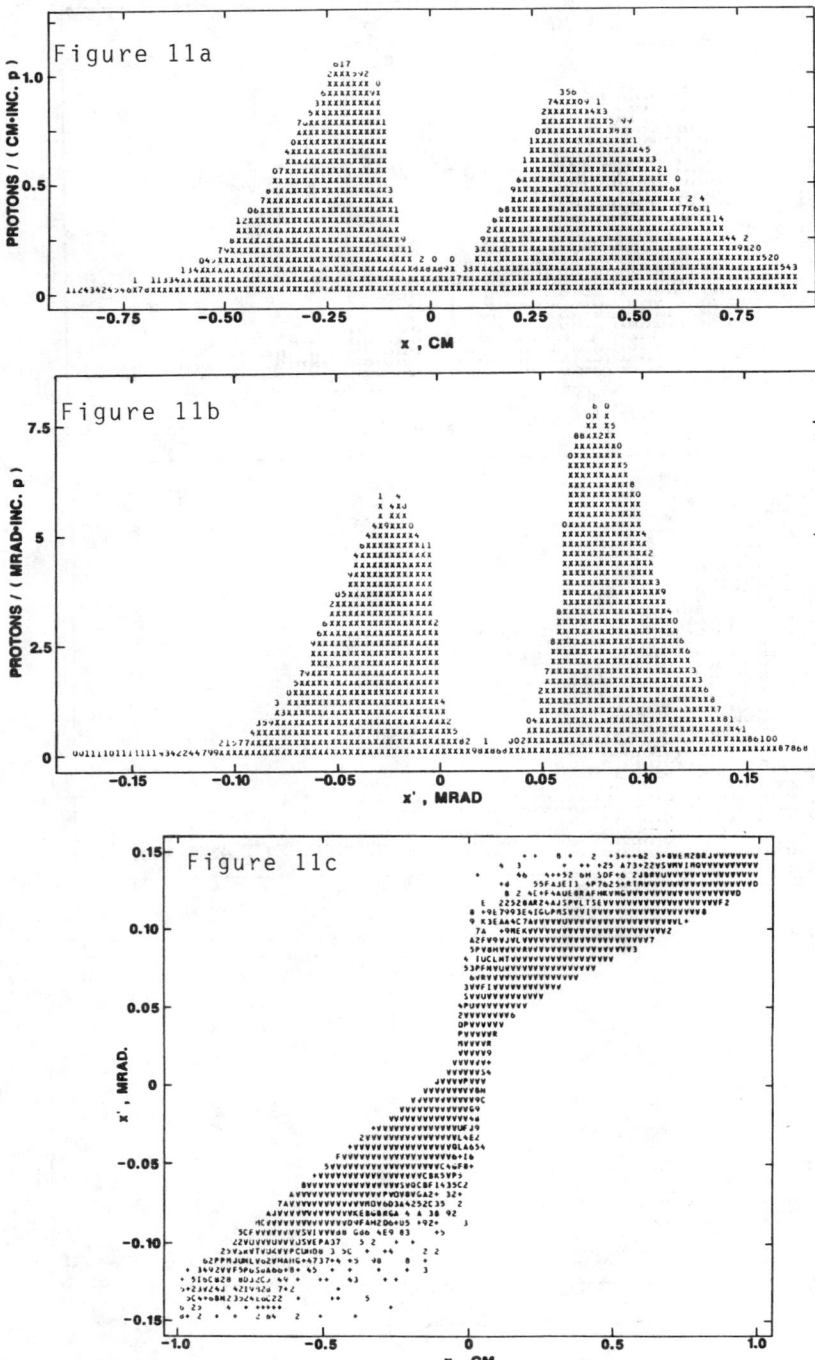

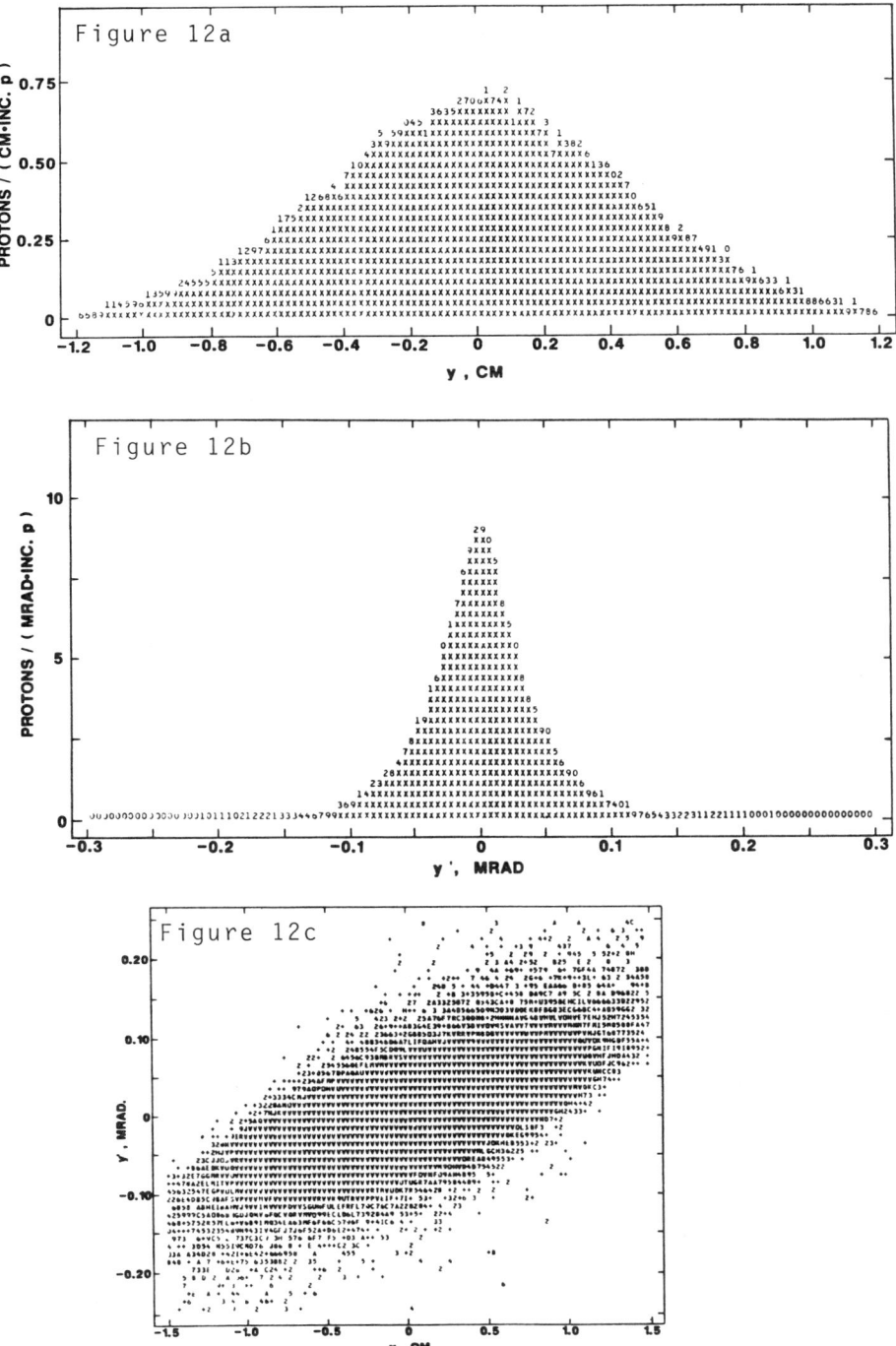

5 PARTICLE TRACKING

Particle tracking divides into two categories; short-range and long-range. Short-range pertains to propagation of inelastics within the first few magnets downstream of the extraction septa and within condensed materials anywhere. Particles receive a kick at regularly spaced intervals (typically ~5 cm) along their trajectory. The incremental displacement due to the field during this step is ignored. The magnitude of the kick is appropriately reduced when stepping across the end faces of a magnet. Magnetic fields are obtained by interpolation from field maps except in the region interior to the coil. Within this region ideal fields (dipole or quadrupole) are used to facilitate comparison with analytic calculations; the effect of nonlinearities within the beam pipe is negligible for the short trajectories involved.

Long-range (significant fraction of a turn) tracking uses a conventional kick algorithm. A particle is propagated to the midpoint of each magnetic element using only linear fields, and an angular deflection calculated from the nonlinear field components is delivered to the particle, which is then propagated through linear fields to the end of the magnet. Correction and adjustment magnets are represented by a kick only. The fields are constructed from the design values of its multipole expansion. Upon striking a boundary, the fields revert to those used for short-range tracking, as in the preceding paragraph.

Tune and chromaticity adjustments are reflected in the appropriate settings of the trim quadrupoles and sextupoles. Similarly, the fields used for extraction quadrupoles and octopoles are based on the settings used in practice.

The x,y coordinates of the particle are explicitly calculated at the entrance, midpoint, and exit of each magnet. If this indicates that the particle is outside the physical aperture, its parameters are noted on a file and the particle is removed from further tracking.

The closed-orbit distortions are incorporated into the program by offsetting the particles with respect to the magnets. The offsets are calculated by interpolating linearly between the quadrupole locations where the beam position is actually measured.

6 ENERGY DEPOSITION IN MAGNETS

6.1 Elastic Transport Through Magnet

The file containing the magnet aperture failures is read by a set of programs each of which includes a description of the detailed geometry and fields in the vicinity of a hotspot of interest located on the first half-turn, or of the Lambertson magnets at A0. If the event occurs in the vicinity of a particular hotspot, the precise coordinates where the proton enters the vacuum chamber wall of a magnet are determined. Because of the small angles of the protons striking the wall and the small radial distances between the inner wall and the superconducting coil, the transport of the scattered beam particles in the magnet requires some care. First the particles are traced through the magnet using the same program as for the elastic part of the beam-on-septum simulation. This elastic part concludes typically with a nuclear interaction and the information on the particle at this point is in turn recorded on file. Occasionally the particle is reflected back into the aperture, but in this work this has been found to be of negligible importance everywhere.

6.2 Energy Deposition

The nuclear interactions of the scattered beam or the inelastics from D0 are the input to a regular CASIM calculation of energy deposition in the magnets. The hadron cascade plus the electromagnetic cascades that develop from π^0 decay are traced through a reasonable geometric representation of the magnet which includes a description of the magnetic fields in the aperture as well as in the rest of the magnet (see Section 5). Both the CASIM code[3] for hadron showers and AEGIS code[10] for electromagnetic showers, which serves here as a CASIM subroutine, are well documented elsewhere.

For a given magnet design, magnetic field, and beam-loss time structure, the temperature rise is directly proportional to the deposited energy density. In the case of fast spill the occurrence of a quench is equivalent to exceeding some given energy density, ρ_E^{max}. It is clear that this energy density is the average over some macroscopic volume but less clear what its dimensions should be. It seems reasonable to choose the volume dimension along each coordinate such that little variation in ρ_E^{max} is expected over a distance comparable to its extent. This volume is typically much smaller than can be accommodated by the Monte Carlo calculation, and some care is therefore needed to estimate ρ_E^{max}. The method used here starts from the commonly generated Monte Carlo output, $\rho_E(r,\phi,z)$, viz. (statistically valid) energy densities averaged over a set of volume bins with dimensions Δr, $\Delta \phi$, Δz which are too large for a direct determination of ρ_E^{max} in accordance with the above criterion. For each magnet ρ_E^{max} is then determined from an interpolation scheme which brings a certain amount of a priori knowledge, about the spatial distribution of ρ_E in general, to bear on the problem.

The volume bins cover the superconducting coils and vacuum chamber wall. The latter is included since ρ_E^{max} is expected to occur at the smallest radius of the superconducting coils, r_c, and some information on ρ_E for $r < r_c$ is clearly desirable. The fact that the beam pipe and coils are close in density and in atomic properties facilitates the interpolation. The ρ_E are determined for either three or four radial bins (one covering the beampipe, the others the superconducting coils), seven azimuthal bins, and from one to five z-bins. The azimuthal bins are unequal and are adjusted in size to accomodate the beam loss spot size. Then ρ_E^{max} is determined by step-wise fitting the ρ_E to a simple function of r, ϕ, and z. At each step energy conservation is imposed by integrating the fitting function over the volume of the bins and equating it to the total energy content of the bins. This constraint is imposed because the total energy deposited in all bins, or in a given subset, is the direct result of the calculation and hence the most statistically reliable result for that volume.

The radial dependence is assumed to be of the form

$$\rho_E(r,\phi,z) = (A/r)\exp(BR)\cdot F(\phi,z)$$

where r is the radial coordinate and R is the thickness of the superconducting coil (or equivalent) between r and the axis. Here, ϕ and z are held constant. Given A and B from the fit, $\rho_E(r_c,\phi,z)$ yields the maximum ρ_E for a given ϕ and z. The ϕ-dependence is next fitted to the Gaussian form

$$\rho_E(r_c,\phi,z) = C\cdot\exp[D(\phi-\phi_0)^2]\cdot F(z) .$$

In most cases ϕ_0 can be set a priori equal to zero (or to π). (A noteworthy exception is the case with a vertical 4-bump, particularly near D0.) In practice, for those cases where ϕ_0 should be zero (or π) it makes little difference whether one forces it or not. The value of ϕ_0 determined by the results (i.e. the centroid of the distribution in ϕ) always agrees well with its a priori value. Clearly $\rho_E(r_c,\phi_0,z)$ has its maximum at $\phi = \phi_0$ and these maxima are next fitted to a simple quadratic in z:

$$\rho_E(r_c,\phi_0,z) = F + Gz + Hz^2 ,$$

and $\rho_E^{max} = \rho_E(r_c,\phi_0,z_m)$ where $z_m = -G/2H$ if $H < 0$ and if it is located within the magnet. Otherwise z_m lies at the front or back end of the magnet. A statistical error analysis on ρ_E for individual bins is performed routinely as part of the Monte Carlo. The error on ρ_E^{max}, σ_m, is obtained from the usual propagation of error formula:

$$\sigma_m^2 = \Sigma\ (\partial\rho_E^{max}/\partial\rho_E)^2\sigma_\rho^2$$

where the σ_ρ refer to the errors of the ρ_E calculated for the individual bins. The partial derivatives are evaluated numerically, i.e. by repeating the entire fitting procedure, changing one ρ_E at a time by a small $\Delta\rho_E$. The partial derivatives are then approximated by $\Delta\rho_E^{max}/\Delta\rho_E$. The same procedure is applied at each intermediate stage of the fitting.

In addition to ρ_E^{max} in the coils, the program computes the total energy deposition in the magnet and in each of its major components. This permits predictions of the total heat load imposed by beam losses on the cryogenic system.

7 BEAM LOSS MONITOR RESPONSE

7.1 Response Function

The BLM characteristics and their placement around the ring are discussed in Section 2. Knowing the output of a BLM near the location where a quench occurs has obvious operational value. Predictions of the BLM output of the type outlined for ρ_E in the coils likewise provide valuable information. In principle, e.g., they can be used to establish "geometric" factors for different hotspots which relate BLM output to quench level.

The energy deposition routines in CASIM specifically address the problem of estimating ρ_E at or reasonably close to its maximum. Radiation problems at large radii (e.g. biological shielding) are typically analyzed in terms of star (i.e. nuclear interaction) densities plus an assumed equilibrium spectrum to convert to dose.[11] This appears of doubtful validity in predicting BLM response. The problem arises because CASIM treats low energy particles very crudely, whereas in this case it is typically low energy neutrons (of a few MeV) that are responsible for the bulk of the BLM dose. The most straightforward solution is to couple CASIM to a low energy neutron code[12] and to combine the calculated ρ_E's. But this approach is not without problems and is avoided in favor of a more empirical procedure.

First a representative BLM was tested in a neutron beam of known spectrum and intensity at the Fermilab Neutron Therapy Facility. This beam is not unlike the radiation environment which prevails near a loss point at the Tevatron, especially for that part of the environment most difficult to calculate with CASIM. Results show that BLM readings agree well with other monitoring devices and are rather insensitive to the presence of steel slabs (1/4 to 1 in. thick) placed directly upstream.

Next the test results are analyzed on the basis of a simple model of low energy neutron interactions within or near the BLM. The model consists of a simple set of assumptions about particle emission (evaporation particles and photons) following neutron-nucleus collisions and how these particles lose energy in the BLM. It is based on experiment,[13] low energy transport calculations,[14] simulations of neutron evaporation,[15] calculated nuclear reaction thresholds, and the enforcement of an energy balance. However, the model retains some abritrariness and is not expected to be accurate. An important ingredient of the model is its parameterization of the effective absorption cross section as a function of neutron energy (here "effective" means properly averaged over the materials of the BLM and nearby components). A key parameter is the ratio of the maximum cross section (assumed to occur in the few-MeV region) to the geometric (or high energy) cross section. This parameter is explicitly chosen to bring about agreement of the model with the BLM response observed in the tests. The value of ~3, determined in this way, is quite reasonable for medium weight nuclei.[16]

7.2 Calculation of BLM Response

The above procedure yields a BLM response curve for low energy (<60-MeV) neutrons. Charged hadrons are treated analogously. Hadrons above this energy and the electron and photon components of electromagnetic showers follow standard CASIM rules.

The BLM response calculation starts with reading a file prepared by the energy deposition computation in the superconducting coils. For a given magnet string associated with a hotspot, this file lists all nuclear interactions occurring in the string. From each such interaction a "recording" particle is generated with momentum and angle roughly proportional to the differential production cross section. This particle is traced through the magnet string undergoing elastic scattering and energy losses (if applicable) while its nuclear interactions are included in an average way, i.e. the weight of the particle is exponentially reduced with distance traversed. (If the particle is a π^0, the usual AEGIS routine is performed.)

For the purpose of estimating the energy deposition of the particle, the lateral dimensions of the magnets are extended in all directions by a hypothetical 5-cm layer of argon gas. Particles traversing this argon layer record their energy deposition in relatively large volume bins, typically 5 × 10 cm for the x,y dimensions and 150 cm in the z-direction. The large volume bins boost statistics with little loss in accuracy, since only slow variation of ρ_E with location is predicted at large radii. One advantage of covering the magnet exterior in this way is that exact placement of the monitor need not be anticipated and the calculation may indicate preferred locations where BLM response is most sensitive to beam loss.

8 RESULTS

The results are separated into two parts: short-range losses that occur in the vicinity of the septum (inelastics) and those that are transported deep into the accelerator structure (elastics).

8.1 Inelastics

In the absence of shielding between septum and downstream superconducting magnets, secondaries produced by inelastic proton interactions in the septum will, in turn, interact in the magnets. These secondaries belong to one of three components: (i) high-energy ("leading") protons which remain in the aperture for some distance before being swept onto the inside wall, (ii) energetic neutrals, mainly γ from π^0, which intercept the outside wall of the beampipe at a spot aligned with the septum but broadened by production and scattering in the septum, (iii) charged secondaries which either strike the front face of the first magnet or are bent into the first few meters of magnet. Earlier calculations[3] indicated that, without shielding, quenching of the elements immediately downstream of the septum is inevitable. These calculations also showed that quenching can be avoided by a 4-bump of conventional magnets (though perhaps not in the limit of full design energy and intensity), which was consequently adopted in the Tevatron design. Qualitatively, the 4-bump absorbs components (ii) and (iii) but relatively little of (i). In the absence of a firm design at the time, these earlier simulations lacked detailed predictive power though the results did establish the feasibility of fast extraction. The present study includes a closer look at the 4-bump.

Space limitations and other practical considerations suggest the use of Main Ring B2 dipoles. Since an achromatic bend is clearly desirable, the most opportune placement of the septum is between the first and second magnet. The orientation of the bump is investigated in some detail. Three cases are analyzed: (i) radially-in, (ii) vertical, and (iii) radially-out bends. Figure 13 intercompares the maximum energy deposition in the coils of each superconducting element in the first cell downstream (inelastic losses beyond this are inconsequential) for a "worst case" (see Section 4.1) phase-space beam of 1000-GeV/c protons incident on the septum. The radially-in bend is seen to produce a peak around the fourth and fifth dipoles, the radially-out bend peaks in the first two quadrupoles, and the vertical bend has its maximum somewhere in between. These differences occur because a radially-out bend sweeps off-momentum positive particles to the inside of the beampipe whereas a radially-in bend sweeps them to the outside. Therefore, following a radially-in bend these particles travel a longer distance before being deposited on the inside wall by the main accelerator guide field. The vertical bend produces no horizontal sweeping and is in this sense an intermediate case. The vertical bend has a certain advantage in that the bump field is at right angles to the guide field thereby introducing extra broadening of the "spot" size where these particles intersect the inside wall. Also in this case the azimuthal variation of ρ_E does not peak in the median plane. In spite of the high ρ_E^{max} in the first quadrupole doublet, the radially-out bend is the choice because the particles causing these high levels are removable by collimation, which is virtually impossible for the particles striking the dipoles further along the string. The effects of collimation are further discussed below.

Figure 14 is a scatter plot showing the correlation between momentum and penetration, defined as the z-distance where the particle leaves the beampipe, for the case of 800-GeV/c incident protons, presented separately for each charge type as well as for the total. The magnetic elements are shown schematically across the top. The only particles reaching the superconducting dipoles in significant numbers are positives above about 600 GeV/c. A modest increase of this threshold momentum does not significantly cut the energy flow into the superconducting dipoles. This means that B2 magnets are well suited for the bump since use of conventional magnets precludes a significantly higher magnetic field and space limitations preclude significantly longer magnets. Figure 15 shows the rejection efficiency of the bump by comparing x,y plots of all secondaries at the downstream end of the septum with those reaching the SC dipoles. For these beam and septum conditions only 7% of the secondaries produced in the septum strike the superconducting dipoles.

The spatial character of the energy deposition in a superconducting magnet in the vicinity of D0 is illustrated in Fig. 16, using the third dipole as an example. The calculation is for a "worst case" phase-space beam of 1000-GeV/c incident protons with a radially-out bump, 66-in. collimator, and thick-walled pipe

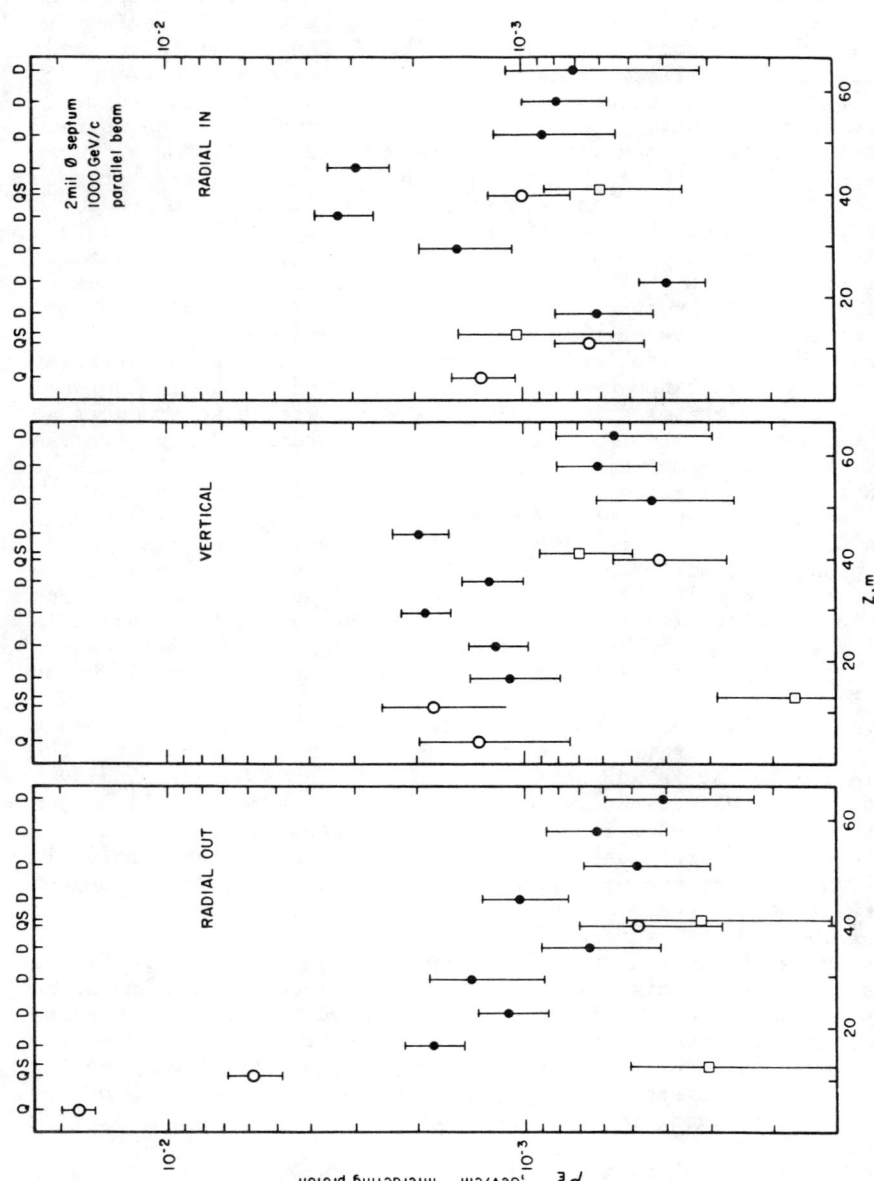

Figure 13 Maximum energy deposition density in each superconducting element at the start of D-sector for different 4-bump orientations.

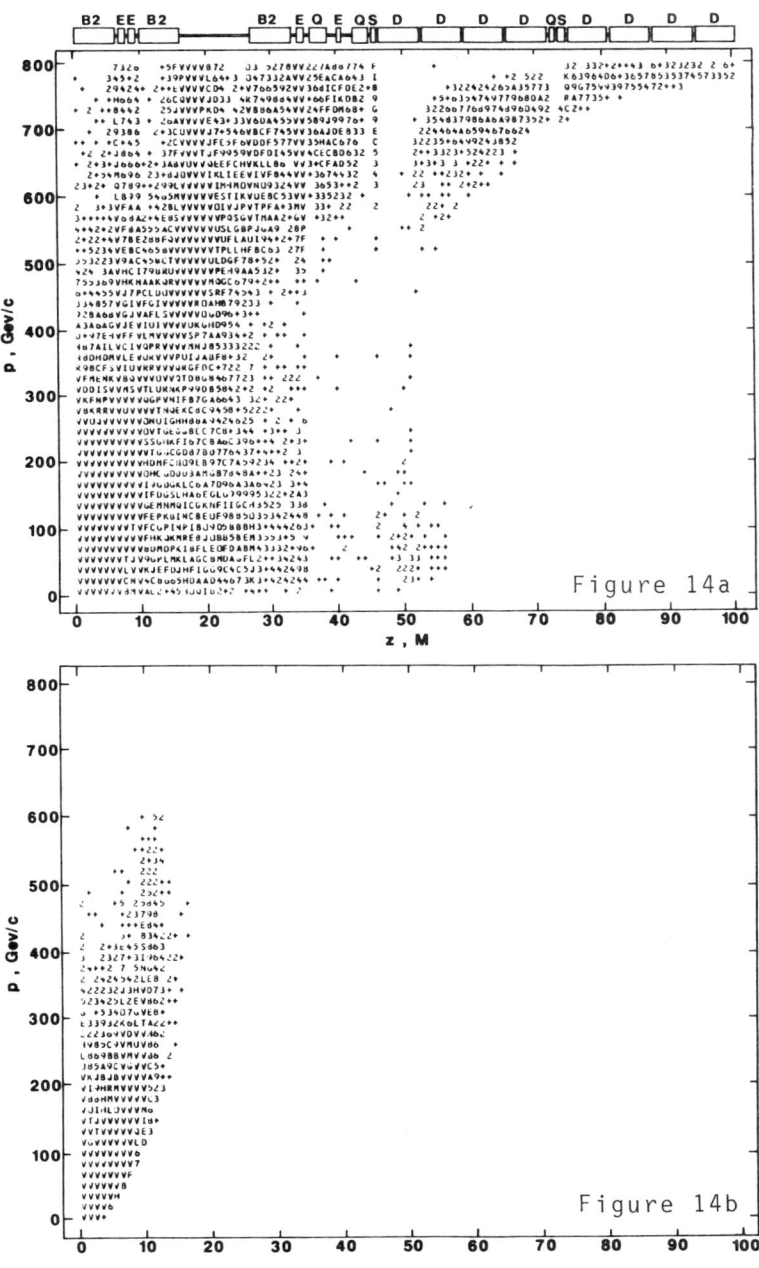

Figure 14 Scatter plot showing correlation between momentum and distance of penetration into 4-bump and superconducting magnets for (a) all particles, (b) positives, (c) negatives, and (d) neutrals.

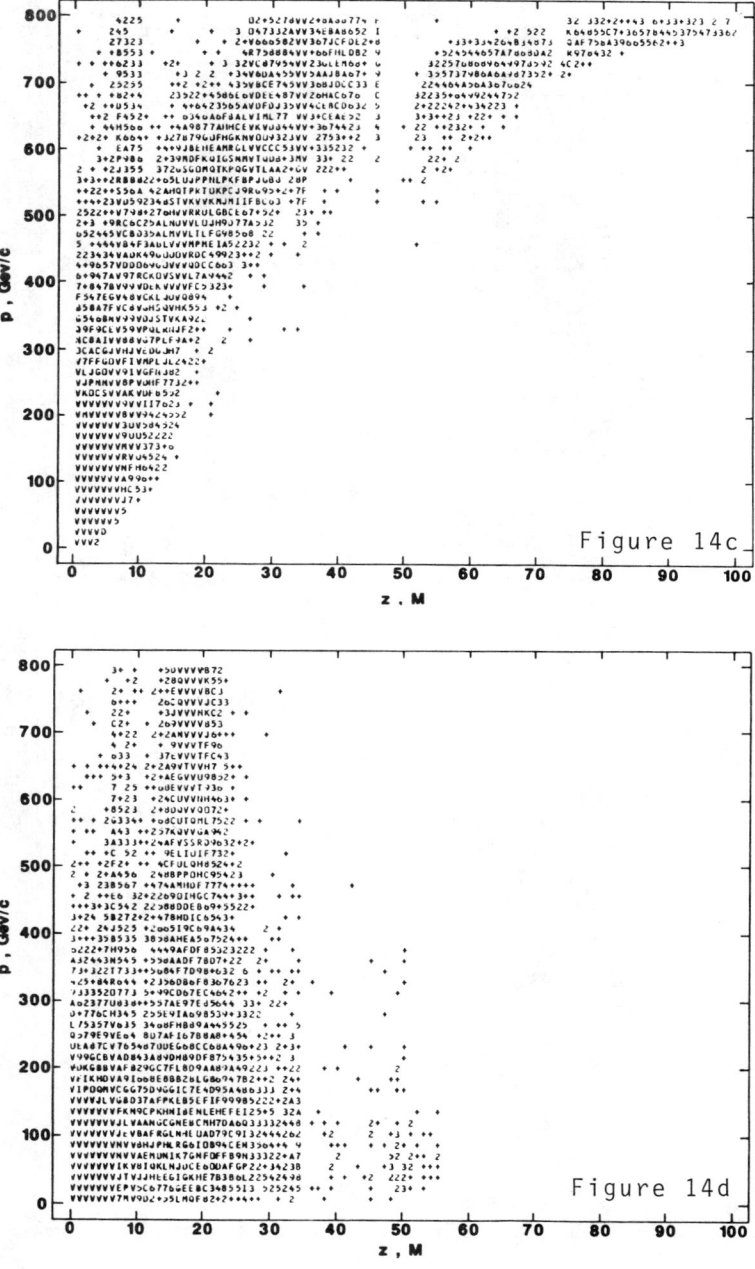

Figure 14 (continued)

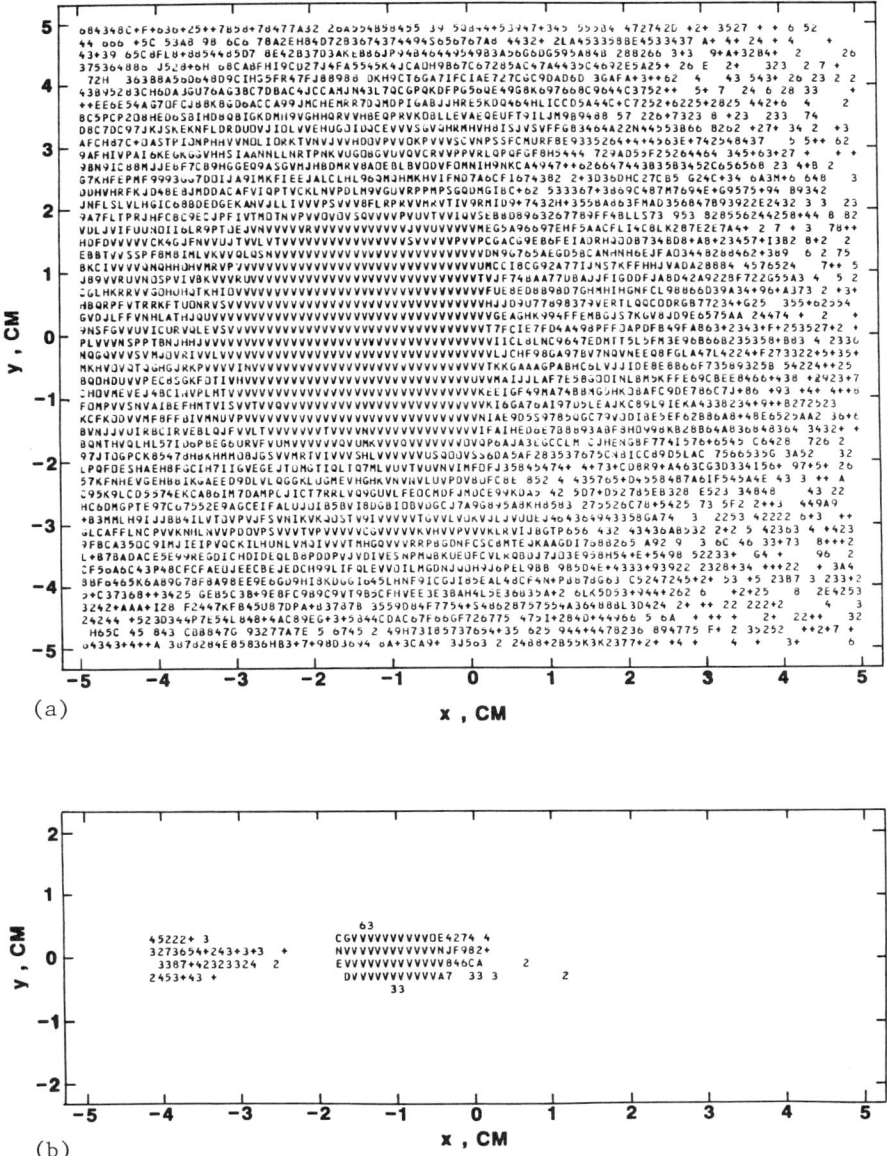

Figure 15 Scatter plot in x-y space showing all inelastically produced particles (a) at downstream end of septum and (b) at the start of the superconducting dipoles.

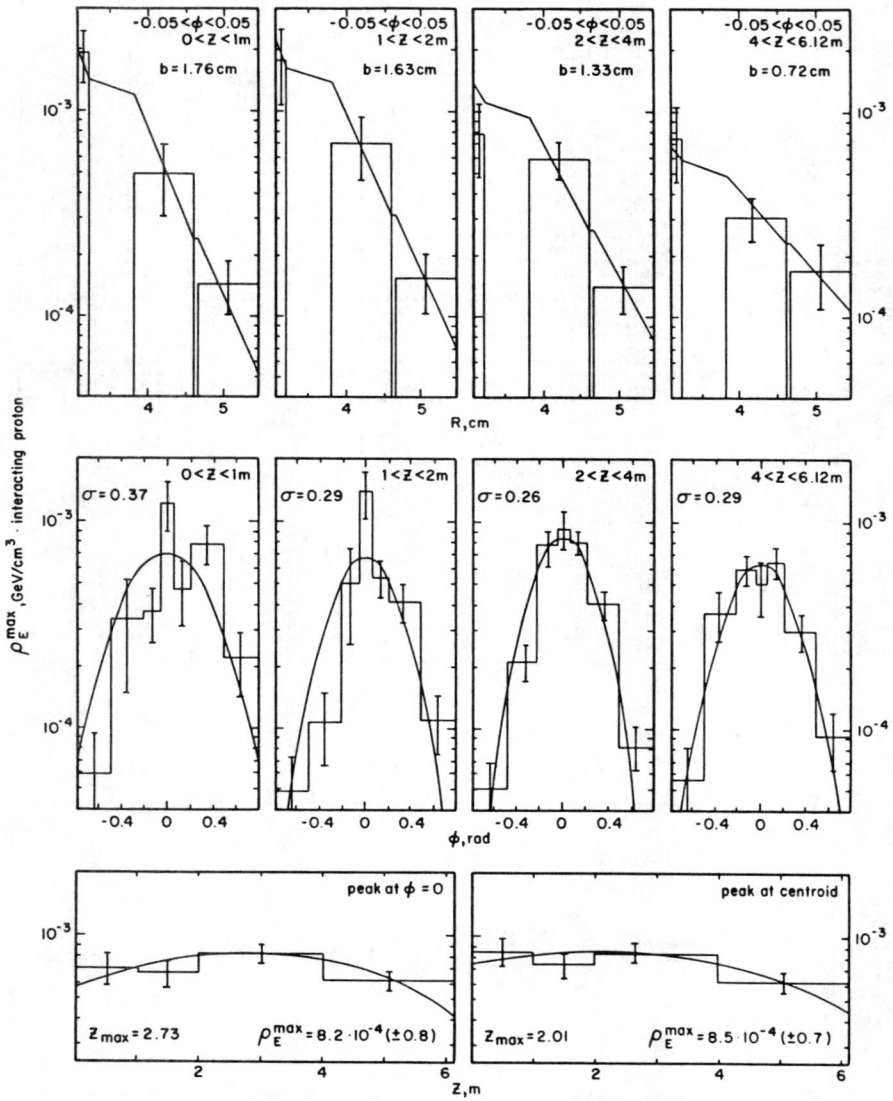

Figure 16 Calculated energy deposition in third superconducting dipole downstream of D0. Top: radial fitting procedure for azimuthal bins containing expected peak ρ_E. Middle: azimuthal fitting of radial maxima. Bottom: longitudinal fitting of azimuthal maxima.

present. Figure 16 also demonstrates the three-stage fitting procedure outlined in Section 6.2. The upper row shows, for each of four z-bins, the calculated energy deposition in the azimuthal bin which includes $\phi = 0$ of the beampipe wall and of the two superconducting coils (as histograms), along with a curve representing the fit to the modified exponential. The maximum energy deposition in the coils, as determined from the fits for each ϕ location, is then shown as histograms in the middle row as function of ϕ for each of the four z-bins. It is compared with Gaussian fits (assuming the peak occurs at $\phi = 0$, i.e., radially inside in the median plane). The bottom row presents ρ_E^{max} as determined from the azimuthal fits again as histograms along with quadratic fits assuming the ϕ-distribution peaks either at zero or at its centroid. The overall ρ_E^{max} as obtained from the quadratic fits is seen to be quite insensitive to the ϕ-fit procedure. Given the statistical uncertainty, the fits represent the underlying histograms quite well.

Energy deposition levels of, e.g., those encountered in Fig. 16 are potentially troublesome when both energy and intensity of the accelerator approach their design values. At the time of these calculations, before commissioning of the superconducting accelerator, a number of protection schemes (in addition to the bump) against radiation-induced quenching were analyzed. (Figure 16 with the presence of a collimator and thick-walled pipe, is of this kind.) Figure 17 summarizes the results of these calculations for a "worst case" phase-space beam of 1000 GeV/c and a radially-out bump. Each plot represents a different protection scheme roughly in order of effectiveness. Each point in a given plot corresponds to the maximum energy deposition in the coils of the magnet (see Fig. 16). The abcissa marks the distance from the end of the last bump magnet.

Figure 17a shows the unprotected case, i.e. only the bump is present. Figure 17b demonstrates the effect of collimation: a stainless steel collimator 66 in. long is placed directly upstream of the last bump magnet with its inner edge at 2 mm from the extracted beam. A dramatic reduction (by about two orders of magnitude) results for the quadrupoles but the dipoles are almost unchanged, i.e. the extracted beam and the high energy secondaries are insufficiently separated for the collimator to be effective. The effect of shielding the front face of the first quadrupole with a 32-in.-long, thick-walled pipe is shown in Fig. 17c. The outer radius of the pipe (7 cm) is sufficient to shadow the superconducting coils completely, and ρ_E^{max} is reduced by roughly a factor of five in the quadrupoles. Figure 17d combines both collimator and thick-walled pipe. (Results for the third dipole of this case are shown in some detail in Fig. 16.) ρ_E^{max} is essentially the same as the collimator-only case of Fig. 17b. To reduce ρ_E^{max} in the dipoles the use of inserts (or, equivalently, thicker beampipes) is investigated. Figures 17e-g show results obtained when the thickness of the 0.032-in. stainless steel beampipe is increased by a factor of two, four, and five, respectively. In six of the eight dipoles ρ_E^{max} is significantly reduced, but for each of the two dipoles which follow a quadrupole

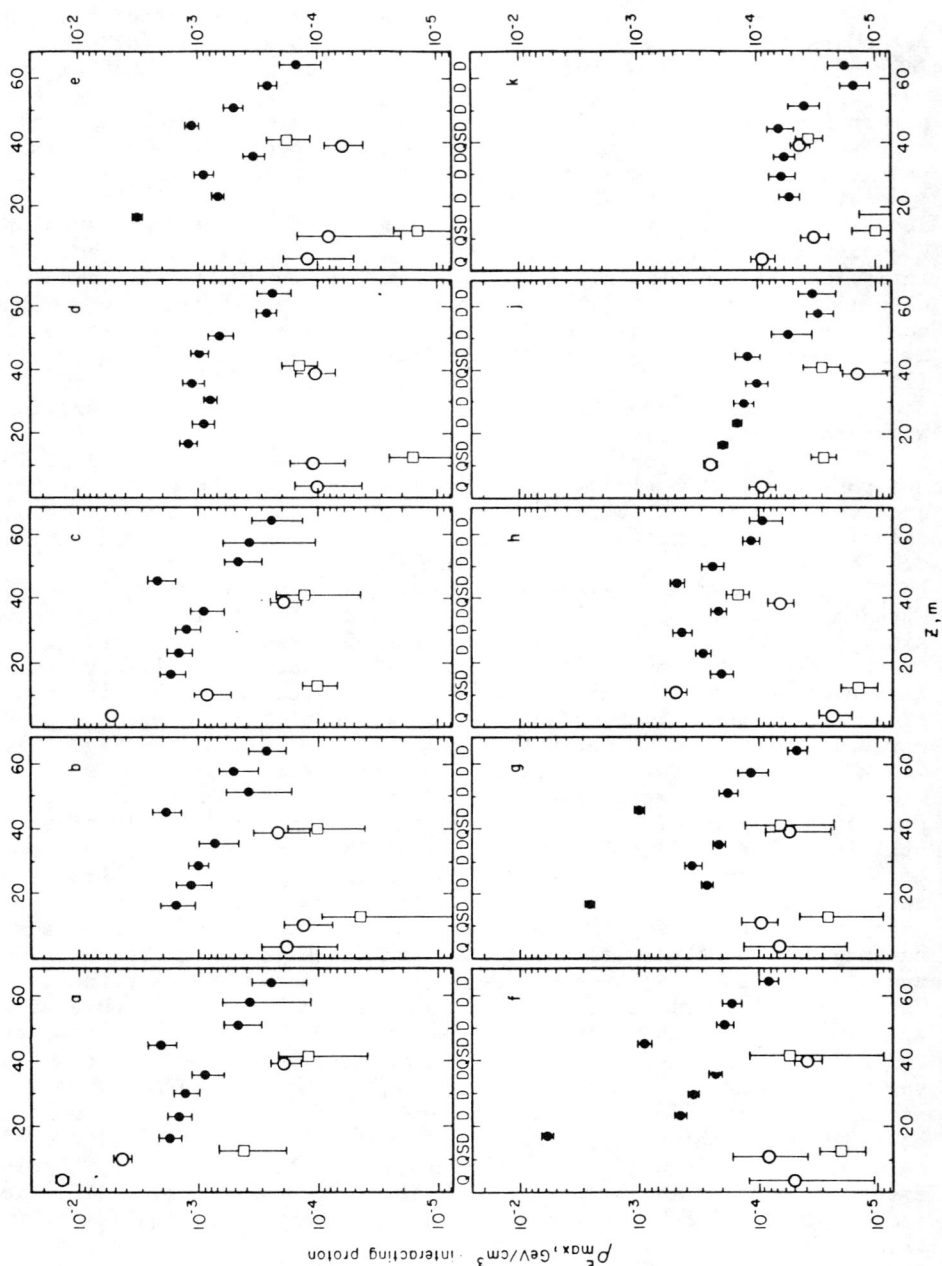

Figure 17 Effect of various protection schemes (see text) on maximum energy density in the superconducting magnets downstream of D0.

(and spoolpiece) there is a marked increase. This is due to an abrupt change in beampipe cross section at the spoolpiece-dipole interface resulting in large losses there. By inserting stainless steel plugs of uniform inner radius in all elements, ρ_E^{max} in these two magnets is reduced to levels comparable to the other six. This is demonstrated for plugs of inner radii of 2.9 cm and 2.5 cm respectively in Figs. 17h,j. Figure 17k shows the effect of higher density of the inserts when the steel of Fig. 17j is replaced with tungsten.

In addition to the 4-bump the only protective measure actually installed in the Tevatron is the 66-in. collimator, corresponding to the situation in Fig. 17b. The installation of aperture inserts poses (as yet unanalyzed) mechanical and cryogenic problems and may also affect magnetic field quality. During two six-month-long running periods with beam energy up to 800 GeV/c, no beam-induced quenches occurred in the D0 vicinity. This experience does not contradict the calculated results (for 1000 GeV/c) here. No detailed simulations of the inelastics at 800 GeV/c have been performed to compare with observations (e.g. absence of quenches, response of beam loss monitors around D0). It is therefore possible that, as both the intensity and energy of the machine are raised, some of the above protective measures need yet be implemented.

8.2 Elastics

The elastically scattered particles emerging from the septum differ relatively little from non-interacting beam particles. Typical angular spreads are shown in Figs. 11b and 12b. The momentum distribution for the case of an 800-GeV/c realistic phase-space beam is presented in Fig. 18. From inspection of these graphs it is clear that such particles tend to remain in the machine aperture for long distances, up to several turns, though they leave eventually since the septum lies outside the stable phase space during extraction.

Figure 19 presents a calculated phase-space distribution of the scattered beam at D17, the first high dispersion point in the lattice downstream of the septum. The tails of this distribution show a strong dependence of phase-space density on position. This is typical of many locations around the ring. Experimental information on these tails is obtained from BLM readings at a given place in the ring by varying the beam position near that location using a closed dipole 3-bump. Figure 20 compares measured beam loss versus orbit position at F28 with predicted values, as calculated from the projected phase-space distribution. There is good agreement, especially in view of the large dynamic range in beam loss covered. The beam displacement is limited to avoid quenching.

Ring-wide losses are studied by tracking a sample of elastically scattered particles, resulting from 10^5 protons incident on the septum, through the machine lattice (see Section 5) until all are lost either by extraction or by striking the beampipe. The latter are recorded on a file to serve as input for energy deposition calculations. These "hit" distributions also provide a

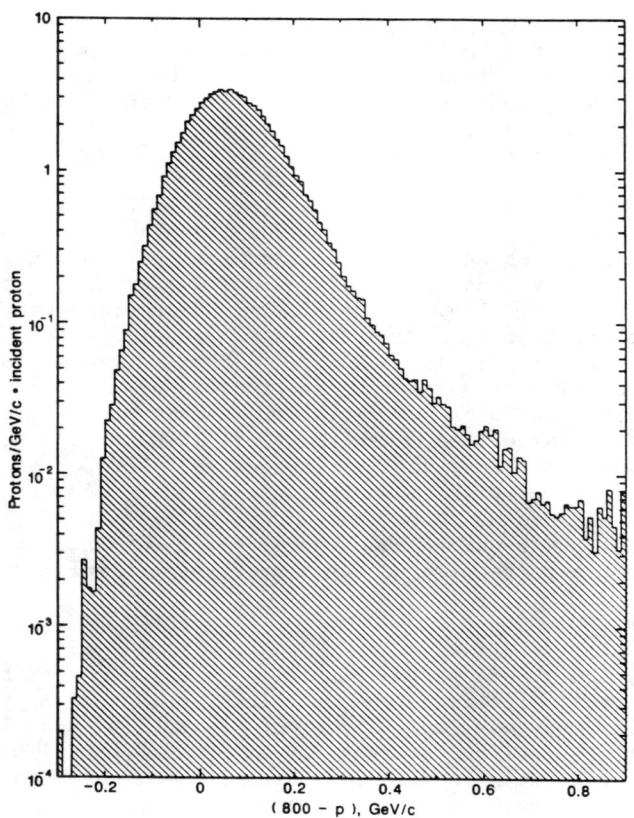

Figure 18 Momentum spectrum of protons elastically scattered in the electrostatic septum.

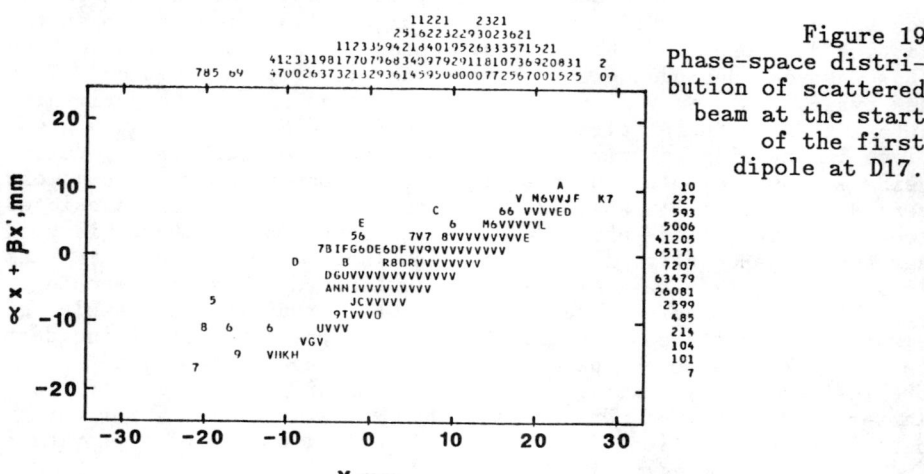

Figure 19 Phase-space distribution of scattered beam at the start of the first dipole at D17.

convenient overview of the energy flow of the losses. This is summarized in Table I, which shows that essentially all particles leave the machine within three turns. Removal by extraction (58.1%) and by inelastic interaction at D0 (22.9%) are the main outcomes. Losses in the superconducting elements are confined to D, E, and F sectors, predominantly on the first turn, and amount to about 10%.

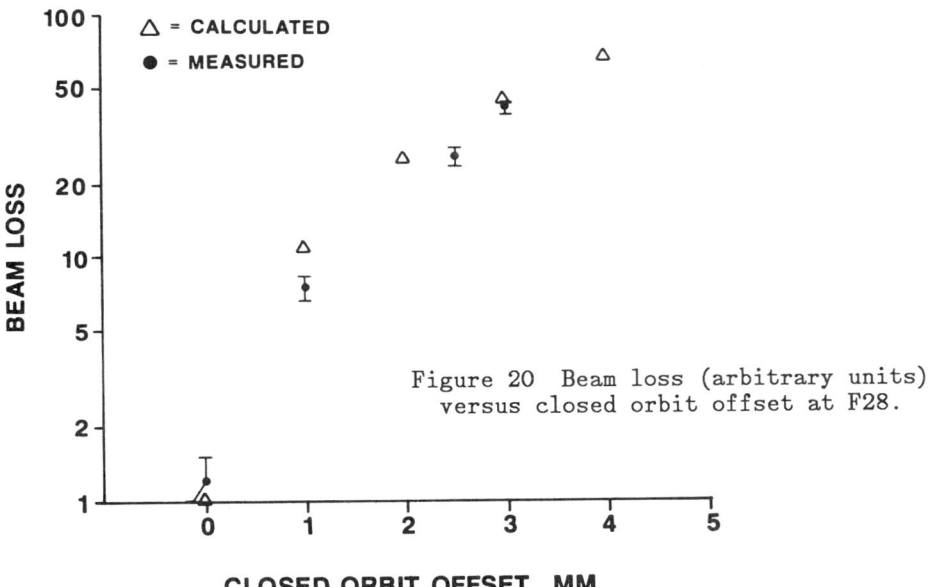

Figure 20 Beam loss (arbitrary units) versus closed orbit offset at F28.

Table I Particle Loss Distribution

	Location	Losses (%)
First Turn:	D0 (inelastics)	22.9
	D sector	3.2
	E sector	3.6
	F sector	2.2
	A0	3.8
	Extracted	32.4
Second Turn:	No losses	
Third Turn:	D0	3.1
	E sector	0.9
	F sector	0.3
	A0	1.6
	Extracted	25.7

A set of scatter plots of first-turn elastics traversing the extraction channel is shown in Fig. 21. The "notch" observed in the x, x' distibutions downstream of the septum (see e.g. Fig. 11) is clearly visible and shadows the magnetic septum. The majority of particles remaining in the machine aperture on the first turn enter the extraction channel two turns later.

BLM recordings of a typical loss distribution between D and A sectors for a fast extraction cycle are shown in Fig. 22. Between D17 and F49 there are 14 locations with significant losses, separated by regions which are virtually loss-free. In Fig. 23 these same data are compared with calculated hit distributions which are normalized to the data so as to yield the same loss integrated over all locations. The calculation predicts significant losses in 12 of the 14 locations and in no case predicts a loss where none occurs. In the remaining two locations, the beam position is within 1 mm of scraping the vacuum chamber wall. However, it is also clear that generally the predictions do not correspond very well in magnitude to the observed values.

One obvious reason is the shortcut of comparing (even after normalization) hits with beam loss monitor response, but this cannot account for large differences, except perhaps at the atypical F49 loss location (see below). The main source of disagreement seems to stem from a lack of sufficiently precise information on the position of the beam with respect to the beampipe. As mentioned in Section 5, the magnetic tracking includes empirical orbit distortions obtained from beam position detectors located at the quadrupoles. At any given location, the best information on beam position derives from such measurements at the neighboring quadrupoles. Table II combines the uncertainties of these measurements with estimated mechanical tolerances of magnet

Table II Beam Position Tolerances (mm)

BPM resolution	0.16
BPM to quad alignment	0.6
Quad to dipole alignment	0.4
Dipole to beam tube alignment	0.8
Total	1.2

alignment to arrive at an overall uncertainty of beam position with respect to the beampipe. The result is an rms value of 1.2 mm. The phase space of Fig. 20 predicts variations in hit distributions of up to an order of magnitude over this range. The error bars on the predicted values in Fig. 23 correspond to the change in number of hits resulting from a ±1 mm variation in magnet position. The number of particles striking the beampipe at a given location is also affected by the alignment of upstream magnets, especially those at loss points immediately upstream. Uncertainties due to this "shadowing" are even harder to assess. The hit distribution shows little or no sensitivity to the initial momentum distribution of the circulating beam, relative alignment of the septum modules, or geometric detail of the dipole interfaces. The loss distribution is

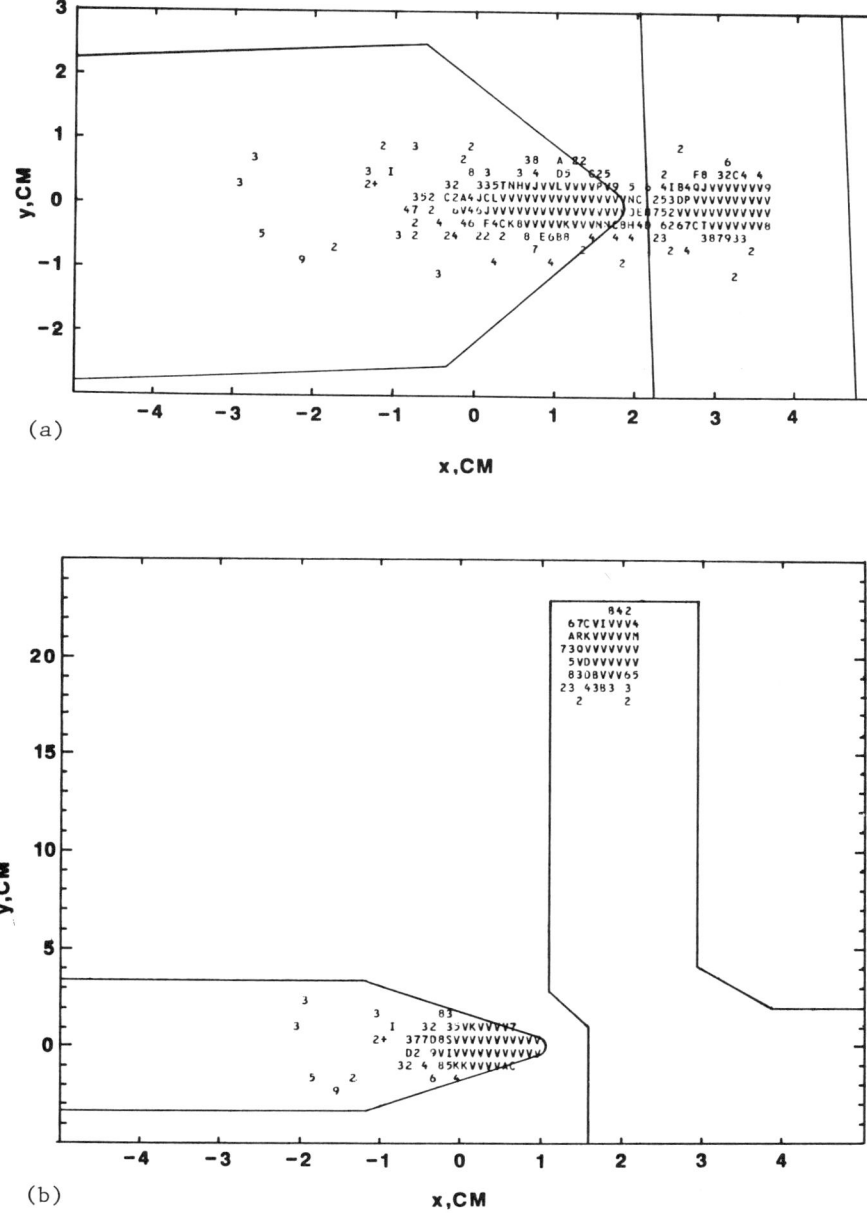

Figure 21 x-y plot of scattered beam: (a) at the start of the extraction channel on the first turn; (b) at the end of the channel.

affected significantly when the magnetic fields associated with the extraction process are turned off. This increases the stable phase-space area in the machine so that fewer particles depart the aperture in the bending arcs but are lost instead in the region of the extraction channel.

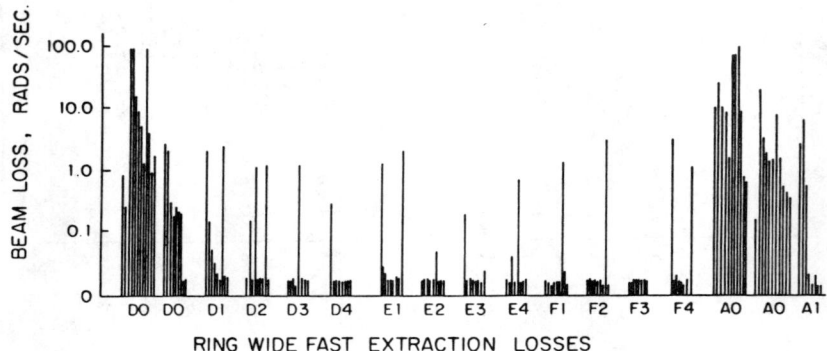

Figure 22 Typical loss pattern between D and A sectors during fast extraction as measured by BLM system.

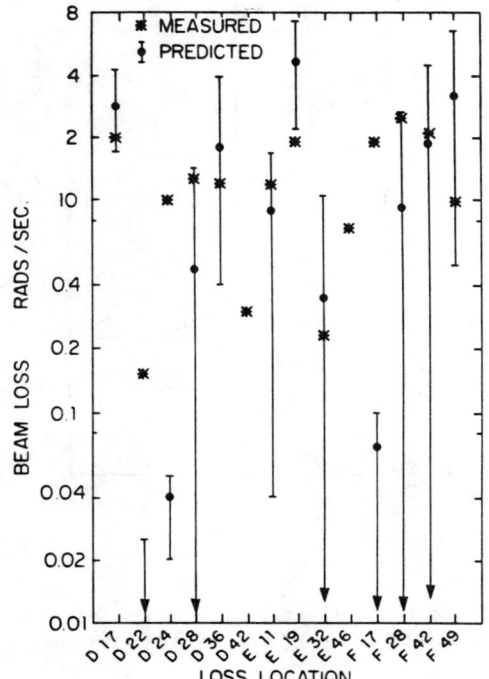

Figure 23
BLM readings of Fig. 22 compared with total hits computed at each location and normalized to same integrated loss.

The large errors inherent in predicting beam loss discourage attempts at ab initio calculation of either energy deposition or BLM response at the typical loss point. Progress is made by incorporating the (measured and calculated) BLM response into the ρ_E^{max} calculations, at the expense of what ideally is an independent test of the calculation. The ratio of the maximum energy deposition in the superconducting coils to the energy deposition in the BLM does not vary much with the total number of hits and its associated geometric sensitivity. Figure 24 shows longitudinal hit distributions in the last dipole of F28 for orbits with offsets of 0.5, 1.0 and 1.5 mm. Although the number of hits changes by a factor of six over this range, the ratio of predicted ρ_E^{max} to BLM response varies by only 30%. This ratio, along with an observed BLM reading, can thus serve to "measure" ρ_E^{max}. Variations at the 30% level are not necessarily significant in the present context, but complete agreement between predicted ρ_E^{max} and BLM response is not expected. Figure 24 shows that as the magnet is moved into the beam, the hit distibution broadens significantly, and also that ρ_E^{max} changes little between the 1-mm and 1.5-mm magnet displacements. The latter means that the broadening of the hit distribution starts exceeding the spread (in ρ_E) of the typical individual shower at the inner radius of the superconducting coils, so that ρ_E^{max} will fail to grow proportionally to the number of hits. The BLM, located at a point where the shower spread is much larger, maintains proportionality somewhat longer.

The method to measure ρ_E^{max} outlined above is applied to two deliberate beam-induced magnet quenches at locations F28 and F49. F28 represents a typical major loss point with beam striking the downstream end of the last dipole in the half cell. At F49 the high β value of the lattice confines beam loss entirely to the second quadrupole of the straight section doublet. This is one of the few places in the ring (and the only major one) where the larger aperture quads experience beam loss and hence represents a radically different geometry from F28. The experimental procedure is to move the beam position at the quench location so as to enhance the losses there and then to raise the intensity by ~10% increments until the element quenches.

The calculated radial, azimuthal, and longitudinal characteristics of the energy deposition at each location are shown in Figs. 25 and 26. Comparing these results with their counterpart for inelastics (Fig. 16), shows a much narrower azimuthal distribution with a steeper radial dependence. This narrower distribution means that, for comparable losses, energy densities (and hence temperature rise) are much larger for the elastics. This is borne out by operational experience, where magnets close to D0 are observed to tolerate much larger losses (as recorded by the BLMs) without quenching than those in the bending arcs. The calculated maximum energy densities in the superconducting coils corresponding to quench threshold are 8.8±5.1 mJ/g at F28 and 6.5±3.8 mJ/g at F49. This compares to an estimated 4.5 mJ/g from Fig. 13-1 of the Tevatron Design Report[1] as derived from tests on prototype accelerator magnets in external beamlines and interpreted

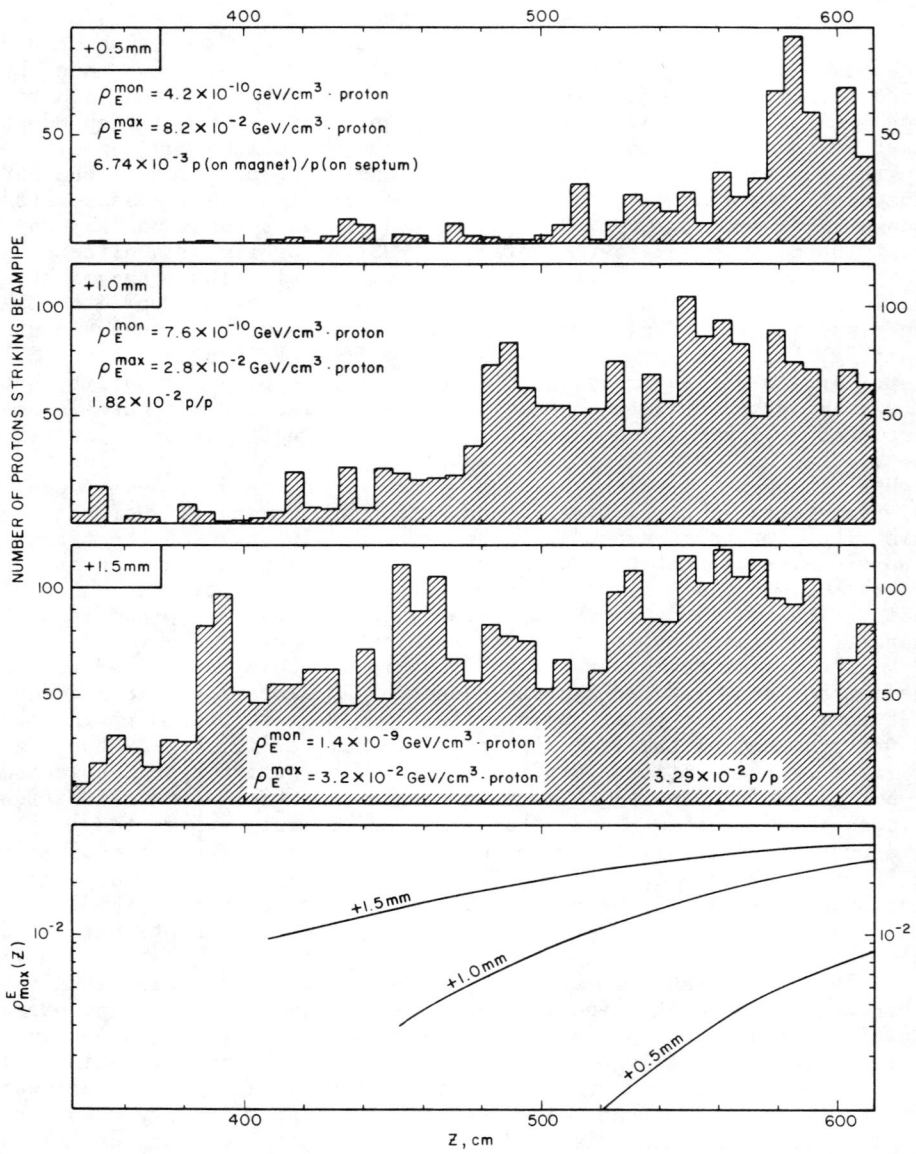

Figure 24 Number of hits as a function of z (distance along magnet), calculated for last dipole at F28 (top three graphs) and for 0.5, 1.0, and 1.5 mm beam offsets. Maximum energy density as a function of z, for same beam offsets.

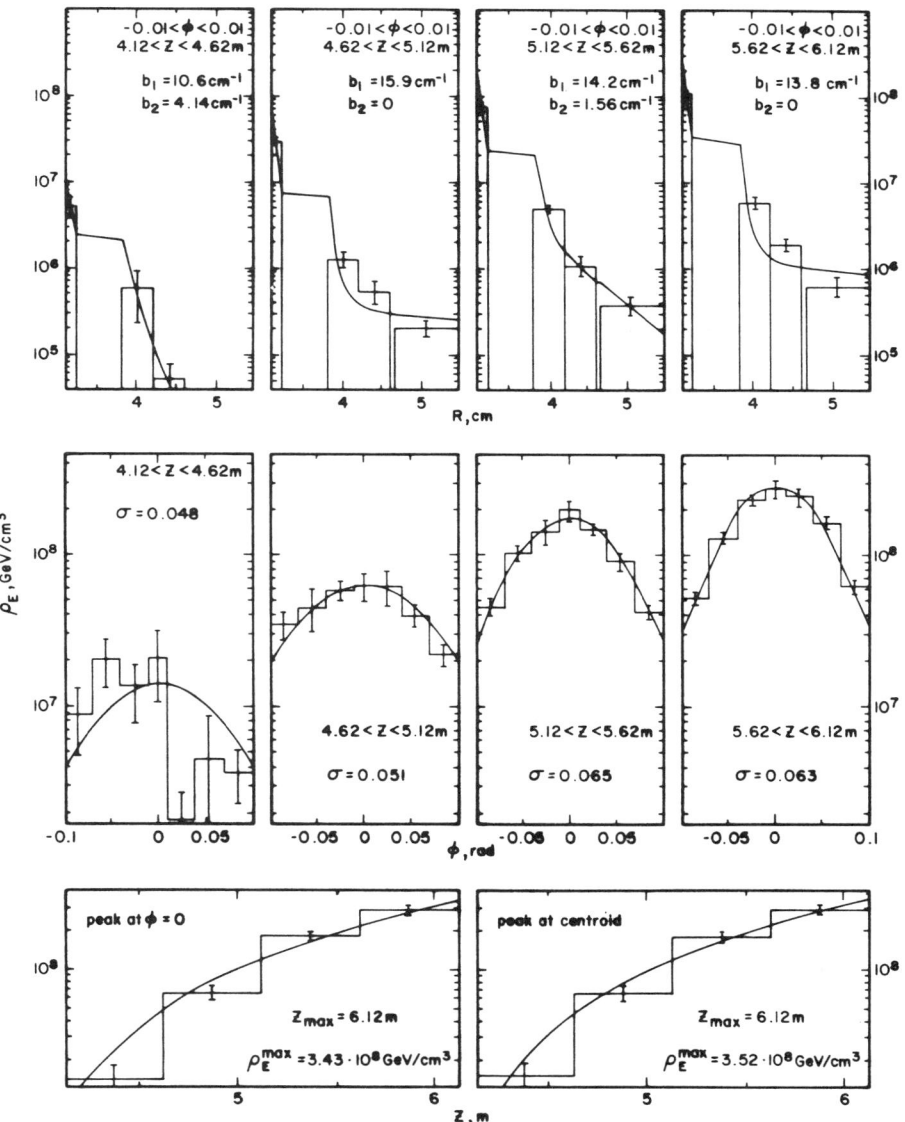

Figure 25 Calculated energy deposition in fourth dipole at F28. Top: radial fitting procedure for azimuthal bins containing expected peak ρ_E. Middle: azimuthal fitting of radial maxima. Bottom: longitudinal fitting of azimuthal maxima.

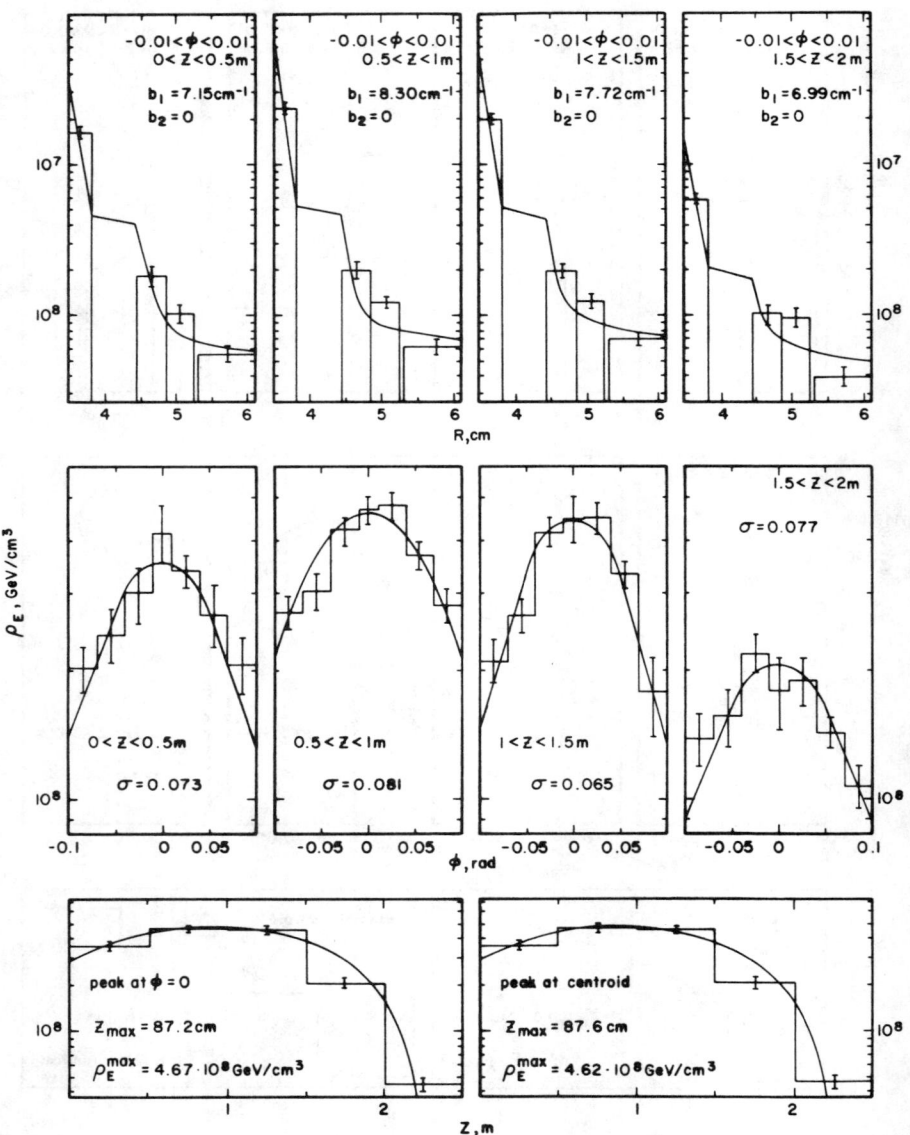

Figure 26 Calculated energy deposition in second quadrupole at F49. Top: radial fitting procedure for azimuthal bins containing expected peak ρ_E. Middle: azimuthal fitting of radial maxima. Bottom: longitudinal fitting of azimuthal maxima.

with the help of CASIM Monte Carlo calculations. Since dipoles run closer to the short sample limit than quadrupoles, a higher quench threshold is expected at F49 than at F28. Other than the considerable calculational uncertainty there is no obvious reason why this is not observed.

The stated rms errors of the quench thresholds themselves carry considerable uncertainty. They are arrived at by combining in quadrature a number of estimated errors associated with the derivation of the quench thresholds: Increasing beam intensity in ~10% steps leads to a 5% error in the quench level. The combined (ρ_E^{max} and BLM) statistical uncertainties of the calculations are evaluated in the Monte Carlo procedure and amount to ~20%. Based on comparisons with target heating experiments,[8] systematic errors associated with the ρ_E^{max} calculation are ~15%. The uncertainty of the ratio of ρ_E^{max} to BLM response varies somewhat with beam position and this is assumed to contribute 30%. The dominant source of error appears to be the systematic error associated with the calculation of BLM response. Section 7 describes how the BLM energy deposition is dominated by low energy neutrons. The treatment of these neutrons and their energy losses in the BLM structure is mostly empirical and as yet uncorroborated by any other experience. Rather arbitrarily, an error of 50% is assigned to this procedure.

9 CONCLUSIONS

The combination of hadron/electromagnetic cascade plus elastic scattering codes with accelerator tracking routines appears to be quite useful in attacking problems of the type encountered here. Such calculations can be valuable tools in design work. Indeed, preliminary versions established that radiation-induced quenching could, with proper precautions, be overcome and that therefore intense beams could be extracted from the Tevatron. This has obviously been shown to be true. More generally, there has been no demonstrable contradiction between experience and observation at the Tevatron and any of the more detailed predictions of the type reported here. Likewise these calculations prove useful when it comes to evaluating competing designs as, e.g., the analysis of the bump orientation in Section 8.1, or the various solutions to protect the superconducting dipoles downstream of D0.

While truly quantitative comparisons seem elusive, at least the reasons for this condition are well understood. But even the results of ab initio calculations are sufficiently close to the mark to merit attention. The comparison between ring-wide losses and calculated number of hits nearby illustrates this point. When other information is brought to bear on the problem, agreement with observations becomes at least semi-quantitative, as witnessed by the evaluation of quench thresholds at F28 and F49. In spite of a rather cavalier approach to the calculation of BLM response, results of the two calculations along with the value of the Tevatron Design Report[1] all fall within a factor of two of each other, thereby lending encouragement to further use of these techniques.

Finally, the calculations performed here show that the underlying hypothesis of a limiting energy deposition density, above which a radiation-induced quench is expected to occur, is valid. Information about heat transport and quench propagation in a magnet coil is, obviously, very useful. But this study shows that for the radiation-induced quenching problem this information can be condensed into the specific limiting value and that this simplifying and labor-saving assumption may be applied to magnets of reasonably similar design.

We wish to thank H. Edwards for numerous discussions and suggestions on this subject. S. Childress likewise contributed his insights during the course of this work. S. Snowdon provided convenient field maps of the various magnets. M. Awschalom and W. Freeman lent their time and expertise to the BLM calibration. C. Rad participated in the early stages of this study.

10 REFERENCES

1. A Report on the Design of the FNAL Superconducting Accelerator, Fermilab (1979).
2. A. VanGinneken, Fermilab-FN-272 (1975).
3. H. Edwards, C. Rode, and J. McCarthy, IEEE Trans. Magn. $\underline{1}$, 666 (1977); B. Cox, P. O. Mazur, and A. VanGinneken, Fermilab-TM-828A (1978); H. Edwards, S. Mori, and A. VanGinneken, Fermilab-UPC-30 (1978); UPC-40 (1979).
4. R. Hagedorn, Suppl. Nuovo Cim. $\underline{3}$, 147 (1965); R. Hagedorn and J. Ranft, ibid. $\underline{6}$, 169 (1968); H. Grote, R. Hagedorn, and J. Ranft, Atlas of Particle Spectra, CERN (1970).
5. J. Ranft, TUL-36, Karl Marx Univ., Leipzig, DDR (1970).
6. J. V. Allaby et al., CERN-70-12 (1970).
7. A. VanGinneken, Fermilab-FN-260 (1974).
8. M. Awschalom et al., Nucl. Inst. Meth. $\underline{131}$, 235 (1975); $\underline{138}$, 521 (1976).
9. A. VanGinneken, Phys. Rev. D $\underline{37}$, 3292 (1988).
10. A. VanGinneken, Fermilab-FN-309 (1978).
11. A. VanGinneken and M. Awschalom, High Energy Particle Interactions in Large Targets, Fermilab (1975).
12. E.g., a code such as MORSE, E. A. Straker et al., ORNL 4585 (1970).
13. D. I. Garber and R. R. Kinsey, Neutron Cross Sections Vol. II, BNL-325 (1976).
14. W. S. Snyder, in Protection against Neutron Radiation, p. 46, NCRP Report 38, Washington DC (1971); R. G. Alsmiller, Jr., ibid. p.86.
15. I. Dostrovsky et al., Phys. Rev. $\underline{111}$, 1659 (1958).
16. S. J. Lindenbaum, Ann. Rev. Nucl. Sci. $\underline{11}$, 213 (1961).

DESIGN AND OPERATION OF THE QUENCH PROTECTION SYSTEM FOR THE FERMILAB TEVATRON

Philip S. Martin
Fermi National Accelerator Laboratory
P. O. Box 500, Batavia, IL 60510

TABLE OF CONTENTS

1	Introduction..	2074
2	Quench Characteristics of Superconductors....................	2074
	2.1 Quench Threshold..	2076
	2.2 Quench Propagation--Theory..............................	2076
	2.3 Quench Propagation--Experimental Results................	2077
3	Temperature Increase in the Adiabatic Limit..................	2077
	3.1 Comparison with Experimental Results....................	2079
4	Design of the Fermilab Quench Protection System..............	2080
	4.1 Layout of the Tevatron Superconducting Accelerator......	2080
	4.2 Quench Protection System................................	2084
	4.3 Quench Detection System.................................	2086
5	Operational Experience at the Tevatron.......................	2090
	5.1 Failures During Operation...............................	2090
	5.2 Other Aspects of Operating a Superconducting Accelerator...	2093
6	Summary..	2094
7	References...	2095

Chapter 11
DESIGN AND OPERATION OF THE QUENCH PROTECTION SYSTEM FOR THE FERMILAB TEVATRON

Philip S. Martin
P. O. Box 500, Batavia, IL 60510

1 INTRODUCTION

The operation of a superconducting accelerator, in addition to cryogenic requirements, introduces a new complexity not present in a conventional accelerator. A method is required for protecting the magnets from possible overheating or overvoltage conditions in the event that some magnets quench, that is, are elevated in temperature so that they are no longer superconducting. The development of that system is the topic of this chapter.

Any quench protection system has two very important ingredients. First, it must be designed with sufficient integrity to remain functional even under abnormal circumstances. The magnets must be protected during power failures, for example. Quenches involving a large number of components can also be hazardous because of the redistribution of voltages during the quench. Some of the system integrity can be achieved through redundancy. Frequent testing of critical elements of the system also assures the overall integrity. Second, the quench protection system must protect against damage from quenches regardless of their location or the excitation current at the time. It is not sufficient to protect just the magnet coils; either the leads between magnets must be fully stabilized or the quench protection system must protect them.

The next section presents a brief discussion of the basic properties of superconductors and the phenomenon of quench propagation. A more complete discussion can be found[1-3] elsewhere.

2 QUENCH CHARACTERISTICS OF SUPERCONDUCTORS

For a given superconductor, there is a surface in the space defined by temperature, magnetic field, and current density which marks the boundary between the superconducting and normal states. In accelerator magnets, in which large quantities of superconductor are required, cost constraints force the magnet designers to minimize the amount of superconductor, and therefore the magnets operate as close to the boundary as possible. It is often convenient to reduce the surface in three-dimensional space to a set of curves of current density versus temperature, with each curve representing the critical current density as a function of temperature at a fixed magnetic field. For any particular magnet design there is a transfer function relating the excitation current and the resulting mgnetic field. This reduces the set of curves to a single curve (Fig. 1), which gives the maximum operating current as a function of temperature. At some particular operating current density, there is similarly some critical temperature above which the superconductor becomes resistive. The superconductor used in

© 1989 American Institute of Physics

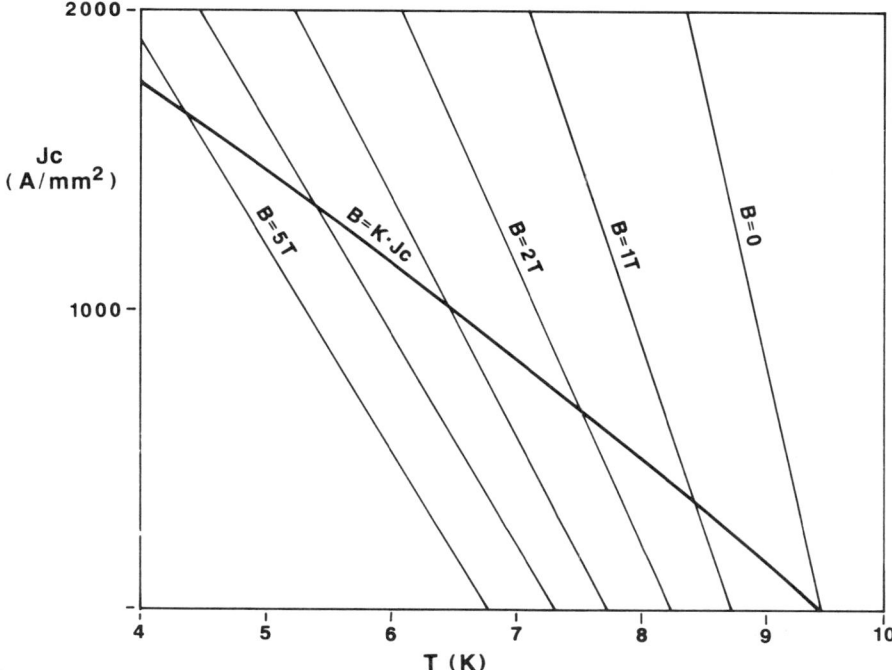

Figure 1 Curves of critical current density vs. temperature for fixed magnetic fields and for a magnetic field proportional to J_c.

the Tevatron was specified to have a critical current density of 1800 A/mm^2 at 5 T and 4.2 K, for the individual strands. Some degradation occurs in the cabling and winding process. In recent years substantial improvement has been made in the quality of superconductor, and current densities >2500 A/mm^2 are now obtainable, although industry has not yet produced superconductor with such a current density on a large scale.

The transition from the superconducting state to the normal state might occur through elevation of the conductor to a temperature above the critical temperature, perhaps due to beam losses, or it might occur at a particular temperature, if the current and field are increased beyond the critical current I_c. The magnetic field seen by the superconductor is not the same as that in the beam aperture. The conductors see an entire range of fields from zero up to a peak field which might exceed that in the beampipe. Therefore, the concepts of critical current and critical temperature are local properties.

The Fermilab magnets are constructed of superconductor consisting of 8-micron filaments of niobium-titanium imbedded in a copper matrix; 23 strands, 0.068 cm in diameter and containing 2100 filaments each, are formed into a keystoned cable.[4] Although much of the discussion here applies specifically to this cable, it can be easily generalized to apply to magnets composed of other types of superconductor and other sizes of cable.

The copper matrix serves two important functions. First, it stabilizes the superconductor by providing an alternative current path, allowing small regions of the superconductor to become resistive for short periods and then recover as they are cooled back to the superconducting state by heat exchange with the environment. Second, in the event a larger section of superconductor becomes normal and cannot recover, it prevents overheating. Niobium-titanium, like other superconductors, is a rather poor conductor once it has exceeded the critical temperature, and the ohmic heating in the absence of the copper would quickly destroy the magnet.

2.1 Quench Threshold

Imagine a superconducting magnet, operating at some current and temperature T_b less than the critical temperature T_c (at that current and magnetic field). The suprconductor will make a transition to the normal (resistive) state if the temperature is raised above T_c. The energy required to raise the temperature from T_b to T_c is determined by the enthalpy of the constituents. For metals at low temperatures, the specific heat varies as T^3 (plus a linear term which is comparable in magnitude at 5 K). For copper, niobium, and titanium, the specific heats are in the range 1 to 5 mJ/cm^3 K. Cables like the Fermilab cable contain about 10% open area between the strands. That space is filled with liquid helium, whose specific heat is far larger than the specific heat of the conductor itself. The details of how the helium participates is not clear; furthermore, it may well depend on the process of interest. For example, fast energy deposition, due to beam loss, is certainly different from slow processes such as eddy-current heating. In the case of quench propagation, discussed in the next section, it seems reasonable to assume that the helium in the normal region has been vaporized. It then expands and displaces the liquid helium surrounding the still superconducting cable. Even under these assumptions, the enthalpy of the gaseous helium is an order of magnitude higher than that of the cable constituents. Experiments have demonstrated[5] that the energy required to initiate a quench varies roughly as 1/I over the range 0.1 to 0.9 I_c. At 0.9 I_c, the energy required is about 4 mJ.

2.2 Quench Propagation--Theory

If some small region of superconducting cable has been elevated in temperature so that it becomes resistive, the region will expand (or contract) according to the balance between the power generated by the resistive section and the heat absorbed by the liquid helium. The heat equation in one dimension is

$$\frac{d}{dx}\left(k \frac{dt}{dx}\right) - c \frac{dT}{dt} - \frac{hP}{A}\left(T - T_b\right) + J^2 \rho = 0$$

where k is the thermal conductivity and c the specific heat of the superconductor, h the coefficient of heat transfer, P the cooled perimeter, A the area of the superconductor, T_b the helium (bath) temperature, and $J^2\rho$ the power density in the normal zone. By transforming to a coordinate system moving with velocity v, one obtains an equation for the motion of the normal zone. The solution to this equation generally is obtained after making a number of assumptions. The thermal conductivity and specific heat are taken to be constants, and the temperature of the normal region is assumed to reach a thermal equilibrium. These assumptions are over-simplifications, but they allow a solution to the equation. That solution takes the form

$$V = \frac{J}{c}\sqrt{\frac{\rho k}{(T_c - T_b)}} .$$

The formula predicts a velocity that grows slightly faster than linearly with current because of the current dependence of T_c. The T_c term also introduces a B-field dependence. As mentioned earlier, the specific heat of the helium in the Fermilab type of cable dominates that of the metallic constituents. Thus, knowledge of details of how the helium participates is necessary if one wishes to calculate quench velocities, either analytically or numerically. In the latter case, one need not make such simplifications as constant specific heats and thermal conductivities.

2.3 Quench Propagation--Experimental Results

The design of the Fermilab quench protection system relies in part on a set of measurements[6,7] made on a short (17-cm) sample of superconductor mounted in a "hairpin" fixture (Fig. 2). A heater was located at one end, and two voltage taps were installed for monitoring the resistance of the sample. A typical voltage signal, after induction of a quench with the heater, is shown in Fig. 3. The voltage growth has two distinct parts. The first part, the very near linear rise in voltage, is due to the propagation along the sample. Measurements at a variety of currents and with different magnetic fields yielded a formula for the quench velocity

$$V = 0.36I^2(1 + 0.077B^2)$$

where I is the current in kiloamperes, B the field in Tesla, and the velocity is given in meters/sec. This formulation holds for currents from 1 to 4 kA.

3 TEMPERATURE INCREASE IN THE ADIABATIC LIMIT

Once a section of conductor has quenched, the Joule heating will start increasing its temperature. If we neglect any heat transfer, the heat equation (setting k and h to zero) becomes simply

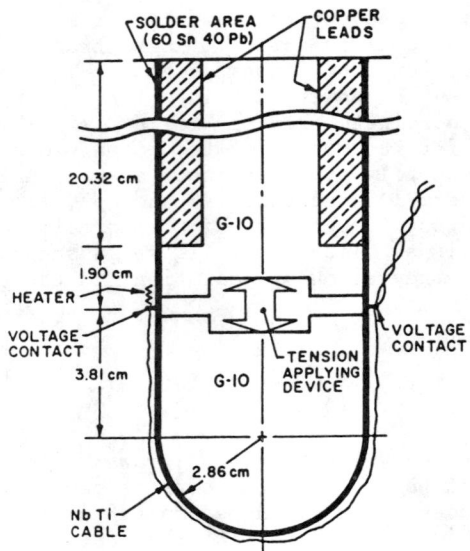

Figure 2 Hairpin fixture used for studying quench velocity and conductor heating.

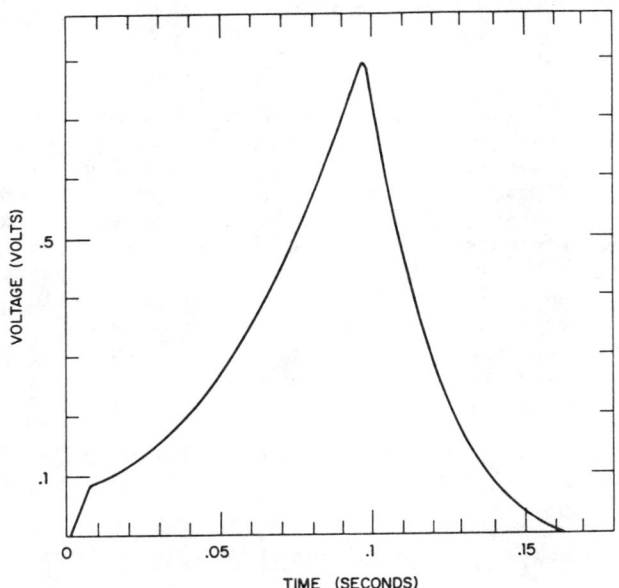

Figure 3 Typical voltage pattern observed in hairpin fixture. The power supply was turned off at 0.095 sec.

$$c \frac{dT}{dt} = J^2 \rho .$$

Let us work in terms of a unit volume of conductor. Then

$$J^2 \rho = I^2 \rho \Big/ \left(A^2 \frac{r}{r+1} \right)$$

where r is the ratio (by volume) of copper to superconductor and ρ is the resistivity of the copper matrix. Taking the resistivity to the other side and integrating, one obtains

$$A^2 \frac{r}{r+1} \int_{T_o}^{T} \frac{c}{\rho} \, dT = \int_0^t I^2 dt .$$

The integral on the right is often referred to as the MIITs (mega-amp-squared-seconds). In the adiabatic limit, the number of MIITs determines the peak temperature in the superconductor. Since we have neglected heat transfer, this is a worst-case calculation. The peak temperature will be somewhat less than that predicted by this formula.

3.1 Comparison with Experimental Results

Measurements[8] of the resistance vs. temperature for samples of superconductor are shown in Fig. 4. The residual resistivity ratio (RRR), i.e. the ratio of resistivity at 273 K to that at 4 K, is ~40 for cables from several different sources. The residual resistivity is a result of impurities in the copper. Above about 40 K the resistance of the sample can be used as a thermometer. The second part of the voltage growth in Fig. 3 is due to the increase in temperature (and resistance) of the conductor. These data have been replotted in Fig. 5 for several different currents. Changing variables, instead of voltage vs. time, we plot temperature vs. MIITs in Fig. 6.

The Fermilab magnets are interconnected with solder, and half of the strands are coated with solder to achieve a balance between proper current-sharing and eddy currents. Therefore, the melting point of solder (450 K) has been selected as the maximum temperature to be allowed during a quench. From Fig. 5 we see that this corresponds to roughly 7 MIITs at 4 kA. That is, within one-half second after a quench begins, the current must be reduced to essentially zero. The selection of 450 K is obviously somewhat arbitrary. Other limits worth noting are (i) degradation of insulation (Kapton, Mylar) begins around 600 K, and (ii) degradation of superconductor occurs above 800 K. Damage to materials like Kapton, however, depends on duration of the temperature level and on presence of oxygen in the environment.

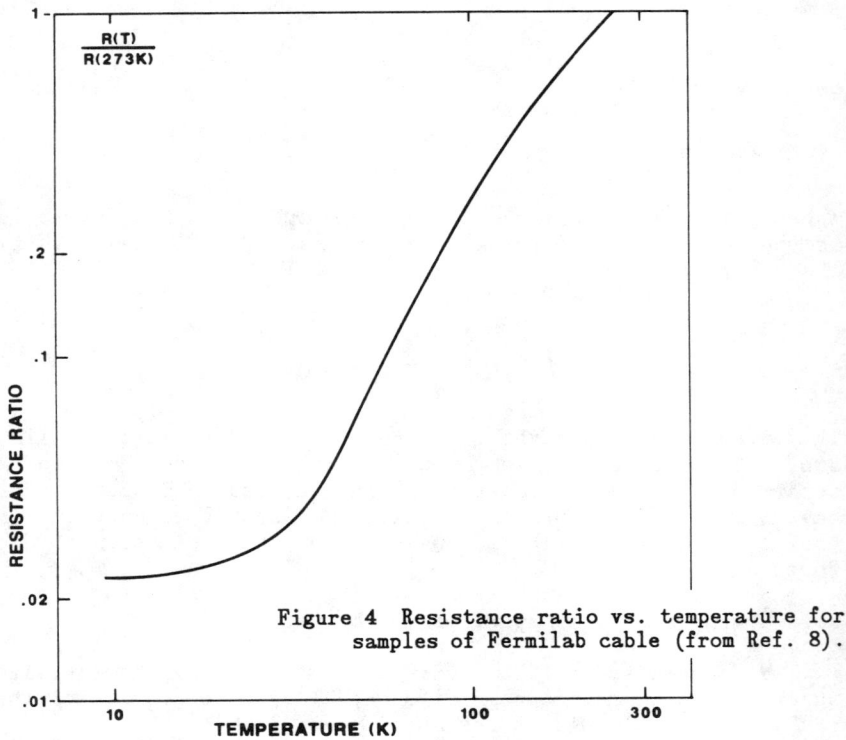

Figure 4 Resistance ratio vs. temperature for samples of Fermilab cable (from Ref. 8).

4 DESIGN OF THE FERMILAB QUENCH PROTECTION SYSTEM

Quench protection implies primarily the removal of the current from a quenching magnet before it overheats. To accomplish this requires two activities: (i) detecting the quench before it's too late, and (ii) taking the necessary action. First let us discuss the configuration of the Tevatron magnets.

4.1 Layout of the Tevatron Superconducting Accelerator

The magnets in the Tevatron are configured as shown in Fig. 7. The magnets and power supplies form a single series circuit consisting of an "upper" and "lower" bus connected at the B0 straight section. Each power supply is capable of ramping to 4500 A at 1 kV. One supply (A2) is the current regulating supply, and is capable of dc operation. The other eleven supplies operate in voltage regulation mode. The uniform spacing of the power supplies minimizes the peak voltage to ground. A 0.25-ohm "dump" resistor is also located at each power supply. When a quench or other fault is detected, the power supplies are turned off and the switches that

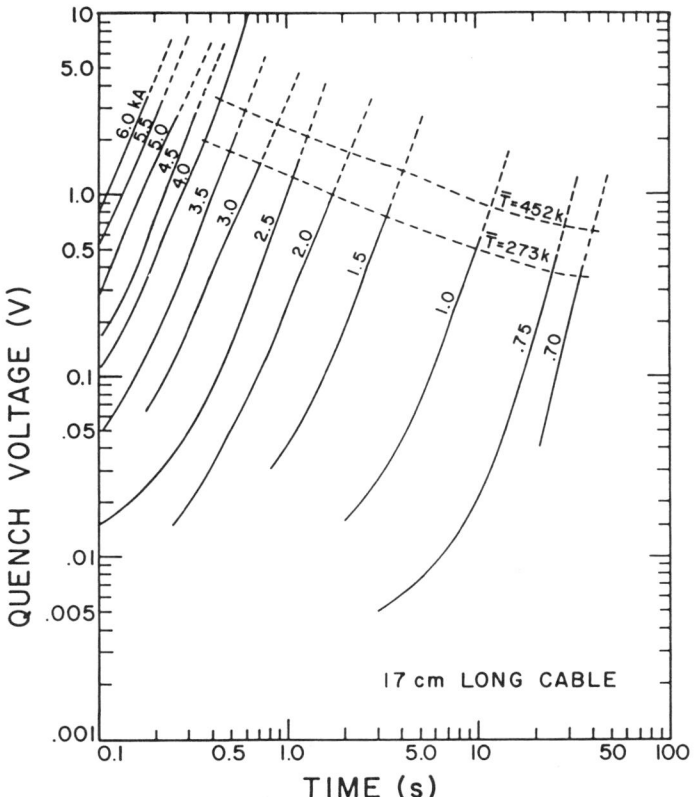

Figure 5 Voltage vs. time from hairpin measurements at different currents.

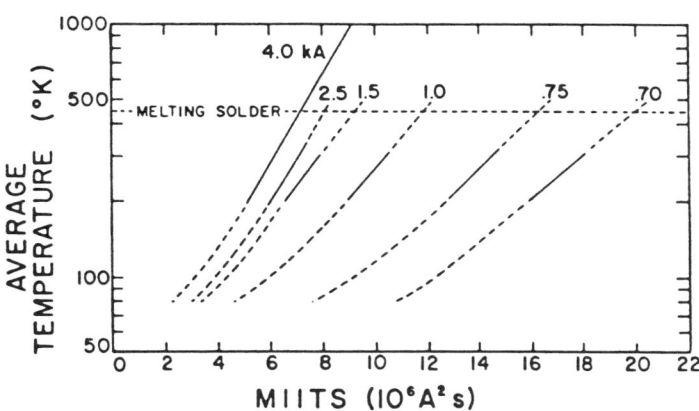

Figure 6 Temperature vs. MIITs.

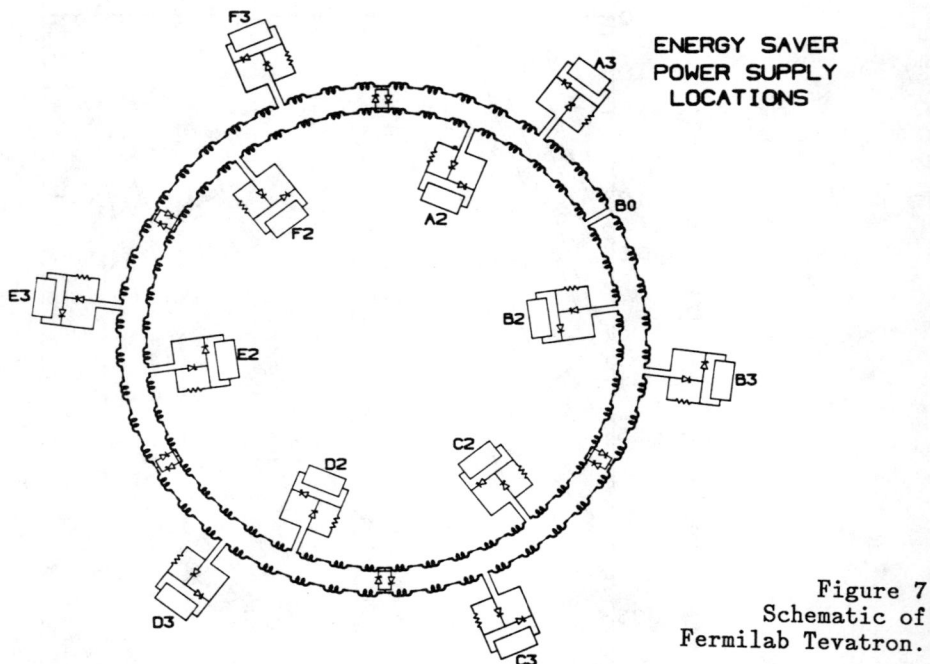

Figure 7
Schematic of
Fermilab Tevatron.

normally conduct current around the dump resistors are opened. The current in the bus then decays with a 12-sec L/R time constant. The dump switch is an SCR backed up by a dc contactor. A schematic of a power supply-dump rack is shown in Fig. 8.

The normal operating conditions of the Tevatron permit voltages up to 1 kV with a single ground fault. To maintain this during a dump requires an inductive cancellation of the voltage provided by the dump resistors. When magnets quench, their inductance is removed from the circuit, and if many magnets quench, substantial redistributions of voltage can occur. At each straight section (other than B0) the upper and lower busses are connected with a bipolar thyristor switch that is closed just before opening the dump switches. This decouples the sectors from one another during the dump, and prevents potentially destructive voltages from arising should many magnets quench simultaneously.

Figure 9 shows the layout of a typical lattice cell of the Tevatron. It consists of eight dipoles and two quadrupoles arranged in an "alternating bus" configuration. Each magnet is a four-pole device, with two leads on each end. The inductance of each magnet is concentrated either on the upper bus (TC type dipole, F quadrupole) or on the lower bus (TB dipole, D quadrupole). The inductance of a typical "half-cell," that is, of one bus of one cell, is about 0.2 H. The nominal ramp rate of the Tevatron is about 250 A/s, so the voltages observed are around 50 V.

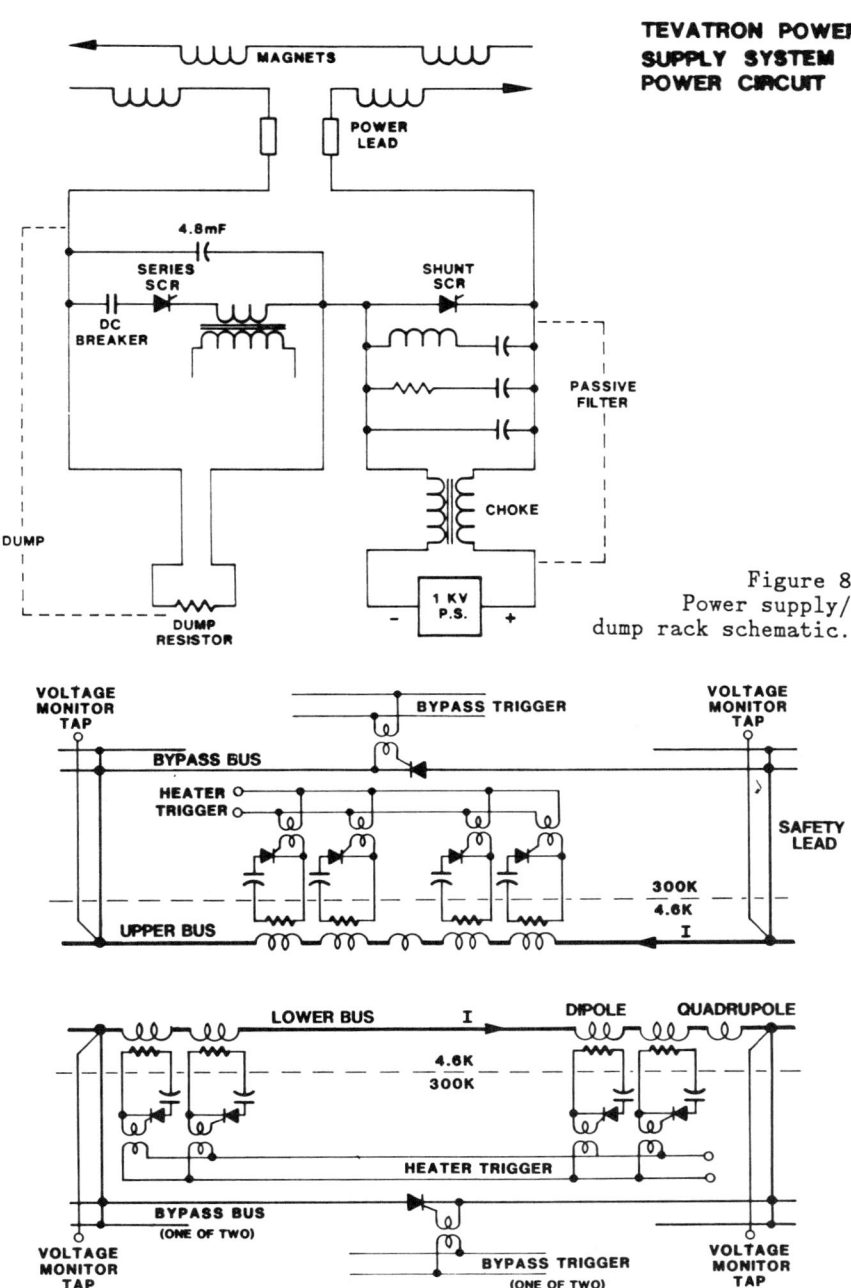

Figure 8 Power supply/dump rack schematic.

Figure 9 Typical cell layout.

4.2 Quench Protection System

The Fermilab quench protection system is an active one; it requires prompt detection of a quench and active components to remove the current from the quenching magnets. The heart of the system is the Quench Protection Monitor (QPM), a microprocessor that monitors the voltages across the cells at a 60-Hz rate. If a quench is detected, heater firing units (HFUs) are discharged into all the dipoles in the cell that is quenching. This drives a large fraction of the cell normal, resulting in a large resistance. The stored energy of the half-cell (2 MJ at full current), rather than being absorbed in the area of the quench origin, will be absorbed by all the coils in the half-cell. At the same time, quench bypass switches (QBSs) are gated on to allow current to bypass the cell. The switches and cables carrying the bypass current are at room temperature and are connected to the superconducting bus through "safety leads" located in the spoolpieces on either end of the cell.

The rate at which current is bypassed depends on the quench resistance and the inductance of the half-cell. Shortly after the heaters are fired, the dump resistors are switched into the circuit. This reverse-biases the QBSs, and no current can be bypassed until the resistive voltage reaches the voltage due to the dump, about 50 V. The current in the cell then begins to decay according to the equation

$$L \frac{dI}{dt} + IR(t) = V_{qbs}$$

where the resistance and SCR drop in the bypass circuit determine the term on the right. The time dependence of R is shown to emphasize that this is not a simple LR circuit with an exponential decay. R is increasing very rapidly with time as the heater-induced quench propagates and the coil heats up. The maximum dI/dt is >10 kA/s, which implies voltages >2 kV. Because the resistive and inductive voltages overlap in space, there is a cancellation, and only modest voltages are observed within a quenching cell (around 200 V across an element). The term V_{qbs} is small and also time dependent. The resistance of the bypass circuit is on the order of 0.010 ohm. That small resistance is useful; it allows one to determine the bypass and magnet currents as a function of time, and from that, the MIITs. The magnet current during a quench is shown in Fig. 10.

The critical elements of the quench protection system are powered from an Uninterruptible Power Supply, which uses storage batteries as a buffer between the ac line power and its output. The UPS continues to supply power to the QPM, HFU logic, and QBS controllers during a power outage.

The heater firing units use capacitors as a storage device; each unit deposits 650 J into the heater strip in the magnet. The heater strip is a thin stainless steel strip located between the superconductor and the collars, electrically isolated from both of

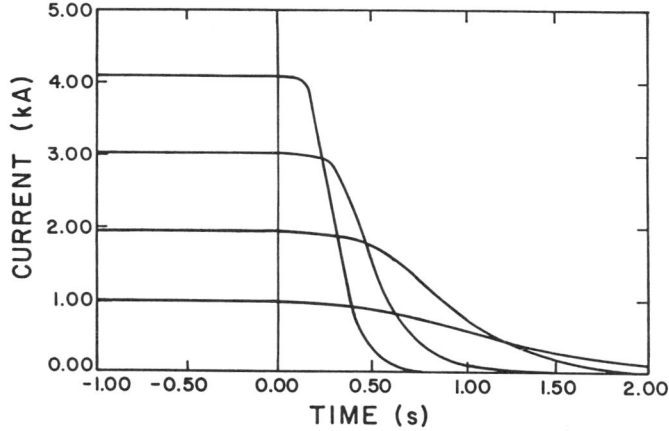

Figure 10
Magnet current for quenches at different currents.

course. In addition, the capacitor charging circuit is transformer-coupled, so that the heater strip is floating. This avoids the heater strips being a potential ground fault for the coils. The insulation between the heater and the coil results in a time delay before the quench is actually initiated. An additional time delay is due to the discharge RC time constant of the heater circuit. This is not a serious problem, however, since the energy stored was determined by the requirement that the heaters work at low currents, where the time scales are relatively long. At high current, only a small fraction of the total energy is required to induce the quench. The HFUs are checked as part of a start-up procedure whenever the Tevatron has been off for >4 hr. The test includes a measurement of the discharge time constant, which insures that the energy is being deposited into the magnets.

The quench bypass circuits consist of two independently controlled thyristors, plus a self-firing circuit that will turn the QBS on once the voltage across the cell reaches 200 V. The "safety leads" that connect the superconducting bus to the room-temperature QBS, as well as the QBS itself, are not intended to carry current for extended periods. Like the superconductor itself, the safety lead and QBS have a MIIT limit, but a much higher one. It is the protection of these elements that dictated the L/R dump time constant of 12 sec. The QBSs, like all semiconductors, are radiation sensitive, and have therefore been shielded by placement in holes bored in the tunnel wall. Each of the QBS circuits is checked separately during the start-up checkout.

The "straight section shorts" -- the devices for clamping the bus-to-bus voltage in each of the straight sections, as mentioned earlier -- are very similar to a normal QBS. The only differences are (i) the self-firing circuit triggers at 1 kV, and (ii) the device is activated only when both QBS circuits are triggered. The latter requirement prevents the straight section shorts from conducting during the tests of the normal QBS system. The 1-kV self-firing level was dictated by the fact that there is only one current-regulating power supply. When the current is first being brought up, the full voltage of that supply can appear across the closest straight section. Also, it is desirable to be able to run the Tevatron with one or more power supplies out of the circuit, which raises the bus-to-bus voltage in the straight sections.

4.3 Quench Detection System

The major component of the Fermilab quench detection system is the Quench Protection Monitors (QPMs). The QPM is responsible for monitoring the voltages across the half-cells, determining whether a quench is occurring, and if so, firing the appropriate HFUs. It must also communicate to TECAR, the Tevatron Excitation Controller and Regulator, that a quench has occurred. TECAR in turn, through the other QPMs, turns off the power supplies, activates the dump switches, and triggers the QBS controllers. The QPM also communicates to the refrigerator microprocessor which cells have quenched so that cool-down can begin promptly.

The half-cell voltage monitoring is accomplished by using voltage-to-frequency converters (VFCs) located in the service buildings near the QPMs. Some care is required in matching resistances and cable capacitances (Fig. 11) to minimize sensitivity to common-mode V and dV/dt. The cables to the magnets in the tunnel are ~200 m long, and one cable generally serves as the positive

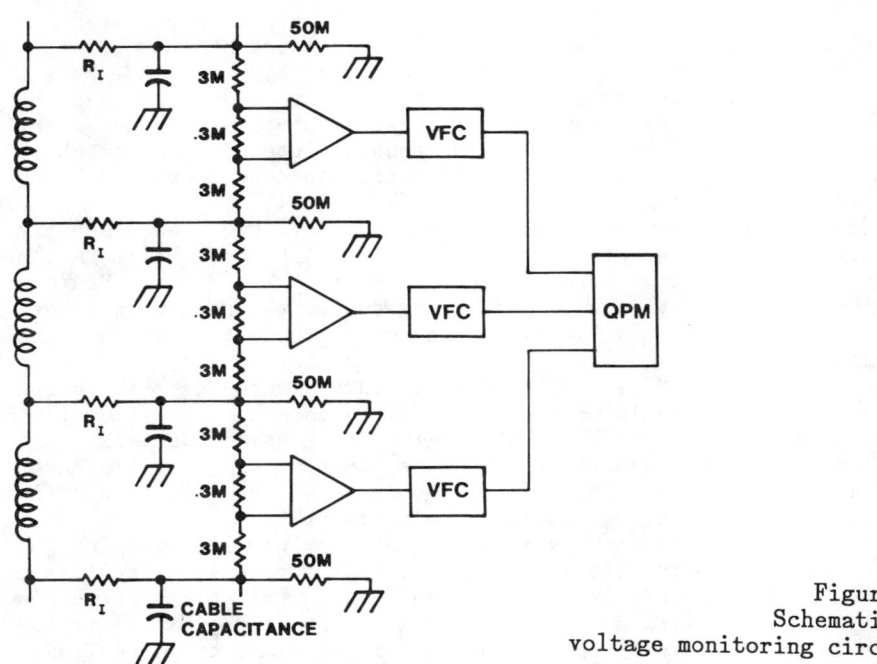

Figure 11
Schematic of voltage monitoring circuit.

input for one VFC and the negative input for its neighbor. This requires a small correction in software. The 50-MΩ resistors that tie one end to ground result in a quench signal if one of the cables is open (e.g. disconnected). The scalers in the QPM, which read the VFC output, can be "auto-zeroed" from the Main Control Room, removing the effect of drift in the VFCs.

The ability of the QPM to turn off the power supplies is critical to the overall protection of the system. As noted above, the primary means is via TECAR. This communication occurs over a dedicated link, whose function is transmitting data between TECAR, the 24 QPMs, and the 12 power supplies around the ring. The power supplies receive the appropriate current or voltage program, and the QPMs receive I and dI/dt information. The link communication is backed up by a second control loop which broadcasts two 100-kHz pulse trains, one positive and one negative. TECAR can clamp either one or both polarities, inhibiting further propagation of the pulses, which in turn causes, in the first case, the power supplies to turn off, and in the second, the dump switches to open. If the QPM-TECAR communication fails, the QPM will clamp both pulses.

The determination that a quench has occurred is based on what we call the "resistive voltage," the difference between the measured voltage across a half-cell and the expected inductive voltage due to L dI/dt. When the resistive voltage exceeds the tolerance, then that half-cell is considered to be quenching. What should the tolerance be? Tests on a string of magnets at the B-12 facility at Fermilab determined that once the HFUs have been fired, a certain time, or MIITs, is required before the current decays away. Those data are shown in Fig. 12. From the earlier hairpin measurements, we know that 7 MIITs is the limit at 4 kA. The difference determines how many MIITs can occur before detection. The hairpin data for voltage vs. time can be extrapolated to longer pieces of

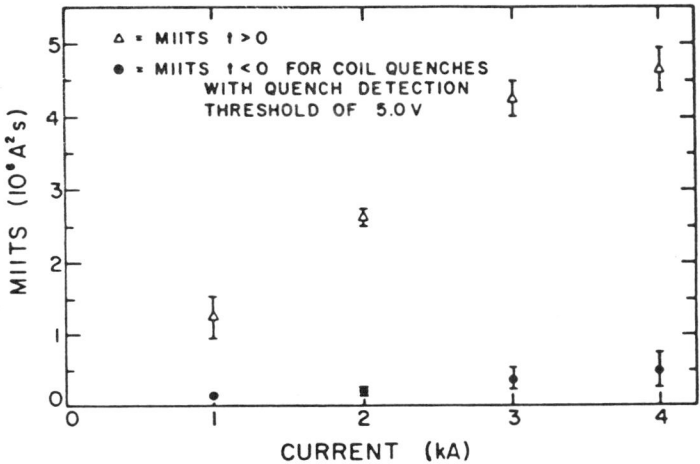

Figure 12   MIITs after firing heaters (open triangles).

cable by a summation procedure, giving the voltage as a function of time for an arbitrarily long piece. The combination of that calculation with the MIITs allowable specifies that the quench must be detected at a level of 0.5 V at 4 kA. This detection level is required in order to protect against quenches that start in the single conductor in low-field regions outside the coil. A higher threshold could be used if one had to protect only the conductor inside the coils, where the propagation is faster, and where the turn-to-turn propagation amplifies the voltage growth. At lower currents, the tolerance increases somewhat, but not very dramatically. The required level at 1 kA is ~3 V. The 0.5-V level is used as the quench detection tolerance in the Fermilab QPM system if the current is >600 A. Below that level, we use a higher tolerance (10 V) which prmits special procedures (such as testing the QBS system) to be performed. The time scales for reaching 3 V at 600 A are so long that a quench must propagate into a magnet coil, regardless of where the quench originates. Once it reaches a coil, many turns contribute to the resistive voltage, and the 10-V level is safe.

In the early development of the quench protection system, the dI/dt signal used in calculating the resistive voltage was generated by analog differentiation of the current signal from a transductor. That method has a number of drawbacks. First, the differentiation process is noisy; as it became clear that 0.5 V was the required tolerance, that noise became a concern. Second, the transductor dI/dt signal was derived in only six locations around the ring, once in each sector. To transmit that signal around to the other QPMs would imply that the quench detection algorithm would require a failsafe transmission system. Failure of the transmission system, or of the transductor dI/dt hardware, would result in a large section of the ring being quenched needlessly. The quench detection algorithm was therefore changed to utilize a "relative" dI/dt. The QPM monitors the voltages across each half-cell within its domain (either 8 or 10 half-cells, depending on the QPM location). Since the upper bus and lower bus half-cells are far apart electrically and may see transients arriving at different times, the QPM treats them as logically distinct. For each (upper and lower) bus, the relative dI/dt is determined by dividing the sum of the measured voltages by the sum of the inductances. This change was made after considerable deliberation. There was concern over the possibility of quenches growing simultaneously in the different half-cells and therefore not being detected. Examination of quenches showed, however, that once the quenches have propagated into the coils, the differences soon exceed 0.5 V. Furthermore, the different half-cells are in different cryogenic circuits, and each circuit has a temperature gradient along it. Earlier papers[9,10] give a more detailed account of the voltage monitoring circuitry, voltage and dV/dt common-mode problems, transient response, and other subtleties that enter into the quench protection algorithm. Figure 13 is a block diagram of the QPM system, showing the interconnections within the system and to the other systems in the accelerator.

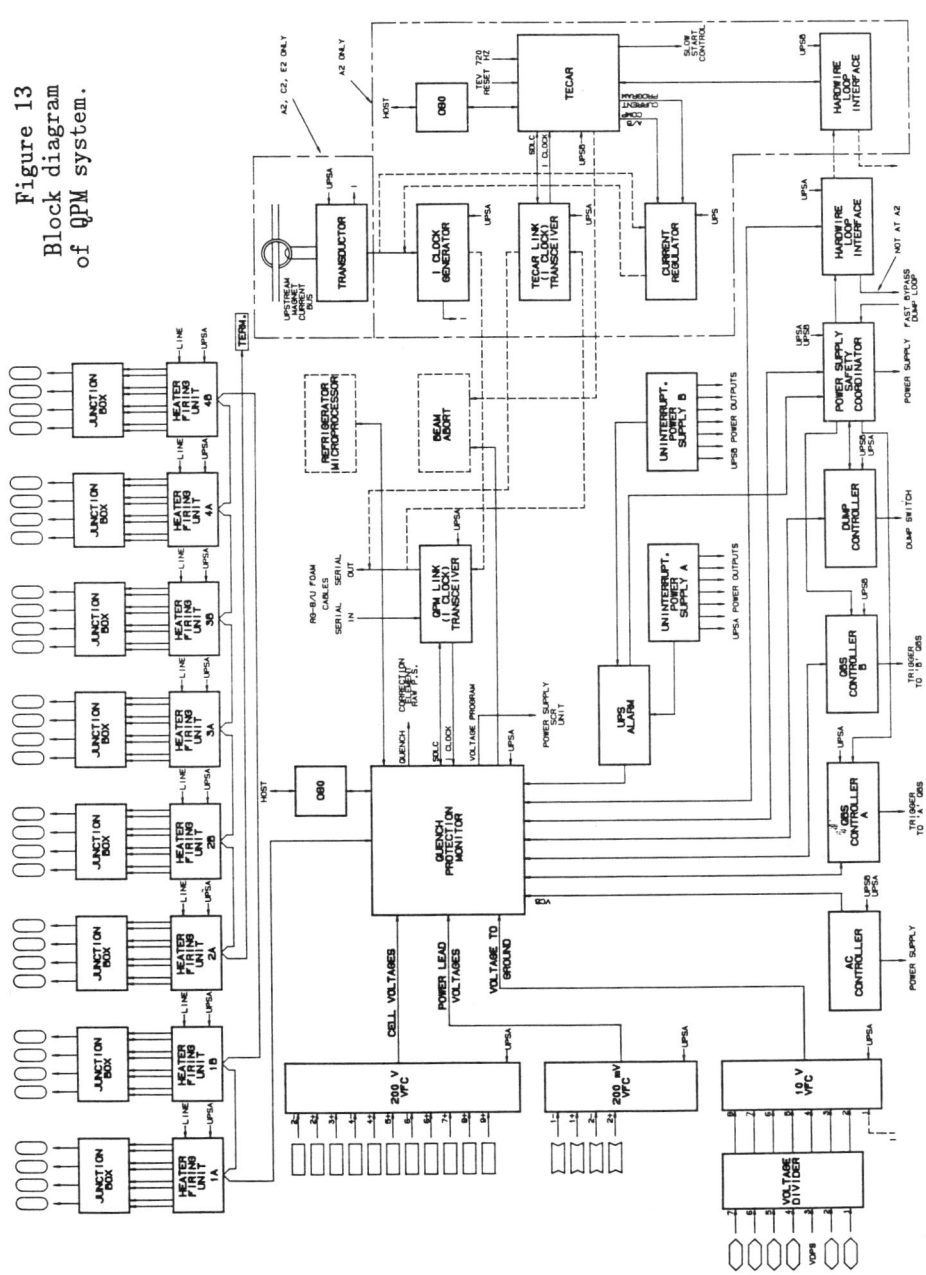

Figure 13
Block diagram of QPM system.

In addition to its active role in quench protection, the QPM has a passive role in the form of data gathering. Each QPM maintains a circular buffer containing analog data and status information; this buffer is stopped whenever something "abnormal" happens, e.g. quench or other event causing the system to deexcite. This circular buffer, which stores data from one second before the event to five seconds after, at a 60-Hz rate, has been invaluable, especially in resolving cause-and-effect questions.

5 OPERATIONAL EXPERIENCE

At the time of this writing, there has been considerable experience in operating the Tevatron, primarily in the "fixed-target" mode, but also in the "collider" mode, in which the Tevatron stores beam for many hours. The accelerator has operated with increasing reliability, but not without some problems along the way.

5.1 Failures During Operation

Only once did the quench protection system fail leaving the magnet system not completely protected. The event was in fact a double fault, one in data transmission and one is software. As the Tevatron was first commissioned, a number of problems were encountered with the power supply turn-on sequence. As a result, one section of QPM software essentially said: it's OK if there is some status wrong, provided the current is <20 A. This situation, combined with a faulty current transmission that indicated a large but <u>negative</u> current, resulted in the failure described below. The situation has been remedied by (i) taking the absolute value of the current and requiring it to be <20 A, (ii) requiring the currents measured separately in the six sectors to agree within 50 A, and (iii) masking off only certain "safe" status bits. Additional checks on dI/dt are also made which further guarantee the system integrity.

The failure occurred when an injection kicker presumably misfired, resulting in beam loss in three separate cells, one each in E1, E2, and E3. The quenches in those three cells were detected and the heaters were fired, but the power supplies were not turned off nor were the QBS gates fired. The self-firing circuits in the QBSs fired, allowing current to bypass the quenched magnets. The power supplies kept ramping for several minutes, overheating the safety leads until the insulation failed and a ground fault tripped the power supplies. Five spoolpieces were damaged and had to be removed from the tunnel for repair. It is ironic that the only quench protection failure to date was a failure mode to which passive quench protection methods are also susceptible, and it underscores the requirement for a failsafe quench detection system (and control of the power supplies and dump switches) even if the magnets themselves are passively protected.

Near the end of the first fixed-target run (mid-1984), problems began developing in the magnets themselves. As noted earlier, the dipoles come in two types, the TB and TC. Slight differences in their construction result in the TC magnets containing a section of cable roughly one foot long from the point it leaves the collared coil assembly to the interface to the next magnet. The Lorentz force from the fringe field at the end of the magnet produces flexing of the cable as the current is ramped up and down. Individual strands began breaking, and the ends of the broken strands were very likely to create ground faults or bus-to-bus shorts. A single ground fault is a rather benign event: the power supplies are turned off but the dumps remain out of the circuit. The current decays away very slowly (several hundred seconds). If a quench occurs, the dump resistors are switched into the circuit and the current comes down quickly. The voltage-to-ground distribution is shifted and larger than normal voltages generally appear, but the system can tolerate these voltages and again nothing terrible happens, although the faulty component will have to be located and replaced.

The voltages normally seen by the magnets are <500 V. The voltage on any particular component depends on its location in the ring. Each magnet has a capacitance to ground of about 60 nanofarads; when a ground fault develops, the charge stored in the capacitance of the adjacent magnets is discharged into the fault. Consequently, ground faults that develop suddenly at high field generally result in a quench signal due to the equal but opposite currents (and dI/dt) into the ground fault.

The fact that ground faults cause quenches (a quench-like signal, that is) turns out to be useful. Ground faults are often difficult to locate, particularly when the joints between magnets cannot be opened up easily. For sparking-type discharges, ground faults can be located to within a few components by looking for differences in arrival time at different locations, even with the magnets superconducting. High-impedance faults are most easily identified by warming the magnets up to 20 K or so, so that the voltage drop of the leakage current across the now-resistive magnet can be sensed. In both cases, tunnel access is required to make the measurements. The situation is obviously much more difficult when the ground fault occurs only at high current. Safety considerations preclude tunnel access when the system is powered to high currents.

Quenches, on the other hand, are relatively easy to locate. The initial quench location is determined by the QPM to be on the upper or lower bus of one cell. By installing a special QPM, known as PDQ, which performs monitoring functions only, it is possible to subdivide the cell into much smaller increments. Every dipole and every spoolpiece has voltage taps that can be monitored. In a typical cell, the only device that does not have a voltage tap is the quadrupole, and a few special devices do not have taps. Thus PDQ can be installed to look across every one or two devices. By stopping its circular buffer at the same time the quench occurs, the quenching component can be identified. (For quenches very close to the interface between magnets, there was originally some confusion in identification, since PDQ really looks not across one device but from the point where the voltage tap is soldered to the superconductor in one magnet to the similar point in the next magnet. As the geometry was better understood, it became possible to identify the location with better accuracy. Quenches in leads, as opposed to coils, have a slower growth, and can therefore be distinguished. If the quench propagates across the interface, one can also determine which end of the magnet is quenching.) The utility of PDQ was first established during the initial Tevatron commissioning. A magnet with a turn-to-turn short and three poorly soldered splices had to be located.

Bus-to-bus shorts, or multiple ground faults, are another matter. The current can take an alternative path so that the quench protection is basically defeated. As a result, such failures are generally more catastrophic; melted superconductor and cryostats with holes burned in them are the most common results. In this case, locating the fault is very easy, but the insulating vacuum and superinsulation are filled with helium. Leak-checking after replacing the faulty component becomes difficult to impossible because of the contamination. A total of five TC dipoles failed during operation between March 26 and July 9, 1984. Two were single ground faults, three involved bus-to-bus shorts. During the 1984 summer shutdown, all TC magnets were repaired by opening the cryostats and tying the two leads together to prevent motion. Broken strands were found in many dipoles; if more than six were broken, a new section of cable was spliced in place. This was done to five dipoles in the course of the repair.

The problems with the TC dipoles all occurred at the upstream end; at the downstream end, where the voltage taps and heater firing connections are made, the leads were tied together already. However, in 1985, one TC dipole, in which the leads had not been tied on the downstream end, also failed in a similar manner.

The strand breakage itself is an interesting situation. Since the cable is in low field, particularly right at the interface where the breakage was occurring, its critical current is much higher. As a result, the increase in current density in the remaining strands does not exceed the critical current if only a few strands are broken, so the system can operate under those conditions. But should a quench occur anywhere in the cell, the heaters will fire and the quench will propagate to the region of broken strands. The temperature increase in the cable is related to the MIITs divided by the area squared. In the bus-to-bus short failures, the short typically appeared well after the quench, and may well have been due to overheating from this effect. The subsequent melting of superconductor may have been inevitable and may have occurred without the shorts.

A TB dipole failed during an 800-GeV quench three days before the scheduled end of the 1985 fixed-target run. The failure occurred in the leads at the upstream end of the magnet, about 1.5 in. from where they exit the coil. A bus-to-bus short and cryostat rupture resulted. A deep score mark was evident in the conductor at the point of the failure, which was inside a G-10 block securing the leads. The damage must have occurred during construction, since the area was covered by G-10. What is most puzzling is the timing: a few days earlier, in an effort to locate the weakest magnets in each sector, that area had been ramped to 900 GeV without quenching. The cell containing the magnet that later failed quenched at 910 GeV without damage, and operation continued at 800 GeV for several days.

Finally, a failure occurred in January 1985, in which a power supply transformer shorted from primary to secondary, raising the bus potential to 13.8 kV, far in excess of its rating. The damage was limited to F-sector, where several devices had to be replaced.

The point of this rather lengthy exposé on magnet failures can best be summarized as follows: the ends of the magnets, in addition to being difficult and costly in terms of cryostat construction, have also been the source of virtually all the magnet failures to date. Designers of future superconducting magnets should pay attention to the electrical details of the ends, as well as to simplifying the cryostat.

5.2 Other Aspects of Operating a Superconducting Accelerator

Quenches were the largest source of unscheduled downtime during 1985. Almost 90% of the quenches were due to beam loss; about half the remainder can be attributed to the Quench Protection System. The QPM-related quenches arise from spontaneous HFU discharges, faulty or drifting VFCs, or QPM failures (in which case all HFUs in the building are fired, so that the magnets are protected). The HFU problem has been the most aggravating, and that system is being redesigned. Another QPM-related problem is the inability to distinguish real quenches from "false" quenches. The latter can occur in association with power supply regulation problems that place fast transients on the bus.

Repeated quenches in the same cell pose a special problem. They occur at certain regions of the ring: near the injection, extraction, and abort channels, where beam loss is most likely. Kicker timing problems, for example, can cause local beam loss. The problem with repeated quenches is that the safety leads which bypass current heat up, and the cooldown time is much longer than the time required for the cryogenic system to recover. To shorten the cooldown, vapor cooling was installed on those spools in regions most likely to quench. Running at 800 GeV, the problem has not been too severe; on a few occasions, after three quenches, the accelerator was shut down to avoid potential overheating should a subsequent quench occur. The overheating problem becomes more severe as the energy of the machine is increased, and vapor cooling is being installed on all leads. The vapor cooling decreases the cooldown time by roughly a factor of two, and some future problems are anticipated. It is desirable to limit the flow of the cooling to avoid forming large icebergs. Even with adjustable flow (e.g. refrigeration-type flow controllers, as now used on the main power leads and the already modified safety leads) the question remains -- how does one insure that there is flow?

With the beginning of commissioning the Tevatron as a Collider, it became evident that the beam lifetime was not limited by the slow losses due to the quality of the vacuum, or to collisions at the anticipated luminosity, but rather, the lifetime was limited by sudden loss of the entire beam as one device or another caused an abort. The QPM system was among those causing aborts. The beam must be aborted any time the system is deexcited. Changes were made in the QPM response with the primary objective of avoiding the need to abort. Many of these changes have had the further benefit of easing the operation for fixed-target physics as well. The most notable example is the response to the refrigeration status becoming "bad." The original response was to deexcite immediately, thereby aborting the beam. Now, the response is to inhibit the next ramp from occurring. In fixed-target mode, this allows the beam to be extracted to the experimental areas and allows the refrigeration system an extra chance to recover. If it does not recover, the accelerator idles at 400 A until it does. This sequence is operationally much easier than turning off and back on again. In the collider mode, the scarcity of antiprotons dictates that the machine continue operation without interruption. It makes little difference whether the beam is lost because of a quench or because of deexcitation. The time required for quench recovery is short enough that it is better to give the refrigerator a chance to recover. A similar response is taken for (i) a single HFU channel showing faulty status or (ii) the circular buffer being stopped.

6 SUMMARY

This chapter has discussed the quench protection system only for the main Tevatron system. In addition, there are four independent low-β quadrupole circuits in B0, each with its own QPM, and three systems in Switchyard. These "auxiliary" systems required the development of somewhat different quench protection algorithms and hardware, but the QPMs themselves are basically the same as the 24 in the Tevatron.

The QPM system is indeed a complicated system. It must monitor and protect the superconducting magnets without introducing potential ground faults. It must be intimately linked to the power supply system. It must have sufficient redundancy that single-device failures do not jeopardize the system's ability to protect the magnets. It must have sufficient sensitivity to detect quenches under many conditions, from low current to high, inside coils and in leads, while ramping or running dc; and yet have sufficient immunity to noise and transient conditions to distinguish them from real quenches. Many people at Fermilab have struggled to accomplish these goals. The QPM system is now a fairly stable system, in contrast to the situation just two years ago, when the software was still rapidly evolving. With the improvements to the HFU system, and streamlining of the checkout procedure, it should prove more and more "user-friendly."

7 REFERENCES

1. M. Wilson, Superconducting Magnets, Clarendon Press, Oxford, 1983.
2. R. Palmer and A. Tollestrup, Ann. Rev. Nucl. Part. Sci. 34, 247 (1984).
3. A. Tollestrup, AIP Conf. Proc. No. 87, 699 (1982).
4. A Report on the Design of the Fermilab Superconducting Accelerator, May 1979.
5. W. Sampson, M. Garber and A. Ghosh, IEEE Trans. Nucl. Sci. NS-28, 3245 (1981).
6. W. Fowler, M. Kuchnir, R. Flora and R. Remsbottom, IEEE Trans. Magn. MAG-17, 925 (1981).
7. K. Koepke, P. Martin and M. Kuchnir, IEEE Trans. Magn. MAG-19, 696 (1983).
8. M. Kuchnir and J. Tague, Fermilab TM-679 (unpublished), 1976.
9. G. Tool, R. Flora, P. Martin and D. Wolff, IEEE Trans. Nucl. Sci. NS-30, 2889 (1983).
10. P. Martin, R. Flora, G. Tool and D. Wolff, Fermilab TM-1134 (unpublished), 1982.

PERSPECTIVES

Science in the Age of Accelerators

MARTIN L. PERL

Stanford Linear Accelerator Center
Stanford, California 94305

TABLE OF CONTENTS

1. Private Science, Public Science 2099
2. Before Accelerators 2101
3. The End of the Amateur 2104
4. The Thirties and Forties 2105
5. Physics Leaves Its Private World 2109
6. Large Particle Colliders: New Physics, New Technology,
   Great Engineering Feats 2111
7. Acknowledgements 2115
8. References 2115

Science in the Age of Accelerators[*][†][‡]

MARTIN L. PERL

*Stanford Linear Accelerator Center
Stanford, California 94305*

1. Private Science, Public Science

Accelerators have brought the particle physicist to work and live in three worlds: the private world of science, the public world of science, and the world of large accelerators. Our private world is our apparatus, our data, our theories, our colleagues, our journals, our meetings, and above all our understanding of elementary particles. There are more intimate areas in that private world, the childhood toys and dreams that led us into physics. There is a connection between building huge accelerators and Erector sets and Meccanos and ham radios. There is a connection, sometimes a painful one, between childhood reading about lone science heros: Pasteur, Madame Curie, Einstein; and then growing up to be part of a group building or using a huge accelerator.

The public world of science is how society sees us, how we want to be seen in newspapers and on TV, how we interact with governments, and most important, how governments support science. Since the 1940's most of us in basic research have not been able to avoid the public world, even if we wanted to stay in our private world. Public money is needed to study agriculture as well as atoms, libidos as well as leptons. If the apparatus is table size, if the laboratory is room size, with a little obliqueness the dependence on public money can be ignored. The builders and users of large accelerators, of large telescopes, of space rockets and satellites cannot ignore their dependence.

At the Spring 1987 Meeting[1] of the American Physical Society I used many slides and two screens to visually trace the intertwining of

---

[*] This work was supported by the Department of Energy, contract DE-AC03-76SF00515.
[†] Written version of talk presented at the American Physical Society Meeting; Arlington, Virginia; April, 1987.
[‡] Figures in this paper without attribution are from SLAC archives or the private collection of the author.

© 1989 American Institute of Physics

the private and public worlds of science with the coming of the age of particle accelerators. I was not trying to do the history or sociology or politics of accelerators. Rather I was illustrating some of the themes laid out historically in Fig. 1. (During my talk, Fig. 1 was always projected, here the reader will have to refer back to it).

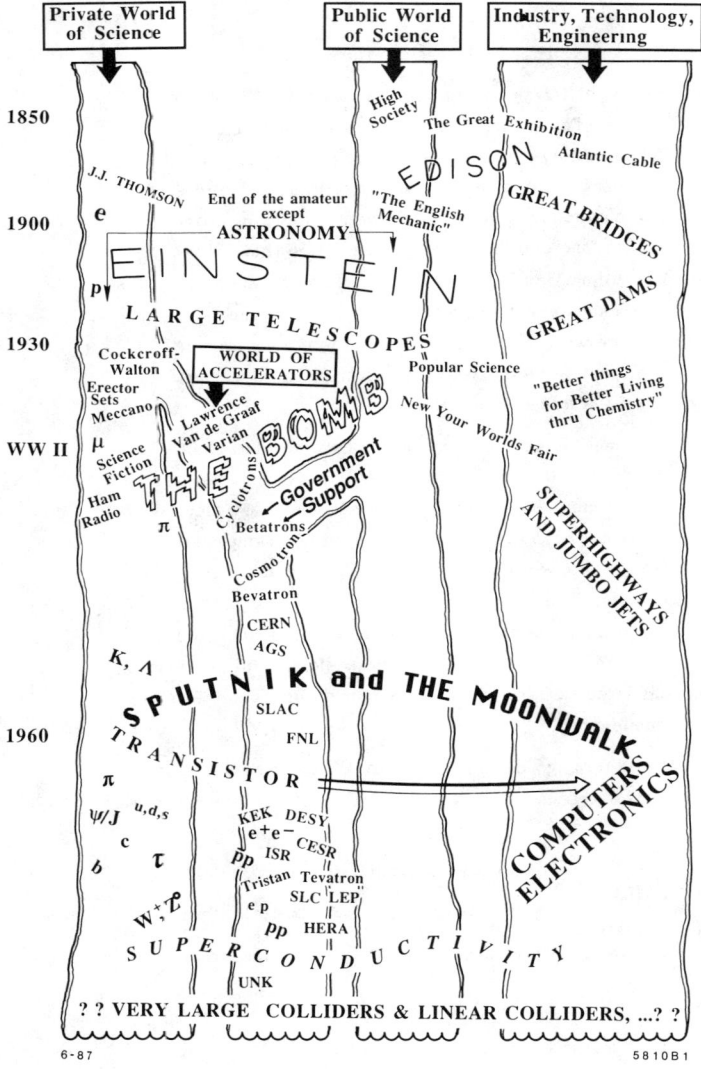

Fig. 1.

There is not space here for all the pictures I used; I retain the unfamiliar images. The reader knows the familiar ones: Rutherford in the Cavendish Laboratory standing under a sign reading "TALK SOFTLY PLEASE"[2] or Livingston and Lawrence in front of the 37-inch cyclotron.[3] These and other familiar images I used came from Refs. 2, 3, and 4, of which the most entrancing is *The Particle Explosion* by Close, Marten and Sutton.[2]

2. Before Accelerators

Before accelerators, J. J. Thomson's cathode ray tube apparatus (Fig. 2) was completely in the physicist's private world. The ideal apparatus, needing only a table and a glassblower, to identify the electron. Not so easy. Thomson writes, "It was only when the vacuum was a good one that the deflection [of the cathode rays] took place." Vacuum problems ninety years ago. In the same article Thomson asks " ... what are these particles? Are they atoms, or molecules, or matter in a still finer state of subdivision?"

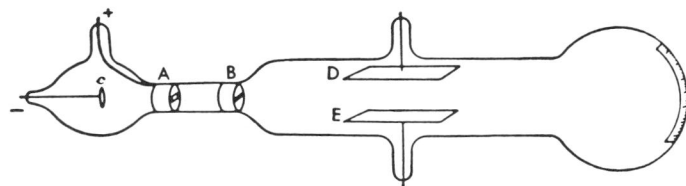

Fig. 2. From J. J. Thomson, Phil. Mag. **44**, 293 (1897).

Society with a capital S (High Society in Fig. 1) was interested in physics as culture and intellectual diversion. To the rest of society physics was hidden, remote. The submarine telegraphic cable (Fig. 3) is my

Fig. 3. Laying the Dover-to-Calais submarine cable in 1850.

metaphor. The public interest is in the enterprise and danger in laying the cable; it is in the wonder of connecting islands and continents. A great engineering feat. Hidden in all this is our physicist hero, Kelvin, and his theory of telegraphic signaling[5] and his idea of a stranded cable.

It is not pure science, but it is great engineering feats which catch the interest and enthusiasm of masses of people in the nineteenth and twentieth centuries: the Atlantic Cable, railroads, large steamships, great bridges (Figs. 4 and 5), great dams. Scientific apparatus can also be

Fig. 4. The Forth Bridge, near Edinburgh, Scotland, completed in 1890. The first large bridge using the cantilever and central girder principle.

Fig. 5. The Brooklyn Bridge, New York, U.S.A., completed in 1883. One of the first large suspension bridges.

great engineering structures. First came the large telescopes, then space rockets and satellites, now huge particle colliders. I will return to this idea later because particle colliders as engineering feats can have special affection from the public and special support from governments. This brings benefits and dangers to particle physics.

It is not pure science, but it is new and visible technology which catches the interest and enthusiasm of masses of people. From the Great Exhibition of 1851 — the Crystal Palace —in London (Fig. 6) to the Trylon and Perisphere of the 1939 New York Worlds Fair (Fig. 6), new

Fig. 6. Two Worlds Fairs: *Above*: The Great Exhibition of 1851 in London usually called the Crystal Palace. *Right*: The Trylon and Perisphere of the 1939 New York Worlds Fair. From a colored postcard.

and future technology has brought the crowds. The Great Exhibition was arranged in four departments: Raw Materials, Machinery, Manufacturers, and Fine Arts. Science is buried in technology.[6] Worlds Fairs fail these days because we are so immediately immersed in new technology.

In North America, Edison was and still is the great symbol of new and visible technology. During the reign of Edison, our private physics world moved on with Maxwell and Hertz and Lorentz and Planck, but Edison was a thousand times more famous. Only Einstein crossed the fame barrier out of our private world. His name stretched across the private and public worlds of physics (Fig. 1).

3. The End of the Amateur

In the last decades of the nineteenth century, a new gulf appeared between the private and public worlds of science. As sciences developed, amateurs could no longer contribute or even fully understand. The usual example is the ninth edition of the Encyclopedia Britannica (1889), the last edition whose physics articles were useful to the professional and to the amateur. After that we have our Handbuch der Physik, and the encyclopedias stay with the public world.

An example I like is *The English Mechanic*, a combined do-it-yourself and amateur scientist magazine (Fig. 7). Building a steam car is an im-

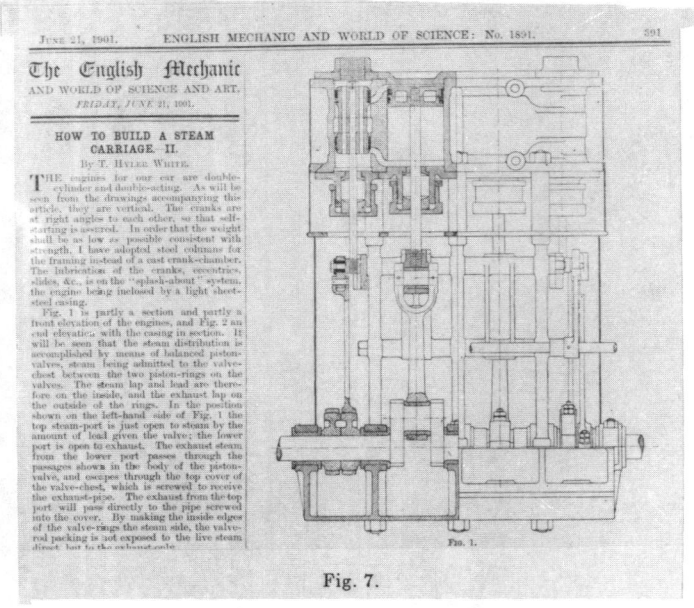

Fig. 7.

Fig. 8. *Left*: The 1864, 80-cm-diameter, reflecting telescope of Foucault at Marseilles incorporating the innovations of a silver-on-glass surface and a parabolic figure. *Above*: The last of the great refracting telescopes, the 40 inch diameter at the Yerkes Observatory, installed in 1897.

pressive hobby, but the inside contents of *The English Mechanic* are more impressive. In this issue there is a summary of a lecture by Dewar on liquid and solid hydrogen; a note on the Curie's work on induced radioactivity; the positions of two new variable stars are given; and there are dozens of queries from readers on subjects ranging from using ammonia for renovating felt hats to using the formula

$$\int_0^{2\pi} \int_{r=0}^{r=6} AB d\theta \, r dr \ .$$

There are no amateur science magazines or amateurs like that today. Except in astronomy. That lucky science has its subject in full view, still has crucial contributions from amateurs, and has apparatus which are also engineering feats. A hundred years ago telescopes were already impressive structures (Fig. 8).

4. The Thirties and Forties

The 1930's and 1940's represent the childhood of the age of accelerators in two ways. First, there are the early accelerators and their inventors: cyclotrons and betatrons and linear accelerators; Cockcroft and Walton and Lawrence and Van de Graff and Wideroe. Familiar names and images. Second, the thirties and forties were the childhood years of the physicists who have since dominated the building and use of large accelerators. That generation is retiring, or will soon retire, from the private world of science. What were our images of physics?

I think our images were quite different from the childhood images of physicists born after the Bomb or Sputnik or the Moonwalk. Before the bomb, physics was a very, very small and private world. In the thirties in the United States the most visible new technology and the public science was chemistry represented by the slogan of the Dupont Company — "Better Things for Better Living Through Chemistry". We, at least the accelerator builders and experimenters, came to physics mostly indirectly through Erector sets and Meccanos (Figs. 9 and 10) and ham radio. Our

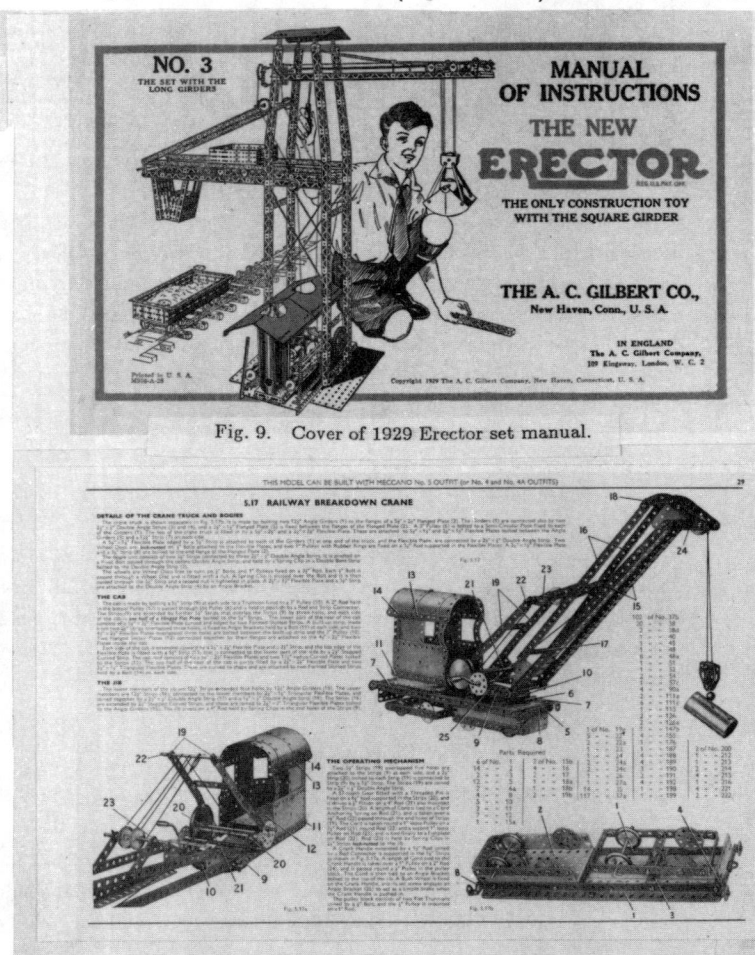

Fig. 9. Cover of 1929 Erector set manual.

Fig. 10. Page of instructions from a 1930's Meccano manual.

reading was the science and hobby magazines (Fig. 11) which were compounded of futuristic technology, science projects usually too complicated for our skill or pocket money (Fig. 12), and occasional perpetual motion (Fig. 13). Popular science magazines have degenerated since the nineteenth century, there was no perpetual motion in the *English Mechanic* because the editors knew the first law of thermodynamics.

Fig. 11. *Left*: Cover of December 1930 Everyday Mechanics featuring the rotor force idea which was popular for futuristic ships and airplanes in the 1930's. The editor Gernsbach pioneered hobby electronics magazines and science fiction magazines in the 1920's. *Right*: Cover of February, 1934 Modern Mechanix.

Fig. 12. A project from the December 1930 issue of Everday Mechanics.

New Rail Car Runs on Air-Electric Perpetual Drive

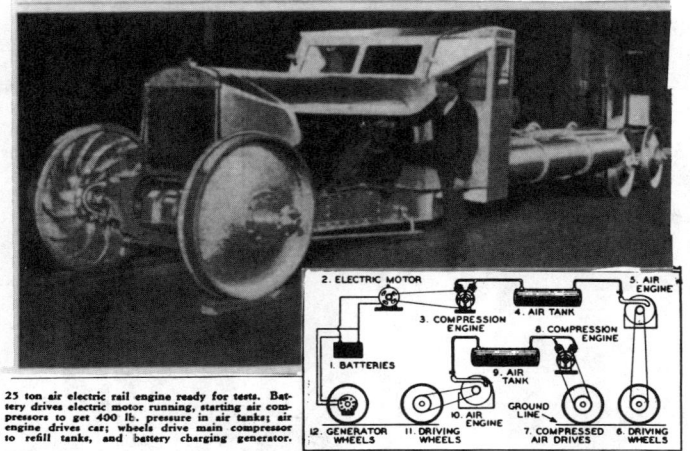

25 ton air electric rail engine ready for tests. Battery drives electric motor running, starting air compressors to get 400 lb. pressure in air tanks; air engine drives car; wheels drive main compressor to refill tanks, and battery charging generator.

FROM coast to coast by rail in 24 hours, traveling literally on air—that is what W. E. Boyette of Atlanta, Georgia, claims for his invention, a railroad engine that runs almost entirely on air.

Air for fuel—speeds of up to 125 miles an hour on rails—low transportation costs —these are possibilities conjured by Boyette's air electric car. After being started by batteries, the car needs only air to keep it running—a close approach to perpetual motion.

Fig. 13. A perpetual motion proposal from the February 1934 issue of Modern Mechanix.

We knew about a few physics greats: Kepler, Newton, Madame Curie if we went to the movies, and, of course, Einstein. But not Bohr or Schrodinger or Fermi or Michelson or Hahn and Strassman. There were great engineering projects going on in the thirties. Boulder Dam (Fig. 14), huge battleships, ocean liners, the China Clipper, and the Twentieth Century Limited entranced us. But we knew that wasn't science. Physics and chemistry and biology and astronomy were science; the problem was, "Could you make a living doing science?"

Fig. 14. *Right*: Boulder Dam across the Colorado River in the U.S.A. completed in 1935. *Left*: A tunnel used during the construction of Boulder Dam.

5. Physics Leaves Its Private World

After World War II the public discovered physics or rather discovered physicists. Not many physicists and not much physics, mostly it was the atom bomb and the hydrogen bomb and nuclear reactors and a little radar that caught the public. But this was enough for all of physics to cross the Rubicon into the public world of science: public interest, public scrutiny, public money.

There was only modest public interest or public scrutiny because we were still not associated with the new technologies and feats of engineering: superhighways, jumbo jets, Sputnik, and an astronaut walking on the moon. The newspapers and TV talked of rocket scientists but we knew they meant rocket engineers. Then the transistor appeared, and the newspapers and TV said that the transistor is physics. We had arrived.

The images are familiar now and I move quickly.

With the building of the Cosmotron and the Bevatron, with the construction of large alternating gradient synchrotrons, with the establishment of CERN, DESY, the Rutherford Laboratory, SLAC (Fig. 15), Fermilab (Fig. 16), government support begins to flow steadily into the world of accelerators (Fig. 1). The builders and users of accelerators now live in three worlds.

We particle physicists are not alone in the necessity of living in three worlds. The same thing happens to space science, to plasma physics, and eventually to material science with its need for high intensity neutron and photon sources. Only the astronomers stay lucky — still able to get

Fig. 15. A view of SLAC showing the new Collider Hall for the SLC in the lower left corner. Photograph by Joe Faust.

some of their telescopes from the world of private wealth. Although not the Hubble Space Telescope.

With public money, following the accelerator pioneers of the thirties and forties, following the dreams started with Erector sets and *Modern Mechanix*, we found there were two kinds of neutrinos (now three) and the proton was made of quarks. We found the ψ/J particles, the τ heavy lepton, the heavy b quark, the gluon that carries the strong force, and the $W^\pm$ and Z^0 that carry the weak force. It has been a splendid, an amazing, twenty-five years. I'm sorry that these discoveries have been given the awful and dull name "standard model". We have come so far in answering Thomson's question " ... or matter in a still finer form of subdivision?", our work deserves a better name.

Fig. 16. A view of Fermilab showing the experimental areas and Accelerator Complex, from the 1986 Annual Report of the Fermi National Accelerator Laboratory.

6. Large Particle Colliders: New Physics, New Technology, Great Engineering Feats

As we plunged forward in our private world of quarks and leptons and intermediate bosons, our accelerator world moved closer and closer to the public world of science, and began to spill over into the industrial world (Fig. 1). In the past, connections between the accelerator world and the industrial world were fitful. We hungrily used some of their new technology — solid state electronics and computers. The passing from electron-positron storage rings to synchrotron light sources has begun to provide important applied research tools to industry. But mostly we kept to ourselves, except for the civil construction involved in building accelerators and accelerator laboratories: SLAC (Fig. 15), Fermilab (Fig. 16), and LEP (Fig. 17).

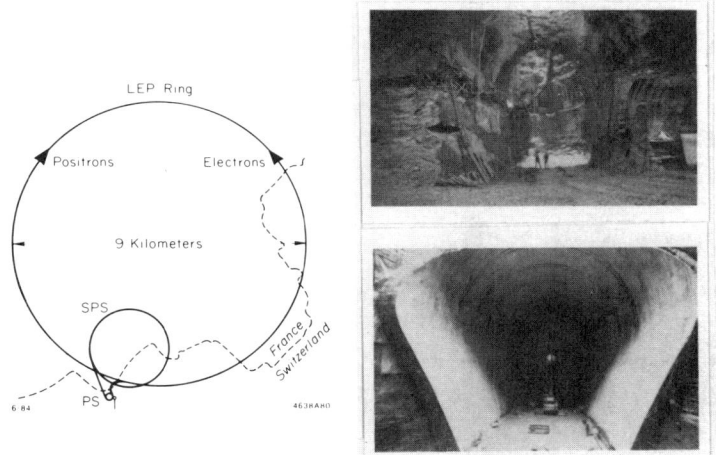

Fig. 17. *Left*: The LEP electron-positron collider under construction at CERN. *Upper Right*: The cave for the DELPHI experiment. *Lower Right*: The cave for the L3 experiment. It is interesting to observe the similarities between these photographic images and the Boulder Dam tunnel in Fig. 14. The photographs are from the CERN Courier, December 1986.

We can no longer keep to ourselves; as accelerators get bigger the civil construction to be done by the industrial world grows bigger. We might still build our own accelerator technical components, but we now have obligations to use industry for the sake of industrial development and for the sale of national economies.

We can no longer keep to ourselves. We need the spiritual and cultural and financial support of the public world of science. The public can provide that support because they continue to be interested and excited by great engineering feats, and that is what our accelerators have become. New and visible technology also interest the public. Our new technology is still esoteric: superconducting magnets (Fig. 18) and linear colliders (Figs. 19 and 20), but it is futuristic technology, and that is interesting to lots of people. There is also a fascination with the contrast between, on one side, the very small objects we study and the precision of some of our devices and, on the other side, the huge size of our tunnels and interaction regions and detectors.

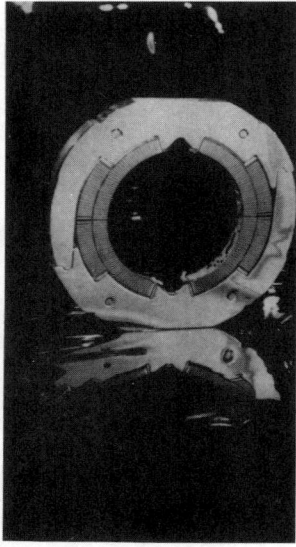

Fig. 18. A cross section of the superconducting magnet coil used in the Tevatron proton-antiproton collider. From the 1986 Annual Report of the Fermi National Accelerator Laboratory.

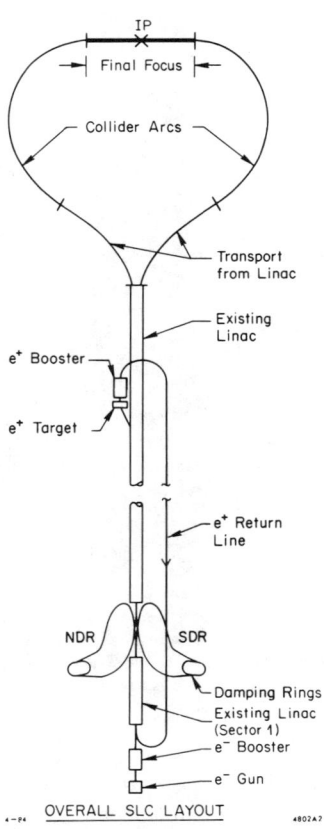

Fig. 19. Schematic of the principles of operation of the SLAC Linear Collider now being commissioned.

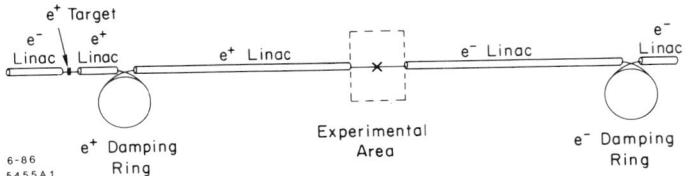

Fig. 20. Diagram for a future linear collider from J. Rees, SLAC-PUB-4037 (1986).

As a demonstration my final two images (Fig. 21) are from a newspaper. Not the New York Times or the Washington Post, but the San Francisco Examiner[7] — a newspaper with an average mixture of national affairs, crime, local politics, and sports in its pages. The science reporter was given the space to do this article because the editors knew that large accelerators are news. Accelerators are news primarily because they are great engineering achievements, secondly because they incorporate highly visible new technology, thirdly, and this is a distant third, because we use them to learn more about the fundamental nature of matter. This is my thesis and my conclusion.

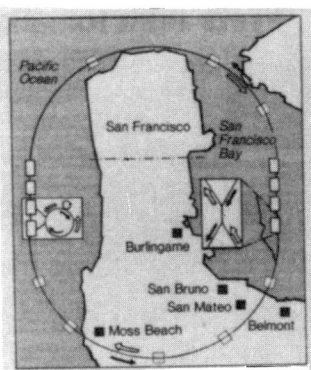

Fig. 21. Figures from articles in the San Francisco Examiner of April 19, 1987. The captions are from the articles.

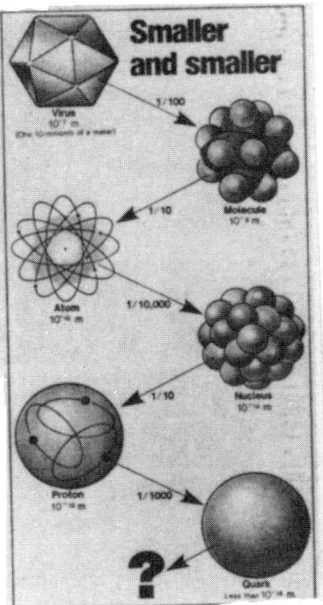

I end with two warnings. The engineering images in this talk were of successful buildings and machines, there were no pictures of the Tacoma Narrows bridge, of the Challenger, of Chernobyl. If we are to build and use successfully the huge new accelerators, we must follow the principles of good engineering as well as good physics. We must know our technology well, we must design carefully — better we must overdesign, we must construct for strength and reliability and durability. If we can't get the public support to build our accelerators truly and well, we had better be honest with the public and tell them we can't do it. We must not fail with the huge accelerators we propose.

My second warning comes from the private world of science — in that dark country where we cannot know what is ahead in the physics of elementary particles. When Roebling designed the Brooklyn Bridge in the 1870's, he could promise that the bridge would take people from Brooklyn to Manhattan and back. It still does. We cannot promise that the next accelerator will take us to the Higgs particle or to the theory of everything or to the next heavy lepton, or even to the top quark.

It is difficult to avoid promises when science gets discussed and displayed in newspapers, on TV, and in government hearings. Witness the new high-temperature superconductors (Fig. 1). However if the promises of these superconductors are not kept, the public world will soon forget. There will be little harm to material sciences or solid state physics. If our huge accelerators fail, our promises will not be so easily forgotten.

7. Acknowledgements

I am greatly indebted to Melvin Month for suggesting this topic and encouraging me to prepare this paper. I wish to thank Lydia Beers for preparing and assembling this manuscript and Sylvia McBride for drawing Fig. 1.

I apologize to my colleagues who are not from the United States or Great Britain for mostly using historical images from those two regions; those are the images I grew up with, feel closest to, and know best.

8. References

1. Martin L. Perl, Bull. Am. Phys. Soc. **32**, 1100 (1987).
2. F. Close, M. Marten, and C. Sutton, *The Particle Explosion* (Oxford Univ. Press, Oxford, 1987); the photograph of Rutherford is on page 40.
3. E. Segrè, *From X-Rays to Quarks*, W. H. Freeman, San Francisco, 1980); the photograph of Livingston and Lawrence is on page 232.
4. D. Varian, *The Inventor and the Pilot* (Pacific Books, Palo Alto, 1983).
5. E. Whittaker, *A History of the Theories of Aether and Electricity* (Harper, New York, 1960), Vol. 1, page 228.
6. Facsimile of *The Art-Journal Illustrated Catalogue of the Industries of All Nations* (Crown Publishers, 1970), section entitled "The Science of the Exhibition".
7. San Francisco Examiner, April 19, 1987.

ACCELERATOR PROJECTS, WORLDWIDE

Lee C. Teng
Fermi National Accelerator Laboratory
Batavia, Illinois 60510

TABLE OF CONTENTS

High Energy Colliders.. 2117
High Intensity, Medium Energy Accelerators....................................... 2123
Synchrotron Radiation Storage Rings... 2125

ACCELERATOR PROJECTS, WORLDWIDE

Lee C. Teng
Fermi National Accelerator Laboratory
Batavia, Illinois 60510

At any given time one always has the feeling that the funding for basic research is tight and inadequate. But an overall examination of the current accelerator projects in the world at all stages of proposal, construction, and operation proves to be rather reassuring. The result of this survey is presented here in tabulated form, and the rationale and utility of these projects are discussed and compared.

Accelerator projects can be broadly classified in the following four categories:

1. High Energy Colliders - The pursuit of high energies or small dimensions for particle physics studies is the initial and basic motivation for the development of accelerators. The high energy frontier is now covered exclusively by colliders.

2. High Intensity, Medium Energy Accelerators - Machines of this category are useful for both particle and nuclear physics research.

3. Synchrotron Radiation Storage Rings - These machines have mushroomed during the past decade into a large and important category of accelerators. The synchrotron radiation from an electron beam travelling inside the dipole magnets or the undulators in a storage ring yields VUV and X-rays of unprecedented brilliance and brightness for studies of atoms, molecules, and condensed matters, and for industrial and medical applications. Most of these storage rings are in the energy range of 0.7 GeV to 7 GeV.

4. Low Energy Medical, Industrial and Research Accelerators - Large numbers of low energy accelerators of all types are used for atomic and nuclear research and for applications in industry and medicine.

In this paper we will describe principally projects of the first two categories.

High Energy Colliders

All recent projects aimed at high energies are colliders. We limit the discussion here to machines having actual or projected completion dates between 1985 and 1995. Geographically these are shown in the world map in Fig. 1. Hadron colliders are underlined, lepton colliders are not underlined, and the mixed collider HERA is underlined with a dotted line. This map contains all colliders that are either in construction or in operation, and two that are approved and are expected to be funded for construction in FY 1989. Altogether nine projects are included within the time period specified.

To compare the utility of hadron colliders with lepton colliders we need the concept of "reach" introduced by Llewellyn Smith. This concept is best illustrated by an examination of the energy dependence of high energy reaction cross-sections. This is given in Table 1 side-by-side for hadrons and leptons.

© 1989 American Institute of Physics

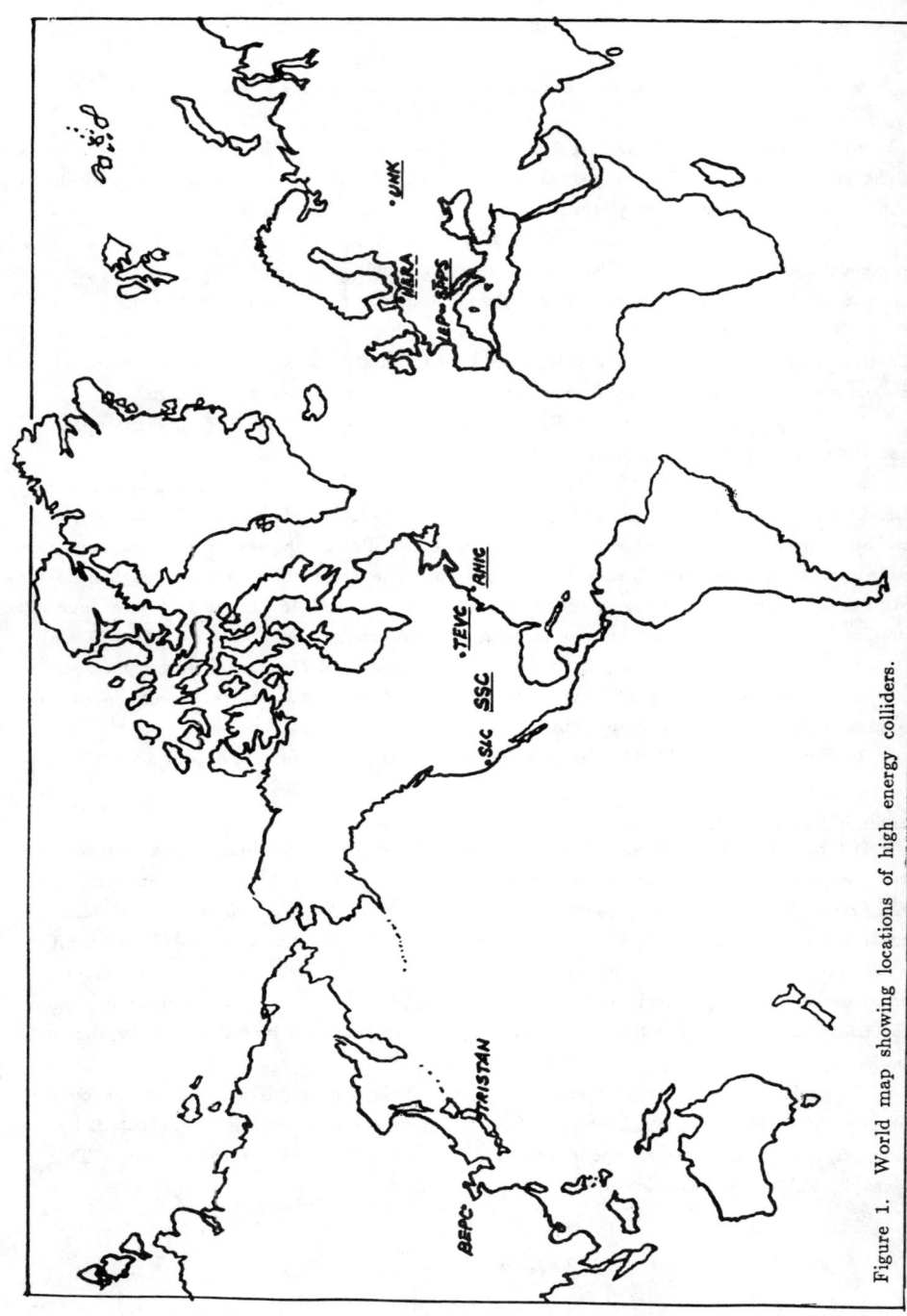

Figure 1. World map showing locations of high energy colliders.

Table 1. High Energy Reaction Cross-section, Luminosity, and "Reach"

	HADRON	LEPTON
Cross-section	$\sigma \propto \frac{1}{s} f\left(\frac{M}{\sqrt{s}}\right)$ $\simeq \frac{1}{s}\left(\frac{M}{\sqrt{s}}\right)^{-6} = \frac{s^2}{M^6}$ M = mass-scale reached	$\sigma \propto \frac{1}{s}$ s = square of C of M energy
Luminosity required	$\mathscr{L} \propto \frac{1}{\sigma} \propto \frac{M^6}{s^2}$	$\mathscr{L} \propto \frac{1}{\sigma} \propto s$ $= \left(\frac{\sqrt{s}}{0.1}\right)^2 \times 10^{31}$
"Reach"	$M \propto s^{1/3} \mathscr{L}^{1/6}$ $< \left(\frac{\sqrt{s}}{20}\right)^{2/3} \left(\frac{\mathscr{L}}{10^{33}}\right)^{1/6} \times 3.8$	$M < \sqrt{s}$

Note: M and $\sqrt{s}$ in TeV, $\mathscr{L}$ in $cm^{-2} sec^{-1}$

For lepton interactions at high energies the cross-sections vary roughly as s^{-1} where s is the square of the center-of-mass energy, and are independent of the mass-scale of the produced particle(s). The necessary luminosity of a lepton collider is, therefore, proportional to s. The coefficient of proportionality is clearly a soft parameter dependent sensitively on the design of the detector and the patience of the experimenter. The value given in Table 1 is only a ball-park number scaled from the parameters of LEP, $\mathscr{L} = 10^{31} cm^{-2} sec^{-1}$ at $\sqrt{s} = 0.1$ TeV.

The fact that the cross-sections are independent of the mass of the produced particle implies that the highest mass-scale M that can be reached (the "reach") is given simply by $\sqrt{s}$. This is the distinguishing characteristic of a lepton collider.

For hadron interactions the situation is quite different. First, the compositeness of hadrons results in energies of collisions between quarks and gluons much lower than the energy of the incident hadrons. Secondly, because of the strong interaction, low mass particles are produced in great multiplicities, thereby reducing the probability for production of high mass particles. Thus, in addition to the s^{-1} dependence, the cross-section is a sharply decreasing function of the ratio of the mass-scale reached, M, to the center-of-mass energy, $\sqrt{s}$. At high energies, available data indicate that $(M/\sqrt{s})^{-6}$ is a fair description of this function,. Thus, the "reach" depends on both the center-of-mass energy and the luminosity. The proportionality coefficient given in Table 1 is again a ball-park number obtained from the discovery of Z^0 and $W^{\pm}$ on $S\bar{P}PS$.

Table 2. High Energy Colliders with First Operation Between 1985 and 1995

Name [Laboratory]	Particles	Energy in GeV	Luminosity in cm^{-2}sec^{-1}	"Reach" in GeV	Year of 1st Operation	Circumf. in m [Field in T]	Cost in M$
In Operation							
S$\bar{\text{P}}$PS [CERN]	$\bar{\text{p}}$, p	270 + 270 315 + 315 450 + 450	4 × 10^{29} (4 × 10^{30})	100 (150) 130 (190)	1981 1985 (1987)	6912 [1.8]	150*
TEV. COLL. [FERMILAB]	$\bar{\text{p}}$, p	800 + 800 900 + 900 (1000 + 1000)	1 × 10^{29} (1 × 10^{30})	160 (260)	1985 1987 (1989)	6283 [4.4]	450
TRISTAN [KEK]	e$^+$, e$^-$	25 + 25 (30 + 30)	2 × 10^{30} (2 × 10^{31})	50 (60)	1986 (1989)	3018 [0.24 (0.29)]	500 [8 × 10^{10}¥]
In Construction							
SLC [SLAC]	e$^+$, e$^-$	50 + 50	6 × 10^{27}	100	1987	Linear	115.4*
BEPC [IHEP-PRC]	e$^+$, e$^-$	2.8 + 2.8	1.7 × 10^{31}	5.6	1988	238 [0.9]	80 [2.4 × 10^8 Yuan]
LEP [CERN]	e$^+$, e$^-$	55 + 55 (100 + 100)	1.6 × 10^{31}	110 (200)	1989	26659 [0.06(0.11)]	750 [1.2 × 10^9 SF]
HERA [DESY]	p, e$^\pm$	820 + 30	1.5 × 10^{31}	130	1990	6336 [4.5, 0.19]	425 [7.8 × 10^8 DM]
UNK [IHEP-USSR]	p, p	3000 + 400 3000 + 3000	10^{32}	590 1200	1992	20772 [5.0, 1.0]	~ 1000

Table 2. High Energy Colliders with First Operation Between 1985 and 1995 (continued)

Name [Laboratory]	Particles	Energy in GeV	Luminosity in cm^{-2}sec^{-1}	"Reach" in GeV	Year of 1st Operation	Circumf. in m [Field in T]	Cost in M$
Proposed, Not Yet Funded							
RHIC [BNL]	Heavy ions 2 × 100 GeV/u [Au + Au]		4.4×10^{26}	-	1994	3834 [3.5]	350
SSC [-]	p, p	20,000 + 20,000	10^{33}	6000	1996	82944 [6.6]	3200
LHC [CERN]	p, p	8000 + 8000	1.4×10^{33}	3500	-	26650 [10.0]	-
	p, e*	8000 + 50	2.7×10^{32}	520		[0.06,10.0]	
CLIC [CERN]	e^+, e^-	1000 + 1000	10^{33}	2000	-	2 × 12500 Linear	-
VLEPP [INP]	e^+, e^-	150 + 150 500 + 500	10^{32}	300 1000	-	2 × 1500 2 × 5000 Linear	-

Note: 1. Numbers in parenthesis are future upgrades.
2. * indicates partial cost only.

The "reach" gives a fair comparison between the capabilities of lepton and hadron colliders. However, for a given lepton collider one should check that the luminosity is adequate to make it a useful machine at all.

In Table 2 we list three colliders in operation; five colliders in construction the first of which, SLC, is now in the commissioning stage; and five proposed colliders of which the first two have been approved for funding in FY 1989. These projects are listed in the order of their actual or anticipated year of first operation.

Several interesting observations deserve mentioning:

1. The list is quite long and rather impressive. If all these projects stay on schedule we will be commissioning new facilities in the period of 1985-1995 at roughly a uniform rate of one per year.

2. Detectors are not included in the Cost entries. These entries are only rough approximations because of the rapidly changing currency exchange and inflation rates and in some cases, the inaccessibility of exact and reliable data. They nevertheless give a rough idea of the magnitude of the efforts involved. One notes that with the two approved proposals included the total cost of all the entries amounts to almost $7 billion, a very impressive sum.

3. One can get a "unit cost" by dividing the Cost by the "Reach." This is given in Table 3, which shows clearly that the cost per "reach" is a monotomically decreasing function of the "reach." This is presumably a demonstration of the principle of "economy of scale."

4. Except for $\overline{SPPS}$, all hadron colliders use superconducting magnets as indicated by Field entries of > 2T.

Table 3. Unit Cost of Colliders in Order of "Reach"

Collider	"Reach" in GeV	Cost in M$	Unit cost in M$/GeV"reach"
BEPC	5.6	80	14
TRISTAN	50 (60)	500	10
SLC	100	115*	1.15*
$\overline{SPPS}$	100 (150)	150*	1.5*
LEP	110 (200)	750	7
TEV. COLL.	160 (260)	450	2.8
HERA	130	425	2.4
UNK	590, 1200	~ 1000	~ 1.6
SSC	6000	3200	0.5
RHIC	-	350	-

Note: * indicates partial cost

5. Despite the disadvantages, all high "reach" machines are hadron colliders. Even the rather futuristic and, so far, not quite ready-for-construction linear lepton colliders CLIC and VLEPP do not come close to the hadron collider SSC in "reach." Among hadron colliders the p̄p option is limited in "reach" by the achievable luminosity. It thus appears that at least for the present, the highest "reach" is obtained by pp colliders. However, this high "reach" is derived only from immense size and cost of the facility.

High Intensity, Medium Energy Accelerators

All so-called high energy phenomena also influence events at low energies, although in most cases the effects are greatly reduced in magnitude. This is even true for the search of some high mass particle whose existence can nevertheless be inferred from effects due to virtual processes on interactions at energies much below its production threshold. To detect these minute effects one needs to perform precision experiments. In addition to yielding information at high mass-scales, precision experiments can reveal new phenomena through studies of rare or forbidden events, violations of symmetry principles, etc. Because of the absence of strong interaction, precision lepton scattering experiments are further useful for probing the fine structures of nucleii and hadrons. In fact, since it appears that colliders with "reaches" much beyond, say, 10 TeV will be too costly to build, at least based on present day technology, it is likely that precision experiments will be the only approach to studying phenomena at extremely high energies. A good example and a case in hand is the proton decay experiment which, performed at < 1 GeV (decay energy), is supposed to test for the symmetry of strong/electroweak interactions at the grand unification mass of 10^{15} GeV.

For precision experiments we need high luminosities, and high luminosities are obtained by high intensity beams striking high density fixed material targets. A 100-μA beam with a cross-sectional area of 1 cm^2 striking a 1-mole target gives a luminosity of $\sim 4 \times 10^{38}$ cm^{-2}sec^{-1}. Both hadron and lepton high intensity accelerators have been proposed, but only one of these machines, CEBAF, a continuous electron beam recirculating linac, has been approved and is now in construction. Most of these designs provide beams at several intermediate steps of energy to enhance their usefulness and rely upon the copious production of secondary and tertiary particles to provide beams of different particle species.

The concept of studying high energy phenomena by doing precision experiments at lower energies was discussed as early as the latter part of 1950's by physicists at MURA (Midwestern Universities Research Association). But the "high intensities" in those days were not very high (a few μA) and the "high energies" were obtainable at not excessively high cost and at not much lower luminosity. "High energies" today are getting to be rather difficult and expensive to come by.

In addition to CEBAF, four proposals for hadron facilities of this category are listed in Table 4. It is understandable that these proposals do not have the appeal of the high energy colliders, but the costs are substantially lower. Hopefully at least one of the high intensity hadron facilities proposed will someday be built.

Table 4. High Intensity, Medium Energy Accelerators

Name [Laboratory]	Particles	Energy in GeV	Current in µA	Cost in M$	Year of Completion	Accelerator type
CEBAF [CEBAF]	e⁻	0.5 - 4.0	200 [Continuous]	216	1992	Recirculating supercon. linac
AHF [LANL]	p	2 15 60	500 25 25	500	-	Linac 6Hz synchrotron 6Hz synchrotron
TRIUMF II [TRIUMF]	p	3 30	100 100	350 [Cn$ 4×10^8]	-	50 Hz synch. + accum. 10 Hz synch. + accum. and stretcher
EHF [-]	p	9 30	100 100	450	-	25 Hz synchrotron 12.5 Hz synch. + accum. and stretcher
JHF [KEK]	p Heavy ions p	2 3.2 1.3 GeV/u 30	200 100 - -	-	-	50 Hz synchrotron 0.5Hz stretcher/synch. [same] -

Synchrotron Radiation Storage Rings

We list in Table 5 the electron (or positron) storage rings used for synchrotron radiation in the world by country. Only their energies, locations and status are given. Because of the very large number of these facilities there may be omissions in the table, but hopefully, they are not major ones.

Instead of high energy or high intensity as for the first two categories, the challenge in accelerator science and technology for these machines is the low beam emittance desired to maximize the brilliance of the synchrotron radiation emitted, and the maintenance of ultra-high vacuum in the presence of outgassing by the very intense synchrotron radiation. As VUV and X-ray sources, these machines yield brilliances many orders-of-magnitude higher than those obtainable from all other types of sources. The costs of these projects range from some $10M for the lowest energy to about $400M for the highest energy facilities.

We will not make further comments on this important category of accelerator projects except to point out the recent outcropping of the subcategory of industrial synchrotron radiation sources indicated by asterisks, *, in the table. These are storage rings in the energy range of 1/2 to 1 GeV and used for the manufacturing of VLSI chips by the method of X-ray lithography. It is expected that with X-ray lithography, feature resolutions of 1/4 μm or better can be achieved. At the present, using optical lithography the feature details are limited to ~ 1 μm. Rough estimates of the storage rings required if all VSLI chips are produced by synchrotron X-ray lithography give numbers in the hundreds. Similar conclusions on the magnitude of needs can also be drawn for medical applications of synchrotron radiation such as angiography. Thus, we can expect further and more extensive mushrooming of accelerator projects of this category in the near future.

Table 5. Synchrotron Radiation Storage Rings

Country	Name (Location)	Energy in GeV	Status+
Brazil	LNRS (Campinas)	2 - 3	C
China	BEPC (Beijing)	2.8	C
	HESYRL (Hefei)	0.8	C
	TLS (Hsinchu)	1.3	C
France	ESRF (Grenoble)	6	C
	ACO (Orsay)	0.54	O
	Super ACO (Orsay)	0.8	C
	DCI (Orsay)	1.8	O
Germany, Federal Republic	BESSY (Berlin)	0.75	O
	BESSY II (Berlin)	1.5 - 2.0	P
	* COSY (Berlin)	0.56	C
	Synchrotron (Bonn)	2.5	O
	ELSA (Bonn)	2.3	C
	DELTA (Dortmund)	-	P
	DORIS II (Hamburg)	3.7	O
	WILMA (Hamburg)	1 - 2	P
	* (Karlsruhe)		C

Table 5. Synchrotron Radiation Storage Rings (continued)

Country	Name (Location)	Energy in GeV	Status+
India	Indore I (Bhabha)	0.8	C
	Indore II (Bhabha)	-	P
Italy	ADONE (Frascati)	1.5	O
	New Ring (Trieste)	1 - 2	C
Japan	KSRS (Osaka)	5 - 6	P
	UVSOR (Okazaki)	0.6	O
	SOR Ring (Tokyo)	0.38	O
	TERAS (Tsukuba)	0.6	O
	NIJI-1 (Tsukuba)	0.3	O
	Photon Factory (Tsukuba)	2.5	O
	TRISTAN (Tsukuba)	30	O
	Accumulator Ring (Tsukuba)	6 - 8	O
	Super SOR (Tsukuba)	1.0	P
	* (Hitachi)	1.0	C
	, (NIT)		C
	* (Sumitomo Heavy Ind.)		C
	* (Sumitomo Electric)		C
Sweden	MAX (Lund)	0.55	O
	* (Scanditronix)	0.5	P
U.K.	SRS (Daresbury)	2.0	O
	* (Oxford Instrument)		C
U.S.A.	APS (Argonne)	7	P
	ALS (Berkeley)	1 - 2	C
	SURF II (Gaithersberg)	0.28	O
	CESR (Ithaca)	4.7 - 5.6	O
	SPEAR (Stanford)	3.5	O
	PEP (Stanford)	15	O
	Tantalus (Stoughton)	0.24	O
	Aladdin (Stoughton)	1.0	O
	VUV Ring (Brookhaven)	0.75	O
	X-ray Ring (Brookhaven)	2.5	O
	* (Brookhaven)		P
USSR	Siberia I (Moscow)	0.4	O
	Siberia II (Moscow)	2.0	C
	VEPP-2M (Novosibirsk)	0.67	O
	VEPP-3 (Novosibirsk)	2	O
	VEPP-4 (Novosibirsk)	5	O
	V3P (Novosibirsk)	-	P
	ARUS (Yerevan)	6	O

Note: + for the Status column: O = in operation, C = in construction, and P = in proposal.

* denotes industrial machines in the energy range of 0.5 - 1.0 GeV. They are generally without names.

Technology and Particle Beams

Donald L. Hartill
Cornell University

Table of Contents

1. Introduction 2128

2. Materials Modification 2130

3. Synchrotron Radiation 2134

   References 2137

Technology and Particle Beams

Donald L. Hartill
Cornell University

Introduction

Particle beams have been developed over the past sixty years by physicists in order to do scattering and production experiments in high energy physics. The earliest beams were the alpha-particle beams from natural radioactive sources used by Rutherford to perform scattering experiments that led to the discovery that the atomic nucleus was very small in diameter compared to atomic diameters. Over this period the technology has ranged from MeV energy proton beams produced by dc and electrostatic generators to modern colliders which collide beams of antiprotons with protons at beam energies of 900 GeV. To contain these energetic beams large-scale superconducting magnets had to be developed and are now in routine use at Fermilab. The proposed SSC which consists of two superconducting magnet rings with protons circulating in opposite directions will provide collisions between 20-TeV protons. The scale of these machines has gone from tens of centimeters to 100 kilometers for the circumference of the SSC in the space of fifty years. The field of particle physics is at the most fundamental level of scientific research since it studies the forces and particle constituents of nature's particles that are the building blocks of matter as we know it. This field has an extensive literature and for the rest of this overview only the technological applications of particle beams will be considered.

One of the more interesting uses of particle beams was the use of cosmic-ray muons by L. Alverez and his collaborators in the early 1970's to "X-ray" one of the giant pyramids in Egypt to look for previously undiscovered burial chambers. The technique involved placing an array of spark chambers in a vault under the pyramid and counting the number of muons as a function of angle. An increase of muon flux in a certain direction would indicate a lower density in that direction and hence a room. No rooms were found. Another historically significant application of particle beams has been the

© 1989 American Institute of Physics

certification of works of art by this same group using a device called the cyclotrino developed by J. Welch[1] which is a very sensitive mass spectrometer. With this technique very small samples of paint can be characterized by their isotopic content, thereby determining that they were the same as used by the artist in question with a high degree of reliablility.

The use of particle beams in medicine has had profound effects on both diagnosis and therapy of diseases. It is difficult to imagine a world without X rays and their modern implementation in the form of CAT scanners for diagnosing internal problems ranging from broken bones to locating tumors. X rays and radioactive sources producing intense gamma-ray beams have been effective therapy tools for selectively killing malignant tumor cells in cancer treatment. During the past thirty years neutron, proton, and pion beams have been added to X rays and gamma rays as specific therapy tools which can produce local damage only in the tumor. There is a consortium of physicists[2] associated with FNAL which has designed and is constructing a small proton synchrotron to be used only for radiation therapy at the Medical School at Loma Linda University.

Besides applications in medicine, materials modification represents the largest technological application of particle beams. These activities range from electron-beam welding and ultra-high-vacuum melting of materials to the construction of three-dimensional VLSI electronic circuits. It is now possible to fabricate layered structures with a scale size of 100 Å and a sharpness of a few Å with one layer of Fe^{57} followed by a layer of Fe^{56} followed by .... Particle beams can then be used to characterize these structures by using Rutherford backscattering to see that they were fabricated as planned.

High energy electron storage rings produce a byproduct that has proved to be extremely useful. Synchrotron radiation has grown from its first experimental verification by P. Hartman et al.[3] at Cornell in the early 1950's to a product that is now being marketed commercially by Oxford Instruments[4] and at least one Japanese manufacturer. The Advanced Light Source under construction at LBL and the Advanced Photon Source presently under design at Argonne represent very large commitments by the scientific community and indicate the importance of high brightness photon sources in the vuv and X-ray energy regions to scientific research. The commercial interest is in providing sources for X-ray lithography which are vital to the production of integrated circuits with feature sizes in the submicron region. There will also be a large demand for tunable X-ray beams for heart blood flow diagnostics following the pioneering experiments carried out at SSRL at SLAC over the past two years. Presently, this form of diagnosis represents a multibillion dollar per year activity, and the present procedures involve substantial risk to the patient.

Other technological applications involve the simulation of beam dynamics in accelerators and colliders. Modern RF power sources depend on a detailed understanding of the interaction of intense beams with their surroundings. In the future, it is likely that particle beams will play an even more important role in our technology as materials requirements become more sophisticated and elaborate.

Materials Modification

Electron-beam welding of materials is now a routine industrial process. Aircraft parts fabricated from titanium can be welded into complex forms with the intrinsic strength of the base metal. Commercial vendors now supply superconducting RF cavities made from high purity niobium which was first vacuum melted with electron beams; then the cavity parts, fabricated by forming the niobium sheet, are electron-beam welded into the final cavity assemblies.

Surface treatment by ion beams can dramatically change the wear resistance of surfaces. Orthopedic hip and knee joints[5] fabricated from titanium and treated by nitrogen ion implantation with doses in the range of 10^{18} cm^{-2} have a wear resistance nearly a factor of 10^3 higher than the pure metal. This gives a greater than twenty-year expected life for the joints. Figure 1 illustrates this surface wear improvement for two different types of steel which are commonly used in mechanical devices.

Ion implantation in semiconductor devices is probably the largest commercial application of particle beams. Essentially every semiconductor device in modern electronics has gone through at least one ion implant step in its fabrication. The techniques vary from a simple ion implant to modify the carrier type and density in the high purity silicon to forming three-dimensional electronic devices. The simplest of the latter category is to use doses of 200-KeV oxygen ions at the level of 2×10^{18} cm^{-2} with a substrate temperature of 600 °C followed by a 1300 °C high temperature anneal for 6

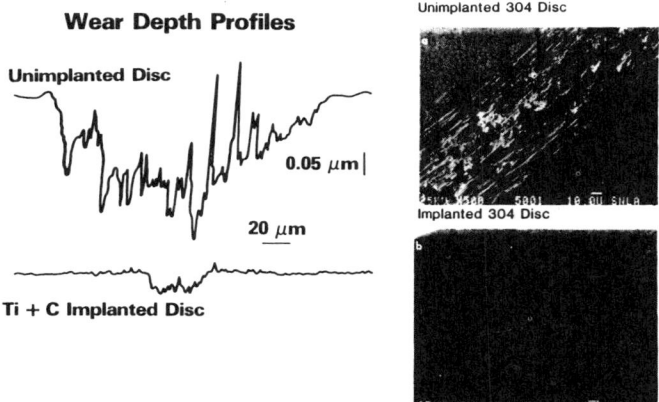

Figure 1. Wear depth profiles and surface view of steels after 1000 wear cycles in unlubricated pin-on-disk tests. Results from disks implanted with Ti and C at fluences of 2×10^{17} cm^{-2} to produce an amorphous surface alloy with Ti and C concentrations of approximately 20 at.% are compared with those from unimplanted disks. This implantation treatment results in significant reduction of the wear and friction (not shown). Wear tracks are shown for a hard bearing steel (440C), but the scanning electron micrographs illustrate similar improvements in a softer steel (304 stainless). (After D. M. Follstaedt et al., Appl. Phys. Lett. 43 [1983] p. 358.)

hours to form a buried SiO_2 layer. This yields a silicon on insulator (SOI) structure that is very useful for constructing fast radiation hard devices. Three-dimensional structures are beginning to appear in the form of optoelectronic heterostructure devices in which both field effect transistors and light emitting diodes are fabricated by laying down layers of AlGaAs and GaAs on a GaAs substrate by molecular-beam epitaxy and then separately doping each of the layers with tightly focused Be and Si ion beams of 60 and 160 KeV to form the structures in each of the layers. Fast self-aligning bipolar devices that currently hold the commercial electronic speed record are fabricated with similar techniques to form the layers but using conventional masked ion-beam implantation to perform the doping operation. Reference 6 provides an overview of these techniques and their likely future evolution.

By using ion beams it is now possible to fabricate very elaborate layered structures. These range from ion-beam-assisted deposition of thin films to manufacturing isotopically pure layers of the same chemical species to form very selective X-ray filters for very high energy resolution X-ray scattering experiments. A typical system consists of an ion source that accelerates the ions to 35 KeV followed by careful momentum analysis and beam collimation. After several stages of differential pumping to maintain surface cleanliness of the sample, the beam passes through an electrostatic lens to form a uniform spot and is then decelerated to 65 eV before striking the sample so that the ions are only deposited on the surface. It is straight forward to produce 100-Å-thick layers in this manner with better than 20Å sharpness between layers. Figure 2 is an example of one of these samples and also illustrates another very important use of particle beams in materials science, namely the characterization of material structures by scattering experiments. Rutherford backscattering with energy analysis of the scattered particle can be used to show the isotopic composition of the material as a function of depth. As indicated in the figure, the analysis technique can resolve features on at least the 20-Å scale.

The ion-beam requirements for materials modification and characterization have a wide range. Very uniform and high current densities with good energy definition are required for production line ion implanters while submicron size beams with moderate currents and fast writing rates are needed for three-dimensional circuit fabrication. A typical Eaton[7] production line ion implanter produces a 1-cm^2 beam of 180-KeV O^+ ions with better than 1% current density uniformity with a total current of approximately 100 mA. One of these machines can process 60 10-cm-diameter Si wafers per shift with a uniform dose of 2×10^{18} cm^{-2}. Typical accelerating voltage stability requirements in critical ion-beam epitaxy processing are in the range of 10 ppm. Beam uniformity and energy definition are the critical parameters in most applications.

Future directions for development of particle beams will be higher currents to speed up processing, and as structures become more complex higher energies will be required to provide deeper implants especially for some of the SOI technologies. More

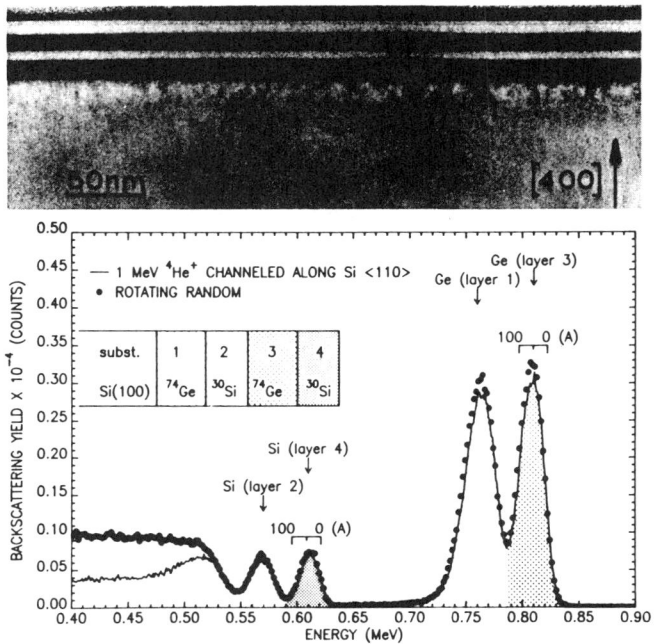

Figure 2. Ion-beam deposition using 65-eV ions to form a Ge/Si multilayer device with a periodicity of 10 nm on Si(100) at room temperature. The cross-section TEM (top) illustrates the sharpness of the Si/Ge interfaces (< 0.35 nm). The Rutherford backscattering spectra (bottom) show the modulation of the composition as a function of depth. (After B. R. Appleton et al., MRS Bulletin, Vol. 12, No. 2 [Feb. 1987] p. 52.)

precise beam definition and faster writing rates will be required to make 3D electronic devices a commercially viable fabrication technique. Higher energy resolution beams and better energy resolution detectors combined with better scattering-angle resolution will enable more precise localization of smaller features in materials.

Synchroton Radiation

Within two years synchrotron radiation sources will enter the manufacturing environment to provide X rays for lithography. The move to ever shorter wavelength radiation for lithography is driven by the need to construct smaller feature size electronic devices. For submicron devices, X rays in the 5 to 10-KeV energy region are a good match to the optics requirements for the lithography. Again to limit processing time in manufacturing these devices, high flux beams with excellent spatial definition are required. Synchrotron radiation from high energy electrons passing through high field dipole magnets satisfies these requirements. Several manufacturers, among them Oxford Instruments[4], are now designing and building compact electron storage rings to be used as dedicated X-ray sources for lithography on submicron feature size integrated circuit production lines. The storage ring being designed by Oxford Instruments is being purchased by IBM for use in their IC production facility.

Medical diagnostics is an area where synchrotron radiation sources will play an increasingly important role. Determining the flow characteristics of the arteries supplying blood to the heart muscle is presently accomplished by a medical procedure that has a probability of several percent of causing very severe side effects and a 1% death rate. By using a beam from a wiggler magnet passing through a fast tunable monochromator centered on the K absorption edge for iodine, and using fast digital subtraction techniques, it is possible to obtain the flow pattern without the insertion of the catheter into the heart artery to periodically inject iodine to measure the flow. It is the insertion of the catheter that produces the high risk. The wiggler beam has sufficient intensity to make the subtraction technique possible on the time scales necessary to determine the flow. At the moment, just this diagnostic procedure represents more than one billion dollars in business per year. There are other diagnostic procedures that can also benefit from both monochromatic X-ray sources and from the very high fluxes possible using wiggler magnets installed in compact electron storage rings.

The intense short pulses of X rays available from wiggler magnets installed in high energy electron-positron storage rings operating with short bunch length particle beams have made possible detailed studies of Si surface solidification after melting with an intense laser pulse[7]. Intense monochromatic beams made by passing a wiggler beam through a monochromator make possible the determination of the structure of very complex organic molecules before beam heating destroys the sample crystal. Recently the determination of the molecular structure of the common cold virus by M. Rossman[9] and his collaborators from Purdue using such a monochromatic beam at the CHESS facility at Cornell has received national attention.

The need for intense monochromatic X-ray beams with very small spot sizes has led to the development of the undulator[10,11] which like the wiggler is an insertion device for electron storage rings. The electron beam requirements are much more stringent since the electron beam emittance determines the sharpness of the X-ray energy spectral lines. Figure 3a shows the calculated spectrum from an undulator presently under construction to be used with CESR operating at 5-GeV beam energy with a low emittance lattice (horizontal emittance = 6 x 10^{-8} m rad). The undulator is a 61-pole device with a 3.3-cm periodicity and uses NdFeB permanent magnet material to produce the magnetic field. It is a prototype for the Argonne Advanced Photon Source. Figure 3b shows the spectrum from the same undulator using the design horizontal emittance (= 7.3 x 10^{-9} m rad) for the Advanced Photon Source operating at 6-GeV

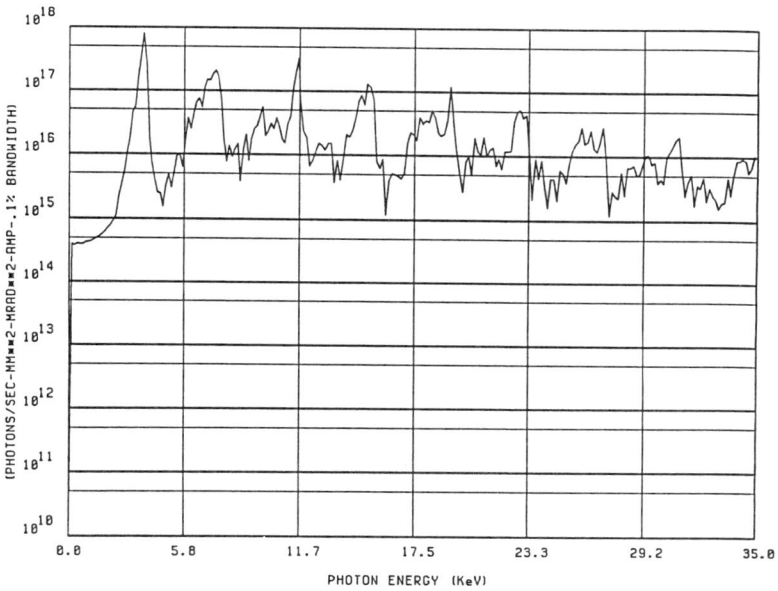

(a) Horizontal Emittance = 6 x 10^{-8} m rad

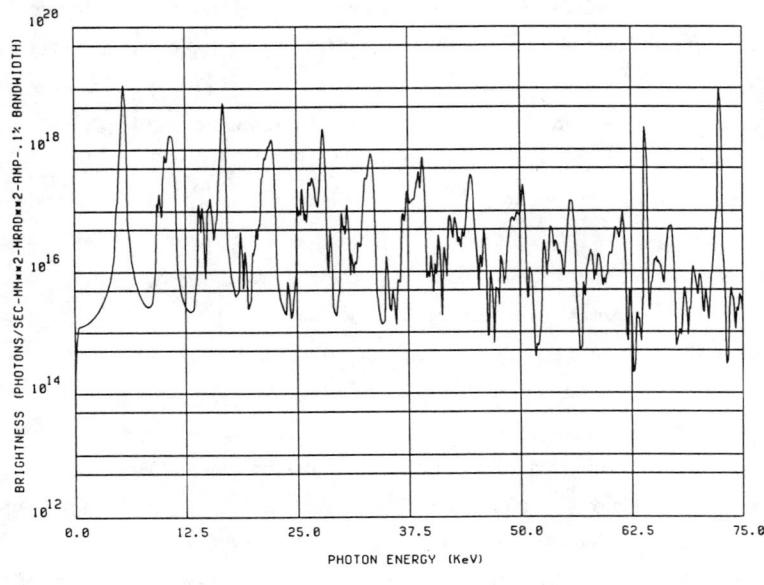

(b) Horizontal Emittance = 7.3×10^{-9} m rad

Figure 3. Full simulation including the electron beam emittance of the X-ray flux from a 61-pole undulator being constructed by Spectratech for Argonne to be used at CHESS at Cornell as a prototype for undulators for the Advanced Photon Source. The top graph shows the results for CESR operating in its low emittance configuration with a horizontal emittance of 6×10^{-8} m rad. The bottom graph gives the expected undulator performance using the horizontal emittance (= 7.3×10^{-9} m rad) projected for the Advanced Photon Source.

beam energy. As can be seen from the two spectra there is a substantial improvement in peak to valley ratios for the lower emittance beam. The roughness of the spectra is a result of the limited number of electrons (400 for the CESR case and 100 for the APS case) used in the simulation.

The future for synchrotron radiation applications is bright and the new facilities coming on-line in the early 1990's will be very rich in their scientific output. The technological applications of both particle beams and X-ray beams will play an increasingly important role in improving our technology that is often taken for granted in our quest for better science. Many other important areas (neutral beams, plasmas, neutron beams for materials research, etc.) using particle beams have been omitted from this short overview because of lack of time and space. Particle beams indeed have a very wide impact on our technology.

References

1. J. Welch et al., Nucl. Instr. and Methods B5, No. 2, 233, Nov. 84.
2. P. Livdahl et al., CERN Courier 22, Oct. 87, and 5, Dec. 86.
3. D. H. Tomboulian and P. Hartman, Phys. Rev. 102, 1423 (1956).
4. B. Feder, The New York Times, D6, Aug. 19, 1987.
5. S. Picraux and P. Peercy, MRS Bulletin, Vol. 12, No. 2, 22, Feb. 87.
6. Advanced Processing of Electronic Materials in the United States and Japan, National Academy Press, 1986. See also MRS Bulletin, Vol. 122, No. 2.
7. Eaton Corporation, 108 Cherry Hill Drive, Beverly, MA. 617-927-5840.
8. D. E. Mills et al., Phys. Rev. Lett. 48, 337 (1982).
9. M. Rossman et al., Nature, Vol. 317, 145-153, (1985).
10. S. Krinsky, IEEE Trans. Nuc. Sci. NS-30, 3078 (1983).
11. G. Brown et al. Nucl. Inst. and Methods 208, 65 (1983).

SYNCHROTRON RADIATION

A GENERAL OVERVIEW AND A REVIEW OF STORAGE RINGS, RESEARCH FACILITIES, AND INSERTION DEVICES

Herman Winick
Stanford Synchrotron Radiation Laboratory
Stanford, CA 94309

TABLE OF CONTENTS

1	Introduction and Overview	2139
2	Basic Nature of Synchrotron Radiation	2140
3	Comparison with Other Sources	2140
4	Range of Applications of Synchrotron Radiation	2144
5	Historical Development (1898-1970)	2147
6	Higher Energy Rings and Research Facilities	2149
7	Dedicated Synchrotron Radiation Sources	2153
8	Insertion Devices (Wiggler and Undulator Magnets)	2155
9	Properties of the Radiation from Bending Magnets, Wigglers, and Undulators	2155
	9.1 Bending Magnets	2156
	9.2 Wiggler Magnets	2156
	9.3 Undulators	2160
	9.4 Permanent Magnet Technology	2165
10	Third-Generation Storage Rings	2168
	10.1 Dynamic Aperture Requirements	2171
	10.2 Lattice Design	2172
	10.3 Lifetime and Stability Considerations	2174
11	Future Storage Rings	2175
12	Bibliography	2178
13	References	2179

ABSTRACT

Synchrotron radiation, the electromagnetic radiation given off by electrons in circular motion, is revolutionizing many branches of science and technology by offering beams of vacuum ultraviolet light and x rays of immense flux and brightness. In the past decade there has been an explosion of interest in these applications leading to increased exploitation of existing rings and activity to construct new research facilities based on advanced storage rings and insertion device sources. Applications include basic and applied research in biology, chemistry, medicine, and physics plus many areas of technology. In this article we present a general overview of the field of synchrotron radiation research, its history, the present status and future prospects of storage rings and research facilities, and the development of wiggler and undulator insertion devices as sources of synchrotron radiation.

SYNCHROTRON RADIATION*

A GENERAL OVERVIEW AND A REVIEW OF STORAGE RINGS, RESEARCH FACILITIES, AND INSERTION DEVICES

Herman Winick
Stanford Synchrotron Radiation Laboratory
Stanford, CA 94309

1   INTRODUCTION AND OVERVIEW

The explosive growth of activity and interest in synchrotron radiation as a scientific and technological tool is an unprecedented phenomenon. It is all the more remarkable considering that the radiation was originally regarded largely as an unfortunate drain on the energy of electrons in circular accelerators, a nuisance and an unavoidable waste product which made it difficult and costly for high energy physicists to push circular electron machines to higher and higher energy.

Today the impact of synchrotron radiation on basic and applied research in physics, chemistry, biology, and their numerous sub-fields has led to intense activity around the world for fuller exploitation of existing electron storage rings and construction of more storage ring sources of this radiation. This interest is further fueled by the expectation that synchrotron radiation will, via the x-ray lithographic process, enable large-scale production of higher density integrated circuits (about six companies are now constructing rings for this purpose) and the possibility that it can be used in less invasive procedures for coronary angiography.

The new rings now in construction range from superconducting devices for x-ray lithography that could fit into a large living room to rings of 1-km circumference that will provide extremely high brightness and coherent power to about 100 simultaneously operational experimental stations, serving thousands of scientists each year. New beam lines on existing rings are often based on wiggler and undulator sources with conventional, superconducting, and permanent magnet designs. Recent demonstrations indicate great interest

---

*Work supported by U.S. Dept. of Energy, Office of Basic Energy Sciences; contract DE-AC03-82-ER13000. This article is based on lectures at the Beijing Symposium/Workshop on Synchrotron Radiation, May 27 to June 7, 1988, whose Proceedings have been published by Gordon & Breach in *Applications of Synchrotron Radiation* (ISBN 2-88124-698-2). This material is used with the permission of the publishers.

© 1989 American Institute of Physics

exploiting the capabilities and potential of very high energy colliding beam rings, particularly when they are operated at lower energy in low emittance modes as very high brightness sources.

From humble beginnings as parasites on accelerators built and operated for high energy physics research, more than 5000 synchrotron radiation users in 8 countries now have access to more than 20 storage rings that are partly or fully dedicated to synchrotron radiation research. About six new basic research facilities are in construction including major centers with construction costs of about $500 million. The rapid buildup of capability is a direct result of the extreme pressure from a broad range of scientists and technologists from universities, government laboratories, and industry who need more beam time and even higher intensity to pursue their work.

2 BASIC NATURE OF SYNCHROTRON RADIATION

Synchrotron radiation is generated by the acceleration of charged particles, a basic mechanism for the generation of electromagnetic radiation. For example, the oscillation (and hence acceleration) of electric currents in antennas at radio frequencies produces radio waves. When the charges move in circular orbits (and hence undergo centripetal acceleration) at velocities close to the velocity of light, relativistic effects become important and the electromagnetic radiation emitted is called synchrotron radiation. Particularly for very light charged particles such as electrons, this radiation has rather spectacular properties; it is extremely intense, highly polarized, tightly collimated, sharply pulsed, partly coherent, and is emitted over a broad spectral range. It is produced copiously by electron beams curving in circular arcs in the bending magnets of high energy electron storage rings and even more copiously when these electrons wiggle or undulate while coursing through the alternating magnetic fields of insertion devices that may be placed between storage ring bending magnets.

3 COMPARISON WITH OTHER SOURCES

In the VUV and x-ray parts of the electromagnetic spectrum (see Figure 1) synchrotron radiation is by far the most powerful source of continuum (i.e., broadband) radiation. In flux (number of photons emitted per second within a

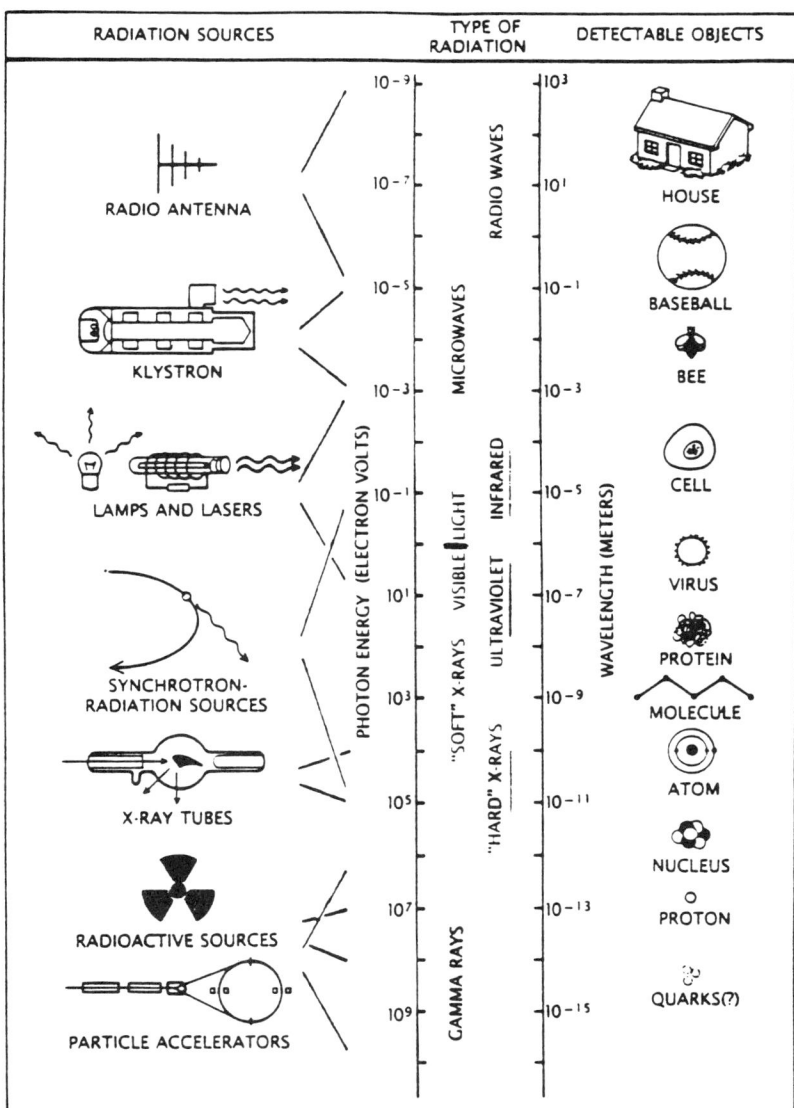

SPECTRUM OF ELECTROMAGNETIC RADIATION represents the range of photon wavelengths or energies that are most scientifically useful. The spectrum can be covered only by a multiplicity of radiation sources. Radiation employed to detect an object must have a wavelength that is equal to or less than the object's dimensions. Synchrotron-radiation sources provide exceedingly bright radiation in the ultraviolet and X-ray regions, which span the wavelengths suited to study atomic and molecular structures.

Figure 1. From Winick.[B8]

specified energy range) synchrotron radiation now exceeds other continuum sources in this spectral range, such as the bremsstrahlung continuum of conventional electron impact x-ray generators, by 3 to 6 orders of magnitude. In brightness (flux per unit source size per unit solid angle) the factor is now about 10 orders of magnitude (see Figure 2). In addition, the time structure, polarization, collimation and coherence properties of synchrotron radiation each offer unique capabilities in the VUV and x-ray spectral region. Just one of these properties would make a radiation source useful for experiments. The combination of all of them makes synchrotron radiation a remarkably capable source for a broad range of applications in basic research, technology, and medicine.

Why have improvements in source capability in the VUV and x-ray spectral regions been so import-

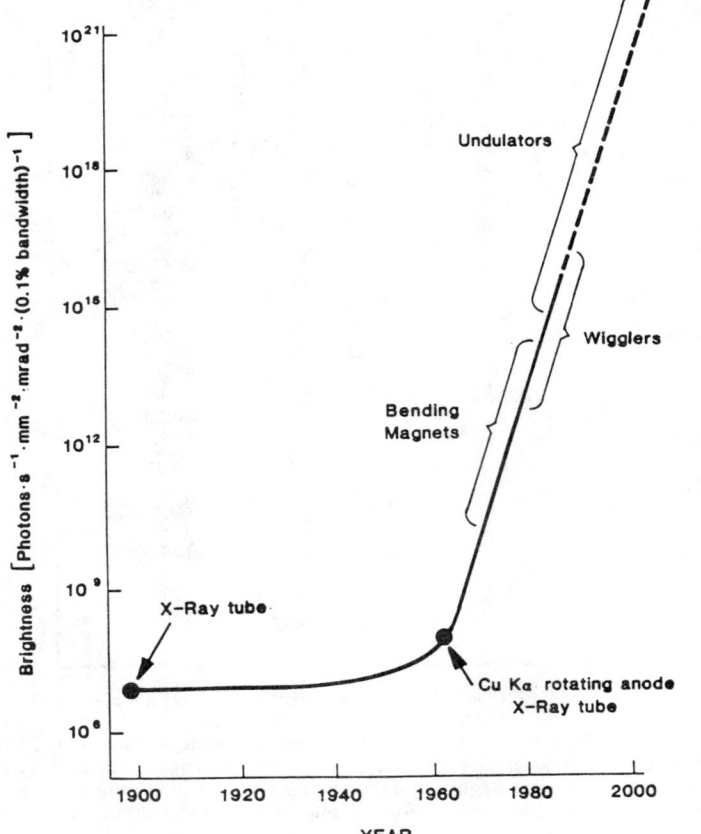

Figure 2. Growth of x-ray brightness with use of synchrotron radiation from bending magnets, wigglers, and undulators.

ant? Comparison with sources and applications in a more familiar part of the electromagnetic spectrum, visible light, may help explain this. Visible light, particularly as produced by a tunable laser, is a powerful structural and spectroscopic tool. Lasers offer pulsed, collimated, monochromatic beams of visible light with extremely high flux and brightness. As a tool for the determination of form or structure, however, visible light, even aided by a microscope, can only reveal features only of the order of the wavelength of the light, i.e. about 300 nanometers. Attempts to "see" smaller objects (e.g. atoms or molecules) with visible light are thwarted by diffraction effects that become more severe as the ratio of the size of the object to the wavelength of the light decreases.

Diffraction effects also determine the minimum size, and hence the number, of elements that can be packed into an integrated circuit chip when visible light photolithography is used in the manufacturing process. Electron beam technology can be used to make masks with features smaller than the wavelength of visible light. However, the replication of these mask patterns onto photosensitive layers on silicon substrates by lithographic contact printing is "diffraction limited" by the wavelength of the light used. With visible light, only features larger than about 500 nanometers can be replicated with high fidelity. With shorter wavelength light (soft x-rays with a wavelength of 1 to 2 nanometers seem optimal) features of the order of 50 nanometers or less have been replicated.

Thus to probe or replicate smaller objects or structures with light, we must use shorter wavelengths. Until the advent of synchrotron radiation, the strength of sources, particularly tunable sources, in these shorter wavelength domains was very weak compared with sources of visible light, severely restricting the applications.

Visible light, particularly monochromatic but tunable visible light such as produced by a tunable laser, is also a powerful tool for the spectroscopic study of electronic structure, chemical bonding, and other properties of matter. With a tunable laser the investigator adjusts the source to the desired wavelength and makes direct use of the narrow wavelength band emitted by the laser to "tune in" to specific electronic transitions, such as those involving the rather weakly bound valence electrons. But many important transitions can be excited only by photons with higher energy than is now available from lasers. In particular, the transitions involving more tightly bound core electrons require photons in the VUV and x-ray parts of the spectrum.

4 RANGE OF APPLICATIONS OF SYNCHROTRON RADIATION

With a broadband VUV and x-ray source such as synchrotron radiation, a grating or crystal monochromator is used to select a narrow photon wavelength band, typically with a spread of the order of 0.1% to 0.01%. Thus narrow band, tunable radiation is readily available at the output of the monochromator.

In the VUV and soft x-ray parts of the spectrum, the intense narrow-band polarized tunable radiation is used, via the photoelectric effect, to study the electronic properties of materials, including systems such as adsorbed layers on metals and metal-semiconductor interfaces. The analysis of the energy and angular distribution of photoelectrons from such systems has led to basic understanding of many surface and interface effects and has provided information important to the development of semiconductor devices.

The tunability of synchrotron radiation has proven to be an extremely important feature. For example, it makes it possible to exploit the fact that each element exhibits a sharp increase in absorption at certain wavelengths, called absorption edges, corresponding to photon energies sufficient to overcome the binding of electrons in particular atomic energy levels (see Figure 3).

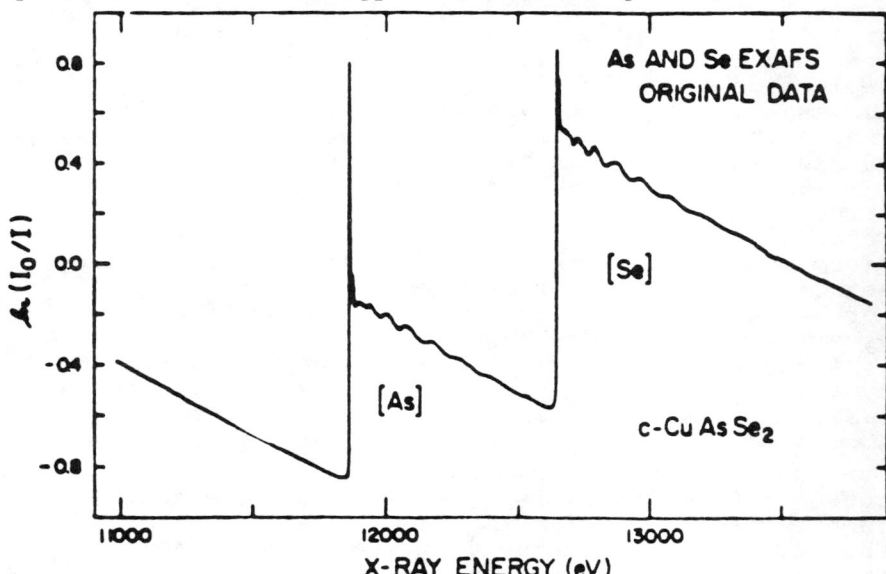

Figure 3. Example of the variation of x-ray absorption with photon energy, showing absorption edges of different elements. From Hunter.[66]

The ability to "tune in" synchrotron radiation to the precise energy of absorption edges in the VUV and x-ray parts of the spectrum has led to very important new techniques and capabilities.

In the anomalous scattering technique, measurements of photon scattering at two or three wavelengths very near an absorption edge are used to obtain unique local or long-range structural information. In the extended x-ray absorption fine structure (EXAFS) technique, information about the local atomic environment around a particular elemental constituent of a complex material is extracted from measurements of x-ray absorption made in a continuous scan of the photon energy from just below, to several hundred volts above, an absorption edge of that element. (See Figure 3.)

These techniques have opened up entirely new possibilities for scientists to learn about atomic arrangements in many condensed matter systems such as catalysts, amorphous materials, surface layers, two-dimensional films, and many others. The investigator can follow very small changes in these atomic arrangements as, for example, a biological enzyme goes through its function or a high temperature superconducting material undergoes a phase change. In many cases the element of interest is present in very low concentrations, e.g. being in a surface monolayer, or being one of many atoms in a very heavy molecule as in many enzymes. Very high flux is needed to obtain the data required to determine the atomic arrangements around these dilute atomic constituents. Very high brightness is required to study small samples such as sub-millimeter protein crystals. Applications such as these, which are limited by available flux and brightness, have paced the development of insertion devices and new dedicated rings.

In yet another example of the use of absorption edges, medical diagnostic radiography, the sensitivity to contrast agents such as iodine can be enhanced by two or more orders of magnitude by taking the difference between a pair of exposures made with monochromatic x-ray beams, one with x-ray energy just above the absorption edge and the other just below. This enhanced sensitivity makes it possible to detect restrictions in the coronary arteries with a much smaller iodine concentration than that now needed with conventional sources. The concentration now required is achieved by direct injection of iodine-containing contrast agent into the coronary arteries via a catheter threaded through a major artery. This procedure produces excellent quality radiographs; however, it is an invasive procedure with significant associated risks. It should be possible to achieve the lower concentration of iodine required in the dual energy subtraction technique without an interarterial catheter, and hence with less risk.

The procedure using synchrotron radiation requires very high x-ray flux (to enable short exposures so that heart motion does not cause blurring) at high photon energy (the K absorption edge of iodine is at 33 keV). A line beam of radiation (15 cm wide but only about 0.5 mm high) is used, and exposures at photon energies above and below the absorption edge are taken within a few milliseconds. The patient is moved through this beam so that the heart is scanned in about 1 to 2 seconds. See Figure 4 for the arrangement used at Stanford. The flux, energy and beam geometry requirements for the angiography application can be met with wigglers on rings above about 2.5 GeV and possibly with the bending magnets of higher energy rings. Angiography programs are underway at synchrotron radiation laboratories in Germany, Japan, the USA, and the USSR, with tests on human subjects recently started at Stanford.

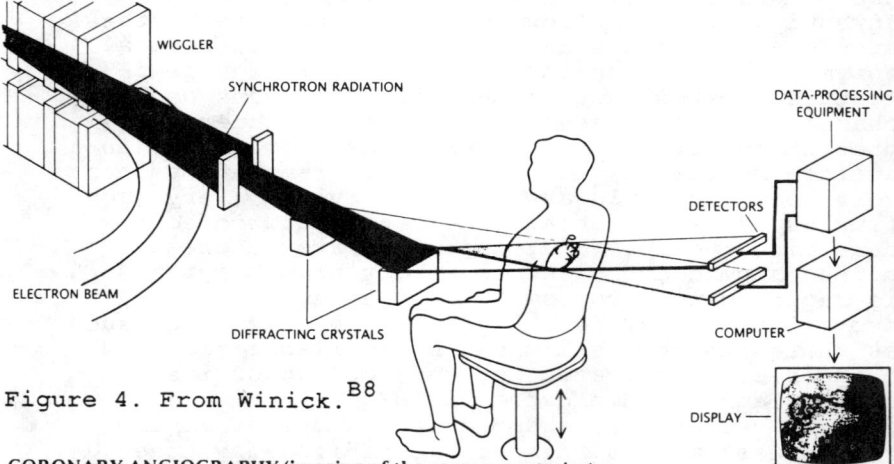

Figure 4. From Winick.[B8]

CORONARY ANGIOGRAPHY (imaging of the coronary arteries) is more sensitive and potentially less risky when synchrotron radiation is used instead of conventional X-ray sources. An iodine-containing liquid has been injected into the patient's blood. Special diffracting crystals convert synchrotron radiation into beams with photon energies just above and just below an absorption edge of iodine. The two beams converge at the patient's heart and then diverge to strike separate detectors. The higher-energy beam is absorbed nearly as much as the lower-energy beam by soft tissue and bone, but the iodine absorbs much more of the higher-energy beam. Taking the difference between the images generated by the two detectors in effect "subtracts out" most of the patient's body tissues and bone, leaving an image in which the coronary arteries and other blood vessels are enhanced. Conventional angiography, because it is less sensitive, requires that the iodine be infused directly into the arteries.

Some of the work described above could be pursued with weaker sources of broadband radiation, such as the bremsstrahlung continuum from a conventional x-ray generator. Such sources were used before the advent of synchrotron radiation and are still being used. In fact, the revitalization of x-ray science due to synchrotron radiation has resulted in an increase in the number of conventional x-ray generators in use around the world. Many groups use these sources for some work while using synchrotron radiation for work requiring higher intensity. Improvements in conventional x-ray generators and in the instrumentation used with them (such as large curved crystals to collect and focus more photon flux) have extended their capabilities, but the low continuum intensity that these sources provide still limits the range of work that can be done.

As an example, consider the impact of synchrotron radiation on the development of the EXAFS technique. With conventional x-ray generators, days or weeks were required to accumulate a spectrum that can be obtained in a few minutes (in some cases a few seconds) with synchrotron radiation sources. The consequences are as follows.
a) Many more samples can be measured.
b) The effects of different preparation techniques and the role of variables such as temperature and pressure can be readily studied.
c) More dilute species can be studied, including samples in which the element of interest is present in only millimolar concentrations or as a fractional surface monolayer.
d) Structural changes during chemical reactions can be followed in real time, and other time-dependent effects can be followed.

The result has been a veritable explosion of EXAFS work in such fields as biochemistry, materials science, and catalytic chemistry, limited partly by the number of stations and operating hours at present synchrotron radiation facilities. The variety and precision of the experimental data have led to improvement in EXAFS theory and analysis techniques leading in turn to the ability to extract more accurate structural information. In short, synchrotron radiation has transformed the EXAFS technique from an interesting curiosity to a broadly applicable tool of analytical chemistry.

5 HISTORICAL DEVELOPMENT (1898-1970)

Although what is now called synchrotron radiation was first directly observed (accidentally) at the General Electric 70-MeV synchrotron in 1947, the theoretical consideration of the radiation by charges in circular motion goes back to 1898 to the work of Lienard.[1] Further theoretical work was done by Schott,[2] Jassinsky,[3] Kerst,[4] Pomeranchuk et al.,[5,6] and others through 1946. Blewett[7] was one of the first to be concerned with the effects of the radiation on the operation of electron accelerators and observed effects on the electron orbit in 1945. Lea[8] provides a more detailed history of these early developments.

The observation of the radiation in 1947 and with construction of electron accelerators in the postwar period led to renewed interest in synchrotron radiation. Comprehensive theoretical treatments were presented by Sokolov et al.[9] and by Schwinger[10] starting in the late 1940s. With these works the theory was fully developed so that accurate predictions could be made regarding the intensity, spectral and angular distributions, polarization, etc. The theory has been reviewed by Jackson[11].

The first experimental investigations of the properties of the radiation were carried out in the late 1940s by Pollack et al.[12] using the General Electric 70-MeV synchrotron. In the 1950s studies of synchrotron radiation were extended by several groups[13] using the 250-MeV synchrotron at the Lebedev Institute in Moscow; by Corson[14] and by Tomboulian et al.[15] using the Cornell 300-MeV synchrotron, and by Codling and Madden[16] using the 180-MeV synchrotron at the National Bureau of Standards (NBS) in Washington, DC. In the mid-1960s Haensel et al.[17] were the first to utilize radiation from a multi-GeV accelerator, the 6-GeV synchrotron in Hamburg. These investigations verified the basic theoretical predictions and provided much useful data and experience in the use of the radiation.

Opportunities to use the radiation increased as more electron synchrotrons were built. Twenty years ago perhaps 50 scientists around the world utilized the radiation from about a dozen synchrotrons that were built and operated primarily for high energy physics research. Ten years ago the number of synchrotrons had decreased, but about a dozen storage rings had synchrotron radiation programs, almost all parasitic to the high energy physics programs, and the total number of users was perhaps 500. A storage ring maintains a current of particles for a long time, typically several hours. Storage rings were developed for high energy physics colliding-beam studies for which, in most cases, counter-rotating beams of electrons and positrons are stored in the same ring.

Today only a few synchrotrons remain, and more than 20 storage rings are used as fully or partly dedicated light sources, by more than 5000 scientists. (See Table I.) The switch to storage rings was initiated in 1968 when Rowe et al. at the University of Wisconsin adapted the 0.24-GeV Tantalus ring[18] for synchrotron radiation research after it was retired from its original use in studies aimed at future high energy physics machines. Although its low electron energy limited its spectral range to photon energies less than about 200 eV, the stable intensity, constant spectrum, and good vacuum enabled Tantalus users to perform experiments not possible with cyclic synchrotrons. This experience led, over the next few years, to use of the radiation from many colliding-beam rings and also, in 1970, to design of the first storage ring intended only as a light source, the 0.4-GeV SOR-Ring at the University of Tokyo, carried out mostly by a group of solid state physicists led by Sasaki.[19]

6 HIGHER ENERGY RINGS AND RESEARCH FACILITIES

Because the energy radiated per turn increases as the fourth power of the electron energy in a ring of fixed radius, it rapidly becomes unmanageable as the ring energy is increased. At a given electron energy the radiation produced by the bending magnets can be reduced only by decreasing the magnetic field, i.e. increasing the bending radius. Therefore rings have had to grow in size and cost as high energy physicists have sought to push the frontier of their field. For example, for colliding-beam storage rings above about one GeV, each doubling of ring energy results in about a factor of four increase in construction cost due to the larger size of the ring and the need to supply more radiofrequency power to compensate for the increase in synchrotron radiation as the electron energy increases. Nevertheless, the energy of colliding beam storage rings has grown by about a factor of 100 over the past 25 years, paced by the desire of elementary particle physicists to work at the high energy frontier to explore new phenomena and create new particles. The amount of energy lost to synchrotron radiation is minimized by the use of weaker bending magnets, resulting in larger radius rings. For example the PEP storage ring operates at 15 GeV with a bending field of 0.3 Tesla, and the LEP ring now in construction in Geneva will operate at 50 GeV and only 0.054 Tesla. LEP has a circumference of 27 km.

TABLE I STORAGE RING SYNCHROTRON RADIATION SOURCES (MAY, 1988)

LOCATION	RING (LAB)	ELEC. EN. (GeV)	NOTES
BRAZIL			
Campinas	LNLS	2.0	Dedicated*
CHINA			
Beijing	BEPC (IHEP)	2.2-2.8	Partly Dedicated
Hefei	HESYRL (USTC)	0.8	Dedicated*
ENGLAND			
Daresbury	SRS (Daresbury)	2.0	Dedicated
FRANCE			
Orsay	ACO (LURE)	0.54	Dedicated
	DCI (LURE)	1.8	Dedicated
	SuperACO (LURE)	0.8	Dedicated
Grenoble	ESRF	6.0	Dedicated*
GERMANY			
Bonn	ELSA	3.5	Partly Dedicated
Dortmund	DELTA	1.5	Design/FEL Use
Hamburg	DORIS II (HASYLAB)	3.5-5.5	Partly Dedicated
West Berlin	BESSY	0.8	Dedicated
	BESSY II	1.5-2	Design/Dedicated
INDIA			
Indore	BARC	0.45	Design/Dedicated
ITALY			
Frascati	ADONE (LNF)	1.5	Partly Dedicated
Trieste	Sincrotrone Trieste	1.5-2	Dedicated*

TABLE I (continued)

JAPAN
Okasaki	UVSOR (IMS)	0.6	Dedicated
Kansai area	6 Gev Ring	6.0	Design/Dedicated
Tokyo	SOR (ISSP)	0.4	Dedicated
Tsukuba	TERAS (ETL)	0.6	Dedicated
Tsukuba	Photon Factory (KEK)	2.5	Dedicated
	Accumulator Ring (KEK)	6-8	Partly Dedicated
	Tristan Main Ring (KEK)	25-30	Planned Use

SWEDEN
Lund	Max (LTH)	0.55	Dedicated

TAIWAN
Hsinchu	SRRC (Synch.Rad.Res.Ctr.)	1.3	Dedicated*

USA
Argonne, IL	APS (ANL)	7.0	Design/Dedicated
Berkeley, CA	ALS (LBL)	1.5	Dedicated*
Gaithersburg, MD	SURF II (NBS)	0.28	Dedicated
Ithaca, NY	CESR (CHESS)	5.5-8	Partly Dedicated
Stanford, CA	SPEAR (SSRL)	4.0	Partly Dedicated
	PEP (SSRL)	5-15	Partly Dedicated
Stoughton, WI	Aladdin (SRC)	0.8-1.0	Dedicated
Upton, NY	NSLS I (BNL)	0.75	Dedicated
	NSLS II (BNL)	2.5	Dedicated

USSR
Karkhov	N-100 (KPI)	0.10	Dedicated
Moscow	Siberia I (Kurchatov)	0.45	Dedicated
	Siberia II (Kurchatov)	2.5	Dedicated*
Novosibirsk	VEPP-2M (INP)	0.7	Partly Dedicated
	VEPP-3 (INP)	2.2	Partly Dedicated
	VEPP-4 (INP)	5-7	Partly Dedicated

* In construction as of 5/88

High energy is not, however, the frontier for the biochemist trying to understand the mechanism by which very few atoms govern the complex actions of enzymes, or for the surface scientist exploring the electronic effect of sub-monolayer concentrations on semiconductor surfaces. For them, and for most users of synchrotron radiation, the frontier is source intensity. The more photons within a given narrow energy interval that can be directed onto a small sample per second, the more subtle are the effects that can be studied. The energy or wavelength of the radiation required for these and other studies using electromagnetic radiation is dictated by factors such as electron binding energies and atomic spacings. Thus, most present applications of synchrotron radiation utilize photons in the VUV, soft x-ray, and x-ray parts of the spectrum, extending from about 5 eV to 50 keV.

The bending magnets even of low energy rings (0.5 to 1 GeV) produce high flux in the lower part of this range (up to about 2 keV), and 3 to 5-GeV rings extend to 50 keV. Thus, even the first small rings built for high energy physics were of interest for synchrotron radiation research, and parasitic programs were started on many of them. As the high energy physics interest shifted to higher energy rings, these programs could expand, and usually some operation dedicated to synchrotron radiation research would become available. In some cases the ring would become completely dedicated. This pattern would be repeated as the next generation of even higher energy colliders were built.

Thus, the early 70s saw the beginning of use of the radiation from several colliding-beam storage rings: ACO (0.54 GeV) in Orsay, France,[20] the CEA (3.5 GeV) in Cambridge, MA,[21] and VEPP-2M (0.67 GeV) in Novosibirsk, USSR[22]. When the CEA was shut down in 1973, a small parasitic program (the Stanford Synchrotron Radiation Project, SSRP[23]) was started at Stanford University using the SPEAR 2.5-GeV (later raised to 4-GeV) collider at SLAC. By 1979 this project had become a laboratory (SSRL) and had grown to 9 experimental stations serving about 300 users with half of the SPEAR operations time dedicated to single-beam operation for synchrotron radiation research.

By about 1980 synchrotron radiation research programs had started and grown rapidly on other large colliders: ADONE[24] (1.5 GeV) at Frascati, Italy, DCI[25] (1.8 GeV) at Orsay, VEPP-3[22] (2.2 GeV) at Novosibirsk, DORIS[26] (5.3 GeV) at Hamburg, Germany, and CESR[27] (5.5 GeV) at Cornell University. By then even higher energy colliders were in construction: VEPP-4 (6 GeV), Petra (23

GeV), and PEP (16 GeV). The shifting of high
energy physics interest to these newer rings made
it possible to expand the synchrotron radiation
programs on smaller rings and also to consider use
of the larger rings. Programs have been started on
VEPP-4[22] and PEP[28] and also on the 6 to 8-GeV
Accumulator Ring (AR)[29] in Japan.[29] Plans are underway to use the Tristan main ring (25 to 30-GeV),
which began colliding-beam physics operation in
1987. The availability of these rings for synchrotron radiation research may be expected to
increase as even higher energy colliders (e.g. the
LEP 50 to 80-GeV ring in Geneva) start operation.
As explained later, the high energy rings offer
possibilities of operation at less than their full
energy in low emittance modes that are better
optimized for producing very high brightness
radiation from undulators.

The use of so many rings for both high energy
physics and synchrotron radiation research has
resulted in important discoveries in both fields.
The synchrotron radiation community owes a great
deal to the high energy physics community for
providing the first sources and for assistance in
adapting their rings and making them available for
synchrotron radiation research. Shared use of
research facilities involves inevitable conflict
between the different research communities. Careful planning and procedures are needed to resolve
these conflicts as the synchrotron radiation activities develop from purely parasitic efforts to
partly dedicated programs, usually leading to eventual full dedication of the ring to synchrotron
radiation research.

7   DEDICATED SYNCHROTRON RADIATION SOURCES

The experience in using these colliding-beam
rings as light sources and the growing number of
users competing for beam time led to the design and
construction of new storage rings to be used as
fully dedicated radiation sources. In the mid to
late 1970s construction began on several fully
dedicated light source storage rings including
major national research facilities in England (the
2-GeV SRS[30] at Daresbury), Germany (the 0.8-GeV
BESSY[31] in Berlin), Japan (the 2.5-GeV Photon
Factory[32] in Tsukuba), and the USA (the 1.0-GeV
Aladdin ring[33] at the University of Wisconsin and
the 0.75- and 2.5-GeV NSLS rings[34] at Brookhaven
National Laboratory).

Several of these rings were designed to
achieve a smaller electron beam emittance in order
to enhance the brightness of the radiation. Emit-

tance is the product of the transverse size and transverse opening angle of the beam. The recognition in 1976, by Ken Green and Rena Chasman at Brookhaven National Laboratory, of the importance of low electron beam emittance to achieving high brightness synchrotron radiation was a major step in the development of optimized synchrotron radiation sources.

Although some of these rings experienced difficulties and delays in reaching operational status, all of them are now reliable operational facilities serving a rapidly growing user community of scientists. Some serve more than 1000 users and are at or near saturation in the number of experimental stations that they can accommodate. They offer excellent experimental conditions, due in no small part to their not being used also for colliding-beam studies.

Besides these facilities, smaller dedicated rings have been completed in Japan (TERAS[35] at the Electrotechnical Laboratory in Tsukuba and UVSOR[36] at the Institute for Molecular Science in Okazaki, both at 0.6 GeV), Sweden (MAX[37] in Lund, 0.58 GeV), and the USSR (Siberia I[38] at the Kurchatov Institute in Moscow, 0.45 GeV). In the People's Republic of China HESYRL,[39] a 0.8-GeV dedicated light source, is in construction in Hefei and BEPC,[40] a 2.8-GeV colliding-beam and light source ring, has recently been completed in Beijing. A 0.45-GeV dedicated ring project has recently been started in Indore, India. Most of these new dedicated rings have been designed to accommodate many bending-magnet beam lines, plus a few from insertion devices.

Besides developing the radiation sources, each of these laboratories is also developing insertion devices, beam lines, and associated instrumentation to utilize the radiation in experiments. Mirrors, monochromators, sample chambers, and detectors are required, as well as protection systems to guard against radiation exposure and vacuum leaks, along with systems to stabilize and steer the narrow photon beams to the small experimental samples. Many facilities offer considerable technical staff support for users and provide facilities such as biochemistry laboratories and darkrooms. This wide range of instrumentation is described in reports of the major laboratories and in the proceedings of conferences on synchrotron radiation instrumentation such as the one in Tsukuba, Japan, in August 1988. See also the bibliography to this report and reviews given by Kulipanov[38] and Winick.[41]

8 INSERTION DEVICES (WIGGLER AND UNDULATOR MAGNETS)

Wiggler and undulator magnets are periodic magnetic structures that can be inserted between the bending magnets of storage rings to provide beams of radiation with higher flux and brightness than those obtained with the ring bending magnets. By deflecting the beam in alternate directions they produce radiation with no net deflection. They can therefore be designed with higher or lower field than the ring bending magnets and with period lengths limited only by technology. They can be designed so that most or all of the radiation produced over their length can be accepted by the beam port and delivered to one or several experiments. Thus, it is not surprising that insertion devices can provide much higher flux and brightness than bending magnets, which bend the beam continuously in one direction. Typically only a few centimeters of electron path length in a bending magnet radiate into a single experimental system whereas a wiggler or undulator can be several meters long.

9 PROPERTIES OF THE RADIATION FROM BENDING MAGNETS, WIGGLERS, AND UNDULATORS

Electrons in circular motion at low velocity emit radiation in a non-directional pattern, as shown in Figure 5 (top). At velocities approaching

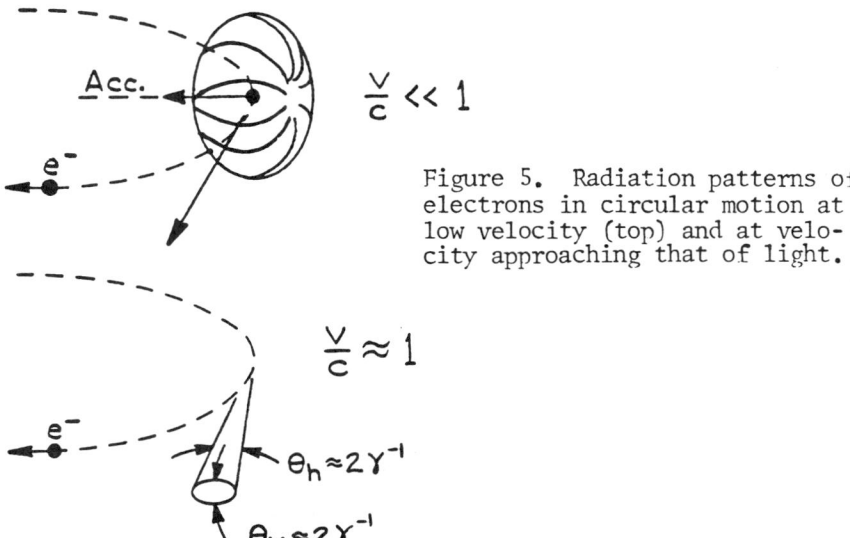

Figure 5. Radiation patterns of electrons in circular motion at low velocity (top) and at velocity approaching that of light.

the velocity of light the radiated power increases dramatically and the pattern is folded forward into a cone with a full opening angle of approximately $2\gamma^{-1} = 2mc^2/E$, as shown in Figure 5 (bottom), where mc^2 is the rest mass energy of the electron (0.51 MeV) and E is the total energy. The opening angle is only 1 milliradian for an electron energy of 1 GeV and correspondingly smaller at higher electron energy. Thus synchrotron radiation has intrinsically high brightness. This geometrical property is exploited differently when the source is a bending magnet, a wiggler, or an undulator.

The total power radiated is given by

$$P(kW) = 1.267 E^2 (GeV) I(A) <B^2(T)> L(m)$$

where $<B^2(T)>$ is the magnetic field in Tesla averaged over the length L in meters, allowing for constant fields (bending magnets) or alternating fields (wiggler and undulator magnets). For sinusoidal fields $<B^2>$ is replaced by $B_o^2/2$, where B_o is the peak field.

For more information about the properties of the radiation (spectral distribution, polarization, time structure) see the review articles listed in the bibliography. Here we concentrate on the distinctions between the radiation produced by bending magnets, wigglers, and undulators with the aim of understanding how wigglers and undulators are designed for different purposes and how storage rings are optimized for them.

9.1. Bending Magnets

Bending magnets produce a large horizontal fan of continuum radiation, much larger than γ^{-1} and much larger than the acceptance of one experiment; see Figure 6 (top). The vertical opening angle remains small. Collecting optics can increase the flux delivered to an experiment by accepting larger horizontal angles, but the brightness is not increased. The smooth spectrum (see Figure 7) is characterized by the critical energy given by

$$\varepsilon_c (keV) = 0.665 B(T) E^2 (GeV).$$

This is the half-power point of the spectrum.

9.2 Wiggler Magnets

A wiggler magnet is a device with several poles, which, when inserted into a straight section of a storage ring, produces one or more oscillations of the electron beam but no net deflection or displacement of the beam. The

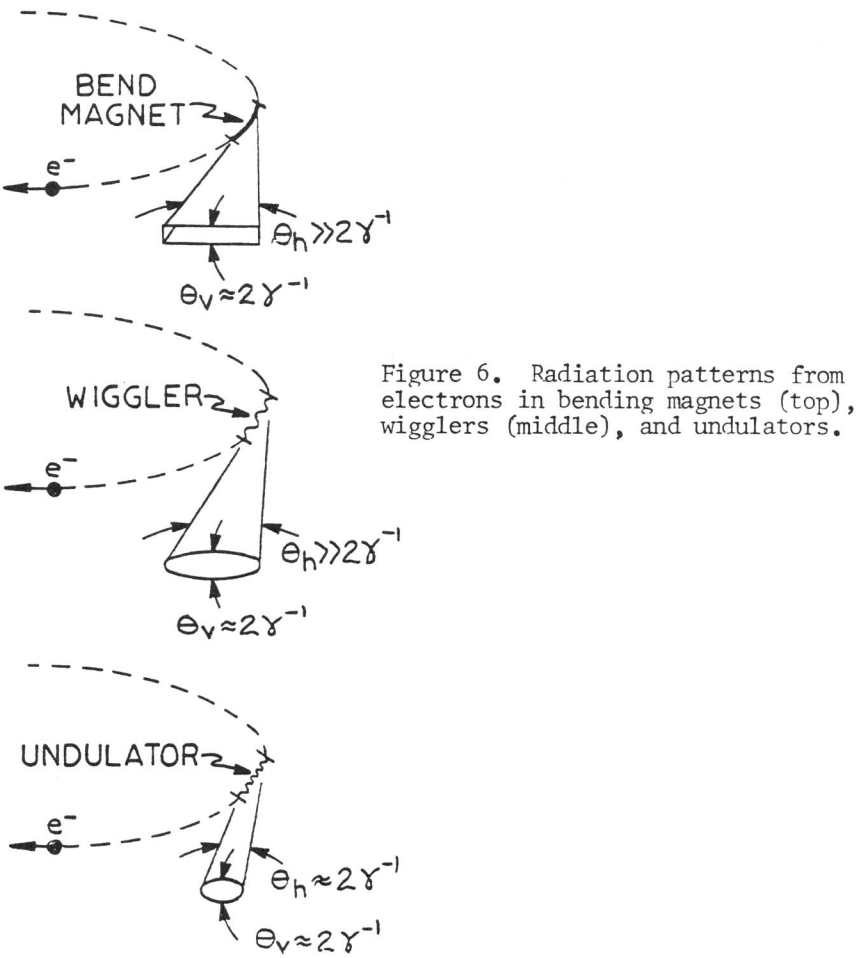

Figure 6. Radiation patterns from electrons in bending magnets (top), wigglers (middle), and undulators.

deflection angle in each pole is large compared with γ^{-1} (see Figure 6 middle.) The spectrum produced by a wiggler, particularly a wiggler with many poles, does have some structure, especially at low photon energies. This structure is due to interference effects and is discussed more in the next section on undulators. At the higher photon energies for which the wiggler is most often used, the structure is usually very small and the spectrum approaches the smooth continuum characteristic of bending magnets.

If the amplitude of the electron beam oscillation is smaller than the electron beam dimension, the wiggler can usually be regarded as a linear succession of bending magnet sources producing a spectrum with a critical energy that is determined by the magnetic field and with an intensity enhanced by the number of poles (see Figure 7). If the amplitude of the electron beam oscillation is large compared with the beam dimension, the source points are no longer co-linear and the geometric pattern becomes more complicated.

Although wiggler magnets had been used for high energy physics purposes,[42] the first wiggler to be used as a radiation source for experiments[43] was installed in the SPEAR storage ring in October 1978 and shown to be compatible with injection and operation of the ring in both single- and colliding-beam modes. This was a 1.2-meter-long electromagnet with 6 poles and a peak field of 1.8 Tesla. With the completion of the beam line in March 1979 it was verified that the expected enhancement of the radiation was achieved, and a new round of experimentation with higher flux and brightness commenced.

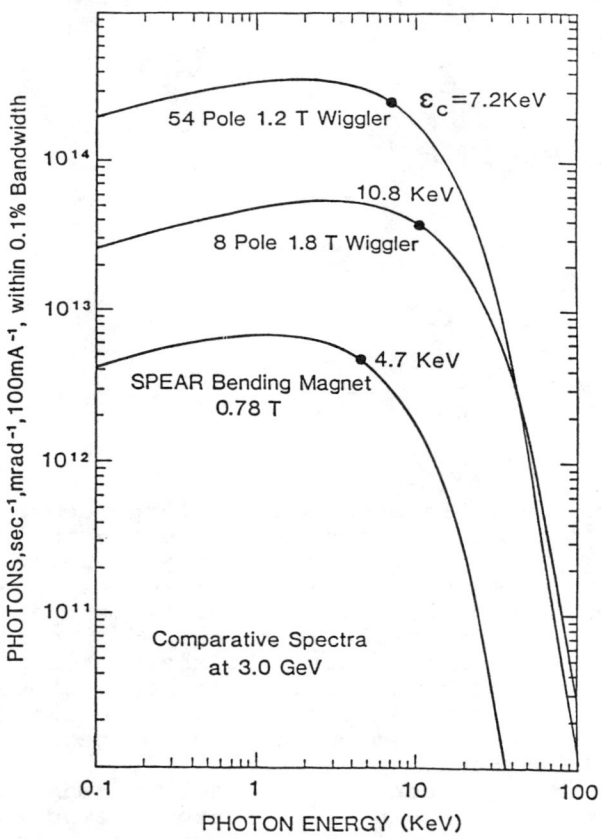

Figure 7. Spectral distribution of radiation from the bending magnets of the SPEAR storage ring at 3 GeV and two wigglers now in operation in SPEAR.

Wiggler magnets can produce beams with a wide range of spectral and geometric properties. The simplest wiggler consists of one strong central pole, which is the main source of radiation, and two weaker, opposite polarity end poles. Together these end poles cancel the deflection of the central pole. In effect the device has two equivalent full poles which produce a single oscillation of the electron beam. The central pole generally has a higher magnetic field than the ring bending magnets, resulting in a hardening of the spectrum. Such single-oscillation wigglers, also called wavelength shifters, with superconducting fields up to about 5 Tesla are in operation at the Daresbury SRS and the Photon Factory. The fan of radiation from wavelength shifters is usually very large and the beam is easily shared by several experimental stations. For example, the SRS wiggler (see Figure 8) fills a beam pipe which accepts a total of 64 milliradians and serves seven simultaneously operating stations.

In Novosibirsk a 3.5-Tesla superconducting wiggler with 20 poles has been used on the VEPP-3 storage ring. A 4-pole, 5-Tesla superconducting wiggler has recently been implemented in the DCI ring, and a 6-pole 5-Tesla wiggler will be installed in the NSLS x-ray ring. (The number of

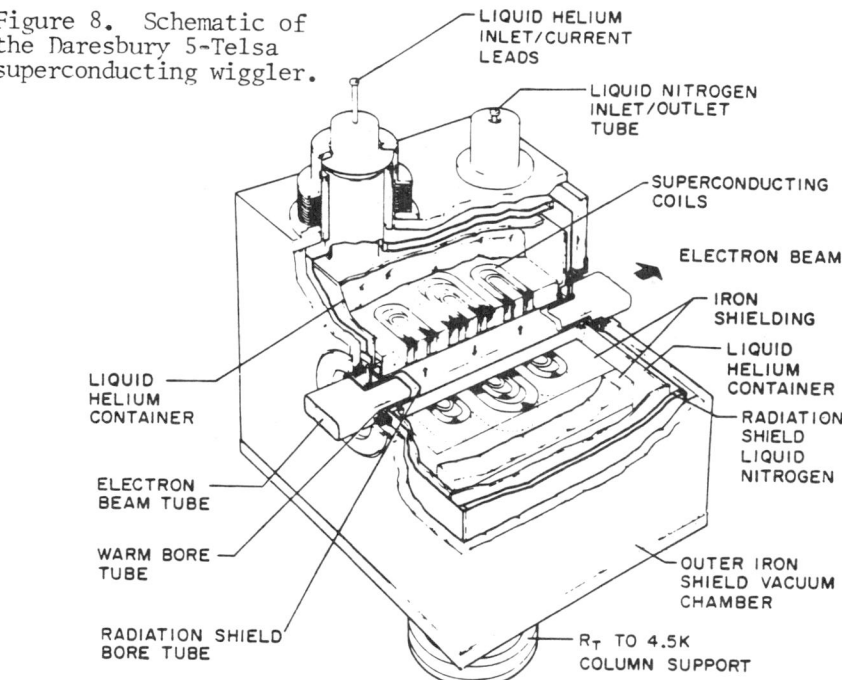

Figure 8. Schematic of the Daresbury 5-Telsa superconducting wiggler.

poles stated is the number of equivalent full
poles; the actual number of poles is one more.)
Six-pole 1.4 to 1.9-Tesla electromagnet wigglers
are in operation at Frascati and Cornell, and two
8-pole, 1.9-Tesla electromagnet wigglers are used
at Stanford (see Figure 9).

In 1983 a 2-meter-long, 54-pole permanent
magnet wiggler[44] (see Figure 10) with a peak field
of 1.2 Tesla began operation at SSRL. The system,
produced in a collaboration between SSRL, LBL, and
the Exxon Corporation, incorporates a unique,
flexible vacuum chamber (see Figure 11) that
provides for the larger aperture required for the
electron beam during injection and permits closing
the chamber, and hence also the external magnet
gap, down to the smaller values that are
permissable after the beam is stored. The many
short poles produce a very intense and narrow beam,
about 2 milliradians wide. Since then other
multipole permanent magnet wigglers have been
implemented at SSRL, BESSY, the Photon Factory, and
other rings, and are planned for other facilities.

Most wigglers employ vertical magnetic fields
so that they can make use of the smaller aperture
requirement for easier achievement of high magnetic
field. The resultant radiation is highly polarized
in the horizontal plane, as it is also with bending
magnet beams. The Photon Factory superconducting
wiggler,[45] however, employs a horizontal magnetic
field, causing the electron beam to execute a
vertical oscillation and producing a unique
vertically polarized synchrotron radiation beam.

Wigglers now in use produce total radiated
power up to several kilowatts and power density on
beam line components up to about 10 kW/cm^2. New
designs for beam line components have been[46] developed to handle these large thermal loads. More
powerful wigglers on existing rings and on proposed
new rings will present even more severe thermal
problems. Clearly we are approaching fundamental
thermal limits on wiggler beams, i.e., we can
produce higher power and power density than can be
handled on beam line components. Undulators reduce
these thermal problems because of their
quasi-monochromatic spectrum; this is one reason
for the increasing interest in undulator sources.

9.3 Undulators

An undulator magnet is a device with N periods
of alternating magnetic field (i.e. 2N effective
poles with alternating polarity) in which each pole
produces an angular deflection of the order of γ^{-1},
the natural emission angle of synchrotron radiation
(see Figure 6, bottom). Thus the intrinsic high

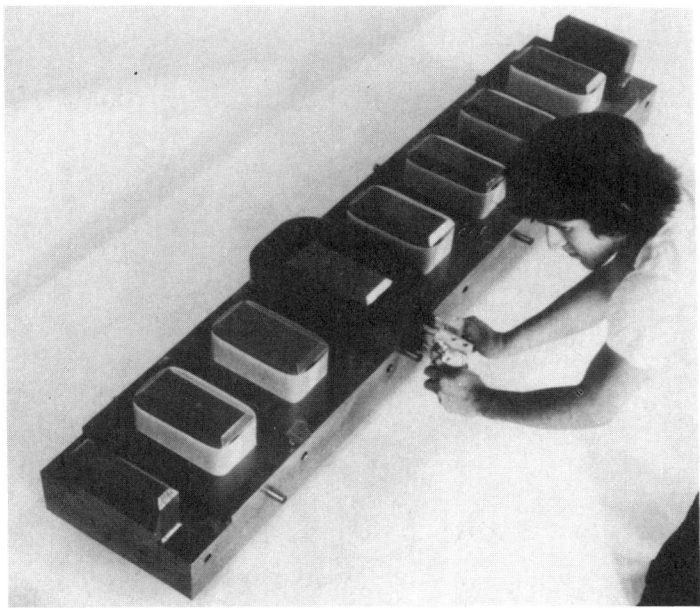

Figure 9. Lower half of SSRL 8-pole, 1.9-Tesla electro-magnet wiggler. Only one coil is shown. There are 9 actual poles, but the end poles are effectively half poles in their contribution to the field integral.

Figure 10. Lower half of LBL/Exxon/SSRL 54-pole permanent magnet hybrid wiggler during magnetic measurements.

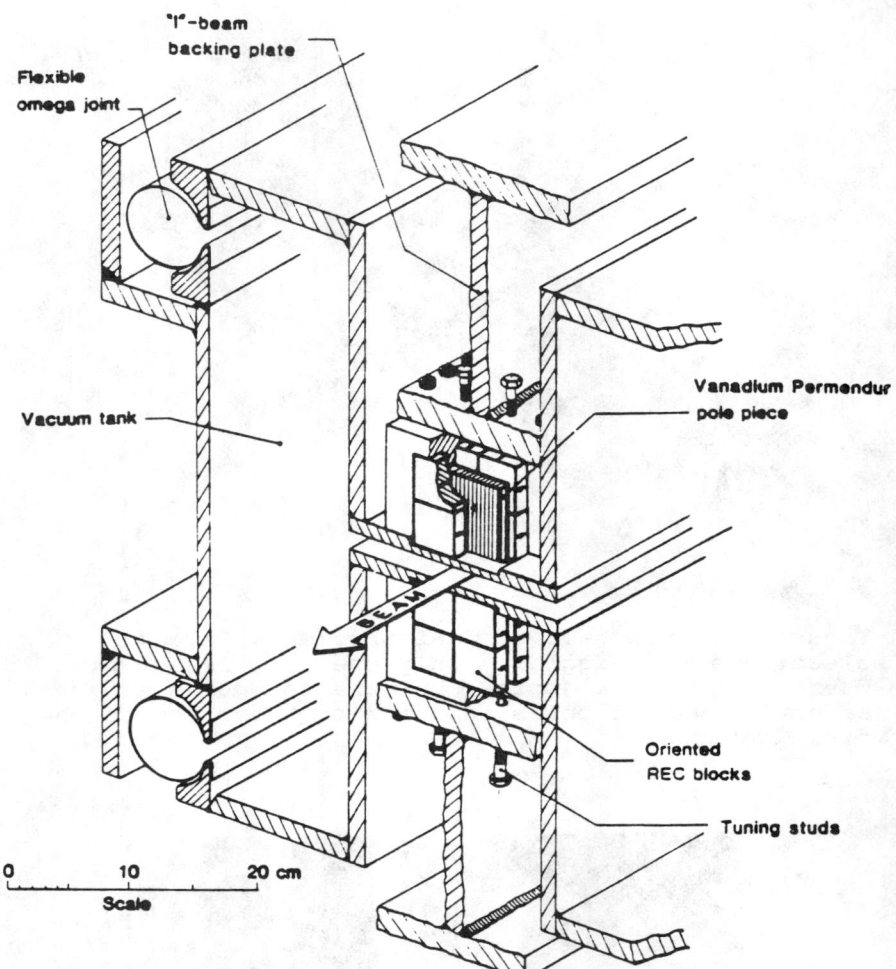

Figure 11. Schematic of the 54-pole wiggler, showing the arrangement of pole pieces, permanent magnet material, and variable gap vacuum chamber.

brightness of the radiation is preserved and enhanced. Furthermore, interference effects in the radiation produced at many essentially co-linear source points result in a modified spectrum[47-49] (see Figures 12 and 13 for examples) with tunable quasi-monochromatic peaks at wavelengths given by

$$\lambda_n = \lambda_u \{1 + K^2/2 + \gamma^2\theta^2\}/(2n\gamma^2)$$

where λ_u is the period length of the magnetic field, n is the harmonic number (n = 1 is the fundamental), $K = 0.934B(T)\lambda_u(cm)$, and θ is the observation angle. The wavelength of the light can be

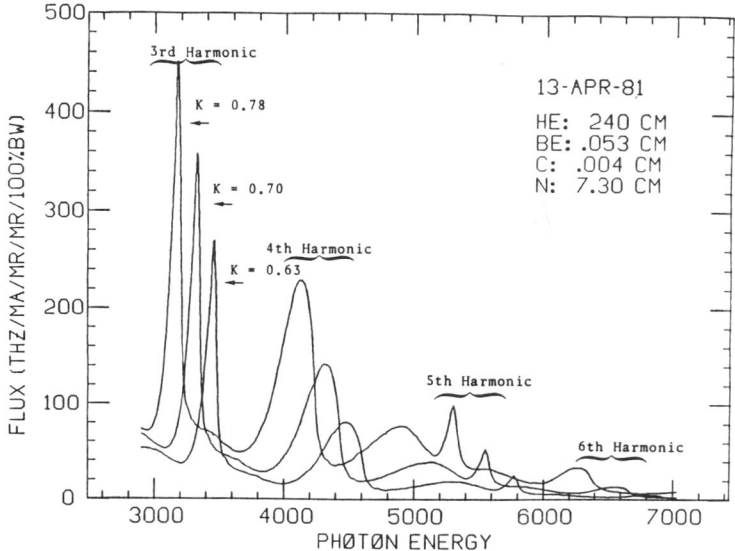

Figure 12. Third to sixth harmonic part of the first undulator brightness spectra taken at SSRL at 3 GeV. Corrections have been made for attenuation as shown on the figure. Even at rather large emittance (450 nanometer-radians) rather pronounced peaks are evident. The undulator used is shown in Figure 15.

most conveniently varied by changing the magnetic field and hence the value of K.

For K << 1, harmonics are very weak and the spectrum consists essentially of one peak, the fundamental. As K increases towards 1, the fundamental moves to longer wavelength, the power in the fundamental increases (reaching a maximum at about K = 1.2), and more harmonics appear. For K >> 1, many closely spaced harmonics appear. Since each harmonic has a finite width, the many closely spaced harmonics eventually blend into the continuum characteristic of bending magnets and wigglers. For K > ~ 4 to 5, the device is generally considered a wiggler, although there is still structure in the spectrum, particularly around the very low energy fundamental peak.

K is also given by $\gamma\delta$, where 2δ is the full angular deflection of the electron beam. Thus undulators (devices for which K ~ 1) produce a very narrow photon beam and wigglers (devices with K >> 1) produce wider beams.

The angular divergence of the electron beam at the undulator source point in the storage ring must also be considered. If it is comparable with or larger than γ^{-1}, the interference effects are

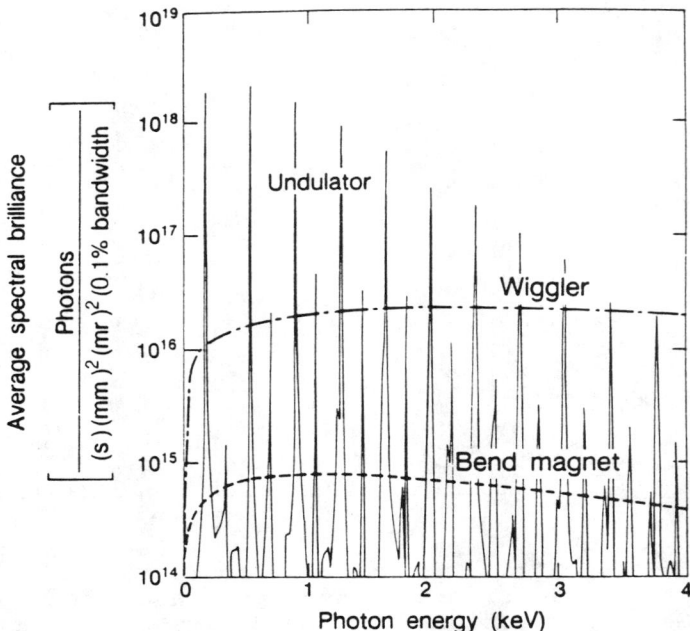

Figure 13. Comparison of bending magnet, wiggler, and undulator spectra calculated for the 1.5-GeV Advanced Light Source at LBL. The peaks are much more pronounced than in the previous figure because of the much lower emittance (7 nanometer-radians). Courtesy Kwang-Je Kim, LBL.

smeared out. If the electron beam angular divergence is of the order of $\gamma^{-1}N^{-1/2}$, the on-axis brightness produced by an undulator can increase as the square of the number of undulator periods. Thus, in new low emittance rings that satisfy this condition, it should be possible to achieve 3 to 4 orders of magnitude enhancement in brightness (compared with ring bending magnets) from undulators with about 100 periods. This is the main driving force behind the design of many future rings.

The above discussion shows that the designer of an undulator generally seeks to achieve a short period (to reach short wavelengths and also to maximize the number of periods in a given straight section length) together with a peak magnetic field high enough to reach a K-value of about 1 (for high brightness in the fundamental and perhaps a few harmonics). High fields are most easily obtained by reducing the gap. The minimum gap of the magnet is determined by the aperture required by the electron beam. Although full vertical apertures of about 1 cm are adequate for a stored beam in most multi-GeV rings, larger apertures are often needed for injection and tune-up. Flexible vacuum chambers[44] offer an attractive solution to this need.

Undulator radiation was studied theoretically and experimentally with linac beams in the early 1950s[50] and by several groups[51] in the Soviet Union using synchrotrons in the 1970s. These were essentially "proof-of-principle" experiments that produced microwaves and visible light respectively, but were not intended as radiation sources for experiments. The first undulators intended as radiation sources for experiments were used at SSRL and in Novosibirsk around 1979 and are described later. Now undulators are have also been used in other rings (Photon Factory, SRS, BESSY, TERAS, NSLS, and others).

Most undulators are planar arrays with magnetic fields alternating in one direction. It is also possible to use helical fields, or a pair of planar arrays with different field directions, to produce elliptical, circular, and other polarizations.[52]

The successful use of insertion device sources had a major impact on SSRL and other laboratories. Shortly after the first wiggler was used, work was cancelled at SSRL on two additional bending magnet lines so that funds could be redirected to the construction of another wiggler line. After the first permanent magnet undulator was used, plans were formulated for additional permanent magnet wigglers and undulators for SPEAR and PEP. At present seven insertion device beam lines are operational at SSRL and several more are being planned.

The demonstration that wiggler and undulator magnets were extremely capable sources, compatible with storage ring operation, provided strong impetus to other labs to develop insertion devices for their rings. At present a total of about 20 insertion device beam lines are in use as radiation sources around the world and more are in construction.

9.4 Permanent Magnet Technology

Permanent magnet technology is making it possible to achieve shorter period devices than can readily be achieved with conventional electromagnets and, in some cases, superconducting magnets. In conventional electromagnets the power density in the coils increases rapidly as the period goes down, eventually making it impossible to cool the coils. In the absence of coils, permanent magnets can be scaled down to very short periods. The peak field decreases exponentially as the ratio of the gap to the period. This sets a practical minimum period of about one to two times the gap. The minimum gap is determined by the aperture requirement for the electron beam.

The suggestion to use permanent magnets, particularly using rare-earth/cobalt material, for short period undulators was made independently around 1979 by Halbach at LBL and by Kulipanov and Vinokurov at Novosibirsk, leading to the construction of the first such devices in both laboratories at about the same time. The LBL device[53] (see Figures 14 and 15), designed and constructed in collaboration with SSRL, was a 2-meter-long, 30-period magnet that was first used in the SPEAR storage ring in late 1980 and used for the first experiments in 1981.

The first permanent magnet insertion devices were made with no iron, as shown in Figure 14. The peak, on-axis magnetic field in such a device is given approximately by

$$B_o = 2B_r [\exp(-\pi g/\lambda_u)] \frac{\sin(\pi/M)}{\pi/M} [1 - \exp(-2\pi h/\lambda_u)]$$

where g is the full magnet gap, λ_u is the period length, M is the number of blocks per period, and h is the height of the blocks. It is most convenient to build such a device with four blocks per period ($M = 4$) and with blocks of square cross section ($h = \lambda_u/4$). In this case, for $B_r = 0.9T$ (which is readily available) the above equation reduces to

$$B_o(T) = 1.28 \exp(-\pi g/\lambda_u).$$

Figure 14. Schematic drawing of a pure permanent magnet undulator based on concepts developed by K. Halbach, LBL.

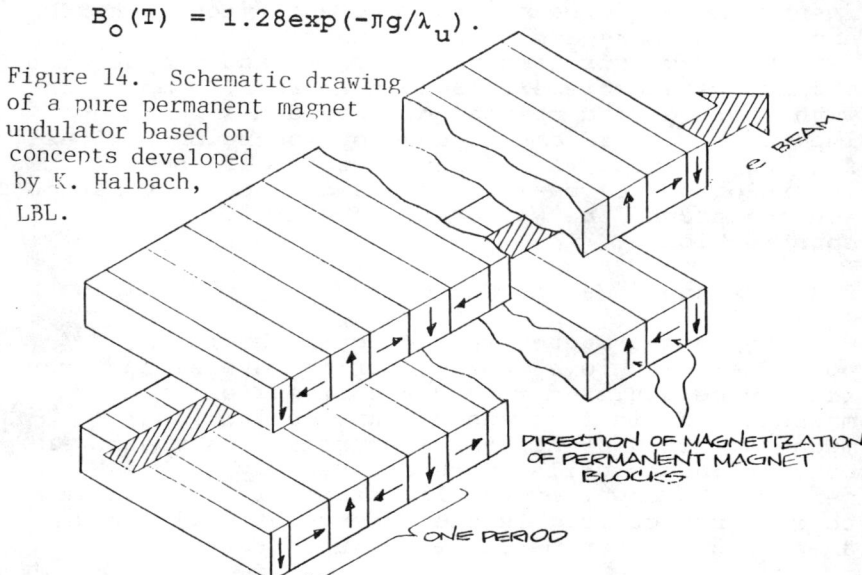

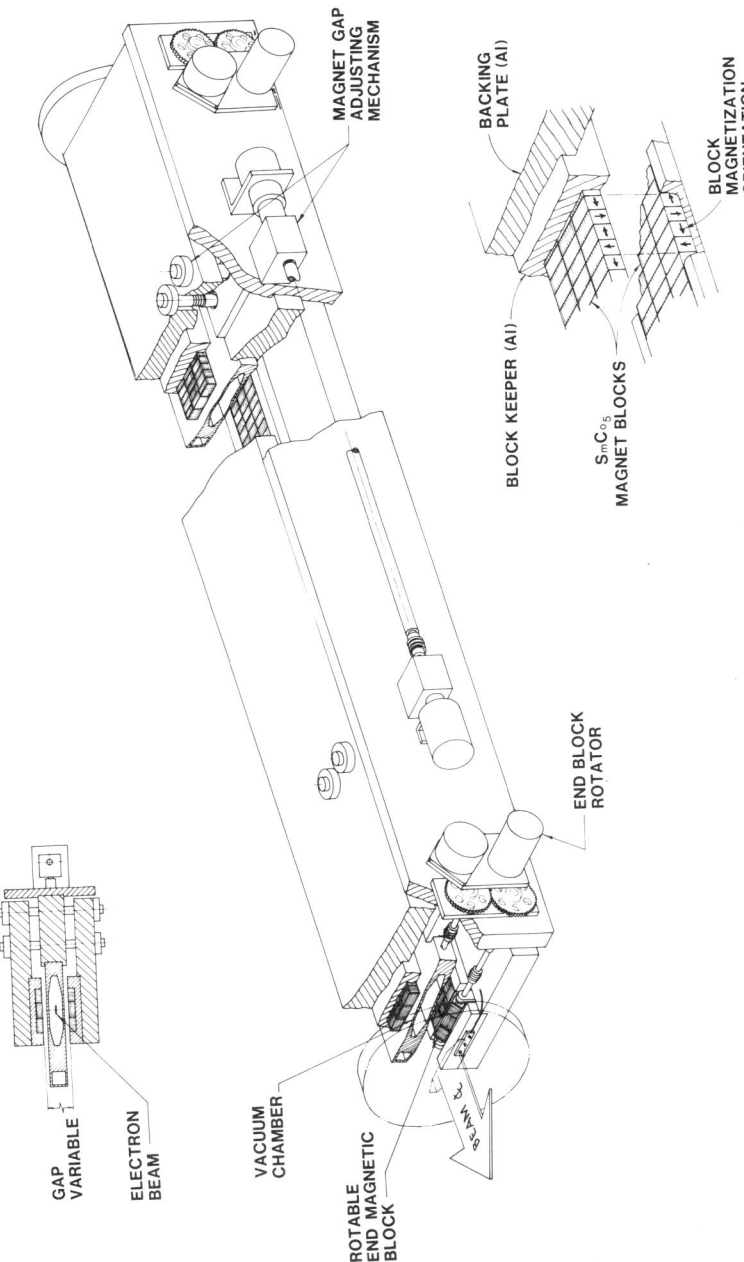

Figure 15. The LBL/SSRL permanent magnet undulator with 6.1-cm period and 30 periods. Courtesy of Egon Hoyer, LBL.

Halbach[54,55] gives more general expressions for the field.

For devices containing iron, also called hybrid devices, the peak field on-axis is given by

$$B_o = a \exp[(-g/\lambda_u)(b - cg/\lambda_u)]$$

where
- B_r = 0.9 or 1.1 Tesla
- a = 3.33   3.44 Tesla
- b = 5.47   5.08
- c = 1.8   1.54

The first column is typical for samarium-cobalt magnetic material and the second for neodymium-iron material.

Many reports on the technology and performance of undulator magnets are in the proceedings of conferences on synchrotron radiation instrumentation and of special conferences (e.g. Refs. 56, 57). The field is developing rapidly with many new ideas such as wedge-pole devices[58] to achieve higher fields with permanent magnets, and micropole undulators[59] to achieve shorter periods and higher photon energy.

10 THIRD-GENERATION STORAGE RINGS

Present rings cannot achieve the full performance potential of insertion devices, particularly the high brightness that can be produced by undulators. This has led to an intense worldwide activity to design and construct the next generation of storage rings, better optimized for the use of wiggler and undulator magnets as sources of synchrotron radiation. The first of these, the 0.8-GeV Super-ACO[60] ring in Orsay, France, began operation in early 1987. Several others have recently been authorized for construction including the 1.3-GeV Taiwan Light Source, the 6-GeV European Synchrotron Radiation Facility (ESRF)[61] in Grenoble, the 1.5-GeV Advanced Light Source (ALS)[62] at Berkeley, a 1.5-GeV ring in Trieste,[63] and a 2.5-GeV ring in Moscow. The Advanced Photon Source[64] (a 7-GeV ring, 1 km in circumference) and BESSY II (1.5 GeV) are in advanced stages of design, and construction approval is expected soon. A 1 to 3-GeV ring is planned in Brazil. See Figures 16 and 17 for examples.

These rings will have many straight sections to accommodate wiggler and undulator insertions and they will have electron beam properties optimized to extend their performance significantly. In particular, the stored electron beams in these rings will have small transverse size and small

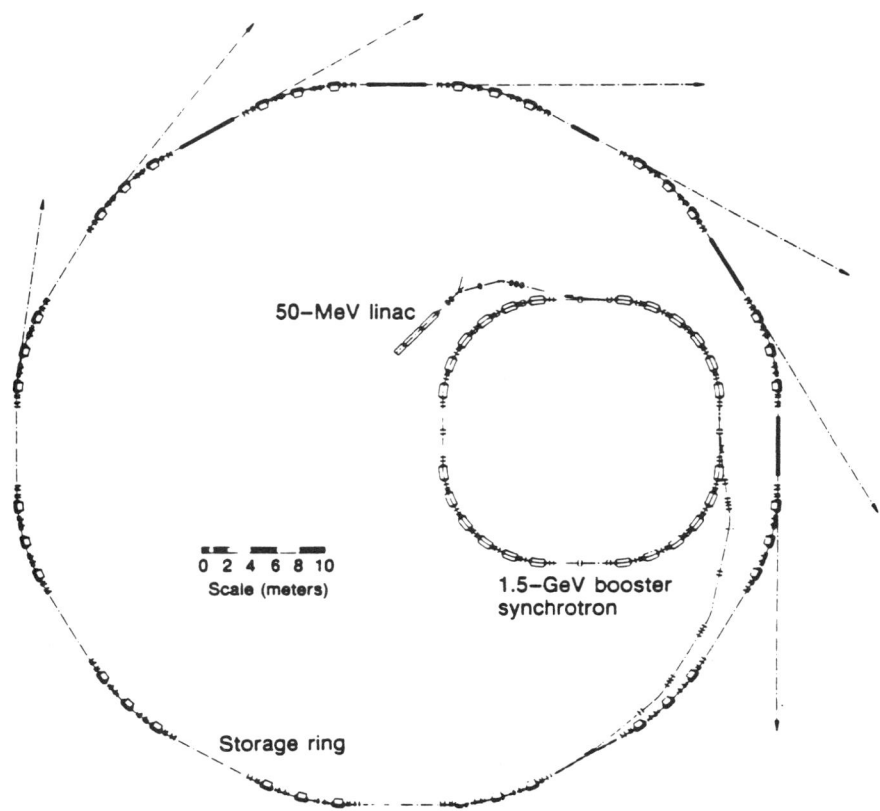

Figure 16. Layout of the LBL 1.5-GeV Advanced Light Source with 12 straight sections, 11 of which can accommodate insertion devices. The circumference of the ring if 196.8 meters. From LBL Conceptual Design Report.

angular divergence. Hence these new rings may be described as low emittance rings optimized for insertion devices. We use the product of source size and divergence, the emittance, to characterize the beams in these rings because emittance is a constant around a ring whereas the size and divergence vary. The bending magnets of these rings will also serve as radiation sources, and their brightness will be higher than that of the bending magnets of existing rings because of the lower emittance.

The synchrotron radiation community has great expectations that these new sources will open the way to important new science. Indeed, our experience with synchrotron radiation over the past two decades has shown that each order of magnitude

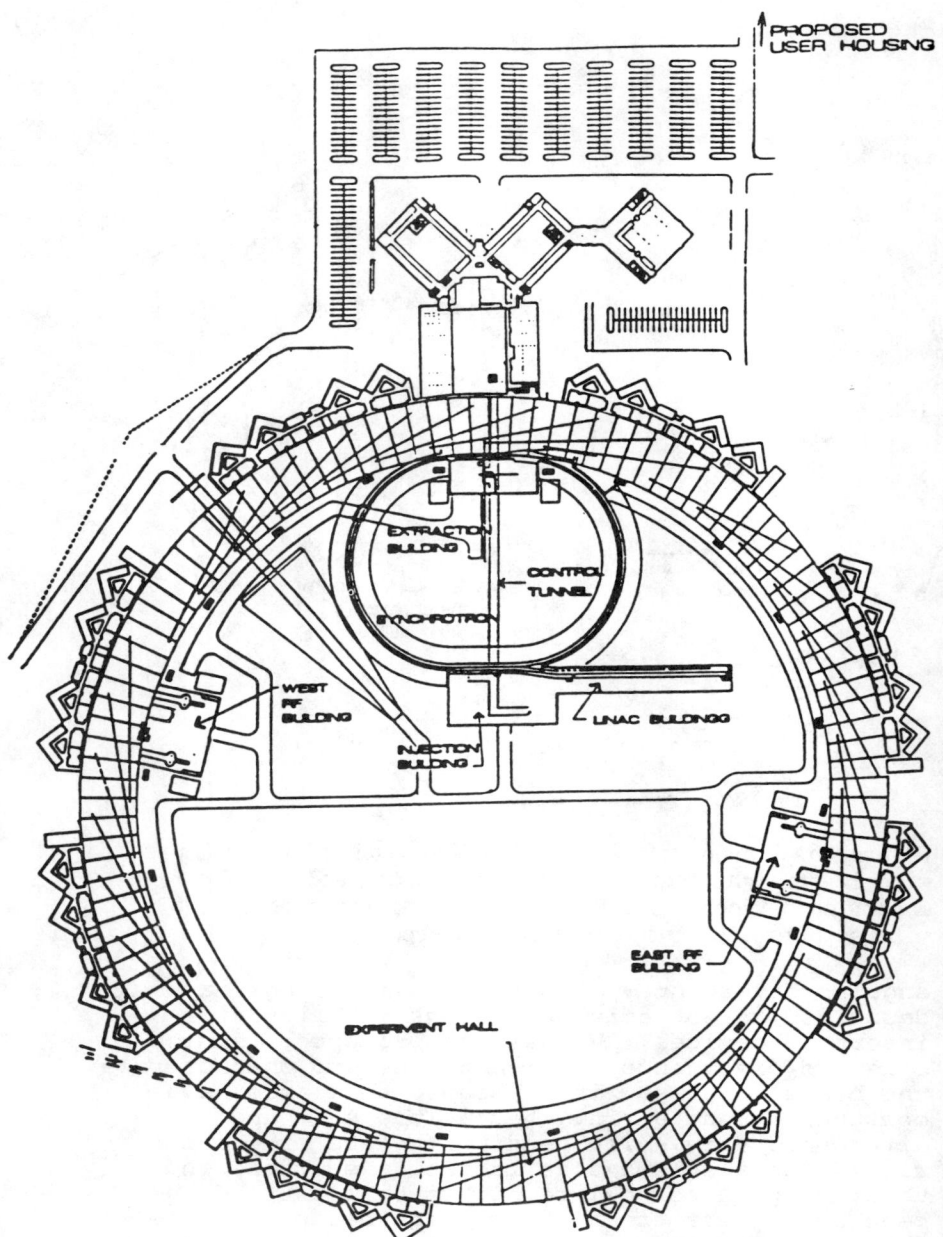

Figure 17. Layout of the Argonne National Laboratory 7-GeV Advanced Photon Source with 40 straight sections, 34 of which can accommodate insertion devices. The circumference of the ring is 1060 meters. From APS Conceptual Design Report.

or so improvement in flux or brightness has resulted in new applications, in many cases unimagined until the source was available. Recent examples include the use of the coherence properties of VUV and soft x-ray synchrotron radiation to make holograms in this new spectral region, and the use of x-ray scattering as a probe of the magnetic properties of materials and in the study of phase transitions in two-dimensional systems such as surface monolayers, lipid bilayers, and liquid crystal films.

The main technical challenge facing the designer of these rings is to achieve an emittance low enough to optimize the performance of undulators with many periods (perhaps 100 or more) while providing adequate dynamic aperture. Emittance is the product of the transverse beam size and angular divergence and is usually measured in nanometer-radians. Dynamic aperture is the "good field" aperture, i.e. the region over which particle oscillations are stable, and is usually measured in millimeters.

The emittance of the beam in an electron storage ring is determined by an equilibrium between the excitation of transverse oscillations (called betatron oscillations) due to the quantum nature of the radiation process, and the damping of these transverse oscillations due to the energy lost to synchrotron radiation being restored by rf cavities that impart momentum only in the longitudinal direction. The resultant emittance depends on the detailed arrangement of bending and focussing magnets that make up the so-called magnet lattice of the ring.

The goal of the third-generation rings now being designed and constructed is to achieve a brilliance of 10^{18} to 10^{19} photons/(s-mm^2-mrad^2) within a 0.1% bandwidth. This can be reached with 5-meter-long undulators in rings operating at about 100 mA with an emittance of 5 to 10 nanometer-radians. For a given ring configuration the emittance scales quadratically with the electron energy.

10.1 Dynamic Aperture Requirements

Stored beams are usually Gaussian in shape. Although individual particles in the Gaussian tails rapidly coalesce with the central beam core because of synchrotron radiation damping, quantum excitation constantly repopulates these tails. Thus, although almost all the particles are contained within +/- 2 or 3 standard deviations, it is necessary to preserve +/- 6 or 7 standard deviations in order to have a good "quantum" lifetime. Since the transverse beam size in a low emittance ring is < 1 mm, this dynamic aperture requirement is easily met.

Even larger dynamic aperture is required to accommodate large amplitude oscillations present during injection and due to Coulomb scattering of stored particles after injection. Large stored current in a synchrotron radiation source is usually achieved by repetitive off-axis injection of electrons until the desired current is accumulated. The freshly injected beam executes large amplitude (about 10 mm or more) oscillations about the already stored beam, coalescing with the small core of the stored beam as the oscillations decrease because of synchrotron radiation damping.

Coulomb scattering of stored beam particles by the residual gas in the ring, even at pressures in the nanotorr range, adds a halo around the Gaussian beam profile which extends well beyond the +/- 6 to 7 standard deviations needed for a good "quantum" lifetime. Although particles leave this halo and return to the beam core as their large amplitude oscillations are decreased by synchrotron radiation damping, there is a steady repopulation due to scattering of particles in the beam core. Thus, to preserve a long lifetime in the presence of Coulomb scattering, the dynamic aperture must be larger than that required for good "quantum" lifetime by an amount that depends on the details of the ring lattice and pressure. The time constant for damping is typically about 5 to 10 msec at full energy and varies inversely as the third power of the electron energy.

A generous dynamic aperture is very desirable because the nonlinear fields of insertion devices cause some reduction in dynamic aperture.

10.2 Lattice Design

Since emittance grows in bending magnets, the desired low emittance is achieved in lattices that use relatively short bending magnets separated by quadrupole lenses. The fact that the emittance increases as the third power of the bending angle leads to rings with many repeating cells and a high degree of symmetry, suitable for accommodating many insertion devices.

For a given cell length the emittance can be reduced by increasing the strength of the quadrupole magnets, which results in stronger focussing and a higher frequency for betatron oscillations or a higher value of the so-called betatron oscillation tune. Quadrupole magnets are the analogues of glass lenses in an optical system. Simple glass lenses have chromatic aberrations, i.e. different focal properties for different colors (or

different wavelengths or photon energies), due to
the variation of index of refraction with photon
energy. Similarly, the ensemble of electrons in a
stored beam has a natural energy spread (due to
statistical fluctuations in the emission of
synchrotron radiation quanta), and the quadrupole
lenses provide different focussing and hence
different tune values for different energy
electrons. This energy dependence of the betatron
oscillation tune is called the chromaticity and is
analogous to the chromatic aberrations in an
optical system for visible light.

The many strong quadrupoles needed to obtain a
low emittance cause the chromaticity to be large.
For the ring to store the desired large current,
this chromaticity must be corrected. This can be
done with sextupole magnets, devices that have six
poles and produce a magnetic field that is zero on
the axis but rises rapidly in a parabolic fashion
off the axis. The nonlinear character of sextupole
fields results in restriction of the range of
stable particle oscillations, i.e. a reduction in
dynamic aperture. Particular care must therefore
be taken in the design of the sextupole correction
system to maintain a large dynamic aperture.

Over the past few years several different
types of lattice have been developed to the point
that they appear to provide large dynamic aperture
and low emittance. The main tool for evaluating
lattices is computer tracking, whereby hundreds of
particles with different oscillation amplitudes are
followed for thousands of turns around a computer
model of the ring to determine the fraction that
become unstable. The effects of misalignment and
field errors are also studied this way. In
general, the larger the dynamic aperture, the more
tolerant the lattice is to alignment and field
errors, and the easier it will be to commission and
operate the ring. It is important that tracking
studies also take into account the effects on the
beam of the many insertion devices, including their
nonlinear fields, since these also affect the
electron beam optics and the dynamic aperture.

Besides computer studies for future rings,
there has also been encouraging operational
experience with a few low emittance rings. The
0.8-GeV Super-ACO ring at Orsay, the first
synchrotron radiation source optimized for
insertion devices, started operation in early 1987
with a moderately low design emittance of 40
nanometer-radians. The 16-GeV PEP colliding-beam
ring has been operated at 7.1 GeV in a low
emittance mode with a calculated emittance of about
6 nanometer-radians. The experience on PEP is
described in more detail later in this report.

10.3 Lifetime and Stability Considerations

For the photon beams from a low emittance ring to have the maximum experimental capability and to realize the highest useful brilliance, the electron beam must be very stable and reproducible in position (to within a fraction of the beam size) and it must have a long lifetime, preferably > 10 hr.

In a well designed ring the beam lifetime is largely determined by the pressure in the vacuum chamber, which in turn is largely determined by the heating and outgassing effects of synchrotron radiation striking the surfaces of specially designed absorbers and other surfaces. When the current in a single bunch is large there can also be heating effects due to the interaction of the stored beam with its environment (so-called higher-mode or parasitic losses). At worst these heating effects cause component failure in addition to gas evolution and a reduction in lifetime. Parasitic losses are minimized by designing chambers with smooth walls and transitions. The pressure at the beam location is minimized by using high speed distributed pumps (such as NEG pumps) and designs incorporating features such as antechambers that keep the gas evolved by the radiation away from the stored beam.

A variety of single and multi-bunch instabilities can limit the stored current below the desired value. The limit can be raised by installing equipment in the ring such as higher frequency rf cavities and feedback systems that use electron-beam position monitors to sense deviations of individual electron bunches and then act on the beam to correct for coherent oscillations of these bunches.

High photon beam brilliance can be achieved only if very tight tolerances on magnet alignment and stability are met. Ground vibration can cause small amplitude (a few micrometers) motion of quadrupole lenses resulting in photon beams vibrating by 100 micrometers or more at experimental stations located 50 m or more from the source. The required short- and long-term stability of magnet supports will be difficult to achieve. Active feedback systems, which use photon beam position monitors to sense fast vibrations and slower drifts and then act on the electron beam to correct for these effects, will be needed to keep millimeter-sized photon beams locked in position to an accuracy of about 20 micrometers over a distance of 50 to 100 m.

Also of concern are the effects of positive ions produced by interactions of the electron beam with the residual gas in the vacuum chamber, even at operating pressures of a few nanotorr. These positive ions (as well as minute dust particles, which can also become positively charged by synchrotron radiation in the vacuum chamber) are attracted by the negatively charged electron beam. They can collect and become trapped in the vicinity of the beam, causing increased scattering and uncontrolled focussing effects. These, in turn, can limit the amount of current that can be accumulated and can also lead to reduction in stored beam lifetime, resonances, and abrupt beam loss.

These problems can be reduced by installing ion clearing electrodes and by leaving a gap in the fill pattern of the ring so that ions can drift out of the beam. The problem can be eliminated if positrons are used rather than electrons. Thus the most capable injector for a synchrotron radiation source is an accelerator capable of producing positrons at the operating energy of the ring. Any positively charged objects in the vacuum chamber are repelled by the positron beam and cannot collect in the beam. Injection at the operating energy eliminates the need to ramp the energy of the beam in the storage ring. Ramping of the ring energy can cause problems with reproducibility of the orbit due to hysteresis effects.

11 FUTURE STORAGE RINGS

Consideration is now being given to radiation sources with even lower emittance than the 5 to 10-nanometer-radian levels being planned for the next-generation rings now in construction. The lowest electron beam emittance that appears to be useful in a light source is numerically about one tenth the wavelength of the light to be used; a lower one would not give improved photon beams because of fundamental diffraction limits. Thus a "diffraction-limited" electron beam for producing 0.1-nanometer x-rays (12 keV) would have an emittance of about 0.01 nanometer, about 500 times less than the emittances planned for the next-generation sources. Achieving such a low emittance, if possible at all, will take decades. The advances along the way, however, will result in a continual increase in the brightness and coherent power levels of synchrotron radiation beams.

A possible way to achieve very low emittance involves the use of so-called damping wigglers,[65] wiggler magnets designed and located in the ring so that the radiation they produce causes more damping than excitation of oscillations. Significant reduction of the emittance may require tens or even hundreds of meters of such damping wigglers. This approach may be a good way to obtain extremely low emittance, because, unlike the approach of increasing the focussing strength, it should not reduce the dynamic aperture.

Another possibility is to inject the entire beam in one shot, using on-axis injection to avoid the aperture requirements for off-axis injection and accumulation. In such a case, stronger focussing might be used, even if the dynamic aperture is reduced. Extremely good vacuum would also be needed in such a ring to maintain a good lifetime with small dynamic aperture. Injection of a large current in one shot could be accomplished by transferring an already accumulated, stored, and damped beam from another storage ring.

High energy colliding-beam storage rings such as PEP (16 GeV) and Tristan (25 to 30 GeV) offer good opportunities to achieve very low emittance. Since emittance decreases as the square of the electron energy, by operating these rings at 1/2 to 1/3 their maximum energy the emittance is reduced by a factor of 4 to 9. Low emittance optics can be used to further reduce the emittance by about a factor of 4. Also, these rings have long, dispersion-free straight sections suitable for damping wigglers to reduce the emittance by another factor of 5 or more. By using these techniques it should be possible to achieve an emittance of less than one nanometer-radian in these rings.

The large circumference (see Figure 18) and weak bending fields of PEP have already made it possible to operate this ring at very low emittance even without damping wigglers. During a 12-day run in December 1987 PEP was operated at 7.1 GeV in a mode for which the calculated emittance is 6.4 nanometer-radians and briefly in a mode with increased horizontal damping and a calculated emittance of 3.3 nanometer-radians. Preliminary analysis of the results indicates that these emittance values were achieved. X-ray spectra taken (see Figure 19) exhibit pronounced fundamental and harmonic peaks, as expected at such low emittance. The beam line used for this work, which includes an x-ray pinhole camera, is shown in Figure 20.

Because of these capabilities, PEP and other high energy colliders offer early opportunities to perform experiments with high brightness x-rays and can serve to provide experience in ring operation and in the development of insertion devices and beam-line instrumentation even before the new low emittance rings come on line. For example, the difficult problems of beam steering and stability

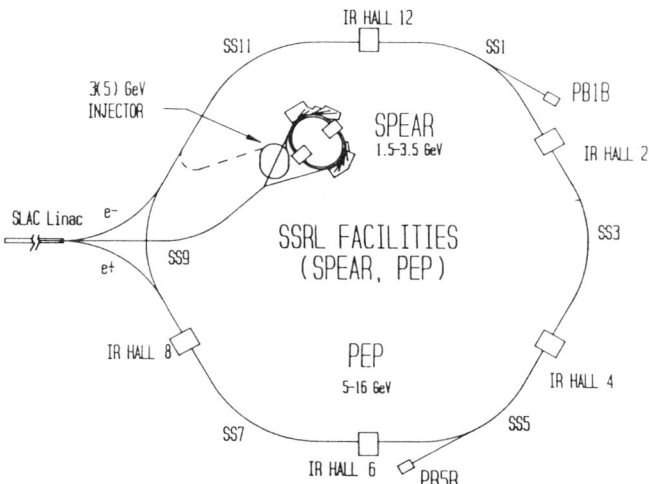

Figure 18. Layout of the PEP and SPEAR storage rings, showing the synchrotron radiation beam lines and a 3-GeV synchrotron injector now under construction for SPEAR and capable of being upgraded to 5 GeV for injection into PEP.

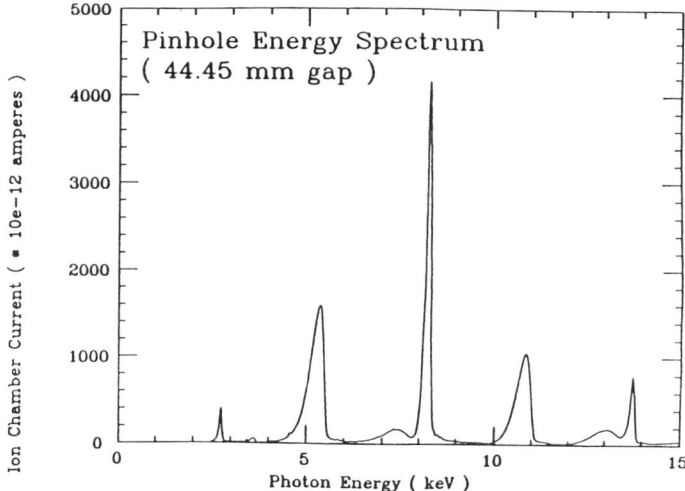

Figure 19. Spectral brightness of the radiation from a 26-period undulator operating on PEP at 7.1 GeV with an emittance of 6 nanometer-radians. The period length is 77 mm and there are 26 periods. The raw, uncorrected experimental data are shown, i.e. the ion chamber current during a scan of the monochromator. The fundamental peak at about 2.8 keV is severely attenuated by the 0.5-mm-thick beryllium window in the beam line. The sharpness of the peaks is due to the low emittance. Compare with the spectrum taken on SPEAR (Figure 12) at much higher emittance.

in low emittance rings and the development of beam-line optics that can withstand the intense thermal loading associated with the high brightness, high energy beams can be pursued and the solutions tested on rings like PEP.

The very long straight sections (up to 117 m in PEP) built for colliding-beam purposes can also serve to accomodate damping wigglers, long undulators, bypasses, and other special equipment, giving these rings even greater potential for extensions of performance. These rings could therefore be prototypes of future high performance rings, offering higher brightness and coherent power, approaching diffraction limits at shorter wavelenghts.

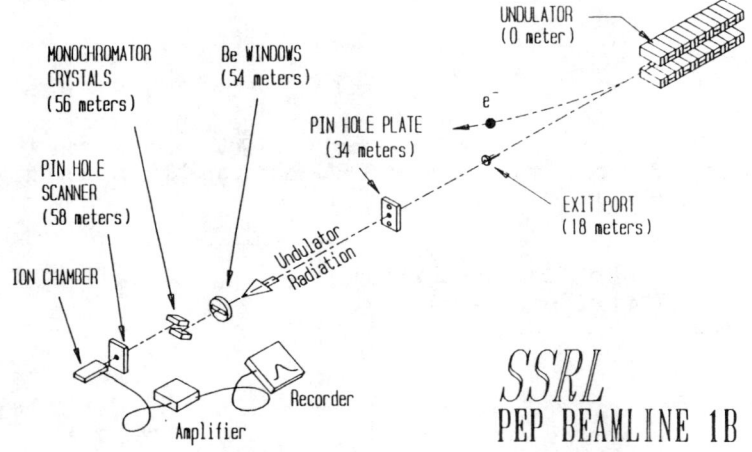

Figure 20. Schematic drawing of the PEP beam line used to take the previous spectrum. The pinhole plate is removable.

12 BIBLIOGRAPHY

B1. Handbook on Synchrotron Radiation, Vol. I, E.-E. Koch, Ed., Vol. II, G. Marr, Ed., North Holland, Amsterdam, 1983, 1987.

B2. A. Bienenstock and H. Winick, Synchrotron Radiation research - An Overview, Physics Today $\underline{36}$ (6) 48 (1983).

B3. Synchrotron Radiation Research, H. Winick and S. Doniach, Ed., Plenum, New York, 1980.

B4. Synchrotron Radiation, Techniques and Applications, C. Kunz, Ed., Springer, Berlin, 1979.

B5. Proc. Int. Conf. on X-Ray and VUV Sync. Rad. Instr., Nucl. Instr. Meth. $\underline{152}$ (1978); $\underline{208}$ (1983); $\underline{A246}$ (1986).

B6. Proc. U.S. Nat. Conf. on Synch. Rad. Instr., Nucl. Instr. Meth. $\underline{172}$ (1980); $\underline{195}$ (1982); $\underline{222}$ (1984).
B7. Proc. USSR Nat. Conf. on Sych. Rad., Nucl. Instr. Meth. $\underline{A261}$ (1987).
B8. H. Winick, Synchrotron Radiation, Scientific American $\underline{225}$ (11), 88-99 (1987).

13 REFERENCES

1. A. Lienard, L'Eclairage Elect. $\underline{16}$, 5 (1898).
2. G.A. Schott, Ann.Phys.(Leipzig) $\underline{24}$, 635 (1907); Electromagnetic Radiation, Ch. 7, 8, Cambridge U. Press, London, 1912.
3. W.W. Jassinsky, J. Exp. Theor. Phys. (USSR) $\underline{5}$, 983 (1935); Arch. Electrotech. (Berlin) $\underline{30}$, 590 (1936).
4. D. Kerst, Phys. Rev. $\underline{60}$, 47 (1941).
5. D. Ivanenko and I. Pomeranchuk, Dokl. Akad. Nauk. SSSR $\underline{44}$, 315 (1944); Phys. Rev. $\underline{65}$, 343 (1944).
6. L. Arzimovitch and I. Pomeranchuk, J. Phys. (Moscow) $\underline{9}$, 267 (1945); J. Exp. Theor. Phys. (USSR) $\underline{16}$, 379 (1946).
7. J.P. Blewett, Phys. Rev. $\underline{69}$, 87 (1946); Nucl. Instr. Meth. $\underline{A266}$, 1-9 (1988).
8. K.R. Lea, Phys. Rep. (Phys. Lett. C) $\underline{43}$, 337-75 (1978).
9. D. Ivanenko and A.A. Sokolov, Dokl. Akad. Nauk SSSR $\underline{59}$, 1551 (1948); A.A. Sokolov, N.P. Klepikov, and I.M. Ternov, Ibid. $\underline{89}$, 665 (1953); A.A. Sokolov and I.M. Ternov, Sov. Phys. JETP $\underline{1}$, 227-30 (1955); $\underline{4}$, 396-400 (1957); Sov. Phys. Dokl. $\underline{8}$, 1203-5 (1964); Synchrotron Radiation, Pergamon, New York, 1968.
10. J. Schwinger, Phys. Rev. $\underline{70}$, 798 (1946); $\underline{75}$, 1912-25 (1949); Proc. Natl. Acad. Sci. USA $\underline{40}$, 132 (1954).
11. J.D. Jackson, Classical Electrodynamics, p. 848, Wiley, New York, 1975.
12. F.R. Elder, A.M. Gurewitsch, R.V. Langmuir, and H.D. Pollack, Phys. Rev. $\underline{71}$, 829-30 (1947); J. Appl. Phys. $\underline{18}$, 810 (1947); F.R. Elder, R.V. Langmuir, and H.C. Pollack, Phys. Rev. $\underline{74}$, 52 (1948).
13. I.M. Ado and P.A. Cherenkov, Sov. Phys. Dokl. $\underline{1}$, 517-19 (1956); F.A. Korolev, V.S. Markov, E.M. Akimov, and O.F. Kulikov, Ibid., 568; F.A. Korolev and O.F. Kulikov, Opt. Spectrosc. (USSR) $\underline{8}$, 1-3 (1960); F.A. Korolev, A.G. Ershov, and O.F. Kulikov, Sov. Phys. Dokl. $\underline{5}$, 1011 (1961); F.A. Korolev, O.F. Kulikov, and A.S. Yarov, Sov. Phys. JETP $\underline{43}$, 1653 (1962).

14. D.A. Corson, Phys. Rev. 86, 1052-3 (1952); 90, 748-52 (1953).
15. P.L. Hartman and D.H. Tomboulian, Phys. Rev. 87, 233 (1952); D.H. Tomboulian, USAEC NP-5803 (1955); D.H. Tomboulian and P.L. Hartman, Phys. Rev. 102, 1423-47 (1956); D.E. Bedo and D.H. Tomboulian, J. Appl. Phys. 29, 804-9 (1958); P.L. Hartman, Nucl. Instr. Meth. 195, 1-6 (1982).
16. K. Codling and R.P. Madden, Phys. Rev. Lett. 10, 516-18 (1963); 12, 106-8 (1964); J. Opt. Soc. Am. 54, 268 (1964); J. Appl. Phys. 36, 830-7 (1965).
17. G. Bathov, E. Freytag, and R. Haensel, J. Appl. Phys. 37, 3449 (1966).
18. E.M. Rowe and F.E. Mills, Part. Accel. 4, 211-27 (1973).
19. T. Miyahara, H. Kitamura, S. Sato, M. Watanabe, S. Mitani, E. Ishiguro, T. Fukushima, T. Ishii, Shigeo Yamaguchi, M. Endo, Y. Iguchi, H. Tsujikawa, T. Sugiura, T. Katayama, T. Yamakawa, Seitaro Yamaguchi, and T. Sasaki, Part. Accel. 7, 163 (1976).
20. P. Dagneaux, C. Depautex, P. Dhez, J. Durup, Y. Farge, R. Fourme, P.-M. Guyon, P. Jaegle, S. Leach, R. Lopez-Delgado, G. Morel, R. Pinchaux, P. Thiry, C. Vermeil, and F. Wuilleumier, Ann. Phys. (NY) 9, 9-65 (1975); P.-M. Goyon, C. Depautex, and G. Morel, Rev. Sci. Instr. 47, 1347 (1976).
21. H. Winick, IEEE Trans. Nucl. Sci. NS-20, 984-8 (1973).
22. G.N. Kulipanov and A.N. Skrinskii, Usp. Fiz. Nauk. 122, 369-418 (1977); Engl. transl., Sov. Phys. Usp. 20, 559-86 (1977).
23. H. Winick, Proc. 9th Int. Conf. on High Energy Accelerators, Stanford, pp. 685-8 (1974); A. Baer, R. Gaxiola, A. Golde, F. Johnson, B. Salsburg, H. Winick, M. Baldwin, N. Dean, J. Harris, E. Hoyt, B. Humphrey, and J. Jurow, IEEE Trans. Nucl. Sci. NS-22, 1794-7 (1975); J. Cerino, A. Golde, J. Hastings, I. Lindau, B. Salsburg, H. Winick, M. Lee, P. Morton, and A. Garren, Ibid. NS-24, 1003-5 (1977).
24. E. Burattini, A. Reale, E. Bernieri, N. Cavallo, A. Morone, M.R. Masullo, R. Rinzivillo, G. Dalba, P. Fornasini, and C. Mencuccini, Nucl. Instr. Meth. 208, 91-96 (1983).
25. C. Depautex and F. Wuilleumier, Nucl. Instr. Meth. 195, 37-48 (1982).
26. E.E. Koch, C. Kunz, and E.W. Weiner, Optik (Stuttgart) 45, 395-410 (1976).
27. D. Mills, Nucl. Instr. Meth. 195, 29-33 (1982).

28. G. Brown, Nucl. Instr. Meth. A246, 149-53 (1986).
29. G. Hirikoshi and Y. Kimura, Proc. 1987 Part. Accel. Conf., IEEE Cat. No. 87CH2387-9, pp.34-8.
30. D.J. Thompson, Nucl. Instr. Meth. 177, 27 (1980).
31. D. Einfeld and G. Mülhaupt, Nucl. Instr. Meth. 172, 55 (1980).
32. K. Huke, Nucl. Instr. and Meth. 177, 1 (1980).
33. E.M. Rowe and P. Woodruff, Nucl. Instr. Meth. 172, 9-12 (1980).
34. A. van Steenbergen, Nucl. Instr. Meth. 177, 53 (1980).
35. T. Tomimasu, T. Noguchi, S. Sugiyama, T. Yamazaki, T. Mikado, and M. Chiwaki, IEEE Trans. Nucl. Sci. NS-30, 3133 (1983).
36. M. Watanabe, T. Kasuga, H. Yonehara, A. Uchida, K. Sakai, O. Matsudo, T. Kinoshita, M. Hasumoto, J. Yamazaki, E. Nakamura, H. Yamamoto, K. Takami, T. Katayama, K. Yoshida, M. Kihara, and G. Saxon, Proc. 5th Symp. on Accel. Sci. and Tech., KEK, Tsukuba, Japan, Sept. 1984, pp. 15-17.
37. M. Eriksson, Nucl. Instr. Meth. 196, 331 (1982).
38. G.M. Kulipanov, Nucl. Instr. Meth. A261, 1-8 (1987).
39. Z. Bao, Nucl. Instr. Meth. A246, 18-20 (1986).
40. D. Xian, Nucl. Instr. Meth. A266, 77-81(1988).
41. H. Winick, Nucl. Instr. Meth. A261, 9-17 (1987).
42. A. Hofmann, R. Little, J.M. Paterson, K.W. Robinson, G.-A. Voss, and H. Winick, Proc. 6th Int. Conf. on High Energy Accel., CEAL 2000 (Sept. 1967), pp. 123-9.
43. M. Berndt, W. Brunk, R. Cronin, D. Jensen, A. King, J. Spencer, T. Taylor, and H. Winick, IEEE Trans. Nucl. Sci. NS-26, 3812-15 (1979).
44. C. Bahr et al., Nucl. Instr. Meth. 208, 117-26 (1983).
45. K. Huke and T. Yamakawa, Nucl. Instr. Meth. 177, 253-7 (1980).
46. R. Avery, Nucl. Instr. Meth. 222, 146-58 (1984).
47. S. Krinsky, IEEE Trans. Nucl. Sci. NS-30, 3078-83 (1983).
48. A. Hofmann, Nucl. Instr. Meth. 152, 17-21 (1978).
49. H. Winick, G. Brown, K. Halbach, and J. Harris, Physics Today 34, 50-63 (May 1981); Nucl. Instr. Meth. 208, 65-77 (1983).
50. H. Motz, J. Appl. Phys. 22, 527-35 (1951); H. Motz, W. Thon, and R.N. Whitehurst, J. Appl. Phys. 24, 826-33 (1953).

51. D. Alferov, Yu.A. Bashmakov, and E.G. Bessonov, Sov. Phys. Tech. Phys. $\underline{18}$, 1336-9 (1974); D.F. Alferov, Yu.A. Bashmakov, K.A. Belovintsev, E.G. Bessonov, and P.A. Cherenkov, JETP Lett. $\underline{26}$, 385-8 (1977); A.N. Didenko, A.V. Kozhevnikov, A. F. Medvevev, and M.M. Nikitin, Pis'ma Sh. Tekh. Fiz $\underline{4}$, 625-9 (1978); Eng. Trans. Sov. Tech. Phys. Lett. $\underline{4}$, 277-8 (1978).
52. M.B. Moiseev, M.M. Nikitin, and N.I.Fedosov, Izvestiya VUZ. Fizika $\underline{3}$, 76-80 (1978); Eng. Trans. Sov. Phys. J. $\underline{21}$, 332-5 (1978); K.-J. Kim, Nucl. Instr. Meth. $\underline{222}$, 11-13 (1984); H. Onuki, N. Saito, and T. Saito, Appl. Phys. Lett. $\underline{52}$ (3), 173-5 (1988).
53. K. Halbach, J. Chin, E. Hoyer, H. Winick, R. Cronin, and J. Yang, IEEE Trans. Nucl. Sci. $\underline{NS-28}$, 3136-8 (1981).
54. K. Halbach, Nucl. Instr. Meth. $\underline{169}$, 1 (1980); Jour. de Phys. $\underline{44}$, C1-211 (1983).
55. K. Halbach, Proc. of Workshop on PEP as a Synch. Rad. Source, R. Coisson and H. Winick, Eds., Oct. 1987, p. 514 (available from SSRL).
56. H. Winick and T. Knight, Eds., Wiggler Workshop, SSRL Report 77/05.
57. R. Tatchyn and I. Lindau, Eds., Int. Conf. on Insertion Devices for Sych. Rad. Sources, Proc. SPIE $\underline{582}$ (1985).
58. D.C. Quimby and A.L. Pindroh, Rev. Sci. Instr. $\underline{58}$, 339-45 (1987).
59. R. Tatchyn and P. Csonka, Proc. Adriatico Res. Conf. on Undulator Magnets for Synch. Rad. Research and Free Electron Lasers, ICTP, Trieste, Italy, June 1987 (to be published).
60. P.C. Marin, Nucl. Instr. Meth. $\underline{A266}$, 18-23 (1988).
61. R. Haensel, Nucl. Instr. Meth. $\underline{A266}$, 68-73 (1988).
62. F.B. Selph, Nucl. Instr. Meth. $\underline{A266}$, 44-58 (1988).
63. M. Cornacchia, M. Campagna, L. Fonda, G. Margaritondo, C. Pellegrini, R. Rosei, and C. Rubbia, Nucl. Instr. Meth. $\underline{A266}$, 38-43 (1988).
64. G.K. Shenoy and D.E. Moncton, Nucl. Instr. Meth. $\underline{A266}$, 38-43 (1988).
65. H. Wiedemann, Nucl. Instr. Meth. $\underline{A266}$, 24-31 (1988); Proc. 1987 Part. Accel. Conf., IEEE Cat. No. 87CH2387-9, pp. 395-9.
66. S. Hunter, PhD Thesis, Stanford U., 1977; SSRL Report 77/04.

INDUCTION ACCELERATORS AND FREE-ELECTRON LASERS AT LLNL*

Richard J. Briggs

Lawrence Livermore National Laboratory

Livermore, California 94550

TABLE OF CONTENTS

Part A: Overview of Linear Induction Accelerators	2185
Physical Principles of Induction Linac Operation	2185
High-Brightness Electron Beams	2187
High-Average-Power Technology	2190
Application of Induction Accelerator Technology to High-Gradient Accelerators	2192
Part B: Overview of FEL Development with Induction Linacs at LLNL	2196
Introduction	2196
General Description of Induction-Linac Powered FEL Amplifiers	2196
FEL Amplifier Physics	2198
High Brightness and High Average Power	2200
FEL Experiments at LLNL	2200
FEL Technology and Fusion Power	2203
FEL Technology and High Power Accelerators	2204
Summary	2206
Acknowledgment	2206
References	2206

INDUCTION ACCELERATORS AND FREE-ELECTRON LASERS AT LLNL[*]

Richard J. Briggs

Lawrence Livermore National Laboratory

Livermore, California 94550

ABSTRACT

Linear induction accelerators have been developed to produce pulses of charged particles at voltages exceeding the capabilities of single-stage, diode-type accelerators and at currents too high for rf accelerators. In principle, one can accelerate charged particles to arbitrarily high voltages using a multi-stage induction machine. The advent of magnetic pulse power systems makes sustained operation at high repetition rates practical, and high-average-power capability is very likely to open up many new applications of induction machines.

In part A of this paper, we survey the U.S. induction linac technology, emphasizing electron machines. We also give a simplified description of how induction machines couple energy to the electron beam to illustrate many general issues that designers of high-brightness and high-average-power induction linacs must consider. We give an example of the application of induction accelerator technology to the relativistic klystron, a power source for high-gradient accelerators.

In part B we address the application of LIAs to free-electron lasers. The multikiloampere peak currents available from linear induction accelerators make high-gain, free-electron laser amplifier configurations feasible. High extraction efficiencies in a single pass of the electron beam are possible if the wiggler parameters are appropriately "tapered," as recently demonstrated at millimeter wavelengths on the 4-MeV ELF facility. Key issues involved in extending the technology to shorter wavelengths and higher average powers are described. Current FEL experiments at LLNL are discussed.

---

[*] Work performed jointly under the auspices of the U.S. Department of Energy by Lawrence Livermore National Laboratory under contract W-7405-ENG-48, for the Strategic Defense Initiative Organization and the U.S. Army Strategic Defense Command.

© 1989 American Institute of Physics

PART A: OVERVIEW OF LINEAR INDUCTION ACCELERATORS

Physical Principles of Induction Linac Operation

A linear induction accelerator (LIA) can be thought of as a series of 1:1 transformers in which the electron beam acts as the secondary. A key feature of this configuration is the absence of a "voltage" on any cable or structure that exceeds the voltage supplied to a single accelerator module driven by the pulse source. The electrons, in effect, "integrate" the axial electric field in the vacuum beam pipe to achieve a final energy of n times the module voltage (for n modules.)

These ideas can be more clearly understood by considering the sketch in Fig. 1 of a geometry similar to that of the accelerator modules used at Lawrence Livermore National Laboratory (LLNL). A voltage pulse is supplied to the accelerator module by coaxial cable transmission lines (driven from two sides in a balanced mode to avoid deflection forces on the electron beam). A cylindrical core of ferromagnetic material (for example, ferrite) located in the cavity as shown presents a very high impedance to the drive transmission lines at their junction point with the cavity. With no electron beam present, the "voltage" ($\int E \cdot dz$) impressed across the accelerator gap is equal to the transmission line voltage at the junction point with the cavity. This statement is a good approximation only when the pulse length is much longer than the transit time of

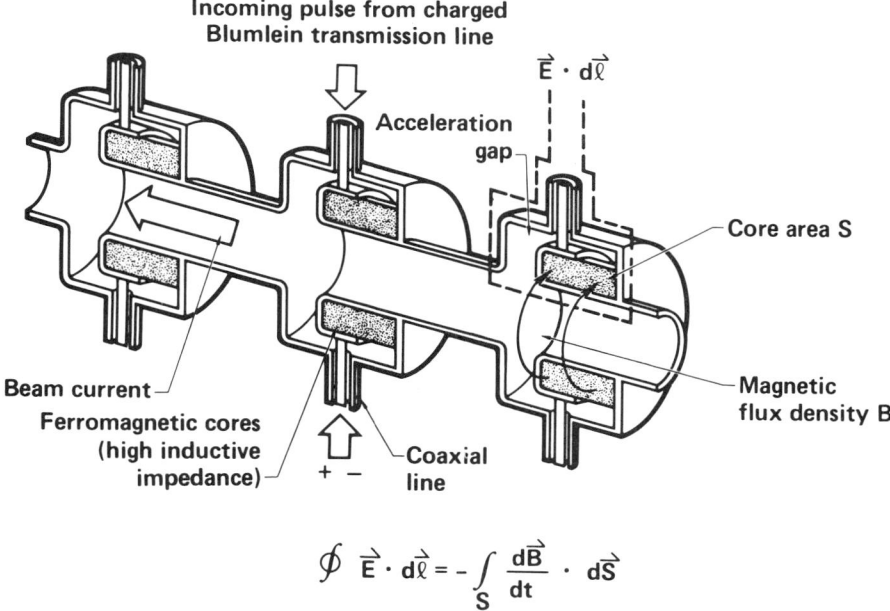

Figure 1. The magnetic induction module.

electromagnetic waves throughout the cylindrical radial line structure (~1 ns typically), so the electromagnetic fields can be treated in a quasistatic approximation. Capacitive and inductive effects of the gap, coaxial leads, and other components can particularly affect the electromagnetic field distribution during the rise and fall times of the voltage (beam current) pulse. Note that the electric field in the vicinity of the gap is quasistatic in shape (see sketch in Fig. 1), and there is no coupling of adjacent modules as long as a beam pipe of reasonable length (> pipe diameter) separates the acceleration gaps.

The impedance of the ferromagnetic core will be high for only a limited time, of course, and the "volt-sec" capability of the material determines the <u>core area</u>, S, for a given module voltage and pulse length. Actually, for ferrite-type modules, the manner in which the electromagnetic fields propagate through the ferromagnetic material cannot be ignored, and the ferrite region is often not accurately represented by lumped circuit models. This feature is not crucial for zero-order modeling since the role of the ferrite is to present a sufficiently large impedance to the drive lines, which it does in cases of interest [since $(\mu/\varepsilon)^{1/2}$ is large compared to the free-space impedance].

In the presence of an electron beam pulse proceeding down the axis of the accelerator, as illustrated in Fig. 1, a return current in the wall will flow up the gap and "load" the driver transmission line. Once again, this simple picture applies when the current pulse is relatively long (e.g., 50 ns) compared to the transit time of electromagnetic waves up the gap (~1 ns).

The pulses are generated with a pulse forming line (PFL); an elementary schematic of such a system is illustrated in Fig. 2. The output of the PFL is applied to the transmission line (in actuality we use a balanced pair of lines to drive the cell.) The transmission lines are long enough to provide "transit time isolation" of the PFL/switch and the cell, that is, the cable transmission time is longer than the pulse length. In this case, the simple equivalent circuit of the drive system shown in Fig. 2 is applicable, where $V_o(t)$ is the pulse waveform supplied to the transmission line by the PFL. The

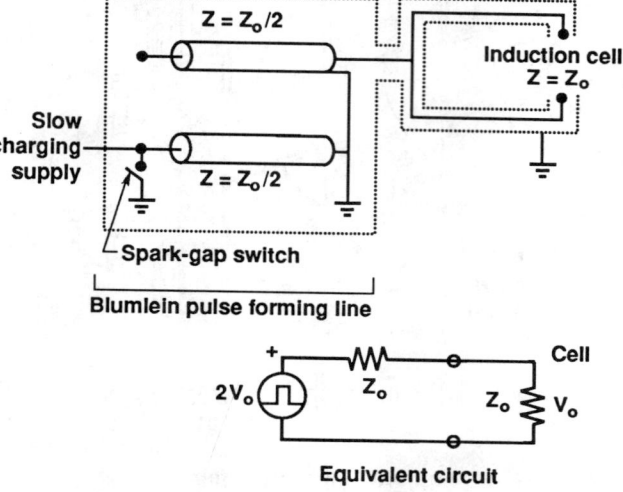

Figure 2. Simplified picture of an accelerator cell driver.

ideal "square wave" shown in the figure is, of course, in practice modified by switch inductances and other factors, which can limit the minimum pulse lengths that have enough of a "flat top" on the waveform to be useful in induction accelerators.

From all these considerations, we can deduce the circuit schematic shown in Fig. 3. The beam current load on the transmission line is accurately represented by a current source because the current is not dependent on the voltage of that stage as it would be in a diode region. External compensation circuits at the transmission line output are often used to help flatten the acceleration voltage pulse and to absorb energy from the transmission lines when the beam is absent (prevent "ringing" of the energy on the cables). In practice, resistors are used on our 50-MeV induction accelerator to absorb half the drive power with a 10-kA beam.

Many observations are readily apparent from this circuit schematic. For example, for optimum efficiency, the transmission line impedance should be matched to the beam impedance, $Z_o = V_o/I_B$ (in the absence of resistor compensation). Since pulse power system efficiencies can be quite high (up to 60–70% with the latest magnetic modulator systems), induction machines can have good overall efficiencies. Parameter choices can compromise this potential for high efficiency; in particular, with long pulse lengths, compensation for voltage "droop" from finite ferrite inductance in the compensation circuitry will waste some drive current.

This simple schematic also shows the difficulty in avoiding beam energy variation in the head and tail of a heavily loaded (relatively efficient) induction machine, where I_B is changing. As a consequence, beam transport systems in induction machines must often accommodate a relatively broad energy variation through the electron beam pulse. Difficulties with our 50-MeV machine, for example, have often been traceable to the problems of handling a time-varying energy on the beam head.

High-Brightness Electron Beams

Over the past two decades, LLNL and the Lawrence Berkeley Laboratory have collaborated on the development of several LIAs. These include the Astron (6 MeV,

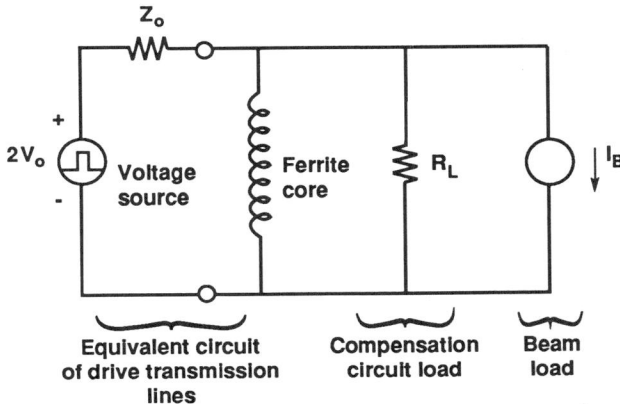

Figure 3. Simplified schematic of induction unit.

0.5 kA, 300 ns), the ERA (4 MeV, 1.5 kA, 30 ns), the Experimental Test Accelerator (ETA) (5 MeV, 10 kA, 30 ns), and the Advanced Test Accelerator (ATA) (50 MeV, 10 kA, 70 ns). When these accelerators were developed, the primary requirement was to achieve high beam currents with relatively little concern for beam emittance. The free-electron laser (FEL) application for LIAs, in contrast, places a very high premium on beam brightness and on high-average-power operation. The technology for these requirements was developed on our Accelerator Research Center (ARC) accelerator (2 MeV, 0.8 kA, 50 ns, and multikilohertz repetition-rate capability). The brightness of an electron beam (the beam's density per unit 4-volume of phase space, V_4) is defined as

$$J = I/(\gamma\beta)^2 V_4 , \tag{1a}$$

$$J[A/(m\text{-rad})^2] = \Psi\, I/(\pi\varepsilon_n)^2 , \tag{1b}$$

where I is the beam current, ε_n is the normalized edge beam emittance ($\beta\gamma RR'$), and Ψ is a shape factor (for uniform density ellipsoids in phase space, determination of edge emittance specifies $\Psi = 2$). Using conservation of momenta ($m_o v_{\text{(source)}} = \gamma m_o v_\perp$) and $\beta = v_z/c$, we can describe the lower bound on beam emittance as it leaves a "perfect" (smooth and uniformly emitting) cathode plane of temperature T_e and radius R:

$$\begin{aligned}\varepsilon_n &= \beta\gamma RR' = \beta\gamma R\frac{v_\perp}{v_z} = R\frac{v_{\text{(source)}}}{c} , \\ &= 0.2 R_{\text{meters}}\sqrt{T_e(\text{eV})} .\end{aligned} \tag{2}$$

Obviously, the "perfect" cathode source for high brightness requires a low effective temperature and a high extraction-current density (small radius). Researchers in our injector development program, in which the goal is to develop a 3-kA beam with normalized brightness of 2×10^9 A/(m-rad)^2 to meet our FEL requirements, have explored various cathode technologies to quantify these operating parameters. Of equal importance, we are trying to determine how requirements of continuous high power and high-repetition-rate operation compromise the performance of various cathodes. In Table 1, we briefly summarize our findings. As is apparent from Table 1, we are concentrating on the dispenser cathode primarily because it is best able to generate the high-brightness beams and satisfy the reliability requirements of high-average-power, high-repetition-rate operation.

The operating parameters of a high-current/high-brightness injector are determined by the following two requirements, which incorporate the criteria set by preservation of the normalized cathode emittance in the subsequent transport and acceleration processes:

1. The extraction electric field E (= V/d, where V is the injector voltage and d is the cathode-to-anode gap spacing) has a practical upper limit if one is to prevent unwanted emission from noncathode surfaces. Even modest electron emissions from such regions, when combined with those originating from the cathode, drastically increase emittance and make management of stray power very difficult in high-average-power operation. Our initial high-average-power tests at full repetition rates indicate that, for satisfactory operation, E must be limited to 120 kV/cm.

Table 1. Study of cathodes for high-brightness and high-average-power operation.

Cathode type	Description	Summary
Plasma surface discharge	Many small sites of surface electrical discharge; energy per discharge site is low and controllable	• Good current density (>20 A/cm^2) and uniformity • Rugged and reliable repetition-rate operation • Poor beam emittance due to granularity and high effective temperature of emitting plasma sheath
Field emission	Surface ionization and/or whisker vaporization forms cathode plasma	• Good current density • Dense plasma closure of anode-cathode gap limits repetition rate • Emittance and uniformity uncertain • Degrades under high-average-power operation
Occluded, gas-enhanced field	Velvet cloth as large surface area gas source with fibers giving field enhancement	• Good current density (50 A/cm^2) • Uniformity of emission degrades with lifetime • Beam brightness few $\times 10^5$ • Plasma closure limits repetition rate
Thermionic emitters	Low work function triple-oxide coatings on heated surfaces	• Limited current density (~10 A/cm^2), but with good uniformity; beam emittance unknown • Taxing vacuum requirements and highly susceptible to poisoning • Thermal stresses weaken bond of oxide to supporting substrate • Not mechanically reliable under high-power operation
LaB$_6$	Low work function material heated to very high temperatures (2400 K)	• Good current density (50 A/cm^2) and uniformity, emittance unknown • Very delicate to thermal stresses • High-average-power operation untested
Dispenser	Low work function barium-impregnated surfaced heated elevated temperature (~1100 K)	• Good current density (20 A/cm^2) and uniformity • Beam brightness > few $\times 10^6$ • Reliable operation at full repetition rate • Taxing vacuum requirements
Photo-emission	Laser-irradiated surface	• Untested, known to have good current density and uniformity, but total current and charge (surface area and purse duration) are limited • Testing for high-repetition rate operation requires major investment in laser development

2. Increases in beam emittance during transport stem from the coupling of space-charge fields to the beam-particle trajectories. Beams with uniform profiles minimize this coupling. For uniform cathode emission, the ratio of cathode radius, a, to gap spacing, d, appears to have a limiting value, which, from extensive numerical simulation, we estimate to be $\leq 1/3$. These simulations seek to optimize beam parameters of cathode radius, gap spacing, anode bore size (that is, the effective size of the hole in the anode through which the beam is extracted), the z-variation in transport axial magnetic field (to avoid angular momentum, the axial magnetic field is zero on the cathode surface and then increases to full focusing strength), the injector voltage, and geometric shapes of all cathode-anode-electrode components.

The total beam current is determined by $I = \pi a^2 j$, where j is the current density (assumed to be uniform according to requirement 2). Note that

$$j = K(\gamma) \frac{V^{3/2}}{d^2}, \tag{3}$$

where K is the perveance, which includes relativistic and anode-depression corrections. To explicitly display the limiting parameters, we write the current as

$$I = \left[\pi K \left(\frac{a}{d}\right)^2 E^{3/2} \right] d^{3/2}. \tag{4}$$

This result numerically implies that if 3 kA is the desired beam current, gaps of 25 cm driven at 3 MeV are required. Using dispenser cathodes, the ARC facility operates at high average powers with the following performance parameters:
- Beam current 0.8 kA.
- Injector voltage 0.8 MeV.
- Post acceleration 1.2 MeV.
- Beam brightness measured at 2.0 MeV was 1.3×10^9 A/(m-rad)^2.

Future research will be directed toward developing higher voltage (3 MeV) injectors, which will allow increased beam current. Work will continue at the ARC facility and on our new, upgraded ETA accelerator (ETA-II). ETA-II began operation in November, 1987, with a 1-kA, 7-MeV, 50-ns electron beam of high brightness [1.3×10^9 A/(m-rad)^2] at a 1-Hz repetition rate. An eventual goal is 3 kA at 2×10^9 A/(m-rad)^2 brightness at repetition rates up to 5 kHz.. The purpose of ETA-II is to integrate all aspects (physics and engineering) of high-average-power, high-brightness electron beam accelerators into one test facility.

High-Average-Power Technology

The generation of intense, relativistic electron beams requires low-impedance, pulse-compression devices to provide the short (70-ns) acceleration voltage pulse. The spark-gap-based system has served us well on both the ETA and ATA accelerators. However, even during the construction of the ATA accelerator, we recognized that the limitations of spark gaps in repetition rates (duty factor) and reliability would not allow us to apply this technology to high-average-power systems.

During construction of ATA, we worked to find a longer term replacement for the spark-gap-driven Blumlein that could be used on the new device. This effort was accelerated when we began the FEL program. FEL applications place very stringent requirements on the power-conditioning systems. Those requirements drive us toward higher repetition rates, higher efficiency, better energy regulation during the acceleration pulse, and higher accuracy in alignment of components.

To generate the required 10-GW acceleration pulse for near-term systems, we are using the well-tested magnetic pulse compressor, MAG-1-D. Although this pulse-compression scheme has been around for a number of years, only recent advances in magnetic materials and technology have allowed us to use it to generate nanosecond pulses with efficiencies exceeding 90%. This technique uses the large changes in permeability exhibited by saturating ferromagnetic materials to produce large changes in impedance. We use state-of-the-art switches (thyratrons, silicon-controlled rectifiers) to initiate the pulse at the 100-MW level and to compress it to the 10-GW levels required by the accelerator. Because the compressor uses strictly passive devices (inductors and capacitors), when properly designed its lifetime exceeds 10^9 cycles.

The current magnetic-drive system is shown in Fig. 4. It consists of an intermediate store switch chassis (ISSC), a precompression stage, a step-up transformer, the first stage of compression, and the final stage of compression. The final stage delivers energy from the PFL to the accelerator in the form of a 70-ns pulse at 125 kV through two 4-Ω lines. The various waveforms in the MAG-1-D are shown in Fig. 5.

This system has been used during the past two years to generate the high voltages required by the ARC injector for studies of high-brightness, high-current beams. The beam from the injector was accelerated through two ten-cell modules (Fig. 6) for a total energy of 2–3 MeV at the point where the brightness measurements were performed.

The stringent peak-power and repetition-rate requirements are stressing the capabilities of current state-of-the-art thyratrons. To date we have achieved 2.5-kHz repetition rates at nearly full-power output. To reach our goal of 5 kHz, we are experimenting with the "Gatling-gun" mode of operation, where two ISSCs are fired in sequence at 2.5 kHz. We will apply this idea in our new ETA-II high-average-power accelerator.

The magnetic power systems described here should make linear induction accelerators available for many applications in the future, including

- E-beam drivers for microwave and millimeter wave sources.
- Fusion plasma electron cyclotron heating current drive with a 1–2-mm-wavelength FEL.
- E-beam driver for infrared to visible-wavelength FELs.
- The two-beam accelerator.
- The relativistic klystron.
- E-beam driver for collective accelerators.
- Radiation processing.

Using induction accelerators for radiation processing is, in one sense, the least obvious application to consider since a high-peak-current capability is not required. Nonetheless, the "rugged" nature of this solid-state pulse power technology, the system simplicity (for example, modest vacuum requirements) made possible because of the very short duration of the high-voltage pulses, and the relatively low cost per watt do warrant serious examination of its applicability, even for applications requiring low peak currents.

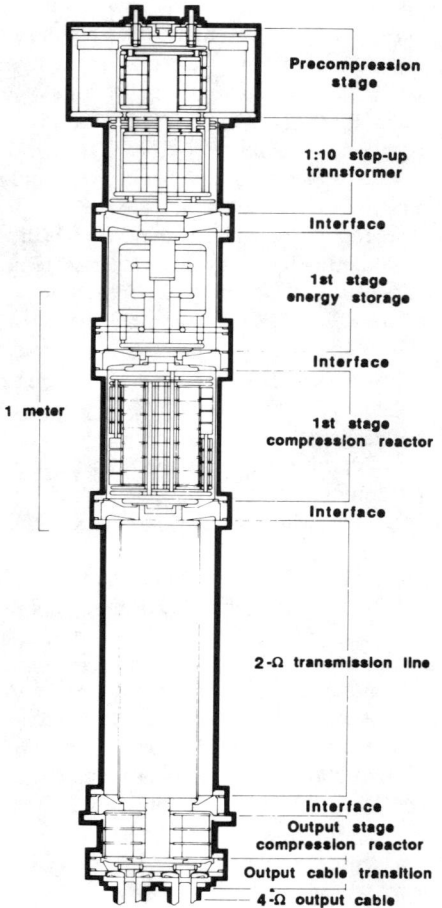

Figure 4. MAG-1-D power compressor.

Application of Induction Accelerator Technology to High-Gradient Accelerators

We would like to discuss, as one example of the applications mentioned above, recent research at LLNL on the relativistic klystron. This work is a collaborative effort with the Stanford Linear Accelerator Center and the Lawrence Berkeley Laboratory.[1]

Relativistic klystrons are being developed as a power source for high-gradient accelerator applications which include large linear electron-positron colliders, compact accelerators, and FEL sources. These applications require a new generation of high-gradient accelerators. Conceptual designs for large linear electron colliders for research at the frontier of particle physics, for example, call for center-of-mass energies of 1–2

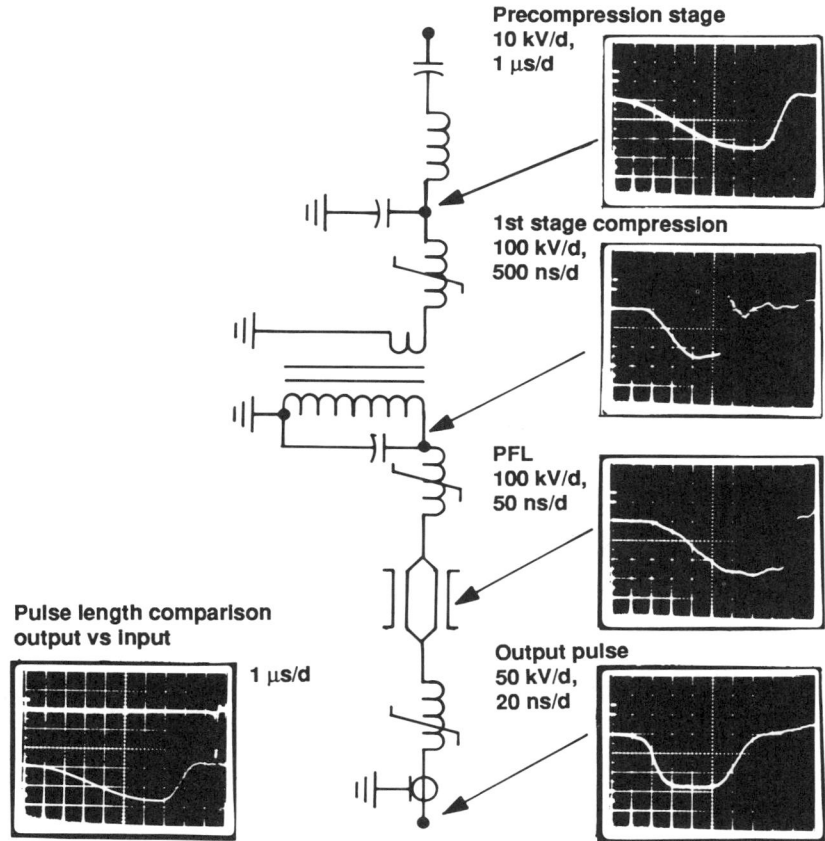

Figure 5. MAG-1-D voltage waveforms.

TeV. Accelerating gradients of 150–200 MV/m are desired in order to keep the accelerator length within acceptable limits. Frequencies of 11–17 GHz are desired in order to keep peak power requirements and beam loading reasonably small. The peak power necessary to drive a traveling wave structure in the desired frequency range with the desired gradient is of order 1 GW/m with a pulse length of 50–100 ns.

Pulsed beams of such high peak power can be obtained using the technologies of magnetic pulse compression and induction acceleration (see Fig. 7). Beam pulses of 1 kA current and 50–100 ns duration are routinely accelerated to several MeV at LLNL. These beams contain several gigawatts of peak power, as we have discussed earlier.

A. M. Sessler and S. S. Yu, following a suggestion by W. K. H. Panofsky, proposed a direct method for energy extraction from an electron beam by bunching a relativistic beam and passing it through extraction cavities. They suggested that if only part of the beam energy were extracted, the beam could be reaccelerated and energy

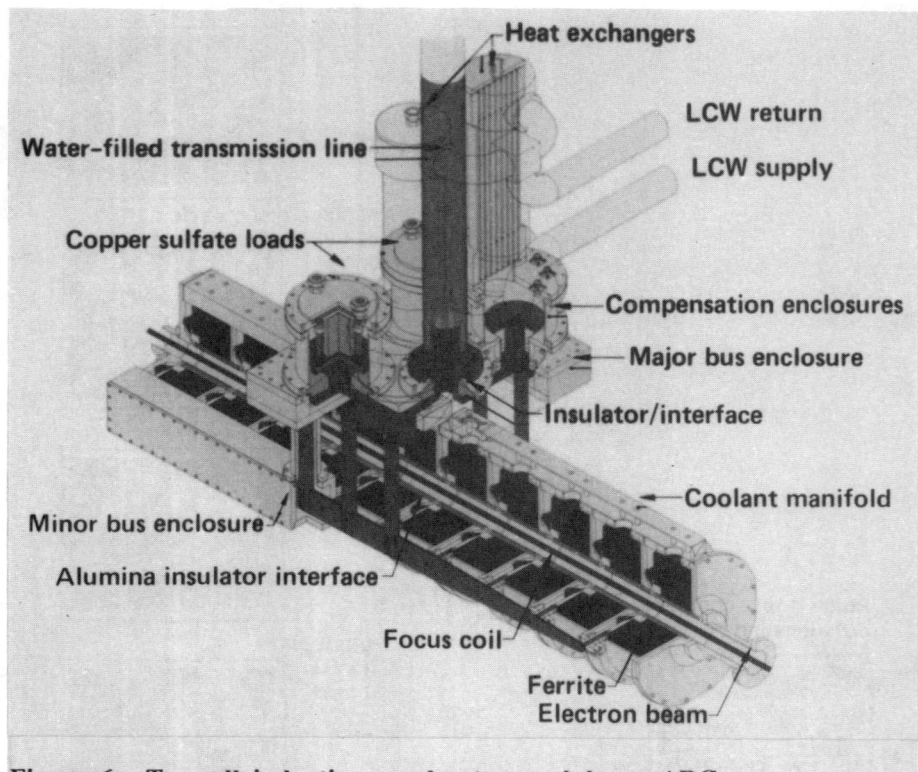

Figure 6. Ten-cell induction accelerator module on ARC.

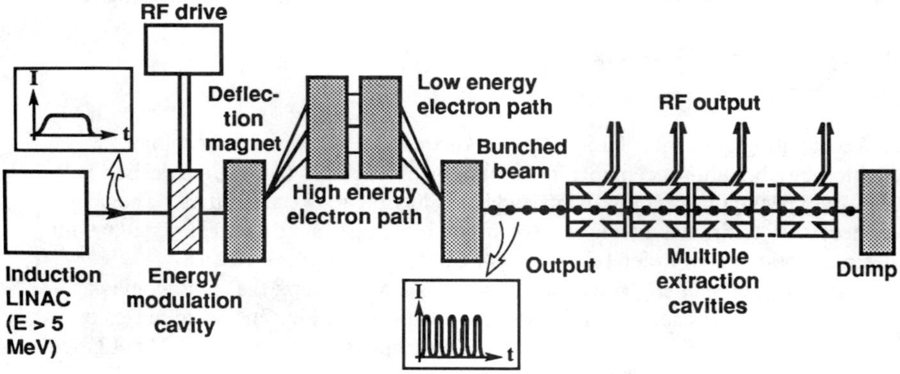

Figure 7. Schematic of a relativistic klystron.

again could be extracted. Repeated reacceleration and extraction was the concept they called a "relativistic klystron two-beam accelerator" (see Fig. 8).[2] The idea of a relativistic klystron, however, is not limited to the two-beam accelerator concept. Relativistic klystrons can be imagined which span the range from a 1-GW device powering 1 m of accelerator, to a 10-GW device powering 10 m, to a two-beam device extending several kilometers.

Our first experiments have been done at our ARC facility using as a gun an induction accelerator designed to produce 1-kA currents with 1.2-MeV kinetic energy for up to 75-ns duration. Three different klystrons have been tested with this injector. They have operated at 8.6 and 11.4 GHz.

In one experiment we achieved a peak output power of 200 MW at 11.4 GHz from one of our relativistic klystrons. We used a 930-kV, 420-A electron beam from the induction injector. The 200-MW peak power was delivered by this klystron to a 26-cm-long high gradient accelerator structure and corresponds to a longitudinal accelerating gradient of 140 MV/m. We are currently trying to understand the rf pulse shapes that were produced, as the rf flat top was much shorter than the beam current pulse, and this shape was not predicted by our numerical simulation. Experiments are continuing on different cavity geometries.

Our goal is to develop a high power (500 MW), short wavelength (2.6 cm) relativistic klystron driven by an electron beam from an induction accelerator at an energy greater than 1 MeV.

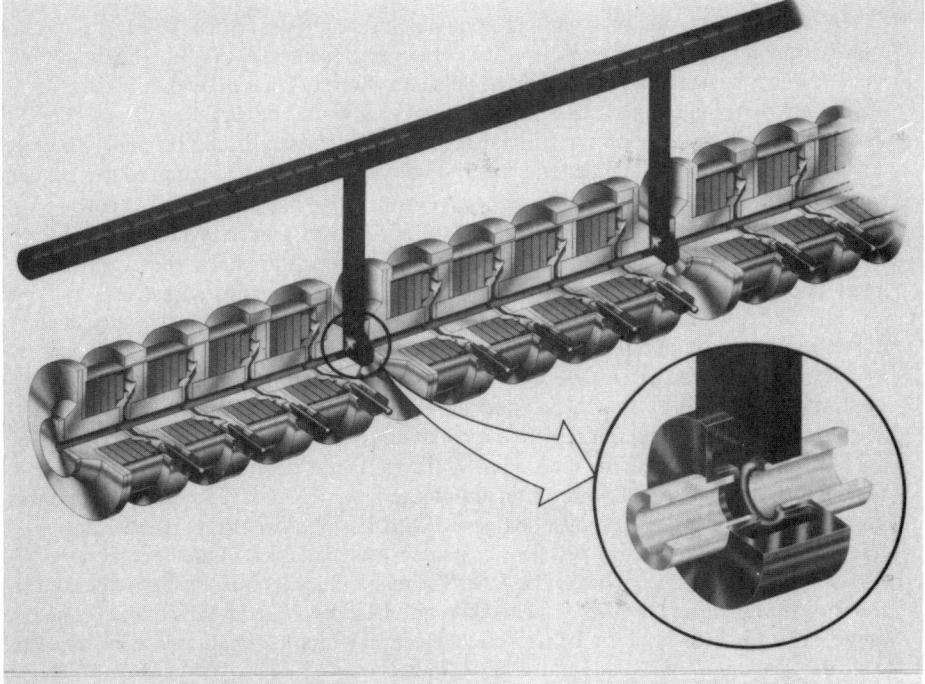

Figure 8. Relativistic klystron version of a two-beam accelerator.

PART B: OVERVIEW OF FEL DEVELOPMENT WITH INDUCTION LINACS AT LLNL

Introduction

Induction linacs are capable of producing multikiloampere peak currents. This capability has stimulated the investigation of single-pass free-electron laser (FEL) amplifiers with very high gain and high conversion efficiency. Experiments on the tapered wiggler concept for achieving high conversion efficiency met with dramatic success on the 35-GHz microwave experiments with the Electron Laser Facility (ELF) at LLNL.[3] Current efforts are focused on extending these results to much shorter wavelengths with the 50-MeV PALADIN experiments on our Advanced Test Accelerator (ATA).

An overview of the induction linac-based FEL master oscillator-power amplifier (MOPA) is given in the next section. The fundamental physics issues involved in the tapered wiggler operation are described in the following section, with particular emphasis on explaining the requirements of small electron-beam energy spread and high brightness for efficient energy conversion.

General Description of Induction-Linac Powered FEL Amplifiers

A schematic showing the various elements of a single-pass MOPA FEL configuration driven by an induction accelerator is presented in Fig. 9. The induction accelerator is basically a linear series of pulse transformers that individually give an increment of voltage to the electrons as they pass down the axis of the system. The pulse lengths of the acceleration voltage are generally chosen by compromising between the desire to minimize the magnetic material volume (volt-sec) and the need to maintain an adequate flat top on the acceleration waveform, as we have discussed in part A. These compromises result in pulse lengths of 50–70 ns in current systems. High average power is obtained by operating at high pulse-repetition rates; the magnetic modulator system is the technology that will enable us to meet this objective.

The electron-beam pulses from the accelerator are sent through the wiggler, then discarded in a beam dump. A drive laser (or microwave source) sends input pulses in synchronism with the driving electron-beam pulses. The front section of the wiggler has uniform properties, acting as a linear amplifier (a "preamp") for the input laser. The electromagnetic power increases exponentially with distance up to the point where the electrons are "trapped" by the wave—at this point the electron beam is strongly bunched in "pancakes," periodic at the input signal wavelength (Fig. 10). Beyond this point, the wiggler properties must be tapered with distance to extract a significant fraction of electron-beam energy (tens of percent), as explained in the next section.

Radiation from the "pancake bunches" of relativistic electrons is highly peaked in the forward direction because of the usual relativistic dipole radiation pattern (cone of angle $\sim 1/\gamma$) and also because the individual electron's dipole radiation is also summed over the "pancake" of radial width $d \gg \lambda$, thereby acting like a "forward-fire" phased-array antenna. The overall result is that the FEL gain pattern is highly peaked in the forward direction. Many of the usual limitations on the maximum gain per module do

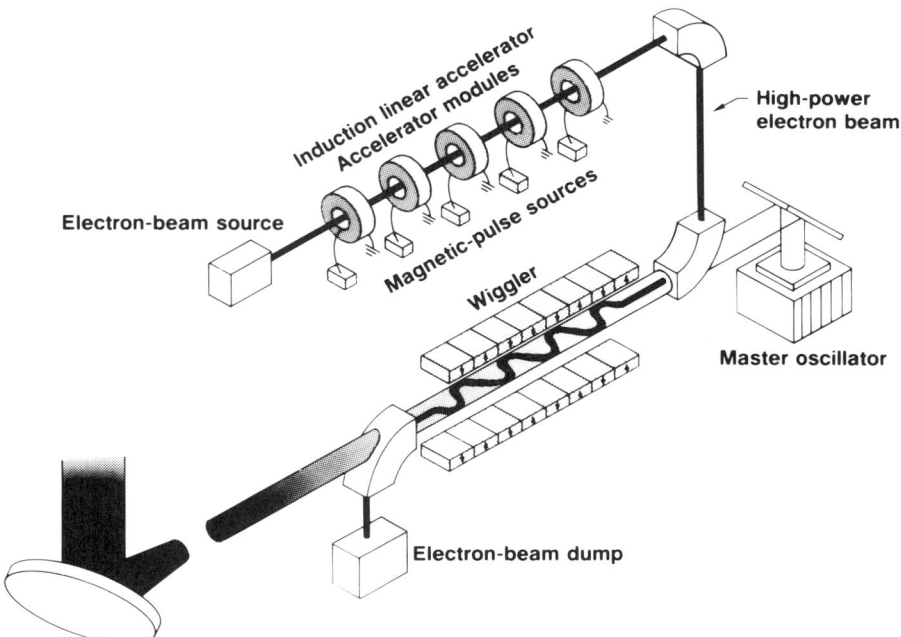

Figure 9. Component technologies for induction linac FEL amplifiers.

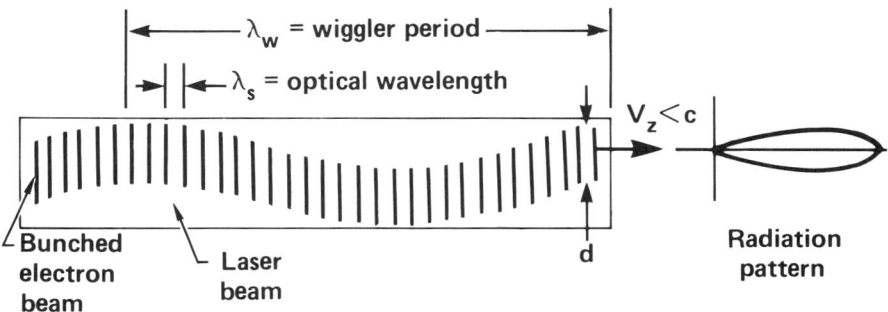

- Electron bunches "slip" one optical period per wiggler wavelength relative to light beam

- "Slices" of laser and electron beam $N\lambda_s$ long interact independently (N = number of wiggler periods)

- Radiation of bunched beam is strongly peaked in forward direction (like a "phased array" of radiating dipoles of width $d \gg \lambda_s$)

Figure 10. Schematic of FEL amplifier operation.

not apply to the FEL, because oscillations from "spurious modes" with transverse gain are absent. The pulsed nature of the driving electron beam also simplifies greatly the handling of reflections in FEL amplifiers with very high gain. The experimentally demonstrated capability for 80-dB small-signal gain in the millimeter wavelength experiment (ELF) without spurious oscillations is noteworthy in this context. (We should note that growth of sideband frequencies that couple to the axial oscillations of electrons trapped by the wave is one of the few spurious modes of potential significance in this laser system.)

The optical beam output from the FEL amplifier can be expanded to a low enough power density for beam transport and beam directors to handle, even with the very high average powers that can be achieved at high repetition rates. In microwave systems it is not necessary to have output windows (in applications such as plasma heating or driving rf accelerators) because sensitive components like the electron cathode are physically well isolated from the output waveguide.

The fundamental characteristics of induction-linac-driven amplifiers are well suited for applications requiring very high peak and/or average power outputs from a single aperture. Electron beams of high power can be generated quite efficiently, and the "gain medium" consisting of electrons, dipole wiggler fields, and a vacuum pipe presents very few limitations on power or power density in the gain region.

FEL Amplifier Physics

The proper operation of an FEL amplifier requires that one maintain a precise relationship between the longitudinal velocity of the electron beam and the wavelengths of both the electromagnetic radiation and the wiggler's magnetic field. This relation is given by

$$\gamma_{\parallel} = \left(\frac{\lambda_w}{2\lambda_s}\right)^{1/2} \pm \delta , \qquad (5)$$

where λ_s is the wavelength of the electromagnetic wave to be amplified, λ_w is the period of the wiggler, and $\gamma_{\parallel}$ is defined by

$$\gamma_{\parallel}^2 = \frac{1}{1 - v_{\parallel}^2/c^2} , \qquad (6)$$

with $v_{\parallel}$ the longitudinal (axial) velocity of the electrons. The quantity δ represents the accuracy to which the electron's longitudinal energy must be held if a significant transfer of energy from the electron to the electromagnetic wave is to take place. Limits on δ depend on both the intensity and the wavelength of the light to be amplified, and these limits can range from a few percent at microwave wavelengths to a few tens of percent at optical wavelengths.

If an amplifier is to be efficient, a very high quality electron beam is required so that all electrons within the beam satisfy Eq. (5). This not only implies careful control of the beam energy, but, in addition, mandates careful control of the beam's emittance

(ε_n). One can estimate the emittance requirement by relating the emittance to the dispersion in axial velocities, which can be written as

$$\frac{\Delta\gamma_\parallel}{\gamma_\parallel} = \frac{b_w \varepsilon_n}{4(1 + a_w^2)}, \tag{7}$$

where $b_w = eBmc$, with B the peak wiggler magnetic field, and $a_w = \frac{b_w}{\sqrt{2}k_w}$, the rms dimensionless vector potential with k_w the wiggler wave number.

Once the electron beam requirements are satisfied, one is left with the problem of extracting large amounts of energy from the electron beam (several tens of percent) under the constraints of Eq. (5). Obviously if the longitudinal velocity, $v_\parallel$, changes substantially as a result of electron beam energy loss, according to Eq. (5), the interaction will cease, and the amplifier will saturate before significant energy extraction has occurred. A method for circumventing this limitation was first described by Kroll, Morton, and Rosenbluth[4]; it is called the tapered wiggler.

The tapered wiggler concept relies on having the wiggler properties vary as the electrons slow down, so that the resonance condition can be preserved while extracting large amounts of energy. The simple realization of this concept would be to decrease λ_w proportional to $\gamma_\parallel^2$, but operationally it is difficult to build wigglers with the requisite short periods at the downstream end. If one recognizes that

$$\gamma_\parallel = \frac{\gamma}{\left[1 + 1/2(b_w/k_w)^2\right]^{1/2}}, \tag{8}$$

however, one can see that the electron's total energy (γ) can be reduced without altering its parallel energy (thus maintaining resonance) by simultaneously reducing the wiggler's magnetic field, b_w. This is the approach taken in the design of our wigglers at LLNL.

ELF, a microwave FEL, was specifically designed to test the tapered wiggler concept. The ELF experiments showed that untapered efficiencies of 5–6% could be increased to 40% by appropriately tapering the profile of $b_w(z)$, in good agreement with the modeling. Also ELF has shown high exponential gain (in the small-signal regime) in accordance with theoretical predictions. Utilizing tapered wigglers, these experiments demonstrated high gain (greater than 42 dB) even when operating at high power levels, i.e., one can make the transition from small signal gain to large signal gain in a single device.

ELF is unable to test certain aspects critical to the operation of an optical FEL. First, one must change the method of transporting or focusing the electron beam in the wiggler. Quadrupoles (as used on ELF) would cause a loss of efficiency. Curved pole pieces are essential for operation of a linearly polarized wiggler and are incorporated on the 10.6-μm wavelength PALADIN experiment at ATA.

Second, ELF is a microwave amplifier operated in a waveguide; therefore, it is unable to address the questions of optical guiding and optical mode control that are being addressed on the PALADIN experiments.

In summary, ELF has demonstrated the validity of the tapered wiggler concept. The PALADIN experiments currently under way are intended to show that those concepts are also valid at optical wavelengths.

High Brightness and High Average Power

The FEL application using linear induction accelerators places a very high premium on electron beam brightness and on high-average-power operation. A discussion of both these topics has been given in part A, "Overview of Linear Induction Accelerators."

FEL Experiments at LLNL

Our first FEL experiment, completed about two years ago, involved a 4-MeV, high power, pulsed electron beam from our ETA accelerator. The master oscillator was a 35-GHz microwave source coupled to the electron beam at the entrance to the wiggler. Figure 11 shows the 3-m wiggler with the waveguide and external quadrupole magnet coils for guiding the electron beam. Some results from this experiment are given in Fig. 12. The main conclusions from these experiments were that the feasibility of the basic FEL concept described above was confirmed, and that our theoretical models were accurately predicting performance of these high-gain amplifiers. Further discussions of results from these experiments are given in Ref. 3.

The extension of FEL amplifier operation to the optical regimes requires higher voltage accelerators, and wigglers of many tens of meters in length. In addition, higher quality electron beams must be generated and transported to the wiggler. In general, all of the technologies became stressed as shorter wavelengths were approached.

Figure 11. The ELF 3-m wiggler.

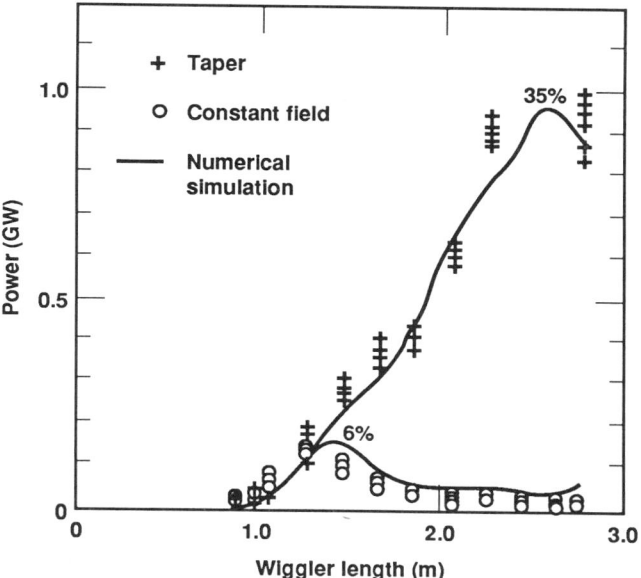

Figure 12. Results from the ELF experiment demonstrate the importance of a tapered wiggler.

The PALADIN experiment, now installed at our Advanced Test Accelerator (ATA), was designed to examine FEL physics at 10.6 μm in a long (25-m) wiggler (see Figs. 13 and 14). This experiment tests the performance of a single-pass wiggler in the regimes of both small and large signal gain. It will give opportunities to validate theoretical models of harmonic generation and will allow examination of the stability of the FEL amplifier against parasitic mechanisms such as sideband instability. We have measured the small signal gain with the first 5 m of the wiggler. At 15 m we have measured the gain of the amplifier and have demonstrated gain guiding. With a 14-kW input signal, and a 45-MeV electron beam, we have measured a gain of 27 dB (a factor of 500 between the input and output radiation signals). We have also observed gain guiding of the radiation, where diffraction losses are masked by the high gain of the FEL amplifier. The radiation signal in this case takes on the shape of the gain medium (electron beam). The signal is not saturated in this experiment; however, we have demonstrated saturation in the amplifier by increasing the input signal to 5 MW. The amplifier then saturates at roughly halfway through the wiggler. Numerical calculations have been successful in modeling these results. Currently we are experimenting with the full 25-m wiggler with the goal of measuring high gain extraction, optical guiding, and the spatial mode purity of the laser beam. Comparison with theoretical predictions is a key part of the experiment.

We are also planning a microwave FEL experiment on our new ETA-II accelerator, which will ultimately lead to heating of plasma in a tokamak fusion machine at LLNL.

Figure 13. Five-meter section of the PALADIN wiggler.

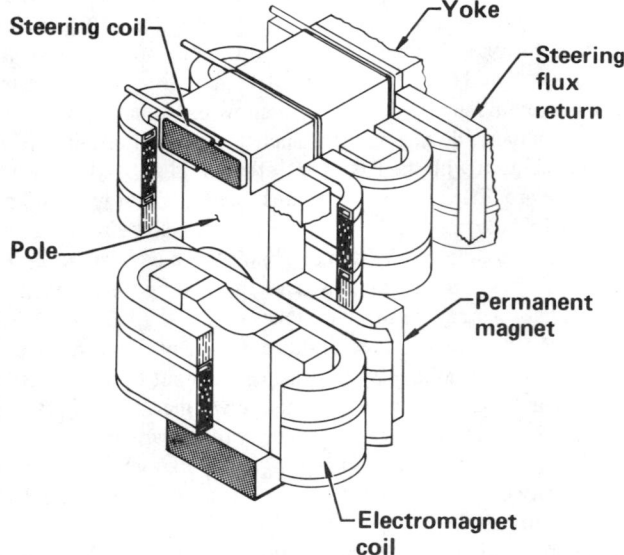

Figure 14. Cutaway of PALADIN wiggler showing poles, main electromagnetic windings, steering coils, and permanent magnet additions to the main pole pieces.

FEL Technology and Fusion Power

The desire to heat plasmas with high power microwaves at millimeter and submillimeter wavelengths may be fulfilled by the free electron laser. The FEL appears ideally suited for this application because it is capable of producing extremely high power and because it has a virtually limitless frequency range. Plasma heating is accomplished by tuning the FEL output frequency to a resonant frequency (cyclotron or twice cyclotron) of the plasma electrons. In a dense plasma, the resonant electrons will absorb the energy of the microwave field and rapidly transfer it to the other plasma particles. Besides having important implications for fusion-power technology, the application of the FEL to electron-cyclotron heating (ECH) may have an impact in other areas of plasma-physics research. The potential for using microwaves to heat plasmas has long been known but has been thwarted in scaling to higher magnetic fields and higher powers by the technology limits of the microwave-generating (gyrotron) tubes. As the frequency and power-per-unit requirements have grown, the tubes designed to heat plasma have become increasingly difficult and expensive to build. The advantage of FELs is that the high power per unit is an inherent feature of their design.

Experiments will be carried out at LLNL to demonstrate that FEL-generated microwaves can be used for both electron heating and current drive in tokamak plasmas. These experiments will address two issues simultaneously: determining how pulsing of the high-intensity field affects power absorption, and understanding the physics of the plasma response to high-power microwave heating.

For our planned experiments, we will use the new ETA II facility at LLNL and the Alcator-C tokamak, recently moved to LLNL from the Massachusetts Institute of Technology. The entire endeavor is called the Microwave Tokamak Experiment (MTX). We have begun the tokamak checkout and will begin microwave system checkout during early 1989. Heating experiments will follow.

A sketch of the facility (Fig. 15) shows the ECH experimental area containing the MTX and the ETA II, which contains the high-energy (7- to 10-MeV) electron-beam accelerator and the FEL wiggler. The electrical and electronic support equipment is not shown.

The MTX will be located just beyond the concrete shield that houses the ETA II. The output end of the FEL will then be situated approximately 19 m from the MTX. Power from the FEL will be transmitted to the tokamak through a quasi-optical, microwave transmission system, which would consist of a network of 0.6-m-diam ducts enclosing four mirrors that focus and direct the FEL microwave beam. The beam will be focused to enter the tokamak vacuum vessel through a narrow (4-cm) port.

Success in these experiments will demonstrate the usefulness of the FEL technology for driving and sustaining reactor plasmas. Good results will also provide a physics base for assessing the FEL's utility in devices for generating fusion power. If this technology fulfills its potential for driving current, controlling instabilities, and improving plasma confinement, ECH could become the method of choice for heating plasmas to ignition and for sustaining steady state plasma current in tokamaks.

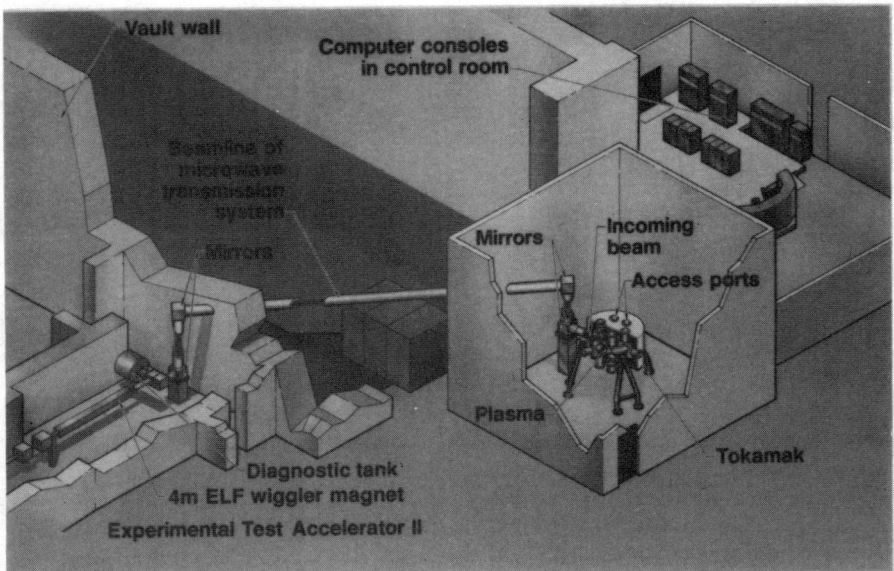

Figure 15. Sketch of the facility for the Microwave Tokamak Experiment (MTX). On the right is the tokamak machine in which the electron-cyclotron heating experiments will be carried out. On the left is the ETA II, containing the electron-beam accelerator and FEL wiggler.

FEL Technology and High Power Accelerators

A final application I wish to discuss is the application of FEL and induction accelerator technology to high power accelerators for high energy physics experiments.

As an alternative approach to the necessary high acceleration gradient via the relativistic klystron, discussed in Part A, we are proceeding with the concept of the two-beam accelerator (TBA), originally proposed by A.M. Sessler in 1982.[5] In concept, the TBA employs a main high-energy-beam linear accelerator driven by a FEL which runs parallel to it and serves as the power source for acceleration. The main accelerator, like that used at the Stanford Linear Accelerator Center (SLAC), is a disk-loaded waveguide powered by microwaves. The innovative feature of the TBA is that it uses an FEL, rather than klystrons or gyrotrons, as the source of the microwaves. With a TBA, power would be tapped off periodically along the FEL wiggler and fed across to the main accelerator. The FEL is designed so that its microwave power increase per unit length is equal to the average power extracted per unit length.

The advantage of employing the FEL—besides its relative simplicity—is its unique, inherent ability to generate economically very high power at very high frequencies. The simplicity of the FEL precludes the need for the thousands of individual microwave generators called for in conventional accelerator designs.

Although the advantage of operating at higher microwave frequencies has long been recognized, there have been no suitable high-power sources at about 1-cm wavelengths until the recent development of gyrotrons and FELs.

Gyrotrons are still being developed and may eventually prove to be a practical power source for some linear accelerators. However, their maximum power output at 1-cm wavelength seems likely to remain below 200 MW. Consequently, a 1-TeV machine would require thousands of such power sources, all properly locked in phase. Obviously, such a device would be impractical.

Our current theoretical and experimental work on the TBA is based on the FEL's ability to supply in excess of 100 MW of average power at 1-cm wavelength. The TBA (Fig. 16) consists of a high-gradient, electron-beam-accelerator structure (HGS) periodically coupled to an FEL as a source of high-power microwave energy.

It is interesting that there is an inverse relationship between the basic functional concepts of the FEL and of the HGS: whereas the microwave field of the FEL wiggler obtains its drive power from an electron beam, the electron beam of the HGS obtains its drive power from a microwave field. This relationship has been likened to the operating principle of the transformer, wherein the microwave field is analogous to the transformer's magnetic coupling field.

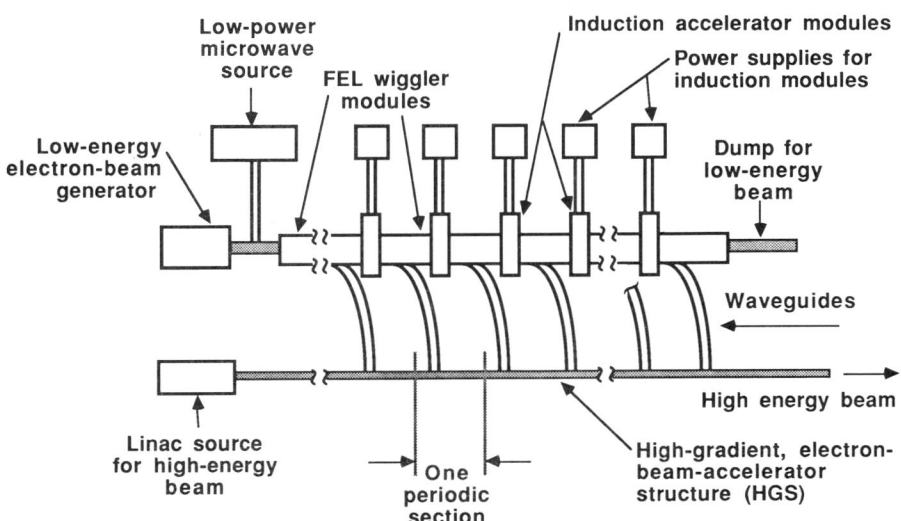

Figure 16. Diagram of the TBA concept. The TBA consists of a high-gradient, electron-beam accelerator structure (HGS) with a FEL wiggler running parallel to it and serving as its microwave-accelerating power source.

Although the FEL electron beam loses energy in the process of generating the high-power microwave field, the energy can be replenished by induction accelerator units placed periodically along the length of the FEL wiggler. The exact arrangement of the FEL driver to the high gradient accelerator, whether continuous along its length or modular, is still being evaluated.

Future research on the application of FELs in accelerators will be three pronged. First, we must optimize the design for the FEL's beam reacceleration cavity. Its overall beamline insertion length must probably be held to a few centimeters to avoid seriously degrading the TBA's high average acceleration gradient. Moreover, the microwave power loss in crossing the reacceleration gap should only be a few percent. Initial measurements of gap loss indicate that special focusing or guiding will be required to keep power loss down at an acceptable level.

Second, we must improve the luminosity of the beam. The desired high-energy beam luminosity is difficult to achieve when accelerating single electron bunches. The situation is mitigated by the acceleration of bursts of multiple bunches. This mode of operation has yet to be fully analyzed and optimized.

Third, we must address the issues of phase stability and control, perhaps the largest outstanding TBA design challenge. Analytical studies of the sensitivity of microwave phase to errors in operating parameters are proceeding at Lawrence Berkeley Laboratory. We require a phase-stabilizing scheme that is automatic and nearly instantaneous in response. We are continuing the search for a practical solution.

Summary

The current state of high-repetition-rate induction machine technology and of tapered wigglers is sufficient for near-term, high-average-power applications in the microwave and millimeter wavelength regimes. The additional physics demonstrations and developments needed for scaling the technology to shorter wavelengths is verification of the tapered wiggler operation in the optical wavelength regime, where "guiding" effects of the radiation are predicted to occur, and the generation and acceleration of very-high-brightness electron beams at multikiloampere current levels. Success in these two developments should open up a wide range of applications of this high-power FEL technology from millimeter to visible wavelengths.

ACKNOWLEGMENT

The author would like to thank Samuel F. Eccles of the Beam Research Program, LLNL, for assembling and editing the material for this article.

REFERENCES

1. Allen, M. A., et al., "Relativistic Klystron Research and Development," European Particle Accelerator Conference, Rome, Italy, June 7–11, 1988.
2. Sessler, A. M. and Yu, S. S., "Relativistic Klystron Two-Beam Accelerator," Phys. Rev. Lett. 58, 2439 (1987).
3. Orzechowski, T. J., et al., "High Efficiency Extraction of Microwave Radiation from a Tapered Wiggler Free Electron Laser," Phys. Rev. Lett. 57, 17 (1986).

4. Kroll, N. M., Morton, P. L., and Rosenbluth, M. N., "Free-Electron Lasers With Variable Parameter Wigglers," IEEE J. Quantum Electron. QE-17 (1981); Prosnitz, D., Szoke, A., and Neil, V. K., "High-Gain, Free-Electron Laser Amplifiers: Design Considerations and Simulation," Phys. Rev. A $\underline{24}$, 3 (1981).
5. Sessler, A. M., "The Free Electron Laser as a Power Source for a High-Gradient Accelerating Structure," in Laser Acceleration of Particles (AIP Conf. Proc. No. 91, American Institute of Physics, New York, 1982), pp. 154-159.

PROGRESS ON NEXT GENERATION LINEAR COLLIDERS[*]

RONALD D. RUTH

Stanford Linear Accelerator Center
Stanford University, Stanford, CA 94305

TABLE OF CONTENTS

1. Introduction 2209
2. Parameters 2212
3. Damping Rings 2214
4. Bunch Compression and Pre-Acceleration 2216
5. Linac 2217
   5.1 Structures 2217
   5.2 RF Power Sources 2218
      5.2.1 RF Pulse Compression 2218
      5.2.2 The Relativistic Klystron 2218
   5.3 Emittance Preservation 2220
      5.3.1 Chromatic Effects 2220
      5.3.2 Transverse Wakefields and BNS Damping 2221
      5.3.3 Jitter 2221
      5.3.4 Coupling 2222
6. Final Focus 2222
   6.1 Final Focus Optics and Tolerances 2222
   6.2 Beam-Beam Effects 2223
7. Multibunch Effects 2224
8. Outlook 2224
9. References 2225

---

[*] Work supported by the Department of Energy, contract DE–AC03–76SF00515.

PROGRESS ON NEXT GENERATION LINEAR COLLIDERS

RONALD D. RUTH

Stanford Linear Accelerator Center
Stanford University, Stanford, California 94309

1. INTRODUCTION

The purpose of this paper is to review progress in the U.S. towards a next generation linear collider. During 1988, there were three workshops held on linear colliders: 1.) "Physics of Linear Colliders," in Capri, Italy, June 14–18, 1988; 2.) Snowmass 88 (Linear Collider subsection) June 27–July 15, 1988; and 3.) SLAC International Workshop on Next Generation Linear Colliders, Nov. 28–Dec. 9, 1988. To obtain detailed current information, the reader is directed to Refs. 1-3 which are the proceedings of each of the workshops. In addition, the Snowmass proceedings for the linear collider working group are collected in Ref. 4. This paper will concentrate on U.S. efforts and will draw heavily from Refs. 3 and 4.

There is also much work ongoing in other parts of the world. The Soviet Union is planning a linear collider at Serpukov which is being designed at Novosibirsk. CERN is working on CLIC (CERN Linear Collider). Finally, KEK is actively engaged in linear collider research towards a JLC (Japanese Linear Collider). Much of this work is covered in Refs. 1 and 3.

In this paper, I focus on reviewing the issues and progress on a next generation linear collider with the general parameters shown in Table 1. The energy range is dictated by physics with a mass reach well beyond LEP, although somewhat short of SSC. The luminosity is that required to obtain $10^3 - 10^4$ units of R_0 per year. The length is consistent with a site on Stanford land with collisions occurring on the SLAC site. The power was determined by economic considerations. Finally, the technology was limited by the desire to have a next generation linear collider before the next century.

Table 1. General parameters.

Energy	0.5 – 1.0 TeV in center-of-mass.
Luminosity	$10^{33} - 10^{34}$ cm^{-2} sec^{-1},
Length	Each Linac $\lesssim 3$ km.
Power	$\lesssim 100$ MW per Linac.
Technology	Must be realizable by 1990–92.

The basic configuration of such a linear collider is shown in Fig. 1. The beam is accelerated by an injector linac and then injected into a damping ring which damps the emittance of the beam and provides the beam with appropriate intensity and repetition rate. After extraction, the bunch must be compressed in length twice in order to achieve the short bunches suitable for the linac and final focus. The linac is used to accelerate the beams to high energy while maintaining the emittance. Finally, the final focus is used to focus the beams to a small spot for collision. This must yield a luminosity with tolerable beam-beam effects (disruption and beamstrahlung) and must also provide a reasonably background-free environment for the detector.

Before proceeding to a detailed discussion of the linear collider subsystem by subsystem, it is useful to discuss generally the overall results of the past year's activities. Perhaps one of the most important developments is the increased interest in an Intermediate Linear Collider (ILC) with an energy of 0.5 TeV in the center-of-mass. This is a factor of two below the TeV Linear Collider (TLC) and thus would require a factor of four less peak power provided that the machines were the same length. One can imagine designing an ILC which would be upgradable in energy by the addition of RF power and minor modifications to the final focus system.

If we begin the discussion of an ILC or TLC at the lower energy end, the damping ring and bunch compressor designs seem relatively straightforward with, however, somewhat tighter tolerances than usual. The main linac will probably have a structure similar to SLAC, except at 4–6 times the frequency. The irises will have slots coupled to radial waveguides to damp the transverse and longitudinal higher-order modes. This makes possible the use of multiple bunches per RF fill, which increases the luminosity by a factor of 10 for "free."

There is no definite power source as yet. The recent demonstration of binary pulse compression at SLAC has focused attention on more conventional approaches to long-pulse power production. Low power, low loss tests of RF pulse compression are continuing at SLAC and initial results look very promising. There are plans to build a high power klystron at SLAC to feed the RF pulse compressor, and there are many new ideas for power sources which would drive RF pulse compressors. The relativistic klystron results have been somewhat discouraging, but much as been learned about the problems associated with these high current, high energy beams.

Once the power source problem is solved, we are still left with the luminosity problem. These two aspects are only partially decoupled due to the use of many bunches (a batch) per RF fill. To obtain the luminosity, we must preserve the emittance of the beam throughout the linac. This means tighter tolerances on vertical magnet alignment than are presently achieved. The final focus demagnifies the beam to obtain a very flat beam at the final focus. The chromatic correction

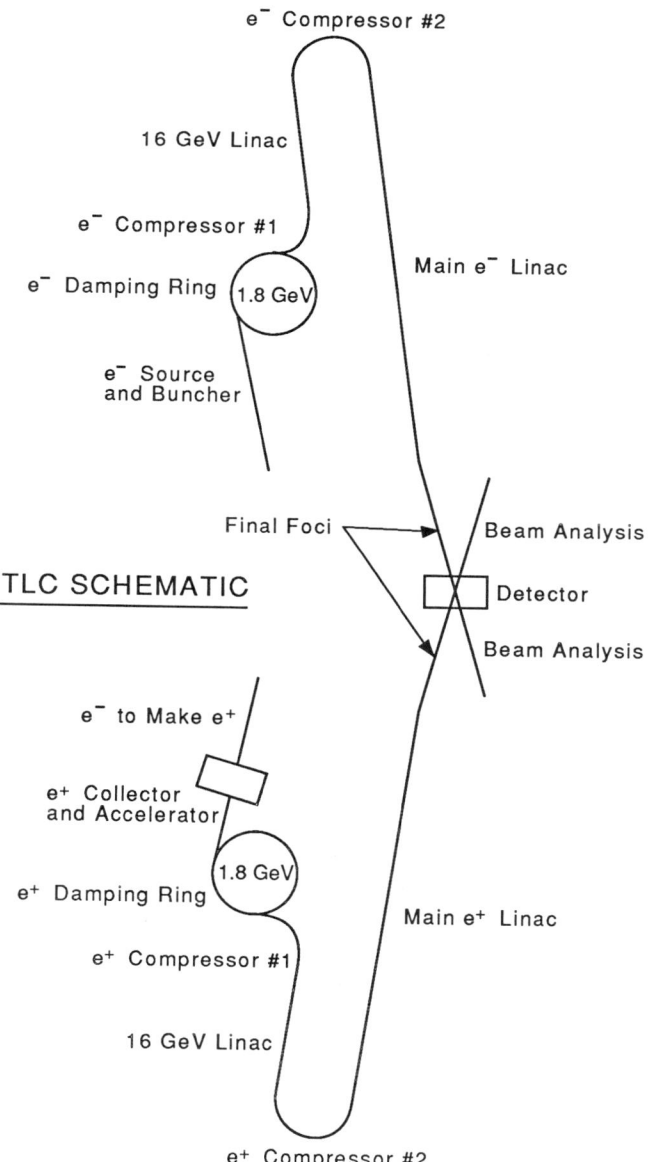

Fig. 1. Schematic layout of the TLC. The angles shown are exaggerated.

for this is quite delicate, and tolerances are tight. Finally, we must measure the beam size at the interaction point in order to tune the final focus. Many of these problems can be addressed via a model final focus at a lower energy. Towards this end, there is presently work ongoing at SLAC to create a Final Focus Test Beam in order to test flat beam final focus optics, measurement techniques, alignment techniques, etc. This would use the 50-GeV SLC beam straight ahead into the old C-line at SLAC.

During the SLAC Workshop in December 1988 following Snowmass, there was one important discovery which should be emphasized here. Beamstrahlung photons create e^+e^- pairs upon interacting with the opposing bunch. One particle of the pair is deflected strongly by the field of the bunch. This, in turn, can cause serious background problems. This will be discussed more thoroughly in the later sections of this paper.

In the next sections, I first discuss parameters briefly and then discuss damping rings. The basic principles of bunch compression are treated in the next section. In the section on the linac, there are three subsections. First I discuss RF structures and power sources, and then I move to a discussion of emittance preservation in the linac. This is followed by a discussion of the final focus and beam-beam effects. Finally, I introduce some of the issues for multibunch effects.

2. PARAMETERS

Linear colliders are being considered for accelerators ranging from B-factories to Z-factories, up to a TeV Linear Collider. In Ref. 5, R. Palmer explores the change in the design of linear colliders as a function of energy given that one is always trying to maximize luminosity, but always respecting the limit on wall plug power shown in Table 1. A very wide range of energies is considered, and this leads to widely differing designs. In particular, one sees that the optimized RF frequency tends to decrease at lower energy while the repetition rate increases.

In addition, due to the interest in the ILC, Palmer considers two possible options for an ILC, both of which would be upgradable to a TLC with additional length/power sources. Perhaps the most attractive option is the low gradient ILC which has a physical layout identical to TLC but has one-half the acceleration gradient. The parameters for ILC and TLC are compared in Table 2.[5]

There is also an addendum to Ref. 5 which discusses the problem of e^+e^- pair creation at the interaction point by beamstrahlung photons interacting with the oncoming bunch. Palmer finds that by using his idea of 'crab crossing' it is possible to collide beams with a very large crossing angle. In this way, with the help of solenoidal guide fields, the deflected e^+ or e^- can exit through a large aperture hole adjacent to the incoming quadrupole. This means that the parameter sets

Table 2. Parameters for TLC and ILC.

		Low grad ILC	High grad ILC	TLC
General				
CM energy	TeV	.5	.5	1
luminosity 10^{33}	$cm^{-2} sec^{-1}$	1.5	2.9	6.2
RF wavelength	cm	1.75	1.75	1.75
repetition rate	kHz	.36	.36	.36
accel gradient	MV/m	93	186	186
number bunches		10	10	10
particles/bunch	10^{10}	.7	1.4	1.4
wall power	MW	52	103	210
length	Km	7.3	3.7	7.3
Damping Ring				
emittance ϵ_x/ϵ_y		100	100	100
emittance $\gamma\epsilon_x$	μm	3.5	6.0	7.0
emittance $\gamma\epsilon_z$	m	.04	.04	.04
bunch spacing	m	.2	.2	.2
RF				
pulse length	ns	60	60	60
peak power/length	MW/m	146	580	580
total RF energy	KJ	51	103	210
Linac				
loading η	%	2.5	2.5	2.5
iris radius a	mm	3.5	3.5	3.5
section length	m	1.6	1.6	1.6
Linac tolerances				
alignment	μm	20	35	30
vibration	μm	.009	.017	.012
Final focus				
β_y^*	mm	.1	.12	.11
crossing angle	mrad	4.2	6.1	3.8
free length	m	.36	.43	.7
Intersection				
σ_y	nm	2.7	3.9	2.8
σ_x/σ_y		132	132	132
σ_z	μm	70	70	70
disruption D		5	5	5
lum enhance H		1.6	1.6	1.6
beamstrahlung δ	%	2	4	11
$\Delta p/p$ physics	%	.7	1.1	3.2

shown here will have to be modified somewhat to include various changes, but the basic parameters still will be rather similar to those given in Table 2.

3. DAMPING RINGS

In Ref. 6, T. Raubenheimer *et al.* discuss many of the basic design considerations for the damping ring. The basic parameters of the TLC damping ring are shown in Table 3 where they are compared to those of the SLC. The key differences are the decrease of the horizontal emittance by an order of magnitude, the increase of the repetition rate, and the requirement of $\epsilon_x/\epsilon_y = 100$. Although asymmetrical emittances have been measured in the SLC damping ring, they are not required for SLC operation.

The desired repetition rate is obtained by having many batches of bunches in the ring. Each batch of 10 bunches is extracted on one kicker pulse and accelerated on one RF fill in the linac. The remaining batches are left in the ring to continue damping while an additional batch is injected to replace the extracted one. The threshold current refers to the threshold for the "microwave instability" or "turbulent bunch lengthening."[7]

The basic layout of a possible damping ring is shown in Fig. 2. Notice that there are several insertions which contain wigglers. In order to obtain the high repetition rate, it is necessary to decrease the damping time by the addition of wigglers in straight sections.

Table 3. Basic parameters of the SLC and TLC damping rings.

	TLC	SLC
Energy	1 ~ 2 GeV	1.15 Gev
Emittance, $\gamma\epsilon_x$	3.0 μmrad	36 μmrad
Emittance, $\gamma\epsilon_y$	30 nmrad	500 nmrad
Repetition rate	360 Hz	180 Hz
Bunch length	4 mm	5 mm[7]
Threshold Current	batches of 10 bunches of 2×10^{10}	1.5×10^{10} [7]

In Tables 4 and 5, you see the basic parameters for the ring. The lattice is combined function which allows the partition of the damping times to trade horizontal damping time for longitudinal. The RF frequency for this example is necessarily 1.4 GHz since the bunch spacing in this example is about 20 cm. The threshold impedance $(Z/n)_t$ is that for the microwave instability. It is quite small due to the small momentum compaction factor, but is only about a factor of three below that obtained in the SLC damping rings.

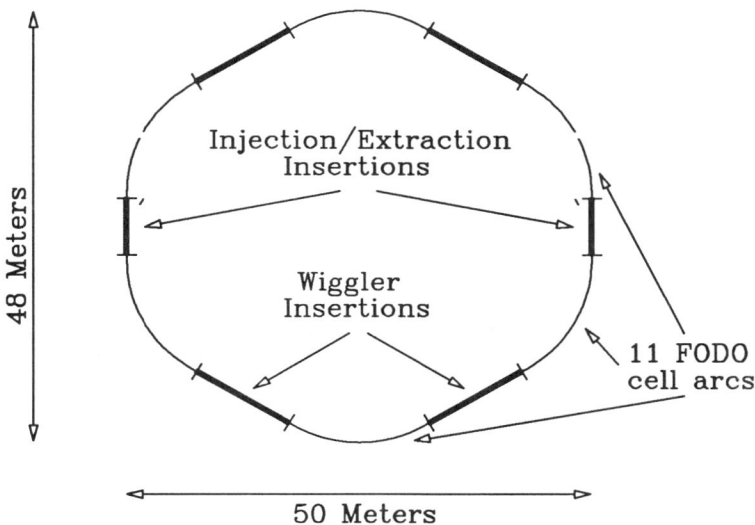

Fig. 2. Schematic of the TLC damping ring

Table 4. TLC damping ring parameters.

Energy	$E_0 = 1.8$ GeV
Length	$L = 155.1$ meters
Momentum compaction	$\alpha = 0.00120$
Tunes	$\nu_x = 24.37, \quad \nu_y = 11.27$
RF frequency	$f_{RF} = 1.4$ GHz
Current	10 batches of 10 bunches of $2 \times 10^{10} e^+/e^-$

Another key aspect of the TLC design is the small vertical emittance. The design calls for an emittance ratio $\epsilon_x/\epsilon_y = 100$. This size emittance ratio is quite common in $e^{\pm}$ storage rings. However, the tolerances for obtaining such a small vertical beam size are proportional to this size. In Ref. 6, those tolerances which are related to maintaining the emittance ratio are calculated. The tolerances presented in Sec. 5 of the paper are in the 100-μm range and could be improved by adding correction skew quadrupoles in the ring.

Table 5. TLC damping ring parameters.

	Wigglers Off	Wigglers On
Natural $\gamma\epsilon_x$	2.46 μmrad	2.00 μmrad
$\gamma\epsilon_x$ w/ intrabeam	3.33 μmrad	2.74 μmrad
Damping, τ_x	3.88 ms	2.50 ms
Damping, τ_y	9.19 ms	3.98 ms
Rep. rate, f_{rep}	155 Hz	360 Hz
Damp. partition, J_x	2.37	1.59
Energy spread, σ_ϵ	0.00128	0.00104
Radiation/turn, U_0	203 keV	468 keV
Bunch length, σ_z	5.6 mm	5.2 mm
Synch. tune, ν_s	0.0068	0.0058
$(Z/n)_t$	$\mathcal{F} \times 0.32\Omega$	$\mathcal{F} \times 0.20\Omega$
Natural chrom., ξ_x	-28.35	-28.07
Natural chrom., ξ_y	-25.10	-22.27

4. BUNCH COMPRESSION AND PRE-ACCELERATION

In order to obtain the very short bunches necessary for the linac, it is necessary to perform at least two bunch compressions after the damping ring. Designs for bunch compression are presented in Ref. 8. A bunch length of about 50 μm in the linac puts a tight constraint on the longitudinal emittance of the damping ring. In addition, during the bunch compressions, it is necessary to keep the energy spread small to avoid the dilution of the transverse emittance. If we assume that we can transport 1% energy spread without diluting either transverse emittance, then at least two bunch compressions are needed. For example, if we consider a 1.8-GeV damping ring with energy spread $\Delta E/E = 10^{-3}$ and a bunch length of 5 mm, the two compressions are shown in Table 6. The first one decreases the bunch length by an order of magnitude. This is followed by a pre-acceleration section to decrease the relative energy spread in the beam by an order of magnitude. One must avoid an increase of energy spread due to the cosine of the RF wave (and also due to beam loading). If this pre-acceleration is done at the present SLAC frequency and if the bunch current is as shown in Table 2, then the additional energy spread induced is about 5×10^{-4}. Neglecting this small increase, the next bunch compression happens at 18 GeV and serves to reduce the bunch length to about 50 μm. This is suitable for injection into the high frequency, high gradient structure.

The two designs shown in Ref. 8 are for bunch compressors which have small bending angles. However, 180° bends which do the same job have also been designed.

Table 6. Bunch compression.

E	$\Delta E/E$	σ_z	Compress $\rightarrow$	$\Delta E/E$	σ_z
1.8 GeV	10^{-3}	5 mm	Compress $\rightarrow$	10^{-2}	0.5 mm
[pre-acceleration at long wavelength, $\lambda = 10.5$ cm]					
18 GeV	10^{-3}	0.5 mm	Compress $\rightarrow$	10^{-2}	50 μm

5. LINAC

The linac is envisioned to be similar to the SLAC disk-loaded structure with a frequency at least four times the present SLAC frequency. The example shown in Table 2 is for six times the present SLAC frequency. The irises in the design are relatively larger to reduce transverse wakefields. The structure may have other modifications to damp long-range transverse wakefields. This would be driven by a power source capable of about 600 MW/m for the TLC or about 150 MW/m in the case of the ILC. In the case of the low gradient ILC, one can imagine an upgrade consisting of the addition of power sources. This section is divided into three subsections. In the first subsection we discuss structures, the second deals with RF power sources, and finally the third treats emittance preservation in the linac.

5.1 STRUCTURES

Since the gradients range from 100 MV/m to 200 MV/m, the first question that arises is RF breakdown. This question is treated in Ref. 9. In this paper G. Loew and J. Wang present results from many experiments at various frequencies. If the scaling laws thus obtained are extrapolated to 11.4 and 17.1 GHz, the breakdown limited surface fields obtained are 660 and 807 MV/m, respectively. To convert this to effective accelerating gradient, a reduction factor of 2.5 is typically used.

In both cases, the accelerating gradient is above the 200 MeV/m used for the TLC design in Table 2. However, the measurements also indicated significant "dark currents" generated by captured field-emitted electrons. The question of the effects of dark current on loading and beam dynamics is not yet resolved and needs further study.

As mentioned in the Introduction, in order to make efficient use of the RF power and to achieve high luminosity, it seems essential to accelerate a train of bunches with each fill of the RF structure. This leads to two problems: (1) the energy of the bunches in the train must be controlled and (2) the transverse stability of the bunch train must be ensured. Both of these problems are helped greatly by damping higher modes (both transverse and longitudinal) in the RF structure. In Ref. 10, R. Palmer describes a technique of using slotted irises coupled to radial waveguides to damp these modes: Q's as low as 10–20 have been measured in model structures. This encouraging evidence has led to a development program at SLAC to do more detailed studies of slotted structures. The beam dynamics consequences of damping the higher modes is explored in the section on Multibunch Effects.

5.2 RF Power Sources

Before discussing results on power sources, it is useful to contrast and compare two basic approaches, RF pulse compression and the relativistic klystron.

5.2.1 RF Pulse Compression

In Fig. 3(a), you see illustrated the basic principle of RF pulse compression. A long modulator pulse is converted by a high power, 'semi-conventional' klystron or some other power source into RF power with the same pulse width. This RF pulse is then compressed by cleverly slicing the pulse using phase shifts and 3-db hybrids and re-routing the portions through delay lines so that they add up at the end to a high peak power but for a small pulse width. This scheme was invented by D. Farkas at SLAC and is presently under experimental investigation.[11] With a factor of 16 in pulse compression, the TLC would require a 50-MW klystron with a 1-μsec pulse length for each meter of the accelerator while the ILC would require a 50-MW klystron for each four meters of structure.

In Ref. 12, P. Wilson describes RF pulse compression in some detail including estimates of efficiencies. There is an experimental test ongoing at SLAC which seeks to test a low loss, low power system. Initial results of this test have been very encouraging. A 100-MW, 11.4-GHz, "semi-conventional" klystron is presently being constructed at SLAC to perform high power tests of pulse compression.

5.2.2 The Relativistic Klystron

In Fig. 3(b), you see the principle of the relativistic klystron illustrated. In this case, the pulse compression happens *before* the creation of RF. This technique makes use of the pulsed power work done at LLNL in which magnetic compressors are used to drive induction linacs to produce multi-MeV e^- beams with kiloampere currents for pulses of about 50 nsec. These e^- beams contain gigawatts of power. The object, then, is to bunch the beam at the RF frequency to extract a significant fraction of this power. This can be done either by velocity modulation or by dispersive magnetic "chicanes." After bunching, the beam is passed by an RF extraction cavity which extracts RF power from the beam.

RF POWER SOURCE DEVELOPMENT

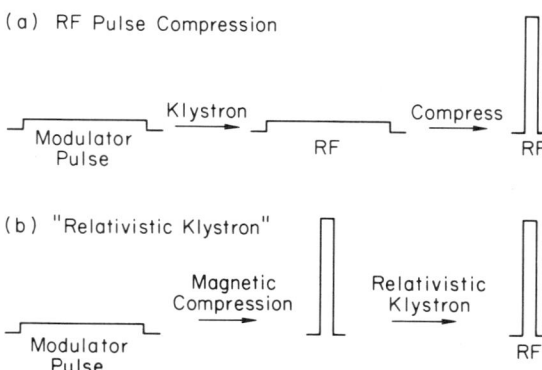

Fig. 3a. Illustration of RF pulse compression.
3b. Illustration of the relativistic klystron with magnetic compression.

In Ref. 13, four experiments on relativistic klystrons are described. These are the result of a SLAC–LLNL–LBL collaboration which makes use of the ARC facility (e^- beams 1.2 MeV and $\lesssim 1$ KA) at LLNL. Thus far the record peak power for any of the devices tested is 200 MW; however, in this case, the RF envelope was noticeably shortened. The highest power obtained with a wide RF pulse was about 80 MW. The most serious problem encountered in the experiment is the pulse shortening phenomenon; recent experiments suggest that this is caused by loading due to anomalous charged particle currents. A second serious problem is poor beam transmission. Finally, this RF power has been used to drive a 26-cm travelling wave structure at 11.4 GHz. The peak power of 200 MW corresponds to a local acceleration gradient of 140 MV/m. Work is continuing on this experiment.

Another interesting possible RF source is the cluster klystron. In Ref. 14, R. Palmer and R. Miller describe a multiple beam array of "klystrinos" which when coupled together can give impressive results. By dividing a single beam into many beams shielded from each other, the problems of space charge are effectively eliminated. This source could be used as a driver for RF pulse compression. Alternatively, with the addition of a grid and an oil-filled transmission line for energy storage, the device could directly produce short RF pulses. Thus far, there has been no experimentation, but calculations and cost estimates are encouraging.

Rather than separating the beam into separate beams, it is also possible to consider ribbon beam geometries. One possibility, the Gigatron, is presented in

Ref. 15. This device makes use of the lasertron concept to produce a bunched beam directly at the cathode. Field emitting arrays are used for the cathode while a ribbon beam geometry is envisioned to control space-charge effects. This device is another candidate for RF pulse compression and has an impressive efficiency on paper. Experimental tests are presently being prepared.

To conclude this section, it seems that if high power tests of RF pulse compression show positive results, there are several candidates to provide the long pulse input RF. Such an RF source combined with RF pulse compression would be a possible power source for an ILC or TLC which could be realized in the near future. All other possibilities seem somewhat more remote and need much more R&D.

5.3   EMITTANCE PRESERVATION

During the process of acceleration, we must take care not to dilute the emittance of the beam. There are several effects which can lead to emittance dilution. In the next few subsections, we discuss a few of the most important effects.

5.3.1 Chromatic Effects

The filamentation of the central trajectory in a linac can cause dilution of the effective emittance of the beam. If we first consider a coherent betatron oscillation down the linac, then to be absolutely safe, we must require that it be small compared to the beam size. If the spread in betatron phase advance is not too large, then this tolerance is increased to perhaps twice the beam size for the cases shown in Table 2.

The chromatic effect of a corrected trajectory is rather different. In this case, it is the distance between an error and a corrector which matters, and the effects partially cancel yielding a growth $\propto \sqrt{N_{quad}}$. This yields a tolerance on magnet misalignment the order of 20 to 30 times the beam size in the linac (about 30 μm) for the cases in Table 2. This is also the tolerance on BPM measurements. If the phase advance of the linac or some subsection is not too large, then this yields a linear correlation of position with momentum (dispersion) which can, in principle, be corrected since it does not vary in time. Therefore, it may be possible to have looser tolerances if such correction is provided.

5.3.2 Transverse Wakefields and BNS Damping

The wakefield left by the head of a bunch of particles, if it is offset in the structure, deflects the tail. If the transverse oscillations of the head and tail have the same wave number, the tail is driven on resonance. This leads to growth of the tail of the bunch.[16] This effect can be controlled by a technique called BNS damping.[17] The bunch is given a head-to-tail energy correlation so that the tail is at lower energy. The offset of the head by an amount $\hat{x}$ induces a deflecting force on the tail away from the axis. The tail, however, feels an additional force $\Delta K \hat{x}$, where ΔK is the difference in focusing strength. These two forces can be arranged to cancel, thereby keeping the coherence of the bunch as a whole. For the designs shown in Table 2, the spread in energy for BNS damping is $\sim \pm.3\%$. This correlation can be accomplished by moving the bunch slightly on the RF wave to obtain a linear variation across the bunch.

Recently, BNS damping has been tested at the SLAC linac with great success. It is now part of normal operating procedure.

5.3.3 Jitter

In order to maintain collisions at the interaction point, the bunch must not move very much from pulse to pulse. Since the optics of the final focus also demagnify this jitter, the tolerance is always set by the local beam divergence compared to the variation of some angular kick. The jitter tolerance on the damping ring kicker is thus related to the divergence of the beam at that point. This is discussed in Ref. 6. At the injection point to the linac, the offset caused by this jitter must be small compared to the local beam size.

If all the quadrupoles in the linac are vibrating in a random way, the effects accumulate down the linac and the orbit offset grows $\propto \sqrt{N_{quad}}$. This sets the tolerance on the random motion of quadrupoles to be much smaller than the beam size. In the examples in Table 2, the random jitter tolerances are $\simeq 0.01$ μm. On the other hand, tolerances for correlated effects are an order of magnitude less severe. In either case, this size motion from pulse to pulse is unlikely due to the large repetition rate of the collider. More gradual motion, which is larger, can be corrected with feedback.

Jitter in RF kicks can cause similar effects. These effects can be reduced by reducing the DC component of the RF kick by eliminating asymmetries in couplers and by careful alignment of structures.

5.3.4 Coupling

Finally, we discuss coupling of the horizontal and vertical emittance. The beam size ratio in the linac is 10:1. The tolerance on random rotations for a flat beam is given by

$$\Theta_{rms} << \frac{\sigma_y}{\sigma_x} \frac{1}{\sqrt{2N_q}} \quad .$$

For the examples shown in Table 2, the right-hand side is about 3 mrad. This seems quite straight forward. If the errors are not random, larger rotations can indeed result; however, because the beam size is so small, the effects are very linear. This means that skew quadrupoles can be used effectively as correction elements. Certainly, in the final focus, skew quads will be an integral part of the tuning procedure to obtain flat beams.

6. FINAL FOCUS

The final focus, as described in the parameters in Table 2, is a flat beam final focus with a crossing angle. The purpose of the flat beam is to increase the luminosity while controlling beamstrahlung and disruption. The crossing angle is to allow different size apertures for the incoming and outgoing beam. Another invention, "crab-wise crossing", discussed in Ref. 5, allows a much larger crossing angle than the diagonal angle of the bunch. As discussed in Ref. 5 and in Ref. 18, this type of geometry may now be essential due to the production of e^+e^- pairs by beamstrahlung photons in the field of the bunches.

6.1 FINAL FOCUS OPTICS AND TOLERANCES

The first job in the final focus is to demagnify the beam to provide a small spot for collision. The design for such a system is presented in Ref. 19 by K. Oide. This is a flat beam final focus which achieves the parameters shown in Table 2 for vertical and horizontal beam size. The vertical size is limited by a fundamental constraint "the Oide limit" due to the synchrotron radiation in the final doublet coupled to the chromatic effect of a quadrupole. The quadrupole gradients necessary are very high and in Oide's design are obtained by conventional iron magnets with 1-mm pole-to-pole distance. Tolerances are very tight in such a final focus. The most restrictive vibration tolerance is on the final doublet which must be stable pulse-to-pulse to about 1 nm.

Since vibration of the final doublet is the most serious problem, it is considered in some detail in Ref. 20. In this paper, it is shown that passive vibration isolation seems to be more than adequate to handle the vibrations above 10 Hz at the high frequency end. For low frequencies, W. Ash suggests an interferometric feedback system to control motion to about 1 μm. Beam steering feedback can then be used to control slow variations in the 1-nm to 1-μm region.

6.2 BEAM-BEAM EFFECTS

When a small bunch of electrons collides with a small bunch of positrons, the fields of one bunch focus the other, causing disruption. Since the opposing particles are strongly bent, they also emit radiation called beamstrahlung. These are the two basic beam-beam effects. The disruption enhances the luminosity by a small amount while the beamstrahlung causes significant energy loss during collision and increases the effective momentum spread for physics. (See Table 2.) These issues are discussed in more detail in Ref. 18.

In addition, there are several other important effects which should be mentioned here. If the beams are offset relative to each other, a kink instability develops. This effect actually causes the luminosity to be less sensitive to offsets because the beams attract each other and collide anyway. There is also a multi-bunch kink instability which is more serious since it can cause the trailing bunches to miss each other entirely. This places restrictions on the product of the vertical and horizontal disruption per bunch.

The final section of Ref. 18 is an addendum added after the SLAC Workshop in Dec. 1988. As mentioned earlier in the Introduction, it was discovered that the beamstrahlung photons pair-produce in the coherent field of the bunch. The corresponding incoherent process has been known for some time, but its importance has only just been realized.[21] The problem is that low energy e^+e^- pairs are produced in an extremely strong field which then deflects the charge of the appropriate sign while confining the other. This leads to large angular kicks, as mentioned earlier in Section 2.

These stray particles can lead to more background problems, which must be addressed by further interaction point design. In Ref. 5, it is suggested that crab-crossing combined with large crossing angles and solenoidal fields would allow one to channel these electrons out through a large exit hole to a beam dump. This idea looks promising but needs much more study.

The measurement of the final spot size is an extremely important, but as yet unsolved, problem. From SLC experience, it is probably possible to use beam-beam effects to minimize spot sizes. However, for the initial tune-up of the final focus, a single-beam method is almost essential. There was some initial work done at the workshop in June 1988 in Capri, Italy, which was also reported at the SLAC workshop.[22] In addition, preliminary results were presented at the SLAC workshop on the use of beamstrahlung from an ionized gas jet.[23] Although this looks promising, there is still much work to be done.

7. MULTIBUNCH EFFECTS

As mentioned earlier, in order to efficiently extract energy from the RF to obtain high luminosity, it is essential to have many bunches per RF fill. This, however, leads to transverse beam breakup. The invention of damped structures discussed in Section 5.1 helps but does not completely solve the problem for the linac. It is also necessary to tune the frequency of the first dipole mode of the accelerating structure. This is discussed in Ref. 24 where the problem of multibunching is traced all the way through the linear collider subsystem by subsystem. Damped accelerating cavities are required for the main linac and the damping rings, while other systems can get by with very strong focusing. Thus, from the transverse point of view, stability seems possible.

In addition, it is necessary to control the energy spread from bunch to bunch very precisely ($\Delta E/E \lesssim 10^{-3}$). This can be accomplished by injecting the bunches before the RF structure is full to match the extraction of energy by the bunches to the incoming energy as the structure fills. This leads to tight tolerances on phase and amplitude of the RF, as well as tight control of the pulse-to-pulse number of particles in a batch of bunches.[25] However, the benefits of multibunching seem to far outweigh any difficulties they impose due to the order of magnitude increase in luminosity.

8. OUTLOOK

During the past few years, there has been tremendous progress towards a next generation linear collider. We now have a much clearer picture of how to obtain both the energy and luminosity required. An important development this past year was the increased interest in an ILC, that is, a linear collider with 0.5 TeV in the CM which would be upgradable to 1.0 TeV with additional power sources. Since there is a factor of four difference in the peak power required for the ILC vs. the TLC, the initial power source looks much easier to do. We will probably see the development of a power source and structure during the next couple of years. This would yield the energy of the collider; what about the luminosity?

Designs of damping rings, bunch compressors and focus systems will continue. Studies of BNS damping in the linac and emittance dilution will continue both experimentally with the SLAC linac and theoretically for the next generation high-frequency linac. However, to really understand tolerances, new measurement techniques, and final focus optics, it is probably essential to build a scale model final focus at SLC energy. This is being planned at SLAC (Final Focus Test Beam).

One key aspect of all linear collider design is background control. With the discovery of the swarm of e^+e^- pairs produced at the interaction point, there now needs to be detailed study of interaction point design to control backgrounds.

To conclude, it looks like we are on the path towards a next generation linear collider and with proper funding of R&D over the next few years we may see a proposal in the early 1990's.

REFERENCES

1. *Proceedings of the Workshop on Physics of Linear Colliders*, Capri, Italy, June 1988.
2. *Proceedings of the Summer Study Physics in the 1990's*, Snowmass, Colorado, July 1988.
3. *Proceedings of the International Workshop on Next Generation Linear Colliders*, SLAC, Stanford, CA, Dec. 1988, SLAC-Report-335.
4. Linear Collider Working Group Reports From Snowmass '88, Ed. R. D. Ruth, SLAC-Report-334.
5. R. B. Palmer, *Energy Scaling, Crab Crossing and the Pair Problem*, SLAC-PUB-4707 and in Refs. 2 and 4.
6. T. O. Raubenheimer, L. Z. Rivkin and R. D. Ruth, *Damping Ring Designs for a TeV Linear Collider*, SLAC-PUB-4808 and in Refs. 2 and 4.
7. L. Z. Rivkin et al., *Bunch Lengthening in the SLC Damping Ring*, SLAC-PUB-4645 (1988).
8. S. A. Kheifets, R. D. Ruth, J. J. Murray and T. H. Fieguth, *Bunch Compression for the TLC. Preliminary Design*, SLAC-PUB-4802 and in Refs. 2 and 4.
9. G. A. Loew and J. W. Wang, *RF Breakdown and Field Emission*, SLAC-PUB-4647 and in Refs. 2 and 4.
10. R. B. Palmer, *Damped Accelerator Cavities*, SLAC-PUB-4542 and in Refs. 2 and 4.
11. Z. D. Farkas, *IEEE Transcripts on Microwave Theory and Techniques*, **MTT-34**, No. 10 (1986) 1036, and also SLAC/AP-59.
12. P. B. Wilson, *RF Pulse Compression and Alternative RF Sources*, SLAC-PUB-4803 and in Refs. 2 and 4.
13. M. A. Allen et al., *Relativistic Klystron Research for Linear Colliders*, SLAC-PUB-4733 and in Refs. 2 and 4.
14. R. B. Palmer and R. Miller, *A Cluster Klystron*, SLAC-PUB-4706 and in Refs. 2 and 4.
15. H. M. Bizek et al., *A Microwave Power Driver for Linac Colliders: Gigatron*, in Refs. 2 and 4.
16. A. Chao, B. Richter and C. Yao, *Nucl. Instr. Meth.* **178**, 1 (1980).

17. V. Balakin, A. Novokhatsky and V. Smirnov, *Proceedings of the 12th International Conference on High Energy Accelerators*, Fermilab (1983), p. 119.
18. P. Chen, *Disruption, Beamstrahlung, and Beamstrahlung Pair Creation*, SLAC-PUB-4822 and in Refs. 2 and 4.
19. K. Oide, *Final Focus System for TLC*, SLAC-PUB-4806 and in Refs. 2 and 4.
20. W. W. Ash, *Final Focus Supports for a TeV Linear Collider*, SLAC-PUB-4782 and in Refs. 2 and 4.
21. M. S. Zolotarev, E. A. Kuraev and V. G. Serbo, *Estimates of Electromagnetic Background Processes for the VLEPP Project*, Inst. Yadernoi Fiziki, Preprint 81-63, 1981; English Translation SLAC TRANS-0227, 1987.
22. J. Norem, presented at the *International Workshop on Next Generation Linear Colliders*, in preparation.
23. D. Burke, P. Chen, M. Hildreth and R. Ruth, *A Plasma Beam Size Monitor*, work in progress.
24. K. A. Thompson and R. D. Ruth, *Multibunch Instabilities in Subsystems of 0.5 and 1.0 TeV Linear Colliders*, SLAC-PUB-4800 and in Refs. 2 and 4.
25. R. D. Ruth, *Multibunch Energy Compensation*, SLAC–PUB–4541 and in Ref. 1.

THE SSC PROJECT

John Peoples

SSC Central Design Group
Lawrence Berkeley Laboratory, Berkeley, California 94720 USA

Table of Contents

Abstract: The SSC Project	2228
Introduction	2228
Essential Parameters of the SSC	2229
Conceptual Design of the Injector	2229
Conceptual Design of the Collider Rings	2230
Magnet Development Status	2233
Site Selection and Conventional Facilities	2238
A Possible Timetable for the SSC	2238
References	2239

Lecture given at the 1st European Particle Accelerator Conference,
June 7–11, 1988, Rome, Italy

THE SSC PROJECT

John Peoples

SSC Central Design Group*
Lawrence Berkeley Laboratory, Berkeley, California 94720 USA

Abstract: The SSC Project

The SSC is a 20-TeV, proton-proton collider proposed for construction by the U.S. Department of Energy (DOE). Completion is planned for the mid-1990s. A technical description of the accelerator is given along with a report on the status of the project.

Introduction

It is fitting that I remind this audience that a more ambitious proposal was made by Enrico Fermi more than thirty years ago. As a graduate student I had heard of his proposal for an earth-circling accelerator as part of the folklore that is passed on from teacher to student. While preparing this talk, I decided to track down just what the proposal was. Thanks to the memory of one of my colleagues, who heard Professor Fermi speak, and to the diligence of the History of Accelerators Project at Fermilab and the J. Regenstein Library, I was fortunate enough to receive a copy of Professor Fermi's typewritten notes for the farewell address that he gave to the American Physical Society on 29 January 1954. Tradition required that the outgoing president of the American Physical Society address the Society at the New York meeting when the new president was inaugurated. Professor Fermi chose as the title of his talk, "What Can We Learn with High Energy Accelerators?" He reminded the audience of what had been learned about elementary particles in the last few decades. He noted that there were too many so-called elementary particles. In spite of the stupendous number of names there was a tantalizing vista. Fermi asked rhetorically "What to do?" His answer was to clamor for higher and higher energy. He extrapolated to 1994; for that date he proposed a machine with a radius of 8000 km, a field of 20000 gauss, and an energy of 5×10^{15} eV (5×10^3 TeV in our contemporary units). He even had a cost estimate: $170 billion.

Although Professor Fermi's machine was a fixed-target machine, I am sure that he would have added another ring if the practicality of colliding beams were then known. Two such rings would give 10^4 TeV in the center of mass. Clearly, the Fermitron would be an international laboratory, since its magnets would circle the earth 1600 km above the surface. As a fixed-target machine, the center-of-mass energy proposed by Professor Fermi was 3 TeV. We are not quite on his ambitious schedule, although the Tevatron has already reached 1.8 TeV in the center of mass. The LHC and SSC will exceed that energy, although a bit later than 1994.

---

*Operated by Universities Research Association, Inc., for the U.S. Department of Energy.

Essential Parameters of the SSC

The Superconducting Super Collider, the SSC, is a proposed high-energy, high-luminosity, pp collider.[1] The basic parameters, which are by now familiar to many people, are given in Table 1.

The 40-TeV center-of-mass energy will be more than adequate to produce experimentally observable collisions of quarks and gluons, the point-like constituents of the protons, at center-of-mass energies in excess of 3 TeV. This should bring us to the threshold of understanding some, if not all, of the many parameters in the standard model.

In this paper I will discuss a few aspects of the R&D program. Let me begin with the Conceptual Design, then move to the magnet systems, and finally review the status of the site selection.

Conceptual Design of the Injector

The Conceptual Design Report[1] (CDR) for the SSC was published in March 1986. It provided a very detailed description of the SSC. No matter how detailed such a description is, however, it is not static; time allows the development of small improvements. While the basic elements of the SSC have remained unchanged since the CDR was completed, a number of refinements have been made and will continue to be made to the design. I will take note of some of them today.[2] Let me begin with the injector. Some of its basic parameters are given in Table 2.

The Linac accelerates H$^-$ ions to a momentum of 1.22 GeV/c. These ions are injected into a rapid-cycling, 10-Hz synchrotron, the low-energy booster (LEB), where they are stripped and accelerated to 8.45 GeV/c. The 8.45-GeV/c protons are transferred to the medium-energy booster (MEB), a second synchrotron built with conventional magnets. Five LEB cycles are needed to fill the MEB. After the MEB is filled, the protons are accelerated to 100 GeV and are then transferred to the high-energy booster (HEB), a synchrotron made with superconducting

Table 1. SSC Parameters — Collider Rings

Circumference of each ring	83.631 km
Interaction Region (IR)	6 (4 initially configured)
Beam Energy	20 TeV
Peak luminosity (at $t = 0$), $\mathcal{L}_p^{\wedge}$	10^{33} cm^{-2}sec^{-1}
Bunch Intensity at $\mathcal{L}_p^{\wedge}$	8×10^9
Number of bunches at $\mathcal{L}_p^{\wedge}$	15,456
β^* at low-β IR	0.5 m
Normalized betatron emittance	1 mm-mr
Nominal beam storage time	24 hr

Table 2. SSC Injector Parameters

	Linac	LEB	MEB	HEB
Injected particle	H^-	H^-	H^+	H^+
Injection momentum (GeV/c)	0	1.22	8.45	100
Extraction momentum (GeV/c)	1.22	8.45	100	1000
Circumference/length (m)	125.0	342.8	1751.5	5336
f_{rf} at extraction (Mhz)	1263.2	62.0	63.0	63.0
Average current (mA)	3.9	100	95.0	95.0
Normalized rms transverse emittance (mm-mr)	0.45	0.75	0.83	0.91
Longitudinal rms emittance/bunch (eV-sec)	0.012×10^{-3}	1.8×10^{-3}	1.8×10^{-3}	35.0×10^{-3}
Cycle time (sec)	0.1	0.1	4.0	60.0

magnets. After three MEB cycles, when the HEB is filled, the protons are accelerated to 100 GeV/c and then transferred to one of the collider rings. Sixteen HEB cycles, which take a total of 16 minutes, are needed to fill each collider ring. After one is filled, the polarity of the HEB is reversed and the second ring is filled. The total filling time for both rings is expected to be less than one hour.

During the past two and a half years, the lattices of the LEB, the MEB, and the HEB have each undergone some change as the designers have sought to make them easier to tune and operate, to reduce the dispersion, and to reduce the sensitivity to the alignment errors. Parameters, such as the tunes, will undoubtedly continue to change slightly as the design is refined further. It is intended to complete the refinement of the lattice design of each of these accelerators next year, following selection of a site. Prototype development of the injector components could begin soon thereafter. Since each of these injector accelerators has a close counterpart in operation today, the need to build prototype components is not nearly as urgent as it is for the collider rings.

Conceptual Design of the Collider Rings

The collider rings consist of two arcs, each containing 144 cells 228.5-m long. Each half-cell contains six 16.54-m 6.6-T dipoles, one 230-T/M quadrupole, and a correction element package. The correction element packages are used to control global properties of the ring such as tune and chromaticity, as well as local beam orbit distortions. In addition, each dipole is intended to have its own small correction package, the bore-tube corrector, which can be thought of as compensation for the unwanted multipoles of the host dipole. The arcs are joined by two clusters of four modules, designated "near" and "far"; within each module there is a straight section.

Two of the IRs in the near cluster are intended to be high-luminosity intersections for large, general-purpose detectors. There will be a 340-m free space centered around the collision point, and β^* will be equal to 0.5 m. The luminosity in these locations is intended to be in excess of 10^{33} cm^{-2}sec^{-1}. The two remaining straight sections in the near cluster will be used for injection, abort, rf, and other accelerator systems requiring space in the straight sections. Two of the IRs in the far cluster are intended to be medium-β IRs with a free space of 234 m centered around the collision point. The luminosity of these interaction regions is expected to be 5×10^{31} cm^{-2}sec^{-1}. The remaining two straight sections in the far cluster have been set aside for future interaction regions.

Because the cost of the SSC is dominated by the cost of the collider rings, they have received the most attention. The largest single cost of each collider ring is the cost of its superconducting magnets. Since their cost is roughly proportional to the diameter of the coils, and since the unwanted multipoles of order n are inversely proportional to the diameter to the $(n+1/2)$ power, there is a strong tension between cost and useful aperture. After an extensive analysis of particle orbits, the inner diameter of the inner coil was chosen to be 40 mm.[3] Subsequent work, as will be noted later, indicates that this diameter provides an adequate dynamic aperture for the beam. Since the CDR was completed, the phase advance/cell of the collider lattice has been increased from 60° to 90°, in order to improve the aperture-versus-cost optimization and to improve the off-momentum beam behavior.[2] The designs of the dispersion suppressor and of the straight sections have also been changed.[3] The current values of the collider ring parameters are summarized in Table 3.

Table 3. Collider Ring Parameters

Circumference	83.631 km
Arc cell length	228.5 m
Half-cell composition	6 dipoles, 1 quadrupole 1 corrector package
Number of cells/arc	144
Phase advance/cell	90°
Horizontal and vertical betatron tune	95.285, 95.265
Dipole field (length*)	6.613 T (16.54 m)
Quadrupole gradient (length*)	228.7 T/m (3.64 m)
Bore tube inner diameter	32.23 mm
Linear aperture (radius)	~ 10.7 ± 2.0 mm
Dynamic aperture (radius)	15.0 ± 0.6 mm

*magnetic length

The available good field aperture has been defined in terms of the linear aperture, defined to be the aperture for which the rms variation with time of the Courant-Snyder invariant amplitude, W, is less than 6.4 percent. W is given by

$$W = \left(\frac{x^2 + (\alpha x + \beta x')^2}{\beta}\right)^{1/2}.$$

It approaches zero as the betatron amplitude of a particle approaches zero. The dynamic aperture is defined as the largest betatron amplitude that remains bounded after an arbitrarily large number of turns. For machines built in the past, the dynamic aperture has usually been larger than the physical aperture of the vacuum chamber and, in those cases, has been only of academic interest. This is not the case for the SSC, since the unwanted multipole fields of the dipole fields limit the dynamic aperture to a value that is less than the bore-tube diameter. The required linear aperture (radius) is estimated to be about 5 mm.

Because the operational performance of the SSC will depend on whether the linear aperture can be achieved with the proposed magnet design, a great deal of effort has gone into theoretical modeling of the particle orbits through tracking programs and analytical calculations. The various methods of calculation give the same results for the size of the linear and dynamic aperture. It is worth noting that these calculations showed that if nothing further were done to reduce the effect of these multipoles, the linear aperture would only be 5 mm for the multipole specifications that have been subsequently adopted for the 16.54-m collider dipoles. In addition, the results of tracking programs have been checked against beam behavior in the Tevatron in experiment E778. In this experiment, the sextupole strength of a number of sextupoles was increased while the tune and chromaticity were kept fixed, in order to simulate the conditions that a beam would encounter in the SSC. The preliminary results reported at this conference are very encouraging, since it appears that the theoretical prediction of the size of the linear aperture agrees with the observed linear aperture.[4] The understanding is not complete, since the predicted dynamic aperture is 20–30 percent larger than the observed dynamic aperture. One possible reason for the discrepancy is that the simulation was obtained in a tracking for a relatively small number of revolutions (500). This work will continue.

As noted earlier, if the systematic multipole fields of the dipoles were corrected only once per half-cell and the random multipole fields were left uncompensated for, then the linear aperture would be considerably smaller than 10 mm. The multipole field of the greatest significance is the sextupole, which has particularly large systematic and random contributions that must be compensated for more often than once per half-cell. Two schemes to neutralize these unwanted multipoles are under consideration. Both have been developed since the CDR was completed. In the first scheme, each dipole has a corrector mounted on the bore tube; each corrector has a sextupole and decapole winding.[5] Since these windings can be individually powered, each can be assigned to a bin according to the sign and magnitude of the random part of a particular multipole of the host dipole. By having up to seven bins for each multipole, and by connecting all of the bore tube corrector windings in each bin in series and then exciting each circuit separately, the random multipole contribution per dipole can be decreased by a factor of five. Thus the unwanted random multipole fields of the dipole can be compensated for locally.[6] The unwanted systematic multipole fields of the dipoles can be compensated for by incrementing the currents in all windings of a given multipole by the same current. This magnet-by-magnet compensation can be achieved either by placing a short magnet on the bore tube at one end of each magnet or by placing thin windings on the outer surface of the bore tube. The latter approach is favored at the moment, and the design is based on it.

The second scheme, due to Neuffer, uses a single additional set of correctors located in the middle of each half-cell.[7] The standard correctors located next to the quadrupoles in each half-cell, together with an additional correction element placed in the middle of each cell, are binned in accordance with the multipole content of the six dipoles in each half-cell. Studies have shown that this second scheme can provide sufficient linear aperture, provided that the sextupole error is below a reasonable level. Beyond this level, it was suggested that octupoles could be added to correct the non-linearities introduced by the strong sextupole correction elements. It appears that both schemes can do the job. The results of the calculations and experiments to date give one confidence that the relatively strong multipoles of the SSC dipoles can be accommodated.

Magnet Development Status

An R&D program to develop a suitable dipole for the collider was initiated by the SSC Central Design Group (CDG) four years ago. Since that time, the CDG has directed an active program of building model magnets at Brookhaven National Laboratory, Fermilab, and Lawrence Berkeley Laboratory. In 1985, the CDG adopted the basic magnet design of a two-layer coil with an approximate $\cos \theta$ current distribution, a non-magnetic collar to limit mechanical motion of the coil perpendicular to the beam direction, and an iron yoke within the cryostat. As can be seen in Figure 1, a cross section of the collared coil, there are four inert copper wedges per quadrant interspersed with the coil windings. The locations of these wedges make it possible to reduce all of the unwanted multipoles, except for the sextupole, b_2, to the specified values. At low fields b_2 is determined by persistent currents in the conductor, and at high fields it is determined by the saturation of the iron yoke. The specification for the random variation of the multipole fields was determined by scaling from the Tevatron and CBA dipoles. To date, the random variations have been less than those tolerances.

Because it is intended that the iron yoke provide 2.2 T of the 6.6-T field, the radial separation between the outer surface of the outer coil and the inner surface of the yoke is only 15 mm. Although this leads to a magnet with a stronger field for a given number of ampere turns, the collar alone does not have sufficient rigidity to limit the mechanical motion of the coil perpendicular to the beam to 50 μ when the coil is excited to 6500 A. Its expansion must be constrained by the iron yoke if the coil is to reach its design current without quenching. Initially, it was intended that the collared coil be unconstrained by the yoke, but we now recognize that the yoke plays a critical role in limiting the coil motion. In order to clamp the coil in this way the temperature of the yoke must be held at 4.35 K.

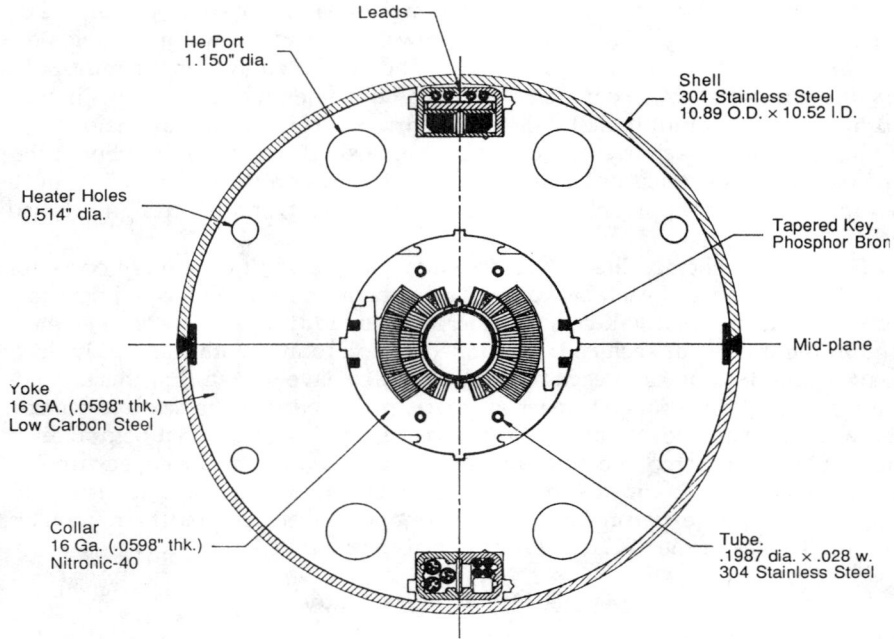

Figure 1. SSC dipole cross section, with BNL C358A coil.

While there are some operational disadvantages with this choice—the iron yoke is the dominant contribution to the mass that must be cooled down—it makes it possible to design a very efficient cryostat and support system. Some of the systems requirements for the collider dipole are summarized in Table 4.

As noted earlier, magnet development is being carried out at three national laboratories under the direction of the CDG. Brookhaven has been responsible for the initial design and subsequent fabrication of all cold masses for the 16.5-m prototypes. Fermilab has been responsible for the design of the cryostat and has installed the 16.5-m cold masses in cryostats. LBL, together with industry, has been responsible for the development of the superconducting wire and cable. To date, all cold masses for 16.5-m full-scale prototypes have been built at BNL and then shipped to Fermilab for installation in a cryostat. Tests of the 16.5-m magnets have been done at Fermilab. In addition to the fabrication and testing of prototype 16.5-m dipoles, BNL and LBL have fabricated and tested short models at their laboratories. During the past year, all of the laboratories have helped to identify and solve problems associated with the 16.5-m prototypes by building and testing short magnets.

The cable performance, the cryostat performance, and nearly all aspects of the cold mass performance meet the requirements of the collider rings. One critical aspect of the first five full-length prototypes that was unsatisfactory was the large number of training quenches required to achieve the 6.6-T design field.

Table 4. Dipole Systems Requirements

Magnetic properties			
Peak magnet field at 4.35 K			6.613 T
Transfer function at 1 TeV			1.0309 T/kA
Transfer function at 20 TeV			1.0147 T/kA
Magnet length		16.54 m	
Random variations of magnetic multipoles (rms)			
b_2	(a_2)	2.0	(0.6)
b_4	(a_4)	0.7	(0.2)
b_6	(a_6)	0.2	(0.1)
b_8	(a_8)	0.1	(0.1)
Heat leak budget/dipole			
at 80 K	27.0 W		
at 20 K	3.3 W		
at 4.35 K	0.32 W		
Synchrotron radiation power deposited in the bore tube dipole			2.34 W
Critical mechanical dimensions			
Slot length		17.34 m	
Bore tube inner diameter			32.26 mm
Vacuum vessel outer diameter			60.96 cm

Three of the first five magnets did not reach the design field of 6.6 T after twenty or more training quenches; two did after about twelve quenches. The sixth magnet, the most recent magnet to be built, reached a plateau above 6500 A at 4.35 K—which is consistent with predictions based on the short sample limit of the cable—without any training quenches.[8] The magnet reached 7675 A, well in excess of the design current, when operated at 3.2 K. The cable used in this magnet was purchased more than two years ago when cable of SSC quality was not being produced regularly, and its short sample current falls somewhat below SSC specifications. Since the cable now being produced meets the SSC specification, magnets made with the new cable should exceed the 6504-A design field at 4.35 K by several hundred amperes. A plot of the quench current versus quench number for magnet DD0012, the most recent magnet to be tested, is shown in Figure 2.

The improved performance is a result of significantly improved mechanical clamping of the coil, azimuthally and radially. During the past year we have carefully reviewed the possible causes of conductor motions. Two stood out: motion of conductors within the body of the magnet perpendicular to the beam direction, and motion of the conductors in the ends. The former caused quenches in the body of the magnet, and the latter caused quenches in the ends of the magnet. With the instrumentation that has been in place since the fifth long magnet, it has been possible to determine the half-turn of the coil in which the quench started. The longitudinal position within the half-turn can be located to within a few tens of centimeters. This instrumentation has allowed us to

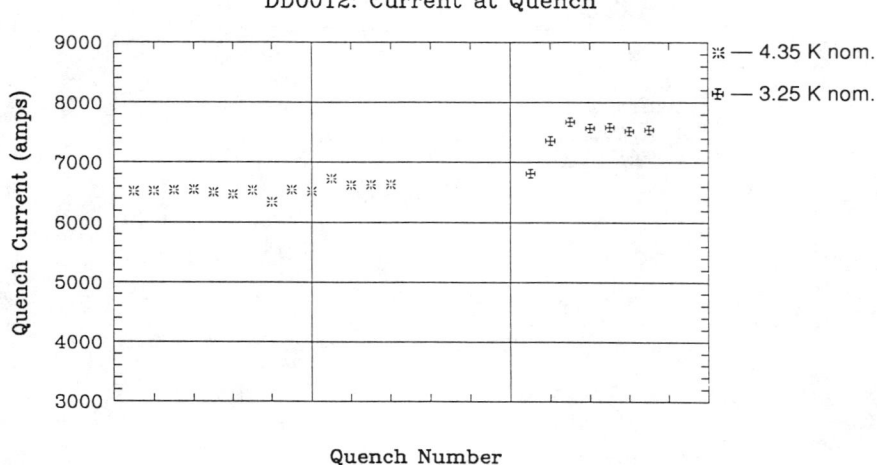

Figure 2. Plot of quench current vs. quench number for DD0012.

pinpoint defects in the magnet construction. A year ago, it was established that the curing fixtures expanded under pressure during the curing process. This caused a larger than allowable variation in the coil size at certain points along the 16.5-m length of the coil. It is believed that the improved coil-clamping scheme introduced in the sixth magnet made it possible to achieve the desired preload in spite of the coil size variation.

Further analysis of both magnet performance and finite-element models of the magnets should remove any remaining uncertainties. In the meantime, the curing fixtures are being rebuilt. It is anticipated that the new fixtures will reduce the coil size variation by a factor of three, within the acceptable range. These fixtures will be available to wind coils early in 1989.

One might wonder whether magnet DD0012 was the result of a favorable fluctuation in dimensions. An examination of the performance of the 1.8-m magnets built and tested at BNL during the past year shows that this is not the case. Table 5 shows the performance of 1.8-m magnets built at BNL during the past year.

The yoke blocks of DSS6R, DSS10, and DSS11 press against the collars of the collared coil and thus limit the expansion of the collared coil in the horizontal direction. Since this was achieved by placing shims between the collared coil and the yoke, there is a nominal 0.15-mm gap between the yoke blocks. As the magnet is excited, the two yoke halves are attracted toward one another, thus creating a vertical force along the collared coil proportional to I^2. This force increases the mechanical clamping of the coil. While in principle the yoke blocks do not bear against the collars in DSS6, this could not be established because the clearance between the yoke blocks and collars was not well defined. The clamping scheme used in DSS6R, DSS10, and DSS11 was also used for DD0012. Magnet DSS9 has an aluminum collar. Since there is an interference at the midplane

Table 5. Performance of recent 1.8-m model dipoles*

1.8-m Model	I on first quench	No. of quenches to reach I_Q at **	I_Q (T)
DSS6	6215 A	3	6460 A (4.49 K))
DSS6R	6491 A	0	6470 A (4.50 K))
DSS9*	6130 A	1	6800 A (4.46 K)
DSS10	5416 A	4	6510 A (4.48 K))
DSS11	6372 A	1	6690 A (4.36 K))

* DSS9 has an NC-9 cross section with aluminum collars; all other models have a C358A cross section with stainless-steel collars.

** I_Q is typically observed to be 2–3 percent higher than the measured short sample current of the cable.

between the collared coil and the yoke blocks at assembly, the expansion of the collared coil in the horizontal direction (along the midplane) is limited by the yokes. In this respect the collared coil constraint of DSS9 is similar to DSS6R.

It is worth noting that all of these magnets exceeded 7 T when tested at 4 K. The difference in the plateau quench current I_Q is due to the widely differing copper-to-superconductor ratio of the superconducting cable. Magnet DSS9 reached a field of 7.8 T at 3.3 K.

Work was begun several years ago at LBL on a 1-m dipole with aluminum rather than stainless-steel collars. The intention of this second design is to exploit the favorable difference in thermal contraction between the aluminum collars and the Cu/NbTi coils. The aluminum collars are expected to shrink more than the coils when cooled from 300 K to 4 K. The large decrease in the coil preload at the collar pole, inherent in magnets made with stainless-steel collars, should thus be reduced. During cooldown the loss in preload of the stainless-steel collared coil amounts to 2000 psi.

In addition to the single 1.8-m magnet of this design built at BNL, a large number of 1-m models of this design have been built and tested at LBL. As a rule, they have reached the design field of 6.6 T at 4.4 K with three or fewer training quenches. In addition, when cooled to 1.8 K these magnets have reached a field of nearly 9 T with very few training quenches. In a joint effort, LBL and BNL built two full-scale 16.5-m prototypes of this design at BNL during the past nine months. They will be tested sometime this summer at Fermilab after being installed in cryostats.

Relative motion of the conductors in the ends of the inner coil has also contributed to excessive training. The ends of the inner coil are difficult to fabricate because of their very small diameter. Small voids between the conductors that collapse during excitation have been difficult to eliminate. The ends of the much larger outer coil, comparable to the inner coil of a Tevatron dipole or a HERA dipole, have not been a problem. Improvements, each a matter of small details, have been steadily made to the fabrication technique. As of this writing, training quenches are no longer occurring in the ends. Nevertheless, we expect to introduce an internal support to the ends this fall, thereby further increasing their strength.

During the past year, a great deal of effort has been spent trying to understand why the long magnets did not perform as well as the short magnets. After all, a long magnet has the same ends and the same two-dimensional cross section. We have found that the small differences in construction of the short and long magnets have led to important differences in the control of the coil dimensions. While our understanding is not complete, we expect to be able to build magnets meeting all of the basic systems requirements once we have completed the evaluation of the six magnets under construction and the upgrade of the tooling and fixtures. This work should be complete early in 1989.

Site Selection and Conventional Facilities

In September 1987 the Department of Energy (DOE) received 36 proposals that were responsive to its site criteria: the site had to be within the United States and the group submitting the proposal had to be able to convey title to the land needed for the SSC. At the request of DOE, these proposals were reviewed by a panel appointed by the National Academies of Sciences and Engineering. The panel recommended eight sites as "best qualified" to the DOE in December of 1987. After completing its own review, the DOE accepted these recommendations and announced the list of eight best qualified sites in January of 1988. Subsequently, the site in New York State was withdrawn from further consideration by the Governor of New York. The remaining seven sites, in Arizona, Colorado, Illinois, Michigan, North Carolina, Tennessee, and Texas, are shown in Figure 3.

On 10 November 1988 Secretary of Energy John S. Herrington selected the location proposed by the State of Texas as the preferred site for the SSC.

A Possible Timetable for the SSC

For the SSC, the past is certain and the future is dreams. The following schedule represents the aspirations of SSC/CDG and the DOE:

Preferred Site Designated	November 1988
Final Site Choice	January 1989
Construction Complete	1996
Commissioning Commences	1996

While 1996 is two years later than the date that Professor Fermi proposed 34 years ago as he retired as President of the American Physical Society, we should hardly be embarrassed. The reality of exceeding 20 TeV in the center of mass is nowhere near as speculative as it was in 1954, for it is now within the realm of the possible.

Figure 3. SSC sites proposed and recommended as best qualified.

References

1. *Conceptual Design of the Superconducting Super Collider*, SSC-SR-1020, edited by J. D. Jackson (March 1986).
2. A. A. Garren and D. E. Johnson, "Status of the SSC Lattice Design," D. Neuffer, "Lumped Correction of the Multipole Content of the SSC", J. M. Peterson and E. Forest, "Correction of Random Multipole Errors in the SSC," Proceedings of the 1st European Particle Conference 7–11, 1988, Rome, Italy.
3. *Optimization of the Cell Lattice Parameters for the SSC*, SSC-SR-1024 (October 1986).
4. N. Merminga, "An Experimental Study of the SSC Magnet Aperture Criterion;" J. M. Peterson, "Dynamic Aperture Measurements at the Tevatron," Proceedings of the 1st European Particle Conference 7–11, 1988, Rome, Italy.
5. CDR, "Correction Coils," Sec. 5.2.7, p. 290.
6. R. Talman, Private Communication (October 1987).
7. D. Neuffer, "Lumped Correction of the Multipole Content of the SSC," Proceedings of the 1st European Particle Conference 7–11, 1988, Rome, Italy.
8. It is actually about 100 A higher than predicted; this is well within the accuracy of the short sample measurement combined with the extrapolation procedure.

PARAMETER CHOICES FOR THE SSC

J.M. Peterson
SSC Central Design Group

TABLE OF CONTENTS

1 Introduction 2241

2 Primary Parameters of the SSC 2242

    2.1 Ring Size 2242
    2.2 Bunched Beam 2242
    2.3 Luminosity-Related Parameters 2243
    2.4 Limiting Effects 2246
        2.4.1 Synchrotron Radiation Power 2247
        2.4.2 Beam-Beam Tune Shifts 2247
        2.4.3 Beam Loss due to Proton-Proton Collisions . 2248
        2.4.4 Beam Instabilities 2248

3 Parameter Optimization 2248

    3.1 Parameter Interrelationships 2248
    3.2 Effects of Varying β^* and ϵ_N 2249
    3.3 Effect of Varying S_B 2250
    3.4 Crossing Angle 2251

4 Primary Parameter Summary 2253

5 References 2254

PARAMETER CHOICES FOR THE SSC

J.M. Peterson
SSC Central Design Group

1 INTRODUCTION: BASIC REQUIREMENTS OF THE SSC

The design of the Superconducting Super Collider has followed from the basic requirement that it produce observable reactions among the elementary constituents of matter at as high an energy as now possible. The basic design parameters are that <u>proton</u> beams of <u>20 TeV</u> be brought into <u>collision</u> at <u>six</u> interaction points in such a way as to produce a luminosity of 10^{33} cm^{-2} s^{-1} per collision point.[1,2,3]

A <u>collider</u> arrangement is chosen to reach the highest possible energy. A 20-TeV proton collider provides 40 TeV of available energy in the center of mass, whereas a 20-TeV fixed-target proton machine has only 0.194 TeV in the center of mass of a proton-nucleon collision.

The energy of <u>20 TeV</u> for each beam is determined by the particle physics to be investigated. High in priority will be new particles with masses in the TeV range that are produced by parton-parton (quark, anti-quark, gluon) collisions. Since each parton carries on the average only a third or so of the proton beam energy, 20 TeV is a practical minimum for investigating this physics domain.

A <u>proton-proton</u> collider, rather than a proton-antiproton collider, is chosen because of the greater intensity that can be obtained with protons. Although the antiproton, because of its antiquark composition, would be inherently advantageous in producing some new particles, this small advantage is outweighed by the greater (x10) luminosity available in proton-proton machines.

The <u>luminosity</u> and <u>number of interaction points</u> are chosen on the basis of the expected reaction rates, the pertinent background rates, and the decay modes and the general observability of the several basic particles that will be searched for. Among the quarry are the Higgs particle, supersymmetric particles, and possibly new gauge particles with masses in the TeV range. A luminosity of 10^{33} cm^{-2} s^{-1} can explore a mass range for a Higgs particle up to 1 TeV, whereas at 10^{32} only the range of up to 1/3 TeV can be covered. To be effective in discovering a Higgs particle, the criterion used is a minimum of 20 such events per year. The criteria for discovering other particles are different depending on their decay modes and the expected backgrounds. To discover new gauge particles, 10^3 events per year are thought to be required, and for gluinos 10^4 per year. In addition, it would be desirable, if practicable, to search for events of even greater rarity -- events so rare that a luminosity of 10^{34} would be needed.[3]

© 1989 American Institute of Physics

Thus the basic parameters of the SSC follow from the particle physics to be explored. The SSC will consist of two colliding storage rings for 20-TeV protons with up to six collision points and provide an average luminosity of up to 10^{33} cm^{-2} s^{-1} and, if feasible, provide for an eventual upgrade to 10^{34} cm^{-2} s^{-1}.

2 PRIMARY PARAMETERS OF THE SSC

Given the few basic requirements and the goals of the SSC, the form and structure of the machine follow in a fairly straightforward manner. Inherent in the design process is the need to minimize the overall cost. We shall limit our considerations to the primary parameters of the collider -- i.e., the relatively small set of principal parameters that define and describe the machine, its components, its layout, and the essential details of operation.

Let us begin by simply listing and briefly discussing the primary parameters. Later we will explore their interrelationships as we look for an optimum set.

2.1 Ring Size

The scale of the machine is set by the bending radius ρ of the dipole magnets.[1] The total circumference, $2\pi R$, is dominated by the total length of the dipoles, $2\pi\rho$. The straight sections for experiments and injection, the quadrupole lengths, and the many, many small intermagnet drift spaces all add up to only about 25% of the total circumference. Superconducting magnets using Nb-Ti superconductor in a cosine-theta winding configuration at a dipole field of 6.6 tesla have been chosen as a practical and optimum technology. At this field strength the 20-TeV magnetic radius is 10.1 km. Higher-field superconductors (e.g. Nb-Sn) exist, but their technologies need further development and at present they do not seem as cost effective as Nb-Ti. Superferric magnets also have been seriously considered -- that is, magnets using Nb-Ti superconductors but with the field shape determined by an iron pole rather than by conductor placement as in the cosine-theta type of magnet. Although the magnet unit cost for superferric magnets was less, the overall optimum (including tunnel cost) favored the cosine-theta design.

2.2 Bunched Beam

The Intersecting Storage Ring (ISR) at CERN used continuous, dc beams of protons. However, the use of bunched beams can produce the same luminosity with a smaller average beam, because a proton in one beam has a thicker "target" if it crosses a bunched beam. In the SSC this is a dominant consideration because the average beam level is limited by the cryogenic heat load produced by synchrotron radiation from 20-TeV protons. Low average current is desirable also to minimize beam-gas scattering background, beam stored energy, residual radioactivity in scrapers and beam dumps, and irradiation of electronics in the tunnel. The disadvantages of using a bunched beam are a higher peak beam current (which encourages collective beam instabilities) and also a low collision duty cycle (which can lead to the confusion caused by multiple reactions per detector resolving time).

Let us note for later use that the ratio of peak beam current i_b to average beam current I_b is

$$i_b/I_b = S_B/\sqrt{2\pi}\,\sigma_z \tag{1a}$$

where S_B is the bunch spacing and σ_z the rms bunch length.

The beam currents can, of course, be expressed in terms of the number of protons per bunch N_B, and the total number of protons per ring N_T:

$$I_b = eN_B\beta_r c/S_B = eN_T\beta_r c/(2\pi R) \tag{1b}$$

where $\beta_r c$ is the proton velocity and e is the electronic charge.

2.3 Luminosity-Related Parameters

The luminosity L is defined as the number of beam reactions per unit reaction cross section per second. For round beams colliding head-on with N_B protons per bunch f times per second,

$$L = N_B^2 f/4\pi\sigma^2 \tag{2}$$

where σ is the rms beam radius in a Gaussian density distribution, and the collision frequency $f = \beta_r c/S_B$.

It is convenient here to express the beam radius in terms of the beam emittance ϵ and the value of the lattice betatron-amplitude function β^* at the interaction point:

$$\sigma^2 = \epsilon\beta^* . \tag{3}$$

This substitution is useful because normally the emittance of a proton beam varies as

$$\epsilon = \epsilon_N/\beta_r\gamma_r \tag{4}$$

where the constant ϵ_N, called the normalized emittance, characterizes the phase-space area occupied by the beam and in most proton machines can, in principle, be an invariant of the acceleration process. The $\beta_r\gamma_r$ factor is the usual relativistic product proportional to proton momentum. In the SSC, β_r is so close to unity that we can safely use γ_r in place of $\beta_r\gamma_r$. At the collision energy of 20 TeV, $\gamma_r \simeq \beta_r\gamma_r = 2.13\times10^4$.

Using these relations we can rewrite the luminosity in terms of the machine parameters N_B, S_B, β^*, and ϵ_N, which we wish to optimize:

$$L = \frac{N_B^2 \gamma_r c}{4\pi\epsilon_N \beta^* S_B} \, . \tag{5}$$

Head-on collisions at the interaction points unfortunately lead to other collisions at spacings of $S_B/2$ until the two beams are separated. To avoid these unwanted collisions, the beams in the SSC cross at an angle in the range of 75 to 150 microradians. This geometry leads to a reduction in luminosity by the factor R_L:

$$R_L = \left[1 + \left(\frac{\alpha\, \sigma_z}{2\, \sigma}\right)^2\right]^{-1/2} \tag{6}$$

where α is the total crossing angle. This reduction factor of course increases slightly the required beam current for a given luminosity. However, it is a complication that can be conveniently postponed without serious error. We shall temporarily ignore this crossing-angle reduction factor so that the principal relationships between the parameters can more easily be recognized.

Another constraint is that there not be too many proton-proton reactions per bunch crossing, because multiple reactions generally cause confusion and ambiguity in the analysis of the observed reactions. The average number of reactions per bunch crossing $<n>$ is

$$<n> = L\, \Sigma_{inel}\, S_B/c \tag{7}$$

where Σ_{inel} is the proton-proton inelastic cross section. (Elastic events are not detected because the elastic deflections generally stay within the dynamic aperture.) At 20 TeV the inelastic cross section is estimated to be about 90 mb. Since in Eq. (7) the average luminosity L is fixed as a basic requirement (10^{33} cm^{-2} s^{-1}), the only parameter available for controlling $<n>$ is the bunch spacing S_B. Setting $<n>$ equal to 1 in Eq. (7), we find the corresponding value of bunch spacing to be 3.3 meters. Although such a spacing can be easily arranged in the storage ring, the corresponding recovery time in the detector system (11 nsec) is a difficult requirement.

The probability that exactly k interactions occur per bunch crossing when the average number is $<n>$ is given by a Poisson distribution. The corresponding effective luminosity L_k is

$$L_k = L <n>^k \exp(-<n>)/k! \, . \tag{8}$$

Thus for experiments requiring one and only one interaction per bunch crossing, the effective luminosity L_1 is

$$L_1 = L <n> \exp(-<n>) \, . \tag{9}$$

Thus for given L and Σ_{inel}, L_1 is a function only of S_B and is plotted in Figure 1. L_1 is a maximum of 0.368 L at $<n>$ = 1.

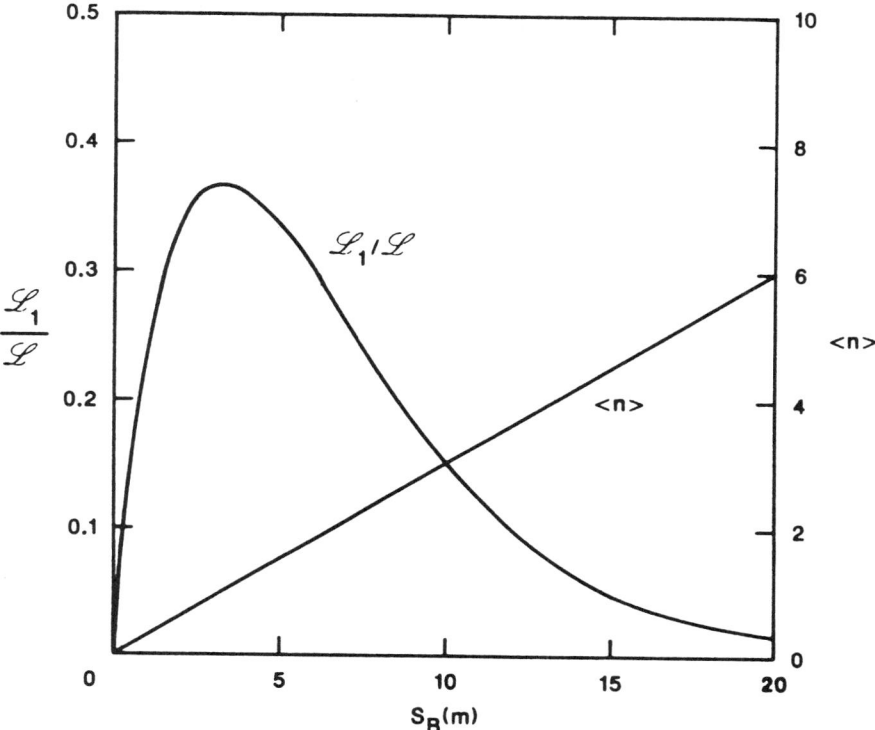

Figure 1. The average number of interactions per bunch crossing <n>, and the single-event luminosity relative to the total luminosity L_1/L, as functions of bunch spacing S_B.

However, by choosing S_B to be somewhat higher (for reasons to be discussed) the penalty in L_1 is seen to be tolerable.

The discussion to this point has considered an interaction region having a luminosity of 10^{33} cm^{-2} s^{-1}. The planned SSC configuration (see Figure 2) has eight lattice insertions, of which two are interaction regions designed to have this luminosity of 10^{33} (and are referred to as high-luminosity I.R.'s) but two are designed to have a luminosity about 20 times less (medium-luminosity I.R.'s) but have a much longer free space on each side of the interaction point. The other four lattice insertions will be built as utility straight sections, two of which will accommodate the injection, beam-abort, and radiofrequency-acceleration functions, and the other two will be reserved for future development as crossing I.R.'s.

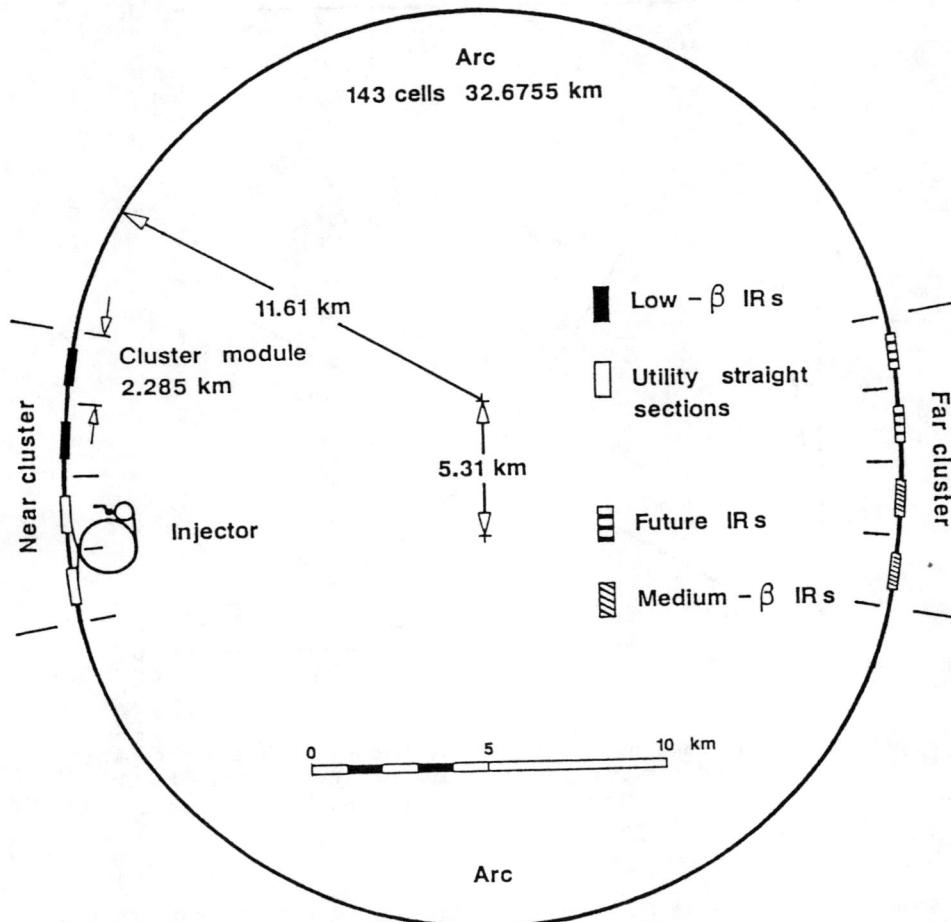

Figure 2. Layout of the Collider ring showing the eight lattice insertions arranged in two clusters. The Near Cluster on the left has two utility straight sections for the injection and abort systems and two low-beta, high-luminosity interaction regions. The Far Cluster on the right has two medium-beta, medium-luminosity interaction regions and the two other insertions configured initially as utility straight sections but reserved for future development as interaction regions.

2.4 Limiting Effects

There are several limitations to the parameter range that can be realistically considered.[1] The limiting effects that must be considered are synchrotron radiation, beam-beam tune shifts, beam loss due to proton-proton collisions, and beam instabilities.

2.4.1 Synchrotron Radiation Power

The total synchrotron radiation power per ring P is given by

$$P = \frac{377e^2c^2\gamma_r^4 N_T}{3(2\pi R)\rho}.$$ (10a)

At 20 TeV, 6.6 tesla (ρ = 10.1 km), and a circumference of 83.6 km, the synchrotron power per ring is

$$P \text{ (kW)} = 7.1 \times 10^{-14} N_T.$$ (10b)

This power is absorbed at 4° K and is the dominant cryogenic heat load in the SSC. It thus represents a major limitation for the design.

2.4.2 Beam-Beam Tune Shifts

As the two beams collide head-on, the particles in one beam are defocused by the electromagnetic field of the other. This highly nonlinear focusing is characterized by a "head-on beam-beam tune shift,"

$$\Delta v_{HO} = N_B r_p / 4\pi \varepsilon_N$$ (11)

where r_p is the classical proton radius (1.53×10^{-18} m). Too large a value of Δv_{HO} causes motion of individual beam particles to become unstable. A Δv_{HO} per crossing of about 0.004 (and a cumulative total tune spread of 0.024 for six crossing points) is an experimentally determined upper limit in the SPS collider at CERN. A similar limit is expected for the SSC.

In addition, there is a secondary, "long-range" beam-beam tune shift due to the electromagnetic interaction as the bunches in one beam come close to the counter-streaming bunches in the opposing beam in the free space on each side of the interaction points. If there were not the finite crossing angle α, these close encounters would, of course, be head-on collisions. There are a large number of such close encounters between the two beams, because they occur at intervals of $S_B/2$ and the free space before the beams are separated is about 80 meters from the interaction point in the high-luminosity interaction regions.

This long-range interaction between passing bunches produces a distortion in the equilibrium orbits and a shift in the single-particle tune. The magnitude of this "long-range beam-beam tune shift" is approximately

$$\Delta v_{LR} = N_B r_p n_{LR} / (2\pi \gamma_r \beta^* \alpha^2)$$ (12)

where n_{LR} is the number of long-range encounters in the interaction region. For typical SSC parameters, the long-range tune shift is about twice the head-on tune shift in the high-luminosity interaction areas. However, the nonlinear effects in the long-range interaction are less severe than in the head-on interaction so that a larger long-range tune shift is acceptable. [In the medium-luminosity interaction areas, the long-range tune shift is much smaller than in the high-luminosity areas because of the β* dependence shown in Eq. (12), whereas the head-on tune shift is about the same.]

2.4.3 Beam Loss due to Proton-Proton Collisions

The particle loss rate due to proton-proton collisions at the interaction point is a significant contribution to the beam lifetime. The partial beam lifetime due to this loss is

$$T_{pp} = N_T/\dot{N}_T = N_T/n_L L \Sigma_{inel} \tag{13}$$

where n_L is the effective number of interaction regions that utilize the full luminosity L. For the SSC, n_L is ~2.1. For the nominal SSC parameters, this partial lifetime is not a strong limitation, but it would become significant if the luminosity were raised to 10^{34} cm^{-2} sec^{-1}.

2.4.4 Beam instabilities

Beam instabilities can produce another limitation to the amount of beam that can be stored in the collider. There are many types of instabilities. Some depend on the peak current (or N_B) and some depend on the average current (or N_T). Some involve transverse degrees of freedom and some longitudinal. All of them must be carefully considered in the design of an accelerator. However, this subject is too large and involved to be adequately treated in this short article. It turns out that for the SSC parameters, the lowest threshold for a peak-current instability is a factor of about six higher than the nominal beam current. The lowest threshold for an average-current-dependent instability is exceeded, but these multibunch instabilities are relatively slow and will be controlled by feedback systems. Higher average currents, if required, could exceed the threshold of the next higher multibunch mode and thus require a second set of feedback systems.

3 PARAMETER OPTIMIZATION[1]

3.1 Parameter Interrelationships

The basic interrelationships between the primary parameters are expressed in the equations of Section 2. Let us explore those interrelationships assuming L = 10^{33} cm^2 s^{-1} at 20 TeV per beam (which are basic requirements), a circumference of 86.3 km, and a magnetic field of 6.6 tesla (at which ρ = 10.1 km) (which are determined from lattice and superconductor-magnet considerations).

Then it is convenient to express $\langle n \rangle$, N_B, N_T, i_b, I_b, L_1, P, T_{pp}, $\Delta\nu_{HO}$, and $\Delta\nu_{LR}$ as functions of the independent variables β^* (in meters), ε_N (in 10^{-6} rad-meters), and S_B (in meters):

$$\langle n \rangle = 0.30 S_B$$

$$N_B\ [10^{10}] = 0.44(\varepsilon_N \beta^*)^{1/2}\ S_B^{1/2}$$

$$N_T\ [10^{14}] = 3.71(\varepsilon_N \beta^*)^{1/2}\ S_B^{-1/2}$$

$$i_B\ [A] = 1.25(\varepsilon_N \beta^*)^{1/2}\ S_B^{1/2}$$

$$I_B\ [mA] = 213(\varepsilon_N \beta^*)^{1/2}\ S_B^{-1/2}$$

$$L_1\ [10^{33} cm^{-2} s^{-1}] = 0.30 S_B\ e^{-0.3 S_B}$$

$$P\ [kW] = 26.3(\varepsilon_N \beta^*)^{1/2}\ S_B^{-1/2}$$

$$T_{pp}\ [d] = 23[\varepsilon_N \beta^*]^{1/2}\ S_B^{-1/2}$$

$$\Delta\nu_{HO}\ [10^{-3}] = 0.54(\beta^*/\varepsilon_N)^{1/2}\ S_B^{1/2} \tag{14}$$

The expressions above have been evaluated for the case of zero crossing angles but are approximately correct for the range of possible crossing angle (up to 150 microradians), the corrections being around 10%. However, since the long-range tune shift does depend strongly on the crossing angle [Eq. (12)], we shall express its dependence on β^*, ε_N, and S_B for the nominal crossing angle $\alpha = 75$ microradians and a close-encounter distance of 80 meters on each side of the interaction point:

$$\Delta\nu_{LR}[10^{-3}] = 2.9(\varepsilon_N/\beta^*)^{1/2}\ S_B^{-1/2}\ . \tag{15}$$

3.2 Effects of Varying β^* and ε_N

Most of the quantities in Eqs. (14) and (15) depend on the product $(\varepsilon_N \beta^*)^{1/2}$, which is the rms beam width at the interaction point. However, the two beam-beam tune shifts depend on the ratio $(\beta^*/\varepsilon_N)^{1/2}$, and in opposite ways. Reducing the beam size at the interaction point reduces the beam current needed for the required luminosity and so is generally beneficial. Thus we can consider the smallest values of ε_N and β^* that can reasonably be obtained and then determine whether such choices produce any difficulties. The minimum value of normalized emittance ε_N is determined principally by achievable ion-source brightness. Sources are available that can satisfy the SSC requirements with ε_N somewhat less than 1×10^{-6} radian-meter. Choosing $\varepsilon_N = 1 \times 10^{-6}$ is conservative and allows the possibility of a later upgrade. The minimum practical value of β^* for the SSC is about 0.5 meter and is set by the maximum quadrupole-gradient and aperture considerations. (A lower value of β^* may be possible through the use of lower temperature or special superconducting material in the strong quadrupole triplets near the interaction points or by decreasing the experimental free space.)

The three quantities that are not improved by minimizing the product $\epsilon_N \beta^*$ are the partial beam lifetime T_{pp} and the head-on and long-range tune shifts. However, with $\epsilon_N = 10^{-6}$ radian-meter and $\beta^* = 0.5$ meter, these quantities are all at acceptable levels (over a range of values for the bunch spacing s_B, to be considered next).

3.3 Effect of Varying S_B

Having fixed the design values of ϵ_N and β^*, we must select S_B by optimizing among the conflicting set of relationships expressed in Eqs. (14) and (15).

In addition to these relationships, S_B is subject also to certain quantization conditions, namely (a) the RF condition that the SSC circumference be an integral multiple of S_B and (b) that every interaction-point spacing be an integral multiple of $S_B/2$ in order that collisions occur at all possible interaction points. [For the SSC geometrical arrangement, condition (b) also satisfies condition (a).] However, these quantization conditions are not restrictive. S_B has many values that satisfy them so that we can treat it as a continuous variable in the optimization and later pick the quantized value closest to the optimum.

The relationships of Eqs. (14) and (15) are illustrated in Figures 1, 3, and 4. As mentioned in Section 2.3, Figure 1 shows that the single-event luminosity L_1 is a maximum at $S_B = 3.3$ meters. However, as shown in Figures 3 and 4, larger values of S_B are favored by considerations of synchrotron-radiation power, long-range tune shift, total beam stored energy, average-current-sensitive instabilities, and beam lifetime due to collisions. In addition, the detector recovery time requirement (11 nsec at $S_B = 3.3$ meters) is relaxed by going to larger S_B, and the bandwidth requirements for feedback systems are similarly softened.

The two factors favoring shorter values of S_B are seen in Figure 3 to be the head-on tune shift and the number of protons per bunch (and peak-current-sensitive beam instabilities).

From these considerations a value of S_B near 5 meters seems to be an overall optimum. Relative to the parameters at $S_B = 3.3$ meters, several important considerations are significantly improved while the single-event luminosity is reduced from its maximum by only about 10%. The resultant head-on tune shift and the number of protons per bunch, although appreciably larger than at $S_B = 3.3$ meters, are still quite tolerable. Finally, the quantization conditions set the exact value of the bunch spacing as 5.078 meters in the current SSC design, which uses a lattice with 90° betatron phase shift per cell and has a circumference of 83.631 km.

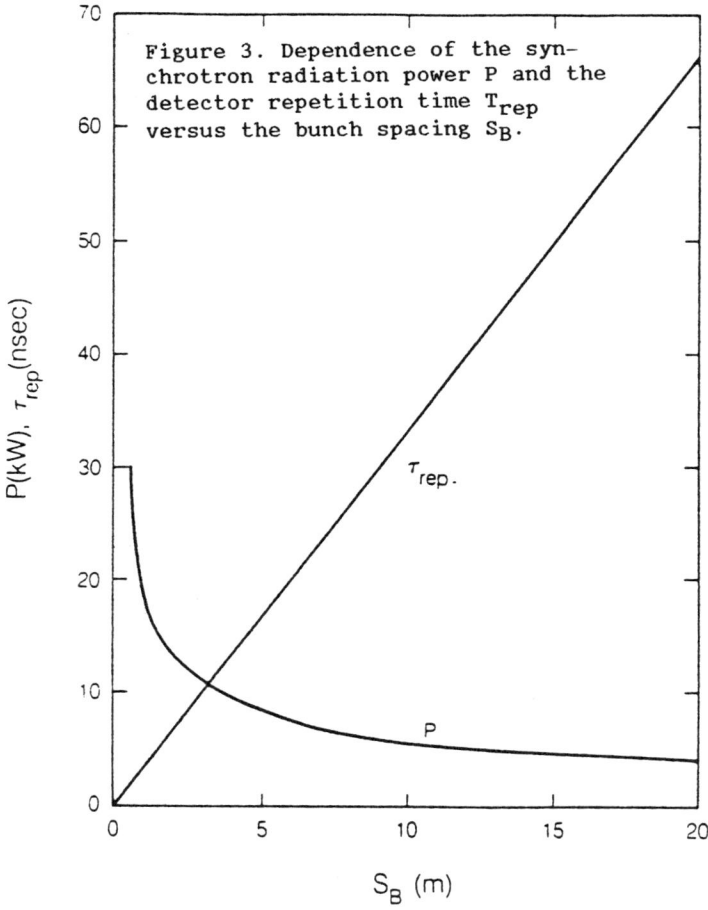

Figure 3. Dependence of the synchrotron radiation power P and the detector repetition time T_{rep} versus the bunch spacing S_B.

3.4 Crossing Angle

For completeness we should include a short discussion of the two parameters that we have mentioned but so far have largely neglected in order to simplify the discussion. The crossing angle α and the rms bunch length σ_z both weakly affect the luminosity reduction factor R_L [Eq. (6)], and α strongly affects the long-range tune shift [Eq. (12)].

A finite crossing angle is necessary in order to enjoy the benefits of a small bunch spacing without having many collision points in the experimental straight sections. To minimize the consequent reduction in luminosity, the product $\alpha\sigma_z$ should be no larger than the rms beam width σ at the collision point, which favors a small crossing angle. The long-range tune shift, on the other hand, favors a large crossing angle.

A rms bunch length of about 7 cm follows from the rms energy spread ($\sigma_E/E = 1.5 \times 10^{-4}$) needed at injection to safely avoid the

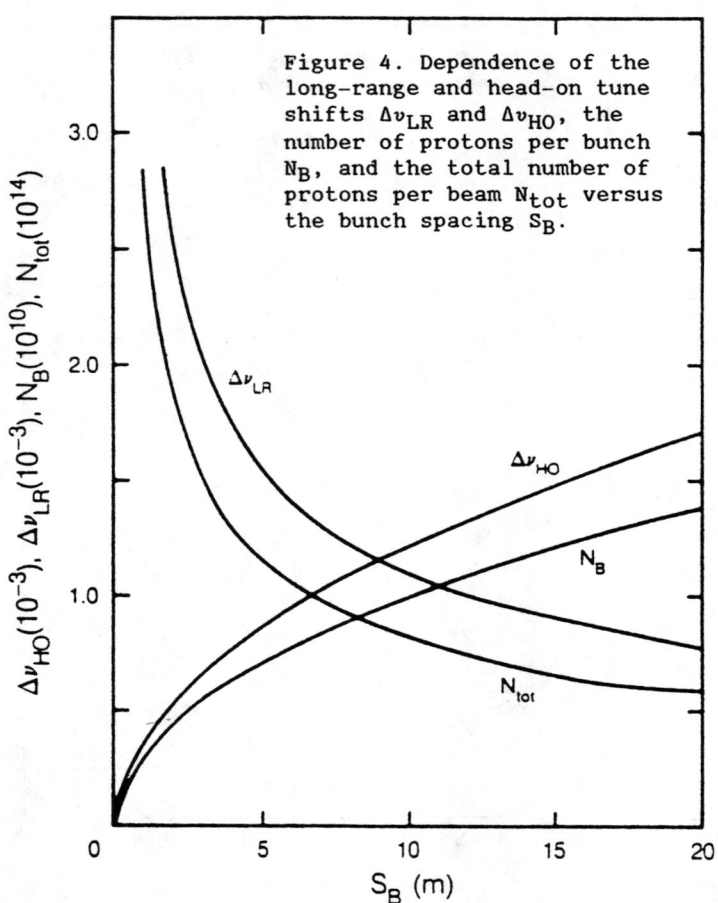

Figure 4. Dependence of the long-range and head-on tune shifts $\Delta\nu_{LR}$ and $\Delta\nu_{HO}$, the number of protons per bunch N_B, and the total number of protons per beam N_{tot} versus the bunch spacing S_B.

single-bunch instabilities. These two parameters are tied together through the properties of the RF bucket.

A further small complication is the effect of the crossing angle on the head-on tune shift. This effect is different in the two planes: Eqs. (11) then is rewritten as

$$\Delta\nu_{HO} = \frac{N_B r_p}{4\pi\varepsilon_N}\left[\frac{2R_L^2}{1+R_L}\right] \quad \text{in the crossing plane}$$

and

$$\Delta\nu_{HO} = \frac{N_B r_p}{4\pi\varepsilon_N}\left[\frac{2R_L}{1+R_L}\right] \quad \text{in the non-crossing plane.} \qquad (16)$$

Consideration of these various crossing-angle effects has resulted in the choice of α = 75 microradians, which allows both an acceptable luminosity reduction factor (R_L = 0.88) and acceptable beam-beam tune shifts.

A final small correction is required to allow for gaps in the beam. Gaps are required to allow for the rise time of the kicker magnet when the beam must be ejected (aborted) to avoid damage that would result if the beam energy (about 400 megajoules) were to strike the walls of the vacuum chamber. In addition to the 3-microsecond abort gap, 335 other smaller gaps are required for the rise times of several injection kicker magnets in the injector chain to provide the 336 beam packets from the Low Energy Booster that fill each ring of the supercollider. In all, these gaps produce a filling factor of 0.938.

4 PRIMARY PARAMETER SUMMARY

The primary parameter set is now complete: Choosing β* = 0.5 m, ϵ_N = 10^{-6} rad-m, S_B = 5.08 m, α = 75 μrad, and σ_z = 6.8 cm, and using Eqs. (14) and (15), we find the rest of the parameter set (without all the crossing-angle and filling-factor effects) as shown in the first column of Table I. For comparison are listed also the values with all the fine corrections included.

Table I. The derived SSC primary parameters without [Eqs. (14) and (15)] and with crossing-angle and filling-factor corrections for the case S_B = 5.08 m, ϵ_N = 10^{-6} rad-m, β* = 0.5 m, α = 75 μrad, σ_z = 6.8 cm.

	Without	With Corrections		
$\langle n \rangle$	1.5	1.7		reactions/crossing
L_1	0.33×10^{33}	0.31×10^{33}	$cm^{-2} s^{-1}$	1-event luminosity
N_B	7.0×10^9	8.0×10^9		protons/bunch
N_T	1.16×10^{10}	1.24×10^{14}		protons/ring
i_B	2.0	2.7 A		pk. beam current
I_B	67	71 mA		av. beam current
P	8.3	8.8 kW		synch. power
R_L	1.0	0.88		lum. reduction factor
T_{pp}	7.2	7.5 days		collision lifetime
Δv_{HO}	0.86×10^{-3}	0.81×10^{-3}		in crossing plane
Δv_{HO}	0.86×10^{-3}	0.92×10^{-3}		in noncrossing plane
Δv_{LR}	1.8×10^{-3}	2.0×10^{-3}		long-range tune shift
σ	4.8×10^{-6}	4.8×10^{-6} m		beam size, collision

These parameters refer to the high-luminosity interaction regions.

For the medium-luminosity interaction regions, the parameters (with corrections) that are different are the following:

β^*	10.0 m
R_L	0.99
L	5.6×10^{31} cm^{-2} s^{-1}
$<n>$	0.10
L_1	4.9×10^{30} cm^{-2} s^{-1}
Δv_{HO}	0.97×10^{-3} (both planes)
Δv_{LR}	2.1×10^{-4}
σ	21.7×10^{-6} m

It should be noted that there was considerable flexibility in arriving at this set of parameters. Should any of these values prove difficult or expensive to achieve, there is reasonable latitude in choosing other equivalent parameter sets that also can satisfy the primary requirements of the supercollider.

5 REFERENCES

1. SSC-SR-2020, Conceptual Design of the Superconducting Super Collider, March 1986.

2. R. Donaldson and J.G. Marfin, Editors, Proceedings of the 1984 Summer Study on the Design and Utilization of the Superconducting Super Collider, July 1984.

3. R. Donaldson and J. Marx, Editors, Proceedings of the 1986 Summer Study on the Physics of the Superconducting Super Collider, July 1986.

SUPERCOLLIDER PHYSICS

CHRIS QUIGG

SSC Central Design Group
Lawrence Berkeley Laboratory, Berkeley, California 94720 USA

Table of Contents

Where We Stand	2257
Why We Believe in Quarks	2258
Gauge Theories of the Fundamental Interactions	2265
Quantum Chromodynamics	2268
Electroweak Theory	2270
Unification of Fundamental Interactions	2272
How Far We Have to Go	2273
The Next Step	2275
The Superconducting Super Collider	2277
Acknowledgments	2279

Lecture given November 16, 1987, at Collège de France, Paris, as
"La Physique à 40 TeV au Supercollisionneur Proton-Proton"
in the 1987 Bernard Gregory Lecture Series

SUPERCOLLIDER PHYSICS

CHRIS QUIGG

SSC Central Design Group[*]
Lawrence Berkeley Laboratory, Berkeley, California 94720 USA

Over the past two decades, the age-old struggle to describe and comprehend the nature of elementary particles and forces has been rewarded by a radically new and simple picture of Nature. This progress has, in large measure, been stimulated by experimental results from particle accelerators, instruments of the kind given to us by Bernard Gregory and his colleagues. We have learned that all matter in its diverse forms is assembled from a few building blocks called quarks and leptons. All known natural phenomena can be described in terms of a few fundamental forces acting among these basic constituents. The new insights embodied in the Standard Model of elementary-particle physics not only provide a framework for describing and understanding the world around us, but also elucidate the first instants after the creation of the universe.

However, as our conception of matter has become better established, we have become increasingly aware of its shortcomings. Although the Standard Model represents a great step toward a complete understanding of the structure of matter, the 1970s brought not only multifarious experimental support for that theoretical framework, but also the realization that it is incomplete. We do not believe that within the energy range now available there can be enough clues for us to piece together a comprehensive theory of the nature of matter. This conviction—together with our awareness of the value of exploration, our appreciation of the countless instances when unexpected, puzzling, and illuminating new observations have come out of accelerator experiments—motivates our quest for higher energies.

My first order of business will be to summarize the microscopic description of matter to which we have come in the past twenty years. I will review the evidence for quarks and leptons as fundamental constituents and explain the strategy of gauge theories of the fundamental interactions. Next, I will discuss questions the Standard Model raises but cannot answer. We can define a frontier where our current understanding ceases to make sense, where the clues that will lead us to a more satisfying description of Nature will have to be found. That frontier lies at energies of about 10^{12} electron volts for collisions among the fundamental constituents. The instrument of choice for reaching this new energy scale is a high-energy, high-luminosity, proton-proton collider. I will conclude by summarizing the status of the American project known as the Superconducting Super Collider, or SSC.

---

[*]Operated by Universities Research Association, Inc., for the U.S. Department of Energy.

Where We Stand

The Standard Model, shown schematically in Figure 1, has emerged from the identification of leptons and quarks as fundamental constituents and the development of gauge theories of the strong, weak, and electromagnetic interactions. Because gauge theories provide a common mathematical framework, we see before us the prospect of a unification of these three interactions and the hope of a simpler and more comprehensive description of Nature.

The leptons experience weak and electromagnetic—but not strong— interactions. We know of six species: the electron, muon, and tau, all electrically charged, and their neutral partners, the neutrinos. The interactions of the electron neutrino and muon neutrino have been observed directly. We know a great deal about the interactions and properties of the tau neutrino and know that it must be a sequential neutrino, distinct from the electron neutrino and antineutrino or the muon neutrino and antineutrino. However, no experiment has been carried out in which a beam of tau neutrinos is produced, penetrates a large column of matter, and interacts in a target to produce a tau. This "three-neutrino experiment" would provide the final demonstration that the tau neutrino exists, but the outcome is not in doubt.

All the leptons are spin-$\frac{1}{2}$, pointlike, Dirac particles, structureless at the current limits of our resolution, about 10^{-16} cm. Much is known about their static properties such as magnetic moments and masses. It is noteworthy that the

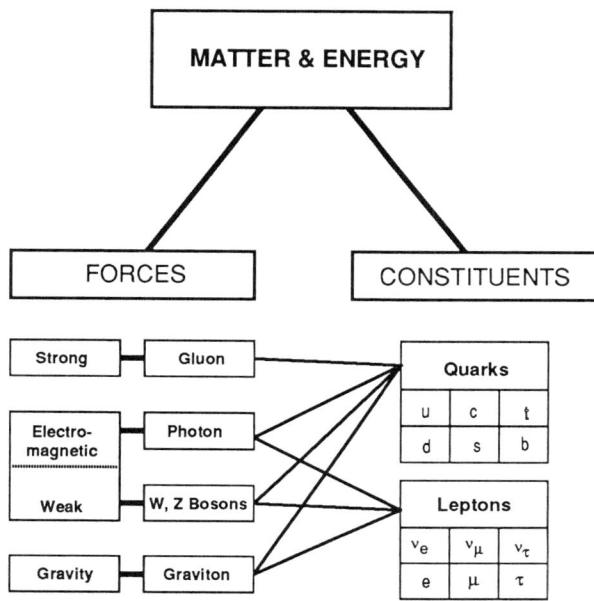

Figure 1: The Standard Model of Elementary-Particle Physics.

leptons fall into weak-interaction families

$$\begin{pmatrix} \nu_e \\ e \end{pmatrix}_L, \begin{pmatrix} \nu_\mu \\ \mu \end{pmatrix}_L, \text{ and } \begin{pmatrix} \nu_\tau \\ \tau \end{pmatrix}_L, \tag{1}$$

where the subscript L signals that the charged-current interaction is left-handed.

The other great class of particles we can study in the laboratory consists of the hadrons, which participate in the strong as well as the weak interactions. There are many hundreds of species, including the proton, neutron, pion, kaon, and Δ-resonance. The hadrons comprise both fermions, which have half-integral spins, and bosons, which have integral spins. All are composite, with sizes on the order of 10^{-13} cm in radius. The hadrons may be exceedingly stable, like the proton, whose lifetime of more than 10^{31} years considerably exceeds the age of the universe, or quite ephemeral, like the Δ, whose lifetime is about 10^{-24} second.

Order is brought to this diverse collection of strongly interacting particles by the idea that hadrons are made up of quarks, which, like the leptons, are spin-$\frac{1}{2}$, pointlike particles, smaller than about 10^{-16} cm in radius. Unlike the leptons, the quarks are not seen directly in the laboratory. It is therefore valuable to spend a few moments recalling the variety of experimental evidence that supports the quark model of the hadrons.

Why We Believe in Quarks

The original motivation for the quark model came from the spectroscopy of the strongly interacting particles developed in the early 1960s. The unitary symmetry group $SU(3)$ was found to be a good classification symmetry for the hadrons then known. In contrast to the experience with the rotation group for angular momentum, or with $SU(2)$ for isospin, it was observed that only a few of the low-lying representations of $SU(3)$ were populated with hadrons. The mesons, hadrons of integer spin like the pion, appeared in families of one or eight members, but not of three, or six, or twenty-seven. The baryons, particles of half-integer spin like the proton, appeared only in families of one, eight, or ten members.

This pattern can be reproduced if we say that the hadrons are built up out of a fundamental triplet of quarks, an isospin doublet called up and down, and an isoscalar strange quark that carries -1 unit of strangeness:

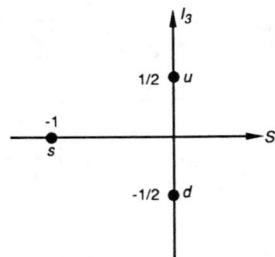

If we make the rule that mesons are composed of a quark and antiquark, $(q\bar{q})$, then according to the arithmetic of $SU(3)$, $3 \otimes 3^* = 1 \oplus 8$, so that mesons should occur in one- and eight-member families. Similarly, if baryons are composed of three quarks, (qqq), then the rules of $SU(3)$ yield $3 \otimes 3 \otimes 3 = 1 \oplus 8 \oplus 8 \oplus 10$, so that baryons should occur in one-, eight-, and ten-member families. It remains to be understood why quarks have only been observed in these combinations.

The first evidence that quarks are real, rather than merely mnemonic devices, came from experiments carried out in the late 1960s at the Stanford Linear Accelerator Center. A beam of high-energy electrons is scattered from a target, as shown in Figure 2. You observe the direction and momentum of the scattered electron and, if you wish, observe something about the recoil particle or particles. The goal of the experiment is to measure the cross section as a function of the angle and energy of the scattered electron and to understand what that reveals about the internal structure of the target.

To appreciate the significance of the SLAC experiment it is useful to look at its historical antecedent. Take as a target a carbon nucleus, scatter electrons from it, and require that the carbon nucleus remain intact after the collision. In this way we are studying the reaction

$$\text{electron + carbon nucleus} \rightarrow \text{electron + carbon nucleus}. \quad (2)$$

If you hit the carbon nucleus very hard, it is likely to fly apart, because it is a loosely bound collection of protons and neutrons (or perhaps of alpha particles). By requiring that the nucleus remain intact you are selecting a very rare occurrence. The rate at which this elastic scattering process occurs decreases rapidly as the momentum transferred to the target increases. This is illustrated in Figure 3(a).

On the other hand, if you relax the constraint that the carbon nucleus remain intact and simply observe the scattered electron without regard to the details of the recoil system, then you find that the cross section is almost independent of how sharp a blow is delivered. This is indicated by the broken line in Figure 3(a). We interpret this constant cross section as evidence that the electrons are being scattered from small charged objects within the nucleus, constituents that are structureless at this resolution. Those are the protons, as may be verified by examining the debris from the shattered carbon nucleus.

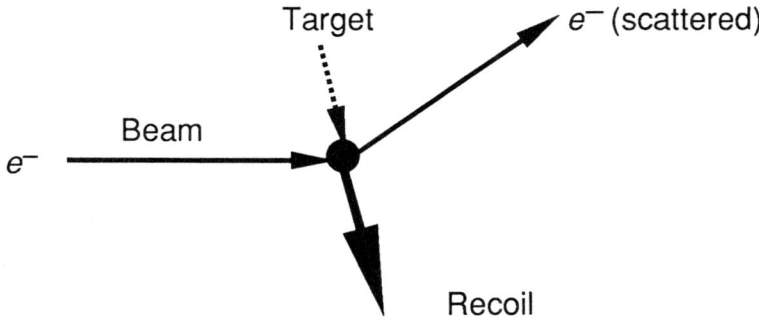

Figure 2: Kinematics of electron scattering.

It is a natural step to repeat these measurements with the proton as target and with beams of increasing energy. The results are indicated in Figure 3(b), where the scale of momentum transfer has been increased by two orders of magnitude. When inspected at this much higher resolution, the proton no longer behaves like a point particle. The structure of the proton is reflected in the fact that the rate for the reaction

$$\text{electron} + \text{proton} \to \text{electron} + \text{proton} \tag{3}$$

falls off rapidly. The proton tends to become excited or to produce new particles when hit hard.

If we relax the constraint that the proton survive the collision intact, we find that, just as for the scattering of an electron from a nucleus, there is a contribution to the "inelastic" cross section that is essentially independent of how hard the proton is hit. Just as we inferred from the earlier experiments with nuclear targets the presence of charged, structureless objects—the protons—within the nucleus, it is tempting to conclude that there are charged, structureless objects within the proton itself. This role is naturally played by the quarks.

Unlike the protons within the nucleus, the quarks within a proton cannot readily be liberated. Our knowledge of the properties of quarks, like our belief in the physical reality of quarks itself, therefore rests on indirect evidence. What can be said about the properties of the quarks?

- From the observation that there are three quarks in a baryon we conclude that each quark has baryon number 1/3.
- The quark charges are given by the Gell-Mann–Nishijima formula as

$$Q = I_3 + \frac{1}{2}(B+S) = \begin{cases} 2/3 & u \\ -1/3 & d \\ -1/3 & s \end{cases}, \tag{4}$$

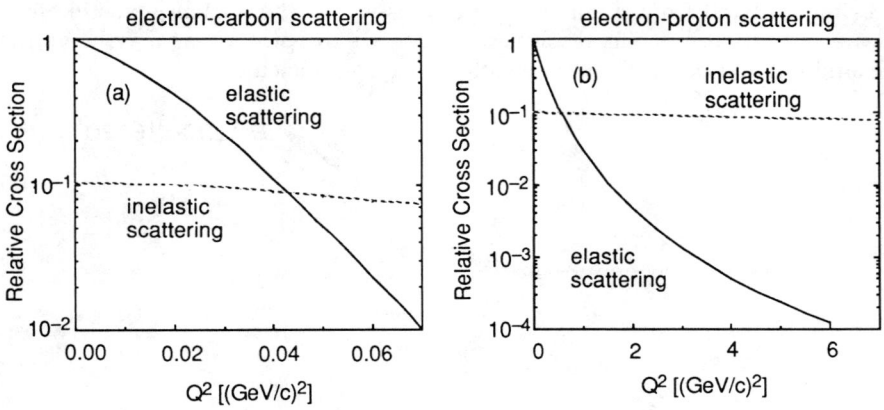

Figure 3: Elastic and inelastic cross sections for (a) electron-carbon scattering; (b) electron-proton scattering.

where B denotes baryon number and S strangeness. These assignments can be tested in several ways. We can note at once that they reproduce the charges (2,1,0,–1) of the baryons. Another test is to look at the decay rates for spin-one particles made of a quark and an antiquark, the so-called vector mesons. These particles decay (rarely) into electron-positron pairs. In the quark model, this decay occurs when the quark and the antiquark annihilate into a virtual photon that subsequently disintegrates, according to the laws of quantum electrodynamics, into the electron-positron pair. The rate at which this leptonic decay occurs is proportional to two parameters: the square of the electric charge carried by the quark and antiquark, and the probability for the quark and the antiquark to get together and annihilate, measured by the square of the bound-state wave function at zero separation, $|\psi(0)|^2$. Taking the wave functions to be similar for all the light vector mesons, ρ, ω, and ϕ, we may test the charge assignments of the quarks that make them up by comparing the leptonic decay rates. The relative rates are just as expected.

• Quarks have spin-$\frac{1}{2}$. The multiplet structure of the hadrons exhibits precisely the pattern given by the quark model. For the mesons, the order of levels follows the sequence

$$J^{PC} = \begin{array}{cc} 0^{-+}\ 1^{--} & 0^{++}\ 1^{+-}\ 2^{++} \\ L=0 & L=1 \end{array} \cdots . \tag{5}$$

This pattern is evident for the "light mesons" made up of up, down, and strange quarks, but it is exhibited most clearly by the bound states of two heavy quarks. I show in Figure 4(a) the spectrum of states composed of a charmed quark and antiquark; the states composed of the fifth quark, called bottom or beauty, and its antiquark are shown in Figure 4(b). The order of levels is just what we expect for bound states of a spin-$\frac{1}{2}$ particle and antiparticle. The analysis of baryon spectra is more involved. It can be summarized in the statement that the $SU(6)=SU(3)_{flavor} \otimes SU(2)_{spin}$ classification of baryons, which hangs upon the spin-$\frac{1}{2}$ nature of the quarks, is highly successful.

Independent evidence for the spin of the quarks—and for the reality of the quarks—comes from the reaction

$$\text{electron + positron} \rightarrow \text{hadrons (mostly pions)} . \tag{6}$$

In the quark-parton model, the electron and positron annihilate into a virtual photon that disintegrates according to the rules of quantum electrodynamics into a quark-antiquark pair. The quark and antiquark are not observed directly. They materialize, by a process not yet completely understood, as hadrons collimated along the directions of the quark and antiquark. The resulting two-jet events are commonplace at high energies; one is shown in Figure 5. The colliding beams are perpendicular to the plane of the page and the projection shown is an end-on view of the cylindrical detector, about two meters in diameter. Charged particles are indicated by their curved tracks in a solenoidal magnetic field, and neutrals are detected by the deposition of energy in the calorimeter cells. The trajectories of neutrals are reconstructed as dotted lines. With this indication that the directions of the quark and the antiquark in the semifinal state may be inferred from the jet directions, we may interpret the angular distribution of the jets with

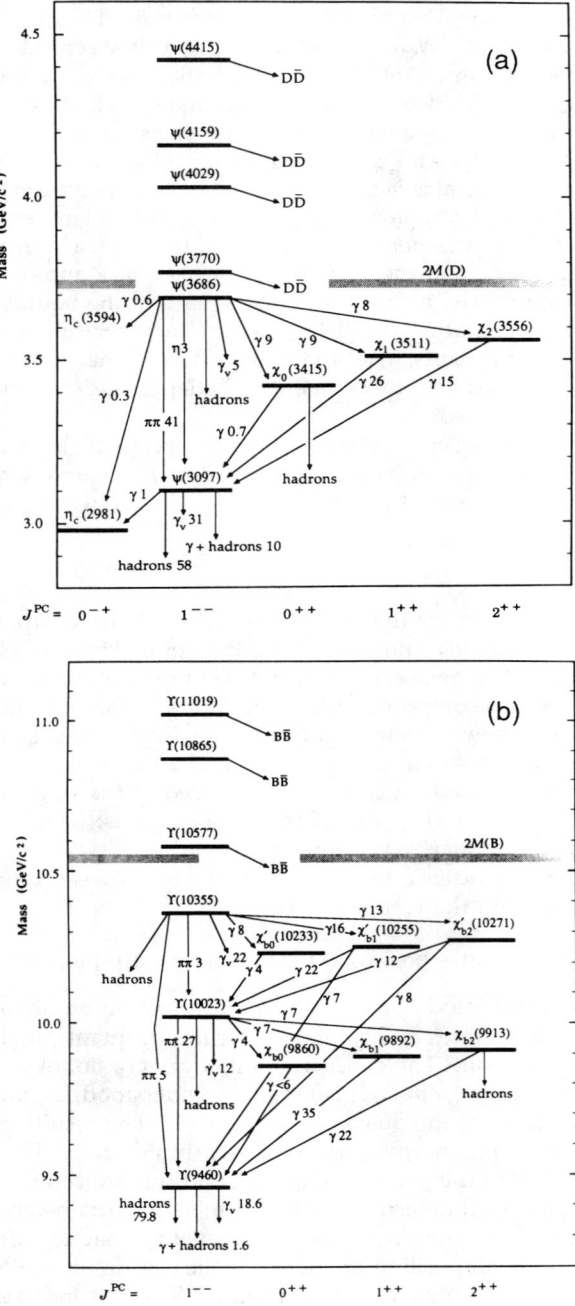

Figure 4: (a) The charmonium states (first observed in 1974);
(b) the upsilon family (first observed in 1977).

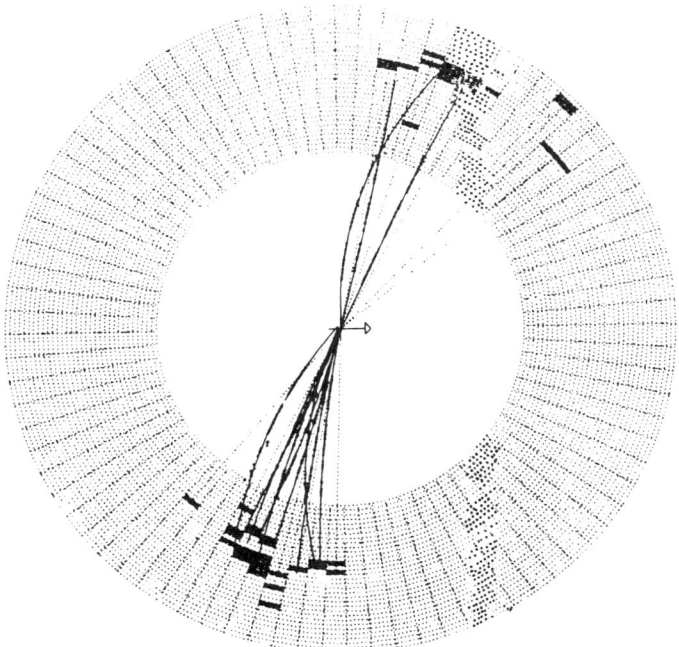

Figure 5: A two-jet event produced in 30-GeV electron-positron annihilations in the JADE Detector at the PETRA storage ring in Hamburg.

respect to the beam axis as the angular distribution of the quark-antiquark pair. The measured distributions are given by

$$\frac{dN}{d(\cos\theta)} \propto 1 + \cos^2\theta, \tag{7}$$

characteristic of the decay of the $J^P = 1^-$ photon into a pair of spin-$\frac{1}{2}$ particles.

This summarizes some of the evidence for the existence of quarks and for their detailed properties. The quark model reproduces much of what we know about the spectrum and interactions of hadrons, but in the form we have described so far it is not completely consistent. In building models of physical phenomena, it has paid off through the years to respect the grand principles that have great force and wide applicability. One such is the Pauli exclusion principle, which is a reliable guide to the construction of the periodic table of the elements. We can derive the exclusion principle, or more precisely the spin-statistics connection, from quantum field theory, so it should apply to quarks as well as to electrons.

The problem is that, for particles such as the Δ^{++} and the Ω^-, the quark model does not respect the Pauli principle. The Δ^{++}, for example, is a pion-nucleon resonance with a mass of about 1232 MeV/c^2. In the quark model it is a uuu state with spin = 3/2 and isospin = 3/2, a member of the ground-state supermultiplet in which all pairs of quarks are in relative s-waves. Thus it is apparently a symmetric state of three identical fermions. Unless we are prepared to suspend the rules of quantum mechanics or to give up the quark model, it is necessary to invoke a new, three-valued, hidden degree of freedom in terms of which the Δ^{++} wave function may be antisymmetrized. This new degree of freedom is called color, and quarks are given the labels red, green, and blue. Any hadron will therefore be colorless: a baryon will be a color-singlet mixture of red, green, and blue quarks, and a meson will be a color-singlet mixture of red-antired, green-antigreen, and blue-antiblue quark-antiquark pairs.

Described in this way, the introduction of color seems arbitrary and artificial. However, a number of observables are sensitive to the number of distinct quark species. Subsequent measurements of these quantities have given strong support to the color hypothesis, which has become the foundation of our understanding of the strong interaction.

We have already seen that the cross section for the inclusive production of hadrons in electron-positron annihilations is described by the elementary process

$$\text{electron} + \text{positron} \rightarrow \text{quark} + \text{antiquark} , \tag{8}$$

where the quark and antiquark materialize with unit probability into the observed hadron jets. At a particular energy, the ratio

$$R \equiv \frac{\sigma(e^+e^- \rightarrow \text{hadrons})}{\sigma(e^+e^- \rightarrow \mu^+\mu^-)} \tag{9}$$

is then simply given as

$$R = \sum_{\text{flavors}} e_q^2 . \tag{10}$$

At the highest energies explored, pairs of up, down, strange, charmed, and bottom quarks are kinematically accessible. In the absence of hadronic color we would therefore expect

$$R = e_u^2 + e_d^2 + e_s^2 + e_c^2 + e_b^2 = \frac{11}{9} , \tag{11}$$

but if each quark flavor exists in three distinct colors, we should have

$$R = 3(e_u^2 + e_d^2 + e_s^2 + e_c^2 + e_b^2) = \frac{11}{3} . \tag{12}$$

The colored-quark prediction is in excellent agreement with the data summarized in Figure 6. These data show, by the way, that the mass of the top quark (whose existence is required for the consistency of the electroweak theory and implied by the characteristics of b-quark decay) must exceed about 26 GeV/c^2. Proton-antiproton collider experiments extend the lower bound to about 40 GeV/c^2.

This completes our survey of the fundamental constituents. Let us now turn our attention to the fundamental interactions and to the gauge theories that describe them.

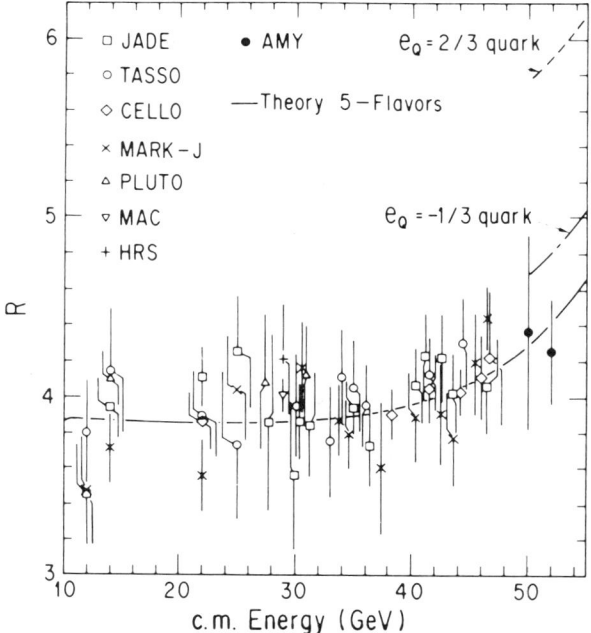

Figure 6: The ratio $R \equiv \sigma(e^+e^- \to \text{hadrons})/\sigma(e^+e^- \to \mu^+\mu^-)$.

Gauge Theories of the Fundamental Interactions

Quantum electrodynamics is in many ways the very model of a successful physical theory. It is renormalizable, calculable in perturbation theory, and in good agreement with experiment. The nine-digit accord between calculation and measurement for the gyromagnetic ratio of the muon sets a standard for other theories to equal. QED has now been subsumed in the Weinberg-Salam theory of the weak and electromagnetic interactions. Though tested only at the level of a few percent—the onset of sensitivity to loop corrections—the electroweak theory has many experimental successes: the prediction of neutral weak currents and charm, and the prediction of the existence and properties of the intermediate bosons $W^\pm$ and Z^0 found in the proton-antiproton collider experiments at CERN in 1983. At the level on which we have been able to test it, the electroweak theory provides a quantitative description of all electroweak phenomena.

The theory of the strong interactions, called quantum chromodynamics, is based on the color symmetry of the quarks. It provides insight into the systematics of hadron structure and hadronic interactions, and it predicted the existence of gluons, the carriers of the strong force. Because the strong interactions are, in most circumstances, strong, it has not yet been possible to derive from QCD many precise, testable predictions. A few quantitative successes, in which experiment agrees with theoretical predictions reliable within a few tens of percent, are known in the regime in which perturbation theory applies.

This is not the place to treat comprehensively theories of the fundamental interactions. However, it is appropriate to review the strategy for formulating a gauge theory, and to summarize the basic hypotheses that underlie our understanding of the strong, weak, and electromagnetic interactions.

Building a gauge theory is easier done than said. I shall first outline the steps and then carry out the construction of a theory in a simple example. First, we recognize conservation laws, or, equivalently, we notice symmetries in Nature, and build equations of physics that respect the symmetries in question. Having accomplished that, we try to impose the symmetry in a stricter form. When the new requirement is imposed, the equations of physics from which we began will have to be modified to accommodate the stricter form of the symmetry. This can be done in a mathematically consistent way only by introducing new interactions and new particles to carry those interactions.

To see what this outline means, suppose that we knew the Schrödinger equation, but not the laws of electrodynamics. Would it be possible to derive Maxwell's equations from a symmetry principle? The answer is yes! It is instructive to trace the steps in detail.

A quantum-mechanical state is described by a complex Schrödinger wave function $\psi(x)$. Quantum-mechanical observables involve inner products—expectation values of Hermitian operators—of the form

$$\langle O \rangle = \int \psi^* O \psi, \tag{13}$$

which are unchanged by a global phase rotation:

$$\psi(x) \to e^{i\theta} \psi(x). \tag{14}$$

In other words, the absolute phase of the wave function cannot be measured and is a matter of convention. Relative phases between wave functions, as measured in interference experiments, are unaffected by such a global rotation. To emphasize the role of symmetry, we may say that ordinary quantum mechanics is invariant if we change our convention for zero phase uniformly everywhere in space. This is illustrated in Figure 7. Let each arrow represent the convention for a complex number with zero phase, i.e., a real number, at different positions in space—say at each seat in the lecture hall. With all its arrows pointing to the right (east), Figure 7(a) indicates the convention we have been trained to consider natural. A new convention for zero phase, directed toward the southeast, is displayed in Figure 7(b). What we have just seen is that the predictions of quantum mechanics do not depend on the orientation of the arrows.

This raises the question: Are we free to choose one phase convention in Paris and another in Geneva, or indeed a different convention at each point in space, as indicated in Figure 8? Differently stated, can quantum mechanics be formulated to be invariant under local (position-dependent) phase rotations

$$\psi(x) \to \psi'(x) = e^{i\alpha(x)} \psi(x) \ ? \tag{15}$$

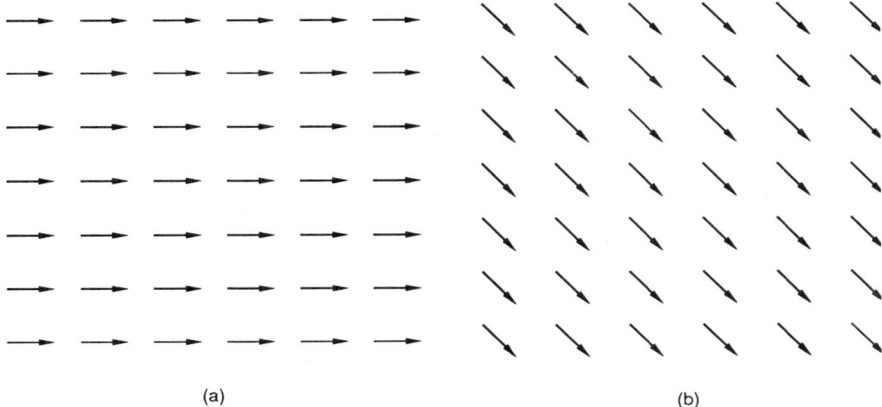

Figure 7: Arrows represent the convention for a complex number of zero phase at different points in space. (a) Original convention for zero phase; (b) new convention for zero phase.

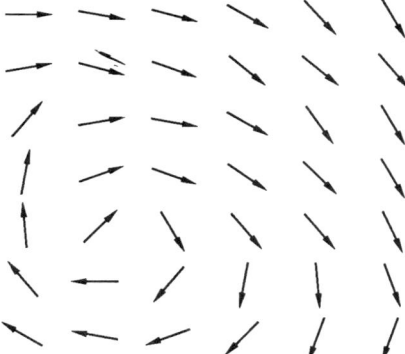

Figure 8: Arrows indicate a different convention for a complex number of zero phase at each point in space.

We shall see that this can be accomplished, but at the price of introducing an interaction. In this example we shall construct that interaction to be electromagnetism.

Quantum mechanical equations of motion, such as the Schrödinger equation, always involve derivatives of the wave function ψ, as do many observables involving energy and momentum. Under local phase rotations, derivatives transform as

$$\nabla\psi(x) \to e^{i\,\alpha(x)}\left[\nabla\psi(x) + i\left(\nabla\alpha(x)\right)\psi(x)\right] , \qquad (16)$$

which involves more than a mere phase change. The additional gradient-of-phase term spoils local phase invariance. Local phase invariance may be achieved, however, if the equations of motion and the observables involving derivatives are modified by the introduction of the electromagnetic field $\mathbf{A}(x)$. If the gradient is everywhere replaced by the gauge-covariant derivative

$$\mathcal{D} \equiv \nabla - ie\mathbf{A} , \qquad (17)$$

where e is the charge in natural units of the particle described by $\psi(x)$ and the field $\mathbf{A}(x)$ transforms under local phase rotations as

$$\mathbf{A}(x) \to \mathbf{A}'(x) \equiv \mathbf{A}(x) + \frac{1}{e}\nabla\alpha(x) , \qquad (18)$$

it is easily verified that under local phase rotations

$$\mathcal{D}\psi(x) \to e^{i\alpha(x)}\mathcal{D}\psi(x) . \qquad (19)$$

Consequently, quantities such as $\psi^*\mathcal{D}\psi$ are invariant under local phase transformations. The required transformation law for the electromagnetic vector potential is precisely the form of a gauge transformation in electrodynamics. Moreover, the gauge-covariant derivative corresponds to the familiar replacement $p \to p - e\mathbf{A}$ for the momentum of a charged particle in the presence of an electromagnetic field. Thus the form of the coupling ($\mathcal{D}\psi$) between the electromagnetic field and matter is suggested by local phase invariance.

Let us summarize the general consequences of the symmetry approach to interactions. Through Noether's theorem, a (continuous) global symmetry implies (and frequently is recognized because of) the existence of a conserved current, the electromagnetic current in our example. A local gauge symmetry requires, in addition, the introduction of a massless vector gauge field $\mathbf{A}$ and prescribes the form of the interaction between matter and the gauge field in the form known as minimal coupling. Electromagnetism, or QED, is the gauge theory based on the group of phase transformations, i.e., on the Abelian group $U(1)$. The generalization to non-Abelian symmetries can be made, so that the construction of a gauge theory can be carried out for any continuous symmetry.

Quantum Chromodynamics

The simplest non-Abelian theory to describe, though regrettably not to solve, is the theory of strong interactions based on color symmetry. QCD is motivated by the observation that color is what distinguishes the quarks from the leptons. Since quarks experience the strong interactions but leptons do not, it is natural to regard color as a strong-interaction charge and to take the symmetry among red, green, and blue quarks as a local gauge symmetry. The unitary symmetry group $SU(3)$ is an apt choice for the color symmetry because, having a complex fundamental representation, it makes a distinction between quarks and antiquarks. We would not want to construct a theory in which the existence of quark-antiquark bound states, the mesons, would imply that there should be quark-quark bound states, which are not observed.

With the choice of $SU(3)_{color}$ as the gauge group, the quark-antiquark interaction is mediated by vector gluons, which transform as an octet under the color symmetry. If physical states must be colorless, which would explain why free mesons and baryons exist but free quarks have not been observed, free gluons will not be seen. Our evidence for the existence of gluons, like that for the existence of quarks, is therefore indirect. It is basically of two sorts. First, energy-momentum sum rules in lepton-nucleon scattering indicate that the partons that interact electromagnetically or weakly, namely the quarks, carry only about half the momentum of a nucleon. Something else, electrically neutral and inert with respect to the weak interactions, must carry the remainder. This is a role for which the gluons are ideally suited. Second, at center-of-momentum energies exceeding about 17 GeV, a fraction of hadronic events produced in electron-positron annihilations display a three-jet structure instead of the familiar two-jet ($e^+e^- \to q\bar{q}$) structure. A typical event of this type is shown in Figure 9. This is interpreted as evidence for the process

$$e^+e^- \to q\bar{q} + \text{gluon}, \qquad (20)$$

in which the gluon is radiated from the outgoing quark in a hadronic analog of electromagnetic bremsstrahlung.

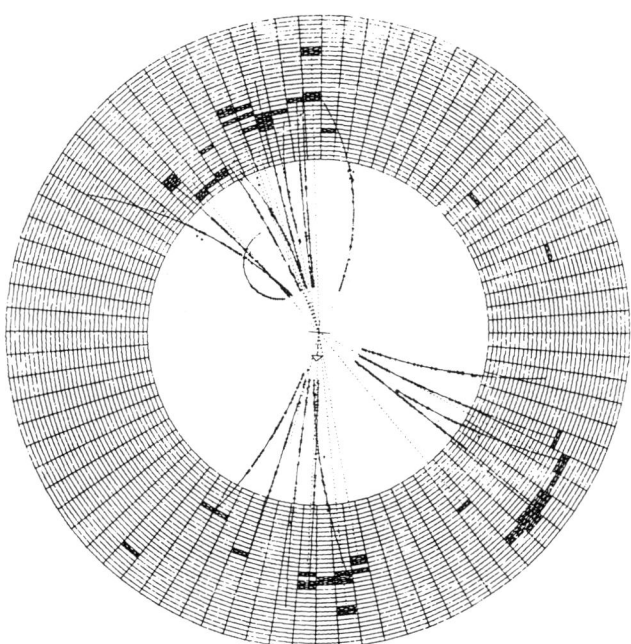

Figure 9: Three-jet event observed in the reaction $e^+e^- \to$ hadrons at 31 GeV, in the JADE detector at PETRA.

Electroweak Theory

The construction of the electroweak theory is somewhat more involved, so I shall describe it only very schematically. We begin from the observation that the leptons (electrons, muons, taus, and their neutrinos) seem to live in distinct families

$$\begin{pmatrix} \nu_e \\ e \end{pmatrix}_L, \begin{pmatrix} \nu_\mu \\ \mu \end{pmatrix}_L, \text{ and } \begin{pmatrix} \nu_\tau \\ \tau \end{pmatrix}_L. \tag{21}$$

The familiar weak interactions change one family member into another but do not cross family lines. This suggests using the weak-isospin symmetry among neutral and charged leptons (or among up-like and down-like quarks) together with a phase invariance, as in electromagnetism, for the gauge symmetry. A theory can be constructed along these lines using the gauge group $SU(2)_L \otimes U(1)_Y$, where Y denotes the weak hypercharge. The resulting force particles are the photon and the two charged intermediate bosons W^+ and W^-, all of which are expected from the classical phenomenology of the weak and electromagnetic interactions, plus a new neutral intermediate boson, the Z^0.

Although the $SU(2)_L \otimes U(1)_Y$ gauge theory contains the familiar charged weak current, it cannot describe the weak interactions without an important modification. It has been known for many years from the absence of pronounced form-factor effects in β–decay that the weak interactions are of very short range. The gauge bosons that mediate the weak interaction cannot be massless, for massless particles mediate forces of infinite range. How can the requirement that the intermediate bosons be massive be reconciled with the appealing constraint of local gauge invariance, which leads naturally to massless gauge bosons?

The answer lies in the phenomenon of spontaneous symmetry breaking, whereby a physical system need not manifest all the symmetries of the interaction that gives rise to it. The most familiar example of a spontaneously broken symmetry is the spontaneous magnetization of soft iron. Electromagnetism is symmetric under spatial rotations; there is no preferred direction in space. This rotation-invariance is manifested by the fact that a lump of warm iron looks the same from all directions. If we regard the piece of iron as a collection of microscopic magnets, the micromagnets are in disorder, pointing randomly in all directions. However, when cooled below the Curie temperature, iron magnetizes spontaneously. The magnetic interaction among the micromagnets overcomes thermal agitation and causes an alignment: the selection of a preferred axis. Now, only rotations about the axis of magnetization leave the lump of iron unchanged in appearance. The full rotation-invariance of electromagnetism is hidden, or spontaneously broken. The hidden symmetry can be recovered by repeating the thermal cycle many times. Each time the iron is cooled, it will magnetize spontaneously, but each time along a different direction. The fact that every direction is equally probable shows that the underlying interaction—electromagnetism—does not have a preferred direction, and so is rotation-invariant.

The most apt analogy for the hiding of the electroweak symmetry is found in superconductivity. In the Ginzburg-Landau description of the superconducting phase transition, a superconducting material is regarded as a collection of two kinds of charge carriers: normal resistive conductors and superconductors. Above the critical temperature for the onset of superconductivity, T_c, the free energy of the substance is supposed to be an increasing function of the density of superconductors. The state of minimum energy, the vacuum state, then corresponds to a purely resistive flow, with no superconductors active. Below the critical temperature, the free energy is minimum when the density of superconductors is nonzero. These two cases are illustrated in Figure 10. If we now consider the behavior of the free energy in an applied magnetic field, we find that, below the critical temperature, the photon acquires a mass within the superconducting material. This is the origin in the Ginzburg-Landau model of the Meissner effect, the exclusion of a magnetic field from a superconductor. More to the point, for our purposes, it shows how a symmetry-hiding phase transition can lead to a massive gauge boson.

To give masses to the intermediate bosons of the weak interaction, we take advantage of a relativistic generalization of the Ginzburg-Landau phase transition known as the Higgs mechanism. We introduce elementary auxiliary scalar fields, with gauge-invariant interactions among themselves and with the fermions and gauge bosons of the electroweak theory. We then arrange their self-interactions so that the vacuum state corresponds to a broken-symmetry solution. As a result, the W and Z bosons acquire masses, as auxiliary scalars assume the role of the missing third (longitudinal) degrees of freedom of what had been massless gauge bosons. The quarks and leptons acquire masses as well, from their Yukawa interactions with the scalars. Finally, there remains as a vestige of the spontaneous breaking of the gauge symmetry a massive, spin-zero particle, the Higgs boson.

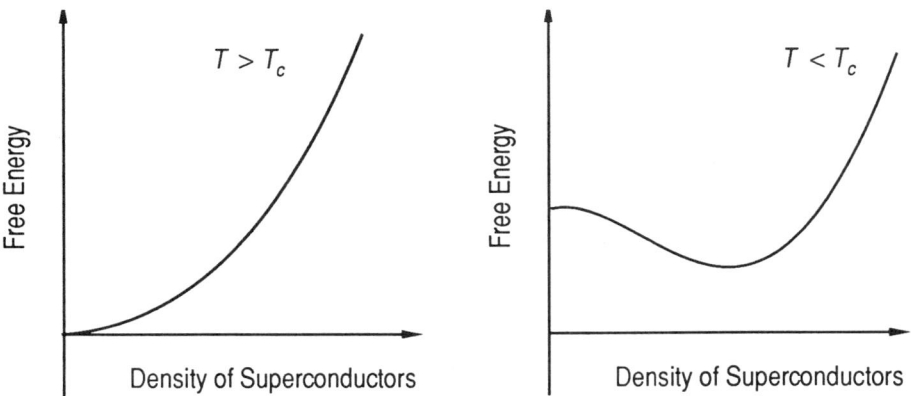

Figure 10: Ginzburg-Landau description of the superconducting phase transition.

The principal consequences of electroweak unification are these: (i) The strengths of the weak and electromagnetic interactions become equal at short distances. (ii) The couplings of the Z^0 to matter are prescribed. (iii) The masses of the gauge bosons $W^\pm$ and Z^0 are predicted. It is remarkable that the resulting theory has been tested at distances ranging from about 10^{-16} cm to about 4×10^{10} cm, especially when we consider that classical electrodynamics has its roots in the tabletop experiments that gave us Coulomb's law. These basic ideas were modified in response to the quantum effects observed in atomic experiments. High-energy physics experiments continued the extension toward still shorter distances and both inspired and tested the unification of the weak and electromagnetic interactions. At distances longer than the scale of common experience, electrodynamics—in the form of the statement that the photon is massless—has been tested in measurements of the magnetic fields of the planets. With additional dynamical assumptions, the observed stability of the Magellanic clouds provides evidence that the photon is massless over distances of about 10^{22} cm.

Unification of the Fundamental Interactions

The theories that describe the fundamental interactions have a number of central elements in common. All are renormalizable field theories, calculable in perturbation theory. All are based on symmetry principles, which suggests the prospect of further unification. In fact, a hint that unification of the strong and electroweak interactions may be required comes from the electroweak theory itself. Applied only to leptons, or only to quarks, the electroweak theory is not mathematically self-consistent: triangle anomalies destroy the renormalizability of the theory. A self-consistent theory requires, for each weak-isospin doublet of leptons, a color-triplet weak-isospin doublet of quarks. The idea that quarks and leptons both are required suggests that they may be related, a notion encouraged by the similarity of their properties. This, in turn, suggests that there should be extended families of quarks and leptons and, in consequence, symmetries that may transform quarks into leptons.

A second hint that the strong, weak, and electromagnetic interactions might be unified arises from the calculated evolution of the coupling constants shown in Figure 11. Coupling constants in quantum field theory depend on the momentum scale at which they are defined. The QCD (strong interaction) coupling decreases at high momentum scales, or short distances. This is the celebrated property of asymptotic freedom, which implies the reliability of perturbation theory in short-distance strong-interaction processes. The couplings associated with weak isospin and the weak-hypercharge phase symmetry also evolve. It is highly suggestive that the three couplings, which differ considerably in strength in low-energy phenomena, evolve toward a common value at an energy near 10^{15} GeV.

As we have noted in the introduction to our discussion of the gauge theories of the fundamental interactions, different theories have survived different degrees of confrontation with experiment. QED has been tested most rigorously, while tests of unified theories of the strong, weak, and electromagnetic interactions remain largely at the level of "yes-or-no" questions. However,

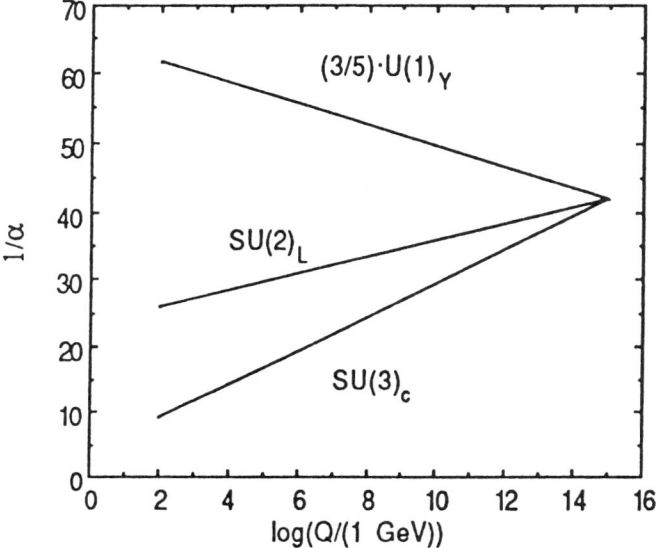

Figure 11: Evolution of the running coupling constants in leading logarithmic approximation in the SU(5) model. Three fermion generations are assumed.

it is very important that for the theories that make up the Standard Model—the $SU(3)_{color} \otimes SU(2)_L \otimes U(1)_Y$ gauge theory of the strong, weak, and electromagnetic interactions involving three generations of color-triplet quarks ($u, d, s, c, b, [t]$) and color-singlet leptons ($e, \nu_e, \mu, \nu_\mu, \tau, \nu_\tau$)—there are no experimental embarrassments. No reliable data contradict the underpinnings or disagree with a credible prediction. Many predictions await sharpening or detailed experimental tests.

How Far We Have to Go

The Standard Model has an appealing simplicity and an impressive generality. The picture at which we have arrived has a pleasing degree of coherence and holds the promise of deeper understanding—in the form of a further unification of the interactions—still to come. This is an accomplishment deserving great respect, but if we have come impressively far in the past two decades, we still have far to go. The very success of the $SU(3)_{color} \otimes SU(2)_L \otimes U(1)_Y$ model prompts new questions: Why does it work? Can it be complete? Where will it fail? As we shall see, the Standard Model itself hints that the frontier of our ignorance lies at about 1 TeV for collisions among the fundamental constituents. In more general terms, the success of the Standard Model suggests that a significant step beyond present-day energies is needed to see breakdowns of our current understanding.

Beyond these generalities, there are many specific questions that the Standard Model raises but cannot answer:

- Although the Higgs mechanism shows how masses could be given to the quarks and leptons, the Standard Model offers no particular insight into the pattern of fermion masses or into the mixing angles that describe transitions that cross quark family lines.
- We have evidence from the requirement that the electroweak theory be anomaly-free, hence renormalizable, that quarks and leptons must occur together in generations. The idea of generations is supported by the explanation of the equality of proton and positron charges in unified theories, but we do not know why generations repeat or how many there are.
- These issues may be summarized in the complaint that there is too much arbitrariness in the Standard Model, a surfeit of parameters. How many parameters is too many?

	3	coupling parameters (α_s, α_{EM}, $\sin^2\theta_W$)
	6	quark masses
	3	Cabibbo-Kobayashi-Maskawa angles
	1	CP-violating phase
	2	parameters of the Higgs potential
	3	charged-lepton masses
	1	vacuum phase
for a total of	19	arbitrary parameters.

The situation is not improved by the unification of the strong, weak, and electromagnetic interactions. Unification imposes constraints among some parameters, but new parameters arise to describe the spontaneous breakdown of the unifying group into $SU(3)_{color} \otimes SU(2)_L \otimes U(1)_Y$.

- CP violation is parametrized, but not explained, by the Standard Model.
- Gravitation is omitted.
- The most serious structural problem is associated with the scalar, or Higgs, sector of the electroweak theory. The scalar sector is responsible for breaking the $SU(2)_L \otimes U(1)_Y$ electroweak symmetry down to $U(1)_{EM}$. Yet the dynamical nature of the spontaneous symmetry breaking is the least understood aspect of the theory. Indeed, the instability of the masses of elementary scalars in interacting field theory gives reason to suspect that the model may, in the end, be inconsistent. At a more operational level, the masses of the $W^{\pm}$ and Z^0 are specified by the theory, but the mass of the Higgs boson is constrained only to lie in the range

$$7 \, \text{GeV}/c^2 \lesssim M_H \lesssim 1 \, \text{TeV}/c^2 \, . \tag{22}$$

The lower bound is strictly valid only in the simplest version of the Standard Model with one elementary Higgs doublet, and it depends upon the mass of the top quark. The upper bound is reasonably model-independent. If the Higgs boson mass exceeds $1 \, \text{TeV}/c^2$, weak interactions must become strong on the TeV scale.

The problem of the scalar sector is exacerbated in unified theories of the strong, weak, and electromagnetic interactions. Several families of Higgs bosons are required to accomplish the breakdown from the unifying group G to the low-energy $SU(3)_{color} \otimes U(1)_{EM}$ symmetry we observe. Moreover, the breaking of the "electronuclear" symmetry must occur in two steps,

$$G \to SU(3)_{color} \otimes SU(2)_L \otimes U(1)_Y \tag{23}$$

and

$$SU(3)_{color} \otimes SU(2)_L \otimes U(1)_Y \to SU(3)_{color} \otimes U(1)_{EM} , \tag{24}$$

at scales separated by many orders of magnitude.

• The large number of quarks and leptons makes it natural to ask whether these fermionic constituents are truly elementary or composites of some more fundamental building block.

• Finally, we may ask what is the origin of the gauge symmetries themselves, why are the weak interactions left-handed, and whether there are new fundamental interactions to be discovered.

This list has inspired imaginative conjectures departing in many directions from the Standard Model paradigm. These have important implications that cannot yet be tested. Although theoretical speculation and synthesis is valuable and necessary, we cannot long advance without new observations. The experimental clues needed to answer our questions can come from several sources, including experiments at high-energy accelerators, experiments at low-energy accelerators and nuclear reactors, nonaccelerator experiments, and deductions from astrophysical observations.

The low-energy and passive experiments provide indirect access to high mass scales through their sensitivity to rare processes. However, according to our present knowledge of particle physics and our past experience, there is no substitute for experiments that probe directly at high energies. Satisfying responses to our complaints about the Standard Model will require experiments at the highest-energy accelerators.

The Next Step

Many of the questions we must confront are beyond the reach of accelerators in operation or under construction. Progress toward a more comprehensive understanding of the nature of matter will depend upon our ability to study phenomena at higher energies, or, equivalently, on shorter scales of distance. What energies must we reach, and what sort of new instruments will be required?

Illuminating the physics of electroweak symmetry breaking is perhaps the most sharply posed assignment for the next generation of accelerators. Unitarity arguments have shown us that new phenomena are to be found in the weak interactions at energies not much larger than 1 TeV for collisions among the fundamental constituents. The same scale is selected by the conjectural extensions to the Standard Model that offer potentially more satisfying descriptions of the scalar sector and thereby of the mechanism for spontaneous symmetry breaking.

One of these ideas involves introducing a complete new set of elementary particles that differ by one-half unit in spin from the known quarks, leptons, and gauge bosons. These new particles are consequences of a postulated supersymmetry, which relates particles of integral and half-integral spin. Supersymmetry would stabilize the mass of the Higgs boson at a value below 1 TeV/c^2 and would require the masses of the supersymmetric partners of the known particles to have masses less than about 1 TeV/c^2. No experimental evidence for superpartners has yet been found.

A second possible solution to the Higgs problem is based on the idea that the Higgs boson is not an elementary particle at all, but is in reality a composite object made up of elementary constituents analogous to the quarks and leptons. This scheme is reminiscent of the Bardeen-Cooper-Schrieffer theory of superconductivity, just as the Standard Model resembles the Ginzburg-Landau description. Although they would resemble the usual quarks and leptons, the new constituents would be subject to a new type of strong interaction, often called technicolor, which would confine them within about 10^{-17} cm. Such new forces could yield new phenomena as rich and diverse as the conventional strong interactions, but on an energy scale a thousand times greater—around 1 TeV. The new phenomena would include a rich spectrum of (technicolor-singlet) bound states, akin to the known hadrons. Again, there is no experimental evidence yet for these particles.

Both general arguments, such as unitarity considerations, and specific conjectures for resolutions to the problem of the scalar sector select 1 TeV as an energy scale on which new phenomena crucial to our understanding of the fundamental interactions must occur. The dynamical origin of electroweak symmetry breaking is but one of the important issues that define the frontier of elementary-particle physics. Because of its immediacy and fundamental importance, this issue helps set the requirements for future accelerators and detectors. Of course, when designing an instrument that will serve our field for a quarter of a century or more, we must ensure that the new device will open a large new territory to investigation.

Either an electron-positron collider with beams of 1–3 TeV or a superconducting proton-(anti)proton collider with beams of about 5–20 TeV would allow an exploration of the TeV regime for hard collisions. The higher beam energy required for protons simply reflects the fact that the proton's energy is shared among its quark and gluon constituents. That partitioning of energy among the constituents has been thoroughly studied in experiments on deeply inelastic scattering of leptons from nucleons, so the rate of collisions among constituents of various energies may be calculated with some confidence. Any accelerator to explore the 1-TeV scale must make available a high collision rate, because the hard-scattering (pointlike) cross sections of central interest scale as E_{cm}^{-2}.

The scientific opportunities presented by both the electron-positron and proton-(anti)proton alternatives are attractive and somewhat complementary. The hadron machine reaches to higher energy and provides a wider variety of constituent collisions, which allow for a greater diversity of phenomena. The simple initial state of the electron-positron machine represents a considerable measurement advantage. However, experiments at the CERN and Fermilab proton-antiproton colliders show that hard collisions at very high energies are relatively easy to identify. Because the essential technology is in hand for the hadron collider, it is the instrument of choice for the first exploration of the TeV regime. Multi-TeV hadron colliders are under active study in the Soviet Union (UNK), in Western Europe (LHC), and in the United States (SSC).

The Superconducting Super Collider

The high-energy physics community in the United States has embarked on the design and eventual construction of a high-energy, high-luminosity, superconducting proton-proton collider to explore the 1-TeV scale. The SSC will produce collisions of 20-TeV proton beams at a luminosity of up to 10^{33} cm^{-2}sec^{-1}, yielding more than 10^8 interactions per second.

Superconducting magnets are chosen for two reasons. First, they make possible confining fields three times as strong as those available with conventional iron magnets. More important, the electrical power requirements for a superconducting machine are far smaller than for a conventional accelerator. A conventional version of the SSC would consume 4 GW of electrical power; the SSC's average consumption will be about forty times smaller.

Even with the more intense fields made possible by superconducting magnets, the SSC will be an instrument of impressive size. In the engineering units appropriate to the problem, the bending radius is related to beam momentum and confining field by

$$\rho = \left(\frac{10}{3} \text{km}\right) \cdot \left(\frac{p}{\text{TeV}/c}\right) \Big/ \left(\frac{B}{\text{tesla}}\right). \tag{25}$$

For a 20-TeV/c beam in a confining field of 6.6 T, the SSC design field, the implied radius of curvature is $\rho \approx 10$ km. With allowance for the straight sections accommodating experimental areas, acceleration gear, etc., the circumference of the SSC will be about 84 km.

The proposed layout of the Supercollider is shown in Figure 12. Interaction halls cluster on the two gently curved sides of the collider ring. In this perspective the near cluster incorporates the injector complex, the radio-frequency accelerating system, beam absorbers, and two of the six interaction halls. The far cluster adds four more interaction halls, two of which are reserved for development after research begins. The schematic enlargement of an interaction hall shows a detector surrounding the point at which two beams collide; the cross section to the right shows the position of the two superconducting magnet rings in the tunnel. The two independent rings for the proton beams will sit one atop the other, 70 cm apart.

The superconducting magnets essential for guiding the protons around the rings are made of two coils arranged to approximate a $\cos\theta$ current distribution. The inner coil has an inside diameter of 4 cm. The coils are wound of cable made from composite strands containing thousands of filaments—each about 6μm in diameter—of a niobium-titanium alloy embedded in a copper matrix. Interlocking stainless steel or aluminum collars surrounded by an iron yoke hold the cables in place. This 17-meter-long package, called the "cold mass," is sealed in its own cryostat, where it can be maintained at an operating temperature of 4.35 K.

The 3840 dipole magnets in each collider ring have a peak operating field of 6.6 teslas, which corresponds to a current of 6504 amperes. Figure 13 shows a cross section of the SSC dipole. Substantial increases in the current-carrying capacity of superconducting cable have resulted from a collaboration among industry, university researchers, and the U.S. national laboratories focused on the development of improved superconductor for the SSC. The SSC specification for

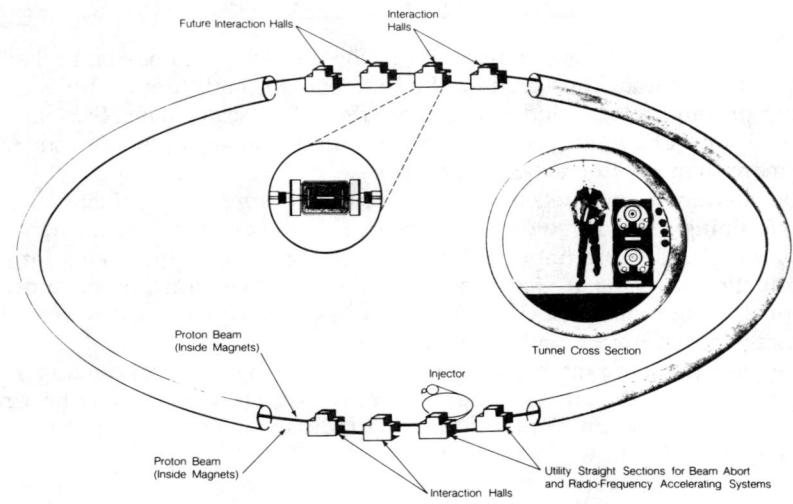

Figure 12: Collider ring layout envisaged in the SSC Conceptual Design Report.

the critical current density in niobium-titanium strand, J_c = 2750 A/mm^2 at 4.2 K and 5 T, represents a 50 percent improvement over the conductor used in the Tevatron. Material meeting this demanding specification is now routinely received in production quantities.

When might the SSC be in operation? An important milestone was passed in March 1986, when the SSC Central Design Group completed a Conceptual Design Report for the SSC. Every major system had been thought through, and a detailed cost estimate had been made. Because a location has not been selected for the Supercollider, the Conceptual Design was not adapted for any specific site. During the summer of 1986, the Department of Energy and independent experts validated the cost and technical feasibility of the machine described in the Conceptual Design Report. President Reagan endorsed the SSC as a national goal in January 1987. In April of that year, the Department of Energy began a site search that has led to a short list of seven "Best Qualified Sites" in the states of Arizona, Colorado, Illinois, Michigan, North Carolina, Tennessee, and Texas. On November 10, 1988, Secretary of Energy John S. Herrington designated Waxahachie, Texas, as the preferred site for the SSC laboratory. A Record of Decision, following completion of an environmental review, is expected early in 1989. From the time the site is available, sometime in 1989, it will take about seven and a half years to build the machine. We hope to commence experimentation with the SSC by 1996.

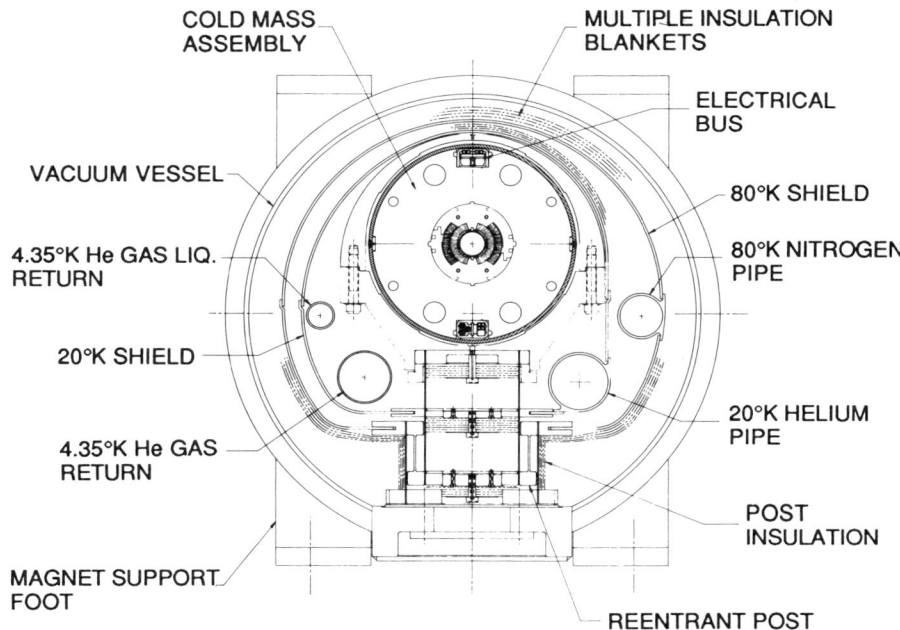

Figure 13: Cross section of the 6.6-T superconducting dipole magnet for the SSC.

We believe that the SSC can foster a new level of international cooperation in particle physics. As a frontier research instrument, the Supercollider will certainly attract to its experimental program many of the best particle physicists from around the world. This, is of course, traditional in our field, but we may hope for more: active international collaborations established early enough to allow significant foreign participation in the design and construction of the SSC and its detectors, and not just in the performance of experiments.

The advances of the past decade have brought us tantalizingly close to a profound new conception of the most basic constituents of matter and their interactions. The simpler and more comprehensive understanding we have gained organizes current knowledge and locates the horizon of particle physics at energies of trillions of electron volts, and the horizon of cosmology at about a millionth of a billionth of a second after the moment of creation. Important answers will be found with the Supercollider: from it we await new discoveries about the unification of the forces of nature and the patterns of the fundamental constituents of matter.

Acknowledgments

It is a pleasure to express my appreciation for the warm hospitality of the Collège de France, and to thank the Bernard Gregory Memorial Fund for making possible this cycle of lectures in Paris and Geneva. I thank my colleagues Estia Eichten, Ian Hinchliffe, and Ken Lane for their contributions to my understanding of supercollider physics. Kate Metropolis made helpful comments on the manuscript.

PARTICLE ACCELERATORS AND DISCOVERIES IN ELEMENTARY PARTICLE PHYSICS

Lawrence W. Jones
Randall Laboratory of Physics, University of Michigan
Ann Arbor, Michigan 48109-1120

Abstract

Some discoveries in elementary particle physics are recounted from personal and historical perspective with particular reference to their interaction with particle accelerators. The particular examples chosen include the ψ/J, the Υ, and the study of nucleon constituents with inelastic electron scattering. Precurser experiments are cited together with the better known discoveries.

PARTICLE ACCELERATORS AND DISCOVERIES IN ELEMENTARY PARTICLE PHYSICS

Lawrence W. Jones[*]
Randall Laboratory of Physics, University of Michigan
Ann Arbor, Michigan 48109-1120

Dr. Month asked me to prepare a talk on high energy physics discoveries and their relationship to particle accelerators. No particular time period was specified; however, high energy physics as a field is less than forty years old. This historical subject was a challenge for me as there was clearly far too much to cover in a comprehensive manner in just one hour. Therefore, I was forced to pick and choose. I am not an historian of physics, and I have not made a systematic study of the recent history of particle physics. Rather I am a physicist who is now becoming old enough to have "been there" when some of these discoveries were made. I have used as one source for this talk the material contained in a report prepared for a larger document "Physics in the 1980's" edited by Dr. Brinkman. Martin Perl of Stanford has chaired a subpanel on elementary particle physics of that task force, on which I served.

I will take two topics from the broad subject matter of particle physics and focus on them: the quark model of hadrons and the quark-lepton interactions as they relate to nucleon structure functions. I will not reach back earlier than the 1960's and I will not try to bring my report up to those most recent experiments which have been the subject of many recent papers, seminars, and lectures. Let me make one warning here: this lecture may be much too elementary for most of you. It is based on the assumption that many of you may be from other fields, or you have been so immersed in accelerator science that you may not have had the opportunity of studying particle physics in depth. We learn by studying history, therefore the objective of a lecture of this character is to observe patterns and to learn from the mistakes and successes of the past in order to assist us in most effectively studying our field and making advances in the future. One thing that has interested me is to observe "precursors" to particular discoveries, that is, to note where there were experiments which preceded the definitive experimental work but in retrospect reported evidence for the same phenomena later announced. I will also depart somewhat from Dr. Month's assignment by identifying instances where accelerators were not used in making important discoveries.

Let me first turn to the subject of the quark model of hadrons. Briefly recapitulating history, the neutron was dicovered in 1932 and the concept of isotopic spin as applied to nuclear states first appeared in the 1930's. In 1946 the charged pion, or pi-meson, was observed in cosmic rays; only a little later the neutral pion was

---

[*] Supported in part by the U.S. National Science Foundation.

identified at particle accelerators. The nucleon 3/2, 3/2 isobar state, the Δ (1238 MeV), was first observed by Fermi and his collaborators in the early 1950's, and the strange particles, the Λ and K-meson, were found in cosmic rays about 1948. So already by the early 1950's there came to be known a large number of meson and baryon states. The Brookhaven Cosmotron first turned on in 1952 and the 6-GeV Bevatron at Berkeley came into operation in 1955. 1954 was also approximately the time the bubble chamber was developed as a tool for the study of particle physics, and following the application of the bubble chamber to the beams from the Bevatron and Cosmotron, there developed the virtual explosion in the discovery of particle resonances.

The proliferation of meson and baryon states was one stimulus to the development of the quark model. Another independent stimulus was provided by the experiments on elastic electron-proton scattering. In the mid-1950's also, Robert Hofstadter, working with the Stanford electron linear accelerator (at that time with an energy of about 180 MeV) studied the angular distribution of the elastically scattered electrons from protons. This was the 1950's analog to the classical Rutherford scattering experiment of alpha particles on gold. The consequence of Hofstadter's experiment was that the scattering distribution of the electrons did not follow the distribution expected for a point charge scattering center. Let me reproduce below the summary of the relevant kinematics and related formula.

Consider the Feynman diagram below (time advances to the right in the figure).

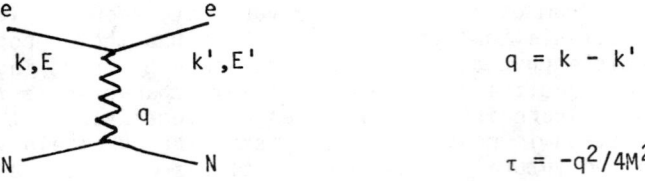

$q = k - k'$

$\tau = -q^2/4M^2$

The differential angular distribution of the scattered electron is given by

$$\frac{d\sigma}{d\Omega} = \frac{\alpha^2}{4E^2 \sin^4 \theta/4} \left(\frac{E'}{E}\right) \left[\frac{G_E^2 + \tau G_M^2}{1 + \tau} \cos^2 \frac{\theta}{2} + 2\tau G_M^2 \sin^2 \frac{\theta}{2}\right]$$

where

$$G_E(q^2) = \frac{1}{\left(1 - \frac{q^2}{0.71}\right)^2} \quad \text{and} \quad G_E = \frac{G_M}{\mu}.$$

The results are that (in the non-relativistic limit)

$$\rho(r) \propto e^{-r/b} \quad ; \quad b^2 = 1/0.71.$$

The consequence of these experiments gave the value for the RMS radius of the proton of about 0.81 fermi.[1] These data were interpreted by Chew and his collaborators at Berkeley as evidence for a vector meson particle in the nucleon structure, that is, the electron was perceived as radiating a virtual gamma ray which coupled to a vector particle in the field of the nucleon (the neutron or the proton). This vector particle was predicted to be a combination or a mixture of an isoscalar or an isovector particle. Later we came to know these as the ρ and the ω mesons. The mass of the ρ was predicted by Chew from the electron scattering experiments to be about 450 MeV, and indeed several experiments set out to look for this vector meson.[2]

I am particularly aware of this bit of history because at Berkeley, Martin Perl and I, working with graduate students, set out to look for such a particle in 1960 by directing a beam of negative pions on a small hydrogen target and looking at the energy and angular distribution of recoiling protons.[3] In order to deduce the mass of the scattered meson state we ran this experiment through a range of incidence pion energies that would be sensitive to the 450-MeV final state meson, but we deliberately avoided extending our incident energies so high that we were above the associated production (Λ-K) threshold, because we were concerned that we would not be able to resolve the background from decaying lambdas, and the interpretation of the data would have then been ambiguous. Of course not long afterward, Walker and his collaborators, using a small hydrogen bubble chamber, discovered the ρ at about 760 MeV,[4] whereas our mass search had only extended up to about 695 MeV. We thus learned a lesson to never take the theorists too seriously.

Subsequently, of course, the ω^0 and the η^0 meson were found at Berkeley; anti-proton annihilation produced invariant mass states that were identified with various resonances in about 1961. In about 1962 the f^0 was observed, later the ϕ, K^*, the A^1 and others were found as well as the baryon family of particles: the Λ^*s, Σs, Ξs, and a whole host of nucleon isobars.[5] The vast majority of these particles were discovered using liquid hydrogen bubble chambers. Now the bubble chambers did not place intensive or excessive demands on operators of accelerators. They did require clean beams and the requirement on accelerator intensity was primarily to provide adequate fluxes at higher energies of negative kaons and anti-protons. The other requirement (which was easy to fulfill) was fast extraction. The bubble chamber should have a beam pass through it in a time interval no more than tens or hundreds of microseconds prior to flashing the light (to record a photograph). The K^- separation from π^- at high energies was a challenging task which was accomplished by electromagnetic separators, devices which are not studied particularly these days. They used a very simple concept of crossed electric and magnetic fields and of course the optics of strong focusing lenses. Much of this progress was due to Luis Alvarez who, in the early days of the Bevatron, insisted that the detectors to be used with the new generation of accelerators warranted an investment that was significant, i.e., it should be in some sense proportional to the effort invested in the accelerators. Thus, the Berkeley 72-inch hydrogen bubble chamber was an enormously large and expensive detector in the context of those years and represented an investment on the order of magnitude of 10% of the investment of the Bevatron itself, where by "investment" I could equally well consider either money or people or laboratory resources.

In the early 1960's the quark model as a framework for understanding the rapidly multiplying numbers of mesons and baryons was put forth independently by Murray Gell-Mann[6] and by George Zweig.[7] As you may know, Gell-Mann's work was published promptly, whereas Zweig's original paper is a CERN report, and (as I have been told) a report which the CERN theoretical group did not feel should be submitted for publication. In any case, the concept here was that all known particles, which then included strange and nonstrange mesons and baryons, might be composed of three fundamental objects or "quarks" (Zweig called them aces). We now refer to these quarks as the up, down, and strange quarks (the u, d, and s) with a one-third or two-thirds integral charge, one-third baryon number, and half-integral spin. Gell-Mann held from the outset that these might be only mathematical constructs and not physical objects. On the other hand, other theorists felt very strongly that if quarks had meaning in the model, they should be physical objects which could be produced and detected in high-energy collisions. Richard Dalitz was a primary proponent of this point of view. One of the turning points in the development of the quark model was the prediction that there was an undiscovered member of a baryon decouplet whose other nine members were the Δ's, the Σ^*'s, and the Ξ^*'s, as shown in the illustration (Figure 1). The missing member was the Ω^-, the baryon composed of three strange quarks with a spin of 3/2 but in only the negative charge state.[8] In 1964 this particle was in fact discovered at Brookhaven, again using a hydrogen bubble chamber and a 5-GeV/c K^- beam.[9] The event was identified as a consequence of the observation of two kaons in the final state; on close examination the event contained the cascade of an Ω^- going to a Ξ^0 which in turn decayed to a Λ^0 which then decayed to a π^- and a proton, all within the bubble chamber.

Now my first example of a precursor is to reference here a 1973 paper by Luis Alvarez entitled "Certification of Three Old Cosmic Ray Emulsion Events as Omega Minus Decays and Interactions".[10] I quote here verbatim the abstract of that paper.

"In the preaccelerator years when large stacks of emulsion were exposed to cosmic rays at high altitude, three events were found in which K^- mesons were emitted from slowly moving particles. The Ω^- is the

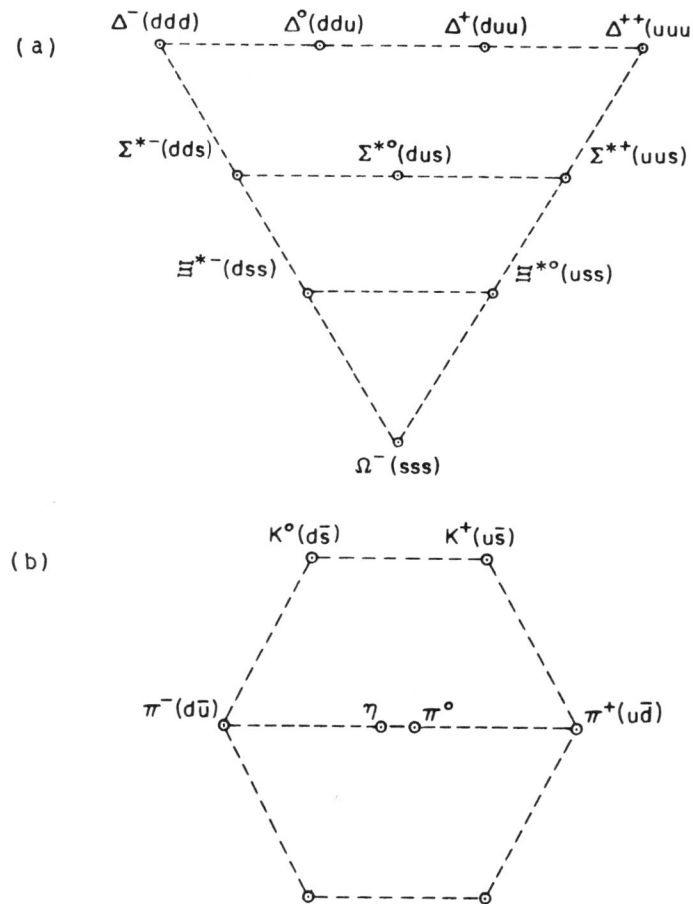

Fig. 1. The Ω^- was predicted by Gell-Mann and was subsequently discovered in the bubble chamber experiments. Figure (a) shows how the delta, sigma-star, xi-star, and omaga family of hadrons are made out of three quarks; (b) shows how the meson family which contains the pion and kaon are made of a quark and an antiquark. The positive pion, π^+, and the positive kaon, K^+, have different properties because the π^+ consists of an up (u) quark and a down antiquark ($\bar{d}$) while the K^+ consists of an up quark (u) and a strange antiquark ($\bar{s}$). The η and π^0 are made of combinations of $u\bar{u}$, $d\bar{d}$, and $s\bar{s}$ quarks.

only presently known particle that can give rise to a K⁻ when moving at nonrelativistic speed, but none of the three events has until now been clearly identified as an Ω^-. One of the cosmic ray events (Eisenberg, 1954) has been incorrectly intrepreted as an Ω^- decaying in flight; it is now shown to be an interaction in flight of an Ω^- with a silver nucleus. The second event is a clear cut example of an Ω^- decaying in orbit bound to an emulsion nucleus. The third event is quite complicated, but can be unambiguously attributed to the decay of an Ω^- atomically bound to an N^{14} nucleus followed by a collision of the daughter Λ with the N^{14}, in which the compound system then fragments into Λ C^{13} + p + n. The mass of the Ω^- determined by each of the last two events (Fry et al., 1955) agrees closely with the mean of all bubble chamber events."

The clinching arguments for the validity of the quark model really came with the discovery in 1974 of the Ψ or J particle. This remarkable period in particle physics included the discovery at the Brookhaven National Laboratory of the J by a group directed by Samuel Ting. The J particle was produced in proton-nucleon collisions from the Brookhaven AGS and detected through its decay into an electron-positron pair.[11] Nearly simultaneously, there was the observation at the SPEAR storage ring at SLAC of electron-positron annihilation into the Ψ particle, in turn producing lepton pairs or hadron pairs in its decay.[12] This experiment was led by a group under the direction of Burton Richter. Among the properties of Ψ/J particle were the fact that it was more massive than any other previously observed hadronic state of baryon number zero. It was about 1000 times more stable, i.e., displayed a narrower width than any massive mesonic state might be expected to from previous experience, and its quantum numbers appeared to be the same as the ϕ^0, the ω^0, and the η^0 mesons. (See Figure 2.)

Some background remarks are appropriate here. At Stanford in the 1960's Gerald K. O'Neill had led a group in constructing a pair of 500-MeV electron storage rings to provide electron-electron collisions for the first time. SPEAR was built as an electron-positron colliding beam ring under Richter's direction. Other e^+e^- rings were built at Frascati in Italy, at Orsay in France, and at Novosibirsk in the USSR. In the late 1960's and earlier 1970's the electron-positron annihilation to produce vector mesons such as the ρ^0, ω^0, and ϕ^0 had already been studied. However, the 1974 observation of the Ψ as a very narrow resonance in the e^+e^- annihilation cross section was the first breakthrough discovery of the colliding beam technique. Subsequently, a rich spectroscopy developed in the Ψ system where the singlet and triplet p states corresponding quite identically to the spectral states of positronium were observed and studied. Of course the Ψ has since been understood as the vector meson whose components are the charm quark and its anti-quark. (See Figure 3.)

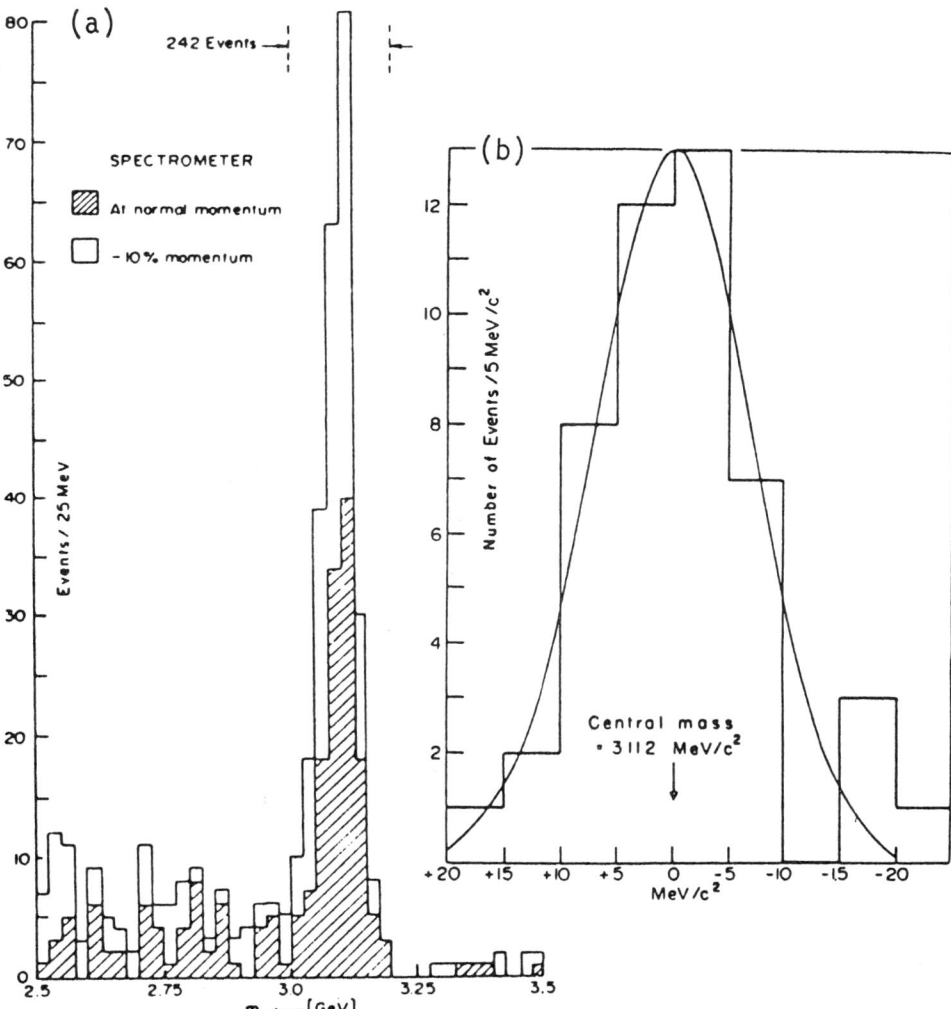

Fig. 2. Data from reference 11 on e^+e^- mass spectrum from the Brookhaven AGS; this was the experiment that first identified the J-meson in strong interactions. (a) Mass spectrum for events in the mass range $2.5 < m_{ee} < 3.5$ GeV/c. The shaded events correspond to those taken at the normal magnet setting, while the unshaded ones correspond to the spectrometer magnet setting at 10% lower than normal value. (b) The measurement of the width of the J. The width is shown to be less than 5 MeV.

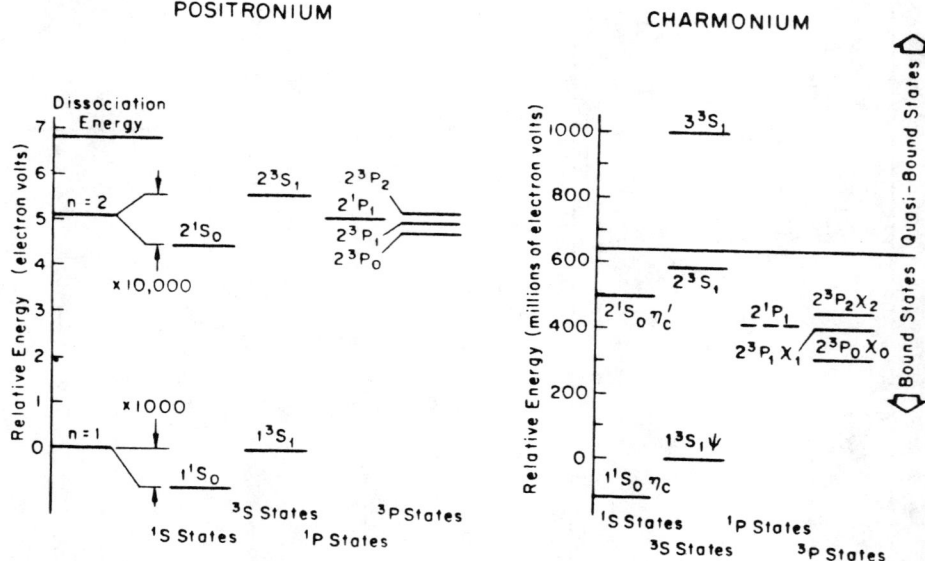

Fig. 3. The spectrum of energy states is similar in positronium and charmonium, but the scale of the energy differences in charmonium is greater by a factor of roughly 100 million. All energies are given with reference to the 1^3S^1 state. At 6.8 electron volts positronium dissociates. At 633 MeV above the energy of the ψ charmonium becomes quasi-bound because it can decay into D^0 and $\overline{D}^0$ mesons.

The experiment at Brookhaven under Ting's direction was very different and totally unrelated to the Stanford work. Ting had earlier studied electron-positron final states of vector mesons working with the 6-GeV electron synchrotron and later with the DORIS electron-positron storage ring at DESY in West Germany. In his Brookhaven experiment he directed a 30-GeV proton beam at a target and looked at the invariant mass spectrum of electron-positron pairs produced. This required a double-armed spectrometer with magnetic analysis, precise particle tracking, and special Cherenkov counters in each leg. He observed a remarkable sharp spike in the invariant mass of the electron-positron pairs resulting from these collisions and identified this particle as the J meson. Of course, it was later recognized that the Ψ and the J are identically the same state.

Once again it is interesting to look at precursors or experiments which anticipated these results. In 1970 Leon Lederman and collaborators undertook an experiment to study the invariant mass spectrum of muon pairs produced by protons on a uranium target at the Brookhaven AGS.[13] Their experiment differed from Ting's in that the muon energies were determined by a coarse range measurement using stepped absorbers, and the muon angles by a modest number of counter telescopes, so that the resolution in invariant mass of the dimuon pairs was rather modest. They reported no clear evidence of new states although their data does show a shoulder at the mass that later proved to be the invariant mass of the ψ/J state. The conclusion of the authors in their paper was that, "As seen both in the mass spectrum and the resulting cross section $d\sigma/dm$ there is no forcing evidence of any structure." This paper was the first report of the process which later became understood by Drell and Yan and studied as the Drell-Yan process.[14] However, Lederman's group was not confident of the physics of the process and was uncertain whether a departure from a smooth curve warranted interpretation as a new state. As a pioneering work this earlier dilepton experiment was totally successful; however, I am certain that the authors have since kicked themselves for missing the dramatic discovery which was later made by Ting's group as a consequence of much finer resolution. (See Figure 4.)

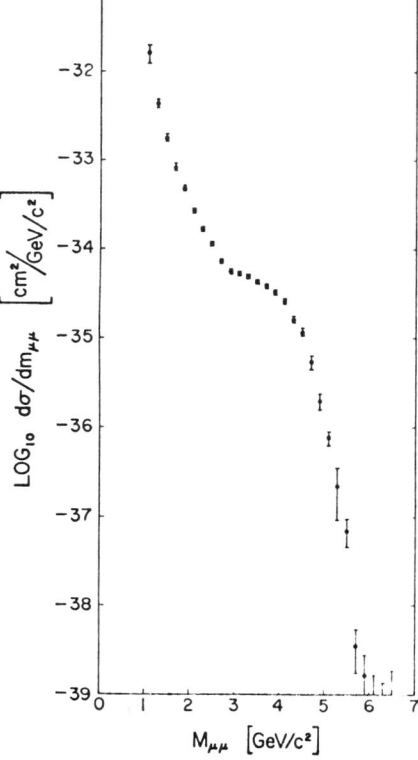

Fig. 4. Cross section versus dimuon invariant mass from reference 13. This experiment clearly observed the ψ/J but lacked sufficient energy resolution and knowledge of the Drell-Yan process to identify it as a new object.

Subsequently, of course, the experiments at Stanford and other e^+e^- colliders observed the D meson, the D^*, the F, etc.; all mesons containing the charm quark. The study of these charmed mesons was also carried out at hadron machines, although there it proved much more difficult. There were many, many experiments at Fermilab looking for the D meson through the mid-1970's, none of which was particularly successful. However, again there was a precursor. In 1970 a Japanese group, flying emulsion stacks in aircraft, reported the observation of short-lived particles produced in high energy proton collisions in emulsions. In particular, a paper presented in 1971 by Dr. Niu at the 12th International Cosmic Ray Conference in Hobart, Tasmania, reported a very interesting event, where two short-lived particles were produced in the same collision.[15] One of the particles decayed in part into a neutral particle which gave rise to two electromagnetic showers with a separation and energies such that the neutral particle was clearly identified as π^0. The lifetime of the particles appeared to be of the order of 10^{-13} seconds and in retrospect it is quite clear that what Niu had observed was a pair of D mesons produced in the cosmic-ray collision. Although the incident energy of the cosmic ray was very much higher than the Tevatron beam energy, the analysis of the final state did not permit particle identification (other than the π^0) so that a determination of the invariant masses was not possible. Niu reported several other interesting events, most of which contained single short-lived final-state particles. Although the cosmic-ray literature is not read as commonly by particle physicists and although Niu's knowledge of the particles he reported was necessarily limited, in retrospect I believe we must credit him and his group with the first documented observation of a charmed meson. (See Figure 5.)

The great success of the experiments of Ting and Richter stimulated further work on studies of the Drell-Yan process and of e^+e^- collisions. Having found one heavy quark, the search was on to find yet more massive states. Lederman's group at Fermilab set up a spectrometer system to look for high mass states decaying into muon pairs or electron-positron pairs over the invariant mass range of 2-1/2 to 20 GeV using 400-GeV protons on a beryllium target. In a Physical Review Letter published in 1976, Lederman's group reported a clustering of 12 electron-positron pairs with an invariant mass between 5.8 and 6.2 GeV which they said suggested that the data contained a new resonance at about 6 GeV.[16] However, they stopped short of making a dramatic claim for a new particle. The word of this indicated resonance spread rapidly and others were quick to follow.

Another Fermilab group working in the meson area set out to confirm this discovery and in a 1976 publication claimed confirming evidence. The abstract from their paper states:[17]

"In a simple search for muon pairs directly produced in proton-nucleon collisions at 300 GeV performed with two range telescopes looking at a beam dump we observed the ψ (3.1GeV) and have an indication for a structure around $m_{\mu\mu}^2 = 36$ GeV^2."

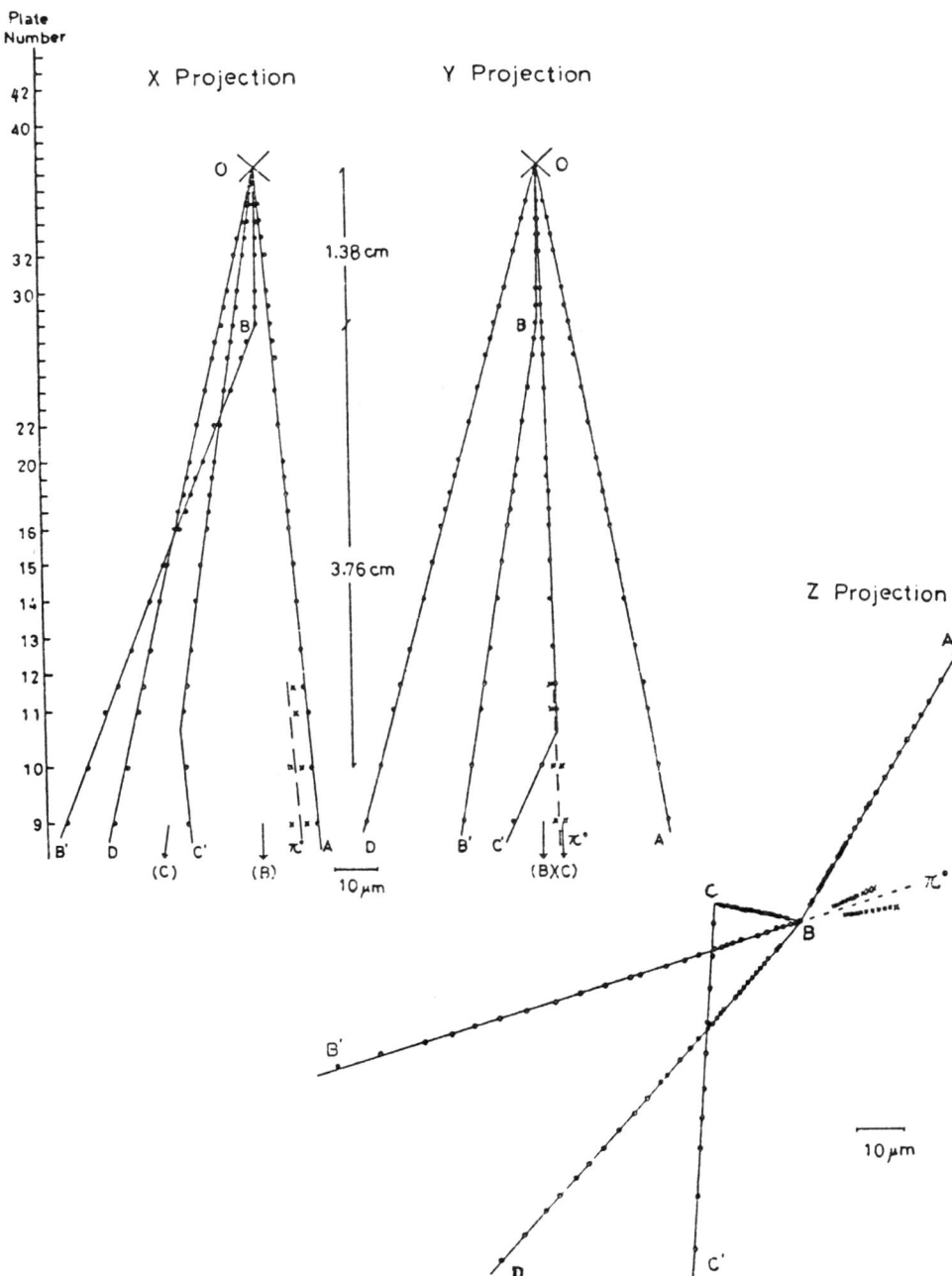

Fig. 5. Reconstruction of an emulsion event from a cosmic-ray exposure by Niu et al. (reference 15) with first observed example of associated production of charmed particles.

Their evidence was statistically no better than Lederman's and their mass resolution somewhat worse. Subsequent experiments, of course, failed to confirm the 6-GeV state which Lederman had already tentatively dubbed the upsilon. In 1977, Lederman's group in looking at muon pairs did a considerably more sensitive study and clearly observed the state now universally recognized as the upsilon (T), the state with a mass of 9.4 GeV decaying into lepton pairs.[18] The electron-positron colliders, as well as further experiments with hadron accelerators have now established the spectroscopy of the upsilon and there is no question that it is a resonance of the b and $\bar{b}$ quark, hence a new quark had been discovered. (See Figure 6.)

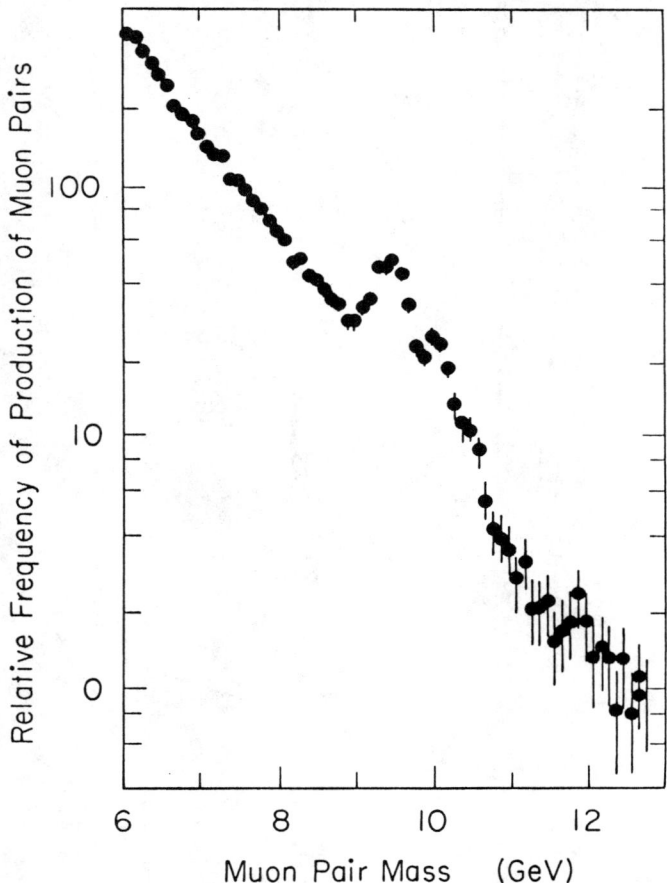

Fig. 6. The upsilon was discovered in 1977 by studying the production of muon pairs or electron pairs in proton collisions. Here the relative frequency of production of muon pairs is shown to decrease as the muon-pair mass increases. The bump in the curve at 9-10 GeV is due to the upsilon.

This discussion of new states of matter would not be complete without reference to the experiments at Stanford in 1975 which resulted in the discovery of the τ lepton. In studying data from the Mark I detector at SPEAR, Martin Perl and his group observed some final states which consisted only of an electron and a muon but with considerable missing energy and missing momentum.[19] These they concluded were compatible with the interpretation that a pair of massive leptons was produced, each of which decayed leptonically to an electron or a muon and neutrinos. The events only occurred above a threshold of about 3.5 GeV total energy and the τ mass was subsequently confirmed to be 1784 MeV. Consequently, the three generations of matter as they are now understood can be represented as shown on Table I. The top quark is generally conceded to exist although its mass is still uncertain. The τ neutrino has not been directly observed but it is also generally believed to exist.

TABLE I. Our present knowledge of the lepton and quark families of particles.

Generation	Particle	Charge	Mass
1	electron (e)	-1	0.51 MeV
1	electron neutrino (ν_e)	0	less than 50 eV
2	muon (μ)	-1	106 MeV = 0.106 GeV
2	muon neutrino (ν_μ)	0	less than 0.5 MeV
3	tau (τ)	-1	1784 MeV = 1.784 GeV
3	tau neutrino* (ν_τ)	0	less than 160 MeV = 0.160 GeV

*indirect evidence

Generation	Particle	Charge	Mass
1	up (u)	+2/3	about 300 MeV = 0.3 GeV
1	down (d)	-1/3	about 300 MeV = 0.3 GeV
2	charm (c)	+2/3	about 1500 MeV = 1.5 GeV
2	strange (s)	-1/3	about 500 MeV = 0.5 GeV
3	top (t)†	+2/3	>50 GeV
3	bottom (b)	-1/3	about 5,000 MeV = 5.0 GeV

†Preliminary evidence

It is amusing in this context to consider the complexity of matter as it has been understood by man down through the ages. A philosophical simplification was made by the Greeks when they believed that all of nature could be understood in terms of earth, air, fire, and water; just four fundamental constituents. However, the alchemists of the middle ages found that a great degree of complexity was necessary; they identified many pure substances such as sulfur, mercury, salt, and so forth which could not be fitted into the earth, air, fire, and water scheme easily. A simplification came in the 19th century with the atomic theory of matter and the understanding that the larger number of chemical compounds were made of a somewhat smaller number of elements. This understanding reached its peak in a philosophic sense with the Mendeleev Periodic Table and the understanding of the 92 elements. Toward the end of the 19th century the electron was discovered and in the early 20th century the nuclear atom was understood. For a time there was the appealing concept that all of the elements might be composed of a cloud of electrons of varying numbers corresponding to the chemical element and a heavier nucleus which might be composed of only protons and electrons. However, this model had difficulties with the uncertainty principle and spin; and the discovery of the neutron greatly improved our understanding of the nucleus but added a new particle to our stable. The complexity of nature became greater with the discovery of mesons and during the 1950's became very confused with the strange particles, hyperons, and the proliferation of meson and baryon states. The simplification brought about by the quark model in the 1960's made it appear that perhaps we only needed a few constituents after all. Now these few constituents have increased to the particles of Table I, and we now have six leptons and six kinds of quarks, each of the quarks to come in three colors and each particle with its anti-particle. The fascinating philosophical question at this point is: is this the end? Are these indeed the fundamental constituents of nature; no more and no less? Or might there be yet heavier quarks and heavier leptons; might there be four or five generations? On the other hand, is it possible that all quarks are in turn composed of more fundamental, simpler constituents and only the different configuration or quantum states of these constituents determines the quark species? Of course, only time and further exploration will answer that question. (See Figure 7.)

I feel it appropriate to remark briefly here that we have searched very diligently for free quarks and it is useful to include here a measure of the limits of the production of free quarks that have been established from various accelerator experiments over the years.[20] The concept of quark confinement is now confortably accommodated by the contemporary theoretical understandings of quantum chromodynamics. Nevertheless, we should recognize that it was forced on theorists earlier when evidence for free quarks simply was not forthcoming from experiments. The positive evidence for free quarks had come from the Fairbank experiments at Stanford[21] and it is my understanding that these may now be open to some question. (See Figure 8.)

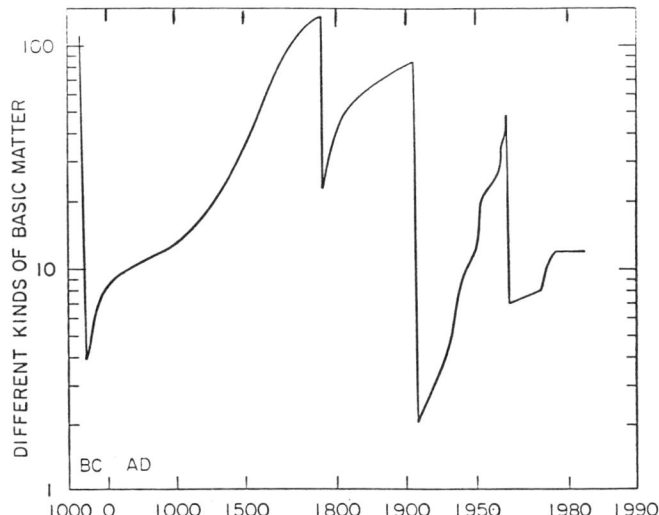

Fig. 7. Mankind has always tried to explain the world as made up of a limited number of different kinds of basic matter. Until a thousand years ago, the basic types of matter were considered to be earth, air, fire, water, et cetera. About 1900 the basic types of matter were thought to be the almost 100 different chemical elements. At present we believe there are about a dozen types of basic matter, namely the leptons and the quarks.

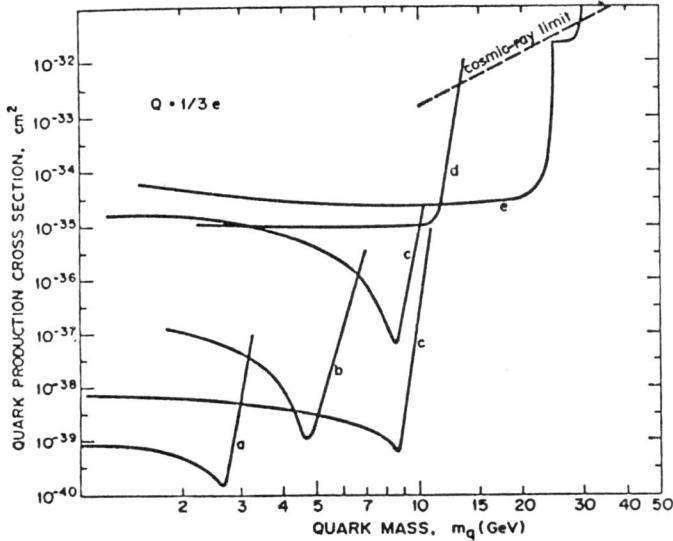

Fig. 8. The upper-limit cross section for production of quarks of charge $-1/3e$ as a function of quark mass from experiments at various particle accelerators and cosmic rays. The different experiments are indicated by letters: (a) CERN 28-GeV proton synchrotron; (b) Serpukov 70-GeV proton synchrotron; (c) and (d) Fermi National Accelerator Laboratory synchrotron; (e) CERN intersecting storage rings.

Let me now turn to the second thread in my discussion: the concept of deep inelastic lepton scattering. The finite size of the proton as revealed by electron scattering in the 1950's was one of the hints that protons might have an underlying structure and that the proton might not be a fundamental particle. Later, the inelastic scattering of electrons on protons, where the proton was broken up into a number of hadrons, resulted in explicit evidence that the constituents of the protons might be point particles.

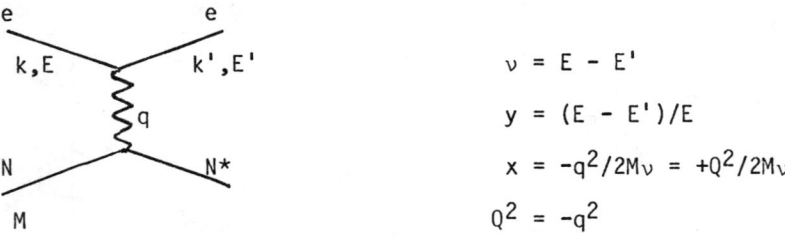

$$\nu = E - E'$$
$$y = (E - E')/E$$
$$x = -q^2/2M\nu = +Q^2/2M\nu$$
$$Q^2 = -q^2$$

The double-differential cross sections, $d^2\sigma/dE'd\Omega$, can be expressed in terms of structure functions, W_1 and W_2, and the related F_1 and F_2, which characterize the quark distributions in the nucleon:

$$\frac{d2\sigma}{dE'd\Omega} = \frac{\alpha^2}{4E^2\sin^4\theta/2} \; [W_2(\nu,q^2)\cos^2 \tfrac{\theta}{2} + 2W_1(\nu,q^2)\sin^2 \tfrac{\theta}{2}] \; .$$

Experimentally, νW_2 is seen to be almost independent of q^2, which suggested in about 1968 that the proton was constructed of point-like constituents.

Currently we have three sources of information on the form factor of protons and neutrons, that is, on the distribution of quark momenta in the proton and neutron. The oldest data come from the Stanford Linear Accelerator experiments on electron-nucleon deep inelastic scattering.[22] This is probably the most important noncollider physics to come from SLAC. Muon beams at Fermilab and CERN have also been used to determine form factors by deep inelastic scattering of muons, where the interpretation of physics is really quite similar to the electron case.[23] Neutrino beams at Fermilab and CERN have been used as well to probe the form factor, although in this case not through the exchange of a virtual photon but through the exchange of an intermediate vector boson.[24] The neutrino experiments determine the form factor in a rather separate way but the results, in fact, totally confirm the leptonic experiments. It is worth noting that these neutrino experiments at the hadron accelerators have placed the greatest demands on the proton beam intensity of any part of the experimental programs at CERN and Fermilab. As the experiments have become better it is observed that the strict scaling is not maintained and that the form factor determination does depend weakly on q^2. This departure from scaling is now understood in terms of quantum chromodynamics. (See Figure 9.)

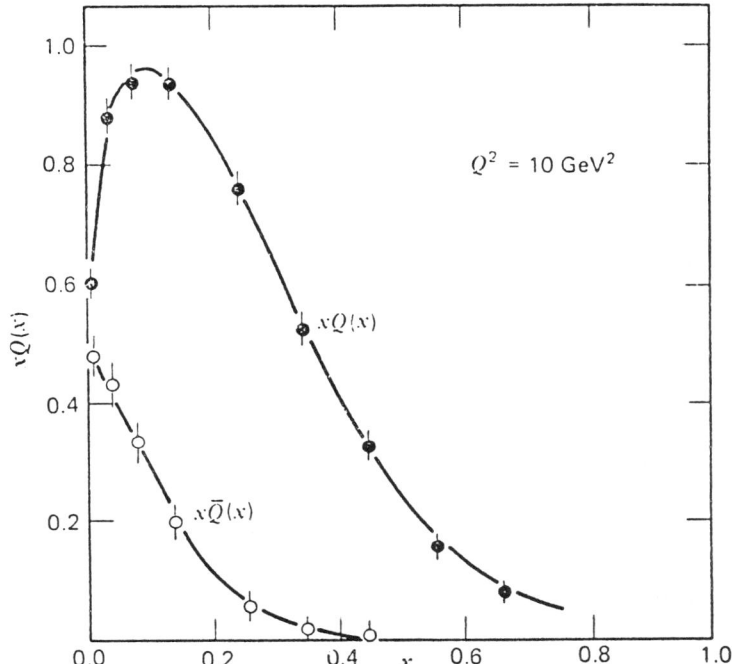

Fig. 9. Quark and antiquark momentum distributions in a nucleon as measured at CERN and the Fermi Laboratory. The experiments reveal that only about half the proton's momentum is carried by quarks. We have associated the remainder with the gluon constituents.

I might summarize briefly some of the spinoffs from the lepton-nucleon deep inelastic experiments beyond the simple determination of form factors. First, there is a scaling violation.[22,23,24] This can be understood by remarking that the quarks with large values of Feynman-x radiate gluons and this softens their scattering contribution.* A second important set of data has clearly identified the electroweak mixing through the parity nonconservation in electron-deuteron scattering using polarized electron beams.[25] A third interesting set of data has come from muon-nucleon scattering, where it is observed that the form factor of quarks determined from muons scattering on an iron nucleus is different from that from scattering on free hydrogen.[26] One might have expected that the distribution of the quark velocities in a heavy nucleus might be broader than in hydrogen due to folding of the form factor for a free proton together with the Fermi motion of that proton in the nucleus. In fact, it is observed that the form

---

*The variable x defined by Feynman is the fraction of the proton's momentum carried by the constituent quark or nucleon.

factor is somewhat narrower, as might be understood simplistically from quantum mechanics of a larger confining box and a broader determination of position or a greater uncertainty of position corresponding to tighter distribution in momentum. This EMC effect (as it's called from the European Muon Collaboration) is still a subject of intense study and is very interesting to the nuclear physicists as well as to particle physicists. The Drell-Yan process can also be used to study form factors, and indeed an experiment in which I was involved at Fermilab using proton collisions with tungsten made a very careful determination of form factors from the study of muon pairs.[27] The upsilon was also observed in this experiment although less clearly than by the Lederman group because of a solid iron spectrometer and the consequent fuzzing of the muon momentum resolution from multiple scattering.

Let me summarize briefly. To be sure, we have only scratched the surface in my simplistic historical summary. We have ignored all hadron dynamics, all of the beautiful studies of the K-meson system with CP violation, etc., and we have ignored almost all of the excitment of electroweak mixing and the other work which had led up to the understanding and later discovery of the intermediate vector bosons. I hope that I have given enough to convey several points, however. One, to emphasize that the threads of our knowledge are interpenetrating, there is a unity in particle physics,and so very often experiments directed at one area find interesting application in understanding other areas. Second, accelerators are used in a wide variety of modes often unanticipated when the machines themselves were built. Third, energy is usually, although not always, important; it has often been cited that the Ting experiment which discovered the J-particle might have been done fifteen years earlier at the AGS and also could have been done more easily and earlier than 1974 at the Fermilab higher energy accelerator. Fourth, intensity is sometimes critical, although the ingenuity of the experimenters, and the reliability and stability of the accelerator, are perhaps more important. Finally, there are often precursors for discoveries. These precursors generally lack either statistics or resolution or a sufficient theoretical understanding of the background phenomenon to have made the definitive experimental statements. And indeed, there are also occasional mistakes. I hope these perspectives might be interesting and might find some application in our thinking about future experimental physics and particle accelerators.

REFERENCES

1. R. Hofstadter, Rev. Mod. Physics $\underline{58}$, 214 (1956): Annu. Rev. Nucl. Sci. $\underline{7}$, 231 (1957).

2. W.E. Frazer and J.R. Fulco, Phys. Rev. $\underline{117}$, 1609 (1960).

3. C.C. Peck, L.W. Jones, and M.L. Perl, Phys. Rev. $\underline{126}$, 1836 (1962).

4. A.R. Erwin, R. March, W.D. Walker, and E. West, Phys. Rev. Lett. $\underline{6}$, 628 (1961).

5. See, for example, Rev. Mod. Phys. 56, 2, II (1984). "Review of Particle Properties."

6. M. Gell-Mann, Phys. Lett. 8, 214 (1964).

7. G. Zweig, CERN-TH 412 (1964) (unpublished).

8. M. Gell-Mann in *Proceedings of the International Conference on High-Energy Nuclear Physics*, Geneva, 1962, p. 805.

9. V.E. Barnes et al., Phys. Rev. Lett. 12, 204 (1964).

10. L.W. Alvarez, Phys. Rev. D8, 702 (1973).

11. J.J. Aubert et al., Phys. Rev. Lett. 33, 1404 (1974).

12. J.E. Augustin et al., Phys. Rev. Lett. 33, 1406 (1974).

13. J.H. Christenson et al., Phys. Rev. Lett. 25, 1523 (1970).

14. S. Drell and T.-M. Yan, Phys. Rev. Lett. 25, 316 (1970).

15. K.C. Niu, E. Mikumo, and Y. Maeda in *Proceedings of the 12th International Cosmic Ray Conference*, Hobart, Tasmania (1971); K.C. Niu, E. Mikumo, and Y. Maeda, Prog. Theor. Phys. 46, 1644 (1971).

16. D.C. Hom et al., Phys. Rev. Lett. 36, 1236 (1976).

17. D. Eartly, F. Giacomelli and K. Pretzl, Phys. Rev. Lett. 36, 1355 (1976).

18. S. Herb et al., Phys. Rev. Lett. 35, 252 (1977).

19. M.L. Perl et al., Phys. Rev. Lett. 39, 1489 (1975).

20. L.W. Jones, Rev. Mod. Phys. 49, 717 (1977).

21. G.S. LaRue, J.D. Phillips, and W.M. Fairbank, Phys. Rev. Lett. 46, 967 (1981).

22. A. Bodek et al., Phys. Rev. D20, 1471 (1979).

23. A.R. Clark et al., Phys. Rev. Lett. 51, 1826 (1983).

24. H. Abramowicz et al., Zeit. Phys. C17, 283 (1983).

25. C. Prescott et al., Phys. Lett. 84B, 524 (1979).

26. J.J. Aubert, Phys. Lett. 123B, 275 (1983); A. Bodek et al., Phys. Rev. Lett. 50, 1431 (1983).

27. S.R. Smith et al., Phys. Rev. Lett. 51, 743 (1983).

FEDERAL SCIENCE POLICY

Harold P. Hanson
Executive Director, Committee on Science, Space, and Technology
U.S. House of Representatives, Washington, D. C.

TABLE OF CONTENTS

1. Introduction 2301
2. How Congress Does Its Work 2301
3. Policy Studies 2302
4. Science Policy Study 2304
5. Technology Policy Study 2306
6. Budget Update -- 1987 2309
7. Budget Update -- 1989 2310

FEDERAL SCIENCE POLICY

Harold P. Hanson
Executive Director, Committee on Science, Space, and Technology
U.S. House of Representatives, Washington, D. C.

1. INTRODUCTION

I am pleased and honored to have been invited to talk with you this afternoon. As Chairman Bjorken indicates, I am the anthropomorphic embodiment of the ergodic theorem of statistical mechanics which assures us that if you live long enough, everything possible will have happened to you, perhaps several times.

I have visited FERMILAB before, and it is a pleasure to return as part of the Accelerator School Program. I am not a nuclear or particle physicist, but I've had many peripheral associations with accelerators -- starting with the war days at Wisconsin when Ray Herb's Van de Graaff machine went to the Manhattan Project; Ray Herb went to do radar research at the MIT Radlab; and I went to be a mine-and-torpedo ensign in the U.S. Navy.

There is another scientific concept that is relevant this afternoon. The Variational Principle for speech-making tells us that one approaches the optimum by keeping the task as short as possible, consistent with the constraint of transmitting a coherent message. A "least time" theorem, so to speak. However, I don't want to sit down quite yet.

I have several items that I want to discuss briefly, but I think I've prepared my remarks carefully enough that they won't make you late for the banquet.

As Ceasar observed about Gaul, my talk is divided into three parts. First, I will give a short course, a very short course, in the political science of science. Next, I shall spend most of my time discussing the two policy studies in which our Committee on Science, Space, and Technology is engaged. Naturally they are a Science Policy Study and a Technology Policy Study. Finally, I want to give you a brief update on the battle of the budget.

2. HOW CONGRESS DOES ITS WORK

First, the short course. The Congress does its work through committees. In the House of Representatives there are 22 standing committees including such things as Armed Services, Agriculture, Commerce, and so on. Unlike university committees, Congressional committees actually do something. I happen to be the Executive Director of the staff of the Science, Space, and Technology Committee of the House of Representatives. We have authorizing responsibility for all non-military and non-biomedical research that the Federal Government supports. Additionally, our oversight responsibility extends to <u>all</u> phases of governmentally-funded research.

Most of the work of any Congressional committee is carried out by subcommittees. I often make a comparison of the functioning of a college to the workings of the Committee and its subcommittees. The subcommittees are like college academic departments.

In both instances, that is where the action is. The subcommittee staff directors are like department chairmen. And the good ones are equally as mean, selfish, and aggressive as their academic counterparts. By analogy, I function as the dean of this special college -- and if you've ever figured out what a dean does for a college, that's what I do for the Committee.

We are past the time of the year when the Congress should have funded the various agencies. Our Committee, but not the Congress, has completed its work on the budgets for NSF, NASA, the Bureau of Standards, the Environmental Protection Agency, NOAA, and the FAA. But it has not completed the work on the research activities of the Department of Energy. More about that later.

You will note that I said our Committee authorizes funding. There are two levels above us that have greater control, or at least a different kind of control. First there is the Appropriations Committee which may or may not appropriate the funding our Committee authorizes. And in some instances where we are not specific or timely, the Appropriations Committee has been known to shape and modify our authorization -- which we don't like very much.

This year, however, the level above the Appropriations level is being felt more than ever. In general, the Budget Committee may or may not create a budget which conforms to the appropriations or authorizations. Further, within broad categories, they too may shape or modify our priorities. In principle, there is an orderly sequence to all this. In practice there is always a bit of creative tension. This year, confusion reigns.

3. POLICY STUDIES

Authorizations by our Committee amount to the annual examination of the trees. In 1985, we thought it was time to give the forest a forty-year checkup, in terms of science policy.

In looking back to World War II, we have asked what changes have occurred in science policy, whether the changes have been good or bad, and what change is called for now to meet the decades ahead.

That study, begun during the last Congress, is currently in its completion stages. It was generally anticipated that the Science Policy Study would be a precursor for a Technology Policy Study and this has been the way matters developed.

But before considering any details about the individual studies, I begin my part two with a pair of pragmatic observations. First, absent anything else, the budget, as an act of the Congress, is itself a policy statement. It creates law. It reflects activities, priorities, and directions. However, as Bismarck said, laws are like sausages, you should not watch them being made. It is not necessarily a pretty sight to watch the budget being put together. If there is no guiding, stabilizing, informing principle contributing to the development of the budget, then it is liable to be a flawed and distorted document, reflecting an uncertain and undefined policy, and yet having a dominating influence.

In recent years, in which the logical budget processes of the

Congress have been left progressively more and more in disarray, this has become, I believe, a major problem.

The extent to which it has become a problem is only partially indicated by the necessity for the Congress to resort to the newly rejuvenated Gramm-Rudman-Hollings Bill. You know that this bill represented the meat-axe approach to deficit cutting. It was pretty much blunted by the Supreme Court, but Congress acted to make it conform to the Constitution. Now it will take over if the targets aren't met by the Continuing Resolution. If all this doesn't work, the deficit grows. If they do work, science will probably be hurt, along with other discretionary items.

My second pragmatic point is one that was brought home to me by a Polish researcher who came by recently to interview me about the Science Policy Study. While she seemed to be, for the most part, very careful not to inject her own attitudes into the questions, it became evident to me that in her view the reason for America's international leadership was that administrators and responsible officials were quite capable of interpreting policy as they saw fit. In Poland, no one would dare not follow the strictest interpretation of state policy. In the United States, in many cases, policy does not seem to exist, and where it does, there may be as many interpretations as there are administrators to carry out policy.

Not denying that there is much truth in these contrasting views that (1) the allotment of dollars is more important than any other documental pronouncement of policy and (2) policy, where it exists, is often best ignored, it still is a fact that over the long trek the impact of a considered policy, generally agreed to, will be a dominating force.

In terms of lip service, it is likely that there never has been a better time for science and technology. Nearly all of Washington tries to outdo itself in proclaiming that science and technology will be the ultimate saving grace for this nation as it competes in the world marketplace. So, in the sense of Charles Dickens, these are the best of times and the worst of times.

Our Committee has authorized the Superconducting Super Collider with all the science and politics that goes with a machine of that magnitude. Virtually every state is deeply involved in some aspect of that competition. We like to think that this machine gives promise of giving us information about the first moments of time and what the "edge" of the universe is like. President Reagan even said, "Today, physicists peering into the infinitely small realms of subatomic particles find reaffirmations of religious faith." I find that intriguing. More concretely, perhaps, it will probably cost about five billion dollars to build and, when operating, will sustain the economy of a small but very affluent town.

We have not brought the SSC to the Floor as yet, and you can well imagine why.

The Space Station will be receiving its first true test of support and Congressional will. It has been the vision, of the nation and of Congress, that we would have a Space Station which would be built and serviced by the Shuttle.

The far-too-optimistic P.R. pipe-dream timetable called for 1992, 500 years after Columbus. That was never realistic, and now it is completely out of the question. 1996 seems more probable. Some tough queries are being posed in the aftermath of the recent horrible accident. We will see issues thought out and fought out again even though they were presumed to be settled decades ago. There will be some new questions such as how the Shuttle can be counted on for the 40 to 50 trips needed to erect the Station.

Add to this that we are designing new, exotic materials and medicines at an incredible pace. High-temperature superconductivity is on its way, perhaps not as quickly as some have predicted, but it's on its way. Also we are seeing a vast build up of the military, but of greatest interest to science and engineering is the simultaneous precipitous quantum jump in funding of research by the Department of Defense. But the thirty billion Mr. Carlucci is scaling down will have an impact -- probably a major one in the R&D area.

We are seeing the impact of the actions of Jeremy Rifkin and his colleagues. As you know, the Rifkinites are the sophisticated Luddites of the modern scientific world who, we should admit, may possibly have a measure of philosophical justification on their side, and who certainly will have more than a measure of hysterical popular support.

Lurking in the background is the spectre of that bogeyman, the deficit, and the influence that this will have on that portion of the nation's budget that is considered to be discretional; that is, not already committed or already spent. We who are scientists will be deluding ourselves if we think that all the truly good and important research and scientific studies that we are ready to do right now, will be done. Even this rich and powerful nation apparently does not have the financial wherewithal, although I hold with those who believe we cannot afford _not_ to move ahead vigorously in science and technology.

4. SCIENCE POLICY STUDY

It is against this backdrop of uncertainty and excitement and optimism that the Science Committee undertook a review of science policy.

Forty years ago, the basic course of American science policy was set by a report commissioned by President Roosevelt while the Second World War was still being fought. <u>Science: The Endless Frontier,</u> was developed by a renowned researcher, Vannevar Bush, who had been Dean of Engineering at MIT, and the wartime Office of Scientific Research and Development. Bush's report proposed that the Federal Government take a new role in the postwar years, by engaging in the financial support of basic and applied research in the nation's universities. The "Bush Report," as it has come to be known, has since that time served as the basic rationale for the Federal Government's funding of a wide-ranging, ever-growing program of support for research -- primarily in universities, but in government laboratories as well.

We have now asked whether the policies laid down in the Bush Report, as they have been amplified and modified or even ignored in the intervening years, are fully adequate to the new environment in which the government/science relationship must function.

For the most part, the inquiry was conducted by inviting a number of hearing-witnesses to address the broad questions posed in the study agenda that we prepared, and then to comment further as they wished on issues germane to the study. The distinguished experts who have appeared before us came from diverse research and administrative positions in academia, government, non-profit institutions, foundations, and private industry. They brought different perspectives to the inquiry.

The formal aspects of the Science Policy Study have been completed for some time now. We held about 60 hearings. We have heard from about 250 witnesses representing agencies, organizations, and societies, as well as themselves. However, we are far behind schedule in producing a finished document. It is like the dissertation that always takes considerably longer to complete than was originally envisaged. But it will be completed.

Before I leave the Science Policy Study, per se, let me touch briefly upon several topics to point out some dilemmas and difficulties.

Naturally, thus far we have heard a great deal of special pleading for causes in which most Congressmen already believe. Consequently, the report will have some elements that are predictable and non-controversial. But if this study is going to have any impact, it will have to take a stand on some tough issues.

I think you can see a spectrum of problems ranging from the easy to the exceedingly difficult. At the tractable end of the range I would say that we have the following: (1) clearly the government must continue its support of science because of the relationship to the nation's economic health and military strength; (2) since nationally significant problems are cross-disciplinary, education should recognize this fact and adapt to it; funding should also be available for cross-disciplinary proposals; and (3) there should be continuity and predictability of funding to get maximum effectiveness in the utilization of the dollars available.

Even though for the most part we could get by with platitudes in addressing these particular questions, they do start already to get a bit sticky. Science should be supported, but at what rate? What fraction of the GNP? How much is enough?

Further, continuity and predictability of funding diminish the flexibility of the government in reacting to new challenges and new opportunities. Also the mechanisms that have been suggested for securing continuity and predictability may well imply a lessened role for the hallowed peer review process.

In the more difficult category I would put the problem of peer review, itself. Over the years there has been a somewhat precarious balance between Congressional responsibilities and prerogatives on the one hand, and the scholarly judgment of scientific peers, on the other.

There are anguished claims that the balance is being upset, which may even be true. The Committee will probably come down squarely on both sides of the fence, which may be a bit hard on the anatomy.

Also in the difficult range is the question of big science and little science. It is quite easy to say, as some are doing, that we must have both -- but what does that mean? The fact is that even in affluent times they tend to impact on one another. They shouldn't, but they do. On the other hand, incremental money that becomes available for a big project, would likely not be available for a small one. If our Policy Study is to be of any value I think we must take a stance as to priorities. Or, at least, how to set priorities.

Perhaps the most difficult problem may turn out to be the role of the military (and I don't mean just Star Wars) in supporting basic research. On both sides of this issue we have Congressmen who feel strongly -- and there may be no compromise possible. Is there a policy to be set? Certainly the situation is there and it must be addressed, but defining a general course of action may be beyond the capacity of the Committee.

Back to the study, per se. When we finish the Study, when the policies have been adopted about which there is consensus, when some of the questions have been answered, when the points still at issue have been delineated and defined, what will we have?

That remains to be seen. The possibility exists that we have labored in vain. Circumstances, which may be the pressures of the economy and politics, may force immediate divergence from the guidelines we develop. This is conceivable, but it is our firm intention to produce a viable, visionary product that will stand the nation in good stead as we move into the future. And we'll do our very best.

5. TECHNOLOGY POLICY STUDY

Now what about technology policy? Having learned a bit (in several ways) from the Science Policy work, this study is forging ahead and I wouldn't be surprised if it doesn't come to final fruition as soon as the Science Policy Study. In thinking about a technology policy, we note that the year 1987 marks the 200th Anniversary of the Constitution of the United States of America. It is a time of pride and of patriotic reflection. We are blessed to be living in the most affluent, the most free, and the most universally-envied nation of the world. Our nation stands as a symbol of freedom, social mobility, and growth throughout the world.

Although the American system still offers much more than any other society, not every prospect pleases. Many of our major industries have declined. Other industries are kept functional through direct government intervention. Some segments of the economy are seeking protection from foreign competitors who are striving to take over international markets. Meanwhile the national debt grows to more than two trillion dollars so that the annual interest on our debt is at 15% of the federal budget, or roughly $150 billion. Thus there seem to be portents of a national future that could be less affluent, less respected, and less influential in the community of nations.

As a nation we do not know with certainty the seriousness of our circumstances. Further, we do not know how much of our perceived plight may be due to poor management and poor planning, and how much may be due to unavoidable flow of history. A cold-eyed assessment can help us correct what is amenable to correction, and adapt to what is inevitable.

In this historic 100th Congress, we have initiated a thorough inquiry of our present technological position in the international community, and we hope to offer measures which may be taken to optimize our future.

This effort by our Committee is by no means isolated or unique. There is a plethora of forums, caucuses, initiatives, symposiums, and studies all designed to analyze and seek out solutions to the problems concerned with our nation's competitive posture.

What is unique is that the Committee on Science, Space, and Technology has both the legislative jurisdiction and the responsibility to do something about the situation. During the last quarter-century the Committee has had jurisdiction over much of the nation's federally-funded civilian scientific and technological development, and this remains our responsibility. Thus, this Committee can occupy a central role in the national examination of the United States as a technological innovator, a manufacturer, and a trader in high and low technology products.

The significance of technology as one of the pillars in the American economic system is well recognized. The country owes much of its high standard of living to its technological preeminence. However, a number of indicators suggest that the U.S. may be living off past technological successes and investments. U.S. industry is now facing, for the first time in many years, technologically equal and well organized foreign competition backed by concentrated fiscal resources, comprehensive industrial policies, and protected home markets.

The setbacks suffered by the U.S. as a competitive nation cannot be blamed solely on macro-economic factors, the over-valued dollar, high wages, or unfair trade practices by its trading partners. Foreign competitors, in many cases, have managed their technology for the long term, have paid more attention to quality, and have been better able to grasp and adapt to opportunities in a technologically competitive world. They have paid attention to modern manufacturing techniques and processes and have produced and designed higher quality products. Many of those products have been better suited to the international market and even to the U.S. market.

By undertaking this Study, the Committee on Science, Space, and Technology and its Technology Policy Task Force will take the lead in examining the longterm problems and structural change that may be needed, and will attempt to establish a framework for legislative initiatives in the coming years. At the same time, when carrying out the Study it is important to keep in mind that the nation's continued ability to compete industrially is both a basic responsibility and a right of the private sector.

The concept of a federal role in the direct support of scientific _advancement_ for the nation emerged at the end of World War II, particularly through the rationale outlined in Vannevar Bush's Science: The Endless Frontier. This support has grown over the ensuing four decades. Today, it is acknowledged and accepted that the Federal Government has responsibility to provide significant support for the nation's _scientific_ research.

The proper role for the Federal Government in the process of fostering _technology development_ is less well understood, defined, or accepted. Yet it is technology, in one form or another, rather than science, that most clearly affects the lives of the nation's citizens. It is by the creation and diverse utilization of technology that, historically, we have measured the progress of civilization. It is primarily through the development and diffusion of technology that the flow of modern business and commerce takes place.

America does not have a formal and articulated policy for stimulating and guiding technology development. We do have numerous laws, as well as many ingrained patterns and practices, that govern the way that we envision, develop, and utilize technology. These guidelines and influences, viewed collectively, form an _an hoc_ national policy for technology. Given the questions about the performance of the United States as a competitor in the domestic and international marketplace, it seems timely to examine this _ad hoc_ policy for its strengths and weaknesses; it seems necessary to ask whether a formal, articulated policy is needed to help us better compete in the changing and complex world economy, and it seems critical to identify the real problems and propose practical solutions. A visionary Congress should initiate dynamic public policy to ensure the development and maintenance of national strategies for maximum utilization of our nation's knowledge and expertise in science and technology.

The objective of the Technology Policy Study is to identify basic problems and make recommendations on policy for technology that represent a cohesive and consistent national approach in support of technology development and utilization for the nation. The Study includes the investigation, collection of data, synthesis, and development of findings related to:

A. The responsibility of the Federal Government for technology support and development.
B. The influence that Federal Government budget mechanisms and procedures have on our nation's long-term ability to develop and apply technology.
C. The need for increased cooperation among the various levels of government (federal, state, local), the private sector, and our educational institutions to foster the utilization of science and engineering data for technology development.
D. The preservation of key technologies, industries, and skills that are required for the constant evolution and growth of society.
E. The necessity to improve the competence and skills of our human resources by better education and training, especially through increased emphasis on science, engineering, and related technical

disciplines.
F. The impact of government policies and regulations on the ability of the private sector to develop and utilize technology for the efficient production of goods and services.
G. The need for developing new mechanisms to aid in the transfer of technologies and expertise from the federal sector to the private sector in order to take full advantage of the large federal investment in science and technology.
H. The influence of the private sector's business strategies on the way that business responds to technological advancement.
I. The ways in which state and local governments contribute to or inhibit technological innovation and development.
J. The implications of foreign involvement in U.S. research and development.

So much for the Technology Policy Study.

6. **BUDGET UPDATE -- 1987**

Now for that brief update on the battle of the budget that I promised. I will speak only on two items. NSF and the SSC which I think are probably the items of greatest moment.

In a speech I made early in the year I said that we were dealing with a good news/bad news story that was not a joke. The good news was that the Administration's proposed science budget was very gratifying to scientists. The bad news was that we wouldn't get the Administration's budget. For a brief heady moment I thought I was wrong and that the NSF was due to get an auspicious start in its program to double its budget in five years by getting a 17% increase this year. I am now more pessimistic again.

A current rumor is that the House Appropriations Bill will be slashed some $50 million from their first mark. The history was that OMB recommended a $250 million increase. Appropriations came up with only $100 million. Several things happened, including our threatening to go the Floor with a full authorization. Appropriations restored another $100 million so they were down only $50 million from the President's figure. Now, however, as I have said, this may be cut another $50 so the NSF increase will be only 10% rather than the generous 17% asked for by OMB.

I say I am fearful because (1) there is a problem with the total budget, (2) NSF and NASA are in the same budget category, Function 250, under the same Appropriations subcommittee, and Jake Garn of the Senate will not permit NASA to be shortchanged, and (3) the Senate has not passed out of subcommittee even one appropriations bill, so we will be funded by a great continuing resolution that will be produced at the last moment under chaos and under duress.

The other specific I want to share with you is the situation for the SSC. I think you've heard and read many times of Al Trivelpiece's valiant efforts which culminated in the President's call for embarking on this program. The House Appropriations Committee picked up the gauntlet to the tune of $35 million for Fiscal '88. The weevil in the boll was that none of it was for construction. Some construction money ($10 million is the canonical number) would be the signal that we are playing for keeps. Now Appropriations was ahead of us. As I said we have not done an authorization budget for the Department of Energy. So we had a number of options including doing nothing, because there is no possibility anything will come out of the Senate. Nevertheless, just before the recess, and after a great deal of politicking that I won't burden you with, we did submit a separate SSC bill.

At the press conference, Mr. Roe said, "The Superconducting Super Collider is the culmination of decades of research on the fundamental nature of matter that began with the discovery of the basic constituents of the atom at the beginning of the century. This device would accelerate beams of protons to energies 20 times greater than the most powerful particle accelerator in existence. At such prodigious energies, scientists will, for the first time in history, observe the fundamental particles of nature that existed at the beginning of creation. From such knowledge, mankind would gain a better understanding, not only of the world as it existed in the past but, perhaps, as we can expect it to evolve in the future.

"The benefits to be gained from such knowledge, as well as the ultimate benefits derivable from the exotic and highly advanced technologies which are required to support this machine, cannot be fully estimated. History, however, shows us that as we improve our understanding of the fundamental laws of nature, we are better able to put this information to great practical use in many diverse areas such as new materials, medical research and treatment, and other fundamental technologies essential to the improvement of man's quality of life. Although the cost of this machine, estimated by the Department of Energy to amount to some $4.5 billion, would be the largest investment made by this country in a scientific experiment, the fundamental knowledge gained and long-term technological benefits more than justify the cost.

"Today, with an overwhelming number of cosponsors, I introduced a bill to authorize the Superconducting Super Collider project in the Department of Energy. I am strongly supportive of this project which I believe must go forward for the U.S. to retain scientific preeminence in the 1990's and beyond. Thus, I am pleased at this convincing demonstration of broad Congressional support."

This decision should be good news to most of you here. We still have to see how the political battle will go.

On that note, I shall conclude my presentation, while wishing the very best for all participants in this U.S. Particle Accelerator School. I trust I haven't violated the minimality condition too badly.

7. BUDGET UPDATE: 1989

There is an inexorable rhythm and repetitiveness to the federal budget process that approaches the semester system...or housework. The Congress has received President Reagan's budget for 1990 which is characteristically unrealistic but which offers high ground from which to defend most science programs. In the meantime the NSF received a 9.7% increase for 1989 which certainly slows the doubling rate but is about twice the percentage increase that other science programs received. An increase of 13.9% is suggested for the NSF in 1990. The SSC and the Space Station were kept viable for 1989, and Mr. Reagan has $2.1 billion scheduled for the Space Station and $.25 billion for the SSC for 1990. The Congress will now work its will.

AIP Conference Proceedings

		L.C. Number	ISBN
No. 1	Feedback and Dynamic Control of Plasmas – 1970	70-141596	0-88318-100-2
No. 2	Particles and Fields – 1971 (Rochester)	71-184662	0-88318-101-0
No. 3	Thermal Expansion – 1971 (Corning)	72-76970	0-88318-102-9
No. 4	Superconductivity in d- and f-Band Metals (Rochester, 1971)	74-18879	0-88318-103-7
No. 5	Magnetism and Magnetic Materials – 1971 (2 parts) (Chicago)	59-2468	0-88318-104-5
No. 6	Particle Physics (Irvine, 1971)	72-81239	0-88318-105-3
No. 7	Exploring the History of Nuclear Physics – 1972	72-81883	0-88318-106-1
No. 8	Experimental Meson Spectroscopy –1972	72-88226	0-88318-107-X
No. 9	Cyclotrons – 1972 (Vancouver)	72-92798	0-88318-108-8
No. 10	Magnetism and Magnetic Materials – 1972	72-623469	0-88318-109-6
No. 11	Transport Phenomena – 1973 (Brown University Conference)	73-80682	0-88318-110-X
No. 12	Experiments on High Energy Particle Collisions – 1973 (Vanderbilt Conference)	73-81705	0-88318-111-8
No. 13	π-π Scattering – 1973 (Tallahassee Conference)	73-81704	0-88318-112-6
No. 14	Particles and Fields – 1973 (APS/DPF Berkeley)	73-91923	0-88318-113-4
No. 15	High Energy Collisions – 1973 (Stony Brook)	73-92324	0-88318-114-2
No. 16	Causality and Physical Theories (Wayne State University, 1973)	73-93420	0-88318-115-0
No. 17	Thermal Expansion – 1973 (Lake of the Ozarks)	73-94415	0-88318-116-9
No. 18	Magnetism and Magnetic Materials – 1973 (2 parts) (Boston)	59-2468	0-88318-117-7
No. 19	Physics and the Energy Problem – 1974 (APS Chicago)	73-94416	0-88318-118-5
No. 20	Tetrahedrally Bonded Amorphous Semiconductors (Yorktown Heights, 1974)	74-80145	0-88318-119-3
No. 21	Experimental Meson Spectroscopy – 1974 (Boston)	74-82628	0-88318-120-7
No. 22	Neutrinos – 1974 (Philadelphia)	74-82413	0-88318-121-5
No. 23	Particles and Fields – 1974 (APS/DPF Williamsburg)	74-27575	0-88318-122-3
No. 24	Magnetism and Magnetic Materials – 1974 (20th Annual Conference, San Francisco)	75-2647	0-88318-123-1
No. 25	Efficient Use of Energy (The APS Studies on the Technical Aspects of the More Efficient Use of Energy)	75-18227	0-88318-124-X

No. 26	High-Energy Physics and Nuclear Structure – 1975 (Santa Fe and Los Alamos)	75-26411	0-88318-125-8
No. 27	Topics in Statistical Mechanics and Biophysics: A Memorial to Julius L. Jackson (Wayne State University, 1975)	75-36309	0-88318-126-6
No. 28	Physics and Our World: A Symposium in Honor of Victor F. Weisskopf (M.I.T., 1974)	76-7207	0-88318-127-4
No. 29	Magnetism and Magnetic Materials – 1975 (21st Annual Conference, Philadelphia)	76-10931	0-88318-128-2
No. 30	Particle Searches and Discoveries – 1976 (Vanderbilt Conference)	76-19949	0-88318-129-0
No. 31	Structure and Excitations of Amorphous Solids (Williamsburg, VA, 1976)	76-22279	0-88318-130-4
No. 32	Materials Technology – 1976 (APS New York Meeting)	76-27967	0-88318-131-2
No. 33	Meson-Nuclear Physics – 1976 (Carnegie-Mellon Conference)	76-26811	0-88318-132-0
No. 34	Magnetism and Magnetic Materials – 1976 (Joint MMM-Intermag Conference, Pittsburgh)	76-47106	0-88318-133-9
No. 35	High Energy Physics with Polarized Beams and Targets (Argonne, 1976)	76-50181	0-88318-134-7
No. 36	Momentum Wave Functions – 1976 (Indiana University)	77-82145	0-88318-135-5
No. 37	Weak Interaction Physics – 1977 (Indiana University)	77-83344	0-88318-136-3
No. 38	Workshop on New Directions in Mossbauer Spectroscopy (Argonne, 1977)	77-90635	0-88318-137-1
No. 39	Physics Careers, Employment and Education (Penn State, 1977)	77-94053	0-88318-138-X
No. 40	Electrical Transport and Optical Properties of Inhomogeneous Media (Ohio State University, 1977)	78-54319	0-88318-139-8
No. 41	Nucleon-Nucleon Interactions – 1977 (Vancouver)	78-54249	0-88318-140-1
No. 42	Higher Energy Polarized Proton Beams (Ann Arbor, 1977)	78-55682	0-88318-141-X
No. 43	Particles and Fields – 1977 (APS/DPF, Argonne)	78-55683	0-88318-142-8
No. 44	Future Trends in Superconductive Electronics (Charlottesville, 1978)	77-9240	0-88318-143-6
No. 45	New Results in High Energy Physics – 1978 (Vanderbilt Conference)	78-67196	0-88318-144-4
No. 46	Topics in Nonlinear Dynamics (La Jolla Institute)	78-57870	0-88318-145-2
No. 47	Clustering Aspects of Nuclear Structure and Nuclear Reactions (Winnepeg, 1978)	78-64942	0-88318-146-0
No. 48	Current Trends in the Theory of Fields (Tallahassee, 1978)	78-72948	0-88318-147-9

No. 49	Cosmic Rays and Particle Physics – 1978 (Bartol Conference)	79-50489	0-88318-148-7
No. 50	Laser-Solid Interactions and Laser Processing – 1978 (Boston)	79-51564	0-88318-149-5
No. 51	High Energy Physics with Polarized Beams and Polarized Targets (Argonne, 1978)	79-64565	0-88318-150-9
No. 52	Long-Distance Neutrino Detection – 1978 (C.L. Cowan Memorial Symposium)	79-52078	0-88318-151-7
No. 53	Modulated Structures – 1979 (Kailua Kona, Hawaii)	79-53846	0-88318-152-5
No. 54	Meson-Nuclear Physics – 1979 (Houston)	79-53978	0-88318-153-3
No. 55	Quantum Chromodynamics (La Jolla, 1978)	79-54969	0-88318-154-1
No. 56	Particle Acceleration Mechanisms in Astrophysics (La Jolla, 1979)	79-55844	0-88318-155-X
No. 57	Nonlinear Dynamics and the Beam-Beam Interaction (Brookhaven, 1979)	79-57341	0-88318-156-8
No. 58	Inhomogeneous Superconductors – 1979 (Berkeley Springs, W.V.)	79-57620	0-88318-157-6
No. 59	Particles and Fields – 1979 (APS/DPF Montreal)	80-66631	0-88318-158-4
No. 60	History of the ZGS (Argonne, 1979)	80-67694	0-88318-159-2
No. 61	Aspects of the Kinetics and Dynamics of Surface Reactions (La Jolla Institute, 1979)	80-68004	0-88318-160-6
No. 62	High Energy e^+e^- Interactions (Vanderbilt, 1980)	80-53377	0-88318-161-4
No. 63	Supernovae Spectra (La Jolla, 1980)	80-70019	0-88318-162-2
No. 64	Laboratory EXAFS Facilities – 1980 (Univ. of Washington)	80-70579	0-88318-163-0
No. 65	Optics in Four Dimensions – 1980 (ICO, Ensenada)	80-70771	0-88318-164-9
No. 66	Physics in the Automotive Industry – 1980 (APS/AAPT Topical Conference)	80-70987	0-88318-165-7
No. 67	Experimental Meson Spectroscopy – 1980 (Sixth International Conference, Brookhaven)	80-71123	0-88318-166-5
No. 68	High Energy Physics – 1980 (XX International Conference, Madison)	81-65032	0-88318-167-3
No. 69	Polarization Phenomena in Nuclear Physics – 1980 (Fifth International Symposium, Santa Fe)	81-65107	0-88318-168-1
No. 70	Chemistry and Physics of Coal Utilization – 1980 (APS, Morgantown)	81-65106	0-88318-169-X
No. 71	Group Theory and its Applications in Physics – 1980 (Latin American School of Physics, Mexico City)	81-66132	0-88318-170-3
No. 72	Weak Interactions as a Probe of Unification (Virginia Polytechnic Institute – 1980)	81-67184	0-88318-171-1
No. 73	Tetrahedrally Bonded Amorphous Semiconductors (Carefree, Arizona, 1981)	81-67419	0-88318-172-X

No. 74	Perturbative Quantum Chromodynamics (Tallahassee, 1981)	81-70372	0-88318-173-8
No. 75	Low Energy X-Ray Diagnostics – 1981 (Monterey)	81-69841	0-88318-174-6
No. 76	Nonlinear Properties of Internal Waves (La Jolla Institute, 1981)	81-71062	0-88318-175-4
No. 77	Gamma Ray Transients and Related Astrophysical Phenomena (La Jolla Institute, 1981)	81-71543	0-88318-176-2
No. 78	Shock Waves in Condensed Matter – 1981 (Menlo Park)	82-70014	0-88318-177-0
No. 79	Pion Production and Absorption in Nuclei – 1981 (Indiana University Cyclotron Facility)	82-70678	0-88318-178-9
No. 80	Polarized Proton Ion Sources (Ann Arbor, 1981)	82-71025	0-88318-179-7
No. 81	Particles and Fields –1981: Testing the Standard Model (APS/DPF, Santa Cruz)	82-71156	0-88318-180-0
No. 82	Interpretation of Climate and Photochemical Models, Ozone and Temperature Measurements (La Jolla Institute, 1981)	82-71345	0-88318-181-9
No. 83	The Galactic Center (Cal. Inst. of Tech., 1982)	82-71635	0-88318-182-7
No. 84	Physics in the Steel Industry (APS/AISI, Lehigh University, 1981)	82-72033	0-88318-183-5
No. 85	Proton-Antiproton Collider Physics –1981 (Madison, Wisconsin)	82-72141	0-88318-184-3
No. 86	Momentum Wave Functions – 1982 (Adelaide, Australia)	82-72375	0-88318-185-1
No. 87	Physics of High Energy Particle Accelerators (Fermilab Summer School, 1981)	82-72421	0-88318-186-X
No. 88	Mathematical Methods in Hydrodynamics and Integrability in Dynamical Systems (La Jolla Institute, 1981)	82-72462	0-88318-187-8
No. 89	Neutron Scattering – 1981 (Argonne National Laboratory)	82-73094	0-88318-188-6
No. 90	Laser Techniques for Extreme Ultraviolt Spectroscopy (Boulder, 1982)	82-73205	0-88318-189-4
No. 91	Laser Acceleration of Particles (Los Alamos, 1982)	82-73361	0-88318-190-8
No. 92	The State of Particle Accelerators and High Energy Physics (Fermilab, 1981)	82-73861	0-88318-191-6
No. 93	Novel Results in Particle Physics (Vanderbilt, 1982)	82-73954	0-88318-192-4
No. 94	X-Ray and Atomic Inner-Shell Physics – 1982 (International Conference, U. of Oregon)	82-74075	0-88318-193-2
No. 95	High Energy Spin Physics – 1982 (Brookhaven National Laboratory)	83-70154	0-88318-194-0
No. 96	Science Underground (Los Alamos, 1982)	83-70377	0-88318-195-9

No. 97	The Interaction Between Medium Energy Nucleons in Nuclei – 1982 (Indiana University)	83-70649	0-88318-196-7
No. 98	Particles and Fields – 1982 (APS/DPF University of Maryland)	83-70807	0-88318-197-5
No. 99	Neutrino Mass and Gauge Structure of Weak Interactions (Telemark, 1982)	83-71072	0-88318-198-3
No. 100	Excimer Lasers – 1983 (OSA, Lake Tahoe, Nevada)	83-71437	0-88318-199-1
No. 101	Positron-Electron Pairs in Astrophysics (Goddard Space Flight Center, 1983)	83-71926	0-88318-200-9
No. 102	Intense Medium Energy Sources of Strangeness (UC-Sant Cruz, 1983)	83-72261	0-88318-201-7
No. 103	Quantum Fluids and Solids – 1983 (Sanibel Island, Florida)	83-72440	0-88318-202-5
No. 104	Physics, Technology and the Nuclear Arms Race (APS Baltimore –1983)	83-72533	0-88318-203-3
No. 105	Physics of High Energy Particle Accelerators (SLAC Summer School, 1982)	83-72986	0-88318-304-8
No. 106	Predictability of Fluid Motions (La Jolla Institute, 1983)	83-73641	0-88318-305-6
No. 107	Physics and Chemistry of Porous Media (Schlumberger-Doll Research, 1983)	83-73640	0-88318-306-4
No. 108	The Time Projection Chamber (TRIUMF, Vancouver, 1983)	83-83445	0-88318-307-2
No. 109	Random Walks and Their Applications in the Physical and Biological Sciences (NBS/La Jolla Institute, 1982)	84-70208	0-88318-308-0
No. 110	Hadron Substructure in Nuclear Physics (Indiana University, 1983)	84-70165	0-88318-309-9
No. 111	Production and Neutralization of Negative Ions and Beams (3rd Int'l Symposium, Brookhaven, 1983)	84-70379	0-88318-310-2
No. 112	Particles and Fields – 1983 (APS/DPF, Blacksburg, VA)	84-70378	0-88318-311-0
No. 113	Experimental Meson Spectroscopy – 1983 (Seventh International Conference, Brookhaven)	84-70910	0-88318-312-9
No. 114	Low Energy Tests of Conservation Laws in Particle Physics (Blacksburg, VA, 1983)	84-71157	0-88318-313-7
No. 115	High Energy Transients in Astrophysics (Santa Cruz, CA, 1983)	84-71205	0-88318-314-5
No. 116	Problems in Unification and Supergravity (La Jolla Institute, 1983)	84-71246	0-88318-315-3
No. 117	Polarized Proton Ion Sources (TRIUMF, Vancouver, 1983)	84-71235	0-88318-316-1

No. 118	Free Electron Generation of Extreme Ultraviolet Coherent Radiation (Brookhaven/OSA, 1983)	84-71539	0-88318-317-X
No. 119	Laser Techniques in the Extreme Ultraviolet (OSA, Boulder, Colorado, 1984)	84-72128	0-88318-318-8
No. 120	Optical Effects in Amorphous Semiconductors (Snowbird, Utah, 1984)	84-72419	0-88318-319-6
No. 121	High Energy e^+e^- Interactions (Vanderbilt, 1984)	84-72632	0-88318-320-X
No. 122	The Physics of VLSI (Xerox, Palo Alto, 1984)	84-72729	0-88318-321-8
No. 123	Intersections Between Particle and Nuclear Physics (Steamboat Springs, 1984)	84-72790	0-88318-322-6
No. 124	Neutron-Nucleus Collisions – A Probe of Nuclear Structure (Burr Oak State Park - 1984)	84-73216	0-88318-323-4
No. 125	Capture Gamma-Ray Spectroscopy and Related Topics – 1984 (Internat. Symposium, Knoxville)	84-73303	0-88318-324-2
No. 126	Solar Neutrinos and Neutrino Astronomy (Homestake, 1984)	84-63143	0-88318-325-0
No. 127	Physics of High Energy Particle Accelerators (BNL/SUNY Summer School, 1983)	85-70057	0-88318-326-9
No. 128	Nuclear Physics with Stored, Cooled Beams (McCormick's Creek State Park, Indiana, 1984)	85-71167	0-88318-327-7
No. 129	Radiofrequency Plasma Heating (Sixth Topical Conference, Callaway Gardens, GA, 1985)	85-48027	0-88318-328-5
No. 130	Laser Acceleration of Particles (Malibu, California, 1985)	85-48028	0-88318-329-3
No. 131	Workshop on Polarized ^{3}He Beams and Targets (Princeton, New Jersey, 1984)	85-48026	0-88318-330-7
No. 132	Hadron Spectroscopy–1985 (International Conference, Univ. of Maryland)	85-72537	0-88318-331-5
No. 133	Hadronic Probes and Nuclear Interactions (Arizona State University, 1985)	85-72638	0-88318-332-3
No. 134	The State of High Energy Physics (BNL/SUNY Summer School, 1983)	85-73170	0-88318-333-1
No. 135	Energy Sources: Conservation and Renewables (APS, Washington, DC, 1985)	85-73019	0-88318-334-X
No. 136	Atomic Theory Workshop on Relativistic and QED Effects in Heavy Atoms	85-73790	0-88318-335-8
No. 137	Polymer-Flow Interaction (La Jolla Institute, 1985)	85-73915	0-88318-336-6
No. 138	Frontiers in Electronic Materials and Processing (Houston, TX, 1985)	86-70108	0-88318-337-4
No. 139	High-Current, High-Brightness, and High-Duty Factor Ion Injectors (La Jolla Institute, 1985)	86-70245	0-88318-338-2

No. 140	Boron-Rich Solids (Albuquerque, NM, 1985)	86-70246	0-88318-339-0
No. 141	Gamma-Ray Bursts (Stanford, CA, 1984)	86-70761	0-88318-340-4
No. 142	Nuclear Structure at High Spin, Excitation, and Momentum Transfer (Indiana University, 1985)	86-70837	0-88318-341-2
No. 143	Mexican School of Particles and Fields (Oaxtepec, México, 1984)	86-81187	0-88318-342-0
No. 144	Magnetospheric Phenomena in Astrophysics (Los Alamos, 1984)	86-71149	0-88318-343-9
No. 145	Polarized Beams at SSC & Polarized Antiprotons (Ann Arbor, MI & Bodega Bay, CA, 1985)	86-71343	0-88318-344-7
No. 146	Advances in Laser Science–I (Dallas, TX, 1985)	86-71536	0-88318-345-5
No. 147	Short Wavelength Coherent Radiation: Generation and Applications (Monterey, CA, 1986)	86-71674	0-88318-346-3
No. 148	Space Colonization: Technology and The Liberal Arts (Geneva, NY, 1985)	86-71675	0-88318-347-1
No. 149	Physics and Chemistry of Protective Coatings (Universal City, CA, 1985)	86-72019	0-88318-348-X
No. 150	Intersections Between Particle and Nuclear Physics (Lake Louise, Canada, 1986)	86-72018	0-88318-349-8
No. 151	Neural Networks for Computing (Snowbird, UT, 1986)	86-72481	0-88318-351-X
No. 152	Heavy Ion Inertial Fusion (Washington, DC, 1986)	86-73185	0-88318-352-8
No. 153	Physics of Particle Accelerators (SLAC Summer School, 1985) (Fermilab Summer School, 1984)	87-70103	0-88318-353-6
No. 154	Physics and Chemistry of Porous Media—II (Ridge Field, CT, 1986)	83-73640	0-88318-354-4
No. 155	The Galactic Center: Proceedings of the Symposium Honoring C. H. Townes (Berkeley, CA, 1986)	86-73186	0-88318-355-2
No. 156	Advanced Accelerator Concepts (Madison, WI, 1986)	87-70635	0-88318-358-0
No. 157	Stability of Amorphous Silicon Alloy Materials and Devices (Palo Alto, CA, 1987)	87-70990	0-88318-359-9
No. 158	Production and Neutralization of Negative Ions and Beams (Brookhaven, NY, 1986)	87-71695	0-88318-358-7

No. 159	Applications of Radio-Frequency Power to Plasma: Seventh Topical Conference (Kissimmee, FL, 1987)	87-71812	0-88318-359-5
No. 160	Advances in Laser Science–II (Seattle, WA, 1986)	87-71962	0-88318-360-9
No. 161	Electron Scattering in Nuclear and Particle Science: In Commemoration of the 35th Anniversary of the Lyman-Hanson-Scott Experiment (Urbana, IL, 1986)	87-72403	0-88318-361-7
No. 162	Few-Body Systems and Multiparticle Dynamics (Crystal City, VA, 1987)	87-72594	0-88318-362-5
No. 163	Pion–Nucleus Physics: Future Directions and New Facilities at LAMPF (Los Alamos, NM, 1987)	87-72961	0-88318-363-3
No. 164	Nuclei Far from Stability: Fifth International Conference (Rosseau Lake, ON, 1987)	87-73214	0-88318-364-1
No. 165	Thin Film Processing and Characterization of High-Temperature Superconductors	87-73420	0-88318-365-X
No. 166	Photovoltaic Safety (Denver, CO, 1988)	88-42854	0-88318-366-8
No. 167	Deposition and Growth: Limits for Microelectronics (Anaheim, CA, 1987)	88-71432	0-88318-367-6
No. 168	Atomic Processes in Plasmas (Santa Fe, NM, 1987)	88-71273	0-88318-368-4
No. 169	Modern Physics in America: A Michelson-Morley Centennial Symposium (Cleveland, OH, 1987)	88-71348	0-88318-369-2
No. 170	Nuclear Spectroscopy of Astrophysical Sources (Washington, D.C., 1987)	88-71625	0-88318-370-6
No. 171	Vacuum Design of Advanced and Compact Synchrotron Light Sources (Upton, NY, 1988)	88-71824	0-88318-371-4
No. 172	Advances in Laser Science–III: Proceedings of the International Laser Science Conference (Atlantic City, NJ, 1987)	88-71879	0-88318-372-2
No. 173	Cooperative Networks in Physics Education (Oaxtepec, Mexico 1987)	88-72091	0-88318-373-0
No. 174	Radio Wave Scattering in the Interstellar Medium (San Diego, CA 1988)	88-72092	0-88318-374-9
No. 175	Non-neutral Plasma Physics (Washington, DC 1988)	88-72275	0-88318-375-7

No. 176	Intersections Between Particle Land Nuclear Physics (Third International Conference) (Rockport, ME 1988)	88-62535	0-88318-376-5
No. 177	Linear Accelerator and Beam Optics Codes (La Jolla, CA 1988)	88-46074	0-88318-377-3
No. 178	Nuclear Arms Technologies in the 1990s (Washington, DC 1988)	88-83262	0-88318-378-1
No. 179	The Michelson Era in American Science: 1870–1930 (Cleveland, OH 1987)	88-83369	0-88318-379-X
No. 180	Frontiers in Science: International Symposium (Urbana, IL, 1987)	88-83526	0-88318-380-3
No. 181	Muon-Catalyzed Fusion (Sanibel Island, FL, 1988)	88-83636	0-88318-381-1
No. 182	High T_c Superconducting Thin Films, Devices, and Application (Atlanta, GA 1988)	88-03947	0-88318-382-X
No. 183	Cosmic Abundances of Matter (Minneapolis, MN 1988)	89-80147	0-88318-383-8

JUL 1 2 1989